MOLECULAR ELECTRONIC-STRUCTURE THEORY

MOLECULAR ELECTRONIC-STRUCTURE THEORY

Trygve Helgaker
Department of Chemistry, University of Oslo, Norway

Poul Jørgensen
Department of Chemistry, University of Aarhus, Denmark

Jeppe Olsen
Department of Chemistry, University of Aarhus, Denmark

JOHN WILEY & SONS, LTD
Chichester • New York • Weinheim • Brisbane • Singapore • Toronto

Copyright © 2000 by   John Wiley & Sons Ltd, The Atrium, Southern Gate, Chichester,
West Sussex PO19 8SQ, England

Telephone   (+44) 1243 779777

Email (for orders and customer service enquiries): cs-books@wiley.co.uk
Visit our Home Page on www.wileyeurope.com or www.wiley.com

Reprinted June 2002, November 2002, May 2004
Reprinted as paperback November 2012

Other Wiley Editorial Offices

John Wiley & Sons Inc., 111 River Street, Hoboken, NJ 07030, USA

Jossey-Bass, 989 Market Street, San Francisco, CA 94103-1741, USA

Wiley-VCH Verlag GmbH, Boschstr. 12, D-69469 Weinheim, Germany

John Wiley & Sons Australia Ltd, 33 Park Road, Milton, Queensland 4064, Australia

John Wiley & Sons (Asia) Pte Ltd, 2 Clementi Loop #02-01, Jin Xing Distripark, Singapore
129809

John Wiley & Sons Canada Ltd, 22 Worcester Road, Etobicoke, Ontario, Canada M9W 1L1

British Library Cataloguing in Publication Data

A catalogue record for this book is available from the British Library

PB ISBN: 9781118531471

Typeset in 10/12pt Times by Laser Words, (India) Ltd

CONTENTS

PREFACE

Quantum chemistry has emerged as an important tool for investigating a wide range of problems in chemistry and molecular physics. With the recent development of computational methods and more powerful computers, it has become possible to solve chemical problems that only a few years ago seemed for ever beyond the reach of a rigorous quantum-mechanical treatment. Today quantum-mechanical methods are routinely applied to problems related to molecular structure and reactivity, and spectroscopic parameters calculated quantum-mechanically are often useful in the interpretation of spectroscopic measurements. With the development and distribution of sophisticated program packages, advanced computational electronic-structure theory has become a practical tool for nonspecialists at universities and in industry.

In view of the increasing importance of computational electronic-structure theory in chemical research, it is somewhat surprising that no comprehensive, up-to-date, technical monograph is available on this subject. This book is an attempt to fill this gap. It covers all the important aspects of modern *ab initio* nonrelativistic wave function-based molecular electronic-structure theory – providing sufficient in-depth background material to enable the reader to appreciate the physical motivation behind the approximations made at the different levels of theory and also to understand the technical machinery needed for efficient implementations on modern computers.

To justify the adoption of the various approximations, this book makes extensive use of numerical examples, designed to illustrate the strengths and shortcomings of each method treated. Statements about the usefulness or deficiencies of the various methods are supported by actual examples and not merely model calculations, making the arguments more forceful. The accuracy and applicability of the various methods have been established from extensive calculations on a sample of molecules followed by a statistical analysis. The purpose of this book is thus not only to motivate approximations and to explain techniques, but also to assess the reliability and accuracy of the various computational methods in quantum chemistry.

The important working equations of computational electronic-structure theory are derived in detail. The level of detail attempted is such that, from the equations presented in this book, the reader should be able to write a computer program without too much difficulty. Thus, all the important aspects of computations are treated: the evaluation of molecular integrals, the parametrization and optimization of the wave function, and the analysis of the results. A detailed description of the contents of the book is given in the Overview.

Some areas of computational electronic-structure theory are not treated in this book. All methods discussed are strictly *ab initio*. Semi-empirical methods are not treated nor is density-functional theory discussed; all techniques discussed involve directly or indirectly the calculation of a wave function. Energy derivatives are not covered, even though these play a prominent role in the evaluation of molecular properties and in the optimization of geometries. Relativistic theory is likewise not treated. In short, the focus is on techniques for solving the nonrelativistic molecular

Born–Oppenheimer electronic Schrödinger equation *ab initio* and on the usefulness and reliability of the solutions.

The presentation of the material relies heavily on the techniques of second quantization. However, no previous knowledge of second quantization is assumed. Indeed, Chapter 1 provides a self-contained introduction to this subject at a level appropriate for the remainder of the text. It is the authors' belief that the beauty and usefulness of second quantization will become apparent as the reader tackles the remainder of the material presented in this book.

Except for the techniques of second quantization, the presentation in this book makes no use of techniques that should not be familiar to any student of quantum chemistry after one or two years of undergraduate study. At the end of each chapter, problems and exercises are provided, complete with hints and solutions. The reader is strongly urged to work through these exercises as a means of deepening and consolidating the understanding of the material presented in each chapter.

The material presented in this book has – in different stages of preparation – been used as teaching material at the *Quantum Chemistry and Molecular Properties* summer schools, organized by the authors biennially since 1990. In addition, it has been used by the authors for teaching one-year courses in computational quantum chemistry in their own departments. As such, it has been subjected to the repeated scrutiny of a large number of students. Their critical comments have been a great help in the preparation of the final manuscript.

Thousands of calculations constitute the background material for this book. We are, in particular, indebted to our colleagues Keld L. Bak, Jürgen Gauss, Asger Halkier, Wim Klopper and Henrik Koch for making these calculations possible. Most of the calculations were carried out with the DALTON program, while ACES II was used especially for spin-unrestricted calculations, for the CCSDT calculations and for geometry optimizations. Many of the DALTON and ACES II calculations were performed by Keld L. Bak and Asger Halkier. Wim Klopper performed all the explicitly correlated calculations, modifying his codes to fit our needs; Jürgen Gauss helped us with the ACES II program, making a number of modifications needed for our nonstandard and seemingly exorbitant calculations; and Henrik Koch made it possible to carry out many of the nonstandard coupled-cluster calculations with the DALTON program. We are also indebted to Jacek Kobus, Leif Laaksonen, Dage Sundholm and Pekka Pyykkö for providing us with their numerical diatomic Hartree–Fock program. The CASPT2 calculations were carried out with the MOLCAS program, the CI and higher-order perturbation calculations with the LUCIA program, while the numerical atomic calculations were carried out with the LUCAS program.

A large number of colleagues and students have provided helpful comments and suggestions for improvements. For this we thank Lars K. Andersen, Alexander Auer, Keld L. Bak, Vebjørn Bakken, Anders Bernhardsson, Thomas Bondo, Ove Christiansen, Attila G. Csaszar, Pål Dahle, Knut Fægri, Jürgen Gauss, Kasper Hald, Asger Halkier, Nicholas C. Handy, Christof Hättig, Hanne Heiberg, Alf C. Hennum, Morten Hohwy, Kristian S. Hvid, Michal Jaszunski, Hans Jørgen Aa. Jensen, Wim Klopper, Jacek Kobus, Henrik Koch, Helena Larsen, Jan Linderberg, Jon Lærdahl, Per-Åke Malmqvist, Kurt V. Mikkelsen, Chris Mohn, Thomas Nyman, Jens Oddershede, Tina D. Poulsen, Antonio Rizzo, Bjørn Roos, Torgeir Ruden, Kenneth Ruud, Trond Saue, Jack Simons, Dage Sundholm, Peter Szalay, Peter R. Taylor, Danny L. Yeager and Hans Ågren.

Most of this manuscript was prepared during numerous stays at the Marinebiological Laboratory of Aarhus University, located at Rønbjerg near Limfjorden in Denmark. We would like to thank the staff of the laboratory – Jens T. Christensen and Ingelise Mortensen – for providing us with an enjoyable and productive setting. Hanne Kirkegaard typed parts of the numerous drafts. We thank her for her competent typing and for never giving up hope in a seemingly never-ending flood of modifications and corrections. Finally, we are grateful to our wives – Barbara, Lise and Jette – for patiently letting us go on with our work. Without their love and support, we would not have finished this book, which is therefore dedicated to them.

Trygve Helgaker, Poul Jørgensen and Jeppe Olsen
Rønbjerg, March 1999

OVERVIEW

Chapters 1–3 introduce second quantization, emphasizing those aspects of the theory that are useful for molecular electronic-structure theory. In **Chapter 1**, second quantization is introduced in the spin-orbital basis, and we show how first-quantization operators and states are represented in the language of second quantization. Next, in **Chapter 2**, we make spin adaptations of such operators and states, introducing spin tensor operators and configuration state functions. Finally, in **Chapter 3**, we discuss unitary transformations and, in particular, their nonredundant formulation in terms of exponentials of matrices and operators. Of particular importance is the exponential parametrization of unitary orbital transformations, used in the subsequent chapters of the book.

In Chapters 4 and 5, we turn our attention away from the tools of second quantization towards wave functions. First, in **Chapter 4**, we discuss important general characteristics of the exact electronic wave function such as antisymmetry, size-extensivity, stationarity, and the cusps that arise from the singularities of the Hamiltonian. Ideally, we would like our approximate wave functions to inherit most of these properties; in practice, compromises must be made between what is desirable and what is practical as illustrated in **Chapter 5**, which introduces the standard models of electronic-structure theory, presenting the basic theory and employing numerical examples to illustrate the usefulness and shortcomings of the different methods.

Chapters 6–8 are concerned with one-electron basis functions and the expansion of many-electron wave functions in such functions. In **Chapter 6**, we discuss the analytical form of the one-electron basis functions – the universal angular part and the different exponential and Gaussian radial parts, applying these to one-electron systems in order to illustrate their merits and shortcomings. In **Chapter 7**, we consider the expansion of the two-electron helium system in products of one-electron functions. In particular, we study the slow convergence of such expansions, arising from the inability of orbital products to tackle head-on the problems associated with short-range interactions and singularities in the Hamiltonian. We here also develop techniques for the extrapolation of the energy to the basis-set limit, later used for many-electron systems. Next, having discussed one- and two-electron systems in Chapters 6 and 7, we proceed in **Chapter 8** to the construction of basis sets for many-electron systems, giving a detailed account of many of the standard Gaussian basis sets of electronic-structure theory – in particular, the correlation-consistent basis sets extensively used in this book. Their convergence is examined by carrying out extrapolations to the basis-set limit, comparing with explicitly correlated calculations.

Chapter 9 discusses the evaluation of molecular integrals over Gaussian atomic functions. We here cover all integrals needed for the evaluation of molecular electronic energies, deriving the working equations in detail. Many techniques have been developed for the calculation of molecular integrals; the more important are treated here: the Obara–Saika scheme, the McMurchie–Davidson scheme, and the Rys-polynomial scheme. However, for large systems, it is better to use multipole expansions to describe the long-range interactions. In this chapter, we develop the multipole method for the individual two-electron integrals; then, we discuss how the multipole method can be organized for the calculation of Coulomb interactions in large systems at a cost that scales linearly with the size of the system.

The next five chapters are each devoted to the study of one particular computational model of *ab initio* electronic-structure theory; **Chapter 10** is devoted to the Hartree–Fock model. Important topics discussed are: the parametrization of the wave function, stationary conditions, the calculation of the electronic gradient, first- and second-order methods of optimization, the self-consistent field method, direct (integral-driven) techniques, canonical orbitals, Koopmans' theorem, and size-extensivity. Also discussed is the direct optimization of the one-electron density, in which the construction of molecular orbitals is avoided, as required for calculations whose cost scales linearly with the size of the system.

In **Chapter 11**, we treat configuration-interaction (CI) theory, concentrating on the full CI wave function and certain classes of truncated CI wave functions. The simplicity of the CI model allows for efficient methods of optimization, as discussed in this chapter. However, we also consider the chief shortcomings of the CI method – namely, the lack of compactness in the description and the loss of size-extensivity that occurs upon truncation of the CI expansion.

In **Chapter 12**, we study the related multiconfigurational self-consistent field (MCSCF) method, in which a simultaneous optimization of orbitals and CI coefficients is attempted. Although the MCSCF method is incapable of providing accurate energies and wave functions, it is a flexible model, well suited to the study of chemical reactions and excited states. This chapter concentrates on techniques of optimization, a difficult problem in MCSCF theory because of the simultaneous optimization of orbitals and CI coefficients.

Chapter 13 discusses coupled-cluster theory. Important concepts such as connected and disconnected clusters, the exponential ansatz, and size-extensivity are discussed; the linked and unlinked equations of coupled-cluster theory are compared and the optimization of the wave function is described. Brueckner theory and orbital-optimized coupled-cluster theory are also discussed, as are the coupled-cluster variational Lagrangian and the equation-of-motion coupled-cluster model. A large section is devoted to the coupled-cluster singles-and-doubles (CCSD) model, whose working equations are derived in detail. A discussion of a spin-restricted open-shell formalism concludes the chapter.

Chapter 14 presents a discussion of perturbation theory in electronic-structure theory. Rayleigh–Schrödinger theory is first treated, covering topics such as Wigner's $2n + 1$ rule and the Hylleraas functional. Next, we consider an important special case of Rayleigh–Schrödinger theory – Møller–Plesset theory – deriving explicit expressions for the energies and wave functions. A convenient Lagrangian formulation of Møller–Plesset theory, in which only size-extensive terms occur, is then developed. The question of convergence is also examined, in particular the conditions under which the Møller–Plesset series is expected to diverge. This chapter concludes with a discussion of hybrid methods: the CCSD(T) model, in which the CCSD energy is perturbatively corrected for the lack of connected triples; and multiconfigurational perturbation theory, in which MCSCF energies are corrected for the effects of dynamical correlation.

In **Chapter 15**, we conclude our exposition of molecular electronic structure theory by applying the standard models of quantum chemistry to the calculation of a number of molecular properties: equilibrium structures, dipole moments, electronic energies, atomization energies, reaction enthalpies, and conformational energies. The performance of each model is examined, comparing with the other models and with experimental measurements. The large number of molecular systems considered enables us to carry out a statistical analysis of the different methods, making the investigations more forceful and reliable than would otherwise be the case. These studies demonstrate that rigorous electronic-structure theory has now advanced to such a level of sophistication that an accuracy comparable to or surpassing that of experimental measurements is routinely obtained for molecules containing five to ten first-row atoms.

PROGRAMS USED IN THE
PREPARATION OF THIS BOOK

DALTON: T. Helgaker, H. J. Aa. Jensen, P. Jørgensen, J. Olsen, K. Ruud, H. Ågren, T. Andersen, K. L. Bak, V. Bakken, O. Christiansen, P. Dahle, E. K. Dalskov, T. Enevoldsen, B. Fernandez, H. Heiberg, H. Hettema, D. Jonsson, S. Kirpekar, R. Kobayashi, H. Koch, K. V. Mikkelsen, P. Norman, M. J. Packer, T. Saue, P. R. Taylor and O. Vahtras, DALTON, an *ab initio* electronic structure program, Release 1.0, 1997.

DIRCCR12: J. Noga and W. Klopper, DIRCCR12, an explicitly correlated coupled-cluster program.

SORE: W. Klopper, SORE, a second-order R12-energy program.

ACES II: J. F. Stanton, J. Gauss, J. D. Watts, W. J. Lauderdale, R. J. Bartlett, ACES II, Quantum Theory Project, University of Florida, Gainesville, Florida, 1992. See also J. F. Stanton, J. Gauss, J. D. Watts, W. J. Lauderdale, R. J. Bartlett, *Int. J. Quantum Chem.* **S26**, 879 (1992). ACES II uses the VMOL integral and VPROPS property integral programs written by J. Almlöf and P. R. Taylor, a modified version of the integral derivative program ABACUS written by T. Helgaker, H. J. Aa. Jensen, P. Jørgensen, J. Olsen, and P. R. Taylor, and a geometry optimization and vibrational analysis package written by J. F. Stanton and D. E. Bernholdt.

LUCIA: J. Olsen, LUCIA, a CI program.

LUCAS: J. Olsen, LUCAS, an atomic-structure program.

2D: J. Kobus, L. Laaksonen, D. Sundholm, 2D, a numerical Hartree–Fock program for diatomic molecules.

MOLCAS4: K. Andersson, M. R. A. Blomberg, M. P. Fülscher, G. Karlström, R. Lindh, P.-Å. Malmqvist, P. Neogrady, J. Olsen, B. O. Roos, A. J. Sadlej, M. Schütz, L. Seijo, L. Serrano-Andrés, P. E. M. Siegbahn, P.-O. Widmark, MOLCAS Version 4, Lund University, Sweden (1997).

1 SECOND QUANTIZATION

In the standard formulation of quantum mechanics, observables are represented by operators and states by functions. In the language of second quantization, the wave functions are also expressed in terms of operators – the creation and annihilation operators working on the vacuum state. The antisymmetry of the electronic wave function follows from the algebra of these operators and requires no special attention. From the creation and annihilation operators, we construct also the standard operators of first-quantization quantum mechanics such as the Hamiltonian operator. This unified description of states and operators in terms of a single set of elementary creation and annihilation operators reduces much of the formal manipulation of quantum mechanics to algebra, allowing important relationships to be developed in an elegant manner. In this chapter, we develop the formalism of second quantization, laying the foundation for our subsequent treatment of molecular electronic structure.

1.1 The Fock space

Let $\{\phi_P(\mathbf{x})\}$ be a basis of M orthonormal *spin orbitals*, where the coordinates $\mathbf{x}$ represent collectively the spatial coordinates $\mathbf{r}$ and the spin coordinate σ of the electron. A *Slater determinant* is an antisymmetrized product of one or more spin orbitals. For example, a normalized Slater determinant for N electrons may be written as

$$|\phi_{P_1}\phi_{P_2}\cdots\phi_{P_N}| = \frac{1}{\sqrt{N!}} \begin{vmatrix} \phi_{P_1}(\mathbf{x}_1) & \phi_{P_2}(\mathbf{x}_1) & \cdots & \phi_{P_N}(\mathbf{x}_1) \\ \phi_{P_1}(\mathbf{x}_2) & \phi_{P_2}(\mathbf{x}_2) & \cdots & \phi_{P_N}(\mathbf{x}_2) \\ \vdots & \vdots & \ddots & \vdots \\ \phi_{P_1}(\mathbf{x}_N) & \phi_{P_2}(\mathbf{x}_N) & \cdots & \phi_{P_N}(\mathbf{x}_N) \end{vmatrix} \tag{1.1.1}$$

We now introduce an abstract linear vector space – the *Fock space* – where each determinant is represented by an *occupation-number (ON) vector* $|\mathbf{k}\rangle$,

$$|\mathbf{k}\rangle = |k_1, k_2, \ldots, k_M\rangle, k_P = \begin{cases} 1 & \phi_P \text{ occupied} \\ 0 & \phi_P \text{ unoccupied} \end{cases} \tag{1.1.2}$$

Thus, the *occupation number* k_P is 1 if ϕ_P is present in the determinant and 0 if it is absent. For an orthonormal set of spin orbitals, we define the *inner product* between two ON vectors $|\mathbf{k}\rangle$ and $|\mathbf{m}\rangle$ as

$$\langle\mathbf{k}|\mathbf{m}\rangle = \delta_{\mathbf{k},\mathbf{m}} = \prod_{P=1}^{M}\delta_{k_P m_P} \tag{1.1.3}$$

This definition is consistent with the overlap between two Slater determinants containing the same number of electrons. However, the extension of (1.1.3) to have a well-defined but zero overlap between states with different electron numbers is a special feature of the Fock-space formulation of quantum mechanics that allows for a unified description of systems with variable numbers of electrons.

In a given spin-orbital basis, there is a one-to-one mapping between the Slater determinants with spin orbitals in canonical order and the ON vectors in the Fock space. Much of the terminology for Slater determinants is therefore used for ON vectors as well. Still, the ON vectors are not Slater determinants – unlike the Slater determinants, the ON vectors have no spatial structure but are just basis vectors in an abstract vector space. This Fock space can be manipulated as an ordinary inner-product vector space. For example, for two general vectors or states in the Fock space

$$|\mathbf{c}\rangle = \sum_{\mathbf{k}} c_{\mathbf{k}} |\mathbf{k}\rangle \tag{1.1.4}$$

$$|\mathbf{d}\rangle = \sum_{\mathbf{k}} d_{\mathbf{k}} |\mathbf{k}\rangle \tag{1.1.5}$$

the inner product is given by

$$\langle \mathbf{c} | \mathbf{d} \rangle = \sum_{\mathbf{km}} c_{\mathbf{k}}^* \langle \mathbf{k} | \mathbf{m} \rangle d_{\mathbf{m}} = \sum_{\mathbf{k}} c_{\mathbf{k}}^* d_{\mathbf{k}} \tag{1.1.6}$$

The resolution of the identity likewise may be written in the usual manner as

$$1 = \sum_{\mathbf{k}} |\mathbf{k}\rangle \langle \mathbf{k}| \tag{1.1.7}$$

where the summation is over the full set of ON vectors for all numbers of electrons.

The ON vectors in (1.1.2) constitute an orthonormal basis in the 2^M-dimensional Fock space $F(M)$. This Fock space may be decomposed as a direct sum of subspaces $F(M, N)$

$$F(M) = F(M, 0) \oplus F(M, 1) \oplus \cdots \oplus F(M, M) \tag{1.1.8}$$

where $F(M, N)$ contains all ON vectors obtained by distributing N electrons among the M spin orbitals – that is, all ON vectors for which the sum of the occupation numbers is N:

$$N = \sum_{P=1}^{M} k_P \tag{1.1.9}$$

The subspace $F(M, 0)$, which consists of ON vectors with no electrons, contains a single vector – the true *vacuum state*

$$|\text{vac}\rangle = |0_1, 0_2, \ldots, 0_M\rangle \tag{1.1.10}$$

which, according to (1.1.3), is normalized to unity:

$$\langle \text{vac} | \text{vac} \rangle = 1 \tag{1.1.11}$$

Approximations to an exact N-electron wave function are expressed in terms of vectors in the Fock subspace $F(M, N)$ of dimension equal to the binomial coefficient $\binom{M}{N}$.

1.2 Creation and annihilation operators

In second quantization, all operators and states can be constructed from a set of elementary creation and annihilation operators. In this section we introduce these operators and explore their basic algebraic properties.

1.2.1 CREATION OPERATORS

The M elementary *creation operators* are defined by the relations

$$a_P^\dagger |k_1, k_2, \ldots, 0_P, \ldots, k_M\rangle = \Gamma_P^{\mathbf{k}} |k_1, k_2, \ldots, 1_P, \ldots, k_M\rangle \tag{1.2.1}$$

$$a_P^\dagger |k_1, k_2, \ldots, 1_P, \ldots, k_M\rangle = 0 \tag{1.2.2}$$

where

$$\Gamma_P^{\mathbf{k}} = \prod_{Q=1}^{P-1} (-1)^{k_Q} \tag{1.2.3}$$

The phase factor $\Gamma_P^{\mathbf{k}}$ is equal to $+1$ if there are an even number of electrons in the spin orbitals $Q < P$ (i.e. to the left of P in the ON vector) and equal to -1 if there are an odd number of electrons in these spin orbitals. As we shall see, this factor is necessary to obtain a representation of wave functions and operators consistent with first quantization. The requirement (1.2.2) that $a_P^\dagger$ produces zero when it operates on a vector with $k_P = 1$ is in agreement with the fact that a Slater determinant vanishes if a spin orbital appears twice.

The spin orbitals that are unoccupied in an ON vector (1.1.2) may be identified from the specification of the occupied spin orbitals. The explicit reference to the unoccupied spin orbitals may be avoided altogether by expressing the ON vector as a string of creation operators in the canonical order (i.e. in the same order as in the ON vector) working on the vacuum state:

$$|\mathbf{k}\rangle = \left[\prod_{P=1}^{M} (a_P^\dagger)^{k_P} \right] |\text{vac}\rangle \tag{1.2.4}$$

We shall later see that the phase factor (1.2.3) is automatically kept track of by the anticommutation relations of the creation operators in (1.2.4), making any reference to this factor unnecessary.

The properties of the creation operators can be deduced from the relations (1.2.1) and (1.2.2), which we here combine in a single defining equation:

$$a_P^\dagger |\mathbf{k}\rangle = \delta_{k_P 0} \Gamma_P^{\mathbf{k}} |k_1, \ldots, 1_P, \ldots, k_M\rangle \tag{1.2.5}$$

Operating twice with $a_P^\dagger$ on an ON vector, we obtain from (1.2.5)

$$a_P^\dagger a_P^\dagger |\mathbf{k}\rangle = a_P^\dagger \delta_{k_P 0} \Gamma_P^{\mathbf{k}} |k_1, \ldots, 1_P, \ldots, k_M\rangle = 0 \tag{1.2.6}$$

Since the product $a_P^\dagger a_P^\dagger$ gives zero when applied to any vector, it must be identical to the zero operator:

$$a_P^\dagger a_P^\dagger = 0 \tag{1.2.7}$$

For $P \neq Q$, the operators $a_P^\dagger$ and $a_Q^\dagger$ may act on an ON vector in two ways. For $Q > P$, we obtain

$$a_P^\dagger a_Q^\dagger |\ldots, k_P, \ldots, k_Q, \ldots\rangle = a_P^\dagger \delta_{k_Q 0} \Gamma_Q^{\mathbf{k}} |\ldots, k_P, \ldots, 1_Q, \ldots\rangle$$

$$= \delta_{k_P 0} \delta_{k_Q 0} \Gamma_P^{\mathbf{k}} \Gamma_Q^{\mathbf{k}} |\ldots, 1_P, \ldots, 1_Q, \ldots\rangle \tag{1.2.8}$$

where the phase factor for P is unaffected by the application of $a_Q^\dagger$ since P appears before Q in the ON vector. Reversing the order of the creation operators, we obtain

$$a_Q^\dagger a_P^\dagger |\ldots, k_P, \ldots, k_Q, \ldots\rangle = a_Q^\dagger \delta_{k_P 0} \Gamma_P^{\mathbf{k}} |\ldots, 1_P, \ldots, k_Q, \ldots\rangle$$

$$= \delta_{k_P 0} \delta_{k_Q 0} \Gamma_P^{\mathbf{k}} (-\Gamma_Q^{\mathbf{k}}) |\ldots, 1_P, \ldots, 1_Q, \ldots\rangle \qquad (1.2.9)$$

The factor $-\Gamma_Q^{\mathbf{k}}$ arises since $a_P^\dagger |\mathbf{k}\rangle$ – if it does not vanish – contains one more electron before spin orbital Q than does $|\mathbf{k}\rangle$. Adding together (1.2.8) and (1.2.9), we obtain

$$(a_P^\dagger a_Q^\dagger + a_Q^\dagger a_P^\dagger)|\mathbf{k}\rangle = 0 \qquad (1.2.10)$$

for $Q > P$. Substitution of dummy indices shows that this equation is valid also for $Q < P$. Finally, from (1.2.7) we see that it is true also for $P = Q$. Since $|\mathbf{k}\rangle$ is an arbitrary ON vector, we conclude that the anticommutation relation

$$a_P^\dagger a_Q^\dagger + a_Q^\dagger a_P^\dagger = [a_P^\dagger, a_Q^\dagger]_+ = 0 \qquad (1.2.11)$$

holds for any pair of creation operators.

1.2.2 ANNIHILATION OPERATORS

Having introduced the creation operators $a_P^\dagger$ of second quantization, we now proceed to the study of their Hermitian adjoints a_P. We shall see that the creation operators and their adjoints are distinct operators and consequently that these operators are not self-adjoint (Hermitian).

The properties of the adjoint or conjugate operators a_P can be inferred from those of the creation operators. Thus, from (1.2.11) the adjoint operators are seen to satisfy the anticommutation relation

$$a_P a_Q + a_Q a_P = [a_P, a_Q]_+ = 0 \qquad (1.2.12)$$

To determine the action of a_P on an ON vector $|\mathbf{k}\rangle$, we invoke the resolution of the identity (1.1.7):

$$a_P |\mathbf{k}\rangle = \sum_{\mathbf{m}} |\mathbf{m}\rangle \langle \mathbf{m}|a_P|\mathbf{k}\rangle \qquad (1.2.13)$$

The matrix element in this expression may be written as

$$\langle \mathbf{m}|a_P|\mathbf{k}\rangle = \langle \mathbf{k}|a_P^\dagger|\mathbf{m}\rangle^* = \begin{cases} \delta_{m_P 0} \Gamma_P^{\mathbf{m}} & \text{if } k_Q = m_Q + \delta_{QP} \\ 0 & \text{otherwise} \end{cases} \qquad (1.2.14)$$

where we have used (1.2.5). From the definition of $\Gamma_P^{\mathbf{k}}$ in (1.2.3) and from $k_Q = m_Q + \delta_{QP}$, we see that $\Gamma_P^{\mathbf{m}} = \Gamma_P^{\mathbf{k}}$. Equation (1.2.14) may therefore be written as

$$\langle \mathbf{m}|a_P|\mathbf{k}\rangle = \begin{cases} \delta_{k_P 1} \Gamma_P^{\mathbf{k}} & \text{if } m_Q = k_Q - \delta_{QP} \\ 0 & \text{otherwise} \end{cases} \qquad (1.2.15)$$

Hence, only one term in (1.2.13) survives and we conclude

$$a_P |\mathbf{k}\rangle = \delta_{k_P 1} \Gamma_P^{\mathbf{k}} |k_1, \ldots, 0_P, \ldots, k_M\rangle \qquad (1.2.16)$$

The operator a_P reduces k_P from 1 to 0 if spin orbital P is occupied and it gives 0 if the spin orbital is unoccupied. It is therefore called an electron *annihilation operator*. An interesting special case of (1.2.16) is

$$a_P|\text{vac}\rangle = 0 \qquad (1.2.17)$$

which states that there are no electrons to be destroyed in the vacuum state.

1.2.3 ANTICOMMUTATION RELATIONS

We have seen that the creation operators anticommute among themselves (1.2.11) and that the same is true for the annihilation operators (1.2.12). We shall now establish the commutation relations between creation and annihilation operators. Combining (1.2.5) and (1.2.16), we obtain

$$a_P^\dagger a_P|\mathbf{k}\rangle = \delta_{k_P 1}|\mathbf{k}\rangle \qquad (1.2.18)$$

$$a_P a_P^\dagger|\mathbf{k}\rangle = \delta_{k_P 0}|\mathbf{k}\rangle \qquad (1.2.19)$$

The phase factors cancel since they appear twice. Adding these equations together, we arrive at the following expression:

$$(a_P^\dagger a_P + a_P a_P^\dagger)|\mathbf{k}\rangle = (\delta_{k_P 1} + \delta_{k_P 0})|\mathbf{k}\rangle = |\mathbf{k}\rangle \qquad (1.2.20)$$

for any ON vector $|\mathbf{k}\rangle$. The operator $a_P^\dagger a_P + a_P a_P^\dagger$ is therefore equal to the identity operator

$$a_P^\dagger a_P + a_P a_P^\dagger = 1 \qquad (1.2.21)$$

For $P > Q$, we obtain

$$a_P^\dagger a_Q|\mathbf{k}\rangle = -\delta_{k_P 0}\delta_{k_Q 1}\Gamma_P^{\mathbf{k}}\Gamma_Q^{\mathbf{k}}|k_1, \ldots, 0_Q, \ldots, 1_P, \ldots, k_M\rangle, \quad P > Q \qquad (1.2.22)$$

$$a_Q a_P^\dagger|\mathbf{k}\rangle = \delta_{k_P 0}\delta_{k_Q 1}\Gamma_P^{\mathbf{k}}\Gamma_Q^{\mathbf{k}}|k_1, \ldots, 0_Q, \ldots, 1_P, \ldots, k_M\rangle, \quad P > Q \qquad (1.2.23)$$

where the minus sign arises since, in $a_Q|\mathbf{k}\rangle$, the number of occupied spin orbitals to the left of spin orbital P has been reduced by one. Adding these two equations together, we obtain

$$(a_P^\dagger a_Q + a_Q a_P^\dagger)|\mathbf{k}\rangle = 0, \quad P > Q \qquad (1.2.24)$$

Since $|\mathbf{k}\rangle$ is an arbitrary ON vector, we have the operator identity

$$a_P^\dagger a_Q + a_Q a_P^\dagger = 0, \quad P > Q \qquad (1.2.25)$$

The case $P < Q$ is obtained by taking the conjugate of this equation and renaming the dummy indices. Combination of (1.2.21) and (1.2.25) shows that, for all P and Q,

$$a_P^\dagger a_Q + a_Q a_P^\dagger = [a_P^\dagger, a_Q]_+ = \delta_{PQ} \qquad (1.2.26)$$

The *anticommutation relations* (1.2.11), (1.2.12), and (1.2.26) constitute the fundamental properties of the creation and annihilation operators. In view of their importance, they are here collected and listed in full:

$$[a_P^\dagger, a_Q^\dagger]_+ = 0 \qquad (1.2.27)$$

$$[a_P, a_Q]_+ = 0 \qquad (1.2.28)$$

$$[a_P^\dagger, a_Q]_+ = \delta_{PQ} \qquad (1.2.29)$$

From these simple relations, all other algebraic properties of the second-quantization formalism follow.

1.3 Number-conserving operators

The creation and annihilation operators introduced in Section 1.2 change the number of particles in a state and therefore couple ON vectors belonging to different subspaces $F(M, N)$. We now turn to operators that conserve the particle number and thus couple ON vectors in the same subspace.

1.3.1 OCCUPATION-NUMBER OPERATORS

We first introduce the *occupation-number (ON) operators* as

$$\hat{N}_P = a_P^\dagger a_P \tag{1.3.1}$$

The ON operator $\hat{N}_P$ counts the number of electrons in spin orbital P:

$$\hat{N}_P|\mathbf{k}\rangle = a_P^\dagger a_P|\mathbf{k}\rangle = \delta_{k_P 1}|\mathbf{k}\rangle = k_P|\mathbf{k}\rangle \tag{1.3.2}$$

Here we have used (1.2.18). The ON operators are Hermitian

$$\hat{N}_P^\dagger = (a_P^\dagger a_P)^\dagger = a_P^\dagger a_P = \hat{N}_P \tag{1.3.3}$$

and commute among themselves

$$\hat{N}_P\hat{N}_Q|\mathbf{k}\rangle = k_P k_Q|\mathbf{k}\rangle = k_Q k_P|\mathbf{k}\rangle = \hat{N}_Q\hat{N}_P|\mathbf{k}\rangle \tag{1.3.4}$$

The ON vectors are thus the simultaneous eigenvectors of the commuting set of Hermitian operators $\hat{N}_P$. Moreover, the set of ON operators is complete in the sense that there is a one-to-one mapping between the ON vectors in the Fock space and the eigenvalues of the ON operators. The eigenvalues of the ON operators characterize the ON vectors completely, consistent with the introduction of the ON vectors as an orthonormal basis for the Fock space.

In the spin-orbital basis, the ON operators are projection operators since, in addition to being Hermitian (1.3.3), they are also idempotent:

$$\hat{N}_P^2 = a_P^\dagger a_P a_P^\dagger a_P = a_P^\dagger(1 - a_P^\dagger a_P)a_P = a_P^\dagger a_P = \hat{N}_P \tag{1.3.5}$$

Here we have used the anticommutators (1.2.29) and (1.2.28) in that order. Applied to a linear combination of ON vectors (1.1.4), the operator $\hat{N}_P$ leaves unaffected vectors where ϕ_P is occupied and annihilates all others:

$$\hat{N}_P|\mathbf{c}\rangle = \sum_\mathbf{k} c_\mathbf{k}\hat{N}_P|\mathbf{k}\rangle = \sum_\mathbf{k} k_P c_\mathbf{k}|\mathbf{k}\rangle \tag{1.3.6}$$

Note that this property of the ON operators holds only in the spin-orbital basis where the occupations are either zero or one.

Using the basic anticommutation relations of creation and annihilation operators, we obtain for the commutators of the ON operators with the creation operators

$$\begin{aligned}
[\hat{N}_P, a_Q^\dagger] &= a_P^\dagger a_P a_Q^\dagger - a_Q^\dagger a_P^\dagger a_P \\
&= a_P^\dagger(\delta_{PQ} - a_Q^\dagger a_P) - a_Q^\dagger a_P^\dagger a_P \\
&= a_P^\dagger \delta_{PQ} - a_P^\dagger a_Q^\dagger a_P - a_Q^\dagger a_P^\dagger a_P \\
&= a_P^\dagger \delta_{PQ} + a_Q^\dagger a_P^\dagger a_P - a_Q^\dagger a_P^\dagger a_P
\end{aligned} \tag{1.3.7}$$

and therefore

$$[\hat{N}_P, a_Q^\dagger] = \delta_{PQ} a_Q^\dagger \tag{1.3.8}$$

Taking the conjugate of this equation, we obtain the corresponding commutator with the annihilation operator

$$[\hat{N}_P, a_Q] = -\delta_{PQ} a_Q \tag{1.3.9}$$

From these commutators, we may also conclude that, for an arbitrary string $\hat{X}$ of creation and annihilation operators such as

$$\hat{X} = a_P^\dagger a_Q a_P a_R^\dagger a_S \tag{1.3.10}$$

the commutators with the ON operators become

$$[\hat{N}_P, \hat{X}] = N_P^X \hat{X} \tag{1.3.11}$$

where N_P^X is the number of times $a_P^\dagger$ occurs in $\hat{X}$ minus the number of times a_P occurs in the same string. To arrive at relation (1.3.11), we have used the commutator expansion (1.8.5) of Section 1.8.

1.3.2 THE NUMBER OPERATOR

Adding together all ON operators in the Fock space, we obtain the Hermitian operator

$$\hat{N} = \sum_{P=1}^{M} a_P^\dagger a_P \tag{1.3.12}$$

which returns the number of electrons in an ON vector

$$\hat{N}|\mathbf{k}\rangle = \sum_{P=1}^{M} k_P |\mathbf{k}\rangle = N|\mathbf{k}\rangle \tag{1.3.13}$$

and therefore is known as the *particle-number operator* or simply the *number operator*. From (1.3.11), we see that the commutator of the number operator with an arbitrary string of operators is given by

$$[\hat{N}, \hat{X}] = N^X \hat{X} \tag{1.3.14}$$

where N^X is the excess of creation operators over annihilation operators in the string. In particular, we find that the number operator commutes with any string $\hat{T}$ that contains an equal number of creation and annihilation operators. Such strings are called *number-conserving*, since they conserve the number of particles in any vector:

$$\hat{N}\hat{T}|\mathbf{k}\rangle = \hat{T}\hat{N}|\mathbf{k}\rangle = \hat{T}N|\mathbf{k}\rangle = N\hat{T}|\mathbf{k}\rangle \tag{1.3.15}$$

In general, the application of the string $\hat{X}$ to a Fock-space vector increases the number of electrons by N^X.

1.3.3 EXCITATION OPERATORS

Apart from the particle-number operators (1.3.12), the simplest number-conserving operators are the elementary *excitation operators* $a_P^\dagger a_Q$, for which we shall occasionally use the notation

$$\hat{X}_Q^P = a_P^\dagger a_Q \tag{1.3.16}$$

Applied to an ON vector, these operators give (see Exercise 1.1)

$$a_P^\dagger a_Q |\mathbf{k}\rangle = \varepsilon_{PQ} \Gamma_P^{\mathbf{k}} \Gamma_Q^{\mathbf{k}} (1 - k_P + \delta_{PQ}) k_Q \left| \begin{matrix} k_P \to 1 \\ k_Q \to \delta_{PQ} \end{matrix} \right\rangle \tag{1.3.17}$$

where

$$\varepsilon_{PQ} = \begin{cases} 1 & P \le Q \\ -1 & P > Q \end{cases} \tag{1.3.18}$$

and where the ket on the right-hand side of (1.3.17) is an ON vector with the same occupation numbers as $|\mathbf{k}\rangle$ except as indicated for spin orbitals P and Q. Equation (1.3.17) shows that $a_P^\dagger a_Q$ excites an electron from spin orbital Q to spin orbital P, thus turning $|\mathbf{k}\rangle$ into another ON vector in the same subspace $F(M, N)$. In fact, each ON vector in $F(M, N)$ can be obtained from any other ON vector in the same subspace by applying a sequence of excitation operators $a_P^\dagger a_Q$. The application of a single such operator yields a single excitation, two operators give a double excitation, and so on. The 'diagonal' excitation operators $a_P^\dagger a_P$ correspond to the occupation-number operators (1.3.1).

In Box 1.1, we summarize the fundamentals of the second-quantization formalism. In Section 1.4, we proceed to discuss the second-quantization representation of standard first-quantization operators such as the electronic Hamiltonian.

Box 1.1   The fundamentals of second quantization

Occupation-number vectors:	$\lvert\mathbf{k}\rangle = \lvert k_1, k_2, \ldots, k_M\rangle, \quad k_P = 0, 1$
Inner products:	$\langle\mathbf{k}\lvert\mathbf{m}\rangle = \prod\limits_{P=1}^{M} \delta_{k_P m_P}$
Creation operators:	$a_P^\dagger \lvert\mathbf{k}\rangle = \delta_{k_P 0} \Gamma_P^{\mathbf{k}} \lvert k_1, \ldots, 1_P, \ldots, k_M\rangle$
	$\Gamma_P^{\mathbf{k}} = \prod\limits_{Q=1}^{P-1} (-1)^{k_Q}$
Annihilation operators:	$a_P \lvert\mathbf{k}\rangle = \delta_{k_P 1} \Gamma_P^{\mathbf{k}} \lvert k_1, \ldots, 0_P, \ldots, k_M\rangle$
Anticommutation relations:	$a_P^\dagger a_Q + a_Q a_P^\dagger = \delta_{PQ}$
	$a_P^\dagger a_Q^\dagger + a_Q^\dagger a_P^\dagger = 0$
	$a_P a_Q + a_Q a_P = 0$
Occupation-number operators:	$\hat{N}_P \lvert\mathbf{k}\rangle = a_P^\dagger a_P \lvert\mathbf{k}\rangle = k_P \lvert\mathbf{k}\rangle$
Particle-number operator:	$\hat{N} \lvert\mathbf{k}\rangle = \sum\limits_{P=1}^{M} \hat{N}_P \lvert\mathbf{k}\rangle = \sum\limits_{P=1}^{M} k_P \lvert\mathbf{k}\rangle$
Excitation operators:	$\hat{X}_Q^P = a_P^\dagger a_Q$
Vacuum state:	$\langle\text{vac}\lvert\text{vac}\rangle = 1$
	$a_P \lvert\text{vac}\rangle = 0$

1.4 The representation of one- and two-electron operators

Expectation values correspond to observables and should therefore be independent of the representation given to the operators and the states. Since expectation values may be expressed as sums of matrix elements of operators, we require the matrix element of a second-quantization operator between two ON vectors to be equal to its counterpart in first quantization. An operator in the Fock space can thus be constructed by requiring its matrix elements between ON vectors to be equal to the corresponding matrix elements between Slater determinants of the first-quantization operator.

Before proceeding to determine the form of the operators in second quantization, we recall that the matrix elements between Slater determinants depend on the spatial form of the spin orbitals. Since the ON vectors are independent of the spatial form of spin orbitals, we conclude that the second-quantization operators – in contrast to their first-quantization counterparts – must depend on the spatial form of the spin orbitals.

First-quantization operators conserve the number of electrons. Following the discussion in Section 1.3, such operators are in the Fock space represented by linear combinations of operators that contain an equal number of creation and annihilation operators. The explicit form of these number-conserving operators depends on whether the first-quantized operator is a one-electron operator or a two-electron operator. One-electron operators are discussed in Section 1.4.1 and two-electron operators in Section 1.4.2. Finally, in Section 1.4.3 we consider the second-quantization representation of the electronic Hamiltonian operator.

1.4.1 ONE-ELECTRON OPERATORS

In first quantization, *one-electron operators* are written as

$$f^c = \sum_{i=1}^{N} f^c(\mathbf{x}_i) \tag{1.4.1}$$

where the summation is over the N electrons of the system. Superscript c indicates that we are working in the coordinate representation of first quantization. Since each term in the operator (1.4.1) involves a single electron, this operator gives a vanishing matrix element whenever the Slater determinants differ in more than one pair of spin orbitals. The second-quantization analogue of (1.4.1) therefore has the structure

$$\hat{f} = \sum_{PQ} f_{PQ} a_P^\dagger a_Q \tag{1.4.2}$$

since the excitation operators $a_P^\dagger a_Q$ shift a single electron in an ON vector. The summation is over all pairs of spin orbitals to secure the highest possible flexibility in the description. The order of the creation and annihilation operators in each term ensures that the one-electron operator $\hat{f}$ produces zero when it works on the vacuum state.

To determine the numerical parameters f_{PQ} in (1.4.2), we evaluate the matrix elements of $\hat{f}$ between two ON vectors and compare with the usual *Slater–Condon rules* for matrix elements between determinants [1]. For one-electron operators there are three distinct cases.

1. The ON vectors are identical:

$$\langle \mathbf{k} | \hat{f} | \mathbf{k} \rangle = \sum_P f_{PP} \langle \mathbf{k} | a_P^\dagger a_P | \mathbf{k} \rangle = \sum_P k_P f_{PP} \tag{1.4.3}$$

2. The ON vectors differ in one pair of occupation numbers:

$$|\mathbf{k}_1\rangle = |k_1, \ldots, 0_I, \ldots, 1_J, \ldots, k_M\rangle \qquad (1.4.4)$$

$$|\mathbf{k}_2\rangle = |k_1, \ldots, 1_I, \ldots, 0_J, \ldots, k_M\rangle \qquad (1.4.5)$$

$$\langle \mathbf{k}_2 | \hat{f} | \mathbf{k}_1 \rangle = \Gamma_I^{\mathbf{k}_2} \Gamma_J^{\mathbf{k}_1} f_{IJ} \qquad (1.4.6)$$

3. The ON vectors differ in more than one pair of occupation numbers:

$$\langle \mathbf{k}_2 | \hat{f} | \mathbf{k}_1 \rangle = 0 \qquad (1.4.7)$$

In these expressions, we have used indices I and J for the spin orbitals with different occupations in the bra and ket vectors.

Let us see how the above results are obtained. For the diagonal element

$$\langle \mathbf{k} | \hat{f} | \mathbf{k} \rangle = \sum_{PQ} f_{PQ} \langle \mathbf{k} | a_P^\dagger a_Q | \mathbf{k} \rangle \qquad (1.4.8)$$

we note from the orthogonality of ON vectors that nonzero contributions can only arise when

$$a_P | \mathbf{k} \rangle = \pm a_Q | \mathbf{k} \rangle \qquad (1.4.9)$$

which occurs for $P = Q$ only. This observation gives (1.4.3). For the ON vectors $|\mathbf{k}_1\rangle$ and $|\mathbf{k}_2\rangle$ in (1.4.6), the transition-matrix element

$$\langle \mathbf{k}_2 | \hat{f} | \mathbf{k}_1 \rangle = \sum_{PQ} f_{PQ} \langle \mathbf{k}_2 | a_P^\dagger a_Q | \mathbf{k}_1 \rangle \qquad (1.4.10)$$

has nonvanishing contributions only if

$$a_P | \mathbf{k}_2 \rangle = \pm a_Q | \mathbf{k}_1 \rangle \qquad (1.4.11)$$

This requirement is only fulfilled for $P = I$ and $Q = J$. Substitution of these values in (1.4.10) gives (1.4.6). Finally, the matrix element (1.4.7) vanishes trivially since the one-electron operator (1.4.2) can change only one pair of occupation numbers.

Comparing with the Slater–Condon rules for one-electron operators, we note that the second-quantization matrix elements (1.4.3), (1.4.6) and (1.4.7) will agree with their first-quantization counterparts if we make the identification

$$f_{PQ} = \int \phi_P^*(\mathbf{x}) f^c(\mathbf{x}) \phi_Q(\mathbf{x}) \, d\mathbf{x} \qquad (1.4.12)$$

The recipe for constructing a second-quantization representation of a one-electron operator is therefore to use (1.4.2) with the integrals (1.4.12). For real spin orbitals, the integrals exhibit the following permutational symmetry

$$f_{PQ} = f_{QP} \qquad \text{(real spin orbitals)} \qquad (1.4.13)$$

In conclusion, we note that the phase factors (1.2.3) were necessary to reproduce the Slater–Condon rules for matrix elements between Slater determinants.

1.4.2 TWO-ELECTRON OPERATORS

In first quantization, *two-electron operators* such as the electronic-repulsion operator are given by the expression

$$g^c = \tfrac{1}{2} \sum_{i \neq j} g^c(\mathbf{x}_i, \mathbf{x}_j) \tag{1.4.14}$$

Other examples of two-electron operators are the two-electron part of the spin–orbit operator and the mass-polarization operator. A two-electron operator gives nonvanishing matrix elements between Slater determinants if the determinants contain at least two electrons and if they differ in the occupations of at most two pairs of electrons. The second-quantization representation of a two-electron operator therefore has the structure

$$\hat{g} = \tfrac{1}{2} \sum_{PQRS} g_{PQRS} a_P^\dagger a_R^\dagger a_S a_Q \tag{1.4.15}$$

The annihilation operators appear to the right of the creation operators in order to ensure that $\hat{g}$ gives zero when it works on an ON vector with less than two electrons. The factor of one-half in (1.4.15) is conventional. Anticommuting the creation and annihilation operators and renaming the dummy indices, we obtain

$$\sum_{PQRS} g_{PQRS} a_P^\dagger a_R^\dagger a_S a_Q = \sum_{PQRS} g_{PQRS} a_R^\dagger a_P^\dagger a_Q a_S = \sum_{PQRS} g_{RSPQ} a_P^\dagger a_R^\dagger a_S a_Q \tag{1.4.16}$$

The parameters g_{PQRS} may therefore be chosen in a symmetric fashion

$$g_{PQRS} = g_{RSPQ} \tag{1.4.17}$$

The numerical values of the parameters g_{PQRS} may be determined by evaluating the matrix element of $\hat{g}$ between two ON vectors and setting the result equal to the matrix element between the corresponding Slater determinants. There are four cases.

1. The ON vectors are identical:

$$\langle \mathbf{k} | \hat{g} | \mathbf{k} \rangle = \tfrac{1}{2} \sum_{PQRS} g_{PQRS} \langle \mathbf{k} | a_P^\dagger a_R^\dagger a_S a_Q | \mathbf{k} \rangle = \tfrac{1}{2} \sum_{PR} k_P k_R (g_{PPRR} - g_{PRRP}) \tag{1.4.18}$$

2. The ON vectors differ in one pair of occupation numbers:

$$|\mathbf{k}_1\rangle = |k_1, \ldots, 0_I, \ldots, 1_J, \ldots, k_M\rangle \tag{1.4.19}$$

$$|\mathbf{k}_2\rangle = |k_1, \ldots, 1_I, \ldots, 0_J, \ldots, k_M\rangle \tag{1.4.20}$$

$$\langle \mathbf{k}_2 | \hat{g} | \mathbf{k}_1 \rangle = \Gamma_I^{\mathbf{k}_2} \Gamma_J^{\mathbf{k}_1} \sum_R k_R (g_{IJRR} - g_{IRRJ}) \tag{1.4.21}$$

3. The ON vectors differ in two pairs of occupation numbers:

$$|\mathbf{k}_1\rangle = |k_1, \ldots, 0_I, \ldots, 0_J, \ldots, 1_K, \ldots, 1_L, \ldots, k_M\rangle \tag{1.4.22}$$

$$|\mathbf{k}_2\rangle = |k_1, \ldots, 1_I, \ldots, 1_J, \ldots, 0_K, \ldots, 0_L, \ldots, k_M\rangle \tag{1.4.23}$$

where $I < J$ and $K < L$

$$\langle \mathbf{k}_2 | \hat{g} | \mathbf{k}_1 \rangle = \Gamma_I^{\mathbf{k}_2} \Gamma_J^{\mathbf{k}_2} \Gamma_K^{\mathbf{k}_1} \Gamma_L^{\mathbf{k}_1} (g_{IKJL} - g_{ILJK}) \tag{1.4.24}$$

4. The ON vectors differ in more than two pairs of occupation numbers:

$$\langle \mathbf{k}_2 | \hat{g} | \mathbf{k}_1 \rangle = 0 \tag{1.4.25}$$

We have in these expressions used the indices I, J, K and L for the spin orbitals with different occupations in the bra and ket vectors.

Let us consider the derivation of these matrix elements in some detail. The diagonal element

$$\langle \mathbf{k} | \hat{g} | \mathbf{k} \rangle = \frac{1}{2} \sum_{PQRS} \langle \mathbf{k} | a_P^\dagger a_R^\dagger a_S a_Q | \mathbf{k} \rangle g_{PQRS} \tag{1.4.26}$$

has nonvanishing contributions if

$$a_R a_P | \mathbf{k} \rangle = \pm a_S a_Q | \mathbf{k} \rangle \tag{1.4.27}$$

This condition holds in two different cases

$$P = Q \text{ and } R = S$$

$$P = S \text{ and } Q = R \tag{1.4.28}$$

If both sets of relations are fulfilled, then the expectation value of the creation and annihilation operators vanishes. We may therefore write the diagonal matrix element in the form

$$\langle \mathbf{k} | \hat{g} | \mathbf{k} \rangle = \frac{1}{2} \sum_{PR} \langle \mathbf{k} | a_P^\dagger a_R^\dagger a_R a_P | \mathbf{k} \rangle g_{PPRR} + \frac{1}{2} \sum_{PR} \langle \mathbf{k} | a_P^\dagger a_R^\dagger a_P a_R | \mathbf{k} \rangle g_{PRRP}$$

$$= \frac{1}{2} \sum_{PR} \langle \mathbf{k} | a_P^\dagger a_R^\dagger a_R a_P | \mathbf{k} \rangle (g_{PPRR} - g_{PRRP}) \tag{1.4.29}$$

From the definition of the ON operators (1.3.1) and the commutator (1.3.9), we obtain

$$a_P^\dagger a_R^\dagger a_R a_P = a_P^\dagger \hat{N}_R a_P = -\delta_{PR} \hat{N}_P + \hat{N}_P \hat{N}_R \tag{1.4.30}$$

The diagonal element (1.4.29) therefore becomes

$$\langle \mathbf{k} | \hat{g} | \mathbf{k} \rangle = \frac{1}{2} \sum_{PR} \langle \mathbf{k} | - \delta_{PR} \hat{N}_P + \hat{N}_P \hat{N}_R | \mathbf{k} \rangle (g_{PPRR} - g_{PRRP})$$

$$= \frac{1}{2} \sum_{PR} k_P k_R (g_{PPRR} - g_{PRRP}) \tag{1.4.31}$$

Next, we consider the case where the ON vectors (1.4.19) and (1.4.20) differ in the occupation numbers of one pair of spin orbitals. The matrix element

$$\langle \mathbf{k}_2 | \hat{g} | \mathbf{k}_1 \rangle = \frac{1}{2} \sum_{PQRS} \langle \mathbf{k}_2 | a_P^\dagger a_R^\dagger a_S a_Q | \mathbf{k}_1 \rangle g_{PQRS} \tag{1.4.32}$$

has nonvanishing contributions if

$$a_R a_P | \mathbf{k}_2 \rangle = \pm a_S a_Q | \mathbf{k}_1 \rangle \tag{1.4.33}$$

This condition holds in four different cases:

$$P = I, \ Q = J \text{ and } R = S$$

$$P = I, \ S = J \text{ and } R = Q$$

$$R = I, \ Q = J \text{ and } P = S$$

$$R = I, \ S = J \text{ and } P = Q \tag{1.4.34}$$

Since the matrix element vanishes if several of the above sets of relations hold, we obtain

$$\langle \mathbf{k}_2 | \hat{g} | \mathbf{k}_1 \rangle = \tfrac{1}{2} \sum_R \langle \mathbf{k}_2 | a_I^\dagger a_R^\dagger a_R a_J | \mathbf{k}_1 \rangle g_{IJRR} + \tfrac{1}{2} \sum_R \langle \mathbf{k}_2 | a_I^\dagger a_R^\dagger a_J a_R | \mathbf{k}_1 \rangle g_{IRRJ}$$

$$+ \tfrac{1}{2} \sum_P \langle \mathbf{k}_2 | a_P^\dagger a_I^\dagger a_P a_J | \mathbf{k}_1 \rangle g_{PJIP} + \tfrac{1}{2} \sum_P \langle \mathbf{k}_2 | a_P^\dagger a_I^\dagger a_J a_P | \mathbf{k}_1 \rangle g_{PPIJ} \qquad (1.4.35)$$

Invoking the permutational symmetry (1.4.17) and the elementary anticommutation relations, we arrive at the final expression (1.4.21)

$$\langle \mathbf{k}_2 | \hat{g} | \mathbf{k}_1 \rangle = \sum_R \langle \mathbf{k}_2 | a_I^\dagger a_R^\dagger a_R a_J | \mathbf{k}_1 \rangle (g_{IJRR} - g_{IRRJ})$$

$$= \sum_R \langle \mathbf{k}_2 | a_I^\dagger a_J a_R^\dagger a_R | \mathbf{k}_1 \rangle (g_{IJRR} - g_{IRRJ})$$

$$= \Gamma_I^{\mathbf{k}_2} \Gamma_J^{\mathbf{k}_1} \sum_R k_R (g_{IJRR} - g_{IRRJ}) \qquad (1.4.36)$$

It does not matter whether the occupation numbers in this expression refer to $|\mathbf{k}_1\rangle$ or $|\mathbf{k}_2\rangle$ since the contributions vanish whenever the occupations differ. The matrix element between ON vectors differing in two pairs of occupations (1.4.24) can be treated in the same way and is left as an exercise.

The two-electron second-quantization matrix elements (1.4.18), (1.4.21), (1.4.24) and (1.4.25) become identical to the corresponding first-quantization elements obtained from the Slater–Condon rules if we choose

$$g_{PQRS} = \iint \phi_P^*(\mathbf{x}_1) \phi_R^*(\mathbf{x}_2) g^c(\mathbf{x}_1, \mathbf{x}_2) \phi_Q(\mathbf{x}_1) \phi_S(\mathbf{x}_2) \, d\mathbf{x}_1 \, d\mathbf{x}_2 \qquad (1.4.37)$$

The recipe for constructing a two-electron second-quantization operator is therefore given by expressions (1.4.15) and (1.4.37). For any interaction between identical particles, the operator $g^c(\mathbf{x}_1, \mathbf{x}_2)$ is symmetric in $\mathbf{x}_1$ and $\mathbf{x}_2$. The integrals (1.4.37) therefore automatically exhibit the permutational symmetry in (1.4.17). We also note the following useful permutational symmetries for real spin orbitals:

$$g_{PQRS} = g_{QPRS} = g_{PQSR} = g_{QPSR} \qquad \text{(real spin orbitals)} \qquad (1.4.38)$$

Thus, for real spin orbitals there are a total of eight permutational symmetries (1.4.17) and (1.4.38) present in the two-electron integrals, whereas for complex spin orbitals there is only one such symmetry (1.4.17).

1.4.3 THE MOLECULAR ELECTRONIC HAMILTONIAN

Combining the results of Sections 1.4.1 and 1.4.2, we may now construct the full second-quantization representation of the electronic Hamiltonian operator in the Born–Oppenheimer approximation. Although not strictly needed for the development of the second-quantization theory in this chapter, we present the detailed form of this operator as an example of the construction of operators in second quantization. In the absence of external fields, the second-quantization nonrelativistic and spin-free *molecular electronic Hamiltonian* is given by

$$\hat{H} = \sum_{PQ} h_{PQ} a_P^\dagger a_Q + \tfrac{1}{2} \sum_{PQRS} g_{PQRS} a_P^\dagger a_R^\dagger a_S a_Q + h_{\text{nuc}} \qquad (1.4.39)$$

where in atomic units (which are always used in this book unless otherwise stated)

$$h_{PQ} = \int \phi_P^*(\mathbf{x}) \left(-\frac{1}{2}\nabla^2 - \sum_I \frac{Z_I}{r_I} \right) \phi_Q(\mathbf{x}) \, d\mathbf{x} \qquad (1.4.40)$$

$$g_{PQRS} = \int\int \frac{\phi_P^*(\mathbf{x}_1)\phi_R^*(\mathbf{x}_2)\phi_Q(\mathbf{x}_1)\phi_S(\mathbf{x}_2)}{r_{12}} \, d\mathbf{x}_1 \, d\mathbf{x}_2 \qquad (1.4.41)$$

$$h_{\text{nuc}} = \frac{1}{2} \sum_{I \neq J} \frac{Z_I Z_J}{R_{IJ}} \qquad (1.4.42)$$

Here the Z_I are the nuclear charges, r_I the electron–nuclear separations, r_{12} the electron–electron separation and R_{IJ} the internuclear separations. The summations are over all nuclei. The scalar term (1.4.42) represents the nuclear-repulsion energy – it is simply added to the Hamiltonian and makes the same contribution to matrix elements as in first quantization since the inner product of two ON vectors is identical to the overlap of the determinants. The molecular one- and two-electron integrals (1.4.40) and (1.4.41) may be calculated using the techniques described in Chapter 9.

The form of the second-quantization Hamiltonian (1.4.39) may be interpreted in the following way. Applied to an electronic state, the Hamiltonian produces a linear combination of the original state with states generated by single and double electron excitations from this state. With each such excitation, there is an associated 'amplitude' h_{PQ} or g_{PQRS}, which represents the probability of this event happening. These probability amplitudes are calculated from the spin orbitals and the one- and two-electron operators according to (1.4.40) and (1.4.41).

1.5 Products of operators in second quantization

In the preceding section, we constructed second-quantization operators for one- and two-electron operators in such a way that the same matrix elements and hence the same expectation values are obtained in the first and second quantizations. Since the expectation values are the only observables in quantum mechanics, we have arrived at a new representation of many-electron systems with the same physical contents as the standard first-quantization representation. In the present section, we examine this new tool in greater detail by comparing the first- and second-quantization representations of operators. In particular, we show that, for operator products $A^c B^c = P^c$, the second-quantization representation of P^c may differ from the product of the second-quantization representations of A^c and B^c unless a complete basis is used.

1.5.1 OPERATOR PRODUCTS

Let A^c and B^c be two one-electron operators in first quantization

$$A^c = \sum_i A^c(\mathbf{x}_i) \qquad (1.5.1)$$

$$B^c = \sum_i B^c(\mathbf{x}_i) \qquad (1.5.2)$$

and let $\hat{A}$ and $\hat{B}$ be the corresponding second-quantization representations

$$\hat{A} = \sum_{PQ} A_{PQ} a_P^\dagger a_Q \qquad (1.5.3)$$

$$\hat{B} = \sum_{PQ} B_{PQ} a_P^\dagger a_Q \tag{1.5.4}$$

From the construction of the second-quantization operators, it is clear that the first-quantization operator $aA^c + bB^c$, where a and b are numbers, is represented by $a\hat{A} + b\hat{B}$. The standard relations

$$\hat{A}(\hat{B}\hat{C}) = (\hat{A}\hat{B})\hat{C} \tag{1.5.5}$$

$$(\hat{A}\hat{B})^\dagger = \hat{B}^\dagger \hat{A}^\dagger \tag{1.5.6}$$

for linear operators in a linear vector space are also valid.

We now consider the representation of *operator products*. The product of the two first-quantization operators $A^c B^c$ can be separated into one- and two-electron parts

$$P^c = A^c B^c = O^c + T^c \tag{1.5.7}$$

where

$$O^c = \sum_i A^c(\mathbf{x}_i) B^c(\mathbf{x}_i) \tag{1.5.8}$$

$$T^c = \tfrac{1}{2} \sum_{i \neq j} [A^c(\mathbf{x}_i) B^c(\mathbf{x}_j) + A^c(\mathbf{x}_j) B^c(\mathbf{x}_i)] \tag{1.5.9}$$

The two-electron operator is written so that it is symmetric with respect to permutation of the particle indices. This symmetrization is necessary since we know only how to generate the second-quantization representation of two-electron operators that are symmetric in the particles.

The second-quantization representation $\hat{P}$ of P^c is the sum of the second-quantization representations of O^c and T^c:

$$\hat{P} = \sum_{PQ} O_{PQ} a_P^\dagger a_Q + \tfrac{1}{2} \sum_{PQRS} T_{PQRS} a_P^\dagger a_R^\dagger a_S a_Q \tag{1.5.10}$$

where

$$O_{PQ} = \int \phi_P^*(\mathbf{x}) A^c(\mathbf{x}) B^c(\mathbf{x}) \phi_Q(\mathbf{x}) \, d\mathbf{x} \tag{1.5.11}$$

$$T_{PQRS} = \int\int \phi_P^*(\mathbf{x}) \phi_R^*(\mathbf{x}') [A^c(\mathbf{x}) B^c(\mathbf{x}') + A^c(\mathbf{x}') B^c(\mathbf{x})] \phi_Q(\mathbf{x}) \phi_S(\mathbf{x}') \, d\mathbf{x} \, d\mathbf{x}'$$
$$= A_{PQ} B_{RS} + A_{RS} B_{PQ} \tag{1.5.12}$$

Inserting this expression for the two-electron parameters in the two-electron part of $\hat{P}$, we obtain

$$\hat{T} = \tfrac{1}{2} \sum_{PQRS} (A_{PQ} B_{RS} + A_{RS} B_{PQ}) a_P^\dagger a_R^\dagger a_S a_Q = \sum_{PQRS} A_{PQ} B_{RS} a_P^\dagger a_R^\dagger a_S a_Q \tag{1.5.13}$$

by substitution of dummy indices. Using the anticommutation relations, we may rewrite this expression as

$$\hat{T} = \sum_{PQRS} A_{PQ}B_{RS}(a_P^\dagger a_Q a_R^\dagger a_S - \delta_{RQ}a_P^\dagger a_S)$$

$$= \left(\sum_{PQ} A_{PQ}a_P^\dagger a_Q\right)\left(\sum_{RS} B_{RS}a_R^\dagger a_S\right) - \sum_{PS}\left(\sum_R A_{PR}B_{RS}\right)a_P^\dagger a_S$$

$$= \hat{A}\hat{B} - \sum_{PQ}\left(\sum_R A_{PR}B_{RQ}\right)a_P^\dagger a_Q \tag{1.5.14}$$

Inserting (1.5.14) in (1.5.10), we finally arrive at the expression

$$\hat{P} = \hat{A}\hat{B} + \sum_{PQ}\left(O_{PQ} - \sum_R A_{PR}B_{RQ}\right)a_P^\dagger a_Q \tag{1.5.15}$$

which shows that the second-quantization representation of $A^c B^c$ is in general not equal to the product of the representations of A^c and B^c.

We shall now demonstrate that the last term in (1.5.15) vanishes for a complete basis. We use the *Dirac delta function* $\delta(\mathbf{x} - \mathbf{x}')$, defined by the relationship [2,3]

$$\int f(\mathbf{x})\delta(\mathbf{x} - \mathbf{x}')\,d\mathbf{x} = f(\mathbf{x}') \tag{1.5.16}$$

For a complete one-electron basis, the delta function may be written in the form

$$\delta(\mathbf{x} - \mathbf{x}') = \sum_{P=1}^{\infty} \phi_P(\mathbf{x})\phi_P^*(\mathbf{x}') \qquad \text{(complete basis)} \tag{1.5.17}$$

Assuming that (1.5.17) holds, we may now write

$$O_{PQ} = \int \phi_P^*(\mathbf{x})A^c(\mathbf{x})B^c(\mathbf{x})\phi_Q(\mathbf{x})\,d\mathbf{x}$$

$$= \int\int \phi_P^*(\mathbf{x})A^c(\mathbf{x})\delta(\mathbf{x} - \mathbf{x}')B^c(\mathbf{x}')\phi_Q(\mathbf{x}')\,d\mathbf{x}\,d\mathbf{x}'$$

$$= \int\int \phi_P^*(\mathbf{x})A^c(\mathbf{x})\left[\sum_{R=1}^{\infty}\phi_R(\mathbf{x})\phi_R^*(\mathbf{x}')\right]B^c(\mathbf{x}')\phi_Q(\mathbf{x}')\,d\mathbf{x}\,d\mathbf{x}'$$

$$= \sum_{R=1}^{\infty} A_{PR}B_{RQ} \qquad \text{(complete basis)} \tag{1.5.18}$$

and the last term in (1.5.15) vanishes. We therefore have for a complete one-electron basis

$$P^c = A^c B^c \Rightarrow \hat{P} = \hat{A}\hat{B} \qquad \text{(complete basis)} \tag{1.5.19}$$

but for finite basis sets this expression does not hold.

The second-quantization operators are projections of the exact operators onto a basis of spin orbitals. For an incomplete basis, the second-quantization representation of an operator product therefore depends on when the projection is made. For a complete basis, however, the representation is exact and independent of when the projection is made.

1.5.2 THE CANONICAL COMMUTATORS

The previous discussion suggests that commutation relations that hold for operators in first quantization, do not necessarily hold for their second-quantization counterparts in a finite basis. Consider the *canonical commutators*

$$[r_\alpha^c, p_\beta^c] = i\delta_{\alpha\beta}^c \tag{1.5.20}$$

where we have contributions from each of the N electrons

$$r_\alpha^c = \sum_{i=1}^N r_\alpha^c(i) \tag{1.5.21}$$

$$p_\alpha^c = \sum_{i=1}^N p_\alpha^c(i) \tag{1.5.22}$$

$$\delta_{\alpha\beta}^c = \sum_{i=1}^N \delta_{\alpha\beta}(i) \tag{1.5.23}$$

and Greek letters denote Cartesian directions. The relationship (1.5.20) holds exactly for first-quantization operators. Note carefully that the operator (1.5.23) depends on the number of electrons and is not the usual Kronecker delta. The second-quantization representations of the position and momentum operators are

$$\hat{r}_\alpha = \sum_{PQ} [r_\alpha^c]_{PQ} a_P^\dagger a_Q \tag{1.5.24}$$

$$\hat{p}_\alpha = \sum_{PQ} [p_\alpha^c]_{PQ} a_P^\dagger a_Q \tag{1.5.25}$$

and the commutator of these operators becomes

$$[\hat{r}_\alpha, \hat{p}_\beta] = \sum_{PQRS} [r_\alpha^c]_{PQ}[p_\beta^c]_{RS}[a_P^\dagger a_Q, a_R^\dagger a_S] \tag{1.5.26}$$

In these expressions, square brackets around a first-quantization operator represent the one-electron integral of this operator in the given basis. This somewhat cumbersome notation is adopted for this discussion to make the dependence of the integrals on the first-quantization operators explicit.

In Section 1.8, the commutator between the two excitation operators is shown to be

$$[a_P^\dagger a_Q, a_R^\dagger a_S] = \delta_{RQ} a_P^\dagger a_S - \delta_{PS} a_R^\dagger a_Q \tag{1.5.27}$$

and the commutator (1.5.26) therefore reduces to

$$[\hat{r}_\alpha, \hat{p}_\beta] = \sum_{PQ} a_P^\dagger a_Q \sum_R ([r_\alpha^c]_{PR}[p_\beta^c]_{RQ} - [p_\beta^c]_{PR}[r_\alpha^c]_{RQ}) \tag{1.5.28}$$

For a complete basis, we may use (1.5.18) and arrive at the following simplifications:

$$\sum_{R=1}^\infty [r_\alpha^c]_{PR}[p_\beta^c]_{RQ} = [r_\alpha^c p_\beta^c]_{PQ} \quad \text{(complete basis)} \tag{1.5.29}$$

$$\sum_{R=1}^{\infty} [p_{\beta}^{c}]_{PR} [r_{\alpha}^{c}]_{RQ} = [p_{\beta}^{c} r_{\alpha}^{c}]_{PQ} \quad \text{(complete basis)} \tag{1.5.30}$$

The second-quantization canonical commutator therefore becomes proportional to the number operator in the limit of a complete basis:

$$[\hat{r}_{\alpha}, \hat{p}_{\beta}] = \sum_{P,Q=1}^{\infty} [[r_{\alpha}^{c}, p_{\beta}^{c}]]_{PQ} a_{P}^{\dagger} a_{Q} = i \sum_{P,Q=1}^{\infty} [\delta_{\alpha\beta}^{c}]_{PQ} a_{P}^{\dagger} a_{Q} = i\delta_{\alpha\beta} \hat{N} \quad \text{(complete basis)} \tag{1.5.31}$$

This expression should be compared with its first-quantization counterpart (1.5.20). For finite basis sets, the second-quantization canonical commutator turns into a general one-electron operator (1.5.28).

1.6 First- and second-quantization operators compared

In Box 1.2, we summarize some of the characteristics of operators in the first and second quantizations. The dependence on the spin-orbital basis is different in the two representations. In first quantization, the Slater determinants depend on the spin-orbital basis whereas the operators are independent of the spin orbitals. In the second-quantization formalism, the ON vectors are basis vectors in a linear vector space and contain no reference to the spin-orbital basis. Instead, the reference to the spin-orbital basis is made in the operators. We also note that, whereas the first-quantization operators depend explicitly on the number of electrons, no such dependence is found in the second-quantization operators.

Box 1.2   First- and second-quantization operators compared

First quantization	Second quantization
→ one-electron operator: $$\sum_{i} f^{c}(\mathbf{x}_i)$$	→ one-electron operator: $$\sum_{PQ} f_{PQ} a_{P}^{\dagger} a_{Q}$$
→ two-electron operator: $$\frac{1}{2} \sum_{i \neq j} g^{c}(\mathbf{x}_i, \mathbf{x}_j)$$	→ two-electron operator: $$\frac{1}{2} \sum_{PQRS} g_{PQRS} a_{P}^{\dagger} a_{R}^{\dagger} a_{S} a_{Q}$$
→ operators are independent of the spin-orbital basis	→ operators depend on the spin-orbital basis
→ operators depend on the number of electrons	→ operators are independent of the number of electrons
→ exact operators	→ projected operators

The fact that the second-quantization operators are projections of the exact operators onto the spin-orbital basis means that a second-quantization operator times an ON vector is just another vector in the Fock space. By contrast, a first-quantization operator times a determinant cannot be expanded as a sum of Slater determinants in a finite basis. This fact often goes unnoticed since, in the first quantization, we usually work directly with matrix elements.

The projected nature of the second-quantization operators has many ramifications. For example, relations that hold for exact operators such as the canonical commutation properties of the coordinate and momentum operators do not necessarily hold for projected operators. Similarly, the projected coordinate operator does not commute with the projected Coulomb repulsion operator. It should be emphasized, however, that these problems are not peculiar to second quantization but arise whenever a finite basis is employed. They also arise in first quantization, but not until the matrix elements are evaluated.

Second quantization treats operators and wave functions in a unified way – they are all expressed in terms of the elementary creation and annihilation operators. This property of the second-quantization formalism can, for example, be exploited to express modifications to the wave function as changes in the operators. To illustrate the unified description of states and operators afforded by second quantization, we note that any ON vector may be written compactly as a string of creation operators working on the vacuum state (1.2.4)

$$|\mathbf{k}\rangle = \hat{X}_{\mathbf{k}}|\text{vac}\rangle = \left[\prod_{P=1}^{M}(a_P^\dagger)^{k_P}\right]|\text{vac}\rangle \tag{1.6.1}$$

Matrix elements may therefore be viewed as the *vacuum expectation value* of an operator

$$\langle\mathbf{k}|\hat{O}|\mathbf{m}\rangle = \langle\text{vac}|\hat{X}_{\mathbf{k}}^\dagger\hat{O}\hat{X}_{\mathbf{m}}|\text{vac}\rangle \tag{1.6.2}$$

and expectation values become linear combinations of vacuum expectation values. The unified description of states and operators in terms of the elementary creation and annihilation operators enables us to carry out most of our manipulations algebraically based on the anticommutation relations of these operators. Thus, the antisymmetry of the electronic wave function follows automatically from the algebra of the elementary operators without the need to keep track of phase factors.

1.7 Density matrices

Having considered the representation of states and operators in second quantization, let us now turn our attention to expectation values. As in first quantization, the evaluation of expectation values is carried out by means of density matrices [4]. Consider a general one- and two-electron Hermitian operator in the spin-orbital basis

$$\hat{\Omega} = \sum_{PQ}\Omega_{PQ}a_P^\dagger a_Q + \frac{1}{2}\sum_{PQRS}\Omega_{PQRS}a_P^\dagger a_R^\dagger a_S a_Q + \Omega_0 \tag{1.7.1}$$

The *expectation value* of this operator with respect to a normalized reference state $|0\rangle$ written as a linear combination of ON vectors,

$$|0\rangle = \sum_{\mathbf{k}} c_{\mathbf{k}}|\mathbf{k}\rangle \tag{1.7.2}$$

$$\langle 0|0\rangle = 1 \tag{1.7.3}$$

is given by the expression

$$\langle 0|\hat{\Omega}|0\rangle = \sum_{PQ} \overline{D}_{PQ} \Omega_{PQ} + \tfrac{1}{2} \sum_{PQRS} \overline{d}_{PQRS} \Omega_{PQRS} + \Omega_0 \tag{1.7.4}$$

where we have introduced the matrix elements

$$\overline{D}_{PQ} = \langle 0|a_P^\dagger a_Q|0\rangle \tag{1.7.5}$$

$$\overline{d}_{PQRS} = \langle 0|a_P^\dagger a_R^\dagger a_S a_Q|0\rangle \tag{1.7.6}$$

Clearly, all information that is required about the wave function (1.7.2) for the evaluation of expectation values (1.7.4) is embodied in the quantities (1.7.5) and (1.7.6) called the *one- and two-electron density-matrix elements*, respectively. Overbars are used for the spin-orbital densities to distinguish these from those that will be introduced in Chapter 2 for the orbital basis. Since the density elements play such an important role in electronic-structure theory, it is appropriate here to examine their properties within the framework of second quantization.

1.7.1   THE ONE-ELECTRON DENSITY MATRIX

The densities (1.7.5) constitute the elements of an $M \times M$ Hermitian matrix – the *one-electron spin-orbital density matrix* $\overline{\mathbf{D}}$ – since the following relation is satisfied:

$$\overline{D}_{PQ}^* = \langle 0|a_P^\dagger a_Q|0\rangle^* = \langle 0|a_Q^\dagger a_P|0\rangle = \overline{D}_{QP} \tag{1.7.7}$$

For real wave functions, the matrix is symmetric:

$$\overline{D}_{PQ} = \overline{D}_{QP} \quad \text{(real wave functions)} \tag{1.7.8}$$

The one-electron density matrix is positive semidefinite since its elements are either trivially equal to zero or inner products of states in the subspace $F(M, N-1)$. The diagonal elements of the spin-orbital density matrix are the expectation values of the occupation-number operators (1.3.1) in $F(M, N)$ and are referred to as the *occupation numbers* $\overline{\omega}_P$ of the electronic state:

$$\overline{\omega}_P = \overline{D}_{PP} = \langle 0|\hat{N}_P|0\rangle \tag{1.7.9}$$

This terminology is appropriate since the diagonal elements of $\overline{\mathbf{D}}$ reduce to the usual occupation numbers k_P in (1.3.2) whenever the reference state is an eigenfunction of the ON operators – that is, when the reference state is an ON vector:

$$\langle \mathbf{k}|\hat{N}_P|\mathbf{k}\rangle = k_P \tag{1.7.10}$$

Since the ON operators are projectors (1.3.5), we may write the occupation numbers in the form

$$\overline{\omega}_P = \langle 0|\hat{N}_P \hat{N}_P|0\rangle \tag{1.7.11}$$

where the projected electronic state is given by

$$\hat{N}_P|0\rangle = \sum_{\mathbf{k}} k_P c_{\mathbf{k}}|\mathbf{k}\rangle \tag{1.7.12}$$

The occupation numbers $\overline{\omega}_P$ may now be written in the form

$$\overline{\omega}_P = \sum_{\mathbf{k}} k_P |c_{\mathbf{k}}|^2 \tag{1.7.13}$$

and interpreted as the squared norm of the part of the reference state where spin orbital ϕ_P is occupied in each ON vector. The occupation numbers $\overline{\omega}_P$ thus serve as indicators of the importance of the spin orbitals in the electronic state.

The expansion coefficients satisfy the normalization condition

$$\sum_{\mathbf{k}} |c_{\mathbf{k}}|^2 = 1 \qquad (1.7.14)$$

Recalling that the occupation numbers k_P of an ON vector are zero or one, we conclude that the occupation numbers $\overline{\omega}_P$ of an electronic state (1.7.13) are real numbers between zero and one – zero for spin orbitals that are unoccupied in all ON vectors, one for spin orbitals that are occupied in all ON vectors, and nonintegral for spin orbitals that are occupied in some but not all ON vectors:

$$0 \leq \overline{\omega}_P \leq 1 \qquad (1.7.15)$$

We also note that the sum of the occupation numbers (i.e. the trace of the density matrix) is equal to the total number of electrons in the system:

$$\text{Tr}\overline{\mathbf{D}} = \sum_P \overline{\omega}_P = \sum_P \langle 0|\hat{N}_P|0\rangle = \langle 0|\hat{N}|0\rangle = N \qquad (1.7.16)$$

Here we have used the definition of the particle-number operator (1.3.12).

For a state consisting of a single ON vector, the one-electron spin-orbital density matrix has a simple diagonal structure:

$$\overline{D}_{PQ}^{\mathbf{k}} = \langle \mathbf{k}|a_P^\dagger a_Q|\mathbf{k}\rangle = \delta_{PQ}k_P \qquad (1.7.17)$$

By contrast, for an electronic state containing several ON vectors, the density matrix is not diagonal. Applying the Schwarz inequality in the $(N-1)$-electron space, we obtain

$$|\langle 0|a_P^\dagger a_Q|0\rangle|^2 \leq \langle 0|a_P^\dagger a_P|0\rangle \langle 0|a_Q^\dagger a_Q|0\rangle \qquad (1.7.18)$$

which gives us an upper bound to the magnitude of the elements of the spin-orbital density matrix equal to the geometric mean of the occupation numbers:

$$|\overline{D}_{PQ}| \leq \sqrt{\overline{\omega}_P \overline{\omega}_Q} \qquad (1.7.19)$$

Of course, since $\overline{\mathbf{D}}$ is a Hermitian matrix, we may eliminate the off-diagonal elements (1.7.19) completely by diagonalization with a unitary matrix:

$$\overline{\mathbf{D}} = \mathbf{U}\overline{\boldsymbol{\eta}}\mathbf{U}^\dagger \qquad (1.7.20)$$

The eigenvalues are real numbers $0 \leq \overline{\eta}_P \leq 1$, known as *the natural-orbital occupation numbers*. The sum of the natural-orbital occupation numbers is again equal to the number of electrons in the system. From the eigenvectors $\mathbf{U}$ of the density matrix, we obtain a new set of spin orbitals called the *natural spin orbitals* of the system. However, we defer the discussion of unitary orbital transformations to Chapter 3.

1.7.2 THE TWO-ELECTRON DENSITY MATRIX

We now turn our attention to the *two-electron density matrix*. We begin by noting that the two-electron density-matrix elements (1.7.6) are not all independent because of the anticommutation relations between the creation operators and between the annihilation operators:

$$\bar{d}_{PQRS} = -\bar{d}_{RQPS} = -\bar{d}_{PSRQ} = \bar{d}_{RSPQ} \qquad (1.7.21)$$

The following elements are therefore zero in accordance with the Pauli principle:

$$\bar{d}_{PQPS} = \bar{d}_{PQRQ} = \bar{d}_{PQPQ} = 0 \qquad (1.7.22)$$

To avoid these redundancies in our representation, we introduce the two-electron density matrix $\mathbf{\bar{T}}$ with elements given by

$$\bar{T}_{PQ,RS} = \langle 0|a_P^\dagger a_Q^\dagger a_S a_R|0\rangle, \qquad P > Q, R > S \qquad (1.7.23)$$

There are $M(M-1)/2$ rows and columns in this matrix with composite indices PQ such that $P > Q$. The elements of $\mathbf{\bar{T}}$ constitute a subset of the two-electron density elements (1.7.6) and differ from these by a reordering of the middle indices:

$$\bar{T}_{PQ,RS} = \bar{d}_{PRQS}, \qquad P > Q, R > S \qquad (1.7.24)$$

The reason for introducing this reordering is that it allows us to examine the two-electron density matrix by analogy with the discussion of the one-electron density in Section 1.7.1. Thus, as in the one-electron case, we note that the two-electron density matrix $\mathbf{\bar{T}}$ is Hermitian

$$\bar{T}_{PQ,RS}^* = \langle 0|a_P^\dagger a_Q^\dagger a_S a_R|0\rangle^* = \langle 0|a_R^\dagger a_S^\dagger a_Q a_P|0\rangle = \bar{T}_{RS,PQ} \qquad (1.7.25)$$

and therefore symmetric for real wave functions

$$\bar{T}_{PQ,RS} = \bar{T}_{RS,PQ} \qquad \text{(real wave functions)} \qquad (1.7.26)$$

Also, the two-electron density matrix $\mathbf{\bar{T}}$ is positive semidefinite since its elements are either trivially equal to zero or inner products of states in $F(M, N-2)$.

We recall that the diagonal elements of $\mathbf{\bar{D}}$ correspond to expectation values of ON operators (1.7.9) and are interpreted as the occupation numbers of the spin orbitals. We now examine the diagonal elements of the two-electron density matrix:

$$\bar{T}_{PQ,PQ} = \langle 0|a_P^\dagger a_Q^\dagger a_Q a_P|0\rangle \qquad (1.7.27)$$

Since $P > Q$, we may anticommute a_P from the fourth to the second position, introduce ON operators, and arrive at the following expression analogous to (1.7.9):

$$\bar{\omega}_{PQ} = \bar{T}_{PQ,PQ} = \langle 0|\hat{N}_P \hat{N}_Q|0\rangle \qquad (1.7.28)$$

Since the ON operators are projectors, we may now interpret the diagonal elements $\bar{\omega}_{PQ}$ as *simultaneous occupations of pairs of spin orbitals (pair occupations)*, noting that $\bar{\omega}_{PQ}$ represents the squared amplitude of the part of the wave function where spin orbitals ϕ_P and ϕ_Q are simultaneously occupied:

$$\hat{N}_P \hat{N}_Q|0\rangle = \sum_\mathbf{k} k_P k_Q c_\mathbf{k}|\mathbf{k}\rangle \qquad (1.7.29)$$

The norm of the wave function is successively reduced by the repeated application of ON projectors; compare (1.7.12) and (1.7.29). This observation agrees with the expectation that the simultaneous occupation of a given spin-orbital pair cannot exceed those of the individual spin orbitals:

$$0 \leq \overline{\omega}_{PQ} \leq \min(\overline{\omega}_P, \overline{\omega}_Q) \leq 1 \tag{1.7.30}$$

The reader may wish to verify that a weaker upper bound to the pair occupations

$$\overline{\omega}_{PQ} \leq \sqrt{\overline{\omega}_P \overline{\omega}_Q} \tag{1.7.31}$$

is arrived at by the application of the Schwarz inequality to (1.7.28). From the trace of the two-electron density matrix

$$\mathrm{Tr}\,\overline{\mathbf{T}} = \sum_{P>Q} \langle 0|\hat{N}_P\hat{N}_Q|0\rangle = \tfrac{1}{2}\sum_{PQ} \langle 0|\hat{N}_P\hat{N}_Q|0\rangle - \tfrac{1}{2}\sum_{P} \langle 0|\hat{N}_P|0\rangle$$

$$= \tfrac{1}{2}\langle 0|\hat{N}^2 - \hat{N}|0\rangle = \tfrac{1}{2}N(N-1) \tag{1.7.32}$$

we find that the sum of all pair occupations $\overline{\omega}_{PQ}$ is equal to the number of electron pairs in the system in the same way that the sum of all single occupations $\overline{\omega}_P$ is equal to the number of electrons in the system (1.7.16). We may summarize these results by stating that the one-electron density matrix probes the individual occupancies of the spin orbitals and describes how the N electrons are distributed among the M spin orbitals, whereas the two-electron density matrix probes the simultaneous occupations of the spin orbitals and describes how the $N(N-1)/2$ electron pairs are distributed among the $M(M-1)/2$ spin-orbital pairs.

For a state containing a single ON vector, the two-electron density matrix has a particularly simple diagonal structure with the following elements

$$\overline{T}^{\mathbf{k}}_{PQ,RS} = \langle \mathbf{k}|a_P^\dagger a_Q^\dagger a_S a_R|\mathbf{k}\rangle = \delta_{PR}\delta_{QS}k_P k_Q \tag{1.7.33}$$

recalling the conditions $P > Q$ and $R > S$. Indeed, for such electronic states, the two-electron density matrix $\overline{T}$ may be constructed directly from the one-electron density matrix

$$\overline{T}^{\mathbf{k}}_{PQ,RS} = \overline{D}^{\mathbf{k}}_{PR}\overline{D}^{\mathbf{k}}_{QS} \tag{1.7.34}$$

and likewise the expectation value of any one- and two-electron operator may be obtained directly from the one-electron density matrix. This observation is consistent with our picture of ON vectors (i.e. determinants) as representing an uncorrelated description of the electronic system where the simultaneous occupations of pairs of spin orbitals are just the products of the individual occupations.

For a general electronic state, containing more than one ON vector and providing a correlated treatment of the electronic system, the two-electron density matrix is in general not diagonal and cannot be generated directly from the one-electron density elements. As in the one-electron case (1.7.19), we may invoke the Schwarz inequality to establish an upper bound to the magnitude of the off-diagonal elements

$$|\overline{T}_{PQ,RS}| \leq \sqrt{\overline{\omega}_{PQ}\overline{\omega}_{RS}} \tag{1.7.35}$$

and we may in principle diagonalize $\overline{\mathbf{T}}$ to obtain a more compact representation of the two-electron density, but this is seldom done in practice.

1.7.3 DENSITY MATRICES IN SPIN-ORBITAL AND COORDINATE REPRESENTATIONS

The density matrices we have discussed so far in this section have all been given in the spin-orbital representation. We shall now see how these matrices are represented in the coordinate representation of quantum mechanics. Since we have not here developed a second-quantization formalism appropriate for the coordinate representation, we shall draw on the equivalence between matrix elements in the first and second quantizations to establish the relationship between density matrices in the spin-orbital and coordinate representations.

We recall that, in the coordinate representation of first quantization, we may write the expectation value of any one-electron operator in the following form

$$\langle \Psi | f^c | \Psi \rangle = \int [f^c(\mathbf{x}_1)\gamma_1(\mathbf{x}_1, \mathbf{x}_1')]_{\mathbf{x}_1'=\mathbf{x}_1} \, d\mathbf{x}_1 \tag{1.7.36}$$

in terms of the *first-order reduced density matrix*

$$\gamma_1(\mathbf{x}_1, \mathbf{x}_1') = N \int \Psi(\mathbf{x}_1, \mathbf{x}_2, \ldots, \mathbf{x}_N)\Psi^*(\mathbf{x}_1', \mathbf{x}_2, \ldots, \mathbf{x}_N) \, d\mathbf{x}_2 \cdots d\mathbf{x}_N \tag{1.7.37}$$

The density matrix in the spin-orbital representation was introduced in second quantization for the evaluation of one-electron expectation values in the following form

$$\langle 0 | \hat{f} | 0 \rangle = \sum_{PQ} \overline{D}_{PQ} f_{PQ} \tag{1.7.38}$$

where the integrals are those given in (1.4.12):

$$f_{PQ} = \int \phi_P^*(\mathbf{x}_1) f^c(\mathbf{x}_1) \phi_Q(\mathbf{x}_1) \, d\mathbf{x}_1 \tag{1.7.39}$$

Combining (1.7.38) and (1.7.39), we obtain

$$\langle 0 | \hat{f} | 0 \rangle = \sum_{PQ} \overline{D}_{PQ} \int \phi_P^*(\mathbf{x}_1) f^c(\mathbf{x}_1) \phi_Q(\mathbf{x}_1) \, d\mathbf{x}_1$$

$$= \int \left[f^c(\mathbf{x}_1) \sum_{PQ} \overline{D}_{PQ} \phi_P^*(\mathbf{x}_1') \phi_Q(\mathbf{x}_1) \right]_{\mathbf{x}_1'=\mathbf{x}_1} \, d\mathbf{x}_1 \tag{1.7.40}$$

and we are therefore able to make the identification

$$\gamma_1(\mathbf{x}_1, \mathbf{x}_1') = \sum_{PQ} \overline{D}_{PQ} \phi_P^*(\mathbf{x}_1') \phi_Q(\mathbf{x}_1) \tag{1.7.41}$$

involving the spin-orbital density matrix $\overline{\mathbf{D}}$ and the first-order reduced density matrix $\gamma_1(\mathbf{x}_1, \mathbf{x}_1')$. Since the spin orbitals are orthonormal, we obtain the following expression after multiplication by spin orbitals and integration

$$\langle 0 | a_P^\dagger a_Q | 0 \rangle = \int \phi_Q^*(\mathbf{x}_1) \gamma_1(\mathbf{x}_1, \mathbf{x}_1') \phi_P(\mathbf{x}_1') \, d\mathbf{x}_1 \, d\mathbf{x}_1' \tag{1.7.42}$$

where we have used (1.7.5). This equation displays the relationship between the density matrices in the spin-orbital representation and in the coordinate representation. We may establish similar relationships for the two-electron densities. Thus, introducing the *second-order reduced density matrix*

$$\gamma_2(\mathbf{x}_1, \mathbf{x}_2, \mathbf{x}_1', \mathbf{x}_2') = \frac{N(N-1)}{2} \int \Psi(\mathbf{x}_1, \mathbf{x}_2, \mathbf{x}_3, \ldots, \mathbf{x}_N)\Psi^*(\mathbf{x}_1', \mathbf{x}_2', \mathbf{x}_3, \ldots, \mathbf{x}_N) \, d\mathbf{x}_3 \cdots d\mathbf{x}_N$$

$$\tag{1.7.43}$$

we obtain the following relationships

$$\gamma_2(\mathbf{x}_1, \mathbf{x}_2, \mathbf{x}_1', \mathbf{x}_2') = \tfrac{1}{2} \sum_{PQRS} \bar{d}_{PQRS} \phi_P^*(\mathbf{x}_1') \phi_Q(\mathbf{x}_1) \phi_R^*(\mathbf{x}_2') \phi_S(\mathbf{x}_2) \tag{1.7.44}$$

$$\langle 0 | a_P^\dagger a_R^\dagger a_S a_Q | 0 \rangle = 2 \int \phi_Q^*(\mathbf{x}_1) \phi_S^*(\mathbf{x}_2) \gamma_2(\mathbf{x}_1, \mathbf{x}_2, \mathbf{x}_1', \mathbf{x}_2') \phi_P(\mathbf{x}_1') \phi_R(\mathbf{x}_2') \, \mathrm{d}\mathbf{x}_1 \, \mathrm{d}\mathbf{x}_2 \, \mathrm{d}\mathbf{x}_1' \, \mathrm{d}\mathbf{x}_2'$$
$$\tag{1.7.45}$$

analogous to (1.7.41) and (1.7.42). The simple algebraic definitions for the densities in second quantization (as expectation values of excitation operators) should be contrasted with the more complicated expression in terms of the reduced densities.

1.8 Commutators and anticommutators

In the manipulation of operators and matrix elements in second quantization, the *commutator*

$$[\hat{A}, \hat{B}] = \hat{A}\hat{B} - \hat{B}\hat{A} \tag{1.8.1}$$

and the *anticommutator*

$$[\hat{A}, \hat{B}]_+ = \hat{A}\hat{B} + \hat{B}\hat{A} \tag{1.8.2}$$

of two operators are often encountered. The elementary creation and annihilation operators satisfy the anticommutation relations (1.2.27)–(1.2.29). Referring to these basic relations, it is usually possible to simplify the commutators and anticommutators between strings of elementary operators considerably. Since manipulations in second quantization frequently involve complicated operator strings, it is important to establish a good strategy for the evaluation of commutators and anticommutators of such strings.

Before considering the evaluation and simplification of commutators and anticommutators, it is useful to introduce the concepts of operator rank and rank reduction. The *(particle) rank* of a string of creation and annihilation operators is simply the number of elementary operators divided by 2. For example, the rank of a creation operator is 1/2 and the rank of an ON operator is 1. *Rank reduction* is said to occur when the rank of a commutator or anticommutator is lower than the combined rank of the operators commuted or anticommuted. Consider the basic anticommutation relation

$$a_P^\dagger a_P + a_P a_P^\dagger = 1 \tag{1.8.3}$$

Here the combined rank of the creation and annihilation operators is 1 whereas the rank of the anticommutator itself is 0. Anticommutation thus reduces the rank by 1.

Rank reduction is desirable since it simplifies the expression. It would clearly be useful if we were able to distinguish at a glance those commutators and anticommutators that reduce the operator rank from those that do not. The following simple rule is sufficient for this purpose: *Rank reduction follows upon anticommutation of two strings of half-integral rank and upon commutation of all other strings.* In Exercise 1.2, rank reduction is demonstrated for a commutator of two strings of integral rank. The remaining cases may be proved in the same manner. The basic anticommutation relations of the elementary operators (1.2.27)–(1.2.29) are the prototypes of rank reduction for strings of half-integral rank.

We now return to the evaluation of commutators and anticommutators. One useful strategy for the evaluation of such expressions is based on their linear expansion in simpler commutators or anticommutators according to the operator identities

$$[\hat{A}, \hat{B}_1\hat{B}_2] = [\hat{A}, \hat{B}_1]\hat{B}_2 + \hat{B}_1[\hat{A}, \hat{B}_2] \tag{1.8.4}$$

$$[\hat{A}, \hat{B}_1 \cdots \hat{B}_n] = \sum_{k=1}^{n} \hat{B}_1 \cdots \hat{B}_{k-1}[\hat{A}, \hat{B}_k]\hat{B}_{k+1} \cdots \hat{B}_n \tag{1.8.5}$$

$$[\hat{A}, \hat{B}_1\hat{B}_2] = [\hat{A}, \hat{B}_1]_+\hat{B}_2 - \hat{B}_1[\hat{A}, \hat{B}_2]_+ \tag{1.8.6}$$

$$[\hat{A}, \hat{B}_1 \cdots \hat{B}_n] = \sum_{k=1}^{n} (-1)^{k-1}\hat{B}_1 \cdots [\hat{A}, \hat{B}_k]_+ \cdots \hat{B}_n \quad (n \text{ even}) \tag{1.8.7}$$

$$[\hat{A}, \hat{B}_1\hat{B}_2]_+ = [\hat{A}, \hat{B}_1]\hat{B}_2 + \hat{B}_1[\hat{A}, \hat{B}_2]_+ = [\hat{A}, \hat{B}_1]_+\hat{B}_2 - \hat{B}_1[\hat{A}, \hat{B}_2] \tag{1.8.8}$$

$$[\hat{A}, \hat{B}_1 \cdots \hat{B}_n]_+ = \sum_{k=1}^{n} (-1)^{k-1}\hat{B}_1 \cdots [\hat{A}, \hat{B}_k]_+ \cdots \hat{B}_n \quad (n \text{ odd}) \tag{1.8.9}$$

Note that the commutators and anticommutators on the right-hand side of these expressions contain fewer operators than does the commutator or the anticommutator on the left-hand side. For proofs, see Exercise 1.3. In deciding what identity to apply in a given case, we follow the 'principle of rank reduction' – that is, we try to expand the expression in commutators or anticommutators that, to the greatest extent possible, exhibit rank reduction.

Let us consider the simplest nontrivial commutator – $[a_P^\dagger, a_Q^\dagger a_R]$. Moving one of the operators in $a_Q^\dagger a_R$ outside the commutator according to the identities (1.8.4) or (1.8.6), we are left either with commutators or with anticommutators of two elementary operators – strings of half-integral rank. In order to reduce the rank and thus the number of operators, we choose the anticommutator expansion (1.8.6):

$$[a_P^\dagger, a_Q^\dagger a_R] = [a_P^\dagger, a_Q^\dagger]_+ a_R - a_Q^\dagger [a_P^\dagger, a_R]_+ \tag{1.8.10}$$

The basic anticommutation relations can now be invoked to give

$$[a_P^\dagger, a_Q^\dagger a_R] = -\delta_{PR} a_Q^\dagger \tag{1.8.11}$$

Proceeding in the same manner, we also obtain

$$[a_P, a_Q^\dagger a_R] = \delta_{PQ} a_R \tag{1.8.12}$$

but we note that this expression is perhaps more easily obtained by conjugating (1.8.11).

The commutator relations (1.8.11) and (1.8.12) may now be used to simplify more complicated commutators. Invoking (1.8.4), we obtain for the commutator between two excitation operators

$$[a_P^\dagger a_Q, a_R^\dagger a_S] = [a_P^\dagger, a_R^\dagger a_S]a_Q + a_P^\dagger[a_Q, a_R^\dagger a_S] \tag{1.8.13}$$

This commutator is expanded in commutators rather than anticommutators since one of the operators on the right-hand side contains two elementary operators. Inserting (1.8.11) and (1.8.12) in this expression, we obtain

$$[a_P^\dagger a_Q, a_R^\dagger a_S] = \delta_{QR} a_P^\dagger a_S - \delta_{PS} a_R^\dagger a_Q \tag{1.8.14}$$

We can now proceed to even more complicated commutators. The proof of the following relationship is left as an exercise:

$$[a_P^\dagger a_Q, a_R^\dagger a_S a_M^\dagger a_N] = \delta_{QR} a_P^\dagger a_S a_M^\dagger a_N - \delta_{PS} a_R^\dagger a_Q a_M^\dagger a_N + \delta_{QM} a_R^\dagger a_S a_P^\dagger a_N - \delta_{PN} a_R^\dagger a_S a_M^\dagger a_Q \tag{1.8.15}$$

The preceding examples should suffice to illustrate the usefulness of the expansions (1.8.4)–(1.8.9) for evaluating commutators and anticommutators involving elementary operators.

Let us finally consider *nested commutators*. Nested commutators may be simplified by the same techniques as the single commutators, thus giving rise to rank reductions greater than 1. For example, the following *double commutator* is easily evaluated using (1.8.14):

$$[a_P^\dagger a_Q, [a_R^\dagger a_S, a_M^\dagger a_N]] = \delta_{SM}\delta_{QR}a_P^\dagger a_N - \delta_{SM}\delta_{PN}a_R^\dagger a_Q - \delta_{RN}\delta_{QM}a_P^\dagger a_S + \delta_{RN}\delta_{PS}a_M^\dagger a_Q \quad (1.8.16)$$

The particle rank is reduced by 2. In manipulating and simplifying nested commutators, the *Jacobi identity* is often useful:

$$[\hat{A}, [\hat{B}, \hat{C}]] + [\hat{C}, [\hat{A}, \hat{B}]] + [\hat{B}, [\hat{C}, \hat{A}]] = 0 \quad (1.8.17)$$

This identity is easily verified by expansion. Note that the Jacobi identity is different from the expansions (1.8.4)–(1.8.9) in that it relates expressions of the same rank and structure.

1.9 Nonorthogonal spin orbitals

Our discussion so far has been concerned with the development of the second-quantization formalism for an orthonormal basis. Occasionally, however, we shall find it more convenient to work with spin orbitals that are not orthogonal. We therefore extend the formalism of second quantization to deal with such spin orbitals, drawing heavily on the development in the preceding sections.

Consider a set of *nonorthogonal spin orbitals* with overlap

$$S_{PQ} = \int \phi_P^*(\mathbf{x})\phi_Q(\mathbf{x})\,d\mathbf{x} \quad (1.9.1)$$

A Fock space for these spin orbitals can now be constructed as an abstract vector space using ON vectors as basis vectors in much the same way as for orthonormal spin orbitals. The inner product of the Fock space is defined such that, for vectors with the same number of electrons, it is equal to the overlap between the corresponding Slater determinants. For vectors with different particle numbers, the inner product is zero. The inner product is thus given by

$$\langle \mathbf{k}|\mathbf{m}\rangle = \delta_{N_k N_m} \det \mathbf{S}^{\mathbf{km}} \quad (1.9.2)$$

where N_k and N_m are the numbers of electrons in the ON vectors. $\mathbf{S}^{\mathbf{km}}$ is the matrix of overlap integrals between the spin orbitals occupied in the ON vectors and $\det \mathbf{S}^{\mathbf{km}}$ represents the determinant of this matrix.

1.9.1 CREATION AND ANNIHILATION OPERATORS

The creation operators $a_P^\dagger$ for nonorthogonal spin orbitals are defined in the same way as for orthonormal spin orbitals (1.2.5). As for orthonormal spin orbitals, the anticommutation relations of the creation operators and the properties of their Hermitian adjoints (the annihilation operators) may be deduced from the definition of the creation operators and from the inner product (1.9.2). However, it is easier to proceed in the following manner. We introduce an auxiliary set of symmetrically orthonormalized spin orbitals

$$\tilde{\phi} = \phi \mathbf{S}^{-1/2} \quad (1.9.3)$$

such that

$$\tilde{S}_{PQ} = \int \tilde{\phi}_P^*(\mathbf{x})\tilde{\phi}_Q(\mathbf{x})\,d\mathbf{x} = [\mathbf{S}^{-1/2}\mathbf{S}\mathbf{S}^{-1/2}]_{PQ} = \delta_{PQ} \tag{1.9.4}$$

For these auxiliary spin orbitals, we next introduce a set of operators that satisfy the usual anti-commutation relations for orthonormal spin orbitals. As discussed in Section 3.2, the relationship between the old and new creation operators is similar to the relationship between the old and new orbitals (1.9.3) and becomes

$$\tilde{a}_P^\dagger = \sum_Q a_Q^\dagger [\mathbf{S}^{-1/2}]_{QP} \tag{1.9.5}$$

$$a_P^\dagger = \sum_Q \tilde{a}_Q^\dagger [\mathbf{S}^{1/2}]_{QP} \tag{1.9.6}$$

The relationships between the annihilation operators are found by conjugating these expressions

$$\tilde{a}_P = \sum_Q a_Q [\mathbf{S}^{-1/2}]_{PQ} \tag{1.9.7}$$

$$a_P = \sum_Q \tilde{a}_Q [\mathbf{S}^{1/2}]_{PQ} \tag{1.9.8}$$

We may now evaluate the anticommutators for the nonorthogonal creation operators

$$[a_P^\dagger, a_Q^\dagger]_+ = \sum_{RS}[\tilde{a}_R^\dagger[\mathbf{S}^{1/2}]_{RP}, \tilde{a}_S^\dagger[\mathbf{S}^{1/2}]_{SQ}]_+ = \sum_{RS}[\mathbf{S}^{1/2}]_{RP}[\mathbf{S}^{1/2}]_{SQ}[\tilde{a}_R^\dagger, \tilde{a}_S^\dagger]_+ = 0 \tag{1.9.9}$$

and obtain the annihilation commutation relations by taking the Hermitian conjugate of this anti-commutator

$$[a_P, a_Q]_+ = 0 \tag{1.9.10}$$

Clearly, the nonorthogonality does not affect the anticommutators between two creation operators and between two annihilation operators. By contrast, the anticommutator between a creation operator and an annihilation operator becomes

$$[a_P^\dagger, a_Q]_+ = \sum_{RS}[\tilde{a}_R^\dagger[\mathbf{S}^{1/2}]_{RP}, \tilde{a}_S[\mathbf{S}^{1/2}]_{QS}]_+ = \sum_{RS}[\mathbf{S}^{1/2}]_{RP}[\mathbf{S}^{1/2}]_{QS}[\tilde{a}_R^\dagger, \tilde{a}_S]_+$$

$$= \sum_{RS}[\mathbf{S}^{1/2}]_{RP}[\mathbf{S}^{1/2}]_{QS}\delta_{RS} = \sum_R[\mathbf{S}^{1/2}]_{QR}[\mathbf{S}^{1/2}]_{RP} \tag{1.9.11}$$

We thus have the following anticommutator for nonorthogonal creation and annihilation operators

$$[a_P^\dagger, a_Q]_+ = S_{QP} \tag{1.9.12}$$

which in general is nonzero for all pairs of operators. For orthonormal spin orbitals, this expression reduces to the standard anticommutator (1.2.29).

Since the nonorthogonal annihilation operators are linear combinations of orthonormal annihilation operators, we note that an annihilation operator times the vacuum state vanishes as for orthonormal spin orbitals:

$$a_P|\text{vac}\rangle = 0 \tag{1.9.13}$$

The effect of an annihilation operator a_P on an N-particle ON vector can therefore be written as

$$a_P|\mathbf{k}\rangle = \left[a_P, \prod_{Q=1}^{M} (a_Q^\dagger)^{k_Q} \right]_{(+)} |\text{vac}\rangle \tag{1.9.14}$$

where, in order to reduce the particle rank as discussed in Section 1.8, we introduce a commutator for even N and an anticommutator for odd N. Using (1.8.7) or (1.8.9), we now expand the (anti)commutator in (1.9.14) as a linear combination of elementary anticommutators:

$$a_P|\mathbf{k}\rangle = \sum_{Q=1}^{M} k_Q \Gamma_Q^{\mathbf{k}} (a_1^\dagger)^{k_1} \cdots [a_P, (a_Q^\dagger)^{k_Q}]_+ \cdots (a_M^\dagger)^{k_M} |\text{vac}\rangle$$

$$= \sum_{Q=1}^{M} k_Q \Gamma_Q^{\mathbf{k}} S_{PQ} |k_1, \ldots, 0_Q, \ldots, k_M\rangle \tag{1.9.15}$$

Thus, an annihilation operator working on a single ON vector generates a linear combination of ON vectors, each obtained by removing one electron from the original vector. Again, for an orthonormal basis, (1.9.15) reduces to the usual one-term expression (1.2.16).

It is noteworthy that the corresponding expression for a nonorthogonal creation operator working on an ON vector is identical to the expression for orthonormal spin orbitals (1.2.5)

$$a_P^\dagger|\mathbf{k}\rangle = (1 - k_P) \Gamma_P^{\mathbf{k}} |k_1, \ldots, 1_P, \ldots, k_M\rangle \tag{1.9.16}$$

since the nonorthogonal creation operators anticommute among themselves.

1.9.2 ONE- AND TWO-ELECTRON OPERATORS

The second-quantization one- and two-electron operators in a nonorthogonal basis may be obtained from the corresponding operators in the auxiliary orthonormal basis discussed in Section 1.4. The one-electron operator is given by

$$\hat{f} = \sum_{PQ} \tilde{f}_{PQ} \tilde{a}_P^\dagger \tilde{a}_Q = \sum_{PQ} [\mathbf{S}^{-1/2} \tilde{\mathbf{f}} \mathbf{S}^{-1/2}]_{PQ} a_P^\dagger a_Q \tag{1.9.17}$$

where the first expression is in the auxiliary basis, and the second expression is obtained by expanding the creation and annihilation operators according to (1.9.5) and (1.9.7). The integrals in the auxiliary basis are related to the integrals in the nonorthogonal basis by the equation

$$\tilde{f}_{PQ} = \int \tilde{\phi}_P^*(\mathbf{x}) f(\mathbf{x}) \tilde{\phi}_Q(\mathbf{x}) \, d\mathbf{x} = [\mathbf{S}^{-1/2} \mathbf{f} \mathbf{S}^{-1/2}]_{PQ} \tag{1.9.18}$$

Inserting this expression in (1.9.17), we obtain the final expression for one-electron operators in a nonorthogonal basis:

$$\hat{f} = \sum_{PQ} [\mathbf{S}^{-1} \mathbf{f} \mathbf{S}^{-1}]_{PQ} a_P^\dagger a_Q \tag{1.9.19}$$

Two-electron operators may be treated in the same way. We obtain

$$\hat{g} = \frac{1}{2} \sum_{PQRS} \left(\sum_{IJKL} [\mathbf{S}^{-1}]_{PI} [\mathbf{S}^{-1}]_{JQ} [\mathbf{S}^{-1}]_{RK} [\mathbf{S}^{-1}]_{LS} \, g_{IJKL} \right) a_P^\dagger a_R^\dagger a_S a_Q \tag{1.9.20}$$

where we have used the indices I, J, K and L for the inner summations. Clearly, the one-electron and two-electron operators both reduce to the standard expression for orthonormal spin orbitals when the overlap matrix becomes the identity matrix.

We now consider the effect of a one-electron operator $\hat{f}$ on an ON vector. Combining equations (1.9.15) and (1.9.16), we obtain

$$\hat{f}|\mathbf{k}\rangle = \sum_{PQ}[\mathbf{S}^{-1}\mathbf{f}\mathbf{S}^{-1}]_{PQ}a_P^\dagger a_Q|\mathbf{k}\rangle$$

$$= \sum_{PQR}[\mathbf{S}^{-1}\mathbf{f}\mathbf{S}^{-1}]_{PQ}S_{QR}\varepsilon_{PR}\Gamma_P^{\mathbf{k}}\Gamma_R^{\mathbf{k}}(1 - k_P + \delta_{RP})k_R \left|\begin{matrix} k_P \to 1 \\ k_R \to \delta_{PR} \end{matrix}\right\rangle$$

$$= \sum_{PQ}[\mathbf{S}^{-1}\mathbf{f}]_{PQ}\varepsilon_{PQ}\Gamma_P^{\mathbf{k}}\Gamma_Q^{\mathbf{k}}(1 - k_P + \delta_{PQ})k_Q \left|\begin{matrix} k_P \to 1 \\ k_Q \to \delta_{PQ} \end{matrix}\right\rangle \qquad (1.9.21)$$

where ε_{PQ} is defined as in (1.3.18). This expression should be compared with (1.3.17), which gives the result of a single-excitation operator for orthonormal spin orbitals. Although the derivation of the effect of a one-electron operator on an ON vector is more complicated in the nonorthogonal case, the final expressions look much the same since in (1.9.21) we were able to eliminate one summation index by modifying the integrals. Note, however, that the integrals in the final expression of (1.9.21) are not symmetric in the indices.

1.9.3 BIORTHOGONAL OPERATORS

It is possible to arrive at the results of Section 1.9.2 in a different way. We note that all complications due to nonorthogonality arise from the anticommutator (1.9.12). We therefore introduce a transformed set of annihilation operators $\bar{a}_P$ that satisfy the anticommutation relation

$$a_P^\dagger \bar{a}_Q + \bar{a}_Q a_P^\dagger = \delta_{PQ} \qquad (1.9.22)$$

To satisfy this relation, the annihilation operators are chosen as

$$\bar{a}_P = \sum_Q a_Q[\mathbf{S}^{-1}]_{PQ} \qquad (1.9.23)$$

That these operators indeed satisfy the anticommutation relation (1.9.22) can be verified by substitution and use of (1.9.12). The operators $a_P^\dagger \bar{a}_Q$ behave just like excitation operators in an orthonormal basis:

$$a_P^\dagger \bar{a}_Q|\mathbf{k}\rangle = \varepsilon_{PQ}\Gamma_P^{\mathbf{k}}\Gamma_Q^{\mathbf{k}}(1 - k_P + \delta_{PQ})k_Q \left|\begin{matrix} k_P \to 1 \\ k_Q \to \delta_{PQ} \end{matrix}\right\rangle \qquad (1.9.24)$$

A general one-electron operator can now be expanded in the excitation operators $a_P^\dagger \bar{a}_Q$

$$\hat{f} = \sum_{PQ}[\mathbf{S}^{-1}\mathbf{f}\mathbf{S}^{-1}]_{PQ}a_P^\dagger a_Q = \sum_{PQ}[\mathbf{S}^{-1}\mathbf{f}]_{PQ}a_P^\dagger \bar{a}_Q \qquad (1.9.25)$$

and we thus obtain (1.9.21) directly from (1.9.24) and (1.9.25). The annihilation operators $\bar{a}_Q$ are said to be *biorthogonal* to the creation operators $a_P^\dagger$ since they fulfill the relation (1.9.22) characteristic of orthonormal spin orbitals [5]. Note, however, that the creation operator $a_P^\dagger$ is not the adjoint of $\bar{a}_P$.

References

[1]   E. U. Condon and G. H. Shortley, *The Theory of Atomic Spectra*, Cambridge University Press, 1967.
[2]   P. A. M. Dirac, *The Principles of Quantum Mechanics*, 4th edn, Oxford University Press, 1958.
[3]   R. D. Richtmyer, *Principles of Advanced Mathematical Physics*, Vol. 1, Springer-Verlag, 1978.
[4]   P.-O. Löwdin, *Phys. Rev.* **97,** 1474 (1955).
[5]   M. Moshinsky and T. H. Seligman, *Ann. Phys. (NY)* **66**, 311 (1971).

Further reading

J. Avery, *Creation and Annihilation Operators*, McGraw-Hill, 1976.
E. R. Davidson, *Reduced Density Matrices in Quantum Chemistry*, Academic Press, 1976.
P. Jørgensen and J. Simons, *Second Quantization-Based Methods in Quantum Chemistry*, Academic Press, 1981.
W. Kutzelnigg, Quantum chemistry in Fock space, in D. Mukherjee (ed.), *Aspects of Many-Body Effects in Molecules and Extended Systems*, Lecture Notes in Chemistry, Vol. 50, Springer-Verlag, 1989, p. 35.
J. Linderberg and Y. Öhrn, *Propagators in Quantum Chemistry*, Academic Press, 1973.
H. C. Longuet-Higgins, Second quantization in the electronic theory of molecules, in P.-O. Löwdin (ed.), *Quantum Theory of Atoms, Molecules, and the Solid State*, Academic Press, 1966, p. 105.
P. R. Surján, *Second Quantized Approach to Quantum Chemistry*, Springer-Verlag, 1989.

Exercises

EXERCISE 1.1

Show that the effect of an excitation operator on an ON vector can be written as

$$a_P^\dagger a_Q |\mathbf{k}\rangle = \varepsilon_{PQ} \Gamma_P^\mathbf{k} \Gamma_Q^\mathbf{k} (1 - k_P + \delta_{PQ}) k_Q \left| \begin{matrix} k_P \to 1 \\ k_Q \to \delta_{PQ} \end{matrix} \right\rangle \tag{1E.1.1}$$

where

$$\varepsilon_{PQ} = \begin{cases} 1 & P \le Q \\ -1 & P > Q \end{cases} \tag{1E.1.2}$$

and where the ket on the right-hand side of (1E.1.1) is an ON vector with the same occupation numbers as $|\mathbf{k}\rangle$ except as indicated for spin orbitals P and Q.

EXERCISE 1.2

Let $\hat{I}_n$ and $\hat{I}_m$ be strings that contain n and m elementary (creation or annihilation) operators, respectively. Show by induction that, for even n and m, $[\hat{I}_n, \hat{I}_m]$ can be reduced to a sum of strings, each of which contains at most $n + m - 2$ elementary operators.

EXERCISE 1.3

Verify the following commutation and anticommutation relations:

1. $[\hat{A}, \hat{B}_1 \hat{B}_2] = [\hat{A}, \hat{B}_1] \hat{B}_2 + \hat{B}_1 [\hat{A}, \hat{B}_2]$ \hfill (1E.3.1)

2. $[\hat{A}, \hat{B}_1 \cdots \hat{B}_n] = \sum_{k=1}^{n} \hat{B}_1 \cdots \hat{B}_{k-1} [\hat{A}, \hat{B}_k] \hat{B}_{k+1} \cdots \hat{B}_n$ \hfill (1E.3.2)

3. $[\hat{A}, \hat{B}_1 \hat{B}_2] = [\hat{A}, \hat{B}_1]_+ \hat{B}_2 - \hat{B}_1 [\hat{A}, \hat{B}_2]_+$ \hfill (1E.3.3)

4. $[\hat{A}, \hat{B}_1 \cdots \hat{B}_n] = \sum_{k=1}^{n} (-1)^{k-1} \hat{B}_1 \cdots [\hat{A}, \hat{B}_k]_+ \cdots \hat{B}_n$ (n even)   (1E.3.4)

5. $[\hat{A}, \hat{B}_1\hat{B}_2]_+ = [\hat{A}, \hat{B}_1]\hat{B}_2 + \hat{B}_1[\hat{A}, \hat{B}_2]_+ = [\hat{A}, \hat{B}_1]_+\hat{B}_2 - \hat{B}_1[\hat{A}, \hat{B}_2]$ (1E.3.5)

6. $[\hat{A}, \hat{B}_1 \cdots \hat{B}_n]_+ = \sum_{k=1}^{n} (-1)^{k-1} \hat{B}_1 \cdots [\hat{A}, \hat{B}_k]_+ \cdots \hat{B}_n$ (n odd)   (1E.3.6)

EXERCISE 1.4

Let $\hat{\kappa}$ and $\hat{f}$ be the one-electron operators

$$\hat{\kappa} = \sum_{PQ} \kappa_{PQ} a_P^\dagger a_Q \tag{1E.4.1}$$

$$\hat{f} = \sum_{PQ} f_{PQ} a_P^\dagger a_Q \tag{1E.4.2}$$

Show that the commutator $[\hat{\kappa}, \hat{f}]$ can be written as a one-electron operator

$$[\hat{\kappa}, \hat{f}] = \sum_{PQ} f_{PQ}^\kappa a_P^\dagger a_Q \tag{1E.4.3}$$

with the modified integrals

$$f_{PQ}^\kappa = \sum_{R} (\kappa_{PR} f_{RQ} - f_{PR}\kappa_{RQ}) \tag{1E.4.4}$$

Solutions

SOLUTION 1.1

We consider first the case $P > Q$. Using (1.2.16) and (1.2.5), we obtain

$$\begin{aligned}
a_P^\dagger a_Q |\mathbf{k}\rangle &= a_P^\dagger \delta_{k_Q 1} \Gamma_Q^{\mathbf{k}} |k_1, \cdots, 0_Q, \cdots, k_P, \cdots, k_M\rangle \\
&= \delta_{k_P 0} \delta_{k_Q 1} \Gamma_Q^{\mathbf{k}} \Gamma_P^{\mathbf{k}} \varepsilon_{PQ} |k_1, \cdots, 0_Q, \cdots, 1_P, \cdots, k_M\rangle \\
&= \varepsilon_{PQ} \Gamma_Q^{\mathbf{k}} \Gamma_P^{\mathbf{k}} (1 - k_P) k_Q \left| \begin{matrix} k_P \to 1 \\ k_Q \to 0 \end{matrix} \right\rangle
\end{aligned} \tag{1S.1.1}$$

in agreement with (1E.1.1). The case $P < Q$ differs from (1S.1.1) only in the interpretation of ε_{PQ}. Finally, the case $P = Q$ is covered by (1.3.2).

SOLUTION 1.2

Assume that the relation

$$[\hat{I}_k, \hat{I}_l] = \sum_i \hat{I}_{kl}^i \tag{1S.2.1}$$

where each $\hat{I}_{kl}^i$ contains at most $k + l - 2$ elementary operators, holds for $m = k$ and $n = l$ where $k \geq 2$ and $l \geq 2$. Introducing

$$\hat{I}_{k+2} = \hat{I}_k \hat{b} \hat{c} \tag{1S.2.2}$$

where $\hat{b}$ and $\hat{c}$ are elementary operators, we obtain

$$[\hat{I}_{k+2}, \hat{I}_l] = [\hat{I}_k \hat{b}\hat{c}, \hat{I}_l] = [\hat{I}_k, \hat{I}_l]\hat{b}\hat{c} + \hat{I}_k[\hat{b}\hat{c}, \hat{I}_l] \tag{1S.2.3}$$

By assumption, both terms on the right-hand side contain at most $k + l$ elementary operators. Thus, if the assumption holds for $m = k$ and $n = l$, it holds also for $m = k + 2$ and $n = l$. By symmetry of k and l in (1S.2.1), the assumption then holds for $m = k$ and $n = l + 2$ as well. The proof is completed by noting that the assumption holds for $k = 2$ and $l = 2$, as is easily verified – for example:

$$[a_P^\dagger a_Q^\dagger, a_R a_S] = \delta_{QR} a_P^\dagger a_S - \delta_{PR} a_Q^\dagger a_S + \delta_{QS} a_R a_P^\dagger - \delta_{PS} a_R a_Q^\dagger \tag{1S.2.4}$$

SOLUTION 1.3

1. Relation (1E.3.1) is verified by expanding both sides and comparing terms.
2. Assume that (1E.3.2) holds for $n = m$. Using (1E.3.1), we then obtain for $n = m + 1$:

$$[\hat{A}, \hat{B}_1 \cdots \hat{B}_m \hat{B}_{m+1}] = [\hat{A}, \hat{B}_1 \cdots \hat{B}_m]\hat{B}_{m+1} + \hat{B}_1 \cdots \hat{B}_m[\hat{A}, \hat{B}_{m+1}]$$

$$= \left(\sum_{k=1}^{m} \hat{B}_1 \cdots \hat{B}_{k-1}[\hat{A}, \hat{B}_k]\hat{B}_{k+1} \cdots \hat{B}_m \right) \hat{B}_{m+1} + \hat{B}_1 \cdots \hat{B}_m[\hat{A}, \hat{B}_{m+1}]$$

$$= \sum_{k=1}^{m+1} \hat{B}_1 \cdots \hat{B}_{k-1}[\hat{A}, \hat{B}_k]\hat{B}_{k+1} \cdots \hat{B}_m \hat{B}_{m+1} \tag{1S.3.1}$$

Since (1E.3.2) holds for $n = 2$, the induction is complete.
3. Relation (1E.3.3) is verified by expanding both sides and comparing terms.
4. To demonstrate (1E.3.4), we first collect the $\hat{B}_i$ in $n/2$ pairs: $(\hat{B}_1 \hat{B}_2)(\hat{B}_3 \hat{B}_4) \cdots$. Next, we apply (1E.3.2) to obtain an expansion over the $n/2$ pairs and finally use (1E.3.3) to resolve the two contributions from each pair.
5. Relation (1E.3.5) is verified by expanding both sides and comparing terms.
6. Assume that (1E.3.6) holds for $n = m$ (m odd). Using (1E.3.5), we obtain

$$[\hat{A}, \hat{B}_1 \cdots \hat{B}_m \hat{B}_{m+1} \hat{B}_{m+2}]_+ = [\hat{A}, \hat{B}_1 \cdots \hat{B}_m]_+ \hat{B}_{m+1} \hat{B}_{m+2} - \hat{B}_1 \cdots \hat{B}_m[\hat{A}, \hat{B}_{m+1} \hat{B}_{m+2}] \tag{1S.3.2}$$

Since

$$[\hat{A}, \hat{B}_{m+1} \hat{B}_{m+2}] = [\hat{A}, \hat{B}_{m+1}]_+ \hat{B}_{m+2} - \hat{B}_{m+1}[\hat{A}, \hat{B}_{m+2}]_+ \tag{1S.3.3}$$

it is easily seen from (1S.3.2) that (1E.3.6) is valid for $n = m + 2$. Since it is valid for $n = 1$, we have proved (1E.3.6).

SOLUTION 1.4

Inserting the operators in the commutator and expanding, we obtain

$$[\hat{\kappa}, \hat{f}] = \sum_{PQRS} \kappa_{PQ} f_{RS}[a_P^\dagger a_Q, a_R^\dagger a_S] = \sum_{PQRS} \kappa_{PQ} f_{RS}(\delta_{QR} a_P^\dagger a_S - \delta_{PS} a_R^\dagger a_Q)$$

$$= \sum_{PQ} \left(\sum_R f_{RQ} \kappa_{PR} - f_{PR} \kappa_{RQ} \right) a_P^\dagger a_Q = \sum_{PQ} f_{PQ}^\kappa a_P^\dagger a_Q \tag{1S.4.1}$$

2 SPIN IN SECOND QUANTIZATION

In the formalism of second quantization as presented in the previous chapter, there is no reference to electron spin – the intrinsic angular momentum of the electron. In nonrelativistic theory, many important simplifications follow by taking spin explicitly into account. In the present chapter, we develop the theory of second quantization further so as to allow for an explicit description of electron spin. Although no fundamentally new concepts of the second-quantization formalism are introduced, the results obtained here are essential for an efficient description of molecular electronic systems in the nonrelativistic limit.

2.1 Spin functions

The spin orbitals introduced in Chapter 1 are functions of three continuous spatial coordinates $\mathbf{r}$ and one discrete spin coordinate m_s. The spin coordinate takes on only two values, representing the two allowed values of the projected spin angular momentum of the electron: $m_s = -\frac{1}{2}$ and $m_s = \frac{1}{2}$. The spin space is accordingly spanned by two functions, which are taken to be the eigenfunctions $\alpha(m_s)$ and $\beta(m_s)$ of the projected spin angular-momentum operator

$$S_z^c \alpha(m_s) = \tfrac{1}{2}\alpha(m_s) \tag{2.1.1}$$

$$S_z^c \beta(m_s) = -\tfrac{1}{2}\beta(m_s) \tag{2.1.2}$$

These spin functions – which we shall generically denote by σ, τ, μ and ν – are eigenfunctions of the total-spin angular-momentum operator as well with quantum number $s = \frac{1}{2}$

$$(S^c)^2 \sigma(m_s) = \tfrac{3}{4}\sigma(m_s) \tag{2.1.3}$$

in accordance with the general theory of angular momentum in quantum mechanics. The functional form of the spin functions is given by the equations

$$\alpha\left(\tfrac{1}{2}\right) = 1, \qquad \alpha\left(-\tfrac{1}{2}\right) = 0 \tag{2.1.4}$$

$$\beta\left(\tfrac{1}{2}\right) = 0, \qquad \beta\left(-\tfrac{1}{2}\right) = 1 \tag{2.1.5}$$

Completeness of the spin basis leads to the following resolution of the identity

$$\sum_\sigma \sigma^*(m_s)\sigma(m_s') = \delta_{m_s, m_s'} \tag{2.1.6}$$

as may be verified from relations (2.1.4) and (2.1.5). It should be noted that we have for convenience written the first-quantization spin operators as operating on the spin function of a single electron. The generalization to an N-electron system is simple and requires no comment except to note that, whereas S_z^c is a true one-electron operator, the operator for the total spin (2.1.3) is a two-electron operator in the sense that it is a linear combination of terms involving two electrons although no physical interactions occur.

We shall occasionally find it convenient to use for the discrete spin functions the same notation as for continuous spatial functions, interpreting integration in spin space as summation over the two discrete values of m_s:

$$\int f(m_s) \, dm_s = \sum_{m_s} f(m_s) \tag{2.1.7}$$

Thus, we may write the orthonormality conditions of the spin functions in the form

$$\int \sigma^*(m_s) \tau(m_s) \, dm_s = \delta_{\sigma\tau} \tag{2.1.8}$$

Like the resolution of the identity (2.1.6), this relationship is easily verified by reference to (2.1.4) and (2.1.5).

The spin-orbital space is spanned by the direct product of a basis for the orbital space and a basis for the spin space. Thus, a general spin orbital may be written as

$$\phi_P(\mathbf{r}, m_s) = \phi_P^\alpha(\mathbf{r}) \alpha(m_s) + \phi_P^\beta(\mathbf{r}) \beta(m_s) \tag{2.1.9}$$

In nonrelativistic theory, it is common to use spin orbitals of the more restricted form

$$\phi_{p\sigma}(\mathbf{r}, m_s) = \phi_p(\mathbf{r}) \sigma(m_s) \tag{2.1.10}$$

so that a given spin orbital consists of an orbital part multiplied by a spin eigenfunction. This simple product form is acceptable since the nonrelativistic Hamiltonian operator does not involve spin and thus cannot couple the spatial and spin parts of the spin orbitals. We note that spin orbitals (2.1.10) with the same orbital parts but different spins are orthogonal.

We shall in this book use lower-case indices for orbitals, reserving upper-case indices for spin orbitals. For spin orbitals $\phi_{p\sigma}$ of the form (2.1.10), we shall usually employ a composite index, where the first index p refers to the orbital part and the second index σ to the spin part. The total number of orbitals is denoted by n. Thus, for a basis of n orbitals, there are a total of $M = 2n$ independent spin orbitals of the form (2.1.10). With the necessary elaboration of notation to accommodate composite indices of the spin orbitals, the theory of second quantization presented in Chapter 1 holds unchanged in the product basis (2.1.10). For example, the anticommutator between creation and annihilation operators (1.2.29) now becomes

$$[a_{p\sigma}^\dagger, a_{q\tau}]_+ = \delta_{p\sigma, q\tau} = \delta_{pq} \delta_{\sigma\tau} \tag{2.1.11}$$

where for example $a_{p\sigma}^\dagger$ is the creation operator associated with the product spin orbital $\phi_{p\sigma}$.

2.2 Operators in the orbital basis

Quantum-mechanical operators may be classified according to how they affect the orbital and spin parts of wave functions. Thus, we classify operators as *spin-free* or *spinless* if they work in ordinary space only without affecting the spin part of a function. Conversely, an operator that works in spin space only, without affecting the spatial part of a function, is termed a *pure spin operator* or simply a *spin operator*. Finally, an operator is *mixed* if it affects both the spatial and spin parts of a function. We shall in this section investigate how each of these three classes of operators is represented in second quantization.

2.2.1 SPIN-FREE OPERATORS

Let us first consider one-electron operators. Following the general discussion in Section 1.4.1, a spin-free one-electron operator of the form

$$f^c = \sum_{i=1}^{N} f^c(\mathbf{r}_i) \tag{2.2.1}$$

may in the spin-orbital basis be written as

$$\hat{f} = \sum_{p\sigma q\tau} f_{p\sigma,q\tau} a_{p\sigma}^\dagger a_{q\tau} \tag{2.2.2}$$

The integrals entering the second-quantization operator $\hat{f}$ vanish for opposite spins since the first-quantization operator f^c is spin-free:

$$f_{p\sigma,q\tau} = \int \phi_p^*(\mathbf{r})\sigma^*(m_s)f^c(\mathbf{r})\phi_q(\mathbf{r})\tau(m_s)\,d\mathbf{r}\,dm_s$$

$$= \delta_{\sigma\tau} \int \phi_p^*(\mathbf{r})f^c(\mathbf{r})\phi_q(\mathbf{r})\,d\mathbf{r} = f_{pq}\delta_{\sigma\tau} \tag{2.2.3}$$

Here we use the notation

$$f_{pq} = \int \phi_p^*(\mathbf{r})f^c(\mathbf{r})\phi_q(\mathbf{r})\,d\mathbf{r} \tag{2.2.4}$$

for the integrals over spatial coordinates and note that these integrals display the usual Hermitian permutational symmetry:

$$f_{pq} = f_{qp}^* \tag{2.2.5}$$

The second-quantization representation of the spin-free one-electron operator (2.2.1) now becomes

$$\hat{f} = \sum_{pq} f_{pq}E_{pq} \tag{2.2.6}$$

where we have introduced the *singlet excitation operators*

$$E_{pq} = a_{p\alpha}^\dagger a_{q\alpha} + a_{p\beta}^\dagger a_{q\beta} \tag{2.2.7}$$

as a linear combination of the spin-orbital excitation operators of Section 1.3.3. The singlet excitation operator is discussed in detail in Sections 2.3.4 and 2.3.5.

We now turn our attention to spinless two-electron operators. According to the discussion in Section 1.4.2, the second-quantization representation of a general spin-free two-electron operator of the form

$$g^c = \frac{1}{2}\sum_{i\neq j} g^c(\mathbf{r}_i, \mathbf{r}_j) \tag{2.2.8}$$

is given by

$$\hat{g} = \tfrac{1}{2} \sum_{\substack{pqrs \\ \sigma\tau\mu\nu}} g_{p\sigma,q\tau,r\mu,s\nu} a^\dagger_{p\sigma} a^\dagger_{r\mu} a_{s\nu} a_{q\tau} \tag{2.2.9}$$

Most of the terms in this operator vanish because of the orthogonality of the spin functions

$$g_{p\sigma,q\tau,r\mu,s\nu} = \iint \phi^*_p(\mathbf{r}_1)\sigma^*(m_1)\phi^*_r(\mathbf{r}_2)\mu^*(m_2)g^c(\mathbf{r}_1,\mathbf{r}_2)\phi_q(\mathbf{r}_1)\tau(m_1)\phi_s(\mathbf{r}_2)\nu(m_2)\,\mathrm{d}\mathbf{r}_1\,\mathrm{d}m_1\,\mathrm{d}\mathbf{r}_2\,\mathrm{d}m_2$$

$$= g_{pqrs}\delta_{\sigma\tau}\delta_{\mu\nu} \tag{2.2.10}$$

where we have introduced two-electron integrals in ordinary space

$$g_{pqrs} = \iint \phi^*_p(\mathbf{r}_1)\phi^*_r(\mathbf{r}_2)g^c(\mathbf{r}_1,\mathbf{r}_2)\phi_q(\mathbf{r}_1)\phi_s(\mathbf{r}_2)\,\mathrm{d}\mathbf{r}_1\,\mathrm{d}\mathbf{r}_2 \tag{2.2.11}$$

Let us consider the permutational symmetries of these integrals. The symmetry

$$g_{pqrs} = g_{rspq} \tag{2.2.12}$$

follows from the symmetry of the interaction operator in (2.2.8) and is always present in the integrals. The remaining symmetries are different for real and complex orbitals. For complex orbitals, we have the Hermitian symmetry

$$g_{pqrs} = g^*_{qpsr} \quad \text{(complex orbitals)} \tag{2.2.13}$$

whereas for real orbitals we have the following permutational symmetries

$$g_{pqrs} = g_{qprs} = g_{pqsr} = g_{qpsr} \quad \text{(real orbitals)} \tag{2.2.14}$$

For real orbitals, therefore, there are a total of eight permutational symmetries present in the integrals, obtained by combining (2.2.12) with (2.2.14).

Inserting the integrals (2.2.10) in (2.2.9), the second-quantization representation of a spin-free two-electron operator can be written as

$$\hat{g} = \tfrac{1}{2} \sum_{pqrs} g_{pqrs} \sum_{\sigma\tau} a^\dagger_{p\sigma} a^\dagger_{r\tau} a_{s\tau} a_{q\sigma}$$

$$= \tfrac{1}{2} \sum_{pqrs} g_{pqrs}(E_{pq}E_{rs} - \delta_{qr}E_{ps})$$

$$= \tfrac{1}{2} \sum_{pqrs} g_{pqrs} e_{pqrs} \tag{2.2.15}$$

where for convenience we have introduced the *two-electron excitation operator*

$$e_{pqrs} = E_{pq}E_{rs} - \delta_{qr}E_{ps} = \sum_{\sigma\tau} a^\dagger_{p\sigma} a^\dagger_{r\tau} a_{s\tau} a_{q\sigma} \tag{2.2.16}$$

Note the permutational symmetry

$$e_{pqrs} = e_{rspq} \tag{2.2.17}$$

which follows directly from the last expression in (2.2.16). There are no permutational symmetries analogous to (2.2.14) for the two-electron excitation operator.

We are now in a position to write up the second-quantization representation of the nonrelativistic and spin-free molecular electronic Hamiltonian in the orbital basis:

$$\hat{H} = \sum_{pq} h_{pq} E_{pq} + \frac{1}{2} \sum_{pqrs} g_{pqrs} e_{pqrs} + h_{\text{nuc}} \tag{2.2.18}$$

This expression should be compared with the operator in the spin-orbital basis (1.4.39), where each summation index runs over twice the number of orbitals. The one- and two-electron integrals in (2.2.18) are the same as those in (1.4.40) and (1.4.41) except that the integrations are over the spatial coordinates only:

$$h_{pq} = \int \phi_p^*(\mathbf{r}) \left(-\frac{1}{2}\nabla^2 - \sum_I \frac{Z_I}{r_I} \right) \phi_q(\mathbf{r}) \, d\mathbf{r} \tag{2.2.19}$$

$$g_{pqrs} = \int\int \frac{\phi_p^*(\mathbf{r}_1)\phi_r^*(\mathbf{r}_2)\phi_q(\mathbf{r}_1)\phi_s(\mathbf{r}_2)}{r_{12}} \, d\mathbf{r}_1 \, d\mathbf{r}_2 \tag{2.2.20}$$

The scalar nuclear-repulsion term h_{nuc} in (2.2.18) was defined in (1.4.42).

2.2.2 SPIN OPERATORS

We now consider the representation of first-quantization operators f^c that work in spin space only. The associated second-quantization operators may be written in the general form

$$\hat{f} = \sum_{p\sigma q\tau} \int \phi_p^*(\mathbf{r})\sigma^*(m_s) f^c(m_s)\phi_q(\mathbf{r})\tau(m_s) \, d\mathbf{r} \, dm_s \, a_{p\sigma}^\dagger a_{q\tau}$$

$$= \sum_{\sigma\tau} \int \sigma^*(m_s) f^c(m_s)\tau(m_s) \, dm_s \sum_p a_{p\sigma}^\dagger a_{p\tau} \tag{2.2.21}$$

Three important examples of pure spin operators are the *raising and lowering operators* S_+^c and S_-^c (also known as the *step-up and step-down operators* or as the *shift operators*) and the operator for the z component of the spin angular momentum S_z^c. From the effect of these operators on the spin functions (again assuming a one-particle state)

$$S_+^c \beta = \alpha, \qquad S_+^c \alpha = 0 \tag{2.2.22}$$

$$S_-^c \beta = 0, \qquad S_-^c \alpha = \beta \tag{2.2.23}$$

$$S_z^c \beta = -\tfrac{1}{2}\beta, \qquad S_z^c \alpha = +\tfrac{1}{2}\alpha \tag{2.2.24}$$

we obtain the following matrix elements

$$[S_+^c]_{p\sigma,q\tau} = \int \phi_p^*(\mathbf{r})\sigma^*(m_s) S_+^c(m_s)\phi_q(\mathbf{r})\tau(m_s) \, d\mathbf{r} \, dm_s = \delta_{pq}\delta_{\sigma\alpha}\delta_{\tau\beta} \tag{2.2.25}$$

$$[S_-^c]_{p\sigma,q\tau} = \int \phi_p^*(\mathbf{r})\sigma^*(m_s) S_-^c(m_s)\phi_q(\mathbf{r})\tau(m_s) \, d\mathbf{r} \, dm_s = \delta_{pq}\delta_{\sigma\beta}\delta_{\tau\alpha} \tag{2.2.26}$$

$$[S_z^c]_{p\sigma,q\tau} = \int \phi_p^*(\mathbf{r})\sigma^*(m_s) S_z^c(m_s)\phi_q(\mathbf{r})\tau(m_s) \, d\mathbf{r} \, dm_s = \tfrac{1}{2}\delta_{pq}\delta_{\sigma\tau}(\delta_{\sigma\alpha} - \delta_{\sigma\beta}) \tag{2.2.27}$$

using the same notation for the matrix elements as in Section 1.5.2. Inserting the integrals (2.2.25)–(2.2.27) in the operator (2.2.21), we arrive at the following expressions for the basic spin operators:

$$\hat{S}_+ = \sum_p a_{p\alpha}^{\dagger} a_{p\beta} \tag{2.2.28}$$

$$\hat{S}_- = \sum_p a_{p\beta}^{\dagger} a_{p\alpha} \tag{2.2.29}$$

$$\hat{S}_z = \tfrac{1}{2} \sum_p (a_{p\alpha}^{\dagger} a_{p\alpha} - a_{p\beta}^{\dagger} a_{p\beta}) \tag{2.2.30}$$

The second-quantization lowering operator is readily seen to be the Hermitian adjoint of the raising operator:

$$\hat{S}_+^{\dagger} = \sum_p (a_{p\alpha}^{\dagger} a_{p\beta})^{\dagger} = \sum_p a_{p\beta}^{\dagger} a_{p\alpha} = \hat{S}_- \tag{2.2.31}$$

For the operators for the x and y components of the spin angular momentum

$$S_x^c = \frac{1}{2}(S_+^c + S_-^c) \tag{2.2.32}$$

$$S_y^c = \frac{1}{2i}(S_+^c - S_-^c) \tag{2.2.33}$$

we obtain from (2.2.28) and (2.2.29) the following expressions for their second-quantization counterparts

$$\hat{S}_x = \frac{1}{2} \sum_p (a_{p\alpha}^{\dagger} a_{p\beta} + a_{p\beta}^{\dagger} a_{p\alpha}) \tag{2.2.34}$$

$$\hat{S}_y = \frac{1}{2i} \sum_p (a_{p\alpha}^{\dagger} a_{p\beta} - a_{p\beta}^{\dagger} a_{p\alpha}) \tag{2.2.35}$$

Finally, from the three components of the spin angular momentum, we may proceed to calculate the operator for the total spin

$$\hat{S}^2 = \hat{S}_x^2 + \hat{S}_y^2 + \hat{S}_z^2 \tag{2.2.36}$$

Since this operator contains products of one-electron operators, it is a two-electron operator.

Our spin basis is – in contrast to the orbital basis – complete. Therefore, for pure spin operators we have none of the problems associated with the representation of product operators discussed in Section 1.5, and the usual first-quantization commutation relations hold also for the second-quantization spin operators. For example, we may easily verify that the commutator between the second-quantization raising and lowering operators is the same as in first quantization:

$$[\hat{S}_+, \hat{S}_-] = \left[\sum_p a_{p\alpha}^{\dagger} a_{p\beta}, \sum_q a_{q\beta}^{\dagger} a_{q\alpha}\right] = \sum_p (a_{p\alpha}^{\dagger} a_{p\alpha} - a_{p\beta}^{\dagger} a_{p\beta}) = 2\hat{S}_z \tag{2.2.37}$$

Here we have used the commutator (1.8.14) and also the fact that creation and annihilation operators for different orbitals anticommute. It has already been noted that the total-spin operator

(2.2.36) is a two-electron operator and therefore somewhat cumbersome to manipulate in second quantization. However, the explicit form of this operator is seldom needed since we may instead employ the standard operator identities

$$\hat{S}^2 = \hat{S}_+\hat{S}_- + \hat{S}_z(\hat{S}_z - 1) = \hat{S}_-\hat{S}_+ + \hat{S}_z(\hat{S}_z + 1) \tag{2.2.38}$$

which are verified by substitution of the expressions for the shift operators in (2.2.36).

2.2.3 MIXED OPERATORS

A number of first-quantization operators such as the fine-structure and hyperfine-structure operators affect both the spatial and spin parts of the wave function. As an example, we here consider the effective spin–orbit interaction operator

$$V_{SO}^c = \sum_{i=1}^{N} V_{SO}^c(\mathbf{r}_i, m_{si}) = \sum_{i=1}^{N} \xi(r_i)\boldsymbol{\ell}^c(i) \cdot \mathbf{S}^c(i) \tag{2.2.39}$$

often employed to describe spin–orbit effects in heavy atoms. In this expression, $\xi(r_i)$ is a radial function where r_i is the distance from electron i to the nucleus, $\boldsymbol{\ell}^c(i)$ is the orbital angular-momentum operator for electron i, and $\mathbf{S}^c(i)$ is the spin angular-momentum operator for this electron. Proceeding in the usual manner, we obtain the second-quantization representation of the effective spin–orbit operator as

$$\hat{V}_{SO} = \sum_{\substack{pq \\ \sigma\tau}} \int \phi_p^*(\mathbf{r})\sigma^*(m_s)V_{SO}^c(\mathbf{r}, m_s)\phi_q(\mathbf{r})\tau(m_s)\, d\mathbf{r}\, dm_s\, a_{p\sigma}^\dagger a_{q\tau}$$

$$= \sum_{pq}(V_{pq}^x \hat{T}_{pq}^x + V_{pq}^y \hat{T}_{pq}^y + V_{pq}^z \hat{T}_{pq}^z) \tag{2.2.40}$$

where (with μ representing x, y and z)

$$V_{pq}^\mu = \int \phi_p^*(\mathbf{r})\xi(r)\ell_\mu^c\phi_q(\mathbf{r})\, d\mathbf{r} \tag{2.2.41}$$

and where the Cartesian components of the *triplet excitation operators* are given by

$$\hat{T}_{pq}^x = \frac{1}{2}(a_{p\alpha}^\dagger a_{q\beta} + a_{p\beta}^\dagger a_{q\alpha}) \tag{2.2.42}$$

$$\hat{T}_{pq}^y = \frac{1}{2i}(a_{p\alpha}^\dagger a_{q\beta} - a_{p\beta}^\dagger a_{q\alpha}) \tag{2.2.43}$$

$$\hat{T}_{pq}^z = \frac{1}{2}(a_{p\alpha}^\dagger a_{q\alpha} - a_{p\beta}^\dagger a_{q\beta}) \tag{2.2.44}$$

The triplet excitation operators are discussed in more detail in Section 2.3.4.

An alternative form of the effective spin–orbit operator is obtained by expressing (2.2.39) in terms of the shift operators

$$V_{SO}^c = \sum_{i=1}^{N} \xi(r_i)\left[\tfrac{1}{2}\ell_+^c(i)S_-^c(i) + \tfrac{1}{2}\ell_-^c(i)S_+^c(i) + \ell_z^c(i)S_z^c(i)\right] \tag{2.2.45}$$

where the orbital-shift operators are given by

$$\ell^c_\pm(i) = \ell^c_x(i) \pm i\ell^c_y(i) \tag{2.2.46}$$

The second-quantization representation of the effective spin–orbit operator now becomes

$$\hat{V}_{SO} = \sum_{pq}[V^+_{pq}a^\dagger_{p\beta}a_{q\alpha} + V^z_{pq}(a^\dagger_{p\alpha}a_{q\alpha} - a^\dagger_{p\beta}a_{q\beta}) + V^-_{pq}a^\dagger_{p\alpha}a_{q\beta}] \tag{2.2.47}$$

where (with μ representing $+$, $-$ or z):

$$V^\mu_{pq} = \tfrac{1}{2}\int \phi^*_p(\mathbf{r})\xi(r)\ell^c_\mu\phi_q(\mathbf{r})\,d\mathbf{r} \tag{2.2.48}$$

From the expression for the spin–orbit operator (2.2.47), we note that the second-quantization representation of a mixed (spin and space) operator depends on both the spin of the electron and the functional form of the orbitals (2.2.48). For comparison, the pure spin operators in Section 2.2.2 are independent of the functional form of the orbitals, whereas the spin-free operators in Section 2.2.1 depend on the orbitals but have the same amplitudes (integrals) for alpha and beta spins. Mixed spin operators are treated in Exercises 2.1 and 2.2.

2.3 Spin tensor operators

The incorporation of spin in second quantization leads to operators with different spin symmetry properties as demonstrated in Section 2.2. Thus, spin-free interactions are represented by operators that are totally symmetric in spin space and thus expressed in terms of orbital excitation operators that affect alpha and beta electrons equally, whereas pure spin interactions are represented by excitation operators that affect alpha and beta electrons differently. For the efficient and transparent manipulation of these operators, we shall apply the standard machinery of group theory. More specifically, we shall adopt the theory of tensor operators for angular momentum in quantum mechanics and develop a useful set of tools for the construction and classification of states and operators with definite spin symmetry properties.

2.3.1 SPIN TENSOR OPERATORS

A *spin tensor operator* of integral or half-integral rank S is a set of $2S + 1$ operators $\hat{T}^{S,M}$ where M runs from $-S$ to S in unit increments and which fulfills the relations [1]

$$[\hat{S}_\pm, \hat{T}^{S,M}] = \sqrt{S(S+1) - M(M\pm1)}\,\hat{T}^{S,M\pm1} \tag{2.3.1}$$

$$[\hat{S}_z, \hat{T}^{S,M}] = M\hat{T}^{S,M} \tag{2.3.2}$$

where we assume that

$$\hat{T}^{S,S+1} = \hat{T}^{S,-S-1} = 0 \tag{2.3.3}$$

A tensor operator working on the vacuum state generates a set of spin eigenfunctions with total and projected spins S and M (provided the tensor operator does not annihilate the vacuum state). We may prove this assertion in the following way. Since the second-quantization spin-component operators $\hat{S}_{\pm}$ and $\hat{S}_z$ produce zero when working on the vacuum state, we obtain from (2.3.1) and (2.3.2)

$$\hat{S}_{\pm}\hat{T}^{S,M}|\text{vac}\rangle = \sqrt{S(S+1) - M(M \pm 1)}\,\hat{T}^{S,M\pm 1}|\text{vac}\rangle \qquad (2.3.4)$$

$$\hat{S}_z\hat{T}^{S,M}|\text{vac}\rangle = M\hat{T}^{S,M}|\text{vac}\rangle \qquad (2.3.5)$$

These are the defining equations for a *spin tensor state* of rank S. The tensor state is obviously an eigenfunction of the projected spin (2.3.5). To determine the effect of the total-spin operator on the tensor state, we combine relations (2.3.4) and (2.3.5) with the expression (2.2.38) for the spin operator:

$$\hat{S}^2\hat{T}^{S,M}|\text{vac}\rangle = [\hat{S}_-\hat{S}_+ + \hat{S}_z(\hat{S}_z + 1)]\hat{T}^{S,M}|\text{vac}\rangle = S(S+1)\hat{T}^{S,M}|\text{vac}\rangle \qquad (2.3.6)$$

We conclude that $\hat{T}^{S,M}|\text{vac}\rangle$ – provided that it does not vanish – represents a tensor state with spin eigenvalues S and M. Because of the close relationship between spin tensor operators and spin eigenfunctions, the terminology for spin functions is often used for spin tensor operators as well. Thus, a spin tensor operator $\hat{T}^{S,M}$ with $S = 0$ is referred to as a *singlet operator*, $S = \frac{1}{2}$ gives a *doublet operator*, $S = 1$ a *triplet*, and so on.

One important observation that should be made about the spin tensor operators is that any singlet operator $\hat{T}^{0,0}$ commutes with both the shift operators $\hat{S}_{\pm}$ and the spin-projection operator $\hat{S}_z$; see (2.3.1) and (2.3.2). It therefore follows that singlet operators also commute with $\hat{S}^2$ since this operator may be expressed in terms of the shift operators and the spin-projection operator (2.2.38):

$$[\hat{S}_z, \hat{T}^{0,0}] = 0 \qquad (2.3.7)$$

$$[\hat{S}^2, \hat{T}^{0,0}] = 0 \qquad (2.3.8)$$

These commutators will be used on several occasions in our development of second quantization for electronic systems.

The commutator of two Hermitian operators is an anti-Hermitian operator. From (2.3.2), we can therefore conclude that spin tensor operators are not in general Hermitian. Indeed, the only possible exception to this rule are the operators $\hat{T}^{S,0}$ where $M = 0$, which may or may not be Hermitian. It is therefore of some interest to examine the Hermitian adjoints of the spin tensor operators. Taking the conjugate of the relations (2.3.1) and (2.3.2), we obtain:

$$[\hat{S}_{\mp}, (\hat{T}^{S,M})^{\dagger}] = -\sqrt{S(S+1) - M(M \pm 1)}\,(\hat{T}^{S,M\pm 1})^{\dagger} \qquad (2.3.9)$$

$$[\hat{S}_z, (\hat{T}^{S,M})^{\dagger}] = -M(\hat{T}^{S,M})^{\dagger} \qquad (2.3.10)$$

Comparing these expressions with (2.3.1) and (2.3.2), we note that the following operators constitute a tensor operator

$$\hat{U}^{S,M} = (-1)^{S+M}(\hat{T}^{S,-M})^{\dagger} \qquad (2.3.11)$$

where the phase factor is required to maintain the correct relative signs among the components. The factor $(-1)^S$ is necessary to ensure that the phase factor is also ± 1 for half-integral values of M. Conjugation of a tensor operator thus always yields a new tensor operator (2.3.11) where the order of the projections is reversed.

Spin tensor operators play an important role in the second-quantization treatment of electronic systems since they may be used to generate states with definite spin properties. In the remainder of this section, we shall examine some important examples of singlet, doublet and triplet tensor operators. Later, in Section 2.6, we shall see how spin tensor operators may be employed to generate many-electron states with definite total and projected spins.

2.3.2 CREATION AND ANNIHILATION OPERATORS

As an example of doublet operators, we first consider the creation operators $\{a_{p\alpha}^\dagger, a_{p\beta}^\dagger\}$. From the form of the spin functions (2.2.28)–(2.2.30), we may verify the following relations by substitution for $m_s = \frac{1}{2}$ and $m_s = -\frac{1}{2}$:

$$[\hat{S}_\pm, a_{pm_s}^\dagger] = \sqrt{\frac{3}{4} - m_s(m_s \pm 1)}\, a_{p,m_s\pm1}^\dagger \tag{2.3.12}$$

$$[\hat{S}_z, a_{pm_s}^\dagger] = m_s a_{pm_s}^\dagger \tag{2.3.13}$$

Comparing these expressions with (2.3.1) and (2.3.2), we conclude that the creation operators $\{a_{p\alpha}^\dagger, a_{p\beta}^\dagger\}$ constitute a doublet tensor operator. Applied to the vacuum state, these operators generate one-electron doublet states. In accordance with the general relation (2.3.11), we obtain a new doublet by conjugation and reordering of $\{a_{p\alpha}^\dagger, a_{p\beta}^\dagger\}$. We find that the pair $\{-a_{p\beta}, a_{p\alpha}\}$ constitutes a doublet tensor:

$$b_{p\alpha} = (-1)^{1/2+1/2}(a_{p\beta}^\dagger)^\dagger = -a_{p\beta} \tag{2.3.14}$$

$$b_{p\beta} = (-1)^{1/2-1/2}(a_{p\alpha}^\dagger)^\dagger = a_{p\alpha} \tag{2.3.15}$$

The beta spin operator $a_{p\beta}$ corresponds to the component $M = \frac{1}{2}$ since it removes an electron with spin projection $-\frac{1}{2}$, thereby increasing the spin projection of the system.

2.3.3 TWO-BODY CREATION OPERATORS

For strings containing two or more elementary operators, it is possible to construct more than one tensor operator. We shall in Section 2.6.7 present a general method for the construction of tensor operators from strings of elementary operators. At present, we note that, by coupling the two doublet operators $\{a_{p\alpha}^\dagger, a_{p\beta}^\dagger\}$ and $\{a_{q\alpha}^\dagger, a_{q\beta}^\dagger\}$, it is possible to generate a *singlet two-body creation operator*

$$\hat{Q}_{pq}^{0,0} = \frac{1}{\sqrt{2}}(a_{p\alpha}^\dagger a_{q\beta}^\dagger - a_{p\beta}^\dagger a_{q\alpha}^\dagger) \tag{2.3.16}$$

as well as a *triplet two-body creation operator*

$$\hat{Q}_{pq}^{1,1} = a_{p\alpha}^\dagger a_{q\alpha}^\dagger \tag{2.3.17}$$

$$\hat{Q}_{pq}^{1,0} = \frac{1}{\sqrt{2}}(a_{p\alpha}^\dagger a_{q\beta}^\dagger + a_{p\beta}^\dagger a_{q\alpha}^\dagger) \tag{2.3.18}$$

$$\hat{Q}_{pq}^{1,-1} = a_{p\beta}^\dagger a_{q\beta}^\dagger \tag{2.3.19}$$

The tensor properties of these operators are verified by substitution in (2.3.1) and (2.3.2). The defining relations (2.3.1) and (2.3.2) allow components of a given tensor operator to be scaled by a common factor. The scaling factor given here for the two-body creation operators is conventional.

We use the term *two-body* rather than *two-electron* for the creation operators, reserving the latter term for number-conserving rank-2 operators such as the two-electron part of the Hamiltonian. In the same manner, we sometimes refer to the elementary operators in Section 2.3.2 as the *one-body* creation and annihilation operators.

The singlet two-body creation operator (2.3.16) is symmetric in the indices p and q whereas the triplet operators are antisymmetric. For $p = q$, the singlet two-body creation operator becomes

$$\hat{Q}^{0,0}_{pp} = \sqrt{2} a^\dagger_{p\alpha} a^\dagger_{p\beta} \tag{2.3.20}$$

whereas the triplet operator vanishes in accordance with the Pauli principle. Applied to the vacuum state, the operator (2.3.20) creates a two-electron closed-shell singlet state assuming that ϕ_p is nondegenerate. For $p \neq q$, the operator $\hat{Q}^{0,0}_{pq}$ creates an open-shell singlet state and $\hat{Q}^{1,0}_{pq}$ generates the associated triplet state with zero projected spin.

2.3.4 EXCITATION OPERATORS

In Section 2.3.3, we coupled two doublet creation operators and obtained in this way singlet and triplet two-body creation operators. In the same manner, we now couple a creation-operator doublet with an annihilation-operator doublet in order to generate excitation operators of singlet and triplet spin symmetries.

Building on our results for the two-body creation operators in Section 2.3.3, we replace $\{a^\dagger_{q\alpha}, a^\dagger_{q\beta}\}$ by $\{-a_{q\beta}, a_{q\alpha}\}$ in (2.3.16)–(2.3.19). We thus obtain the *singlet excitation operator*

$$\hat{S}^{0,0}_{pq} = \frac{1}{\sqrt{2}}(a^\dagger_{p\alpha} a_{q\alpha} + a^\dagger_{p\beta} a_{q\beta}) \tag{2.3.21}$$

and the three components of the *triplet excitation operator*

$$\hat{T}^{1,1}_{pq} = -a^\dagger_{p\alpha} a_{q\beta} \tag{2.3.22}$$

$$\hat{T}^{1,0}_{pq} = \frac{1}{\sqrt{2}}(a^\dagger_{p\alpha} a_{q\alpha} - a^\dagger_{p\beta} a_{q\beta}) \tag{2.3.23}$$

$$\hat{T}^{1,-1}_{pq} = a^\dagger_{p\beta} a_{q\alpha} \tag{2.3.24}$$

The scaling factor chosen here agrees with common usage. Note that, in contrast to the two-body creation operators of Section 2.3.3, there is no permutational symmetry in the indices p and q. Thus, both the singlet and the triplet excitation operators are nonzero for $p = q$. This lack of permutational symmetry is expected since an excitation from ϕ_p to ϕ_q is different from an excitation from ϕ_q to ϕ_p. Instead, the excitation operators exhibit the symmetries

$$\hat{S}^{0,0\dagger}_{pq} = \hat{S}^{0,0}_{qp} \tag{2.3.25}$$

$$\hat{T}^{S,M\dagger}_{pq} = (-1)^M \hat{T}^{S,-M}_{qp} \tag{2.3.26}$$

which are easily verified by conjugation. Thus, applied to an excitation operator, the conjugation relation for tensor operators (2.3.11) returns the associated 'de-excitation' operator. The triplet operators in the spin-tensor form (2.3.22)–(2.3.24) are related to the Cartesian triplet operators (2.2.42)–(2.2.44) by a simple linear transformation

$$(\hat{T}^x_{pq}, \hat{T}^y_{pq}, \hat{T}^z_{pq}) = (\hat{T}^{1,1}_{pq}, \hat{T}^{1,-1}_{pq}, \hat{T}^{1,0}_{pq}) \begin{pmatrix} -\dfrac{1}{2} & -\dfrac{1}{2i} & 0 \\[2mm] \dfrac{1}{2} & -\dfrac{1}{2i} & 0 \\[2mm] 0 & 0 & \dfrac{1}{\sqrt{2}} \end{pmatrix} \tag{2.3.27}$$

Note that the Cartesian components satisfy the same simple conjugation symmetry

$$\hat{T}^{\mu\dagger}_{pq} = \hat{T}^{\mu}_{qp} \tag{2.3.28}$$

as does the singlet excitation operator (2.3.25), with no mixing of the Cartesian components upon conjugation.

Except for a scaling factor, the singlet excitation operator (2.3.21) is identical to the orbital excitation operator in (2.2.7):

$$E_{pq} = \sqrt{2}\hat{S}^{0,0}_{pq} = a^{\dagger}_{p\alpha}a_{q\alpha} + a^{\dagger}_{p\beta}a_{q\beta} \tag{2.3.29}$$

The spin-free one- and two-electron operators (2.2.6) and (2.2.15) are thus expressed entirely in terms of the singlet excitation operator and we may, for example, write the electronic Hamiltonian (2.2.18) in the form

$$\hat{H} = \sum_{pq} h_{pq}E_{pq} + \tfrac{1}{2}\sum_{pqrs} g_{pqrs}(E_{pq}E_{rs} - \delta_{qr}E_{ps}) + h_{\text{nuc}} \tag{2.3.30}$$

whereas the triplet operators (2.3.22)–(2.3.24) are needed for the pure spin operators and the mixed operators. For example, we may write the spin-component operators in the form

$$\hat{S}_{+} = -\sum_{p} \hat{T}^{1,1}_{pp} \tag{2.3.31}$$

$$\hat{S}_{z} = \frac{1}{\sqrt{2}}\sum_{p} \hat{T}^{1,0}_{pp} \tag{2.3.32}$$

$$\hat{S}_{-} = \sum_{p} \hat{T}^{1,-1}_{pp} \tag{2.3.33}$$

and the mixed spin–orbit operator (2.2.47) may be written as

$$\hat{V}_{\text{SO}} = \sum_{pq}(V^{+}_{pq}\hat{T}^{1,-1}_{pq} + \sqrt{2}V^{z}_{pq}\hat{T}^{1,0}_{pq} - V^{-}_{pq}\hat{T}^{1,1}_{pq}) \tag{2.3.34}$$

Comparing (2.3.34) and (2.2.40), we note that the three components of the spin–orbit operator are treated alike in the Cartesian form (2.2.40) but differently in the spin-tensor form (2.3.34). The spin-tensor representation (2.3.34), on the other hand, separates the spin–orbit operator into three terms, each of which produces a well-defined change in the spin projection. From the discussion in this section, we see that the singlet and triplet excitation operators (in Cartesian or spin-tensor form) allow for a compact representation of the second-quantization operators in the orbital basis. The coupling of more than two elementary operators to strings or linear combinations of strings that transform as irreducible spin tensor operators is described in Section 2.6.7.

2.3.5 SINGLET EXCITATION OPERATORS

The singlet excitation operators play an important role in the second-quantization treatment of molecular electronic structure. More generally, they are known as the generators of the unitary group, satisfying the same commutation relations

$$[E_{mn}, E_{pq}] = E_{mq}\delta_{pn} - E_{pn}\delta_{mq} \tag{2.3.35}$$

as the $n \times n$ matrices $\mathbf{E}_{pq}$ where 1 appears in element pq and 0 elsewhere (sometimes referred to as the *matrix units*). We also note the following commutators that are useful for evaluating many of the expressions that arise in second quantization:

$$[E_{mn}, a^\dagger_{p\sigma}] = \delta_{np}a^\dagger_{m\sigma} \tag{2.3.36}$$

$$[E_{mn}, a_{p\sigma}] = -\delta_{mp}a_{n\sigma} \tag{2.3.37}$$

$$[E_{mn}, e_{pqrs}] = \delta_{pn}e_{mqrs} - \delta_{mq}e_{pnrs} + \delta_{rn}e_{pqms} - \delta_{ms}e_{pqrn} \tag{2.3.38}$$

Rank reduction occurs in all cases since the singlet excitation operator is a rank-1 operator. These commutators are proved in Exercise 2.3.

To gain some familiarity with the singlet excitation operator, we shall consider its effect on a simple two-electron system. By substitution and expansion, the following commutator is seen to hold for the singlet excitation operator and the two-body creation operators:

$$[E_{pq}, \hat{Q}^{S,M}_{rs}] = \hat{Q}^{S,M}_{ps}\delta_{qr} + \hat{Q}^{S,M}_{rp}\delta_{sq} \tag{2.3.39}$$

We consider the excitation of electrons from orbital ϕ_q to an orbital ϕ_p different from ϕ_q. Both orbitals are assumed to be nondegenerate. The initial closed-shell state is given by

$$\hat{Q}^{0,0}_{qq}|\text{vac}\rangle = \sqrt{2}a^\dagger_{q\alpha}a^\dagger_{q\beta}|\text{vac}\rangle \tag{2.3.40}$$

Applying the excitation operator to this state, we arrive at an open-shell singlet state

$$E_{pq}\hat{Q}^{0,0}_{qq}|\text{vac}\rangle = [E_{pq}, \hat{Q}^{0,0}_{qq}]|\text{vac}\rangle = (\hat{Q}^{0,0}_{pq}\delta_{qq} + \hat{Q}^{0,0}_{qp}\delta_{qq})|\text{vac}\rangle = 2\hat{Q}^{0,0}_{pq}|\text{vac}\rangle \tag{2.3.41}$$

with one electron in ϕ_p and one electron in ϕ_q:

$$2\hat{Q}^{0,0}_{pq}|\text{vac}\rangle = \sqrt{2}\,(a^\dagger_{p\alpha}a^\dagger_{q\beta} - a^\dagger_{p\beta}a^\dagger_{q\alpha})|\text{vac}\rangle \tag{2.3.42}$$

In the derivation (2.3.41), we used the commutator (2.3.39) and the permutational symmetry of the singlet two-body creation operator. Next, we apply the same excitation operator once more:

$$E_{pq}2\hat{Q}^{0,0}_{pq}|\text{vac}\rangle = 2[E_{pq}, \hat{Q}^{0,0}_{pq}]|\text{vac}\rangle = 2(\hat{Q}^{0,0}_{pq}\delta_{pq} + \hat{Q}^{0,0}_{pp}\delta_{qq})|\text{vac}\rangle = 2\hat{Q}^{0,0}_{pp}|\text{vac}\rangle \tag{2.3.43}$$

and arrive at the final closed-shell doubly excited state with both electrons in ϕ_p. Further application of the same excitation operator annihilates the state since there are no electrons left to be excited from ϕ_q.

2.4 Spin properties of determinants

Having considered in Section 2.3 the spin-tensor properties of the basic operators in second quantization, we now turn our attention to the determinants. In particular, we shall establish under what conditions the determinants are eigenfunctions of the total and projected spins. Knowledge of the spin properties of determinants is essential since the determinants are the basic building blocks of N-electron wave functions.

At this point, a comment on terminology is in order. In Chapter 1, we used the term *ON vector* for second-quantization vectors that correspond to Slater determinants in first quantization. The unusual term *ON vector* was employed in order to make a clear distinction between the first- and second-quantization representations of the same object and to emphasize the separate and independent structure of second quantization. We shall from now on use the conventional term *Slater determinant* or simply *determinant* for ON vectors, bearing in mind that determinants in second quantization are just vectors with elements representing spin-orbital occupations.

2.4.1   GENERAL CONSIDERATIONS

The exact wave function in Fock space is taken to be an eigenfunction of the nonrelativistic Hamiltonian (2.2.18):

$$\hat{H}|0\rangle = E|0\rangle \tag{2.4.1}$$

Since this Hamiltonian is a singlet operator (see Section 2.3.4), it commutes with the operators for the total and projected spins:

$$[\hat{S}^2, \hat{H}] = 0 \tag{2.4.2}$$

$$[\hat{S}_z, \hat{H}] = 0 \tag{2.4.3}$$

The Hamiltonian $\hat{H}$ therefore possesses a common set of eigenfunctions with $\hat{S}^2$ and $\hat{S}_z$ and we shall assume that the exact wave function (2.4.1) is a spin eigenfunction with quantum numbers S and M, respectively. Consequently, when calculating an approximation to this state, we shall often find it convenient to restrict the optimization to the part of the Fock space that is spanned by spin eigenfunctions with quantum numbers S and M. It is therefore important to examine the spin properties of determinants.

Slater determinants are in general not eigenfunctions of the Hamiltonian (2.2.18) but are instead the nondegenerate eigenfunctions of the spin-orbital ON operators:

$$\hat{N}_{p\sigma}|\mathbf{k}\rangle = k_{p\sigma}|\mathbf{k}\rangle \tag{2.4.4}$$

We should therefore be able to establish the general spin properties of determinants by examining the commutators between the spin operators and the spin-orbital ON operators. We note that the spin-orbital ON operators commute with the spin-projection operator:

$$[\hat{S}_z, \hat{N}_{p\sigma}] = 0 \tag{2.4.5}$$

This relation follows from the observation that the spin-projection operator (2.2.30) is a linear combination of spin-orbital ON operators. Thus, since the ON operators commute among themselves (1.3.4), they must also commute with the spin-projection operator. From the commutation relations (2.4.5) and from the observation that there are no degeneracies among the spin-orbital occupation numbers (2.4.4), we conclude that the Slater determinants are eigenfunctions of the projected spin:

$$\hat{S}_z|\mathbf{k}\rangle = M|\mathbf{k}\rangle \tag{2.4.6}$$

We shall determine the precise eigenvalues of the projected spin in Section 2.4.2.

We now proceed to consider the total spin of Slater determinants. Using the relation (1.8.14), we obtain the following commutators between the spin-orbital ON operators and the shift operators:

$$[\hat{S}_+, \hat{N}_{p\sigma}] = a_{p\alpha}^\dagger a_{p\sigma} \delta_{\beta\sigma} - a_{p\sigma}^\dagger a_{p\beta} \delta_{\alpha\sigma} \tag{2.4.7}$$

$$[\hat{S}_-, \hat{N}_{p\sigma}] = a_{p\beta}^\dagger a_{p\sigma} \delta_{\alpha\sigma} - a_{p\sigma}^\dagger a_{p\alpha} \delta_{\beta\sigma} \tag{2.4.8}$$

Evaluating the commutators with $\hat{S}^2$ in the form (2.2.38), we find upon expansion:

$$[\hat{S}^2, \hat{N}_{p\alpha}] = \hat{S}_+ a_{p\beta}^\dagger a_{p\alpha} - a_{p\alpha}^\dagger a_{p\beta} \hat{S}_- \tag{2.4.9}$$

$$[\hat{S}^2, \hat{N}_{p\beta}] = a_{p\alpha}^\dagger a_{p\beta} \hat{S}_- - \hat{S}_+ a_{p\beta}^\dagger a_{p\alpha} \tag{2.4.10}$$

Clearly, the spin-orbital ON operators do not commute with the operator for total spin. In general, therefore, we do not expect Slater determinants to be eigenfunctions of the total spin. Of course, operators that do not commute may still possess common eigenfunctions in special cases. Indeed, in Section 2.4.3, we shall see that there are important classes of determinants that are also spin eigenfunctions.

Although Slater determinants are not by themselves spin eigenfunctions, it is possible to determine spin eigenfunctions as simple linear combinations of determinants. A clue to the procedure for generating spin-adapted determinants is obtained from the observation that both the total- and projected-spin operators commute with the sum of the ON operators for alpha and beta spins:

$$[\hat{S}_z, \hat{N}_{p\alpha} + \hat{N}_{p\beta}] = 0 \tag{2.4.11}$$

$$[\hat{S}^2, \hat{N}_{p\alpha} + \hat{N}_{p\beta}] = 0 \tag{2.4.12}$$

Expression (2.4.12) may be obtained by combining the commutators (2.4.9) and (2.4.10). Relations (2.4.11) and (2.4.12) also follow directly from the observation that $\hat{N}_{p\alpha} + \hat{N}_{p\beta}$ are singlet operators. Spin-adapted functions can therefore be constructed as the simultaneous eigenfunctions of $\hat{S}_z$, $\hat{S}^2$ and $\hat{N}_{p\alpha} + \hat{N}_{p\beta}$. This approach is explored in Sections 2.5 and 2.6.

2.4.2 SPIN PROJECTION OF DETERMINANTS

We assume that the determinant is written as a product of alpha creation operators, *an alpha string*, times a product of beta creation operators, *a beta string*, working on the vacuum state:

$$|\mathbf{k}_\alpha \mathbf{k}_\beta\rangle = \left[\prod_{p=1}^n (a_{p\alpha}^\dagger)^{k_{p\alpha}}\right]\left[\prod_{p=1}^n (a_{p\beta}^\dagger)^{k_{p\beta}}\right]|\text{vac}\rangle \tag{2.4.13}$$

The operators corresponding to the doubly occupied orbitals can be moved out in front of the other creation operators. The determinant then appears as a product of three strings – a core string, an alpha string and a beta string:

$$|\mathbf{k}_\alpha \mathbf{k}_\beta\rangle = \hat{A}_c^\dagger \hat{A}_\alpha^\dagger \hat{A}_\beta^\dagger |\text{vac}\rangle \tag{2.4.14}$$

The core string is given by

$$\hat{A}_c^\dagger = \text{sgn} \prod_{p=1}^n (a_{p\alpha}^\dagger a_{p\beta}^\dagger)^{k_{p\alpha} k_{p\beta}} \tag{2.4.15}$$

where the sign factor sgn depends on the number of transpositions needed to extract the core. The alpha and beta strings are given by

$$\hat{A}_\alpha^\dagger = \prod_{p=1}^n (a_{p\alpha}^\dagger)^{k_{p\alpha} - k_{p\alpha}k_{p\beta}} \tag{2.4.16}$$

$$\hat{A}_\beta^\dagger = \prod_{p=1}^n (a_{p\beta}^\dagger)^{k_{p\beta} - k_{p\alpha}k_{p\beta}} \tag{2.4.17}$$

where only singly occupied orbitals contribute. The spin-projection eigenvalue equation (2.4.6) may now be written in the form

$$\hat{S}_z \hat{A}_c^\dagger \hat{A}_\alpha^\dagger \hat{A}_\beta^\dagger |\text{vac}\rangle = [\hat{S}_z, \hat{A}_c^\dagger \hat{A}_\alpha^\dagger \hat{A}_\beta^\dagger]|\text{vac}\rangle \tag{2.4.18}$$

To obtain the eigenvalue of the projected-spin operator, we first calculate the commutators between $\hat{S}_z$ and the strings (2.4.15)–(2.4.17). Using (2.3.13), these commutators become

$$[\hat{S}_z, \hat{A}_c^\dagger] = 0 \tag{2.4.19}$$

$$[\hat{S}_z, \hat{A}_\alpha^\dagger] = \tfrac{1}{2} n_\alpha \hat{A}_\alpha^\dagger \tag{2.4.20}$$

$$[\hat{S}_z, \hat{A}_\beta^\dagger] = -\tfrac{1}{2} n_\beta \hat{A}_\beta^\dagger \tag{2.4.21}$$

where n_α and n_β are the numbers of orbitals in the alpha and beta strings (i.e. the numbers of unpaired alpha and beta electrons):

$$n_\alpha = \sum_{p=1}^n k_{p\alpha}(1 - k_{p\beta}) \tag{2.4.22}$$

$$n_\beta = \sum_{p=1}^n k_{p\beta}(1 - k_{p\alpha}) \tag{2.4.23}$$

The commutator in the eigenvalue equation (2.4.18) now becomes

$$[\hat{S}_z, \hat{A}_c^\dagger \hat{A}_\alpha^\dagger \hat{A}_\beta^\dagger] = \tfrac{1}{2}(n_\alpha - n_\beta)\hat{A}_c^\dagger \hat{A}_\alpha^\dagger \hat{A}_\beta^\dagger \tag{2.4.24}$$

which implies that all determinants are projected-spin eigenfunctions

$$\hat{S}_z |\mathbf{k}_\alpha \mathbf{k}_\beta\rangle = \tfrac{1}{2}(n_\alpha - n_\beta)|\mathbf{k}_\alpha \mathbf{k}_\beta\rangle \tag{2.4.25}$$

as asserted in Section 2.4.1. As expected, the projected spin is easily obtained from the excess of unpaired alpha electrons over unpaired beta electrons.

2.4.3 TOTAL SPIN OF DETERMINANTS

We now proceed to determine the effect of the operator of the total spin on the determinants:

$$\hat{S}^2 |\mathbf{k}_\alpha \mathbf{k}_\beta\rangle = [\hat{S}_- \hat{S}_+ + \hat{S}_z(\hat{S}_z + 1)]|\mathbf{k}_\alpha \mathbf{k}_\beta\rangle \tag{2.4.26}$$

For this purpose, we shall need the commutators between the operator strings (2.4.15)–(2.4.17) and the shift operators. The commutators involving the raising operator are given by

$$[\hat{S}_+, \hat{A}_c^\dagger] = 0 \tag{2.4.27}$$

$$[\hat{S}_+, \hat{A}_\alpha^\dagger] = 0 \tag{2.4.28}$$

$$[\hat{S}_+, \hat{A}_\beta^\dagger] = \sum_{p=1}^{n} k_{p\beta}(1 - k_{p\alpha}) \hat{A}_\beta^\dagger (a_{p\beta}^\dagger \rightarrow a_{p\alpha}^\dagger) \tag{2.4.29}$$

The effect of the raising operator on a beta string is thus to produce a linear combination of n_β strings, each of which is obtained from the original string by replacing one beta creation operator by the corresponding alpha operator as indicated in (2.4.29). In the same notation, the commutators with the lowering operator are found to be

$$[\hat{S}_-, \hat{A}_c^\dagger] = 0 \tag{2.4.30}$$

$$[\hat{S}_-, \hat{A}_\alpha^\dagger] = \sum_{p=1}^{n} k_{p\alpha}(1 - k_{p\beta}) \hat{A}_\alpha^\dagger (a_{p\alpha}^\dagger \rightarrow a_{p\beta}^\dagger) \tag{2.4.31}$$

$$[\hat{S}_-, \hat{A}_\beta^\dagger] = 0 \tag{2.4.32}$$

The effect of the lowering operator on the alpha string is to generate a linear combination of n_α strings, each obtained by lowering the spin projection of one orbital in the alpha string.

To evaluate (2.4.26), we first apply the raising operator to the determinants. The raising operator generates a linear combination of determinants, as seen from the expression

$$\hat{S}_+ |\mathbf{k}_\alpha \mathbf{k}_\beta\rangle = \tilde{A}_c^\dagger \hat{A}_\alpha^\dagger [\hat{S}_+, \hat{A}_\beta^\dagger]|\text{vac}\rangle \tag{2.4.33}$$

Application of the lowering operator now yields

$$\hat{S}_- \hat{S}_+ |\mathbf{k}_\alpha \mathbf{k}_\beta\rangle = \hat{A}_c^\dagger [\hat{S}_-, \hat{A}_\alpha^\dagger][\hat{S}_+, \hat{A}_\beta^\dagger]|\text{vac}\rangle + \hat{A}_c^\dagger \hat{A}_\alpha^\dagger [\hat{S}_-, [\hat{S}_+, \hat{A}_\beta^\dagger]]|\text{vac}\rangle \tag{2.4.34}$$

Invoking the Jacobi identity (1.8.17), the double commutator may be written as

$$[\hat{S}_-, [\hat{S}_+, \hat{A}_\beta^\dagger]] = [\hat{A}_\beta^\dagger, [\hat{S}_+, \hat{S}_-]] = 2[\hat{A}_\beta^\dagger, \hat{S}_z] = n_\beta \hat{A}_\beta^\dagger \tag{2.4.35}$$

where we also have used (2.2.37) and (2.4.21). Inserting this expression in (2.4.34), we obtain

$$\hat{S}_- \hat{S}_+ |\mathbf{k}_\alpha \mathbf{k}_\beta\rangle = \hat{A}_c^\dagger [\hat{S}_-, \hat{A}_\alpha^\dagger][\hat{S}_+, \hat{A}_\beta^\dagger]|\text{vac}\rangle + n_\beta |\mathbf{k}_\alpha \mathbf{k}_\beta\rangle \tag{2.4.36}$$

The effect of the spin-projection operator in (2.4.26) is easily obtained from (2.4.24):

$$\hat{S}_z(\hat{S}_z + 1)|\mathbf{k}_\alpha \mathbf{k}_\beta\rangle = \tfrac{1}{4}(n_\alpha - n_\beta)(n_\alpha - n_\beta + 2)|\mathbf{k}_\alpha \mathbf{k}_\beta\rangle \tag{2.4.37}$$

Combining the last two results, we obtain the following expression for the effect of the total-spin operator on the determinants

$$\hat{S}^2 |\mathbf{k}_\alpha \mathbf{k}_\beta\rangle = \tfrac{1}{4}[(n_\alpha - n_\beta)^2 + 2(n_\alpha + n_\beta)]|\mathbf{k}_\alpha \mathbf{k}_\beta\rangle + \hat{A}_c^\dagger [\hat{S}_-, \hat{A}_\alpha^\dagger][\hat{S}_+, \hat{A}_\beta^\dagger]|\text{vac}\rangle \tag{2.4.38}$$

which confirms that determinants are in general not eigenfunctions of the total spin. The commutator of the lowering operator with the alpha string and the commutator of the raising operator with the beta string create linear combinations of determinants where the spins of two orbitals have been flipped. However, if one of the commutators vanishes, the determinant becomes an eigenfunction of the total-spin operator.

There are two important cases where Slater determinants are eigenfunctions of the total spin. In *closed-shell systems*, all orbitals are doubly occupied, and the alpha and beta strings become the identity operators. The commutators in (2.4.38) are then zero and the determinant is an eigenfunction of the total spin with eigenvalue zero (since both n_α and n_β are zero). In *high-spin states*, all singly occupied orbitals have the same spin. The string for the opposite spin reduces to the identity operator and the term containing the commutators in (2.4.38) becomes zero. Equation (2.4.38) then reduces to the eigenvalue equation for $\hat{S}^2$

$$\hat{S}^2|\mathbf{k}_\alpha\mathbf{k}_\beta(n_\beta = 0)\rangle = \frac{n_\alpha}{2}\left(\frac{n_\alpha}{2} + 1\right)|\mathbf{k}_\alpha\mathbf{k}_\beta(n_\beta = 0)\rangle \qquad (2.4.39)$$

$$\hat{S}^2|\mathbf{k}_\alpha(n_\alpha = 0)\mathbf{k}_\beta\rangle = \frac{n_\beta}{2}\left(\frac{n_\beta}{2} + 1\right)|\mathbf{k}_\alpha(n_\alpha = 0)\mathbf{k}_\beta\rangle \qquad (2.4.40)$$

where the core string is not referenced in the kets. The determinant $|\mathbf{k}_\alpha\mathbf{k}_\beta(n_\beta = 0)\rangle$ is thus a spin eigenfunction with total spin $n_\alpha/2$ and spin projection $n_\alpha/2$. Similarly, the determinant $|\mathbf{k}_\alpha(n_\alpha = 0)\mathbf{k}_\beta\rangle$ is an eigenfunction with total spin $n_\beta/2$ and projection $-n_\beta/2$.

2.5 Configuration state functions

As discussed in Section 2.4, the exact nonrelativistic wave function is an eigenfunction of the total and projected spins but Slater determinants are eigenfunctions of the projected spin only. Since it is often advantageous to employ a basis of spin eigenfunctions in approximate calculations, we shall in this section see how we can set up a basis of spin eigenfunctions by taking linear combinations of Slater determinants.

The lack of spin symmetry in the determinants arises since the spin-orbital ON operators – of which the determinants are eigenfunctions – do not commute with the operator for the total spin; see (2.4.9) and (2.4.10). By contrast, the *orbital ON operators*

$$\hat{N}_p^o = \hat{N}_{p\alpha} + \hat{N}_{p\beta} \qquad (2.5.1)$$

are singlet operators and commute with the operators for the projected and total spins:

$$[\hat{N}_p^o, \hat{S}_z] = 0 \qquad (2.5.2)$$

$$[\hat{N}_p^o, \hat{S}^2] = 0 \qquad (2.5.3)$$

We can therefore set up a basis of functions that are simultaneously eigenfunctions of the orbital ON operators as well as the operators for the projected and total spins. Such spin-adapted functions are called *configuration state functions (CSFs)* [2].

Let us investigate the eigenfunctions of the orbital ON operators in greater detail. We first note that Slater determinants are eigenfunctions of the orbital ON operators:

$$\hat{N}_p^o|\mathbf{k}\rangle = (k_{p\alpha} + k_{p\beta})|\mathbf{k}\rangle = k_p^o|\mathbf{k}\rangle \qquad (2.5.4)$$

Next, we note that different determinants may have the same orbital occupation numbers but different spin-orbital occupation numbers since an orbital occupation equal to 1 may represent either the occupation of an alpha spin orbital or the occupation of a beta spin orbital by an unpaired electron. The set of all determinants with the same orbital occupation numbers but different spin-orbital occupation numbers is said to constitute an *orbital configuration*. In a sense, an orbital configuration contains all determinants that are degenerate with respect to the orbital ON operators. In constructing a basis of CSFs, we combine the degenerate determinants belonging to the same orbital configuration. The construction of CSFs is considered in Section 2.6.

The validity of the CSF approach follows immediately from the fact that the orbital ON operators commute with the spin operators. However, it is instructive to verify directly that the operator for the total spin does not couple determinants belonging to different orbital configurations. Taking the matrix element of the commutator (2.5.3) between two different determinants and invoking (2.5.4), we obtain

$$(m_p^o - k_p^o)\langle \mathbf{m}|\hat{S}^2|\mathbf{k}\rangle = 0 \tag{2.5.5}$$

If the determinants belong to different configurations, there must be at least one orbital p for which m_p^o and k_p^o differ, and the matrix element of the total-spin operator between such determinants must therefore be equal to zero.

The principal advantage of CSFs is that their use imposes the correct spin symmetry on the approximate wave function. As an additional benefit, the use of CSFs leads to a shorter expansion of the wave function since, for a fixed spin projection $M \geq 0$, the number of CSFs with a given total spin $S = M$ is always less than or equal to the number of determinants. Let us be more specific and compare in a given orbital configuration the number of determinants of spin projection M with the number of CSFs of the same projection M and total spin $S = M$. We begin by counting the number of determinants with projection M in a given orbital configuration.

An orbital configuration is characterized by the number of unpaired electrons – that is, by the number of orbital occupations equal to 1. For a given determinant in this configuration, the total number of unpaired electrons is equal to the number of unpaired alpha electrons n_α plus the number of unpaired beta electrons n_β:

$$N_{\text{open}} = n_\alpha + n_\beta \tag{2.5.6}$$

According to (2.4.25), the spin projection of this determinant is given by

$$M = \tfrac{1}{2}(n_\alpha - n_\beta) \tag{2.5.7}$$

Combining (2.5.6) and (2.5.7), we find that, for a fixed spin projection M, the number of alpha electrons and also the number of beta electrons are the same in all determinants:

$$n_\alpha = \tfrac{1}{2}N_{\text{open}} + M \tag{2.5.8}$$

$$n_\beta = \tfrac{1}{2}N_{\text{open}} - M \tag{2.5.9}$$

In a given orbital configuration, the number of determinants with spin projection M is therefore equal to the number of distinct ways we may distribute the n_α alpha electrons or alternatively the n_β beta electrons among the N_{open} spin orbitals:

$$N_M^d = \binom{N_{\text{open}}}{\tfrac{1}{2}N_{\text{open}} \pm M} \tag{2.5.10}$$

In this expression, both signs yield the same number of determinants, as is easily verified. The total number of determinants for all spin projections is equal to $2^{N_{\text{open}}}$.

To arrive at the number of CSFs with spin projection M and total spin $S = M$, we first note that the number of determinants N_M^d with spin projection M is equal to the number of CSFs with spin projection M and total spin $S \geq M$:

$$N_M^d = \sum_{S=M}^{S_{high}} N_{SM}^c \tag{2.5.11}$$

Here N_{SM}^c is the number of CSFs with total spin S and spin projection M and S_{high} is the highest total spin allowed in the orbital configuration:

$$S_{high} = \tfrac{1}{2} N_{open} \tag{2.5.12}$$

Likewise, the number of determinants N_{M+1}^d with spin projection $M+1$ is equal to the number of CSFs with spin projection $M+1$ and spin $S \geq M+1$:

$$N_{M+1}^d = \sum_{S=M+1}^{S_{high}} N_{S,M+1}^c = \sum_{S=M+1}^{S_{high}} N_{SM}^c \tag{2.5.13}$$

To obtain the last expression, we have used the fact that CSFs with total spin $S \geq M+1$ have the same number of components with projection M as with projection $M+1$. Subtracting (2.5.13) from (2.5.11), we obtain

$$N_{S=M,M}^c = N_M^d - N_{M+1}^d \tag{2.5.14}$$

which gives the number of CSFs with spin projection M and total spin $S = M$. Inserting (2.5.10) in this expression, we obtain, after some simple algebra,

$$N_{S,M=S}^c = \frac{2S+1}{S_{high} + S + 1} \binom{N_{open}}{\tfrac{1}{2}N_{open} - S} \tag{2.5.15}$$

We follow here the convention that the binomial coefficients vanish if any of the factorials contain negative numbers or half integers. The ratio between the number of CSFs of total spin S and projection $M = S$ and the number of determinants of the same projection is therefore

$$\frac{N_{S,M=S}^c}{N_{M=S}^d} = \frac{2S+1}{S_{high} + S + 1} \tag{2.5.16}$$

For singlets, there are thus $S_{high} + 1$ determinants for each CSF. For configurations with a large number of unpaired electrons and a low total spin, the number of CSFs is significantly smaller than the number of determinants.

In conclusion, it should be emphasized that the Slater determinants and the CSFs constitute two alternative sets of basis functions from which we may proceed to calculate approximate wave functions. The determinants are simpler than the CSFs but lead to longer expansions of the wave functions and may sometimes also give solutions that are not true spin eigenfunctions. The CSFs are more complicated – each CSF contains several determinants – but provide a more compact representation of the electronic system with the correct spin symmetry imposed.

2.6   The genealogical coupling scheme

We give in this section an introduction to the construction of CSFs and more generally to the construction of spin tensor operators. We shall employ the *genealogical coupling scheme*, where the final CSF for N electrons is arrived at in a sequence of N steps [2]. At each step, a new electron is introduced and coupled to those already present. We thus arrive at the final CSF through a sequence of $N-1$ intermediate CSFs, each of which represents a spin eigenfunction.

The first spin orbital corresponds to a doublet spin eigenfunction. When the second spin orbital is introduced, it may be coupled to the first one to yield either a singlet or a triplet spin eigenfunction. These two intermediate CSFs for two electrons may then be coupled to a third spin orbital, giving rise to two three-electron doublet CSFs and one quartet. This process is continued until all N electrons have been coupled to yield a final CSF of the desired spin symmetry.

In the construction of CSFs, there is no need to consider the doubly occupied orbitals explicitly – the core string $\hat{A}_c^\dagger$ is a singlet operator and it is the same for all determinants that belong to the same orbital configuration; see the discussion in Section 2.5. We shall therefore be concerned only with those orbitals that are singly occupied. There are a total of N_{open} such orbitals (2.5.6), but for ease of notation we shall use N rather than N_{open} to denote the number of singly occupied orbitals. We also note that, although the present discussion is completely general with respect to the coupling of electron spin, we do not concern ourselves here with the coupling of orbital angular momentum.

2.6.1 REPRESENTATIONS OF DETERMINANTS AND CSFs

A genealogical coupling may be represented by a vector $\mathbf{T}$ of length N where each element T_i indicates the total spin resulting from the spin coupling of the first i electrons. For $N = 3$, the two doublets are represented by the vectors $|^1\mathbf{T}\rangle^c = \left|\frac{1}{2}, 1, \frac{1}{2}\right\rangle^c$ and $|^2\mathbf{T}\rangle^c = \left|\frac{1}{2}, 0, \frac{1}{2}\right\rangle^c$. Superscript c is used to distinguish CSFs from determinants, for which superscript d will be used. The final spin is given by the last entry. Alternatively, the intermediate spin couplings can be specified relative to one another in a vector $\mathbf{t}$ with elements $t_i = T_i - T_{i-1}$ for $i > 1$ and $t_1 = T_1$. The two doublets now become $|^1\mathbf{t}\rangle^c = |+, +, -\rangle^c$ and $|^2\mathbf{t}\rangle^c = |+, -, +\rangle^c$, where for brevity only signs are displayed. In this notation, the final spin is equal to the sum of the elements. The genealogical spin functions can be visualized by drawing straight lines between the points (i, T_i) on a graph. In Figure 2.1, all possible spin couplings are shown for up to 12 unpaired electrons. The number of distinct spin couplings for each value of N and S is also indicated.

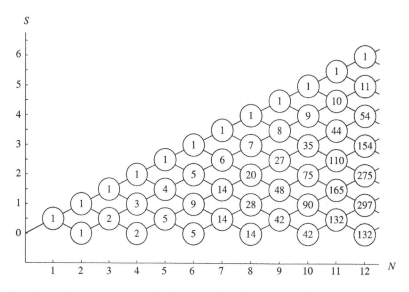

Fig. 2.1. Branching diagram for up to 12 electrons. N is the number of electrons and S the total spin.

We also need a systematic way to represent the Slater determinants belonging to a given configuration. Following the scheme for CSFs, we represent each determinant by a vector $\mathbf{P}$, where the element P_i gives the total spin projection of the first i spin orbitals. Alternatively, the determinant may be represented by a vector $\mathbf{p}$, where the element p_i is equal to the spin projection of spin orbital i (i.e. $-\frac{1}{2}$ or $+\frac{1}{2}$). Three determinants arise when three spin orbitals are combined to give $M = \frac{1}{2}$. In the first scheme, these three determinants are represented by $|\frac{1}{2}, 1, \frac{1}{2}\rangle^{\mathrm{d}}$, $|\frac{1}{2}, 0, \frac{1}{2}\rangle^{\mathrm{d}}$ and $|-\frac{1}{2}, 0, \frac{1}{2}\rangle^{\mathrm{d}}$. In the second scheme, they are represented by $|+, +, -\rangle^{\mathrm{d}}$, $|+, -, +\rangle^{\mathrm{d}}$ and $|-, +, +\rangle^{\mathrm{d}}$. As in the 'relative' notation for CSFs, we display only the signs of the spin projections in the vectors. As a mnemonic for memorizing the conventions for $\mathbf{T}$ and $\mathbf{P}$, the reader may think of $\mathbf{T}$ as standing for 'total spin' and $\mathbf{P}$ for 'projected spin'.

2.6.2 GENEALOGICAL COUPLING

We now consider the expansion of a given N-electron genealogical CSF with total spin S and projected spin M in determinants with spin projection M belonging to the same orbital configuration:

$$|\mathbf{t}\rangle^{\mathrm{c}} = \sum_i d_i |{}^i\mathbf{p}\rangle^{\mathrm{d}} \tag{2.6.1}$$

Each determinant may be written as a product of creation operators working on the closed shell of the spin-paired core electrons:

$$|\mathbf{p}\rangle^{\mathrm{d}} = a_{1p_1}^\dagger a_{2p_2}^\dagger \cdots a_{Np_N}^\dagger |\mathrm{cs}\rangle \tag{2.6.2}$$

Here we enumerate the singly occupied orbitals by integers from 1 to N. The spin projection of each spin orbital is obtained from the corresponding element of $\mathbf{p}$. To determine the coefficients of the expansion (2.6.1), it is convenient to rewrite the CSF as a tensor operator working on the closed shell of core electrons:

$$|\mathbf{t}\rangle^{\mathrm{c}} = \hat{O}_N^{S,M}(\mathbf{t})|\mathrm{cs}\rangle \tag{2.6.3}$$

The operator $\hat{O}_N^{S,M}(\mathbf{t})$ generates a normalized N-electron spin eigenfunction as specified by the genealogical vector $\mathbf{t}$. Since CSFs may be written as linear combinations of Slater determinants, the tensor operator in (2.6.3) must be a linear combination of operator strings each containing N creation operators. The tensor operator therefore commutes with creation operators for even N and anticommutes for odd N.

In the genealogical scheme, we envisage each N-electron state $|\mathbf{t}\rangle^{\mathrm{c}}$ as arising by a coupling of the one-body alpha and beta creation operators with a spin-adapted $(N-1)$-electron state:

$$\hat{O}_N^{S,M}(\mathbf{t})|\mathrm{cs}\rangle = \sum_{\sigma=-1/2}^{1/2} C_{t_N,\sigma}^{S,M} \hat{O}_{N-1}^{S-t_N,M-\sigma}(\mathbf{t}) a_{N\sigma}^\dagger |\mathrm{cs}\rangle \tag{2.6.4}$$

In general, both alpha and beta spin orbitals contribute to this coupling. We therefore have a sum over two $(N-1)$-electron states in (2.6.4), one with spin projection $M - \frac{1}{2}$ (coupled to an alpha electron) and one with projection $M + \frac{1}{2}$ (coupled to a beta electron). The total spin is $T_{N-1} = S - t_N$ for each $(N-1)$-electron state, where t_N is the last element in the genealogical vector. The coupling coefficients in (2.6.4) depend on the total and projected spins of the coupled state S and M and also on the spins of the creation operator σ and t_N, both of which may take on the values $+\frac{1}{2}$ and $-\frac{1}{2}$. For high-spin CSFs, the term $\sigma = -\frac{1}{2}$ vanishes in (2.6.4) for $M = S$, whereas the term $\sigma = \frac{1}{2}$ vanishes for $M = -S$.

The *genealogical coupling coefficients* $C_{t_N,\sigma}^{S,M}$ in (2.6.4) are determined from the requirement that $\hat{O}_N^{S,M}(\mathbf{t})$ should be a normalized tensor operator given that the corresponding $(N-1)$-electron operators on the right-hand side of (2.6.4) are normalized tensor operators. The coefficients $C_{t_N,\sigma}^{S,M}$ correspond to the *vector-coupling coefficients* or the *Clebsch–Gordan coefficients* that are used for coupling two states with spins $S - t_N$ and $\frac{1}{2}$ so as to obtain a new state with total spin S and spin projection M [1]. Their values are tabulated in standard textbooks. In Exercise 2.5, the coupling coefficients are obtained as:

$$C_{1/2,\sigma}^{S,M} = \sqrt{\frac{S + 2\sigma M}{2S}} \tag{2.6.5}$$

$$C_{-1/2,\sigma}^{S,M} = -2\sigma \sqrt{\frac{S + 1 - 2\sigma M}{2(S+1)}} \tag{2.6.6}$$

Note that the genealogical coupling coefficients are independent of the number of electrons.

Once our N-electron CSF has been reduced to a linear combination of one or two $(N-1)$-electron spin eigenfunctions each multiplied by a creation operator, we may go one step further and expand each $(N-1)$-electron state in two $(N-2)$-electron spin eigenfunctions as dictated by the penultimate element t_{N-1} in the genealogical coupling vector $\mathbf{t}$. After $N - 1$ such steps, we arrive at an expansion in terms of determinants with projected spin M. In this way, we are led directly to an expansion of CSFs in Slater determinants where the coefficients are products of the genealogical coupling coefficients in (2.6.5) and (2.6.6).

2.6.3 COUPLING COEFFICIENTS

In Section 2.6.2, we presented an N-step procedure for reducing any given CSF to a linear combination of Slater determinants (2.6.1). In practice, it is easier to calculate the coupling coefficients directly from inner products [3,4]. Since the Slater determinants are orthonormal, the expansion coefficient d_i for the determinant $|^i\mathbf{p}\rangle^\mathrm{d}$ may be extracted from the expansion by multiplying (2.6.1) from the left by the associated bra determinant:

$$d_i = {}^\mathrm{d}\langle {}^i\mathbf{p}|\mathbf{t}\rangle^\mathrm{c} \tag{2.6.7}$$

For ease of notation, we discard the superscript in $^i\mathbf{p}$ and consider the product $^\mathrm{d}\langle \mathbf{p}|\mathbf{t}\rangle^\mathrm{c}$ with an unspecified Slater determinant. Writing the determinant as a product of creation operators (2.6.2) and the CSF by means of a spin tensor operator (2.6.3), we obtain

$$^\mathrm{d}\langle \mathbf{p}|\mathbf{t}\rangle^\mathrm{c} = \langle \mathrm{cs}|a_{Np_N} \cdots a_{1p_1} \hat{O}_N^{T_N,P_N}(\mathbf{t})|\mathrm{cs}\rangle \tag{2.6.8}$$

where we have used the fact that the total and projected spins of the final CSF are given by T_N and P_N, respectively. We may now introduce (2.6.4) and obtain

$$
\begin{aligned}
^\mathrm{d}\langle \mathbf{p}|\mathbf{t}\rangle^\mathrm{c} &= \sum_\sigma C_{t_N,\sigma}^{T_N,P_N} \langle \mathrm{cs}|a_{Np_N} \cdots a_{1p_1} \hat{O}_{N-1}^{T_N-t_N,P_N-\sigma}(\mathbf{t}) a_{N\sigma}^\dagger|\mathrm{cs}\rangle \\
&= C_{t_N,p_N}^{T_N,P_N} \langle \mathrm{cs}|a_{N-1,p_{N-1}} \cdots a_{1p_1} a_{Np_N} a_{Np_N}^\dagger \hat{O}_{N-1}^{T_N-t_N,P_N-p_N}(\mathbf{t})|\mathrm{cs}\rangle \\
&= C_{t_N,p_N}^{T_N,P_N} \langle \mathrm{cs}|a_{N-1,p_{N-1}} \cdots a_{1p_1} \hat{O}_{N-1}^{T_{N-1},P_{N-1}}(\mathbf{t})|\mathrm{cs}\rangle
\end{aligned} \tag{2.6.9}
$$

In going from the first to the second expression, we recognize that only one of the two states contributes since the creation operator $a_{N\sigma}^\dagger$ introduced in the expansion must match one of the annihilation operators from the determinant. This fixes σ to have the value p_N. To obtain the last expression, we eliminate the creation and annihilation operators for spin orbital N and express the superscripts on the tensor operator in terms of T_{N-1} and P_{N-1}. Comparing (2.6.8) and (2.6.9), we note that this process may be repeated to give the final result

$$^{\mathrm{d}}\langle \mathbf{p}|\mathbf{t}\rangle^{\mathrm{c}} = \prod_{n=1}^{N} C_{t_n, p_n}^{T_n, P_n} \tag{2.6.10}$$

where it is understood that the coupling coefficients (2.6.5) and (2.6.6) vanish unless the condition $|P_n| \leq T_n$ is satisfied. The coefficients with $|P_n| > T_n$ vanish since in the expansion of $|\mathbf{T}\rangle^{\mathrm{c}}$ only determinants $|\mathbf{P}\rangle^{\mathrm{d}}$ with $|P_n| \leq T_n$ contribute, as in each intermediate step the magnitude of the spin projection must be smaller than the total spin.

2.6.4   AN EXAMPLE: THREE ELECTRONS IN THREE ORBITALS

To illustrate the genealogical coupling scheme, we shall consider the genealogical CSFs with spin projection $\frac{1}{2}$ that may be generated from a distribution of three electrons among three orbitals. Such an orbital configuration contains two doublets and a quartet, as may easily be verified using (2.5.15). The genealogical doublets are $|\frac{1}{2}, 1, \frac{1}{2}\rangle^{\mathrm{c}}$ and $|\frac{1}{2}, 0, \frac{1}{2}\rangle^{\mathrm{c}}$ and the quartet is $|\frac{3}{2}, 1, \frac{3}{2}\rangle^{\mathrm{c}}$. The Slater determinants with projected spin $\frac{1}{2}$ are $|\frac{1}{2}, 1, \frac{1}{2}\rangle^{\mathrm{d}}$, $|\frac{1}{2}, 0, \frac{1}{2}\rangle^{\mathrm{d}}$ and $|-\frac{1}{2}, 0, \frac{1}{2}\rangle^{\mathrm{d}}$. Our task is to expand the three genealogical CSFs in these three determinants.

Starting with the first doublet, we note that all three determinants contribute since the intermediate spin projections are always numerically smaller than the intermediate spins. Their expansion coefficients are determined as

$$^{\mathrm{d}}\langle \tfrac{1}{2}, 1, \tfrac{1}{2} | \tfrac{1}{2}, 1, \tfrac{1}{2}\rangle^{\mathrm{c}} = C_{1/2,1/2}^{1/2,1/2} \, C_{1/2,1/2}^{1,1} \, C_{-1/2,-1/2}^{1/2,1/2} = 1 \times 1 \times \sqrt{\tfrac{2}{3}} = \sqrt{\tfrac{2}{3}} \tag{2.6.11}$$

$$^{\mathrm{d}}\langle \tfrac{1}{2}, 0, \tfrac{1}{2} | \tfrac{1}{2}, 1, \tfrac{1}{2}\rangle^{\mathrm{c}} = C_{1/2,1/2}^{1/2,1/2} \, C_{1/2,-1/2}^{1,0} \, C_{-1/2,1/2}^{1/2,1/2} = 1 \times \sqrt{\tfrac{1}{2}} \times \left(-\sqrt{\tfrac{1}{3}}\right) = -\sqrt{\tfrac{1}{6}} \tag{2.6.12}$$

$$^{\mathrm{d}}\langle -\tfrac{1}{2}, 0, \tfrac{1}{2} | \tfrac{1}{2}, 1, \tfrac{1}{2}\rangle^{\mathrm{c}} = C_{1/2,-1/2}^{1/2,-1/2} \, C_{1/2,1/2}^{1,0} \, C_{-1/2,1/2}^{1/2,1/2} = 1 \times \sqrt{\tfrac{1}{2}} \times \left(-\sqrt{\tfrac{1}{3}}\right) = -\sqrt{\tfrac{1}{6}} \tag{2.6.13}$$

The expansion of the CSF therefore becomes

$$|\tfrac{1}{2}, 1, \tfrac{1}{2}\rangle^{\mathrm{c}} = \sqrt{\tfrac{2}{3}} \, |\tfrac{1}{2}, 1, \tfrac{1}{2}\rangle^{\mathrm{d}} - \tfrac{1}{\sqrt{6}} \, |\tfrac{1}{2}, 0, \tfrac{1}{2}\rangle^{\mathrm{d}} - \tfrac{1}{\sqrt{6}} \, |-\tfrac{1}{2}, 0, \tfrac{1}{2}\rangle^{\mathrm{d}} \tag{2.6.14}$$

For the CSF $|\frac{1}{2}, 0, \frac{1}{2}\rangle^{\mathrm{c}}$, we note that the Slater determinant $|\frac{1}{2}, 1, \frac{1}{2}\rangle^{\mathrm{d}}$ cannot contribute since, in the second coupling step, the intermediate spin projection would exceed the total spin. Carrying out the calculations, we arrive at the expansion (see Exercise 2.6)

$$|\tfrac{1}{2}, 0, \tfrac{1}{2}\rangle^{\mathrm{c}} = \tfrac{1}{\sqrt{2}} \, |\tfrac{1}{2}, 0, \tfrac{1}{2}\rangle^{\mathrm{d}} - \tfrac{1}{\sqrt{2}} \, |-\tfrac{1}{2}, 0, \tfrac{1}{2}\rangle^{\mathrm{d}} \tag{2.6.15}$$

Using the orthonormality of the Slater determinants, it is easy to verify that the doublets (2.6.14) and (2.6.15) are normalized and orthogonal

$$^{\mathrm{c}}\langle \tfrac{1}{2}, 0, \tfrac{1}{2} | \tfrac{1}{2}, 0, \tfrac{1}{2}\rangle^{\mathrm{c}} = {}^{\mathrm{c}}\langle \tfrac{1}{2}, 1, \tfrac{1}{2} | \tfrac{1}{2}, 1, \tfrac{1}{2}\rangle^{\mathrm{c}} = 1 \tag{2.6.16}$$

$$^c\langle \tfrac{1}{2}, 0, \tfrac{1}{2} \,|\, \tfrac{1}{2}, 1, \tfrac{1}{2} \rangle^c = 0 \tag{2.6.17}$$

Finally, the quartet CSF with the spin projection $\tfrac{1}{2}$ may be generated genealogically, yielding the expansion (see Exercise 2.6)

$$|\tfrac{1}{2}, 1, \tfrac{3}{2}\rangle^c = \tfrac{1}{\sqrt{3}} |\tfrac{1}{2}, 1, \tfrac{1}{2}\rangle^d + \tfrac{1}{\sqrt{3}} |\tfrac{1}{2}, 0, \tfrac{1}{2}\rangle^d + \tfrac{1}{\sqrt{3}} |-\tfrac{1}{2}, 0, \tfrac{1}{2}\rangle^d \tag{2.6.18}$$

From the orthonormality of the determinants, the normalization of this quartet is easily verified as is the orthogonality to the doublets, although orthogonality of course follows directly from the different spin symmetries of the doublets and the quartet. We have thus succeeded in constructing the three genealogical CSFs that can generated by distributing three electrons among three orbitals; see Figure 2.1.

2.6.5 COMPLETENESS AND ORTHONORMALITY

The three CSFs generated by the genealogical scheme in Section 2.6.4 constitute a complete and orthonormal set of states for three unpaired electrons with $M = \tfrac{1}{2}$. Completeness follows since there are only three determinants with $M = \tfrac{1}{2}$ and orthonormality was verified by explicit calculation of inner products. More generally, it can be shown that the genealogical scheme always generates a complete and orthonormal set of spin eigenfunctions for a given orbital configuration. The first part of the proof is given in Exercise 2.7, where it is demonstrated that the number of distinct genealogical couplings is indeed given by (2.5.15). The second part of the proof is given here, where we demonstrate that the genealogical CSFs are orthonormal and therefore linearly independent.

We proceed by induction, assuming that the $(N-1)$-electron CSFs $\hat{O}^{S,M}_{N-1}(\mathbf{t})|\mathrm{cs}\rangle$ constitute an orthonormal set of spin eigenfunctions. This is trivially true for one-electron CSFs. Next, we calculate the overlap between two arbitrarily chosen N-electron CSFs $\hat{O}^{S,M}_N(\mathbf{u})|\mathrm{cs}\rangle$ and $\hat{O}^{S,M}_N(\mathbf{v})|\mathrm{cs}\rangle$. Expanding these CSFs according to (2.6.4), we obtain

$$\hat{O}^{S,M}_N(\mathbf{u}) = \sum_{\sigma=-1/2}^{1/2} C^{S,M}_{u_N,\sigma} \hat{O}^{S-u_N,M-\sigma}_{N-1}(\mathbf{u}) a^\dagger_{N\sigma} \tag{2.6.19}$$

$$\hat{O}^{S,M}_N(\mathbf{v}) = \sum_{\tau=-1/2}^{1/2} C^{S,M}_{v_N,\tau} \hat{O}^{S-v_N,M-\tau}_{N-1}(\mathbf{v}) a^\dagger_{N\tau} \tag{2.6.20}$$

Inserting these expansions in the dot product, we obtain

$$^c\langle \mathrm{cs}|\hat{O}^{S,M}_N(\mathbf{u})^\dagger \hat{O}^{S,M}_N(\mathbf{v})|\mathrm{cs}\rangle^c = \sum_{\sigma\tau} C^{S,M}_{u_N,\sigma} C^{S,M}_{v_N,\tau} \,^c\langle \mathrm{cs}|a_{N\sigma} \hat{O}^{S-u_N,M-\sigma}_{N-1}(\mathbf{u})^\dagger \hat{O}^{S-v_N,M-\tau}_{N-1}(\mathbf{v}) a^\dagger_{N\tau}|\mathrm{cs}\rangle^c$$

$$= \sum_{\sigma\tau} C^{S,M}_{u_N,\sigma} C^{S,M}_{v_N,\tau} \,^c\langle \mathrm{cs}|\hat{O}^{S-u_N,M-\sigma}_{N-1}(\mathbf{u})^\dagger a_{N\sigma} a^\dagger_{N\tau} \hat{O}^{S-v_N,M-\tau}_{N-1}(\mathbf{v})|\mathrm{cs}\rangle^c$$

$$= \sum_{\sigma} C^{S,M}_{u_N,\sigma} C^{S,M}_{v_N,\sigma} \,^c\langle \mathrm{cs}|\hat{O}^{S-u_N,M-\sigma}_{N-1}(\mathbf{u})^\dagger \hat{O}^{S-v_N,M-\sigma}_{N-1}(\mathbf{v})|\mathrm{cs}\rangle^c \tag{2.6.21}$$

The second expression follows from the first since the creation operator to the right commutes or anticommutes with the second spin tensor operator, and likewise the annihilation operator to the left commutes or anticommutes with the first tensor operator. The last expression of (2.6.21) – which follows from the anticommutation of creation and annihilation operators – involves the inner product of two $(N-1)$-electron operators, which are orthonormal by assumption. We may now write the inner product in the form

$$^{\mathrm{c}}\langle \mathrm{cs}|\hat{O}_N^{S,M}(\mathbf{u})^\dagger \hat{O}_N^{S,M}(\mathbf{v})|\mathrm{cs}\rangle^{\mathrm{c}} = \sum_\sigma C_{u_N,\sigma}^{S,M} C_{v_N,\sigma}^{S,M} \delta_{u_N,v_N} \prod_{k=1}^{N-1} \delta_{u_k,v_k} \qquad (2.6.22)$$

The first Kronecker delta arises since different u_N and v_N lead to different spins in the $(N-1)$-electron tensor operators. The remaining Kronecker deltas arise since the $(N-1)$-electron tensor operators are orthogonal by assumption if any of the pairs u_k and v_k are different, although the final spin symmetries are identical. We now write (2.6.22) in the more compact form

$$^{\mathrm{c}}\langle \mathrm{cs}|\hat{O}_N^{S,M}(\mathbf{u})^\dagger \hat{O}_N^{S,M}(\mathbf{v})|\mathrm{cs}\rangle^{\mathrm{c}} = \delta_{\mathbf{u},\mathbf{v}} \sum_\sigma (C_{u_N,\sigma}^{S,M})^2 \qquad (2.6.23)$$

which clearly displays the orthogonality of the spin tensor operators of different genealogies. Normalization of the genealogical CSFs is easily verified. Thus, for $u_N = \frac{1}{2}$ normalization follows by substitution of (2.6.5) in (2.6.23)

$$\sum_\sigma (C_{1/2,\sigma}^{S,M})^2 = (C_{1/2,1/2}^{S,M})^2 + (C_{1/2,-1/2}^{S,M})^2 = \frac{S+M}{2S} + \frac{S-M}{2S} = 1 \qquad (2.6.24)$$

and for $u_N = -\frac{1}{2}$ it follows by substitution of (2.6.6) in (2.6.23).

2.6.6 TRANSFORMATIONS BETWEEN DETERMINANT AND CSF BASES

We now have two different and useful sets of orthonormal basis functions for the Fock space at our disposal – Slater determinants and CSFs. Neither set of functions is superior to the other in all respects – the CSFs exhibit the correct spin symmetry and lead to more compact expansions of the wave function but are more complicated than the Slater determinants, which in many situations are far easier to manipulate than the CSFs. Clearly, there will be situations where we would like to transform our representation of the wave function from one basis to the other. Such transformations are discussed in the present subsection.

For this purpose, we label the Slater determinants $|m, i, j\rangle^{\mathrm{d}}$, where j specifies the type of orbital configuration as characterized by the number of unpaired electrons, i counts the orbital configurations of each type and m specifies the determinants in each configuration. Similarly, the CSFs are labelled $|n, i, j\rangle^{\mathrm{c}}$, where i and j are the same indices as in the determinant basis and n counts the CSFs in each orbital configuration. Slater determinants and CSFs belonging to different configurations are orthogonal:

$$^{\mathrm{d}}\langle m, i, j|n, i', j'\rangle^{\mathrm{c}} = d_{mn}^j \delta_{i,i'} \delta_{j,j'} \qquad (2.6.25)$$

We note that the overlap depends on the type of orbital configuration j (as specified by the number of unpaired electrons) and also on the particular determinant m and CSF n that are chosen within this configuration type. However, the overlaps are the same for all orbital configurations with the same number of unpaired electrons; see (2.6.10).

We first assume that the wave function has been expanded in CSFs:

$$|\mathbf{C}\rangle = \sum_{nij} C_{nij}^{\mathrm{c}} |n, i, j\rangle^{\mathrm{c}} \qquad (2.6.26)$$

To transform to the determinant basis, we invoke the resolution of the identity and obtain

$$|\mathbf{C}\rangle = \sum_{nij} C^c_{nij} \sum_m |m, i, j\rangle^d \,^d\langle m, i, j|n, i, j\rangle^c$$

$$= \sum_{mnij} d^j_{mn} C^c_{nij} |m, i, j\rangle^d = \sum_{mij} C^d_{mij} |m, i, j\rangle^d \tag{2.6.27}$$

where the expansion coefficients are given by

$$C^d_{mij} = \sum_n d^j_{mn} C^c_{nij} \tag{2.6.28}$$

The transformation back to the CSF basis may be carried out in the same way, this time invoking the resolution of the identity within each configuration:

$$|\mathbf{C}\rangle = \sum_{mij} C^d_{mij} |m, i, j\rangle^d = \sum_{mij} C^d_{mij} \sum_n |n, i, j\rangle^c \,^c\langle n, i, j|m, i, j\rangle^d$$

$$= \sum_{nij} C^c_{nij} |n, i, j\rangle^c \tag{2.6.29}$$

where

$$C^c_{nij} = \sum_m d^j_{mn} C^d_{mij} \tag{2.6.30}$$

The transformations between the CSF and determinant bases can thus be carried out as matrix multiplications (2.6.28) and (2.6.30). Since only CSFs and determinants belonging to the same configurations are mixed, the number of operations required for this transformation is considerably less than that needed for a full transformation of the expanded wave function.

2.6.7 GENEALOGICAL COUPLING OF OPERATORS

In the discussion of the genealogical scheme for spin coupling, we have so far considered only the construction of spin eigenfunctions by the linear combination of determinants belonging to the same orbital configuration. The genealogical scheme may, however, also be applied to the construction of spin tensor operators. Indeed, within the framework of second quantization, we may regard the determinants (2.6.2) as strings of operators working on the vacuum state

$$|\mathbf{p}\rangle^d = \hat{X}(\mathbf{p})|\text{vac}\rangle \tag{2.6.31}$$

From this point of view, then, the construction of CSFs from determinants (2.6.1) may be viewed as the construction of a spin tensor operator from strings of creation operators

$$\hat{O}^{S,M}_N(\mathbf{t}) = \sum_i d_i \hat{X}(^i\mathbf{p}) \tag{2.6.32}$$

carried out in N intermediate steps

$$\hat{O}^{S,M}_N(\mathbf{t}) = C^{S,M}_{t_N, 1/2} \hat{O}^{S-t_N, M-1/2}_{N-1}(\mathbf{t}) a^\dagger_{N\alpha} + C^{S,M}_{t_N, -1/2} \hat{O}^{S-t_N, M+1/2}_{N-1}(\mathbf{t}) a^\dagger_{N\beta} \tag{2.6.33}$$

where the coupling coefficients are given by (2.6.5) and (2.6.6). As for CSFs, we may generate the expansion coefficients for the strings in (2.6.32) directly from the coupling coefficients using (2.6.10).

This approach to the construction of a tensor operator is not sufficiently general, however, since the strings in (2.6.32) contain creation operators only. Clearly, for the construction of a general tensor operator, we must extend the scheme in (2.6.32) and (2.6.33) so as to allow for the coupling of both creation and annihilation operators. Fortunately, it turns out that this extension is trivial. Thus, our development of the genealogical scheme for coupling of determinants was based strictly on the tensor and anticommutation relations of the creation operators. These properties are exactly the same for the doublet operator $\{-a_{p\beta}, a_{p\alpha}\}$ as for $\{a_{p\alpha}^\dagger, a_{p\beta}^\dagger\}$. Therefore, at any step in the genealogical coupling of creation operators, we may replace $\{a_{p\alpha}^\dagger, a_{p\beta}^\dagger\}$ by $\{-a_{p\beta}, a_{p\alpha}\}$ and obtain instead a general tensor operator containing annihilation as well as creation operators:

$$\hat{O}_N^{S,M}(\mathbf{t}) = C_{t_N,-1/2}^{S,M} \hat{O}_{N-1}^{S-t_N,M+1/2}(\mathbf{t})a_{N\alpha} - C_{t_N,1/2}^{S,M} \hat{O}_{N-1}^{S-t_N,M-1/2}(\mathbf{t})a_{N\beta} \qquad (2.6.34)$$

Thus, in the construction of a spin tensor operator, we employ (2.6.33) for creation operators and (2.6.34) for annihilation operators, assuming that the number and relative ordering of the creation and annihilation operators are fixed.

As an example of the genealogical scheme for tensor operators, we consider the construction of a one-electron excitation operator of triplet symmetry $S = 1$ and $M = 0$. Because of the requirement of zero spin projection, only two strings will contribute:

$$\hat{O}_2^{1,0}\left(\tfrac{1}{2}, 1\right) = C_{1/2,1/2}^{1/2,1/2} C_{1/2,-1/2}^{1,0} a_{p\alpha}^\dagger a_{q\alpha} + C_{1/2,-1/2}^{1/2,-1/2} C_{1/2,1/2}^{1,0} a_{p\beta}^\dagger (-a_{q\beta}) \qquad (2.6.35)$$

We recall that the spin projection of the annihilation operators $a_{q\alpha}$ and $a_{q\beta}$ is $-\frac{1}{2}$ and $\frac{1}{2}$, respectively. Evaluating the coupling coefficients, we arrive at the operator

$$\hat{O}_2^{1,0}\left(\tfrac{1}{2}, 1\right) = \frac{1}{\sqrt{2}}\left(a_{p\alpha}^\dagger a_{q\alpha} - a_{p\beta}^\dagger a_{q\beta}\right) \qquad (2.6.36)$$

which is identical to the triplet excitation operator (2.3.23). In Exercise 2.8, we derive the two remaining components of the triplet operator and the corresponding singlet; in Exercise 2.9, we discuss the singlet spin coupling of double-excitation operators.

2.7 Density matrices

In Section 1.7, we discussed density matrices in the spin-orbital basis. In the orbital basis, the density matrices are conveniently defined in terms of the spin tensor excitation operators of Section 2.3.4 rather than in terms of the spin-orbital excitation operators of Section 1.3.3. We shall in this section consider the properties of density matrices in the orbital basis, relating their properties to the elementary spin-orbital densities. The exposition in this section follows closely that of Section 1.7.

2.7.1 ORBITAL-DENSITY MATRICES

We consider the expectation value of a Hermitian operator of singlet spin symmetry

$$\hat{\Omega} = \sum_{pq} \Omega_{pq} E_{pq} + \tfrac{1}{2}\sum_{pqrs} \Omega_{pqrs} e_{pqrs} + \Omega_0 \qquad (2.7.1)$$

with respect to some normalized reference state

$$|0\rangle = \sum_{\mathbf{k}} c_{\mathbf{k}} |\mathbf{k}\rangle \tag{2.7.2}$$

$$\langle 0|0\rangle = 1 \tag{2.7.3}$$

which we assume is an eigenfunction of the projected spin

$$\hat{S}_z |0\rangle = M |0\rangle \tag{2.7.4}$$

The function may be an eigenfunction also of the total-spin operator, although this property is not assumed in the development of this section.

Evaluation of the expectation value of $\hat{\Omega}$ with respect to the reference state (2.7.2) gives the expression

$$\langle 0|\hat{\Omega}|0\rangle = \sum_{pq} D_{pq} \Omega_{pq} + \tfrac{1}{2} \sum_{pqrs} d_{pqrs} \Omega_{pqrs} + \Omega_0 \tag{2.7.5}$$

where we have introduced the *one- and two-electron orbital density-matrix elements*

$$D_{pq} = \langle 0|E_{pq}|0\rangle \tag{2.7.6}$$

$$d_{pqrs} = \langle 0|e_{pqrs}|0\rangle = \langle 0|E_{pq}E_{rs} - \delta_{rq}E_{ps}|0\rangle = \sum_{\sigma\tau} \langle 0|a_{p\sigma}^\dagger a_{r\tau}^\dagger a_{s\tau} a_{q\sigma}|0\rangle \tag{2.7.7}$$

These expressions should be compared with (1.7.5) and (1.7.6) in the spin-orbital basis. Expanding the orbital densities in the spin-orbital densities following the definitions of the excitation operators (2.2.7) and (2.2.16), we obtain

$$D_{pq} = \overline{D}_{p\alpha,q\alpha} + \overline{D}_{p\beta,q\beta} \tag{2.7.8}$$

$$d_{pqrs} = \overline{d}_{p\alpha,q\alpha,r\alpha,s\alpha} + \overline{d}_{p\beta,q\beta,r\beta,s\beta} + \overline{d}_{p\alpha,q\alpha,r\beta,s\beta} + \overline{d}_{p\beta,q\beta,r\alpha,s\alpha} \tag{2.7.9}$$

recalling from Section 1.7 that overbars are used for density matrices in the spin-orbital basis. Using these expressions, we may determine the properties of the orbital densities from those of the spin-orbital densities. In particular, we note that the Hermitian and permutational symmetries are the same except that the two-electron orbital density matrix exhibits no symmetry with respect to a permutation of the first and third indices and with respect to a permutation of the second and fourth indices. The full set of permutational symmetries are therefore

$$D_{pq} = D_{qp}^* \tag{2.7.10}$$

$$d_{pqrs} = d_{rspq} = d_{qpsr}^* = d_{srqp}^* \tag{2.7.11}$$

The loss of the permutational symmetries (upper-case indices used for spin orbitals)

$$\overline{d}_{PQRS} = -\overline{d}_{RQPS} \tag{2.7.12}$$

$$\overline{d}_{PQRS} = -\overline{d}_{PSRQ} \tag{2.7.13}$$

occurs because of the orthogonality of the spin functions, which upon integration (2.2.10) forces the first two orbitals and the last two orbitals of the densities to be associated with identical spin functions (2.7.9).

In Section 1.7, we interpreted the diagonal elements of density matrices in the spin-orbital basis as occupation numbers. A similar interpretation may be given to the diagonal elements of the density matrices in the orbital basis. We thus introduce the following *orbital occupation numbers*

$$\omega_p = D_{pp} \tag{2.7.14}$$

$$\omega_{pq} = d_{ppqq} \tag{2.7.15}$$

by analogy with the spin-orbital basis. The quantities ω_p are interpreted as the occupation numbers of a single orbital, whereas ω_{pq} represent the simultaneous (pair) occupations of two orbitals. Expanding the densities according to (2.7.8) and (2.7.9), we obtain the following relationships between the orbital and spin-orbital occupation numbers:

$$\omega_p = \overline{\omega}_{p\alpha} + \overline{\omega}_{p\beta} \tag{2.7.16}$$

$$\omega_{pq} = \overline{\omega}_{p\alpha,q\alpha} + \overline{\omega}_{p\beta,q\beta} + \overline{\omega}_{p\alpha,q\beta} + \overline{\omega}_{p\beta,q\alpha} \tag{2.7.17}$$

Referring to the inequalities (1.7.15) and (1.7.30), we find that the orbital occupation numbers are real numbers restricted to the following intervals:

$$0 \leq \omega_p \leq 2 \tag{2.7.18}$$

$$0 \leq \omega_{pq} \leq 2(2 - \delta_{pq}) \tag{2.7.19}$$

In obtaining (2.7.19), it should be recalled that (1.7.30) was derived under the assumption $p\sigma > q\tau$. The first two terms in (2.7.17) therefore cannot be obtained from (1.7.30) for $p = q$. However, both terms are easily seen to be zero when generalized to $p = q$ and (2.7.19) is then straightforwardly obtained.

The orbital occupations (2.7.18) may exceed 1 since each orbital may be occupied by two electrons of opposite spins. For two distinct orbitals, the pair occupations (2.7.19) may be as large as 4, representing a situation where both orbitals are doubly occupied in all determinants. The pair occupations are defined for a single orbital $p = q$ since two electrons may simultaneously occupy the same orbital. Thus, $\omega_{pp} = d_{pppp}$ measures the simultaneous occupation of electrons in orbitals $\phi_{p\alpha}$ and $\phi_{p\beta}$ and is equal to 2 if both orbitals are occupied in each determinant.

A particularly useful set of occupation numbers is obtained by diagonalizing the one-electron density matrix with a unitary matrix:

$$\mathbf{D} = \mathbf{U}\boldsymbol{\eta}\mathbf{U}^\dagger \tag{2.7.20}$$

The eigenvalues η_p are here the *natural-orbital occupation numbers* and from the eigenvectors $\mathbf{U}$ we may generate a set of orbitals known as the *natural orbitals* of the system. The natural-orbital occupation numbers in the orbital basis are completely analogous to those in the spin-orbital basis (1.7.20) except that the occupation numbers are now real numbers in the range $0 \leq \eta_p \leq 2$. The sum of the natural-orbital occupation numbers is equal to the number of electrons in the system.

2.7.2 SPIN-DENSITY MATRICES

From the orbital density matrices considered in Section 2.7.1, we may calculate expectation values of singlet operators. For triplet operators such as the Fermi contact operator, a different set of density matrices is needed. Consider the evaluation of the expectation value for a one-electron triplet operator of the general form

$$\hat{\Omega}^{\mathrm{T}} = \sum_{pq}(\Omega_{pq}^{1,-1}\hat{T}_{pq}^{1,1} + \Omega_{pq}^{1,0}\hat{T}_{pq}^{1,0} + \Omega_{pq}^{1,1}\hat{T}_{pq}^{1,-1}) \tag{2.7.21}$$

Taking the expectation value of (2.7.21) with respect to the reference function, we obtain

$$\langle 0|\hat{\Omega}^{\mathrm{T}}|0\rangle = \sum_{pq} \Omega_{pq}^{1,0} \langle 0|\hat{T}_{pq}^{1,0}|0\rangle \tag{2.7.22}$$

The expectation values of the remaining two terms in (2.7.21) vanish from spin-symmetry considerations since the reference wave function is an eigenfunction of the spin-projection operator (2.7.4):

$$\langle 0|\hat{T}_{pq}^{1,1}|0\rangle = -\langle 0|a_{p\alpha}^{\dagger}a_{q\beta}|0\rangle = 0 \tag{2.7.23}$$

$$\langle 0|\hat{T}_{pq}^{1,-1}|0\rangle = \langle 0|a_{p\beta}^{\dagger}a_{q\alpha}|0\rangle = 0 \tag{2.7.24}$$

We now introduce the *spin-density matrix* with elements

$$D_{pq}^{\mathrm{T}} = \frac{1}{\sqrt{2}}\langle 0|\hat{T}_{pq}^{1,0}|0\rangle = \frac{1}{2}\langle 0|a_{p\alpha}^{\dagger}a_{q\alpha} - a_{p\beta}^{\dagger}a_{q\beta}|0\rangle \tag{2.7.25}$$

that may be related to the spin-orbital densities in the following way:

$$D_{pq}^{\mathrm{T}} = \tfrac{1}{2}(\overline{D}_{p\alpha,q\alpha} - \overline{D}_{p\beta,q\beta}) \tag{2.7.26}$$

The spin-density matrix thus measures the excess of the density of alpha electrons over beta electrons. Similarly, the spin occupation numbers

$$\omega_p^{\mathrm{T}} = \tfrac{1}{2}(\overline{\omega}_{p\alpha} - \overline{\omega}_{p\beta}) \tag{2.7.27}$$

measure the excess of alpha electrons over beta electrons in ϕ_p. The trace of the spin-density matrix is equal to the total spin projection of the wave function

$$\mathrm{Tr}\,\mathbf{D}^{\mathrm{T}} = \frac{1}{\sqrt{2}}\sum_{p}\langle 0|\hat{T}_{pp}^{1,0}|0\rangle = \langle 0|\hat{S}_z|0\rangle = M \tag{2.7.28}$$

which should be compared with the trace of the orbital density matrix $\mathbf{D}$, which is equal to the total number of electrons in the system.

2.7.3 DENSITY FUNCTIONS

In our discussion, we have so far examined the electron density in the spin-orbital and orbital spaces. Let us now consider the electron density in ordinary space. Of particular interest are the expectation values of operators that probe the presence of electrons at particular points in space. Thus, the one-electron first-quantization operator in the form of a linear combination of Dirac delta functions

$$\delta^{\mathrm{c}}(\mathbf{r}) = \sum_{i=1}^{N} \delta^{\mathrm{c}}(\mathbf{r} - \mathbf{r}_i) \tag{2.7.29}$$

probes the presence of a single electron at $\mathbf{r}$, whereas the two-electron operator

$$\delta^{\mathrm{c}}(\mathbf{r}_1, \mathbf{r}_2) = \tfrac{1}{2}\sum_{i\neq j} \delta^{\mathrm{c}}(\mathbf{r}_1 - \mathbf{r}_i)\delta^{\mathrm{c}}(\mathbf{r}_2 - \mathbf{r}_j) \tag{2.7.30}$$

probes the simultaneous presence of an electron at $\mathbf{r}_1$ and an electron at $\mathbf{r}_2$. Constructing the second-quantization counterparts of these operators in the usual manner, we obtain

$$\hat{\delta}(\mathbf{r}) = \sum_{pq} \phi_p^*(\mathbf{r})\phi_q(\mathbf{r})E_{pq} \tag{2.7.31}$$

$$\hat{\delta}(\mathbf{r}_1, \mathbf{r}_2) = \frac{1}{2} \sum_{pqrs} \phi_p^*(\mathbf{r}_1)\phi_q(\mathbf{r}_1)\phi_r^*(\mathbf{r}_2)\phi_s(\mathbf{r}_2)e_{pqrs} \tag{2.7.32}$$

where we have invoked the definition (1.5.16). Taking the expectation values of these operators, we obtain the *electron-density functions*

$$\rho(\mathbf{r}) = \langle 0|\hat{\delta}(\mathbf{r})|0\rangle = \sum_{pq} D_{pq}\phi_p^*(\mathbf{r})\phi_q(\mathbf{r}) \tag{2.7.33}$$

$$\rho(\mathbf{r}_1, \mathbf{r}_2) = \langle 0|\hat{\delta}(\mathbf{r}_1, \mathbf{r}_2)|0\rangle = \frac{1}{2} \sum_{pqrs} d_{pqrs}\phi_p^*(\mathbf{r}_1)\phi_q(\mathbf{r}_1)\phi_r^*(\mathbf{r}_2)\phi_s(\mathbf{r}_2) \tag{2.7.34}$$

where D_{pq} and d_{pqrs} are elements of the one- and two-electron density matrices.

Calculating the expectation values of (2.7.29) and (2.7.30) in first quantization, we find that these physical densities correspond to the diagonal elements of the first- and second-order reduced density matrices (1.7.37) and (1.7.43) integrated over the spin coordinates

$$\rho(\mathbf{r}_1) = \int \gamma_1(\mathbf{x}_1, \mathbf{x}_1)\,d\sigma_1 \tag{2.7.35}$$

$$\rho(\mathbf{r}_1, \mathbf{r}_2) = \int \gamma_2(\mathbf{x}_1, \mathbf{x}_2, \mathbf{x}_1, \mathbf{x}_2)\,d\sigma_1\,d\sigma_2 \tag{2.7.36}$$

These expressions may also be arrived at by integrating the reduced density matrices (1.7.41) and (1.7.44) over spin, introducing the orbital densities (2.7.8) and (2.7.9), and finally invoking (2.7.33) and (2.7.34). The one-electron function (2.7.35) gives the density of the electronic charge in ordinary space, whereas the pair function (2.7.36) gives information about the correlated distribution of electron pairs in space.

It is also possible to set up an operator that returns the excess of alpha electrons over beta electrons at a particular point in space. In first quantization, such an operator may be written in the form

$$\delta_S^c(\mathbf{r}) = \sum_{i=1}^{N} \delta^c(\mathbf{r} - \mathbf{r}_i)S_z^c(i) \tag{2.7.37}$$

The second-quantization representation of this operator becomes

$$\hat{\delta}_S(\mathbf{r}) = \frac{1}{\sqrt{2}} \sum_{pq} \phi_p^*(\mathbf{r})\phi_q(\mathbf{r})\hat{T}_{pq}^{1,0} \tag{2.7.38}$$

and the *spin-density function* is therefore given by

$$\rho_S(\mathbf{r}) = \langle 0|\hat{\delta}_S(\mathbf{r})|0\rangle = \sum_{pq} D_{pq}^{\mathrm{T}}\phi_p^*(\mathbf{r})\phi_q(\mathbf{r}) \tag{2.7.39}$$

where the triplet densities are given by (2.7.26) analogous to (2.7.33) for the total density. Again, we may relate this density to the reduced density matrix (1.7.37) introduced in Section 1.7.3.

We obtain the following expression

$$\rho_S(\mathbf{r}) = \tfrac{1}{2}[\gamma_1(\mathbf{r}, \alpha; \mathbf{r}, \alpha) - \gamma_1(\mathbf{r}, \beta; \mathbf{r}, \beta)] \tag{2.7.40}$$

as is readily verified by calculating the expectation value of (2.7.37) in first quantization.

References

[1]   D. M. Brink and G. R. Satchler, *Angular Momentum*, 2nd edn, Clarendon Press, 1968.
[2]   R. Pauncz, *Spin Eigenfunctions*, Plenum Press, 1979.
[3]   J. E. Grabenstetter, T. J. Tseng and F. Grein, *Int. J. Quantum Chem.* **10**, 143 (1976).
[4]   W. Duch, *GRMS or Graphical Representation of Model Spaces I: Basics*, Lecture Notes in Chemistry, Vol. 42, Springer-Verlag, 1986.

Further reading

E. R. Davidson, *Reduced Density Matrices in Quantum Chemistry*, Academic Press, 1976.
W. Duch, *GRMS or Graphical Representation of Model Spaces I: Basics*, Lecture Notes in Chemistry, Vol. 42, Springer-Verlag, 1986.
R. Pauncz, *Spin Eigenfunctions*, Plenum Press, 1979.

Exercises

EXERCISE 2.1

One of the operators that describe the interaction between nuclear magnetic dipoles and the electrons is the *Fermi contact operator*

$$H_{\text{FC}}^{\text{c}} = \sum_A \sum_i \gamma_A \delta(\mathbf{r}_i - \mathbf{R}_A) \mathbf{S}_i^{\text{c}} \cdot \mathbf{I}_A \tag{2E.1.1}$$

where $\mathbf{S}_i^{\text{c}}$ is the spin of electron i, $\mathbf{R}_A$ the position of nucleus A, $\mathbf{I}_A$ the spin of nucleus A, and γ_A its magnetogyric ratio. Determine the second-quantization representation of H_{FC}^{c}.

EXERCISE 2.2

In this exercise, we consider the relationship between two relativistic operators: the *two-electron Darwin operator* and the *spin–spin contact operator*:

$$H_{\text{2D}}^{\text{c}} = -\frac{\pi}{c^2} \sum_{i>j} \delta(\mathbf{r}_i - \mathbf{r}_j) \tag{2E.2.1}$$

$$H_{\text{SSC}}^{\text{c}} = -\frac{8\pi}{3c^2} \sum_{i>j} \delta(\mathbf{r}_i - \mathbf{r}_j) \mathbf{S}_i^{\text{c}} \cdot \mathbf{S}_j^{\text{c}} \tag{2E.2.2}$$

1. Show that the second-quantization representation of the two-electron Darwin operator becomes

$$\hat{H}_{\text{2D}} = \sum_{pqrs} S_{pqrs} a_{p\alpha}^\dagger a_{r\beta}^\dagger a_{s\beta} a_{q\alpha} \tag{2E.2.3}$$

where

$$S_{pqrs} = -\frac{\pi}{c^2} \int \phi_p^*(\mathbf{r}) \phi_q(\mathbf{r}) \phi_r^*(\mathbf{r}) \phi_s(\mathbf{r}) \, d\mathbf{r} \tag{2E.2.4}$$

2. Show that, in second quantization, the two-electron Darwin and spin–spin contact operators are related as

$$\hat{H}_{SSC} = -2\hat{H}_{2D} \tag{2E.2.5}$$

EXERCISE 2.3

Verify the following commutation relations

1. $[E_{mn}, a_{p\sigma}^\dagger] = \delta_{np} a_{m\sigma}^\dagger$ (2E.3.1)

2. $[E_{mn}, a_{p\sigma}] = -\delta_{mp} a_{n\sigma}$ (2E.3.2)

3. $[E_{mn}, E_{pq}] = E_{mq}\delta_{pn} - E_{pn}\delta_{mq}$ (2E.3.3)

4. $[E_{mn}, e_{pqrs}] = \delta_{pn}e_{mqrs} - \delta_{mq}e_{pnrs} + \delta_{rn}e_{pqms} - \delta_{ms}e_{pqrn}$ (2E.3.4)

EXERCISE 2.4

Let $\hat{\kappa}$ be a one-electron operator

$$\hat{\kappa} = \sum_{mn} \kappa_{mn} E_{mn} \tag{2E.4.1}$$

and $\hat{H}$ the electronic Hamiltonian

$$\hat{H} = \sum_{pq} h_{pq} E_{pq} + \frac{1}{2} \sum_{pqrs} g_{pqrs} e_{pqrs} + h_{\text{nuc}} \tag{2E.4.2}$$

The orbitals are assumed to be real so that the two-electron integrals have the permutational symmetries (2.2.14) and similarly for the one-electron integrals.

1. Show that the commutator $[\hat{\kappa}, \hat{H}]$ can be written as

$$[\hat{\kappa}, \hat{H}] = \sum_{pq} h_{pq}^\kappa E_{pq} + \frac{1}{2} \sum_{pqrs} g_{pqrs}^\kappa e_{pqrs} \tag{2E.4.3}$$

in terms of the *one-index transformed integrals*

$$h_{pq}^\kappa = \sum_m (\kappa_{pm} h_{mq} - \kappa_{mq} h_{pm}) \tag{2E.4.4}$$

$$g_{pqrs}^\kappa = \sum_m (\kappa_{pm} g_{mqrs} - \kappa_{mq} g_{pmrs} + \kappa_{rm} g_{pqms} - \kappa_{ms} g_{pqrm}) \tag{2E.4.5}$$

2. Show that g_{pqrs}^κ is symmetric with respect to the permutation of the two electrons:

$$g_{pqrs}^\kappa = g_{rspq}^\kappa \tag{2E.4.6}$$

3. Assume that κ_{mn} is antisymmetric, $\kappa_{mn} = -\kappa_{nm}$. Show that h_{pq}^{κ} and g_{pqrs}^{κ} then satisfy the same symmetry relations as h_{pq} and g_{pqrs}:

$$h_{pq}^{\kappa} = h_{qp}^{\kappa} \tag{2E.4.7}$$

$$g_{pqrs}^{\kappa} = g_{qprs}^{\kappa} = g_{pqsr}^{\kappa} = g_{qpsr}^{\kappa} \tag{2E.4.8}$$

EXERCISE 2.5

1. Assume that $\hat{O}_{N-1}^{S-1/2,M\pm1/2}$ are components of a normalized spin tensor operator of rank $S - \frac{1}{2}$. Show that the operator

$$\hat{O}_N^{S,M} = C_{1/2,1/2}^{S,M} \hat{O}_{N-1}^{S-1/2,M-1/2} a_{n\alpha}^{\dagger} + C_{1/2,-1/2}^{S,M} \hat{O}_{N-1}^{S-1/2,M+1/2} a_{n\beta}^{\dagger} \tag{2E.5.1}$$

becomes a normalized spin tensor operator of rank S if the coefficients are chosen as

$$C_{1/2,\pm1/2}^{S,M} = \sqrt{\frac{S \pm M}{2S}} \tag{2E.5.2}$$

2. Next, assume that $\hat{O}_{N-1}^{S+1/2,M\pm1/2}$ are components of a normalized tensor operator of rank $S + 1/2$. Show that the operator

$$\hat{O}_N^{S,M} = C_{-1/2,1/2}^{S,M} \hat{O}_{N-1}^{S+1/2,M-1/2} a_{n\alpha}^{\dagger} + C_{-1/2,-1/2}^{S,M} \hat{O}_{N-1}^{S+1/2,M+1/2} a_{n\beta}^{\dagger} \tag{2E.5.3}$$

becomes a normalized spin tensor operator of rank S if the coefficients are chosen as

$$C_{-1/2,\pm1/2}^{S,M} = \mp\sqrt{\frac{S + 1 \mp M}{2(S + 1)}} \tag{2E.5.4}$$

EXERCISE 2.6

Consider a configuration with three singly occupied orbitals. For $M_s = \frac{1}{2}$, determine the determinantal expansions of the doublet with spin coupling $\left|\frac{1}{2}, 0, \frac{1}{2}\right\rangle^c$ and of the quartet with spin coupling $\left|\frac{1}{2}, 1, \frac{3}{2}\right\rangle^c$.

EXERCISE 2.7

Show that the number of distinct genealogical spin couplings with N_{open} unpaired electrons and spin S is given as

$$N_{S,N_{\text{open}}}^c = \binom{N_{\text{open}}}{\frac{1}{2}N_{\text{open}} - S} - \binom{N_{\text{open}}}{\frac{1}{2}N_{\text{open}} - S - 1} \tag{2E.7.1}$$

with the convention that $\binom{n}{m}$ is zero for $m < 0$ or $n < m$. This number is identical to the number of CSFs as given by (2.5.15).

EXERCISE 2.8

In this exercise, we use the genealogical scheme to couple the tensor operators $\{a_{p\alpha}^{\dagger}, a_{p\beta}^{\dagger}\}$ and $\{-a_{q\beta}, a_{q\alpha}\}$ to obtain spin-adapted one-electron operators.

1. Show that the singlet operator becomes

$$\hat{S}_{pq}^{0,0} = \frac{1}{\sqrt{2}}(a_{p\alpha}^{\dagger}a_{q\alpha} + a_{p\beta}^{\dagger}a_{q\beta}) \tag{2E.8.1}$$

2. Show that the three components of the triplet operator become

$$\hat{T}_{pq}^{1,1} = -a_{p\alpha}^{\dagger}a_{q\beta} \tag{2E.8.2}$$

$$\hat{T}_{pq}^{1,0} = \frac{1}{\sqrt{2}}(a_{p\alpha}^{\dagger}a_{q\alpha} - a_{p\beta}^{\dagger}a_{q\beta}) \tag{2E.8.3}$$

$$\hat{T}_{pq}^{1,-1} = a_{p\beta}^{\dagger}a_{q\alpha} \tag{2E.8.4}$$

EXERCISE 2.9

Consider the genealogical coupling of $\{a_{p\alpha}^{\dagger}, a_{p\beta}^{\dagger}\}$, $\{a_{r\alpha}^{\dagger}, a_{r\beta}^{\dagger}\}$, $\{-a_{q\beta}, a_{q\alpha}\}$ and $\{-a_{s\beta}, a_{s\alpha}\}$ to spin-adapted two-electron excitation operators. We assume that p, q, r, s are all different and couple the elementary operators in the order creation–annihilation–creation–annihilation. The two singlet operators become $\hat{O}_4^{0,0}\left(\frac{1}{2}, 0, \frac{1}{2}, 0\right)$ and $\hat{O}_4^{0,0}\left(\frac{1}{2}, 1, \frac{1}{2}, 0\right)$, where the numbers in parentheses describe the intermediate spin couplings.

1. Construct the two-electron singlet operator $\hat{O}_4^{0,0}\left(\frac{1}{2}, 0, \frac{1}{2}, 0\right)$ and show that it may be expressed in terms of one-electron singlet operators in the following manner:

$$\hat{O}_4^{0,0}\left(\tfrac{1}{2}, 0, \tfrac{1}{2}, 0\right) = \hat{S}_{pq}^{0,0}\hat{S}_{rs}^{0,0} = \tfrac{1}{2}E_{pq}E_{rs} \tag{2E.9.1}$$

2. Construct the two-electron singlet operator $\hat{O}_4^{0,0}\left(\frac{1}{2}, 1, \frac{1}{2}, 0\right)$ and express it in terms of the one-electron triplet operators:

$$\hat{O}_4^{0,0}\left(\tfrac{1}{2}, 1, \tfrac{1}{2}, 0\right) = \frac{1}{\sqrt{3}}(\hat{T}_{pq}^{1,1}\hat{T}_{rs}^{1,-1} - \hat{T}_{pq}^{1,0}\hat{T}_{rs}^{1,0} + \hat{T}_{pq}^{1,-1}\hat{T}_{rs}^{1,1}) \tag{2E.9.2}$$

3. Show that (2E.9.2) can be expressed in terms of singlet one-electron operators as

$$\hat{O}_4^{0,0}\left(\tfrac{1}{2}, 1, \tfrac{1}{2}, 0\right) = \frac{1}{\sqrt{3}}\left(E_{ps}E_{rq} + \tfrac{1}{2}E_{pq}E_{rs}\right) \tag{2E.9.3}$$

4. The operators $\hat{O}_4^{0,0}\left(\frac{1}{2}, 0, \frac{1}{2}, 0\right)$ and $\hat{O}_4^{0,0}\left(\frac{1}{2}, 1, \frac{1}{2}, 0\right)$ can be unitarily transformed in the following manner:

$$(\hat{O}_4^{0,0}(1), \hat{O}_4^{0,0}(-1)) = \left(\hat{O}_4^{0,0}\left(\tfrac{1}{2}, 0, \tfrac{1}{2}, 0\right), \hat{O}_4^{0,0}\left(\tfrac{1}{2}, 1, \tfrac{1}{2}, 0\right)\right)\begin{pmatrix} \cos\theta & -\sin\theta \\ \sin\theta & \cos\theta \end{pmatrix} \tag{2E.9.4}$$

Show that it is possible to select θ such that the transformed operators become the more symmetrical pair

$$\hat{O}_4^{0,0}(S) = \frac{S}{2\sqrt{2} - S}(E_{pq}E_{rs} + SE_{ps}E_{rq}) \tag{2E.9.5}$$

where $S = \pm 1$.

EXERCISE 2.10

Consider the spin tensor operator $\hat{T}^{S,M}$. Show that

$$\hat{R} = \sum_{M=-S}^{S} (-1)^M \hat{T}^{S,M} \hat{T}^{S,-M} \tag{2E.10.1}$$

is a singlet spin tensor operator.

EXERCISE 2.11

Use the properties of spin tensor operators to verify the following relationship between the matrix elements of a singlet operator $\hat{O}^{0,0}$

$$\langle S, M | \hat{O}^{0,0} | S, M \rangle = \langle S, M+1 | \hat{O}^{0,0} | S, M+1 \rangle \tag{2E.11.1}$$

for $M < S$. This result shows that the diagonal matrix elements of a singlet operator are independent of the spin projection M.

Solutions

SOLUTION 2.1

The second-quantization representation of (2E.1.1) may be written in the form:

$$
\hat{H}_{FC} = \sum_A \gamma_A \sum_{p\sigma q\tau} \int \phi_p^*(\mathbf{r}) \sigma(m_s) \delta(\mathbf{r} - \mathbf{R}_A) \mathbf{S}^c \cdot \mathbf{I}_A \phi_q(\mathbf{r}) \tau(m_s) \, d\mathbf{r} \, dm_s \, a_{p\sigma}^\dagger a_{q\tau}
$$

$$
= \sum_A \gamma_A \sum_{p\sigma q\tau} \phi_p^*(\mathbf{R}_A) \phi_q(\mathbf{R}_A) \int \sigma(m_s) \mathbf{S}^c \cdot \mathbf{I}_A \tau(m_s) \, dm_s \, a_{p\sigma}^\dagger a_{q\tau} \tag{2S.1.1}
$$

Using (2.2.32) and (2.2.33), the dot product in (2S.1.1) becomes

$$\mathbf{S}^c \cdot \mathbf{I}_A = \frac{1}{2}(S_+^c + S_-^c) I_{Ax} + \frac{1}{2i}(S_+^c - S_-^c) I_{Ay} + S_z^c I_{Az} \tag{2S.1.2}$$

and (2S.1.1) may be expressed as

$$\hat{H}_{FC} = \sum_A \gamma_A \sum_{pq} \phi_p^*(\mathbf{R}_A) \phi_q(\mathbf{R}_A)$$

$$\times \left[\frac{1}{2} I_{Ax}(a_{p\alpha}^\dagger a_{q\beta} + a_{p\beta}^\dagger a_{q\alpha}) + \frac{1}{2i} I_{Ay}(a_{p\alpha}^\dagger a_{q\beta} - a_{p\beta}^\dagger a_{q\alpha}) + \frac{1}{2} I_{Az}(a_{p\alpha}^\dagger a_{q\alpha} - a_{p\beta}^\dagger a_{q\beta}) \right] \tag{2S.1.3}$$

where we have used relations (2.2.22)–(2.2.24). Finally, introducing the Cartesian triplet excitation operator (2.2.42)–(2.2.44), we obtain for the Fermi contact operator:

$$\hat{H}_{FC} = \sum_A \gamma_A \sum_{pq} \phi_p^*(\mathbf{R}_A) \phi_q(\mathbf{R}_A)(I_{Ax} \hat{T}_{pq}^x + I_{Ay} \hat{T}_{pq}^y + I_{Az} \hat{T}_{pq}^z) \tag{2S.1.4}$$

SOLUTION 2.2

1. Writing the spin-free two-electron Darwin operator in the more symmetric form

$$H^c_{2D} = -\frac{\pi}{2c^2} \sum_{i \neq j} \delta(\mathbf{r}_i - \mathbf{r}_j) \tag{2S.2.1}$$

we find that its second-quantization representation becomes

$$\hat{H}_{2D} = \frac{1}{2} \sum_{\substack{pqrs \\ \sigma\tau}} S_{pqrs} a^\dagger_{p\sigma} a^\dagger_{r\tau} a_{s\tau} a_{q\sigma} \tag{2S.2.2}$$

where the two-electron integrals are given by

$$S_{pqrs} = -\frac{\pi}{c^2} \iint \phi^*_p(\mathbf{r}_1)\phi_q(\mathbf{r}_1)\delta(\mathbf{r}_1 - \mathbf{r}_2)\phi^*_r(\mathbf{r}_2)\phi_s(\mathbf{r}_2)\,\mathrm{d}\mathbf{r}_1\,\mathrm{d}\mathbf{r}_2$$

$$= -\frac{\pi}{c^2} \int \phi^*_p(\mathbf{r})\phi_q(\mathbf{r})\phi^*_r(\mathbf{r})\phi_s(\mathbf{r})\,\mathrm{d}\mathbf{r} \tag{2S.2.3}$$

Consider the terms in (2S.2.2) with $\sigma = \tau = \alpha$. From the anticommutation relations of the creation operators, we obtain

$$\frac{1}{2} \sum_{pqrs} S_{pqrs} a^\dagger_{p\alpha} a^\dagger_{r\alpha} a_{s\alpha} a_{q\alpha} = \frac{1}{2} \sum_{\substack{p>r \\ qs}} S_{pqrs} a^\dagger_{p\alpha} a^\dagger_{r\alpha} a_{s\alpha} a_{q\alpha} + \frac{1}{2} \sum_{\substack{p<r \\ qs}} S_{pqrs} a^\dagger_{p\alpha} a^\dagger_{r\alpha} a_{s\alpha} a_{q\alpha}$$

$$= \frac{1}{2} \sum_{\substack{p>r \\ qs}} S_{pqrs} a^\dagger_{p\alpha} a^\dagger_{r\alpha} a_{s\alpha} a_{q\alpha} + \frac{1}{2} \sum_{\substack{p>r \\ qs}} S_{rqps} a^\dagger_{r\alpha} a^\dagger_{p\alpha} a_{s\alpha} a_{q\alpha} = 0 \tag{2S.2.4}$$

where, to obtain the second equality, we have renamed the indices p and r and used the special permutational symmetry of the two-electron integrals

$$S_{pqrs} = S_{rqps} = S_{psrq} = S_{rspq} \tag{2S.2.5}$$

The terms $\sigma = \tau = \beta$ vanish in the same manner and (2S.2.2) becomes

$$\hat{H}_{2D} = \frac{1}{2} \sum_{pqrs} S_{pqrs}(a^\dagger_{p\alpha} a^\dagger_{r\beta} a_{s\beta} a_{q\alpha} + a^\dagger_{p\beta} a^\dagger_{r\alpha} a_{s\alpha} a_{q\beta}) = \sum_{pqrs} S_{pqrs} a^\dagger_{p\alpha} a^\dagger_{r\beta} a_{s\beta} a_{q\alpha} \tag{2S.2.6}$$

where again we have used the permutational symmetries (2S.2.5).

2. We first rewrite the spin-dependent two-electron spin–spin contact operator in the form

$$H^c_{SSC} = -\frac{4\pi}{3c^2} \sum_{i \neq j} \delta(\mathbf{r}_i - \mathbf{r}_j) \left(\tfrac{1}{2}S^c_{i+}S^c_{j-} + \tfrac{1}{2}S^c_{i-}S^c_{j+} + S_{iz}S_{jz}\right) \tag{2S.2.7}$$

Following the general procedure for constructing two-electron operators in the spin-orbital basis, we obtain

$$\hat{H}_{SSC} = \frac{1}{2} \sum_{poq\tau r\mu sv} [H^c_{SSC}]_{poq\tau r\mu sv} a^\dagger_{p\sigma} a^\dagger_{r\mu} a_{sv} a_{q\tau} \tag{2S.2.8}$$

where the integrals are given by

$$[H^c_{SSC}]_{poq\tau r\mu sv} = \tfrac{4}{3} S_{pqrs} \iint \sigma^*(m_1)\mu^*(m_2)(S^c_{1+}S^c_{2-} + S^c_{1-}S^c_{2+} + 2S_{1z}S_{2z})\tau(m_1)v(m_2)\,\mathrm{d}m_1\,\mathrm{d}m_2 \tag{2S.2.9}$$

where the spatial factor S_{pqrs} is defined in (2E.2.4). From the properties of the spin operators, we find that only the following integrals are nonzero:

$$[H_{SCC}]_{p\alpha q\alpha r\alpha s\alpha} = [H_{SCC}]_{p\beta q\beta r\beta s\beta} = \frac{2}{3}S_{pqrs} \tag{2S.2.10}$$

$$[H_{SCC}]_{p\alpha q\alpha r\beta s\beta} = [H_{SCC}]_{p\beta q\beta r\alpha s\alpha} = -\frac{2}{3}S_{pqrs} \tag{2S.2.11}$$

$$[H_{SCC}]_{p\alpha q\beta r\beta s\alpha} = [H_{SCC}]_{p\beta q\alpha r\alpha s\beta} = \frac{4}{3}S_{pqrs} \tag{2S.2.12}$$

The contributions from the (2S.2.10) integrals to (2S.2.8) vanish for the same reason as for the two-electron Darwin operator. The second-quantization representation of the two-electron spin–spin contact term then becomes

$$\hat{H}_{SSC} = -\frac{1}{3}\sum_{pqrs} S_{pqrs}(a_{p\alpha}^{\dagger}a_{r\beta}^{\dagger}a_{s\beta}a_{q\alpha} + a_{p\beta}^{\dagger}a_{r\alpha}^{\dagger}a_{s\alpha}a_{q\beta})$$

$$+ \frac{2}{3}\sum_{pqrs} S_{pqrs}(a_{p\alpha}^{\dagger}a_{r\beta}^{\dagger}a_{s\alpha}a_{q\beta} + a_{p\beta}^{\dagger}a_{r\alpha}^{\dagger}a_{s\beta}a_{q\alpha})$$

$$= -2\sum_{pqrs} S_{pqrs}a_{p\alpha}^{\dagger}a_{r\beta}^{\dagger}a_{s\beta}a_{q\alpha} = -2\hat{H}_{2D} \tag{2S.2.13}$$

To obtain the last equality, we have used the permutational symmetry of the integrals in (2S.2.5).

SOLUTION 2.3

1. Expand the excitation operator and use the commutator (1.8.11):

$$[E_{mn}, a_{p\sigma}^{\dagger}] = \sum_{\tau} \left[a_{m\tau}^{\dagger}a_{n\tau}, a_{p\sigma}^{\dagger}\right] = \delta_{np}a_{m\sigma}^{\dagger} \tag{2S.3.1}$$

2. Expand the excitation operator and use the commutator (1.8.12):

$$[E_{mn}, a_{p\sigma}] = \sum_{\tau} \left[a_{m\tau}^{\dagger}a_{n\tau}, a_{p\sigma}\right] = -\delta_{mp}a_{n\sigma} \tag{2S.3.2}$$

3. Expand the excitation operators and use (1.8.11) and (1.8.12):

$$[E_{mn}, E_{pq}] = \sum_{\sigma\tau} \left[a_{m\sigma}^{\dagger}a_{n\sigma}, a_{p\tau}^{\dagger}a_{q\tau}\right]$$

$$= \sum_{\sigma\tau} \left(\left[a_{m\sigma}^{\dagger}a_{n\sigma}, a_{p\tau}^{\dagger}\right]a_{q\tau} + a_{p\tau}^{\dagger}\left[a_{m\sigma}^{\dagger}a_{n\sigma}, a_{q\tau}\right]\right)$$

$$= \sum_{\sigma\tau} \left(\delta_{np}\delta_{\sigma\tau}a_{m\sigma}^{\dagger}a_{q\tau} - \delta_{qm}\delta_{\sigma\tau}a_{p\tau}^{\dagger}a_{n\sigma}\right)$$

$$= \delta_{np}E_{mq} - \delta_{qm}E_{pn} \tag{2S.3.3}$$

4. The strategy is to expand the two-electron excitation operators in the commutator (2E.3.4), reduce the rank by evaluating the commutators (2E.3.3), and express the result in terms of the two-electron operators. Inserting (2.2.16) in (2E.3.4), we obtain

$$[E_{mn}, e_{pqrs}] = [E_{mn}, E_{pq}E_{rs} - \delta_{qr}E_{ps}]$$

$$= E_{pq}[E_{mn}, E_{rs}] + [E_{mn}, E_{pq}]E_{rs} - \delta_{qr}[E_{mn}, E_{ps}] \tag{2S.3.4}$$

The commutators are now evaluated using (2E.3.3), giving

$$[E_{mn}, e_{pqrs}] = \delta_{nr} E_{pq} E_{ms} - \delta_{ms} E_{pq} E_{rn} + \delta_{np} E_{mq} E_{rs} - \delta_{mq} E_{pn} E_{rs}$$
$$- \delta_{qr} \delta_{np} E_{ms} + \delta_{qr} \delta_{ms} E_{pn} \tag{2S.3.5}$$

Next, we replace all products of excitation operators by the two-electron excitation operators according to (2.2.16). For example, in the first term in (2S.3.5), we make the substitution

$$E_{pq} E_{ms} = e_{pqms} + \delta_{qm} E_{ps} \tag{2S.3.6}$$

All terms involving one-electron excitation operators such as E_{ps} in (2S.3.6) cancel, yielding (2E.3.4).

SOLUTION 2.4

1. Using (2.3.35), we obtain for the one-electron part:

$$[\hat{\kappa}, \hat{h}] = \sum_{pqmn} \kappa_{mn} h_{pq} [E_{mn}, E_{pq}]$$

$$= \sum_{pqmn} \kappa_{mn} h_{pq} (E_{mq} \delta_{pn} - E_{pn} \delta_{mq})$$

$$= \sum_{pqm} (\kappa_{pm} h_{mq} - \kappa_{mq} h_{pm}) E_{pq} \tag{2S.4.1}$$

Next, using (2.3.38) for the two-electron part, we obtain

$$[\hat{\kappa}, \hat{g}] = \tfrac{1}{2} \sum_{\substack{pqrs \\ mn}} \kappa_{mn} g_{pqrs} [E_{mn}, e_{pqrs}]$$

$$= \tfrac{1}{2} \sum_{\substack{pqrs \\ mn}} \kappa_{mn} g_{pqrs} (\delta_{pn} e_{mqrs} - \delta_{mq} e_{pnrs} + \delta_{rn} e_{pqms} - \delta_{ms} e_{pqrn})$$

$$= \tfrac{1}{2} \sum_{\substack{pqrs \\ m}} (\kappa_{pm} g_{mqrs} - \kappa_{mq} g_{pmrs} + \kappa_{rm} g_{pqms} - \kappa_{ms} g_{pqrm}) e_{pqrs} \tag{2S.4.2}$$

2. The particle symmetry (2E.4.6) follows directly from the corresponding symmetry of the untransformed integrals:

$$g^{\kappa}_{rspq} = \sum_{m} (\kappa_{rm} g_{mspq} - \kappa_{ms} g_{rmpq} + \kappa_{pm} g_{rsmq} - \kappa_{mq} g_{rspm})$$

$$= \sum_{m} (\kappa_{rm} g_{pqms} - \kappa_{ms} g_{pqrm} + \kappa_{pm} g_{mqrs} - \kappa_{mq} g_{pmrs})$$

$$= g^{\kappa}_{pqrs} \tag{2S.4.3}$$

3. The permutational symmetry (2E.4.7) follows from the symmetry of h_{pq} and the antisymmetry of κ_{mn}:

$$h^{\kappa}_{qp} = \sum_{m} (\kappa_{qm} h_{mp} - \kappa_{mp} h_{qm}) = \sum_{m} (-\kappa_{mq} h_{pm} + \kappa_{pm} h_{mq}) = h^{\kappa}_{pq} \tag{2S.4.4}$$

The first permutational symmetry in (2E.4.8) follows from the corresponding symmetry of g_{pqrs} and the antisymmetry of κ_{mn}:

$$g_{qprs}^{\kappa} = \sum_m (\kappa_{qm}g_{mprs} - \kappa_{mp}g_{qmrs} + \kappa_{rm}g_{qpms} - \kappa_{ms}g_{qprm})$$

$$= \sum_m (-\kappa_{mq}g_{pmrs} + \kappa_{pm}g_{mqrs} + \kappa_{rm}g_{pqms} - \kappa_{ms}g_{pqrm})$$

$$= g_{pqrs}^{\kappa} \qquad (2S.4.5)$$

The remaining permutational symmetries in (2E.4.8) may be verified in the same manner.

SOLUTION 2.5

1. Applying (2.3.1) twice, we find that the spin tensor operator $\hat{O}_N^{S,M}$ must satisfy the condition

$$[\hat{S}_-, [\hat{S}_+, \hat{O}_N^{S,M}]] = A_{0,0}^2 \hat{O}_N^{S,M} \qquad (2S.5.1)$$

where we have used the short-hand notation

$$A_{s,m} = \sqrt{(S-s)(S-s+1) - (M-m)(M-m+1)} \qquad (2S.5.2)$$

We shall use (2S.5.1) to establish the ratio between the two coefficients in (2E.5.2). Evaluating the inner commutator in (2S.5.1), we obtain

$$[\hat{S}_+, \hat{O}_N^{S,M}] = (A_{1/2,1/2} C_{1/2,1/2}^{S,M} + C_{1/2,-1/2}^{S,M}) \hat{O}_{N-1}^{S-1/2,M+1/2} a_{n\alpha}^\dagger$$
$$+ A_{1/2,-1/2} C_{1/2,-1/2}^{S,M} \hat{O}_{N-1}^{S-1/2,M+3/2} a_{n\beta}^\dagger \qquad (2S.5.3)$$

Next, evaluating the outer commutator, we arrive at the expression

$$[\hat{S}_-, [\hat{S}_+, \hat{O}_N^{S,M}]] = A_{1/2,1/2}(A_{1/2,1/2} C_{1/2,1/2}^{S,M} + C_{1/2,-1/2}^{S,M}) \hat{O}_{N-1}^{S-1/2,M-1/2} a_{n\alpha}^\dagger$$
$$+ [A_{1/2,1/2} C_{1/2,1/2}^{S,M} + (1 + A_{1/2,-1/2}^2) C_{1/2,-1/2}^{S,M}] \hat{O}_{N-1}^{S-1/2,M+1/2} a_{n\beta}^\dagger \qquad (2S.5.4)$$

Substituting this into (2S.5.1) and equating the coefficients of the same operators on the two sides, we obtain the equations:

$$A_{1/2,1/2}^2 C_{1/2,1/2}^{S,M} + A_{1/2,1/2} C_{1/2,-1/2}^{S,M} = A_{0,0}^2 C_{1/2,1/2}^{S,M} \qquad (2S.5.5)$$

$$A_{1/2,1/2} C_{1/2,1/2}^{S,M} + (1 + A_{1/2,-1/2}^2) C_{1/2,-1/2}^{S,M} = A_{0,0}^2 C_{1/2,-1/2}^{S,M} \qquad (2S.5.6)$$

Either equation may be used to express $C_{1/2,1/2}^{S,M}$ in terms of $C_{1/2,-1/2}^{S,M}$, yielding

$$C_{1/2,1/2}^{S,M} = C_{1/2,-1/2}^{S,M} \sqrt{\frac{S+M}{S-M}} \qquad (2S.5.7)$$

To establish the absolute magnitudes of the coefficients, we invoke the normalization condition

$$\langle \text{vac}|(\hat{O}_N^{S,M})^\dagger \hat{O}_N^{S,M}|\text{vac}\rangle = (C_{1/2,1/2}^{S,M})^2 + (C_{1/2,-1/2}^{S,M})^2 = 1 \qquad (2S.5.8)$$

which, combined with (2S.5.7), yields

$$(C^{S,M}_{1/2,-1/2})^2 \left(\frac{S+M}{S-M} + 1 \right) = 1 \qquad (2S.5.9)$$

With the phase factor equal to $+1$, this equation yields the coefficient $C^{S,M}_{1/2,-1/2}$ in (2E.5.2). $C^{S,M}_{1/2,1/2}$ is obtained from (2S.5.7).

2. The coefficients in (2E.5.3) are determined in the same manner as in Exercise 2.5.1. Evaluating the commutator in

$$[\hat{S}_-, [\hat{S}_+, \hat{O}^{S,M}_N]] = A^2_{0,0} \hat{O}^{S,M}_N \qquad (2S.5.10)$$

with the operator in the form (2E.5.3), we obtain

$$[\hat{S}_-, [\hat{S}_+, \hat{O}^{S,M}_N]] = A_{-1/2,1/2}(A_{-1/2,1/2}C^{S,M}_{-1/2,1/2} + C^{S,M}_{-1/2,-1/2})\hat{O}^{S+1/2,M-1/2}_{N-1}a^\dagger_{n\alpha}$$

$$+ [A_{-1/2,1/2}C^{S,M}_{-1/2,1/2} + (1 + A^2_{-1/2,-1/2})C^{S,M}_{-1/2,-1/2}]\hat{O}^{S+1/2,M+1/2}_{N-1}a^\dagger_{n\beta} \qquad (2S.5.11)$$

Inserting this into (2S.5.10) and equating the coefficients in front of $\hat{O}^{S+1/2,M-1/2}_{N-1}a^\dagger_{n\alpha}$ on the two sides, we obtain the condition:

$$A^2_{-1/2,1/2}C^{S,M}_{-1/2,1/2} + A_{-1/2,1/2}C^{S,M}_{-1/2,-1/2} = A^2_{0,0}C^{S,M}_{-1/2,1/2} \qquad (2S.5.12)$$

Solving for $C^{S,M}_{-1/2,1/2}$, we obtain the relation:

$$C^{S,M}_{-1/2,1/2} = -\sqrt{\frac{S+1-M}{S+1+M}}C^{S,M}_{-1/2,-1/2} \qquad (2S.5.13)$$

Next, invoking the normalization condition on the spin tensor operator (2E.5.3), we obtain

$$\left(1 + \frac{S+1-M}{S+1+M} \right)(C^{S,M}_{-1/2,-1/2})^2 = 1 \qquad (2S.5.14)$$

From (2S.5.13) and (2S.5.14), we arrive at (2E.5.4), having made an arbitrary choice of phase factor.

SOLUTION 2.6

There are three determinants with $M = \frac{1}{2}$: $|\frac{1}{2}, 1, \frac{1}{2}\rangle^d$, $|\frac{1}{2}, 0, \frac{1}{2}\rangle^d$ and $|-\frac{1}{2}, 0, \frac{1}{2}\rangle^d$. The doublet $|\frac{1}{2}, 0, \frac{1}{2}\rangle^c$ has no contribution from $|\frac{1}{2}, 1, \frac{1}{2}\rangle^d$ since this determinant has spin projection one after the second coupling, while the CSF has a total spin zero after the second coupling:

$$^d\langle \frac{1}{2}, 1, \frac{1}{2} | \frac{1}{2}, 0, \frac{1}{2} \rangle^c = 0 \qquad (2S.6.1)$$

The coefficients for the remaining determinants in $|\frac{1}{2}, 0, \frac{1}{2}\rangle^c$ are

$$^d\langle \frac{1}{2}, 0, \frac{1}{2} | \frac{1}{2}, 0, \frac{1}{2} \rangle^c = C^{1/2,1/2}_{1/2,1/2}\, C^{0,0}_{-1/2,-1/2}\, C^{1/2,1/2}_{1/2,1/2} = 1 \times \frac{1}{\sqrt{2}} \times 1 = \frac{1}{\sqrt{2}} \qquad (2S.6.2)$$

$$^d\langle -\frac{1}{2}, 0, \frac{1}{2} | \frac{1}{2}, 0, \frac{1}{2} \rangle^c = C^{1/2,-1/2}_{1/2,-1/2}\, C^{0,0}_{-1/2,1/2}\, C^{1/2,1/2}_{1/2,1/2} = 1 \times \frac{-1}{\sqrt{2}} \times 1 = -\frac{1}{\sqrt{2}} \qquad (2S.6.3)$$

leading to the expansion (2.6.15):

$$\left|\tfrac{1}{2}, 0, \tfrac{1}{2}\right\rangle^{\text{c}} = \tfrac{1}{\sqrt{2}} \left|\tfrac{1}{2}, 0, \tfrac{1}{2}\right\rangle^{\text{d}} - \tfrac{1}{\sqrt{2}} \left|-\tfrac{1}{2}, 0, \tfrac{1}{2}\right\rangle^{\text{d}} \tag{2S.6.4}$$

The expansion coefficients for the quartet state $\left|\tfrac{1}{2}, 1, \tfrac{3}{2}\right\rangle^{\text{c}}$ are given by

$$^{\text{d}}\left\langle \tfrac{1}{2}, 1, \tfrac{1}{2} \middle| \tfrac{1}{2}, 1, \tfrac{3}{2}\right\rangle^{\text{c}} = C_{1/2,1/2}^{1/2,1/2}\, C_{1/2,1/2}^{1,1}\, C_{1/2,-1/2}^{3/2,1/2} = 1 \times 1 \times \sqrt{\tfrac{1}{3}} = \tfrac{1}{\sqrt{3}} \tag{2S.6.5}$$

$$^{\text{d}}\left\langle \tfrac{1}{2}, 0, \tfrac{1}{2} \middle| \tfrac{1}{2}, 1, \tfrac{3}{2}\right\rangle^{\text{c}} = C_{1/2,1/2}^{1/2,1/2}\, C_{1/2,-1/2}^{1,0}\, C_{1/2,1/2}^{3/2,1/2} = 1 \times \sqrt{\tfrac{1}{2}} \times \sqrt{\tfrac{2}{3}} = \tfrac{1}{\sqrt{3}} \tag{2S.6.6}$$

$$^{\text{d}}\left\langle -\tfrac{1}{2}, 0, \tfrac{1}{2} \middle| \tfrac{1}{2}, 1, \tfrac{3}{2}\right\rangle^{\text{c}} = C_{1/2,-1/2}^{1/2,-1/2}\, C_{1/2,1/2}^{1,0}\, C_{1/2,1/2}^{3/2,1/2} = 1 \times \sqrt{\tfrac{1}{2}} \times \sqrt{\tfrac{2}{3}} = \tfrac{1}{\sqrt{3}} \tag{2S.6.7}$$

yielding the CSF in (2.6.18):

$$\left|\tfrac{1}{2}, 1, \tfrac{3}{2}\right\rangle^{\text{c}} = \tfrac{1}{\sqrt{3}} \left|\tfrac{1}{2}, 1, \tfrac{1}{2}\right\rangle^{\text{d}} + \tfrac{1}{\sqrt{3}} \left|\tfrac{1}{2}, 0, \tfrac{1}{2}\right\rangle^{\text{d}} + \tfrac{1}{\sqrt{3}} \left|-\tfrac{1}{2}, 0, \tfrac{1}{2}\right\rangle^{\text{d}} \tag{2S.6.8}$$

SOLUTION 2.7

Assuming that (2E.7.1) is valid for $N_{\text{open}} = k$, we shall prove that it is valid for $N_{\text{open}} = k + 1$:

$$N_{S,k+1}^{\text{c}} = \binom{k+1}{\tfrac{1}{2}(k+1-2S)} - \binom{k+1}{\tfrac{1}{2}(k+1-2S)-1} \tag{2S.7.1}$$

We consider three cases separately. First, the highest spin $S = (k+1)/2$ is obtained from coupling with the highest k-electron spin function and occurs exactly once, in agreement with (2S.7.1):

$$N_{(k+1)/2,k+1}^{\text{c}} = \binom{k+1}{0} - \binom{k+1}{-1} = 1 \tag{2S.7.2}$$

Next, for $0 < S < (k+1)/2$, there is one spin coupling for each k-electron spin function with spin $S + \tfrac{1}{2}$ or $S - \tfrac{1}{2}$:

$$N_{S,k+1}^{\text{c}} = N_{S-1/2,k}^{\text{c}} + N_{S+1/2,k}^{\text{c}} = \binom{k}{\tfrac{1}{2}(k-2S+1)} - \binom{k}{\tfrac{1}{2}(k-2S-1)-1} \tag{2S.7.3}$$

From the identity

$$\binom{n}{m} + \binom{n}{m+1} = \binom{n+1}{m+1} \tag{2S.7.4}$$

we can rewrite the coefficients in (2S.7.3) as

$$\binom{k}{\tfrac{1}{2}(k-2S+1)} = \binom{k+1}{\tfrac{1}{2}(k-2S+1)} - \binom{k}{\tfrac{1}{2}(k-2S-1)} \tag{2S.7.5}$$

$$\binom{k}{\tfrac{1}{2}(k-2S-1)-1} = \binom{k+1}{\tfrac{1}{2}(k-2S-1)} - \binom{k}{\tfrac{1}{2}(k-2S-1)} \tag{2S.7.6}$$

Hence

$$N^c_{S,k+1} = \binom{k+1}{\frac{1}{2}(k+1-2S)} - \binom{k+1}{\frac{1}{2}(k+1-2S)-1} \tag{2S.7.7}$$

We conclude that (2S.7.1) is valid for all $S > 0$.

Finally, $S = 0$ occurs only when $k + 1$ is an even number. The number of singlets is then equal to the number of doublets with k orbitals:

$$N^c_{0,k+1} = N^c_{1/2,k} = \binom{k}{\frac{1}{2}(k-1)} - \binom{k}{\frac{1}{2}(k-1)-1}$$

$$= \binom{k+1}{\frac{1}{2}(k+1)} - \binom{k+1}{\frac{1}{2}(k+1)-1} \tag{2S.7.8}$$

To obtain the last equality, we have used (2S.7.4) and

$$\binom{n}{m} = \binom{n}{n-m} \tag{2S.7.9}$$

From (2S.7.2), (2S.7.7), and (2S.7.8) we see that (2S.7.1) is valid for $N_{\text{open}} = k + 1$ if it is valid for $N_{\text{open}} = k$. Since it is valid for $k = 2$ (one singlet and one triplet), it is valid in general.

SOLUTION 2.8

1. Genealogical coupling of the singlet operator:

$$\hat{S}^{0,0}_{pq} = \hat{O}^{0,0}_2 \left(\tfrac{1}{2}, 0\right) = C^{1/2,1/2}_{1/2,1/2} C^{0,0}_{-1/2,-1/2} a^\dagger_{p\alpha} a_{q\alpha} - C^{1/2,-1/2}_{1/2,-1/2} C^{0,0}_{-1/2,1/2} a^\dagger_{p\beta} a_{q\beta}$$

$$= \frac{1}{\sqrt{2}} (a^\dagger_{p\alpha} a_{q\alpha} + a^\dagger_{p\beta} a_{q\beta}) \tag{2S.8.1}$$

2. Genealogical coupling of the triplet operator:

$$\hat{T}^{1,1}_{pq} = \hat{O}^{1,1}_2 \left(\tfrac{1}{2}, 1\right) = -C^{1/2,1/2}_{1/2,1/2} C^{1,1}_{1/2,1/2} a^\dagger_{p\alpha} a_{q\beta} = -a^\dagger_{p\alpha} a_{q\beta} \tag{2S.8.2}$$

$$\hat{T}^{1,0}_{pq} = \hat{O}^{1,0}_2 \left(\tfrac{1}{2}, 1\right) = C^{1/2,1/2}_{1/2,1/2} C^{1,0}_{1/2,-1/2} a^\dagger_{p\alpha} a_{q\alpha} - C^{1/2,-1/2}_{1/2,-1/2} C^{1,0}_{1/2,1/2} a^\dagger_{p\beta} a_{q\beta} \tag{2S.8.3}$$

$$= \frac{1}{\sqrt{2}} (a^\dagger_{p\alpha} a_{q\alpha} - a^\dagger_{p\beta} a_{q\beta})$$

$$\hat{T}^{1,-1}_{pq} = \hat{O}^{1,-1}_2 \left(\tfrac{1}{2}, 1\right) = C^{1/2,-1/2}_{1/2,-1/2} C^{1,-1}_{1/2,-1/2} a^\dagger_{p\beta} a_{q\alpha} = a^\dagger_{p\beta} a_{q\alpha} \tag{2S.8.4}$$

SOLUTION 2.9

1. Four operator strings contribute to $\hat{O}^{0,0}_4 \left(\tfrac{1}{2}, 0, \tfrac{1}{2}, 0\right)$:

$$\hat{O}^{0,0}_4 \left(\tfrac{1}{2}, 0, \tfrac{1}{2}, 0\right) = C^{1/2,1/2}_{1/2,1/2} C^{0,0}_{-1/2,-1/2} C^{1/2,1/2}_{1/2,1/2} C^{0,0}_{-1/2,-1/2} a^\dagger_{p\alpha} a_{q\alpha} a^\dagger_{r\alpha} a_{s\alpha}$$

$$- C^{1/2,1/2}_{1/2,1/2} C^{0,0}_{-1/2,-1/2} C^{1/2,-1/2}_{1/2,-1/2} C^{0,0}_{-1/2,1/2} a^\dagger_{p\alpha} a_{q\alpha} a^\dagger_{r\beta} a_{s\beta}$$

$$- C^{1/2,-1/2}_{1/2,-1/2} C^{0,0}_{-1/2,1/2} C^{1/2,1/2}_{1/2,1/2} C^{0,0}_{-1/2,-1/2} a^\dagger_{p\beta} a_{q\beta} a^\dagger_{r\alpha} a_{s\alpha}$$

$$+ C^{1/2,-1/2}_{1/2,-1/2} C^{0,0}_{-1/2,1/2} C^{1/2,-1/2}_{1/2,-1/2} C^{0,0}_{-1/2,1/2} a^\dagger_{p\beta} a_{q\beta} a^\dagger_{r\beta} a_{s\beta} \tag{2S.9.1}$$

Evaluating the coefficients and expressing the result in terms of the singlet excitation operators (2.3.21) and (2.3.29), we obtain:

$$\hat{O}_4^{0,0}\left(\tfrac{1}{2},0,\tfrac{1}{2},0\right) = \tfrac{1}{2}a_{p\alpha}^\dagger a_{q\alpha}a_{r\alpha}^\dagger a_{s\alpha} + \tfrac{1}{2}a_{p\alpha}^\dagger a_{q\alpha}a_{r\beta}^\dagger a_{s\beta} + \tfrac{1}{2}a_{p\beta}^\dagger a_{q\beta}a_{r\alpha}^\dagger a_{s\alpha} + \tfrac{1}{2}a_{p\beta}^\dagger a_{q\beta}a_{r\beta}^\dagger a_{s\beta}$$

$$= \hat{S}_{pq}^{0,0}\hat{S}_{rs}^{0,0} = \tfrac{1}{2}E_{pq}E_{rs} \tag{2S.9.2}$$

2. Six operator strings contribute to $\hat{O}_4^{0,0}\left(\tfrac{1}{2},1,\tfrac{1}{2},0\right)$:

$$\hat{O}_4^{0,0}\left(\tfrac{1}{2},1,\tfrac{1}{2},0\right) = C_{1/2,1/2}^{1/2,1/2}\,C_{1/2,-1/2}^{1,0}\,C_{-1/2,1/2}^{1/2,1/2}\,C_{-1/2,-1/2}^{0,0}a_{p\alpha}^\dagger a_{q\alpha}a_{r\alpha}^\dagger a_{s\alpha}$$

$$- C_{1/2,1/2}^{1/2,1/2}\,C_{1/2,-1/2}^{1,0}\,C_{-1/2,-1/2}^{1/2,-1/2}\,C_{-1/2,1/2}^{0,0}a_{p\alpha}^\dagger a_{q\alpha}a_{r\beta}^\dagger a_{s\beta}$$

$$- C_{1/2,-1/2}^{1/2,-1/2}\,C_{1/2,1/2}^{1,0}\,C_{-1/2,1/2}^{1/2,1/2}\,C_{-1/2,-1/2}^{0,0}a_{p\beta}^\dagger a_{q\beta}a_{r\alpha}^\dagger a_{s\alpha}$$

$$+ C_{1/2,-1/2}^{1/2,-1/2}\,C_{1/2,1/2}^{1,0}\,C_{-1/2,-1/2}^{1/2,-1/2}\,C_{-1/2,1/2}^{0,0}a_{p\beta}^\dagger a_{q\beta}a_{r\beta}^\dagger a_{s\beta}$$

$$- C_{1/2,1/2}^{1/2,1/2}\,C_{1/2,1/2}^{1,1}\,C_{-1/2,-1/2}^{1/2,1/2}\,C_{-1/2,-1/2}^{0,0}a_{p\alpha}^\dagger a_{q\beta}a_{r\beta}^\dagger a_{s\alpha}$$

$$- C_{1/2,-1/2}^{1/2,-1/2}\,C_{1/2,-1/2}^{1,-1}\,C_{-1/2,1/2}^{1/2,-1/2}\,C_{-1/2,1/2}^{0,0}a_{p\beta}^\dagger a_{q\alpha}a_{r\alpha}^\dagger a_{s\beta} \tag{2S.9.3}$$

Evaluating the coefficients and expressing the result in terms of the one-electron triplet excitation operator (2.3.22)–(2.3.24), we obtain

$$\hat{O}_4^{0,0}\left(\tfrac{1}{2},1,\tfrac{1}{2},0\right) = -\frac{1}{2\sqrt{3}}a_{p\alpha}^\dagger a_{q\alpha}a_{r\alpha}^\dagger a_{s\alpha} + \frac{1}{2\sqrt{3}}a_{p\alpha}^\dagger a_{q\alpha}a_{r\beta}^\dagger a_{s\beta} + \frac{1}{2\sqrt{3}}a_{p\beta}^\dagger a_{q\beta}a_{r\alpha}^\dagger a_{s\alpha}$$

$$- \frac{1}{2\sqrt{3}}a_{p\beta}^\dagger a_{q\beta}a_{r\beta}^\dagger a_{s\beta} - \frac{1}{\sqrt{3}}a_{p\alpha}^\dagger a_{q\beta}a_{r\beta}^\dagger a_{s\alpha} - \frac{1}{\sqrt{3}}a_{p\beta}^\dagger a_{q\alpha}a_{r\alpha}^\dagger a_{s\beta}$$

$$= \frac{1}{\sqrt{3}}(\hat{T}_{pq}^{1,1}\hat{T}_{rs}^{1,-1} - \hat{T}_{pq}^{1,0}\hat{T}_{rs}^{1,0} + \hat{T}_{pq}^{1,-1}\hat{T}_{rs}^{1,1}) \tag{2S.9.4}$$

3. By adding and subtracting $(2\sqrt{3})^{-1}a_{p\alpha}^\dagger a_{q\alpha}a_{r\alpha}^\dagger a_{s\alpha}$ and $(2\sqrt{3})^{-1}a_{p\beta}^\dagger a_{q\beta}a_{r\beta}^\dagger a_{s\beta}$ and interchanging positions of the annihilation operators in (2S.9.4), we obtain

$$\hat{O}_4^{0,0}\left(\tfrac{1}{2},1,\tfrac{1}{2},0\right) = \frac{1}{2\sqrt{3}}(a_{p\alpha}^\dagger a_{q\alpha} + a_{p\beta}^\dagger a_{q\beta})(a_{r\alpha}^\dagger a_{s\alpha} + a_{r\beta}^\dagger a_{s\beta})$$

$$+ \frac{1}{\sqrt{3}}(a_{p\alpha}^\dagger a_{s\alpha} + a_{p\beta}^\dagger a_{s\beta})(a_{r\alpha}^\dagger a_{q\alpha} + a_{r\beta}^\dagger a_{q\beta})$$

$$= \frac{1}{\sqrt{3}}(2S_{ps}^{0,0}S_{rq}^{0,0} + S_{pq}^{0,0}S_{rs}^{0,0})$$

$$= \frac{1}{\sqrt{3}}\left(E_{ps}E_{rq} + \frac{1}{2}E_{pq}E_{rs}\right) \tag{2S.9.5}$$

4. The transformed and untransformed excitation operators are related as

$$\frac{1}{2}(E_{pq}E_{rs} + E_{ps}E_{rq}) = \frac{1}{2}\left(\cos\theta + \frac{\sin\theta}{\sqrt{3}}\right)E_{pq}E_{rs} + \frac{\sin\theta}{\sqrt{3}}E_{ps}E_{rq} \tag{2S.9.6}$$

$$-\frac{1}{2\sqrt{3}}(E_{pq}E_{rs} - E_{ps}E_{rq}) = \frac{1}{2}\left(\frac{\cos\theta}{\sqrt{3}} - \sin\theta\right)E_{pq}E_{rs} + \frac{\cos\theta}{\sqrt{3}}E_{ps}E_{rq} \tag{2S.9.7}$$

which has the solution

$$\theta = \frac{\pi}{3} \tag{2S.9.8}$$

SOLUTION 2.10

We must verify that the operator (2E.10.1) satisfies conditions (2.3.1) and (2.3.2). For $\hat{S}_+$, condition (2.3.1) gives

$$
\begin{aligned}
[\hat{S}_+, \hat{R}] &= \sum_{M=-S}^{S} (-1)^M ([\hat{S}_+, \hat{T}^{S,M}]\hat{T}^{S,-M} + \hat{T}^{S,M}[\hat{S}_+, \hat{T}^{S,-M}]) \\
&= \sum_{M=-S}^{S-1} (-1)^M \sqrt{S(S+1) - M(M+1)}\, \hat{T}^{S,M+1}\hat{T}^{S,-M} \\
&\quad - \sum_{M=-S+1}^{S} (-1)^{M-1} \sqrt{S(S+1) - M(M-1)}\, \hat{T}^{S,M}\hat{T}^{S,-M+1} \\
&= 0 \tag{2S.10.1}
\end{aligned}
$$

where the cancellation of all terms follows when the substitution $P = M - 1$ is made in the last summation. The condition for $\hat{S}_-$ may be verified in a similar manner. Finally, condition (2.3.2) is easily verified:

$$
\begin{aligned}
[\hat{S}_z, \hat{R}] &= \sum_{M=-S}^{S} (-1)^M ([\hat{S}_z, \hat{T}^{S,M}]\hat{T}^{S,-M} + \hat{T}^{S,M}[\hat{S}_z, \hat{T}^{S,-M}]) \\
&= \sum_{M=-S}^{S} (-1)^M (M\hat{T}^{S,M}\hat{T}^{S,-M} - M\hat{T}^{S,M}\hat{T}^{S,-M}) = 0 \tag{2S.10.2}
\end{aligned}
$$

Since $\hat{R}$ satisfies conditions (2.3.1) and (2.3.2) with the right-hand sides equal to zero, it must be a singlet tensor operator.

SOLUTION 2.11

Applying (2.3.4) and the relation (2.2.31), we obtain

$$
\langle S, M+1|\hat{O}^{0,0}|S, M+1\rangle = \frac{\langle S, M|\hat{S}_-\hat{O}^{0,0}\hat{S}_+|S, M\rangle}{S(S+1) - M(M+1)} = \frac{\langle S, M|\hat{O}^{0,0}\hat{S}_-\hat{S}_+|S, M\rangle}{S(S+1) - M(M+1)} \tag{2S.11.1}
$$

where we have also made use of the fact that the shift operators commute with singlet operators (2.3.1). Applying (2.3.4) twice – first for the raising operator and then for the lowering operator on the right-hand side of this equation – we arrive at (2E.11.1).

3 ORBITAL ROTATIONS

The Fock space as introduced in Chapter 1 is defined in terms of a set of orthonormal spin orbitals. In many situations – for example, during the optimization of an electronic state or in the calculation of the response of an electronic state to an external perturbation – it becomes necessary to carry out transformations between different sets of orthonormal spin orbitals. In this chapter, we consider the unitary transformations of creation and annihilation operators and of Fock-space states that are generated by such transformations of the underlying spin-orbital basis. In particular, we shall see how, in second quantization, the unitary transformations can be conveniently carried out by the exponential of an anti-Hermitian operator, written as a linear combination of excitation operators.

The exponential parametrization of a unitary operator is independent in the sense that there are no restrictions on the allowed values of the numerical parameters in the operator – any choice of numerical parameters gives rise to a bona fide unitary operator. In many situations, however, we would like to carry out restricted spin-orbital and orbital rotations in order to preserve, for example, the spin symmetries of the electronic state. Such constrained transformations are also considered in this chapter, which contains an analysis of the symmetry properties of unitary orbital-rotation operators in second quantization. We begin, however, our exposition of spin-orbital and orbital rotations in second quantization with a discussion of unitary matrices and matrix exponentials.

3.1 Unitary transformations and matrix exponentials

A *unitary matrix* $\mathbf{U}$ is characterized by the relation

$$\mathbf{U}^\dagger \mathbf{U} = \mathbf{U}\mathbf{U}^\dagger = \mathbf{1} \tag{3.1.1}$$

where $\mathbf{U}^\dagger$ is the adjoint of $\mathbf{U}$ (obtained by transposition followed by complex conjugation). Our task in this section is to develop a parametrization of unitary matrices such that: any unitary matrix can be generated from this parametrization; only unitary matrices may be generated; and all parameters are independent. The elements of $\mathbf{U}$ cannot be used for this purpose since they are coupled by the unitary relation (3.1.1). Instead, we shall develop an independent (i.e. unconstrained) parametrization by writing the unitary matrix in terms of an exponential matrix. However, before considering the exponential unitary parametrization in Section 3.1.2, we shall first introduce and discuss matrix exponentials in Section 3.1.1.

3.1.1 MATRIX EXPONENTIALS

We define the *exponential of a matrix* $\mathbf{A}$ as

$$\exp(\mathbf{A}) = \sum_{n=0}^{\infty} \frac{\mathbf{A}^n}{n!} \tag{3.1.2}$$

where $\mathbf{A}^0$ is the identity matrix. The mathematically inclined reader may wish to prove that the sum in (3.1.2) converges for any finite-dimensional matrix. The following relations for matrix exponentials follow from definition (3.1.2):

$$\exp(-\mathbf{A})\exp(\mathbf{A}) = \mathbf{1} \tag{3.1.3}$$

$$\exp(\mathbf{A})^{\dagger} = \exp(\mathbf{A}^{\dagger}) \tag{3.1.4}$$

$$\mathbf{B}\exp(\mathbf{A})\mathbf{B}^{-1} = \exp(\mathbf{B}\mathbf{A}\mathbf{B}^{-1}) \tag{3.1.5}$$

$$\exp(\mathbf{A} + \mathbf{B}) = \exp(\mathbf{A})\exp(\mathbf{B}) \Leftarrow [\mathbf{A}, \mathbf{B}] = \mathbf{0} \tag{3.1.6}$$

$$\exp(-\mathbf{A})\mathbf{B}\exp(\mathbf{A}) = \mathbf{B} + [\mathbf{B}, \mathbf{A}] + \frac{1}{2!}[[\mathbf{B}, \mathbf{A}], \mathbf{A}] + \frac{1}{3!}[[[\mathbf{B}, \mathbf{A}], \mathbf{A}], \mathbf{A}] + \cdots \tag{3.1.7}$$

Substituting $-\mathbf{A}$ for $\mathbf{A}$ in (3.1.7), we obtain the equivalent expansion

$$\exp(\mathbf{A})\mathbf{B}\exp(-\mathbf{A}) = \mathbf{B} + [\mathbf{A}, \mathbf{B}] + \frac{1}{2!}[\mathbf{A}, [\mathbf{A}, \mathbf{B}]] + \frac{1}{3!}[\mathbf{A}, [\mathbf{A}, [\mathbf{A}, \mathbf{B}]]] + \cdots \tag{3.1.8}$$

Expressions (3.1.7) and (3.1.8) are two realizations of the *Baker–Campbell–Hausdorff (BCH) expansion*. For proofs of the identities (3.1.3)–(3.1.7), see Exercise 3.1.

3.1.2 EXPONENTIAL REPRESENTATIONS OF UNITARY MATRICES

We shall now demonstrate that any unitary matrix $\mathbf{U}$ can be written as the matrix exponential of an anti-Hermitian matrix:

$$\mathbf{U} = \exp(\mathbf{X}), \qquad \mathbf{X}^{\dagger} = -\mathbf{X} \tag{3.1.9}$$

We first show that, for any unitary matrix $\mathbf{U}$, we can always find an anti-Hermitian matrix $\mathbf{X}$ such that (3.1.9) is satisfied. For this purpose, we recall that the spectral theorem states that any unitary matrix can be diagonalized as

$$\mathbf{U} = \mathbf{V}\boldsymbol{\varepsilon}\mathbf{V}^{\dagger} \tag{3.1.10}$$

where $\mathbf{V}$ is unitary and $\boldsymbol{\varepsilon}$ is a complex diagonal matrix, the elements of which may be written in the form

$$\varepsilon_k = \exp(i\delta_k) \tag{3.1.11}$$

where the parameters δ_k are real numbers. Introducing the diagonal matrix $\boldsymbol{\delta}$ with elements δ_k, we obtain using (3.1.5)

$$\mathbf{U} = \mathbf{V}\exp(i\boldsymbol{\delta})\mathbf{V}^{\dagger} = \exp(i\mathbf{V}\boldsymbol{\delta}\mathbf{V}^{\dagger}) \tag{3.1.12}$$

Since $i\mathbf{V}\boldsymbol{\delta}\mathbf{V}^{\dagger}$ is anti-Hermitian, we have shown that any unitary matrix can be written in the exponential form (3.1.9).

We have now satisfied the first of the three requirements for the unitary parametrization stated in the introduction to Section 3.1. To satisfy the second requirement, we note that, for any anti-Hermitian matrix $\mathbf{X}$, the exponential $\exp(\mathbf{X})$ is always unitary since, from the relation $\mathbf{X} = -\mathbf{X}^\dagger$, it follows that

$$\exp(\mathbf{X})^\dagger \exp(\mathbf{X}) = \exp(-\mathbf{X})\exp(\mathbf{X}) = \mathbf{1} \tag{3.1.13}$$

Finally, we note that the third requirement for the parametrization is also satisfied since anti-Hermitian matrices are trivially represented by a set of independent parameters. We may, for instance, take the matrix elements at the diagonal and below the diagonal as the independent ones and generate the remaining elements of the matrix from the anti-Hermitian condition $X^*_{pq} = -X_{qp}$. Note that the diagonal elements of an anti-Hermitian matrix are pure imaginary and that the off-diagonal elements are complex.

3.1.3 SPECIAL UNITARY MATRICES

The exponential parametrization of a unitary matrix in (3.1.9) is a general one, applicable under all circumstances. We shall now consider more special forms of unitary matrices. We begin by writing the anti-Hermitian matrix $\mathbf{X}$ in the form

$$\mathbf{X} = \frac{\mathbf{X} - \mathbf{X}^T}{2} + i\frac{\mathbf{X} + \mathbf{X}^T}{2i} \tag{3.1.14}$$

where $\mathbf{X}^T$ is the transpose of $\mathbf{X}$. Any anti-Hermitian matrix may therefore be uniquely decomposed as

$$\mathbf{X} = {}^R\mathbf{X} + i({}^D\mathbf{X} + {}^I\mathbf{X}) \tag{3.1.15}$$

where ${}^R\mathbf{X}$ is a real antisymmetric (skew-symmetric) matrix

$$^R\mathbf{X} = \frac{\mathbf{X} - \mathbf{X}^T}{2} \tag{3.1.16}$$

and where ${}^D\mathbf{X}$ is a real diagonal matrix and ${}^I\mathbf{X}$ a real symmetric matrix with vanishing diagonal elements

$$^D\mathbf{X} + {}^I\mathbf{X} = \frac{\mathbf{X} + \mathbf{X}^T}{2i} \tag{3.1.17}$$

The unitary matrix (3.1.9) can thus be written in the form

$$\mathbf{U} = \exp(i{}^D\mathbf{X} + {}^R\mathbf{X} + i{}^I\mathbf{X}) \tag{3.1.18}$$

in terms of real symmetric and skew-symmetric matrices. If the diagonal matrix ${}^D\mathbf{X}$ is set equal to zero, the resulting unitary matrix has a determinant equal to 1:

$$\det[\exp({}^R\mathbf{X} + i{}^I\mathbf{X})] = \exp[\mathrm{Tr}({}^R\mathbf{X} + i{}^I\mathbf{X})] = 1 \tag{3.1.19}$$

We have here used the identity (proved in Exercise 3.2)

$$\det[\exp(\mathbf{A})] = \exp(\mathrm{Tr}\,\mathbf{A}) \tag{3.1.20}$$

and the fact that ${}^R\mathbf{X} + i{}^I\mathbf{X}$ has vanishing diagonal elements and thus zero trace. The unitary matrices of the form

$$\mathbf{O} = \exp({}^R\mathbf{X} + i{}^I\mathbf{X}) \tag{3.1.21}$$

therefore have determinants equal to 1 and are examples of *special unitary matrices*, which in general are defined by (3.1.9) for a traceless matrix **X**.

It can be demonstrated that any unitary matrix can be written as a unitary diagonal matrix times a special unitary matrix [1]

$$\mathbf{U} = \exp(i^D\mathbf{X} + {}^R\mathbf{X} + i^I\mathbf{X}) = \exp(i^D\tilde{\mathbf{X}})\exp({}^R\tilde{\mathbf{X}} + i^I\tilde{\mathbf{X}}) \tag{3.1.22}$$

where in general, for a given matrix **U**, the real matrices $^D\tilde{\mathbf{X}}$, $^R\tilde{\mathbf{X}}$ and $^I\tilde{\mathbf{X}}$ are different from the real matrices $^D\mathbf{X}$, $^R\mathbf{X}$ and $^I\mathbf{X}$. Equation (3.1.22) thus gives two alternative exponential parametrizations of unitary matrices. The diagonal unitary matrix $\exp(i^D\tilde{\mathbf{X}})$ introduces *complex phase shifts*, which are usually redundant in our work, and only the special unitary matrices (3.1.21) are then needed.

3.1.4 ORTHOGONAL MATRICES

In many cases, we are interested in unitary transformations among real rather than complex vectors. Such transformations are known as *orthogonal* and the associated *real orthogonal matrices* satisfy the relation

$$\mathbf{Q}^T\mathbf{Q} = \mathbf{Q}\mathbf{Q}^T = \mathbf{1} \tag{3.1.23}$$

analogous to (3.1.1). From (3.1.22), we note that any orthogonal matrix may be written in the exponential form

$$\mathbf{Q} = \mathbf{d}\exp({}^R\mathbf{X}) \tag{3.1.24}$$

where **d** is a diagonal matrix with elements ± 1 and $^R\mathbf{X}$ is a real antisymmetric matrix. Omitting the phase factors, we obtain the set of *special orthogonal matrices*

$$\mathbf{R} = \exp({}^R\mathbf{X}) \tag{3.1.25}$$

Special orthogonal matrices are often useful, for example, for the rotation of orbitals.

3.1.5 EVALUATION OF MATRIX EXPONENTIALS

The expansion of the exponential matrix (3.1.2) is rapidly convergent and may often be used for the evaluation of unitary matrices, especially if the anti-Hermitian matrix **X** has a small norm and high accuracy is not required. An alternative strategy is to diagonalize **X**:

$$\mathbf{X} = i\mathbf{V}\boldsymbol{\delta}\mathbf{V}^\dagger, \qquad \mathbf{V}^\dagger\mathbf{V} = \mathbf{1} \tag{3.1.26}$$

where $\boldsymbol{\delta}$ is a real diagonal matrix (noting that the eigenvalues of the Hermitian matrix $i\mathbf{X}$ are real) and then use (3.1.5) to rewrite the exponential as

$$\mathbf{U} = \exp(\mathbf{X}) = \exp(i\mathbf{V}\boldsymbol{\delta}\mathbf{V}^\dagger) = \mathbf{V}\exp(i\boldsymbol{\delta})\mathbf{V}^\dagger \tag{3.1.27}$$

The exponential of the pure imaginary diagonal matrix $i\boldsymbol{\delta}$ with complex elements $i\delta_k$ is easily calculated. The final matrix **U** is then obtained after a few matrix multiplications.

The same method may be used for special orthogonal matrices (3.1.25). Diagonalization of the real antisymmetric matrix $^R\mathbf{X}$ gives

$${}^R\mathbf{X} = i\mathbf{V}\boldsymbol{\tau}\mathbf{V}^\dagger, \qquad \mathbf{V}^\dagger\mathbf{V} = \mathbf{1} \tag{3.1.28}$$

where the eigenvalues $i\tau_k$ in the diagonal matrix $i\tau$ are pure imaginary or zero, and $\mathbf{V}$ is a complex unitary matrix. The orthogonal matrix $\mathbf{R}$ is now obtained from the expression

$$\mathbf{R} = \exp(^R\mathbf{X}) = \exp(i\mathbf{V}\tau\mathbf{V}^\dagger) = \mathbf{V}\exp(i\tau)\mathbf{V}^\dagger \tag{3.1.29}$$

Note that, even though $\mathbf{R}$ and $^R\mathbf{X}$ are both real, the evaluation of (3.1.29) involves complex arithmetic since $\mathbf{V}$ is complex and $i\tau$ imaginary.

To see how the real orthogonal matrix $\mathbf{R}$ may be obtained using only real arithmetic, we note that the square of $^R\mathbf{X}$ may be diagonalized by an orthogonal matrix

$$^R\mathbf{X}^2 = -\mathbf{W}\tau^2\mathbf{W}^T, \qquad \mathbf{W}^T\mathbf{W} = \mathbf{1} \tag{3.1.30}$$

Since $^R\mathbf{X}^2$ is real and symmetric, its eigenvalues and eigenvectors are real. In fact, the eigenvalues are nonpositive as they may be obtained also by squaring the imaginary eigenvalues of $^R\mathbf{X}$. To arrive at an expression for $\mathbf{R}$, we now expand the exponential (3.1.25) in a Taylor series

$$\mathbf{R} = \exp(^R\mathbf{X}) = \sum_{n=0}^{\infty} \frac{1}{(2n)!} {}^R\mathbf{X}^{2n} + \sum_{n=0}^{\infty} \frac{1}{(2n+1)!} {}^R\mathbf{X}^{2n}\,{}^R\mathbf{X}$$

$$= \mathbf{W}\cos(\tau)\mathbf{W}^T + \mathbf{W}\tau^{-1}\sin(\tau)\mathbf{W}^T\,{}^R\mathbf{X} \tag{3.1.31}$$

where we have used (3.1.30) as well as the Taylor expansions for $\cos(\tau)$ and $\sin(\tau)$:

$$\cos(\tau) = \sum_{n=0}^{\infty} \frac{(-1)^n}{(2n)!} \tau^{2n} \tag{3.1.32}$$

$$\sin(\tau) = \sum_{n=0}^{\infty} \frac{(-1)^n}{(2n+1)!} \tau^{2n+1} \tag{3.1.33}$$

Expression (3.1.31) involves only real arithmetic and may be used instead of the complex expression (3.1.29). Care should be taken to avoid vanishing eigenvalues in (3.1.31). In the two-dimensional case, unitary matrices may be given a simple analytical form; see Exercise 3.3.

3.1.6 NONUNITARY TRANSFORMATIONS

Let us consider transformations that preserve the Hermitian nonsingular *metric matrix* $\mathbf{S}$:

$$\mathbf{W}^\dagger\mathbf{S}\mathbf{W} = \mathbf{S} \tag{3.1.34}$$

To proceed, we first introduce the general nonintegral power $\mathbf{A}^y$ of a nonsingular matrix

$$\mathbf{A} = \mathbf{U}\alpha\mathbf{U}^\dagger \tag{3.1.35}$$

as

$$\mathbf{A}^y = \mathbf{U}\alpha^y\mathbf{U}^\dagger \tag{3.1.36}$$

By substitution, it is now easily verified that $\mathbf{W}$ may be written in terms of a unitary matrix $\mathbf{U}$ as

$$\mathbf{W} = \mathbf{S}^{-1/2}\mathbf{U}\mathbf{S}^{1/2} \tag{3.1.37}$$

Parametrizing $\mathbf{U}$ in terms of the anti-Hermitian matrix $\mathbf{X}$

$$\mathbf{U} = \exp{(\mathbf{S}^{1/2}\mathbf{X}\mathbf{S}^{1/2})} \tag{3.1.38}$$

we obtain, upon substitution in (3.1.37), the expression

$$\mathbf{W} = \exp(\mathbf{X}\mathbf{S}) \tag{3.1.39}$$

as the generalization of the exponential parametrization to metric matrices $\mathbf{S}$ different from $\mathbf{1}$. Substituting (3.1.39) into (3.1.34) and taking the inverse, we obtain the two expressions

$$\exp(-\mathbf{S}\mathbf{X})\mathbf{S}\exp(\mathbf{X}\mathbf{S}) = \mathbf{S} \tag{3.1.40}$$

$$\exp(-\mathbf{X}\mathbf{S})\mathbf{S}^{-1}\exp(\mathbf{S}\mathbf{X}) = \mathbf{S}^{-1} \tag{3.1.41}$$

which are the nonunitary generalizations of (3.1.3).

Let us now consider the BCH expansions for the nonunitary transformations induced by $\exp(\mathbf{X}\mathbf{S})$ and $\exp(\mathbf{S}\mathbf{X})$. In the notation

$$\tilde{\mathbf{M}} = \mathbf{S}^{1/2}\mathbf{M}\mathbf{S}^{1/2} \tag{3.1.42}$$

we obtain

$$\exp(-\mathbf{X}\mathbf{S})\mathbf{A}\exp(\mathbf{S}\mathbf{X}) = \mathbf{S}^{-1/2}\exp(-\tilde{\mathbf{X}})\tilde{\mathbf{A}}\exp(\tilde{\mathbf{X}})\mathbf{S}^{-1/2} \tag{3.1.43}$$

Since $\tilde{\mathbf{X}}$ is anti-Hermitian, $\exp(-\tilde{\mathbf{X}})\tilde{\mathbf{A}}\exp(\tilde{\mathbf{X}})$ represents a unitary transformation. Carrying out a standard BCH expansion, we obtain

$$\exp(-\mathbf{X}\mathbf{S})\mathbf{A}\exp(\mathbf{S}\mathbf{X}) = \mathbf{S}^{-1/2}\tilde{\mathbf{A}}\mathbf{S}^{-1/2} + \mathbf{S}^{-1/2}[\tilde{\mathbf{A}}, \tilde{\mathbf{X}}]\mathbf{S}^{-1/2} + \tfrac{1}{2}\mathbf{S}^{-1/2}[[\tilde{\mathbf{A}}, \tilde{\mathbf{X}}], \tilde{\mathbf{X}}]\mathbf{S}^{-1/2} + \cdots \tag{3.1.44}$$

which may be written in the expanded form

$$\exp(-\mathbf{X}\mathbf{S})\mathbf{A}\exp(\mathbf{S}\mathbf{X}) = \mathbf{A} + \mathbf{A}\mathbf{S}\mathbf{X} - \mathbf{X}\mathbf{S}\mathbf{A} + \tfrac{1}{2}\mathbf{A}\mathbf{S}\mathbf{X}\mathbf{S}\mathbf{X} - \mathbf{X}\mathbf{S}\mathbf{A}\mathbf{S}\mathbf{X} + \tfrac{1}{2}\mathbf{X}\mathbf{S}\mathbf{X}\mathbf{S}\mathbf{A} + \cdots \tag{3.1.45}$$

This expansion is identical to a conventional BCH expansion except that the metric $\mathbf{S}$ has been inserted between each pair of adjacent matrices. Introducing the S commutators

$$[\mathbf{A}, \mathbf{B}]_S = \mathbf{A}\mathbf{S}\mathbf{B} - \mathbf{B}\mathbf{S}\mathbf{A} \tag{3.1.46}$$

we find that the transformation (3.1.44) may be expressed compactly as *an asymmetric BCH expansion*

$$\exp(-\mathbf{X}\mathbf{S})\mathbf{A}\exp(\mathbf{S}\mathbf{X}) = \mathbf{A} + [\mathbf{A}, \mathbf{X}]_S + \tfrac{1}{2}[[\mathbf{A}, \mathbf{X}]_S, \mathbf{X}]_S + \cdots \tag{3.1.47}$$

Note that the inverse metric $\mathbf{S}^{-1}$ commutes with all matrices $\mathbf{B}$ in the sense that

$$[\mathbf{S}^{-1}, \mathbf{B}]_S = \mathbf{0} \tag{3.1.48}$$

in accordance with the relation (3.1.41).

Let us finally consider the transformation $\exp(-\mathbf{S}\mathbf{X})\mathbf{A}\exp(\mathbf{X}\mathbf{S})$, which we rewrite as

$$\exp(-\mathbf{S}\mathbf{X})\mathbf{A}\exp(\mathbf{X}\mathbf{S}) = \mathbf{S}\exp(-\mathbf{X}\mathbf{S})\mathbf{S}^{-1}\mathbf{A}\mathbf{S}^{-1}\exp(\mathbf{S}\mathbf{X})\mathbf{S} \tag{3.1.49}$$

Invoking (3.1.47), we obtain the asymmetric BCH expansion

$$\exp(-\mathbf{S}\mathbf{X})\mathbf{A}\exp(\mathbf{X}\mathbf{S}) = \mathbf{A} + \mathbf{S}[\mathbf{S}^{-1}\mathbf{A}\mathbf{S}^{-1}, \mathbf{X}]_S\mathbf{S} + \tfrac{1}{2}\mathbf{S}[[\mathbf{S}^{-1}\mathbf{A}\mathbf{S}^{-1}, \mathbf{X}]_S, \mathbf{X}]_S\mathbf{S} + \cdots \tag{3.1.50}$$

or, in expanded form,

$$\exp(-\mathbf{SX})\mathbf{A}\exp(\mathbf{XS}) = \mathbf{A} + \mathbf{AXS} - \mathbf{SXA} + \tfrac{1}{2}\mathbf{AXSXS} - \mathbf{SXAXS} + \tfrac{1}{2}\mathbf{SXSXA} + \cdots \quad (3.1.51)$$

Formally, this expansion may be obtained from the BCH expansion of $\exp(-\mathbf{X})\mathbf{A}\exp(\mathbf{X})$ by padding $\mathbf{X}$ with $\mathbf{S}$ everywhere except next to $\mathbf{A}$. The theory of nonunitary transformations is used in the development of density-based Hartree–Fock theory in Section 10.7.

3.2 Unitary spin-orbital transformations

It is often necessary to consider states where the occupation numbers refer to a set of orthonormal spin orbitals $\tilde{\phi}_P$ obtained from another set ϕ_P by a unitary transformation:

$$\tilde{\phi}_P = \sum_Q \phi_Q U_{QP} \quad (3.2.1)$$

According to the discussion in Section 3.1, the unitary matrix $\mathbf{U}$ may be written in terms of an anti-Hermitian matrix $\boldsymbol{\kappa}$ as

$$\mathbf{U} = \exp(-\boldsymbol{\kappa}), \qquad \boldsymbol{\kappa}^\dagger = -\boldsymbol{\kappa} \quad (3.2.2)$$

where the minus sign in the exponential is conventional.

Let $a_P^\dagger$ and a_P be the elementary creation and annihilation operators associated with the untransformed spin orbitals ϕ_P, and let $|0\rangle$ be any state in Fock space expressed in terms of the untransformed elementary operators. We shall in this section demonstrate that the elementary operators $\tilde{a}_P^\dagger$ and $\tilde{a}_P$ and state $|\tilde{0}\rangle$ generated by the unitary transformation (3.2.1) and (3.2.2) can be expressed in terms of the untransformed operators and states as

$$\tilde{a}_P^\dagger = \exp(-\hat{\kappa})a_P^\dagger \exp(\hat{\kappa}) \quad (3.2.3)$$

$$\tilde{a}_P = \exp(-\hat{\kappa})a_P \exp(\hat{\kappa}) \quad (3.2.4)$$

$$|\tilde{0}\rangle = \exp(-\hat{\kappa})|0\rangle \quad (3.2.5)$$

We have here introduced the anti-Hermitian operator

$$\hat{\kappa} = \sum_{PQ} \kappa_{PQ} a_P^\dagger a_Q, \qquad \hat{\kappa}^\dagger = -\hat{\kappa} \quad (3.2.6)$$

where the summation is over all pairs of creation and annihilation operators (i.e. over all excitation operators) and the parameters κ_{PQ} are the elements of the anti-Hermitian matrix $\boldsymbol{\kappa}$ in (3.2.2).

The operator exponential in expressions (3.2.3)–(3.2.5) is defined by analogy with the exponential of a matrix (3.1.2)

$$\exp(\hat{\kappa}) = \sum_{n=0}^{\infty} \frac{\hat{\kappa}^n}{n!} \quad (3.2.7)$$

and the rules (3.1.3)–(3.1.8) – established for matrices in Section 3.1.1 – are easily seen to hold for operators as well. As an example of the application of these rules to operators, we note that the unitary property of the exponential of an anti-Hermitian operator $\hat{\kappa}$ follows directly from the relations (3.1.3) and (3.1.4)

$$[\exp(\hat{\kappa})]^\dagger \exp(\hat{\kappa}) = \exp(-\hat{\kappa})\exp(\hat{\kappa}) = 1 \quad (3.2.8)$$

by analogy with (3.1.13) for matrix exponentials.

Expressions (3.2.3)–(3.2.5) are important since they allow us to express transformations between operators and states belonging to different sets of spin orbitals in a compact manner, greatly simplifying many of the algebraic manipulations in second quantization. In the remainder of this section, we shall prove these relations.

3.2.1 UNITARY MATRIX EXPANSIONS OF CREATION AND ANNIHILATION OPERATORS

Let us begin by considering the relationship between spin-orbital transformations in first quantization and the transformations of creation operators in second quantization. Assume that we have constructed two Fock spaces, one from the set of first-quantization spin orbitals ϕ_P and another from the transformed set $\tilde{\phi}_P$ in (3.2.1). What is the relationship between the second-quantization creation operators of the two Fock spaces? In the original spin-orbital basis, an N-electron Slater determinant in first quantization

$$\Phi = |\phi_{P_1}\phi_{P_2}\cdots\phi_{P_N}| \tag{3.2.9}$$

is in Fock space represented by the ON vector

$$|\mathbf{k}\rangle = a_{P_1}^\dagger a_{P_2}^\dagger \cdots a_{P_N}^\dagger |\text{vac}\rangle \tag{3.2.10}$$

assuming some canonical ordering of the spin orbitals. In the transformed spin-orbital basis, the transformed Slater determinant

$$\tilde{\Phi} = |\tilde{\phi}_{P_1}\tilde{\phi}_{P_2}\cdots\tilde{\phi}_{P_N}| \tag{3.2.11}$$

is similarly represented by the ON vector

$$|\tilde{\mathbf{k}}\rangle = \tilde{a}_{P_1}^\dagger \tilde{a}_{P_2}^\dagger \cdots \tilde{a}_{P_N}^\dagger |\text{vac}\rangle \tag{3.2.12}$$

obtained by applying a string of transformed creation operators to the vacuum state. We wish to determine the transformed creation operators in terms of the untransformed ones. Expanding the spin orbitals in (3.2.11) according to (3.2.1), we arrive at the expression

$$\tilde{\Phi} = \sum_{Q_1,Q_2,\dots,Q_N} U_{Q_1P_1}\cdots U_{Q_NP_N}|\phi_{Q_1}\cdots\phi_{Q_N}| \tag{3.2.13}$$

with a second-quantization representation given by

$$|\tilde{\mathbf{k}}\rangle = \sum_{Q_1,Q_2,\dots,Q_N} U_{Q_1P_1}\cdots U_{Q_NP_N}a_{Q_1}^\dagger \cdots a_{Q_N}^\dagger |\text{vac}\rangle \tag{3.2.14}$$

Combining (3.2.12) and (3.2.14), we obtain

$$\tilde{a}_{P_1}^\dagger \cdots \tilde{a}_{P_N}^\dagger |\text{vac}\rangle = \sum_{Q_1,Q_2,\dots,Q_N} U_{Q_1P_1}\cdots U_{Q_NP_N}a_{Q_1}^\dagger \cdots a_{Q_N}^\dagger |\text{vac}\rangle \tag{3.2.15}$$

which means that the one-to-one mapping between Slater determinants in first quantization and ON vectors in second quantization is preserved provided the creation operators transform in the same way as the spin orbitals

$$\tilde{a}_P^\dagger = \sum_Q a_Q^\dagger U_{QP} = \sum_Q a_Q^\dagger [\exp(-\kappa)]_{QP} \tag{3.2.16}$$

where we have also used (3.2.2). For the annihilation operators, we likewise obtain

$$\tilde{a}_P = \sum_Q a_Q U_{QP}^* = \sum_Q a_Q [\exp(-\kappa)]_{QP}^* \tag{3.2.17}$$

by taking the Hermitian conjugate of (3.2.16). The unitarily transformed creation and annihilation operators (3.2.16) and (3.2.17) satisfy the usual anticommutation relations (1.2.27)–(1.2.29), as is readily verified.

3.2.2 EXPONENTIAL UNITARY TRANSFORMATIONS OF THE ELEMENTARY OPERATORS

Having expanded the transformed creation and annihilation operators (3.2.16) and (3.2.17) in the original ones, let us now develop a more convenient representation of the transformed operators. Consider the operators

$$\bar{a}_P^\dagger = \exp(-\hat{\kappa}) a_P^\dagger \exp(\hat{\kappa}) \tag{3.2.18}$$

$$\bar{a}_P = \exp(-\hat{\kappa}) a_P \exp(\hat{\kappa}) \tag{3.2.19}$$

where the anti-Hermitian operator $\hat{\kappa}$ is given by (3.2.6). Using the rules in Section 3.1.1, the operators (3.2.18) and (3.2.19) are easily seen to be the Hermitian conjugate of each other and to satisfy the usual anticommutation relations for creation and annihilation operators.

We shall now demonstrate that the operators $\bar{a}_P^\dagger$ in (3.2.18) are identical to the transformed creation operators $\tilde{a}_p^\dagger$ in (3.2.16), thereby establishing the relationship (3.2.3). The corresponding expression for the annihilation operators (3.2.4) then follows by taking the Hermitian conjugate of (3.2.3). Expanding $\bar{a}_P^\dagger$ using the BCH expression (3.1.7), we obtain

$$\bar{a}_P^\dagger = a_P^\dagger + [a_P^\dagger, \hat{\kappa}] + \frac{1}{2!}[[a_P^\dagger, \hat{\kappa}], \hat{\kappa}] + \cdots \tag{3.2.20}$$

The first two commutators in this expansion may be expanded as

$$[a_P^\dagger, \hat{\kappa}] = -\sum_Q a_Q^\dagger \kappa_{QP} \tag{3.2.21}$$

$$[[a_P^\dagger, \hat{\kappa}], \hat{\kappa}] = \sum_Q a_Q^\dagger [\kappa^2]_{QP} \tag{3.2.22}$$

and, for the n-fold nested commutator, we obtain

$$[\ldots[[a_P^\dagger, \hat{\kappa}], \hat{\kappa}], \ldots] = (-1)^n \sum_Q a_Q^\dagger [\kappa^n]_{QP} \tag{3.2.23}$$

We may therefore write the BCH expansion (3.2.20) in the form

$$\bar{a}_P^\dagger = \sum_Q a_Q^\dagger \left\{ \delta_{QP} - \kappa_{QP} + \cdots + \frac{(-1)^n}{n!}[\kappa^n]_{QP} + \cdots \right\}$$

$$= \sum_Q a_Q^\dagger \sum_n \frac{(-1)^n}{n!}[\kappa^n]_{QP} = \sum_Q a_Q^\dagger [\exp(-\kappa)]_{QP} \tag{3.2.24}$$

which shows that the operators (3.2.18) are identical to (3.2.16), demonstrating the validity of (3.2.3) for the creation operators.

3.2.3 EXPONENTIAL UNITARY TRANSFORMATIONS OF STATES IN FOCK SPACE

The usefulness of the exponential operator $\exp(-\hat{\kappa})$ becomes clear when we consider the transformed ON vector (3.2.12). Inserting the expression for the transformed creation operators (3.2.3) into (3.2.12), we obtain

$$|\tilde{\mathbf{k}}\rangle = \left[\prod_P (\tilde{a}_P^\dagger)^{k_P}\right] |\mathrm{vac}\rangle = \exp(-\hat{\kappa}) \left[\prod_P (a_P^\dagger)^{k_P}\right] \exp(\hat{\kappa})|\mathrm{vac}\rangle \tag{3.2.25}$$

since the pairs of unitary operators appearing between the creation operators cancel according to (3.2.8). Further simplifications follow from the observation that, for $\hat{\kappa}$ of the form (3.2.6),

$$\hat{\kappa}|\mathrm{vac}\rangle = 0 \Rightarrow \exp(\hat{\kappa})|\mathrm{vac}\rangle = |\mathrm{vac}\rangle \tag{3.2.26}$$

which allows us to express the transformed ON vector in the form

$$|\tilde{\mathbf{k}}\rangle = \exp(-\hat{\kappa})|\mathbf{k}\rangle \tag{3.2.27}$$

A similar expression must hold for any linear combination of ON vectors – that is, for any state in Fock space. We have thus established equation (3.2.5): applied to any electronic state $|0\rangle$, the operator $\exp(-\hat{\kappa})$ generates a new state $|\tilde{0}\rangle$ where all the spin orbitals have been unitarily transformed according to (3.2.1).

3.3 Symmetry-restricted unitary transformations

The unitary transformations discussed in Section 3.2 are completely general, allowing us to express any unitarily transformed operator and state in terms of a set of independent parameters. However, in many situations less general transformations are required owing to the presence of special symmetries in the electronic system. A more general discussion of symmetry restrictions is given in Chapter 4. Here we anticipate this development by considering *symmetry-restricted* or *symmetry-constrained* unitary operators $\exp(-\hat{\kappa})$.

3.3.1 THE NEED FOR SYMMETRY RESTRICTIONS

Let us assume that $|0\rangle$ is an approximate state associated with the electronic Hamiltonian $\hat{H}$ and that $|0\rangle$ possesses definite spin and spatial symmetries. If we wish to determine a set of parameters κ_{pq} in $\exp(-\hat{\kappa})$ that transforms the approximate state $|0\rangle$ into a new state $|\tilde{0}\rangle$ with the same spin and spatial symmetries

$$|\tilde{0}\rangle = \exp(-\hat{\kappa})|0\rangle \tag{3.3.1}$$

then we must impose constraints on the spin-orbital variations. To see the need for such constraints, we note that the transformation from $|0\rangle$ to $|\tilde{0}\rangle$ is expressed in terms of a set of *nonlinear* spin-orbital rotation parameters:

$$|\tilde{0}\rangle = \left(1 - \hat{\kappa} + \frac{1}{2!}\hat{\kappa}^2 - \frac{1}{3!}\hat{\kappa}^3 + \cdots\right)|0\rangle \tag{3.3.2}$$

As discussed in Section 4.4, the conservation of spin and spatial symmetries in the optimized wave function is then guaranteed only if the symmetry restrictions are explicitly imposed on the parametrization. In other words, the spin and spatial symmetries are conserved only if $\hat{\kappa}$ contains only those spin-orbital excitation operators that transform as the totally symmetric representation of the Hamiltonian $\hat{H}$. If nontotally symmetric rotations are allowed, then $\hat{\kappa}|0\rangle$ will not have the symmetry of $|0\rangle$ and the symmetries of $|\tilde{0}\rangle$ and $|0\rangle$ will be different.

When a perturbation $\hat{V}$ is applied to the system, the Hamiltonian becomes $\hat{H} + \hat{V}$. The allowed variations then become those that transform according to the totally symmetric representation of $\hat{H} + \hat{V}$ rather than of $\hat{H}$. For example, if we consider the ground state of the oxygen molecule and if $\hat{V}$ corresponds to an electric field perpendicular to the internuclear axis, then the allowed spin-orbital excitations are represented by spin-conserving orbital excitation operators that are totally symmetric in the C_{2v} point group. In other words, the allowed variations are described by operators E_{pq} where the direct product of the irreducible representations of the orbitals ϕ_p and ϕ_q is totally symmetric in C_{2v}.

These examples should suffice to illustrate the need for *symmetry-constrained spin-orbital rotations*. Let us summarize the most common situations where restricted rotations are required:

1. *Real and imaginary rotations*: Field-free nonrelativistic wave functions may be constructed from real orbitals and only real rotations are needed for optimizing the wave functions. On the other hand, in the presence of perturbations such as an external magnetic field, imaginary rotations are required to describe the perturbed state.

2. *Spin-adapted rotations*: The spin-free nonrelativistic Hamiltonian commutes with the total and projected spin operators. We are therefore usually interested only in wave functions with well-defined spin quantum numbers. Such functions may be generated from spin tensor operators that are totally symmetric in spin space. For optimizations, we need consider only singlet operators since these are the only ones that conserve the spin of the wave function. Spin perturbations, on the other hand, may mix spin eigenstates and require the inclusion also of triplet rotations.

3. *Point-group symmetry-adapted rotations*: Since the exact wave function transforms as one of the irreducible representations of the Hamiltonian point group, it is convenient to work in a representation of symmetry-adapted orbitals. In the course of the optimization, we need then consider only totally symmetric rotations since these preserve the symmetry of the wave function. If symmetry-breaking geometrical distortions are applied, nontotally symmetric rotations must be considered as well.

The use of restricted orbital rotations not only ensures that the approximate electronic wave function has the desired spin and spatial symmetries. As an additional benefit, the computational cost is lowered by reducing the number of free parameters.

3.3.2 SYMMETRY RESTRICTIONS IN THE SPIN-ORBITAL BASIS

Having seen the need for restricted spin-orbital rotations, let us first see how the antisymmetric $\hat{\kappa}$ operator (3.2.6)

$$\hat{\kappa} = \sum_{PQ} \kappa_{PQ} a_P^\dagger a_Q \tag{3.3.3}$$

appearing in the rotated operators and states (3.2.3)–(3.2.5) may be separated into real and imaginary parts; spin symmetries are considered in Section 3.3.3. In the spin-orbital basis, the operator (3.3.3) may conveniently be rewritten as

$$\hat{\kappa} = \sum_P \kappa_{PP} a_P^\dagger a_P + \sum_{P \neq Q} \kappa_{PQ} a_P^\dagger a_Q$$

$$= i \sum_P {}^I\kappa_{PP} a_P^\dagger a_P + i \sum_{P>Q} {}^I\kappa_{PQ}(a_P^\dagger a_Q + a_Q^\dagger a_P) + \sum_{P>Q} {}^R\kappa_{PQ}(a_P^\dagger a_Q - a_Q^\dagger a_P) \qquad (3.3.4)$$

where the real and imaginary parts of the spin-orbital rotation parameters are given by

$$^R\kappa_{PQ} = \text{Re } \kappa_{PQ} = \frac{\kappa_{PQ} - \kappa_{QP}}{2} \qquad (3.3.5)$$

$$^I\kappa_{PQ} = \text{Im } \kappa_{PQ} = \frac{\kappa_{PQ} + \kappa_{QP}}{2i} \qquad (3.3.6)$$

Since $\hat{\kappa}$ in (3.3.4) must be anti-Hermitian, the number of independent spin-orbital excitation parameters κ_{PQ} in a basis of M spin orbitals becomes

$$M + \frac{M^2 - M}{2} + \frac{M^2 - M}{2} = M^2 \qquad (3.3.7)$$

For the optimization of M real spin orbitals, it is sufficient to consider the real part of $\hat{\kappa}$ and ignore the phase factors. The resulting spin-orbital operator

$$^R\hat{\kappa} = \sum_{P>Q} {}^R\kappa_{PQ}(a_P^\dagger a_Q - a_Q^\dagger a_P) \qquad (3.3.8)$$

contains only $M(M-1)/2$ parameters.

3.3.3 SYMMETRY RESTRICTIONS IN THE ORBITAL BASIS

To allow for spin-symmetry constraints in the rotation operator (3.3.3), we express $\hat{\kappa}$ in the orbital basis and introduce spin tensor operators. We first consider the contributions to (3.3.3) from excitation operators that are diagonal in the orbital indices

$$\sum_{p\sigma\tau} \kappa_{p\sigma p\tau} a_{p\sigma}^\dagger a_{p\tau} = i \sum_p ({}^I\kappa_{p\alpha p\alpha} a_{p\alpha}^\dagger a_{p\alpha} + {}^I\kappa_{p\beta p\beta} a_{p\beta}^\dagger a_{p\beta}) + \sum_p (\kappa_{p\alpha p\beta} a_{p\alpha}^\dagger a_{p\beta} + \kappa_{p\beta p\alpha} a_{p\beta}^\dagger a_{p\alpha})$$

$$(3.3.9)$$

where $^R\kappa_{p\alpha p\alpha}$ and $^R\kappa_{p\beta p\beta}$ vanish since $\hat{\kappa}$ is anti-Hermitian. We may write (3.3.9) also in terms of parameters that refer to the irreducible orbital excitation operators (2.3.29) and (2.3.22)–(2.3.24). Equation (3.3.9) now becomes

$$\sum_{p\sigma\tau} \kappa_{p\sigma p\tau} a_{p\sigma}^\dagger a_{p\tau} = i \sum_p {}^I\kappa_{pp}^{0,0} E_{pp} + i \sum_p {}^I\kappa_{pp}^{1,0} \hat{T}_{pp}^{1,0} + \sum_p \kappa_{pp}^{1,-1} \hat{T}_{pp}^{1,-1} + \sum_p \kappa_{pp}^{1,1} \hat{T}_{pp}^{1,1} \qquad (3.3.10)$$

where

$$^I\kappa_{pp}^{0,0} = \frac{1}{2}({}^I\kappa_{p\alpha p\alpha} + {}^I\kappa_{p\beta p\beta}) \qquad (3.3.11)$$

$$^I\kappa_{pp}^{1,0} = \frac{1}{\sqrt{2}}(^I\kappa_{p\alpha p\alpha} - {}^I\kappa_{p\beta p\beta}) \tag{3.3.12}$$

$$\kappa_{pp}^{1,1} = -\kappa_{p\alpha p\beta} \tag{3.3.13}$$

$$\kappa_{pp}^{1,-1} = \kappa_{p\beta p\alpha} \tag{3.3.14}$$

Treating the nondiagonal orbital excitation operators in the same way, we arrive at the following expression for $\hat{\kappa}$:

$$\hat{\kappa} = \sum_p \kappa_{pp}^{0,0} E_{pp} + \sum_{M=-1}^{1}\sum_p \kappa_{pp}^{1,M}\hat{T}_{pp}^{1,M} + \sum_{p\neq q}\kappa_{pq}^{0,0} E_{pq} + \sum_{M=-1}^{1}\sum_{p\neq q}\kappa_{pq}^{1,M}\hat{T}_{pq}^{1,M} \tag{3.3.15}$$

Here we have isolated the diagonal elements but made no separation of real and imaginary parts. All parameters are complex except $\kappa_{pp}^{0,0}$ and $\kappa_{pp}^{1,0}$, which have only imaginary components.

If we insist on using tensor operators as well as on separating real and imaginary rotations, the antisymmetry of $\hat{\kappa}$ imposes the following constraints on the orbital excitation parameters

$$^R\kappa_{pq}^{0,0} = -{}^R\kappa_{qp}^{0,0}; \quad {}^I\kappa_{pq}^{0,0} = {}^I\kappa_{qp}^{0,0} \tag{3.3.16}$$

$$^R\kappa_{pq}^{1,0} = -{}^R\kappa_{qp}^{1,0}; \quad {}^I\kappa_{pq}^{1,0} = {}^I\kappa_{qp}^{1,0} \tag{3.3.17}$$

$$^R\kappa_{pq}^{1,-1} = {}^R\kappa_{qp}^{1,1}; \quad {}^I\kappa_{pq}^{1,-1} = -{}^I\kappa_{qp}^{1,1} \tag{3.3.18}$$

The operator $\hat{\kappa}$ then can be expressed as

$$\hat{\kappa} = i\sum_p {}^I\kappa_{pp}^{0,0} E_{pp} + i\sum_p {}^I\kappa_{pp}^{1,0}\hat{T}_{pp}^{1,0} + i\sum_{p>q}{}^I\kappa_{pq}^{0,0}(E_{pq} + E_{qp}) + \sum_{p>q}{}^R\kappa_{pq}^{0,0}(E_{pq} - E_{qp})$$

$$+ i\sum_{p>q}{}^I\kappa_{pq}^{1,0}(\hat{T}_{pq}^{1,0} + \hat{T}_{qp}^{1,0}) + \sum_{p>q}{}^R\kappa_{pq}^{1,0}(\hat{T}_{pq}^{1,0} - \hat{T}_{qp}^{1,0})$$

$$+ i\sum_{pq}{}^I\kappa_{pq}^{1,1}(\hat{T}_{pq}^{1,1} - \hat{T}_{qp}^{1,-1}) + \sum_{pq}{}^R\kappa_{pq}^{1,1}(\hat{T}_{pq}^{1,1} + \hat{T}_{qp}^{1,-1}) \tag{3.3.19}$$

This expression contains

$$2n + 4\frac{n^2 - n}{2} + n^2 + n^2 = 4n^2 \tag{3.3.20}$$

parameters, where n is the number of orbitals. The number of parameters (3.3.20) is identical to M^2 obtained using (3.3.7) for $M = 2n$ spin orbitals.

Alternatively, we may write $\hat{\kappa}$ in terms of the Cartesian triplet operators (2.3.27). Since the Cartesian components obey the simple conjugation relation (2.3.28), we now obtain the following more symmetric form for a general anti-Hermitian operator $\hat{\kappa}$:

$$\hat{\kappa} = i\sum_p \left[{}^I\kappa_{pp}^{0,0} E_{pp} + \sum_\mu {}^I\kappa_{pp}^\mu \hat{T}_{pp}^\mu\right] + i\sum_{p>q}\left[{}^I\kappa_{pq}^{0,0}(E_{pq} + E_{qp}) + \sum_\mu {}^I\kappa_{pq}^\mu(\hat{T}_{pq}^\mu + \hat{T}_{qp}^\mu)\right]$$

$$+ \sum_{p>q}\left[{}^R\kappa_{pq}^{0,0}(E_{pq} - E_{qp}) + \sum_\mu {}^R\kappa_{pq}^\mu(\hat{T}_{pq}^\mu - \hat{T}_{qp}^\mu)\right] \tag{3.3.21}$$

where the inner summations are over the three Cartesian components. The spherical and Cartesian forms (3.3.19) and (3.3.21) are equivalent, containing the same number of rotational parameters $4n^2$. The exact relation between these parameters is seldom needed and is not given here.

In contrast to the spin-orbital rotation operator in (3.3.4), we have in (3.3.19) and (3.3.21) introduced irreducible tensor components in spin space and at the same time made a clear distinction between real and imaginary rotational parameters. This procedure makes it easy to select the components of $\hat{\kappa}$ that are necessary for a particular task. For example, in optimizations of real orbitals, we may neglect all complex rotations and rotations that mix spin states as well as the complex phase factors. The orbital-rotation operator then reduces to

$$^{\mathrm{R}}\hat{\kappa}^{0,0} = \sum_{p>q} {}^{\mathrm{R}}\kappa_{pq}^{0,0}(E_{pq} - E_{qp}) \tag{3.3.22}$$

which contains only $n(n-1)/2$ real parameters. The triplet and imaginary parts of (3.3.19) and (3.3.21) are needed for perturbations that mix spin and involve imaginary operators.

3.4   The logarithmic matrix function

In this section, we introduce the logarithm, $\log \mathbf{A}$, of a nonsingular matrix $\mathbf{A}$, required to satisfy the relationship

$$\exp \log(\mathbf{A}) = \mathbf{A} \tag{3.4.1}$$

To simplify our discussion of $\log \mathbf{A}$, we restrict the domain of the logarithmic matrix function to nonsingular matrices $\mathbf{A}$ that can be diagonalized

$$\mathbf{A} = \mathbf{Z}\boldsymbol{\alpha}\mathbf{Z}^{-1} \tag{3.4.2}$$

where $\boldsymbol{\alpha}$ is a diagonal matrix containing real or complex elements. For the special cases of Hermitian, anti-Hermitian, and unitary matrices $\mathbf{A}$, the diagonalization (3.4.2) may be carried out by means of a unitary matrix $\mathbf{Z}$.

3.4.1   DEFINITION OF THE LOGARITHMIC MATRIX FUNCTION

For a nonsingular matrix $\mathbf{A}$ that can be diagonalized as in (3.4.2), we define the *logarithmic matrix function* $\log \mathbf{A}$ as

$$\log \mathbf{A} = \mathbf{Z} \log(\boldsymbol{\alpha})\mathbf{Z}^{-1} \tag{3.4.3}$$

This definition satisfies (3.4.1) since

$$\exp \log \mathbf{A} = \exp[\mathbf{Z} \log(\boldsymbol{\alpha})\mathbf{Z}^{-1}] = \mathbf{Z} \exp[\log(\boldsymbol{\alpha})]\mathbf{Z}^{-1} = \mathbf{Z}\boldsymbol{\alpha}\mathbf{Z}^{-1} = \mathbf{A} \tag{3.4.4}$$

The evaluation of $\log \mathbf{A}$ is thus carried out in terms of the complex scalar logarithmic function. For a complex number z, the logarithm is obtained from the polar representation

$$z = |z| \exp[\mathrm{i}(\phi + 2\pi n)] \tag{3.4.5}$$

where n is an integer [2]. However, since the argument of a complex number z is fixed only to within a multiple of 2π, the logarithm of z becomes a multiple-valued function

$$\log z = \log |z| + \mathrm{i}(\phi + 2\pi n) \tag{3.4.6}$$

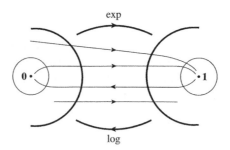

Fig. 3.1. The relationship between the exponential and logarithmic matrix functions. The domains for one-to-one mappings in the neighbourhoods of **0** and **1** are indicated by circles.

where $\log |z|$ is the standard real logarithmic function. A single-valued logarithmic function is obtained by restricting ϕ to the interval $-\pi < \phi \le \pi$. With this restriction, the logarithmic matrix function is uniquely defined by (3.4.3).

It is natural to enquire whether the logarithm is the inverse of the exponential in some domain of the matrices (3.4.2). The relation between exponential and logarithmic functions is illustrated in Figure 3.1. The exponential maps an infinite number of matrices with eigenvalues differing by multiples of $2\pi i$ onto the same matrix. The exponential therefore does not have an inverse function over the entire domain of matrices of the type (3.4.2). In the neighbourhood of **0**, however, the exponential provides a one-to-one mapping onto the neighbourhood of **1**, as is readily appreciated from the Taylor expansion

$$\exp \mathbf{A} \approx \mathbf{1} + \mathbf{A} \tag{3.4.7}$$

The logarithmic matrix function $\log \mathbf{A}$ is thus the true inverse function of the exponential $\exp \mathbf{A}$ in the neighbourhood of **1**, mapping matrices in the neighbourhood of **1** onto matrices in the neighbourhood of **0**.

3.4.2 EXPANSION OF THE LOGARITHMIC MATRIX FUNCTION

We now consider the Taylor expansion of the logarithmic matrix function $\log \mathbf{A}$ in the neighbourhood of **1** and write $\mathbf{A} \approx \mathbf{1}$ in (3.4.2) in the form

$$\mathbf{A} = \mathbf{1} + \mathbf{S} = \mathbf{Z}(\mathbf{1} + \sigma)\mathbf{Z}^{-1} \tag{3.4.8}$$

where σ is a diagonal matrix containing the eigenvalues of $\mathbf{S}$ (which has the same eigenvectors as $\mathbf{A}$). We assume that the eigenvalues of $\mathbf{S}$ are smaller than unity: $|\sigma_i| < 1$. Using the scalar Taylor expansion

$$\log(1 + x) = -\sum_{n=1}^{\infty} \frac{(-1)^n}{n} x^n, \qquad |x| < 1 \tag{3.4.9}$$

we may write equation (3.4.3) in the form

$$\log(\mathbf{1} + \mathbf{S}) = -\sum_{n=1}^{\infty} \frac{(-1)^n}{n} \mathbf{Z} \sigma^n \mathbf{Z}^{-1} \tag{3.4.10}$$

Thus, the logarithmic matrix function may be expanded as

$$\log(\mathbf{1} + \mathbf{S}) = -\sum_{n=1}^{\infty} \frac{(-1)^n}{n} \mathbf{S}^n \tag{3.4.11}$$

provided the eigenvalues of $\mathbf{S}$ are all smaller than 1 in magnitude.

3.4.3 PROPERTIES OF THE LOGARITHMIC MATRIX FUNCTION

The following properties of the logarithmic matrix function are proved in Exercise 3.6:

$$(\log \mathbf{A})^\dagger = \log(\mathbf{A}^\dagger) \tag{3.4.12}$$

$$\log(\mathbf{A}^n) = n \log \mathbf{A} \tag{3.4.13}$$

$$\log(\mathbf{AB}) = \log \mathbf{A} + \log \mathbf{B} \Leftarrow [\log \mathbf{A}, \log \mathbf{B}] = \mathbf{0} \tag{3.4.14}$$

$$(\log \mathbf{U})^\dagger = -\log \mathbf{U} \Leftarrow \mathbf{U}^\dagger \mathbf{U} = \mathbf{1} \tag{3.4.15}$$

Equation (3.4.12) holds for all matrices and is the analogue of (3.1.4) for matrix exponentials. Relationships (3.4.13) and (3.4.14) are similar to those of the scalar logarithmic function. Equation (3.4.13) holds for all values of n (not necessarily integers) when matrix powers are defined as

$$\mathbf{A}^n = \mathbf{Z}\boldsymbol{\alpha}^n \mathbf{Z}^\dagger \tag{3.4.16}$$

In Exercise 3.6, the proof given for (3.4.14) is valid only for matrices belonging to the domain where the logarithmic and exponential matrix functions are inverse functions. Finally, (3.4.15) – which follows from (3.4.12) and (3.4.13) – shows that the logarithmic function maps unitary matrices onto anti-Hermitian matrices as expected from the fact that the exponential function maps anti-Hermitian matrices onto unitary matrices.

References

[1]   J. Linderberg and Y. Öhrn, *Int. J. Quantum Chem.* **12**, 161 (1977).
[2]   H. A. Pristley, *Introduction to Complex Analysis*, Clarendon Press, 1985.

Further reading

M. L. Curtis, *Matrix Groups*, Springer-Verlag, 1979.
G. H. Golub and C. F. Van Loan, *Matrix Computations*, 2nd edn, Johns Hopkins University Press, 1989.

Exercises

EXERCISE 3.1

Verify the following relations for matrix exponentials:

1. $\exp(\mathbf{A})^\dagger = \exp(\mathbf{A}^\dagger)$ $\tag{3E.1.1}$

2. $\mathbf{B}\exp(\mathbf{A})\mathbf{B}^{-1} = \exp(\mathbf{BAB}^{-1})$ $\tag{3E.1.2}$

3. $\exp(\mathbf{A} + \mathbf{B}) = \exp(\mathbf{A})\exp(\mathbf{B}) \Leftarrow [\mathbf{A}, \mathbf{B}] = \mathbf{0}$ (3E.1.3)

4. $\exp(-\mathbf{A})\exp(\mathbf{A}) = \mathbf{1}$ (3E.1.4)

5. $\dfrac{d}{d\lambda}\exp(\lambda\mathbf{A}) = \mathbf{A}\exp(\lambda\mathbf{A}) = \exp(\lambda\mathbf{A})\mathbf{A}$ (3E.1.5)

6. $\exp(-\mathbf{A})\mathbf{B}\exp(\mathbf{A}) = \mathbf{B} + [\mathbf{B}, \mathbf{A}] + \dfrac{1}{2}[[\mathbf{B}, \mathbf{A}], \mathbf{A}] + \dfrac{1}{3!}[[[\mathbf{B}, \mathbf{A}], \mathbf{A}], \mathbf{A}] + \cdots$ (3E.1.6)

Hint: Consider the Taylor expansion of $\exp(-\lambda\mathbf{A})\mathbf{B}\exp(\lambda\mathbf{A})$ about $\lambda = 0$.

EXERCISE 3.2

Let $\mathbf{D}$ be a diagonal matrix with diagonal elements d_i:

$$\mathbf{D} = \text{diag}(d_1, d_2, \ldots, d_n)$$ (3E.2.1)

1. Show that the exponential of the diagonal matrix is given by:

$$\exp(\mathbf{D}) = \text{diag}(\exp(d_1), \exp(d_2), \ldots, \exp(d_n))$$ (3E.2.2)

2. Let $\mathbf{A}$ be a matrix that can be diagonalized:

$$\mathbf{A} = \mathbf{X}\mathbf{D}\mathbf{X}^{-1}$$ (3E.2.3)

Verify the following relation for the determinant of the exponential of $\mathbf{A}$:

$$\det(\exp \mathbf{A}) = \exp(\text{Tr }\mathbf{A})$$ (3E.2.4)

EXERCISE 3.3

Determine closed-form expressions for $\exp(\boldsymbol{\kappa})$ for the following anti-Hermitian matrices $\boldsymbol{\kappa}$:

1. a general 2×2 anti-Hermitian matrix

$$\boldsymbol{\kappa} = \begin{pmatrix} i\delta_1 & \alpha \\ -\alpha^* & i\delta_2 \end{pmatrix}$$ (3E.3.1)

where α is complex and where δ_1 and δ_2 are real. *Hint:* Decompose $\boldsymbol{\kappa}$ as

$$\boldsymbol{\kappa} = \begin{pmatrix} \dfrac{i}{2}(\delta_1 + \delta_2) & 0 \\ 0 & \dfrac{i}{2}(\delta_1 + \delta_2) \end{pmatrix} + \begin{pmatrix} \dfrac{i}{2}(\delta_1 - \delta_2) & \alpha \\ -\alpha^* & -\dfrac{i}{2}(\delta_1 - \delta_2) \end{pmatrix}$$ (3E.3.2)

2. a complex 2×2 anti-Hermitian matrix with zero diagonal elements:

$$\boldsymbol{\kappa} = \begin{pmatrix} 0 & \alpha \\ -\alpha^* & 0 \end{pmatrix}$$ (3E.3.3)

3. a real 2×2 antisymmetric matrix:

$$\boldsymbol{\kappa} = \begin{pmatrix} 0 & \lambda \\ -\lambda & 0 \end{pmatrix}$$ (3E.3.4)

EXERCISE 3.4

In this exercise, we consider the *spin-flip operators* – that is, unitary operators that transform alpha operators into beta operators and visa versa. Although we shall investigate only the effect of the spin-flip operators on the creation operators, we note that the corresponding results for the annihilation operators are readily obtained by taking the adjoints of the relations for the creation operators.

1. Consider the unitary operators

$$\hat{V}_x = \exp(i\pi \hat{S}_x) \tag{3E.4.1}$$

$$\hat{V}_y = \exp(i\pi \hat{S}_y) \tag{3E.4.2}$$

Show that these operators are spin-flip operators in the sense that

$$\hat{V}_x a_{p\alpha}^\dagger \hat{V}_x^\dagger = i a_{p\beta}^\dagger \tag{3E.4.3}$$

$$\hat{V}_x a_{p\beta}^\dagger \hat{V}_x^\dagger = i a_{p\alpha}^\dagger \tag{3E.4.4}$$

and

$$\hat{V}_y a_{p\alpha}^\dagger \hat{V}_y^\dagger = -a_{p\beta}^\dagger \tag{3E.4.5}$$

$$\hat{V}_y a_{p\beta}^\dagger \hat{V}_y^\dagger = a_{p\alpha}^\dagger \tag{3E.4.6}$$

2. Demonstrate that the unitary operator

$$\hat{W} = \exp\left(-\frac{i\pi}{2}\hat{N}\right) \tag{3E.4.7}$$

changes the phase of the creation operators in the following manner:

$$\hat{W} a_{p\sigma}^\dagger \hat{W}^\dagger = -i a_{p\sigma}^\dagger \tag{3E.4.8}$$

3. Verify that the unitary operator

$$\hat{U}_x = \exp\left[i\pi\left(\hat{S}_x - \tfrac{1}{2}\hat{N}\right)\right] \tag{3E.4.9}$$

carries out clean spin flips:

$$\hat{U}_x a_{p\alpha}^\dagger \hat{U}_x^\dagger = a_{p\beta}^\dagger \tag{3E.4.10}$$

$$\hat{U}_x a_{p\beta}^\dagger \hat{U}_x^\dagger = a_{p\alpha}^\dagger \tag{3E.4.11}$$

Determine the effect of applying $\hat{U}_x$ to an ON vector $|\mathbf{k}\rangle$. This operator will be used in Section 11.8.6, in our discussion of direct configuration-interaction methods.

4. The *time-reversal operator*, which changes the direction of time in quantum mechanics, is an antiunitary operator $\hat{T}$ that can be written as

$$\hat{T} = \hat{K}\hat{U}_y \tag{3E.4.12}$$

where $\hat{K}$ is the complex-conjugation operator and $\hat{U}_y$ the unitary spin operator that transforms the spin orbitals as

$$\hat{U}_y a_{p\alpha}^\dagger \hat{U}_y^\dagger = i a_{p\beta}^\dagger \tag{3E.4.13}$$

$$\hat{U}_y a_{p\beta}^\dagger \hat{U}_y^\dagger = -i a_{p\alpha}^\dagger \tag{3E.4.14}$$

Show that these rotations are obtained with the unitary operator

$$\hat{U}_y = \exp\left[i\pi \left(\hat{S}_y - \tfrac{1}{2}\hat{N} \right) \right] \tag{3E.4.15}$$

EXERCISE 3.5

Consider the orbital-rotation operator

$$\hat{\kappa} = \sum_{P>Q} \kappa_{PQ} (a_P^\dagger a_Q - a_Q^\dagger a_P) \tag{3E.5.1}$$

where κ_{PQ} are real parameters. We shall derive a general expression for the first derivatives of $\exp(-\hat{\kappa})$ with respect to κ_{PQ}, valid also for $\kappa_{PQ} \neq 0$.

1. Verify the relation

$$\exp[-(\hat{\kappa} + \delta\hat{\kappa})] - \exp(-\hat{\kappa}) = \int_0^1 \frac{d}{dz} \exp[-(\hat{\kappa} + \delta\hat{\kappa})z] \exp(\hat{\kappa}z)\, dz\, \exp(-\hat{\kappa}) \tag{3E.5.2}$$

2. Using (3E.5.2), write the first-order variation of $\exp(-\hat{\kappa})$ with respect to $\delta\hat{\kappa}$ as

$$\delta \exp(-\hat{\kappa}) = \sum_{n=0}^\infty \frac{(-1)^{n+1}}{(n+1)!} \underbrace{[\hat{\kappa}, [\hat{\kappa}, \ldots [\hat{\kappa}, \delta\hat{\kappa}] \ldots]]}_{n \text{ commutators}} \exp(-\hat{\kappa}) \tag{3E.5.3}$$

We shall consider this expression further in Exercise 3.5.5, having first developed a convenient expression for the first-order variation $\delta\hat{\kappa}$.

3. Show that the real antisymmetric matrix $\boldsymbol{\kappa}$ with elements

$$[\boldsymbol{\kappa}]_{PQ} = \begin{cases} \kappa_{PQ} & P \geq Q \\ -\kappa_{PQ} & P < Q \end{cases} \tag{3E.5.4}$$

can be diagonalized as

$$\boldsymbol{\kappa} = -i\mathbf{U}\boldsymbol{\lambda}\mathbf{U}^\dagger \tag{3E.5.5}$$

where $\mathbf{U}$ is a unitary matrix and $\boldsymbol{\lambda}$ a real diagonal matrix.

4. The eigenvectors $\mathbf{U}$ in (3E.5.5) define a basis

$$\bar{a}_P^\dagger = \sum_R a_R^\dagger U_{RP} \tag{3E.5.6}$$

Show that, in this eigenvector basis, the orbital-rotation operator and its first-order variation may be expressed as:

$$\hat{\kappa} = -i \sum_P \lambda_P \bar{a}_P^\dagger \bar{a}_P \tag{3E.5.7}$$

$$\delta\hat{\kappa} = \sum_{PQ} [\mathbf{U}^\dagger \delta\boldsymbol{\kappa} \mathbf{U}]_{PQ} \bar{a}_P^\dagger \bar{a}_Q \tag{3E.5.8}$$

5. Using the expansions (3E.5.7) and (3E.5.8), rewrite (3E.5.3) as

$$\delta \exp(-\hat{\kappa}) = - \sum_{PQ} \frac{\exp[i(\lambda_P - \lambda_Q)] - 1}{i(\lambda_P - \lambda_Q)} [\mathbf{U}^\dagger \delta \boldsymbol{\kappa} \mathbf{U}]_{PQ} \, \bar{a}_P^\dagger \bar{a}_Q \exp(-\hat{\kappa}) \qquad (3E.5.9)$$

6. Show that the first derivatives of the orbital-rotation operator are given by

$$\frac{\partial \exp(-\hat{\kappa})}{\partial \kappa_{PQ}} = - \sum_{RS} \frac{\exp[i(\lambda_R - \lambda_S)] - 1}{i(\lambda_R - \lambda_S)} (U_{PR}^* U_{QS} - U_{QR}^* U_{PS}) \bar{a}_R^\dagger \bar{a}_S \exp(-\hat{\kappa}) \qquad (3E.5.10)$$

This expression is valid for all values of κ_{PQ}, not just for $\kappa_{PQ} = 0$.

EXERCISE 3.6

In this exercise, we consider some properties of the logarithmic matrix function $\log \mathbf{A}$.

1. Prove the following relations for the logarithmic matrix function:

$$(\log \mathbf{A})^\dagger = \log(\mathbf{A}^\dagger) \qquad (3E.6.1)$$

$$\log(\mathbf{A}^n) = n \log \mathbf{A} \qquad (3E.6.2)$$

2. Assume that $\log \mathbf{A}$ and $\log \mathbf{B}$ commute and that $\mathbf{A}$, $\mathbf{B}$ and $\mathbf{AB}$ all belong to the domain where the logarithm is the inverse of the exponential. Verify the relation

$$\log(\mathbf{AB}) = \log \mathbf{A} + \log \mathbf{B} \Leftarrow [\log \mathbf{A}, \log \mathbf{B}] = \mathbf{0} \qquad (3E.6.3)$$

3. Assume that $\mathbf{U}$ is a unitary matrix. Prove the relation

$$(\log \mathbf{U})^\dagger = - \log \mathbf{U} \Leftarrow \mathbf{U}^\dagger \mathbf{U} = \mathbf{1} \qquad (3E.6.4)$$

Solutions

SOLUTION 3.1

1. Introduce the Taylor expansion of the exponential:

$$\exp(\mathbf{A})^\dagger = \left(\sum_{n=0}^{\infty} \frac{1}{n!} \mathbf{A}^n \right)^\dagger = \sum_{n=0}^{\infty} \frac{1}{n!} (\mathbf{A}^\dagger)^n = \exp(\mathbf{A}^\dagger) \qquad (3S.1.1)$$

2. Introducing the Taylor expansion of the exponential and using the identity

$$\mathbf{B} \mathbf{A}^n \mathbf{B}^{-1} = (\mathbf{B} \mathbf{A} \mathbf{B}^{-1})^n \qquad (3S.1.2)$$

(easily proved by expansion), we obtain

$$\mathbf{B} \exp(\mathbf{A}) \mathbf{B}^{-1} = \mathbf{B} \left(\sum_{n=0}^{\infty} \frac{1}{n!} \mathbf{A}^n \right) \mathbf{B}^{-1} = \sum_{n=0}^{\infty} \frac{1}{n!} (\mathbf{B} \mathbf{A} \mathbf{B}^{-1})^n = \exp(\mathbf{B} \mathbf{A} \mathbf{B}^{-1}) \qquad (3S.1.3)$$

3. Invoking the binomial expansion for commuting matrices

$$(\mathbf{A} + \mathbf{B})^n = \sum_{i=0}^{n} \frac{n!}{i!(n-i)!} \mathbf{A}^i \mathbf{B}^{n-i} \tag{3S.1.4}$$

we obtain

$$\exp(\mathbf{A}) \exp(\mathbf{B}) = \sum_{i=0}^{\infty} \sum_{j=0}^{\infty} \frac{1}{i!j!} \mathbf{A}^i \mathbf{B}^j = \sum_{n=0}^{\infty} \frac{1}{n!} \sum_{i=0}^{n} \frac{n!}{i!(n-i)!} \mathbf{A}^i \mathbf{B}^{n-i}$$

$$= \sum_{n=0}^{\infty} \frac{1}{n!} (\mathbf{A} + \mathbf{B})^n = \exp(\mathbf{A} + \mathbf{B}) \tag{3S.1.5}$$

4. Use (3E.1.3) with $\mathbf{B} = -\mathbf{A}$.
5. Differentiating the Taylor expansion of the exponential, we obtain

$$\frac{d}{d\lambda} \exp(\lambda \mathbf{A}) = \sum_{n=1}^{\infty} \frac{1}{(n-1)!} \lambda^{n-1} \mathbf{A}^n = \mathbf{A} \sum_{n=0}^{\infty} \frac{1}{n!} \lambda^n \mathbf{A}^n = \mathbf{A} \exp(\lambda \mathbf{A}) \tag{3S.1.6}$$

6. We consider the Taylor expansion of the function

$$f(\lambda) = \exp(-\lambda \mathbf{A}) \mathbf{B} \exp(\lambda \mathbf{A}) \tag{3S.1.7}$$

around $\lambda = 0$. Applying (3E.1.5), we obtain the following expressions for the first and second derivatives of $f(\lambda)$:

$$\frac{df(\lambda)}{d\lambda} = \exp(-\lambda \mathbf{A})[\mathbf{B}, \mathbf{A}] \exp(\lambda \mathbf{A}) \tag{3S.1.8}$$

$$\frac{d^2 f(\lambda)}{d\lambda^2} = \exp(-\lambda \mathbf{A})[[\mathbf{B}, \mathbf{A}], \mathbf{A}] \exp(\lambda \mathbf{A}) \tag{3S.1.9}$$

The general expression is therefore given by:

$$\frac{d^n f(\lambda)}{d\lambda^n} = \exp(-\lambda \mathbf{A})[[\ldots [\mathbf{B}, \mathbf{A}] \ldots, \mathbf{A}], \mathbf{A}] \exp(\lambda \mathbf{A}) \tag{3S.1.10}$$

From the Taylor expansion of $f(\lambda)$ about $\lambda = 0$, we obtain (3E.1.6) by setting $\lambda = 1$.

SOLUTION 3.2

1. Expanding the exponential, we obtain

$$\exp(\mathbf{D}) = \sum_{k=0}^{\infty} \frac{1}{k!} \mathbf{D}^k \tag{3S.2.1}$$

Since the kth power of the diagonal matrix $\mathbf{D}$ is a diagonal matrix with elements d_i^k

$$\mathbf{D}^k = \text{diag}(d_1^k, d_2^k, \ldots, d_n^k) \qquad (3S.2.2)$$

we may write (3S.2.1) in the form

$$\exp(\mathbf{D}) = \text{diag}\left(\sum_{k=0}^{\infty} \frac{1}{k!} d_1^k, \sum_{k=0}^{\infty} \frac{1}{k!} d_2^k, \ldots, \sum_{k=0}^{\infty} \frac{1}{k!} d_n^k\right)$$
$$= \text{diag}(\exp(d_1), \exp(d_2), \ldots, \exp(d_n)) \qquad (3S.2.3)$$

2. Using (3.1.5), the exponential of the matrix $\mathbf{A}$ may be written in the form

$$\exp(\mathbf{A}) = \exp(\mathbf{XDX}^{-1}) = \mathbf{X}\exp(\mathbf{D})\mathbf{X}^{-1} \qquad (3S.2.4)$$

Taking the determinant of this matrix and using the elementary properties of determinants and traces, we obtain:

$$\det[\exp(\mathbf{A})] = \det(\mathbf{X})\det[\exp(\mathbf{D})]\det(\mathbf{X}^{-1}) = \det[\exp(\mathbf{D})] = \prod_{i=1}^{n}\exp(d_i)$$

$$= \exp(\text{Tr }\mathbf{D}) = \exp(\text{Tr }\mathbf{DX}^{-1}\mathbf{X}) = \exp(\text{Tr }\mathbf{XDX}^{-1}) = \exp(\text{Tr }\mathbf{A}) \quad (3S.2.5)$$

SOLUTION 3.3

1. Introducing the notation

$$\delta_{\pm} = \frac{\delta_1 \pm \delta_2}{2} \qquad (3S.3.1)$$

and decomposing the matrix (3E.3.1) according to (3E.3.2), we obtain

$$\exp\begin{pmatrix} i\delta_1 & \alpha \\ -\alpha^* & i\delta_2 \end{pmatrix} = \exp\left[\begin{pmatrix} i\delta_+ & 0 \\ 0 & i\delta_+ \end{pmatrix} + \begin{pmatrix} i\delta_- & \alpha \\ -\alpha^* & -i\delta_- \end{pmatrix}\right]$$
$$= \exp\begin{pmatrix} i\delta_+ & 0 \\ 0 & i\delta_+ \end{pmatrix}\exp\begin{pmatrix} i\delta_- & \alpha \\ -\alpha^* & -i\delta_- \end{pmatrix}$$
$$= \exp(i\delta_+)\exp\begin{pmatrix} i\delta_- & \alpha \\ -\alpha^* & -i\delta_- \end{pmatrix} \qquad (3S.3.2)$$

where we have used (3.1.6). Noting that the square of the 2×2 matrix

$$\mathbf{A} = \begin{pmatrix} i\delta_- & \alpha \\ -\alpha^* & -i\delta_- \end{pmatrix} \qquad (3S.3.3)$$

is given by

$$\mathbf{A}^2 = -a^2 \begin{pmatrix} 1 & 0 \\ 0 & 1 \end{pmatrix} \qquad (3S.3.4)$$

$$a = \sqrt{\delta_-^2 + |\alpha|^2} \qquad (3S.3.5)$$

we obtain the following closed-form expression for its exponential:

$$
\exp(\mathbf{A}) = \sum_{n=0}^{\infty} \frac{1}{(2n)!} \mathbf{A}^{2n} + \sum_{n=0}^{\infty} \frac{1}{(2n+1)!} \mathbf{A}^{2n} \mathbf{A}
$$

$$
= \sum_{n=0}^{\infty} \frac{(-1)^n}{(2n)!} a^{2n} \begin{pmatrix} 1 & 0 \\ 0 & 1 \end{pmatrix} + \frac{1}{a} \sum_{n=0}^{\infty} \frac{(-1)^n}{(2n+1)!} a^{2n+1} \mathbf{A}
$$

$$
= \cos a \begin{pmatrix} 1 & 0 \\ 0 & 1 \end{pmatrix} + \frac{\sin a}{a} \begin{pmatrix} i\delta_- & \alpha \\ -\alpha^* & -i\delta_- \end{pmatrix} \tag{3S.3.6}
$$

Substituting (3S.3.6) in (3S.3.2), we arrive at the desired expression for the exponential of a general anti-Hermitian matrix:

$$
\exp \begin{pmatrix} i\delta_1 & \alpha \\ -\alpha^* & i\delta_2 \end{pmatrix} = \exp(i\delta_+) \begin{pmatrix} \cos a + \dfrac{\sin a}{a} i\delta_- & \dfrac{\sin a}{a} \alpha \\ -\dfrac{\sin a}{a} \alpha^* & \cos a - \dfrac{\sin a}{a} i\delta_- \end{pmatrix} \tag{3S.3.7}
$$

The reader may wish to verify that this matrix is indeed unitary.

2. Setting δ_1 and δ_2 equal to zero in (3S.3.7), we obtain

$$
\exp \begin{pmatrix} 0 & \alpha \\ -\alpha^* & 0 \end{pmatrix} = \begin{pmatrix} \cos |\alpha| & \sin |\alpha| \dfrac{\alpha}{|\alpha|} \\ -\sin |\alpha| \dfrac{\alpha^*}{|\alpha|} & \cos |\alpha| \end{pmatrix} \tag{3S.3.8}
$$

3. Setting the imaginary part of α equal to zero in (3S.3.8), we arrive at the following simple expression for the exponential of a real antisymmetric matrix:

$$
\exp \begin{pmatrix} 0 & \lambda \\ -\lambda & 0 \end{pmatrix} = \begin{pmatrix} \cos \lambda & \sin \lambda \\ -\sin \lambda & \cos \lambda \end{pmatrix} \tag{3S.3.9}
$$

SOLUTION 3.4

1. To demonstrate (3E.4.3)–(3E.4.6), we shall evaluate the BCH expansions of the unitary transformations. For this purpose, we must first evaluate the commutators of the creation operators with $\hat{S}_x$ and $\hat{S}_y$. Expressing $\hat{S}_x$ and $\hat{S}_y$ in terms of the spin-shift operators

$$
\hat{S}_x = \frac{1}{2}(\hat{S}_+ + \hat{S}_-) \tag{3S.4.1}
$$

$$
\hat{S}_y = \frac{i}{2}(\hat{S}_- - \hat{S}_+) \tag{3S.4.2}
$$

the single and double commutators are seen to be given by

$$
[\hat{S}_x, a_{p\alpha}^\dagger] = \frac{1}{2} a_{p\beta}^\dagger \tag{3S.4.3}
$$

$$
[\hat{S}_x, [\hat{S}_x, a_{p\alpha}^\dagger]] = \frac{1}{4} a_{p\alpha}^\dagger \tag{3S.4.4}
$$

and

$$[\hat{S}_y, a^\dagger_{p\alpha}] = \frac{\mathrm{i}}{2}a^\dagger_{p\beta} \tag{3S.4.5}$$

$$[\hat{S}_y, [\hat{S}_y, a^\dagger_{p\alpha}]] = \frac{1}{4}a^\dagger_{p\alpha} \tag{3S.4.6}$$

In general, therefore, the commutators in the BCH expansions of (3E.4.3)–(3E.4.6) are very simple. Separating the expansion of (3E.4.3) into terms involving even and odd numbers of nested commutators, we obtain

$$\hat{V}_x a^\dagger_{p\alpha} \hat{V}^\dagger_x = \sum_{n=0}^{\infty} \frac{(-1)^n \pi^{2n}}{(2n)!} \underbrace{[\hat{S}_x, [\hat{S}_x, \ldots [\hat{S}_x, a^\dagger_{p\alpha}] \ldots]]}_{2n \text{ nested commutators}}$$

$$+ \mathrm{i} \sum_{n=0}^{\infty} \frac{(-1)^n \pi^{2n+1}}{(2n+1)!} \underbrace{[\hat{S}_x, [\hat{S}_x, \ldots [\hat{S}_x, a^\dagger_{p\alpha}] \ldots]]}_{2n+1 \text{ nested commutators}} \tag{3S.4.7}$$

Inserting the expressions for the commutators generalized from (3S.4.3) and (3S.4.4), we obtain

$$\hat{V}_x a^\dagger_{p\alpha} \hat{V}^\dagger_x = \sum_{n=0}^{\infty} \frac{(-1)^n}{(2n)!} \left(\frac{\pi}{2}\right)^{2n} a^\dagger_{p\alpha} + \mathrm{i} \sum_{n=0}^{\infty} \frac{(-1)^n}{(2n+1)!} \left(\frac{\pi}{2}\right)^{2n+1} a^\dagger_{p\beta}$$

$$= \cos\left(\frac{\pi}{2}\right) a^\dagger_{p\alpha} + \mathrm{i} \sin\left(\frac{\pi}{2}\right) a^\dagger_{p\beta} = \mathrm{i}a^\dagger_{p\beta} \tag{3S.4.8}$$

which is the desired result (3E.4.3). The corresponding relation (3E.4.5) for $\hat{V}_y$ is obtained in the same manner, using (3S.4.5) and (3S.4.6) in place of (3S.4.3) and (3S.4.4). The relation (3E.4.4) may be demonstrated in the same manner but is more easily obtained from (3E.4.3) by evaluating the commutator of both sides with $\hat{S}_x$:

$$[\hat{S}_x, \hat{V}_x a^\dagger_{p\alpha} \hat{V}^\dagger_x] = [\hat{S}_x, \mathrm{i}a^\dagger_{p\beta}] \tag{3S.4.9}$$

Since $\hat{S}_x$ commutes with $\hat{V}_x$, we obtain for the left-hand side

$$[\hat{S}_x, \hat{V}_x a^\dagger_{p\alpha} \hat{V}^\dagger_x] = \hat{V}_x [\hat{S}_x, a^\dagger_{p\alpha}] \hat{V}^\dagger_x = \tfrac{1}{2} \hat{V}_x a^\dagger_{p\beta} \hat{V}^\dagger_x \tag{3S.4.10}$$

using (3S.4.3). For the right-hand side, we obtain

$$[\hat{S}_x, \mathrm{i}a^\dagger_{p\beta}] = \tfrac{1}{2}\mathrm{i}a^\dagger_{p\alpha} \tag{3S.4.11}$$

Substituting (3S.4.10) and (3S.4.11) into (3S.4.9), we arrive at (3E.4.4). The relation (3E.4.6) may be proved in the same manner.

2. Noting the commutation relation

$$[\hat{N}, a^\dagger_{p\sigma}] = a^\dagger_{p\sigma} \tag{3S.4.12}$$

as a special case of (1.3.14), we obtain for the BCH expansion of (3E.4.8):

$$\hat{W} a^\dagger_{p\sigma} \hat{W}^\dagger = \left[\sum_{n=0}^{\infty} \frac{1}{n!}\left(-\frac{\mathrm{i}\pi}{2}\right)^n\right] a^\dagger_{p\sigma} = \exp\left(-\frac{\mathrm{i}\pi}{2}\right) a^\dagger_{p\sigma} = -\mathrm{i}a^\dagger_{p\sigma} \tag{3S.4.13}$$

3. Since $\hat{S}_x$ and $\hat{N}$ commute, we may express (3E.4.9) as the product of $\hat{W}$ and $\hat{V}_x$:

$$\hat{U}_x = \exp\left[i\pi\left(\hat{S}_x - \frac{1}{2}\hat{N}\right)\right] = \exp\left(-\frac{i\pi}{2}\hat{N}\right)\exp(i\pi\hat{S}_x) = \hat{W}\hat{V}_x \tag{3S.4.14}$$

Applying (3E.4.3) and then (3E.4.8), we obtain (3E.4.10)

$$\hat{U}_x a_{p\alpha}^\dagger \hat{U}_x^\dagger = \hat{W}\hat{V}_x a_{p\alpha}^\dagger \hat{V}_x^\dagger \hat{W}^\dagger = i\hat{W}a_{p\beta}^\dagger \hat{W}^\dagger = a_{p\beta}^\dagger \tag{3S.4.15}$$

and similarly for (3E.4.11). When applied to an ON vector, $\hat{U}_x$ changes the spin projection of all creation operators – for example

$$\hat{U}_x a_{p\alpha}^\dagger a_{q\beta}^\dagger |\text{vac}\rangle = \hat{U}_x a_{p\alpha}^\dagger \hat{U}_x^\dagger \hat{U}_x a_{q\beta}^\dagger \hat{U}_x^\dagger \hat{U}_x |\text{vac}\rangle$$

$$= [\hat{U}_x a_{p\alpha}^\dagger \hat{U}_x^\dagger][\hat{U}_x a_{q\beta}^\dagger \hat{U}_x^\dagger]\hat{U}_x |\text{vac}\rangle = a_{p\beta}^\dagger a_{q\alpha}^\dagger |\text{vac}\rangle \tag{3S.4.16}$$

since the vacuum state is unaffected by the unitary transformation:

$$\hat{U}_x |\text{vac}\rangle = |\text{vac}\rangle \tag{3S.4.17}$$

4. Expressing (3E.4.15) in the product form

$$\hat{U}_y = \hat{W}\hat{V}_y \tag{3S.4.18}$$

the relations (3E.4.13) and (3E.4.14) follow in the same manner as (3E.4.10) and (3E.4.11) for $\hat{U}_x$ in Exercise 3.4.3.

SOLUTION 3.5

1. Rearranging, we find

$$\exp[-(\hat{\kappa} + \delta\hat{\kappa})] - \exp(-\hat{\kappa}) = \{\exp[-(\hat{\kappa} + \delta\hat{\kappa})]\exp(\hat{\kappa}) - 1\}\exp(-\hat{\kappa})$$

$$= \exp[-(\hat{\kappa} + \delta\hat{\kappa})z]\exp(\hat{\kappa}z)|_0^1 \exp(-\hat{\kappa})$$

$$= \int_0^1 \frac{d}{dz}\exp[-(\hat{\kappa} + \delta\hat{\kappa})z]\exp(\hat{\kappa}z)\,dz\exp(-\hat{\kappa}) \tag{3S.5.1}$$

2. Carrying out the differentiation on the right-hand side of (3E.5.2) and keeping terms to first order in $\delta\hat{\kappa}$, we obtain:

$$\delta\exp(-\hat{\kappa}) = -\int_0^1 \exp(-\hat{\kappa}z)\delta\hat{\kappa}\exp(\hat{\kappa}z)\,dz\exp(-\hat{\kappa}) \tag{3S.5.2}$$

We now expand $\exp(-\hat{\kappa}z)\delta\hat{\kappa}\exp(\hat{\kappa}z)$ in a BCH series:

$$\delta\exp(-\hat{\kappa}) = -\int_0^1 \sum_{n=0}^\infty \frac{(-z)^n}{n!}\,dz\,\underbrace{[\hat{\kappa},[\hat{\kappa},\ldots[\hat{\kappa},\delta\hat{\kappa}]\ldots]]}_{n\text{ commutators}}\exp(-\hat{\kappa})$$

$$= \sum_{n=0}^\infty \frac{(-1)^{n+1}}{(n+1)!}\underbrace{[\hat{\kappa},[\hat{\kappa},\ldots[\hat{\kappa},\delta\hat{\kappa}]\ldots]]}_{n\text{ commutators}}\exp(-\hat{\kappa}) \tag{3S.5.3}$$

3. Since κ is antisymmetric, $i\kappa$ is Hermitian and may be diagonalized by a unitary matrix $\mathbf{U}$ to yield a real diagonal matrix λ:

$$i\kappa = \mathbf{U}\lambda\mathbf{U}^\dagger \tag{3S.5.4}$$

From this result, (3E.5.5) follows trivially.

4. From (3E.5.6), it follows that we may express the original elementary operators in terms of the transformed operators as

$$a_P^\dagger = \sum_R \bar{a}_R^\dagger U_{PR}^* \tag{3S.5.5}$$

$$a_P = \sum_R \bar{a}_R U_{PR} \tag{3S.5.6}$$

Inserting these expansions in (3E.5.1), we obtain an expression for the orbital-rotation operator in terms of the transformed elementary operators

$$\hat{\kappa} = \sum_{P>Q} \kappa_{PQ}(a_P^\dagger a_Q - a_Q^\dagger a_P) = \sum_{PQ} \kappa_{PQ} a_P^\dagger a_Q$$

$$= \sum_{PQ} \kappa_{PQ} \sum_{RS} \bar{a}_R^\dagger U_{PR}^* \bar{a}_S U_{QS} = \sum_{RS} [\mathbf{U}^\dagger\kappa\mathbf{U}]_{RS} \bar{a}_R^\dagger \bar{a}_S \tag{3S.5.7}$$

where (3E.5.4) has been used to express $\hat{\kappa}$ as a free summation. The expression for $\hat{\kappa}$ in (3E.5.7) now follows by application of (3E.5.5), whereas the expression for $\delta\hat{\kappa}$ in (3E.5.8) is obtained by replacing κ_{PQ} with $\delta\kappa_{PQ}$.

5. To evaluate (3E.5.3), we must determine the nested commutators. From (3E.5.7) and (3E.5.8), we obtain for the first commutator:

$$[\hat{\kappa}, \delta\hat{\kappa}] = -i\sum_{PQR} \lambda_P [\mathbf{U}^\dagger\delta\kappa\mathbf{U}]_{QR}[\bar{a}_P^\dagger\bar{a}_P, \bar{a}_Q^\dagger\bar{a}_R] = -i\sum_{PQ}(\lambda_P - \lambda_Q)[\mathbf{U}^\dagger\delta\kappa\mathbf{U}]_{PQ}\bar{a}_P^\dagger\bar{a}_Q \tag{3S.5.8}$$

Comparing this with the expression for $\delta\hat{\kappa}$ in (3E.5.8), we conclude that the general expression for the nested commutators in (3E.5.3) is given by

$$\underbrace{[\hat{\kappa}, [\hat{\kappa}, \dots [\hat{\kappa}, \delta\hat{\kappa}]\dots]]}_{n \text{ commutators}} = (-i)^n \sum_{PQ}(\lambda_P - \lambda_Q)^n [\mathbf{U}^\dagger\delta\kappa\mathbf{U}]_{PQ}\bar{a}_P^\dagger\bar{a}_Q \tag{3S.5.9}$$

We now arrive at (3E.5.9) by inserting this expression in (3E.5.3) and carrying out some manipulations:

$$\delta\exp(-\hat{\kappa}) = i\sum_{PQ}\sum_{n=0}^\infty \frac{i^{n+1}}{(n+1)!}(\lambda_P - \lambda_Q)^n [\mathbf{U}^\dagger\delta\kappa\mathbf{U}]_{PQ}\bar{a}_P^\dagger\bar{a}_Q \exp(-\hat{\kappa})$$

$$= \sum_{PQ} \frac{i}{\lambda_P - \lambda_Q}\sum_{n=1}^\infty \frac{[i(\lambda_P - \lambda_Q)]^n}{n!}[\mathbf{U}^\dagger\delta\kappa\mathbf{U}]_{PQ}\bar{a}_P^\dagger\bar{a}_Q \exp(-\hat{\kappa})$$

$$= -\sum_{PQ} \frac{\exp[i(\lambda_P - \lambda_Q)] - 1}{i(\lambda_P - \lambda_Q)}[\mathbf{U}^\dagger\delta\kappa\mathbf{U}]_{PQ}\bar{a}_P^\dagger\bar{a}_Q \exp(-\hat{\kappa}) \tag{3S.5.10}$$

6. From (3E.5.9), we obtain directly (3E.5.10):

$$\frac{\partial \exp(-\hat{\kappa})}{\partial \kappa_{PQ}} = -\sum_{RS} \frac{\exp[i(\lambda_R - \lambda_S)] - 1}{i(\lambda_R - \lambda_S)} [\mathbf{U}^\dagger (\partial \boldsymbol{\kappa} / \partial \kappa_{PQ}) \mathbf{U}]_{RS} \bar{a}_R^\dagger \bar{a}_S \exp(-\hat{\kappa})$$

$$= -\sum_{RS} \frac{\exp[i(\lambda_R - \lambda_S)] - 1}{i(\lambda_R - \lambda_S)} (U_{PR}^* U_{QS} - U_{QR}^* U_{PS}) \bar{a}_R^\dagger \bar{a}_S \exp(-\hat{\kappa}) \qquad (3S.5.11)$$

SOLUTION 3.6

1. To verify (3E.6.1) and (3E.6.2) for $\log \mathbf{A}$, we first note the relations

$$\mathbf{A}^\dagger = (\mathbf{Z}^{-1})^\dagger \boldsymbol{\alpha}^* \mathbf{Z}^\dagger \qquad (3S.6.1)$$

$$\mathbf{A}^n = \mathbf{Z} \boldsymbol{\alpha}^n \mathbf{Z}^{-1} \qquad (3S.6.2)$$

which follow from (3.4.2). We now obtain directly:

$$(\log \mathbf{A})^\dagger = [\mathbf{Z} \log(\boldsymbol{\alpha}) \mathbf{Z}^{-1}]^\dagger = (\mathbf{Z}^{-1})^\dagger \log(\boldsymbol{\alpha}^*) \mathbf{Z}^\dagger = \log(\mathbf{A}^\dagger) \qquad (3S.6.3)$$

$$\log(\mathbf{A}^n) = \log(\mathbf{Z} \boldsymbol{\alpha}^n \mathbf{Z}^{-1}) = \mathbf{Z} \log(\boldsymbol{\alpha}^n) \mathbf{Z}^{-1} = n \mathbf{Z} \log(\boldsymbol{\alpha}) \mathbf{Z}^{-1} = n \log(\mathbf{A}) \qquad (3S.6.4)$$

2. Since $\mathbf{A}$, $\mathbf{B}$ and $\mathbf{AB}$ belong to the domain where the exponential matrix function is the inverse of the logarithmic matrix function, we obtain

$$\mathbf{AB} = [\exp(\log \mathbf{A})][\exp(\log \mathbf{B})] = \exp(\log \mathbf{A} + \log \mathbf{B}) \qquad (3S.6.5)$$

where we have applied (3.1.6), assuming that $\log \mathbf{A}$ and $\log \mathbf{B}$ commute. Taking the logarithm, we arrive at (3E.6.3).

3. Using (3E.6.1) and then (3E.6.2), we obtain for a unitary matrix $\mathbf{U}$

$$(\log \mathbf{U})^\dagger = \log(\mathbf{U}^\dagger) = \log(\mathbf{U}^{-1}) = -\log \mathbf{U} \qquad (3S.6.6)$$

4

EXACT AND APPROXIMATE WAVE FUNCTIONS

We have now developed the basic tools needed for the construction of approximate electronic wave functions and for analysing the different approaches to molecular electronic-structure calculations. However, before we embark on this project, it is useful to consider in general terms some of the requirements we would like to place on any approximate wave function. We therefore discuss in this chapter the relationship between exact wave functions and approximate wave functions, with emphasis on topics such as size-extensivity, the variation principle and symmetry restrictions.

We first survey, in Section 4.1, the more important characteristic properties of the exact solution to the time-independent Schrödinger equation for a molecular electronic system and relate these characteristics to those of approximate wave functions. More detailed treatments of some of these topics follow in the subsequent sections: the variation principle in Section 4.2, size-extensivity in Section 4.3 and symmetry constraints in Section 4.4.

4.1 Characteristics of the exact wave function

The time-independent molecular electronic Schrödinger equation for an N-electron system with the Hamiltonian H in the coordinate representation reads

$$H\Psi = E\Psi \tag{4.1.1}$$

where the exact eigenstate Ψ is a function of the space and spin coordinates of the N electrons and E is the associated total electronic energy. In practice, it is exceedingly difficult to solve (4.1.1) since the wave function describes the correlated motion of N interacting particles. Indeed, analytic solutions to the Schrödinger equation can be found only for a few simple one-electron systems such as the one-electron hydrogen atom and the H_2^+ molecule. For more complicated systems, containing more than one electron, approximations must be introduced.

One popular strategy – which forms the basis for the methods examined in this book – is to seek an approximate solution to the Schrödinger equation (4.1.1) as a linear combination of Slater determinants constructed from a set of orthonormal orbitals:

$$|0\rangle = \sum_i c_i |i\rangle \tag{4.1.2}$$

In the limit of a complete set of orbitals, we may arrive at the *exact* solution to the Schrödinger equation in the form (4.1.2) but the determinantal expansion then becomes infinite. In practice, we must resort to truncated expansions and thus be satisfied with approximate solutions to the electronic Schrödinger equation.

Approximations should not be made in a haphazard manner. Rather we should seek to retain in our wave function as many symmetries and properties of the exact solution as possible. Indeed, some of the characteristics of the exact wave function are so important that we should try to incorporate them at each level of theory, and a few are so fundamental that they are introduced into our models without thought. We here list some of these properties and symmetries of the exact wave function that may guide us in the construction of models. We note that many of these characteristics or symmetries of the exact solution to the Schrödinger equation may be written in the form of some supplementary eigenvalue equation

$$A\Psi = a\Psi \qquad (4.1.3)$$

where A commutes with the molecular electronic Hamiltonian.

1. The exact state Ψ is a function of the space and spin coordinates of N electrons. The approximate state $|0\rangle$ should therefore be an eigenfunction of the number operator (1.3.12) with an eigenvalue equal to the number of electrons:

$$\hat{N}|0\rangle = N|0\rangle \qquad (4.1.4)$$

This requirement is of course trivially satisfied by writing the wave function as a linear combination of Slater determinants belonging to the subspace $F(M, N)$ of the Fock space; see Section 1.1.

2. The exact wave function is *antisymmetric* with respect to the permutation of any pair of electrons

$$P_{ij}\Psi = -\Psi \qquad (4.1.5)$$

In second quantization, the Pauli antisymmetry principle is incorporated through the algebraic properties of the creation and annihilation operators as discussed in Chapter 1. We note that, in density-functional theory (which bypasses the construction of the wave function and concentrates on the electron density), the fulfilment of the N-representability condition on the density represents a less trivial problem. A density is said to be *N-representable* if it can be derived from an antisymmetric wave function for N particles [1].

3. For a bound state, the exact wave function is *square-integrable* and hence normalizable

$$\langle\Psi|\Psi\rangle = 1 \qquad (4.1.6)$$

This boundary condition on the Schrödinger equation may be satisfied provided the approximate wave function is expanded in a set of normalized orbitals. In practice, the orthonormal orbitals from which the determinants are constructed are expanded in a finite set of Gaussian functions (and sometimes Slater-type functions) as discussed in Chapter 6. The asymptotic form of these functions ensures that the wave function is square-integrable.

4. The exact wave function is *variational* in the sense that, for all possible variations $\delta\Psi$ that are orthogonal to the wave function, the energy is stable [2]:

$$\langle\delta\Psi|\Psi\rangle = 0 \Rightarrow \langle\delta\Psi|H|\Psi\rangle = 0 \qquad (4.1.7)$$

It is desirable to incorporate this property in approximate calculations, for several reasons. First, it ensures that the calculated energy represents an upper bound to the true ground-state energy. This observation forms the basis for a systematic procedure for improving the quality of any approximate variational wave function: the ground-state energy is obtained by minimizing the expectation value of the Hamiltonian with respect to a set of variational parameters and improved descriptions are arrived at by systematically extending the variational space. Second, the variational property simplifies the calculation of many molecular properties. Unfortunately, it is not always easy to reconcile the variational requirement with the size-extensivity requirement to be discussed shortly. The variation principle is examined in Section 4.2.

5. The exact wave function is *size-extensive* in the sense that, for a system containing noninteracting subsystems, the total energy is equal to the sum of the energies of the individual systems. More specifically, if the total Hamiltonian can be written as a sum of M noninteracting Hamiltonians

$$H_T = \sum_{i=1}^{M} H_i \tag{4.1.8}$$

then the energy of the total system E_T as obtained from the equation

$$H_T \Psi_T = E_T \Psi_T \tag{4.1.9}$$

is equal to the sum of the energies E_i

$$E_T = \sum_{i=1}^{M} E_i \tag{4.1.10}$$

obtained as the eigenvalues of the Schrödinger equations of the individual systems:

$$H_i \Psi_i = E_i \Psi_i \tag{4.1.11}$$

In approximate calculations, we should thus require that the energy obtained by applying a particular computational scheme to the supersystem is equal to the sum of the energies obtained by applying the same scheme separately to each noninteracting subsystem. This size-extensivity condition forces us to consider carefully the parametrization of wave functions in the Fock space and we shall find that some popular methods do not satisfy this condition. Size-extensivity is discussed in Section 4.3.

6. In nonrelativistic theory, the exact stationary states are eigenfunctions of the total and projected spin operators

$$S^2 \Psi = S(S + 1)\Psi \tag{4.1.12}$$

$$S_z \Psi = M \Psi \tag{4.1.13}$$

and we may wish to impose these *spin symmetries* on the approximate wave functions as well. As discussed in Chapter 2, the correct spin behaviour may be enforced by expanding the wave functions in symmetry-adapted linear combinations of determinants (CSFs).

7. Within the Born–Oppenheimer approximation, the exact stationary states form a basis for an irreducible representation of the molecular point group. We may enforce the same *spatial symmetry* on the approximate state by expanding the wave function in determinants constructed from a set of symmetry-adapted orbitals. For atoms, in particular, the use of point-group

symmetry ensures that the wave function is an eigenfunction of the orbital angular-momentum operators:

$$L^2 \Psi = L(L+1)\Psi \tag{4.1.14}$$

$$L_z \Psi = M_L \Psi \tag{4.1.15}$$

For diatomics and other linear molecules, only (4.1.15) is satisfied, whereas, for nonlinear molecules, the wave function is an eigenfunction of the discrete symmetry operations R of the (finite) molecular point group. It should be emphasized that the enforcement of spin and point-group symmetries on approximate wave functions constitutes a restriction on the wave function, which in variational ground-state calculations may raise the electronic energy above what would be obtained in an unrestricted treatment. Symmetry restrictions are discussed in Section 4.4.

8. Owing to the presence of the Coulomb potential, the molecular electronic Hamiltonian becomes singular when two electrons coincide in space. To balance this singularity, the exact wave function exhibits a characteristic nondifferentiable behaviour for coinciding electrons, giving rise to the *electronic Coulomb cusp condition* [3]

$$\lim_{r_{ij} \to 0} \left(\frac{\partial \Psi}{\partial r_{ij}} \right)_{\text{ave}} = \frac{1}{2} \Psi(r_{ij} = 0) \tag{4.1.16}$$

(spherical averaging implied) discussed in Section 7.2. This nondifferential behaviour of the wave function is not easily modelled by determinantal expansions. Indeed, the difficulties encountered in the description of the electronic cusp represent an important obstacle to the accurate calculation of electronic wave functions. A condition similar to (4.1.16) arises when electrons coincide with point-charge nuclei (*the nuclear cusp condition*). In most situations, the nuclear cusp is less of a problem than the electronic Coulomb cusp since it is more easily treated within the orbital model and since it affects mostly the calculation of molecular properties related to the electronic densities at the nuclei. In calculations that do not employ point-charge nuclei, the nuclear cusp condition does not arise, although the correct behaviour at the nuclei may still be difficult to model.

9. Another analytic result concerns the asymptotic behaviour at large distances from the molecule. It may be shown that, at large distances, the electron density decays exponentially

$$\rho(r) \approx \exp(-2^{3/2}\sqrt{I}r) \quad \text{(large } r) \tag{4.1.17}$$

where I is the first ionization potential of the molecule [4]. The a priori incorporation of the correct *long-range exponential decay* is impossible since the ionization potential is unknown at the outset of the calculation. The correct decay can be ensured only by a flexible description of the variational space in the outer regions of the molecular system.

10. The exact wave function changes in a characteristic way under *gauge transformations* of the potentials associated with electromagnetic fields, thereby ensuring that all molecular properties that may be calculated or extracted from the wave function are unaffected by such transformations [5]. It is desirable to incorporate the same *gauge invariance* in the calculation of properties from approximate wave functions so as to make the calculations unambiguous and well defined.

Although not exhaustive, the above list gives some indication of what characteristics of the exact wave function we may wish to incorporate in our approximate wave function to ensure that it gives a reasonable description of the electronic system. Some of these characteristics, such as square-integrability, the correct number of particles and the Pauli principle, are imposed without much thought. Other characteristics – for example, size-extensivity and in particular the cusp condition – may be more difficult to impose but still desirable. Still others, such as the nuclear cusp condition, the gauge invariance and the exponential decay, are of interest only in special situations. In the remainder of this chapter, we shall discuss some of these properties at greater length and in particular concern ourselves with the variation principle, size-extensivity and symmetry restrictions; the cusp condition is studied in Chapter 7. However, before we proceed, we note that, in this book, we are concerned only with solutions to the electronic problem within the *Born–Oppenheimer approximation* – treating the nuclei as stationary sources of electrostatic and possibly magnetostatic fields rather than as true particles.

4.2 The variation principle

We consider in this section the variation principle in molecular electronic-structure theory. Having established the particular relationship between the Schrödinger equation and the variational condition that constitutes the variation principle, we proceed to examine the variation method as a computational tool in quantum chemistry, paying special attention to the application of the variation method to linearly expanded wave functions. Next, we examine two important theorems of quantum chemistry – the Hellmann–Feynman theorem and the molecular virial theorem – both of which are closely associated with the variational condition for exact and approximate wave functions. We conclude this section by presenting a mathematical device for recasting any electronic energy function in a variational form so as to benefit to the greatest extent possible from the simplifications associated with the fulfilment of the variational condition.

4.2.1 THE VARIATION PRINCIPLE

The solution of the time-independent Schrödinger equation

$$\hat{H}|0\rangle = E_0|0\rangle \tag{4.2.1}$$

may be recast in the form of a variation principle for the energy written as an expectation value

$$E[\tilde{0}] = \frac{\langle \tilde{0}|\hat{H}|\tilde{0}\rangle}{\langle \tilde{0}|\tilde{0}\rangle} \tag{4.2.2}$$

where $|\tilde{0}\rangle$ is some approximation to the eigenstate in (4.2.1). We use square brackets here to indicate that the functional depends on the form of the wave function rather than on a set of parameters. In most applications, the wave function is described in terms of a finite set of parameters and the usual notation for functions will then be used.

To establish the one-to-one relationship between the stationary points of the energy functional $E[\tilde{0}]$ and the solutions to the Schrödinger equation, we first assume that $|0\rangle$ represents a solution to the Schrödinger equation (4.2.1) and that $|\delta\rangle$ is an allowed variation

$$|\tilde{0}\rangle = |0\rangle + |\delta\rangle \tag{4.2.3}$$

Inserting this expression in the energy functional (4.2.2) and expanding in orders of $|\delta\rangle$ around $|0\rangle$, we obtain

$$E[0 + \delta] = \frac{\langle 0|\hat{H}|0\rangle + \langle 0|\hat{H}|\delta\rangle + \langle \delta|\hat{H}|0\rangle + \langle \delta|\hat{H}|\delta\rangle}{\langle 0|0\rangle + \langle 0|\delta\rangle + \langle \delta|0\rangle + \langle \delta|\delta\rangle}$$

$$= E_0 + \langle 0|\hat{H} - E_0|\delta\rangle + \langle \delta|\hat{H} - E_0|0\rangle + O(\delta^2) = E_0 + O(\delta^2) \qquad (4.2.4)$$

where we have used the relation

$$\frac{1}{1+x} = 1 - x + O(x^2) \qquad (4.2.5)$$

The first-order variation in the energy functional $E[\tilde{0}]$ therefore vanishes whenever $|\tilde{0}\rangle$ corresponds to one of the eigenstates $|0\rangle$, thus demonstrating that the eigenstates of the Schrödinger equation represent stationary points of the energy functional.

Conversely, to show that all stationary points of the energy functional represent eigenstates of the Schrödinger equation, let $|0\rangle$ be a stationary point of $E[\tilde{0}]$. For the variation $|\delta\rangle$, we obtain by expanding the energy functional around the stationary point in the same way as in (4.2.4)

$$\langle 0|\hat{H} - E[0]|\delta\rangle + \langle \delta|\hat{H} - E[0]|0\rangle = 0 \qquad (4.2.6)$$

and, for the variation $i|\delta\rangle$, we find

$$\langle 0|\hat{H} - E[0]|\delta\rangle - \langle \delta|\hat{H} - E[0]|0\rangle = 0 \qquad (4.2.7)$$

Combination of these two equations yields

$$\langle \delta|\hat{H} - E[0]|0\rangle = 0 \qquad (4.2.8)$$

and, since this relation holds for an arbitrary variation $|\delta\rangle$, we arrive at the eigenvalue equation

$$\hat{H}|0\rangle = E[0]|0\rangle \qquad (4.2.9)$$

which shows that each stationary point $E[0]$ of the energy functional $E[\tilde{0}]$ also represents a solution $|0\rangle$ to the Schrödinger equation with eigenvalue $E[0]$. We have thus established *the variation principle*: the solution of the Schrödinger equation (4.2.1) is equivalent to a variational optimization of the energy functional (4.2.2).

4.2.2 THE VARIATION METHOD

The variation principle provides us with a simple and powerful procedure for generating approximate wave functions: *the variation method*. For some proposed model or ansatz for the wave function, we express the electronic state $|\mathbf{C}\rangle$ in terms of a finite set of numerical parameters $\mathbf{C}$, and the 'best' values of $\mathbf{C}$ are deemed to be those that correspond to the stationary points of the energy function

$$E(\mathbf{C}) = \frac{\langle \mathbf{C}|\hat{H}|\mathbf{C}\rangle}{\langle \mathbf{C}|\mathbf{C}\rangle} \qquad (4.2.10)$$

The stationary points of $E(\mathbf{C})$ represent the approximate electronic states $|\mathbf{C}\rangle$ and the values of $E(\mathbf{C})$ at the stationary points are the approximate energies. We shall discuss the meaning of the term 'best' approximation later.

The exact wave function $|0\rangle$ corresponds to a stationary point of the expectation value of the Hamiltonian. The expectation value of the Hamiltonian for *any* approximate wave function $|0\rangle + |\delta\rangle$ is therefore correct to second order in the error $|\delta\rangle$ and it follows that the energy calculated as an *expectation value* is more accurate than the wave function itself (for sufficiently small errors). This result is important as it shows that small contributions to the wave function may be neglected without affecting the calculated energy significantly. In Exercise 4.1, we show how the error in the energy can be estimated from the norm of the gradient.

4.2.3 LINEAR EXPANSIONS AND EIGENVALUE EQUATIONS

A particularly simple realization of the variation method arises if we make a linear ansatz for the wave function, expanding the approximate electronic state in an m-dimensional set of normalized antisymmetric N-electron functions (e.g. Slater determinants or CSFs):

$$|\mathbf{C}\rangle = \sum_{i=1}^{m} C_i |i\rangle \tag{4.2.11}$$

We here assume that the wave function is real. The energy function for this state depends on the numerical parameters C_i and is given by

$$E(\mathbf{C}) = \frac{\langle \mathbf{C}|\hat{H}|\mathbf{C}\rangle}{\langle \mathbf{C}|\mathbf{C}\rangle} \tag{4.2.12}$$

where $\hat{H}$ is assumed to be a real Hermitian operator. In order to locate and characterize the stationary points of this function, we need its first and second derivatives with respect to the variational parameters:

$$E_i^{(1)}(\mathbf{C}) = \frac{\partial E(\mathbf{C})}{\partial C_i} \tag{4.2.13}$$

$$E_{ij}^{(2)}(\mathbf{C}) = \frac{\partial^2 E(\mathbf{C})}{\partial C_i \partial C_j} \tag{4.2.14}$$

The first derivatives are the elements of a vector called the *electronic gradient* $\mathbf{E}^{(1)}$ and the second derivatives form a matrix known as the *electronic Hessian* $\mathbf{E}^{(2)}$. To obtain the elements of the gradient and the Hessian, it is convenient to rewrite (4.2.12) in the form

$$E(\mathbf{C})\langle \mathbf{C}|\mathbf{C}\rangle = \langle \mathbf{C}|\hat{H}|\mathbf{C}\rangle \tag{4.2.15}$$

Differentiating, we obtain the following expressions for a real wave function:

$$E_i^{(1)}(\mathbf{C})\langle \mathbf{C}|\mathbf{C}\rangle + 2E(\mathbf{C})\langle i|\mathbf{C}\rangle = 2\langle i|\hat{H}|\mathbf{C}\rangle \tag{4.2.16}$$

$$E_{ij}^{(2)}(\mathbf{C})\langle \mathbf{C}|\mathbf{C}\rangle + 2E_i^{(1)}(\mathbf{C})\langle j|\mathbf{C}\rangle + 2E_j^{(1)}(\mathbf{C})\langle i|\mathbf{C}\rangle + 2E(\mathbf{C})\langle i|j\rangle = 2\langle i|\hat{H}|j\rangle \tag{4.2.17}$$

The gradient and Hessian elements are therefore given by

$$E_i^{(1)}(\mathbf{C}) = 2\frac{\langle i|\hat{H}|\mathbf{C}\rangle - E(\mathbf{C})\langle i|\mathbf{C}\rangle}{\langle \mathbf{C}|\mathbf{C}\rangle} \tag{4.2.18}$$

$$E_{ij}^{(2)}(\mathbf{C}) = 2\frac{\langle i|\hat{H}|j\rangle - E(\mathbf{C})\langle i|j\rangle}{\langle \mathbf{C}|\mathbf{C}\rangle} - 2E_i^{(1)}(\mathbf{C})\frac{\langle j|\mathbf{C}\rangle}{\langle \mathbf{C}|\mathbf{C}\rangle} - 2E_j^{(1)}(\mathbf{C})\frac{\langle i|\mathbf{C}\rangle}{\langle \mathbf{C}|\mathbf{C}\rangle} \tag{4.2.19}$$

Returning to the variational problem, we note that the conditions for the stationary points

$$E_i^{(1)}(\mathbf{C}) = 0 \tag{4.2.20}$$

are now equivalent to the requirement

$$\langle i|\hat{H}|\mathbf{C}\rangle = E(\mathbf{C})\langle i|\mathbf{C}\rangle \tag{4.2.21}$$

which, in matrix notation, may be written as

$$\mathbf{HC} = E(\mathbf{C})\mathbf{SC} \tag{4.2.22}$$

where the elements of the Hamiltonian matrix $\mathbf{H}$ and the overlap matrix $\mathbf{S}$ are given by

$$H_{ij} = \langle i|\hat{H}|j\rangle \tag{4.2.23}$$

$$S_{ij} = \langle i|j\rangle \tag{4.2.24}$$

Assuming that the antisymmetric N-electron functions $|i\rangle$ are orthonormal

$$\langle i|j\rangle = \delta_{ij} \tag{4.2.25}$$

we obtain a standard m-dimensional eigenvalue problem of linear algebra

$$\mathbf{HC} = E(\mathbf{C})\mathbf{C} \tag{4.2.26}$$

Since $\mathbf{H}$ is Hermitian, this equation has exactly m orthonormal solutions

$$\mathbf{C}_K = \begin{pmatrix} C_{1K} \\ C_{2K} \\ \vdots \\ C_{mK} \end{pmatrix}, \qquad \mathbf{C}_K^{\mathrm{T}}\mathbf{C}_L = \delta_{KL} \tag{4.2.27}$$

with the associated real eigenvalues

$$E_K = E(\mathbf{C}_K), \qquad E_1 \le E_2 \le \cdots \le E_m \tag{4.2.28}$$

which may or may not be distinct. The eigenvectors $\mathbf{C}_K$ in (4.2.27) represent the approximate wave functions

$$|K\rangle = \sum_{i=1}^{m} C_{iK}|i\rangle \tag{4.2.29}$$

and the associated eigenvalues E_K in (4.2.28) represent the approximate energies.

To characterize the stationary points, we note that the Hessian elements (4.2.19) at these points are given by

$$^{K}E_{ij}^{(2)} = E_{ij}^{(2)}(\mathbf{C}_K) = 2(\langle i|\hat{H}|j\rangle - E_K\langle i|j\rangle) \tag{4.2.30}$$

where we have used the fact that the gradient vanishes at the stationary point and that the eigenvectors are normalized. Further simplifications are obtained if the Hessian is expressed in the basis of the eigenvectors of the Hamiltonian:

$$^{K}E_{MN}^{(2)} = 2(\langle M|\hat{H}|N\rangle - E_K\langle M|N\rangle) = 2(E_M - E_K)\delta_{MN} \tag{4.2.31}$$

Thus, the eigenvalues of the Hessian at the stationary points are given by

$$^K E_{MM}^{(2)} = 2(E_M - E_K) \tag{4.2.32}$$

and correspond to the excitation energies (multiplied by 2) for the transition from $|K\rangle$ to $|M\rangle$. We note that the Hessian is singular since $^K E_{KK}^{(2)} = 0$ and that the Kth electronic state has exactly $K - 1$ negative eigenvalues since, in the absence of degeneracies, $E_M < E_K$ for $M < K$. Thus, in the space orthogonal to $\mathbf{C}_K$, the first solution to (4.2.26) corresponds to a minimum, the second corresponds to a first-order saddle point, the third to a second-order saddle point, and so on.

To summarize, the linear ansatz for the wave function (4.2.11) leads, in an orthonormal basis, to an eigenvalue equation for the approximate wave function (4.2.26). In an m-dimensional vector space, this equation has exactly m solutions, which we associate with the ground state, the first excited state, and so on. In the limit of a complete expansion, the exact electronic states are recovered. Thus, it should be possible to improve our approximate solutions to the Schrödinger equation in a controlled manner, by systematically extending the basis of N-electron functions. Convergence can be monitored by examining the behaviour of the eigenvalue spectrum as the vector space is extended.

4.2.4   UPPER BOUNDS AND THE HYLLERAAS–UNDHEIM THEOREM

Consider two orthonormal sets of N-electron basis functions $S' = \{|I'\rangle\}$ and $S'' = \{|I''\rangle\}$ where S' is a subset of S''. The eigenvalue equations in the two basis sets are given by

$$\mathbf{H}'\mathbf{C}_K' = E_K'\mathbf{C}_K' \tag{4.2.33}$$

$$\mathbf{H}''\mathbf{C}_K'' = E_K''\mathbf{C}_K'' \tag{4.2.34}$$

and we assume that the eigenvalues are sorted in ascending order. Since S' is contained in S'', we may express the eigenvectors of S' exactly in terms of the eigenvectors of S''

$$|K'\rangle = \sum_L a_{KL}|L''\rangle \tag{4.2.35}$$

where the normalization of $|K'\rangle$ implies the condition

$$\sum_L |a_{KL}|^2 = 1 \tag{4.2.36}$$

The first energy in the smallest basis can now be written as

$$E_1' = \sum_{KL} a_{1K}^* \langle K''|\hat{H}|L''\rangle a_{1L} = \sum_K |a_{1K}|^2 E_K'' \geq E_1'' \sum_K |a_{1K}|^2 = E_1'' \tag{4.2.37}$$

which shows that the lowest eigenvalue in S' represents an upper bound to the lowest eigenvalue in S'':

$$E_1'' \leq E_1' \tag{4.2.38}$$

Therefore, in the linear variation method, the lowest eigenvalue will always drop as the variational space is extended, providing us with a highly systematic way of approaching the exact ground state.

The result (4.2.38) is just a special case of a more general result embodied in *Cauchy's interlace theorem* [6]. According to this theorem, which is proved in Exercise 4.2, the eigenvalues of two linear variational spaces $S' \subset S''$ are related to one another in the following manner

$$E_K'' \leq E_K' \leq E_{K+\delta}'' \tag{4.2.39}$$

where δ is the difference in the dimensions of the two spaces:

$$\delta = \dim S'' - \dim S' \tag{4.2.40}$$

It is therefore not only the lowest eigenvalue that is lowered as the linear variational space increases – the Kth eigenvalue of the extended space S'' is always less than or equal to the Kth eigenvalue of the original space S'. In particular, in any linear vector space, the m lowest eigenvalues of the Hamiltonian matrix $\mathbf{H}$ provide rigorous upper bounds to the m lowest eigenvalues of the exact solutions to the Schrödinger equation. As a special case of Cauchy's interlace theorem, we obtain for $\delta = 1$ *the Hylleraas–Undheim theorem* [7]

$$E_1'' \leq E_1' \leq E_2'' \leq E_2' \leq \cdots \leq E_m'' \leq E_m' \leq E_{m+1}'' \tag{4.2.41}$$

which relates the eigenvalues of two variational spaces $S' \subset S''$, one of which contains one more basis function than the other space.

To illustrate Cauchy's interlace theorem, we shall consider linear variational calculations of the five lowest $^1\Sigma^+$ electronic energies of the BH molecule, all calculated at the ground-state equilibrium geometry. Although the computational details need not concern us here, we note that the variational calculations have been carried out in the Fock space of the canonical Hartree–Fock orbitals of the $^1\Sigma^+$ ground state, generated in the d-aug-cc-pVDZ basis (see Section 8.3). Moreover, a total of six linear variational spaces are considered. The smallest space is denoted by S (for singles) and contains, in addition to the Hartree–Fock state, all configurations generated by single excitations out of the Hartree–Fock configuration. Next, the SD (singles–doubles) space is obtained from S by also including all double excitations, the SDT (singles–doubles–triples) space is obtained by including the triple excitations and the SDTQ (singles–doubles–triples–quadruples) space by including the quadruple excitations. The final variational space contains the full set of six-electron states in the Fock space and is denoted by FCI (full configuration interaction). For a description of the computational models, see Chapter 5; for a discussion of one-electron basis sets, see Chapter 8.

In Figure 4.1, we have plotted, for each variational space, the five lowest $^1\Sigma^+$ eigenvalues of the BH molecule. In agreement with Cauchy's interlace theorem, all energies drop as the variational space increases. However, since the dimension of the variational space increases by more than 1 as we go from one space to the next, the Hylleraas–Undheim theorem (4.2.41) does not apply. The energy E_K obtained in one space is therefore in general not a lower bound to the energy E_{K+1} obtained in the next space – for example, from an inspection of Figure 4.1, we see that $E_2(S) > E_3(SD)$.

We conclude that the linear variational expansion of the wave function provides us with a high degree of control over our calculations. From a sequence of variational calculations where the vector space is systematically extended, we may generate a sequence of energies that converge monotonically from above towards the exact energies of the Schrödinger equation. Also, whenever the electronic energies are calculated by applying the variation principle to a wave function expanded in an m-dimensional vector space, we may unambiguously identify the m eigenvalues as approximations to the lowest m exact energies.

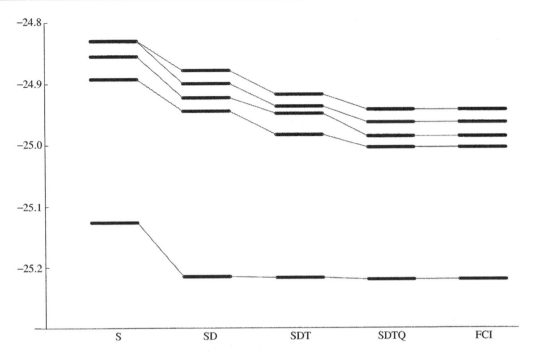

Fig. 4.1. The five lowest totally symmetric $^1\Sigma^+$ energies (in E_h) of the BH molecule calculated in the d-aug-cc-pVDZ basis at the molecular ground-state equilibrium bond distance of $2.3289a_0$.

4.2.5 NONLINEAR EXPANSIONS

The discussion of the variation method has so far been restricted to the linear variation method. For more general models, where the variational parameters appear in a nonlinear fashion, variational energies will still be upper bounds to the exact lowest energy, but the variationally determined excited energies will no longer necessarily represent upper bounds to the exact energies of the excited states.

To investigate in more detail the relationship between linearly and nonlinearly optimized energies, consider a nonlinearly determined state $|\mathbf{C}\rangle$ in Fock space. The nature of this state need not concern us here, but we may, for example, take it to represent a variationally optimized Hartree–Fock approximation to the ground state. This nonlinearly optimized state may now be expanded in a set of orthonormal eigenvectors $|K\rangle$:

$$|\mathbf{C}\rangle = \sum_K C_K |K\rangle \qquad (4.2.42)$$

The vectors $|K\rangle$ may be the exact eigenvectors of the Hamiltonian ordered as $E_{K+1} > E_K$ or the Hamiltonian eigenvectors of the Fock space containing $|\mathbf{C}\rangle$, ordered in the same manner. Assuming that $|\mathbf{C}\rangle$ is normalized and that its energy is written as an expectation value of the Hamiltonian, we obtain

$$E_{\mathbf{C}} = \langle \mathbf{C} | \hat{H} | \mathbf{C} \rangle = \sum_K C_K^2 \langle K | \hat{H} | K \rangle = \sum_K C_K^2 E_K \qquad (4.2.43)$$

We can now relate E_C to one particular Hamiltonian eigenvalue E_I in the following manner

$$E_C = E_I + \sum_K C_K^2 (E_K - E_I) \tag{4.2.44}$$

The energy E_C is less than, equal to or greater than the selected eigenvalue E_I, depending on which of the following relations are satisfied

$$-\sum_{K<I} C_K^2 (E_K - E_I) > \sum_{K>I} C_K^2 (E_K - E_I) \tag{4.2.45}$$

$$-\sum_{K<I} C_K^2 (E_K - E_I) = \sum_{K>I} C_K^2 (E_K - E_I) \tag{4.2.46}$$

$$-\sum_{K<I} C_K^2 (E_K - E_I) < \sum_{K>I} C_K^2 (E_K - E_I) \tag{4.2.47}$$

Clearly, if $|C\rangle$ is orthogonal to all states $|K\rangle$ of energy lower than E_I, then E_C will be an upper bound to E_I. For example, E_C will always be an upper bound to the exact ground-state energy E_1. Moreover, if the nonlinearly optimized state $|C\rangle$ has been determined such that it is orthogonal to the ground-state eigenfunction $|1\rangle$, then E_C will also be an upper bound to the first excited state, and so on.

In general, for a given set of nonlinearly optimized approximations to several of the lowest states, nothing can be said about their relationship to the exact energies except that they all represent upper bounds to the exact ground state. However, we may always consider these nonlinearly optimized electronic states as basis functions of a (presumably small) linear vector space. By solving the corresponding linear variation problem in this vector space, we obtain a set of eigenvalues that, in accordance with Cauchy's interlace theorem, represent upper bounds to the exact energies.

As an illustration, consider the two lowest ^1S states of the helium atom. Nonlinear Hartree–Fock optimizations give -2.861627 E_h for the $1s^2$ ground state and -2.169652 E_h for the $1s2s$ excited state generated in the t-aug-cc-pVTZ basis. Both energies provide upper bounds to the exact ground-state energy (-2.903724 E_h). However, since these approximate energies have been obtained by a nonlinear procedure, they do not represent lower or upper bounds to the exact energy of the first excited state of the same symmetry (-2.145974 E_h). On the contrary, in this particular case, both approximate energies turn out to be lower than the exact excited energy.

Let us now construct two approximate electronic states of ^1S symmetry in the basis of the Hartree–Fock approximations to the $1s^2$ and $1s2s$ states. In this basis, the electronic states are written in the following manner

$$|C\rangle = C_{1s^2} |1s^2\rangle + C_{1s2s} |1s2s\rangle \tag{4.2.48}$$

Minimization of the energy with respect to the two linear parameters leads to the following two-dimensional eigenvalue problem

$$\begin{pmatrix} \langle 1s^2|\hat{H}|1s^2\rangle & \langle 1s^2|\hat{H}|1s2s\rangle \\ \langle 1s2s|\hat{H}|1s^2\rangle & \langle 1s2s|\hat{H}|1s2s\rangle \end{pmatrix} \begin{pmatrix} C_{1s^2} \\ C_{1s2s} \end{pmatrix} = E \begin{pmatrix} \langle 1s^2|1s^2\rangle & \langle 1s^2|1s2s\rangle \\ \langle 1s2s|1s^2\rangle & \langle 1s2s|1s2s\rangle \end{pmatrix} \begin{pmatrix} C_{1s^2} \\ C_{1s2s} \end{pmatrix} \tag{4.2.49}$$

The diagonal elements are the energies and norms of the two Hartree–Fock states and are trivially obtained from the separate Hartree–Fock calculations. The off-diagonal matrix elements involve

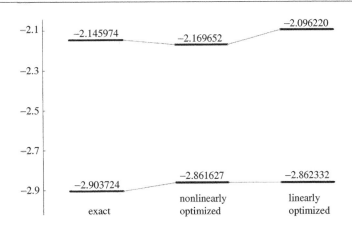

Fig. 4.2. Exact and approximate energies for the ground state and the first excited state of ^1S symmetry of the helium atom (in E_h). The nonlinearly optimized energies correspond to two separately optimized Hartree–Fock energies for the $1s^2$ and $1s2s$ configurations; the linearly optimized energies are those obtained by solving a 2×2 eigenvalue problem in the basis of the same two Hartree–Fock states.

two Hartree–Fock states of different (nonorthogonal) orbitals and must be evaluated using the special technique discussed in Section 11.9. The eigenvalue problem then becomes

$$\begin{pmatrix} -2.861627 & -0.825295 \\ -0.825295 & -2.169652 \end{pmatrix} \begin{pmatrix} C_{1s^2} \\ C_{1s2s} \end{pmatrix} = E \begin{pmatrix} 1.000000 & 0.280609 \\ 0.280609 & 1.000000 \end{pmatrix} \begin{pmatrix} C_{1s^2} \\ C_{1s2s} \end{pmatrix} \tag{4.2.50}$$

Solving the eigenvalue problem, we arrive at the energies -2.862332 and -2.096220 E_h, which represent rigorous upper bounds to the exact energies of the two lowest ^1S states of the helium atom. The relationships among the approximate and exact energies of the helium atom are illustrated in Figure 4.2.

The solutions obtained by applying the variation method to a nonlinear wave function are thus not orthogonal to one another and do not provide us with a diagonal representation of the Hamiltonian. In effect, there is no simple relationship among the different variational solutions in the nonlinear case – even the number of distinct variational solutions is usually unknown. Indeed, many of the stationary points of the energy function cannot be associated with any particular electronic state and should be rejected as unphysical. As a result, each calculated stationary point should be scrutinized to confirm that it indeed represents an acceptable solution to the Schrödinger equation – for example, by examining the number of negative eigenvalues of the electronic Hessian. These difficulties, which arise from the nonlinear parametrization, should be contrasted with the transparency of the linear variation method, where the solutions constitute an orthonormal basis in a linear m-dimensional vector space and where each solution may be unambiguously associated with a well-defined physical state.

4.2.6 THE HELLMANN–FEYNMAN THEOREM

Many of the theorems for exact wave functions hold also for approximate wave functions that are variationally determined. We consider here the *Hellmann–Feynman theorem*, which states that the first-order change in the energy due to a perturbation may be calculated as the expectation value of the perturbation operator V [8]. This result is easily derived:

$$\frac{dE(\alpha)}{d\alpha}\bigg|_{\alpha=0} = \frac{\partial}{\partial\alpha}\frac{\langle\Psi_\alpha|H+\alpha V|\Psi_\alpha\rangle}{\langle\Psi_\alpha|\Psi_\alpha\rangle}\bigg|_{\alpha=0}$$

$$= 2\mathrm{Re}\left\langle\frac{\partial\Psi_\alpha}{\partial\alpha}\bigg|_{\alpha=0}\bigg|H-E(0)\bigg|\Psi\right\rangle + \langle\Psi|V|\Psi\rangle$$

$$= \langle\Psi|V|\Psi\rangle \tag{4.2.51}$$

Here $|\Psi_\alpha\rangle$ is the wave function associated with $H+\alpha V$ and $|\Psi\rangle$ is the normalized, unperturbed wave function at $\alpha=0$. As an example, consider a molecule with the nuclei initially located at $\mathbf{R}_I$. Displacing the nuclei to $\mathbf{R}_I + \boldsymbol{\alpha}_I$, we obtain the Hamiltonian

$$H(\mathbf{R}_I + \boldsymbol{\alpha}_I) = -\frac{1}{2}\sum_i \nabla_i^2 + \sum_{i>j}\frac{1}{|\mathbf{r}_i - \mathbf{r}_j|} - \sum_{iI}\frac{Z_I}{|\mathbf{r}_i - \mathbf{R}_I - \boldsymbol{\alpha}_I|} + \sum_{I>J}\frac{Z_I Z_J}{|\mathbf{R}_I - \mathbf{R}_J + \boldsymbol{\alpha}_I - \boldsymbol{\alpha}_J|}$$

$$= H(\mathbf{R}_I) - \sum_{iI}\frac{Z_I(\mathbf{r}_i - \mathbf{R}_I)\cdot\boldsymbol{\alpha}_I}{|\mathbf{r}_i - \mathbf{R}_I|^3} - \sum_{I\neq J}\frac{Z_I Z_J(\mathbf{R}_I - \mathbf{R}_J)\cdot\boldsymbol{\alpha}_I}{|\mathbf{R}_I - \mathbf{R}_J|^3} + O(\boldsymbol{\alpha}^2) \tag{4.2.52}$$

Also the wave function is modified when the nuclei are displaced. However, the Hellmann–Feynman theorem shows that the contribution from the changes in the wave function vanishes for first derivatives:

$$\frac{dE}{d\boldsymbol{\alpha}_I} = -\left\langle\Psi\bigg|\sum_i\frac{Z_I(\mathbf{r}_i - \mathbf{R}_I)}{|\mathbf{r}_i - \mathbf{R}_I|^3}\bigg|\Psi\right\rangle - \sum_{J\neq I}\frac{Z_I Z_J(\mathbf{R}_I - \mathbf{R}_J)}{|\mathbf{R}_I - \mathbf{R}_J|^3} \tag{4.2.53}$$

The Hellmann–Feynman theorem thus greatly simplifies the evaluation of the first derivatives of variational energies.

The Hellmann–Feynman theorem also holds for approximate wave functions provided they are optimized with respect to the changes induced by the perturbation:

$$|\Psi\rangle \to |\Psi\rangle + \alpha\left|\frac{\partial\Psi}{\partial\alpha}\right\rangle \tag{4.2.54}$$

This point sometimes requires careful consideration. For example, in molecular calculations, we usually construct our wave functions from a finite set of analytical functions attached to the atomic nuclei. Consequently, whenever the geometry is distorted, we also change the basis set in terms of which our electronic wave function is expanded. We are thus faced with a *restricted but variable variational space*, and a wave function optimized at one particular geometry cannot be accurately represented in terms of the basis set associated with a different geometry. From this observation it follows that a geometrical distortion of the molecule will introduce changes in the wave function that cannot be described by the variational parameters associated with the undistorted geometry. As a result, the unperturbed electronic energy is not variational with respect to these changes and the conditions of the Hellmann–Feynman theorem are not satisfied. It should be emphasized that this situation holds for all wave functions expanded in finite and variable basis sets, even for those that are fully optimized within the variational space at each geometry. In general, we cannot invoke the Hellmann–Feynman theorem for any wave function optimized within a restricted but variable variational space.

The Hellmann–Feynman theorem as presented above is associated with the coordinate representation of quantum mechanics. In the picture provided by second quantization, the situation is different since the one-electron basis functions appear only in the Hamiltonian operator and do not enter the calculations in any other way. Consequently, any changes that are imposed on the basis set modify only the electronic Hamiltonian and not the wave function directly. In fact, the formalism of second quantization allows us to treat fixed basis sets and variable basis sets on the same footing by isolating all complications associated with variable sets in the Hamiltonian. As a consequence, variable basis sets do not affect the conditions of the Hellmann–Feynman theorem, which may therefore also be invoked in such cases. Although this circumstance does not lead to any computational savings, it does represent a conceptual simplification in the sense that it allows us to manipulate wave functions expanded in finite basis sets in the same manner as we manipulate exact, fully variational functions [9].

4.2.7 THE MOLECULAR ELECTRONIC VIRIAL THEOREM

The exact molecular electronic energy is variational with respect to an arbitrary change in the wave function. In particular, therefore, it is variational with respect to the following uniform scaling of the electron coordinates:

$$\Psi(\mathbf{r}_i) \to \Psi_\alpha(\mathbf{r}_i) = \alpha^{3N/2}\Psi(\alpha\mathbf{r}_i) \tag{4.2.55}$$

Here α is the scaling factor and N the number of electrons in the system. The overall scaling factor $\alpha^{3N/2}$ has been incorporated so as to preserve normalization of the *scaled wave function* Ψ_α (see Exercise 4.3):

$$\langle\Psi_\alpha|\Psi_\alpha\rangle = 1 \tag{4.2.56}$$

We may now write the stationary condition on the energy in the form

$$\frac{\partial}{\partial\alpha}\langle\Psi_\alpha|H(\mathbf{R})|\Psi_\alpha\rangle\bigg|_{\alpha=1} = 0 \tag{4.2.57}$$

where $H(\mathbf{R})$ is the electronic Hamiltonian

$$H(\mathbf{R}) = T + V(\mathbf{R}) \tag{4.2.58}$$

$$T = -\frac{1}{2}\sum_i \nabla_i^2 \tag{4.2.59}$$

$$V(\mathbf{R}) = \sum_{i>j}\frac{1}{|\mathbf{r}_i - \mathbf{r}_j|} - \sum_{iI}\frac{Z_I}{|\mathbf{r}_i - \mathbf{R}_I|} + \sum_{I>J}\frac{Z_I Z_J}{|\mathbf{R}_I - \mathbf{R}_J|} \tag{4.2.60}$$

Note that the undisturbed system is represented here by $\alpha = 1$ rather than $\alpha = 0$ as in the discussion of the Hellmann–Feynman theorem. The Hamiltonian (4.2.58) depends parametrically on the nuclear positions $\mathbf{R}_I$, collectively denoted by $\mathbf{R}$. Before evaluating the variational condition (4.2.57), we note the following identities (see Exercise 4.3)

$$\langle\Psi_\alpha|T|\Psi_\alpha\rangle = \alpha^2\langle\Psi|T|\Psi\rangle \tag{4.2.61}$$

$$\langle\Psi_\alpha|V(\mathbf{R})|\Psi_\alpha\rangle = \alpha\langle\Psi|V(\alpha\mathbf{R})|\Psi\rangle \tag{4.2.62}$$

The variational condition may therefore be written in the form

$$\frac{\partial}{\partial\alpha}[\alpha^2\langle\Psi|T|\Psi\rangle + \alpha\langle\Psi|V(\alpha\mathbf{R})|\Psi\rangle]\bigg|_{\alpha=1} = 0 \tag{4.2.63}$$

Carrying out the differentiation, we obtain

$$2\langle\Psi|T|\Psi\rangle + \langle\Psi|V(\mathbf{R})|\Psi\rangle = -\left\langle\Psi\left|\frac{dV(\alpha\mathbf{R})}{d\alpha}\right|_{\alpha=1}\right|\Psi\right\rangle \tag{4.2.64}$$

which constitutes the quantum-mechanical *virial theorem* for a field-free nonrelativistic electronic molecular system [10]. It holds for the exact electronic wave function and also for any approximate wave function that is variational with respect to a uniform scaling of the electronic coordinates. It does not hold for an approximate wave function expanded in an incomplete basis unless the scaling parameter α has been explicitly optimized. However, both sides of (4.2.64) are easily evaluated. A comparison of these two terms may then serve as an indication of the quality of the one-electron basis employed in the construction of the wave function.

The molecular electronic virial theorem is usually written in a slightly different fashion. According to the Hellmann–Feynman theorem, we have the following relationship between the derivative of the total electronic energy and the derivative of the potential-energy operator:

$$\left.\frac{dE(\alpha\mathbf{R})}{d\alpha}\right|_{\alpha=1} = \left\langle\Psi\left|\frac{dV(\alpha\mathbf{R})}{d\alpha}\right|_{\alpha=1}\right|\Psi\right\rangle \tag{4.2.65}$$

The virial theorem may therefore be written in the form

$$2\langle\Psi|T|\Psi\rangle + \langle\Psi|V(\mathbf{R})|\Psi\rangle = -\left.\frac{dE(\alpha\mathbf{R})}{d\alpha}\right|_{\alpha=1} \tag{4.2.66}$$

Thus, according to the molecular electronic virial theorem, the classical force with respect to a uniform scaling of the nuclear framework is equal to twice the electronic kinetic energy plus the potential energy. Whenever this force vanishes, the kinetic energy is equal to minus one-half of the potential energy. Most important, the scaling force vanishes at molecular equilibrium geometries $\mathbf{R}_e$ and also for atoms:

$$\langle\Psi|T|\Psi\rangle|_{\mathbf{R}=\mathbf{R}_e} = -\tfrac{1}{2}\langle\Psi|V(\mathbf{R}_e)|\Psi\rangle|_{\mathbf{R}=\mathbf{R}_e} \tag{4.2.67}$$

Combining the virial theorem at equilibrium with the expression for the total electronic energy

$$E(\mathbf{R}) = \langle\Psi|T|\Psi\rangle + \langle\Psi|V(\mathbf{R})|\Psi\rangle \tag{4.2.68}$$

we arrive at the following simple relationships for the exact wave function:

$$\langle\Psi|T|\Psi\rangle|_{\mathbf{R}=\mathbf{R}_e} = -E(\mathbf{R}_e) \tag{4.2.69}$$

$$\langle\Psi|V(\mathbf{R}_e)|\Psi\rangle|_{\mathbf{R}=\mathbf{R}_e} = 2E(\mathbf{R}_e) \tag{4.2.70}$$

These expressions also hold for any approximate wave function provided it is variational with respect to a uniform scaling of the nuclear as well as the electronic coordinates. The requirement that the electronic wave function should be variationally optimized with respect to a uniform scaling of the nuclear coordinates arises since otherwise the Hellmann–Feynman theorem (4.2.65) cannot be invoked. The requirements for fulfilment of (4.2.66) are thus more stringent than for (4.2.64), which assumes stability only with respect to scaling of the electronic coordinates. On the other hand, (4.2.64) is less useful than (4.2.66) in the sense that, for an approximate wave function that does not satisfy the conditions of the Hellmann-Feynman theorem, its right-hand side does not vanish at physically interesting geometries. An interesting corollary of the virial theorem in the form (4.2.69) is that no stationary points of positive energy (relative to the infinite separation of all particles) can exist since, at such a point, the kinetic energy would have to be negative.

The above expressions for the molecular electronic virial theorem have been given in terms of the force with respect to a uniform scaling of the nuclear framework. This scaling force may easily be related to the classical Cartesian forces on the nuclei

$$\mathbf{F}_I(\mathbf{R}) = -\frac{dE(\mathbf{R})}{d\mathbf{R}_I} \tag{4.2.71}$$

by invoking the chain rule

$$\frac{dE(\alpha\mathbf{R})}{d\alpha}\bigg|_{\alpha=1} = \sum_I \frac{d(\alpha\mathbf{R}_I)}{d\alpha}\bigg|_{\alpha=1} \cdot \frac{dE(\alpha\mathbf{R})}{d(\alpha\mathbf{R}_I)}\bigg|_{\alpha=1} = -\sum_I \mathbf{R}_I \cdot \mathbf{F}_I(\mathbf{R}) \tag{4.2.72}$$

Combining (4.2.72) with (4.2.66), we may now turn the molecular electronic virial theorem around and regard it as setting up a condition on the Cartesian forces acting on the nuclei:

$$\sum_I \mathbf{R}_I \cdot \mathbf{F}_I(\mathbf{R}) = 2\langle\Psi|T|\Psi\rangle + \langle\Psi|V(\mathbf{R})|\Psi\rangle \tag{4.2.73}$$

There are other conditions on these forces, related to the translational and rotational invariance of the electronic energy (which are always exactly fulfilled also for approximate wave functions). In particular, in a diatomic molecule, there are three translational conditions – two rotational conditions and one condition provided by the electronic virial theorem. Taken together, these conditions determine the diatomic nuclear force field completely. The only nonvanishing force acts along the molecular axis and may be obtained directly from the kinetic and potential energies if the wave function is fully variational with respect to a scaling of both the electronic and the nuclear coordinates.

Before concluding our discussion of the virial theorem, we note that, for molecular wave functions describing nuclear as well as electronic degrees of freedom, the virial theorem takes a simpler form

$$2\langle\Psi_{\text{mol}}|T_{\text{mol}}|\Psi_{\text{mol}}\rangle + \langle\Psi_{\text{mol}}|V_{\text{mol}}|\Psi_{\text{mol}}\rangle = 0 \tag{4.2.74}$$

This simplification occurs since, in this case, the potential-energy operator scales in a simple fashion characteristic of the Coulomb potential when the coordinates of *all* particles (electrons and nuclei) are uniformly scaled. It is the failure of the electron–nuclear potential-energy operator (4.2.62) to scale in a simple fashion in response to the electron coordinate scaling that introduces the scaling-force term in the molecular electronic virial theorem.

4.2.8 VARIATIONAL REFORMULATION OF NONVARIATIONAL ENERGIES

Although many computational techniques rely on the use of the variation principle at some early stage to provide a zero-order wave function, they do not always provide us with final energies that are variationally determined (i.e. stationary) and we cannot then invoke the Hellmann–Feynman theorem to simplify the calculation of molecular properties. Indeed, it would appear that, in order to calculate first-order properties for such nonvariational wave functions, we would need to calculate the first-order response of the wave function to the perturbation of interest, making the calculation of properties from nonvariational wave functions rather cumbersome.

Fortunately, the additional complications associated with nonvariational energies are not as severe as we might first suspect since, in most cases, it is rather easy to modify the energy function of a nonvariational wave function in such a way that the optimized energy becomes stationary with respect to the variables of this new function. In this variational formulation, then, the conditions of the Hellmann–Feynman theorem are indeed fulfilled and molecular properties may be calculated by a procedure that is essentially the same as for variational wave functions.

To see how this reformulation may be accomplished, we shall consider the Hellmann–Feynman theorem from a somewhat more general point of view than in Section 4.2.6, assuming nothing about the structure of the energy expression $E(\alpha, \lambda)$ except that it depends on two sets of parameters – the parameter α that describes the perturbation and the parameters λ that describe the wave function. The parameter α may for example represent an external field or a geometrical distortion; the parameters λ may represent the orbitals or the expansion in N-particle space. We need not concern ourselves here with the structure or nature of these parameters but note that the optimized electronic energy $E(\alpha)$ at α is obtained by substituting the optimized set of parameters λ^* in the energy function $E(\alpha, \lambda)$:

$$E(\alpha) = E(\alpha, \lambda^*) \tag{4.2.75}$$

At each value of α, λ^* is obtained as the solution to some set of equations

$$\mathbf{f}(\alpha, \lambda^*) = \mathbf{0} \qquad \text{(at each } \alpha\text{)} \tag{4.2.76}$$

For variational wave functions, this condition corresponds to the stationary requirement

$$\left.\frac{\partial E_{\text{var}}(\alpha, \lambda)}{\partial \lambda}\right|_* = \mathbf{0} \qquad \text{(at each } \alpha\text{)} \tag{4.2.77}$$

but for nonvariational wave functions this is not so. Asterisk indicates that the derivatives are evaluated at λ^*.

Now consider the evaluation of the derivative of $E(\alpha)$ with respect to α. In terms of partial derivatives, we obtain

$$\frac{dE(\alpha)}{d\alpha} = \left.\frac{dE(\alpha, \lambda)}{d\alpha}\right|_* = \left.\frac{\partial E(\alpha, \lambda)}{\partial \alpha}\right|_* + \left.\frac{\partial E(\alpha, \lambda)}{\partial \lambda}\right|_* \cdot \left.\frac{\partial \lambda}{\partial \alpha}\right|_* \tag{4.2.78}$$

For variational wave functions, the last term vanishes since (4.2.77) is satisfied. We then obtain the following expression

$$\frac{dE_{\text{var}}(\alpha)}{d\alpha} = \left.\frac{\partial E_{\text{var}}(\alpha, \lambda)}{\partial \alpha}\right|_* \tag{4.2.79}$$

which tells us that the first-order total derivative is obtained directly from the corresponding partial derivative. Thus, if the variational energy corresponds to an expectation value

$$E_{\text{var}}(\alpha, \lambda) = \langle \lambda | H(\alpha) | \lambda \rangle \tag{4.2.80}$$

and the perturbed system is described by the Hamiltonian

$$H(\alpha) = H + \alpha V \tag{4.2.81}$$

we recover the usual expression

$$\left.\frac{dE_{\text{var}}(\alpha)}{d\alpha}\right|_{\alpha=0} = \langle \lambda^* | V | \lambda^* \rangle \tag{4.2.82}$$

in accordance with the Hellmann–Feynman theorem and also with first-order perturbation theory.

Returning to nonvariational energies, we can no longer use the simple expression (4.2.79) since the variational condition (4.2.77) does not hold. To make up for this loss of computational and conceptual convenience, we shall try to replace our original nonvariational function $E(\alpha, \lambda)$ by a new variational function $L(\alpha, \lambda, \bar{\lambda})$ with a stationary point $\{\lambda^*, \bar{\lambda}^*\}$ that satisfies the nonvariational condition (4.2.76) and whose value at this point $L(\alpha, \lambda^*, \bar{\lambda}^*)$ corresponds to the optimized energy $E(\alpha, \lambda^*)$:

$$E(\alpha) = E(\alpha, \lambda^*) = L(\alpha, \lambda^*, \bar{\lambda}^*) \qquad (4.2.83)$$

The variational function $L(\alpha, \lambda, \bar{\lambda})$ is a function of the auxiliary set of parameters $\bar{\lambda}$ as well as the original set λ. To arrive at this function, we use *Lagrange's method of undetermined multipliers* and regard the original electronic energy $E(\alpha, \lambda)$ as variationally optimized subject to the constraints that the variational parameters λ satisfy (4.2.76) at each value of α. Formally, therefore, the electronic energy may be viewed as obtained by an unconstrained optimization of the *Lagrangian*

$$L(\alpha, \lambda, \bar{\lambda}) = E(\alpha, \lambda) + \bar{\lambda} \cdot \mathbf{f}(\alpha, \lambda) \qquad (4.2.84)$$

where both the original wave-function parameters λ and the associated *Lagrange multipliers* $\bar{\lambda}$ are determined variationally [11, 12]. The variational conditions on the Lagrangian are

$$\frac{\partial L(\alpha, \lambda, \bar{\lambda})}{\partial \lambda} = \frac{\partial E(\alpha, \lambda)}{\partial \lambda} + \bar{\lambda}^{\mathrm{T}} \frac{\partial \mathbf{f}(\alpha, \lambda)}{\partial \lambda} = \mathbf{0} \qquad \text{(at each } \alpha\text{)} \qquad (4.2.85)$$

$$\frac{\partial L(\alpha, \lambda, \bar{\lambda})}{\partial \bar{\lambda}} = \mathbf{f}(\alpha, \lambda) = \mathbf{0} \qquad \text{(at each } \alpha\text{)} \qquad (4.2.86)$$

Thus, the variational conditions with respect to the wave-function parameters λ determine the Lagrange multipliers, and the variational conditions with respect to the multipliers determine the original nonvariational parameters by reducing to the original nonvariational conditions (4.2.76). Note that $E(\alpha, \lambda) \equiv L(\alpha, \lambda, \mathbf{0})$ and $E(\alpha, \lambda^*) \equiv L(\alpha, \lambda^*, \bar{\lambda})$, but for all other values of their arguments these two functions differ. Also note that the stationary point of the Lagrangian (4.2.84) is not a minimum, even for electronic ground states. Indeed, since the Lagrangian is linear in the multipliers, all stationary points are saddle points.

We have succeeded in replacing the original nonvariational energy function by a variational one. The price we pay for this reformulation is the need to introduce an auxiliary set of parameters – the Lagrange multipliers. These parameters enter the function linearly (4.2.84) and are rather easy to incorporate, requiring only the evaluation of the partial derivatives of $E(\alpha, \lambda)$ and $\mathbf{f}(\alpha, \lambda)$ with respect to λ and the solution of a set of linear equations. Although the Lagrangian may sometimes be useful for optimizing the energy (since its variational property ensures that the error in the energy is quadratic in the errors in λ^* and $\bar{\lambda}^*$), it is mostly used to simplify the calculation of properties. Thus, invoking the variational property of the Lagrangian, we obtain the following expression

$$\frac{\mathrm{d}E(\alpha)}{\mathrm{d}\alpha} = \frac{\mathrm{d}L(\alpha, \lambda, \bar{\lambda})}{\mathrm{d}\alpha}\bigg|_* = \frac{\partial L(\alpha, \lambda, \bar{\lambda})}{\partial \alpha}\bigg|_* = \frac{\partial E(\alpha, \lambda)}{\partial \alpha}\bigg|_* + \bar{\lambda}^* \cdot \frac{\partial \mathbf{f}(\alpha, \lambda)}{\partial \alpha}\bigg|_* \qquad (4.2.87)$$

which does not involve the response of the wave function to the perturbation [13].

The usefulness of the variational reformulation depends on the relative costs of setting up the Lagrangian (i.e. of obtaining the Lagrange multipliers) and of calculating the perturbed wave functions. Often the cost of calculating the Lagrangian is small, making the Lagrangian approach the preferred one. The Lagrangian reformulation is useful for other purposes than calculating derivatives, however. Thus, it forms the basis for a variational formulation of coupled-cluster theory considered in Section 13.5, and it provides a convenient framework for studying perturbation theory as demonstrated in Sections 14.1 and 14.3. In addition, the variational Lagrangian provides the starting point for a treatment of time-dependent properties for nonvariational wave functions.

4.2.9 THE VARIATION PRINCIPLE SUMMARIZED

The variation principle and its application in the form of the variation method provide a computational framework with a number of appealing properties – the energy obtained by minimization of the expectation value of the Hamiltonian represents an upper bound to the exact ground-state energy, and in the linear variation method we also obtain upper bounds to the energies of excited states. As additional benefits, the error in the calculated wave function affects the energy to second order only and the calculation of molecular properties is particularly simple, as illustrated by the Hellmann–Feynman theorem. Indeed, the variation method represents the single most important computational technique in quantum chemistry. All standard computational procedures rely in some way or other on the variation principle, either directly for the construction of the wave function or indirectly for providing a zero-order wave function.

However, absolute energies *per se* are usually not important since we are mostly interested in energy differences, such as dissociation energies and excitation energies. Therefore, the variational energy criterion should be used with some caution. We should always try to construct our wave functions so as to give a balanced description of the electronic systems for all situations of interest. For example, in studies of excitation energies our model should give a balanced treatment of both electronic states, in studies of electric properties it should give a balanced description of the electronic system in the presence and in the absence of the external field, and in studies of reactivity it should give a balanced treatment of the electronic structure for all important internuclear configurations. If the wave function favours one particular situation, our calculations may turn out to be meaningless.

4.3 Size-extensivity

In quantum chemistry, we often need to compare energies of different systems. For example, the atomization energy is obtained by subtracting the energy of the molecule from those of its atoms. If we are to obtain reliable results, the energies of the molecule and its fragments must be calculated with comparable accuracy. More generally, our methods should be such that we may apply them to large and small systems alike and expect to obtain results of the same accuracy. Obviously, terms such as 'comparable accuracy' are vague. Fortunately, there are situations where we may give a very precise meaning to this term – namely, for systems comprising two or more noninteracting subsystems. In such cases, our methods should generate the same energy for the system irrespective of whether we have carried out the calculations for each subsystem separately or for all subsystems simultaneously. We summarize this requirement by stating that the calculations and, more generally, the computational methods should be *size-extensive*. On size-extensivity and related concepts, see [14, 15].

4.3.1 SIZE-EXTENSIVITY OF EXACT WAVE FUNCTIONS

Consider a system of two noninteracting molecular fragments A and B. For such a system, the Hamiltonian operator may be written in the form

$$\hat{H}_{AB} = \hat{H}_A + \hat{H}_B \qquad (4.3.1)$$

where $\hat{H}_A$ is associated with fragment A and $\hat{H}_B$ with fragment B. The elementary operators of $\hat{H}_A$ anticommute with those of $\hat{H}_B$ since the associated orbitals are orthogonal (see (1.9.12)):

$$[a_{P_A}^\dagger, a_{Q_B}]_+ = 0 \tag{4.3.2}$$

Strictly speaking, the Hamiltonian (4.3.1) is an idealized operator that cannot be realized in practice, but – in the limit of an infinite separation between the fragments – it provides an exact representation of the true operator.

The exact solution for the combined system represented by the Hamiltonian (4.3.1) satisfies the Schrödinger equation

$$\hat{H}_{AB}|AB\rangle = E_{AB}|AB\rangle \tag{4.3.3}$$

We shall consider how this solution is related to the exact solutions for the fragments

$$\hat{H}_A|A\rangle = E_A|A\rangle \tag{4.3.4}$$

$$\hat{H}_B|B\rangle = E_B|B\rangle \tag{4.3.5}$$

within the framework of second quantization. In particular, introducing the *wave operators* for the *fragment wave functions*

$$|A\rangle = \hat{\psi}_A|vac\rangle \tag{4.3.6}$$

$$|B\rangle = \hat{\psi}_B|vac\rangle \tag{4.3.7}$$

we shall demonstrate that the exact *compound wave function* may be written in the multiplicative form

$$|AB\rangle = \hat{\psi}_{AB}|vac\rangle = \hat{\psi}_A\hat{\psi}_B|vac\rangle \tag{4.3.8}$$

and that the exact energy for the compound system may be written in the additive form

$$E_{AB} = E_A + E_B \tag{4.3.9}$$

The relations (4.3.8) and (4.3.9) constitute the requirements for *size-extensivity*. The wave function is said to be *multiplicatively separable* and the energy *additively separable*.

Before discussing size-extensivity, we consider the commutation relations between two operators $\hat{O}_A$ and $\hat{O}_B$ associated with fragments A and B, respectively. We first assume that both operators represent a single string of elementary operators and recall that the elementary operators of $\hat{O}_A$ anticommute with those of $\hat{O}_B$. The operators $\hat{O}_A$ and $\hat{O}_B$ therefore anticommute if an odd number of transpositions of elementary operators is required to reorder $\hat{O}_A$ and $\hat{O}_B$ and otherwise commute. The number of transpositions is equal to the number of elementary operators in $\hat{O}_A$ times the number of elementary operators in $\hat{O}_B$. Therefore, $\hat{O}_A$ and $\hat{O}_B$ anticommute if both operators contain an odd number of elementary operators and otherwise commute:

$$[\hat{O}_A, \hat{O}_B]_+ = 0 \qquad \text{(both operators of half-integral rank)} \tag{4.3.10}$$

$$[\hat{O}_A, \hat{O}_B] = 0 \qquad \text{(one or both operators of integral rank)} \tag{4.3.11}$$

This result holds also when $\hat{O}_A$ and $\hat{O}_B$ are linear combinations of strings (each of the same rank) and agrees with the rank-reduction rule given in Section 1.8, which states that rank reduction occurs upon anticommutation of two operators of half-integral rank and upon commutation of all other operators. In particular, we note that the Hamiltonian for a given fragment is an operator of integral rank and therefore commutes with any operator associated with a different, noninteracting fragment.

Since the wave functions are linear combinations of Slater determinants, we may write the wave operators for the fragments as linear combinations of strings of creation operators $\hat{\psi}_{iA}$ and $\hat{\psi}_{iB}$ associated with systems A and B, respectively:

$$\hat{\psi}_A = \sum_i C_{iA} \hat{\psi}_{iA} \tag{4.3.12}$$

$$\hat{\psi}_B = \sum_i C_{iB} \hat{\psi}_{iB} \tag{4.3.13}$$

The wave operator and the Hamiltonian associated with different, noninteracting fragments therefore commute. We also note that, in accordance with (4.3.10) and (4.3.11),

$$\hat{\psi}_A \hat{\psi}_B = \pm \hat{\psi}_B \hat{\psi}_A \tag{4.3.14}$$

where the minus sign applies when both states contain an odd number of electrons. In contrast to the situation in first quantization, no special antisymmetrization operator is required for the description of product wave functions (4.3.8) – the correct permutational symmetry is imposed by the algebra of the elementary operators as displayed in, for example, (4.3.14).

Having established the commutation relations of operators associated with noninteracting fragments, we examine the action of the Hamiltonian (4.3.1) on the compound wave function in the multiplicatively separable form (4.3.8):

$$\begin{aligned}
\hat{H}_{AB} \hat{\psi}_{AB} |\text{vac}\rangle &= (\hat{H}_A + \hat{H}_B) \hat{\psi}_A \hat{\psi}_B |\text{vac}\rangle \\
&= \pm \hat{\psi}_B \hat{H}_A \hat{\psi}_A |\text{vac}\rangle + \hat{\psi}_A \hat{H}_B \hat{\psi}_B |\text{vac}\rangle \\
&= \pm \hat{\psi}_B E_A \hat{\psi}_A |\text{vac}\rangle + \hat{\psi}_A E_B \hat{\psi}_B |\text{vac}\rangle \\
&= (E_A + E_B) \hat{\psi}_A \hat{\psi}_B |\text{vac}\rangle
\end{aligned} \tag{4.3.15}$$

Here we have used the commutation relations (4.3.10) and (4.3.11) in addition to the fragment eigenvalue equations (4.3.4) and (4.3.5). Comparing (4.3.3) and (4.3.15), we see that the exact compound wave function is multiplicatively separable and that the exact energy is additively separable.

Of course, the exact state cannot be anything but size-extensive and the demonstration given in this subsection may therefore seem somewhat pedantic. However, the discussion of size-extensivity for exact states prepares us for the discussion of size-extensivity for approximate wave functions in Sections 4.3.2–4.3.4. Moreover, the demonstration of size-extensivity given here is valid also for the exact solution in a finite-dimensional Fock space, not just the exact state in a complete basis.

It should be understood that size-extensivity alone does not ensure that a molecule and its dissociation products are described with the same accuracy. Most size-extensive methods, for example, describe exactly (i.e. within a given one-electron basis) the products of the $H_4 \rightarrow 4H$ dissociation, whereas the parent system is described only approximately in the same basis. Size-extensivity ensures only that the same energy is obtained, irrespective of whether this energy is obtained from calculations on the individual fragments or from a single calculation on the whole system.

4.3.2 SIZE-EXTENSIVITY OF LINEAR VARIATIONAL WAVE FUNCTIONS

Size-extensivity holds trivially for exact wave functions. For approximate wave functions, however, size-extensivity is not always observed. We now examine size-extensivity for the linear variational model of Section 4.2.3. We shall find that, for this simple model, size-extensivity may be imposed by a careful construction of the variational space for the compound wave function. For ease of presentation, we assume that all wave functions are real.

We begin our discussion of size-extensivity by considering the fragments. The fragment wave functions are written in the form

$$|\mathbf{c}_A\rangle = \sum_i c_{iA}|i_A\rangle = \sum_i c_{iA}\hat{\psi}_{iA}|\text{vac}\rangle \tag{4.3.16}$$

$$|\mathbf{c}_B\rangle = \sum_i c_{iB}|i_B\rangle = \sum_i c_{iB}\hat{\psi}_{iB}|\text{vac}\rangle \tag{4.3.17}$$

where the summations are over a complete or incomplete set of determinants, and $\mathbf{c}_A$ and $\mathbf{c}_B$ are the variational coefficients. The energies are no longer eigenvalues of the Hamiltonian as in Section 4.3.1 but are calculated as expectation values

$$E_A(\mathbf{c}_A) = \frac{\langle \mathbf{c}_A|\hat{H}_A|\mathbf{c}_A\rangle}{\langle \mathbf{c}_A|\mathbf{c}_A\rangle} \tag{4.3.18}$$

$$E_B(\mathbf{c}_B) = \frac{\langle \mathbf{c}_B|\hat{H}_B|\mathbf{c}_B\rangle}{\langle \mathbf{c}_B|\mathbf{c}_B\rangle} \tag{4.3.19}$$

and the optimized variational wave functions

$$|A\rangle = \sum_i C_{iA}|i_A\rangle = \hat{\psi}_A|\text{vac}\rangle \tag{4.3.20}$$

$$|B\rangle = \sum_i C_{iB}|i_B\rangle = \hat{\psi}_B|\text{vac}\rangle \tag{4.3.21}$$

satisfy the equations

$$\left.\frac{\partial E_A(\mathbf{c}_A)}{\partial c_{iA}}\right|_* = 2\frac{\langle i_A|\hat{H}_A - E_A|A\rangle}{\langle A|A\rangle} = 0 \tag{4.3.22}$$

$$\left.\frac{\partial E_B(\mathbf{c}_B)}{\partial c_{iB}}\right|_* = 2\frac{\langle i_B|\hat{H}_B - E_B|B\rangle}{\langle B|B\rangle} = 0 \tag{4.3.23}$$

where asterisks indicate that the derivatives are evaluated at the optimal values $\mathbf{C}_A$ and $\mathbf{C}_B$ of the coefficients $\mathbf{c}_A$ and $\mathbf{c}_B$. The optimized energies are given by

$$E_A = \frac{\langle A|\hat{H}_A|A\rangle}{\langle A|A\rangle} \tag{4.3.24}$$

$$E_B = \frac{\langle B|\hat{H}_B|B\rangle}{\langle B|B\rangle} \tag{4.3.25}$$

in terms of the optimized states (4.3.20) and (4.3.21).

Let us now consider the compound system. We assume that the compound wave function is written as a linear combination of determinants in the *direct-product space* of the fragment spaces

$$|\mathbf{c}_{AB}\rangle = \sum_{ij} c_{ij}|ij\rangle = \sum_{ij} c_{ij}\hat{\psi}_{iA}\hat{\psi}_{jB}|\text{vac}\rangle \tag{4.3.26}$$

and that the variational solution is obtained by optimizing the energy function

$$E_{AB}(\mathbf{c}_{AB}) = \frac{\langle \mathbf{c}_{AB} | \hat{H}_{AB} | \mathbf{c}_{AB} \rangle}{\langle \mathbf{c}_{AB} | \mathbf{c}_{AB} \rangle} \tag{4.3.27}$$

with respect to the coefficients. The optimized compound wave function is written as

$$|AB\rangle = \sum_{ij} C_{ij} |ij\rangle \tag{4.3.28}$$

and satisfies the variational conditions

$$\left. \frac{\partial E_{AB}(\mathbf{c}_{AB})}{\partial c_{ij}} \right|_* = 2 \frac{\langle ij | \hat{H}_{AB} - E_{AB} | AB \rangle}{\langle AB | AB \rangle} = 0 \tag{4.3.29}$$

where the energy of the compound wave function is given by

$$E_{AB} = \frac{\langle AB | \hat{H}_{AB} | AB \rangle}{\langle AB | AB \rangle} \tag{4.3.30}$$

The compound wave function is thus determined in exactly the same manner as the fragment wave functions but in the direct-product space of the fragment spaces. We shall now demonstrate that, for noninteracting systems A and B, the variationally optimized energy of the compound system in the direct-product space is equal to the sum of the energies of the fragments

$$E_{AB} = E_A + E_B \tag{4.3.31}$$

and that the variational wave function is the product of the variational wave functions of the fragments

$$|AB\rangle = \hat{\psi}_A \hat{\psi}_B |vac\rangle \tag{4.3.32}$$

The relations (4.3.31) and (4.3.32) are the conditions for size-extensivity.

We first establish the additive separability of the expectation value for the product wave function (4.3.31). Introducing the wave operators, we obtain

$$\begin{aligned}
\langle AB | \hat{H}_A | AB \rangle &= \langle vac | \hat{\psi}_B^\dagger \hat{\psi}_A^\dagger \hat{H}_A \hat{\psi}_A \hat{\psi}_B | vac \rangle \\
&= \langle vac | \hat{\psi}_A^\dagger \hat{H}_A \hat{\psi}_A \hat{\psi}_B^\dagger \hat{\psi}_B | vac \rangle \\
&= \langle vac | \hat{\psi}_A^\dagger \hat{H}_A \hat{\psi}_A | vac \rangle \langle vac | \hat{\psi}_B^\dagger \hat{\psi}_B | vac \rangle \\
&= \langle A | \hat{H}_A | A \rangle \langle B | B \rangle
\end{aligned} \tag{4.3.33}$$

where we have used (4.3.10) and (4.3.11). In addition, we have invoked the resolution of the identity in vacuum space: the Hermitian operator $\hat{\psi}_B^\dagger \hat{\psi}_B$ is number-conserving since it is a linear combination of strings that contain an equal number of creation and annihilation operators. The state $\hat{\psi}_B^\dagger \hat{\psi}_B | vac \rangle$ therefore contains no electrons and the resolution of the identity in (4.3.33) involves the vacuum state only. Proceeding in the same manner, we also find

$$\langle AB | \hat{H}_B | AB \rangle = \langle A | A \rangle \langle B | \hat{H}_B | B \rangle \tag{4.3.34}$$

$$\langle AB | AB \rangle = \langle A | A \rangle \langle B | B \rangle \tag{4.3.35}$$

and we are led to the following expression for the expectation value of the total Hamiltonian:

$$\frac{\langle AB|\hat{H}_{AB}|AB\rangle}{\langle AB|AB\rangle} = \frac{\langle A|\hat{H}_A|A\rangle}{\langle A|A\rangle} + \frac{\langle B|\hat{H}_B|B\rangle}{\langle B|B\rangle} \qquad (4.3.36)$$

which establishes (4.3.31) for expectation values of product wave functions.

Let us now consider the variational condition (4.3.29) for the compound wave function (4.3.28). Following the procedure in (4.3.33), we obtain the following matrix elements:

$$\langle ij|\hat{H}_A|AB\rangle = \langle i_A|\hat{H}_A|A\rangle\langle j_B|B\rangle \qquad (4.3.37)$$

$$\langle ij|\hat{H}_B|AB\rangle = \langle i_A|A\rangle\langle j_B|\hat{H}_B|B\rangle \qquad (4.3.38)$$

$$\langle ij|AB\rangle = \langle i_A|A\rangle\langle j_B|B\rangle \qquad (4.3.39)$$

Inserting these expressions in the variational condition (4.3.29), we arrive at the expression

$$\frac{\langle ij|\hat{H}_{AB} - E_{AB}|AB\rangle}{\langle AB|AB\rangle} = \frac{\langle i_A|\hat{H}_A - E_A|A\rangle}{\langle A|A\rangle}\frac{\langle j_B|B\rangle}{\langle B|B\rangle} + \frac{\langle i_A|A\rangle}{\langle A|A\rangle}\frac{\langle j_B|\hat{H}_B - E_B|B\rangle}{\langle B|B\rangle} \qquad (4.3.40)$$

which, in conjunction with (4.3.22) and (4.3.23), establishes (4.3.29) for the product state (4.3.32). We have thus proved that the linear variation method yields size-extensive energies provided the variational space for the compound wave function is equal to the *direct product of the variational spaces for the fragment wave functions*.

For fragment spaces of dimensions m_A and m_B, the product space will be of dimension $m_A m_B$. There are then exactly $m_A m_B$ variational solutions in the product space, representing all possible combinations of the m_A and m_B solutions in the fragment spaces. However, if the variational space for the compound system differs from the direct-product space, then the energies obtained will be different from those obtained in the fragment calculations and the number of solutions will be different as well. For example, if we remove one of the determinants from the direct-product space, then the energy calculated for the compound wave function will be higher than that obtained from the fragment calculations and the number of solutions will be smaller.

The requirement for size-extensivity for linear variational wave functions may appear simple and straightforward to satisfy in practice. In reality, however, the requirement of a direct-product basis for the compound system is an awkward one. First, we note that the number of linear variational parameters in the product space grows exponentially with the number of noninteracting systems. Such an exponential dependence is neither physical nor economical – indeed, in a more suitable parametrization, we would expect the number of variational parameters to scale linearly with the number of noninteracting systems. Second, although it is in principle easy to generate a compound variational space from the fragment spaces, it is usually impossible to go in the opposite direction. Thus, for a given variational space of the compound system, it may upon fragmentation be impossible to 'factorize' the space and generate 'factor' wave functions that ensure size-extensivity.

4.3.3 MATRIX REPRESENTATION OF THE NONINTERACTING EIGENVALUE PROBLEM

It is of some interest to examine the matrix representation of the variational conditions for the fragment states and the compound state. Thus, the variational conditions for the compound system (4.3.29) may be written in the form

$$\mathbf{H}_{AB}\mathbf{C}_{AB} = E_{AB}\mathbf{C}_{AB} \qquad (4.3.41)$$

where the elements of the Hamiltonian matrix in a direct-product basis decouple for noninteracting fragments:

$$[\mathbf{H}_{AB}]_{ij,kl} = \langle ij|\hat{H}_A|kl\rangle + \langle ij|\hat{H}_B|kl\rangle = \langle i_A|\hat{H}_A|k_A\rangle\delta_{j_B,l_B} + \delta_{i_A,k_A}\langle j_B|\hat{H}_B|l_B\rangle \qquad (4.3.42)$$

For a compact representation of the eigenvalue equations for noninteracting systems, we introduce the direct products of matrices and vectors

$$[\mathbf{A}\otimes\mathbf{B}]_{ij,kl} = A_{ik}B_{jl} \qquad (4.3.43)$$

When $\mathbf{A}$ and $\mathbf{B}$ are vectors, the second index in (4.3.43) is absent or equal to unity (A_{i1} and B_{j1}). The Hamiltonian matrix with elements (4.3.42) may now be written in the form

$$\mathbf{H}_{AB} = \mathbf{H}_A \otimes \mathbf{1}_B + \mathbf{1}_A \otimes \mathbf{H}_B \qquad (4.3.44)$$

which clearly reflects the use of a direct-product basis. The Hamiltonian matrix is additively separable and contains two parts – one that involves the Hamiltonian for system A and another that involves the Hamiltonian for system B.

According to the preceding discussion, the solutions to the equations (4.3.41) with a separable Hamiltonian matrix (4.3.44) may be obtained in the form of the direct products

$$\mathbf{C}_{AB} = \mathbf{C}_A \otimes \mathbf{C}_B \qquad (4.3.45)$$

of the solutions to the fragment eigenvalue equations

$$\mathbf{H}_A\mathbf{C}_A = E_A\mathbf{C}_A \qquad (4.3.46)$$

$$\mathbf{H}_B\mathbf{C}_B = E_B\mathbf{C}_B \qquad (4.3.47)$$

To verify the direct-product form of the solutions of (4.3.41), we first note the identity

$$(\mathbf{A}\otimes\mathbf{B})(\mathbf{C}\otimes\mathbf{D}) = (\mathbf{AC})\otimes(\mathbf{BD}) \qquad (4.3.48)$$

which is readily established from the definition (4.3.43) for direct-product matrices. By application of this identity, the following equation follows from (4.3.46) and (4.3.47)

$$(\mathbf{H}_A\otimes\mathbf{1}_B + \mathbf{1}_A\otimes\mathbf{H}_B)(\mathbf{C}_A\otimes\mathbf{C}_B) = (E_A + E_B)(\mathbf{C}_A\otimes\mathbf{C}_B) \qquad (4.3.49)$$

confirming the multiplicative separability of the wave function and the additive separability of the eigenvalues in the direct-product basis. In conclusion, variational calculations on the compound system *in the direct-product space of the fragment spaces* yield wave functions that are multiplicatively separable and energies that are additively separable. Any other choice of variational space for the compound calculation will yield energies different from those of the fragments.

4.3.4 SIZE-EXTENSIVITY OF EXPONENTIAL WAVE FUNCTIONS

The great difficulties faced by the linear model in representing noninteracting fragments in a compact manner suggest that we should consider alternative models for the wave function, where the separability of the wave function is built into the ansatz itself. Specifically, we shall in this subsection advocate the use of *exponential wave functions* as a natural ansatz for separable approximate wave functions.

We begin our discussion by writing the exact wave operators for the noninteracting fragments in the exponential form

$$\hat{\psi}_A(\mathbf{p}_A) = \exp\left(\sum_i p_{iA}\hat{X}_{iA}\right)\hat{A}_A^\dagger \tag{4.3.50}$$

$$\hat{\psi}_B(\mathbf{p}_B) = \exp\left(\sum_i p_{iB}\hat{X}_{iB}\right)\hat{A}_B^\dagger \tag{4.3.51}$$

where $\hat{A}_A^\dagger$ and $\hat{A}_B^\dagger$ are strings of creation operators that, when applied to the vacuum state, generate the reference determinants for fragments A and B. The excitation operators $\hat{X}_{iA}$ and $\hat{X}_{iB}$ excite electrons out of the spin orbitals of $\hat{A}_A^\dagger$ and $\hat{A}_B^\dagger$ and into spin orbitals that are occupied in neither $\hat{A}_A^\dagger$ nor $\hat{A}_B^\dagger$. The operators $\hat{X}_{iA}$ and $\hat{X}_{iB}$ therefore commute. The parameters $\mathbf{p}_A$ and $\mathbf{p}_B$ are chosen such that $\hat{\psi}_A(\mathbf{p}_A)|\text{vac}\rangle$ and $\hat{\psi}_B(\mathbf{p}_B)|\text{vac}\rangle$ are solutions to the Schrödinger equation for the Hamiltonians $\hat{H}_A$ and $\hat{H}_B$, respectively.

From Section 4.3.1, we know that the exact wave operator for the compound system is given by the product of $\hat{\psi}_A$ and $\hat{\psi}_B$. In the exponential representation, this product wave function becomes

$$\hat{\psi}_A(\mathbf{p}_A)\hat{\psi}_B(\mathbf{p}_B) = \left[\exp\left(\sum_i p_{iA}\hat{X}_{iA}\right)\hat{A}_A^\dagger\right]\left[\exp\left(\sum_i p_{iB}\hat{X}_{iB}\right)\hat{A}_B^\dagger\right]$$

$$= \exp\left(\sum_i p_{iA}\hat{X}_{iA} + \sum_i p_{iB}\hat{X}_{iB}\right)\hat{A}_A^\dagger\hat{A}_B^\dagger \tag{4.3.52}$$

Thus, in this particular representation, the compound wave function for the noninteracting fragments is generated by the sum of the excitation operators for the individual systems:

$$\hat{\psi}_A(\mathbf{p}_A)\hat{\psi}_B(\mathbf{p}_B) = \hat{\psi}_{AB}(\mathbf{p}_A, \mathbf{p}_B) \tag{4.3.53}$$

This situation should be contrasted with the linear variation method, where the wave operators for the fragments may be written in the form

$$\hat{\psi}_A(\mathbf{C}_A) = \left(1 + \sum_i C_{iA}\hat{X}_{iA}\right)\hat{A}_A^\dagger \tag{4.3.54}$$

$$\hat{\psi}_B(\mathbf{C}_B) = \left(1 + \sum_i C_{iB}\hat{X}_{iB}\right)\hat{A}_B^\dagger \tag{4.3.55}$$

The product wave function now contains nonzero parameters that refer to both fragments simultaneously

$$\hat{\psi}_A(\mathbf{C}_A)\hat{\psi}_B(\mathbf{C}_B) = \left(1 + \sum_i C_{iA}\hat{X}_{iA} + \sum_i C_{iB}\hat{X}_{iB}\right)\hat{A}_A^\dagger\hat{A}_B^\dagger + \left(\sum_{ij} C_{iA}C_{jB}\hat{X}_{iA}\hat{X}_{jB}\right)\hat{A}_A^\dagger\hat{A}_B^\dagger \tag{4.3.56}$$

The relationship between the fragment and product wave functions is therefore given by

$$\hat{\psi}_A(\mathbf{C}_A)\hat{\psi}_B(\mathbf{C}_B) = \hat{\psi}_{AB}(\mathbf{C}_A, \mathbf{C}_B, \mathbf{C}_{AB} = \mathbf{C}_A \otimes \mathbf{C}_B) \tag{4.3.57}$$

where the nonzero interaction parameters (for the noninteracting fragments) are represented by the direct product $\mathbf{C}_A \otimes \mathbf{C}_B$ of the fragment parameters $\mathbf{C}_A$ and $\mathbf{C}_B$. In the exponential representation (4.3.53), no such spurious interaction parameters occur.

Clearly, we cannot generate all possible wave functions from (4.3.52) since this would require the use of the full direct-product space as previously discussed. On the other hand, the exponential ansatz, with operators and parameters referring only to the individual subsystems, provides exactly those operators that are needed to describe the products of *isolated* systems. Assume, for example, that some choice of operators $\hat{X}_{iA}$ and $\hat{X}_{iB}$ has been made and that the parameters $\mathbf{p}_A$ have been chosen to optimize the energy of fragment A

$$\frac{\partial E_A}{\partial p_{iA}} = \frac{\partial}{\partial p_{iA}} \frac{\langle A|\hat{H}_A|A \rangle}{\langle A|A \rangle} = 0 \qquad (4.3.58)$$

and similarly for fragment B. We can now use the operator $\hat{\psi}_B \hat{\psi}_A$ as the wave operator for the compound system. The product wave function is automatically optimized for noninteracting fragments since

$$\frac{\partial E_{AB}}{\partial p_{iA}} = \frac{\partial}{\partial p_{iA}} \frac{\langle AB|\hat{H}_{AB}|AB \rangle}{\langle AB|AB \rangle} = \frac{\partial}{\partial p_{iA}} \frac{\langle A|\hat{H}_A|A \rangle}{\langle A|A \rangle} + \frac{\partial}{\partial p_{iA}} \frac{\langle B|\hat{H}_B|B \rangle}{\langle B|B \rangle} = 0 \qquad (4.3.59)$$

with a similar result for the derivatives with respect to p_{iB}.

The discussion of the exponential ansatz leads us to the concept of *manifest separability*. Let the fragment wave functions in some model be denoted by $\hat{\psi}_A(\mathbf{p}_A)$ and $\hat{\psi}_B(\mathbf{p}_B)$, where $\mathbf{p}_A$ and $\mathbf{p}_B$ represent the variational parameters. The compound wave function $\hat{\psi}_{AB}(\mathbf{p}_A, \mathbf{p}_B, \mathbf{p}_{AB})$ must contain parameters that refer to each fragment A and B separately as well as parameters $\mathbf{p}_{AB}$ that refer to both fragments simultaneously. The compound wave function and more generally the model itself are said to be *manifestly separable* if the following requirement is fulfilled for all values of $\mathbf{p}_A$ and $\mathbf{p}_B$

$$\hat{\psi}_{AB}(\mathbf{p}_A, \mathbf{p}_B, \mathbf{p}_{AB} = \mathbf{0}) \equiv \hat{\psi}_A(\mathbf{p}_A)\hat{\psi}_B(\mathbf{p}_B) \qquad (4.3.60)$$

Thus, for a manifestly separable compound wave function, the interaction parameters $\mathbf{p}_{AB}$ become 0 and the wave function turns into a product wave function as the interactions between A and B vanish. Manifestly separable wave functions do not require the use of interaction parameters $\mathbf{p}_{AB}$ to describe noninteracting systems size-extensively and the number of variational parameters therefore scales linearly with the number of noninteracting systems. Obviously, the exponential form of the wave function is manifestly separable, whereas the linear expansion suffers from the lack of this property.

So far, we have considered only wave functions that are variationally determined. For technical reasons, many popular methods do not employ the variation principle for the construction of the wave function. As discussed in Section 13.1.4, the exponential ansatz does not lend itself easily to variational treatments and the wave functions are instead generated by a different principle. Nevertheless, we shall see that the exponential ansatz may still be given a size-extensive formulation.

The examples presented in connection with size-extensivity should give some indication of the crucial role that the choice of wave-function model or parametrization plays in quantum chemistry – by a judicious choice of model, we may simplify the calculation and the interpretation of the wave function considerably. An analogy is provided by classical mechanics, where the choice of coordinate system is often crucial for the successful solution of a problem. We shall return to the topic of size-extensivity on several occasions later in this book, in particular in the discussions of the different wave-function models in Chapters 10–14.

4.4 Symmetry constraints

The exact electronic Hamiltonian H usually commutes with a number of additional operators Λ_p and the exact wave functions can then be chosen to be simultaneous eigenfunctions of H and Λ_p. For example, the commutation of H with the spin operators S^2 and S_z implies that the eigenfunctions of H can be chosen as eigenfunctions of the total and projected spins, and the commutation of H with the symmetry operators of the molecular point group implies that we can classify the wave functions according to the irreducible representations of this point group.

Approximate wave functions do not necessarily display the full symmetry of the exact state. Since approximate wave functions are not eigenfunctions of the exact Hamiltonian, the commutation relationships of the exact Hamiltonian have no direct influence on the symmetry of such wave functions. For example, the fact that the Hamiltonian is spinless and commutes with S^2 and S_z does not imply that a variationally determined approximate wave function is also a spin eigenfunction.

Nevertheless, it is often desirable to impose on the approximate solutions the full symmetry of the exact Hamiltonian, thereby simplifying the identification of the approximate states with exact states and minimizing the computational cost by reducing the variational space. As an example, molecular spin orbitals are invariably chosen to have a well-defined spin-projection symmetry. This symmetry is then imparted to the Slater determinants as described in Section 2.4 and therefore also to the wave function. Similarly, the absence of degenerate irreducible representations in Abelian point groups means that all Slater determinants in a symmetry-adapted orbital basis have a well-defined point-group symmetry. For degenerate symmetries – such as the total spin and the spatial symmetries of point groups with degenerate irreducible representations – several Slater determinants must be combined linearly to arrive at a symmetry-adapted many-electron wave function. The linear combination of Slater determinants into spin eigenfunctions was discussed in Sections 2.5 and 2.6.

In the linear variation method, we can obtain conditions that ensure symmetry-adapted solutions even in the absence of symmetry-adapted basis vectors. Consider some symmetry associated with the operator Λ that commutes with the exact Hamiltonian

$$H\Lambda = \Lambda H \tag{4.4.1}$$

In the many-electron space of the linear variation method, the operators H and Λ are represented by the matrices $\mathbf{H}$ and $\mathbf{\Lambda}$ with elements

$$H_{ij} = \langle i|H|j \rangle \tag{4.4.2}$$

$$\Lambda_{ij} = \langle i|\Lambda|j \rangle \tag{4.4.3}$$

The basis vectors $|i\rangle$ may, for example, be a set of Slater determinants or linear combinations of such determinants. The approximate wave functions are the eigenvectors of the Hamiltonian matrix $\mathbf{H}$. These approximate states are therefore eigenvectors of $\mathbf{\Lambda}$ provided that $\mathbf{H}$ and $\mathbf{\Lambda}$ commute.

It is important to note that the commutation of $\mathbf{H}$ and $\mathbf{\Lambda}$ does not in general follow from the commutation of the exact operators (4.4.1), which implies only the following relationship in the many-electron vector basis:

$$\langle i|H\Lambda|j \rangle = \langle i|\Lambda H|j \rangle \tag{4.4.4}$$

However, let us now assume that the many-electron basis is closed under the action of the exact symmetry operator Λ:

$$\Lambda|j\rangle = \sum_k |k\rangle\langle k|\Lambda|j\rangle \tag{4.4.5}$$

The relationship (4.4.4) may now be written as

$$\sum_k \langle i|H|k\rangle\langle k|\Lambda|j\rangle = \sum_k \langle i|\Lambda|k\rangle\langle k|H|j\rangle \tag{4.4.6}$$

implying that $\mathbf{H}$ and $\boldsymbol{\Lambda}$ commute:

$$\mathbf{H}\boldsymbol{\Lambda} = \boldsymbol{\Lambda}\mathbf{H} \tag{4.4.7}$$

The requirement that the variational space is closed under the application of the exact symmetry operator Λ is thus sufficient to ensure the existence of a common set of eigenvectors of $\mathbf{H}$ and $\boldsymbol{\Lambda}$. We may now take the approximate variational wave function

$$|0\rangle = \sum_i C_i|i\rangle \tag{4.4.8}$$

to be a simultaneous eigenvector of $\mathbf{H}$ and $\boldsymbol{\Lambda}$

$$\mathbf{H}\mathbf{C} = E\mathbf{C} \tag{4.4.9}$$

$$\boldsymbol{\Lambda}\mathbf{C} = \lambda\mathbf{C} \tag{4.4.10}$$

Moreover, the approximate wave function (4.4.8) is also an eigenfunction of the exact operator Λ:

$$\Lambda|0\rangle = \sum_i C_i\Lambda|i\rangle = \sum_{ij} |j\rangle\langle j|\Lambda|i\rangle C_i = \sum_j |j\rangle[\boldsymbol{\Lambda}\mathbf{C}]_j = \lambda\sum_j C_j|j\rangle = \lambda|0\rangle \tag{4.4.11}$$

where we have used (4.4.5) and (4.4.10).

Thus, within the *linear* variation method, it is straightforward to design vector spaces that yield eigenvectors with well-defined symmetry properties. Consider, for instance, the operator for total spin S^2. Following the discussion in Section 2.5, $S^2|i\rangle$ is a linear combination of Slater determinants belonging to the same orbital configuration. To ensure that the approximate wave functions are eigenfunctions of S^2, we must then include in the variational space either all or none of the determinants belonging to each orbital configuration.

For *nonlinear* variational wave functions, the stationary conditions are in general not sufficient to ensure the correct symmetry. Instead, we must *impose* the desired symmetry on the wave function by restricting the model for the wave function to allow only functions of the desired symmetry. For example, to transfer an approximate state $|0\rangle$ to a new state $|\tilde{0}\rangle$ of the same spin and spatial symmetry by means of an exponential operator

$$|\tilde{0}\rangle = \exp(-\hat{\kappa})|0\rangle \tag{4.4.12}$$

we must require that $\hat{\kappa}$ contains only orbital excitation operators that transform as the totally symmetric irreducible representation of the Hamiltonian. For real wave functions and a Hamiltonian that is totally symmetric, the operator $^R\hat{\kappa}^{0,0}$ in (3.3.22) must then be used, with the restriction that the excitation operators transform as the totally symmetric representation of the molecular point group. Clearly, the energy obtained in such a symmetry-restricted calculation represents an upper bound to the energy obtained without restrictions. The use of symmetry constraints in nonlinear variational calculations may therefore (but by no means always) raise the energy. This situation is sometimes referred to as the *symmetry dilemma* [16]. For a discussion of the symmetry dilemma, see for example Section 10.10.

References

[1] R. G. Parr and W. Yang, *Density-Functional Theory of Atoms and Molecules*, Oxford University Press, 1989.

[2] S. T. Epstein, *The Variation Method in Quantum Chemistry*, Academic Press, 1974.

[3] T. Kato, *Commun. Pure Appl. Math.* **10**, 151 (1957).

[4] J. Katriel and E. R. Davidson, *Proc. Natl. Acad. Sci. USA* **77**, 4403 (1980).

[5] C. Cohen-Tannoudji, B. Diu and F. Laloë, *Quantum Mechanics*, Wiley, 1977.

[6] B. N. Parlett, *The Symmetric Eigenvalue Problem*, Prentice Hall, 1980.

[7] E. A. Hylleraas and B. Undheim, *Z. Phys.* **65**, 759 (1930) J. K. L. MacDonald, *Phys. Rev.* **43**, 830 (1933).

[8] H. Hellmann, *Einführung in die Quantenchemie*, Franz Deuticke, 1937; R. P. Feynman, *Phys. Rev.* **56**, 340 (1939).

[9] T. U. Helgaker and J. Almlöf, *Int. J. Quantum Chem.* **26**, 275 (1984).

[10] P.-O. Löwdin, *J. Mol. Spectrosc.* **3**, 46 (1959).

[11] T. Helgaker and P. Jørgensen, *Theor. Chim. Acta* **75**, 111 (1989).

[12] T. Helgaker and P. Jørgensen, in S. Wilson and G. H. F. Diercksen (eds), *Methods in Computational Molecular Physics,* Plenum 1992, p.353.

[13] N. C. Handy and H. F. Schaefer III, *J. Chem. Phys.* **81**, 5031 (1984).

[14] R. J. Bartlett and G. D. Purvis, *Int. J. Quantum Chem.* **14**, 561 (1978); *Phys. Scripta* **21**, 255 (1980).

[15] J. A. Pople, J. S. Binkley and R. Seeger, *Int J. Quantum Chem.* **S10**, 1 (1976).

[16] P.-O. Löwdin, *Rev. Mod. Phys.* **35**, 496 (1963).

Further reading

B. T. Sutcliffe, Fundamentals of computational chemistry, in G. H. F. Diercksen, B. T. Sutcliffe and A. Veillard, (eds), *Computational Techniques in Quantum Chemistry and Molecular Physics*, Reidel, 1975, p. 1.

Exercises

EXERCISE 4.1

In Section 4.2.2, it is noted that, if the energy is evaluated as an expectation value, then the error in the energy is quadratic in the error in the wave function. Usually, however, the error in the wave function is unknown. In this exercise, we show how the error in the energy can be estimated from the norm of the gradient.

1. Show that the electronic energy, gradient and Hessian can be expanded around a stationary point as

$$E(\boldsymbol{\lambda}) = E + \tfrac{1}{2}\boldsymbol{\lambda}^{\mathrm{T}}\mathbf{E}^{(2)}\boldsymbol{\lambda} + O(\|\boldsymbol{\lambda}\|^3) \tag{4E.1.1}$$

$$\mathbf{E}^{(1)}(\boldsymbol{\lambda}) = \mathbf{E}^{(2)}\boldsymbol{\lambda} + O(\|\boldsymbol{\lambda}\|^2) \tag{4E.1.2}$$

$$\mathbf{E}^{(2)}(\boldsymbol{\lambda}) = \mathbf{E}^{(2)} + O(\|\boldsymbol{\lambda}\|) \tag{4E.1.3}$$

Here λ is the vector containing the wave-function parameters and $\lambda = \mathbf{0}$ represents the stationary point. For the energy and Hessian at $\lambda = \mathbf{0}$, we use the notation E and $\mathbf{E}^{(2)}$.

2. Use the above expansions to show that the errors in the wave function and in the energy are related to the gradient as

$$\lambda = [\mathbf{E}^{(2)}(\lambda)]^{-1}\mathbf{E}^{(1)}(\lambda) + O(\|\mathbf{E}^{(1)}(\lambda)\|^2) \tag{4E.1.4}$$

$$E(\lambda) - E = \tfrac{1}{2}[\mathbf{E}^{(1)}(\lambda)]^{\mathrm{T}}[\mathbf{E}^{(2)}(\lambda)]^{-1}\mathbf{E}^{(1)}(\lambda) + O(\|\mathbf{E}^{(1)}(\lambda)\|^3) \tag{4E.1.5}$$

If the gradient and Hessian are available, these relations can be used to estimate the errors in the wave function and the energy as illustrated in Exercise 10.4.

EXERCISE 4.2

In this exercise, we examine the relation between the eigenvalues of a Hermitian matrix and its submatrices. Consider the $D \times D$ matrix $\mathbf{M}$ and its $d \times d$ submatrix $\mathbf{m}$:

$$\mathbf{M} = \begin{pmatrix} \mathbf{m} & \mathbf{V} \\ \mathbf{V}^{\dagger} & \mathbf{W} \end{pmatrix} \tag{4E.2.1}$$

We denote the orthonormal eigenvectors of $\mathbf{M}$ by $\mathbf{B}_1, \mathbf{B}_2, \ldots, \mathbf{B}_D$ and the eigenvalues by $E_1 \leq E_2 \leq \cdots \leq E_D$. Likewise, for the eigenvectors and eigenvalues of $\mathbf{m}$, we use the notation $\mathbf{b}_1, \mathbf{b}_2, \ldots, \mathbf{b}_d$ and $e_1 \leq e_2 \leq \cdots \leq e_d$. In the following, we shall regard the eigenvectors $\mathbf{b}_k$ as vectors in the full D-dimensional space, with the last $D - d$ elements set equal to 0. The eigenvectors $\mathbf{b}_k$ may then be expanded in the eigenvectors $\mathbf{B}_K$ in the usual manner:

$$\mathbf{b}_k = \sum_{K=1}^{D} a_K \mathbf{B}_K \tag{4E.2.2}$$

1. Show that the lowest eigenvalue of $\mathbf{m}$ is an upper bound to the lowest eigenvalue of $\mathbf{M}$:

$$E_1 \leq e_1 \tag{4E.2.3}$$

2. Show that

$$\frac{\mathbf{X}^{\dagger}\mathbf{M}\mathbf{X}}{\mathbf{X}^{\dagger}\mathbf{X}} \geq E_i \tag{4E.2.4}$$

if $\mathbf{X}$ is orthogonal to all $\mathbf{B}_K$ with $K < i$.

3. Assume that the square matrix containing $\mathbf{B}_K^{\dagger}\mathbf{b}_k$ with $1 \leq K < i$ and $1 \leq k < i$ is nonsingular. Show that the coefficients a_k in the expansion

$$\mathbf{c}_i = \mathbf{b}_i + \sum_{k=1}^{i-1} a_k \mathbf{b}_k \tag{4E.2.5}$$

can be chosen such that $\mathbf{c}_i$ becomes orthogonal to all $\mathbf{B}_K$ with $K < i$.

4. Using (4E.2.4) and (4E.2.5), show that, for $1 < i \leq d$, the following inequality holds

$$E_i \leq e_i \tag{4E.2.6}$$

if the matrix containing $\mathbf{B}_K^{\dagger}\mathbf{b}_k$ with $1 \leq K < i$ and $1 \leq k < i$ is nonsingular.

5. Show that the highest eigenvalue of $\mathbf{M}$ is an upper bound to the highest eigenvalue of $\mathbf{m}$:

$$e_d \leq E_D \tag{4E.2.7}$$

Also show that, for $1 \leq i < d$, the inequality

$$e_{d-i} \leq E_{D-i} \tag{4E.2.8}$$

holds if the matrix containing $\mathbf{B}_{D-K}^\dagger \mathbf{b}_{d-k}$ with $0 \leq K < i$ and $0 \leq k < i$ is nonsingular.

6. Show that, under the usual assumptions, the above results imply the inequalities

$$E_i \leq e_i \leq E_{i+D-d} \tag{4E.2.9}$$

and that, for the special case of $d = D - 1$, we obtain

$$E_1 \leq e_1 \leq E_2 \leq e_2 \leq E_3 \leq \cdots \leq E_d \leq e_d \leq E_{d+1} \tag{4E.2.10}$$

EXERCISE 4.3

In this exercise, we determine the effect of coordinate scaling on expectation values. Consider the uniform scaling of the electron coordinates

$$\Psi(\mathbf{r}_i) \to \Psi_\alpha(\mathbf{r}_i) = \alpha^{3N/2} \Psi(\alpha \mathbf{r}_i) \tag{4E.3.1}$$

Here α is a real scaling factor and N the number of electrons. We assume that $\Psi(\mathbf{r}_i)$ is normalized.

1. Show that the overlap is unaffected by the scaling:

$$\langle \Psi_\alpha | \Psi_\alpha \rangle = 1 \tag{4E.3.2}$$

2. Show that the expectation values of the kinetic and potential energies transform as

$$\langle \Psi_\alpha | T | \Psi_\alpha \rangle = \alpha^2 \langle \Psi | T | \Psi \rangle \tag{4E.3.3}$$

$$\langle \Psi_\alpha | V(\mathbf{R}) | \Psi_\alpha \rangle = \alpha \langle \Psi | V(\alpha \mathbf{R}) | \Psi \rangle \tag{4E.3.4}$$

where

$$T = -\frac{1}{2} \sum_i \nabla_i^2 \tag{4E.3.5}$$

$$V(\mathbf{R}) = \sum_{i>j} \frac{1}{|\mathbf{r}_i - \mathbf{r}_j|} - \sum_{iI} \frac{Z_I}{|\mathbf{r}_i - \mathbf{R}_I|} + \sum_{I>J} \frac{Z_I Z_J}{|\mathbf{R}_I - \mathbf{R}_J|} \tag{4E.3.6}$$

Solutions

SOLUTION 4.1

1. Expanding $E(\boldsymbol{\lambda})$ around $\boldsymbol{\lambda} = \mathbf{0}$, we obtain

$$E(\boldsymbol{\lambda}) = E + \boldsymbol{\lambda}^\mathrm{T} \mathbf{E}^{(1)} + \tfrac{1}{2} \boldsymbol{\lambda}^\mathrm{T} \mathbf{E}^{(2)} \boldsymbol{\lambda} + O(\|\boldsymbol{\lambda}\|^3) = E + \tfrac{1}{2} \boldsymbol{\lambda}^\mathrm{T} \mathbf{E}^{(2)} \boldsymbol{\lambda} + O(\|\boldsymbol{\lambda}\|^3) \tag{4S.1.1}$$

since the gradient vanishes at the stationary point. Differentiation of $E(\boldsymbol{\lambda})$ with respect to $\boldsymbol{\lambda}$ then gives the expressions for the gradient and the Hessian in (4E.1.2) and (4E.1.3).

2. Solving (4E.1.2) for λ and using (4E.1.3), we obtain (4E.1.4):

$$\lambda = -(\mathbf{E}^{(2)})^{-1}\mathbf{E}^{(1)}(\lambda) + O(\|\mathbf{E}^{(1)}(\lambda)\|^2) = -[\mathbf{E}^{(2)}(\lambda)]^{-1}\mathbf{E}^{(1)}(\lambda) + O(\|\mathbf{E}^{(1)}(\lambda)\|^2) \qquad (4S.1.2)$$

Inserting this expression into (4S.1.1) and using (4E.1.3) once more, we arrive at (4E.1.5).

SOLUTION 4.2

1. Expanding the first eigenvector of $\mathbf{m}$ in the eigenvectors of $\mathbf{M}$, we obtain

$$\mathbf{b}_1 = \sum_{K=1}^{D} a_K \mathbf{B}_K \qquad (4S.2.1)$$

where the coefficients satisfy the normalization condition

$$\sum_{K=1}^{D} |a_K|^2 = 1 \qquad (4S.2.2)$$

The lowest eigenvalue of $\mathbf{m}$ may now be related to lowest eigenvalue of $\mathbf{M}$ as follows:

$$e_1 = \mathbf{b}_1^\dagger \mathbf{m}\mathbf{b}_1 = \sum_{K,L=1}^{D} a_K^* \mathbf{B}_K^\dagger \mathbf{M} \mathbf{B}_L a_L = \sum_{K=1}^{D} |a_K|^2 E_K \geq \sum_{K=1}^{D} |a_K|^2 E_1 = E_1 \qquad (4S.2.3)$$

2. Since $\mathbf{X}$ is orthogonal to all $\mathbf{B}_K$ with $K < i$, we may expand it as follows:

$$\mathbf{X} = \sum_{K=i}^{D} a_K \mathbf{B}_K \qquad (4S.2.4)$$

The expectation value is then bounded as

$$\frac{\mathbf{X}^\dagger \mathbf{M} \mathbf{X}}{\mathbf{X}^\dagger \mathbf{X}} = \frac{\displaystyle\sum_{K,L=i}^{D} a_K^* \mathbf{B}_K^\dagger \mathbf{M} \mathbf{B}_L a_L}{\displaystyle\sum_{K=i}^{D} |a_K|^2} = \frac{\displaystyle\sum_{K=i}^{D} |a_K|^2 E_K}{\displaystyle\sum_{K=i}^{D} |a_K|^2} \geq \frac{\displaystyle\sum_{K=i}^{D} |a_K|^2 E_i}{\displaystyle\sum_{K=i}^{D} |a_K|^2} = E_i \qquad (4S.2.5)$$

3. Inserting the expansion (4E.2.5) into the orthogonality conditions

$$\mathbf{B}_K^\dagger \mathbf{c}_i = 0 \qquad (4S.2.6)$$

for $K < i$ and rearranging, we obtain a linear set of equations

$$\sum_{k=1}^{i-1} Q_{Kk} a_k = -\mathbf{B}_K^\dagger \mathbf{b}_i \qquad (4S.2.7)$$

$$Q_{Kk} = \mathbf{B}_K^\dagger \mathbf{b}_k \qquad (4S.2.8)$$

which has a unique solution since the matrix containing the elements (4S.2.8) is nonsingular.

4. Using (4E.2.4), we obtain

$$E_i \leq \frac{\mathbf{c}_i^\dagger \mathbf{m} \mathbf{c}_i}{\mathbf{c}_i^\dagger \mathbf{c}_i} \tag{4S.2.9}$$

where $\mathbf{c}_i$ with $i > 1$ is given by (4E.2.5), with the coefficients chosen such that it is orthogonal to $\mathbf{B}_K$ for $K < i$. Inserting (4E.2.5) in (4S.2.9), we arrive at (4E.2.6):

$$E_i \leq \frac{e_i + \sum_{k=1}^{i-1} |a_k|^2 e_k}{1 + \sum_{k=1}^{i-1} |a_k|^2} \leq \frac{e_i + \sum_{k=1}^{i-1} |a_k|^2 e_i}{1 + \sum_{k=1}^{i-1} |a_k|^2} = e_i \tag{4S.2.10}$$

5. Equations (4E.2.7) and (4E.2.8) may be obtained from (4E.2.3) and (4E.2.6) by considering the eigenvalues of the matrices $-\mathbf{M}$ and $-\mathbf{m}$.

6. In (4E.2.9), the first inequality corresponds to (4E.2.3) and (4E.2.6), whereas the second is obtained from (4E.2.7) and (4E.2.8), having substituted $d - i$ for i. Equation (4E.2.10) is obtained from (4E.2.9) by setting $d = D - 1$.

SOLUTION 4.3

1. Evaluating the overlap integral for the wave function (4E.3.1), we find

$$\langle \Psi_\alpha | \Psi_\alpha \rangle = \alpha^{3N} \int \cdots \int |\Psi(\alpha \mathbf{r}_i)|^2 \, d\mathbf{r}_1 \cdots d\mathbf{r}_N$$

$$= \int \cdots \int |\Psi(\alpha \mathbf{r}_i)|^2 \, d(\alpha \mathbf{r}_1) \cdots d(\alpha \mathbf{r}_N)$$

$$= \int \cdots \int |\Psi(\mathbf{r}_i')|^2 \, d\mathbf{r}_1' \cdots d\mathbf{r}_N' = 1 \tag{4S.3.1}$$

2. Likewise, for the kinetic-energy integral, we obtain

$$\langle \Psi_\alpha | T | \Psi_\alpha \rangle = \int \cdots \int \Psi^*(\alpha \mathbf{r}_i) \left(-\frac{1}{2} \alpha^2 \sum_{i\mu} \frac{\partial^2}{\partial (\alpha r_{i\mu})^2} \right) \Psi(\alpha \mathbf{r}_i) \, d(\alpha \mathbf{r}_1) \cdots d(\alpha \mathbf{r}_N)$$

$$= \alpha^2 \int \cdots \int \Psi^*(\mathbf{r}_i') \left(-\frac{1}{2} \sum_{i\mu} \frac{\partial^2}{\partial r_{i\mu}'^2} \right) \Psi(\mathbf{r}_i') \, d\mathbf{r}_1' \cdots d\mathbf{r}_N'$$

$$= \alpha^2 \langle \Psi | T | \Psi \rangle \tag{4S.3.2}$$

where the summations are over the three Cartesian components $r_{i\mu}$ of the position vectors $\mathbf{r}_i$ of all electrons. The potential-energy integral may be evaluated in the following manner:

$$\langle \Psi_\alpha | V(\mathbf{R}) | \Psi_\alpha \rangle = \int \cdots \int \Psi^*(\alpha \mathbf{r}_i) \left(\alpha \sum_{i>j} \frac{1}{\alpha |\mathbf{r}_i - \mathbf{r}_j|} - \alpha \sum_{il} \frac{Z_I}{\alpha |\mathbf{r}_i - \mathbf{R}_I|} + \alpha \sum_{I>J} \frac{Z_I Z_J}{\alpha |\mathbf{R}_I - \mathbf{R}_J|} \right)$$

$$\times \Psi(\alpha \mathbf{r}_i) \, d(\alpha \mathbf{r}_1) \cdots d(\alpha \mathbf{r}_N) = \alpha \langle \Psi | V(\alpha \mathbf{R}) | \Psi \rangle \tag{4S.3.3}$$

5 THE STANDARD MODELS

For nearly all systems of chemical interest, the exact solution to the Schrödinger equation cannot be obtained and we must introduce approximations in our solutions. Although unavoidable, these approximations should be introduced with care. In particular, all approximations should be unambiguous and precisely defined. They should be designed such that their inadequacies may be mended in a systematic fashion, so as to yield more and more elaborate solutions until in principle the exact solution is recovered. Thus we speak of different *models, levels of theory* and *hierarchies of approximations*.

The advantages of such an approach are many. First, it allows us to build up experience about the performance and accuracy of each level of approximation. This accumulated experience makes it possible to assess the usefulness, quality and reliability of a given calculation. By exploring different levels of theory, it becomes possible to extrapolate the results towards the exact solution and to make estimates of errors. Also, the use of well-established levels of theory makes the calculations and their results reproducible and unambiguous.

Clearly, the errors introduced by the approximations should be as small as possible. Nevertheless, in most molecular calculations, the absolute errors will still be large compared with energies of chemical interest. Such calculations may nevertheless be meaningful if the errors can be controlled or cancelled in some manner. Thus, our approximations should be designed so as to yield solutions of the same quality for the systems and physical situations to be compared. For example, in dissociation processes, the approximations should give results of similar quality for the reactants and the products. To describe correctly the response of the wave function to external perturbations such as electric and magnetic fields, the approximations should give a balanced treatment of the electronic system in the field and out of the field.

Computational quantum chemistry has developed a few *standard models* for the construction of approximate electronic wave functions. At the simplest level, the wave function is represented by a single Slater determinant. At the most complex level, we represent the true wave function by a variationally determined superposition of all determinants in the N-particle Fock space. Between these extremes, a large number of intermediate models and hierarchies of models have been developed, allowing for descriptions of varying cost and flexibility. None of these models is applicable under all circumstances and none succeeds in incorporating all properties that characterize the exact electronic state as discussed in Chapter 4. However, each model is applicable to the study of a broad range of molecular systems, providing approximate solutions to the Schrödinger equation of known quality and flexibility at a given computational cost.

The purpose of this chapter is to introduce and examine the more important standard models of quantum chemistry, comparing and characterizing the various models in relationship to one another and to the exact solution in Fock space. The exposition is descriptive and qualitative rather than technical. In this way, we hope to develop some rudimentary understanding of the standard models – in particular, concerning the flexibility they offer in the description of the electronic structure and their shortcomings *vis-à-vis* the exact wave function in Fock space. We thus make preparations for the more detailed and technical expositions of the models in Chapters 10–14 and for the final assessment and comparison of the models in Chapter 15.

We begin by considering, in Section 5.1, the relationship between the expansions made in the one-particle space (the orbital space) and in the N-particle space (the Fock space). For an exact representation of the true wave function, both expansions must be complete. The one-particle expansions – which are common to all models – are not considered in detail in this chapter but are examined separately in Chapters 6–8. In the present chapter, we concentrate on the N-particle treatment, which is studied here in isolation from the one-particle treatment, proceeding on the assumption that the one-particle basis is sufficiently flexible to ensure a reasonably accurate representation of the true wave function.

Before examining the standard models in any detail, we consider in Section 5.2 the representation of the electronic structure of the hydrogen molecule in a variational space of two orbitals. The purpose of this simple exercise is to familiarize ourselves with the way in which electronic states are represented by Slater determinants, with emphasis on the interplay between orbitals and configurations and on the description of electron correlation by means of superpositions of configurations.

In Section 5.3, we begin our study of the standard models by presenting, for a given Fock space, exact N-particle treatments of the electronic structures of the hydrogen and water molecules, using truncated but reasonably accurate one-particle expansions. The purpose of these calculations is to serve as benchmarks for the less accurate models introduced in the subsequent sections of this chapter: the Hartree–Fock model in Section 5.4, the multiconfigurational self-consistent field model in Section 5.5, the configuration-interaction model in Section 5.6, the coupled-cluster model in Section 5.7 and finally Møller–Plesset perturbation theory in Section 5.8. For each model, we first give a short introductory description and then go on to present calculations on the hydrogen and water molecules. In comparison with the exact Fock-space treatments of Section 5.3, these calculations serve to illustrate the flexibility and accuracy characteristic of each model.

5.1 One- and N-electron expansions

The molecular electronic wave functions studied in this book are constructed from N-electron Slater determinants, each of which represents an antisymmetric product of one-electron functions (orbitals). Consider the Schrödinger equation for an N-electron system in Fock space:

$$\hat{H}|0\rangle = E|0\rangle \tag{5.1.1}$$

As discussed in Chapters 1 and 2, the spin-free nonrelativistic molecular electronic Hamiltonian

$$\hat{H} = \hat{h} + \hat{g} + h_{\text{nuc}} \tag{5.1.2}$$

contains one- and two-electron operators

$$\hat{h} = \sum_{pq} h_{pq} E_{pq} \tag{5.1.3}$$

$$\hat{g} = \tfrac{1}{2} \sum_{pqrs} g_{pqrs} e_{pqrs} \tag{5.1.4}$$

and the nuclear–nuclear repulsion term h_{nuc}. The operators (5.1.3) and (5.1.4) are expressed in terms of a set of orthonormal *molecular orbitals* (MOs) $\phi_p(\mathbf{r})$. For (5.1.2) to be an exact Hamiltonian and (5.1.1) to be an exact representation of the Schrödinger equation, the MOs must form a complete set of one-electron functions. Usually, these MOs are expanded in a finite set of nonorthogonal *atomic orbitals* (AOs), each of which is a function of the Cartesian coordinates of a single electron:

$$\phi_p(\mathbf{r}) = \sum_\mu C_{\mu p} \chi_\mu(\mathbf{r}) \tag{5.1.5}$$

The set of all AOs employed in a particular calculation is referred to as the AO basis and corresponds usually to a set of simple analytical functions centred on the atomic nuclei as discussed in Chapter 6.

If we express the electronic Hamiltonian (5.1.2) in an MO basis in which the one-electron Hamiltonian matrix is diagonal

$$h_{pq} = \delta_{pq} \varepsilon_p \tag{5.1.6}$$

then the one-electron functions $a_{p\sigma}^\dagger |\text{vac}\rangle$ become eigenfunctions of the one-electron part of the Hamiltonian

$$\hat{h} a_{p\sigma}^\dagger |\text{vac}\rangle = \varepsilon_p a_{p\sigma}^\dagger |\text{vac}\rangle \tag{5.1.7}$$

Furthermore, the N-electron determinants written as products of creation operators

$$|i\rangle = \left(\prod_{p\sigma \in |i\rangle} a_{p\sigma}^\dagger \right) |\text{vac}\rangle \tag{5.1.8}$$

become eigenfunctions of the one-electron part of the Hamiltonian

$$\hat{h}|\text{i}\rangle = E_i |\text{i}\rangle \tag{5.1.9}$$

with eigenvalues equal to the sum of the energies of the occupied MOs:

$$E_i = \sum_{p\sigma \in |i\rangle} \varepsilon_p \tag{5.1.10}$$

Thus, provided the two-electron part of the Hamiltonian may be neglected, we can solve the Schrödinger equation (5.1.1) exactly by diagonalizing the one-electron Hamiltonian (5.1.6) in a complete one-electron basis (5.1.5). The question of completeness in the one-electron space is a subtle one, to which we shall briefly return in Chapter 6.

In practice, the two-electron contributions (5.1.4) are not small and the determinants (5.1.8) therefore do not represent accurate solutions to the Schrödinger equation (5.1.1). However, since the determinants are eigenfunctions of a Hermitian operator (5.1.3) and thus constitute a complete set of N-electron functions, we may express the exact solution to the Schrödinger equation as a superposition of determinants:

$$|0\rangle = \sum_i C_i |i\rangle \tag{5.1.11}$$

To arrive at an exact solution to the Schrödinger equation in the form of (5.1.11), the following requirements must therefore be satisfied:

1. The orbitals (5.1.5) in terms of which the determinants are constructed must form a complete set in one-electron space.
2. In the expansion of the N-electron wave function (5.1.11), we must include all determinants that can be generated from these orbitals.

Failure to fulfil either requirement will introduce errors in the wave function. In this context, it is important to realize that the errors that are introduced by truncating the one-electron space are qualitatively different from those that arise from approximations made in the N-electron description (and vice versa). Consequently, a poor description in either of the two spaces can never be compensated for by an improved description in the other, although a fortuitous cancellation of the errors in the two spaces sometimes may be observed.

Figure 5.1 illustrates how the exact solution to the Schrödinger equation may be approached by systematic improvements in the one- and N-electron descriptions. To understand what accuracy may be expected in a given calculation, it is important to analyse separately the errors arising from the one- and N-electron treatments. The errors in the N-electron treatment, in particular, can be identified by comparison with the exact solution in the same Fock space (i.e. in the same one-electron basis). In this chapter, such comparisons will be made to illustrate the errors characteristic of the different N-electron models; in Chapters 6–8, we shall investigate the errors that arise from truncating the one-electron basis.

Although it is important to distinguish between the errors made in the one- and N-electron treatments, it should be realized that these errors are not entirely independent of each other. For example, the basis-set requirements for a single-determinant wave function are rather different from those for a wave function containing a large number of determinants. It should also be pointed out that the exact solution to the Schrödinger equation can be approached in a manner different from that illustrated in Figure 5.1 – for example, by employing explicitly correlated wave

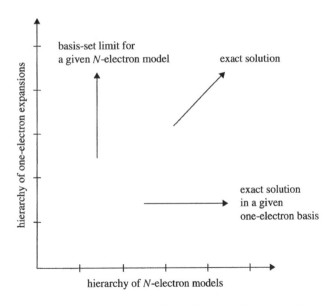

Fig. 5.1. The systematic approach to the exact solution of the Schrödinger equation by successive improvements in the one- and N-electron descriptions.

functions as discussed in Chapter 7. Still, our emphasis in this book is on Fock-space methods, where the wave functions are written as linear combinations of Slater determinants.

5.2 A model system: the hydrogen molecule in a minimal basis

Before considering accurate solutions to the Schrödinger equation, it is useful to examine in some detail the description of the electronic system of the hydrogen molecule by means of a very simple model – the expansion of the wave function in a set of two hydrogenic orbitals centred on the atomic nuclei. Although crude, this model provides some useful insight into the way that we represent electronic systems by means of determinants in Fock space, preparing us for the discussions of the standard models in the subsequent sections.

5.2.1 ONE-ELECTRON BASIS

The normalized AOs are given by

$$1s_A(\mathbf{r}) = \frac{1}{\sqrt{\pi}} \exp(-r_A) \tag{5.2.1}$$

$$1s_B(\mathbf{r}) = \frac{1}{\sqrt{\pi}} \exp(-r_B) \tag{5.2.2}$$

where r_A and r_B are the distances between the electron and nuclei A and B, respectively. From these AOs, we may construct symmetry-adapted orthonormal MOs in the usual manner:

$$\phi_1(\mathbf{r}) = 1\sigma_g(\mathbf{r}) = N_g \left[1s_A(\mathbf{r}) + 1s_B(\mathbf{r}) \right] \tag{5.2.3}$$
$$\phi_2(\mathbf{r}) = 1\sigma_u(\mathbf{r}) = N_u \left[1s_A(\mathbf{r}) - 1s_B(\mathbf{r}) \right] \tag{5.2.4}$$

The normalization constants N_g and N_u are given by

$$N_g = \frac{1}{\sqrt{2(1+S)}} \tag{5.2.5}$$

$$N_u = \frac{1}{\sqrt{2(1-S)}} \tag{5.2.6}$$

where S is the overlap between the AOs:

$$S = \int 1s_A(\mathbf{r}) 1s_B(\mathbf{r}) \, d\mathbf{r} \tag{5.2.7}$$

Upon integration, the overlap integral is found to take the form

$$S = (1 + R + \tfrac{1}{3}R^2) \exp(-R) \tag{5.2.8}$$

where R is the internuclear separation; see Exercise 5.1.

The functional forms of the AOs and MOs along the molecular axis are illustrated in Figure 5.2 for the nuclear separation $R = 1.4a_0$, which corresponds to the true ground-state equilibrium

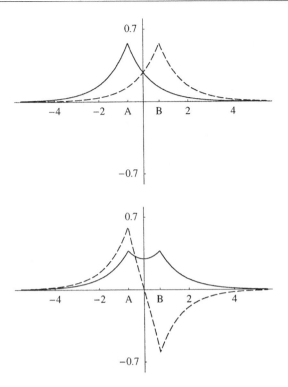

Fig. 5.2. The AOs and MOs of the hydrogen molecule at an internuclear separation of $1.4a_0$. At the top, the $1s_A$ (solid line) and $1s_B$ (dashed line) AOs are depicted. The $1\sigma_g$ (solid line) and $1\sigma_u$ (dashed line) MOs are depicted at the bottom. Atomic units are used.

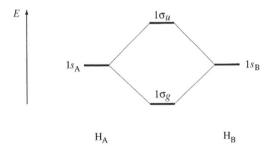

Fig. 5.3. Molecular-orbital energy-level diagram for the hydrogen molecule.

distance of the hydrogen molecule. The nuclei are assumed to be located on the z axis, with the origin at the molecular centre of inversion, and the plots are along the molecular z axis. The AO basis employed here is called *minimal* since there is only one basis function for each orbital occupied in the constituent atomic systems. For a general discussion of atomic basis functions, see Chapter 6.

The energies of the $1\sigma_g$ and $1\sigma_u$ orbitals are depicted in the usual manner in the molecular-orbital energy-level diagram in Figure 5.3. At equilibrium, the $1\sigma_g$ orbital is below the $1\sigma_u$ orbital in energy; in the dissociation limit, the MOs are degenerate.

5.2.2   N-ELECTRON BASIS

From the bonding and antibonding MOs (5.2.3) and (5.2.4), we may generate a total of six determinants by distributing the two electrons among the four spin orbitals in all possible ways allowed for by the Pauli principle. In our calculations, however, we shall employ a basis of CSFs rather than determinants. As discussed in Section 2.5, the CSFs are spin-adapted linear combinations of determinants belonging to the same orbital configuration.

Applying to the vacuum state the two-body creation operators introduced in Section 2.3.3, we obtain two closed-shell CSFs of $^1\Sigma_g^+$ symmetry (the bonding and antibonding configurations)

$$|^1\Sigma_g^+(g^2)\rangle = |1\sigma_g^2\rangle = \frac{1}{\sqrt{2}}\hat{Q}_{11}^{0,0}|\text{vac}\rangle = a_{1\alpha}^\dagger a_{1\beta}^\dagger|\text{vac}\rangle \tag{5.2.9}$$

$$|^1\Sigma_g^+(u^2)\rangle = |1\sigma_u^2\rangle = \frac{1}{\sqrt{2}}\hat{Q}_{22}^{0,0}|\text{vac}\rangle = a_{2\alpha}^\dagger a_{2\beta}^\dagger|\text{vac}\rangle \tag{5.2.10}$$

three spin components of an open-shell CSF of $^3\Sigma_u^+$ symmetry

$$|^3\Sigma_u^+\rangle = \begin{cases} \hat{Q}_{21}^{1,1}|\text{vac}\rangle = a_{2\alpha}^\dagger a_{1\alpha}^\dagger|\text{vac}\rangle \\ \hat{Q}_{21}^{1,0}|\text{vac}\rangle = \frac{1}{\sqrt{2}}(a_{2\alpha}^\dagger a_{1\beta}^\dagger + a_{2\beta}^\dagger a_{1\alpha}^\dagger)|\text{vac}\rangle \\ \hat{Q}_{21}^{1,-1}|\text{vac}\rangle = a_{2\beta}^\dagger a_{1\beta}^\dagger|\text{vac}\rangle \end{cases} \tag{5.2.11}$$

and one open-shell CSF of $^1\Sigma_u^+$ symmetry

$$|^1\Sigma_u^+\rangle = \hat{Q}_{21}^{0,0}|\text{vac}\rangle = \frac{1}{\sqrt{2}}(a_{2\alpha}^\dagger a_{1\beta}^\dagger - a_{2\beta}^\dagger a_{1\alpha}^\dagger)|\text{vac}\rangle \tag{5.2.12}$$

Only states of the same spin and space symmetries are coupled by the totally symmetric electronic Hamiltonian. Thus, the only CSFs that will mix in our calculations are those of gerade symmetry. An arbitrary, normalized state of this symmetry may be written in the following manner

$$|^1\Sigma_g^+(\tau)\rangle = \cos(\tau)|1\sigma_g^2\rangle + \sin(\tau)|1\sigma_u^2\rangle \tag{5.2.13}$$

where τ is the variational parameter. The associated orthogonal state is obtained by adding $\pi/2$ to the variational parameter

$$\left\langle {}^1\Sigma_g^+(\tau)\middle|{}^1\Sigma_g^+\left(\tau + \frac{\pi}{2}\right)\right\rangle = 0 \tag{5.2.14}$$

as is easily verified.

5.2.3   DENSITY MATRICES AND MOLECULAR INTEGRALS

In the remainder of Section 5.2, we shall investigate the description of the H_2 electronic system made possible by the six CSFs generated from the minimal atomic basis. In particular, we shall examine and compare the electronic energies and the density functions associated with the different states generated in this basis.

We recall that, for a normalized electronic state $|0\rangle$, the expectation value of the energy is obtained from the expression

$$E = \sum_{pq} D_{pq} h_{pq} + \frac{1}{2} \sum_{pqrs} d_{pqrs} g_{pqrs} + \frac{1}{R} \tag{5.2.15}$$

and that the electron-density functions of Section 2.7.3 are given by

$$\rho(\mathbf{r}) = \sum_{pq} D_{pq} \phi_p^*(\mathbf{r}) \phi_q(\mathbf{r}) \tag{5.2.16}$$

$$\rho(\mathbf{r}_1, \mathbf{r}_2) = \tfrac{1}{2} \sum_{pqrs} d_{pqrs} \phi_p^*(\mathbf{r}_1) \phi_q(\mathbf{r}_1) \phi_r^*(\mathbf{r}_2) \phi_s(\mathbf{r}_2) \tag{5.2.17}$$

In the minimal H_2 basis, the summations are over only the bonding and antibonding MOs. For the calculation of energy and density functions, we need the integrals h_{pq} and g_{pqrs} and the density-matrix elements D_{pq} and d_{pqrs} in the MO basis.

In Table 5.1, we have listed the density-matrix elements of the CSFs (5.2.9)–(5.2.12) and of the two-configuration state (5.2.13). In Exercise 5.2, these elements are derived from

$$D_{pq} = \langle 0 | E_{pq} | 0 \rangle = \sum_{\sigma} \langle 0 | a_{p\sigma}^\dagger a_{q\sigma} | 0 \rangle \tag{5.2.18}$$

$$d_{pqrs} = \langle 0 | E_{pq} E_{rs} - \delta_{rq} E_{ps} | 0 \rangle = \sum_{\sigma\tau} \langle 0 | a_{p\sigma}^\dagger a_{r\tau}^\dagger a_{s\tau} a_{q\sigma} | 0 \rangle \tag{5.2.19}$$

Whereas the density elements for the CSFs are integers, those for the superposition of CSFs are real numbers, not necessarily integers. The one-electron density matrices in Table 5.1 are all diagonal, with elements corresponding to the natural-orbital occupation numbers of Section 2.7.1.

In Table 5.1, we have also listed the molecular integrals needed for the evaluation of the electronic energy at the experimental equilibrium distance and at infinite separation. These integrals have been obtained using special techniques for many-centre integrals over hydrogenic orbitals and should be taken at face value. In Chapter 9, we shall develop techniques appropriate for

Table 5.1 The density-matrix elements and molecular integrals for the hydrogen molecule in a symmetry-adapted basis of hydrogenic $1s$ functions with exponents 1 (atomic units). Rows containing only zero elements and rows with elements that are related to those of other rows by permutational symmetry are not listed

Indices	Density elements					Integrals	
	$\|1\sigma_g^2\rangle$	$\|1\sigma_u^2\rangle$	$\|^1\Sigma_g^+(\tau)\rangle$	$\|^3\Sigma_u^+\rangle$	$\|^1\Sigma_u^+\rangle$	$R = 1.4a_0$	$R = \infty$
11	2	0	$2\cos^2\tau$	1	1	-1.1856^a	$-1/2^c$
22	0	2	$2\sin^2\tau$	1	1	-0.5737^b	$-1/2^c$
1111	2	0	$2\cos^2\tau$	0	0	0.5660	5/16
2222	0	2	$2\sin^2\tau$	0	0	0.5863	5/16
2211	0	0	0	1	1	0.5564	5/16
2121	0	0	$\sin 2\tau$	0	0	0.1403	5/16
2112	0	0	0	-1	1	0.1403	5/16

[a]Kinetic and nuclear-attraction contributions: $0.4081 - 1.5937 = -1.1856$.
[b]Kinetic and nuclear-attraction contributions: $1.1521 - 1.7258 = -0.5737$.
[c]Kinetic and nuclear-attraction contributions: $\frac{1}{2} - 1 = -\frac{1}{2}$.

the calculation of integrals over the Gaussian-type orbitals used in most molecular calculations. Gaussians are not employed in this section since, in a minimal basis, the hydrogenic orbitals give a more accurate representation of the electronic system.

5.2.4 BONDING AND ANTIBONDING CONFIGURATIONS

We first consider the descriptions provided by the closed-shell CSFs of $^1\Sigma_g^+$ symmetry. From Table 5.1, we obtain the following expressions for the energies of the bonding and antibonding configurations:

$$E(g^2) = 2h_{11} + g_{1111} + \frac{1}{R} \tag{5.2.20}$$

$$E(u^2) = 2h_{22} + g_{2222} + \frac{1}{R} \tag{5.2.21}$$

These expressions may be given a simple physical interpretation. Thus, the first term in (5.2.20) represents the kinetic energies of the two electrons in orbital ϕ_1 as well as the interactions of these electrons with the nuclei. The second term represents the Coulomb repulsion between the charge distributions of the two electrons. Similar considerations apply to (5.2.21).

In Table 5.2, we have given the energies of the bonding and antibonding configurations (as well as of other electronic states to be discussed shortly) as calculated from the densities and integrals in Table 5.1. Also listed are the individual contributions to the electronic energies. The energies of the bonding and antibonding configurations are -1.0909 and 0.1532 E_h, respectively, which should be compared with -1 E_h for two noninteracting ground-state hydrogen atoms. The difference in energies between the bonding and antibonding configurations arises mainly from a difference in the kinetic energies – the potential energies are quite similar and in fact higher for the bonding configuration.

In Figure 5.4, we have plotted the density functions (5.2.16) and (5.2.17) on the molecular axis for the bonding and antibonding configurations. The one-electron densities have been plotted along the z axis – that is, as functions of $(0, 0, z)$. Likewise, the two-electron densities have been

Table 5.2 The electronic energies of the hydrogen molecule at an internuclear separation of $1.4a_0$ in a minimal basis of hydrogenic $1s$ orbitals with unit exponents (E_h)

State	Kinetic	Attraction	Electron repulsion[a]	Nuc. rep.	Total
$^1\Sigma_g^+ 1\sigma_g^2$	0.8162	-3.1874	$0.5660 + 0.0000 = 0.5660$	0.7143	-1.0909
$^1\Sigma_g^+ 1\sigma_u^2$	2.3042	-3.4516	$0.5863 + 0.0000 = 0.5863$	0.7143	0.1532
$^1\Sigma_g^+(\tau_0)$[b]	0.8344	-3.1907	$0.5663 - 0.0308 = 0.5354$	0.7143	-1.1066
$^1\Sigma_g^+(\tau_1)$[c]	2.2860	-3.4484	$0.5861 + 0.0308 = 0.6169$	0.7143	0.1688
$^1\Sigma_g^+(\tau_{cov})$[d]	0.8452	-3.1926	$0.5664 - 0.0388 = 0.5277$	0.7143	-1.1055
$^1\Sigma_g^+(\tau_{ion})$[e]	0.8452	-3.1926	$0.5664 + 0.0388 = 0.6052$	0.7143	-1.0279
$^3\Sigma_u^+$	1.5602	-3.3195	$0.5564 - 0.1403 = 0.4162$	0.7143	-0.6289
$^1\Sigma_u^+$	1.5602	-3.3195	$0.5564 + 0.1403 = 0.6967$	0.7143	-0.3484

[a] The electron-repulsion energy is written as the sum of the classical Coulomb contribution and the exchange and correlation contributions.
[b] The ground state calculated from $\tau_0 = -0.1109$.
[c] The excited state calculated from $\tau_1 = -0.1109 + \pi/2$.
[d] The covalent state calculated from $\tau_{cov} = -0.1400$.
[e] The ionic state calculated from $\tau_{ion} = 0.1400$.

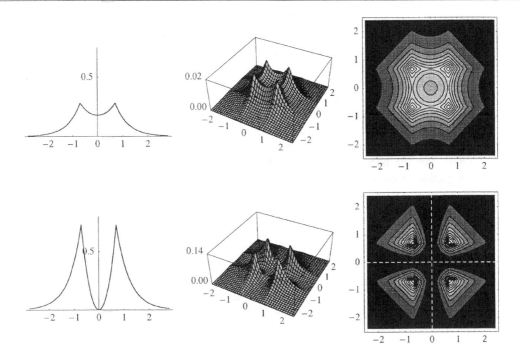

Fig. 5.4. The one- and two-electron density functions of the bonding $|1\sigma_g^2\rangle$ (upper plots) and antibonding $|1\sigma_u^2\rangle$ (lower plots) configurations of the hydrogen molecule on the molecular axis (atomic units). The two-electron densities are represented by surface and contour plots. In the contour plot for the antibonding configuration, the two-electron nodes are represented by dashed lines. The density functions have been calculated in a minimal basis of hydrogenic $1s$ functions with unit exponents.

plotted as functions of $(0, 0, z_1)$ and $(0, 0, z_2)$. Comparing the one-electron densities of the two configurations

$$\rho_{1\sigma_g^2}(\mathbf{r}) = 2\phi_1^2(\mathbf{r}) \tag{5.2.22}$$

$$\rho_{1\sigma_u^2}(\mathbf{r}) = 2\phi_2^2(\mathbf{r}) \tag{5.2.23}$$

we find – as expected from the form of the one-electron functions in Figure 5.2 – that the bonding configuration exhibits a characteristic accumulation of density in the internuclear region. By contrast, the density of the antibonding configuration is depleted in the same region and the wave function has a nodal surface perpendicular to the molecular axis.

As discussed in Section 2.7.3, the two-electron densities represent the probability of simultaneously locating two electrons at two given positions in the molecule. According to (5.2.17), the two-electron density functions of the bonding and antibonding configurations are given by

$$\rho_{1\sigma_g^2}(\mathbf{r}_1, \mathbf{r}_2) = \phi_1^2(\mathbf{r}_1)\phi_1^2(\mathbf{r}_2) \tag{5.2.24}$$

$$\rho_{1\sigma_u^2}(\mathbf{r}_1, \mathbf{r}_2) = \phi_2^2(\mathbf{r}_1)\phi_2^2(\mathbf{r}_2) \tag{5.2.25}$$

which are simply the products of the one-electron densities divided by 4:

$$\rho_{1\sigma_g^2}(\mathbf{r}_1, \mathbf{r}_2) = \tfrac{1}{4}\rho_{1\sigma_g^2}(\mathbf{r}_1)\rho_{1\sigma_g^2}(\mathbf{r}_2) \tag{5.2.26}$$

$$\rho_{1\sigma_u^2}(\mathbf{r}_1, \mathbf{r}_2) = \tfrac{1}{4}\rho_{1\sigma_u^2}(\mathbf{r}_1)\rho_{1\sigma_u^2}(\mathbf{r}_2) \tag{5.2.27}$$

This factorization of the two-electron density functions is apparent in the two-electron plots in Figure 5.4, which represent the probability of simultaneously locating the two electrons at different positions on the molecular axis. Thus, the relative probabilities of locating one electron at different positions on the molecular axis are independent of the whereabouts of the other electron. For example, the probability of locating one electron at nucleus A is identical to the probability of locating the same electron at nucleus B – even when the other electron is known to be located at B.

In the two-electron density functions plotted in Figure 5.4, the nuclear cusps appear as sharp ridges intersecting at the nuclei. The interior of the centre square formed by the nuclear cusps represents the region where both electrons are located between the nuclei and on the molecular axis. For the bonding and antibonding configurations, we observe the same accumulation and depletion of electron density in the internuclear region as for the one-electron density.

5.2.5 SUPERPOSITION OF CONFIGURATIONS

In Section 5.2.4, we found that the simple bonding and antibonding configurations of $^1\Sigma_g^+$ symmetry provide uncorrelated descriptions of the electronic system in the sense that the probability of locating one electron at a given position is unrelated to the position of the other electron. We shall now see that, by mixing the bonding and antibonding configurations, we arrive at a more realistic description of the electronic system, where the motion of the electrons is correlated in such a way that the two-electron density function is no longer a simple product of the one-electron density functions.

From the density-matrix elements in Table 5.1, we find that the energy of the two-configuration state (5.2.13) is given by the expression

$$E_\tau(^1\Sigma_g^+) = \cos^2(\tau)E(g^2) + \sin^2(\tau)E(u^2) + \sin(2\tau)g_{2121} \tag{5.2.28}$$

Whereas the first two terms represent the weighted average of the energies of the bonding (5.2.20) and antibonding (5.2.21) configurations in Table 5.2, the third term represents the interaction between these configurations and is responsible for their mixing.

To arrive at the best approximation to the electronic ground state in the minimal basis, we invoke the variation principle and minimize (5.2.28) with respect to τ. Differentiating the energy with respect to τ and setting the result equal to zero, we obtain:

$$\sin(2\tau)E(g^2) - \sin(2\tau)E(u^2) - 2\cos(2\tau)g_{2121} = 0 \tag{5.2.29}$$

A simple rearrangement gives

$$\tan(2\tau) = \frac{2g_{2121}}{E(g^2) - E(u^2)} \tag{5.2.30}$$

from which we obtain the solutions

$$\tau_n = \frac{1}{2}\arctan\left[\frac{2g_{2121}}{E(g^2) - E(u^2)}\right] + \frac{n\pi}{2} \tag{5.2.31}$$

where n is an integer.

The two-configuration energy (5.2.28) is periodic in τ with period π. We choose τ_0 and τ_1 in (5.2.31) as representing the ground and excited states since

$$-\frac{\pi}{2} \leq \tau_0 \leq 0 \leq \tau_1 \leq \frac{\pi}{2} \tag{5.2.32}$$

implies that the interaction term in (5.2.28) is negative for $\tau_0 < 0$ and positive for $\tau_1 > 0$. Substituting into (5.2.31) the numerical values for the integral and the energies, we obtain $\tau_0 = -0.1109$ and $\tau_1 = -0.1109 + \pi/2$. At the equilibrium bond distance, the ground-state and excited-state wave functions (5.2.13) may therefore be written in the form

$$|^1\Sigma_g^+(\tau_0)\rangle_e = 0.9939|1\sigma_g^2\rangle - 0.1106|1\sigma_u^2\rangle \tag{5.2.33}$$

$$|^1\Sigma_g^+(\tau_1)\rangle_e = 0.1106|1\sigma_g^2\rangle + 0.9939|1\sigma_u^2\rangle \tag{5.2.34}$$

The ground state is dominated by the bonding configuration with a weight of 98.8% (obtained by squaring the expansion coefficient), the antibonding configuration contributing merely 1.2%. These weights are reflected also in the natural-orbital occupation numbers, which are 1.98 for the bonding orbital and 0.02 for the antibonding one; see Table 5.1.

The electronic energies of the optimized states are listed in Table 5.2. The energy of the ground state (5.2.33) is lowered relative to that of the bonding $|1\sigma_g^2\rangle$ configuration by 1.4% and the energy of the excited state (5.2.34) is raised above that of the antibonding $|1\sigma_u^2\rangle$ configuration. These shifts occur because of modifications to the electron-repulsion energy as described by the third term in (5.2.28), the changes in the one-electron energies being somewhat smaller and oppositely directed.

Let us now consider the densities of the two-configuration states (5.2.33) and (5.2.34); see Figure 5.5. The one-electron densities – which now contain contributions from both orbitals – do not differ much from those of the single-configuration states, as expected from the modest 1%

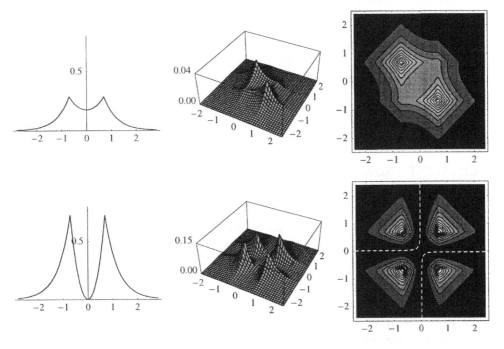

Fig. 5.5. The one- and two-electron density functions of the two-configuration $^1\Sigma_g^+$ ground (upper plots) and excited (lower plots) states of the hydrogen molecule on the molecular axis (atomic units). The two-electron densities are represented by surface and contour plots. In the contour plot for the excited state, the dashed lines represent two-electron nodes. The density functions have been calculated in a minimal basis of hydrogenic $1s$ functions with unit exponents.

change in the occupation numbers. Although not obvious from the plot in Figure 5.5, the one-electron density of the excited state is no longer zero at the origin – the small contribution from the bonding configuration ensures that the density always remains nonzero (but small) in the internuclear region.

Unlike the one-electron density, the two-electron density changes markedly due to the interaction between the bonding and antibonding configurations; compare Figures 5.4 and 5.5. Thus, although the ground-state two-electron density is still confined mostly to the internuclear region, it has now been oriented along the diagonal axis $z_2 = -z_1$, exposing a preference for the ground-state electrons to be located around opposite nuclei. The interaction between the bonding and antibonding configurations has thus endowed the system with a distinct *covalent* character, responsible for the 5% reduction in the electron-repulsion energy in Table 5.2. The energy is reduced since the electrons tend to avoid each other by residing on different nuclei. This covalent description of the electronic system is said to be *correlated*, in contrast to the *uncorrelated* description of the bonding configuration $|1\sigma_g^2\rangle$. For a correlated electronic system, the two-electron density function (5.2.17) cannot be written as a product of the one-electron density functions (5.2.16), as may be verified for the two-configuration wave function by inspection of Table 5.1.

For the excited state in Figure 5.5, the effect of correlation is perhaps less conspicuous than for the ground state. As for the antibonding configuration $|1\sigma_u^2\rangle$ in Figure 5.4, the density is depleted in the internuclear region but it now has a preferred orientation along the diagonal $z_2 = z_1$, reflecting a tendency of the electrons to be located around the same nucleus. This *ionic* character increases the electron repulsion by about 5% relative to the uncorrelated description of the antibonding configuration. Also the nodal structure is different from that of the antibonding configuration – the two-electron nodes avoid the ionic diagonal $z_2 = z_1$ and are restricted to the covalent regions, although this covalent shift is strong only close to the origin.

Our observations so far may be summarized by stating that, whereas the *bonding and antibonding character* of the one-electron density determines the coarse structure of the electronic system such as the arrangement of the electrons relative to the nuclei (accumulation and depletion of electrons in the internuclear region), the *covalent and ionic character* of two-electron density determines the finer details such as the arrangement of the electrons relative to each other (electron correlation).

5.2.6 COVALENT AND IONIC STATES

We have seen that the closed-shell bonding and antibonding configurations each provide an uncorrelated description of the electronic system but that a superposition of these configurations introduces correlation. In the ground state, the electrons tend to be located around opposite nuclei, whereas, in the excited state, there is a tendency for the electrons to be located around the same nucleus. Further insight into the correlation problem may be obtained by isolating the pure *covalent and ionic states*, where the electrons are either always located around opposite nuclei or always located around the same nucleus.

To determine the covalent and ionic states, we expand the bonding and antibonding configurations in terms of the localized AOs $1s_A$ and $1s_B$ rather than in terms of the delocalized MOs $1\sigma_g$ and $1\sigma_u$. We therefore write the creation operators for the $1\sigma_g$ and $1\sigma_u$ orbitals in terms of the operators for the localized $1s_A$ and $1s_B$ orbitals:

$$a_{1\sigma}^\dagger = N_g(a_{A\sigma}^\dagger + a_{B\sigma}^\dagger) \tag{5.2.35}$$

$$a_{2\sigma}^\dagger = N_u(a_{A\sigma}^\dagger - a_{B\sigma}^\dagger) \tag{5.2.36}$$

Here $a_{A\sigma}^\dagger$ and $a_{B\sigma}^\dagger$ are the creation operators for electrons of spin σ in $1s_A$ and $1s_B$, respectively. The $1s_A$ and $1s_B$ orbitals are nonorthogonal, but we recall from Section 1.9 that nonorthogonality affects only the anticommutation relations between creation and annihilation operators (1.9.12) – not the anticommutations between two creation operators and between two annihilation operators.

Substituting the operators (5.2.35) and (5.2.36) into the bonding and antibonding CSFs (5.2.9) and (5.2.10), we obtain

$$|1\sigma_g^2\rangle = \frac{N_g^2}{N_c}|\text{cov}\rangle + \frac{N_g^2}{N_i}|\text{ion}\rangle \tag{5.2.37}$$

$$|1\sigma_u^2\rangle = \frac{N_u^2}{N_i}|\text{ion}\rangle - \frac{N_u^2}{N_c}|\text{cov}\rangle \tag{5.2.38}$$

where the normalized (but nonorthogonal) covalent and ionic states are given by

$$|\text{cov}\rangle = N_c(a_{A\alpha}^\dagger a_{B\beta}^\dagger + a_{B\alpha}^\dagger a_{A\beta}^\dagger)|\text{vac}\rangle \tag{5.2.39}$$

$$|\text{ion}\rangle = N_i(a_{A\alpha}^\dagger a_{A\beta}^\dagger + a_{B\alpha}^\dagger a_{B\beta}^\dagger)|\text{vac}\rangle \tag{5.2.40}$$

Using the anticommutation relations for nonorthogonal orbitals, the reader may verify that the normalization constants are given by

$$N_c = N_i = \frac{1}{\sqrt{2(1 + S^2)}} \tag{5.2.41}$$

Whereas the covalent state (5.2.39) represents an electronic structure with the electrons located always around different nuclei, the ionic state (5.2.40) represents a situation where the electrons are located around the same nucleus.

To determine the expansions of the covalent and ionic states in the bonding and antibonding configurations, we substitute (5.2.37) and (5.2.38) in the two-configuration state (5.2.13):

$$|^1\Sigma_g^+(\tau)\rangle = \frac{N_g^2\cos(\tau) - N_u^2\sin(\tau)}{N_c}|\text{cov}\rangle + \frac{N_g^2\cos(\tau) + N_u^2\sin(\tau)}{N_i}|\text{ion}\rangle \tag{5.2.42}$$

The covalent and ionic states are retrieved from this expression by setting the ionic and covalent coefficients, respectively, equal to zero. We obtain

$$\tau_{\text{cov}} = -\tau_{\text{ion}} = -\arctan\left(\frac{1 - S}{1 + S}\right) \tag{5.2.43}$$

which allows us to write

$$|\text{cov}\rangle = \cos(\tau_{\text{cov}})|1\sigma_g^2\rangle + \sin(\tau_{\text{cov}})|1\sigma_u^2\rangle \tag{5.2.44}$$

$$|\text{ion}\rangle = \cos(\tau_{\text{cov}})|1\sigma_g^2\rangle - \sin(\tau_{\text{cov}})|1\sigma_u^2\rangle \tag{5.2.45}$$

as normalized representations of the covalent and ionic states. At the equilibrium bond distance, we obtain from (5.2.8) $\tau_{\text{cov}} = -0.1400$:

$$|\text{cov}\rangle_e = 0.9902|1\sigma_g^2\rangle - 0.1396|1\sigma_u^2\rangle \tag{5.2.46}$$

$$|\text{ion}\rangle_e = 0.9902|1\sigma_g^2\rangle + 0.1396|1\sigma_u^2\rangle \tag{5.2.47}$$

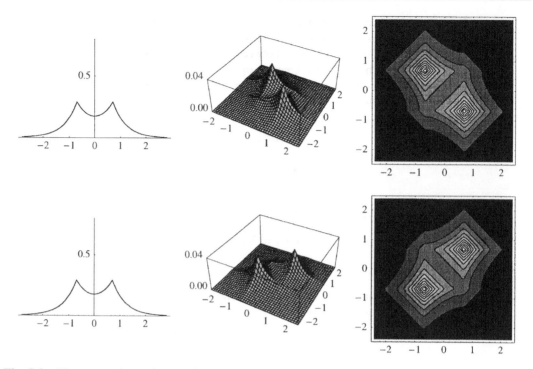

Fig. 5.6. The one- and two-electron density functions of the covalent $|\text{cov}\rangle$ (upper plots) and ionic $|\text{ion}\rangle$ (lower plots) electronic structures of the hydrogen molecule on the molecular axis (atomic units). The two-electron densities are represented by surface and contour plots. The density functions have been calculated in a minimal basis of hydrogenic $1s$ functions with unit exponents.

At equilibrium, therefore, the covalent and ionic states are both dominated by the bonding configuration. In Section 5.2.9, we shall see that the bonding and antibonding configurations contribute equally to the covalent and ionic states at infinite internuclear separation.

The energy of the covalent state is lower than that of the bonding configuration and close to that of the optimized ground state (5.2.33) – see Table 5.2. Indeed, in the covalent state, the lowering of the electron-repulsion energy is greater than in the optimized ground state, but this additional lowering is more than compensated for by an increase in the kinetic energy.

From Figure 5.6, we see that the one- and two-electron densities of the covalent state are very similar to those of the optimized ground state. In particular, the two-electron density is oriented along the covalent diagonal $z_2 = -z_1$. The ionic state has exactly the same one-electron density as does the covalent state but a different two-electron density, oriented along the ionic diagonal $z_2 = z_1$.

5.2.7 OPEN-SHELL STATES

The open-shell states are of ungerade symmetry and all the minimal-basis CSFs have different space and spin symmetries; see Section 5.2.2. The open-shell CSFs therefore represent our final approximate states for the model system. The energies of these states are obtained from the expressions

$$E(^3\Sigma_u^+) = h_{11} + h_{22} + g_{2211} - g_{2112} + \frac{1}{R} \qquad (5.2.48)$$

$$E(^1\Sigma_u^+) = h_{11} + h_{22} + g_{2211} + g_{2112} + \frac{1}{R} \qquad (5.2.49)$$

and are listed in Table 5.2. These expressions may be given a simple physical interpretation. Thus, the first two terms represent the kinetic and nuclear-attraction energies of electrons in orbitals ϕ_1 and ϕ_2 and the third term represents the Coulomb repulsion between the electrons in these orbitals. The fourth term is equal in magnitude but of opposite sign for the two states, reflecting two different modes of correlation: the energy is lowered for the triplet state and raised for the singlet state.

The density functions are plotted in Figure 5.7. The two states have singly occupied bonding and antibonding orbitals and thus the same one-electron density functions. Comparing Figures 5.4 and 5.7, we note that the internuclear one-electron densities of the open-shell states are intermediate between the densities of the closed-shell bonding and antibonding CSFs – in the same way that the energies of the triplet and singlet states (5.2.48) and (5.2.49) are intermediate between those of the ground and excited closed-shell states, see Table 5.2.

The different modes of correlation in the singlet and triplet states are obvious from an inspection of the plots of the two-electron densities in Figure 5.7. The triplet state is covalently correlated and the singlet state ionically. Moreover, whereas the triplet state has a two-electron node running

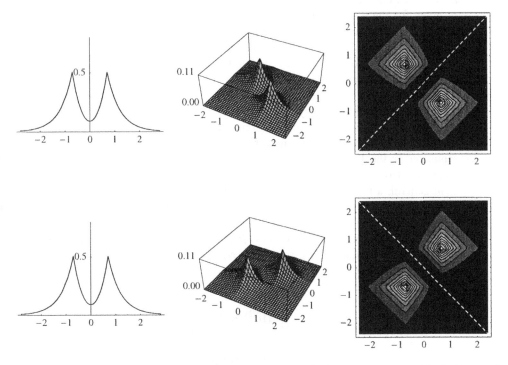

Fig. 5.7. The one- and two-electron density functions of the $^3\Sigma_u^+$ (upper plots) and $^1\Sigma_u^+$ (lower plots) electronic states of the hydrogen molecule on the molecular axis (atomic units). The two-electron densities are represented by surface and contour plots. In the contour plots, the dashed lines represent two-electron nodes. The density functions have been calculated in a minimal basis of hydrogenic $1s$ functions with unit exponents.

along the ionic diagonal, the singlet state has a node along the covalent diagonal. The covalent and ionic characters of these states are confirmed by expanding the CSFs (5.2.11) and (5.2.12) in the localized orbitals (5.2.35) and (5.2.36):

$$|^3\Sigma_u^+\rangle = \begin{cases} 2N_gN_u a_{A\alpha}^\dagger a_{B\alpha}^\dagger |\text{vac}\rangle \\ \sqrt{2}N_gN_u(a_{A\alpha}^\dagger a_{B\beta}^\dagger + a_{A\beta}^\dagger a_{B\alpha}^\dagger)|\text{vac}\rangle \\ 2N_gN_u a_{A\beta}^\dagger a_{B\beta}^\dagger |\text{vac}\rangle \end{cases} \tag{5.2.50}$$

$$|^1\Sigma_u^+\rangle = \sqrt{2}N_gN_u(a_{A\alpha}^\dagger a_{A\beta}^\dagger - a_{B\alpha}^\dagger a_{B\beta}^\dagger)|\text{vac}\rangle \tag{5.2.51}$$

The difference between the repulsion energies of the triplet and singlet states is quite large. Thus, in the triplet state, the covalent rearrangement of the two-electron density reduces the repulsion by 25%; in the singlet state, the repulsion is raised by the same amount (relative to the classical Coulomb contribution). As a result, the electron–electron repulsion energy of the singlet state is as much as 67% higher than that of the triplet state.

5.2.8 ELECTRON CORRELATION

In our discussion of the open-shell states in Section 5.2.7, we used the terms *correlation* and *correlated* in a rather general and perhaps somewhat unusual sense. In normal usage, these terms are reserved for the correlation that occurs upon superposition of CSFs and the description provided by a single CSF is consequently referred to as uncorrelated. Thus, in the conventional terminology, the singlet and triplet CSFs are classified as uncorrelated just like the closed-shell bonding and antibonding configurations. The correlated behaviour of the singlet and triplet states apparent in Figure 5.7 is instead described as an exchange effect since it arises not from a superposition of CSFs but from antisymmetrization of the wave function in accordance with the Pauli principle – that is, from the antisymmetric superposition of spin-orbital products. Sometimes, the term *Fermi correlation* is used for correlation arising from the Pauli antisymmetry principle. However, whatever the physical reasons may be for the correlated behaviour of the electrons – electron repulsion or the Pauli principle – the effect is always to modify the electron-repulsion energy calculated from the electron distribution of the system.

In general, a truly uncorrelated many-particle state is always represented by a product of one-particle functions. Conversely, any superposition of such products represents a state where the motion of the particles is correlated. Nevertheless, a Slater determinant – which represents an antisymmetric superposition of *spin-orbital* products – may in some cases represent a truly uncorrelated electronic state. Thus, in the closed-shell bonding and antibonding configurations of the hydrogen molecule, the Pauli principle is satisfied by an antisymmetrization of the spin part of the wave function

$$|1\sigma_g^2| = \phi_1\phi_1 \frac{1}{\sqrt{2}}(\alpha\beta - \beta\alpha) \tag{5.2.52}$$

$$|1\sigma_u^2| = \phi_2\phi_2 \frac{1}{\sqrt{2}}(\alpha\beta - \beta\alpha) \tag{5.2.53}$$

and the spatial part is therefore in the product form characteristic of an uncorrelated wave function. (We here use the notation (1.1.1) for Slater determinants and follow the convention that the first term in a product is associated with electron 1 and the second term with electron 2.) The superposition of these configurations introduces correlation since the spatial part is no longer in a product form

$$|{}^1\Sigma_g^+(\tau)| = [\cos(\tau)\phi_1\phi_1 + \sin(\tau)\phi_2\phi_2]\frac{1}{\sqrt{2}}(\alpha\beta - \beta\alpha) \tag{5.2.54}$$

as discussed in Sections 5.2.5 and 5.2.6. For the open-shell states in Section 5.2.7, the triplet state combines an antisymmetric orbital part with a symmetric spin part:

$$|{}^3\Sigma_u^+| = \frac{1}{\sqrt{2}}(\phi_1\phi_2 - \phi_2\phi_1)\begin{cases}\alpha\alpha \\ \frac{1}{\sqrt{2}}(\alpha\beta + \beta\alpha) \\ \beta\beta\end{cases} \tag{5.2.55}$$

Thus, whereas the superposition of orbital products in (5.2.54) is a consequence of electron repulsion, the superposition of orbital products in (5.2.55) arises from the antisymmetry requirement. In the open-shell singlet state, the situation is reversed and the orbital part is symmetric

$$|{}^1\Sigma_u^+| = \frac{1}{\sqrt{2}}(\phi_1\phi_2 + \phi_2\phi_1)\frac{1}{\sqrt{2}}(\alpha\beta - \beta\alpha) \tag{5.2.56}$$

leading to an ionically correlated structure. In passing, we note that the nodal structures of the two-electron densities in Figure 5.7 are easily understood from the form of the spatial parts in (5.2.55) and (5.2.56). Thus, whereas the triplet wave function vanishes whenever the coordinates of the electrons coincide

$$\phi_1(z)\phi_2(z) - \phi_2(z)\phi_1(z) = 0 \tag{5.2.57}$$

the singlet wave function vanishes when the electrons are located exactly opposite to each other

$$\phi_1(z)\phi_2(-z) - \phi_2(z)\phi_1(-z) = 0 \tag{5.2.58}$$

recalling that ϕ_1 is gerade and ϕ_2 ungerade.

5.2.9 THE DISSOCIATION LIMIT

Let us finally consider the hydrogen molecule in the dissociation limit. There are two possible outcomes of the dissociation – in the limit of infinite separation and no interactions, the electrons may end up in the same atom (*ionic dissociation*) or they may end up in different atoms (*covalent dissociation*). In the ionic dissociation, the wave function is a closed-shell singlet state; in the covalent case, the electrons may be coupled to yield either a singlet or a triplet.

We now investigate how our model wave functions perform for the dissociated system. In the limit of no interactions, the orbitals and thus the bonding and antibonding configurations (5.2.9) and (5.2.10) become degenerate; see Table 5.3. Because of this degeneracy, we expect the optimal wave function to be a linear combination of these two configurations, with the same weight on each configuration. Indeed, an optimization of the two-configuration wave function (5.2.13) yields two independent solutions, corresponding to the covalent ($\tau = -\pi/4$) and ionic ($\tau = \pi/4$) states:

$$|\text{cov}\rangle_\infty = \frac{1}{\sqrt{2}}|1\sigma_g^2\rangle_\infty - \frac{1}{\sqrt{2}}|1\sigma_u^2\rangle_\infty \tag{5.2.59}$$

$$|\text{ion}\rangle_\infty = \frac{1}{\sqrt{2}}|1\sigma_g^2\rangle_\infty + \frac{1}{\sqrt{2}}|1\sigma_u^2\rangle_\infty \tag{5.2.60}$$

Table 5.3 The electronic energies of the hydrogen molecule at an infinite internuclear separation in a minimal basis of hydrogenic $1s$ orbitals with unit exponents (E_h)

State	Kinetic	Attraction	Electron repulsion[a]	Nuc. rep.	Total
$^1\Sigma_g^+ 1\sigma_g^2$	1	-2	$5/16 + 0 = 5/16$	0	$-11/16$
$^1\Sigma_g^+ 1\sigma_u^2$	1	-2	$5/16 + 0 = 5/16$	0	$-11/16$
$^1\Sigma_g^+ (\tau_0)$[b]	1	-2	$5/16 - 5/16 = 0$	0	-1
$^1\Sigma_g^+ (\tau_1)$[c]	1	-2	$5/16 + 5/16 = 5/8$	0	$-3/8$
$^1\Sigma_g^+ (\tau_{cov})$	1	-2	$5/16 - 5/16 = 0$	0	-1
$^1\Sigma_g^+ (\tau_{ion})$	1	-2	$5/16 + 5/16 = 5/8$	0	$-3/8$
$^3\Sigma_u^+$	1	-2	$5/16 - 5/16 = 0$	0	-1
$^1\Sigma_u^+$	1	-2	$5/16 + 5/16 = 5/8$	0	$-3/8$

[a]The nuclear-repulsion energy has been broken down in two parts: the classical Coulomb contribution and the exchange and correlation contributions.
[b]The ground state calculated from $\tau_0 = -\pi/4$.
[c]The excited state calculated from $\tau_1 = \pi/4$.
[d]The covalent state calculated from $\tau_{cov} = -\pi/4$.
[e]The ionic state calculated from $\tau_{ion} = \pi/4$.

Whereas the covalent state is lower in energy and represents two noninteracting hydrogen atoms in the ground state, the ionic state represents the hydrogen anion. The energies of these states are given by

$$E_{cov}^\infty = h_{AA} + h_{BB} \tag{5.2.61}$$

$$E_{ion}^\infty = \tfrac{1}{2}(2h_{AA} + g_{AAAA}) + \tfrac{1}{2}(2h_{BB} + g_{BBBB}) \tag{5.2.62}$$

in the basis of the AOs (5.2.1) and (5.2.2). To obtain these expressions, we have used the identities

$$h_{11} = h_{22} = \tfrac{1}{2}(h_{AA} + h_{BB}) \tag{5.2.63}$$

$$g_{1111} = g_{2222} = g_{2211} = g_{2121} = \tfrac{1}{4}(g_{AAAA} + g_{BBBB}) \tag{5.2.64}$$

which hold only at infinite separation. Hence, whereas the energy of the covalent state (5.2.61) is just the sum of the energies of the two separate hydrogen atoms in the ^2S ground state

$$E_{cov}^\infty = 2E(H) \tag{5.2.65}$$

the energy of the ionic state (5.2.62) represents the average of the energies of two uncorrelated electrons in the same atom – that is, the energy of the hydrogen anion in the ^1S ground state

$$E_{ion}^\infty = E(H^-) \tag{5.2.66}$$

Let us also consider the energies of the bonding and antibonding configurations at infinite separation. From (5.2.59) and (5.2.60), we obtain

$$|1\sigma_g^2\rangle_\infty = \frac{1}{\sqrt{2}}(|ion\rangle_\infty + |cov\rangle_\infty) \tag{5.2.67}$$

$$|1\sigma_u^2\rangle_\infty = \frac{1}{\sqrt{2}}(|ion\rangle_\infty - |cov\rangle_\infty) \tag{5.2.68}$$

which give the following Hamiltonian expectation values for the degenerate bonding and anti-bonding configurations at infinite separation

$$E(g^2)_\infty = E(u^2)_\infty = E(\mathrm{H}) + \tfrac{1}{2}E(\mathrm{H}^-) \qquad (5.2.69)$$

as may be confirmed by inspection of Table 5.3.

It is noteworthy that, in order to describe the $^1\Sigma_g^+$ states of the hydrogen molecule at infinite separation, we are forced to use a two-configuration wave function even though the final states are completely uncorrelated; see (5.2.59) and (5.2.60). This happens since we insist on expressing the wave function for two noninteracting systems in terms of a set of symmetrized and therefore delocalized MOs (5.2.3) and (5.2.4). Clearly, in this limit, it would be better to expand our wave function in terms of the localized (and in this limit orthonormal) orbitals (5.2.1) and (5.2.2), which would yield single-determinant wave functions directly. In the intermediate region, on the other hand, the situation is less clear. To maintain spin eigenfunctions throughout the dissociation, we must employ a two-configuration wave function based on the use of the CSFs of Section 5.2.2. Alternatively, we may employ determinants that are not spin eigenfunctions, instead representing the wave function in terms of orbitals that, during the dissociation, transform smoothly from delocalized MOs to localized AOs.

In Figure 5.8, we have plotted the potential-energy curves for the H_2 electronic states discussed in this section. The curves are bracketed by those of the optimized two-configuration closed-shell states and the degeneracies in the dissociation limit are clearly illustrated. The optimized ground state has an equilibrium bond length of $1.67a_0$, slightly longer than the $|1\sigma_g^2\rangle$ bond

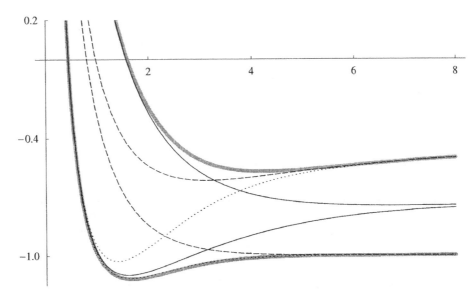

Fig. 5.8. Potential-energy curves for H_2 calculated in a minimal basis of hydrogenic $1s$ functions with unit exponents (atomic units). The closed-shell two-configuration $^1\Sigma_g^+$ ground and excited states are represented by thick grey lines and the dashed lines represent the open-shell $^3\Sigma_u^+$ (lower curve) and $^1\Sigma_u^+$ (upper curve) states. Also depicted are the energy curves of the single-configuration bonding and antibonding states $|1\sigma_g^2\rangle$ and $|\sigma_u^2\rangle$ (thin full lines) as well as of the covalent and ionic states $|\mathrm{cov}\rangle$ and $|\mathrm{ion}\rangle$ (dotted lines).

length of $1.60a_0$. These results should be compared with the experimental ground-state bond length of $1.40a_0$ in Table 15.3.

The triplet open-shell state is dissociative but the singlet is stable with an equilibrium bond distance of $3.11a_0$ – considerably longer than the experimental distance of $2.44a_0$ for this state [1]. Whereas the localized covalent state is close to the ground state at all distances, the ionic state is degenerate with the ground state at short distances and with the excited state in the dissociation limit. We finally note that the open-shell triplet CSF correlates with the triplet covalent state in the dissociation limit, which is degenerate with the closed-shell singlet covalent state. Likewise, the open-shell singlet state correlates with the singlet ionic state, which is degenerate with the closed-shell ionic state.

5.2.10 STATIC AND DYNAMICAL CORRELATION

In the preceding discussion of the electronic structure of the hydrogen molecule, we have seen how the superposition of determinants may serve several purposes. Let us now summarize our reasons for going beyond the single-configuration representation of the electronic wave function.

For the stable hydrogen molecule at the equilibrium geometry, the need to go beyond the single-determinant representation of the electronic wave function arises from a desire to describe the *detailed correlated motion of the electrons* as induced by their instantaneous mutual repulsion. In the molecular dissociation limit, by contrast, the need to go beyond the single-determinant representation arises for a completely different reason, related to the *degeneracy of the bonding and antibonding configurations* (which means that the configurations will interact strongly and cannot be treated in isolation from each other) and unrelated to the repulsion between the electrons. In the intermediate region, therefore, the superposition of determinants serves the double purpose of accounting for the effects of Coulomb repulsion and the near-degeneracy of the configurations. Of course, no clear distinction can be drawn between these two effects (except in the limit of a complete dissociation where the repulsion vanishes). In quantum chemistry, therefore, both effects are attributed to electron correlation since they require a description beyond that provided by a single configuration. Nevertheless, a distinction is often drawn between *dynamical correlation* (arising from the Coulomb repulsion) and *near-degeneracy or static correlation* (arising from near-degeneracies among configurations). Static correlation is sometimes also referred to as *nondynamical correlation*.

5.3 Exact wave functions in Fock space

In the present section, we consider wave functions that, for a given one-electron basis, represent exact solutions to the Schrödinger equation in Fock space. In the subsequent Sections 5.4–5.8, these solutions to the Schrödinger equation will serve as benchmarks for less accurate but more practical models of quantum chemistry.

5.3.1 FULL CONFIGURATION-INTERACTION WAVE FUNCTIONS

For a given one-electron expansion, the exact solution to the Schrödinger equation may be written as a linear combination of all determinants that can be constructed from this one-electron basis in the N-electron Fock space:

$$|\text{FCI}\rangle = \sum_i C_i |i\rangle \tag{5.3.1}$$

The expansion coefficients may be obtained from the variation principle as described in Section 4.2.3 and the solution is called the *full configuration-interaction (FCI)* wave function. The number of determinants in an FCI wave function is given by the expression

$$N_{\text{det}} = \binom{M}{N} = \frac{M!}{N!(M-N)!} \tag{5.3.2}$$

where N is the number of electrons and M the number of spin orbitals. For a tabulation of the number of determinants in the special case of $N = 2M$ and zero spin projection, see Table 11.1 in Chapter 11.

The factorial dependence of the number of Slater determinants on the number of spin orbitals and electrons (5.3.2) makes FCI wave functions intractable for all but the smallest systems. However, in those cases where FCI calculations can be carried out, the results are often very useful since their solutions are exact within the chosen one-electron basis. In comparison with FCI, the errors introduced in the treatment of the N-electron expansion by less accurate wave functions can be identified and examined in isolation from the errors in the one-electron space. In this way, the FCI calculations may serve as useful *benchmarks* for the N-electron treatments of the standard models, exhibiting their strengths and deficiencies in a transparent manner. Still, for the FCI benchmark calculations to be worthwhile, the orbital basis must be sufficiently flexible to provide a reasonably faithful representation of the electronic system – otherwise our conclusions will be based on a distorted representation of the true system, with little relevance for accurate calculations carried out in larger basis sets.

In this section, we shall present two examples of FCI wave functions that, in the subsequent sections, will be used in our discussion of the standard computational models: the hydrogen molecule in the large cc-pVQZ basis and the water molecule in the small cc-pVDZ basis. For a description of these basis sets, see Section 8.3.3.

5.3.2   THE ELECTRONIC GROUND STATE OF THE HYDROGEN MOLECULE

For two-electron systems, the number of Slater determinants in the FCI wave function (5.3.2) increases only as M^2, where M is the number of spin orbitals. For the hydrogen molecule, it is therefore possible to carry out FCI calculations in large basis sets. We here employ the cc-pVQZ basis, which contains a total of 60 AOs and is capable of providing an accurate description of the $X^1\Sigma_g^+$ ground state of the hydrogen molecule. In this basis, there are 552 Slater determinants consistent with $^1\Sigma_g^+$ symmetry in the FCI wave function.

The FCI potential-energy curve for the ground state of H_2 is plotted in Figure 5.9. In Table 5.4, we have – for selected bond distances – listed the electronic energy and the occupation numbers of the most important natural orbitals. For later reference, we have also given the weight (i.e. the squared expansion coefficient) of the Hartree–Fock determinant in the FCI wave function. At the cc-pVQZ level, the equilibrium bond distance is $1.402a_0$ (74.2 pm) and the electronic dissociation energy 173.9 mE$_h$ (456.6 kJ/mol), which may be compared with the experimental equilibrium distance of $1.401a_0$ in Table 15.3 and the experimental dissociation energy of 458.0 kJ/mol in Table 15.20.

Let us analyse the electronic state in terms of the natural-orbital occupation numbers in Table 5.4. At the equilibrium geometry, the occupation number of the $1\sigma_g$ orbital is 1.9643, indicating that a single Slater determinant with a doubly occupied $1\sigma_g$ orbital should provide a reasonably accurate representation of the exact wave function. With an occupation number of 0.0199, the $1\sigma_u$ orbital (which corresponds to the antibonding $1s$ orbital of the minimal basis) is the second most important

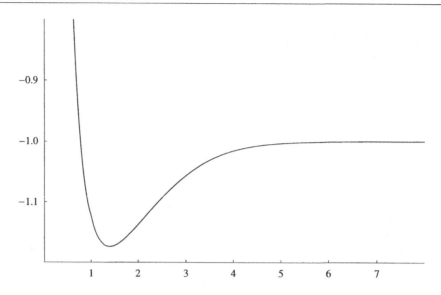

Fig. 5.9. The FCI/cc-pVQZ potential-energy curve of the hydrogen molecule in the $X^1\Sigma_g^+$ ground state (atomic units).

Table 5.4 The FCI/cc-pVQZ energy (E_{FCI}), the weight of the Hartree–Fock determinant (W_{HF}) and the largest natural-orbital occupation numbers of the FCI wave function for the hydrogen molecule at different internuclear separations. The occupations of the dominant natural orbitals have been shaded

	$R = 1.4a_0$	$R = 4.0a_0$	$R = 15.0a_0$
E_{FCI}	-1.173796 E_h	-1.015724 E_h	-0.999891 E_h
W_{HF}	0.9820	0.7445	0.4751
$\eta(1\sigma_g)$	1.9643	1.5162	1.0000
$\eta(2\sigma_g)$	0.0061	0.0015	0.0000
$\eta(3\sigma_g)$	0.0003	0.0000	0.0000
$\eta(1\sigma_u)$	0.0199	0.4804	1.0000
$\eta(2\sigma_u)$	0.0002	0.0000	0.0000
$\eta(1\pi_u)$	0.0043	0.0000	0.0000
$\eta(1\pi_g)$	0.0001	0.0000	0.0000

orbital. It has a nodal plane that coincides with the symmetry plane perpendicular to the molecular axis. As discussed in Section 5.2.5, its occupation increases the probability of locating the electrons around opposite nuclei, often referred to as *left–right correlation*.

From Section 5.2, we recall that it is impossible to tell from its occupation number alone whether the $1\sigma_u$ orbital decreases or increases the probability of having the two electrons close to the same nucleus. Indeed, this information can be obtained only from an inspection of the two-electron density matrix or the wave function itself. The one-electron occupation numbers are nevertheless useful for analysing the electronic state, providing information about the importance of the individual orbitals for the one-electron density.

Although the $1\sigma_u$ orbital, with an occupation number of 0.0199, is by far the most important orbital for describing correlation in the hydrogen molecule, other orbitals – in particular, the two $1\pi_u$ orbitals and the $2\sigma_g$ orbital – also make significant contributions. Thus, the $1\pi_u$ orbitals both have an occupation number of 0.0043 and possess nodal planes that introduce *angular correlation*, increasing the probability of finding the electrons on opposite sides of the molecular axis. The $2\sigma_g$ orbital belongs to the same irreducible representation as the $1\sigma_g$ orbital but possesses a radial node. With an occupation number of 0.0061, this orbital introduces *radial correlation*, increasing the probability of locating the electrons at different distances from the molecular axis. The remaining orbitals have occupations equal to or less than 10^{-4} and are less important for the description of the ground-state electronic system, although their accumulative effect is such that they cannot be ignored for an accurate representation of the electronic structure.

As the atoms separate, the electronic ground state of the hydrogen molecule correlates with the ground-state wave functions of two noninteracting hydrogen atoms coupled to a singlet spin eigenfunction. The importance of *dynamical* correlation decreases and the importance of *static* correlation increases as the interaction between the electrons vanishes and the orbitals $1\sigma_g$ and $1\sigma_u$ become degenerate. At infinite separation, the occupations of $1\sigma_g$ and $1\sigma_u$ are equal to 1 and all others are 0. In this limit, the *exact* wave function is a superposition of two degenerate configurations – the bonding $1\sigma_g^2$ determinant and the antibonding $1\sigma_u^2$ determinant. The purpose of this superposition is not to describe the dynamical correlation that arises from the electronic interaction. Rather, it is mandated by the degeneracy of the bonding and antibonding configurations and is required for a proper description of the two neutral atoms, as discussed in Section 5.2.9.

5.3.3   THE ELECTRONIC GROUND STATE OF THE WATER MOLECULE

Let us review the electronic structure of the water molecule in the simple minimal-basis MO picture. We place the molecule in the yz plane such that the z axis coincides with the molecular axis. The occupied oxygen orbitals $1s$, $2s$, $2p_x$, $2p_y$ and $2p_z$ transform as the irreducible representations A_1, A_1, B_1, B_2 and A_1 of the C_{2v} point group, respectively. In the same point group, the symmetric and antisymmetric hydrogen $1s$ orbitals transform as A_1 and B_2.

The molecular-orbital energy-level diagram for H_2O is reproduced in Figure 5.10. The $1a_1$ orbital is a pure $1s$ oxygen orbital and does not participate in the bonding of the water molecule. In contrast, the valence A_1 orbitals $2a_1$–$4a_1$ and the valence B_2 orbitals $1b_2$ and $2b_2$ are mixed oxygen–hydrogen orbitals that participate in the bonding. The $1b_1$ orbital – which corresponds to the $2p_x$ orbital on oxygen – represents the electronic lone pair in the water molecule.

The FCI $^1A_1$ ground-state wave function of water is determined in a basis of 24 Gaussian orbitals, corresponding to the cc-pVDZ basis. Distributing 10 electrons among 24 orbitals in all possible ways consistent with A_1 spatial symmetry and zero spin projection, we obtain a total of 451 681 246 Slater determinants for the expansion of the wave function. That such a large number of determinants may arise for such a small basis and such a small molecule demonstrates in a spectacular fashion the intractability of FCI wave functions for large systems.

The cc-pVDZ basis contains two AOs for each occupied orbital in the valence region. In addition, there is a set of *polarization functions* – that is, AOs of angular momentum higher than that of the occupied orbitals in the atoms. As demonstrated by the systematic investigations of basis sets and wave functions in Chapter 15, the cc-pVDZ basis provides only a crude description of electron correlation – it is inadequate for high-precision work but sufficient for semi-quantitative investigations. In particular, since it provides a uniformly accurate representation of the whole energy surface of the water molecule, the cc-pVDZ basis is sufficient for our purposes – that is,

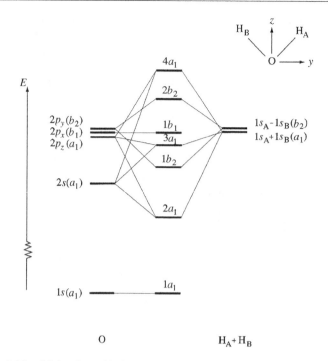

Fig. 5.10. Molecular-orbital energy-level diagram for the water molecule.

to illustrate and compare the behaviour of the different standard models in relation to that of the FCI model.

The experimental equilibrium structure of water is 95.7 pm for the OH bond distance (Table 15.3) and 104.5° for the HOH bond angle (Table 15.8). In our calculations, we have used a slightly different reference geometry with a bond length of $R_{ref} = 1.84345a_0$ (97.6 pm) and a bond angle of 110.565°, corresponding to the equilibrium structure of a CISD wave function in a small double-zeta basis (see Section 5.6 for a discussion of truncated CI wave functions). In Figure 5.11, we have – for the FCI/cc-pVDZ wave function – plotted the C_{2v} potential-energy surface as a function of the OH bond length for a fixed angle of 110.565°. In Table 5.5, the electronic energy and natural-orbital occupation numbers are listed for selected bond distances.

In the equilibrium region, the occupation numbers in Table 5.5 are close to 2 for all orbitals in the configuration $1a_1^2 2a_1^2 1b_2^2 3a_1^2 1b_1^2$, indicating that the wave function is dominated by this particular configuration. The occupations of the remaining 19 orbitals arise from dynamical correlation and are at least two orders of magnitude smaller. The most important of these correlating orbitals are the antibonding valence orbitals $4a_1$ and $2b_2$ (see Figure 5.10), which introduce left–right correlation in the OH bonds.

When we stretch the OH bonds, the $1a_1$, $2a_1$ and $1b_1$ orbitals remain doubly occupied as they transform themselves into the $1s$, $2s$ and $2p_x$ oxygen orbitals. In contrast, the bonding $3a_1$ and $1b_2$ orbitals lose occupations to the antibonding $4a_1$ and $2b_2$ orbitals. (Recall that $3a_1$ and $4a_1$ are linear combinations of $2p_z$ with the symmetric $1s$ hydrogen orbital, and that $1b_2$ and $2b_2$ are combinations of $2p_y$ with the antisymmetric $1s$ hydrogen orbital.) At $2R_{ref}$, the partial occupations of these four orbitals indicate that the dissociation has begun. Moreover, the configuration that dominates at equilibrium has now a weight of only 59%; the remaining part of the wave function

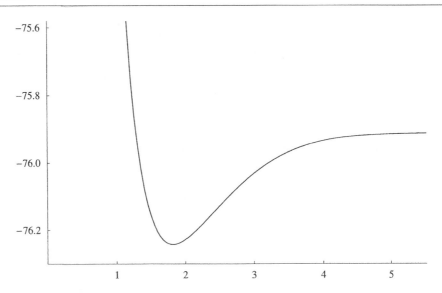

Fig. 5.11. The FCI/cc-pVDZ potential-energy curve of the C_{2v} water molecule in the X^1A_1 ground state for a fixed HOH bond angle of $110.565°$ (atomic units).

contains nine determinants with a total weight of 34% and a very large number of determinants with small weights (less than 1%).

At the internuclear distance of $8R_{\text{ref}}$, the occupation numbers in Table 5.5 indicate that dissociation is complete and that the OH bonds have been broken. Thus, the ^2S ground-state hydrogen atoms are represented by the singly occupied $3a_1$ and $1b_2$ orbitals, which now correspond to the symmetric and antisymmetric combinations of the $1s$ hydrogen orbitals – just like the $1\sigma_g$ and $1\sigma_u$ orbitals of the dissociated hydrogen molecule. Furthermore, the ^3P ground-state oxygen atom is represented by three doubly occupied orbitals ($1s$, $2s$ and $2p_x$) and two orbitals of single occupation ($2p_y$ and $2p_z$). It is noteworthy that the configuration that dominates the bonded molecule makes no contribution in the dissociation limit. For a discussion of this particular point, see Section 10.11.

5.4 The Hartree–Fock approximation

We begin our survey of the standard models of quantum chemistry with a discussion of the Hartree–Fock model – the simplest wave-function model in *ab initio* electronic-structure theory. It serves not only as a useful approximation in its own right, but also constitutes a convenient starting point for other, more accurate models of molecular electronic structure.

5.4.1 THE HARTREE–FOCK MODEL

In the *Hartree–Fock approximation*, the electronic wave function is approximated by a single configuration of spin orbitals (i.e. by a single Slater determinant or by a single space- and spin-adapted CSF) and the energy is optimized with respect to variations of these spin orbitals. Thus, the wave function may be written in the form

$$|\kappa\rangle = \exp(-\hat{\kappa})|0\rangle \tag{5.4.1}$$

Table 5.5 The FCI energy (E_{FCI}), the weight of the Hartree–Fock determinant in the FCI wave function (W_{HF}) and the FCI natural-orbital occupation numbers for the C_{2v} water molecule in the cc-pVDZ basis for various OH bond distances R (atomic units). The HOH bond angle is fixed at 110.565° and $R_{ref} = 1.84345a_0$. Orbitals that are doubly occupied at all geometries are shaded dark grey and orbitals with variable occupancies – which participate directly in the OH bond breaking – are shaded light grey

	$R = R_{ref}$	$R = 2R_{ref}$	$R = 2.5R_{ref}$	$R = 8R_{ref}$
E_{FCI}	-76.241860 E_h	-75.951665 E_h	-75.917991 E_h	-75.910300 E_h
W_{HF}	0.941	0.589	0.337	0.000
$\eta(1a_1)$	1.9999	1.9999	1.9999	1.9999
$\eta(2a_1)$	1.9837	1.9801	1.9806	1.9810
$\eta(3a_1)$	1.9638	1.5538	1.2270	1.0000
$\eta(4a_1)$	0.0278	0.4434	0.7680	0.9949
$\eta(5a_1)$	0.0123	0.0086	0.0084	0.0083
$\eta(6a_1)$	0.0060	0.0057	0.0059	0.0058
$\eta(7a_1)$	0.0045	0.0036	0.0037	0.0037
$\eta(8a_1)$	0.0013	0.0031	0.0032	0.0032
$\eta(9a_1)$	0.0007	0.0006	0.0001	0.0000
$\eta(10a_1)$	0.0005	0.0001	0.0000	0.0000
$\eta(11a_1)$	0.0000	0.0000	0.0000	0.0000
$\eta(1a_2)$	0.0054	0.0037	0.0036	0.0036
$\eta(2a_2)$	0.0008	0.0001	0.0000	0.0000
$\eta(1b_1)$	1.9714	1.9739	1.9758	1.9765
$\eta(2b_1)$	0.0179	0.0169	0.0158	0.0154
$\eta(3b_1)$	0.0044	0.0036	0.0036	0.0036
$\eta(4b_1)$	0.0006	0.0001	0.0000	0.0000
$\eta(1b_2)$	1.9609	1.6242	1.2995	1.0000
$\eta(2b_2)$	0.0290	0.3689	0.6957	0.9949
$\eta(3b_2)$	0.0062	0.0053	0.0058	0.0058
$\eta(4b_2)$	0.0013	0.0034	0.0032	0.0032
$\eta(5b_2)$	0.0008	0.0004	0.0001	0.0000
$\eta(6b_2)$	0.0006	0.0001	0.0000	0.0000
$\eta(7b_2)$	0.0000	0.0000	0.0000	0.0000

where $|0\rangle$ is some reference configuration and $\exp(-\hat{\kappa})$ is an operator that carries out unitary transformations among the spin orbitals. Following the discussion in Section 3.2, the anti-Hermitian operator $\hat{\kappa}$ may in its most general form be written as

$$\hat{\kappa} = \sum_{PQ} \kappa_{PQ} a_P^\dagger a_Q \tag{5.4.2}$$

where κ_{PQ} are the elements of an anti-Hermitian matrix κ and the summation is over all pairs of spin orbitals. The ground-state Hartree–Fock wave function is obtained by minimizing the energy with respect to the orbital-rotation parameters

$$E_{HF} = \min_{\kappa} \langle \kappa | \hat{H} | \kappa \rangle \tag{5.4.3}$$

Since the parameters κ occur nonlinearly in the energy expression, an iterative procedure must be invoked to determine the Hartree–Fock state. For a detailed account of Hartree–Fock theory, see Chapter 10.

5.4.2 THE FOCK OPERATOR AND THE CANONICAL REPRESENTATION

The optimization of the Hartree–Fock wave function may be carried out using any of the standard techniques of numerical analysis. However, for many purposes, it is better to use an alternative scheme, which more directly reflects the physical contents of the Hartree–Fock state. Thus, from the discussion in Section 5.1, we recall that an antisymmetric product of spin orbitals represents a state where each electron behaves as an independent particle (but subject to Fermi correlation as discussed in Section 5.2.8). This observation suggests that the optimal determinant – that is, the Hartree–Fock determinant in (5.4.3) – may be found by solving a set of effective one-electron Schrödinger equations for the spin orbitals. Such an approach is indeed possible: the effective one-electron Schrödinger equations are called the *Hartree–Fock equations* and the associated Hamiltonian is the *Fock operator*

$$\hat{f} = \sum_{PQ} f_{PQ} a_P^\dagger a_Q \tag{5.4.4}$$

where the elements f_{PQ} constitute the *Fock matrix*. In the Fock operator, the one-electron part of the true Hamiltonian (5.1.2) is retained but the two-electron part is replaced by an effective one-electron *Fock potential* $\hat{V}$

$$\hat{f} = \hat{h} + \hat{V} \tag{5.4.5}$$

which in an average sense incorporates the effects of the Coulomb repulsion among the electrons (corrected for Fermi correlation)

$$\hat{V} = \sum_{PQ} V_{PQ} a_P^\dagger a_Q \tag{5.4.6}$$

$$V_{PQ} = \sum_I (g_{PQII} - g_{PIIQ}) \tag{5.4.7}$$

In these expressions, the index I runs over all occupied spin orbitals while P and Q run over the full set of spin orbitals – unoccupied as well as occupied. The Fock potential depends on the form of the occupied spin orbitals. Whereas the first term in (5.4.7) represents the classical Coulomb interaction of the electron with the charge distribution of the occupied spin orbitals, the second *exchange* term is a correction that arises from the antisymmetry of the wave function.

The Fock equations are solved by diagonalizing the Fock matrix. The resulting eigenvectors are called the *canonical spin orbitals* of the system and the *orbital energies* are the eigenvalues of the Fock matrix:

$$f_{PQ} = \delta_{PQ} \varepsilon_P \tag{5.4.8}$$

However, since the Fock matrix is defined in terms of its own eigenvectors, the canonical spin orbitals and the orbital energies can only be obtained by means of an iterative procedure, where the Fock matrix is repeatedly reconstructed and rediagonalized until the spin orbitals generated by its diagonalization become identical to those from which the Fock matrix has been constructed. This iterative procedure is known as the *self-consistent field (SCF) method* and the resulting wave function – which satisfies (5.4.3) – is called the *SCF wave function*.

In the *canonical representation* – that is, in the diagonal representation of the Fock matrix – the electrons occupy the spin orbitals in accordance with the Pauli principle, moving independently of one another in the electrostatic field set up by the stationary nuclei and by the charge distributions of all the remaining electrons, appropriately modified to take into account the effects of Pauli antisymmetry. The orbital energies are the one-electron energies associated with the independent electrons and may consequently be interpreted as the energies required to ionize the system and thus to remove a single electron.

It turns out that the Hartree–Fock state is invariant to unitary transformations among the occupied spin orbitals. The spin orbitals of the Hartree–Fock state are therefore not uniquely determined by the stationary condition (5.4.3) and the canonical spin orbitals are therefore just one possible choice of spin orbitals for the optimized N-particle state. Indeed, any set of energy-optimized spin orbitals decomposes the Fock matrix into two noninteracting blocks – one for the occupied spin orbitals and another for the unoccupied spin orbitals. When these subblocks are diagonalized, the canonical spin orbitals are obtained.

The Hartree–Fock method can be applied to truly large molecules containing several hundred atoms. For such systems, it becomes impossible to construct a set of orthonormal orbitals (5.1.5), much less a set of canonical orbitals. However, as we shall see in Chapter 10, all information about the Hartree–Fock wave function is contained in the one-electron density matrix, which may be expressed directly in the basis of AOs. For large molecules, the density-matrix elements can be optimized by an algorithm whose complexity scales linearly with the size of the system.

5.4.3 RESTRICTED AND UNRESTRICTED HARTREE–FOCK THEORY

In Fock space, the exact solution to the Schrödinger equation is an eigenfunction of the total and projected spins. Often, we would like the approximate solutions to exhibit the same spin symmetries. However, according to the discussion in Section 4.4, it does not follow that the Hartree–Fock wave function – which is not an eigenfunction of the exact Hamiltonian – possesses the same symmetries as the exact solution. In general, therefore, these and other symmetries of the exact state must be *imposed* on the Hartree–Fock solution.

In the *restricted Hartree–Fock (RHF)* approximation, the energy is optimized subject to the condition that the wave function is an eigenfunction of the total and projected spins. Spin adaptation is accomplished by requiring the alpha and beta spin orbitals to have the same spatial parts and by writing the wave function as a CSF rather than as a Slater determinant. In addition, the RHF wave function is required to transform as an irreducible representation of the molecular point group, which is accomplished by requiring the spatial orbitals to transform according to the irreducible representations of the appropriate point group.

In the *unrestricted Hartree–Fock (UHF)* approximation, on the other hand, the wave function is not required to be a spin eigenfunction and different spatial orbitals are used for different spins. Occasionally, the spatial symmetry restrictions are also lifted, and the MOs are then no longer required to transform as irreducible representations of the molecular point group. It should be noted that, in particular for systems close to the equilibrium geometry, the symmetries of the exact state are sometimes present in the UHF state as well even though they have not been imposed during its optimization. In such cases, the UHF and RHF states will coincide.

5.4.4 THE CORRELATION ENERGY

In Section 5.2.8, we introduced the term *electron correlation* for the correlating effects associated with wave functions that go beyond the single-configuration approximation. For a given electronic state, we now define the *correlation energy* as the difference between the exact nonrelativistic energy (i.e. the FCI energy) and the Hartree–Fock energy of the electronic system (both calculated in a complete basis) [2]

$$E_{corr} = E_{exact} - E_{HF} \tag{5.4.9}$$

The correlation energy is usually defined relative to the energy of the RHF (rather than the UHF) wave function. Note that this definition comprises both the dynamical and the nondynamical correlation discussed in Section 5.2.10 but not the Fermi correlation.

The correlation energy in (5.4.9) is defined in terms of a complete one-electron basis. In practice, however, an incomplete basis must be used for the calculation of the correlation energy. The term *correlation energy* is then used more loosely to denote the energy obtained from (5.4.9) in a given one-electron basis. As we shall see for example in Section 8.4.2, the correlation energy usually increases in magnitude with the size of the orbital basis, since a small basis does not have the flexibility required for an accurate representation of correlation effects.

5.4.5   THE GROUND STATE OF THE HYDROGEN MOLECULE

To investigate the usefulness of the Hartree–Fock approximation and to gain some insight into the performance of this important class of wave functions, we shall carry out calculations on the hydrogen and water molecules, using as benchmarks the FCI calculations of Section 5.3. Since FCI calculations are exact in the chosen one-electron basis, such comparisons provide us with an opportunity to examine the Hartree–Fock method in a particularly transparent fashion. On the other hand, since the number of systems for which meaningful FCI calculations can be carried out is limited, a full appreciation of the Hartree–Fock method can be acquired only from experience with a large body of calculations, where comparisons are made with experimental measurements. Such comparisons are presented in Chapter 15. Here, we shall restrict ourselves to comparisons with FCI solutions, examining the Hartree–Fock method as applied first to the hydrogen molecule in the large cc-pVQZ basis and next to the water molecule in the small cc-pVDZ basis.

For the hydrogen molecule, the RHF wave function is a singlet spin eigenfunction with doubly occupied symmetry-adapted $1\sigma_g$ orbitals; the UHF wave function, on the other hand, is not required to be a spin eigenfunction and there are no symmetry restrictions on the orbitals:

$$|RHF\rangle = a^{\dagger}_{1\sigma_g\alpha}a^{\dagger}_{1\sigma_g\beta}|vac\rangle \tag{5.4.10}$$

$$|UHF\rangle = a^{\dagger}_{\phi_1\alpha}a^{\dagger}_{\phi_2\beta}|vac\rangle \tag{5.4.11}$$

In general, the spatial UHF orbitals ϕ_1 and ϕ_2 are different but not orthogonal. From our discussion in Section 5.2, we expect the RHF wave function to perform well for short internuclear separations but poorly for large separations. By contrast, the UHF wave function, whose energy is equal to or lower than the RHF energy at all separations, should dissociate correctly by transforming its orbitals smoothly into the localized $1s_A$ and $1s_B$ AOs. On the other hand, whereas the RHF wave function remains a true singlet at all distances, we have no such expectations for the UHF wave function.

Figure 5.12 confirms these expectations. At short internuclear distances, the RHF and UHF wave functions are identical, providing a reasonable but crude representation of the potential-energy surface. At larger separations, the RHF and UHF wave functions behave differently. While the RHF wave function behaves poorly with respect to the energy, the UHF wave function behaves incorrectly with respect to the spin (yielding a spin intermediate between that of a singlet and a triplet). Thus, outside the bonded region, neither wave function provides a satisfactory representation of the exact wave function.

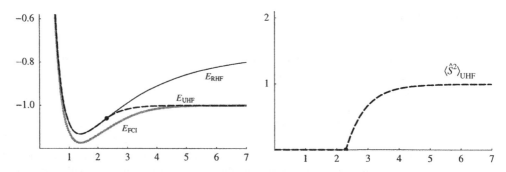

Fig. 5.12. The Hartree–Fock dissociation of the hydrogen molecule (atomic units). On the left, we have plotted the cc-pVQZ dissociation curves for the RHF wave function (solid line), for the UHF wave function (dashed line) and for the FCI wave function (thick grey line). On the right, we have plotted the expectation value of the total spin squared for the UHF wave function.

5.4.6 THE BONDED HYDROGEN MOLECULE

For bond distances shorter than $2.30a_0$, the optimized RHF and UHF wave functions are identical and there is no conflict between the spin-symmetry requirements and energy minimization. In addition, the Hartree–Fock wave function dominates the exact wave function, contributing a weight of 0.98 to the FCI wave function at the equilibrium geometry; see Table 5.4.

The Hartree–Fock equilibrium bond distance of the hydrogen molecule is $1.387a_0$, just 1.1% shorter than the FCI value of $1.402a_0$. In Section 15.3, we shall see that this excellent behaviour with respect to bond distances is often observed for the Hartree–Fock approximation and that Hartree–Fock bond lengths are typically 2–3 pm too short.

At the internuclear separation of $1.40a_0$, the Hartree–Fock energy is $-1.1335\,\mathrm{E_h}$, which should be compared with the FCI energy of $-1.1738\,\mathrm{E_h}$ (see Table 5.6, where the electronic energies are tabulated for several internuclear separations). In the cc-pVQZ basis, the correlation energy is thus $-40.3\,\mathrm{mE_h}$, representing as much as 3.4% of the total energy. Usually, the correlation energy constitutes a much smaller proportion of the total energy, typically 0.5% for molecules containing first-row atoms – see, for example, Table 15.14.

5.4.7 THE RHF DISSOCIATION OF THE HYDROGEN MOLECULE

Having examined the Hartree–Fock wave function at small internuclear separations, we now consider the dissociation process. At separations greater than $2.30a_0$, the UHF and RHF wave functions no longer coincide. Also, the RHF determinant no longer dominates the FCI wave function – its weight is 0.74 at a separation of $4.0a_0$ and 1/2 in the dissociation limit; see Section 5.3.2. Clearly, the quality of the RHF description deteriorates as the nuclei separate.

Let us consider how we might go about calculating the dissociation energy for the hydrogen molecule in the RHF approximation. There are two possible approaches: we may either calculate the energy of the dissociated system using the same RHF wave function as for the bonded system (5.4.10) or obtain the energy of the dissociated system from separate RHF calculations on each atom. The first *supermolecule approach* – which is the only possible one if we wish to describe the system uniformly at all distances – yields according to Table 5.6 a dissociation energy of $414.6\,\mathrm{mE_h}$, more than twice that of the FCI dissociation energy of $173.9\,\mathrm{mE_h}$.

Table 5.6 Total energies of the hydrogen molecule at the FCI, RHF, UHF, two-configuration MCSCF and MP2 levels in the cc-pVQZ basis at various internuclear separations (in E_h)

	$R = 1.4a_0$	$R = 4.0a_0$	$R = 15.0a_0$	$R = 100.0a_0$
E_{FCI}	-1.173796	-1.015724	-0.999891	-0.999891
E_{RHF}	-1.133459	-0.911164	-0.747191	-0.718827
E_{UHF}	-1.133459	-1.002786	-0.999891	-0.999891
E_{MCSCF}	-1.151978	-1.013026	-0.999891	-0.999891
E_{MP2}	-1.166573	-0.968012	-1.169828	-4.034127

In the second *fragment approach,* we first calculate the energy of the isolated hydrogen atom in the cc-pVQZ basis and obtain -0.4999 E_h, close to the exact value of $-\frac{1}{2}$ E_h. From this energy and from the energy of the bonded system, we obtain a dissociation energy of 133.6 mE_h, which represents 77% of the FCI dissociation energy – clearly an improvement on the supermolecule calculation. Obviously, the fragment approach cannot be extended to the calculation of energies at intermediate separations. In short, there is no satisfactory way to calculate the dissociation energy from RHF calculations if the RHF description breaks down during the dissociation.

In Section 5.2.9, we found that the dissociated RHF wave function $|1\sigma_g^2\rangle$ contains equal contributions from covalent and ionic structures. We should therefore be able to reproduce the RHF energy of the dissociated system by using (5.2.69)

$$E_\infty = E(\mathrm{H}) + \tfrac{1}{2}E(\mathrm{H}^-) \tag{5.4.12}$$

where $E(\mathrm{H}^-)$ is the energy of the anionic atom in the ^1S ground state and $E(\mathrm{H})$ that of the neutral atom in the ^2S ground state. In the cc-pVQZ basis, RHF calculations on the neutral and anionic atoms give -0.4999 and -0.4735 E_h, respectively. Substituting these energies into (5.4.12), we obtain for the dissociated system -0.7367 E_h, which is in fact somewhat lower than -0.7188 E_h obtained from a supermolecule calculation at a separation of $100a_0$; see Table 5.6. This difference arises since the $1\sigma_g$ RHF orbital of the supermolecule is not optimal for the fragments. In contrast, in the minimal basis of Section 5.2, (5.4.12) holds exactly since the orbitals are then identical for the molecular and atomic systems; see Table 5.3.

5.4.8 THE UHF DISSOCIATION OF THE HYDROGEN MOLECULE

Let us now examine the UHF description of the dissociation process. As already noted, the UHF wave function provides a crude but qualitatively correct representation of the potential-energy curve of the hydrogen molecule; see Figure 5.12. The UHF curve coincides exactly with the RHF curve for bond distances shorter than $2.30a_0$ and to within 1 mE_h with the FCI curve for distances longer than $5.5a_0$. In the intermediate region – where the electrons are being separated and the bond is being broken – the UHF curve falls below the RHF curve and flattens out towards the FCI curve as the total spin increases rapidly from 0 to 1 at a separation of about $3a_0$.

For the UHF wave function of the dissociated system, the expectation value of the total spin is equal to 1, as is easily verified by writing the dissociated UHF wave function

$$|\mathrm{UHF}\rangle_\infty = a^\dagger_{1s_A\alpha} a^\dagger_{1s_B\beta} |\mathrm{vac}\rangle \tag{5.4.13}$$

in terms of the singlet and triplet creation operators (2.3.16) and (2.3.18) of Section 2.3.3:

$$|\text{UHF}\rangle_\infty = \frac{1}{\sqrt{2}}(Q_{1s_A\alpha 1s_B\beta}^{0,0} + Q_{1s_A\alpha 1s_B\beta}^{1,0})|\text{vac}\rangle \tag{5.4.14}$$

The dissociated UHF wave function is thus an even superposition of singlet and triplet states. Using the relationship (2.3.6), we obtain for the total spin squared

$$\langle \text{UHF}|\hat{S}^2|\text{UHF}\rangle_\infty = 1 \tag{5.4.15}$$

We may also obtain this value from the general expression (2.4.38), noting that, for the UHF wave function, the last term in (2.4.38) does not contribute to the expectation value.

Since the UHF wave function behaves reasonably well at all internuclear distances and in particular in the bonded and dissociated limits, we should obtain fairly accurate dissociation energies from this wave function. In fact, for this system, the UHF dissociation energy of 133.6 mE$_h$ is identical to that obtained from the RHF fragment calculations since the RHF and UHF wave functions are the same, for the fragments as well as for the parent molecule. This binding energy is lower than the FCI binding energy since the uncorrelated UHF wave function treats the fragments (one-electron systems) more accurately than the parent molecule (a two-electron system).

5.4.9 THE GROUND STATE OF THE WATER MOLECULE

In Figure 5.13, we have plotted the RHF and UHF dissociation curves for the C_{2v} water molecule. The behaviour of the Hartree–Fock wave functions is similar to that for the hydrogen molecule but perhaps somewhat poorer. Indeed, a comparison of Tables 5.4 and 5.5 reveals that, as the water atoms separate, the contribution of the RHF determinant to the FCI wave function decreases more rapidly than for the hydrogen molecule. The UHF and RHF solutions coincide up to $2.64a_0$, beyond which the UHF energy falls below the RHF energy. Although the UHF curve soon becomes parallel to the FCI curve, it does not coincide as for the hydrogen molecule since the oxygen atom – unlike the hydrogen atom – is a many-electron system, for which the FCI energy is lower than the UHF energy.

In Figure 5.13, we have indicated the presence of a second RHF solution for bond distances greater than $4.00a_0$, using a dotted line terminated by an arrow. This $^1A_1$ solution – which represents an excited state since its electronic Hessian has one negative eigenvalue – approaches the ground-state RHF wave function closely in an avoided crossing at $4.00a_0$. For shorter bond distances, the calculation of the excited RHF energy becomes difficult. No plot has therefore been attempted to the left of the avoided crossing, although the existence of this solution also at short distances has been indicated by an arrow. Multiple solutions that are close in energy are usually found at high energies and more generally whenever the Hartree–Fock wave function provides an inadequate representation of the electronic system. Their appearance may therefore be taken as an indication that the RHF model is inadequate and is usually accompanied by difficulties in the optimization of the wave function.

Close to the equilibrium geometry, the correlation energy is small – only 0.29% of the total energy; see Table 5.7. We recall that, in the hydrogen molecule, the correlation energy constitutes 3.4% of the total energy. The smaller relative correlation energy in water arises because of the strong attraction between the oxygen nucleus and the electrons, which reduces the total energy considerably without correlating the electrons more strongly. Another reason is the small basis in the water calculation. Thus, in the basis-set limit, the correlation energy in water constitutes 0.5% of the total energy; see Table 15.14. The significantly smaller correlation energy at the cc-pVDZ

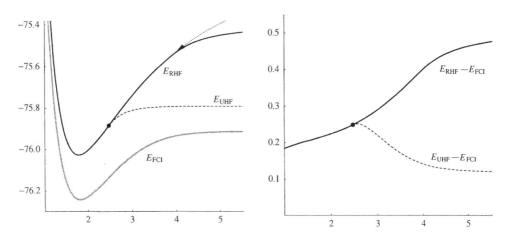

Fig. 5.13. The Hartree–Fock dissociation of the C_{2v} water molecule in the cc-pVDZ basis for a fixed HOH bond angle of $110.565°$ (atomic units). On the left, we have plotted energies on an absolute scale; on the right, we have plotted energies relative to that of the FCI wave function. The FCI curve is plotted in grey, the RHF curves are represented by full black lines and the UHF curves by dashed black lines. In the figure on the left, we have also indicated (with a dotted line) the presence of a second RHF solution of $^1A_1$ symmetry.

Table 5.7 Total cc-pVDZ energies (E_h) for the C_{2v} water molecule at the FCI, RHF, UHF, valence CASSCF and MP2 levels at various internuclear separations R. The HOH bond angle is fixed at $110.565°$ and $R_{ref} = 1.84345a_0$

	$R = R_{ref}$	$R = 2R_{ref}$	$R = 2.5R_{ref}$	$R = 8R_{ref}$
E_{FCI}	-76.241860	-75.951665	-75.917985	-75.910300
E_{RHF}	-76.024039	-75.587711	-75.441244^a	-75.393278^a
E_{UHF}	-76.024039	-75.793668	-75.791198	-75.790723
E_{CASSCF}	-76.077835	-75.818097	-75.791669	-75.786070
E_{MP2}	-76.228729	-75.896935	-75.849661^b	-76.307938^b

[a]The RHF energy corresponds to that of the upper RHF solution in Figure 5.13.
[b]The MP2 energy has been calculated from a reference wave function corresponding to the upper RHF solution in Figure 5.13.

level serves as a reminder of the limited flexibility this basis possesses for the description of electron correlation, and that comparisons made in this basis should be interpreted with caution.

5.4.10 THE DISSOCIATION OF THE WATER MOLECULE

The situation for the binding energy of the water molecule is the same as for the hydrogen molecule. In the supermolecule approach, the binding energy – which is 331.6 mE$_h$ for the FCI wave function – is overestimated by RHF theory (630.8 mE$_h$) and underestimated by UHF theory (233.3 mE$_h$); see Table 5.7. Since the RHF model neglects the antibonding configurations required for correct dissociation, it provides a far better representation of the bonded molecule than of the dissociated system. Thus, the RHF error arises mainly because of the neglect of nondynamical correlation in the dissociated system.

In the UHF picture, there are no near-degeneracy effects and consequently no neglect of nondynamical correlation. Instead, the dissociation energy is underestimated because of the less severe neglect of dynamical correlation. When water dissociates, the OH bonds are broken and the electrons in the bonding orbitals are separated and thus taken from a situation of strong mutual repulsion to one of no interaction. The repulsion experienced by these electrons in the bonded molecule is overestimated by the UHF wave function, which neglects the dynamical correlation that would otherwise reduce the electron repulsion by correlating the electrons covalently in the OH bonds; see Figure 5.13.

As for the hydrogen molecule, we may also calculate the binding energy of water from separate RHF calculations on the dissociation products. Thus, RHF calculations in the cc-pVDZ basis give -74.7875 E_h for the ^3P ground state of oxygen and -0.4993 E_h for the ^2S ground state of hydrogen, leading to a binding energy of 237.9 mE_h. As expected, this binding energy is close to that of the UHF model and the reason for the underestimation relative to FCI is the same as for UHF.

5.4.11 FINAL COMMENTS

For many purposes, the RHF wave function is able to provide results in qualitative agreement with experiment. Examples are energies related to conformational changes such as inversion and rotation barriers, proton affinities and hydration energies. In these cases, no drastic rearrangement of the electronic structure occurs during the reaction or the molecular rearrangement – no electron pairs are broken or formed and the orbitals are only slightly modified. The error in the energy therefore remains more or less constant and cancels as we compare energies of the different species. Molecular equilibrium structures, where differential total energies are compared, also belong to the class of properties that are usually well described by the Hartree–Fock model.

When the RHF description is reasonably accurate, the correlation energy is predominantly dynamical and may be recovered by employing post-Hartree–Fock methods based on the dominance of a single electronic configuration. The simplest such approach is perturbation theory, where the dynamical correlation energy is evaluated to a given order in the perturbation – that is, to a given order in the difference between the true Hamiltonian and the Fock operator. In the remaining sections of this chapter, we shall consider this and other correlated models developed for systems dominated by a single configuration. First, however, we shall examine a model that is not based on the dominance of a single electronic configuration and which is therefore particularly useful for the description of nondynamical correlation – the multiconfigurational self-consistent field method.

5.5 Multiconfigurational self-consistent field theory

As the next standard model of quantum chemistry, we consider here the generalization of the Hartree–Fock wave function to systems dominated by more than one electronic configuration: the *multiconfigurational self-consistent field* (MCSCF) wave function. This flexible model may be useful for describing the electronic structure of bonded molecular systems, in particular for excited states. Perhaps more important, however, is its ability to describe bond breakings and molecular dissociation processes.

5.5.1 THE MULTICONFIGURATIONAL SELF-CONSISTENT FIELD MODEL

In MCSCF theory, the wave function is written as a linear combination of determinants or CSFs, whose expansion coefficients are optimized simultaneously with the MOs according to the variation principle. Thus, the MCSCF wave function may be written in the form

$$|\kappa, \mathbf{C}\rangle = \exp(-\hat{\kappa}) \sum_i C_i |i\rangle \tag{5.5.1}$$

where C_i are the configuration expansion coefficients (normalized to unity) and the operator $\exp(-\hat{\kappa})$ carries out unitary transformations among the spin orbitals in the same way as for Hartree–Fock wave functions. The ground-state MCSCF wave function is obtained by minimizing the energy with respect to the variational parameters:

$$E_{\mathrm{MC}} = \min_{\kappa, \mathbf{C}} \frac{\langle \kappa, \mathbf{C} | \hat{H} | \kappa, \mathbf{C} \rangle}{\langle \kappa, \mathbf{C} | \kappa, \mathbf{C} \rangle} \tag{5.5.2}$$

This model allows for a highly flexible description of the electronic system, where both the one-electron functions (the MOs) and the N-electron function (the configurations) may adapt to the physical situation.

The MCSCF wave function is well suited to studies of systems involving degenerate or nearly degenerate configurations, where static correlation is important. Such situations are usually encountered in the description of reaction processes where chemical bonds are being broken, but sometimes also in ground-state molecular systems at the equilibrium geometry. MCSCF wave functions are therefore indispensable in many situations, although there are often considerable difficulties associated with their optimization. Moreover, it has proved difficult to generalize MCSCF theory in such a way that dynamical correlation effects can be calculated accurately for multiconfigurational electronic systems, although the application of second-order perturbation theory to MCSCF wave functions has been quite successful. A detailed exposition of MCSCF theory is given in Chapter 12.

5.5.2 THE GROUND STATE OF THE HYDROGEN MOLECULE

In most applications, MCSCF theory is best regarded as a generalization of Hartree–Fock theory to systems that are not dominated by a single configuration. For example, the simplest spin-adapted wave function that gives a reasonably accurate and uniform description of the hydrogen molecule at all internuclear separations is

$$|\mathrm{MC}\rangle = C_1 a_{1\sigma_g \alpha}^\dagger a_{1\sigma_g \beta}^\dagger |\mathrm{vac}\rangle + C_2 a_{1\sigma_u \alpha}^\dagger a_{1\sigma_u \beta}^\dagger |\mathrm{vac}\rangle \tag{5.5.3}$$

where the variation principle is invoked to optimize the configuration coefficients as well as the orbitals. In Figure 5.14, we have plotted the potential-energy curve of the hydrogen molecule using the two-configuration MCSCF wave function (5.5.3). Unlike the RHF wave function, the MCSCF wave function gives a balanced description of the whole potential-energy curve, dissociating correctly to the ^2S ground-state hydrogen atoms. Indeed, at the internuclear distance of $100a_0$ in Table 5.6, a separate calculation on the hydrogen atom (in the same basis) gives exactly one-half of the MCSCF energy.

5.5.3 THE SELECTION OF MCSCF CONFIGURATION SPACES

Usually, the greatest difficulty faced in setting up an MCSCF calculation is the selection of configuration space. For the ground-state hydrogen molecule in Section 5.5.2, the choice of configuration space presented no problems – the bonding and antibonding configurations are the only ones that

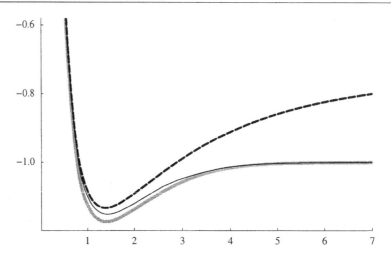

Fig. 5.14. The cc-pVQZ potential-energy curve of the hydrogen molecule for a two-configuration MCSCF wave function (solid line) compared with the potential-energy curves of the FCI wave function (thick grey line) and the RHF wave function (dashed line). Atomic units are used.

contribute appreciably to the wave function in the course of the dissociation. For more complicated molecules, however, the selection of a suitable configuration space can be very difficult – indeed, even for rather small systems, it is often impossible to generate an MCSCF configuration space sufficiently flexible to describe the physical process and yet sufficiently small to be computationally tractable.

The selection of MCSCF configurations is usually not carried out by inspection of the individual configurations. Instead, whole classes of configurations are simultaneously selected according to some general criterion. One successful approach has been to partition the MO space into three subspaces containing the *inactive, active* and *secondary (virtual)* orbitals. Whereas the inactive orbitals are doubly occupied and the virtual orbitals unoccupied in all configurations, the active orbitals have variable occupancies of 0, 1 or 2. The MCSCF expansion is then obtained by distributing the *active electrons* in all possible ways among the active orbitals. In the final optimized state, the active orbitals therefore have nonintegral occupancies between 0 and 2. This method is known as the *complete active space (CAS) SCF method*. Note that a CAS wave function may be regarded as an FCI wave function in a limited but variationally optimized active orbital space: when the active space comprises all the orbitals in the system, the true FCI wave function is recovered; when the active space is empty, the CAS wave function reduces to the Hartree–Fock wave function.

5.5.4 THE GROUND STATE OF THE WATER MOLECULE

As an example of the generation of active spaces, we consider the dissociation of the water molecule into the ground-state atomic fragments. From the molecular-orbital energy-level diagram in Figure 5.10, we conclude that the flexibility needed to describe the breaking of the two OH bonds is secured with an active space containing four electrons and comprising the four MOs $3a_1$, $4a_1$, $1b_2$ and $2b_2$. Such an active space is designated (2,0,0,2), where the numbers of active orbitals in the four irreducible representations of the C_{2v} point group are enclosed in parentheses in the order (A_1, A_2, B_1, B_2). In this *minimal active space* for the dissociation of the water molecule, the three occupied MOs $1a_1$, $2a_1$ and $1b_1$ are left inactive. The adequacy of this active space for the dissociation process is confirmed by an inspection of the FCI occupation numbers in Table 5.5.

To ensure a symmetry-correct dissociation of the water molecule – that is, the dissociation to an oxygen atom in the ^3P ground state and to two hydrogen atoms in the ^2S ground state – the active space must contain the full set of degenerate $2p$ orbitals. Incorporating the $2p_x$ lone-pair orbital of B_1 symmetry, we arrive at the $(2,0,1,2)$ active space, where the only inactive orbitals are the oxygen $1s$ and $2s$ orbitals. In this small active space, six active electrons are distributed among five orbitals in all possible ways consistent with the $^1A_1$ symmetry of the electronic system.

The $2s$ orbital belongs to the valence space and it is therefore expected to interact quite strongly with the $2p$ orbitals. Incorporating this orbital in our active space, we arrive at the *valence active space* $(3,0,1,2)$, with eight electrons distributed among six orbitals in accordance with $^1A_1$ symmetry. If at all possible, valence (or larger) active spaces should be employed in CASSCF calculations since such spaces ensure a reasonably uniform representation of the potential-energy surface as the molecule distorts, but even this rather modest requirement is often impossible to satisfy in practice.

In Figure 5.15, we have plotted the potential-energy curve of the valence $(3,0,1,2)$ CAS wave function as well as the difference between the CAS and FCI energies. The energies at selected bond distances are listed in Table 5.7. Clearly, the $(3,0,1,2)$ CASSCF wave function gives a uniform although somewhat crude representation of the potential-energy curve. Comparing with the RHF energy, we find that the error relative to FCI is reduced from 217.8 to 164.0 mE_h at R_{ref} and from 364.0 to 133.6 mE_h at $2R_{ref}$. For larger internuclear separations, the CAS error decreases somewhat since there are fewer electrons to correlate in the dissociation limit; see Figure 5.15. Indeed, the behaviour of the CAS wave function is here similar to that of the UHF wave function in Figure 5.13 and occurs for the same reason. But, unlike the UHF wave function, the CAS wave function remains a true singlet at all separations. In terms of the correlation energy, the CASSCF wave function recovers 25% at R_{ref} and – because of the very poor performance of RHF theory at large separations – as much as 63% at $2R_{ref}$.

It has already been pointed out that a CAS wave function corresponds to an FCI wave function in a restricted but variationally optimized active space. Indeed, comparing Tables 5.8 and 5.5, we find that the occupation numbers of the $(3,0,1,2)$ CAS wave function are quite similar to those of the FCI wave function. In general, natural-orbital occupation numbers – obtained, for example,

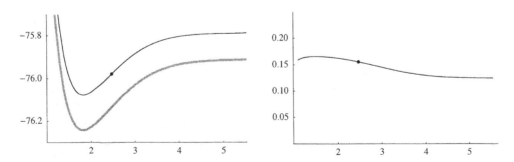

Fig. 5.15. The CASSCF dissociation of the C_{2v} water molecule in the cc-pVDZ basis (atomic units) for a fixed HOH bond angle of 110.565° On the left, we have plotted the potential-energy curves of the valence CASSCF wave function (black line) and the FCI wave function (grey line); on the right, we have plotted the difference between the energies of the valence CAS wave function and the FCI wave function.

Table 5.8 The cc-pVDZ natural-orbital occupation numbers of the valence (3,0,1,2) CASSCF wave function for the C_{2v} water molecule at various internuclear separations R. The HOH bond angle is fixed at $110.565°$ and $R_{ref} = 1.84345a_0$

	$R = R_{ref}$	$R = 2R_{ref}$	$R = 2.5R_{ref}$	$R = 8R_{ref}$
$\eta(1a_1)$	2.0000	2.0000	2.0000	2.0000
$\eta(2a_1)$	1.9991	1.9990	1.9998	2.0000
$\eta(3a_1)$	1.9780	1.5162	1.2053	1.0000
$\eta(4a_1)$	0.0235	0.4866	0.7948	1.0000
$\eta(1b_1)$	1.9994	1.9995	1.9999	2.0000
$\eta(1b_2)$	1.9768	1.5840	1.2715	1.0000
$\eta(2b_2)$	0.0232	0.4147	0.7286	1.0000

from perturbation theory rather than from FCI theory as in this section – are useful in providing information about the importance of the individual orbitals for the description of correlation, thereby guiding the selection of MCSCF orbital spaces: orbitals with occupation numbers close to 2 become inactive, those with occupations close to 0 become secondary, and the remaining ones active.

Gaps in the occupation numbers are often helpful in designing balanced CAS spaces. Thus, in Table 5.5, there is a noticeable gap between the smallest occupation of the valence orbitals (0.0278) and the largest occupation of the nonvalence virtual orbitals (0.0179), indicating that the valence (3,0,1,2) active space is indeed a sensible one. From the FCI occupation numbers at the equilibrium geometry, we conclude that the smallest balanced extension of the configuration space should also include the $5a_1$ and $2b_1$ orbitals. In the resulting (4,0,2,2) active space, there is one weakly occupied *correlating orbital* for each strongly occupied active orbital – a situation often encountered in balanced active spaces. For further improvement, we might employ the larger (6,1,3,3) active space with the $1s$ oxygen orbital still left inactive.

5.5.5 FINAL COMMENTS

The occupation numbers of the FCI natural orbitals – that is, the eigenvalues of the one-electron FCI density matrix – approach zero very slowly; see Table 5.5. As discussed in Chapter 7, for sufficiently small occupations, this slow rate of decay is governed by the difficulties encountered in describing the Coulomb cusp by means of orbital expansions. Since the configuration expansions that can be managed within the framework of CASSCF theory correspond to small active spaces – comprising not many more than ten orbitals and ten electrons – it is in general impossible to recover the dynamical correlation energy by MCSCF wave functions. Consequently, MCSCF wave functions are seldom appropriate for high-accuracy applications. On the other hand, MCSCF wave functions are unique in providing a flexible framework for treating the static correlation that arises from the presence of degenerate or nearly degenerate electronic configurations. In such cases, MCSCF wave functions may give a useful, spin-adapted zero-order description of the electronic system, even though high accuracy cannot be expected. For high accuracy and the treatment of dynamical correlation, additional calculations must be carried out based on the initial MCSCF description – either within the framework of multireference configuration-interaction theory as discussed in Section 5.6.3 or by means of multiconfigurational perturbation theory as discussed in Section 14.7.

5.6   Configuration-interaction theory

We now turn our attention to the first standard model that incorporates the effects of dynamical correlation: the *configuration-interaction (CI) model*. This model, which arises naturally in the MO picture as the superposition of determinants, has been quite successful in molecular electronic applications but has more recently been superseded by the coupled-cluster model as a computational tool of quantum chemistry, at least in the more common areas of application.

5.6.1   THE CONFIGURATION-INTERACTION MODEL

In the CI method, the wave function is constructed as a linear combination of determinants or CSFs

$$|\mathbf{C}\rangle = \sum_i C_i |i\rangle \tag{5.6.1}$$

with the coefficients determined by a variational optimization of the expectation value of the electronic energy – that is, by minimization for the ground state:

$$E_{\mathrm{CI}} = \min_{\mathbf{C}} \frac{\langle \mathbf{C} | \hat{H} | \mathbf{C} \rangle}{\langle \mathbf{C} | \mathbf{C} \rangle} \tag{5.6.2}$$

As described in Section 4.2, this condition is equivalent to a set of eigenvalue equations for the energy and the expansion coefficients

$$\mathbf{HC} = E_{\mathrm{CI}} \mathbf{C} \tag{5.6.3}$$

where $\mathbf{H}$ is the Hamiltonian matrix with elements

$$H_{ij} = \langle i | \hat{H} | j \rangle \tag{5.6.4}$$

and $\mathbf{C}$ is a vector containing the expansion coefficients C_i. Equation (5.6.3) is a standard Hermitian eigenvalue problem of linear algebra. Thus, the construction of the CI wave function may be carried out by diagonalizing the Hamiltonian matrix, but more often iterative techniques are used to extract selected eigenvalues and eigenvectors. A thorough technical discussion of the CI model is given in Chapter 11.

Owing to its formal and conceptual simplicity, the CI method has been extensively and successfully applied in quantum chemistry. The simple structure of the wave function has made it possible to develop schemes for treating a very large number of Slater determinants – indeed, calculations involving more than 10^9 determinants have been reported. In Section 5.3, FCI calculations, comprising all the configurations that may be generated in a given orbital basis, were presented. However, for any but the simplest systems, it is not possible to carry out such FCI calculations and it then becomes necessary to *truncate* the CI expansion so that only a small subset of the full set of the determinants is included. The *truncated CI expansion* should preferably recover a large part of the correlation energy and, for chemical reactions, provide a uniform description of the electronic structure over the whole potential-energy surface. In this section, we consider the truncated CI wave function (as distinct from the intractable FCI wave function) as a practical wave-function model in quantum chemistry.

In CI theory, only the configuration expansion is variationally optimized – the orbitals are generated separately in a preceding Hartree–Fock or MCSCF calculation and are held fixed during the optimization of the configuration expansion. In contrast, the MCSCF wave function of Section 5.5

is obtained by a simultaneous variational optimization of the orbitals and the configuration expansion. The MCSCF wave function is thus more flexible than the corresponding CI wave function. On the other hand, the optimization of the nonlinearly parametrized MCSCF wave function is considerably more difficult than the optimization of the CI wave function. In practice, therefore, MCSCF wave functions are restricted to small configuration expansions – adequate for the description of static correlation but inadequate for the description of dynamical correlation.

5.6.2 SINGLE-REFERENCE CI WAVE FUNCTIONS

The FCI wave function is often dominated by a single *reference configuration*, usually the Hartree–Fock state. It is then convenient to think of the FCI wave function as generated from this reference configuration by the application of a linear combination of spin-orbital excitation operators

$$|\text{FCI}\rangle = \left(1 + \sum_{AI} \hat{X}_I^A + \sum_{A>B,I>J} \hat{X}_{IJ}^{AB} + \cdots \right)|\text{HF}\rangle \tag{5.6.5}$$

where, for example,

$$\hat{X}_I^A|\text{HF}\rangle = C_I^A a_A^\dagger a_I|\text{HF}\rangle \tag{5.6.6}$$

$$\hat{X}_{IJ}^{AB}|\text{HF}\rangle = C_{IJ}^{AB} a_A^\dagger a_B^\dagger a_I a_J|\text{HF}\rangle \tag{5.6.7}$$

Thus, we may characterize the determinants in the FCI expansion as single (S), double (D), triple (T), quadruple (Q), quintuple (5), sextuple (6) and higher excitations relative to the Hartree–Fock state.

In general, it is reasonable to assume that the lower-order excitations in (5.6.5) are more important than those of higher orders. In particular, we expect the doubly excited configurations to dominate since these are the only ones that interact directly with the Hartree–Fock state (singles do not interact according to the Brillouin theorem; see Chapter 10). We should thus be able to set up a *hierarchy* of successively more accurate wave functions by retaining in the expansion (5.6.5) only determinants up to a given level of excitation. First, only the single and double excitations are retained – these are usually the most important (the singles interacting via the doubles). At the next level, we include also the triples. In principle (but seldom in practice), this procedure of *hierarchical truncation* may be continued until the FCI wave function is recovered.

In Table 5.9, we have listed the electronic energies of truncated CI wave functions for the water molecule at R_ref and $2R_\text{ref}$. Since the CI model is variational, the FCI energy is approached monotonically from above. From the weights, we see how the FCI wave function is approached as higher and higher excitations are included in the expansion.

Let us first consider the calculations at R_ref (i.e. close to the equilibrium geometry) in Table 5.9. Although the energy converges monotonically towards the FCI limit, the contributions from the different excitation levels vary considerably. In particular, whereas the contributions drop significantly as we go from an even-order level $2n$ to an odd-order level $2n + 1$, the contributions from levels $2n + 1$ and $2n + 2$ are comparable. For example, at R_ref, the contribution from the doubles is -205.8 mE$_\text{h}$ – considerably larger than those from the triples -3.0 mE$_\text{h}$ and from the quadruples -8.7 mE$_\text{h}$, both of which are significantly larger than that from the quintuples -0.2 mE$_\text{h}$. For a balanced correlation treatment, CI wave functions should therefore be truncated only at even orders. Consequently, the first useful truncated CI wave function – the *CI singles-and-doubles (CISD) wave function* – contains all singles and doubles configurations and recovers 94.5% of the

Table 5.9 The cc-pVDZ energies of truncated CI wave functions relative to the FCI energy (in E_h) for the C_{2v} water molecule at OH separations $R = R_{ref}$ and $R = 2R_{ref}$. The weights W of the CI wave functions in the FCI wave function are also given. The HOH bond angle is fixed at $110.565°$ and $R_{ref} = 1.84345a_0$

	$R = R_{ref}$		$R = 2R_{ref}$	
	$E - E_{FCI}$	W	$E - E_{FCI}$	W
RHF	0.217822	0.941050	0.363954	0.589664
CISD	0.012024	0.998047	0.072015	0.948757
CISDT	0.009043	0.998548	0.056094	0.959086
CISDTQ	0.000327	0.999964	0.005817	0.998756
CISDTQ5	0.000139	0.999985	0.002234	0.999553
CISDTQ56	0.000003	1.000000	0.000074	0.999993

correlation energy, confirming our claim that singly and doubly excited determinants are the most important ones for the description of correlation. While the improvement at the *CISDT level* is modest with only 95.9% of the correlation energy recovered, as much as 99.9% of the correlation energy is recovered at the *CISDTQ level*.

At the stretched geometry $2R_{ref}$ in Table 5.9, the situation is similar to that at the equilibrium geometry except that the convergence towards the FCI limit is slower. Thus, the CISD wave function now recovers only 80.2% of the correlation energy, compared with 94.5% at the equilibrium geometry. Similarly, the Hartree–Fock determinant is less dominant – its weight has dropped from 94% to 59% as the antibonding configurations necessary for describing bond breaking become important. Clearly, the strategy of defining a correlation hierarchy in terms of excitations out of a single reference determinant such as the Hartree–Fock determinant does not provide a uniform description of the dissociation process.

In Figure 5.16, we have plotted the RHF-based CISD, CISDT and CISDTQ energies in the cc-pVDZ basis as functions of the OH bond distance in the C_{2v} water molecule. The improvement on the Hartree–Fock description in Figure 5.13 is quite striking at all levels of truncation. Still, since the CI wave functions are based on the dominance of the RHF determinant, their description deteriorates notably as the OH bonds lengthen, in particular for CISD and CISDT. For long bond distances, the single-reference calculations become meaningless and the potential-energy curves in Figure 5.16 have not been plotted beyond $4.0a_0$.

5.6.3 MULTIREFERENCE CI WAVE FUNCTIONS

To overcome the shortcomings of the single-reference CI method, we introduce CI wave functions based on the idea of a *reference space* comprising more than a single configuration – for example, all configurations that may become important in the region of the energy surface needed to describe a particular reaction. The *multireference CI (MRCI)* wave function is then generated by including in the wave function all configurations belonging to this reference space as well as all excitations up to a given level from each reference configuration. Commonly, all single and double excitations from the reference space are included, resulting in the *multireference singles-and-doubles CI (MRSDCI)* wave function.

The construction of a multireference CI wave function begins with the generation of a set of orbitals and the reference space of configurations. MCSCF calculations are ideally suited to this task. Returning to the water molecule, we would typically include in our reference space the

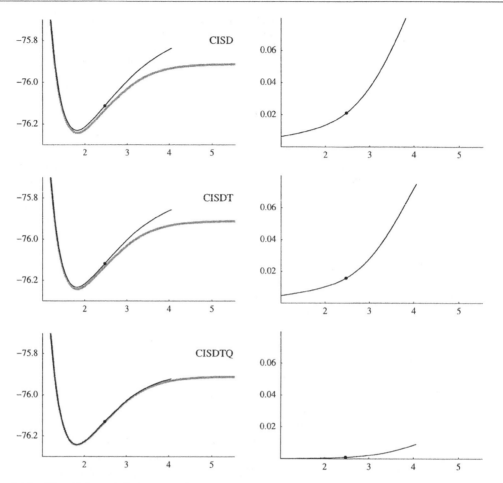

Fig. 5.16. The CI dissociation curves of the C_{2v} water molecule in the cc-pVDZ basis (atomic units) for a fixed HOH bond angle of 110.565°. In the two uppermost figures, we have on the left plotted the dissociation curves for the CISD wave function (black line) and the FCI wave function (grey line); on the right, we have plotted the difference between the CISD and FCI energies. In the middle and lower figures, we have made similar plots for the CISDT and CISDTQ wave functions.

configurations obtained by distributing the eight valence electrons among the six valence orbitals ($2a_1, 3a_1, 4a_1, 1b_1, 1b_2, 2b_2$) in all possible ways consistent with the spin- and space-symmetry restrictions of the system. Thus, we first carry out a preliminary $(3,0,1,2)$ CASSCF calculation as described in Section 5.5.4. The orbitals and configurations from this calculation are then used for the construction of the MRSDCI wave function. The reference space contains all CAS configurations and the remaining configurations of the MRSDCI wave function correspond to all single and double excitations that may be generated from these reference configurations.

The H_2O potential-energy curve of the MRSDCI wave function is plotted in Figure 5.17. This curve should be compared with that of the corresponding CAS wave function in Figure 5.15 and with those of the single-reference CI wave functions in Figure 5.16. Clearly, by combining a truncated CI expansion with a multiconfigurational reference space, we have arrived at a wave function that provides a uniformly accurate approximation to the FCI energy at all internuclear distances.

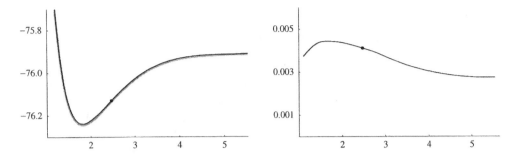

Fig. 5.17. The MRSDCI dissociation of the C_{2v} water molecule in the cc-pVDZ basis (atomic units) for a fixed bond angle of 110.565°. On the left, we have plotted the potential-energy curves of the MRSDCI wave function (black line) and the FCI wave function (grey line); on the right, we have plotted the difference between the energies of the MRSDCI wave function and the FCI wave function.

Table 5.10 The cc-pVDZ energies of RHF, CAS and MRSDCI wave functions relative to FCI (in E_h) for the C_{2v} water molecule at the OH separations $R = R_{ref}$ and $R = 2R_{ref}$. The weight of the Hartree–Fock configuration in the FCI wave function and the combined weight of the CAS reference configurations in the MRSDCI wave function are also given. The HOH bond angle is fixed at 110.565° and $R_{ref} = 1.84345a_0$

| | $R = R_{ref}$ | | $R = 2R_{ref}$ | |
	$E - E_{FCI}$	W	$E - E_{FCI}$	W
RHF	0.217822	0.941[a]	0.363954	0.589[a]
CAS (3,0,1,2)	0.164025	0.962[b]	0.133568	0.967[b]
MRSDCI	0.004425	–	0.003208	–

[a]Weight of the Hartree–Fock configuration in the FCI wave function.
[b]Combined weight of the CAS reference configurations in the MRSDCI wave function.

From Table 5.10, we see that the MRSDCI error is about 40 times smaller than the CASSCF error and in fact comparable with the CISDT error at R_{ref} and the CISDTQ5 error at $2R_{ref}$. Indeed, of all the approximate wave functions considered in this chapter, the MRSDCI wave function gives the consistently most accurate representation of the dissociation surface. It is noteworthy that the combined weight of the CAS reference determinants in the MRSDCI wave function is almost constant across the surface, differing by only 0.5% between R_{ref} and $2R_{ref}$. By contrast, the weight of the Hartree–Fock determinant in the FCI wave function drops dramatically as the bond is stretched.

The MRSDCI model is accurate provided all the important configurations are included in the reference space. Unfortunately, in extended systems, the number of reference configurations that must be included for a uniform description often becomes very large, making the calculations time-consuming if not impossible. To solve this problem, we may use different reference configurations for different regions of the energy surface, omitting those configurations that are unimportant in each particular region. Unfortunately, this procedure introduces unphysical kinks in the energy surface as we go from one region to another. Often, the construction of a suitable MRSDCI wave function represents a great challenge to the user's chemical insight and patience as a large number of exploratory calculations must be carried out to arrive at an appropriate reference space.

5.6.4 FINAL COMMENTS

The CI approach to the many-electron problem suffers from two main disadvantages. First, since the CI model is not manifestly separable, it does not provide size-extensive energies. In Section 4.3.2, we found that size-extensivity follows only when the variational space of the compound system is equal to the direct product of the fragment spaces. This requirement cannot be satisfied for CI wave functions truncated at a fixed excitation level relative to the reference configuration. For example, if two fragments are each described at the CISD level, then a size-extensive treatment of the compound system requires the wave function to be augmented with certain triples and quadruples – namely, those that represent products of single and double excitations in the two subsystems. Such wave functions are not defined within the hierarchy of truncated CI wave functions based on excitation levels.

Second, the CI description of the electronic system is not particularly compact. Thus, even though higher excitations are less important than those of lower orders, their number is very large. As a result, the CI wave function converges slowly with the number of variational parameters (one parameter for each configuration). Partly, this slow convergence is dictated by the difficulties faced in describing the Coulomb cusp by determinantal expansions – a problem shared by all Fock-space models. However, the slow convergence arises also from the use of a linear parametrization in the CI model (5.6.1). In Section 5.7, we shall see how this particular shortcoming of the CI model may be remedied within the Fock-space formulation of electronic-structure theory.

5.7 Coupled-cluster theory

As noted in Section 5.6, the CI hierarchy of wave functions suffers from two serious shortcomings: the lack of size-extensivity and the slow convergence towards the FCI limit. We shall now see how these problems are overcome through the introduction of the coupled-cluster model, which represents a nonlinear but manifestly separable parametrization of the correlated electronic state. The coupled-cluster model constitutes a particularly successful approach to electronic-structure theory, providing for many purposes – often in combination with perturbation theory – the most efficient strategy for the accurate calculation of electronic energies and wave functions.

5.7.1 THE COUPLED-CLUSTER MODEL

The lack of size-extensivity of the truncated CI wave function arises from the use of the linear parametrization (5.6.1). To impose size-extensivity on our description, we recast the linear FCI expansion (5.6.5) in the form of a product wave function:

$$|\text{CC}\rangle = \left[\prod_{AI}(1 + \hat{X}_I^A) \right] \left[\prod_{A>B,I>J}(1 + \hat{X}_{IJ}^{AB}) \right] \cdots |\text{HF}\rangle \qquad (5.7.1)$$

This *coupled-cluster wave function* is manifestly separable (see Section 4.3.4) and differs from the CI wave function (5.6.5) by the presence of terms that are nonlinear in the excitation operators. Since the excitation operators such as (5.6.6) and (5.6.7) commute with one another, there is no problem with the order of the operators in the product wave function (5.7.1).

It should be noted that the CI and coupled-cluster wave functions (5.6.5) and (5.7.1) are entirely equivalent provided all excitations are included in the expressions, differing only in their parametrization. Their nonequivalence becomes apparent only when some of the excitation operators – for example, all excitations higher than doubles – are omitted from the wave function. In particular, whereas size-extensivity is always preserved for the manifestly separable coupled-cluster wave function, it is lost upon truncation of the linearly parametrized CI wave function, as discussed in Section 5.6.4.

The transition from a linear model for the wave function to a product model shifts the emphasis away from excitation levels and excited determinants towards excitations and excitation processes. Thus, applied to some electronic state or configuration $|0\rangle$, each operator in (5.7.1) produces a superposition of the original state and a correction term that represents an excitation from the original state:

$$(1 + \hat{X}_{IJ}^{AB})|0\rangle = |0\rangle + C_{IJ}^{AB} a_A^\dagger a_B^\dagger a_I a_J |0\rangle \tag{5.7.2}$$

The weight of the correction is independent of the original state, depending only on the excitation process itself.

In the coupled-cluster state (5.7.1), a given excited configuration may be generated in several distinct ways, each with its own probability. Consider, for instance, a configuration that is quadruply excited relative to the Hartree–Fock state:

$$\left| \begin{matrix} ABCD \\ IJKL \end{matrix} \right\rangle = a_A^\dagger a_B^\dagger a_C^\dagger a_D^\dagger a_I a_J a_K a_L |\mathrm{HF}\rangle \tag{5.7.3}$$

This configuration may be generated either directly as the result of a concerted simultaneous interaction among four electrons (i.e. by the application of a single quadruple-excitation operator)

$$(1 + \hat{X}_{IJKL}^{ABCD})|\mathrm{HF}\rangle = |\mathrm{HF}\rangle + C_{IJKL}^{ABCD} \left| \begin{matrix} ABCD \\ IJKL \end{matrix} \right\rangle \tag{5.7.4}$$

or indirectly as the result of two independent interactions, each occurring between two electrons (i.e. by the combined application of two distinct double-excitation operators)

$$(1 + \hat{X}_{IJ}^{AB})(1 + \hat{X}_{KL}^{CD})|\mathrm{HF}\rangle = |\mathrm{HF}\rangle + C_{IJ}^{AB} \left| \begin{matrix} AB \\ IJ \end{matrix} \right\rangle + C_{KL}^{CD} \left| \begin{matrix} CD \\ KL \end{matrix} \right\rangle + C_{IJ}^{AB} C_{KL}^{CD} \left| \begin{matrix} ABCD \\ IJKL \end{matrix} \right\rangle \tag{5.7.5}$$

If we include single excitations, other mechanisms become possible as well. With each excitation mechanism, we associate a characteristic probability amplitude. The final electronic state (5.7.1) is thus a linear combination of excited configurations, each with a total weight equal to the combined probabilities of all mechanisms leading to this particular configuration. In this respect, the coupled-cluster wave function differs fundamentally from the CI wave function, where we parametrize the total weight of each excited configuration individually.

5.7.2 THE EXPONENTIAL ANSATZ OF COUPLED-CLUSTER THEORY

The product form (5.7.1) for the wave function is inconvenient for algebraic manipulations. We note that, since

$$\hat{X}_{IJ}^{AB} \hat{X}_{IJ}^{AB} = 0 \tag{5.7.6}$$

we may write

$$1 + \hat{X}_{IJ}^{AB} = 1 + \hat{X}_{IJ}^{AB} + \tfrac{1}{2}\hat{X}_{IJ}^{AB}\hat{X}_{IJ}^{AB} + \cdots = \exp(\hat{X}_{IJ}^{AB}) \tag{5.7.7}$$

and similarly for the other excitations. Inserting (5.7.7) into (5.7.1) and similarly for the other excitations, we arrive at the following expression for the coupled-cluster wave function

$$|CC\rangle = \exp\left(\sum_{AI} t_I^A a_A^\dagger a_I + \sum_{A>B,I>J} t_{IJ}^{AB} a_A^\dagger a_B^\dagger a_I a_J + \cdots\right)|HF\rangle \qquad (5.7.8)$$

where we have conventionally changed the notation for the *excitation amplitudes* or the *coupled-cluster amplitudes* t_I^A and t_{IJ}^{AB}. The coupled-cluster model (5.7.1) therefore corresponds to a nonlinear exponential parametrization of the wave function

$$|CC\rangle = \exp(\hat{T})|HF\rangle \qquad (5.7.9)$$

where the non-Hermitian *cluster operator*

$$\hat{T} = \hat{T}_1 + \hat{T}_2 + \cdots \qquad (5.7.10)$$

contains the single and double excitation operators

$$\hat{T}_1 = \sum_{AI} t_I^A a_A^\dagger a_I \qquad (5.7.11)$$

$$\hat{T}_2 = \frac{1}{4} \sum_{\substack{AB \\ IJ}} t_{IJ}^{AB} a_A^\dagger a_B^\dagger a_I a_J \qquad (5.7.12)$$

as well as all higher-order operators. Expressions (5.7.9)–(5.7.12) constitute the standard *exponential ansatz* of coupled-cluster theory.

As in CI theory, we now introduce a hierarchy of coupled-cluster wave functions by truncating the cluster operator (5.7.10) at different excitation levels. At the simplest level, we obtain the *coupled-cluster singles-and-doubles (CCSD) model*, omitting from the cluster operator all terms that involve higher than single and double excitations. At the next level, we also retain the triples and arrive at the *CCSDT wave function*, and so on.

It should be realized that, because of the product parametrization (5.7.1), any approximate coupled-cluster wave function such as CCSD or CCSDT – obtained by truncating the cluster operator (5.7.10) at a given excitation level – contains contributions from *all* determinants in Fock space (of the correct spin and space symmetries), not just those that arise from a *linear* application of the cluster operator. Thus, we may regard, for example, the CCSD wave function as a particular approximation to the FCI state where *all* configurations have nonzero coefficients, but where the coefficients of the triples and higher configurations are generated from those of the singles and doubles so as to ensure size-extensivity. In the limit of a full cluster operator (5.7.10), we recover the FCI wave function, although we might then as well use the simple linear CI parametrization directly.

Unlike the wave functions considered so far (Hartree–Fock, MCSCF and CI), the coupled-cluster wave function is not optimized according to the variation principle. Instead, the cluster amplitudes are determined by projecting the Schrödinger equation in the form

$$\exp(-\hat{T})\hat{H}\exp(\hat{T})|HF\rangle = E_{CC}|HF\rangle \qquad (5.7.13)$$

against a set of configurations $\langle\mu|$ that spans the space of all states that can be reached by applying the truncated cluster operator $\hat{T}$ linearly to the reference state $|HF\rangle$:

$$\langle\mu|\exp(-\hat{T})\hat{H}\exp(\hat{T})|HF\rangle = 0 \qquad (5.7.14)$$

As an example, the CCSD amplitudes are obtained by projecting (5.7.13) against all singles and doubles configurations and solving the resulting nonlinear equations for the amplitudes. The final coupled-cluster energy is calculated according to the expression

$$E_{CC} = \langle HF| \exp(-\hat{T})\hat{H} \exp(\hat{T})|HF\rangle \qquad (5.7.15)$$

obtained from (5.7.13) by projection onto the Hartree–Fock state.

The reasons for not invoking the variation principle in the optimization of the wave function are given in Chapter 13, which provides a detailed account of coupled-cluster theory. We here only note that the loss of the variational property characteristic of the exact wave function is unfortunate, but only mildly so. Thus, even though the coupled-cluster method does not provide an upper bound to the FCI energy, the energy is usually so accurate that the absence of an upper bound does not matter anyway. Also, because of the Lagrangian method of Section 4.2.8, the complications that arise in connection with the evaluation of molecular properties for the nonvariational coupled-cluster model are of little practical consequence.

5.7.3 THE GROUND STATE OF THE WATER MOLECULE

Let us examine the convergence of the coupled-cluster hierarchy towards the FCI wave function. In Table 5.11, we have – for the water molecule in the cc-pVDZ basis – listed the differences between the (spin- and space-restricted) CCSD, CCSDT and CCSDTQ energies and the FCI energy [3]. Comparing with the CI energies in Table 5.9, we find that the coupled-cluster hierarchy converges faster and more uniformly towards the FCI limit. In particular, there is a significant reduction in the error at each level of truncation – not just at even-order levels as in CI theory.

Clearly, the nonlinear coupled-cluster parametrization is more compact and efficient than the CI parametrization, containing determinants of higher excitations than the corresponding CI wave function. The CCSD wave function, for instance, contains contributions from triple and higher excitations as products of lower-order excitations:

$$|CCSD\rangle = |HF\rangle + \hat{T}_1|HF\rangle + \left(\hat{T}_2 + \tfrac{1}{2}\hat{T}_1^2\right)|HF\rangle + \left(\hat{T}_2\hat{T}_1 + \tfrac{1}{6}\hat{T}_1^3\right)|HF\rangle$$

$$+ \left(\tfrac{1}{2}\hat{T}_2^2 + \tfrac{1}{2}\hat{T}_2\hat{T}_1^2 + \tfrac{1}{24}\hat{T}_1^4\right)|HF\rangle + \cdots \qquad (5.7.16)$$

Apparently, the more important higher-order excitations – in particular, the quadruples – are quite accurately reproduced by the product operators. Thus, at the equilibrium geometry, the CCSD

Table 5.11 The difference between the cc-pVDZ electronic energies of restricted coupled-cluster wave functions and the FCI wave function ($E - E_{FCI}$) of the C_{2v} water molecule for the OH distances R_{ref} and $2R_{ref}$ (in E_h). The HOH bond angle is fixed at $110.565°$ and $R_{ref} = 1.84345a_0$

	$R = R_{ref}$	$R = 2R_{ref}$
RHF	0.217822	0.363954
CCSD	0.003744	0.022032
CCSDT	0.000493	−0.001405
CCSDTQ	0.000019	−0.000446
CCSDTQ5	0.000003	–

model recovers a larger part of the correlation energy than does the CISDT model. Moreover, the CCSDT model recovers almost as much of the correlation energy as does the CISDTQ5 model.

Since the coupled-cluster wave functions in Table 5.11 have been generated from a space- and spin-restricted RHF wave function, their performance deteriorates as the OH bonds lengthen and the RHF contribution to the FCI wave function decreases. Thus, at R_{ref}, the RHF determinant contributes 94% to the FCI wave function and the CCSD error is 3.7 mE$_h$; at $2R_{ref}$, the RHF determinant contributes only 59% and the CCSD error is six times as great. We also note that the nonvariational CCSDT and CCSDTQ energies are *lower* than the FCI energy at $2R_{ref}$, demonstrating that the coupled-cluster energy may converge to the FCI energy from either side.

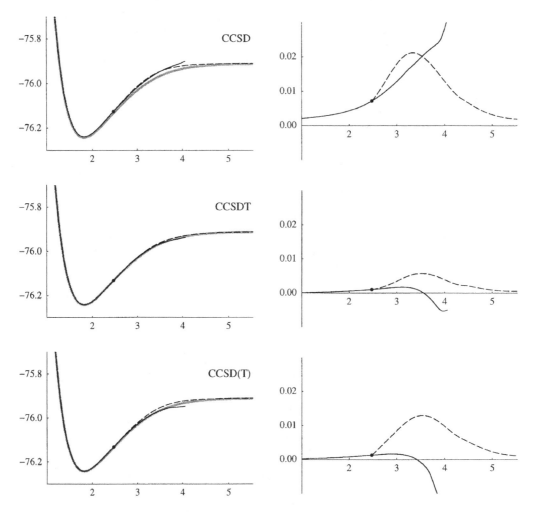

Fig. 5.18. The coupled-cluster dissociation curves of the C_{2v} water molecule in the cc-pVDZ basis (atomic units) for a fixed HOH bond angle of 110.565°. In the two uppermost figures, we have on the left plotted the dissociation curves for the restricted CCSD wave function (full black line), the unrestricted CCSD wave function (dashed black line) and the FCI wave function (grey line); on the right, we have plotted the corresponding differences between the CCSD and FCI energies. In the middle and lower figures, we have made similar plots for the restricted and unrestricted CCSDT and CCSD(T) models.

In Figure 5.18 we have plotted the potential-energy curves of the restricted CCSD and CCSDT wave functions. For short bond distances, the excellent behaviour of both models – in particular, the CCSDT model – is noted. Indeed, in this region, the CCSD and CCSDT wave functions both perform better than the MRSDCI wave function in Figure 5.17 and significantly better than the CI wave functions in Figure 5.16. For longer bond distances their performance deteriorates, although they still easily outperform their CI counterparts (CISD and CISDT).

5.7.4   THE UNRESTRICTED COUPLED-CLUSTER MODEL

Since the UHF wave function dissociates correctly to the neutral atoms (see Section 5.4.10 and, in particular, Figure 5.13), it is of some interest to examine the performance of the unrestricted CCSD and CCSDT models upon dissociation. An inspection of Figure 5.18 reveals that the unrestricted coupled-cluster wave functions dissociate correctly to the neutral atoms, describing these fragments to the same accuracy as the bonded molecule. However, the unrestricted coupled-cluster curves exhibit a characteristic hump (relative to the FCI curve) beyond $2.64a_0$, where the restricted and unrestricted curves separate. It is noteworthy that, unlike the variational UHF energy in Figure 5.13, the unrestricted coupled-cluster energies are *higher* here than the restricted ones. Indeed, this somewhat peculiar behaviour of the unrestricted coupled-cluster energies is another manifestation of the nonvariational character of the coupled-cluster model.

Even though the unrestricted coupled-cluster wave functions are more flexible than their restricted counterparts, their use is less widespread. There are several reasons for this. First, their calculation is several times more expensive than that of the restricted wave functions. Next, although the unrestricted wave functions behave correctly in the dissociation limit, they are not in general spin eigenfunctions and in the intermediate region (where the electrons recouple to form or break chemical bonds) their performance is not as good as for the reactants and for the fragments. In practice, therefore, bond breakings are usually described by means of multiconfigurational methods such as the CASSCF method of Section 5.5 and the MRSDCI method of Section 5.6.

5.7.5   APPROXIMATE TREATMENTS OF TRIPLE EXCITATIONS

For high-precision work, the CCSD model is usually not sufficiently accurate and the CCSDT model is too expensive. Various *hybrid methods*, in which the contribution from the triples amplitudes is estimated by perturbation theory, have therefore been developed to provide descriptions intermediate between CCSD and CCSDT in cost and accuracy. The most popular is the *CCSD(T) method*, whose performance for the dissociation of the water molecule is displayed in Figure 15.18.

For OH distances shorter than $3.5a_0$, the CCSD(T) model works well, giving about 90% of the full CCSDT triples correction. However, the model breaks down at longer distances. Also, the unrestricted CCSD(T) model does not provide a uniform description of the dissociation process, performing not much better than the simpler unrestricted CCSD model in the intermediate region. Hybrid methods are considered in Chapter 14, following a general discussion of perturbation theory.

5.7.6   FINAL COMMENTS

The coupled-cluster model represents a significant improvement on the truncated CI model in that it provides a description of the electronic structure that is both size-extensive and more compact. On the other hand, it has proved difficult to extend the application of coupled-cluster theory to systems that are characterized by degenerate or nearly degenerate electronic configurations. For such systems, MCSCF wave functions and multireference CI wave functions are better suited.

When the Hartree–Fock description is reasonably accurate – as for the stable water molecule – the restricted CCSD model appears to provide a satisfactory representation of the FCI wave function. Although less accurate than CCSDT, it represents a useful compromise between accuracy and cost and can be routinely applied to relatively large molecular systems in a black-box manner. The excellent performance of the CCSDT model is quite remarkable but its cost is so high that it is applicable only to small systems. For most purposes, however, the CCSDT model may be replaced by the CCSD(T) model at little or no loss of accuracy, making it possible to carry out highly accurate calculations on systems containing up to ten atoms.

5.8   Perturbation theory

When the Hartree–Fock wave function provides a reasonably accurate description of the electronic structure, it is tempting to try and improve on it by the application of perturbation theory. Indeed, this approach to the correlation problem has been quite successful in quantum chemistry – in particular, in the form of Møller–Plesset perturbation theory, to which we now turn our attention.

5.8.1   MØLLER–PLESSET PERTURBATION THEORY

In *Møller–Plesset perturbation theory (MPPT)*, the electronic Hamiltonian $\hat{H}$ in (5.1.2) is partitioned as

$$\hat{H} = \hat{f} + \hat{\Phi} + h_{\text{nuc}} \tag{5.8.1}$$

where $\hat{f}$ is the *Fock operator* (5.4.4) of Section 5.4.2, $\hat{\Phi}$ the *fluctuation potential* and h_{nuc} the nuclear–nuclear term. The fluctuation potential represents the difference between the true two-electron Coulomb potential $\hat{g}$ of the Hamiltonian operator (5.1.2) and the effective one-electron Fock potential $\hat{V}$ of the Fock operator (5.4.5):

$$\hat{\Phi} = \hat{H} - \hat{f} - h_{\text{nuc}} = \hat{g} - \hat{V} \tag{5.8.2}$$

In MPPT, the Fock operator represents the zero-order operator and the fluctuation potential the perturbation. The zero-order electronic state is represented by the Hartree–Fock state in the canonical representation

$$\hat{f}|\text{HF}\rangle = \sum_I \varepsilon_I |\text{HF}\rangle \tag{5.8.3}$$

where the summation is over the occupied spin orbitals. The excited zero-order states are spanned by all single, double and higher excitations $|\mu\rangle$ with respect to the Hartree–Fock state.

Applying the standard machinery of perturbation theory, we obtain to second order in the perturbation

$$E_{\text{MP}}^{(0)} = \langle \text{HF}|\hat{f}|\text{HF}\rangle = \sum_I \varepsilon_I \tag{5.8.4}$$

$$E_{\text{MP}}^{(1)} = \langle \text{HF}|\hat{\Phi}|\text{HF}\rangle \tag{5.8.5}$$

$$E_{\text{MP}}^{(2)} = - \sum_{A>B, I>J} \frac{|g_{AIBJ} - g_{AJBI}|^2}{\varepsilon_A + \varepsilon_B - \varepsilon_I - \varepsilon_J} \tag{5.8.6}$$

in the spin-orbital basis. Thus, the Hartree–Fock energy is equal to the sum of the zero- and first-order corrections

$$E_{\text{HF}} = E_{\text{MP}}^{(0)} + E_{\text{MP}}^{(1)} + h_{\text{nuc}} = \langle \text{HF}|\hat{H}|\text{HF} \rangle \tag{5.8.7}$$

and, by adding the second-order correction, we obtain the *second-order Møller–Plesset energy*

$$E_{\text{MP2}} = E_{\text{MP}}^{(0)} + E_{\text{MP}}^{(1)} + E_{\text{MP}}^{(2)} + h_{\text{nuc}} = E_{\text{HF}} - \sum_{A>B,I>J} \frac{|g_{AIBJ} - g_{AJBI}|^2}{\varepsilon_A + \varepsilon_B - \varepsilon_I - \varepsilon_J} \tag{5.8.8}$$

The MP2 model represents a highly successful approach to the correlation problem in quantum chemistry, providing a surprisingly accurate, size-extensive correction at low cost. Higher-order corrections may be derived as well. The MP3 and MP4 corrections, in particular, have found widespread use but represent less successful compromises between cost and accuracy than does the MP2 correction. For a detailed exposition of perturbation theory, we refer to Chapter 14.

5.8.2 THE GROUND STATE OF THE WATER MOLECULE

Let us consider the application of Møller–Plesset theory to the calculation of the dissociation curve of the water molecule. In Figure 5.19, we have plotted the cc-pVDZ potential-energy curves of the restricted and unrestricted Møller–Plesset models at the MP2, MP3 and MP4 levels. Close to the equilibrium geometry, the restricted MP2 energy represents a significant improvement on the Hartree–Fock energy. Referring to the CI energy curves in Figure 5.16, we find that the MP2 energy compares favourably with that of the more expensive CISD wave function, in particular when we recall that the MP2 correlation correction is size-extensive. The comparison with coupled-cluster theory in Figure 5.18 is less favourable.

The restricted Møller–Plesset description of the dissociation process is improved as we go to higher orders in perturbation theory, in particular at the MP4 level. However, as for all methods based on the dominance of a single electronic configuration, the description deteriorates as the OH bonds are stretched, although the MP4 curve is quite satisfactory for bond distances up to $3.5a_0$. For sufficiently large distances, the restricted potential-energy curves diverge at all levels of theory.

The performance of the unrestricted Møller–Plesset theory is perhaps somewhat surprising: even though the unrestricted theory performs well in the dissociation limit, its performance in the intermediate region is altogether unsatisfactory. Thus, the hump apparent in the coupled-cluster dissociation curve in Figure 5.18 is now much more prominent and persists even at the MP4 level. In addition, a kink has appeared where the restricted and unrestricted curves separate. Clearly, unrestricted Møller–Plesset perturbation theory does not provide a uniform description of the dissociation process and does not appear to be an appropriate tool for the study of such processes.

5.8.3 CONVERGENCE OF THE MØLLER–PLESSET PERTURBATION SERIES

Like any perturbation method, the Møller–Plesset series does not converge unconditionally. In Table 5.12, we have listed the restricted Møller–Plesset energies up to order 15 at R_{ref} and $2R_{\text{ref}}$. Although the series converges at the equilibrium bond distance, the convergence is less obvious at $2R_{\text{ref}}$. Clearly, at some point along the dissociation curve, the dominance of the RHF determinant is sufficiently eroded to destroy the convergence completely, although the exact location for the onset of the divergence would be difficult to pinpoint. For a discussion of convergence in Møller–Plesset theory, we refer to Section 14.5.

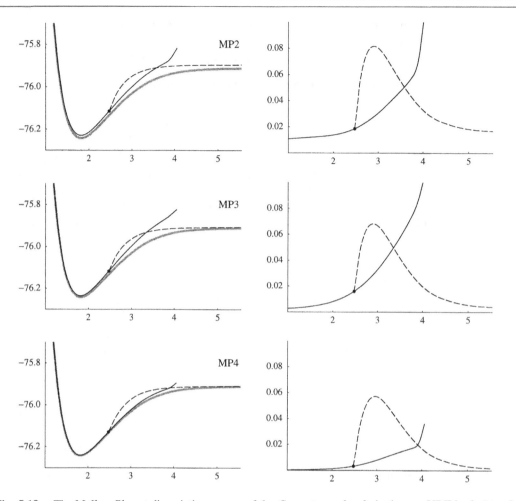

Fig. 5.19. The Møller–Plesset dissociation curves of the C_{2v} water molecule in the cc-pVDZ basis (atomic units) for a fixed HOH bond angle of 110.565°. In the two uppermost figures, we have on the left plotted the restricted MP2 dissociation curve (full black line), the unrestricted MP2 curve (dashed black line) and the FCI curve (grey line); on the right, we have plotted the corresponding differences between the MP2 and FCI energies. In the middle and lower figures, we have made similar plots for the MP3 and MP4 models.

As a minor point of interest, we note that the Møller–Plesset energy in Table 5.12 occasionally falls below that of the FCI wave function. This behaviour should not surprise us since the Møller–Plesset energy (like the coupled-cluster energy) is not variational. Indeed, we have already seen in Figure 5.19 that the unrestricted Møller–Plesset energy may be higher than the corresponding restricted energy, which would never happen for a variational wave function.

5.8.4 THE GROUND STATE OF THE HYDROGEN MOLECULE

As another illustration of the performance and, in particular, the conditional convergence of the Møller–Plesset series, we have in Figure 5.20 plotted the restricted MP2, MP3, MP4 and MP50 dissociation curves for the hydrogen molecule in the cc-pVQZ basis. For comparison, we have also plotted the FCI and RHF dissociation curves in the same basis.

Table 5.12 The differences $(E_{\mathrm{MP}n} - E_{\mathrm{FCI}})$ between the restricted cc-pVDZ Møller–Plesset energies and the FCI energy (in E_h) of the C_{2v} water molecule at the OH separations R_{ref} and $2R_{\mathrm{ref}}$. The HOH bond angle is fixed at $110.565°$ and $R_{\mathrm{ref}} = 1.84345a_0$

	$R = R_{\mathrm{ref}}$	$R = 2R_{\mathrm{ref}}$
RHF	0.217822	0.363954
MP2	0.013131	0.054730
MP3	0.006439	0.069096
MP4	0.001069	0.016046
MP5	0.000511	0.016686
MP6	0.000130	0.004300
MP7	0.000067	0.000626
MP8	0.000014	−0.000475
MP9	0.000011	−0.002065
MP10	−0.000001	−0.001652
MP11	−0.000002	−0.001332
MP12	0.000000	−0.001130
MP13	0.000000	−0.000523
MP14	0.000000	−0.000397
MP15	0.000000	−0.000146

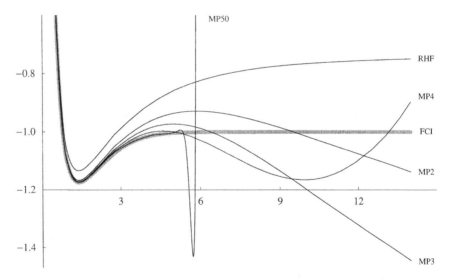

Fig. 5.20. The cc-pVQZ potential-energy curves of the hydrogen molecule at the FCI, RHF, MP2, MP3, MP4 and MP50 levels (atomic units). The oscillations in the MP50 potential-energy curve for bond lengths beyond $6a_0$ are not shown.

For bond distances shorter than $5a_0$, the perturbation series converges and MPPT provides excellent representations of the FCI curve, particularly for distances shorter than $3a_0$. At 5–$6a_0$, however, the MPPT convergence is disrupted, as seen from the pathological behaviour of the MP50 curve. Also, the lower-order curves behave in an unphysical manner for long bond distances, although in a less spectacular fashion. Thus, the MP2, MP3 and MP4 curves bend downwards and intersect the FCI curve at the internuclear separations of 9.5, 6.4 and $5.2a_0$, respectively. At a separation of $13.2a_0$, the MP4 curve rises above the FCI curve again.

5.8.5 FINAL COMMENTS

Møller–Plesset perturbation theory represents a useful approach to the calculation of size-extensive correlation energies for systems dominated by a single electronic configuration. The MP2 model, in particular, represents a successful compromise between computational cost and accuracy. Higher-order corrections may also be calculated, but it should be emphasized that the Møller–Plesset series does not converge unconditionally.

The application of Møller–Plesset theory is limited to systems dominated by a single configuration, making the theory ill suited to the study of near-degeneracy problems. Also, since Møller–Plesset theory is based on the Hartree–Fock description, it is not well suited to the study of excited electronic states. For such problems, *multiconfigurational perturbation theory* has been developed, based on the dominance of an active reference space. Multiconfigurational perturbation theory is discussed in Chapter 14, as part of a detailed exposition of perturbation methods in electronic-structure theory.

References

[1]   K. P. Huber and G.H. Herzberg, *Constants of Diatomic Molecules*, Van Nostrand Reinhold, 1979.
[2]   P.-O. Löwdin, *Adv. Chem. Phys.* **2**, 207 (1959).
[3]   J. Olsen, P. Jørgensen, H. Koch, A. Balkova and R. J. Bartlett, *J. Chem. Phys.* **104**, 8007 (1996).

Further reading

J. Almlöf, Notes on Hartree–Fock theory and related topics, in B. O. Roos (ed.), *Lecture Notes in Quantum Chemistry II*, Lecture Notes in Chemistry Vol. 64, Springer-Verlag, 1994, p. 1.
R. J. Bartlett, Coupled-cluster theory: an overview of recent developments, in D. R. Yarkony (ed.), *Modern Electronic Structure Theory*, World Scientific, 1995, p. 1047.
B. O. Roos, The complete active space self-consistent field method and its applications in electronic structure calculations, *Adv. Chem. Phys.* **69**, 399 (1987).
B. O. Roos, The multiconfigurational (MC) self-consistent field (SCF) theory, in B. O. Roos (ed.), *Lecture Notes in Quantum Chemistry*, Lecture Notes in Chemistry Vol. 58, Springer-Verlag, 1992, p. 177.
P. E. M. Siegbahn, The configuration interaction method, in B. O. Roos (ed.), *Lecture Notes in Quantum Chemistry*, Lecture Notes in Chemistry Vol. 58, Springer-Verlag, 1992, p. 255.
P. R. Taylor, Coupled-cluster methods in quantum chemistry, in B. O. Roos, (ed.), *Lecture Notes in Quantum Chemistry II*, Lecture Notes in Chemistry Vol. 64, Springer-Verlag, 1994, p. 125.
H.-J. Werner, Matrix-formulated direct multiconfiguration self-consistent field and multiconfiguration reference configuration-interaction methods, *Adv. Chem. Phys.* **69**, 1 (1987).

Exercises

EXERCISE 5.1

Calculate the overlap integral

$$S = \int 1s_A(\mathbf{r})1s_B(\mathbf{r})\,d\mathbf{r} \tag{5E.1.1}$$

between two normalized s orbitals centred on nuclei A and B

$$1s_A(\mathbf{r}) = \frac{1}{\sqrt{\pi}} \exp(-r_A) \tag{5E.1.2}$$

$$1s_B(\mathbf{r}) = \frac{1}{\sqrt{\pi}} \exp(-r_B) \tag{5E.1.3}$$

where r_A and r_B are the distances between the electron and nuclei A and B, respectively. Use the confocal elliptical coordinates (μ, ν, ϕ), where μ and ν are defined in terms of r_A and r_B and the internuclear separation R as

$$\mu = \frac{r_A + r_B}{R}, \qquad 1 \leq \mu < \infty \tag{5E.1.4}$$

$$\nu = \frac{r_A - r_B}{R}, \qquad -1 \leq \nu \leq 1 \tag{5E.1.5}$$

and ϕ is the azimuthal angle $0 \leq \phi \leq \pi$. The volume element is given by

$$d\mathbf{r} = \tfrac{1}{8}R^3(\mu^2 - \nu^2)\, d\mu\, d\nu\, d\phi \tag{5E.1.6}$$

EXERCISE 5.2

Calculate the nonvanishing one- and two-electron density-matrix elements

$$D_{pq} = \langle 0|E_{pq}|0\rangle \tag{5E.2.1}$$

$$d_{pqrs} = \langle 0|e_{pqrs}|0\rangle = \sum_{\sigma\tau} \langle 0|a_{p\sigma}^\dagger a_{r\tau}^\dagger a_{s\tau} a_{q\sigma}|0\rangle \tag{5E.2.2}$$

for the following states of the hydrogen molecule in the minimal basis of Section 5.2:
1. the single-configuration singlet states

$$|1\sigma_g^2\rangle = a_{1\alpha}^\dagger a_{1\beta}^\dagger|\text{vac}\rangle \tag{5E.2.3}$$

$$|1\sigma_u^2\rangle = a_{2\alpha}^\dagger a_{2\beta}^\dagger|\text{vac}\rangle \tag{5E.2.4}$$

2. the two-configuration singlet state

$$|{}^1\Sigma_g^+(\tau)\rangle = \cos(\tau)|1\sigma_g^2\rangle + \sin(\tau)|1\sigma_u^2\rangle \tag{5E.2.5}$$

3. the triplet and singlet ungerade states

$$|{}^3\Sigma_u^+\rangle = \frac{1}{\sqrt{2}}(a_{2\alpha}^\dagger a_{1\beta}^\dagger + a_{2\beta}^\dagger a_{1\alpha}^\dagger)|\text{vac}\rangle \tag{5E.2.6}$$

$$|{}^1\Sigma_u^+\rangle = \frac{1}{\sqrt{2}}(a_{2\alpha}^\dagger a_{1\beta}^\dagger - a_{2\beta}^\dagger a_{1\alpha}^\dagger)|\text{vac}\rangle \tag{5E.2.7}$$

Solutions

SOLUTION 5.1

Introducing the confocal elliptical coordinates and carrying out the integration first over the azimuthal angle ϕ and then over ν, we obtain

$$
\begin{aligned}
S &= \frac{1}{\pi} \int \exp[-(r_A + r_B)] \, d\mathbf{r} \\
&= \frac{1}{4} R^3 \int_{-1}^{1} \int_{1}^{\infty} \exp(-R\mu)(\mu^2 - \nu^2) \, d\mu \, d\nu \\
&= \frac{1}{2} R^3 \int_{1}^{\infty} \mu^2 \exp(-R\mu) \, d\mu - \frac{1}{6} R^3 \int_{1}^{\infty} \exp(-R\mu) \, d\mu
\end{aligned}
\tag{5S.1.1}
$$

Repeated partial integration

$$
\int x^m \exp(ax) \, dx = \frac{x^m \exp(ax)}{a} - \frac{m}{a} \int x^{m-1} \exp(ax) \, dx
\tag{5S.1.2}
$$

now gives the following expression for the overlap integral (5E.1.1):

$$
S = \left(1 + R + \tfrac{1}{3} R^2\right) \exp(-R)
\tag{5S.1.3}
$$

SOLUTION 5.2

1. In the state (5E.2.3), there is only one occupied orbital. The nonzero one-electron density-matrix elements must therefore have $p = q = 1$:

$$
D_{11}(1\sigma_g^2) = \langle 1\sigma_g^2 | E_{11} | 1\sigma_g^2 \rangle = 2
\tag{5S.2.1}
$$

Likewise, the nonzero two-electron density-matrix elements have $p = q = r = s = 1$ for this state:

$$
d_{1111}(1\sigma_g^2) = \sum_{\sigma\tau} \langle 1\sigma_g^2 | a_{1\sigma}^\dagger a_{1\tau}^\dagger a_{1\tau} a_{1\sigma} | 1\sigma_g^2 \rangle = 2
\tag{5S.2.2}
$$

By the same arguments, the nonzero density-matrix elements for (5E.2.4) are given by

$$
D_{22}(1\sigma_u^2) = \langle 1\sigma_u^2 | E_{22} | 1\sigma_u^2 \rangle = 2
\tag{5S.2.3}
$$

$$
d_{2222}(1\sigma_u^2) = \sum_{\sigma\tau} \langle 1\sigma_u^2 | a_{2\sigma}^\dagger a_{2\tau}^\dagger a_{2\tau} a_{2\sigma} | 1\sigma_u^2 \rangle = 2
\tag{5S.2.4}
$$

2. To determine the nonzero one- and two-electron density-matrix elements

$$
D_{pq}(\tau) = \langle {}^1\Sigma_g^+(\tau) | E_{pq} | {}^1\Sigma_g^+(\tau) \rangle
\tag{5S.2.5}
$$

$$
d_{pqrs}(\tau) = \langle {}^1\Sigma_g^+(\tau) | e_{pqrs} | {}^1\Sigma_g^+(\tau) \rangle
\tag{5S.2.6}
$$

we first expand these elements according to (5E.2.5):

$$
\begin{aligned}
D_{pq}(\tau) = {}&\cos^2(\tau) D_{pq}(1\sigma_g^2) + \sin^2(\tau) D_{pq}(1\sigma_u^2) \\
&+ \cos(\tau) \sin(\tau) (\langle 1\sigma_g^2 | E_{pq} | 1\sigma_u^2 \rangle + \langle 1\sigma_u^2 | E_{pq} | 1\sigma_g^2 \rangle)
\end{aligned}
\tag{5S.2.7}
$$

$$d_{pqrs}(\tau) = \cos^2(\tau)d_{pqrs}(1\sigma_g^2) + \sin^2(\tau)d_{pqrs}(1\sigma_u^2)$$

$$+\cos(\tau)\sin(\tau)(\langle1\sigma_g^2|e_{pqrs}|1\sigma_u^2\rangle + \langle1\sigma_u^2|e_{pqrs}|1\sigma_g^2\rangle) \tag{5S.2.8}$$

Thus, in addition to the density-matrix elements for $|1\sigma_g^2\rangle$ and $|1\sigma_u^2\rangle$ determined in Exercise 5.2.1, we must consider the transition-density elements. The one-electron transition-density elements in (5S.2.7) vanish since a single-excitation operator cannot connect two determinants that differ by more than a single occupation. By contrast, because of the double-excitation operator present in the two-electron transition-density elements

$$\langle1\sigma_g^2|e_{pqrs}|1\sigma_u^2\rangle = \sum_{\sigma\tau}\langle1\sigma_g^2|a_{po}^\dagger a_{rt}^\dagger a_{st} a_{q\sigma}|1\sigma_u^2\rangle \tag{5S.2.9}$$

there is a nonzero element for $p = r = 1$ and $s = q = 2$

$$\langle1\sigma_g^2|e_{1212}|1\sigma_u^2\rangle = \sum_{\sigma\tau}\langle1\sigma_g^2|a_{1\sigma}^\dagger a_{1\tau}^\dagger a_{2\tau} a_{2\sigma}|1\sigma_u^2\rangle = 2 \tag{5S.2.10}$$

that contributes to $d_{1212}(\tau)$ in (5S.2.8). Likewise, there is a nonzero contribution to $d_{2121}(\tau)$:

$$\langle1\sigma_u^2|e_{2121}|1\sigma_g^2\rangle = \sum_{\sigma\tau}\langle1\sigma_u^2|a_{2\sigma}^\dagger a_{2\tau}^\dagger a_{1\tau} a_{1\sigma}|1\sigma_g^2\rangle = 2 \tag{5S.2.11}$$

From these considerations, we arrive at the following nonvanishing density-matrix elements for (5E.2.5):

$$D_{11}(\tau) = 2\cos^2(\tau) \tag{5S.2.12}$$

$$D_{22}(\tau) = 2\sin^2(\tau) \tag{5S.2.13}$$

$$d_{1111}(\tau) = 2\cos^2(\tau) \tag{5S.2.14}$$

$$d_{2222}(\tau) = 2\sin^2(\tau) \tag{5S.2.15}$$

$$d_{1212}(\tau) = d_{2121}(\tau) = 2\cos(\tau)\sin(\tau) = \sin(2\tau) \tag{5S.2.16}$$

3. Let us first consider the one-electron density matrix. Since (5E.2.6) and (5E.2.7) both contain two singly occupied orbitals, the one-electron density matrices must be diagonal, with the same occupations for the two orbitals:

$$D_{11}(^3\Sigma_u^+) = D_{11}(^1\Sigma_u^+) = 1 \tag{5S.2.17}$$

$$D_{22}(^3\Sigma_u^+) = D_{22}(^1\Sigma_u^+) = 1 \tag{5S.2.18}$$

To determine the two-electron density matrices, we insert (5E.2.6) and (5E.2.7) in (5E.2.2) and obtain upon expansion

$$d_{pqrs}^\pm = \frac{1}{2}\sum_{\sigma\tau}\langle\mathrm{vac}|a_{1\beta}a_{2\alpha}a_{po}^\dagger a_{rt}^\dagger a_{st} a_{q\sigma}a_{2\alpha}^\dagger a_{1\beta}^\dagger|\mathrm{vac}\rangle \pm \frac{1}{2}\sum_{\sigma\tau}\langle\mathrm{vac}|a_{1\alpha}a_{2\beta}a_{po}^\dagger a_{rt}^\dagger a_{st} a_{q\sigma}a_{2\alpha}^\dagger a_{1\beta}^\dagger|\mathrm{vac}\rangle$$

$$+ \frac{1}{2}\sum_{\sigma\tau}\langle\mathrm{vac}|a_{1\alpha}a_{2\beta}a_{po}^\dagger a_{rt}^\dagger a_{st} a_{q\sigma}a_{2\beta}^\dagger a_{1\alpha}^\dagger|\mathrm{vac}\rangle \pm \frac{1}{2}\sum_{\sigma\tau}\langle\mathrm{vac}|a_{1\beta}a_{2\alpha}a_{po}^\dagger a_{rt}^\dagger a_{st} a_{q\sigma}a_{2\beta}^\dagger a_{1\alpha}^\dagger|\mathrm{vac}\rangle$$

$$\tag{5S.2.19}$$

where the plus sign refers to the triplet state and the minus sign to the singlet state. This expression may be simplified, noting that the last two terms are identical to the first two except that the spin labels α and β have been interchanged. Since the summations are symmetric in α and β, we may write (5S.2.19) in the form

$$d_{pqrs}^{\pm} = \sum_{\sigma\tau} \langle \text{vac} | a_{1\beta} a_{2\alpha} a_{p\sigma}^{\dagger} a_{r\tau}^{\dagger} a_{s\tau} a_{q\sigma} a_{2\alpha}^{\dagger} a_{1\beta}^{\dagger} | \text{vac} \rangle \pm \sum_{\sigma\tau} \langle \text{vac} | a_{1\alpha} a_{2\beta} a_{p\sigma}^{\dagger} a_{r\tau}^{\dagger} a_{s\tau} a_{q\sigma} a_{2\alpha}^{\dagger} a_{1\beta}^{\dagger} | \text{vac} \rangle \tag{5S.2.20}$$

To identify the nonzero elements, we first note that p and r must be different and also that q and s must be different. There are then four possibilities for the indices $pqrs$: 1122, 2211, 1221 and 2112. The elements $d_{1122}^{\pm}$ and $d_{2211}^{\pm}$ are equal by particle symmetry, as are $d_{1221}^{\pm}$ and $d_{2112}^{\pm}$. Only the elements $d_{1122}^{\pm}$ and $d_{1221}^{\pm}$ must therefore be determined explicitly. For $d_{1122}^{\pm}$, we obtain

$$d_{1122}^{\pm} = \sum_{\sigma\tau} \langle \text{vac} | a_{1\beta} a_{2\alpha} a_{1\sigma}^{\dagger} a_{2\tau}^{\dagger} a_{2\tau} a_{1\sigma} a_{2\alpha}^{\dagger} a_{1\beta}^{\dagger} | \text{vac} \rangle \pm \sum_{\sigma\tau} \langle \text{vac} | a_{1\alpha} a_{2\beta} a_{1\sigma}^{\dagger} a_{2\tau}^{\dagger} a_{2\tau} a_{1\sigma} a_{2\alpha}^{\dagger} a_{1\beta}^{\dagger} | \text{vac} \rangle \tag{5S.2.21}$$

The second sum does not contribute since σ cannot be equal to both α and β and we obtain

$$d_{1122}^{\pm} = \langle \text{vac} | a_{1\beta} a_{2\alpha} a_{1\beta}^{\dagger} a_{2\alpha}^{\dagger} a_{2\alpha} a_{1\beta} a_{2\alpha}^{\dagger} a_{1\beta}^{\dagger} | \text{vac} \rangle = 1 \tag{5S.2.22}$$

For $d_{1221}^{\pm}$, on the other hand, only the second sum in (5S.2.20) contributes, giving

$$d_{1221}^{\pm} = \pm \langle \text{vac} | a_{1\alpha} a_{2\beta} a_{1\alpha}^{\dagger} a_{2\beta}^{\dagger} a_{1\beta} a_{2\alpha} a_{2\alpha}^{\dagger} a_{1\beta}^{\dagger} | \text{vac} \rangle = \mp 1 \tag{5S.2.23}$$

We may now summarize our results as follows:

$$d_{1122}(^{3}\Sigma_{u}^{+}) = d_{2211}(^{3}\Sigma_{u}^{+}) = d_{1122}(^{1}\Sigma_{u}^{+}) = d_{2211}(^{1}\Sigma_{u}^{+}) = 1 \tag{5S.2.24}$$

$$d_{1221}(^{3}\Sigma_{u}^{+}) = d_{2112}(^{3}\Sigma_{u}^{+}) = -d_{1221}(^{1}\Sigma_{u}^{+}) = -d_{2112}(^{1}\Sigma_{u}^{+}) = -1 \tag{5S.2.25}$$

6 ATOMIC BASIS FUNCTIONS

The standard wave functions of quantum chemistry are all constructed from antisymmetric products of MOs. In most applications, these MOs are generated by expansion in a finite set of simple analytical functions – the atomic basis functions. The choice of basis functions for a molecular calculation is therefore an important one, which ultimately determines the quality of the wave function. In this chapter, the mathematical properties of the atomic basis functions are investigated and their usefulness is explored by carrying out simple expansions of the ground-state orbitals of the carbon atom.

The present chapter is the first of four chapters that are concerned with basis functions and basis sets. In Chapter 7, we go on to consider the expansion of the two-electron helium ground-state wave function in products of atomic basis functions, preparing ourselves for the construction of general molecular basis sets in Chapter 8. The evaluation of integrals over atomic basis functions is discussed in Chapter 9.

6.1 Requirements on one-electron basis functions

Molecular orbitals may be constructed in one of two ways: numerically or algebraically by linear expansion in a set of basis functions. The *numerical approach* offers great flexibility and accuracy, but it is computationally intractable because of the large number of grid points needed except for highly symmetric systems such as atoms [1] and linear molecules [2,3]. For the latter systems, accurate numerical calculations are often very useful in providing benchmarks with which results obtained by the algebraic method can be compared. For polyatomic systems, we are forced to take the *algebraic approach* and expand the MOs $\phi_p(\mathbf{r})$ in a set of simple analytical one-electron functions:

$$\phi_p(\mathbf{r}) = \sum_\mu C_{\mu p} \chi_\mu(\mathbf{r}) \tag{6.1.1}$$

The question then arises as to what functions $\chi_\mu(\mathbf{r})$ are suitable for orbital expansions. Ideally, a suitable set of basis functions should fulfil the following requirements:

1. The basis should be designed such that it allows for an orderly and systematic extension towards completeness with respect to one-electron square-integrable functions.

2. The basis should allow for a rapid convergence to any atomic or molecular electronic state, requiring only a few terms for a reasonably accurate description of molecular electron distributions.

3. The functions should have an analytical form that allows for easy manipulation. In particular, all molecular integrals over these functions should be easy to evaluate. It is also desirable that the basis functions are orthogonal or at least that their nonorthogonality does not present problems related to numerical instability.

In practice, it has proved difficult to construct basis sets that combine all these properties. For example, as we shall see in Section 6.2, it is not difficult to set up a complete orthonormal set of basis functions that may be applied to any molecular system in an unambiguous fashion. The problem is that such a set in general does not converge rapidly to the exact solution and therefore becomes intractable because of the large number of terms that must be included. A useful set of one-electron functions must therefore represent a compromise between the three requirements. In particular, we shall see that the Gaussian basis functions, introduced in Section 6.6, are very successful in this respect, combining reasonably short expansions with a fast integral evaluation.

Before proceeding with the construction of basis sets, let us be more precise about the meaning of completeness. We are here concerned with *square-integrable one-electron functions* $f(\mathbf{r})$ of three spatial coordinates – that is, functions belonging to the Hilbert space $L^2(\mathbb{R}^3)$. (Although we shall not be concerned with the details and niceties of Hilbert-space theory here, we note that a *Hilbert space* is an inner-product space for which every Cauchy sequence converges to an element of that space [4–7].) We wish to approximate our functions $f(\mathbf{r})$ as accurately as possible by linear expansions in basis functions $h_n(\mathbf{r})$ belonging to $L^2(\mathbb{R}^3)$. The basis is said to be *complete* if, for any given $\varepsilon > 0$, we can satisfy the equation

$$\left\| f - \sum_{n=1}^{N} c_n h_n \right\| < \varepsilon \tag{6.1.2}$$

by including a sufficiently large number N of functions in the expansion. Here we use the notation

$$\| f \| = \sqrt{\langle f | f \rangle} \tag{6.1.3}$$

for the norm of a function in Hilbert space. Thus, we are concerned with *completeness in the norm* (6.1.3) rather than with pointwise convergence. We note that the requirement of completeness in the norm does not in general ensure pointwise convergence. For example, complete sets $h_n(\mathbf{r})$ exist in the sense of (6.1.2) for which $h_n(\mathbf{0}) = 0$ for all n. Pointwise convergence to functions that are nonzero at the origin then cannot be obtained. The fact that it is sufficient to have completeness in the norm is a fundamental and nontrivial result for certain classes of partial differential equations such as the Schrödinger equation.

Two more points should be made concerning convergence. First, if a set of one-electron functions is complete in $L^2(\mathbb{R}^3)$, then the set of all possible Slater determinants that can be constructed from these functions constitutes a complete basis in the space of all antisymmetric square-integrable functions. Thus, the use of a complete one-electron basis ensures that any N-electron wave function – the exact wave function, in particular – can be approximated to any accuracy. Second, in an FCI calculation, completeness in the norm (6.1.3) does not guarantee convergence to the exact energy and wave function. The kinetic-energy operator T is a differential operator, which requires completeness with respect to the *Sobolev norms* $\| f \|_{S1}$ with weight $\sqrt{1 + T}$ and $\| f \|_{S2}$ with weight $1 + T$ to ensure convergence to the exact energy and wave function [7,8]. The question of completeness is thus a complicated one, depending, for example, on the form of the Hamiltonian employed in our calculations.

6.2 One- and many-centre expansions

As an introduction to the subject of one-electron basis functions, it is instructive to see how a simple, universal basis may be set up. Since the eigenfunctions of any Hermitian operator constitute a complete orthonormal set (assuming discrete eigenvalues), we may simply take our basis functions to be the eigenfunctions

$$H\psi_n = E_n\psi_n \tag{6.2.1}$$

of a suitable one-electron Hamiltonian

$$H = -\tfrac{1}{2}\nabla^2 + V \tag{6.2.2}$$

It remains only to specify the form of the bounded potential V. Clearly, the potential should be such that the eigenfunctions (6.2.1) can be written in a closed analytical form. Also, since our prescription for setting up the basis should be valid for any molecule, the potential should take the form of a superposition of approximate atomic potentials

$$V = \sum_A V_A \tag{6.2.3}$$

in such a way that V allows for an analytical solution. A suitable, simple choice of V_A would be the three-dimensional *harmonic potentials*

$$V_A = \tfrac{1}{2}k_A r_A^2 \tag{6.2.4}$$

where

$$\mathbf{r}_A = \mathbf{r} - \mathbf{A} \tag{6.2.5}$$

is the position of the electron with respect to nucleus A at $\mathbf{A}$ and k_A is a *force constant* associated with that nucleus. The suitability of the harmonic potential arises from the fact that a superposition of such potentials remains harmonic. Thus, in Exercise 6.1, it is shown that

$$\tfrac{1}{2}\sum_A k_A r_A^2 = V^{(0)} + \tfrac{1}{2}K r_S^2 \tag{6.2.6}$$

where

$$K = \sum_A k_A \tag{6.2.7}$$

$$V^{(0)} = \frac{1}{2K}\sum_{A>B} k_A k_B R_{AB}^2 \tag{6.2.8}$$

$$\mathbf{S} = K^{-1}\sum_A k_A \mathbf{A} \tag{6.2.9}$$

$$\mathbf{R}_{AB} = \mathbf{A} - \mathbf{B} \tag{6.2.10}$$

For a given molecular system, therefore, the basis functions correspond to the eigenfunctions of a harmonic potential centred at $\mathbf{S}$ with force constant K. We have thus succeeded in constructing, for any system of any size, a complete and discrete orthonormal basis of *harmonic-oscillator (HO) functions.* Our prescription is universal in the sense that the basis is completely specified once and for all by assigning a force constant to each atom type.

Unfortunately, the above prescription is not very useful since it leads to slow convergence, in particular for large molecules. The reason for this predicament is quite clear: We have not incorporated much of the physical characteristics of the electronic system in our basis. Presumably, a very large number of functions would be required for an accurate representation of the electronic system, in particular far away from the centre of expansion. Thus, even though we have succeeded in generating a basis that fulfils the first and third requirements in Section 6.1 (integrals over HO functions are easy to evaluate), we have failed to fulfil the second requirement. Clearly, a different approach is called for if we wish to obtain a basis that converges with a modest number of functions. In particular, if we can succeed in capturing some of the physical characteristics of the electronic system and incorporate these in the analytical form of the basis functions, we should be able to reduce the number of functions needed for an accurate description of the system.

We note that, for molecular systems, the atomic electron distributions are largely unaffected by the formation of chemical bonds. We may therefore first concentrate our attention on atomic systems, seeking a set of simple analytical functions suitable for expansions of orbitals in many-electron atoms. Once such functions have been found, we may generalize our procedure to polyatomic molecules by introducing a separate basis of AOs for each atom in the molecule, being careful to include in our basis any additional functions that may be needed to describe the molecular bonding.

This *atoms-in-molecule approach* to the design of basis functions has several advantages. First, we may incorporate in our basis some of the important physical characteristics of the atomic systems, thus reducing the number of functions needed for an accurate description of the electronic system. Second, since the functions are associated with and attached to the nuclei, our representation of the electronic system should be of more or less the same quality for molecules of all sizes and all geometrical configurations. Even so, in a finite basis, the exact same quality can never be achieved across a potential-energy surface, creating subtle problems when the results of calculations carried out at different geometries are compared – for example, in the study of van der Waals complexes in Section 8.5.

Their advantages notwithstanding, there are certain complications associated with the use of many-centre as opposed to one-centre basis sets. First, in one-centre expansions, analytic relationships such as orthogonality and recurrence relations often simplify the calculations; in general, no such relationships exist between functions on different centres. Second, the use of atom-fixed functions complicates the calculation of properties related to geometrical distortions, since we must take into account also the effects of basis-set distortion. Third, the use of many-centre expansions complicates the extension towards completeness. Clearly, we cannot freely extend each atomic basis since this would introduce linear dependencies between functions on different centres. In contrast, the extension of one-centre bases is simple and unambiguous. All things considered, however, many-centre atoms-in-molecule expansions are much more compact and economical than one-centre expansions, in particular for large molecules.

6.3 The one-electron central-field system

Let us consider in general terms the quantum-mechanical treatment of *one-electron systems in a central field*. Such systems are treated in detail in most textbooks on quantum mechanics; see, for example, [9]. Our emphasis here is on the separation of the solution into radial and angular parts. In particular, we shall see that the angular part of the solution is independent of the form of the central field. The angular solution will therefore reappear in all one-electron basis functions considered in this chapter, their differences being only in their radial forms.

For an electron in a central field, the Schrödinger equation may be written as

$$-\tfrac{1}{2}\nabla^2\psi + V(r)\psi = E\psi \tag{6.3.1}$$

where the potential $V(r)$ is *spherically symmetric*, depending only on the separation r of the electron from the origin. Because of the spherical symmetry of the potential, it is convenient to solve the Schrödinger equation (6.3.1) in polar coordinates (r, θ, φ), related to the Cartesian (x, y, z) as follows:

$$x = r\sin\theta\cos\varphi$$
$$y = r\sin\theta\sin\varphi \tag{6.3.2}$$
$$z = r\cos\theta$$

In polar coordinates, the *Laplacian* in the Schrödinger equation (6.3.1) takes the form

$$\nabla^2 = \frac{1}{r^2}\frac{\partial}{\partial r}\left(r^2\frac{\partial}{\partial r}\right) - \frac{L^2}{r^2} \tag{6.3.3}$$

where L^2 is the operator for the total angular momentum about the origin

$$L^2 = -\frac{1}{\sin\theta}\frac{\partial}{\partial\theta}\left(\sin\theta\frac{\partial}{\partial\theta}\right) - \frac{1}{\sin^2\theta}\frac{\partial^2}{\partial\varphi^2} \tag{6.3.4}$$

All angular dependencies are thus isolated in the angular-momentum operator and there is no term in the Hamiltonian that involves differentiation with respect to both the radial and the angular coordinates. We therefore attempt a solution to the Schrödinger equation in the product form

$$\psi(r, \theta, \varphi) = R(r)Y(\theta, \varphi) \tag{6.3.5}$$

where the *radial part* $R(r)$ depends only on the radial coordinate and the *angular part* $Y(\theta, \varphi)$ only on the angular coordinates. In passing, we note that this is not a separation in the usual sense since the part of the Hamiltonian that contains the angular-momentum operator depends also on the inverse square of the radial distance; see (6.3.3). We therefore expect the radial part of the solution (6.3.5) to depend on the angular solution.

Inserting the product form of the wave function (6.3.5) into the Schrödinger equation (6.3.1) with the Laplacian in polar form (6.3.3), we find, after some simple algebra, that the angular solution must be an eigenfunction of the total angular momentum. Since the Laplacian (6.3.3) and hence the Hamiltonian in (6.3.1) commute with the operator for the total angular momentum L^2 and also with the operator for the projected angular momentum L_z

$$L_z = -\mathrm{i}\frac{\partial}{\partial\varphi} \tag{6.3.6}$$

we may choose the angular part of the one-electron wave function ψ to correspond to the usual *spherical harmonics* [10,11]

$$L^2 Y_{lm}(\theta, \varphi) = l(l+1)Y_{lm}(\theta, \varphi) \tag{6.3.7}$$

$$L_z Y_{lm}(\theta, \varphi) = mY_{lm}(\theta, \varphi) \tag{6.3.8}$$

where l runs over all integers $l \geq 0$ and m over all integers $|m| \leq l$. The spherical harmonics form a complete orthonormal set of square-integrable functions on the sphere – that is, in the Hilbert space $L^2(S)$ – and are normalized as

$$\int_0^{2\pi} \int_0^{\pi} Y_{lm}^*(\theta, \varphi) Y_{l'm'}(\theta, \varphi) \sin \theta \, d\theta \, d\varphi = \delta_{ll'} \delta_{mm'} \tag{6.3.9}$$

The spherical harmonics are discussed in detail in Section 6.4.

With the angular part in the form of a spherical harmonic, we find that the radial part may be obtained from the differential equation

$$-\frac{1}{2} \frac{\partial^2 R(r)}{\partial r^2} - \frac{1}{r} \frac{\partial R(r)}{\partial r} + \left[V(r) + \frac{l(l+1)}{2r^2} \right] R(r) = ER(r) \tag{6.3.10}$$

In addition to the (presumably) attractive potential $V(r)$, the radial equation contains a repulsive *centrifugal potential*, which increases with the square of the total angular momentum. The radial part of the solution (6.3.5) therefore depends on the quantum number l for the total angular momentum of the system.

Since the differential equation for the radial part (6.3.10) contains a term linear in the differential operator, it is not in the standard Hermitian form of a one-dimensional Schrödinger equation. To eliminate the linear term, we multiply (6.3.10) from the left by r and obtain, after some rearrangement, a standard Schrödinger equation

$$-\frac{1}{2} \frac{\partial^2 r R_{nl}(r)}{\partial r^2} + \left[V(r) + \frac{l(l+1)}{2r^2} \right] r R_{nl}(r) = E_{nl} r R_{nl}(r) \tag{6.3.11}$$

The radial solutions are here distinguished by two quantum numbers – recognizing that, for each $l \geq 0$, we may solve (6.3.11) and obtain a complete set of orthonormal radial functions $r R_{nl}(r)$ in $L^2(\mathbb{R}^+)$

$$\int_0^{\infty} R_{ml}^*(r) R_{nl}(r) r^2 dr = \delta_{mn} \tag{6.3.12}$$

or, equivalently, a complete orthonormal set $R_{nl}(r)$ in $L^2(\mathbb{R}^+, r^2)$. Note that radial functions of different quantum numbers l are not orthogonal, being eigenfunctions of Hamiltonians with different effective potentials (6.3.11).

In conclusion, the wave function for a one-electron central-field system (6.3.1) may be written as a product of a radial function and a spherical-harmonic angular function

$$\psi_{nlm}(r, \theta, \varphi) = R_{nl}(r) Y_{lm}(\theta, \varphi) \tag{6.3.13}$$

Completeness and orthonormality of ψ_{nlm} in $L^2(\mathbb{R}^3)$ then follow from completeness and orthonormality of the spherical harmonics on the sphere $L^2(S)$ and of the radial functions in $L^2(\mathbb{R}^+, r^2)$ [10,11]. Whereas the angular part is the same for all systems, the radial part is obtained from (6.3.11) and depends on the potential $V(r)$. Important examples are the hydrogenic radial functions associated with the Coulomb potential (see Section 6.5.2) and the HO functions associated with the three-dimensional isotropic harmonic potential (see Section 6.6.1).

Returning to the subject of basis sets, we shall invariably employ basis functions written as products of radial and angular parts (6.3.13). For the radial part $R(r)$, we shall consider a variety of functional forms, all obtained by adapting in some way the simple analytical functions found in model systems – the hydrogenic atom and the three-dimensional HO system, for instance – to the complicated potential experienced by the electrons in a many-electron atomic system. From the discussion in this section, we know that our basis will be complete if the radial functions $R(r)$ are complete in $L^2(\mathbb{R}^+, r^2)$ – that is, in the one-dimensional Hilbert space with inner product

$$\langle f | g \rangle = \int_0^\infty f^*(r) g(r) r^2 \, \mathrm{d}r \tag{6.3.14}$$

However, before we begin the discussion of the radial part of the atomic basis functions, we shall in Section 6.4 consider the universal angular part.

6.4 The angular basis

As discussed in Section 6.3, the angular part of the atomic basis functions is universal – that is, the same for all kinds of basis functions. The purpose of the present section is to consider this angular part in more detail, preparing us for the discussion of radial functions in Sections 6.5 and 6.6.

We begin in Section 6.4.1 by reviewing, without proofs or derivations, the standard properties of the spherical harmonics and their relationship to the associated Legendre polynomials. The closely related solid harmonics are next introduced in Section 6.4.2. In Sections 6.4.3 and 6.4.4, we derive explicit Cartesian expressions for the complex and real solid harmonics, respectively. Finally, in Section 6.4.5, we derive a set of recurrence relations for the real solid harmonics.

6.4.1 THE SPHERICAL HARMONICS

The *spherical harmonic functions* [10,11] are characterized by two quantum numbers

$$L^2 Y_{lm}(\theta, \varphi) = l(l+1) Y_{lm}(\theta, \varphi) \tag{6.4.1}$$

$$L_z Y_{lm}(\theta, \varphi) = m Y_{lm}(\theta, \varphi) \tag{6.4.2}$$

and constitute a complete orthonormal set of functions in $L^2(S)$ (i.e. for the angular degrees of freedom) when the quantum numbers l and m take on the integers $l \geq 0$ and $|m| \leq l$. In Table 6.1, we have listed the normalized spherical harmonics for $l \leq 3$ in standard form. The angular-momentum eigenvalue equations (6.4.1) and (6.4.2) may be verified for these functions using L^2 in the form (6.3.4) and L_z in the form (6.3.6).

Let us consider the general form of the spherical harmonics. In the standard treatment, the spherical harmonics are formed by taking products of the associated Legendre polynomials in $\cos \theta$ and exponentials in $im\varphi$:

$$Y_{lm}(\theta, \varphi) = \sqrt{\frac{2l+1}{4\pi} \frac{(l-m)!}{(l+m)!}} \, P_l^m(\cos \theta) \exp(im\varphi) \tag{6.4.3}$$

The *associated Legendre polynomials* [11,12] may be obtained from the *Rodrigues expression* (in the phase convention of Condon and Shortley)

$$P_l^m(x) = \frac{(-1)^m}{2^l l!} (1 - x^2)^{m/2} \frac{\mathrm{d}^{l+m}}{\mathrm{d}x^{l+m}} (x^2 - 1)^l \tag{6.4.4}$$

Table 6.1 The spherical harmonics $Y_{lm}(\theta,\varphi)$ for $l \le 3$

m \ l	0	1	2	3
3				$-\frac{1}{8}\sqrt{\frac{35}{\pi}}\sin^3\theta\,e^{3i\varphi}$
2			$\frac{1}{4}\sqrt{\frac{15}{2\pi}}\sin^2\theta\,e^{2i\varphi}$	$\frac{1}{4}\sqrt{\frac{105}{2\pi}}\cos\theta\sin^2\theta\,e^{2i\varphi}$
1		$-\frac{1}{2}\sqrt{\frac{3}{2\pi}}\sin\theta\,e^{i\varphi}$	$-\frac{1}{2}\sqrt{\frac{15}{2\pi}}\cos\theta\sin\theta\,e^{i\varphi}$	$-\frac{1}{8}\sqrt{\frac{21}{\pi}}(5\cos^2\theta-1)\sin\theta\,e^{i\varphi}$
0	$\frac{1}{2\sqrt{\pi}}$	$\frac{1}{2}\sqrt{\frac{3}{\pi}}\cos\theta$	$\frac{1}{4}\sqrt{\frac{5}{\pi}}(3\cos^2\theta-1)$	$\frac{1}{4}\sqrt{\frac{7}{\pi}}(5\cos^2\theta-3)\cos\theta$
-1		$\frac{1}{2}\sqrt{\frac{3}{2\pi}}\sin\theta\,e^{-i\varphi}$	$\frac{1}{2}\sqrt{\frac{15}{2\pi}}\cos\theta\sin\theta\,e^{-i\varphi}$	$\frac{1}{8}\sqrt{\frac{21}{\pi}}(5\cos^2\theta-1)\sin\theta\,e^{-i\varphi}$
-2			$\frac{1}{4}\sqrt{\frac{15}{2\pi}}\sin^2\theta\,e^{-2i\varphi}$	$\frac{1}{4}\sqrt{\frac{105}{2\pi}}\cos\theta\sin^2\theta\,e^{-2i\varphi}$
-3				$\frac{1}{8}\sqrt{\frac{35}{\pi}}\sin^3\theta\,e^{-3i\varphi}$

which is valid for integers $l \ge 0$ and $|m| \le l$. The associated Legendre polynomials are here defined for negative as well as positive m, the relationship between these polynomials being

$$P_l^{-m}(x) = (-1)^m \frac{(l-m)!}{(l+m)!} P_l^m(x) \tag{6.4.5}$$

The associated Legendre polynomials are orthogonal on the interval $-1 \le x \le 1$ in the sense that

$$\int_{-1}^{1} P_k^m(x) P_l^m(x)\,\mathrm{d}x = \frac{2}{2l+1}\frac{(l+m)!}{(l-m)!}\delta_{kl} \tag{6.4.6}$$

The reader may wish to verify that the orthonormality (6.3.9) of the spherical harmonics in $L^2(S)$ follows from the orthogonality of the associated Legendre polynomials in $L^2(-1, 1)$.

In Table 6.2, the associated Legendre polynomials are listed for $l \le 3$. We note that, for odd integers m, the associated Legendre polynomials are not true polynomials in x. Rather, they are generalizations of the *Legendre polynomials*

$$P_l(x) = P_l^0(x) \tag{6.4.7}$$

which are true polynomials of degree l. Further generalizations of $P_l^m(x)$ to nonintegral complex m and l exist and are known as the *associated Legendre functions*.

As is usually the case for the standard orthogonal polynomials, different conventions exist. In particular, the associated Legendre polynomials are sometimes defined as

$$\mathcal{P}_l^m(x) = \sqrt{\frac{(l-m)!(l+|m|)!}{(l+m)!(l-|m|)!}}\, P_l^m(x) \tag{6.4.8}$$

This definition is identical to (6.4.4) except for negative m, obeying the relation

$$\mathcal{P}_l^{-m}(x) = (-1)^m \mathcal{P}_l^m(x) \tag{6.4.9}$$

Table 6.2 The associated Legendre polynomials $P_l^m(x)$ for $l \leq 3$

$m \backslash\, l$	0	1	2	3
3				$-15(1-x^2)^{3/2}$
2			$3(1-x^2)$	$15x(1-x^2)$
1		$-\sqrt{1-x^2}$	$-3x\sqrt{1-x^2}$	$-\dfrac{3}{2}(5x^2-1)\sqrt{1-x^2}$
0	1	x	$\dfrac{1}{2}(3x^2-1)$	$\dfrac{1}{2}(5x^2-3)x$
-1		$\dfrac{1}{2}\sqrt{1-x^2}$	$\dfrac{1}{2}x\sqrt{1-x^2}$	$\dfrac{1}{8}(5x^2-1)\sqrt{1-x^2}$
-2			$\dfrac{1}{8}(1-x^2)$	$\dfrac{1}{8}x(1-x^2)$
-3				$\dfrac{1}{48}(1-x^2)^{3/2}$

rather than (6.4.5). Finally, for the spherical harmonics, *Racah's normalization* is sometimes used [13]:

$$C_{lm}(\theta, \varphi) = \sqrt{\frac{4\pi}{2l+1}} Y_{lm}(\theta, \varphi) \tag{6.4.10}$$

In this normalization, $C_{00}(\theta, \varphi)$ is equal to 1 – compare with the spherical harmonics in Table 6.1.

6.4.2 THE SOLID HARMONICS

Henceforth, we shall work almost exclusively with radial functions of the form

$$R_{nl}(r) = r^l \mathcal{R}_{nl}(r) \tag{6.4.11}$$

Associating the radial factor r^l in (6.4.11) with the spherical-harmonic part of the one-electron functions (6.3.13), we may write the one-centre functions in the form

$$\psi_{nlm}(r, \theta, \varphi) = R_{nl}(r)\mathcal{Y}_{lm}(r, \theta, \varphi) \tag{6.4.12}$$

where we have introduced the *solid harmonics* or the *harmonic polynomials* as products of the spherical harmonics with the monomials r^l [13]:

$$\mathcal{Y}_{lm}(\mathbf{r}) = \mathcal{Y}_{lm}(r, \theta, \varphi) = r^l Y_{lm}(\theta, \varphi) \tag{6.4.13}$$

In Section 6.4.3, we shall demonstrate that the solid harmonics may be expressed as

$$\mathcal{Y}_{lm}(\mathbf{r}) = N_{lm}[x + \mathrm{sgn}(m)\,\mathrm{i}y]^{|m|} \sum_{t=0}^{[(l-|m|)/2]} C_t^{l|m|}(x^2+y^2)^t z^{l-2t-|m|} \tag{6.4.14}$$

$$C_t^{lm} = \left(-\frac{1}{4}\right)^t \binom{l-t}{|m|+t}\binom{l}{t} \tag{6.4.15}$$

$$N_{lm} = (-1)^{(m+|m|)/2}\frac{1}{2^{|m|}l!}\sqrt{\frac{2l+1}{4\pi}(l+|m|)!(l-|m|)!} \tag{6.4.16}$$

where the normalization constant agrees with the Condon–Shortley phase convention. In (6.4.14), $[k]$ is the largest integer less than or equal to k, and we have used the function $\text{sgn}(m)$, which is equal to 1 for nonnegative m and equal to -1 for negative m. As is easily verified from (6.4.14)–(6.4.16), the solid harmonics obey the following relations

$$\mathcal{Y}_{lm}(-\mathbf{r}) = (-1)^l \mathcal{Y}_{lm}(\mathbf{r}) \tag{6.4.17}$$

$$\mathcal{Y}_{lm}^*(\mathbf{r}) = (-1)^m \mathcal{Y}_{l,-m}(\mathbf{r}) \tag{6.4.18}$$

which will prove useful in our manipulation of these functions. Since the solid harmonics have been obtained from the spherical harmonics by a simple scaling with r^l, they obey the usual angular-momentum eigenvalue equations (6.4.1) and (6.4.2). This result follows since the operators L^2 in (6.3.4) and L_z in (6.3.6) are independent of r and therefore commute with r^l.

Being eigenfunctions of the angular-momentum operators, the solid harmonics $\mathcal{Y}_{lm}(\mathbf{r})$ are complex functions. In molecules without spherical or axial symmetry, nothing is gained by expanding the wave function in orbitals that are eigenfunctions of the angular-momentum operators. Usually, therefore, we avoid complex arithmetic altogether by taking the real and imaginary parts of the solid harmonics $\mathcal{Y}_{lm}(\mathbf{r})$, noting the relation (6.4.18). The *real-valued solid harmonics* are defined as

$$S_{l0} = \sqrt{\frac{4\pi}{2l+1}} \mathcal{Y}_{l0} \tag{6.4.19}$$

$$S_{lm} = (-1)^m \sqrt{\frac{8\pi}{2l+1}} \, \text{Re} \, \mathcal{Y}_{lm}, \qquad m > 0 \tag{6.4.20}$$

$$S_{l,-m} = (-1)^m \sqrt{\frac{8\pi}{2l+1}} \, \text{Im} \, \mathcal{Y}_{lm}, \qquad m > 0 \tag{6.4.21}$$

and constitute an orthogonal set of functions. To demonstrate their orthogonality, we first introduce the complex solid harmonics in Racah's normalization

$$C_{lm}(\mathbf{r}) = r^l C_{lm}(\theta, \varphi) = \sqrt{\frac{4\pi}{2l+1}} \mathcal{Y}_{lm}(\mathbf{r}) \tag{6.4.22}$$

whose orthogonality follows trivially from the orthogonality of $\mathcal{Y}_{lm}(\mathbf{r})$. Using (6.4.18), we may now express the transformation (6.4.20) and (6.4.21) in the form

$$\begin{pmatrix} S_{lm} \\ S_{l,-m} \end{pmatrix} = \frac{1}{\sqrt{2}} \begin{pmatrix} (-1)^m & 1 \\ -(-1)^m i & i \end{pmatrix} \begin{pmatrix} C_{lm} \\ C_{l,-m} \end{pmatrix}, \qquad m > 0 \tag{6.4.23}$$

where the matrix is unitary, showing that the real solid harmonics $S_{lm}(\mathbf{r})$ are related to the renormalized solid harmonics $C_{lm}(\mathbf{r})$ by a unitary transformation. The lowest-order real solid harmonics are listed in Table 6.3. In Figure 6.1, we have given a pictorial representation of the same functions.

6.4.3 EXPLICIT CARTESIAN EXPRESSIONS FOR THE COMPLEX SOLID HARMONICS

The complex solid harmonics are eigenfunctions of the total angular-momentum operator, which in (6.3.4) is given in polar coordinates. The form of this operator is rather complicated, however, making the solution of the associated eigenvalue problem (in order to arrive at the solid harmonics)

Table 6.3 The real solid harmonics $S_{lm}(\mathbf{r})$ for $l \leq 4$

$m\backslash l$	0	1	2	3	4
4					$\frac{1}{8}\sqrt{35}(x^4 - 6x^2y^2 + y^4)$
3				$\frac{1}{2}\sqrt{\frac{5}{2}}(x^2 - 3y^2)x$	$\frac{1}{2}\sqrt{\frac{35}{2}}(x^2 - 3y^2)xz$
2			$\frac{1}{2}\sqrt{3}(x^2 - y^2)$	$\frac{1}{2}\sqrt{15}(x^2 - y^2)z$	$\frac{1}{4}\sqrt{5}(7z^2 - r^2)(x^2 - y^2)$
1		x	$\sqrt{3}xz$	$\frac{1}{2}\sqrt{\frac{3}{2}}(5z^2 - r^2)x$	$\frac{1}{2}\sqrt{\frac{5}{2}}(7z^2 - 3r^2)xz$
0	1	z	$\frac{1}{2}(3z^2 - r^2)$	$\frac{1}{2}(5z^2 - 3r^2)z$	$\frac{1}{8}(35z^4 - 30z^2r^2 + 3r^4)$
−1		y	$\sqrt{3}yz$	$\frac{1}{2}\sqrt{\frac{3}{2}}(5z^2 - r^2)y$	$\frac{1}{2}\sqrt{\frac{5}{2}}(7z^2 - 3r^2)yz$
−2			$\sqrt{3}xy$	$\sqrt{15}xyz$	$\frac{1}{2}\sqrt{5}(7z^2 - r^2)xy$
−3				$\frac{1}{2}\sqrt{\frac{5}{2}}(3x^2 - y^2)y$	$\frac{1}{2}\sqrt{\frac{35}{2}}(3x^2 - y^2)yz$
−4					$\frac{1}{2}\sqrt{35}(x^2 - y^2)xy$

tedious. We now recall that the angular part of the electronic wave function is the same for all central-field systems (6.3.1). A particularly simple equation, known as *Laplace's equation*, is obtained in the limit of zero potential and energy:

$$\nabla^2 \psi = 0 \qquad (6.4.24)$$

The solid harmonics are easily seen to be solutions to Laplace's equation [14]. Thus, from the expressions (6.3.3) and (6.4.1), we find

$$\nabla^2 r^l Y_{lm} = \left(\frac{1}{r^2} \frac{\partial}{\partial r} r^2 \frac{\partial}{\partial r} - \frac{L^2}{r^2} \right) r^l Y_{lm} = \frac{Y_{lm}}{r^2} \frac{\partial}{\partial r} r^2 \frac{\partial r^l}{\partial r} - \frac{r^l L^2 Y_{lm}}{r^2}$$

$$= [l(l+1) - l(l+1)] r^{l-2} Y_{lm} = 0 \qquad (6.4.25)$$

Moreover, if we have a solution to Laplace's equation in the form $r^l f(\theta, \varphi)$, then the angular part $f(\theta, \varphi)$ must be an eigenfunction of L^2 with eigenvalue $l(l+1)$ – in other words, $f(\theta, \varphi)$ must then be a linear combination of the spherical harmonics Y_{lm} with $-l \leq m \leq l$.

Let us determine, in Cartesian coordinates, the solutions to Laplace's equation that correspond to the solid harmonics for a given angular momentum l. According to the preceding discussion, the solution must be a polynomial in x, y and z of degree l:

$$\psi_l(x, y, z) = N_l \sum_{pq} C^l_{pq}(x + \mathrm{i}y)^p (x - \mathrm{i}y)^q z^{l-p-q} \qquad (6.4.26)$$

To simplify future algebraic manipulations, we have expressed the solution here in terms of the complex variables

$$x \pm \mathrm{i}y = r \sin\theta \exp(\pm \mathrm{i}\varphi) \qquad (6.4.27)$$

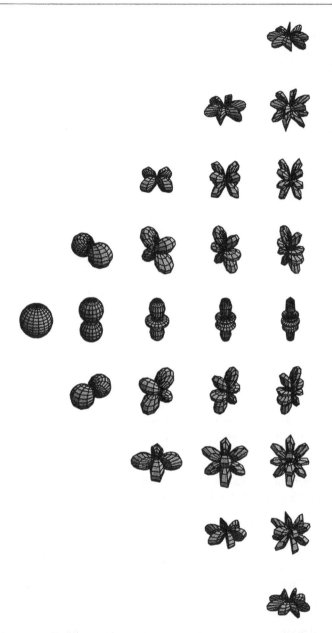

Fig. 6.1. The real solid harmonics $S_{lm}(\mathbf{r})$ for $l \leq 4$. The functions have been arranged as in Table 6.3.

rather than in terms of x and y. In (6.4.26), the summation is over all pairs of indices p and q subject to the condition

$$0 \leq p + q \leq l \tag{6.4.28}$$

Our task is to determine the expansion coefficients in ψ_l in such a way that Laplace's equation is satisfied.

Inserting (6.4.26) in Laplace's equation (6.4.24) and carrying out the differentiation, we obtain, after some simple algebra,

$$4 \sum_{pq} pq C^l_{pq}(x+iy)^{p-1}(x-iy)^{q-1}z^{l-p-q} = -\sum_{pq}(l-p-q)(l-p-q-1)C^l_{pq}$$

$$\times (x+iy)^p(x-iy)^q z^{l-p-q-2} \qquad (6.4.29)$$

In this expression, the left-hand side arises from differentiation of ψ_l with respect to x and y and the right-hand side from differentiation with respect to z. To satisfy Laplace's equation, the coefficients that multiply the same polynomial terms on the two sides of this equation must be identical. We then arrive at the following conditions on the expansion coefficients in the solution to Laplace's equation:

$$C^l_{pq} = -\frac{(l-p-q+2)(l-p-q+1)}{4pq}C^l_{p-1,q-1} \qquad (6.4.30)$$

This expression shows that nontrivial solutions to Laplace's equation exist where all terms vanish except those where the difference between p and q is equal to a fixed (positive, zero or negative) integer

$$m = p - q \qquad (6.4.31)$$

For example, for $m = 1$, we may choose the first coefficient as $C^l_{10} = 1$ and generate the remaining ones by the relation (6.4.30), noting that this process terminates when the recursion arrives at $p + q = l + 1$ or $p + q = l + 2$. A nontrivial solution to Laplace's equation has then been generated. We also note that, for a given l, we may choose m in $2l + 1$ different ways $0 \le |m| \le l$.

Since, for a given solution to Laplace's equation, the indices p and q are not independent, it is convenient to impose this condition explicitly on the solution (6.4.26). Introducing the new summation index

$$t = \frac{p+q-|m|}{2} \qquad (6.4.32)$$

to be used in conjunction with m in place of p and q, we may write the solution to Laplace's equation as

$$\psi_{lm} = N_{lm} \sum_{t=0}^{[(l-|m|)/2]} C^{lm}_t(x+iy)^{(2t+|m|+m)/2}(x-iy)^{(2t+|m|-m)/2}z^{l-2t-|m|} \qquad (6.4.33)$$

where the new coefficients are related to the old ones by

$$C^{lm}_t = C^l_{t+(|m|+m)/2,\,t+(|m|-m)/2} \qquad (6.4.34)$$

and $[k]$ is the largest integer less than or equal to k. The reader should verify that, as we let the index t in (6.4.33) run over all integers

$$0 \le t \le \left[\frac{l-|m|}{2}\right] \qquad (6.4.35)$$

we recover the same terms as when we let p and q run over all pairs of indices (6.4.28) subject to the condition (6.4.31).

Let us investigate in more detail the functional form of the solution to Laplace's equation (6.4.33). Noting the relationship

$$(x + \mathrm{i}y)^{(2t+|m|+m)/2}(x - \mathrm{i}y)^{(2t+|m|-m)/2} = \begin{cases} (x^2 + y^2)^t(x + \mathrm{i}y)^{|m|} & m \geq 0 \\ (x^2 + y^2)^t(x - \mathrm{i}y)^{|m|} & m < 0 \end{cases} \tag{6.4.36}$$

we write ψ_{lm} in the form

$$\psi_{lm} = N_{lm}[x + \mathrm{sgn}(m)\,\mathrm{i}y]^{|m|} \sum_{t=0}^{[(l-|m|)/2]} C_t^{lm}(x^2 + y^2)^t z^{l-2t-|m|} \tag{6.4.37}$$

Next, we introduce polar coordinates using (6.3.2) and (6.4.27) and find that (6.4.37) may be written as

$$\psi_{lm} = N_{lm}r^l \exp(\mathrm{i}m\varphi) \sum_{t=0}^{[(l-|m|)/2]} C_t^{lm} \cos^{l-2t-|m|}\theta \sin^{2t+|m|}\theta \tag{6.4.38}$$

According to the discussion following (6.4.25), the factor multiplying r^l must be a linear combination of the spherical harmonics $Y_{lm'}$ with $-l \leq m' \leq l$. Moreover, from the functional form (6.4.38), we see immediately that ψ_{lm} also satisfies the eigenvalue equation for the z component of the angular momentum (6.3.6) since

$$L_z\psi_{lm} = -\mathrm{i}\frac{\partial\psi_{lm}}{\partial\varphi} = m\psi_{lm} \tag{6.4.39}$$

We have thus established that the solutions (6.4.37) to Laplace's equation represent products of r^l with the spherical harmonics Y_{lm} – that is, they correspond to the solid harmonics $\mathcal{Y}_{lm}$.

It remains to determine the expansion coefficients (6.4.34). Expressing p and q in terms of t and m and substituting the results into (6.4.30), we obtain the recurrence relation

$$C_t^{lm} = -\frac{(l - |m| - 2t + 2)(l - |m| - 2t + 1)}{4t(|m| + t)} C_{t-1}^{lm} \tag{6.4.40}$$

Iterating this relation with the first coefficient chosen for convenience as

$$C_0^{lm} = \binom{l}{|m|} \tag{6.4.41}$$

we arrive at the following explicit expression for the coefficients (see Exercise 6.2)

$$C_t^{lm} = \left(-\frac{1}{4}\right)^t \binom{l-t}{|m|+t}\binom{l}{t} \tag{6.4.42}$$

Except for the normalization constant (see Exercises 6.3 and 6.4), this step completes our derivation of the solid harmonics (6.4.33) from Laplace's equation. Our results are summarized in (6.4.14)–(6.4.16).

6.4.4 EXPLICIT CARTESIAN EXPRESSIONS FOR THE REAL SOLID HARMONICS

Let us consider the evaluation of the real solid harmonics $S_{lm}(\mathbf{r})$ (6.4.19)–(6.4.21) from the complex solid harmonics $\mathcal{Y}_{lm}$ (6.4.14)–(6.4.16). For the polynomial in $x^2 + y^2$ in (6.4.14), we invoke the binomial theorem:

$$(x^2 + y^2)^t = \sum_{u=0}^{t} \binom{t}{u} x^{2(t-u)} y^{2u} \tag{6.4.43}$$

The term $[x + \text{sgn}(m)\, iy]^{|m|}$ can be separated into real and imaginary parts as

$$(x + iy)^{|m|} = \sum_{u=0}^{|m|} \binom{|m|}{u} x^{|m|-u} (iy)^u$$

$$= \sum_{v=0}^{[|m|/2]} \binom{|m|}{2v} x^{|m|-2v} (-1)^v y^{2v} + i \sum_{v=0}^{[|m-1|/2]} \binom{|m|}{2v+1} x^{|m|-2v-1} (-1)^v y^{2v+1} \tag{6.4.44}$$

To highlight the symmetry between the real and imaginary parts, it is convenient to use $v + 1/2$ as the summation index in the imaginary part:

$$\text{Re}\,(x + iy)^{|m|} = \sum_{v=0}^{[|m|/2]} (-1)^v \binom{|m|}{2v} x^{|m|-2v} y^{2v} \tag{6.4.45}$$

$$\text{Im}\,(x + iy)^{|m|} = \sum_{v=1/2}^{[(|m|-1)/2]+1/2} (-1)^{v-1/2} \binom{|m|}{2v} x^{|m|-2v} y^{2v} \tag{6.4.46}$$

Substituting the binomial expansions (6.4.43), (6.4.45) and (6.4.46) in the real and imaginary parts of the complex solid harmonics (6.4.14), we arrive at the following explicit expression for the real solid harmonics:

$$S_{lm} = N_{lm}^S \sum_{t=0}^{[(l-|m|)/2]} \sum_{u=0}^{t} \sum_{v=v_m}^{[|m|/2-v_m]+v_m} C_{tuv}^{lm} x^{2t+|m|-2(u+v)} y^{2(u+v)} z^{l-2t-|m|} \tag{6.4.47}$$

$$C_{tuv}^{lm} = (-1)^{t+v-v_m} \left(\frac{1}{4}\right)^t \binom{l}{t} \binom{l-t}{|m|+t} \binom{t}{u} \binom{|m|}{2v} \tag{6.4.48}$$

$$N_{lm}^S = \frac{1}{2^{|m|} l!} \sqrt{\frac{2(l+|m|)!(l-|m|)!}{2^{\delta_{0m}}}} \tag{6.4.49}$$

where v_m is given by

$$v_m = \begin{cases} 0 & m \geq 0 \\ \frac{1}{2} & m < 0 \end{cases} \tag{6.4.50}$$

Note that all terms in the solid harmonics of order l in (6.4.47) are of degree l exactly. The reader may wish to verify that (6.4.47) reproduces the real solid harmonics in Table 6.3.

6.4.5   RECURRENCE RELATIONS FOR THE REAL SOLID HARMONICS

In Section 6.4.4, we derived an explicit expression for the real solid harmonics. Often, it is more convenient to calculate the solid harmonics by recursion, in particular when the full set of solid harmonics up to a given angular momentum l is needed. In the present subsection, we shall derive a set of recurrence relations for the real solid harmonics of the form

$$\bar{S}_{lm}(\mathbf{r}) = n_{lm} S_{lm}(\mathbf{r}) \tag{6.4.51}$$

The factor n_{lm} has here been introduced since, in our discussion of multipole expansions of the Coulomb integrals in Section 9.13, we shall work with different sets of scaled solid harmonics. Of course, when the solid harmonics are calculated from an explicit expression such as (6.4.47), any scaling factor is trivially incorporated and requires no special attention.

To simplify matters, we shall consider the generation of the solid harmonics $\overline{S}_{lm}$ in two steps. First, we evaluate all functions of the type $\overline{S}_{l,\pm l}$ by *diagonal recursion*, where the quantum numbers l and m are changed simultaneously. Next, the remaining solid harmonics $\overline{S}_{lm}$ are generated by *vertical recursion*, where m is kept fixed and l is incremented. While the diagonal recurrence relations are easily established, the derivation of the vertical recurrences requires more work.

To obtain the diagonal recurrences, we first note from (6.4.19)–(6.4.21) that the solid harmonics $\overline{S}_{l,\pm l}$ may be written in the simple form

$$\overline{S}_{00} = n_{00} \tag{6.4.52}$$

$$\overline{S}_{l,l} = n_{l,l} \frac{\sqrt{2(2l)!}}{2^l l!} \frac{(x+iy)^l + (x-iy)^l}{2} \tag{6.4.53}$$

$$\overline{S}_{l,-l} = n_{l,-l} \frac{\sqrt{2(2l)!}}{2^l l!} \frac{(x+iy)^l - (x-iy)^l}{2i} \tag{6.4.54}$$

Next, elementary algebra shows that these functions satisfy the *diagonal recurrence relations*

$$\overline{S}_{l+1,l+1} = \sqrt{2^{\delta_{l0}} \frac{2l+1}{2l+2}} \left[\frac{n_{l+1,l+1}}{n_{l,l}} x \overline{S}_{l,l} - (1 - \delta_{l0}) \frac{n_{l+1,l+1}}{n_{l,-l}} y \overline{S}_{l,-l} \right] \tag{6.4.55}$$

$$\overline{S}_{l+1,-l-1} = \sqrt{2^{\delta_{l0}} \frac{2l+1}{2l+2}} \left[\frac{n_{l+1,-l-1}}{n_{l,l}} y \overline{S}_{l,l} + (1 - \delta_{l0}) \frac{n_{l+1,-l-1}}{n_{l,-l}} x \overline{S}_{l,-l} \right] \tag{6.4.56}$$

where the Kronecker deltas have been introduced to accommodate the special definition of S_{00} in (6.4.52).

To establish the vertical recursion, we use (6.4.14)–(6.4.16) to write the (scaled) solid harmonics (6.4.19)–(6.4.21) in the form

$$\overline{S}_{lm} = \text{Re} \, (x+iy)^m \sum_{t=0}^{[(l-m)/2]} E_t^{lm} (x^2 + y^2)^t z^{l-2t-m}, \qquad m \geq 0 \tag{6.4.57}$$

$$\overline{S}_{l,-m} = \text{Im} \, (x+iy)^m \sum_{t=0}^{[(l-m)/2]} E_t^{lm} (x^2 + y^2)^t z^{l-2t-m}, \qquad m > 0 \tag{6.4.58}$$

where the special solid harmonics $\overline{S}_{l0}$ are included in (6.4.57) and where the expansion coefficients are given as

$$E_t^{lm} = n_{lm} \left(-\frac{1}{4} \right)^t \frac{\sqrt{(2 - \delta_{m0})(l+m)!(l-m)!}}{2^m (l-2t-m)!(t+m)!t!} \tag{6.4.59}$$

To obtain the desired vertical recurrence relations for the solid harmonics, it is sufficient to determine a recurrence for the expansion coefficients E_t^{lm} of the form

$$E_t^{lm} = \sum_{ij \neq 00} c_{ij}^{lm} E_{t-j}^{l-i,m} \tag{6.4.60}$$

where the c_{ij}^{lm} are independent of t and the summation is over all nonnegative i and j (avoiding $ij = 00$). The condition that the coefficients c_{ij}^{lm} are independent of t determines the number of nonzero terms in (6.4.60). We shall proceed as follows: first, we shall generate three recurrences for E_t^{lm} whose coefficients depend on t; next, these recurrences will be combined to yield a final recurrence relation whose coefficients c_{ij}^{lm} are independent of t.

From the definition of the expansion coefficients E_t^{lm} in (6.4.59), we obtain the first basic recurrence:

$$E_t^{l-1,m} = \frac{l-m-2t}{\sqrt{(l+m)(l-m)}} \frac{n_{l-1,m}}{n_{lm}} E_t^{lm} \tag{6.4.61}$$

This relation is linear in t. To eliminate t from the recursion, we must establish an independent relation that involves t linearly. Decrementing t in (6.4.59), we obtain

$$E_{t-1}^{lm} = \frac{-4t(t+m)}{(l-m-2t+1)(l-m-2t+2)} E_t^{lm} \tag{6.4.62}$$

which contains t in the denominator. This awkward dependence on t is avoided by applying (6.4.61) twice on the left-hand side of (6.4.62):

$$E_{t-1}^{l-2,m} = \frac{-4t(t+m)}{\sqrt{(l+m-1)(l-m-1)(l+m)(l-m)}} \frac{n_{l-2,m}}{n_{lm}} E_t^{lm} \tag{6.4.63}$$

However, since this recurrence is quadratic in t, it cannot be combined linearly with (6.4.61) to eliminate t. Clearly, a second recurrence quadratic in t is needed. Such a recurrence is easily generated by applying (6.4.61) twice:

$$E_t^{l-2,m} = \frac{(l-m-2t)(l-m-2t-1)}{\sqrt{(l+m-1)(l-m-1)(l+m)(l-m)}} \frac{n_{l-2,m}}{n_{lm}} E_t^{lm} \tag{6.4.64}$$

Equipped with the three recurrences (6.4.61), (6.4.63) and (6.4.64), we are now in a position to construct a linear combination of expansion coefficients $E_{t-j}^{l-i,m}$ that satisfies the equation

$$E_t^{lm} = c_{10}^{lm} E_t^{l-1,m} + c_{20}^{lm} E_t^{l-2,m} + c_{21}^{lm} E_{t-1}^{l-2,m} \tag{6.4.65}$$

where the coefficients c_{ij}^{lm} are independent of t.

To fix the coefficients c_{ij}^{lm} in (6.4.65), we substitute (6.4.61), (6.4.63) and (6.4.64) into (6.4.65) and eliminate E_t^{lm} from the resulting expression. This yields an equation that contains the constant 1 on the left-hand side and a quadratic polynomial in t on the right-hand side. Setting this polynomial equal to 1 and its first and second derivatives equal to 0, we obtain

$$c_{10}^{lm} = \frac{2l-1}{\sqrt{(l+m)(l-m)}} \frac{n_{lm}}{n_{l-1,m}} \tag{6.4.66}$$

$$c_{20}^{lm} = c_{21}^{lm} = -\sqrt{\frac{(l+m-1)(l-m-1)}{(l+m)(l-m)}} \frac{n_{lm}}{n_{l-2,m}} \tag{6.4.67}$$

The required recurrence relation (6.4.65) for E_t^{lm} has now been established. To establish the corresponding recurrence for the solid harmonics $\bar{S}_{lm}$, we substitute (6.4.60) into (6.4.57) and (6.4.58). Rearranging indices, we obtain a single equation valid for all m:

$$\bar{S}_{lm} = \sum_{ij} c_{ij}^{lm} z^{i-2j} (x^2+y^2)^j \bar{S}_{l-i,m} \tag{6.4.68}$$

Inserting the coefficients (6.4.66) and (6.4.67) into this expression, we arrive at the final *vertical recurrence relations* for the scaled solid harmonics

$$\overline{S}_{lm} = \frac{(2l-1)}{\sqrt{(l+m)(l-m)}} \frac{n_{lm}}{n_{l-1,m}} z \, \overline{S}_{l-1,m} - \sqrt{\frac{(l+m-1)(l-m-1)}{(l+m)(l-m)}} \frac{n_{lm}}{n_{l-2,m}} r^2 \, \overline{S}_{l-2,m} \quad (6.4.69)$$

where we have used r^2 rather than $x^2 + y^2 + z^2$. In this expression, all solid harmonics $\overline{S}_{lm}$ with $l < 0$ are assumed to be 0.

For ease of reference, we conclude this section by gathering together the diagonal and vertical recurrence relations for the unscaled real solid harmonics:

$$S_{00} = 1 \tag{6.4.70}$$

$$S_{l+1,l+1} = \sqrt{2^{\delta_{l0}} \frac{2l+1}{2l+2}} [x S_{l,l} - (1 - \delta_{l0}) y S_{l,-l}] \tag{6.4.71}$$

$$S_{l+1,-l-1} = \sqrt{2^{\delta_{l0}} \frac{2l+1}{2l+2}} [y S_{l,l} + (1 - \delta_{l0}) x S_{l,-l}] \tag{6.4.72}$$

$$S_{l+1,m} = \frac{(2l+1) z S_{lm} - \sqrt{(l+m)(l-m)} r^2 S_{l-1,m}}{\sqrt{(l+m+1)(l-m+1)}} \tag{6.4.73}$$

Using these two-term recurrences, the full set of solid harmonics to order l may be calculated at a cost that scales as l^2 – that is, linearly in the number of solid harmonics. The corresponding scaled recurrences are needed in Section 9.13, in our discussion of the multipole method for the evaluation of the two-electron integrals.

6.5 Exponential radial functions

Having discussed the angular part of the one-centre basis functions, we now turn our attention to the more difficult problem of representing their radial part. In the present section, we consider basis functions that contain an exponential radial part, in accordance with the solutions to the one-electron hydrogenic problem; a different class of radial functions – the Gaussians, which decay in a less physical but computationally more tractable manner – are introduced in Section 6.6. The basis functions discussed in this chapter are listed in Box 6.1.

We begin the present section by reviewing the properties of the Laguerre polynomials, which will be used frequently both here and in Section 6.6. After a brief discussion of the hydrogenic functions, we go on to consider the more important exponential Laguerre functions and, in particular, their nodeless counterparts – the Slater-type orbitals.

6.5.1 THE LAGUERRE POLYNOMIALS

The *associated* or *generalized Laguerre polynomials* $L_n^\alpha(x)$ [11,12] may be obtained from the Rodrigues expression

$$L_n^\alpha(x) = \frac{\exp(x) x^{-\alpha}}{n!} \frac{d^n}{dx^n} [\exp(-x) x^{n+\alpha}] \tag{6.5.1}$$

and are polynomials of degree n in x for a fixed real α, although only integral and half-integral α are used in this chapter. The associated Laguerre polynomials $L_n^\alpha(x)$ have n (negative or positive) roots, which are shifted upwards with increasing α. For $\alpha > -1$, the roots are always positive.

Box 6.1   One-electron basis functions expressed in terms of the generalized Laguerre polynomials (6.5.1) and the solid harmonics (6.4.13). The functions are not normalized

Hydrogenic functions	$L_{n-l-1}^{2l+1}\left(\dfrac{2Zr}{n}\right)\exp\left(-\dfrac{Zr}{n}\right)\mathcal{Y}_{lm}(\mathbf{r})$	Section 6.5.2 incomplete for fixed Z orthogonal
Laguerre functions	$L_{n-l-1}^{2l+2}(2\zeta r)\exp(-\zeta r)\mathcal{Y}_{lm}(\mathbf{r})$	Sections 6.5.3 and 6.5.4 complete for fixed ζ orthogonal
Slater-type functions	$r^{n-l-1}\exp(-\zeta r)\mathcal{Y}_{lm}(\mathbf{r})$	Sections 6.5.5–6.5.7 complete for fixed ζ nonorthogonal no radial nodes
Harmonic-oscillator functions	$L_{n-l-1}^{l+1/2}(2\alpha r^2)\exp(-\alpha r^2)\mathcal{Y}_{lm}(\mathbf{r})$	Sections 6.6.1–6.6.2 complete for fixed α orthogonal
Gaussian-type functions	$r^{2(n-l-1)}\exp(-\alpha r^2)\mathcal{Y}_{lm}(\mathbf{r})$	Section 6.6.3 complete for fixed α nonorthogonal no radial nodes
	$\exp(-\alpha r^2)\mathcal{Y}_{lm}(\mathbf{r})$	Sections 6.6.4–6.6.5 fixed $n = l + 1$ complete for variable α nonorthogonal no radial nodes

The associated Laguerre polynomials $L_n^\alpha(x)$ with $n > 0$ then constitute a complete orthogonal set of polynomials in $\mathbb{R}^+$ with weight function $x^\alpha\exp(-x)$ [5]

$$\int_0^\infty L_n^\alpha(x)L_m^\alpha(x)x^\alpha\exp(-x)\,\mathrm{d}x = \frac{\Gamma(n+\alpha+1)}{n!}\delta_{mn} \qquad (6.5.2)$$

where $\Gamma(x)$ is the Euler gamma function to be discussed shortly.

The associated Laguerre polynomials are related to the *Laguerre polynomials* $L_n(x)$ by the expression

$$L_n(x) = L_n^0(x) \qquad (6.5.3)$$

The Laguerre polynomials $L_n(x)$ are normalized to unity; see (6.5.2). We note that, for positive integers α, the associated Laguerre polynomials may be obtained from the Laguerre polynomials by differentiation

$$L_n^\alpha(x) = (-1)^\alpha \frac{d^\alpha}{dx^\alpha} L_{n+\alpha}(x) \tag{6.5.4}$$

It is important to realize that different conventions are used for the associated Laguerre polynomials. Thus, the associated Laguerre polynomials are frequently defined as

$$\mathcal{L}_n^\alpha(x) = (-1)^\alpha n! L_{n-\alpha}^\alpha(x) \tag{6.5.5}$$

which differ from (6.5.1) in the phase convention, in the normalization convention and in the index convention. An advantage of the definition (6.5.1) adopted here is that the lower index n determines the number of roots (i.e. the degree of the polynomial) and the upper index α their positions. Examples of the associated Laguerre polynomials are given in Table 6.4.

The orthogonality relation for the associated Laguerre polynomials (6.5.2) contains the *Euler gamma function* $\Gamma(x)$, which is a generalization of the factorial function $n!$ to nonintegral arguments. $\Gamma(x)$ is defined for complex arguments but has poles at zero and at all negative integers [11,12]. For Re $x > 0$, it may be expressed in the form

$$\Gamma(x) = \int_0^\infty \exp(-t) t^{x-1} \, dt \tag{6.5.6}$$

The following special cases for integers and half integers are important:

$$\Gamma(n) = (n-1)!, \quad \text{all integers } n > 0 \tag{6.5.7}$$

$$\Gamma\left(n + \frac{1}{2}\right) = \frac{(2n-1)!!\sqrt{\pi}}{2^n}, \quad \text{all integers } n \tag{6.5.8}$$

Here the *factorial function* is defined for zero and all positive integers

$$n! = \begin{cases} 1 & n = 0 \\ n(n-1)(n-2)\cdots 1 & n > 0 \end{cases} \tag{6.5.9}$$

whereas the *double factorial function* is also defined for odd negative integers

$$n!! = \begin{cases} 1 & n = 0 \\ n(n-2)(n-4)\cdots 2 & \text{even } n > 0 \\ n(n-2)(n-4)\cdots 1 & \text{odd } n > 0 \\ \dfrac{1}{(n+2)(n+4)\cdots 1} & \text{odd } n < 0 \end{cases} \tag{6.5.10}$$

Table 6.4 The associated Laguerre polynomials $L_n^\alpha(x)$ for selected n and α

$\alpha \backslash n$	0	1	2	3
3	1	$4 - x$	$10 - 5x + \frac{1}{2}x^2$	$20 - 15x + 3x^2 - \frac{1}{6}x^3$
2	1	$3 - x$	$6 - 4x + \frac{1}{2}x^2$	$10 - 10x + \frac{5}{2}x^2 - \frac{1}{6}x^3$
1	1	$2 - x$	$3 - 3x + \frac{1}{2}x^2$	$4 - 6x + 2x^2 - \frac{1}{6}x^3$
0	1	$1 - x$	$1 - 2x + \frac{1}{2}x^2$	$1 - 3x + \frac{3}{2}x^2 - \frac{1}{6}x^3$
-1	1	$-x$	$-x + \frac{1}{2}x^2$	$-x + x^2 - \frac{1}{6}x^3$

Note that the factorial is undefined for all negative integers and the double factorial for all even negative integers.

6.5.2 THE HYDROGENIC FUNCTIONS

A natural starting point for the discussion of radial basis functions is the one-electron hydrogenic system, for which the wave function can be written in a closed analytical form. Indeed, we would naively expect the simple hydrogenic eigenstates – the exact solutions for a one-electron atom – to represent an ideal set of functions in terms of which we may expand the more complicated orbitals of many-electron atoms. Somewhat surprisingly, perhaps, the hydrogenic orbitals have certain deficiencies when used as basis functions for many-electron atoms. Nevertheless, from a consideration of the hydrogenic wave functions, we should learn much about the desirable analytical properties of radial atomic basis functions.

In atomic units, the Hamiltonian of a *hydrogenic system* with nuclear charge Z is given by

$$H = -\frac{1}{2}\nabla^2 - \frac{Z}{r} \tag{6.5.11}$$

The eigenstates of this operator consists of a set of bounded states and set of continuum states. In polar coordinates, the normalized *bounded eigenfunctions* may be written in the product form [9]

$$\psi_{nlm}(r, \theta, \varphi) = R_{nl}(r) Y_{lm}(\theta, \varphi) \tag{6.5.12}$$

$$R_{nl}(r) = \left(\frac{2Z}{n}\right)^{3/2} \sqrt{\frac{(n-l-1)!}{2n(n+l)!}} \left(\frac{2Zr}{n}\right)^l L_{n-l-1}^{2l+1}\left(\frac{2Zr}{n}\right) \exp\left(-\frac{Zr}{n}\right) \tag{6.5.13}$$

where *the principal quantum number n* is a positive integer $n > 0$ and *the angular-momentum quantum number $l < n$*. The radial part contains an associated Laguerre polynomial L_{n-l-1}^{2l+1} in $2Zr/n$ and an exponential in $-Zr/n$. The energy of the bounded hydrogenic state ψ_{nlm} is given by

$$E_n = -\frac{Z^2}{2n^2} \tag{6.5.14}$$

The degeneracy of the hydrogenic states of different angular momenta is peculiar to the Coulomb potential and is lifted in a many-electron system. The degeneracy of the one-electron isotropic HO system is also different from that of the hydrogenic system; see Section 6.6.1.

The most notable features of the radial part of the bounded hydrogenic wave function (6.5.13) are the presence of the exponential in $-Zr/n$ (which ensures that the wave function decays exponentially at large distances) and the nuclear cusp for s orbitals (i.e. nonvanishing functional value and derivative at $r = 0$). The Laguerre polynomial introduces $n - l - 1$ radial nodes in the wave function. We note that the orthogonality of the hydrogenic wave functions does not follow from the orthogonality of the Laguerre polynomials (6.5.2) since, in (6.5.13), the arguments both in the polynomial and in the exponential depend on the quantum number n. Rather, the orthogonality of the hydrogenic wave functions follows directly from the fact that these functions are eigenfunctions of the Hermitian operator (6.5.11).

Even though the hydrogenic functions represent the exact solutions for a one-electron Coulombic system, the bounded functions ψ_{nlm} in (6.5.12) are not particularly useful as basis functions for many-electron atomic or molecular systems. First, they do not constitute a complete set by themselves but must be supplemented by the unbounded continuum states, clearly a nuisance in practical calculations. Second, they spread out quickly and become very diffuse because of the presence of the inverse quantum number n^{-1} in the polynomial and in the exponential. The diffuseness of ψ_{nlm} for large n is reflected in the presence of n^2 in the expectation value of r

$$\langle \psi_{nlm} | r | \psi_{nlm} \rangle = \frac{3n^2 - l(l+1)}{2Z} \tag{6.5.15}$$

(see Exercise 6.5). A large number of terms is therefore needed for a flexible description of the core and valence regions of a many-electron atom. To avoid the problems associated with continuum states and the slow convergence in the core and valence regions, we shall in the following consider basis functions that retain the product form of the hydrogenic wave functions but have a more compact exponential radial form.

6.5.3 THE LAGUERRE FUNCTIONS

In an attempt to retain the exponential form of the hydrogenic wave function while avoiding the problems associated with the use of continuum states, we are led to the following class of functions:

$$\chi_{nlm}^{\mathrm{LF}}(r, \theta, \varphi) = R_{nl}^{\mathrm{LF}}(r) Y_{lm}(\theta, \varphi) \tag{6.5.16}$$

$$R_{nl}^{\mathrm{LF}}(r) = (2\zeta)^{3/2} \sqrt{\frac{(n-l-1)!}{(n+l+1)!}} (2\zeta r)^l L_{n-l-1}^{2l+2}(2\zeta r) \exp(-\zeta r) \tag{6.5.17}$$

where the quantum numbers n, l and m take on the same values as in the hydrogenic wave functions. For fixed l and ζ, the radial functions (6.5.17) with $n > l$ constitute a complete orthonormal set in $L^2(\mathbb{R}^+, r^2)$; orthonormality follows from (6.5.2) and completeness from the completeness of the associated Laguerre polynomials $L_{n-l-1}^{2l+2}(2\zeta r)$ with weight function $(2\zeta r)^{2l+2} \exp(-2\zeta r)$; see Section 6.5.1. The functions R_{nl}^{LF} and more generally χ_{nlm}^{LF} are known as the *Laguerre functions*, a term that is also used for the generalization of $L_n^\alpha(x)$ to nonintegral complex indices n and α.

The Laguerre functions (6.5.16) form a complete orthonormal set of functions that combine a fixed-exponent exponential decay with a polynomial in r. They therefore arise quite naturally when searching for a suitable set of one-electron basis functions for atomic applications. Because of their obvious similarity with the hydrogenic wave functions, we shall use for the Laguerre functions the same notation as for the hydrogenic functions, referring, for example, to χ_{200}^{LF} as the 2s Laguerre function and to n as the principal quantum number.

The essential difference between the hydrogenic and Laguerre functions is the absence of the inverse quantum number n^{-1} in the latter functions. The Laguerre functions are therefore considerably more compact than the hydrogenic ones for large n, as seen from the average value of r (see Exercise 6.5)

$$\langle \chi_{nlm}^{\mathrm{LF}} | r | \chi_{nlm}^{\mathrm{LF}} \rangle = \frac{2n+1}{2\zeta} \tag{6.5.18}$$

which depends on n linearly rather than quadratically as in the hydrogenic functions; see (6.5.15). The Laguerre functions are therefore better suited to an accurate description of the electron distribution in the core and valence regions. Note that, for the Laguerre functions, the expectation value of r is independent of the angular momentum l. The hydrogenic and Laguerre functions are compared in Figure 6.2, where we have plotted the *radial distributions* $R_{nl}^2 r^2$ (full lines) and $(R_{nl}^{\mathrm{LF}})^2 r^2$ (dashed lines) for unit exponents. The Laguerre functions exhibit the same nodal structure as the hydrogenic ones but their shape is quite different (except for the 1s functions).

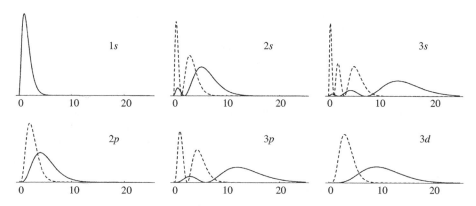

Fig. 6.2. The radial distribution functions of the hydrogenic (full line) and Laguerre (dashed line) functions for $Z = 1$ and $\zeta = 1$ (atomic units). The scales are the same in all plots. The $1s$ functions are identical.

6.5.4 THE CARBON ORBITALS EXPANDED IN LAGUERRE FUNCTIONS

To illustrate the convergence properties of the Laguerre functions, we shall use these functions to expand the numerical Hartree–Fock orbitals of the ^3P ground state of the carbon atom. However, to see how such expansions are obtained, we shall first consider in general terms the expansion of a function $f(x)$ in a set of basis functions $h_i(x)$. Thus, we wish to determine an expansion in the form

$$g(x) = \sum_i c_i h_i(x) \tag{6.5.19}$$

which is the best *least-squares approximation* to $f(x)$ in the sense that $\|f - g\|^2$ is as small as possible. Let $\mathbf{S}$ be the overlap matrix for the basis functions

$$S_{ij} = \langle h_i | h_j \rangle \tag{6.5.20}$$

and let $\mathbf{T}$ denote the overlaps between these functions and the target function

$$T_i = \langle h_i | f \rangle \tag{6.5.21}$$

As shown in Exercise 6.6, the best approximation to $f(x)$ is given by

$$g(x) = \sum_{ij} h_i(x) [\mathbf{S}^{-1}]_{ij} T_j \tag{6.5.22}$$

which involves the inverse of the overlap matrix. For an orthonormal basis, the expansion (6.5.22) may be written in the simpler form

$$g(x) = \sum_i h_i(x) \langle h_i | f \rangle, \qquad \langle h_i | h_j \rangle = \delta_{ij} \tag{6.5.23}$$

Returning now to the radial functions, we recall that the overlap integrals in $L^2(\mathbb{R}^+, r^2)$ are given by

$$\langle f | g \rangle = \int_0^\infty f^*(r) g(r) r^2 \, \mathrm{d}r \tag{6.5.24}$$

and note that we may employ the simple expression (6.5.23) for expansions in the orthonormal Laguerre functions.

In Figure 6.3, we have plotted the radial distribution functions of the $1s$, $2s$ and $2p$ numerical Hartree–Fock orbitals of the carbon ^3P ground state together with three approximate orbitals, expanded in 2, 8 and 15 Laguerre functions with unit exponent. More precisely, the $1s$ and $2s$ carbon orbitals have been expanded in Laguerre functions R_{n0}^{LF} with $1 \le n \le 15$ and the $2p$ orbital in functions R_{n1}^{LF} with $2 \le n \le 16$. We have also plotted the error in the expansions – that is, the norm of the difference between the numerical and expanded orbitals.

The $2s$ and $2p$ orbitals are fairly well reproduced by Laguerre functions with unit exponent. Errors less than 0.1 are obtained in expansions with 6 and 4 terms, respectively, but an accuracy of 0.01 requires as many as 16 and 13 functions. The $1s$ function, on the other hand, is much harder to reproduce. Here 12 functions are needed for an accuracy of 0.1, and 21 functions for an accuracy of 0.01.

It is not difficult to see why the $1s$ orbital is so hard to reproduce. It is very compact, with an expectation value of r of about 0.3. For comparison, the $2s$ and $2p$ expectation values are 1.6 and 1.7. According to (6.5.18), our basis functions with unit exponent have expectation values equal to $n + \frac{1}{2}$ and are thus not well suited for describing the core orbital. We conclude that the Laguerre functions with a fixed exponent are ill suited for describing orbitals with widely different radial distributions and that a large number of such functions would be needed to ensure a uniform description of the core and valence regions of an atomic system.

Presumably, with a more fortunate choice of exponent, we should be able to describe the $1s$ orbital quite accurately with only a few terms. Indeed, with $\zeta = 5.6$, only three terms are needed for an accuracy of 0.01. Similarly, we may improve the convergence for the other two orbitals by

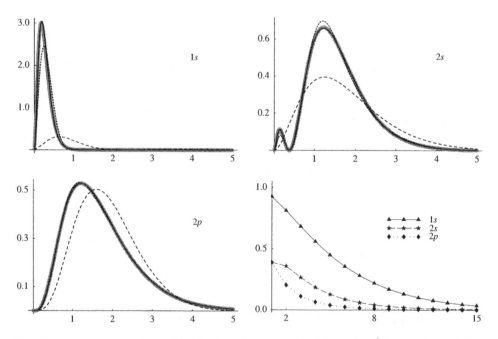

Fig. 6.3. Least-squares expansions of the $1s$, $2s$ and $2p$ orbitals of the carbon ^3P ground state in Laguerre functions with $\zeta = 1$ (atomic units). The radial distribution functions of the carbon orbitals are depicted using thick grey lines. The dashed, dotted and full thin lines correspond to Laguerre expansions containing 2, 8 and 15 terms, respectively. Also plotted are the errors in the expansions against the number of basis functions employed.

adjusting the Laguerre exponents. The best exponents are 2.0 for the $2s$ orbital and 1.6 for $2p$. With these exponents, convergence to 0.01 is reached after six and five terms, respectively.

The conclusions we may draw from this example are quite clear. We cannot hope to expand all the orbitals of an atomic system accurately in a small number of Laguerre functions with a single, fixed exponent. Instead, we must take a more pragmatic approach and introduce functions specifically designed to reproduce as closely as possible the different orbitals of each atom. In this way, we may hope to reproduce the electron distribution of each atom with a small number of terms. Indeed, our example suggests that it may be possible to obtain a crude but qualitatively correct description with no more than one basis function for each occupied AO – if we use basis functions with *variable exponents*.

6.5.5 THE NODELESS SLATER-TYPE ORBITALS

Introducing variable exponents, we note that the use of different exponents destroys the orthogonality of the Laguerre functions. Once this orthogonality has been lost, there is little point in retaining the complicated polynomial form of the Laguerre functions – the nodal structure of the Laguerre functions is dictated by orthogonality alone and does not necessarily resemble that of the carbon functions. In Figure 6.4, for example, we compare the radial distribution of the carbon $2s$ function with that of the $2s$ Laguerre function with exponent $\zeta = 2.52$, chosen as it gives the largest overlap (0.82) with the carbon function. Even though the maxima are close, the radial functions are quite different: the carbon function has a very small density inside the node (0.02); the Laguerre function, by contrast, has a sizeable density in this region (0.35).

In simplifying the polynomial structure of the Laguerre functions, we note that R_{nlm}^{LF} corresponds to an exponential in r times a polynomial of degree $n - 1$ in r. More precisely, we see from (6.5.17) that r occurs to power l in r^l and to powers $n - l - 1$ and lower in the Laguerre polynomial. Thus, a nodeless one-electron function that resembles χ_{nlm}^{LF} and the hydrogenic functions closely for large r is obtained by retaining in the Laguerre function only the term of the highest power of r. This prescription yields the following nonorthogonal set of functions:

$$\chi_{nlm}^{\mathrm{STO}}(r, \theta, \varphi) = R_n^{\mathrm{STO}}(r) Y_{lm}(\theta, \varphi) \tag{6.5.25}$$

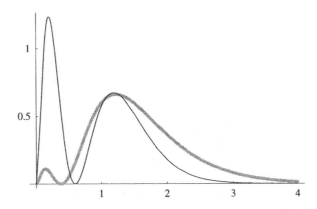

Fig. 6.4. The radial distributions of the $2s$ carbon orbital (thick grey line) and a $2s$ Laguerre function with exponent 2.52 (thin line) (atomic units).

$$R_n^{STO}(r) = \frac{(2\zeta)^{3/2}}{\sqrt{\Gamma(2n+1)}}(2\zeta r)^{n-1}\exp(-\zeta r) \tag{6.5.26}$$

where the normalization constant has been written so as to allow for nonintegral values of n (used in some rare cases). These functions are known as the *Slater-type orbitals (STOs)* [15,16].

For a fixed exponent ζ, the STOs (6.5.25) constitute a complete set of functions when the quantum numbers take on the same values as for the hydrogenic and Laguerre functions. To see this, we note that all Laguerre functions of orders n and lower may be written in terms of STOs of the same orders. For example, we may expand the $1s$, $2s$ and $3s$ Laguerre functions in terms of STOs as follows:

$$\chi_{1s}^{LF} = \chi_{1s}^{STO} \tag{6.5.27}$$

$$\chi_{2s}^{LF} = \sqrt{3}\chi_{1s}^{STO} - 2\chi_{2s}^{STO} \tag{6.5.28}$$

$$\chi_{3s}^{LF} = \sqrt{6}\chi_{1s}^{STO} - 4\sqrt{2}\chi_{2s}^{STO} + \sqrt{15}\chi_{3s}^{STO} \tag{6.5.29}$$

For the STOs, we use the same notation here as for the hydrogenic functions ($1s$, $2s$, $2p$, etc.), bearing in mind that the STOs are nodeless and that the principal quantum number n now refers only to the monomial factor r^{n-1} in (6.5.26). Note the alternating signs in (6.5.28) and (6.5.29) arising from the nodes in the Laguerre functions and that the coefficients are independent of the exponents. Since the sets of STOs and Laguerre functions up to a given order in n are equivalent, our conclusions regarding the convergence of the Laguerre functions (based on our experience with the carbon atom) are valid also for fixed-exponent STOs.

6.5.6 STOs WITH VARIABLE EXPONENTS

Although it is reassuring to know that the set of STOs constitute a complete set of functions for a fixed exponent, our motivation for introducing the STOs was to pave the way for functions with *variable exponents*. Let us examine the radial extent of the STOs. The expectation value of r and the maximum in the radial distribution curve r_{max}^{STO} are given by (see Exercise 6.7)

$$\langle\chi_{nlm}^{STO}|r|\chi_{nlm}^{STO}\rangle = \frac{2n+1}{2\zeta} \tag{6.5.30}$$

$$r_{max}^{STO} = \frac{n}{\zeta} \tag{6.5.31}$$

From these expressions, it is apparent that we may introduce flexibility in the radial part of the one-electron space not only by using functions with different principal quantum numbers n, but also by using functions with different exponents ζ_n. For example, by choosing the exponents $\zeta_n = n^{-1}$ in $1s$ STOs, we may hope to span the same space as we do by introducing higher-order STOs since their radial maxima r_{max}^{STO} are the same. In Figure 6.5, we compare the $1s$, $2s$ and $3s$ STOs with a fixed exponent equal to 1 with the sequence of $1s$ STOs with exponents 1, $\frac{1}{2}$ and $\frac{1}{3}$.

The question as to what sequences of STOs with variable exponents constitute a complete set is a difficult one, which has been analysed in some detail. It has been possible to establish certain conditions that ensure completeness in the limit of infinitely many functions [17]. In particular, it has been shown that the set

$$\chi_{\zeta_{nl}lm}^{STO}(r,\theta,\varphi) = R_{\zeta_{nl}l}^{STO}(r)Y_{lm}(\theta,\varphi) \tag{6.5.32}$$

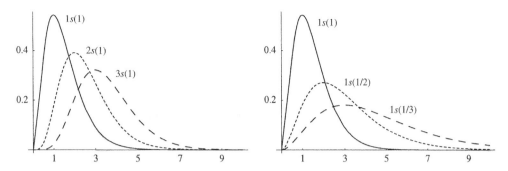

Fig. 6.5. The radial distribution functions for $1s$, $2s$ and $3s$ STOs with exponents $\zeta = 1$ (left) and for $1s$ STOs with exponents $\zeta = 1$, $1/2$ and $1/3$ (right) (atomic units). The notation is $ns(\zeta)$. The radial maxima $r_{\text{max}}^{\text{STO}}$ are the same in the two figures.

$$R_{\zeta_{nl}l}^{\text{STO}}(r) = \frac{(2\zeta_{nl})^{3/2}}{\sqrt{(2l+2)!}}(2\zeta_{nl}r)^l \exp(-\zeta_{nl}r) \tag{6.5.33}$$

is complete (with respect to functions and energies) if, for each l, the sequence of exponents ζ_{nl} has a finite accumulation point $0 < \zeta_l < \infty$. For example, the exponents $1 + n^{-1}$ fulfil this criterion, whereas the sequences n and n^{-1} do not. It is important to realize, however, that this condition is sufficient rather than necessary. Therefore, it may still be the case that the sequences n and n^{-1} generate complete sets. The importance of this criterion is not so much that it guides us in the construction of basis sets but rather that it proves that it is indeed possible to generate complete sets by employing variable exponents.

6.5.7 STO BASIS SETS

We now have at our disposal two distinct techniques for generating complete sets of STOs. We may choose to work with a single, fixed exponent according to (6.5.26), describing the radial space by functions with different n. Such a basis will contain the orbitals $(1s, 2s, \ldots)$, $(2p, 3p, \ldots)$, and so on – all with the same exponent ζ. Alternatively, we may describe the radial space by functions with variable exponents according to (6.5.33). For each angular momentum l, we then employ only the functions of the lowest principal quantum number n, yielding a basis of the type $(1s(\zeta_{1s}), 1s(\zeta_{2s}), \ldots)$, $(2p(\zeta_{1p}), 2p(\zeta_{2p}), \ldots)$, and so on.

In practice, a combined approach has proved more fruitful and the one-electron space is expanded both in terms of the principal quantum number and by means of variable exponents. Thus, for the carbon atom, we introduce a set of $1s$ functions with variable exponents, a set of $2s$ functions with variable exponents, and a set of $2p$ functions with variable exponents. The exponents are chosen to ensure an accurate representation of the wave function.

To illustrate the construction of such a basis set, we consider the representation of the occupied carbon Hartree–Fock orbitals by an STO expansion that is *minimal* in the sense that it contains only one STO for each occupied Hartree–Fock orbital. The expectation values of r for the $1s$, $2s$ and $2p$ numerical Hartree–Fock orbitals are $0.2690a_0$, $1.589a_0$ and $1.715a_0$, respectively. As basis functions, we choose STOs with expectation values (6.5.30) that match those of the Hartree–Fock orbitals. This prescription leads to an STO basis with the following radial forms:

$$R_{1s}^{\text{STO}}(r) = N_{1s} \exp(-5.58r) \tag{6.5.34}$$

$$R_{2s}^{STO}(r) = N_{2s} r \exp(-1.57r) \tag{6.5.35}$$

$$R_{2p}^{STO}(r) = N_{2p} r \exp(-1.46r) \tag{6.5.36}$$

Expanding the radial Hartree–Fock orbitals in these functions, we obtain the following normalized least-squares approximations:

$$\phi_{1s}(r) = 0.998 R_{1s}^{STO}(r) + 0.009 R_{2s}^{STO}(r) \tag{6.5.37}$$

$$\phi_{2s}(r) = -0.231 R_{1s}^{STO}(r) + 1.024 R_{2s}^{STO}(r) \tag{6.5.38}$$

$$\phi_{2p}(r) = R_{2p}^{STO}(r) \tag{6.5.39}$$

Note that, in (6.5.37) and (6.5.38), there is a mixing of the $1s$ and $2s$ STOs – in particular for the carbon $2s$ orbital, where both STOs make an important contribution because of the node. The carbon $1s$ orbital, on the other hand, is dominated by a single STO. The $1s$ and $2s$ functions are not strictly orthogonal since they have been obtained from independent least-squares fits. Finally, since there is only one STO of p symmetry, the $2p$ orbital is represented by a single function. The quality of this basis may be inspected in Figure 6.6.

The excellent agreement with the exact Hartree–Fock orbitals with such a *minimal* or *single-zeta* basis set is quite remarkable. Recall, however, that our basis has been specially designed with these atomic carbon functions in mind. The basis is able to reproduce in detail only the ^3P ground-state orbitals of the carbon atom and has very little flexibility to describe the modifications that these orbitals undergo on formation of chemical bonds or as a result of some external perturbation. For such purposes, larger and more flexible basis sets are needed. Before turning our attention to such sets, we note that a minimal STO basis set can be set up for any atom by following a set of rules given by Slater [15,16]. Using these rules, we obtain an exponent of 5.70 for the carbon $1s$ orbital and 1.625 for the $2s$ and $2p$ orbitals, rather close to the exponents in the expansions (6.5.37)–(6.5.39).

For accurate calculations, we need to go beyond the STO minimal representation. An obvious extension would be to introduce more functions of the same type as those already present in order to improve the radial flexibility. Thus, in the *double-zeta* basis sets, we use two STOs for each occupied AO. The exponents of such sets may, for example, be obtained from atomic Hartree–Fock calculations, optimized to give the lowest energy.

The double-zeta basis sets usually represent a significant improvement on the minimal Hartree–Fock description and may be used for near-quantitative calculations. For high accuracy, still more functions must be added, yielding *extended basis sets*. For example, the numerical Hartree–Fock orbitals referred to in this chapter can be accurately reproduced in a basis of two $1s$

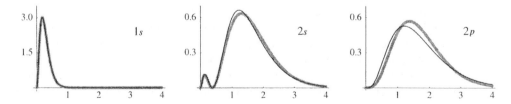

Fig. 6.6. The radial distribution functions of the carbon orbitals (thick grey lines) compared with their minimal STO representations (6.5.37)–(6.5.39) (thin lines) (atomic units).

Table 6.5 The exponents and expansion coefficients of the $1s$, $2s$ and $2p$ orbitals of the carbon atom in the ^3P ground state [18]

STO type	exponents	$1s$	$2s$	$2p$
$1s$ STO	9.2863	0.07657	−0.01196	
	5.4125	0.92604	−0.21041	
$2s$ STO	4.2595	0.00210	−0.13209	
	2.5897	0.00638	0.34624	
	1.5020	0.00167	0.74108	
	1.0311	−0.00073	0.06495	
$2p$ STO	6.3438			0.01090
	2.5873			0.23563
	1.4209			0.57774
	0.9554			0.24756

STOs, four $2s$ STOs and four $2p$ STOs [18]; see Table 6.5. Indeed, in a radial-distribution plot, the numerical Hartree–Fock orbitals and the orbitals in Table 6.5 would be indistinguishable.

At this point, it should be noted that the above extensions of the minimal STO sets – the addition of more basis functions of the same type as those already present – are appropriate only for atomic Hartree–Fock calculations. For correlated calculations and for molecular (rather than atomic) calculations, we must consider also basis functions of symmetries that do not occur in the uncorrelated atomic calculations. The purpose of these functions is twofold: to allow for a distortion of the atomic charge distribution in a molecular environment of lower symmetry; and to allow for a description of electronic correlation. We shall return to this subject in our discussion of Gaussian basis sets in Chapter 8.

The STO basis sets discussed in this subsection have been quite successful for atoms and diatoms, but much less so for molecules other than diatoms since no efficient method has been developed for calculating the many-centre two-electron STO integrals required for molecular calculations. Indeed, this problem has been so difficult that the STOs have been largely abandoned in favour of the Gaussian-type orbitals introduced in the next section, at least for systems more complicated than atoms and diatoms.

6.6 Gaussian radial functions

Except for atoms and perhaps diatoms, modern calculations of electronic structure usually employ basis sets with radial forms that contain the Gaussian distributions $\exp(-\alpha r^2)$ rather than the exponential forms of the STOs $\exp(-\zeta r)$. Since Gaussian distributions resemble the charge distributions of atomic systems less closely than do the STOs, a larger number of Gaussians is needed for an accurate representation of the electronic system. However, the slow convergence of the Gaussians is more than compensated for by the faster evaluation of the molecular integrals, allowing a greater number of functions to be used in a calculation.

The discussion in this section follows closely that for the exponential orbitals in Section 6.5. First, a complete set of orthonormal functions with Gaussian radial forms and a fixed exponent α is introduced. From these orbitals, we next arrive at a more flexible and useful set of nonorthogonal basis functions by simplifying the polynomial part and by introducing variable exponents. As for the exponential functions in Section 6.5, the performance of the Gaussian-based functions is illustrated by carrying out simple expansions of the ^3P ground-state orbitals of the carbon atom. We conclude by reviewing those properties of the Gaussians that simplify the calculation of molecular integrals over such functions.

6.6.1 THE HARMONIC-OSCILLATOR FUNCTIONS IN POLAR COORDINATES

We consider an electron in a spherically symmetric harmonic potential [9]. In polar coordinates, the Hamiltonian of such a *three-dimensional isotropic harmonic oscillator (HO)* is given by

$$H = -\tfrac{1}{2}\nabla^2 + \tfrac{1}{2}(2\alpha)^2 r^2 \tag{6.6.1}$$

where the force constant is equal to $4\alpha^2$. The eigenfunctions of this operator are called the *HO functions* and may be written in the form

$$\chi_{nlm}^{HO}(r, \theta, \varphi) = R_{nl}^{HO}(r)Y_{lm}(\theta, \varphi) \tag{6.6.2}$$

$$R_{nl}^{HO}(r) = \frac{(2\alpha)^{3/4}}{\pi^{1/4}}\sqrt{\frac{2^{n+1}(n-l-1)!}{(2n-1)!!}}(\sqrt{2\alpha}\,r)^l L_{n-l-1}^{l+1/2}(2\alpha r^2)\exp(-\alpha r^2) \tag{6.6.3}$$

where the quantum numbers n, l and m take on the same values as in the hydrogenic functions and the same notation ($1s$, $2s$, $2p$, etc.) is used for the orbitals. We note the similarity with the exponential hydrogenic and Laguerre functions in (6.5.13) and (6.5.17); the essential difference is the use of r^2 rather than r in the associated Laguerre polynomial and in the exponential. Note the half-integral upper index in the associated Laguerre polynomial in (6.6.3) – required for orthogonality of the HO functions according to (6.5.2). Since the lower index of the associated Laguerre polynomial of the HO function is the same as in the hydrogenic and Laguerre functions, their nodal structures are the same, as illustrated in Figure 6.7.

The energy levels of the isotropic HO depend on the quantum numbers n and l in the following manner:

$$E_{nl} = \left[2(n-1)-l+\frac{3}{2}\right]2\alpha \tag{6.6.4}$$

Use of the principal quantum number n for the HO system emphasizes the close relationship of the HO functions (6.6.3) with the Laguerre functions (6.5.17) but complicates the expression for the

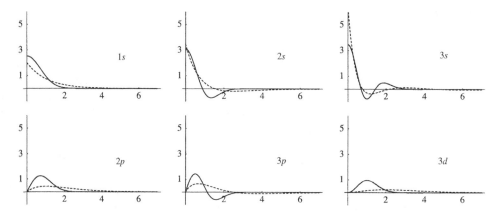

Fig. 6.7. The radial HO functions (full line) and Laguerre functions (dashed line) with unit exponents (atomic units).

energy. In fact, it is more common to express the HO functions in terms of the principal quantum number

$$\bar{n} = 2(n - 1) - l \tag{6.6.5}$$

which determines the energy (6.6.4) completely in the sense that all HO functions with the same principal quantum number $\bar{n}$ are degenerate. For a given angular momentum l, n takes on the values $l + 1, l + 2, l + 3, \ldots$ and the allowed values of $\bar{n}$ are therefore $l, l + 2, l + 4, \ldots$. From the expression for the HO functions (6.6.3), we also note that $\bar{n}$ corresponds directly to the highest power of r in the radial part of the wave function. Consequently, all HO functions of the same highest power of r are degenerate.

Since, according to (6.6.5), $R_{nl}^{\text{HO}}(r)$ and $R_{n-k,l-2k}^{\text{HO}}(r)$ have the same principal quantum number $\bar{n}$, they are degenerate. The first five energy levels of the three-dimensional isotropic HO therefore contain the functions $(1s)$, $(2p)$, $(3d,2s)$, $(4f,3p)$ and $(5g,4d,3s)$ with degeneracies 1, 3, 6, 10 and 15. As shown in Exercise 6.8, the degeneracy of level $\bar{n}$ is given by $(\bar{n} + 1)(\bar{n} + 2)/2$. In Table 6.6, we have listed the real-valued HO functions for the first three energy levels, using the solid harmonics of Table 6.3. These levels should be compared with those of the hydrogenic system, where the first three levels contain the functions $(1s)$, $(2s,2p)$ and $(3s,3p,3d)$ with degeneracies 1, 4 and 9 (i.e. n^2).

6.6.2 THE CARBON ORBITALS EXPANDED IN HO FUNCTIONS

Since the HO functions are the eigenfunctions of a Hermitian operator, they form a complete set of functions and are therefore valid candidates for atomic basis functions. Of course, a different question altogether is their usefulness as basis functions. Since the HO functions do not decay exponentially and do not have a cusp at the origin, we expect that more HO functions than Laguerre functions are required to model the orbitals in atomic and molecular systems accurately.

To investigate the usefulness of the HO functions, we proceed as for the Laguerre functions in Section 6.5.4, expanding the radial $1s$, $2s$ and $2p$ functions of the carbon atom in the HO functions (6.6.2) with a fixed exponent α. The results of such expansions with $\alpha = 1$ are shown in Figure 6.8, where we have used the same number of basis functions as for the Laguerre expansions in Figure 6.3. Clearly, the convergence of the HO functions is slower than that of the Laguerre functions. In fact, for the $1s$ orbital, as many as 30 terms are needed to reduce the error to 0.1, and, for the $2s$ and $2p$ orbitals, we need 10 terms. We note a tendency of the HO functions to

Table 6.6 The unnormalized HO functions in real form for the first three levels

$\bar{n}$	l	Component	Form
0	0	$1s$	$\exp(-\alpha r^2)$
1	1	$2p_x$	$x \exp(-\alpha r^2)$
		$2p_z$	$z \exp(-\alpha r^2)$
		$2p_y$	$y \exp(-\alpha r^2)$
2	2	$3d_{x^2-y^2}$	$(x^2 - y^2) \exp(-\alpha r^2)$
		$3d_{xz}$	$xz \exp(-\alpha r^2)$
		$3d_{z^2}$	$(3z^2 - r^2) \exp(-\alpha r^2)$
		$3d_{yz}$	$yz \exp(-\alpha r^2)$
		$3d_{xy}$	$xy \exp(-\alpha r^2)$
	0	$2s$	$(4\alpha r^2 - 3) \exp(-\alpha r^2)$

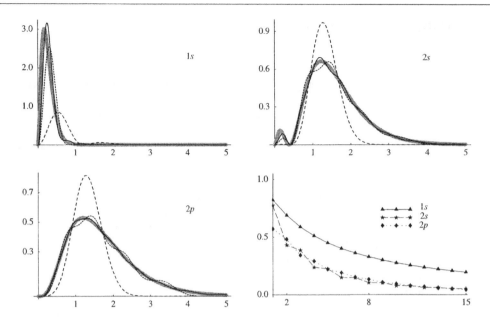

Fig. 6.8. Least-squares expansions of the $1s$, $2s$ and $2p$ orbitals of the ^3P carbon ground state in HO functions with unit exponent (atomic units). The radial distribution functions of the carbon orbitals are depicted using thick grey lines. The dashed, dotted and full thin lines correspond to HO expansions with 2, 8 and 15 functions, respectively. Also plotted are the errors in the expansions.

develop bumpy low-order approximations to the orbitals, in particular in comparison with the smooth convergence of the Laguerre functions.

As for the Laguerre functions in Section 6.5.4, we may adjust the exponents to improve the convergence. With an exponent of 11, the $1s$ HO expansion yields an error less than 0.1 with 3 terms included and less than 0.01 with 21 terms. Long expansions are needed for the $2s$ and $2p$ functions as well. Again, the HO functions converge much more slowly than the corresponding Laguerre functions, even with optimal exponents.

6.6.3 THE NODELESS GAUSSIAN-TYPE ORBITALS

In Section 6.6.2, we found that the HO functions converge more slowly than the Laguerre functions. The need to improve convergence is therefore even more critical than for the exponential functions. Again, we shall attempt to improve the convergence by employing a set of functions with different exponents, abandoning any hope of arriving at a universal single-exponent Gaussian basis for all atoms.

Following the same reasoning as for exponential functions in Section 6.5.5, we note that there is no longer any point in maintaining the nodal structure of the Laguerre polynomials. Instead, we introduce a nodeless set of functions by retaining in each HO function only the monomial term of highest power of r, which according to the discussion in Section 6.6.1 is given by $\bar{n} = 2(n-1) - l$. We then arrive at the following complete set of nonorthogonal basis functions (for normalization, see Exercise 6.9) known as the *spherical-harmonic Gaussian-type orbitals (GTOs)* [19]:

$$\chi_{nlm}^{\text{GTO}}(r, \theta, \varphi) = R_{nl}^{\text{GTO}}(r) Y_{lm}(\theta, \varphi) \tag{6.6.6}$$

$$R_{nl}^{\mathrm{GTO}}(r) = \frac{2(2\alpha)^{3/4}}{\pi^{1/4}}\sqrt{\frac{2^{2n-l-2}}{(4n-2l-3)!!}}(\sqrt{2\alpha}\ r)^{2n-l-2}\exp(-\alpha r^2) \qquad (6.6.7)$$

The term *spherical-harmonic* is used to distinguish these orbitals from the *Cartesian* GTOs introduced in Section 6.6.7. Note that the GTOs of even l contain even powers of r and those of odd l contain odd powers. Thus, whereas the radial $2s$ and $3s$ functions correspond to $r^2\exp(-\alpha r^2)$ and $r^4\exp(-\alpha r^2)$, the radial $2p$ and $3p$ functions are represented by $r\exp(-\alpha r^2)$ and $r^3\exp(-\alpha r^2)$.

Completeness of the single-exponent spherical-harmonic GTOs follows from the fact that the HO functions (which constitute a complete set) may be written as finite linear combinations of GTOs, for example

$$\chi_{1s}^{\mathrm{HO}} = \chi_{1s}^{\mathrm{GTO}} \qquad (6.6.8)$$

$$\chi_{2s}^{\mathrm{HO}} = \sqrt{\frac{3}{2}}\chi_{1s}^{\mathrm{GTO}} - \sqrt{\frac{5}{2}}\chi_{2s}^{\mathrm{GTO}} \qquad (6.6.9)$$

$$\chi_{3s}^{\mathrm{HO}} = \frac{1}{2}\sqrt{\frac{15}{2}}\chi_{1s}^{\mathrm{GTO}} - \frac{5}{\sqrt{2}}\chi_{2s}^{\mathrm{GTO}} + \frac{3}{2}\sqrt{\frac{7}{2}}\chi_{3s}^{\mathrm{GTO}} \qquad (6.6.10)$$

where the nodal structure of the HO functions is reflected in the alternating signs. The situation is analogous to that of the exponential functions in Section 6.5.5, where the nodeless STOs are obtained from the orthonormal Laguerre functions in the same manner.

To compare the nodeless STOs and GTOs, we have in Figure 6.9 plotted the first three STO and GTO functions of s symmetry with unit exponent. We note that the STOs fall off much more slowly than the GTOs, decaying exponentially; that the $1s$ STO has a cusp at the nucleus; and that the GTOs are more localized in space.

6.6.4 THE GTOs WITH VARIABLE EXPONENTS

Equipped with a complete set of simple spherical-harmonic GTOs of the form (6.6.6), we now turn our attention to the description of the radial space by means of variable exponents. We begin by comparing the expansion of the radial part of the one-electron space by means of variable exponents and by means of the principal quantum number n.

In Figure 6.10, we have plotted the $1s$, $2s$ and $3s$ GTOs (6.6.6) with unit exponent as well as the first three $1s$ GTOs with exponent $1/n$. The fixed-exponent functions are more localized in

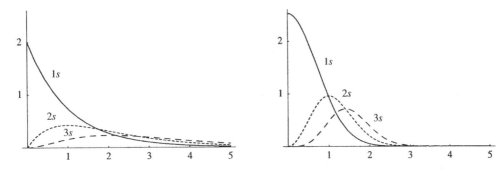

Fig. 6.9. The radial forms $R_{ns}^{\mathrm{STO}}(r)$ of the STOs (left) and $R_{ns}^{\mathrm{GTO}}(r)$ of the GTOs (right) with unit exponents (atomic units).

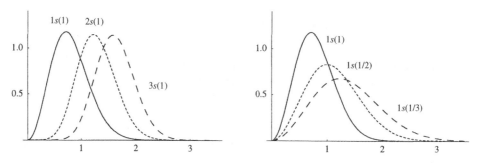

Fig. 6.10. The radial distribution functions in atomic units for the $1s$, $2s$ and $3s$ GTOs with exponent $\alpha = 1$ (left) and for the $1s$ GTOs with exponents $\alpha = 1$, $\frac{1}{2}$ and $\frac{1}{3}$ (right). The notation is $ns(\alpha)$.

space. The spatial extent of the GTOs is given by the formulae (see Exercise 6.9)

$$\langle \chi_{nlm}^{\text{GTO}} | r | \chi_{nlm}^{\text{GTO}} \rangle = \sqrt{\frac{2}{\pi\alpha}} \frac{2^{2n-l-2}(2n-l-1)!}{(4n-2l-3)!!} \approx \sqrt{\frac{2n-l-2}{2\alpha}} \qquad (6.6.11)$$

$$r_{\max}^{\text{GTO}} = \sqrt{\frac{2n-l-1}{2\alpha}} \qquad (6.6.12)$$

where $r_{\max}^{\text{GTO}}$ represents the position of the peak of the radial distribution curve. These expressions should be compared with (6.5.30) and (6.5.31) for the STOs. Whereas the extent of the STOs depends linearly on n and ζ^{-1}, the extent of the GTOs depends linearly on the square roots of n and α^{-1}. Therefore, for a flexible description of the outer regions of the electron distribution, much larger quantum numbers or much smaller exponents are needed for the GTOs than for the STOs.

In practice, the use of GTOs with variable exponents differs somewhat from the procedure for STOs. From Section 6.5.7, we recall that, for STOs, the radial form is described partly by the principal quantum number n and partly by the use of different exponents. For GTOs, on the other hand, we describe the radial space exclusively by means of variable exponents, using for this purpose only spherical-harmonic GTOs with $n = l + 1$. The only powers of r introduced in the GTOs are therefore those associated with the angular-momentum quantum number l of the spherical harmonics. The following set of spherical-harmonic GTOs is thus used:

$$\chi_{\alpha_{nl}lm}^{\text{GTO}} = R_{\alpha_{nl}l}^{\text{GTO}}(r) Y_{lm}(\theta, \varphi) \qquad (6.6.13)$$

$$R_{\alpha_{nl}l}^{\text{GTO}}(r) = \frac{2(2\alpha_{nl})^{3/4}}{\pi^{1/4}} \sqrt{\frac{2^l}{(2l+1)!!}} (\sqrt{2\alpha_{nl}}\, r)^l \exp(-\alpha_{nl}r^2) \qquad (6.6.14)$$

Complex algebra is avoided by rewriting the spherical harmonics in terms of the real solid harmonics of Section 6.4.2 (which then absorb the monomials in r), leaving us with the following simple set of real-valued spherical-harmonic GTOs

$$\chi_{\alpha_{nl}lm}^{\text{GTO}} = N_{\alpha_{nl}lm}^{\text{GTO}} S_{lm} \exp(-\alpha_{nl}r^2) \qquad (6.6.15)$$

where $N_{\alpha_{nl}lm}^{\text{GTO}}$ is the normalization constant.

We must now consider the choice of the orbital exponents. Again, some sufficient but not necessary conditions for completeness have been established [17]. Thus, completeness with respect to wave functions and energies is guaranteed if, for a fixed l, the sequence of exponents α_{nl} has an accumulation point α_l such that $0 < \alpha_l < \infty$. Completeness is ensured also if the set of exponents contains a sequence α_{nl} that decreases monotonically to zero and whose elements sum to infinity:

$$\sum_n \alpha_{nl} = \infty \tag{6.6.16}$$

This condition is satisfied by for example the sequences n^{-1} or $n^{-1/2}$, although these exponents would lead to basis sets with slow convergence and tremendous numerical problems. As for STOs in Section 6.5.6, these criteria are not very helpful in practice since they do not guide us towards basis sets that converge rapidly (since the conditions do not depend in any way on the atoms or orbitals to be expanded). They do tell us, however, that it is possible to construct complete basis sets of the simple form (6.6.15).

6.6.5 THE CARBON ORBITALS EXPANDED IN GTOs

Let us now expand the ^3P ground-state carbon orbitals in a minimal basis of GTOs with exponents chosen so as to match the expectation values of r given in Section 6.5.7. This prescription gives us the following basis functions

$$R_{1s}^{\text{GTO}}(r) = N_{1s} \exp(-8.80 r^2) \tag{6.6.17}$$

$$R_{2s}^{\text{GTO}}(r) = N_{2s} \exp(-0.252 r^2) \tag{6.6.18}$$

$$R_{2p}^{\text{GTO}}(r) = N_{2p} r \exp(-0.385 r^2) \tag{6.6.19}$$

As expected, the exponents are more spread out than in the corresponding STOs (6.5.34)–(6.5.36). The normalized least-squares approximations

$$\phi_{1s}(r) = 0.982 R_{1s}^{\text{GTO}}(r) + 0.077 R_{2s}^{\text{GTO}}(r) \tag{6.6.20}$$

$$\phi_{2s}(r) = -0.266 R_{1s}^{\text{GTO}}(r) + 1.016 R_{2s}^{\text{GTO}}(r) \tag{6.6.21}$$

$$\phi_{2p}(r) = R_{2p}^{\text{GTO}}(r) \tag{6.6.22}$$

are similar to the STO expansions (6.5.37)–(6.5.39). Comparing Figures 6.6 and 6.11, we find that the GTOs perform reasonably well for the s functions but poorly for the $2p$ function, for which only one basis function is present in the basis.

We have finally arrived at a set of basis functions suitable for molecular calculations: the spherical-harmonic GTOs in the form (6.6.15) with variable exponents for the radial part of the

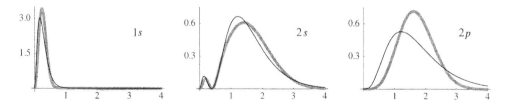

Fig. 6.11.   The radial distribution functions of the carbon orbitals (thick grey lines) in atomic units compared with their minimal GTO representations (6.6.20)–(6.6.22) (thin lines).

one-electron space, combined with solid harmonics for the angular part. Since the GTOs are spatially localized, compact functions, we must be prepared to work with a large number of them to ensure a uniform flexibility throughout the important regions of space. However, as we shall see in Chapter 9, integrals over such functions are easily generated. The reason for the ease with which integrals over GTOs may be obtained is related to two important properties of the Gaussian distributions, to be discussed in Section 6.6.7: their separability in the Cartesian directions and the Gaussian product rule. But before considering these properties, we shall briefly return to the HO functions of Section 6.6.1.

6.6.6 THE HO FUNCTIONS IN CARTESIAN COORDINATES

In our discussion of the HO functions and the GTOs so far, we have worked in polar coordinates in order to separate the functions in radial and angular parts. However, a different symmetry is also present in the Hamiltonian of the three-dimensional isotropic HO. Writing the Hamiltonian (6.6.1) in Cartesian coordinates, we obtain

$$H = -\tfrac{1}{2}\nabla^2 + \tfrac{1}{2}(2\alpha)^2(x^2 + y^2 + z^2) \tag{6.6.23}$$

which is separable in the three Cartesian directions. When the Schrödinger equation is solved with the Hamiltonian in this form, its solution is a product of one-dimensional HO functions [9]

$$\chi_{ijk}^{HO}(x, y, z) = \chi_i^{HO}(x)\chi_j^{HO}(y)\chi_k^{HO}(z) \tag{6.6.24}$$

where the quantum numbers take on the integers $i \geq 0$, $j \geq 0$ and $k \geq 0$ and where $\chi_i^{HO}(x)$ is the one-dimensional HO function in the x direction

$$\chi_i^{HO}(x) = \left(\frac{2\alpha}{\pi}\right)^{1/4}\frac{1}{\sqrt{2^i i!}}H_i(\sqrt{2\alpha}x)\exp(-\alpha x^2) \tag{6.6.25}$$

and similarly for the other two directions. The $H_i(\sqrt{2\alpha}x)$ are the Hermite polynomials to be discussed shortly. The energy of the isotropic three-dimensional HO is still given by (6.6.4)

$$E_{\bar{n}} = \left(\bar{n} + \tfrac{3}{2}\right)2\alpha \tag{6.6.26}$$

but the principal quantum number (6.6.5) is now given by

$$\bar{n} = i + j + k \tag{6.6.27}$$

From this expression, it is easy to verify the degeneracy $(\bar{n}+1)(\bar{n}+2)/2$ of the HO system.

The *Hermite polynomials* of degree n, $H_n(x)$, that appear in the HO functions in Cartesian form fulfil the orthogonality relation [11,12]

$$\int_{-\infty}^{+\infty} H_m(x)H_n(x)\exp(-x^2)\,\mathrm{d}x = \sqrt{\pi}2^n n!\delta_{mn} \tag{6.6.28}$$

and may be generated from the Rodrigues expression

$$H_n(x) = (-1)^n \exp(x^2)\frac{\mathrm{d}^n}{\mathrm{d}x^n}\exp(-x^2) \tag{6.6.29}$$

We note that the Hermite polynomials are either symmetric or antisymmetric with respect to inversion through the origin. The reader should be aware that a different form of the Hermite polynomials is often used, related to (6.6.29) by the expression

$$\mathcal{H}_n(\sqrt{2}x) = H_n(x) \tag{6.6.30}$$

Also, generalizations of the $H_n(x)$ to nonintegral complex n exist, called the Hermite functions. The lowest-degree Hermite polynomials are listed in Table 6.7.

6.6.7 THE CARTESIAN GTOs

The three-dimensional HO functions (6.6.24) possess a rather complicated nodal structure. If we retain only the highest-order terms in the Hermite polynomials, we arrive at the following set of nodeless *Cartesian GTOs*:

$$\chi_{ijk}^{\text{GTO}}(x, y, z) = \chi_i^{\text{GTO}}(x)\chi_j^{\text{GTO}}(y)\chi_k^{\text{GTO}}(z) \tag{6.6.31}$$

where the x component is given by

$$\chi_i^{\text{GTO}}(x) = \left(\frac{2\alpha}{\pi}\right)^{1/4} \sqrt{\frac{(4\alpha)^i}{(2i-1)!!}}\, x^i \exp(-\alpha x^2) \tag{6.6.32}$$

and similarly for the y and z components. Since the HO functions (6.6.24) are finite linear combinations of Cartesian GTOs, we conclude that, for a fixed exponent α, the Cartesian GTOs form a complete set and therefore represent a valid set of one-electron basis functions. We also note that the spherical-harmonic GTOs are simple linear combinations of Cartesian GTOs, where the expansion coefficients may be generated from (6.4.47).

The Cartesian GTOs are particularly well suited to the evaluation of molecular integrals since they are *separable* in the three Cartesian directions and since the product of two or more Cartesian GTOs on different centres may be written as a simple linear combination of Cartesian GTOs (the Gaussian product rule) [19]. Neither of these properties is shared by the exponential functions of Section 6.5, making the evaluation of molecular integrals over STOs much more difficult. Of course, the spherical-harmonic GTOs are not separable either, but integrals over such GTOs may be generated (explicitly or implicitly) as linear combinations of integrals over Cartesian GTOs. To avoid this transformation, the Cartesian GTOs are sometimes used as basis functions in place of the spherical-harmonic GTOs. However, for a given quantum number l, the spherical-harmonic GTOs contain fewer functions and provide a more compact representation of the electronic system than do the Cartesian GTOs. The use of spherical-harmonic GTOs is therefore strongly recommended,

Table 6.7 The Hermite polynomials $H_n(x)$ for $n \leq 5$

n	$H_n(x)$
0	1
1	$2x$
2	$4x^2 - 2$
3	$8x^3 - 12x$
4	$16x^4 - 48x^2 + 12$
5	$32x^5 - 160x^3 + 120x$

in particular since the transformation from Cartesian to spherical-harmonic GTOs, properly implemented, makes the integration over spherical-harmonic GTOs faster than the integration over the corresponding, larger set of Cartesian GTOs.

Let us finally consider the *Gaussian product rule* for a simple product of two *s* functions

$$\chi_A = \exp(-\alpha r_A^2) \tag{6.6.33}$$

$$\chi_B = \exp(-\beta r_B^2) \tag{6.6.34}$$

with the exponents α and β and centred on **A** and **B**. The product of these functions may be expressed in terms of a third Gaussian

$$\chi_{AB} = \exp\left(-\frac{\alpha\beta}{\alpha+\beta}R_{AB}^2\right) \exp[-(\alpha+\beta)r_P^2] \tag{6.6.35}$$

centred on **P** given by

$$\mathbf{P} = \frac{\alpha\mathbf{A} + \beta\mathbf{B}}{\alpha+\beta} \tag{6.6.36}$$

and where R_{AB} is the distance between **A** and **B**. Note the similarity with the transformation of the harmonic potentials in Section 6.2. Using the formula (6.2.6), we may therefore rewrite the product of any number of Gaussian distributions as a single Gaussian. The Gaussian product rule simplifies the calculation of molecular integrals considerably, allowing all the common molecular integrals to be calculated either in a closed analytical form or by means of a one-dimensional numerical integration. An exposition of integration techniques for Gaussian orbitals is given in Chapter 9.

References

[1] C. F. Fischer, *The Hartree–Fock Method for Atoms*, Wiley, 1977.
[2] L. Laaksonen, P. Pyykkö and D. Sundholm, *Comput. Phys. Rept.* **4**, 313 (1986).
[3] E. A. McCullough, Jr, *Comput. Phys. Rept.* **4**, 265 (1986).
[4] E. Prugovecki, *Quantum Mechanics in Hilbert Space*, 2nd edn, Academic Press, 1981.
[5] J. R. Higgins, *Completeness and Basis Properties of Sets of Special Functions*, Cambridge University Press, 1977.
[6] N. I. Akhiezer and I. M. Glazman, *Theory of Linear Operators in Hilbert Space*, Dover, 1993.
[7] R. D. Richtmyer, *Principles of Advanced Mathematical Physics*, Vol. 1, Springer-Verlag, 1978.
[8] B. Klahn and W. A. Bingel. *Theor. Chim. Acta* **44**, 9 (1977).
[9] C. Cohen-Tannoudji, B. Diu and F. Laloë, *Quantum Mechanics*, Wiley, 1977.
[10] D. M. Brink and G. R. Satchler, *Angular Momentum*, 2nd edn, Clarendon Press, 1968.
[11] G. Arfken, *Mathematical Methods for Physicists*, Academic Press, 1970.
[12] M. Abramowitz and I. A. Stegun, *Handbook of Mathematical Functions*, Dover, 1965.
[13] A. J. Stone, *The Theory of Intermolecular Forces*, Clarendon, 1996.
[14] B. L. van der Waerden, *Group Theory and Quantum Mechanics*, Springer-Verlag, 1974.
[15] J. C. Slater, *Phys. Rev.* **36**, 57 (1930).
[16] J. C. Slater, *Quantum Theory of Atomic Structure*, Vol. 1, McGraw–Hill, 1960.
[17] B. Klahn and W. A. Bingel, *Theor. Chim. Acta* **44**, 27 (1977).
[18] E. Clementi, C. C. J. Roothaan and M. Yoshimine, *Phys. Rev.* **127**, 1618 (1962).
[19] S. F. Boys, *Proc. Roy. Soc.* **A200**, 542 (1950).

Further reading

I. Shavitt, The history and evolution of Gaussian basis sets, *Isr. J. Chem.* **33**, 357 (1993).

Exercises

EXERCISE 6.1

Consider the sum of harmonic potentials

$$V = \tfrac{1}{2} \sum_A k_A r_A^2 \tag{6E.1.1}$$

where

$$\mathbf{r}_A = \mathbf{r} - \mathbf{A}$$

and k_A is a force constant associated with centre $\mathbf{A}$. Show that the sum of the harmonic potentials (6E.1.1) can be written as a single harmonic potential

$$V = V^{(0)} + \tfrac{1}{2} K r_S^2 \tag{6E.1.2}$$

with

$$K = \sum_A k_A \tag{6E.1.3}$$

$$V^{(0)} = \frac{1}{2K} \sum_{A>B} k_A k_B R_{AB}^2 \tag{6E.1.4}$$

$$\mathbf{S} = K^{-1} \sum_A k_A \mathbf{A} \tag{6E.1.5}$$

$$\mathbf{R}_{AB} = \mathbf{A} - \mathbf{B} \tag{6E.1.6}$$

EXERCISE 6.2

Verify that the recurrence relation

$$C_t^{lm} = -\frac{(l - |m| - 2t + 2)(l - |m| - 2t + 1)}{4t(|m| + t)} C_{t-1}^{lm} \tag{6E.2.1}$$

for $0 \le t \le [(l - |m|)/2]$, combined with the coefficient

$$C_0^{lm} = \binom{l}{|m|} \tag{6E.2.2}$$

for $t = 0$, gives the following closed-form expression for the coefficients in the solid harmonics:

$$C_t^{lm} = \left(-\frac{1}{4}\right)^t \binom{l-t}{|m|+t} \binom{l}{t} \tag{6E.2.3}$$

EXERCISE 6.3

In this exercise, we introduce the *angular-momentum shift operators*

$$L_+ = L_x + iL_y \tag{6E.3.1}$$

$$L_- = L_x - iL_y \tag{6E.3.2}$$

and show how they can be used to step up and down between the different components of the solid harmonic $\mathcal{Y}_{lm}$ with $-l \le m \le l$.

1. Verify the relations

$$[L_z, L_\pm] = \pm L_\pm \tag{6E.3.3}$$

$$[L^2, L_\pm] = 0 \tag{6E.3.4}$$

$$L_+ L_- = L^2 - L_z(L_z - 1) \tag{6E.3.5}$$

Hint: Use the commutators

$$[L_x, L_y] = iL_z \tag{6E.3.6}$$

$$[L_z, L_x] = iL_y \tag{6E.3.7}$$

$$[L_y, L_z] = iL_x \tag{6E.3.8}$$

2. Verify the relations

$$L_z L_\pm \mathcal{Y}_{lm} = (m \pm 1)L_\pm \mathcal{Y}_{lm} \tag{6E.3.9}$$

$$L^2 L_\pm \mathcal{Y}_{lm} = l(l+1)L_\pm \mathcal{Y}_{lm} \tag{6E.3.10}$$

which show that $L_+ \mathcal{Y}_{lm}$ and $L_- \mathcal{Y}_{lm}$ are proportional to $\mathcal{Y}_{l,m+1}$ and $\mathcal{Y}_{l,m-1}$, respectively.
3. Assume next that the angular part of $\mathcal{Y}_{lm}$ has been normalized over the angular coordinates:

$$\int \left(\frac{\mathcal{Y}_{lm}}{r^l}\right)^* \left(\frac{\mathcal{Y}_{lm}}{r^l}\right) d\Omega = \int_0^{2\pi} \int_0^\pi \frac{\mathcal{Y}_{lm}^*}{r^l} \frac{\mathcal{Y}_{lm}}{r^l} \sin\theta \, d\theta \, d\varphi = 1 \tag{6E.3.11}$$

Show that

$$L_- \mathcal{Y}_{lm} = e^{i\phi_{lm}} \sqrt{l(l+1) - m(m-1)} \mathcal{Y}_{l,m-1} \tag{6E.3.12}$$

$$L_+ \mathcal{Y}_{lm} = e^{-i\phi_{l,m+1}} \sqrt{l(l+1) - m(m+1)} \mathcal{Y}_{l,m+1} \tag{6E.3.13}$$

where the phase ϕ_{lm} is a real number.
4. In the following, we shall assume that, for all m, the $\mathcal{Y}_{lm}$ have been chosen such that $\phi_{lm} = 0$:

$$L_\pm \mathcal{Y}_{lm} = \sqrt{l(l+1) - m(m \pm 1)} \mathcal{Y}_{l,m\pm 1} \tag{6E.3.14}$$

Show that

$$\mathcal{Y}_{lm} = \sqrt{\frac{(l-m)!(l+k)!}{(l+m)!(l-k)!}} L_+^{m-k} \mathcal{Y}_{lk}, \quad m \ge k \tag{6E.3.15}$$

$$\mathcal{Y}_{lm} = \sqrt{\frac{(l+m)!(l-k)!}{(l-m)!(l+k)!}} L_-^{k-m} \mathcal{Y}_{lk}, \quad m \le k \tag{6E.3.16}$$

5. Show that

$$L_z \mathcal{Y}_{lm}^* = -m \mathcal{Y}_{lm}^* \tag{6E.3.17}$$

and use this relation to demonstrate that $\mathcal{Y}_{l0}$ is a constant times a real function.

6. Since $\mathcal{Y}_{l0}$ is a constant times a real function, we may scale all $\mathcal{Y}_{lm}$ by a common factor so that $\mathcal{Y}_{l0}$ becomes real. Show that, if $\mathcal{Y}_{l0}$ is real, then

$$\mathcal{Y}_{lm}^* = (-1)^m \mathcal{Y}_{l,-m} \tag{6E.3.18}$$

Hint: Set $k = 0$ in (6E.3.15) and (6E.3.16).

EXERCISE 6.4

The purpose of the present exercise is to scale the complex solid harmonics $\mathcal{Y}_{lm}$ in such a way that the angular part is normalized to 1, $\mathcal{Y}_{l0}$ is a real function and the components are related as

$$L_\pm \mathcal{Y}_{lm} = \sqrt{l(l+1) - m(m \pm 1)} \mathcal{Y}_{l,m\pm 1} \tag{6E.4.1}$$

The solid harmonics are given by

$$\mathcal{Y}_{lm} = N_{lm}[x + \text{sgn}(m)\mathrm{i}y]^{|m|} \sum_{t=0}^{[(l-|m|)/2]} C_t^{l|m|}(x^2 + y^2)^t z^{l-2t-|m|} \tag{6E.4.2}$$

where

$$C_t^{lm} = \left(-\frac{1}{4}\right)^t \binom{l-t}{|m|+t}\binom{l}{t} \tag{6E.4.3}$$

and N_{lm} is the normalization constant to be determined. Our strategy is first to normalize $\mathcal{Y}_{ll}$, next to generate $\mathcal{Y}_{lm}$ for $m > 0$ by applying the step-down operator and finally to normalize $\mathcal{Y}_{l,-|m|}$ by using the relation (6E.3.18).

1. From (6E.4.2) and (6E.4.3), we obtain for the solid harmonic with $m = l$:

$$\mathcal{Y}_{ll} = N_{ll}(x + \mathrm{i}y)^l \tag{6E.4.4}$$

Show that the norm of the angular part of $\mathcal{Y}_{ll}$ is given by

$$\int \left(\frac{\mathcal{Y}_{ll}}{r^l}\right)^* \left(\frac{\mathcal{Y}_{ll}}{r^l}\right) \mathrm{d}\Omega = \int_0^{2\pi} \int_0^\pi \frac{\mathcal{Y}_{ll}^*}{r^l} \frac{\mathcal{Y}_{ll}}{r^l} \sin\theta \, \mathrm{d}\theta \, \mathrm{d}\varphi = \frac{4\pi}{2l+1} \frac{2^{2l}(l!)^2}{(2l)!} |N_{ll}|^2 \tag{6E.4.5}$$

Hint: A useful integral is

$$\int_{-1}^1 (1 - x^2)^n \mathrm{d}x = 2^{2n+1} \frac{(n!)^2}{(2n+1)!}, \quad n \geq 0 \tag{6E.4.6}$$

Following the usual practice in quantum mechanics, we adopt the phase convention

$$N_{ll} = (-1)^l \sqrt{\frac{(2l+1)(2l)!}{4\pi}} \frac{1}{2^l l!} \tag{6E.4.7}$$

2. Show that, in Cartesian coordinates, the step-down operator becomes

$$L_- = (x - iy)\frac{\partial}{\partial z} - z\left(\frac{\partial}{\partial x} - i\frac{\partial}{\partial y}\right) \tag{6E.4.8}$$

3. Verify the following relation for $m > 0$:

$$L_-\mathcal{Y}_{lm} = N_{lm}(x + iy)^{m-1}\sum_{t=0}^{[(l-m)/2]}C_t^{lm}(l - 2t - m)(x^2 + y^2)^{t+1}z^{l-2t-m-1}$$

$$- N_{lm}(x + iy)^{m-1}\sum_{t=0}^{[(l-m)/2]}C_t^{lm}(2m + 2t)(x^2 + y^2)^t z^{l-2t-m+1} \tag{6E.4.9}$$

4. For $m > 0$, use (6E.4.3) to verify the relations

$$C_t^{l,m-1} = \frac{m + t}{l - m - 2t + 1}C_t^{l,m} \tag{6E.4.10}$$

$$C_t^{l,m-1} = -\frac{l - m - 2t + 2}{4t}C_{t-1}^{lm} \tag{6E.4.11}$$

5. Use (6E.4.10) and (6E.4.11) to rewrite (6E.4.9) in the form

$$L_-\mathcal{Y}_{lm} = -2N_{lm}(l - m + 1)(x+iy)^{m-1}\sum_{t=0}^{[(l-m)/2]}C_t^{l,m-1}(x^2 + y^2)^t z^{l-2t-m+1} + N_{lm}(x + iy)^{m-1}T_{lm} \tag{6E.4.12}$$

where

$$T_{lm} = (l - 2t_{max} - m)C_{t_{max}}^{lm}(x^2 + y^2)^{t_{max}+1}z^{l-2t_{max}-m-1} \tag{6E.4.13}$$

$$t_{max} = \left[\frac{l - m}{2}\right] \tag{6E.4.14}$$

6. Show that (6E.4.13) can be written in the form

$$T_{lm} = -2(l - m + 1)\sum_{t=[(l-m)/2]+1}^{[(l-(m-1))/2]}C_t^{l,m-1}(x^2 + y^2)^t z^{l-2t-m+1} \tag{6E.4.15}$$

7. Verify the following relation for $m > 0$:

$$L_-\mathcal{Y}_{lm} = -2(l - m + 1)\frac{N_{lm}}{N_{l,m-1}}\mathcal{Y}_{l,m-1} \tag{6E.4.16}$$

8. Combine (6E.4.16) with (6E.4.1) to show that the normalization constants are related by

$$\frac{N_{lm}}{N_{l,m-1}} = -\frac{1}{2}\sqrt{\frac{l + m}{l - m + 1}} \tag{6E.4.17}$$

9. Show that

$$N_{lm} = (-1)^m\frac{1}{2^m l!}\sqrt{\frac{2l + 1}{4\pi}(l + m)!(l - m)!}, \quad m \geq 0 \tag{6E.4.18}$$

contains N_{ll} in (6E.4.7) as a special case and that it fulfils (6E.4.17).

10. We have now obtained the normalization constants for $m \geq 0$. Since these constants are real, we see from (6E.4.2) that $\mathcal{Y}_{l0}$ is real as well. We can then obtain the normalization constants for $m < 0$ from (6E.3.18). Show that, in general, the normalization constant can be written as

$$N_{lm} = (-1)^{\frac{m+|m|}{2}} \frac{1}{2^{|m|} l!} \sqrt{\frac{2l+1}{4\pi} (l+|m|)!(l-|m|)!} \tag{6E.4.19}$$

Note that the sign of the normalization constant is positive for $m < 0$ but alternates for $m > 0$.

EXERCISE 6.5

We here examine the radial extent of the hydrogenic radial functions R_{nl} and the Laguerre functions R_{nl}^{LF}. In terms of the associated Laguerre polynomials $L_n^k(x)$, these functions may be expressed as

$$R_{nl}(r) = \left(\frac{2Z}{n}\right)^{3/2} \sqrt{\frac{(n-l-1)!}{2n(n+l)!}} \left(\frac{2Zr}{n}\right)^l L_{n-l-1}^{2l+1}\left(\frac{2Zr}{n}\right) \exp\left(\frac{-Zr}{n}\right) \tag{6E.5.1}$$

$$R_{nl}^{\mathrm{LF}}(r) = (2\zeta)^{3/2} \sqrt{\frac{(n-l-1)!}{(n+l+1)!}} (2\zeta r)^l L_{n-l-1}^{2l+2}(2\zeta r) \exp(-\zeta r) \tag{6E.5.2}$$

1. Calculate the radial extent of the Laguerre functions:

$$\langle r \rangle^{\mathrm{LF}} = \int_0^\infty R_{nl}^{\mathrm{LF}}(r) r R_{nl}^{\mathrm{LF}}(r) r^2 \mathrm{d}r \tag{6E.5.3}$$

Hint: Use the orthogonality and recurrence relations of the Laguerre polynomials:

$$\int_0^\infty L_m^k(x) L_n^k(x) x^k \exp(-x) \, \mathrm{d}x = \frac{(n+k)!}{n!} \delta_{mn} \tag{6E.5.4}$$

$$(n+1)L_{n+1}^k(x) = (2n+k+1-x)L_n^k(x) - (n+k)L_{n-1}^k(x) \tag{6E.5.5}$$

2. Calculate the radial extent of the hydrogenic wave functions:

$$\langle r \rangle^{\mathrm{H}} = \int_0^\infty R_{nl}(r) r R_{nl}(r) r^2 \mathrm{d}r \tag{6E.5.6}$$

EXERCISE 6.6

In the least-squares approximation, a given function $f(x)$ is approximated as

$$g(x) = \sum_i c_i h_i(x) \tag{6E.6.1}$$

where the $h_i(x)$ are a fixed set of functions and the coefficients c_i are chosen to minimize

$$\|f - g\|^2 = \langle f - g | f - g \rangle \tag{6E.6.2}$$

for some given inner product. Show that, for real functions, the coefficients are given as

$$\mathbf{c} = \mathbf{S}^{-1}\mathbf{T} \tag{6E.6.3}$$

where the matrix elements of **S** and **T** are

$$S_{ij} = \langle h_i | h_j \rangle \tag{6E.6.4}$$

$$T_i = \langle h_i | f \rangle \tag{6E.6.5}$$

EXERCISE 6.7

In this exercise, we consider the extent of the radial STOs:

$$R_n^{STO}(r) = \frac{(2\zeta)^{3/2}}{\sqrt{(2n)!}} (2\zeta r)^{n-1} \exp(-\zeta r) \tag{6E.7.1}$$

1. Calculate the expectation value

$$\langle r \rangle^{STO} = \int_0^\infty R_n^{STO}(r) r R_n^{STO}(r) r^2 dr \tag{6E.7.2}$$

Hint: A useful integral is

$$\int_0^\infty x^n \exp(-\alpha x) \, dx = \frac{n!}{\alpha^{n+1}}, \quad n \geq 0, \alpha > 0 \tag{6E.7.3}$$

2. Show that the maximum of the radial distribution function $4\pi r^2 [R_n^{STO}(r)]^2$ is given by

$$r_{max}^{STO} = \frac{n}{\zeta} \tag{6E.7.4}$$

EXERCISE 6.8

The energy of the three-dimensional isotropic HO function $R_{nl}^{HO} Y_{lm}$ is given by

$$E_{\bar{n}} = \left(\bar{n} + \tfrac{3}{2}\right) 2\alpha \tag{6E.8.1}$$

where

$$\bar{n} = 2(n-1) - l \tag{6E.8.2}$$

1. Show that, for a fixed $\bar{n}$, the possible values of (n, l) are $(\bar{n} + 1, \bar{n})$, $(\bar{n}, \bar{n} - 2)$, $\cdots (\bar{n} + 1 - [\bar{n}/2], \bar{n} - 2[\bar{n}/2])$.
2. Show that the degeneracy of the energy level $\bar{n}$ is given by

$$g = \tfrac{1}{2}(\bar{n} + 1)(\bar{n} + 2) \tag{6E.8.3}$$

EXERCISE 6.9

In this exercise, we study the radial form of the GTOs, the most frequently used basis functions in electronic-structure theory:

$$R_{\alpha l}^{GTO}(r) = N_{\alpha l}^{GTO} (\sqrt{2\alpha} \, r)^l \exp(-\alpha r^2) \tag{6E.9.1}$$

The related functions $R_{nl}^{GTO}(r)$ in (6.6.7) may be studied in the same manner, replacing l by $2n - l - 2$ in (6E.9.1).

1. Using the integral

$$\int_0^\infty x^{2n} \exp(-\alpha x^2)\,dx = \frac{(2n-1)!!}{2^{n+1}} \sqrt{\frac{\pi}{\alpha^{2n+1}}}, \quad n \geq 0, \alpha > 0 \tag{6E.9.2}$$

verify that the normalization constant $N_{\alpha l}^{\text{GTO}}$ in (6E.9.1) is given by

$$N_{\alpha l}^{\text{GTO}} = \frac{2(2\alpha)^{3/4}}{\pi^{1/4}} \sqrt{\frac{2^l}{(2l+1)!!}} \tag{6E.9.3}$$

2. Determine the radial overlap between two GTOs:

$$S_{\alpha_1 \alpha_2 l} = \int_0^\infty R_{\alpha_1 l}^{\text{GTO}}(r) R_{\alpha_2 l}^{\text{GTO}}(r) r^2\,dr \tag{6E.9.4}$$

3. Setting

$$\alpha_2 = \beta\alpha_1 \tag{6E.9.5}$$

verify that the overlap $S_{\alpha_1,\beta\alpha_1,l}$ in (6E.9.4) is independent of α_1.

4. Using the integral

$$\int_0^\infty x^{2n+1} \exp(-\alpha x^2)\,dx = \frac{n!}{2\alpha^{n+1}}, \quad n \geq 0, \alpha > 0 \tag{6E.9.6}$$

show that the expectation value of r becomes

$$\langle r \rangle^{\text{GTO}} = \int_0^\infty R_{\alpha l}^{\text{GTO}}(r) r R_{\alpha l}^{\text{GTO}}(r) r^2\,dr = \sqrt{\frac{2}{\pi\alpha}} \frac{2^l (l+1)!}{(2l+1)!!} \tag{6E.9.7}$$

5. Use the Stirling approximation for factorials

$$k! \approx k^{k+1/2} \sqrt{2\pi} e^{-k} \quad \text{(large } k) \tag{6E.9.8}$$

to show that $\langle r \rangle^{\text{GTO}}$ for large l becomes

$$\langle r \rangle^{\text{GTO}} \approx \sqrt{\frac{l}{2\alpha}} \quad \text{(large } l) \tag{6E.9.9}$$

6. Verify that the radial distribution function $4\pi r^2 [R_{\alpha l}^{\text{GTO}}(r)]^2$ has a maximum at

$$r_{\max}^{\text{GTO}} = \sqrt{\frac{l+1}{2\alpha}} \tag{6E.9.10}$$

7. Consider two GTOs of different l, $R_{\alpha_1 l_1}^{\text{GTO}}$ and $R_{\alpha_2 l_2}^{\text{GTO}}$. Determine the relation between α_1 and α_2 so that the radial distributions of these GTOs have the same maximum.

Solutions

SOLUTION 6.1

Expanding r_A in the potential, we obtain

$$V = \tfrac{1}{2} \sum_A k_A r_A^2 = \tfrac{1}{2} \sum_A k_A (\mathbf{r} - \mathbf{A}) \cdot (\mathbf{r} - \mathbf{A})$$

$$= \tfrac{1}{2} \left(\sum_A k_A \right) \mathbf{r} \cdot \mathbf{r} - \sum_A k_A \mathbf{A} \cdot \mathbf{r} + \tfrac{1}{2} \sum_A k_A A^2 \tag{6S.1.1}$$

We now introduce (6E.1.3) and (6E.1.5) and rearrange terms, yielding

$$V = \tfrac{1}{2}K\mathbf{r} \cdot \mathbf{r} - K\mathbf{S} \cdot \mathbf{r} + \tfrac{1}{2}\sum_A k_A A^2$$

$$= \tfrac{1}{2}K(\mathbf{r} - \mathbf{S}) \cdot (\mathbf{r} - \mathbf{S}) + \tfrac{1}{2}\sum_A k_A A^2 - \tfrac{1}{2}KS^2 \qquad (6S.1.2)$$

Finally, we re-expand $\mathbf{S}$ in the last term and obtain

$$V = \frac{1}{2}Kr_S^2 + \frac{1}{2K}\left(\sum_{AB} k_A k_B A^2 - \sum_{AB} k_A k_B \mathbf{A} \cdot \mathbf{B}\right)$$

$$= \frac{1}{2}Kr_S^2 + \frac{1}{2K}\sum_{A>B} k_A k_B (A^2 + B^2 - 2\mathbf{A} \cdot \mathbf{B})$$

$$= \frac{1}{2}Kr_S^2 + \frac{1}{2K}\sum_{A>B} k_A k_B R_{AB}^2 = \frac{1}{2}Kr_S^2 + V^{(0)} \qquad (6S.1.3)$$

where we have introduced (6E.1.4).

SOLUTION 6.2

Expression (6E.2.3) is obviously correct for $t = 0$. Assuming that (6E.2.3) is true for $t - 1$, we obtain

$$C_t^{lm} = -\frac{(l - |m| - 2t + 2)(l - |m| - 2t + 1)}{4t(|m| + t)}C_{t-1}^{lm}$$

$$= \left(-\frac{1}{4}\right)^t \frac{(l - |m| - 2t + 2)(l - |m| - 2t + 1)(l - t + 1)!\,l!}{t(|m| + t)(|m| + t - 1)!(l - |m| - 2t + 2)!(t - 1)!(l - t + 1)!}$$

$$= \left(-\frac{1}{4}\right)^t \frac{(l - t)!\,l!}{(|m| + t)!(l - |m| - 2t)!\,t!(l - t)!}$$

$$= \left(-\frac{1}{4}\right)^t \binom{l - t}{|m| + t}\binom{l}{t} \qquad (6S.2.1)$$

Equation (6E.2.3) has thereby been proved by induction.

SOLUTION 6.3

1. Relation (6E.3.3) follows by inserting (6E.3.1) and (6E.3.2) in the commutator and using the basic commutation relations (6E.3.6)–(6E.3.8):

$$[L_z, L_\pm] = [L_z, L_x] \pm \mathrm{i}[L_z, L_y] = \pm L_\pm \qquad (6S.3.1)$$

Relation (6E.3.4) follows in a similar manner, noting that L^2 commutes with its components L_x, L_y, and L_z – for example:

$$[L^2, L_x] = [L_y^2, L_x] + [L_z^2, L_x]$$

$$= L_y[L_y, L_x] + [L_y, L_x]L_y + L_z[L_z, L_x] + [L_z, L_x]L_z$$

$$= -\mathrm{i}L_y L_z - \mathrm{i}L_z L_y + \mathrm{i}L_z L_y + \mathrm{i}L_y L_z = 0 \qquad (6S.3.2)$$

Relation (6E.3.5) is verified as follows:

$$L_+L_- = L_x^2 + L_y^2 + i[L_y, L_x] = L^2 - L_z(L_z - 1) \tag{6S.3.3}$$

2. Using the commutation relations (6E.3.3) and (6E.3.4), we obtain:

$$L_z L_\pm \mathcal{Y}_{lm} = [L_z, L_\pm]\mathcal{Y}_{lm} + L_\pm L_z \mathcal{Y}_{lm} = (m \pm 1)L_\pm \mathcal{Y}_{lm} \tag{6S.3.4}$$

$$L^2 L_\pm \mathcal{Y}_{lm} = [L^2, L_\pm]\mathcal{Y}_{lm} + L_\pm L^2 \mathcal{Y}_{lm} = l(l+1)L_\pm \mathcal{Y}_{lm} \tag{6S.3.5}$$

3. Since $L_- \mathcal{Y}_{lm}$ is proportional to $\mathcal{Y}_{l,m-1}$, we have

$$L_- \mathcal{Y}_{lm} = C_{lm} \mathcal{Y}_{l,m-1} \tag{6S.3.6}$$

The constant C_{lm} can be obtained from the equation

$$\int \left(\frac{L_- \mathcal{Y}_{lm}}{r^l}\right)^* \left(\frac{L_- \mathcal{Y}_{lm}}{r^l}\right) \, d\Omega = \langle L_- \mathcal{Y}_{lm} | L_- \mathcal{Y}_{lm}\rangle$$

$$= |C_{lm}|^2 \langle \mathcal{Y}_{l,m-1} | \mathcal{Y}_{l,m-1}\rangle = |C_{lm}|^2 \tag{6S.3.7}$$

To evaluate the left-hand side, we use the relation $L_-^\dagger = L_+$ and (6E.3.5):

$$\langle L_- \mathcal{Y}_{lm} | L_- \mathcal{Y}_{lm}\rangle = \langle \mathcal{Y}_{lm} | L_+ L_- | \mathcal{Y}_{lm}\rangle = \langle \mathcal{Y}_{lm} | L^2 - L_z(L_z - 1) | \mathcal{Y}_{lm}\rangle$$

$$= l(l+1) - m(m-1) \tag{6S.3.8}$$

The coefficient C_{lm} can thus be written as

$$C_{lm} = e^{i\phi_{lm}} \sqrt{l(l+1) - m(m-1)} \tag{6S.3.9}$$

in agreement with (6E.3.12). To arrive at (6E.3.13), we invoke (6E.3.12) and obtain

$$L_+ \mathcal{Y}_{lm} = L_+ \frac{e^{-i\phi_{l,m+1}} L_- \mathcal{Y}_{l,m+1}}{\sqrt{l(l+1) - m(m+1)}} = e^{-i\phi_{l,m+1}} \sqrt{l(l+1) - m(m+1)} \mathcal{Y}_{l,m+1} \tag{6S.3.10}$$

where again (6E.3.5) has been used.

4. Relations (6E.3.15) and (6E.3.16) hold trivially for $m = k$. Assume next that (6E.3.15) holds for $m - 1$. Relation (6E.3.15) then follows by induction:

$$\mathcal{Y}_{lm} = \frac{1}{\sqrt{l(l+1) - m(m-1)}} L_+ \mathcal{Y}_{l,m-1}$$

$$= \frac{1}{\sqrt{(l+m)(l-m+1)}} L_+ \left[\sqrt{\frac{(l-m+1)!(l+k)!}{(l+m-1)!(l-k)!}} L_+^{m-1-k} \mathcal{Y}_{lk}\right]$$

$$= \sqrt{\frac{(l-m)!(l+k)!}{(l+m)!(l-k)!}} L_+^{m-k} \mathcal{Y}_{lk} \tag{6S.3.11}$$

Relation (6E.3.16) follows in a similar manner, assuming that it holds for $m + 1$:

$$\mathcal{Y}_{lm} = \frac{1}{\sqrt{l(l+1) - m(m+1)}} L_- \mathcal{Y}_{l,m+1}$$

$$= \frac{1}{\sqrt{(l+m+1)(l-m)}} L_- \left[\sqrt{\frac{(l+m+1)!(l-k)!}{(l-m-1)!(l+k)!}} L_-^{k-m-1} \mathcal{Y}_{lk} \right]$$

$$= \sqrt{\frac{(l+m)!(l-k)!}{(l-m)!(l+k)!}} L_-^{k-m} \mathcal{Y}_{lk} \tag{6S.3.12}$$

5. Since L_z is a pure imaginary operator, we obtain

$$L_z \mathcal{Y}_{lm}^* = -(L_z \mathcal{Y}_{lm})^* = -m \mathcal{Y}_{lm}^* \tag{6S.3.13}$$

The function $\mathcal{Y}_{lm}^*$ is therefore proportional to $\mathcal{Y}_{l,-m}$. As a special case, $\mathcal{Y}_{l0}^*$ and $\mathcal{Y}_{l0}$ are proportional to each other and $\mathcal{Y}_{l0}$ is therefore a constant times a real function.

6. From (6E.3.15) and (6E.3.16), we first note the relations:

$$\mathcal{Y}_{l,|m|} = \sqrt{\frac{(l-|m|)!}{(l+|m|)!}} L_+^{|m|} \mathcal{Y}_{l0} \tag{6S.3.14}$$

$$\mathcal{Y}_{l,-|m|} = \sqrt{\frac{(l-|m|)!}{(l+|m|)!}} L_-^{|m|} \mathcal{Y}_{l0} \tag{6S.3.15}$$

Taking the complex conjugate of (6S.3.15), we now obtain

$$\mathcal{Y}_{l,-|m|}^* = \sqrt{\frac{(l-|m|)!}{(l+|m|)!}} (L_-^{|m|} \mathcal{Y}_{l0})^* = \sqrt{\frac{(l-|m|)!}{(l+|m|)!}} (-L_+)^{|m|} \mathcal{Y}_{l0}^* = (-1)^{|m|} \mathcal{Y}_{l|m|} \tag{6S.3.16}$$

which yields (6E.3.18). This relation – the Condon–Shortley phase convention – is a consequence of choosing $\mathcal{Y}_{l0}$ real and defining the phases of $\mathcal{Y}_{lm}$ for $m \neq 0$ such that (6E.3.14) is fulfilled.

SOLUTION 6.4

1. Using (6E.4.4) and (6.4.27), we find

$$\frac{\mathcal{Y}_{ll}}{r^l} = N_{ll} \left(\frac{x + iy}{r} \right)^l = N_{ll} \sin^l \theta \exp(il\varphi) \tag{6S.4.1}$$

The normalization integral then becomes

$$\int \left(\frac{\mathcal{Y}_{ll}}{r^l} \right)^* \left(\frac{\mathcal{Y}_{ll}}{r^l} \right) d\Omega = |N_{ll}|^2 \int_0^{2\pi} \int_0^{\pi} \sin^{2l} \theta \sin \theta \, d\theta \, d\varphi$$

$$= |N_{ll}|^2 2\pi \int_{-1}^{1} (1 - \cos^2 \theta)^l \, d\cos \theta$$

$$= |N_{ll}|^2 \frac{4\pi}{2l+1} \frac{2^{2l} (l!)^2}{(2l)!} \tag{6S.4.2}$$

2. This expression is obtained by substituting

$$L_x = -i\left(y\frac{\partial}{\partial z} - z\frac{\partial}{\partial y}\right) \tag{6S.4.3}$$

$$L_y = -i\left(z\frac{\partial}{\partial x} - x\frac{\partial}{\partial z}\right) \tag{6S.4.4}$$

in the step-down operator (6E.3.2).

3. Inserting (6E.4.8) in (6E.4.2) and carrying out the differentiation, we obtain

$$L_-\mathscr{Y}_{lm} = N_{lm}L_-(x+iy)^m \sum_{t=0}^{[(l-m)/2]} C_t^{lm}(x^2+y^2)^t z^{l-2t-m}$$

$$= N_{lm}(x-iy)(x+iy)^m \sum_{t=0}^{[(l-m)/2]} C_t^{lm}(x^2+y^2)^t(l-2t-m)z^{l-2t-m-1}$$

$$- N_{lm}2m(x+iy)^{m-1} \sum_{t=0}^{[(l-m)/2]} C_t^{lm}(x^2+y^2)^t z^{l-2t-m+1}$$

$$- N_{lm}(x+iy)^m \sum_{t=0}^{[(l-m)/2]} C_t^{lm}2t(x-iy)(x^2+y^2)^{t-1}z^{l-2t-m+1} \tag{6S.4.5}$$

which gives (6E.4.9).

4. The first relation (6E.4.10) follows by expanding the first binomial coefficient in (6E.4.3):

$$C_t^{l,m-1} = \left(-\frac{1}{4}\right)^t \binom{l-t}{m-1+t}\binom{l}{t}$$

$$= \left(-\frac{1}{4}\right)^t \frac{(l-t)!}{(m-1+t)!(l-m-2t+1)!}\binom{l}{t}$$

$$= \frac{m+t}{l-m-2t+1}\left(-\frac{1}{4}\right)^t \frac{(l-t)!}{(m+t)!(l-m-2t)!}\binom{l}{t}$$

$$= \frac{m+t}{l-m-2t+1}C_t^{lm} \tag{6S.4.6}$$

For the second relation (6E.4.11), both coefficients must be expanded:

$$C_t^{l,m-1} = \left(-\frac{1}{4}\right)^t \binom{l-t}{m-1+t}\binom{l}{t}$$

$$= \left(-\frac{1}{4}\right)^t \frac{(l-t)!l!}{(m-1+t)!(l-m-2t+1)!t!(l-t)!}$$

$$= \left(-\frac{1}{4}\right)\left(-\frac{1}{4}\right)^{t-1} \frac{(l-m-2t+2)(l-t+1)!l!}{t(m-1+t)!(l-m-2t+2)!(t-1)!(l-t+1)!}$$

$$= -\frac{l-m-2t+2}{4t}C_{t-1}^{lm} \tag{6S.4.7}$$

5. The right-hand side of (6E.4.9) contains two summations A_{lm} and B_{lm}

$$L_- \mathcal{Y}_{lm} = N_{lm}(x + iy)^{m-1}(A_{lm} + B_{lm}) \tag{6S.4.8}$$

which we rewrite in the following manner

$$A_{lm} = \sum_{t=0}^{[(l-m)/2]} C_t^{lm}(l - 2t - m)(x^2 + y^2)^{t+1} z^{l-2t-m-1}$$

$$= \sum_{t=1}^{[(l-m)/2]} (l - m - 2t + 2)C_{t-1}^{lm}(x^2 + y^2)^t z^{l-2t-m+1}$$

$$+ C_{t_{max}}^{lm}(l - 2t_{max} - m)(x^2 + y^2)^{t_{max}+1} z^{l-2t_{max}-m-1} \tag{6S.4.9}$$

$$B_{lm} = -\sum_{t=0}^{[(l-m)/2]} C_t^{lm}(2m + 2t)(x^2 + y^2)^t z^{l-2t-m+1}$$

$$= -2m C_0^{lm} z^{l-m+1} - \sum_{t=1}^{[(l-m)/2]} 2(m + t)C_t^{lm}(x^2 + y^2)^t z^{l-2t-m+1} \tag{6S.4.10}$$

where t_{max} is defined in (6E.4.14). Except for the different numerical factors, the rewritten summations in (6S.4.9) and (6S.4.10) are identical and may be combined. Using (6E.4.10) and (6E.4.11), we obtain

$$(l - m - 2t + 2)C_{t-1}^{lm} - 2(m + t)C_t^{lm} = -2(l - m + 1)C_t^{l,m-1} \tag{6S.4.11}$$

$$-2m C_0^{lm} = -2(l - m + 1)C_0^{l,m-1} \tag{6S.4.12}$$

which allows us to combine the summations as

$$A_{lm} + B_{lm} = -2(l - m + 1)\sum_{t=0}^{[(l-m)/2]} C_t^{l,m-1}(x^2 + y^2)^t z^{l-2t-m+1}$$

$$+ (l - 2t_{max} - m)C_{t_{max}}^{lm}(x^2 + y^2)^{t_{max}+1} z^{l-2t_{max}-m-1} \tag{6S.4.13}$$

Insertion of (6S.4.13) in (6S.4.8) gives (6E.4.12).

6. We distinguish between two cases, depending on whether $l - m$ is even or odd:

$$t_{max} = \begin{cases} \dfrac{l - m}{2} & l - m \text{ even} \\ \dfrac{l - m - 1}{2} & l - m \text{ odd} \end{cases} \tag{6S.4.14}$$

If $l - m$ is even, then (6E.4.13) and (6E.4.15) both vanish – the former trivially and the latter because the upper limit is smaller than the lower limit. If $l - m$ is odd, then we may rewrite (6E.4.13) in the form

$$T_{lm} = C_{(l-m-1)/2}^{lm}(x^2 + y^2)^{(l-m+1)/2}$$

$$= -2(l - m + 1)C_{(l-m+1)/2}^{l,m-1}(x^2 + y^2)^{(l-m+1)/2}$$

$$= -2(l - m + 1)\sum_{t=[(l-m)/2]+1}^{[(l-m+1)/2]} C_t^{l,m-1}(x^2 + y^2)^t z^{l-2t-m+1} \tag{6S.4.15}$$

7. Combining (6E.4.12) and (6E.4.15), we obtain

$$L_- \mathscr{Y}_{lm} = -2(l - m + 1)N_{lm}(x + iy)^{m-1} \sum_{t=0}^{[(l-(m-1))/2]} C_t^{l,m-1}(x^2 + y^2)^t z^{l-2t-m+1}$$

$$= -2(l - m + 1)\frac{N_{lm}}{N_{l,m-1}}\mathscr{Y}_{l,m-1} \tag{6S.4.16}$$

where we have used the definition of $\mathscr{Y}_{l,m-1}$ in (6E.4.2).

8. Combining (6E.4.1) with (6E.4.16), we find

$$\frac{N_{lm}}{N_{l,m-1}} = \frac{\sqrt{l(l+1) - m(m-1)}}{-2(l-m+1)} = -\frac{1}{2}\sqrt{\frac{l+m}{l-m+1}} \tag{6S.4.17}$$

9. For $m = l$, we obtain from (6E.4.18)

$$N_{ll} = (-1)^l \frac{1}{2^l l!}\sqrt{\frac{2l+1}{4\pi}(2l)!} \tag{6S.4.18}$$

which is identical to (6E.4.7). Next, using N_{lm} and $N_{l,m-1}$ from (6E.4.18), we obtain

$$\frac{N_{lm}}{N_{l,m-1}} = -\frac{1}{2}\sqrt{\frac{(l+m)!(l-m)!}{(l+m-1)!(l-m+1)!}} = -\frac{1}{2}\sqrt{\frac{l+m}{l-m+1}} \tag{6S.4.19}$$

which is identical to (6E.4.17).

10. From (6E.3.18) and (6E.4.2), we obtain

$$N_{l,-|m|} = (-1)^{|m|}N_{l,|m|} \tag{6S.4.20}$$

which, combined with (6E.4.18), yields the following normalization for $m < 0$:

$$N_{l,-|m|} = \frac{1}{2^{|m|}l!}\sqrt{\frac{2l+1}{4\pi}(l+|m|)!(l-|m|)!} \tag{6S.4.21}$$

Equations (6E.4.18) and (6S.4.21) can be combined to yield (6E.4.19), which is valid for all m.

SOLUTION 6.5

1. Inserting (6E.5.2) in (6E.5.3), we obtain

$$\langle r \rangle^{\mathrm{LF}} = \frac{(n-l-1)!}{(n+l+1)!}\int_0^\infty \exp(-2\zeta r)(2\zeta r)^{2l+3}[L_{n-l-1}^{2l+2}(2\zeta r)]^2 \, dr$$

$$= \frac{(n-l-1)!}{2\zeta(n+l+1)!}\int_0^\infty \exp(-x)x^{2l+2}xL_{n-l-1}^{2l+2}(x)L_{n-l-1}^{2l+2}(x) \, dx \tag{6S.5.1}$$

The recurrence relation (6E.5.5) may be written in the form

$$xL_{n-l-1}^{2l+2}(x) = (2n+1)L_{n-l-1}^{2l+2}(x) - (n+l+1)L_{n-l-2}^{2l+2}(x) - (n-l)L_{n-l}^{2l+2}(x) \tag{6S.5.2}$$

Inserting this relation into (6S.5.1) and invoking the orthogonality relation (6E.5.4), we arrive at the expectation value:

$$\langle r \rangle^{\text{LF}} = \frac{(2n+1)(n-l-1)!}{2\zeta(n+l+1)!} \int_0^\infty \exp(-x) x^{2l+2} L_{n-l-1}^{2l+2}(x) L_{n-l-1}^{2l+2}(x) \, dx = \frac{2n+1}{2\zeta} \quad (6S.5.3)$$

2. Introducing

$$\alpha = \frac{2Z}{n} \quad (6S.5.4)$$

we may write the expectation value as

$$\langle r \rangle^{\text{H}} = \alpha^3 \frac{(n-l-1)!}{2n(n+l)!} \int_0^\infty \exp(-\alpha r)(\alpha r)^{2l} [L_{n-l-1}^{2l+1}(\alpha r)]^2 r^3 \, dr$$

$$= \frac{(n-l-1)!}{2n(n+l)!\alpha} \int_0^\infty \exp(-x) x^{2l+1} x^2 L_{n-l-1}^{2l+1}(x) L_{n-l-1}^{2l+1}(x) \, dx \quad (6S.5.5)$$

Applying (6E.5.5) twice, we obtain

$$x^2 L_{n-l-1}^{2l+1}(x) = [6n^2 - 2l(l+1)] L_{n-l-1}^{2l+1}(x) - 2(2n-1)(n+l) L_{n-l-2}^{2l+1}(x)$$

$$- 2(2n+1)(n-l) L_{n-l}^{2l+1}(x) + (n+l)(n+l-1) L_{n-l-3}^{2l+1}(x)$$

$$+ (n-l)(n-l+1) L_{n-l+1}^{2l+1}(x) \quad (6S.5.6)$$

Insertion of (6S.5.6) into (6S.5.5) and use of (6E.5.4) and (6S.5.4) give us

$$\langle r \rangle^{\text{H}} = \frac{[3n^2 - l(l+1)](n-l-1)!}{\alpha(n+l)!n} \int_0^\infty x^{2l+1} [L_{n-l-1}^{2l+1}(x)]^2 \exp(-x) \, dx$$

$$= \frac{3n^2 - l(l+1)}{2Z} \quad (6S.5.7)$$

SOLUTION 6.6

Expanding the function to be minimized (6E.6.2), we obtain

$$\langle f - g | f - g \rangle = \langle f | f \rangle - 2 \sum_i c_i T_i + \sum_{ij} c_i S_{ij} c_j \quad (6S.6.1)$$

The gradient elements of this function are given by

$$\frac{\partial \langle f - g | f - g \rangle}{\partial c_i} = 2(\mathbf{Sc} - \mathbf{T})_i \quad (6S.6.2)$$

The gradient therefore vanishes if

$$\mathbf{Sc} = \mathbf{T} \quad (6S.6.3)$$

which is equivalent to (6E.6.3). The stationary point corresponds to a minimum since the Hessian 2S is positive definite.

SOLUTION 6.7

1. Using (6E.7.3), we obtain

$$\langle r \rangle^{\text{STO}} = \frac{(2\zeta)^{2n+1}}{(2n)!} \int_0^\infty r^{2n+1} \exp(-2\zeta r)\, dr = \frac{(2\zeta)^{2n+1}(2n+1)!}{(2n)!(2\zeta)^{2n+2}} = \frac{2n+1}{2\zeta} \tag{6S.7.1}$$

2. The derivative of the radial distribution function is given by

$$4\pi \frac{d}{dr} r^2 [R_n^{\text{STO}}(r)]^2 = 4\pi \frac{(2\zeta)^{2n+1}}{(2n)!} \frac{d}{dr} r^{2n} \exp(-2\zeta r)$$

$$= 4\pi \frac{(2\zeta)^{2n+1}}{(2n)!} r^{2n-1} (2n - 2\zeta r) \exp(-2\zeta r) \tag{6S.7.2}$$

which vanishes for

$$r_{\text{max}}^{\text{STO}} = \frac{n}{\zeta} \tag{6S.7.3}$$

SOLUTION 6.8

1. Let us first determine the pair $(n_{\text{max}}, l_{\text{max}})$ where n and l take on the largest values. From (6E.8.2), n_{max} and l_{max} are related by

$$\bar{n} = 2(n_{\text{max}} - 1) - l_{\text{max}} \tag{6S.8.1}$$

Since

$$l_{\text{max}} = n_{\text{max}} - k \tag{6S.8.2}$$

where $k > 0$, we obtain from (6S.8.1)

$$n_{\text{max}} = \bar{n} + 2 - k \tag{6S.8.3}$$

which implies that n_{max} is obtained for $k = 1$:

$$n_{\text{max}} = \bar{n} + 1 \tag{6S.8.4}$$

$$l_{\text{max}} = \bar{n} \tag{6S.8.5}$$

The remaining (n, l) pairs are generated by decrementing n and l by 1 and 2, respectively. This process terminates when l becomes 0 or 1, giving the following smallest values of l and n:

$$l_{\text{min}} = l_{\text{max}} - 2 \left[\frac{l_{\text{max}}}{2}\right] = \bar{n} - 2 \left[\frac{\bar{n}}{2}\right] \tag{6S.8.6}$$

$$n_{\text{min}} = \bar{n} - \left[\frac{\bar{n}}{2}\right] + 1 \tag{6S.8.7}$$

For a given $\bar{n}$, the allowed (n, l) are thus $(\bar{n} + 1, \bar{n})$, $(\bar{n}, \bar{n} - 2), \ldots, (\bar{n} - [\bar{n}/2] + 1, \bar{n} - 2[\bar{n}/2])$.

2. For each allowed l, there are $2l + 1$ states. From (6E.8.2), we obtain

$$l = 2(n - 1) - \bar{n} \tag{6S.8.8}$$

For a given n, the degeneracy is thus given by

$$2l + 1 = 4n - 3 - 2\bar{n} \tag{6S.8.9}$$

and the total degeneracy becomes

$$g = \sum_{n=\bar{n}-[\bar{n}/2]+1}^{\bar{n}+1} (4n - 3 - 2\bar{n}) \tag{6S.8.10}$$

This sum is a simple arithmetic series, which is evaluated by multiplying the number of terms by the mean of the smallest and largest terms:

$$g = \left[\bar{n} + 1 - \left(\bar{n} - \left[\frac{\bar{n}}{2}\right] + 1\right) + 1\right] \frac{(4\bar{n} + 4 - 3 - 2\bar{n}) + (4\bar{n} - 4[\bar{n}/2] + 4 - 3 - 2\bar{n})}{2}$$

$$= \frac{1}{2}\left(2\left[\frac{\bar{n}}{2}\right] + 2\right)\left[2\left(\bar{n} - \left[\frac{\bar{n}}{2}\right]\right) + 1\right] \tag{6S.8.11}$$

As is easily verified, this expression reduces to (6E.8.3), for both even and odd $\bar{n}$.

SOLUTION 6.9

1. From (6E.9.2), we obtain

$$\int_0^\infty \left[R_{\alpha l}^{\text{GTO}}(r)\right]^2 r^2 \, dr = \left(N_{\alpha l}^{\text{GTO}}\right)^2 (2\alpha)^l \int_0^\infty r^{2l+2} \exp(-2\alpha r^2) \, dr$$

$$= (N_{\alpha l}^{\text{GTO}})^2 (2\alpha)^l \frac{(2l+1)!!}{2^{l+2}} \sqrt{\frac{\pi}{(2\alpha)^{2l+3}}} \tag{6S.9.1}$$

The following normalization constant is obtained by setting this integral equal to 1:

$$N_{\alpha l}^{\text{GTO}} = \frac{2(2\alpha)^{3/4}}{\pi^{1/4}} \sqrt{\frac{2^l}{(2l+1)!!}} \tag{6S.9.2}$$

2. Using (6E.9.2), we obtain the overlap integral

$$S_{\alpha_1\alpha_2 l} = \frac{4(4\alpha_1\alpha_2)^{3/4}}{\pi^{1/2}} \frac{2^l(\sqrt{4\alpha_1\alpha_2})^l}{(2l+1)!!} \int_0^\infty r^{2l+2} \exp[-(\alpha_1 + \alpha_2)r^2] \, dr$$

$$= \frac{4(4\alpha_1\alpha_2)^{3/4}}{\pi^{1/2}} \frac{2^l(\sqrt{4\alpha_1\alpha_2})^l}{(2l+1)!!} \frac{(2l+1)!!}{2^{l+2}} \sqrt{\frac{\pi}{(\alpha_1 + \alpha_2)^{2l+3}}} = \left(\frac{\sqrt{4\alpha_1\alpha_2}}{\alpha_1 + \alpha_2}\right)^{l+3/2} \tag{6S.9.3}$$

3. We insert (6E.9.5) in (6S.9.3) and obtain an expression which is independent of α_1:

$$S_{\alpha_1, \alpha_1\beta, l} = \left[\frac{\sqrt{4\alpha_1^2\beta}}{\alpha_1(1 + \beta)}\right]^{l+3/2} = \left(\frac{2\sqrt{\beta}}{1 + \beta}\right)^{l+3/2} \tag{6S.9.4}$$

4. Substituting (6E.9.1) with the normalization constant (6E.9.3) in (6E.9.7), we find

$$\langle r \rangle^{\text{GTO}} = \frac{4(2\alpha)^{3/2}}{\sqrt{\pi}} \frac{2^l}{(2l+1)!!} (2\alpha)^l \int_0^\infty r^{2l+3} \exp(-2\alpha r^2) \, dr$$

$$= \frac{4(2\alpha)^{3/2}}{\sqrt{\pi}} \frac{2^l}{(2l+1)!!} (2\alpha)^l \frac{(l+1)!}{2(2\alpha)^{l+2}} = \sqrt{\frac{2}{\pi\alpha}} \frac{2^l(l+1)!}{(2l+1)!!} \tag{6S.9.5}$$

where (6E.9.6) was used to obtain the integral.

5. We first write the expectation value (6E.9.7) in the form

$$\langle r \rangle^{\mathrm{GTO}} = \sqrt{\frac{2}{\pi\alpha}} \frac{2^{2l}(l+1)!l!}{(2l+1)!} \tag{6S.9.6}$$

Introducing the Stirling approximation (6E.9.8), we now obtain

$$\langle r \rangle^{\mathrm{GTO}} \approx \sqrt{\frac{2}{\pi\alpha}} \frac{\sqrt{2\pi}2^{2l}(l+1)^{l+1+1/2}\mathrm{e}^{-l-1}l^{l+1/2}\mathrm{e}^{-l}}{(2l+1)^{2l+1+1/2}\mathrm{e}^{-2l-1}}$$

$$\approx \frac{2^{2l+1}l^{l+3/2}l^{l+1/2}}{\sqrt{\alpha}(2l)^{2l+3/2}} = \sqrt{\frac{l}{2\alpha}} \tag{6S.9.7}$$

6. Taking the derivative of the radial distribution function

$$4\pi r^2 [R_{\alpha l}^{\mathrm{GTO}}(r)]^2 = 4\pi (N_{\alpha l}^{\mathrm{GTO}})^2 (2\alpha)^l r^{2l+2} \exp(-2\alpha r^2) \tag{6S.9.8}$$

with respect to r, we obtain

$$4\pi \frac{\mathrm{d}}{\mathrm{d}r} r^2 [R_{\alpha l}^{\mathrm{GTO}}(r)]^2 = 4\pi (N_{\alpha l}^{\mathrm{GTO}})^2 (2\alpha)^l r^{2l+1}(2l+2-4\alpha r^2) \exp(-2\alpha r^2) \tag{6S.9.9}$$

which gives a maximum at (6E.9.10).

7. From (6E.9.10), we see that the two GTOs have the same maximum for

$$\sqrt{\frac{l_1+1}{2\alpha_1}} = \sqrt{\frac{l_2+1}{2\alpha_2}} \tag{6S.9.10}$$

which gives the following relation between the exponents:

$$\frac{\alpha_1}{\alpha_2} = \frac{l_1+1}{l_2+1} \tag{6S.9.11}$$

7
SHORT-RANGE INTERACTIONS AND ORBITAL EXPANSIONS

In the standard models of quantum chemistry of Chapter 5, the N-electron wave function is expanded in products of the one-electron basis functions introduced in Chapter 6. In the present chapter, we shall examine the orbital expansion of the many-electron systems by considering a simple two-electron system: the ground-state helium wave function. In this manner, we shall obtain a clearer picture of the possibilities and limitations offered by the orbital expansion for many-electron systems. The information obtained by studying the helium system will later prove useful when we go on to construct molecular basis sets from AOs in Chapter 8.

We begin this chapter by examining the exact helium ground-state wave function. In particular, we consider the helium Coulomb hole in Section 7.1 and the Coulomb cusp in Section 7.2. Next, in Section 7.3, we examine the description of the helium ground state by different approximate models: the conventional CI expansion in Section 7.3.1, the explicitly correlated CI expansion in Section 7.3.2, and the Hylleraas expansion in Section 7.3.3. For each wave function, we consider the convergence of the energy with respect to the number of terms in the expansion, thereby illustrating the slow convergence of the standard CI expansion and the improvements obtained by introducing the electronic distance into the wave function. In the final two sections of this chapter, we consider in some detail two important special cases of CI expansions of the helium ground state: the partial-wave expansion in Section 7.4 and the principal expansion in Section 7.5. The partial-wave expansion is important as its convergence has been carefully analysed theoretically; the principal expansion is important as it is closely related to the standard basis sets developed for correlated calculations on molecular systems in Chapter 8.

7.1 The Coulomb hole

Consider the wave function for the ^1S ground state of the helium atom. In the Hartree–Fock approximation, the two electrons move independently of each other, each in the field generated by the average motion of the electrons. The motion of one electron is therefore unaffected by the instantaneous position of the other, as illustrated in Figure 7.1. In this figure, we have plotted the Hartree–Fock wave function for the helium ground state with one electron fixed at a position $0.5a_0$ from the nucleus, in a plane that contains the nucleus and the fixed electron. For clarity, combined surface and contour plots are presented, with the positions of the fixed electron and the nucleus marked by vertical bars. As expected, the Hartree–Fock contour plot consists of a set of concentric circles that are undistorted by the fixed electron – that is, the wave-function amplitude for one electron depends only on its distance to the nucleus, not on the distance to the other electron.

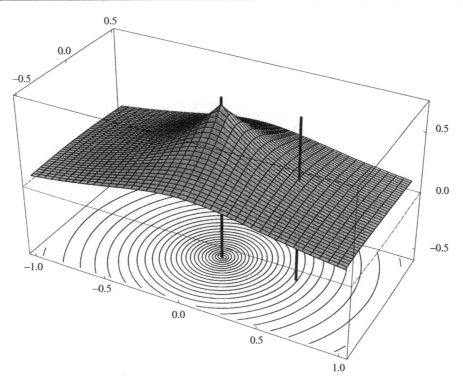

Fig. 7.1. The Hartree–Fock ground-state wave function of the helium atom with one electron fixed at a position $0.5a_0$ away from the nucleus. The wave function is plotted (in atomic units) in a plane that contains the nucleus and the fixed electron, the positions of which are indicated by vertical bars.

Having considered the Hartree–Fock description of the helium atom, let us now turn to the exact wave function for this system. Figure 7.2 shows, for the exact ground-state wave function, a plot similar to that for the Hartree–Fock wave function in Figure 7.1. As seen from the distortion of the concentric contour lines close to the fixed electron, the amplitude of the free electron now depends on its position relative to the nucleus as well as to the fixed electron. Moreover, a careful comparison of Figures 7.1 and 7.2 reveals that the probability of finding the two electrons close to each other is overestimated at the Hartree–Fock level. Still, the description afforded by the Hartree–Fock model is reasonably accurate, differing from that of the exact wave function only in the details.

In order better to bring out the details of the electron–electron interaction in the ground-state helium atom, we have in Figure 7.3 plotted the difference between the exact wave function and the Hartree–Fock wave function. In the exact wave function, the probability amplitude for the free electron is shifted away from the fixed electron, creating a *Coulomb hole* around this electron. The Coulomb hole is wide but rather shallow, with a minimum of -0.068 (relative to the Hartree–Fock description) close to the fixed electron (more precisely, at a distance of $0.49a_0$ from the nucleus). Since the free electron is pushed away from the fixed electron, its amplitude accumulates elsewhere and has a maximum of 0.012 relative to the Hartree–Fock description on the opposite side of the nucleus – at a distance of $0.79a_0$ from the nucleus and $1.29a_0$ from the fixed electron.

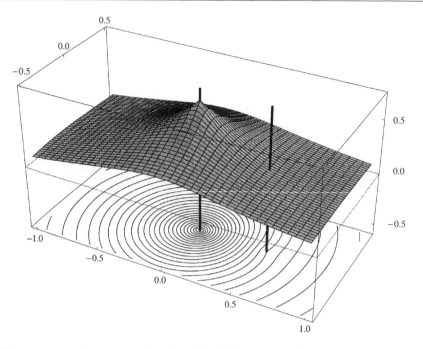

Fig. 7.2. The exact ground-state wave function of the helium atom with one electron fixed at a position $0.5a_0$ away from the nucleus. The wave function is plotted (in atomic units) in a plane that contains the nucleus and the fixed electron, the positions of which are indicated by vertical bars.

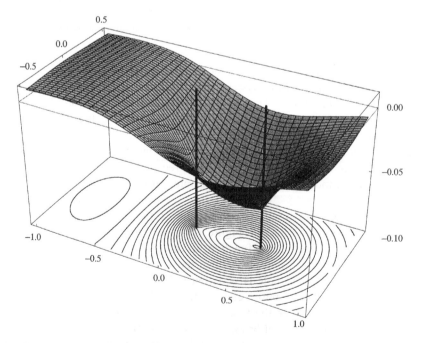

Fig. 7.3. The Coulomb hole in the ground-state helium atom. The plotted function represents the difference between the exact wave function and the Hartree–Fock wave function in a plane that contains the nucleus and the fixed electron, the positions of which are indicated by vertical bars (atomic units).

In the wave-function models that go beyond the Hartree–Fock description, attempts are made at representing – as accurately as possible – the Coulomb hole of the exact wave function. By superposition of antisymmetric products of spin orbitals, it is fairly easy to arrive at a *crude* representation of the Coulomb hole but exceedingly difficult to arrive at an *accurate* description of the short-range interactions – see also the discussion in Section 5.2. In the following, we shall examine this *correlation problem* in greater detail. We begin by investigating the analytical behaviour of the exact wave function for coinciding electrons – that is, the behaviour of the exact wave function in the centre of the Coulomb hole in Figure 7.3.

7.2 The Coulomb cusp

In the present section, we consider the behaviour of the exact wave function for coinciding particles. At such points, the electronic Hamiltonian becomes singular and gives rise to a characteristic cusp in the wave function. To simplify the discussion, we shall again examine the ground state of the helium atom, for which highly accurate approximations to the true wave function are easily generated.

In Cartesian coordinates with the origin at the nucleus, the nonrelativistic electronic Hamiltonian of the helium-like atoms may be written in the usual manner as

$$H = -\frac{1}{2}\nabla_1^2 - \frac{1}{2}\nabla_2^2 - \frac{Z}{|\mathbf{r}_1|} - \frac{Z}{|\mathbf{r}_2|} + \frac{1}{|\mathbf{r}_1 - \mathbf{r}_2|} \tag{7.2.1}$$

where $\mathbf{r}_1$ and $\mathbf{r}_2$ are the coordinates of the two electrons and Z the nuclear charge. This Hamiltonian has singularities in the attraction terms for $\mathbf{r}_1 = \mathbf{0}$ and $\mathbf{r}_2 = \mathbf{0}$ and in the repulsion term for $\mathbf{r}_1 = \mathbf{r}_2$. At these points, the exact solution to the Schrödinger equation

$$H\Psi(\mathbf{r}_1, \mathbf{r}_2) = E\Psi(\mathbf{r}_1, \mathbf{r}_2) \tag{7.2.2}$$

must provide contributions to the product $H\Psi$ that balance the singularities in the Hamiltonian (7.2.1) to ensure that the *local energy*

$$\varepsilon(\mathbf{r}_1, \mathbf{r}_2) = \frac{H\Psi(\mathbf{r}_1, \mathbf{r}_2)}{\Psi(\mathbf{r}_1, \mathbf{r}_2)} \tag{7.2.3}$$

remains constant and equal to the eigenvalue E. The only possible source of such balancing contributions is the kinetic energy. In Figure 7.4, we have plotted the function

$$\Psi_K(\mathbf{r}_1) = -\tfrac{1}{2}\nabla_1^2\Psi\left[\mathbf{r}_1, \left(0, 0, \tfrac{1}{2}a_0\right)\right] \tag{7.2.4}$$

with the second electron fixed at the usual distance of $0.5a_0$ from the nucleus along the z axis. This function, which may be interpreted as the 'local kinetic energy' of the first electron for a fixed position of the second electron, has two sharp peaks – one at the nucleus and the other at the second electron. As expected, the local kinetic energy is positive around the nucleus and negative around the electron, becoming infinite at the points of coalescence so as to balance the singularities in the Hamiltonian. To produce such an infinite local kinetic energy, the wave function must be *nondifferentiable* at the singularities.

To see the implications of the Hamiltonian singularities for the wave function, it is convenient to express the Hamiltonian in a different set of coordinates. Since the ground-state wave function of the helium atom is totally symmetric, it can be expressed in terms of the three radial coordinates

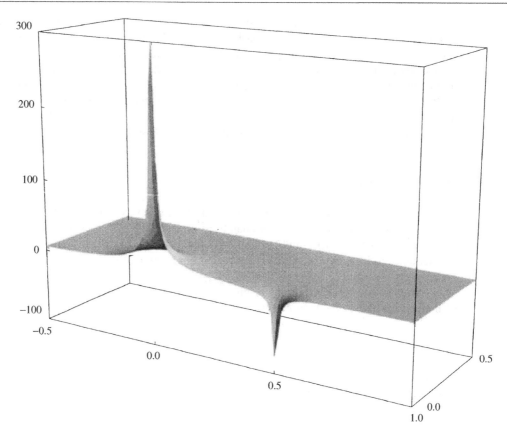

Fig. 7.4. The local kinetic energy (7.2.4) of an electron in the ground state of the helium atom. The local kinetic energy is plotted in a plane that contains the nucleus at the origin and the second electron fixed at a position $0.5a_0$ from the nucleus. Atomic units are used.

r_1, r_2 and r_{12}, where r_1 and r_2 are the distances of the electrons to the nucleus and r_{12} the interelectronic distance. Expressing the kinetic-energy operator in these coordinates, we obtain the following expression for the Hamiltonian [1]

$$H = -\frac{1}{2} \sum_{i=1}^{2} \left(\frac{\partial^2}{\partial r_i^2} + \frac{2}{r_i} \frac{\partial}{\partial r_i} + \frac{2Z}{r_i} \right) - \left(\frac{\partial^2}{\partial r_{12}^2} + \frac{2}{r_{12}} \frac{\partial}{\partial r_{12}} - \frac{1}{r_{12}} \right)$$
$$- \left(\frac{\mathbf{r}_1}{r_1} \cdot \frac{\mathbf{r}_{12}}{r_{12}} \frac{\partial}{\partial r_1} + \frac{\mathbf{r}_2}{r_2} \cdot \frac{\mathbf{r}_{21}}{r_{21}} \frac{\partial}{\partial r_2} \right) \frac{\partial}{\partial r_{12}} \tag{7.2.5}$$

Note that the two one-electron parts of this Hamiltonian each have the same form as the radial Hamiltonian for an electron of zero angular momentum in a spherical potential (6.3.10). The singularities at the nucleus are now seen to be balanced by the kinetic-energy terms proportional to $1/r_i$:

$$\left. \frac{\partial \Psi}{\partial r_i} \right|_{r_i=0} = -Z\Psi(r_i = 0) \tag{7.2.6}$$

Likewise, the terms that multiply $1/r_{12}$ at $r_{12} = 0$ must vanish in $H\Psi$, imposing the additional condition

$$\left. \frac{\partial \Psi}{\partial r_{12}} \right|_{r_{12}=0} = \frac{1}{2} \Psi(r_{12} = 0) \tag{7.2.7}$$

on the wave function. Equation (7.2.6) establishes the behaviour of the ground-state wave function in the vicinity of the nucleus and is known as the *nuclear cusp condition*. Similarly, equation (7.2.7) describes the behaviour of the wave function when the electrons coincide and represents the *(electronic) Coulomb cusp condition* [2,3].

Let us examine the implications of the *nuclear cusp condition* for the first electron, assuming that the wave function does not vanish at $r_1 = 0$ (which holds for the helium ground state). The cusp condition is satisfied if the wave function exhibits an *exponential dependence* on r_1 close to the nucleus:

$$\Psi(r_1, r_2, r_{12}) = (1 - Zr_1 + \cdots)\Psi(0, r_2, r_{12})$$

$$= \exp(-Zr_1)\Psi(0, r_2, r_{12}) \quad \text{(small } r_1) \tag{7.2.8}$$

As discussed in Chapter 6, molecular electronic wave functions are usually expanded in simple analytical functions centred on the atomic nuclei (AOs). Assuming that, close to a given nucleus, the behaviour of the electrons is determined completely by the analytical form of the one-electron basis functions at that nucleus, the nuclear cusp condition imposes constraints on the form of these basis functions close to the nucleus. By employing atomic basis functions that satisfy these constraints, it is possible to construct approximate wave functions that satisfy the nuclear cusp condition exactly. In particular, the STOs of Section 6.5 are compatible with the nuclear cusp condition but not so the GTOs of Section 6.6, which depend on the square of the electronic distance to the atomic centre. In passing, we note that the nuclear cusp condition applies only to point-charge nuclei such as that present in (7.2.1).

The *Coulomb cusp condition* at $r_{12} = 0$ has an even more severe implication for the approximate wave function. Consider the ground-state helium wave function for a collinear arrangement of the nucleus and the two electrons. Expanding the wave function around $r_2 = r_1$ and $r_{12} = 0$, we obtain

$$\Psi(r_1, r_2, r_{12}) = \Psi(r_1, r_1, 0) + (r_2 - r_1) \left. \frac{\partial \Psi}{\partial r_2} \right|_{r_2=r_1} + r_{12} \left. \frac{\partial \Psi}{\partial r_{12}} \right|_{r_2=r_1} + \cdots \tag{7.2.9}$$

which, upon substitution of the cusp condition (7.2.7), gives us

$$\Psi(r_1, r_2, r_{12}) = \Psi(r_1, r_1, 0) + (r_2 - r_1) \left. \frac{\partial \Psi}{\partial r_2} \right|_{r_2=r_1} + \frac{1}{2}|\mathbf{r}_2 - \mathbf{r}_1|\Psi(r_1, r_1, 0) + \cdots \tag{7.2.10}$$

The Coulomb cusp condition therefore leads to a wave function that is continuous but, because of the last term in (7.2.10), not smooth at $r_{12} = 0$. Consequently, the wave function has *discontinuous first derivatives* for coinciding electrons.

To illustrate the nuclear and electronic Coulomb cusp conditions, we have in Figure 7.5 plotted the ground-state helium wave function with one electron fixed at a point $0.5a_0$ from the nucleus. On the left, the wave function is plotted with the free electron restricted to a circle of radius $0.5a_0$ centred at the nucleus (with the fixed electron at the origin of the plot); on the right, the wave function is plotted on the straight line through the nucleus and the fixed electron. The wave function is differentiable everywhere except at the points where the particles coincide.

It should be noted that the cusp conditions (7.2.6) and (7.2.7) have been derived for the totally symmetric singlet ground state of the helium atom. More generally, the cusp conditions should be

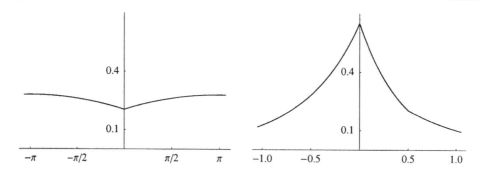

Fig. 7.5. The electronic and nuclear cusps of the ground-state helium atom with one electron fixed at a position $0.5a_0$ from the nucleus (atomic units). On the left, the wave function is plotted on a circle of radius $0.5a_0$ centred at the nucleus; on the right, the wave function is plotted along the axis through the nucleus and the fixed electron.

written in the following manner [3]:

$$\lim_{r_i \to 0} \left(\frac{\partial \Psi}{\partial r_i} \right)_{\text{ave}} = -Z\Psi(r_i = 0) \tag{7.2.11}$$

$$\lim_{r_{ij} \to 0} \left(\frac{\partial \Psi}{\partial r_{ij}} \right)_{\text{ave}} = \frac{1}{2}\Psi(r_{ij} = 0) \tag{7.2.12}$$

where a spherical averaging over all directions is implied. As a further caveat, these conditions hold only for wave functions that do not vanish at the singularities in the Hamiltonian. For wave functions that vanish at the singularities, different conditions apply [4,5].

7.3 Approximate treatments of the ground-state helium atom

Having examined the behaviour of the *exact* wave function in the ground-state helium atom, let us now consider the description of this atom by *approximate* electronic wave functions. We begin by examining the standard expansion of CI theory and then go on to investigate how this model may be amended to allow for a better description of the short-range electronic interactions. We conclude this section with a discussion of the Hylleraas expansion.

7.3.1 CONFIGURATION-INTERACTION EXPANSIONS

In the CI approach, we expand the wave function in Slater determinants – that is, in antisymmetric products of spin orbitals. For the spatial part of the spin orbitals, we shall first employ the (unnormalized) STOs of the form (6.5.25) and (6.5.26)

$$\chi_{nlm}(\mathbf{r}) = r^{n-1} \exp(-\zeta r) Y_{lm}(\theta, \varphi), \qquad n > l \geq |m| \geq 0 \tag{7.3.1}$$

where ζ is the orbital exponent and $Y_{lm}(\theta, \varphi)$ a spherical harmonic, which, in terms of the associated Legendre polynomial $P_l^m(\cos \theta)$, may be written as (6.4.3):

$$Y_{lm}(\theta, \varphi) = \sqrt{\frac{2l + 1}{4\pi} \frac{(l - m)!}{(l + m)!}} P_l^m(\cos \theta) \exp(im\varphi) \tag{7.3.2}$$

In these expressions, $\mathbf{r}$ represents the position of the electron and (r, θ, φ) are the polar coordinates. The spherical harmonics and the Legendre polynomials were introduced in Section 6.4.1 and the STOs in Section 6.5.5. We recall that, for any fixed exponent ζ, the STOs constitute a complete set of one-electron functions; see Section 6.5.5.

We now turn to the CI expansion in STOs (7.3.1) for the totally symmetric singlet ground state of the helium atom. In Exercise 7.1, we invoke the *addition theorem* for the spherical harmonics [6]

$$P_l(\cos\theta_{12}) = \frac{4\pi}{2l+1} \sum_{m=-l}^{l} (-1)^m Y_{l,m}(\theta_1, \varphi_1) Y_{l,-m}(\theta_2, \varphi_2) \tag{7.3.3}$$

where θ_{12} is the angle between the electrons

$$\cos\theta_{12} = \frac{\mathbf{r}_1 \cdot \mathbf{r}_2}{r_1 r_2} \tag{7.3.4}$$

to show that, for a totally symmetric singlet two-electron state, the spatial part of the CI expansion may be written as

$$\Psi^{\text{CI}} = \exp[-\zeta(r_1 + r_2)] \sum_{\substack{l \geq 0 \\ n_1 \geq n_2 \\ n_2 > l}} C_{n_1 n_2 l} (r_1^{n_1-1} r_2^{n_2-1} + r_2^{n_1-1} r_1^{n_2-1}) P_l(\cos\theta_{12}) \tag{7.3.5}$$

in terms of the Legendre polynomials $P_l(\cos\theta_{12}) = P_l^0(\cos\theta_{12})$. In the CI approach, therefore, we employ for our description of the ground-state helium atom three coordinates: the radial coordinates r_1 and r_2 and the relative angle θ_{12}. For a fixed ζ, the CI expansion (7.3.5) is complete – provided the summations are over all principal quantum numbers n_1 and n_2 and over all angular-momentum quantum numbers l.

In Figure 7.6, we have plotted the error in this single-zeta CI energy as more and more terms from (7.3.5) are included in the wave function, truncating the expansion according to the *principal quantum number N*:

$$\Psi_N^{\text{CI}} = \exp[-\zeta(r_1 + r_2)] \sum_{l=0}^{N-1} P_l(\cos\theta_{12}) \sum_{n_1=l+1}^{N} \sum_{n_2=l+1}^{n_1} C_{n_1 n_2 l} (r_1^{n_1-1} r_2^{n_2-1} + r_2^{n_1-1} r_1^{n_2-1}) \tag{7.3.6}$$

We shall refer to this expansion of the helium wave function as the *principal expansion*. The first plotted point corresponds to the expectation value of the $1s^2$ Hartree–Fock state

$$\Psi_1^{\text{CI}} = 2C_{110} \exp[-\zeta(r_1 + r_2)] \tag{7.3.7}$$

where the exponent is variationally optimized. The second point is obtained by variationally optimizing the exponent and the coefficients of the expansion

$$\Psi_2^{\text{CI}} = \exp[-\zeta(r_1 + r_2)][2C_{110} + C_{210}(r_1 + r_2) + 2C_{220}r_1 r_2 + 2C_{221}r_1 r_2 \cos\theta_{12}] \tag{7.3.8}$$

and represents a superposition of the configurations $1s^2$, $1s2s$, $2s^2$ and $2p^2$ – that is, the full set of totally symmetric configurations that may be constructed from the STOs of principal quantum numbers $n_i \leq 2$. As can be seen from Figure 7.6, the CI expansion converges very slowly and it appears difficult to arrive at an error less than 0.1 mE$_h$ in this manner.

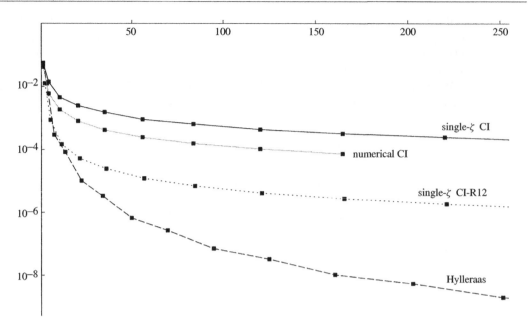

Fig. 7.6. The error in the electronic energy of the ground-state helium atom (E_h). The error is plotted on a logarithmic scale as a function of the number of terms in the expansions.

In Figure 7.6, we have also plotted the energies obtained by a CI expansion where the orbitals are not the simple STOs with a fixed single exponent ζ as in (7.3.6), but rather a set of variationally optimized orbitals $\phi_{nl}(r)$ of the same structure:

$$\Psi_N^{\text{num-CI}} = \sum_{l=0}^{N-1} P_l(\cos\theta_{12}) \sum_{n_1=l+1}^{N} \sum_{n_2=l+1}^{n_1} C_{n_1 n_2 l}[\phi_{n_1 l}(r_1)\phi_{n_2 l}(r_2) + \phi_{n_1 l}(r_2)\phi_{n_2 l}(r_1)] \qquad (7.3.9)$$

The numerical optimization of the orbitals reduces the error in the energy (by a factor of 3–4 at each level) but does not improve substantially on the intrinsically slow convergence of the CI expansion. A more detailed discussion of the convergence of the principal expansion (7.3.9) is given in Section 7.5. In routine molecular calculations, we employ AOs more flexible than the single-zeta STOs in (7.3.6) but less flexible than the numerical orbitals in (7.3.9) – in practice, the results are close to those of the numerical CI expansion in Figure 7.6, see Section 8.4.2.

7.3.2   CORRELATING FUNCTIONS AND EXPLICITLY CORRELATED METHODS

To uncover the underlying reason for the slow convergence of the CI energy, we express the CI expansion (7.3.5) in terms of r_{12} rather than θ_{12}. We first note that an equivalent form of the CI expansion (7.3.5) is given by the expression

$$\Psi^{\text{CI}} = \exp[-\zeta(r_1 + r_2)] \sum_{\substack{l \geq 0 \\ n_1 \geq n_2 \\ n_2 > l}} C_{n_1 n_2 l}(r_1^{n_1-1} r_2^{n_2-1} + r_2^{n_1-1} r_1^{n_2-1}) \cos^l \theta_{12} \qquad (7.3.10)$$

since $P_l(\cos\theta_{12})$ is a polynomial of degree l in $\cos\theta_{12}$. Introducing into (7.3.10) the relation

$$r_{12}^2 = r_1^2 + r_2^2 - 2r_1r_2\cos\theta_{12} \tag{7.3.11}$$

which follows by expansion of $\mathbf{r}_{12}\cdot\mathbf{r}_{12}$, we obtain the equivalent form

$$\Psi^{CI} = \exp[-\zeta(r_1+r_2)]\sum_{\substack{l\geq 0 \\ n_1\geq n_2 \\ n_2 > l}} C_{n_1n_2l}(r_1^{n_1-l-1}r_2^{n_2-l-1} + r_2^{n_1-l-1}r_1^{n_2-l-1})(r_1^2 + r_2^2 - r_{12}^2)^l \tag{7.3.12}$$

where, for convenience, the coefficients $C_{n_1n_2l}$ have been redefined. After a reordering of the terms, this expression leads us to the following simple ansatz for the two-electron CI wave function:

$$\Psi^{CI} = \exp[-\zeta(r_1+r_2)]\sum_{ijk} C_{ijk}(r_1^i r_2^j + r_1^j r_2^i)r_{12}^{2k} \tag{7.3.13}$$

where the summation is over all nonnegative integers. The CI expansion thus contains all possible combinations of powers in r_1, r_2 and r_{12} except for those of odd powers in r_{12}. In particular, no terms linear in r_{12} are present in the CI expansion. The cusp condition (7.2.7) therefore can never be satisfied for CI wave functions – no matter how many terms are included or how the truncation is carried out – since differentiation of (7.3.13) with respect to r_{12} always yields zero

$$\left.\frac{\partial\Psi^{CI}}{\partial r_{12}}\right|_{r_{12}=0} = 0 \tag{7.3.14}$$

when carried out at $r_{12} = 0$.

Clearly, to have nonzero derivatives of the wave function with respect to r_{12} at the singularity $r_{12} = 0$, we must include terms that are linear in r_{12}. Indeed, the following simple modification of the CI wave function

$$\Psi_{r12}^{CI} = \left(1 + \tfrac{1}{2}r_{12}\right)\Psi^{CI} \tag{7.3.15}$$

leads to a function that satisfies the cusp condition (7.2.7) exactly since

$$\left.\frac{\partial\Psi_{r12}^{CI}}{\partial r_{12}}\right|_{r_{12}=0} = \frac{1}{2}\Psi^{CI}(r_{12}=0) = \frac{1}{2}\Psi_{r12}^{CI}(r_{12}=0) \tag{7.3.16}$$

More generally, we may impose the correct Coulomb-cusp behaviour on any determinant-based wave function Φ by multiplying the determinantal expansion Φ by some *correlating function* γ of the form [7]

$$\gamma = 1 + \tfrac{1}{2}\sum_{i>j} r_{ij} \tag{7.3.17}$$

which, as shown in Exercise 7.2, leads to the correct nondifferentiable cusp in the product function $\gamma\Phi$. Methods that employ correlating functions or otherwise make explicit use of the interelectronic coordinates in the wave function are known as *explicitly correlated methods*.

Although the presence of the correlating function imposes the correct cusp behaviour on the wave function, it does not a priori follow that the associated improvements in the energy are significant. To investigate the convergence of the energy for an explicitly correlated wave function, we have in Figure 7.6 plotted the error in the energy of the *CI-R12 wave function* Ψ_N^{R12}, obtained from the standard CI expansion Ψ_N^{CI} by adding a single term, consisting of the dominant Hartree–Fock configuration Ψ_1^{CI} (7.3.7) multiplied by r_{12}:

$$\Psi_N^{R12} = \Psi_N^{CI} + c_{12}r_{12}\Psi_1^{CI} \tag{7.3.18}$$

Otherwise, the CI-R12 wave function is generated in the same manner as the original CI wave function – that is, by variationally optimizing the coefficients and the orbital exponent, the only difference being the additional optimization of c_{12}.

As seen from Figure 7.6, the addition of a single linear r_{12} term to the CI wave function has a remarkable effect on the energy, reducing the error by about 2 orders of magnitude at all levels of truncation. Thus, whereas, for an error less than 1 mE$_h$, the CI energy requires all orbitals with principal quantum numbers $n_i \leq 6$, the CI-R12 energy is converged to within 1 mE$_h$ with $n_i \leq 2$ and to within 0.1 mE$_h$ with $n_i \leq 4$. Still, it appears difficult to converge the CI-R12 energy to within 1 μE$_h$, just as it is difficult to converge the (standard) CI energy below 0.1 mE$_h$. Clearly, for errors of order 1 μE$_h$ and smaller, an even more flexible wave function is needed.

7.3.3 THE HYLLERAAS FUNCTION

To arrive at a wave function capable of providing errors smaller than 1 μE$_h$, we generalize the CI expansion in (7.3.13) to include *all* powers in the interelectronic separation – odd powers as well as even ones. The resulting wave function may be written as

$$\Psi^H = \exp[-\zeta(r_1 + r_2)] \sum_{ijk} C_{ijk}(r_1^i r_2^j + r_2^i r_1^j)r_{12}^k \tag{7.3.19}$$

and is known as the *Hylleraas expansion* or the *Hylleraas function* [1,8]. The Hylleraas function is usually expressed in the following manner:

$$\Psi^H = \exp(-\zeta s) \sum_{ijk} c_{ijk} s^i t^{2j} u^k \tag{7.3.20}$$

in terms of the *Hylleraas coordinates*

$$s = r_1 + r_2 \tag{7.3.21}$$

$$t = r_1 - r_2 \tag{7.3.22}$$

$$u = r_{12} \tag{7.3.23}$$

Note that only terms that are even in t are included in the Hylleraas function, in order to ensure that the spatial wave function is symmetric in the two electrons, as appropriate for the singlet ground state. Conversely, by allowing only odd powers of t in the expansion, an approximation to the triplet $1s2s$ state is obtained instead.

The performance of the Hylleraas function is illustrated in Figure 7.6, where we have plotted the energy (variationally optimized with respect to the exponent and the coefficients at each level) as a function of the number of terms generated by truncating the expansion (7.3.20) according to the condition

$$i + 2j + k \leq N \tag{7.3.24}$$

Clearly, with this ansatz, very high accuracy is attained with only a few terms in the expansion. Indeed, the plots of the Coulomb hole, the kinetic energy and the Coulomb cusp in this chapter have all been obtained from the Hylleraas function with $N = 21$. Unfortunately, the Hylleraas method is applicable only to atomic systems containing few electrons.

7.3.4 CONCLUSIONS

We have seen that, by means of determinantal expansions, it is exceedingly difficult to converge the electronic energy to *chemical accuracy* – that is, to an error of the order of 1 mE_h. For high accuracy, it appears necessary to introduce terms that depend linearly on the interelectronic distance, for example by including correlating functions of the form (7.3.17). Methods that employ a single additional configuration, multiplied linearly by the interelectronic distances as in (7.3.18), are referred to as *R12 methods*. R12 methods may be set up within the usual framework of CI theory, coupled-cluster theory or Møller–Plesset theory, always leading to a significant improvement in the description of the electronic system [9].

The improvements in the wave function notwithstanding, the introduction of correlating functions leads to a rather complex representation of the energy. Whereas a determinantal wave function, written in terms of products of one-electron functions, leads to an energy expression containing at most two-electron integrals, the inclusion of two-electron functions in the wave function gives rise to three- and four-electron integrals in the energy expression. Nevertheless, the use of correlating functions leads to expansions more compact than the standard expansions in Fock space. Ultimately, therefore, the use of correlating functions may provide a more efficient means of generating highly accurate wave functions. Still, the explicitly correlated methods have not yet reached an advanced stage of development and have not yet been universally adopted as a standard computational tool in quantum chemistry.

7.4 The partial-wave expansion of the ground-state helium atom

In accurate CI expansions of the electronic wave function, by far most of the determinants are introduced to describe the short-range correlation of the electrons. If we employ instead a wave function that contains r_{12} linearly, much shorter expansions are sufficient, as demonstrated in Section 7.3. In some sense, therefore, we may regard the standard CI approach as a brave attempt to expand the nondifferentiable interelectronic distance r_{12} in products of one-electron functions. In general, any such attempt will lead to a slow convergence of the wave function and the energy, as we shall discuss in this section and the next.

7.4.1 PARTIAL-WAVE EXPANSION OF THE INTERELECTRONIC DISTANCE

In the Hartree–Fock approximation, the ^1S ground state of the helium atom is represented by a single Slater determinant containing a doubly occupied $1s$ orbital. Following the discussion in Section 7.3, we may impose the correct cusp behaviour on the wave function by introducing a correlating function (7.3.17) containing the interelectronic distance r_{12} explicitly:

$$\Psi(\mathbf{x}_1, \mathbf{x}_2) = \left(1 + \tfrac{1}{2}r_{12}\right) |1s_\alpha 1s_\beta| \tag{7.4.1}$$

However, because of technical difficulties associated with the explicit use of r_{12} in the wave function, we shall instead explore the expansion of r_{12} in a set of one-electron variables. Thus, in the *partial-wave expansion* of the electronic distance, we express r_{12} in terms of the three variables

r_1, r_2 and $\cos\theta_{12}$ as

$$r_{12} = \sum_{l=0}^{\infty} P_l(\cos\theta_{12}) \left(\frac{1}{2l+3} \frac{r_<^{l+2}}{r_>^{l+1}} - \frac{1}{2l-1} \frac{r_<^l}{r_>^{l-1}} \right) \tag{7.4.2}$$

where $r_> = \max(r_1, r_2)$ and $r_< = \min(r_1, r_2)$. In Exercise 7.3, this expression for r_{12} is derived from the *partial-wave expansion* of r_{12}^{-1} [6]:

$$\frac{1}{r_{12}} = \frac{1}{r_>} \sum_{l=0}^{\infty} \left(\frac{r_<}{r_>} \right)^l P_l(\cos\theta_{12}) \tag{7.4.3}$$

Before employing the partial-wave expansion (7.4.2) in the electronic wave function, we note that its convergence is rapid whenever r_1 and r_2 are very different from each other and slow when they are similar. This behaviour is illustrated in Figure 7.7, where, for different truncations $l \leq L$, we have plotted the partial-wave expansions of r_{12} as functions of r_1 for fixed $r_2 = 1$ and $\theta_{12} = 0$ (i.e. with the electrons arranged collinearly with the nucleus). Note, in particular, the difficulties faced by the truncated expansions in approximating r_{12} (the grey line) in the region $r_1 \approx r_2$.

7.4.2 PARTIAL-WAVE EXPANSION OF THE WAVE FUNCTION

To arrive at the partial-wave expansion of the wave function (7.4.1), we insert the expansion of the interelectronic distance (7.4.2) and invoke the addition theorem (7.3.3):

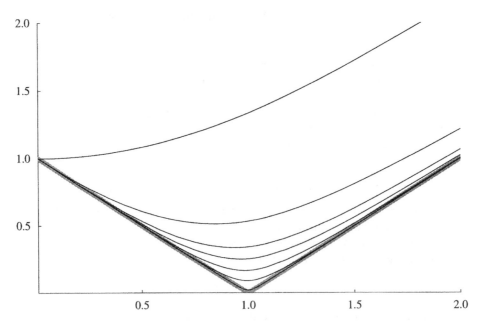

Fig. 7.7. Partial-wave expansions of r_{12} in r_1, r_2 and θ_{12} for $r_2 = 1$ and $\theta_{12} = 0$. The orders of the expansions are $L = 0,1, 2, 3, 5, 10$ and 50. The exact function r_{12} is represented by a thick grey line.

$$\Psi(\mathbf{x}_1, \mathbf{x}_2) = \left[1s(r_1)1s(r_2) + \frac{1}{2}1s(r_1)1s(r_2) \sum_{l=0}^{\infty} \sum_{m=-l}^{l} (-1)^m \frac{4\pi}{2l+1} Y_{l,m}(\theta_1, \varphi_1) Y_{l,-m}(\theta_2, \varphi_2) \right.$$

$$\left. \times \left(\frac{1}{2l+3} \frac{r_<^{l+2}}{r_>^{l+1}} - \frac{1}{2l-1} \frac{r_<^l}{r_>^{l-1}} \right) \right] \frac{1}{\sqrt{2}} [\alpha(1)\beta(2) - \beta(1)\alpha(2)] \tag{7.4.4}$$

The terms containing $r_<$ and $r_>$ are now expanded in a complete set of radial functions $\phi_p(r)$, yielding

$$\Psi(\mathbf{x}_1, \mathbf{x}_2) = \left[1s(r_1)1s(r_2) + \sum_{lm} \sum_{pq} C_{pq}^{lm} \phi_p(r_1) Y_{l,m}(\theta_1, \varphi_1) \phi_q(r_2) Y_{l,-m}(\theta_2, \varphi_2) \right]$$

$$\times \frac{1}{\sqrt{2}} [\alpha(1)\beta(2) - \beta(1)\alpha(2)] \tag{7.4.5}$$

where the coefficients are determined so as to reproduce the expansion (7.4.4) exactly. The partial-wave expansion of $\Psi(\mathbf{x}_1, \mathbf{x}_2)$ is obtained by collecting terms of the same angular momentum l

$$\Psi(\mathbf{x}_1, \mathbf{x}_2) = \left[\sum_{l=0}^{\infty} \psi_l(\mathbf{r}_1, \mathbf{r}_2) \right] \frac{1}{\sqrt{2}} [\alpha(1)\beta(2) - \beta(1)\alpha(2)] \tag{7.4.6}$$

Each partial wave is given by

$$\psi_l(\mathbf{r}_1, \mathbf{r}_2) = \delta_{l0} 1s(r_1)1s(r_2) + \frac{1}{2} 1s(r_1)1s(r_2) P_l(\cos\theta) \left(\frac{1}{2l+3} \frac{r_<^{l+2}}{r_>^{l+1}} - \frac{1}{2l-1} \frac{r_<^l}{r_>^{l-1}} \right)$$

$$= \delta_{l0} 1s(r_1)1s(r_2) + \sum_{\substack{pq \\ m}} C_{pq}^{lm} \phi_{p,l,m}(r_1, \theta_1, \varphi_1) \phi_{q,l,-m}(r_2, \theta_2, \varphi_2) \tag{7.4.7}$$

where we have introduced the spatial orbitals of the form

$$\phi_{plm}(r, \theta, \varphi) = \phi_p(r) Y_{lm}(\theta, \varphi) \tag{7.4.8}$$

Nonzero contributions to the partial waves (7.4.7) occur only for orbitals of the same angular momentum l but opposite projections. From the slow convergence of the partial-wave expansion of r_{12} in Figure 7.7, we conclude that the convergence of the correlating function in the partial-wave expansion of $\Psi(\mathbf{x}_1, \mathbf{x}_2)$ should be slow as well. As for r_{12}, the partial waves $\psi_l(\mathbf{r}_1, \mathbf{r}_2)$ should fall off rapidly for $r_> \gg r_<$ but slowly for $r_> \approx r_<$.

In the determinantal representation, the ground state of the helium atom (7.4.1) may be written in the form

$$\Psi(\mathbf{x}_1, \mathbf{x}_2) = \sum_{l=0}^{\infty} \Psi_l(\mathbf{x}_1, \mathbf{x}_2) \tag{7.4.9}$$

where each partial wave is a linear combination of Slater determinants

$$\Psi_l(\mathbf{x}_1, \mathbf{x}_2) = \delta_{l0} |1s_\alpha 1s_\beta| + \sum_{\substack{pq \\ m}} C_{pq}^{lm} |\phi_{p,l,m,\alpha} \phi_{q,l,-m,\beta}| \tag{7.4.10}$$

with the expansion coefficients determined so as to reproduce the partial-wave expansion of the correlating function in (7.4.1). However, for an accurate representation of the electronic system, we must provide a flexible description of the long-range interactions as well. In standard CI theory, the long-range interactions are described by terms of the same structure as those already present in the partial-wave expansion (7.4.7). The wave function is therefore still written as in (7.4.9) but each partial wave now takes the form

$$\Psi_l(\mathbf{x}_1, \mathbf{x}_2) = \sum_{\substack{pq \\ m}} c_{pq}^{lm} | \phi_{p,l,m,\alpha} \, \phi_{q,l,-m,\beta}| \tag{7.4.11}$$

with the coefficients determined so as to satisfy (as accurately as possible) the Schrödinger equation rather than to reproduce the wave function (7.4.1).

In the partial-wave expansion of the CI wave function (7.4.9) and (7.4.11), the description of the long-range interactions is expected to converge more rapidly than that of the short-range interactions. Ultimately, therefore, the convergence of the partial waves of the CI wave function should be determined by the difficulties faced in describing the short-range interactions of the electronic system. In Sections 7.4.3 and 7.4.4, we shall examine more closely the asymptotic convergence of the partial-wave expansion.

7.4.3   THE ASYMPTOTIC CONVERGENCE OF THE PARTIAL-WAVE EXPANSION

The asymptotic convergence of the partial-wave expansion of the ground-state helium atom has been carefully analysed theoretically. We shall not attempt such an analysis here but instead present some numerical calculations to illustrate the more important theoretical results. For this purpose, we employ a wave function of the form (7.4.9), expanding each partial wave in Slater determinants (7.4.11). First, we calculate the CI energy in a large (and for all practical purposes complete) set of s functions so as to establish the s limit of the energy E_0. Next, the p limit E_1 is established by augmenting the s basis with a large number of p functions. At this point, the CI wave function is a linear combination of determinants containing either two s functions or two p functions. This procedure is continued up to $l \leq L = 8$. In Table 7.1, we have listed the energies E_L and the increments

$$\varepsilon_L = E_L - E_{L-1} \tag{7.4.12}$$

obtained for each CI wave function in this sequence. For our purposes, it is convenient to think of each partial wave l in the exact CI ground-state helium wave function (7.4.9) as contributing ε_l to the total energy.

For low angular momenta, the contributions of the partial waves to the energy fall off rapidly; for high angular momenta, the convergence appears to slow down. Indeed, a careful theoretical analysis of the partial-wave expansion reveals that, in the asymptotic limit, the energy increments behave according to the expression [7,10]

$$\varepsilon_L = e_4 \left(L + \tfrac{1}{2} \right)^{-4} + e_5 \left(L + \tfrac{1}{2} \right)^{-5} + \cdots \tag{7.4.13}$$

To illustrate the partial-wave convergence of the energy of the ground-state helium atom, we have in Figure 7.8 plotted $\log_{10}(-\varepsilon_L)$ against $\log_{10}(L + 1/2)$. Superimposed on this plot is a straight line that represents the function $e_4 (L + 1/2)^{-4}$ with e_4 fitted to the calculated numbers. The calculated points form nearly a straight line, indicating that (7.4.13) is dominated by the first term.

It is of some interest to investigate at what level of truncation L the error in the energy increments coincides with that predicted by the asymptotic behaviour of the short-range interactions. From a

Table 7.1 The convergence of the ground-state energy (in E_h) in the partial-wave expansion of the helium atom. E_L is the CI energy calculated in a basis of AOs of angular momentum $l \le L$

L	E_L	$\varepsilon_L = E_L - E_{L-1}$
0	-2.8790288	
1	-2.9005162	-0.0214874
2	-2.9027668	-0.0022506
3	-2.9033211	-0.0005543
4	-2.9035186	-0.0001975
5	-2.9036058	-0.0000872
6	-2.9036500	-0.0000442
7	-2.9036747	-0.0000247
8	-2.9036897	-0.0000150

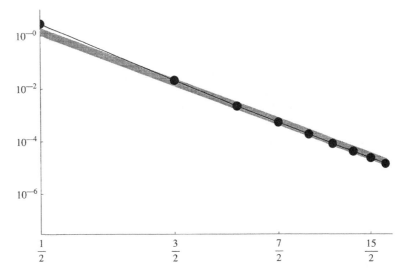

Fig. 7.8. The energy increments $\log_{10}(-\varepsilon_L)$ of the partial-wave expansion of the ground-state helium atom plotted against $\log_{10}(L + 1/2)$. The thick grey line represents the function $e_4 (L + 1/2)^{-4}$ with e_4 fitted to the calculated numbers. Atomic units are used.

theoretical analysis, it may be shown that, in the asymptotic limit, the short-range behaviour of the energy becomes

$$\varepsilon_L = -0.074226 \left(L + \tfrac{1}{2}\right)^{-4} - 0.030989 \left(L + \tfrac{1}{2}\right)^{-5} + \cdots \qquad (7.4.14)$$

In Table 7.2, we have listed the coefficients e_4 and e_5 obtained by fitting (7.4.13) to the energies ε_L in Table 7.1, including in the fit only the increments ε_L with $L \ge l_{\min}$. Comparing the fitted coefficients in Table 7.2 with (7.4.14), we conclude that, in the ground-state helium atom, orbitals higher than d or f in angular momentum are needed mostly for the description of the short-range interactions.

The fitted coefficients in Table 7.2 may be used to calculate the higher-order energy increments from (7.4.13). Using, for example, the coefficients obtained for $l_{\min} = 3$, we are able to reproduce

Table 7.2 The coefficients of (7.4.13) for the ground-state helium atom obtained from two-parameter fits to the calculated ε_L with $L \geq l_{\min}$ (see Table 7.1)

$l_{\min}$	e_4	e_5
1	−0.05781	−0.07646
2	−0.07175	−0.04041
3	−0.07361	−0.03348
4	−0.07433	−0.02994
5	−0.07427	−0.03036

each ε_L with $L \geq 3$ in Table 7.1 to an error less than 1 μE_h. We may also predict the truncation error for $L = 8$ by adding up all the predicted increments for $l > L$ (i.e. all the increments not included in Table 7.1):

$$\Delta E_L = \sum_{l=L+1}^{\infty} \varepsilon_l \tag{7.4.15}$$

The truncation error is predicted to be -34.7 μE_h, giving a total energy of -2.903724 E_h, in perfect agreement with the correct ^1S ground-state helium energy of -2.903724377 E_h [11]. In Section 7.4.4, we shall obtain a simple, closed form for this truncation error.

7.4.4   THE TRUNCATION ERROR OF THE PARTIAL-WAVE EXPANSION

When the partial-wave expansion has been truncated by including only terms with $l \leq L$, the error in the energy is given as

$$\Delta E_L = e_4 \sum_{l=L+1}^{\infty} \left(l + \tfrac{1}{2}\right)^{-4} + e_5 \sum_{l=L+1}^{\infty} \left(l + \tfrac{1}{2}\right)^{-5} + \cdots \tag{7.4.16}$$

To evaluate this error, we replace the summations by integrations

$$\sum_{l=L+1}^{\infty} \left(l + \frac{1}{2}\right)^{-k} \approx \int_{L+1/2}^{\infty} \left(l + \frac{1}{2}\right)^{-k} dl = \frac{1}{k-1}(L+1)^{-(k-1)} \tag{7.4.17}$$

and obtain

$$\Delta E_L \approx \tfrac{1}{3}e_4(L+1)^{-3} + \tfrac{1}{4}e_5(L+1)^{-4} + \cdots \tag{7.4.18}$$

Retaining only the leading term, we may write the truncation error in the following simple way

$$\Delta E_L \approx a(L+1)^{-3} \tag{7.4.19}$$

where the numerical constant a is independent of L. For the ground-state helium atom, therefore, the error in the energy is inversely proportional to the third power of the angular momentum $L + 1$ of the first neglected partial wave.

In Section 7.4.3, we obtained a truncation error $\Delta E_8 = -34.7$ μE_h by summing all the predicted energy increments with $l > 8$ in (7.4.16), in agreement with the true truncation error. Let us compare this number with the truncation error obtained from the first two terms in (7.4.18). Using again the coefficients for $l_{min} = 3$ in Table 7.2, we obtain $\Delta E_8 = -34.9$ μE_h, indicating that the approximation (7.4.17) is a good one. For $L = 4$, the truncation errors obtained from (7.4.16) and (7.4.18) are -205.5 and -209.7 μE_h, respectively, in good agreement with the true truncation error of -205.8 μE_h.

7.5 The principal expansion of the ground-state helium atom

The convergence rate of the partial-wave expansion discussed in Section 7.4 is theoretically interesting but difficult to realize in practice since, for each partial wave, an infinite expansion in orbitals is required. In the CI framework, a more useful approach is to consider a series of calculations where a (reasonably) fast convergence of the correlation energy is obtained in a small orbital space. This is achieved by means of the principal expansion discussed in the present section.

7.5.1 THE PRINCIPAL EXPANSION AND ITS ASYMPTOTIC CONVERGENCE

In the *principal expansion*, all orbitals of principal quantum number $n \leq N$ are included at a given level. The orbitals are thus grouped according to the total number of radial and angular nodes and the principal expansion of order N is obtained by including in the wave function all orbitals with the total number of nodes (angular and radial) less than N. Thus, at each level, we include all orbitals of the same orbital energy. In this manner, we generate the sequence of orbital spaces $(1s)$, $(2s1p)$, $(3s2p1d)$, $(4s3p2d1f)$, and so on. At the first level, we obtain the numerical Hartree–Fock energy. Next, we generate a CI wave function in the $(2s1p)$ space, with the orbitals fully optimized for this particular expansion, and so on. The formal expression for the wave function in the principal expansion was given in (7.3.9). In Figure 7.9, the sets of orbitals included at each level in the principal and partial-wave expansions are compared.

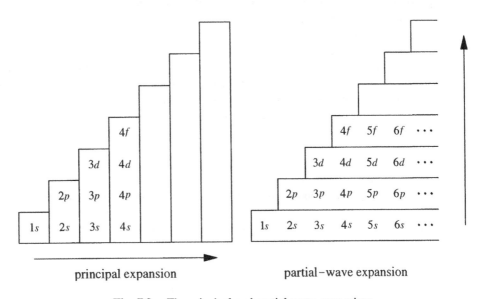

Fig. 7.9. The principal and partial-wave expansions.

Table 7.3 The convergence of the principal expansion of the ground-state energy of the helium atom. E_N is the energy (in E_h) obtained in a basis containing all AOs of principal quantum number N and lower

N	E_N	$\varepsilon_N = E_N - E_{N-1}$
1	$-2.861\,680$	
2	$-2.897\,674$	$-0.035\,994$
3	$-2.901\,841$	$-0.004\,167$
4	$-2.902\,909$	$-0.001\,068$
5	$-2.903\,300$	$-0.000\,391$
6	$-2.903\,476$	$-0.000\,176$
7	$-2.903\,567$	$-0.000\,091$
8	$-2.903\,618$	$-0.000\,051$
9	$-2.903\,650$	$-0.000\,032$

In Table 7.3, the total energies of the principal expansion are listed, together with the energy contributions from each new shell:

$$\varepsilon_N = E_N - E_{N-1} \tag{7.5.1}$$

In Figure 7.6, the curve marked as 'numerical CI' corresponds to the energies obtained by the principal expansion in this table. Comparing the energies in Table 7.3 with those in Table 7.1, we find that, at each level $N = L + 1$, the energy is lower for the partial-wave expansion than for the principal expansion. This difference is expected since the partial-wave expansion $L + 1$ contains – in addition to all the determinants included in the principal expansion N – all other determinants with $l \leq L + 1$. For example, at the first level of the principal expansion we employ a single-determinant Hartree–Fock wave function, whereas at the first level of the partial-wave expansion we employ a CI wave function in the complete set of s orbitals.

Keeping in mind that the partial-wave expansion requires an infinite number of determinants at each level, the differences between the partial-wave and principal expansions in Tables 7.1 and 7.3 are rather small, in particular for high $N = L + 1$. In this sense, the convergence of the principal expansion appears to be satisfactory. In Figure 7.10 we have plotted the energy increments of the principal expansion $\log_{10}(-\varepsilon_N)$ against $\log_{10}\left(N - \frac{1}{2}\right)$ in the same way as for the partial-wave expansion in Figure 7.8. The plotted points lie almost on a straight line with a slope of -4, indicating that the energy contributions may be written in the form

$$\varepsilon_N = c_4 \left(N - \tfrac{1}{2}\right)^{-4} + c_5 \left(N - \tfrac{1}{2}\right)^{-5} + \cdots \tag{7.5.2}$$

with a dominant leading term. Thus, the convergence does not seem to be slower than for the partial-wave expansion, indicating that the principal expansion should form a very useful basis for the treatment of larger electronic systems. As for the partial-wave expansion, the error of the energy of the principal expansion may be written in the form

$$\Delta E_N \approx cN^{-3} \tag{7.5.3}$$

Using this expression, we may predict the CI energy in a complete one-electron basis from the energies obtained for small maximum principal quantum numbers N.

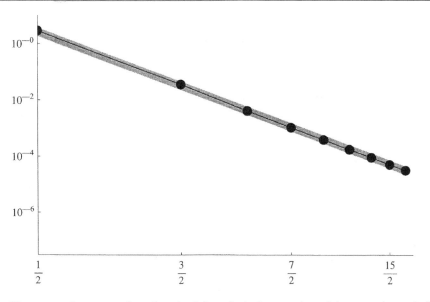

Fig. 7.10. The energy increments $\log_{10}(-\varepsilon_N)$ of the principal expansion of the ground-state helium atom plotted against $\log_{10}(N-1/2)$. The thick grey line represents the function $c_4\,(N-1/2)^{-4}$ where the constant c_4 has been determined by a fit to the calculated numbers. Atomic units are used.

7.5.2 COMPARISON OF THE PARTIAL-WAVE AND PRINCIPAL EXPANSIONS

We have seen that the partial-wave and principal expansions give energy increments that are inversely proportional to the fourth power of L and N, respectively – see (7.4.13) and (7.5.2). This behaviour may be understood by considering the energy contributions from the individual orbitals [12].

It has been found that, for high quantum numbers n and l, the energy recovered by adding the natural orbital ϕ_{nlm} to the wave function is well described by the expression [13]

$$\varepsilon_{nlm} = -An^{-6} \tag{7.5.4}$$

Thus, in this approximation, the contribution to the energy is independent of the angular momentum and inversely proportional to the sixth power of the principal quantum number. We shall now use this expression to predict the energy increments for a given L in the partial-wave expansion and for a given N in the principal expansion, summing the contributions from all orbitals added at each level as illustrated in Figure 7.9.

First, noting that, for each n, there are $2L+1$ orbitals of a given L and replacing the summation over $n > L$ by integration, we obtain for the energy increments of the partial-wave expansion

$$\varepsilon_L = -A(2L+1)\sum_{n=L+1}^{\infty}n^{-6} \approx -A(2L+1)\int_{L+1/2}^{\infty}n^{-6}\,\mathrm{d}n$$

$$= -\tfrac{1}{5}A(2L+1)\left(L+\tfrac{1}{2}\right)^{-5} = -\tfrac{2}{5}A\left(L+\tfrac{1}{2}\right)^{-4} \tag{7.5.5}$$

Next, recalling that there are N^2 orbitals of a given principal quantum number N, we obtain for the energy increments of the principal expansion

$$\varepsilon_N = -AN^2 N^{-6} = -AN^{-4} \approx -A\left(N - \tfrac{1}{2}\right)^{-4} \tag{7.5.6}$$

neglecting terms of higher inverse powers in $N - \tfrac{1}{2}$. Thus, the increments (7.4.13) for the partial-wave expansion and (7.5.2) for the principal expansion both follow from the orbital contribution (7.5.4). Moreover, for high $N = L + 1$, the partial-wave increment should be about 40% of the principal increment. Comparing the increments in Table 7.1 with those in Table 7.3, we find that the true ratio decreases monotonically from 60% for $N = 2$ to 47% for $N = 9$.

In reality, the energy contributions ε_{nlm} are not completely independent of the angular momentum. However, the above argument shows that the observed rates of the partial-wave and principal expansions are compatible with each other, assuming that each orbital of a given principal quantum number contributes the same amount of energy, irrespective of its angular momentum. Moreover, from the assumption (7.5.4), the advantages of the principal expansion over the partial-wave expansion follow directly since, at each level in the principal expansion, we have included all orbitals that have energy contributions of a given magnitude or larger. Indeed, this is the idea behind the correlation-consistent basis sets introduced in Section 8.3.3.

7.5.3 THE COULOMB HOLE IN THE PRINCIPAL EXPANSION

Since the principal expansion appears to be optimal in terms of the number of basis functions employed, it is of some interest to investigate how the description of the Coulomb hole is improved as the principal quantum number increases. In Figure 7.11, we have, for the principal expansions $2 \leq N \leq 5$, plotted the ground-state helium wave function in the same manner as in Figure 7.5, with one electron fixed at a point $0.5a_0$ from the nucleus. In the left-hand plot, the wave function is plotted on a circle of radius $0.5a_0$ centred at the nucleus (with the fixed electron at the origin of the plot); in the right-hand plot, the wave function is plotted along the axis containing the fixed electron and the nucleus. Superimposed on these plots are the numerical Hartree–Fock wave function (dotted lines) and the Hylleraas function (grey lines). The overall shape of the Coulomb hole is surprisingly well described already at low orders; the cusp region, by contrast, is much more difficult to describe and requires high principal quantum numbers for an accurate representation.

7.5.4 CONCLUSIONS

To conclude our discussion of the helium atom, we have found that the convergence of the orbital expansion is excruciatingly slow for energies smaller than $1 \text{ m}E_h = 2.625 \text{ kJ/mol}$. Fortunately, the bulk of the correlation energy is recovered without too much difficulty, making the use of determinantal expansions a practical solution in most cases. Furthermore, the short-range correlation corrections should be rather insensitive to, for example, geometrical distortions and the application of an external field. A small error in the description of the Coulomb cusp should therefore not affect appreciably the prediction of equilibrium geometries and other molecular properties. In cases where it is necessary to obtain energies with an error less than $1 \text{ m}E_h$, we may take advantage of the systematic behaviour of the energy corrections illustrated by (7.4.13) and (7.5.2) and extrapolate the energy to the basis-set limit. Such extrapolations are explored in Section 8.4.

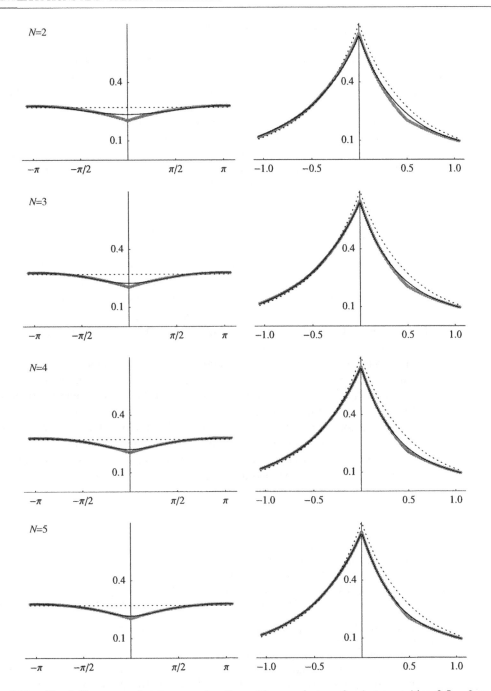

Fig. 7.11. The helium ground-state wave function with one electron fixed at a position $0.5a_0$ from the nucleus for different principal expansions $2 \leq N \leq 5$ (thin black line) compared with the exact (Hylleraas) function (thick grey line) and the Hartree–Fock wave function (dotted line). On the left, the wave function is plotted on a circle of radius $0.5a_0$ centred at the nucleus; on the right, the wave function is plotted along the axis defined by the positions of the nucleus and the fixed electron. Atomic units are used.

7.6 Electron-correlation effects summarized

From the discussions of electron correlation in the hydrogen molecule in Section 5.2 and in the helium atom in the present chapter, a clear picture of the important differences between short-range and long-range correlation should emerge. *Short-range correlation* arises from the singularities in the Hamiltonian and manifests itself in the appearance of cusps in the exact wave function. The behaviour of the electrons near the cusps is exceedingly difficult to describe by means of orbital expansions. By extending the orbital basis and by including more and more determinants in our description, we can systematically reduce the regions where the wave function is inaccurate and thereby improve our description of short-range correlation, but this procedure becomes progressively more difficult for smaller and smaller interelectronic separations. By contrast, the *long-range correlation* of the electrons – as manifested, for example, in the left–right, angular and radial correlation in the hydrogen molecule as discussed in Section 5.3.2 – can usually be adequately described by the superposition of a relatively small number of determinants.

Before closing this chapter, it is appropriate to review the different modes of correlation encountered in our description of electronic systems. *Fermi correlation* arises from the Pauli antisymmetry of the wave function and is taken into account already at the Hartree–Fock level as discussed in Section 5.2.8. *Static correlation* – also known as near-degeneracy or nondynamical correlation – arises from the near-degeneracy of electronic configurations and may at the simplest level be described by MCSCF theory; see Section 5.2.10. *Dynamical correlation*, also introduced in Section 5.2.10, is associated with the instantaneous correlation among the electrons arising from their mutual repulsion and requires for its description a large number of electronic configurations. It is often useful to distinguish between *long-range dynamical correlation* and *short-range dynamical correlation*, the latter related to the singularities in the Hamiltonian operator and giving rise to the Coulomb cusp in the electronic wave function. Dynamical correlation is best accounted for by perturbation theory or by coupled-cluster theory. Long-range dynamical correlation can usually be adequately described by means of a relatively small number of determinants, whereas short-range correlation is exceedingly difficult to account for and is perhaps best treated using explicitly correlated methods, which go beyond the Fock-space description of molecular electronic structure.

References

[1]   E. A. Hylleraas, *Adv. Quantum Chem.* **1**, 1 (1964).
[2]   J. C. Slater, *Phys. Rev.* **31**, 333 (1928).
[3]   T. Kato, *Commun. Pure Appl. Math.* **10**, 151 (1957).
[4]   R. T. Pack and W. Byers Brown, *J. Chem. Phys.* **45**, 556 (1966).
[5]   W. Kutzelnigg and J. D. Morgan III, *J. Chem. Phys.* **96**, 4484 (1992); *J. Chem. Phys.* **97**, 8821 (1992) (erratum).
[6]   G. Arfken, *Mathematical Methods for Physicists*, Academic Press, 1970.
[7]   W. Kutzelnigg, *Theor. Chim. Acta* **68**, 445 (1985).
[8]   E. A. Hylleraas, *Z. Phys.* **54**, 347 (1929).
[9]   W. Klopper, R. Röhse and W. Kutzelnigg, *Chem. Phys. Lett.* **178**, 455 (1991); J. Noga, W. Kutzelnigg and W. Klopper, *Chem. Phys. Lett.* **199**, 497 (1992); W. Klopper and W. Kutzelnigg, *Chem. Phys. Lett.* **134**, 17 (1987).
[10] C. Schwartz, *Phys. Rev.* **126**, 1015 (1962); *Meth. Comput. Phys.* **2**, 241 (1963).
[11] K. Frankowski and C. L. Pekeris, *Phys. Rev.* **146**, 46 (1966).
[12] W. Klopper, K. L. Bak, P. Jørgensen, J. Olsen and T. Helgaker, *J. Phys. B: At. Mol. Opt. Phys.* **32**, R103 (1999).
[13] D. P. Carroll, H. J. Silverstone and R. M. Metzger, *J. Chem. Phys.* **71**, 4142 (1979).

Further reading

N. C. Handy, Correlated wave functions, in G. H. F. Diercksen, B. T. Sutcliffe and A. Veillard (eds), *Computational Techniques in Quantum Chemistry and Molecular Physics*, Reidel, 1975, p. 425.

W. Klopper, $r12$-dependent wave functions, in P. v. R. Schleyer, N. L. Allinger, T. Clark, J. Gasteiger, P. A. Kollman, H. F. Schaefer III and P. R. Schreiner (eds), *Encyclopedia of Computational Chemistry*, Vol. 4, Wiley, 1998, p. 2351.

Exercises

EXERCISE 7.1

In this exercise, we consider the CI expansion of the ^1S ground state of the helium atom in the orbitals

$$\phi_{nlm}(\mathbf{r}) = R_{nl}(r)Y_{lm}(\theta, \varphi) \tag{7E.1.1}$$

A general singlet two-electron state may be expanded as

$$\Psi(\mathbf{x}_1, \mathbf{x}_2) = \sum_{nlm} C_{nlm,nlm} |\phi_{nlm}\alpha \ \phi_{nlm}\beta|$$

$$+ \sum_{(n_1 l_1 m_1) > (n_2 l_2 m_2)} C_{n_1 l_1 m_1, n_2 l_2 m_2} (|\phi_{n_1 l_1 m_1}\alpha \ \phi_{n_2 l_2 m_2}\beta| - |\phi_{n_1 l_1 m_1}\beta \ \phi_{n_2 l_2 m_2}\alpha|) \tag{7E.1.2}$$

where $|\phi_1\phi_2|$ is the two-electron Slater determinant

$$|\phi_1\phi_2| = \frac{1}{\sqrt{2}}[\phi_1(\mathbf{x}_1)\phi_2(\mathbf{x}_2) - \phi_2(\mathbf{x}_1)\phi_1(\mathbf{x}_2)] \tag{7E.1.3}$$

1. Show that the singlet function (7E.1.2) may be written as

$$\Psi(\mathbf{x}_1, \mathbf{x}_2) = \Psi(\mathbf{r}_1, \mathbf{r}_2)\frac{1}{\sqrt{2}}[\alpha(1)\beta(2) - \beta(1)\alpha(2)] \tag{7E.1.4}$$

where

$$\Psi(\mathbf{r}_1, \mathbf{r}_2) = \sum_{nlm} C_{nlm,nlm}\phi_{nlm}(\mathbf{r}_1)\phi_{nlm}(\mathbf{r}_2) + \sum_{(n_1 l_1 m_1) > (n_2 l_2 m_2)} C_{n_1 l_1 m_1, n_2 l_2 m_2}$$

$$\times [\phi_{n_1 l_1 m_1}(\mathbf{r}_1)\phi_{n_2 l_2 m_2}(\mathbf{r}_2) + \phi_{n_2 l_2 m_2}(\mathbf{r}_1)\phi_{n_1 l_1 m_1}(\mathbf{r}_2)] \tag{7E.1.5}$$

The two-electron singlet wave function has now been separated into a symmetric spatial part and an antisymmetric spin part.

2. We shall now develop further the spatial part of the ground-state helium wave function (7E.1.5). For S states, the following restrictions arise from angular-momentum theory:

$$l_1 = l_2 \tag{7E.1.6}$$

$$m_1 = -m_2 \tag{7E.1.7}$$

$$C_{n_1,l,m,n_2,l,-m} = (-1)^m C_{n_1,l,0,n_2,l,0} \tag{7E.1.8}$$

Thus, to form an S state, l_1 and l_2 must be identical and the total projected angular momentum $m_1 + m_2$ must be zero. The relation (7E.1.8) between the coefficients of different m follows from the requirement that, for a given set of quantum numbers n_1, n_2 and l, the sum over all m must give a totally symmetric state. In Exercise 2.10, it is demonstrated that a similar relation among the coefficients in the expansion of a spin tensor operator yields a totally symmetric operator.

Substitute the orbitals (7E.1.1) into (7E.1.5) and then use (7E.1.6)–(7E.1.8) and the spherical-harmonic addition theorem (7.3.3)

$$P_l(\cos\theta_{12}) = \frac{4\pi}{2l+1} \sum_{m=-l}^{l} (-1)^m Y_{lm}(\theta_1, \varphi_1) Y_{l,-m}(\theta_2, \varphi_2) \tag{7E.1.9}$$

to rewrite the spatial part of the ^1S wave function as

$$\Psi(\mathbf{r}_1, \mathbf{r}_2) = \sum_{l \geq 0} P_l(\cos\theta_{12}) \sum_{\substack{n_1 \geq n_2 \\ n_2 > l}} C_{n_1 n_2 l}[R_{n_1 l}(r_1) R_{n_2 l}(r_2) + R_{n_2 l}(r_1) R_{n_1 l}(r_2)] \tag{7E.1.10}$$

where

$$C_{n_1 n_2 l} = \frac{2l+1}{4\pi}(1 + \delta_{n_1 n_2})^{-1} C_{n_1 l 0 n_2 l 0} \tag{7E.1.11}$$

Assuming that the radial functions are those of the (nonorthogonal) STOs with exponent ζ

$$R_{nl}(r) = r^{n-1} \exp(-\zeta r) \tag{7E.1.12}$$

we then arrive at the spatial part of the ^1S wave function (7.3.5):

$$\Psi(\mathbf{r}_1, \mathbf{r}_2) = \exp[-\zeta(r_1 + r_2)] \sum_{\substack{l \geq 0 \\ n_1 \geq n_2 \\ n_2 > l}} C_{n_1 n_2 l}(r_1^{n_1-1} r_2^{n_2-1} + r_2^{n_1-1} r_1^{n_2-1}) P_l(\cos\theta_{12}) \tag{7E.1.13}$$

EXERCISE 7.2

In this exercise, we consider the use of correlating functions – that is, functions that depend only on the interelectronic distances – to describe the Coulomb cusp of many-electron systems. Consider a two-electron system with the Hamiltonian

$$H = T + V + g \tag{7E.2.1}$$

where

$$T = T_1 + T_2 = -\tfrac{1}{2}\nabla_1^2 - \tfrac{1}{2}\nabla_2^2 \tag{7E.2.2}$$

$$V = V_1 + V_2 = -\sum_I \frac{Z_I}{|\mathbf{r}_1 - \mathbf{R}_I|} - \sum_I \frac{Z_I}{|\mathbf{r}_2 - \mathbf{R}_I|} \tag{7E.2.3}$$

$$g = \frac{1}{r_{12}} \tag{7E.2.4}$$

Write the wave function in the form

$$\Psi = \gamma\Phi \tag{7E.2.5}$$

where γ is a continuous *correlating function* that depends only on r_{12}, and Φ is a smooth function (i.e. Φ and its derivatives are continuous) of $\mathbf{r}_1$ and $\mathbf{r}_2$. The functions γ and Φ are both finite – that is, they do not become infinite for any choice of coordinates.

1. Show that the Schrödinger equation

$$H\Psi = E\Psi \tag{7E.2.6}$$

can be written in the form

$$g\gamma\Phi + [T, \gamma]\Phi = (E - V)\gamma\Phi - \gamma T\Phi \tag{7E.2.7}$$

2. Consider (7E.2.7) as $r_{12} \to 0$ and g becomes singular. Assuming that the electrons and nuclei do not coincide, show that the singularity in g must be compensated for by a singularity in $[T, \gamma]$ such that $g\gamma + [T, \gamma]$ remains finite.

3. Show that

$$[T, r_{12}] = -\frac{2}{r_{12}} - \frac{\mathbf{r}_{12}}{r_{12}} \cdot (\nabla_1 - \nabla_2) \tag{7E.2.8}$$

but that, for $n > 1$, $[T, r_{12}^n]$ contains no terms proportional to r_{12}^{-1}.

4. Consider the following expansion of γ about $r_{12} = 0$:

$$\gamma = 1 + \beta r_{12} + O(r_{12}^2) \tag{7E.2.9}$$

Determine the parameter β such that $g\gamma + [T, \gamma]$ becomes nonsingular.

5. For more than two electrons, we may again approximate the wave function by $\gamma\Phi$, where γ depends on the interelectronic coordinates r_{ij} and where Φ is a smooth function of the electronic coordinates $\mathbf{r}_i$. Show that the form

$$\gamma = 1 + \beta \sum_{i>j} r_{ij} \tag{7E.2.10}$$

eliminates the two-electron singularities in the many-electron Schrödinger equation if β is chosen as for the two-electron system in Exercise 7.2.4.

6. Determine which of the following correlating functions γ remove the singularities at $r_{ij} = 0$:

$$\gamma = 1 + \beta \sum_{i>j} r_{ij} \tag{7E.2.11}$$

$$\gamma = \prod_{i>j}(1 + \beta r_{ij}) \tag{7E.2.12}$$

$$\gamma = \exp\left(\beta \sum_{i>j} r_{ij}\right) \tag{7E.2.13}$$

$$\gamma = 1 + \sum_p \alpha_p \exp\left(\beta_p \sum_{i>j} r_{ij}^2\right) \tag{7E.2.14}$$

EXERCISE 7.3

We shall here consider the partial-wave expansion of the interelectronic distance r_{12} in the coordinates r_1, r_2 and θ_{12} (i.e. the angle between the position vectors $\mathbf{r}_1$ and $\mathbf{r}_2$ of the two electrons). Expressing the angular dependence of r_{12} in terms of the Legendre polynomials, we obtain

$$r_{12} = \sum_{l=0}^{\infty} P_l(\cos\theta_{12}) f_l(r_1, r_2) \tag{7E.3.1}$$

where $f_l(r_1, r_2)$ are radial functions to be determined in this exercise.

1. We shall derive our expansion of r_{12} from the partial-wave expansion of r_{12}^{-1}. For this purpose, we need to express r_{12} in terms of r_{12}^{-1} and the three coordinates r_1, r_2 and θ_{12}. Verify the relation

$$r_{12} = \frac{r_1^2 + r_2^2 - 2r_1 r_2 \cos \theta_{12}}{r_{12}} \tag{7E.3.2}$$

2. Consider the partial-wave expansion of r_{12}^{-1}

$$\frac{1}{r_{12}} = \frac{1}{r_>} \sum_{l=0}^{\infty} \left(\frac{r_<}{r_>}\right)^l P_l(\cos\theta_{12}) \tag{7E.3.3}$$

where $r_<$ is the smaller and $r_>$ the greater of r_1 and r_2. Invoking the recurrence relation for the Legendre polynomials

$$(l+1)P_{l+1}(x) = (2l+1)xP_l(x) - lP_{l-1}(x) \tag{7E.3.4}$$

show that the radial functions $f_l(r_1, r_2)$ in (7E.3.1) are given by

$$f_l(r_1, r_2) = \frac{1}{2l+3}\frac{r_<^{l+2}}{r_>^{l+1}} - \frac{1}{2l-1}\frac{r_<^{l}}{r_>^{l-1}} \tag{7E.3.5}$$

3. Express the partial-wave expansion (7E.3.1) in products of one-electron functions, using the spherical-harmonic addition theorem.

4. Let us consider the partial-wave expansion of r_{12} in the special case of $\theta_{12} = 0$. Truncating the expansion at $l = L$ and noting that $P_l(1) = 1$ for all l, we obtain from (7E.3.1)

$$F_L = \sum_{l=0}^{L} f_l(r_1, r_2) \tag{7E.3.6}$$

Show that the truncated expansion may be written in the form

$$F_L = r_{12} + \frac{1}{2L+3}\frac{r_<^{L+2}}{r_>^{L+1}} + \frac{1}{2L+1}\frac{r_<^{L+1}}{r_>^{L}} \tag{7E.3.7}$$

and discuss its convergence for $r_> \gg r_<$ and $r_> \approx r_<$.

Solutions

SOLUTION 7.1

1. The Slater determinants appearing in the first sum in (7E.1.2) may be expanded as

$$|\phi_{nlm}\alpha\ \phi_{nlm}\beta| = \frac{1}{\sqrt{2}}\phi_{nlm}(\mathbf{r}_1)\alpha(1)\phi_{nlm}(\mathbf{r}_2)\beta(2) - \frac{1}{\sqrt{2}}\phi_{nlm}(\mathbf{r}_1)\beta(1)\phi_{nlm}(\mathbf{r}_2)\alpha(2)$$

$$= \phi_{nlm}(\mathbf{r}_1)\phi_{nlm}(\mathbf{r}_2)\frac{1}{\sqrt{2}}[\alpha(1)\beta(2) - \beta(1)\alpha(2)] \tag{7S.1.1}$$

Likewise, the determinants in the second sum in (7E.1.2) may be expanded as

$$|\phi_{n_1 l_1 m_1} \alpha \; \phi_{n_2 l_2 m_2} \beta| - |\phi_{n_1 l_1 m_1} \beta \; \phi_{n_2 l_2 m_2} \alpha| = [\phi_{n_1 l_1 m_1}(\mathbf{r}_1)\phi_{n_2 l_2 m_2}(\mathbf{r}_2) + \phi_{n_2 l_2 m_2}(\mathbf{r}_1)\phi_{n_1 l_1 m_1}(\mathbf{r}_2)]$$

$$\times \frac{1}{\sqrt{2}}[\alpha(1)\beta(2) - \beta(1)\alpha(2)] \qquad (7S.1.2)$$

Inserting these expressions in (7E.1.2) and collecting terms, we arrive at (7E.1.4).

2. Rewriting (7E.1.5) as a free summation

$$\Psi(\mathbf{r}_1, \mathbf{r}_2) = \tfrac{1}{2} \sum_{\substack{n_1 l_1 m_1 \\ n_2 l_2 m_2}} C_{n_1 l_1 m_1 n_2 l_2 m_2}[\phi_{n_1 l_1 m_1}(\mathbf{r}_1)\phi_{n_2 l_2 m_2}(\mathbf{r}_2) + \phi_{n_2 l_2 m_2}(\mathbf{r}_1)\phi_{n_1 l_1 m_1}(\mathbf{r}_2)] \qquad (7S.1.3)$$

and introducing the restrictions (7E.1.6)–(7E.1.8), we arrive at the expansion

$$\Psi(\mathbf{r}_1, \mathbf{r}_2) = \tfrac{1}{2} \sum_{n_1 n_2 l} C_{n_1, l, 0, n_2, l, 0}$$

$$\times \sum_{m=-l}^{l} (-1)^m [\phi_{n_1, l, m}(\mathbf{r}_1)\phi_{n_2, l, -m}(\mathbf{r}_2) + \phi_{n_2, l, -m}(\mathbf{r}_1)\phi_{n_1, l, m}(\mathbf{r}_2)] \qquad (7S.1.4)$$

Next, we invoke the addition theorem (7E.1.9) to obtain

$$\Psi(\mathbf{r}_1, \mathbf{r}_2) = \tfrac{1}{2} \sum_{n_1 n_2 l} C_{n_1, l, 0, n_2, l, 0}$$

$$\times \frac{2l+1}{4\pi} P_l(\cos\theta_{12})[R_{n_1 l}(r_1)R_{n_2 l}(r_2) + R_{n_2 l}(r_1)R_{n_1 l}(r_2)] \qquad (7S.1.5)$$

which may be rewritten to give (7E.1.10) with the coefficients (7E.1.11).

SOLUTION 7.2

1. Inserting (7E.2.1) and (7E.2.5) into the Schrödinger equation (7E.2.6) and rearranging, we obtain

$$(V - E)\gamma\Phi + g\gamma\Phi + T\gamma\Phi = 0 \qquad (7S.2.1)$$

which is equivalent to (7E.2.7), as is readily seen by expanding the commutator.

2. Consider the terms on the right-hand side of (7E.2.7). Since there are no singularities in $\gamma\Phi$, the first term $(V - E)\gamma\Phi$ does not contain any singularities provided (as assumed) that the electrons do not coincide with the nuclei. Next, since Φ is smooth and γ is finite, the second term on the right-hand side likewise does not give rise to any singularities. It thus follows that the left-hand side of (7E.2.7) $g\gamma\Phi + [T, \gamma]\Phi$ must be nonsingular as long as the electrons do not coincide with the nuclei. Since Φ is finite, $g\gamma + [T, \gamma]$ must be nonsingular as well.

3. From elementary vector analysis, we obtain the following expressions for the gradient and the Laplacian of the interelectronic distance:

$$\nabla_1 r_{12} = \left(\frac{\partial r_{12}}{\partial x_1}, \frac{\partial r_{12}}{\partial y_1}, \frac{\partial r_{12}}{\partial z_1} \right) = \left(\frac{x_1 - x_2}{r_{12}}, \frac{y_1 - y_2}{r_{12}}, \frac{z_1 - z_2}{r_{12}} \right) = \frac{\mathbf{r}_{12}}{r_{12}} \qquad (7S.2.2)$$

$$\nabla_2 r_{12} = -\frac{\mathbf{r}_{12}}{r_{12}} \tag{7S.2.3}$$

$$\nabla_1^2 r_{12} = \frac{\partial}{\partial x_1}\left(\frac{x_1 - x_2}{r_{12}}\right) + \frac{\partial}{\partial y_1}\left(\frac{y_1 - y_2}{r_{12}}\right) + \frac{\partial}{\partial z_1}\left(\frac{z_1 - z_2}{r_{12}}\right)$$

$$= \frac{3}{r_{12}} - \frac{(x_1 - x_2)^2}{r_{12}^3} - \frac{(y_1 - y_2)^2}{r_{12}^3} - \frac{(z_1 - z_2)^2}{r_{12}^3} = \frac{2}{r_{12}} \tag{7S.2.4}$$

$$\nabla_2^2 r_{12} = \frac{2}{r_{12}} \tag{7S.2.5}$$

Using these expressions, we obtain for the commutator of T and r_{12}

$$[T, r_{12}] = -\tfrac{1}{2}\nabla_1^2 r_{12} - \tfrac{1}{2}\nabla_2^2 r_{12} - (\nabla_1 r_{12}) \cdot \nabla_1 - (\nabla_2 r_{12}) \cdot \nabla_2$$

$$= -\frac{2}{r_{12}} - \frac{\mathbf{r}_{12}}{r_{12}} \cdot \nabla_1 + \frac{\mathbf{r}_{12}}{r_{12}} \cdot \nabla_2 \tag{7S.2.6}$$

which is equivalent to (7E.2.8). For higher powers of r_{12}, we obtain the gradients and Laplacians

$$\nabla_1 r_{12}^n = n\mathbf{r}_{12} r_{12}^{n-2} \tag{7S.2.7}$$

$$\nabla_2 r_{12}^n = -n\mathbf{r}_{12} r_{12}^{n-2} \tag{7S.2.8}$$

$$\nabla_1^2 r_{12}^n = n(n+1) r_{12}^{n-2} \tag{7S.2.9}$$

$$\nabla_2^2 r_{12}^n = n(n+1) r_{12}^{n-2} \tag{7S.2.10}$$

Therefore, for $n > 1$, $[T, r_{12}^n]$ contains no terms proportional to r_{12}^{-1}.

4. We wish to make $g\gamma + [T, \gamma]$ nonsingular. Introducing the expansion (7E.2.9), we obtain

$$g\gamma + [T, \gamma] = \beta + \frac{1 - 2\beta}{r_{12}} - \beta \frac{\mathbf{r}_{12}}{r_{12}} \cdot (\nabla_1 - \nabla_2) + O(r_{12}) \tag{7S.2.11}$$

The choice $\beta = \tfrac{1}{2}$ therefore gives

$$g\gamma + [T, \gamma] = \frac{1}{2} - \frac{1}{2}\frac{\mathbf{r}_{12}}{r_{12}} \cdot (\nabla_1 - \nabla_2) + O(r_{12}), \qquad \beta = \tfrac{1}{2} \tag{7S.2.12}$$

which is nonsingular, noting that $\mathbf{r}_{12}/r_{12}$ is the unit vector pointing from $\mathbf{r}_2$ to $\mathbf{r}_1$.

5. Consider the singularity at $r_{ij} = 0$ for a given electron pair ij. The many-electron Schrödinger equation can be written as

$$g_{ij}\gamma\Phi + [T_i + T_j, \gamma]\Phi = (E - V)\gamma\Phi - \frac{1}{2}\sum_{kl \neq ij} g_{kl}\gamma\Phi - \sum_{k \neq i,j} T_k\gamma\Phi - \gamma(T_i + T_j)\Phi \tag{7S.2.13}$$

in the notation

$$g_{kl} = \frac{1}{r_{kl}} \tag{7S.2.14}$$

$$T_k = -\tfrac{1}{2}\nabla_k^2 \tag{7S.2.15}$$

Again, the only term that may compensate the singularity of $g_{ij}\gamma\Phi$ at $r_{ij} = 0$ in (7S.2.13) is $[T_i + T_j, \gamma]\Phi$. Expanding γ in r_{kl} around $r_{kl} = 0$, we obtain

$$\gamma = 1 + \sum_{k>l} \beta_{kl} r_{kl} + O(r_{kl}^2) \tag{7S.2.16}$$

The condition that $g_{ij}\gamma + [T_i + T_j, \gamma]$ is nonsingular at $r_{ij} = 0$ then becomes equivalent to the condition that $g_{ij} + \beta_{ij}[T_i + T_j, r_{ij}]$ is nonsingular. Again, the singularities are removed if we choose $\beta_{ij} = \beta = \frac{1}{2}$.

6. Whereas (7E.2.11)–(7E.2.13) have the correct local behaviour, (7E.2.14) behaves incorrectly for coinciding electrons, no matter how many terms are included in the expansion.

SOLUTION 7.3

1. Since

$$r_{12} = \sqrt{(\mathbf{r}_1 - \mathbf{r}_2) \cdot (\mathbf{r}_1 - \mathbf{r}_2)} = \sqrt{r_1^2 + r_2^2 - 2r_1 r_2 \cos\theta_{12}} \tag{7S.3.1}$$

we obtain

$$r_{12} = \frac{r_{12}^2}{r_{12}} = \frac{r_1^2 + r_2^2 - 2r_1 r_2 \cos\theta_{12}}{r_{12}} \tag{7S.3.2}$$

2. Inserting (7E.3.3) in (7E.3.2), we obtain the expansion

$$r_{12} = \frac{r_<^2 + r_>^2 - 2r_< r_> \cos\theta_{12}}{r_>} \sum_{l=0}^{\infty} \left(\frac{r_<}{r_>}\right)^l P_l(\cos\theta_{12}) \tag{7S.3.3}$$

From the recurrence relation (7E.3.4) written in the form

$$\cos\theta_{12} P_l(\cos\theta_{12}) = \frac{l+1}{2l+1} P_{l+1}(\cos\theta_{12}) + \frac{l}{2l+1} P_{l-1}(\cos\theta_{12}) \tag{7S.3.4}$$

we arrive at the identity

$$\sum_{l=0}^{\infty} \left(\frac{r_<}{r_>}\right)^l \cos\theta_{12} P_l(\cos\theta_{12}) = \sum_{l=0}^{\infty} \left[\frac{l}{2l-1}\left(\frac{r_<}{r_>}\right)^{l-1} + \frac{l+1}{2l+3}\left(\frac{r_<}{r_>}\right)^{l+1}\right] P_l(\cos\theta_{12}) \tag{7S.3.5}$$

Using this expression in (7S.3.3), we obtain

$$r_{12} = \sum_{l=0}^{\infty} \left(\frac{r_<^{l+2}}{r_>^{l+1}} + \frac{r_<^l}{r_>^{l-1}} - \frac{2l}{2l-1}\frac{r_<^l}{r_>^{l-1}} - \frac{2l+2}{2l+3}\frac{r_<^{l+2}}{r_>^{l+1}}\right) P_l(\cos\theta_{12})$$

$$= \sum_{l=0}^{\infty} \left(\frac{1}{2l+3}\frac{r_<^{l+2}}{r_>^{l+1}} - \frac{1}{2l-1}\frac{r_<^l}{r_>^{l-1}}\right) P_l(\cos\theta_{12}) \tag{7S.3.6}$$

in accordance with (7E.3.5).

3. Inserting (7E.3.5) for $f_l(r_1, r_2)$ in (7E.3.1) and using the addition theorem (7.3.3) for the Legendre polynomials, we arrive at the following product expansion of r_{12}:

$$r_{12} = \sum_{l=0}^{\infty} \sum_{m=-l}^{l} (-1)^m \frac{4\pi}{2l+1} \left(\frac{1}{2l+3} \frac{r_<^{l+2}}{r_>^{l+1}} - \frac{1}{2l-1} \frac{r_<^l}{r_>^{l-1}} \right) Y_{l,m}(\theta_1, \varphi_1) Y_{l,-m}(\theta_2, \varphi_2) \quad (7S.3.7)$$

4. Inserting (7E.3.5) in (7E.3.6) and rearranging terms (noting that many cancel), we obtain, for $\theta = 0$:

$$\begin{aligned}
F_L &= \sum_{l=0}^{L} \frac{1}{2l+3} \frac{r_<^{l+2}}{r_>^{l+1}} - \sum_{l=0}^{L} \frac{1}{2l-1} \frac{r_<^l}{r_>^{l-1}} \\
&= r_> - r_< + \sum_{l=0}^{L} \frac{1}{2l+3} \frac{r_<^{l+2}}{r_>^{l+1}} - \sum_{l=2}^{L} \frac{1}{2l-1} \frac{r_<^l}{r_>^{l-1}} \\
&= r_{12} + \frac{1}{2L+3} \frac{r_<^{L+2}}{r_>^{L+1}} + \frac{1}{2L+1} \frac{r_<^{L+1}}{r_>^L} \quad (7S.3.8)
\end{aligned}$$

When $r_> \gg r_<$, the series converges rapidly since each increment in L reduces the error by a factor of $r_</r_>$; when $r_> \approx r_<$, the convergence is slow since $(2L+3)^{-1}$ and $(2L+1)^{-1}$ both tend to zero slowly.

8 GAUSSIAN BASIS SETS

Over the years, many different sets of Gaussian orbitals have been proposed for molecular calculations. Indeed, the large number of Gaussian basis sets in routine use bear witness to the fact that the design of molecular basis sets is a very tricky business. Although several good, flexible basis sets have been developed for molecular calculations at the Hartree–Fock and correlated levels, it has not been possible to construct a single, universal molecular basis set that is applicable under all circumstances. Thus, the basis-set requirements for ground states and for excited states may be quite different from each other and the description of an anionic system may require functions that make a negligible contribution to the neutral system. Also, the accurate description of many molecular properties requires the use of functions that are of little or no importance for the unperturbed system. Finally, the different wave-function models may have quite different requirements with regard to the basis set. The requirements for accurate correlated treatments, in particular, are more severe than those for Hartree–Fock calculations.

To gain some understanding of the principles underlying the design of Gaussian basis sets, we shall in this chapter consider a selection of representative basis sets, ranging from small minimal sets, suitable for simple calculations at the Hartree–Fock level, to large sets, appropriate for accurate correlated calculations. The discussion of each basis set should reveal some of the principles behind the construction of molecular basis sets.

The present chapter is divided into five sections. First, the functional form of the basis functions is reviewed in Section 8.1. Next, we discuss Gaussian basis sets for Hartree–Fock calculations in Section 8.2 and for correlated calculations in Section 8.3. The convergence of the basis sets towards the basis-set limit is then examined in Section 8.4. Finally, in Section 8.5, we consider the problems associated with basis-set superposition errors, in particular in calculations of interaction energies.

8.1 Gaussian basis functions

Before we begin our discussion of Gaussian basis sets, let us briefly review the one-electron basis functions studied in Chapter 6. The *complex spherical-harmonic* GTOs are given by

$$\chi_{\alpha l m}^{\text{GTO}}(r, \theta, \varphi) = R_l^{\text{GTO}}(\alpha, r) Y_{lm}(\theta, \varphi) \tag{8.1.1}$$

$$R_l^{\text{GTO}}(\alpha, r) = \frac{2(2\alpha)^{3/4}}{\pi^{1/4}} \sqrt{\frac{2^l}{(2l+1)!!}} \left(\sqrt{2\alpha}\, r\right)^l \exp(-\alpha r^2) \tag{8.1.2}$$

where α is the orbital exponent and $Y_{lm}(\theta, \varphi)$ one of the *spherical harmonics* (6.4.3) introduced in Section 6.4.1. In most molecular applications, we use the *real spherical-harmonic GTOs*

$$\chi_{\alpha lm}^{\text{GTO}}(x, y, z) = N_{\alpha lm}^{\text{GTO}} S_{lm}(x, y, z) \exp(-\alpha r^2) \tag{8.1.3}$$

where $N_{\alpha lm}^{\text{GTO}}$ is the normalization constant and $S_{lm}(x, y, z)$ is one of the *real solid harmonics*, related to the *complex solid harmonics*

$$\mathcal{Y}_{lm}(r, \theta, \varphi) = r^l Y_{lm}(\theta, \varphi) \tag{8.1.4}$$

in the following way

$$S_{l0} = \sqrt{\frac{4\pi}{2l+1}} \mathcal{Y}_{l0} \tag{8.1.5}$$

$$S_{lm} + \mathrm{i} S_{l,-m} = (-1)^m \sqrt{\frac{8\pi}{2l+1}} \mathcal{Y}_{lm}, \qquad m > 0 \tag{8.1.6}$$

The real solid harmonics may be generated from (6.4.47); the first few are illustrated in Figure 6.1 and listed in Table 6.3. We also recall that the STOs are given by the expressions

$$\chi_{nlm}^{\text{STO}}(r, \theta, \varphi) = R_n^{\text{STO}}(\zeta, r) Y_{lm}(\theta, \varphi) \tag{8.1.7}$$

$$R_n^{\text{STO}}(\zeta, r) = \frac{(2\zeta)^{3/2}}{\sqrt{(2n)!}} (2\zeta r)^{n-1} \exp(-\zeta r) \tag{8.1.8}$$

where $n > l$. The use of STOs is restricted mainly to atoms and diatoms since their exponential radial part is much less amenable to polyatomic calculations than the Gaussian radial part of the GTOs. In the following, we shall consider the construction of molecular basis sets from a finite number of real spherical-harmonic GTOs of the form (8.1.3).

8.2 Gaussian basis sets for Hartree–Fock calculations

We consider in this section Gaussian basis sets developed for the calculation of Hartree–Fock wave functions, ranging from the minimal STO-3G basis sets to larger basis sets such as 6-31G**. In the course of our discussion, we shall cover such topics as even-tempered basis functions (Section 8.2.3), contracted functions (Section 8.2.4) and polarization functions (Section 8.2.7).

8.2.1 STO-kG BASIS SETS

In the 1960s, much work was directed towards developing Gaussian basis functions that would serve as one-for-one substitutes for STOs, to be used in minimal or single-zeta calculations at the Hartree–Fock level. Thus, the STOs were retained as the conceptual basis for the calculations, but the difficulties associated with the evaluation of STO integrals were side-stepped by representing each STO as a fixed, linear combination of GTOs. Obviously, such basis sets can provide only a crude description of the electronic structure and the accuracy achieved with such sets is necessarily limited. Nevertheless, such basis sets may be useful for qualitative or preliminary investigations, at least for molecules containing no higher than first-row atoms.

The most commonly used Gaussian expansions of STOs are the *STO-kG functions* of Hehre, Stewart and Pople [1], where the exponents and the expansion coefficients are obtained by a least-squares procedure as described in Section 6.5.4. Thus, we consider k-term GTO expansions of the form

$$R_{nl}^k(\zeta, \mathbf{d}, \boldsymbol{\alpha}, r) = \sum_{i=1}^k d_i R_l^{\text{GTO}}(\zeta^2 \alpha_i, r) \tag{8.2.1}$$

where ζ is the exponent of the STO

$$\chi_{nlm}^{\text{STO}}(\zeta, r, \theta, \varphi) = R_n^{\text{STO}}(\zeta, r) Y_{lm}(\theta, \varphi) \tag{8.2.2}$$

whose radial form we wish to approximate. In the expansion (8.2.1), we have introduced the square of ζ in anticipation of future simplifications. To determine the expansion (8.2.1), we fix the exponents α_i and expansion coefficients d_i by minimizing the error function

$$\Delta_{nl}^k(\zeta, \mathbf{d}, \boldsymbol{\alpha}) = \langle R_n^{\text{STO}}(\zeta) - R_{nl}^k(\zeta, \mathbf{d}, \boldsymbol{\alpha}) | R_n^{\text{STO}}(\zeta) - R_{nl}^k(\zeta, \mathbf{d}, \boldsymbol{\alpha}) \rangle \tag{8.2.3}$$

subject to the condition that the expansion (8.2.1) is normalized.

An important simplification arises from the scaling properties of the STOs and GTOs. Introducing the scaled coordinate

$$\rho = \zeta r \tag{8.2.4}$$

we find that

$$R_n^{\text{STO}}(\zeta, r) = \zeta^{3/2} R_n^{\text{STO}}(1, \rho) \tag{8.2.5}$$

$$R_l^{\text{GTO}}(\zeta^2 \alpha_i, r) = \zeta^{3/2} R_l^{\text{GTO}}(\alpha_i, \rho) \tag{8.2.6}$$

$$R_{nl}^k(\zeta, \mathbf{d}, \boldsymbol{\alpha}, r) = \zeta^{3/2} R_{nl}^k(1, \mathbf{d}, \boldsymbol{\alpha}, \rho) \tag{8.2.7}$$

where we have used (8.1.8), (8.1.2) and (8.2.1). Thus, all overlap integrals are unaffected by a change in ζ

$$\langle f(\zeta) | g(\zeta) \rangle = \langle f(1) | g(1) \rangle \tag{8.2.8}$$

when $f(\zeta)$ and $g(\zeta)$ are any of the functions (8.2.5)–(8.2.7). The error function $\Delta_{nl}^k(\zeta, \mathbf{d}, \boldsymbol{\alpha})$ is therefore independent of ζ. Clearly, we need only determine the coefficients $\mathbf{d}$ and the exponents $\boldsymbol{\alpha}$ for a single exponent $\zeta = 1$; the expansion for a different Slater exponent ζ is then obtained by a universal scaling of the Gaussian exponents $\boldsymbol{\alpha}$ by ζ^2 according to (8.2.1).

The STO-kG set approximates each STO by k GTOs in the least-squares sense. For example, in the STO-3G basis, each function is represented by a linear combination of three Gaussians. To simplify the evaluation of integrals, the Gaussian exponents are constrained to be the same for all orbitals belonging to the same shell. Thus, the $1s$ orbital is expanded in three $1s$ GTOs, whose exponents and coefficients are determined by minimizing the mean square deviation of the K shell

$$\Delta_K^{\text{STO}-kG} = \Delta_{1s}^k(1, \mathbf{d}_{1s}, \boldsymbol{\alpha}_K) \tag{8.2.9}$$

subject to the normalization condition on the $1s$ orbital. Likewise, the $2s$ and $2p$ orbitals are determined by minimizing the mean square deviation of the L shell

$$\Delta_L^{\text{STO}-kG} = \Delta_{2s}^k(1, \mathbf{d}_{2s}, \boldsymbol{\alpha}_L) + \Delta_{2p}^k(1, \mathbf{d}_{2p}, \boldsymbol{\alpha}_L) \tag{8.2.10}$$

subject to the normalization conditions. The final exponents and coefficients of the STO-3G basis are listed in Table 8.1. Note that, even though the radial forms of the $2s$ and $2p$ STOs are identical, their expansion coefficients are different since different GTOs are used in the two cases – the $2s$ function is expanded in $1s$ GTOs and the $2p$ function in $2p$ GTOs.

In Figure 8.1, we compare the $1s$ STO-kG functions with the corresponding $1s$ STO with unit exponent. Clearly, at least three Gaussians are needed for a fair representation of the STO function and more functions are needed for high accuracy. In practice, only STO-3G functions are used. These functions reproduce the STOs satisfactorily for crude calculations; for higher accuracy, different basis functions are in any case needed.

To complete the construction of the STO-kG basis sets, we must specify the Slater exponents ζ to be used for the orbitals. One solution would be to use exponents optimized in separate atomic calculations for each STO-kG set. The resulting exponents are similar for all k and also close to the exponents of Clementi and Raimondi (5.67 for $1s$, 1.61 for $2s$ and 1.57 for $2p$ in the carbon atom), optimized for single STOs in atoms [2]. In practice, for the first-row atoms, the atomic exponents of Clementi and Raimondi are used for the K shell. For the L shell, however, a set of standard molecular exponents (obtained from calculations on small molecules) is used – recognizing that, in a molecular environment, the optimum exponents for the valence shells differ somewhat from their atomic values. For the carbon atom, an exponent of 1.72 is used for the $2s$ and $2p$ orbitals. For the hydrogen $1s$ orbital, the exponent is 1.24. Thus, the hydrogen exponent differs substantially from its atomic value, reflecting a significant contraction of the atomic electron distribution in a molecular environment.

In Figure 8.2, we compare the STO-3G carbon basis with the ^3P ground-state Hartree–Fock orbitals. Note the poor description of the $2p$ orbital and the absence of a radial node in the STO-3G

Table 8.1   The STO-3G basis set for unit Slater exponent

α_K	$\mathbf{d}_{1s}$	α_L	$\mathbf{d}_{2s}$	$\mathbf{d}_{2p}$
2.22766	0.154329	0.994203	−0.0999672	0.155916
0.405771	0.535328	0.231031	0.399513	0.607684
0.109818	0.444635	0.0751386	0.700115	0.391957

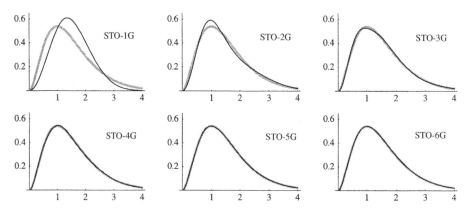

Fig. 8.1.   Radial distribution functions of the 1s STO-kG basis functions (thin lines) superimposed on a $1s$ STO with $\zeta = 1$ (thick grey lines) (atomic units).

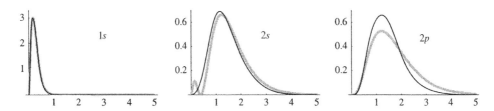

Fig. 8.2. The radial distribution functions of the carbon STO-3G basis functions (thin lines) superimposed on the numerical ^3P ground-state Hartree–Fock orbitals (thick grey lines) (atomic units).

$2s$ orbital. In calculations using this basis, the $2s$ radial node will arise from a linear combination of the STO-3G $1s$ and $2s$ orbitals. The absence of a node in the STO-3G orbital is therefore not a problem. The poor representation of the carbon $2p$ orbital is to be expected since the $2p$ STO orbital gives a poor description of this orbital; see Figure 6.6.

8.2.2 PRIMITIVE EXPANSIONS OF HARTREE–FOCK ORBITALS

For basis sets larger than minimal, the Gaussian expansions are determined by minimization of the total Hartree–Fock energy rather than by least-squares fits to numerical or some other accurate AOs. A set of such energy-optimized exponents and coefficients is given in Table 8.2. This *Huzinaga basis* consists of nine *primitive* $1s$ GTOs in terms of which the $1s$ and $2s$ carbon orbitals are expanded, and five $2p$ GTOs in terms of which the $2p$ carbon orbital is expanded [3]. The resulting basis is referred to as $(9s5p)$. In Figure 8.3, we have plotted the Hartree–Fock $(9s5p)$ orbitals together with the numerical Hartree–Fock orbitals. Clearly, we have made a dramatic improvement on the STO-kG basis sets of Section 8.2.1.

Although the $(9s5p)$ expansions look excellent to the eye, the orbitals are still not converged to the Hartree–Fock limit. In this limit, the ^3P ground-state energy is -37.688619 E_h, whereas the $(9s5p)$ expansion with an energy of -37.685247 E_h is in error by 3.372 mE_h. For comparison, a double-zeta STO expansion is in error by only 1.942 mE_h. Of course, we may improve on our GTO

Table 8.2 The Gaussian $(9s5p)$ exponents and expansion coefficients for the ^3P ground-state carbon orbitals

Exponents	$1s$	$2s$	$2p$
4232.61	0.00122	−0.00026	
634.882	0.00934	−0.00202	
146.097	0.04534	−0.00974	
42.4974	0.15459	−0.03606	
14.1892	0.35867	−0.08938	
5.1477	0.43809	−0.17699	
1.9666	0.14581	−0.05267	
0.4962	0.00199	0.57408	
0.1533	0.00041	0.54768	
18.1557			0.01469
3.9864			0.09150
1.1429			0.30611
0.3594			0.50734
0.1146			0.31735

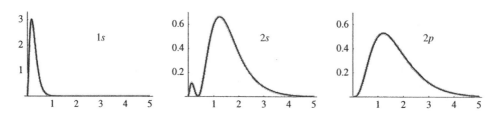

Fig. 8.3. The radial distribution functions of the ^3P ground-state orbitals of the carbon atom (atomic units). The thick grey lines are the numerical Hartree–Fock orbitals, the thin black lines are the Hartree–Fock orbitals expanded in the $(9s5p)$ basis of Huzinaga in Table 8.2.

description by including more functions. Indeed, with an error of 1.295 mE$_h$, the $(10s6p)$ energy turns out to be slightly better than the double-zeta STO energy [3]. Returning briefly to the STO-kG expansions of Section 8.2.2, we find that the STO-3G and STO-6G energies are in error (relative to the Hartree–Fock limit) by as much as 460.0 and 79.6 mE$_h$, respectively. We recall, however, that the purpose of these expansions is to reproduce the radial form of an optimal *single-zeta* STO basis, the ^3P ground-state energy of which is 66.8 mE$_h$ above the Hartree–Fock limit [2].

These examples confirm our previous conclusions – namely that, although it is possible to converge our calculations in terms of GTOs, we must be prepared to use a large number of such functions, several times more than for STOs. We should keep in mind, however, that our discussion here has been concerned only with the convergence of the Hartree–Fock energy of an isolated atom. In correlated molecular calculations, the demands on our basis sets are much more severe – both for GTOs and STOs – since we must then describe the polarization of the charge distribution and also provide an orbital space suitable for recovering correlation effects. It turns out that GTOs are no less suitable than STOs for describing polarization and correlation. The same is true also for the description of excited states and perturbed systems (molecular properties). Therefore, the disadvantages of the GTOs are much less pronounced in accurate, correlated calculations on molecular systems than in Hartree–Fock calculations on the unperturbed ground states of atomic systems.

8.2.3 EVEN-TEMPERED BASIS SETS

The optimization of Gaussian exponents is a highly nonlinear problem with multiple solutions. The determination of exponents for large basis sets is therefore a difficult problem, which must be repeated for each atomic system since no simple scaling relation such as those for STO-kG orbitals exists for the Hartree–Fock orbitals. However, an inspection of the exponents in Table 8.2 reveals a regularity – the ratio between consecutive exponents seems to be fairly constant. In Figure 8.4, we have plotted the logarithms of the s and p exponents of the Huzinaga $(9s5p)$ basis set. The exponents form almost straight lines.

Since the ratio between the exponents is approximately constant, we should be able to obtain a fair representation of the exponents in terms of only two numbers:

$$\alpha_i = \alpha\beta^{i-1} \tag{8.2.11}$$

Using the method of least squares, we obtain the following parameters for the $(9s5p)$ Huzinaga basis set:

$$\alpha_s = 0.1364, \qquad \beta_s = 3.381 \tag{8.2.12}$$

$$\alpha_p = 0.1041, \qquad \beta_p = 3.503 \tag{8.2.13}$$

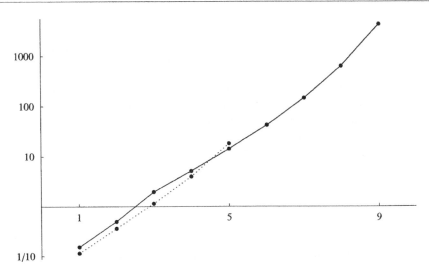

Fig. 8.4. A logarithmic plot of the exponents in the Huzinaga $(9s5p)$ carbon basis. The s exponents are connected by full lines and the p exponents by dotted lines.

Clearly, the $(9s5p)$ set generated by the parameters (8.2.12) and (8.2.13) will yield a somewhat poorer representation of the carbon orbitals than the energy-optimized set in Table 8.2. The important point is, however, that it is possible to generate an acceptable set of Gaussian exponents for the carbon atom by adjusting only four parameters (two for each angular momentum). In view of the many difficulties associated with an unconstrained optimization of the 14 exponents of the $(9s5p)$ basis, this is an attractive alternative.

Basis sets in which the exponents are constrained to be of the form (8.2.11) are referred to as *even-tempered* [4–8]. Even-tempered basis sets are less economical than unconstrained sets in terms of the number of Gaussians required for a given accuracy in the basis set. The difference corresponds roughly to the effect of one s and one p function for first-row atoms, so that an even-tempered $(10s6p)$ set with optimized parameters α and β is roughly equivalent to an unconstrained $(9s5p)$ set with the exponents α_i freely optimized.

Even though even-tempered basis sets are rarely used in practice for first-row atoms, they are important both conceptually and practically. Thus, an even-tempered sequence of exponents may be used as a starting point for an unconstrained optimization of the individual exponents. More important, they allow for a simple generation of sequences of Gaussian basis sets, where, by decreasing β and increasing the number of primitives, one can approach the basis-set limit in a simple and controlled fashion.

Our justification for introducing even-tempered basis sets has been an empirical one, based on the observation that energy-optimized exponents form an approximately geometrical sequence. More insight into the nature of even-tempered basis sets is obtained by considering the overlap between two Gaussians of the same angular momentum (see Exercise 6.9 in Chapter 6):

$$\left\langle \chi^{\text{GTO}}_{\alpha_i lm} | \chi^{\text{GTO}}_{\alpha_j lm} \right\rangle = \left(\frac{\sqrt{4\alpha_i \alpha_j}}{\alpha_i + \alpha_j} \right)^{3/2+l} \qquad (8.2.14)$$

Clearly, the overlap between two Gaussians is determined completely by the ratio between the exponents and, for an even-tempered basis set, the overlap between two consecutive Gaussians becomes

$$\left\langle \chi_{\alpha_i,lm}^{\mathrm{GTO}} \big| \chi_{\alpha_{i+1},lm}^{\mathrm{GTO}} \right\rangle = \left(\frac{2\sqrt{\beta}}{1+\beta} \right)^{3/2+l} \quad \text{(even-tempered basis)} \tag{8.2.15}$$

Hence, in even-tempered basis sets, the overlap between two consecutive Gaussians is constant and determined only by the parameter β. In some sense, this observation implies that the individual exponents in an even-tempered basis are chosen so as to maintain an even distribution of Gaussians in the radial space, not allowing neighbouring functions to overlap by more than a certain amount. Using (8.2.15) with $l = 0$, the reader may verify that, for $\beta = 1.5, 2, 4$ and 8, we obtain the s overlaps 0.9698, 0.9155, 0.7155 and 0.4983.

8.2.4 CONTRACTED GAUSSIANS

We have seen that a relatively large number of Gaussians are needed to represent the ground-state Hartree–Fock orbitals in a satisfactory manner. The large number of Gaussian functions presents a problem, not so much with the evaluation of the integrals as with the subsequent calculation of the wave function. To solve this problem, we note that, upon formation of a chemical bond, the atomic electron distributions do not change very much, in particular in the core regions. Furthermore, the large number of GTOs is needed not so much to describe the changes that occur upon bond formation as to describe the *unperturbed* electron distribution. Presumably, there is usually no need to employ the individual GTOs as basis functions. Instead, we may construct our basis set from fixed linear combinations of GTOs:

$$R_{al}^{\mathrm{CGTO}}(r) = \sum_i d_{ia} R_l^{\mathrm{GTO}}(\alpha_i, r) \tag{8.2.16}$$

Fixed linear combinations such as (8.2.16) are known as *contracted GTOs*, and the individual Gaussians from which the contracted GTOs are constructed are referred to as *primitive GTOs*. The coefficients in (8.2.16) are called the *contraction coefficients*.

It is important to distinguish between *general* and *segmented contractions* of primitive Gaussians. In the general-contraction scheme, we place no restrictions on the contractions and allow all primitive functions of a given angular momentum on an atom to mix freely [9]. A generally contracted basis set for carbon may, for example, consist of the accurate $1s$, $2s$ and $2p$ functions in Table 8.2. In the segmented approach, on the other hand, each primitive function is allowed to contribute to only one contracted orbital. The segmented and general contractions are illustrated in Figure 8.5 for seven primitives contracted to four functions.

The reason for imposing the restriction of segmented contractions is a technical one. It simplifies the evaluation of molecular integrals over contracted functions, reducing the transformation of integrals from the primitive to the contracted basis to a simple addition process where each integral makes only one contribution. In passing, we note that it is always possible to carry out calculations over generally contracted basis sets using programs that are designed only for segmented basis sets: we may simply include each primitive function several times in our basis, one time for each contracted function to which it contributes. Such an approach is, however, inefficient since each primitive integral is calculated several times.

In the remainder of this section, we shall consider only segmented basis sets. Generally contracted basis sets have been developed primarily for the accurate calculation of electron correlation and will be discussed in Section 8.3.

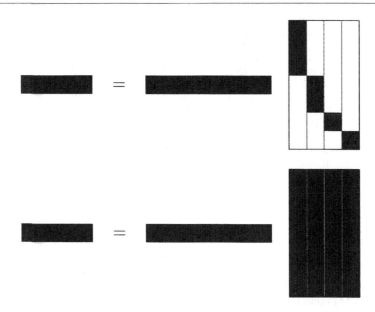

Fig. 8.5. Segmented and general contractions.

8.2.5 SEGMENTED CONTRACTIONS

Let us see how we may generate a segmented basis set from the primitive functions in Table 8.2. We would like to combine these functions into a smaller set of contracted GTOs that reproduces as accurately as possible the Hartree–Fock orbitals and which at the same time has the flexibility to describe the changes that occur in the valence region upon formation of chemical bonds. Clearly, some compromises must be made since an accurate representation of the Hartree–Fock orbitals requires the use of generally contracted GTOs.

A simple way to design a segmented basis set is to start with the orbital coefficients obtained in an uncontracted atomic Hartree–Fock calculation. The contracted functions are then generated by taking fixed linear combinations of the primitive functions with the coefficients taken from the atomic calculations. An optimal contraction scheme (for a given size of the contracted set) is arrived at by systematically trying different combinations of contracted functions in simple molecular calculations, attempting to minimize the impact of the contraction on the calculated energies. As an example, we have in Table 8.3 listed *Dunning's segmented contractions* [10] of the energy-optimized primitive Huzinaga $(9s5p)$ basis [3]. The contraction coefficients correspond to the coefficients in Table 8.2, scaled to make the contracted functions normalized. The number of contracted functions is conventionally given in square brackets. Thus, the second column in Table 8.3 corresponds to a $[3s]$ set contracted from a $(9s)$ primitive set.

As can be seen from Table 8.3, the best segmentations are typically obtained by leaving the outermost primitive functions uncontracted and by combining the innermost functions in fixed linear combinations. Note, however, that the primitive s function with exponent 5.1477 is left uncontracted in all contractions since it contributes strongly to both the $1s$ and the $2s$ orbitals. Also note that, in the $[3s]$ set, the two most diffuse primitives have been contracted using scaled $2s$ coefficients.

Of the Dunning–Huzinaga basis sets, the $[4s2p]$ set has been particularly popular. This *double-zeta (DZ)* basis set, in which there are two contracted functions for each occupied AO, represents

Table 8.3 Dunning's segmented contractions of the Huzinaga carbon $(9s5p)$ basis

Exponents	[3s]	[4s]	[5s]	[2p]	[3p]
4232.61	0.002029	0.002029	0.006228		
634.882	0.015535	0.015535	0.047676		
146.097	0.075411	0.075411	0.231439		
42.4974	0.257121	0.257121	0.789108		
14.1892	0.596555	0.596555	0.791751		
1.9666	0.242517	0.242517	0.321870		
5.1477	1.000000	1.000000	1.000000		
0.4962	0.542048	1.000000	1.000000		
0.1533	0.517121	1.000000	1.000000		
18.1557				0.018534	0.039196
3.9864				0.115442	0.244144
1.1429				0.386206	0.816775
0.3594				0.640089	1.000000
0.1146				1.000000	1.000000

a good balance between the number of primitive functions and the number of contracted functions. The $[4s2p]$ contractions of the $(9s5p)$ set are illustrated in Figure 8.6. The double-zeta character of the basis set is evident: Each Hartree–Fock orbital is bracketed by two nodeless segmented basis functions, thus providing some flexibility to describe the changes that occur upon formation of a chemical bond.

Expanding the numerical Hartree–Fock ^3P ground-state orbitals in the Dunning–Huzinaga $[4s2p]$ basis (in the least-squares sense), we arrive at the following functions:

$$\phi_{1s} = 0.601\chi_{1s} + 0.438\chi_{2s} + 0.005\chi_{3s} + 0.001\chi_{4s} \tag{8.2.17}$$

$$\phi_{2s} = -0.142\chi_{1s} - 0.195\chi_{2s} + 0.562\chi_{3s} + 0.553\chi_{4s} \tag{8.2.18}$$

$$\phi_{2p} = 0.792\chi_{1p} + 0.318\chi_{2p} \tag{8.2.19}$$

Note that the contracted $3s$ and $4s$ functions make negligible contributions to the $1s$ Hartree–Fock orbital but that all four contracted functions make a significant contribution to the $2s$ orbital. From (8.2.18), we see that the two innermost contracted functions are needed to reproduce the nodal structure of the $2s$ function. In Figure 8.7, the orbitals (8.2.17)–(8.2.19) are compared with the numerical Hartree–Fock orbitals.

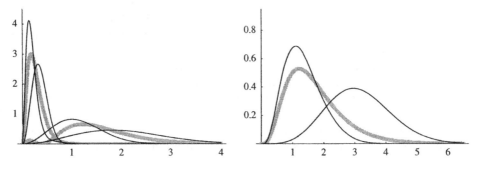

Fig. 8.6. The radial distribution functions of the segmented Dunning–Huzinaga $[4s2p]$ carbon basis (thin lines) superimposed on the numerical ^3P ground-state Hartree–Fock orbitals (thick grey lines) (atomic units).

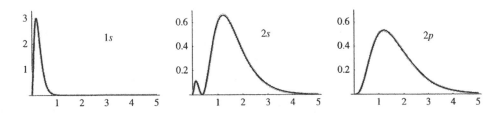

Fig. 8.7. The radial distribution functions of the carbon ^3P ground-state Hartree–Fock orbitals (thick grey lines) expanded in the Dunning–Huzinaga $[4s2p]$ basis (thin lines) (atomic units).

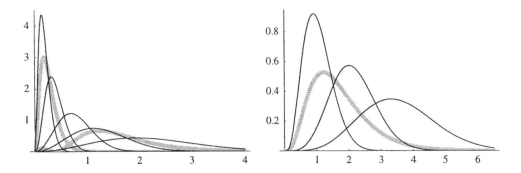

Fig. 8.8. The radial distribution functions of Dunning's segmented $[5s3p]$ contraction of Huzinaga's $(10s6p)$ primitive carbon basis (thin lines) superimposed on the numerical ^3P ground-state Hartree–Fock orbitals (thick grey lines) (atomic units).

For an accuracy higher than that provided by the $[4s2p]$ basis set, we must introduce more contracted functions. Although it is possible to use the $[5s3p]$ contraction in Table 8.3, it is more common to employ a $[5s3p]$ set contracted from a somewhat larger $(10s6p)$ primitive basis [11]. As is apparent from Figure 8.8, the improved flexibility is mainly in the valence or inner valence region.

8.2.6   SIMULTANEOUS OPTIMIZATION OF EXPONENTS AND COEFFICIENTS

The segmented Huzinaga–Dunning sets discussed in Section 8.2.5 are obtained in a two-step procedure. First, a set of primitive functions is obtained by carrying out calculations on atomic systems. Next, a smaller set of contracted functions is generated by taking fixed linear combinations of the primitive atomic functions. At this point, no exponent reoptimization is carried out to ensure that the final segmented basis is optimal for a given number of primitive and contracted functions. Although such a two-step procedure has the advantage of simplicity, it hardly makes optimal use of exponents or coefficients.

A more flexible approach is to optimize the exponents and coefficients simultaneously, obtaining in some sense the best segmented basis set for a fixed number of primitive functions and a fixed number of contracted functions. Several such sets have been designed. Particularly popular are the split-valence *3-21G* [12] and *6-31G* [13] basis sets of Pople and coworkers. (The term *split-valence* indicates that there is a single-zeta representation of the core shell and a double-zeta representation of the valence shell.) In the 6-31G basis, for example, the $1s$ core orbital is described by a single function expanded in six GTOs, whereas the valence $2s$ and $2p$ orbitals are each represented by two contracted functions. Each inner-valence function is expanded in three GTOs and each outer-valence function is represented by a single primitive GTO. As in the STO-3G set, the valence $2s$ and $2p$ orbitals share exponents. This restricted scheme simplifies the nonlinear optimization of the basis set and leads to a faster evaluation of molecular integrals.

In the split-valence sets such as 3-21G and 6-31G, the exponents and coefficients are determined variationally in Hartree–Fock calculations carried out on atomic systems. The resulting exponents and coefficients for the carbon atom are listed in Table 8.4. To make the basis more suitable for molecular applications, the atomic valence exponents in Table 8.4 should be scaled by factors obtained from molecular studies. For the carbon atom, the scaling factors are 1.00^2 and 1.04^2 for the inner and outer valence orbitals, respectively. Thus, the outer valence exponents in Table 8.4 should be multiplied by a factor of 1.0816 for molecular applications. The remaining exponents are unscaled. As expected, the atom most affected by scaling is hydrogen, for which the atomic inner and outer valence orbitals are scaled uniformly by 1.20^2 and 1.15^2, respectively.

The 6-31G carbon basis is displayed in Figure 8.9. Its structure is similar to that of the Huzi-naga–Dunning DZ basis in Figure 8.6, the main difference being that, in the 6-31G basis, a single contracted function is used to represent the $1s$ function.

Table 8.4   The 6-31G carbon basis set. For molecular applications, the outer valence exponents should be uniformly scaled by a factor of $1.04^2 = 1.0816$

Exponents	$1s$	$2s$	$2p$
3047.52	0.00183474		
457.37	0.0140373		
103.949	0.0688426		
29.2102	0.232184		
9.28666	0.467941		
3.16393	0.362312		
7.86827		−0.119332	0.0689991
1.88129		−0.160854	0.316424
0.544249		1.14346	0.744308
0.168714		1.0000	1.0000

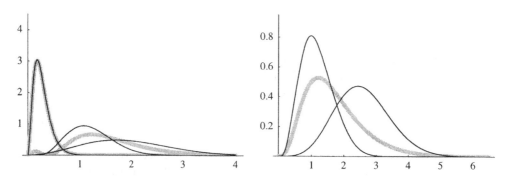

Fig. 8.9.   The radial distribution functions of the 6-31G carbon basis (thin lines) superimposed on the numerical ^3P ground-state Hartree–Fock orbitals (thick grey lines) (atomic units).

8.2.7 POLARIZATION FUNCTIONS

In the basis sets discussed so far, we have included only orbitals of the same symmetries as those of the occupied AOs. However, because of the lower symmetry of molecules compared with that of their constituent atoms, functions of symmetry different from those of the occupied orbitals can contribute to the Hartree–Fock wave function. For instance, in a diatomic molecule, the σ component of the $2p$ orbitals can contribute to the occupied $1\sigma_g$ orbital. These functions of higher angular momentum describe the polarization of the charge distribution upon formation of a chemical bond. To see the polarizing effect of such functions, let $s(\mathbf{A})$ be a $1s$ GTO positioned at $\mathbf{A}$:

$$s(\mathbf{A}) = \exp(-\alpha r_A^2) \tag{8.2.20}$$

A displacement of $s(\mathbf{A})$ by an amount δ_z along the z axis may be written as a Taylor expansion in the displacement:

$$s(\mathbf{A} + \boldsymbol{\delta}_z) = s(\mathbf{A}) + 2\alpha z_A s(\mathbf{A})\delta_z + \cdots \tag{8.2.21}$$

Since a $2p_z$ function at $\mathbf{A}$ is given by

$$p_z(\mathbf{A}) = z_A \exp(-\alpha r_A^2) \tag{8.2.22}$$

we may, to first order in the displacement, write the displaced $1s$ function as a linear combination of the undisplaced $1s$ and $2p_z$ functions:

$$s(\mathbf{A} + \boldsymbol{\delta}_z) = s(\mathbf{A}) + 2\alpha p_z(\mathbf{A})\delta_z + \cdots \tag{8.2.23}$$

Basis functions of angular momentum higher than that of the occupied AOs may thus be used to describe the polarization of the atomic charge distribution in a molecular environment and are called *polarization functions*. The introduction of polarization functions in the basis improves the flexibility in the *angular* part of the wave function. By contrast, going from a minimal to a double-zeta basis, we improve the *radial* flexibility of the basis. In passing, we note that the same flexibility may be achieved by using *floating orbitals* – that is, AOs with positions variationally optimized in space rather than fixed to the atomic nuclei [14,15]. In practice, the usefulness of floating orbitals is restricted by the difficulties encountered in the optimization of their positions and by the fact that, in correlated calculations, higher angular-momentum functions are in any case needed for the description of electron correlation.

Let us consider the numerical values of the exponents for polarization functions. If the polarization functions are to be effective in describing a distortion of the charge density, it is natural to expect that they should contribute most where the charge density of the occupied AOs to be polarized is at its maximum. Thus, a reasonable rule of thumb for choosing polarization exponents for molecular Hartree–Fock calculations is to determine where the radial maximum of the atomic density occurs and to use a polarization-function exponent that gives a similar maximum. Optimization actually leads to somewhat more diffuse functions than this prescription since it is the outer part of the electron density that is most affected by polarization. An alternative, therefore, would be to explore a range of molecules and adjust the polarization-function exponents on the basis of maximum energy lowering. Extensive experience indicates that Hartree–Fock energies are not particularly sensitive to the choice of polarization functions and a wide variety of polarization functions for Hartree–Fock calculations have been proposed.

Let us consider the selection of polarization functions for simple GTOs of the form $1s$, $2p$, $3d$, and so on. The average value of r is given by (6.6.11) with $n = l + 1$ and the position

Table 8.5 The equilibrium geometry of the water molecule optimized with unpolarized and polarized basis sets at the Hartree–Fock level

	6-31G	6-31G*	6-31G**	cc-pV5Z
R_{OH}(pm)	94.96	94.76	94.27	93.96
HOH(°)	111.6	105.6	106.1	106.3

of the maximum in the radial distribution curve by (6.6.12). From these expressions, we see that an uncontracted GTO of angular momentum $l + 1$ has the same expectation value of r or, alternatively, the same position of the maximum as a GTO of angular momentum l and exponent α_l provided its exponent α_{l+1} is given by:

$$\alpha_{l+1}^{\text{ave}} = \left(\frac{2l + 4}{2l + 3} \right)^2 \alpha_l \tag{8.2.24}$$

$$\alpha_{l+1}^{\text{max}} = \frac{l + 2}{l + 1} \alpha_l \tag{8.2.25}$$

Thus, the optimum polarization-function exponent is somewhat larger than the exponent of the function to be polarized, in particular for small quantum numbers – for s orbitals, for example, the expressions (8.2.24) and (8.2.25) give $(16/9)\alpha_0$ and $2\alpha_0$, respectively.

For lighter atoms, the most significant improvement on DZ basis sets is the inclusion of polarization functions. For the Huzinaga–Dunning DZ basis sets, the addition of a single d function yields a *polarized DZ (DZP) basis*. The corresponding set for the hydrogen atom would have a primitive p set added to the original [2s] basis. For the Pople split-valence sets, the addition of a d set on first-row atoms is indicated by an asterisk (6-31G*) and the addition of a d set on first-row atoms and a p set on hydrogen by a double asterisk (6-31G**) [16]. Exponents for the polarization functions have been derived from a variety of sources. Again, the exact numerical values of the polarization exponents are not very critical; in the 6-31G* basis, the same exponent (0.8) is used for all $3d$ polarization functions for first-row atoms.

To see the effect of polarization functions, we have in Table 8.5 listed the equilibrium geometry of the water molecule in the unpolarized 6-31G basis as well as in the polarized 6-31G* and 6-31G** sets. The 6-31G* basis contains a set of d functions with exponent 0.8 on oxygen; in the 6-31G** basis, there is an additional p function with exponent 1.0 on hydrogen. The importance of including polarization functions is evident. For comparison, we have also listed the optimized geometry in a large cc-pV5Z basis (to be described in Section 8.3.3), containing g functions on the hydrogen atoms and h functions on the oxygen atom.

Polarized double-zeta or split-valence basis sets yield Hartree–Fock energies and geometries not far from those obtained with large basis sets. For higher accuracy, larger basis sets such as *triple-zeta (TZ)* or *triple-split valence sets* such as 6-311G provide even better agreement (if augmented with polarization functions) [17]. We defer the discussion of more accurate basis sets to the next section, where we consider basis sets designed for correlated calculations.

8.3   Gaussian basis sets for correlated calculations

So far, we have considered basis sets appropriate for ground-state Hartree–Fock calculations of molecular systems as generated from uncorrelated calculations on atomic systems. For correlated calculations, the basis-set requirements are different and more demanding since we must also provide a *virtual orbital space* capable of recovering a large part of the correlation energy. Clearly, we cannot obtain such basis sets from uncorrelated atomic calculations but must instead consider *correlated* calculations on atomic systems. Techniques for the construction of correlated basis sets are discussed in the present section.

Although an accurate treatment of electron correlation is essential for most molecular applications, it is often sufficient to correlate only the valence electrons – the core electrons are largely unaffected by molecular rearrangements and also by many physical perturbations. Without a great loss of accuracy, we may then significantly reduce the computational cost by not correlating the core electrons. The important separation of *core and valence correlation energies* is considered in Section 8.3.1, before our discussion of the construction and performance of basis sets for correlated calculations.

In Section 8.3.2, we consider the use of *atomic natural orbitals* as basis functions for correlated wave functions. The atomic natural orbitals constitute a conceptually important class of atomic basis functions, useful for systematic investigations of molecular electronic structure at the correlated level. An alternative class of basis functions, perhaps somewhat more useful in practice, is provided by the *correlation-consistent basis sets* discussed in Section 8.3.3 (for valence correlation) and in Section 8.3.4 (for core and valence correlation). The correlation-consistent basis sets are used extensively in the remainder of this book.

8.3.1 CORE AND VALENCE CORRELATION ENERGIES

In atomic and molecular systems, the occupied orbitals can be divided into two broad classes: the core orbitals and the valence orbitals. The *core orbitals* are the innermost orbitals, located close to the atomic nuclei; the *valence orbitals* are the outermost orbitals, spatially more extended than the core orbitals. In the water molecule, for example, the oxygen $1s$ orbital is the core orbital; the $2s$ and $2p$ orbitals on oxygen and the $1s$ orbital on hydrogen constitute the valence orbitals. The atomic valence orbitals overlap considerably at the equilibrium geometry and molecular bonding can be understood as interactions among the electrons occupying these orbitals. The core orbitals, by contrast, are concentrated close to the atomic nuclei and have very small overlaps with the valence orbitals and with the core orbitals on other atoms. Upon molecular bonding, the core orbitals are only slightly modified from their atomic form.

The electronic correlation energy can likewise be partitioned into core and valence contributions. We define the *valence correlation energy* as the correlation energy obtained from calculations where only the valence orbitals are correlated – that is, from calculations where excitations only from the valence orbitals are allowed. The *core correlation energy* is defined as the difference between the *all-electron correlation energy* (obtained by correlating all the electrons) and the valence correlation energy. For the water molecule, the valence correlation energy is obtained by correlating the eight electrons in the valence orbitals, leaving the core orbital *frozen* – that is, doubly occupied in all determinants.

Although, in absolute terms, the core correlation energy may be as large as the valence correlation energy, it can often be neglected in molecular calculations since it is nearly constant across the potential-energy surface and therefore does not affect the calculated equilibrium geometries and dissociation energies significantly. The valence correlation energy, by contrast, is considerably more sensitive to molecular rearrangements and cannot be neglected in accurate calculations.

As an example of the relative importance of the core and valence correlation energies, we consider the ground state of the BH molecule. The Hartree–Fock configuration for this system is

$1\sigma^2 2\sigma^2 3\sigma^2$. The 1σ and 2σ orbitals correspond roughly to the boron $1s$ and $2s$ orbitals, whereas the bonding 3σ orbital is a linear combination of the hydrogen $1s$ orbital and the axial boron $2p$ orbital. The 2σ and 3σ orbitals constitute the valence orbitals and 1σ is the core orbital. On the left in Figure 8.10, we have plotted three potential-energy curves for the BH ground-state system: the uppermost curve represents the Hartree–Fock energy E_{HF}; the next curve represents the energy E_{FCI4} obtained by keeping the core orbital frozen and carrying out an FCI calculation for the four valence electrons; the lower curve represents the energy E_{FCI6} obtained by including all six electrons in the FCI calculation. The electronic energies have been calculated in the cc-pCVTZ basis discussed in Section 8.3.4, which is sufficiently flexible to allow for an accurate description of core and valence correlation energies.

Whereas the Hartree–Fock and the correlated potential-energy curves in Figure 8.10 are noticeably different, the two correlated curves are similar, differing by about the same amount at all internuclear distances. The different geometry dependencies of the core and valence correlation energies are most easily seen from the plot on the right in Figure 8.10, where we have plotted the valence correlation energy (full line) and the core correlation energy (dotted line) separately as functions of the internuclear distance.

For a more quantitative comparison of the energies, we have in Table 8.6, for three internuclear distances of BH, listed the Hartree–Fock energy E_{HF}, the all-electron correlation energy E_{all}^{corr}, the valence correlation energy E_{val}^{corr} and the core correlation energy E_{core}^{corr} calculated as

$$E_{all}^{corr} = E_{FCI6} - E_{HF} \tag{8.3.1}$$

$$E_{val}^{corr} = E_{FCI4} - E_{HF} \tag{8.3.2}$$

$$E_{core}^{corr} = E_{FCI6} - E_{FCI4} \tag{8.3.3}$$

The core correlation energy constitutes about one-third of the all-electron correlation energy, which in turn represents only about 0.6% of the total electronic energy.

When the BH bond is stretched from the equilibrium distance of $2.3289a_0$ [18] to $4.0a_0$, the total electronic energy increases by 77.72 mE$_h$: whereas the Hartree–Fock energy increases by 95.63 mE$_h$, the correlation energy changes by -17.91 mE$_h$. The change in the correlation energy is dominated by the valence correlation energy, which changes by as much as -18.34 mE$_h$; the core contribution increases by 0.43 mE$_h$. The change in the core correlation energy thus represents only 2.4% of the change in the correlation energy and as little as 0.6% of the change in the total

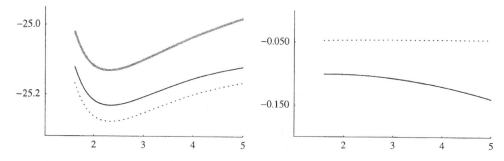

Fig. 8.10. Ground-state potential-energy curves of BH in the cc-pCVTZ basis (in atomic units). On the left, we have plotted the Hartree–Fock energy (thick grey line), the FCI energy with a frozen core (thin full line) and the FCI energy without a frozen core (dotted line). On the right, we have plotted the corresponding valence correlation energy (full line) and core correlation energy (dotted line).

Table 8.6 The Hartree–Fock energy, the all-electron correlation energy (8.3.1), the valence correlation energy (8.3.2), the core correlation energy (8.3.3), the core–valence correlation energy (8.3.4) and the core–core correlation energy (8.3.5) in the ground state of BH (in E_h)

	$R = 2.0a_0$	$R = 2.3289a_0$	$R = 4.0a_0$
E_{HF}	−25.116363	−25.130001	−25.034374
E_{all}^{corr}	−0.146789	−0.147759	−0.165669
E_{val}^{corr}	−0.100445	−0.101723	−0.120064
E_{core}^{corr}	−0.046344	−0.046036	−0.045605
$E_{core-val}^{corr}$	−0.007386	−0.007004	−0.006595
$E_{core-core}^{corr}$	−0.038958	−0.039032	−0.039010

Table 8.7 The equilibrium bond length R_e and the dissociation energy D_e of BH in the electronic ground state calculated in the Hartree–Fock approximation and using FCI wave functions with and without core correlation. The experimental bond length is $2.3289a_0$ [18]. The experimental dissociation energy of 131.0 mE$_h$ is in doubt – see [18] and Exercise 15.1

	$R_e(a_0)$	$D_e(mE_h)$
RHF	2.3080	101.89[a]
FCI (frozen core)	2.3347	132.67
FCI (all electrons)	2.3308	132.90
CICV	2.3302	132.84

[a]The Hartree–Fock energy in the dissociation limit has been obtained from open-shell calculations on the separate atoms.

electronic energy. Finally, when the bond length is shortened from the equilibrium geometry to a distance of $2.0a_0$, the total electronic energy increases by 14.61 mE$_h$, which corresponds to a change of 13.64 mE$_h$ in the Hartree–Fock energy, a small change of 1.28 mE$_h$ in the valence correlation energy and a change of −0.31 mE$_h$ in the core correlation energy.

To illustrate the validity of the *frozen-core approximation* for the calculation of molecular spectroscopic constants, we have in Table 8.7 listed the equilibrium geometry and the dissociation energy for the ground state of BH calculated using the Hartree–Fock wave function and FCI wave functions with and without core correlation. The valence correlation energy alters the equilibrium bond length and in particular the dissociation energy considerably – by $0.0267a_0$ and 30.78 mE$_h$, respectively. The subsequent inclusion of the core correlation energy modifies these results only slightly (i.e. by $-0.0039a_0$ for the bond length and by 0.23 mE$_h$ for the dissociation energy).

The core correlation energy is often partitioned further into two contributions: the *core–core correlation energy* and the *core–valence correlation energy*. Whereas the core–core correlation energy arises from double or higher excitations out of the core orbitals, the core–valence correlation energy arises from single excitations out of the core orbitals coupled with excitations out of the valence orbitals. (In keeping with this terminology, the valence correlation energy is sometimes referred to as the *valence–valence correlation energy*.) For the water molecule, we would calculate the valence–valence correlation energy by not allowing any excitations out of the oxygen $1s$ core orbital, we would include core–valence correlation effects by allowing all excitations that keep the $1s$ core orbital singly occupied, and we would include core–core correlation effects by also allowing double excitations out of the $1s$ core orbital.

Whereas the core–core correlation energy arises from the internal correlation of the electrons in the core region, the core–valence correlation energy arises from the correlation between valence and core electrons. The core–valence correlation energy therefore represents the part of the core correlation energy that is most sensitive to the geometry of the molecule and the chief differential effects of core correlation can thus usually be recovered by including only the core–valence correlation in the calculation (in addition to the valence correlation).

In Table 8.6, we have partitioned the core correlation energy of BH into core–valence and core–core contributions. Denoting the CI energy obtained by including all determinants with at least one electron in the $1s$ core orbital by E_{CICV}, the core–valence and core–core correlation energies are obtained as

$$E_{\text{core–val}}^{\text{corr}} = E_{\text{CICV}} - E_{\text{FCI4}} \tag{8.3.4}$$

$$E_{\text{core–core}}^{\text{corr}} = E_{\text{FCI6}} - E_{\text{CICV}} \tag{8.3.5}$$

We see from the table that, although the core–core correlation energy is numerically larger than the core–valence energy, it is the latter energy that gives the chief differential effects. The core–core correlation energy is particularly insensitive to bond stretching: when the bond is stretched from $2.3289a_0$ to $4.0a_0$, the core–core correlation energy changes by 0.02 mE$_\text{h}$ and the core–valence correlation energy by 0.41mE$_\text{h}$. The dominance of the core–valence correlation energy over the core–core energy is illustrated also in Table 8.7, which gives the equilibrium bond distance and the dissociation energy obtained in the CICV calculations. Clearly, the main core contributions to the bond length and to the dissociation energy are recovered at the core–valence level.

8.3.2 ATOMIC NATURAL ORBITALS

Let us consider how we might construct a suitable set of basis functions for correlated calculations. What is needed for the purpose of carrying out accurate correlated calculations are not only a set of AOs that resemble as closely as possible the occupied orbitals of the atomic systems but also a set of virtual *correlating orbitals* into which the correlated electrons can be excited. An obvious candidate are the canonical orbitals from atomic Hartree–Fock calculations. However, since the lowest virtual canonical orbitals – that is, the canonical orbitals of positive energy close to zero – are very diffuse, they are ill suited for correlating the ground-state electrons (except when the full set of canonical orbitals is used). A better strategy is to generate the correlating orbitals from *correlated* atomic calculations, choosing those orbitals that, in some sense, make the largest contributions to the correlated wave function. This idea leads us to consider the *atomic natural orbitals (ANOs)* – that is, the orbitals obtained by diagonalizing the one-electron density matrix from a correlated atomic calculation – and to select for our basis those correlating orbitals that make an appreciable contribution to the atomic density.

We recall that the *natural orbitals* are the eigenfunctions of the one-electron density matrix and that the associated eigenvalues, known as the *occupation numbers*, represent a measure of the contributions that the different natural orbitals make to the density matrix. Natural orbitals with small occupation numbers can thus be presumed to be less important than those with larger occupations. For a discussion of natural orbitals, see Sections 1.7.1 and 2.7.1.

We now have the following strategy for the generation of basis sets for correlated calculations. First, a Hartree–Fock calculation is carried out on the atomic system in a given primitive Gaussian basis, giving us a set of occupied canonical orbitals as well as a set of virtual canonical orbitals. We retain the occupied canonical orbitals (in order to represent the occupied atomic configuration as closely as possible) but discard the virtual ones. Next, we carry out a correlated atomic calculation in the full canonical basis of the occupied and virtual orbitals and proceed to diagonalize the virtual part of the resulting one-electron density matrix. A basis of correlating orbitals is now obtained by selecting among the virtual natural orbitals those with the largest occupation numbers. Together with the occupied canonical orbitals, the correlating natural orbitals constitute a basis for correlated molecular calculations, known as an *ANO basis* [19,20].

An important advantage of this procedure is that it allows for hierarchical basis-set truncation. For example, for the ^3P state of the carbon atom, we obtain the pattern of occupation numbers in Table 8.8. These numbers were obtained from a four-electron valence CISD calculation involving all single and double excitations from the degenerate Hartree–Fock configurations $2s^2 2p_x^1 2p_y^1$, $2s^2 2p_x^1 2p_z^1$ and $2s^2 2p_y^1 2p_z^1$ in a primitive $(13s8p6d4f2g)$ basis. For occupation numbers larger than 10^{-5}, there is a well-defined hierarchical pattern; by selecting thresholds of 10^{-3}, 10^{-4} and 10^{-5} for inclusion in the basis, we obtain the contracted sets $[3s2p1d]$, $[4s3p2d1f]$ and $[5s4p3d2f1g]$, respectively, where the smaller sets are subsets of the larger ones.

In Figure 8.11, we have depicted the orbitals of the valence carbon $[5s4p3d2f1g]$ ANO basis set. The arrangement of the orbitals is the same as in Table 8.8. The first two orbitals of s symmetry and the first orbital of p symmetry represent the occupied Hartree–Fock orbitals. The correlating orbitals are localized in the same radial regions as the occupied valence orbitals but contain more radial and angular nodes. The nodal structure is similar to that of the hydrogenic functions; see Figure 6.2. The number of nodes thus appears to be a good measure of the importance of a given orbital. For example, the $[3s2p1d]$ set contains all ANOs with at most two nodes, the $[4s3p2d1f]$ set contains all ANOs with at most three nodes, and the $[5s4p3d2f1g]$ set contains all ANOs with at most four nodes. There is thus an appealing simplicity about the valence ANOs – the truncations may be carried out based on the number of nodes, by analogy with the principal expansion of Section 7.5.

Although the ANOs constitute a promising hierarchical set of correlating orbitals, we have not yet investigated their convergence with respect to the total electronic energy. Indeed, for the ANOs to constitute a useful set of correlating orbitals, the electronic energy should converge uniformly towards the energy obtained in the full primitive basis as more and more contracted functions are added. Just as important, most of the correlation energy should be recovered with the smaller sets. In Table 8.9, we have listed the incremental energies ΔE_{nl} recovered by adding successively functions in the order $3s, 3p, 3d, 4s, \ldots, 5f, 5g$. Also listed are the correlation energies ΔE_n recovered by adding all ANOs of the same principal quantum number n.

Table 8.8 The orbital occupation numbers η_{nl} obtained by diagonalizing separately the occupied and virtual parts of the one-electron density matrix, generated in a CISD calculation on the ^3P ground state of the carbon atom using a $(13s8p6d4f2g)$ primitive set

	s	p	d	f	g
η_{1l}	2.000000	–	–	–	–
η_{2l}	1.924675	0.674781	–	–	–
η_{3l}	0.008356	0.004136	0.008834	–	–
η_{4l}	0.000347	0.000331	0.000124	0.000186	–
η_{5l}	0.000021	0.000034	0.000016	0.000011	0.000018

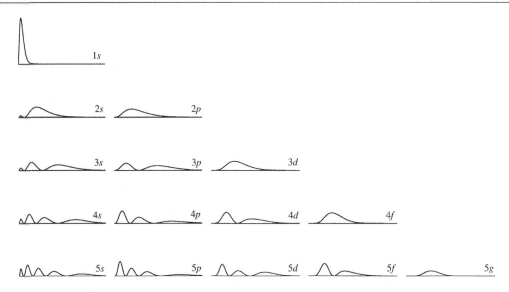

Fig. 8.11. The radial distribution functions of the ANOs for the ground-state carbon atom. The $1s$, $2s$ and $2p$ orbitals correspond to the canonical Hartree–Fock orbitals. The remaining orbitals are valence correlating orbitals, with nodal structures that closely resemble those of the hydrogenic and HO functions. The orbitals are arranged in the same order as in Table 8.8.

Table 8.9 The carbon ground-state correlation energies ΔE_{nl} recovered by adding ANOs in the order $3s, 3p, 3d, 4s, \ldots, 5f, 5g$. ΔE_n is the energy recovered by adding all ANOs of principal quantum number n. All energies are in units of E_h. In the full primitive basis, the Hartree–Fock energy is $-36.688501\ E_h$ and the CISD correlation energy $-98.38\ mE_h$

	s	p	d	f	g	ΔE_n
ΔE_{3l}	-21.570	-19.394	-43.136	$-$	$-$	-84.100
ΔE_{4l}	-0.650	-3.759	-1.833	-4.772	$-$	-11.014
ΔE_{5l}	-0.099	-0.670	-0.421	-0.502	-0.896	-2.588

With the notable exceptions of the $4s$ and $5s$ orbitals, the incremental energies follow the same general pattern as the occupation numbers in Table 8.8. About 85% of the total correlation energy is recovered by the $[3s2p1d]$ basis and as much as 95% of the energy by the $[4s3p2d1f]$ set. It appears that the strategy of selecting correlating orbitals from virtual natural orbitals in order of decreasing occupation number is a sound one, leading to a set of basis functions with excellent convergence characteristics. Note that the incremental energies ΔE_{nl} in Table 8.9 represent the total contributions from all orbitals of the same quantum numbers n and l. If these contributions were divided by the number of orbitals of the same quantum numbers, the similarity with the occupation numbers in Table 8.8 would become even more evident.

As an additional justification for the ANOs, we have calculated, in the full set of canonical Hartree–Fock orbitals but with the occupied orbitals frozen, the MCSCF energies in the valence CISD configuration spaces generated by the ANO sets $[3s2p1d]$, $[4s3p2d1f]$ and $[5s4p3d2f1g]$. The energies $E_{\text{CISD}}^{\text{opt/GTO}}$ obtained in this manner represent the best possible correlation energies in the given orbital basis, providing an interesting set of numbers to compare with the ANO energies; see Table 8.10. The MCSCF and ANO energies are very similar, indicating that the ANO orbitals

Table 8.10 The total energy of the ground-state carbon atom obtained in valence CISD calculations using natural orbitals E_{CISD}, using correlating orbitals optimized in a primitive GTO basis $E_{\text{CISD}}^{\text{opt/GTO}}$ and using numerical correlating orbitals $E_{\text{CISD}}^{\text{opt/num}}$. The energies are given in mE_h relative to the Hartree–Fock energy $E_{\text{HF}} = -37.688501\ E_h$ in the full primitive basis

	$[3s2p1d]$	$[4s3p2d1f]$	$[5s4p3d2f1g]$
$E_{\text{CISD}} - E_{\text{HF}}$	−84.100	−95.114	−97.702
$E_{\text{CISD}}^{\text{opt/GTO}} - E_{\text{HF}}$	−84.209	−95.194	−97.741
$E_{\text{CISD}}^{\text{opt/num}} - E_{\text{HF}}$	−84.220	−95.283	−97.881

are very nearly optimal. As a final test, we also compare $E_{\text{CISD}}^{\text{opt/GTO}}$ with the energies $E_{\text{CISD}}^{\text{opt/num}}$, obtained from a set of fully optimized, numerical correlating orbitals. The latter energies represent the CISD energies obtained by expanding the correlating orbitals in a complete basis. The ANO energies reproduce the complete basis-set energies remarkably well.

An important difference between the ANOs and the basis sets discussed in Section 8.2 is that the ANOs provide us with a natural criterion for selecting basis functions of different symmetries and for increasing the basis set, based on the contribution that each orbital makes to the total atomic one-electron density. Thus, although the ANOs have been constructed with correlated calculations in mind, their design also makes their use appropriate for uncorrelated calculations, resolving some of the ambiguities inherent in the design of basis functions for uncorrelated calculations, in particular with respect to the choice of polarization functions. In this sense, the ANOs represent a significant improvement on the basis sets discussed in Section 8.2.

Some deficiencies and limitations of the ANOs should be noted. First, even the smallest ANO sets are constructed from a large number of primitive orbitals, making their use rather expensive. Second, the ANOs are strongly biased towards reproducing the unperturbed energies of the atomic systems. They do not provide as much flexibility in the valence region as would be desirable in highly accurate molecular calculations. They also lack the flexibility required for accurate calculations of molecular properties, in particular properties that depend on an accurate description of the outer valence region. To overcome these problems, additional flexibility may be introduced by adding more diffuse (uncontracted) primitives to the basis sets and by decontracting some of the diffuse primitives present in the original basis sets [20,21].

8.3.3 CORRELATION-CONSISTENT BASIS SETS

In our discussion of ANOs in Section 8.3.2, we found that the correlating orbitals, selected in order of increasing occupation number, group together according to their principal quantum number and that orbitals belonging to the same group make similar contributions to the energy – providing in some sense a generalization of the principal expansion of the helium wave function in Section 7.5. These observations suggest that it may be worthwhile to try and generate correlating AOs for molecular calculations by relying on the energy criterion alone. By adjusting the exponents of the correlating orbitals so as to maximize their contributions to the correlation energy, we should be able to generate sets of correlating orbitals that are more compact (i.e. contain fewer primitives) than the rather heavily contracted ANOs.

Such an approach is taken in the design of the *correlation-consistent basis sets*, where each correlating orbital is represented by a single primitive Gaussian chosen so as to maximize its contribution to the correlation energy, and where all correlating orbitals that make similar contributions to the correlation energy are added simultaneously [22]. A hierarchy of basis sets may then be set up that is *correlation-consistent* in the sense that each basis set contains all correlating orbitals that lower the energy by comparable amounts as well as all orbitals that lower the energy by larger amounts. It turns out that such a strategy for the design of atomic basis sets gives the same pattern of correlating orbitals as found in the ANOs. The main advantage of the energy criterion over the occupation-number criterion invoked for the ANOs is that it allows us to employ smaller primitive sets.

Let us consider the structure of the correlation-consistent basis sets in more detail. The suite of *correlation-consistent polarized valence basis sets* is denoted by cc-pVXZ, where X is known as the *cardinal number*. The cardinal number is equal to 2 for basis sets of double-zeta quality, 3 for sets of triple-zeta quality, and so on. For a first-row atom, the correlation-consistent polarized valence double-zeta basis set cc-pVDZ with $X = 2$ contains the functions $[3s2p1d]$, the correlation-consistent polarized valence triple-zeta basis cc-pVTZ with $X = 3$ contains the functions $[4s3p2d1f]$, the quadruple set cc-pVQZ contains the functions $[5s4p3d2f1g]$, the quintuple set cc-pV5Z contains the functions $[6s5p4d3f2g1h]$, and so on. It should be emphasized that these basis sets have been designed for the calculation of valence correlation energies only. More flexible correlation-consistent sets are considered in Section 8.3.4.

Each correlation-consistent basis set is built around an accurate representation of the atomic ground-state Hartree–Fock orbitals, obtained by carrying out a general contraction of a primitive Gaussian basis, the size of which increases with the cardinal number – unlike the ANO sets, which all employ the same underlying primitive set. Thus, the cc-pVDZ set contains a $[2s1p]$ contraction of a $(9s4p)$ set for the occupied orbitals as well as a $[1s1p1d]$ set of primitive correlating orbitals. The occupied $[2s1p]$ contraction is determined variationally in an atomic Hartree–Fock calculation and the correlating $[1s1p1d]$ set is then added so as to maximize its contribution to the correlation energy in an atomic valence CISD calculation. It turns out that the optimized s and p exponents are similar to the most diffuse exponents in the primitive set employed for the occupied Hartree–Fock orbitals. To reduce the computational cost, the exponents of the correlating $3s$ and $3p$ orbitals are therefore chosen to be equal to the most diffuse exponents in the primitive set. At the polarized double-zeta level, the underlying primitive set is then $(9s4p1d)$. The radial forms of the cc-pVDZ basis are plotted in Figure 8.12. The occupied $1s$, $2s$ and $2p$ orbitals are faithful representations of the true carbon orbitals with the $2s$ nodal structure well reproduced. By contrast, the correlating $3s$, $3p$ and $3d$ orbitals contain a single primitive function and are nodeless – compare with the corresponding correlating ANOs in Figure 8.11.

The cc-pVDZ basis set contains too few functions to be useful for accurate molecular calculations, although it may be useful for qualitative and semi-quantitative calculations. At the next

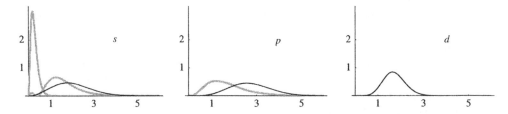

Fig. 8.12. The radial distribution functions of the cc-pVDZ $[3s2p1d]$ basis for carbon (atomic units). The occupied orbitals are depicted by thick grey lines and the correlating orbitals by thin lines.

level, the cc-pVTZ set of composition $[4s3p2d1f]$ is built around a $[2s1p]$ contraction of a $(10s5p)$ primitive set with a $[2s2p2d1f]$ set of orbitals added for correlation; see Figure 8.13. The underlying primitive set is $(10s5p2d1f)$. Comparing Figures 8.12 and 8.13, we note that, for each correlating cc-pVDZ orbital, there is a pair of bracketing cc-pVTZ orbitals of the same symmetry. In addition, a set of f functions is introduced at the triple-zeta level.

For many purposes, the cc-pVTZ is sufficiently flexible for an accurate representation of the electronic wave function. Nevertheless, there are situations where still higher accuracy is required. In such cases, the cc-pVQZ basis in Figure 8.14 and the cc-pV5Z basis in Figure 8.15 may be used. These contracted $[5s4p3d2f1g]$ and $[6s5p4d3f2g1h]$ sets contain the primitive sets $(12s6p3d2f1g)$ and $(14s8p4d3f2g1h)$, respectively.

Correlation-consistent basis sets have been developed for atoms other than those of the first row. In Table 8.11, we summarize the characteristics of the correlation-consistent basis sets for

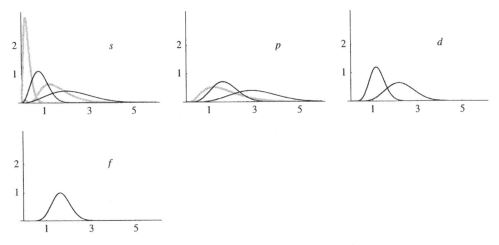

Fig. 8.13. The radial distribution functions of the cc-pVTZ $[4s3p2d1f]$ basis for carbon (atomic units). The occupied orbitals are depicted by thick grey lines and the correlating orbitals by thin lines.

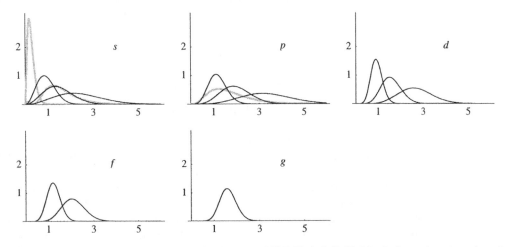

Fig. 8.14. The radial distribution functions of the cc-pVQZ $[5s4p3d2f1g]$ basis for carbon (atomic units). The occupied orbitals are depicted by thick grey lines and the correlating orbitals by thin lines.

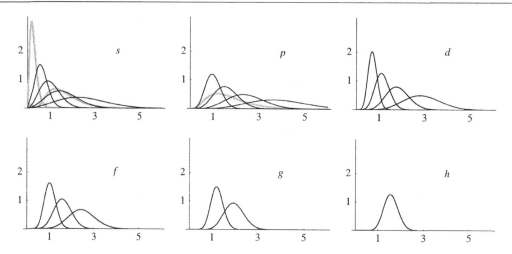

Fig. 8.15. The radial distribution functions of the cc-pV5Z $[6s5p4d3f2g1h]$ basis for carbon (atomic units). The occupied orbitals are depicted by thick grey lines and the correlating orbitals by thin lines.

Table 8.11 The structure of the correlation-consistent polarized valence basis sets

	H–He		B–Ne		Al–Ar	
cc-pVDZ	$[2s1p]$	5	$[3s2p1d]$	14	$[4s3p1d]$	18
cc-pVTZ	$[3s2p1d]$	14	$[4s3p2d1f]$	30	$[5s4p2d1f]$	34
cc-pVQZ	$[4s3p2d1f]$	30	$[5s4p3d2f1g]$	55	$[6s5p3d2f1g]$	59
cc-pV5Z	$[5s4p3d2f1g]$	55	$[6s5p4d3f2g1h]$	91	$[7s6p4d3f2g1h]$	95

the atoms H and He, for the first-row atoms B to Ne [22], and for the second-row atoms Al to Ar [23]. For all atoms, there is a single-zeta representation of the core orbitals and a multiple-zeta representation of the valence orbitals. The basis sets for the second-row atoms are therefore not much bigger than those for the first-row atoms. The reader may verify that the number of contracted functions in the cc-pVXZ sets increases as the third power of the cardinal number according to the expressions

$$N_V^0(X) = \tfrac{1}{3}X\left(X + \tfrac{1}{2}\right)(X + 1) \tag{8.3.6}$$

$$N_V^1(X) = \tfrac{1}{3}(X + 1)\left(X + \tfrac{3}{2}\right)(X + 2) \tag{8.3.7}$$

$$N_V^2(X) = \tfrac{1}{3}(X + 1)\left(X + \tfrac{3}{2}\right)(X + 2) + 4 \tag{8.3.8}$$

In later chapters, we shall see that, for a given molecular system, the computational cost of the more accurate standard models scales as the fourth power of the number of correlating orbitals, indicating that the cost of accurate molecular calculations should scale as X^{12}.

To illustrate the performance of the cc-pVXZ sets, we have, for cardinal numbers $X \leq 5$, calculated the Hartree–Fock and CISD energies of the ^3P ground state of the carbon atom; see Table 8.12. The correlation-energy contributions compare quite favourably with those of the ANOs in Table 8.10, keeping in mind that, for small cardinal numbers, the ANOs are constructed from much larger primitive sets. Thus, from the point of view of energy contributions at least, the use of primitive rather than contracted correlating orbitals appears to be fully justified. Also,

Table 8.12 The Hartree–Fock energy E_{HF} and the CISD correlation energy $E_{CISD} - E_{HF}$ for the ^3P ground state of the carbon atom calculated using correlation-consistent basis sets. All energies are given in atomic units. The Hartree–Fock basis-set limit is $-37.688619\ E_h$ and the estimated CISD correlation-energy basis-set limit is $-99.7 \pm 0.4\ mE_h$

	cc-pVDZ	cc-pVTZ	cc-pVQZ	cc-pV5Z
E_{HF}	−37.682391	−37.686662	−37.688234	−37.688571
$E_{CISD} - E_{HF}$	−0.077613	−0.093252	−0.097280	−0.098545

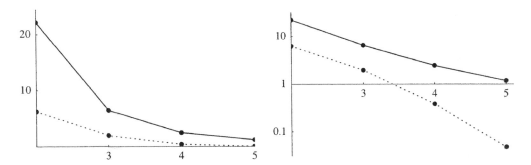

Fig. 8.16. Errors (in mE_h) relative to the basis-set limit in the Hartree–Fock energy (dotted line) and in the valence CISD correlation energy (full line) for the ^3P ground state of the carbon atom calculated using correlation-consistent basis sets cc-pVXZ. On the left, we have plotted the errors on a linear scale as a function of the cardinal number X; on the right, we have plotted the errors on a logarithmic scale.

since the ANO energies in Table 8.10 are close to the energies in a numerical set of correlating orbitals, we may more generally conclude that GTOs are as good as STOs for this particular purpose. In short, there appears to be no compelling reason to employ Slater-type correlating orbitals in molecular calculations.

In Figure 8.16 we have, for the correlation-consistent basis sets, plotted the error in the Hartree–Fock energy relative to the basis-set limit of $-37.688619\ E_h$ as well as the error in the CISD correlation energy relative to an estimated basis-set limit of $-99.7 \pm 0.4\ mE_h$. For a given cardinal number, the error in the Hartree–Fock energy relative to the basis-set limit is considerably smaller than the corresponding error in the correlation energy. There is a significant improvement in both descriptions as we go from the cc-pVDZ level to the cc-pVTZ level, where the errors in the Hartree–Fock and correlation energies are approximately 2 and 6 mE_h, respectively. At the cc-pV5Z level, the error in the correlation energy is approximately 1 mE_h and the Hartree–Fock error is only 0.05 mE_h. Clearly, the energies are quite well converged at this level. It should be noted, however, that these numbers are particular to the carbon atom. For other atoms, the errors may deviate considerably from those in the carbon atom although the overall behaviour will be the same.

The carbon calculations demonstrate that, for atoms at least, the correlation-consistent basis sets perform quite well. As the logarithmic plots in Figure 8.16 indicate, the errors are reduced by several factors with each increment in the cardinal number, although the progression is not uniform – the Hartree–Fock curve bends downwards and the correlation curve upwards as the cardinal number increases. The Hartree–Fock error is thus reduced by a factor of 3.2 going from cc-pVDZ to cc-pVTZ, by a factor of 5.1 going to cc-pVQZ, and by a factor of 8.0 going to cc-pV5Z; the corresponding factors for the correlation energy are 3.4, 2.6 and 2.1. The difficulty experienced in converging the dynamical correlation energy for large cardinal numbers is related to the presence of the Coulomb hole, as discussed in Chapter 7.

The calculations presented here suggest that the correlation-consistent basis sets – which have been specifically designed for correlated calculations – perform quite well also for uncorrelated energies. It should be pointed out, however, that, since the preceding calculations are atomic, only the s and p parts of the correlation-consistent basis sets have been put to test in the Hartree–Fock calculations. A more interesting and sensitive test of the correlation-consistent basis sets would be to carry out calculations on molecular systems. Such calculations are discussed in Section 8.4.

8.3.4 EXTENDED CORRELATION-CONSISTENT BASIS SETS

The correlation-consistent basis sets described in Section 8.3.3 have been designed for one particular purpose: the accurate calculation of valence-correlated wave functions of ground-state neutral systems. The cc-pVXZ basis sets therefore do not have the flexibility required either for the investigation of core correlation discussed in Section 8.3.1 or for the study of anions and excited states with diffuse electron distributions. For such applications, additional AOs must be introduced. In the present subsection, we shall first discuss the *correlation-consistent polarized core–valence sets* cc-pCVXZ [24], where the standard cc-pVXZ sets have been extended for additional flexibility in the core region, and next the *augmented correlation-consistent basis sets* aug-cc-pVXZ and aug-cc-pCVXZ [25], where diffuse functions have been added so as to improve the flexibility in the outer valence region.

For the description of core correlation, functions are needed with radial maxima close to the atomic nuclei. In the core–valence sets cc-pCVXZ [24], the standard valence sets cc-pVXZ are augmented with tight correlating orbitals, with exponents selected to maximize the magnitude of the core correlation energy

$$E_{\text{core}}^{\text{corr}} = E_{\text{CISD}} \text{ (all electrons) } - E_{\text{CISD}} \text{ (frozen core)} \tag{8.3.9}$$

The correlating orbitals are introduced in a correlation-consistent manner – that is, all functions with similar contributions to the correlation energy are introduced simultaneously. This strategy results in a set of $[1s1p]$ core-correlating orbitals for augmentation of the double-zeta set, giving rise to the cc-pCVDZ set of composition $[4s3p1d]$. Proceeding in the same manner for cc-pVTZ, we obtain a set of $[2s2p1d]$ core-correlating orbitals and arrive at the cc-pCVTZ set of composition $[6s5p3d1f]$.

The polarized core–valence correlation-consistent basis sets for first-row atoms are listed in Table 8.13. We note that the same pattern is followed for the core–valence functions of the first-row atoms as for the valence functions of helium in Table 8.11. The number of functions in the core–valence sets may therefore be calculated from the cardinal number as

$$N_{\text{CV}}^1(X) = N_{\text{V}}^1(X) + N_{\text{V}}^1(X-1) - 1 \tag{8.3.10}$$

The core–valence sets for the first-row atoms are seen to be considerably larger than the valence sets and, in the limit of large cardinal numbers, the inclusion of core–valence functions increases the basis set by a factor of 2. The calculation of core-correlation effects is clearly an expensive undertaking, in particular when we consider the small effects that core correlation often has on the calculated molecular properties as discussed in Section 8.3.1.

Table 8.13 The structure of the cc-pVXZ, cc-pCVXZ and aug-cc-pVXZ sets for the first-row atoms. For the cc-pCVXZ and aug-cc-pVXZ sets, only the core-correlating and diffuse orbitals are listed, respectively. N_V, N_{CV} and N_{aug} are the total numbers of contracted orbitals in the basis sets

	cc-pVXZ	N_V	cc-pCVXZ	N_{CV}	aug-cc-pVXZ	N_{aug}
D	$[3s2p1d]$	14	$+1s1p$	18	$+1s1p1d$	23
T	$[4s3p2d1f]$	30	$+2s2p1d$	43	$+1s1p1d1f$	46
Q	$[5s4p3d2f1g]$	55	$+3s3p2d1f$	84	$+1s1p1d1f1g$	80
5	$[6s5p4d3f2g1h]$	91	$+4s4p3d2f1g$	145	$+1s1p1d1f1g1h$	127

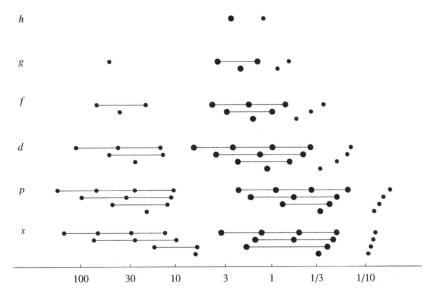

Fig. 8.17. Gaussian exponents of the carbon correlation-consistent basis sets of cardinal numbers 2–5 on a logarithmic scale with tight functions to the left and diffuse functions to the right. The exponents of the valence-correlating orbitals present in the cc-pVXZ root sets are located in the middle and are plotted using larger dots, the exponents of the core-correlating orbitals are located on the left, and the diffuse functions of the augmented aug-cc-p(C)VXZ sets are located on the right. For each angular momentum, we have plotted the exponents for cardinal numbers $X = 2$ at the bottom and $X = 5$ at the top.

In Figure 8.17 we have plotted the Gaussian exponents of the carbon correlation-consistent basis sets on a logarithmic scale. In this plot, the exponents that contribute only to the Hartree–Fock orbitals have not been included. The exponents of the cc-pVXZ basis sets are located in the middle (valence) region and the exponents of the core-correlating orbitals on the left. We note the orderly development of the exponents for d and higher angular-momentum functions. For large cardinal numbers, there is an almost even-tempered distribution of exponents throughout the core and valence regions. For small cardinal numbers, there is a noticeable gap in the core–valence region (except for the somewhat unusual behaviour of the s functions in the cc-pCVTZ basis). The largest spread in the exponents is found in the p sets.

In Table 8.14, we have listed the core–core, core–valence and valence–valence correlation energies of the ground-state carbon atom calculated using the basis sets cc-pVXZ and cc-pCVXZ. There are only small differences in the valence–valence correlation energies for the two basis sets – as we would expect since the cc-pVXZ sets have been designed specifically with the

Table 8.14 The core–core, core–valence and valence–valence correlation energies (in mE_h) for the ^3P ground state of the carbon atom calculated using the cc-p(C)VXZ basis sets

	Core–core	Core–valence	Valence–valence
cc-pVDZ	−0.276	−0.947	−77.613
cc-pVTZ	−4.704	−3.856	−93.252
cc-pVQZ	−17.252	−7.479	−97.280
cc-pV5Z	−20.937	−9.143	−98.545
cc-pCVDZ	−30.297	−4.948	−78.900
cc-pCVTZ	−37.061	−9.509	−94.300
cc-pCVQZ	−39.288	−11.513	−97.736
cc-pCV5Z	−40.052	−11.987	−98.736

valence correlation energy in mind. The core–core correlation energy, by contrast, is very poorly described by the valence sets, even at the quintuple-zeta level. Clearly, in any application where core–core correlation is important, the cc-pVXZ sets are inadequate and the larger and more expensive cc-pCVXZ sets must be used. For the core–valence correlation energy, the situation is slightly different and, for high cardinal numbers, the valence basis sets are able to recover a sizeable amount of the core–valence energy – about 76% at the quintuple-zeta level. In some applications, this amount may be sufficiently large to make the use of the core–valence basis sets unnecessary.

We now turn our attention to the outer, diffuse regions of the electronic system. The valence and core–valence basis sets considered so far are inadequate for the description of the diffuse electron distributions characteristic of anionic systems and excited states. In addition, these sets do not have the flexibility required for a proper description of interactions with external electric fields and hence the accurate calculation of dipole moments and polarizabilities. For such calculations, additional functions must be added in the outer valence region. Within the framework of correlation-consistent basis sets, we proceed by adding primitive functions to the standard cc-pVXZ sets, with exponents adjusted so as optimize the energy of atomic anions. Diffuse functions are added in groups, with one set of functions for each angular momentum present in the root set. This procedure leads to the *augmented correlation-consistent polarized valence basis sets* aug-cc-pVXZ [25], the composition and size of which are listed in Table 8.13. The number of functions in the aug-cc-pVXZ sets may be calculated as

$$N^1_{aug}(X) = N^1_V(X) + (X+1)^2 \tag{8.3.11}$$

and the number of diffuse functions therefore increases quadratically with the cardinal number. The exponents of the diffuse functions (typically 0.1 or smaller) are located on the right in the diagram in Figure 8.17. Sometimes, functions that are even more diffuse are needed; these are included in the doubly and triply augmented d-aug-cc-pVXZ and t-aug-cc-pVXZ basis sets [26]. When both diffuse and tight functions are needed, the aug-cc-pCVXZ sets may be used.

In Table 8.15, we have listed the Hartree–Fock energies and the CISD correlation energies obtained using the cc-pVXZ and aug-cc-pVXZ basis sets for the neutral carbon atom in the ^3P ground state and for the more diffuse anionic carbon atom in the ^4S ground state. Augmentation has little effect on the neutral atom but is essential for the anionic system. We note in passing that it is important to add whole shells of diffuse functions rather than just sp sets. For example, successive additions of diffuse s, p, d and f functions to the cc-pVTZ basis give the incremental correlation-energy contributions of −1.3, −2.4, −2.0 and −1.3 mE_h, respectively, for the carbon

Table 8.15 The Hartree–Fock energy and the CISD valence correlation energy for the ^3P ground state of the carbon atom and for the ^4S ground state of the carbon anion calculated using the cc-pVXZ and aug-cc-pVXZ basis sets (atomic units). The Hartree–Fock basis-set limits for the neutral and anionic atoms are -37.688619 and -37.708844 E_h, respectively. The estimated CISD basis-set limits are -99.7 ± 0.4 mE_h and -120.7 ± 0.4 mE_h, respectively

		(aug-)cc-pVDZ	(aug-)cc-pVTZ	(aug-)cc-pVQZ	(aug-)cc-pV5Z
E_{HF}	cc	-37.682391	-37.686662	-37.688234	-37.688571
	aug-cc	-37.683071	-37.686777	-37.688256	-37.688573
E_{corr}	cc	-0.077613	-0.093252	-0.097280	-0.098545
	aug-cc	-0.081283	-0.094162	-0.097571	-0.098652
E_{HF}^{-}	cc	-37.678694	-37.695083	-37.702274	-37.706499
	aug-cc	-37.703612	-37.707024	-37.708496	-37.708798
E_{corr}^{-}	cc	-0.083526	-0.107474	-0.114742	-0.117654
	aug-cc	-0.099528	-0.114530	-0.118325	-0.119520

anion. Clearly, the diffuse functions of each angular momentum make similar contributions to the correlation energy.

8.4 Basis-set convergence

In the present section, we investigate the basis-set convergence of the uncorrelated Hartree–Fock model and the correlated MP2 and CCSD models. Using a variety of basis sets, we calculate the electronic energies of Ne, N_2 and H_2O at geometries that correspond to the experimental equilibrium structures in Tables 15.3 and 15.8 ($R_{NN} = 109.77$ pm for N_2; $R_{OH} = 95.72$ pm and $\theta_{HOH} = 104.52°$ for H_2O). At the correlated level, some calculations are carried out also for the helium atom and the BH molecule. Although we shall be concerned mainly with the electronic energy, we shall examine also the equilibrium structures of N_2 and H_2O. The calculations in this section should give the reader some indication of the performance of the standard basis sets and their convergence for molecular systems, recalling that the previous examples in this chapter were concerned mostly with atomic calculations.

Because of the different basis-set requirements for correlated and uncorrelated calculations, we examine the convergence of the Hartree–Fock and correlation energies separately, in Sections 8.4.1 and 8.4.2. Next, in Section 8.4.3, the extrapolation of the correlation energy is discussed. Finally, in Section 8.4.4, we consider the binding energy. Basis-set convergence is also studied in Section 8.5 (for van der Waals systems) and in Chapter 15, in the broader context of a critical assessment of the standard models of quantum chemistry.

8.4.1 BASIS-SET CONVERGENCE OF THE HARTREE–FOCK MODEL

The Hartree–Fock energies and structures are listed in Table 8.16. The energies are given relative to the basis-set limit. For Ne and N_2, these limits have been obtained from numerical calculations. For H_2O, we have used the cc-pV6Z energy, which should be within 0.1 mE_h of the basis-set limit. Note that the tabulated energies have all been obtained at the experimental rather than optimized geometries.

The STO-3G errors are very large. The electronic energies are $1-2$ E_h too high and the NN bond distance is overestimated by 40 pm. Although this particular bond may represent a rare case of poor performance (noting that, for the π orbitals, there is only one basis function of each

Table 8.16 The basis-set convergence in Hartree–Fock calculations on Ne, N_2 and H_2O. For the electronic energy, the errors relative to the Hartree–Fock limit at the experimental equilibrium geometry are listed (in mE_h). For the molecular structures of N_2 and H_2O, the calculated bond distances (in pm) and bond angle (in degrees) are listed

Basis set	ΔE_{HF}			Geometrical parameters		
	Ne[a]	N_2^b	H_2O^c	R_{NN}	R_{OH}	θ_{HOH}
STO-3G	1942.57	1497.29	1104.47	146.82	98.94	100.03
6-31G	73.22	125.43	83.40	108.91	94.96	111.55
6-311G	24.54	99.02	58.01	108.60	94.54	111.88
6-31G*	73.22	51.32	58.27	107.81	94.76	105.58
6-31G**	73.22	51.32	44.75	107.81	94.27	106.05
6-311G**	24.54	23.76	20.95	107.03	94.10	105.46
cc-pVDZ	58.32	39.06	40.60	107.73	94.63	104.61
cc-pVTZ	15.23	9.72	10.23	106.71	94.06	106.00
cc-pVQZ	3.62	2.11	2.57	106.56	93.96	106.22
cc-pV5Z	0.32	0.43	0.31	106.54	93.96	106.33
cc-pCVDZ	58.17	38.27	40.20	107.65	94.60	104.64
cc-pCVTZ	15.14	8.79	10.04	106.60	94.05	106.00
cc-pCVQZ	3.52	1.88	2.45	106.55	93.96	106.22
cc-pCV5Z	0.32	0.36	0.30	106.54	93.96	106.33

[a]Numerical Hartree–Fock energy: -128.547094 E_h.
[b]Numerical Hartree–Fock energy: -108.9931881 E_h.
[c]cc-pV6Z Hartree–Fock energy: -76.067401 E_h.

symmetry and therefore no variational flexibility), it is clear that the minimal STO-3G basis is too inflexible to be useful except in preliminary investigations on large systems.

In view of the poor performance of the STO-3G basis, the improvements at the unpolarized split-valence 6-31G level are quite striking. The errors in the energies are reduced by a factor of 10–30 and the bond distances are only 1–2 pm longer than the Hartree–Fock limit. Still, a large error remains in the HOH bond angle, which is now 111.6° – about 5° larger than the Hartree–Fock limit of 106.3°. The bond angle is not improved by going to the larger 6-311G basis, which improves only the atomic valence description. For molecular systems, therefore, some energetic but no structural improvement is observed as we go from 6-31G to 6-311G.

As discussed in Section 8.2.7, the correct description of molecular bonding requires the use of polarization functions. When polarization functions are added to oxygen at the 6-31G* level and finally to the hydrogens at the 6-31G** level, the HOH bond angle becomes 105.6° and 106.1°, respectively. The bond distances are likewise improved by the polarization functions: at the 6-31G** level, the distances are in error by 1.3 pm for the NN bond and by 0.3 pm for the OH bond. The error in the energy is reduced by a factor of 2 when polarization functions are added.

The polarized split-valence sets represent the first level of description where the agreement with the Hartree–Fock limit is reasonable. For quantitative agreement, however, still larger sets are needed. At the 6-311G** level, the error in the electronic energy is reduced by a factor of about 2 (relative to 6-31G**) and the bond distances are now in error by only 0.5 pm for the NN bond and 0.1 pm for the OH bond (relative to the Hartree–Fock limit). The improvement is not uniform, however, since the 6-311G** HOH bond angle is in error by $-0.8°$, compared with the 6-31G** error of $-0.3°$. What is needed is a sequence of basis sets that allows us to approach the Hartree–Fock limit in a more systematic manner. We therefore turn our attention to the correlation-consistent basis sets.

The first point to note about the correlation-consistent basis sets in Table 8.16 is that the convergence is in all cases uniform and systematic – for the energies, for the bond distances, and for the bond angle. Scrutiny of the table reveals that, with each increment in the cardinal number, all errors are reduced by a factor of at least 3 or 4. Clearly, the correlation-consistent basis sets provide a convenient framework for the quantitative study of molecular systems at the Hartree–Fock level. We also note that the results for the cc-pVXZ and cc-pCVXZ basis sets are very similar. Apparently, the molecular core orbitals are quite atom-like and unpolarized by chemical bonding. In Hartree–Fock calculations, therefore, the use of the smaller valence cc-pVXZ sets is recommended.

Concerning the geometrical parameters in Table 8.16, we note that the bond distances (NN and OH) converge (monotonically from above) somewhat faster than the energy (with reduction factors of 7 and more with each increment in the cardinal number). At the cc-pVDZ level, the error in the bond distance is about 1 pm, at the cc-pVTZ level it is reduced to 0.1 pm, and at the cc-pVQZ level the error is only 0.01 pm. The HOH bond angle converges more slowly than the distances, the corresponding errors being $-1.7°$, $-0.3°$ and $-0.1°$. With regard to geometries as well as energies, therefore, we conclude that the cc-pVDZ basis is sufficient for qualitative studies (at the Hartree–Fock level) and that the cc-pVQZ basis gives properties very close to the Hartree–Fock limit.

Comparing the correlation-consistent basis sets with those discussed at the beginning of this subsection, we find that the cc-pVDZ set performs slightly better than the 6-31G** set. These sets contain the same number of contracted functions, but have been constructed differently. In the cc-pVDZ basis, the contraction is general and the exponents different for the s and p functions. By contrast, in the 6-31G** basis, the contraction is segmented and the exponents are shared among the s and p functions; also, the number of primitives is smaller. These restrictions lead to a basis that performs slightly worse than the cc-pVDZ set but gives a faster evaluation of the molecular integrals.

For the correlation-consistent basis sets, the error in an *atomic* ground-state Hartree–Fock energy is determined solely by the quality of the primitive set since the AOs spanned by these primitives are contained (as contracted functions) in the basis. In other words, there are no contraction errors in such energies. The convergence for the neon atom in Table 8.16 therefore reflects the quality of the primitive set and is typical of first-row atoms although, for a given cardinal number, the magnitude of the error increases with the atomic number – compare, for example, the errors for the carbon atom in Table 8.15 and for the neon atom in Table 8.16. Conversely, in *molecular* correlation-consistent calculations, the molecular bonding is described entirely by means of the correlating orbitals. Although introduced to describe atomic correlation, these orbitals also perform well for the description of charge polarization and chemical bonding – as indicated by the errors in the Hartree–Fock energies of N_2 and H_2O in this subsection and more explicitly demonstrated by the errors in the binding energies in Section 8.4.4.

8.4.2 BASIS-SET CONVERGENCE OF CORRELATED MODELS

A particularly useful system for exploring basis-set convergence of correlation energies is the ground-state helium atom, for which the Hylleraas expansion (7.3.19) provides, for all practical purposes, an exact solution to the Schrödinger equation. Indeed, this system was extensively studied in Chapter 7, in our discussion of the short-range electron correlation problem. In the present subsection, we begin our discussion of basis-set convergence by reinvestigating this simple and transparent system so as to benchmark the performance of the standard correlation-consistent basis sets cc-pVXZ, preparing us for the larger electronic systems considered later in this subsection.

In Section 7.5, we investigated the principal expansion of the correlation energy in the helium atom. The numerical orbitals generated by this expansion have the same composition and nodal structure as the correlation-consistent basis sets cc-pVXZ. In the present subsection, we shall refer to the numerical orbitals of the principal expansion as the *numerical correlation-consistent basis sets n-cc-pVXZ*, by analogy with the standard *analytical correlation-consistent basis sets cc-pVXZ*. By comparing the analytical and numerical basis sets, we should obtain some impression of the performance of the standard cc-pVXZ sets and in particular of their deficiencies *vis-à-vis* a fully optimized set of correlating orbitals.

In Table 8.17 the correlation energies obtained with the cc-pVXZ and n-cc-pVXZ sets are listed for $X \leq 5$. For these cardinal numbers, the numerical sets recover 86–99% of the correlation energy, whereas the analytical sets recover 77–99%. Thus, at the double-zeta level, the numerical n-cc-pVDZ set recovers about 8.5% more of the energy than does the corresponding analytical set; for higher cardinal numbers, the differences are 2.6%, 0.7% and 0.2%. The cc-pVXZ sets therefore recover nearly all the correlation energy available in the principal expansion beyond the triple-zeta level.

At the quintuple-zeta level, as much as 99% of the ground-state correlation energy is recovered by the correlation-consistent basis sets. Referring back to Figure 7.6 (where the convergence of the n-cc-pVXZ basis sets is plotted), it is clear that we should not be too enthusiastic about this result since further substantial reductions are almost impossible by means of orbital expansions. Also, we shall soon see that, in molecules containing heavier elements, a smaller proportion of the total correlation energy is recovered for a given cardinal number.

The performance of the correlation-consistent basis sets is further illustrated in Figure 8.18, where, for the analytical and numerical cc-pVXZ and n-cc-pVXZ sets, we have plotted the ground-state helium wave function with one electron fixed at a distance of $0.5a_0$ from the nucleus and with the other electron confined to a circle of radius $0.5a_0$ about the nucleus. For comparison with the correlation-consistent basis sets, the Hylleraas and Hartree–Fock wave functions are plotted as well.

At the double-zeta level, the helium wave function is poorly represented by the correlation-consistent basis sets, in particular in the region of the Coulomb hole, where the amplitude is strongly overestimated. Although the numerical double-zeta basis performs somewhat better than its analytical counterpart, the description of the short-range correlation is still poor. Some improvement is observed at the triple-zeta level, although, inside the Coulomb hole, the description of the correlation-consistent basis sets still differs appreciably from that of the Hylleraas function. On the other hand, the difference between the numerical and analytical descriptions is considerably smaller at the triple-zeta level, being most conspicuous in the long-range region.

At the higher quadruple- and quintuple-zeta levels, the description of the helium ground state appears to be quite accurate everywhere except in the cusp region – for the analytical as well as the

Table 8.17 The correlation energies recovered by the analytical and numerical correlation-consistent basis sets for the ground-state helium atom. The exact correlation energy is -42.044 mE$_h$

	cc-pVXZ		n-cc-pVXZ	
X	E_{corr}(mE$_h$)	E_{corr} (%)	E_{corr}(mE$_h$)	E_{corr} (%)
2	-32.4	77.1	-36.0	85.6
3	-39.1	93.0	-40.2	95.6
4	-40.9	97.3	-41.2	98.0
5	-41.5	98.7	-41.6	98.9

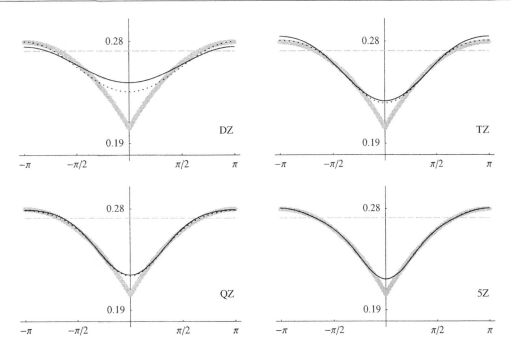

Fig. 8.18. The ground-state helium wave function plotted on a circle of radius $0.5a_0$ about the nucleus with one electron fixed at the origin of the plot. The full and dotted black lines represent the wave functions generated by the analytical and numerical correlation-consistent basis sets cc-pVXZ and n-cc-pVXZ, respectively; the full and dashed grey lines represent the Hylleraas and Hartree–Fock functions, respectively.

numerical basis sets. The only noticeable difference between cc-pVQZ and cc-pV5Z is the depth of the Coulomb hole, which is slightly greater at the quintuple-zeta level. We may thus conclude our discussion of the helium atom by stating that, for large cardinal numbers, the correlation-consistent basis sets provide an almost optimal description of short-range correlation – that is, within the framework of the principal expansion.

To examine the convergence of correlation energies for larger electronic systems, we have calculated the MP2 and CCSD energies of Ne, N_2 and H_2O at the experimental equilibrium geometries for a variety of basis sets; see Table 8.18. In addition, we have listed in Table 8.19 the optimized MP2 geometries of N_2 and H_2O for the same basis sets. We shall first be concerned mainly with the convergence of the correlation energies, returning to the geometries at the end of this subsection.

Table 8.18 contains a set of energies labelled 'extrapolated' as well as a set of energies labelled 'R12'. The R12 energies, which have been obtained from explicitly correlated MP2 and CCSD calculations similar to the CI-R12 calculations in Section 7.3.2, should be close to the basis-set limit [27]. The extrapolated energies in Table 8.18 have been obtained from the quintuple- and sextuple-zeta energies using the extrapolation technique described in Section 8.4.3.

As the basis-set requirements in correlated calculations are more severe than those in uncorrelated calculations, some of the smaller basis sets considered for the Hartree–Fock wave function in Section 8.4.1 are omitted here. In particular, it is pointless to calculate correlation energies using basis sets smaller than polarized valence double-zeta sets. To illustrate this point, we have in Tables 8.18 and 8.19 given the results for one particular basis set that is too small for correlated

Table 8.18 The convergence of the MP2 and CCSD correlation energies (in mE_h) for Ne, N_2 and H_2O at the experimental equilibrium geometries

Electrons correlated	Basis set	Ne		N_2		H_2O	
		MP2	CCSD	MP2	CCSD	MP2	CCSD
valence	6-31G	−113.4	−114.3	−236.4	−225.8	−127.8	−134.4
	6-31G**	−150.3	−152.2	−305.3	−308.3	−194.6	−203.8
	6-311G**	−209.0	−210.6	−326.4	−326.3	−217.4	−224.9
	cc-pVDZ	−185.5	−189.0	−306.3	−309.3	−201.6	−211.2
	cc-pVTZ	−264.3	−266.3	−373.7	−371.9	−261.5	−267.4
	cc-pVQZ	−293.6	−294.7	−398.8	−393.1	−282.8	−286.0
	cc-pV5Z	−306.2	−305.5	−409.1	−400.6	−291.5	−292.4
	cc-pV6Z	−311.8	−309.9	−413.8	−403.7	−295.2	−294.9
	extrapolated	−319.5	−315.9	−420.3	−408.0	−300.3	−298.3
	R12[a]	−320(1)	−316(1)	−421(2)	−408(2)	−300(1)	−298(1)
all	cc-pCVDZ	−228.3	−232.2	−382.7	−387.8	−241.3	−251.8
	cc-pCVTZ	−329.1	−331.4	−477.8	−478.2	−317.5	−324.2
	cc-pCVQZ	−361.5	−362.7	−510.7	−507.1	−342.6	−346.5
	cc-pCV5Z	−374.1	−373.7	−523.1	−516.7	−352.3	−353.9
	cc-pCV6Z	−379.8	−378.2	−528.7	−520.6	−356.4	−356.9
	extrapolated	−387.6	−384.4	−536.4	−526.0	−362.0	−361.0
	R12[a]	−388(1)	−384(1)	−537(2)	−526(2)	−361(1)	−361(2)

[a]Estimated error bars (relative to the basis-set limit) in parentheses

Table 8.19 The convergence of the MP2 molecular geometries for N_2 and H_2O. The bond distances are given in pm and the bond angle in degrees

Electrons correlated	Basis set	R_{NN}	R_{OH}	θ_{HOH}
valence	6-31G	115.43	97.46	109.28
	6-31G**	113.07	96.14	103.83
	6-311G**	111.98	95.78	102.44
	cc-pVDZ	112.99	96.49	101.90
	cc-pVTZ	111.36	95.91	103.52
	c-pVQZ	111.04	95.77	104.01
	cc-pV5Z	110.96	95.79	104.29
all	cc-pCVDZ	112.84	96.42	101.91
	cc-pCVTZ	111.01	95.80	103.63
	cc-pCVQZ	110.78	95.69	104.14
	cc-pCV5Z	110.70	95.70	104.41

calculations – namely, the 6-31G basis set. With this basis, less than one-half of the correlation energy is recovered and, perhaps more important, there is no improvement (relative to the Hartree–Fock description) in the equilibrium geometry, in particular in the bond angle. In short, this basis set does not have the required flexibility to support correlated calculations.

In Tables 8.18 and 8.19, two classes of basis sets are listed: the valence cc-pVXZ sets, designed for correlation of the valence electrons; and the core–valence cc-pCVXZ sets, designed for correlation of all electrons. Accordingly, we have correlated only the valence electrons in the cc-pVXZ calculations, keeping the $1s$ orbital frozen. In the cc-pCVXZ calculations, the full set of electrons has been correlated. Comparing the cc-pVXZ and cc-pCVXZ sequences in Table 8.18, we see that the correlation energies converge to different limits and that, for these particular systems, the valence correlation energy constitutes approximately 80% of the all-electron correlation energy. As an example, the R12 estimate of the CCSD all-electron correlation energy of H_2O is -361 mE_h, whereas the core correlation energy is -63 mE_h, which may be further partitioned into a core–core contribution of -42 mE_h and a core–valence contribution of -21 mE_h.

Comparing the correlation energies in Table 8.18 with the Hartree–Fock energies in Table 8.16, we find that the correlation energies converge considerably more slowly than the Hartree–Fock energies. As an example, the Hartree–Fock energies change by only 2–3 mE_h as we go from cc-pCVQZ to cc-pCV5Z and the remaining errors at this level are only about 0.3 mE_h. By contrast, the MP2 correlation energies change by as much as 9–13 mE_h as we go from cc-pCVQZ to cc-pCV5Z and, more important, the remaining errors are still 7–14 mE_h. The proportion of the total correlation energy recovered by the correlation-consistent basis sets is, on average, 67%, 88%, 95%, 97% and 98% for the cardinal numbers $2 \leq X \leq 6$.

Even though the convergence patterns of the MP2 and CCSD energies in Table 8.18 are similar, a slight difference is noticeable. For the cc-p(C)VDZ sets, the CCSD energies are lower than the MP2 energies. As the cardinal number increases, the situation is reversed and the extrapolated CCSD energies are always higher (by 1–12 mE_h) than the MP2 energies. For easy comparison of all the models considered in this section, we have in Table 8.20 collected the Hartree–Fock, MP2 and CCSD errors for the cc-pCVXZ basis sets. The considerably faster convergence of the Hartree–Fock energy relative to the correlation energies is here evident, as is the similarity of the MP2 and CCSD errors.

Let us now consider the convergence of the molecular equilibrium structures. Comparing the correlated structures in Table 8.19 with the Hartree–Fock structures in Table 8.16, we find that, for the correlated models, the geometries converge somewhat more slowly and that, for small basis sets, their errors are usually slightly larger (relative to the basis-set limit). The HOH bond angle, in particular, is underestimated by more than $2°$ at the MP2 double-zeta level (relative to the basis-set limit). We also note that the NN distance increases by 4–5 pm upon correlation and the OH distance by about 2 pm; the bond angle is reduced by 2–3°. In accordance with our discussion in Section 8.3.1, the neglect of core–core and core–valence correlation has a small but noticeable effect on the geometries. The bond distances contract by 0.1–0.3 pm when the core electrons are correlated, whereas the HOH bond angle increases by about 0.1°. In short, core correlation works in the opposite direction of valence correlation but its effects are an order of magnitude smaller.

In the present subsection we have established that, for molecules containing first-row atoms, the convergence of the correlation energy with the extension of the basis set is slow for errors smaller

Table 8.20 Errors in the calculated Hartree–Fock and correlation energies in mE_h for the cc-pCVXZ basis sets at the experimental equilibrium geometry

	RHF			MP2			CCSD		
	Ne	N_2	H_2O	Ne	N_2	H_2O	Ne	N_2	H_2O
cc-pCVDZ	58	38	40	160	154	120	152	138	109
cc-pCVTZ	15	9	10	59	59	44	53	48	37
cc-pCVQZ	4	2	2	26	26	18	21	19	15
cc-pCV5Z	0	0	0	14	14	9	10	9	7
cc-pCV6Z	0	0	0	8	8	5	6	5	4

than 10–20 mE$_h$. However, although it may be difficult to establish *absolute energies* to an accuracy of 1 mE$_h$, other properties may often be easier to converge to the required accuracy. For example, an inspection of the bond distances in Table 8.19 indicates that a basis-set error of the order of 0.5 pm or less (an accuracy typical of many experiments) is usually achieved at the cc-pVTZ level and that, at the cc-pVQZ level, the basis-set errors are only of the order of 0.1–0.2 pm. Thus, even though the asymptotic convergence of bond distances and other geometrical parameters is indeed slow, it does not necessarily present insurmountable difficulties since sufficient accuracy (to within a few tenths of a picometre and a few tenths of a degree) is often reached with the basis sets cc-pVTZ and cc-pVQZ. Also, since the convergence of the correlation energy (and thus other molecular properties) is systematic, we may in many cases carry out extrapolations to the basis-set limit as a practical alternative to more accurate calculations.

8.4.3 THE ASYMPTOTIC CONVERGENCE OF THE CORRELATION ENERGY

As discussed in Chapter 7, the slow convergence of the correlation energy in the asymptotic region arises from difficulties in the description of the short-range electronic interactions. Clearly, it would be useful if we could reliably extrapolate to the basis-set limit from the energies for low cardinal numbers. Indeed, such extrapolations should be possible since the short-range correlation effects are universal and not sensitive to the details of the molecular structure.

To carry out the extrapolations, some simple analytical model for the convergence of the energy must be assumed. In our discussion of the *principal expansion* of the helium atom in Section 7.5, we found that the error in the FCI energy obtained by neglecting all orbitals of principal quantum number n greater than N is given by the expression

$$\Delta E_N \approx cN^{-3} \tag{8.4.1}$$

For the helium atom, the correlation-consistent sets cc-pVXZ have the same composition as the orbitals of the principal expansion with $N = X$. Identifying the cardinal number X with N in (8.4.1), we then obtain the following simple expression for the error in the cc-pVXZ basis:

$$\Delta E_X \approx AX^{-3} \tag{8.4.2}$$

The relationship between the correlation-consistent energies and the energy in the basis-set limit may now be written as

$$E_{\text{exact}} \approx E_X + AX^{-3} \tag{8.4.3}$$

Using this expression, we shall attempt to extrapolate to the exact correlation energy E_{exact} (within a given wave-function model) from the correlation-consistent energies E_X for small cardinal numbers X.

To apply (8.4.3), only two energies are needed. Assume that, for the cardinal numbers X and Y, we have obtained the energies E_X and E_Y. From these energies, we wish to determine the extrapolated basis-set limit E_{XY}^* and the linear parameter A_{XY} such that the following two equations are satisfied

$$E_{XY}^* = E_X + A_{XY}X^{-3} \tag{8.4.4}$$

$$E_{XY}^* = E_Y + A_{XY}Y^{-3} \tag{8.4.5}$$

Solving for E_{XY}^* and A_{XY}, we obtain the following expressions for the extrapolated correlation energy and the linear parameter A_{XY} [27]:

$$E_{XY}^* = \frac{X^3 E_X - Y^3 E_Y}{X^3 - Y^3} \tag{8.4.6}$$

$$A_{XY} = -\frac{E_X - E_Y}{X^{-3} - Y^{-3}} \tag{8.4.7}$$

The extrapolated energies in Table 8.18 have been obtained in this manner, using the energies calculated with the cc-p(C)V5Z and cc-p(C)V6Z basis sets. These E_{56}^* extrapolations agree nicely with the R12 energies, giving a mean absolute deviation from the R12 values of only 0.4 mE$_h$ and differing by at most 1.0 mE$_h$.

We may also carry out extrapolations using energies obtained with lower cardinal numbers, but the agreement is then less satisfactory. Thus, in the E_{45}^*, E_{34}^* and E_{23}^* extrapolations, we obtain mean absolute deviations relative to R12 of 0.9, 2.0 and 13.5 mE$_h$, respectively; the corresponding maximum deviations are 1.5, 5.0 and 22.5 mE$_h$. Apparently, the double-zeta energies are too far away from the basis-set limit (recovering less than 70% of the correlation energy and making significant contributions to the polarization of the charge distribution) to contain much useful information about the asymptotic behaviour of the correlation energy and should be avoided in the extrapolations.

In Figure 8.19, we have illustrated the asymptotic convergence of the MP2 all-electron correlation energy of N_2. On the left, we have plotted, superimposed on the calculated energies, the fitted expression (8.4.3) as a function of the cardinal number; on the right, we have made the same plot on a logarithmic scale (full line), comparing it with the corresponding errors in the Hartree–Fock energy (dotted line). The slow convergence of the correlation energy towards the basis-set limit is well illustrated. For an error in the correlation energy of less than 1 mE$_h$, a basis set with cardinal number greater than 10 would be required. For the Hartree–Fock energy, which converges exponentially, this accuracy is already achieved at the cc-pCV5Z level. These plots should be compared with Figure 8.16 for the carbon atom, where the errors are smaller but the behaviour otherwise similar.

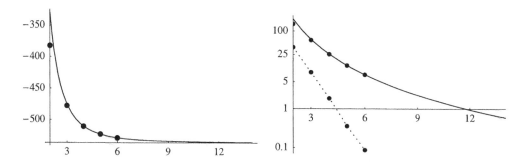

Fig. 8.19. The convergence of the MP2 correlation energy (full line) and the Hartree–Fock energy (dotted line) in mE$_h$ for N_2 calculated using the cc-pCVXZ basis sets. On the left, we have plotted the *correlation energies* superimposed on a fit of the form (8.4.3) with the horizontal axis representing the asymptotic limit of -537 mE$_h$. On the right, we have plotted the *errors in the correlation energy* superimposed on the fitted form (8.4.3) (full line) as well as the errors in the Hartree–Fock energy (dotted line) on a logarithmic scale.

An inspection of the logarithmic curve for the correlation energy in Figure 8.19 might suggest that our extrapolation to the basis-set limit from two low-order energies is somewhat presumptuous and that the behaviour for high cardinal numbers is rather speculative. The main assumptions underlying our extrapolations are the following: the expression for the error of the principal expansion of the helium CI energy (8.4.1) also holds for the correlation-consistent MP2 and CCSD energies in the form (8.4.2); the low-order terms are sufficiently close to the asymptotic limit to ensure a reasonable extrapolation; and chemical bonding effects do not destroy the convergence. Since our extrapolations in Table 8.18 are close to the calculated R12 energies, these assumptions appear to be well founded. Thus, although the convergence of the correlation energy is excruciatingly slow, it is nevertheless smooth (for correlation-consistent basis sets), allowing us to extrapolate to the basis-set limit from the energies calculated for $3 \leq X \leq 6$. Still, it is always wise to be prudent with any extrapolation to the infinite limit since the assumptions made here may not always be satisfied.

As a final illustration of the systematic asymptotic behaviour of the correlation energy, we have in Figure 8.20 plotted the errors in the Hartree–Fock and MP2 energies for Ne, N_2 and H_2O as functions of the inverse number of basis functions n^{-1}. The calculated points form line segments, which, at the Hartree–Fock level, bend and approach zero for small n. The MP2 energies, by contrast, are located on (almost) straight lines, emanating from the origin of the plot, which represents the solution in an infinite basis. This behaviour indicates that the errors in the correlation energies are roughly inversely proportional to the number of basis functions, which may be understood from the observations that, according to (8.4.3), the correlation energies converge asymptotically as X^{-3} and that, according to (8.3.6) and (8.3.7), the number of basis functions in the correlation-consistent basis sets is proportional to X^3.

8.4.4 BASIS-SET CONVERGENCE OF THE BINDING ENERGY

The correlation-consistent basis sets are constructed by carrying out correlated atomic calculations and it therefore does not necessarily follow that these sets are suitable for molecular calculations, where the atomic charge distribution becomes polarized in the presence of the other atoms. However, it has been found that, with the correlation-consistent basis sets, the error in the molecular correlation energy decreases as X^{-3} and it is therefore apparent that molecular binding energies are reasonably well described and do not dominate the errors for large cardinal numbers. In the present subsection, we investigate in more detail how well the correlation-consistent basis sets are able to describe molecular binding.

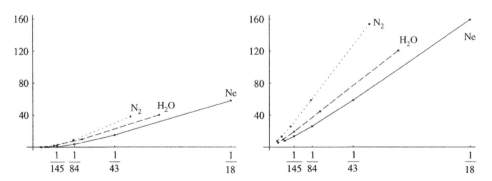

Fig. 8.20. The errors in the Hartree–Fock energy (left) and in the MP2 energy (right) in mE$_h$ plotted as functions of the inverse number of basis functions n^{-1}.

Consider the Hartree–Fock description of the binding energy of N_2 at the equilibrium bond distance of 109.77 pm. In Table 8.21, we have given the Hartree–Fock errors for N_2 and N relative to the numerical Hartree–Fock energies of -108.993188 and -54.400934 E_h, respectively. From these errors, we have calculated the errors in the binding energy as

$$\Delta E_{NN}^{bnd} = 2\Delta E_N^{atm} - \Delta E_{N_2}^{mol} \qquad (8.4.8)$$

By comparing the errors in the binding energy with those in the atomic energies, we obtain an impression of the ability of the correlation-consistent basis sets to describe molecular bonding.

From Table 8.21 we find that the error in the binding energy is similar to the error in the atomic energies – both errors decrease monotonically with the cardinal number in an approximately geometrical fashion. Hence, the correlation-consistent basis sets provide a balanced description of the atomic energies and the binding energy at the Hartree–Fock level. This is illustrated in Figure 8.21, where we have plotted the molecular errors as well as the fragment errors for N_2. The shaded area represents the errors associated with chemical bonding.

Table 8.21 Basis-set errors in the Hartree–Fock energy of N_2 and its constituent atoms (in mE_h). The error in the description of the binding energy is calculated as the difference (8.4.8) between the errors for the molecule and for the constituent atoms. All errors are relative to the numerical Hartree–Fock energies

	$\Delta E_{N_2}^{mol}$	ΔE_N^{atm}	ΔE_{NN}^{bnd}
cc-pVDZ	39.062	12.520	-14.022
cc-pVTZ	9.720	3.576	-2.568
cc-pVQZ	2.107	0.758	-0.591
cc-pV5Z	0.426	0.082	-0.262

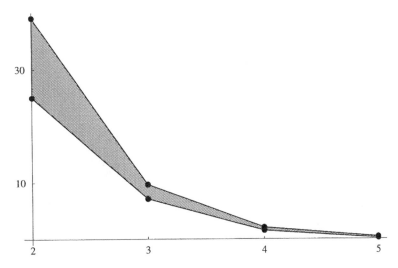

Fig. 8.21. The error in the N_2 Hartree–Fock energy (in mE_h) for the cc-pVXZ basis sets as a function of the cardinal number X. The upper curve represents the error in the molecular energy, whereas the lower curve represents the error in the energy of the atomic fragments.

To examine the errors at the correlated level, we consider the binding energy of BH at the valence FCI level. In Table 8.22, we have listed the errors obtained at the valence FCI/cc-pVXZ levels, with the Hartree–Fock errors included for comparison. The binding errors have been calculated as

$$\Delta E_{\text{BH}}^{\text{bnd}} = \Delta E_{\text{B}}^{\text{atm}} + \Delta E_{\text{H}}^{\text{atm}} - \Delta E_{\text{BH}}^{\text{mol}} \tag{8.4.9}$$

where $\Delta E_{\text{BH}}^{\text{mol}}$ is the error in the molecular energy and $\Delta E_{\text{B}}^{\text{atm}}$ and $\Delta E_{\text{H}}^{\text{atm}}$ the errors in the symmetry-restricted atomic energies. At the uncorrelated level, the errors are relative to the numerical Hartree–Fock energy (-25.131639 E_h for BH and -24.529061 E_h for B); at the correlated level, the errors are relative to the valence FCI basis-set limits (-25.2378 E_h for BH and -24.6026 E_h for B). The binding energies have been calculated at the experimental equilibrium bond distance of $2.3289 a_0$.

The numbers in Table 8.22 show that the errors in the binding energy are similar to the atomic errors, at both the correlated and the uncorrelated levels. As expected, the FCI errors are larger than the Hartree–Fock errors and also decrease more slowly with the cardinal number. This behaviour is illustrated in Figure 8.22, where we have plotted the molecular and fragment errors. The shaded areas represent the errors in the binding energy.

The results for N_2 and BH show that, for a given cardinal number X, the errors in the binding energy are about as large as the errors in the molecular energy for $X + 1$. Binding energies are therefore described to a higher absolute accuracy than are total energies. This observation, combined with the fact that the errors in the molecular and atomic calculations are of the same

Table 8.22 Basis-set errors in the Hartree–Fock and valence FCI energies of BH and its constituent atoms (in mE_h). The errors in the binding energies are calculated as the difference between the errors of the atoms and the molecule according to (8.4.9)

	Hartree–Fock				Valence FCI			
	$\Delta E_{\text{BH}}^{\text{mol}}$	$\Delta E_{\text{B}}^{\text{atm}}$	$\Delta E_{\text{H}}^{\text{atm}}$	$\Delta E_{\text{BH}}^{\text{bnd}}$	$\Delta E_{\text{BH}}^{\text{mol}}$	$\Delta E_{\text{B}}^{\text{atm}}$	$\Delta E_{\text{H}}^{\text{atm}}$	$\Delta E_{\text{BH}}^{\text{bnd}}$
cc-pVDZ	6.307	2.497	0.722	−3.088	22.7	12.5	0.7	−9.5
cc-pVTZ	1.711	0.963	0.190	−0.553	6.7	3.9	0.2	−2.6
cc-pVQZ	0.344	0.163	0.054	−0.127	2.2	1.3	0.1	−0.9
cc-pV5Z	0.083	0.025	0.005	−0.053	1.1	0.6	0.0	−0.5

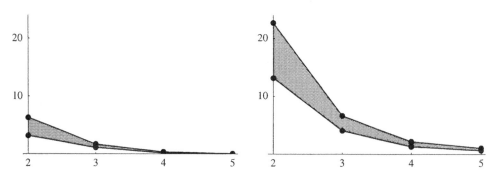

Fig. 8.22. The errors in the Hartree–Fock (left) and valence FCI (right) BH energies in mE_h for the sets cc-pVXZ as functions of the cardinal number X. The upper curves represent the errors in the molecular energy, whereas the lower curves represent the errors in the energy of the atomic fragments.

magnitude, leads us to conclude that the atomic correlating orbitals of the cc-pVXZ sets are also suitable for molecular calculations.

8.5 Basis-set superposition error

In electronic-structure calculations, we are often interested in evaluating the interaction energy of molecular complexes. For two fragments A and B, the interaction energy of the complex AB is given by

$$\Delta E_{AB} = E_{AB} - E_A - E_B \tag{8.5.1}$$

where E_A and E_B are the energies of the fragments and E_{AB} the energy of the complex. In Chapter 4, we showed that, for a balanced description of the fragments and the complex, such interaction energies should be calculated by size-extensive methods. In the present section, we discuss a second problem that arises in the calculation of interaction energies using a finite set of basis functions: the basis-set superposition error.

8.5.1 BASIS-SET SUPERPOSITION ERROR AND THE COUNTERPOISE CORRECTION

A simple approach to the calculation of molecular interaction energies (8.5.1) would be to calculate each of the fragment energies E_A and E_B in its own basis and the energy E_{AB} in the combined basis of the two fragments. However, in finite basis sets, this approach introduces a systematic error in the interaction energy. To see how this error arises, we note that, in the calculation on the complex AB, the description of fragment A is improved by the basis functions on B and visa versa. By contrast, in the calculations on the separate fragments A or B, only the basis belonging to a single fragment is used. In general, the energy of a single fragment in the presence of the basis of the other fragment as well as in its own basis will be lower than the energy calculated only in its own basis. The calculated interaction energy will therefore contain – in addition to the 'pure' interaction energy – a spurious attractive contribution that arises from an improved description of the fragments in the combined basis rather than from a true physical interaction between the fragments. This attractive contribution to the interaction energy, which may arise even in the absence of any genuine interaction between the fragments, is known as the *basis-set superposition error (BSSE)*.

The BSSE arises from an improved description of each fragment in the presence of the basis of the other fragment. An obvious strategy to solve this problem, suggested by Boys and Bernardi [28], is to carry out all calculations – those on the fragments as well as those on the complex – in the same combined atomic basis. In this way, no spurious attraction should arise in the description of the interaction between the fragments. Thus, in the calculation on fragment A, the presence of the *ghost basis* for fragment B serves to counterbalance the effect that basis B has on fragment A in the calculation on the complex AB. The interaction energy calculated in this manner is known as the *counterpoise-corrected interaction energy*

$$\Delta E_{AB}^{CP} = E_{AB} - E_A^{CP} - E_B^{CP} \tag{8.5.2}$$

Here, E_A^{CP} is the counterpoise-corrected energy of fragment A calculated in the full basis of the complex AB (i.e. by including the ghost functions located on system B), and similarly for E_B^{CP}. The *counterpoise correction* to the interaction energy becomes

$$\Delta E_{corr}^{CP} = \Delta E_{AB}^{CP} - \Delta E_{AB} \tag{8.5.3}$$

where ΔE_{AB}^{CP} is calculated according to (8.5.2) and ΔE_{AB} according to (8.5.1). Inserting (8.5.1) and (8.5.2) in the counterpoise correction, we obtain

$$\Delta E_{corr}^{CP} = \left(E_A - E_A^{CP}\right) + \left(E_B - E_B^{CP}\right) \tag{8.5.4}$$

Thus, for variational wave functions, the counterpoise correction is always positive since $E_A > E_A^{CP}$ and $E_B > E_B^{CP}$. For nonvariational wave functions, the correction is not positive by construction, although in practice we expect it to be positive.

We now have at our disposal two methods for the calculation of interaction energies: the standard or naive method, where the energies of the fragments are calculated in the usual manner without ghost functions; and the counterpoise-correction method, where the energies of the fragments are calculated in the presence of ghost functions. In the limit of a complete basis, the two methods give the same results; in a finite basis, different results are obtained. In the following, we shall compare these methods and in particular investigate under what circumstances the simpler standard method may be employed.

Depending on the nature of the interaction, molecular interaction energies vary considerably in magnitude. Typically, the interactions are 100–500 mE_h for covalent bonds, 1–10 mE_h for hydrogen-bonded complexes and 50–500 μE_h for complexes bound by dispersion. The BSSE is present in all cases but is more important for weakly bound van der Waals complexes than for chemically bonded molecules. In the following, we shall investigate the importance of BSSE at the Hartree–Fock and correlated levels for three systems, selected to represent the three cases of interaction listed above: the BH molecule, the water dimer and the neon dimer.

8.5.2 BSSE IN THE NEON DIMER

In the neon dimer, the closed-shell, nonpolar neon atoms are held together by dispersion forces alone. This very weak interaction is a pure correlation effect and cannot be described by the Hartree–Fock model, which, in the limit of a complete basis, gives rise to a purely repulsive interaction between the two atoms. An empirical potential-energy curve for the neon dimer gives an interaction energy of -134 μE_h with a minimum at a separation of $5.84a_0$ [29]. Because of the extreme weakness of this interaction (it constitutes less than a millionth of the total energy of the system), we must carefully consider the effects of BSSE.

Let us first examine the calculations on the individual atoms. In Table 8.23, we have listed the basis-set errors and the counterpoise corrections of a single neon atom as calculated at the Hartree–Fock and valence MP2 levels. For the Hartree–Fock calculations, the errors are relative to the numerical Hartree–Fock energy of -128.547094 E_h; for the MP2 calculations, the reference energy of -128.867 E_h is obtained by adding the R12 estimate of the correlation energy (-0.320 E_h; see Table 8.18) to the numerical Hartree–Fock energy. The counterpoise corrections are calculated with a ghost basis positioned at a distance of $5.84a_0$ from the nucleus. The table contains data for basis sets from cc-pVDZ to t-aug-cc-pV5Z. Moving horizontally, we change the cardinal number; moving vertically, we change the augmentation level.

From the table, we see that the basis-set errors are considerably larger in magnitude than the interaction energy. Moreover, even for the largest basis sets, the counterpoise corrections in the correlated calculations are as large (in magnitude) as the interaction itself. Clearly, care should be exercised to ensure that the interaction energy is calculated in a reliable manner.

For the neon atom, the basis-set error decreases as we go from cc-pVXZ to aug-cc-pVXZ, where it becomes stable with respect to further addition of diffuse functions. The counterpoise correction

Table 8.23 The basis-set errors and the counterpoise corrections (in mE$_h$) of a single neon atom calculated at the Hartree–Fock and valence MP2 levels

		Hartree–Fock				MP2			
		D	T	Q	5	D	T	Q	5
cc-pVXZ	error	58.318	15.232	3.624	0.324	192.8	70.9	30.1	14.2
	CPC	0.040	0.069	0.031	0.003	0.048	0.098	0.057	0.011
aug-cc	error	50.744	13.821	3.338	0.308	163.9	61.3	26.1	12.3
	CPC	0.018	0.022	0.008	0.002	0.061	0.044	0.024	0.012
d-aug-cc	error	50.730	13.806	3.327	0.306	163.5	60.7	25.8	12.2
	CPC	0.077	0.044	0.026	0.005	0.141	0.089	0.061	0.029
t-aug-cc	error	50.725	13.800	3.323	0.305	163.5	60.7	25.8	12.2
	CPC	0.088	0.093	0.063	0.008	0.189	0.187	0.150	0.052

shows no convergence with respect to the addition of diffuse functions and the largest correction occurs at the triply augmented level. On the other hand, the counterpoise correction decreases with the cardinal number, at least beyond the double-zeta level. In the largest basis considered here (i.e. t-aug-cc-pV5Z), the Hartree–Fock correction is 8 µE$_h$ and the MP2 correction 52 µE$_h$, comparable with the interaction energy. The MP2 corrections are in general larger than the Hartree–Fock corrections, usually by a factor of 2 or more.

In Figure 8.23 we have plotted the counterpoise-corrected and raw interaction energies of the neon dimer as a function of the cardinal number at various augmentation levels. The most striking feature is the slow convergence of the raw, uncorrected energies (grey lines). At the counterpoise-corrected level (black lines), the convergence is much more satisfactory: saturation with respect to diffuse functions occurs at the aug-cc-pVXZ level for the Hartree–Fock energy and at the d-aug-cc-pVXZ level for the MP2 energy. With these diffuse functions added, accurate interaction energies are obtained even for relatively small cardinal numbers. From a detailed analysis of the interaction energies plotted in Figure 8.23, we find that the Hartree–Fock and valence MP2 basis-set limits are +95(1) and −82(2) µE$_h$, respectively, at the experimental minimum.

Since the dispersion interaction cannot be recovered at the uncorrelated Hartree–Fock level, the Hartree–Fock potential-energy curve should be repulsive for all internuclear separations. Nevertheless, we note from Figure 8.23 that, for several of the basis sets considered here, the uncorrected

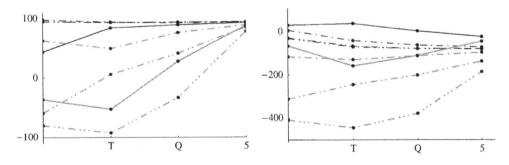

Fig. 8.23. The neon–neon interaction energy (in µE$_h$) at the Hartree–Fock level (left) and the MP2 level (right) at the experimental equilibrium separation of 5.84a_0. The black lines connect the counterpoise-corrected energies; the grey lines are used for the uncorrected energies. Full lines are used for the cc-pVXZ basis sets; for the remaining basis sets, the augmentation level is indicated by the number of consecutive dots.

Hartree–Fock interaction is attractive at the experimental minimum. The origin of this interaction is BSSE, whose spurious stabilization of the neon dimer outweighs the Pauli repulsion for small basis sets. These effects are illustrated on the left in Figure 8.24, where we have plotted the (uncorrected) Hartree–Fock potential-energy curve for various doubly augmented basis sets. As the cardinal number increases, the spurious Hartree–Fock minimum becomes shallower and located further out. We note that the potential-energy curve may contain several such spurious minima. Thus, the d-aug-cc-pVDZ curve has a minimum of -73 μE_h at a separation of $6.1a_0$ and a shallower minimum of -14 μE_h at a separation of $8.5a_0$.

It is not our purpose here to analyse the dispersion interaction of the neon dimer as such. However, to demonstrate the importance of a proper correlation treatment, we have on the right in Figure 8.24 plotted the counterpoise-corrected d-aug-pV5Z neon–neon dispersion energy at the HF, MP2, CCSD and CCSD(T) levels of theory. The agreement with the experimental curve is clearly seen to improve as the N-particle treatment is refined. The estimated basis-set limit for the valence MP2 interaction energy is $-82(2)$ μE_h at the experimental van der Waals equilibrium separation of $5.84a_0$, much smaller than the experimental energy of -134 μE_h. At the valence CCSD level, the estimated interaction energy of $-103(2)$ μE_h is still far away from the true interaction, indicating that higher-order excitations are important. Indeed, with the introduction of the perturbative triples correction at the CCSD(T) level, we obtain an estimated interaction energy of $-132(2)$ μE_h, in close agreement with experiment.

The calculations on the neon dimer demonstrate that, for an accurate calculation of dispersion energies, it is essential to apply the counterpoise correction. Furthermore, sequences of calculations must be carried out, where the cardinal number and the augmentation level are varied to ensure convergence to a particular threshold. Only in this way is it possible to control the massive cancellation of errors in such calculations. Proceeding in this manner, we were able to establish that, for the neon dimer, the d-aug-cc-pV5Z basis provides a valence MP2 energy within 3–4% of the basis-set limit – even though the resulting MP2 interaction energy of about -82 μE_h is more than 300 times smaller than the basis-set error of 24 mE_h at this level. By themselves, the uncorrected raw energies are not particularly useful, but they serve a useful purpose in that they may be compared with the corrected energies and thus help in assessing the accuracy of the

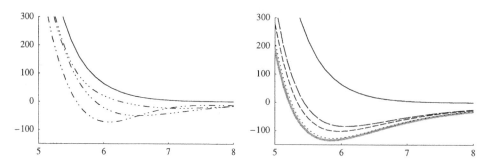

Fig. 8.24. The interaction energy of the neon dimer (in μE_h) plotted as a function of the internuclear separation (in a_0). On the left, we have plotted the interaction energies calculated at the Hartree–Fock level in a complete basis (full line) and using the d-aug-cc-pVXZ sets with $X \leq 4$ (with the number of consecutive dots indicating the cardinal number). No counterpoise correction has been applied to any of the Hartree–Fock curves. On the right, we have plotted the (counterpoise-corrected) interaction energies for different *ab initio* models at the frozen-core d-aug-cc-pV5Z level: the Hartree–Fock model (full line), the MP2 model (longer dashes), the CCSD model (shorter dashes) and the CCSD(T) model (dots). The thick grey line represents the potential-energy curve extracted from experiment [29].

counterpoise-corrected interaction energies. We have also demonstrated the need to employ highly correlated methods to describe dispersion interactions accurately.

8.5.3 BSSE IN THE WATER DIMER

We now turn our attention to hydrogen-bonded van der Waals complexes, where the total interaction is one to two orders of magnitude larger than the dispersion interaction. Whereas complexes bound by dispersion forces require high-level correlation treatments and large basis sets, hydrogen bonds are electrostatic in origin and may be described at less demanding levels of theory. In particular, we shall see that the basis-set requirements for hydrogen-bonded systems are less stringent than for the neon–neon dimer in Section 8.5.2 and that a qualitatively correct description is obtained even at the uncorrelated level.

For our discussion of BSSEs in hydrogen-bonded systems, we shall consider the water dimer. The calculations have been carried out at the experimental geometry of the monomers (see Section 8.4), with their relative orientation optimized at the all-electron CCSD(T)/aug-cc-pVTZ level [30,31]. The optimized water dimer is depicted in Figure 8.25 and has C_s symmetry, with one water molecule and an oxygen atom in the symmetry plane. The calculated oxygen–oxygen separation of 289.5 pm should be compared with the experimental separation of 294.6 pm, based on microwave measurements [32]. The experimental interaction energy is -8.6 ± 1.1 mE$_h$ [33,34].

Let us first examine the monomers. In Table 8.24, we have listed the basis-set errors and the counterpoise corrections of the monomers obtained at the Hartree–Fock and valence MP2 levels. In the Hartree–Fock calculations, we have used for reference the cc-pV6Z energy of -76.067401 E$_h$; the MP2 reference energy of -76.367 E$_h$ is obtained by adding the MP2-R12 estimate of the valence correlation energy (-0.300 E$_h$) to the Hartree–Fock energy. The counterpoise correction is calculated with a ghost basis at the CCSD(T) geometry. Since the two monomers are not equivalent (see Figure 8.25), their counterpoise corrections are different and must be calculated separately. Thus, for each relative orientation of the monomers, three calculations must be carried out in the combined basis of two monomers: one calculation with both monomers present and two calculations with only one monomer present.

From Table 8.24, we note that the experimental water-dimer interaction energy of -8.6 mE$_h$ is of the same order of magnitude as the basis-set error with $X = 3$ for Hartree–Fock and with $X = 5$ for MP2. Moreover, the basis-set error is more than an order of magnitude larger than the counterpoise

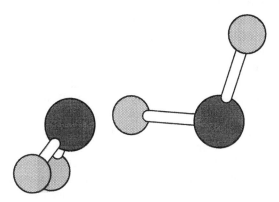

Fig. 8.25. The water dimer. Experimental monomer geometries are used with their relative orientation optimized at the all-electron CCSD(T)/aug-cc-pVTZ level.

Table 8.24 The basis-set energy errors and the counterpoise corrections (in mE_h) for the water monomers of the water dimer calculated at the Hartree–Fock and valence MP2 levels

		Hartree–Fock				Valence MP2			
		D	T	Q	5	D	T	Q	5
cc-pVXZ	error	40.60	10.23	2.57	0.31	139.0	48.8	19.8	8.8
	CPC(A)	2.97	1.10	0.42	0.10	4.84	2.15	0.97	0.34
	CPC(B)	0.36	0.23	0.14	0.05	0.62	0.50	0.31	0.13
aug	error	25.97	6.79	1.40	0.08	106.6	38.4	15.5	7.2
	CPC(A)	0.32	0.08	0.03	0.00	0.95	0.44	0.25	0.13
	CPC(B)	0.08	0.05	0.02	0.00	0.41	0.32	0.13	0.08
d-aug	error	25.56	6.72	1.38	–	105.8	38.0	15.3	–
	CPC(A)	0.34	0.10	0.04	–	0.95	0.49	0.28	–
	CPC(B)	0.14	0.07	0.03	–	0.54	0.42	0.19	–

correction. By contrast, for the neon dimer in Table 8.23, the interaction energy is much smaller than the basis-set error. Based on these observations, we expect accurate calculations of the interaction energy of the water dimer to be considerably easier than for the neon dimer.

For the water dimer, the counterpoise correction decreases uniformly with the cardinal number: at the Hartree–Fock level, the correction is reduced by a factor of 2–3 with each increment in the cardinal number; at the MP2 level, it is larger and reduced by a factor of about 2 with each increment. Diffuse functions are particularly efficient at reducing the counterpoise correction: when a single set of diffuse functions is added, the combined counterpoise correction of the two fragments is reduced by an order of magnitude at the Hartree–Fock level and by a factor of 2–4 at the MP2 level. Apparently, this reduction occurs because, at the aug-cc-pVXZ level, a better saturation is obtained of the outer valence region and the ghost functions therefore have a smaller effect. No improvement is observed when a second set of diffuse functions is added.

In Figure 8.26, we have plotted the counterpoise-corrected and raw interaction energies of the water dimer as functions of the cardinal number for the different augmentation levels. The Hartree–Fock model (on the left) and the MP2 model (on the right) behave differently. Whereas the counterpoise-corrected MP2 model underestimates the correlation contribution to the interaction energy, the counterpoise-corrected Hartree–Fock model overestimates the Hartree–Fock interaction energy. Also, at the cc-pVXZ level, the counterpoise correction improves the description significantly at the Hartree–Fock level but apparently not so at the MP2 level. Figure 8.26 illustrates clearly the improvement observed when diffuse functions are added to the basis set.

Still, a closer look at the MP2 corrections in Figure 8.26 reveals that the counterpoise-corrected energies converge more smoothly than the uncorrected energies. Indeed, unlike for the raw energies, the convergence of the counterpoise-corrected energies is similar to that of the MP2 correlation energy of N_2 in Figure 8.19. An explanation for the behaviour in Figure 8.26 is therefore that, in the raw MP2 energies, there is a partial cancellation of the BSSE (which lowers the energy) and the basis-set correlation error (which raises the energy). This hypothesis is confirmed by Table 8.25, which shows that the two-point extrapolation (8.4.6) works much better for the counterpoise-corrected energies than for the raw energies [31].

The basis-set limit of the MP2 correction to the interaction energy is -2.2 mE_h, as established by the R12-MP2 method. At the cc-pV(DT)Z level, the extrapolation reduces the error in the counterpoise-corrected energy by almost a factor of 8 but increases the error in the uncorrected energy. Moreover, at the cc-pV(TQ)Z level, the counterpoise-corrected energies are converged to within 0.01 mE_h of the basis-set limit, while the uncorrected energy is still in error by 0.09 mE_h.

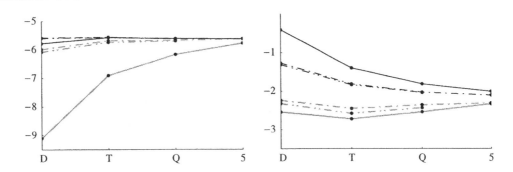

Fig. 8.26. The Hartree–Fock (left) and MP2 (right) contributions to the interaction energy (in mE_h) of the water dimer. The black lines connect the counterpoise-corrected energies; the grey lines are used for the uncorrected energies. Full lines are used for the cc-pVXZ basis sets; for the basis sets with diffuse functions added, the augmentation level is indicated by the number of consecutive dots.

Table 8.25 The raw and counterpoise-corrected MP2 contributions to the interaction energy (in mE_h) of the water dimer. The interaction energies have been calculated at the aug-cc-pVXZ level, with and without the extrapolation (8.4.6) applied. The MP2-R12 numbers have been obtained as described in [31]

	D	T	Q	5	DT	TQ	Q5	R12
Uncorrected	−2.24	−2.46	−2.36	−2.31	−2.55	−2.29	−2.26	−2.20
CP corrected	−1.27	−1.82	−2.03	−2.11	−2.05	−2.19	−2.19	−2.20

Clearly, for a proper description of the interaction energy, the counterpoise correction should be applied at the aug-cc-pVXZ level but only in conjunction with a careful analysis of the convergence of the energy with respect to the cardinal number.

Based on the energies plotted in Figure 8.26, the basis-set limit of the interaction energy of the water dimer is estimated to be −5.6 and −7.8(1) mE_h, respectively, at the Hartree–Fock and valence MP2 levels. At the CCSD and CCSD(T) levels, the limits are −7.5(1) and −7.9(1) mE_h, respectively. The MP2 estimate agrees with experiment (−8.6 ± 1.1 mE_h) and correlation treatments beyond MP2 appear to be less important than for interactions dominated by dispersion.

8.5.4 BSSE IN THE BH MOLECULE

Let us finally consider the importance of the BSSE for the binding energy of the covalent bond in the BH molecule. The binding energy is about 131 mE_h – see [18] and Exercise 15.1. In Table 8.26 we have listed the basis-set errors and the counterpoise corrections for BH, calculated at the Hartree–Fock and valence FCI levels in the cc-pVXZ basis sets at the experimental equilibrium geometry of $2.3289a_0$ [18]. At this geometry, the numerical Hartree–Fock energy of BH is −25.131639 E_h and the basis-set limit of the valence FCI energy is −25.2378 E_h. Diffuse functions are not needed for the description of covalent bonds and have not been added in these calculations.

The counterpoise corrections are larger at the FCI level than at the Hartree–Fock level but are rapidly reduced as the cardinal number increases. Compared with the total binding energy, the counterpoise corrections are very small – at the cc-pVDZ level, for example, the correction is only

Table 8.26 The basis-set errors and counterpoise corrections (in mE_h) at the Hartree–Fock and valence FCI levels for the BH molecule in the cc-pVXZ basis

	Hartree–Fock				Valence FCI			
	D	T	Q	5	D	T	Q	5
Error	6.307	1.706	0.344	0.083	22.7	6.7	2.2	1.1
CPC(B)	0.016	0.007	0.002	0.001	0.391	0.208	0.086	0.041
CPC(H)	0.011	0.006	0.002	0.000	0.011	0.006	0.002	0.000

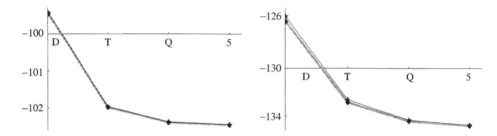

Fig. 8.27. The interaction energy (in mE_h) of the BH molecule at the Hartree–Fock level (left) and at the valence FCI level (right). The black lines connect the counterpoise-corrected energies; grey lines are used for the uncorrected energies. The calculations have been carried out in the cc-pVXZ sets at a separation of $2.3289 a_0$.

0.2% of the Hartree–Fock interaction energy and 0.3% of the FCI energy. At the cc-pV5Z level, the corrections have been reduced by one order of magnitude.

In Figure 8.27, we have plotted the corrected (black lines) and uncorrected (grey lines) interaction energies (i.e. the negative binding energies) of the BH molecule at the Hartree–Fock level (on the left) and at the valence FCI level (on the right). The counterpoise corrections are very small – not only compared with the binding energy but also relative to the changes that occur as the cardinal number increases. The BSSE may therefore be neglected in the description of the covalent bond of BH. The Hartree–Fock and valence FCI basis-set limits of the binding energy (estimated from the numbers on which Figure 8.27 is based) are 103 and 135 mE_h, respectively.

8.5.5 SUMMARY

In this section, we have seen that, although BSSE is present in all calculations of molecular interaction energies, its importance varies considerably. To control the effects of BSSE, systematic sequences of calculations must be carried out with and without the counterpoise correction applied. The BSSE problem is most acute for complexes held together by dispersion forces, where the counterpoise correction may be as large as the interaction itself, and least important for covalent bonds, where it is so small compared with the interaction energy that it may be neglected. For hydrogen-bonded systems, the counterpoise-corrected interaction energy exhibits the characteristically slow X^{-3} basis-set convergence associated with the description of dynamical correlation by orbital-based methods. For such systems, the basis-set limit may be estimated by the two-point extrapolation of Section 8.4.3.

References

[1]   W. J. Hehre, R. F. Stewart and J. A. Pople, *J. Chem. Phys.* **51**, 2657 (1969).
[2]   E. Clementi and D. L. Raimondi, *J. Chem. Phys.* **38**, 2686 (1963); E. Clementi, D. L. Raimondi and W. P. Reinhardt, *J. Chem. Phys.* **47**, 1300 (1967).
[3]   S. Huzinaga, *J. Chem. Phys.* **42**, 1293 (1965).
[4]   C. M. Reeves, *J. Chem. Phys.* **39**, 1 (1963).
[5]   C. M. Reeves and M. C. Harrison, *J. Chem. Phys.* **39**, 11 (1963).
[6]   R. C. Raffenetti, *J. Chem. Phys.* **59**, 5936 (1973).
[7]   R. D. Bardo and K. Ruedenberg, *J. Chem. Phys.* **59**, 5956 (1973).
[8]   M. W. Schmidt and K. Ruedenberg, *J. Chem. Phys.* **71**, 3951 (1979).
[9]   R. C. Raffenetti, *J. Chem. Phys.* **58**, 4452 (1973).
[10] T. H. Dunning, Jr, *J. Chem. Phys.* **53**, 2823 (1970).
[11] T. H. Dunning, Jr, *J. Chem. Phys.* **55**, 716 (1971).
[12] J. S. Binkley, J. A. Pople and W. J. Hehre, *J. Am. Chem. Soc.* **102**, 939 (1980).
[13] W. J. Hehre, R. Ditchfield and J. A. Pople, *J. Chem. Phys.* **56**, 2257 (1972).
[14] A. A. Frost, H. F. Schaefer III (ed.), in *Methods of Electronic Structure Theory*, Plenum, 1977, p. 29.
[15] T. Helgaker and J. Almlöf, *J. Chem. Phys.* **89**, 4889 (1988).
[16] P. C. Hariharan and J. A. Pople, *Theor. Chim. Acta* **28**, 213 (1973).
[17] R. Krishnan, J. S. Binkley, R. Seeger and J. A. Pople, *J. Chem. Phys.* **72**, 650 (1980).
[18] K. P. Huber and G. H. Herzberg, *Constants of Diatomic Molecules*, Van Nostrand Reinhold, 1979.
[19] J. Almlöf and P. R. Taylor, *J. Chem. Phys.* **86**, 4070 (1987).
[20] J. Almlöf and P. R. Taylor, *Adv. Quantum Chem.* **22**, 301 (1991).
[21] P.-O. Widmark, P.-Å. Malmqvist and B. O. Roos, *Theor. Chim. Acta* **77**, 291 (1990).
[22] T. H. Dunning, Jr, *J. Chem. Phys.* **90**, 1007 (1989).
[23] D. E. Woon and T. H. Dunning, Jr, *J. Chem. Phys.* **98**, 1358 (1993).
[24] D. E. Woon and T. H. Dunning, Jr, *J. Chem. Phys.* **103**, 4572 (1995).
[25] R. A. Kendall, T. H. Dunning, Jr and R. J. Harrison, *J. Chem. Phys.* **96**, 6796 (1992).
[26] D. E. Woon and T. H. Dunning, Jr, *J. Chem. Phys.* **100**, 2975 (1994).
[27] A. Halkier, T. Helgaker, P. Jørgensen, W. Klopper, H. Koch, J. Olsen and A. K. Wilson, *Chem. Phys. Lett.* **286**, 243 (1998).
[28] S. F. Boys and F. Bernardi, *Mol. Phys.* **19**, 553 (1970).
[29] R. A. Aziz and M. J. Slaman, *Chem. Phys.* **130**, 187 (1989).
[30] A. Halkier, H. Koch, P. Jørgensen, O. Christiansen, I. M. Beck Nielsen and T. Helgaker, *Theor. Chem. Acc.* **97**, 150 (1997).
[31] A. Halkier, W. Klopper, T. Helgaker, P. Jørgensen and P. R. Taylor, *J. Chem. Phys.* **111**, 9157 (1999).
[32] J. A. Odutola and T. R. Dyke, *J. Chem. Phys.* **72**, 5062 (1980).
[33] L. A. Curtiss, D. J. Frurip and M. Blander, *J. Chem. Phys.* **71**, 2703 (1979).
[34] J. R. Reimers, R. O. Watts and M. L. Klein, *Chem. Phys.* **64**, 95 (1982).

Further reading

T. Helgaker and P. R. Taylor, Gaussian basis sets and molecular integrals, in D. R. Yarkony (ed.), *Modern Electronic Structure Theory*, World Scientific, 1995, p. 725.

E. R. Davidson and D. Feller, Basis set selection for molecular calculations, *Chem. Rev.* **86**, 681 (1986).

T. H. Dunning, Jr, K. A. Peterson and D. E. Woon, Basis sets: correlation consistent sets, in P. v. R. Schleyer, N. L. Allinger, T. Clark, J. Gasteiger, P. A. Kollman, H. F. Schaefer III and P. R. Schreiner (eds), *Encyclopedia of Computational Chemistry*, Vol. 1, Wiley, 1998, p. 88.

F. B. van Duijneveldt, J. G. C. M. van Duijneveldt-van de Rijdt and J. H. van Lenthe, State of the art in counterpoise theory, *Chem. Rev.* **94**, 1873 (1994).

I. Shavitt, The history and evolution of Gaussian basis sets, *Isr. J. Chem.* **33**, 357 (1993).

9 MOLECULAR INTEGRAL EVALUATION

We consider in this chapter techniques for the evaluation of the one- and two-electron molecular integrals needed for the calculation of the standard *ab initio* wave functions and energies. Only integrals over GTOs will be considered, since these are the only basis functions widely used for multi-centre molecular calculations. Having introduced the GTOs in Sections 9.1 and 9.2, we discuss simple one-electron integrals in Sections 9.3–9.6 and Coulomb integrals in Sections 9.7–9.11. After a discussion of the scaling properties of integrals in Section 9.12, we consider the multipole method for the evaluation of two-electron Coulomb integrals in Section 9.13. We conclude with a discussion of the fast multipole method for the evaluation of Coulomb interactions in large systems in Section 9.14. Although the integrals needed for the evaluation of a few basic molecular properties are considered in this chapter, a comprehensive account of property integrals is not given.

9.1 Contracted spherical-harmonic Gaussians

The *one- and two-electron integrals* considered in this chapter may all be written in the form

$$O_{\mu\nu} = \langle \chi_\mu | \hat{O}_1 | \chi_\nu \rangle = \int \chi_\mu(\mathbf{r}) \hat{O}(\mathbf{r}) \chi_\nu(\mathbf{r}) \, d\mathbf{r} \tag{9.1.1}$$

$$O_{\mu\nu\lambda\sigma} = \langle \chi_\mu(1) \chi_\nu(1) | \hat{O}_{12} | \chi_\lambda(2) \chi_\sigma(2) \rangle$$

$$= \int\int \chi_\mu(\mathbf{r}_1) \chi_\nu(\mathbf{r}_1) \hat{O}(\mathbf{r}_1, \mathbf{r}_2) \chi_\lambda(\mathbf{r}_2) \chi_\sigma(\mathbf{r}_2) \, d\mathbf{r}_1 \, d\mathbf{r}_2 \tag{9.1.2}$$

where the integrations are over the full space of Cartesian coordinates of one and two electrons, respectively. We shall return to the operators that appear in these integrals later. In the present section, we shall discuss the form of the AOs $\chi_\mu(\mathbf{r})$ over which the integrals (9.1.1) and (9.1.2) are calculated.

9.1.1 PRIMITIVE CARTESIAN GTOs

The functional form of the AOs was discussed in Chapters 6 and 8. In nearly all calculations on polyatomic systems, the AOs are taken as fixed linear combinations of real-valued *primitive Cartesian GTOs* of the form [1,2]

$$G_{ijk}(\mathbf{r}, a, \mathbf{A}) = x_A^i y_A^j z_A^k \exp(-a r_A^2) \tag{9.1.3}$$

centred on the atomic nuclei of the molecular system. In (9.1.3), the position of the electron is defined relative to a nucleus at position $\mathbf{A}$:

$$\mathbf{r}_A = \mathbf{r} - \mathbf{A} \tag{9.1.4}$$

The vector $\mathbf{r}_A$ has the three Cartesian components x_A, y_A and z_A and the norm r_A. The *Gaussian exponent* a is a real number greater than zero, the numerical value of which is determined in calculations on atomic systems as discussed in Chapter 8. The *Cartesian quantum numbers* i, j and k are integers greater than or equal to zero. The full set of GTOs of a given *total angular-momentum quantum number*

$$l = i + j + k \tag{9.1.5}$$

and of the same exponent a are said to constitute a *shell*. Gaussians with the angular-momentum quantum number $l = i + j + k = 0$ are referred to as *spherical Gaussians* (not to be confused with the *spherical-harmonic Gaussians* discussed in Section 9.1.2). The number of Cartesian GTOs in a shell of total angular-momentum quantum number l is given by the expression

$$N_l^c = \frac{(l+1)(l+2)}{2} \tag{9.1.6}$$

In molecular calculations, we always employ full shells of GTOs – that is, the individual Cartesian components of a given shell are never used alone. As we shall see in the present chapter, the simple analytical form of the primitive Cartesian GTOs (9.1.3) allows for an efficient evaluation of the polyatomic integrals entering the standard Hamiltonian operator (2.2.18).

As discussed in Chapter 8, the primitive Cartesian GTOs (9.1.3) are mostly used in fixed linear combinations $\chi_\mu(\mathbf{r})$. A typical AO thus consists of a linear combination of primitive Cartesian GTOs of the same angular-momentum quantum number l but of different Cartesian quantum numbers i, j and k and of different exponents a. In Section 9.1.2, we shall discuss how Cartesian GTOs of the same l but different i, j and k are combined to yield the real-valued spherical-harmonic GTOs; next, in Section 9.1.3, we shall see how the GTOs of different exponents are combined to yield the final AOs as contracted spherical-harmonic GTOs.

9.1.2   SPHERICAL-HARMONIC GTOs

A real-valued *spherical-harmonic GTO* of quantum numbers l and m, with exponent a and centred on $\mathbf{A}$ is given by the expression

$$G_{lm}(\mathbf{r}, a, \mathbf{A}) = S_{lm}(x_A, y_A, z_A) \exp(-a r_A^2) \tag{9.1.7}$$

where $S_{lm}(\mathbf{r}_A)$ is one of the real solid harmonics introduced in Section 6.4.2 – see in particular equations (6.4.19)–(6.4.21). A shell of spherical-harmonic GTOs contains all GTOs of the same a and l but different $|m| \le l$. The number of spherical-harmonic GTOs that constitute a complete shell is therefore given by the expression

$$N_l^s = 2l + 1 \tag{9.1.8}$$

In a given shell, the number of spherical-harmonic GTOs (1,3,5,7,9, ...) is always less than or equal to the number of Cartesian GTOs (1,3,6,10,15, ...).

As discussed in Section 6.4, the relationship between the spherical-harmonic and Cartesian GTOs is given by the expression

$$G_{lm}(\mathbf{r}, a, \mathbf{A}) = N_{lm}^S \sum_{t=0}^{[(l-|m|)/2]} \sum_{u=0}^{t} \sum_{v=v_m}^{[|m|/2-v_m]+v_m} C_{tuv}^{lm} G_{2t+|m|-2(u+v),2(u+v),l-2t-|m|}(\mathbf{r}, a, \mathbf{A}) \tag{9.1.9}$$

where

$$N_{lm}^S = \frac{1}{2^{|m|}l!}\sqrt{\frac{2(l+|m|)!(l-|m|)!}{2^{\delta_{0m}}}} \tag{9.1.10}$$

$$C_{tuv}^{lm} = (-1)^{t+v-v_m}\left(\frac{1}{4}\right)^t \binom{l}{t}\binom{l-t}{|m|+t}\binom{t}{u}\binom{|m|}{2v} \tag{9.1.11}$$

$$v_m = \begin{cases} 0 & m \geq 0 \\ \frac{1}{2} & m < 0 \end{cases} \tag{9.1.12}$$

The analytical form of the spherical-harmonic GTOs is thus more complicated than that of the Cartesian GTOs – see Table 6.3, which contains the lowest-order real-valued solid harmonics. In particular, the angular part of the spherical-harmonic GTOs is not separable in the Cartesian directions, making the integration over spherical-harmonic GTOs more complicated than the integration over Cartesian GTOs.

The integration techniques discussed in this chapter involve most directly the Cartesian GTOs. The integrals over Cartesian GTOs may subsequently be transformed to integrals over spherical-harmonic GTOs. However, since a relatively large number of Cartesian GTOs contribute to a smaller number of spherical-harmonic GTOs, it is advantageous to carry out the transformation (9.1.9) as early as possible in the evaluation of the molecular integrals. We shall return to this point later in this chapter, in our discussion of two-electron integrals.

9.1.3 CONTRACTED GTOs

The primitive Cartesian Gaussians are combined not only in their angular parts, but also in their radial parts. The final *contracted GTOs* may be written as linear combinations of primitive spherical-harmonic GTOs of different exponents a_v

$$G_{\mu lm}(\mathbf{r}, \mathbf{a}, \mathbf{A}) = \sum_v G_{lm}(\mathbf{r}, a_v, \mathbf{A})d_{v\mu} \tag{9.1.13}$$

where the *contraction coefficients* $d_{v\mu}$ are the same for all the angular components. The number of contracted GTOs is often considerably smaller than the number of primitive GTOs. To reduce the computational cost, the contraction should therefore (like the transformation to solid harmonics) be carried out as early as possible in the integration over contracted GTOs.

As discussed in Section 8.2.4, contractions of primitive GTOs come in two varieties: segmented and general contractions. In *segmented contractions*, each primitive GTO contributes to only one contracted GTO. The contraction (9.1.13) can then be carried out as a simple addition process, in which primitive integrals premultiplied by the appropriate contraction coefficients are added together. In *general contractions*, each primitive GTO contributes to several contracted GTOs. The contraction must then be carried out as a matrix multiplication involving the contraction coefficients and the integrals over primitive GTOs.

9.1.4 COMPUTATIONAL CONSIDERATIONS

As should be apparent from our discussion so far, a large number of integrals over *primitive Cartesian GTOs* (9.1.3) contribute to a smaller number of integrals over *contracted spherical-harmonic GTOs* (9.1.13). This is especially true for the two-electron integrals (9.1.2), since the number of such integrals grows as the fourth power of the number of GTOs. Therefore, although the algorithms discussed in this chapter are mostly presented in terms of primitive Cartesian GTOs, we should as early as possible carry out the transformations (9.1.9) and (9.1.13) so as to reduce the number of quantities to be manipulated further in the integration. Note that the Cartesian-to-spherical-harmonic and primitive-to-contracted transformations (9.1.9) and (9.1.13) are independent of each other and may be carried out in the order that suits the integration best.

Another important observation is that each primitive Cartesian GTO may contribute to several contracted spherical-harmonic GTOs. First, for $l \geq 2$, each Cartesian GTO may contribute to several spherical-harmonic GTOs; second, in the general contraction scheme, each primitive GTO may contribute to several contracted GTOs. To avoid any recalculation of integrals over primitive Cartesian GTOs, all contracted integrals made up from the same primitive integrals should be evaluated simultaneously. In fact, even if the different Cartesian components of the same shell are not combined in the generation of the spherical-harmonic components, it will still be advantageous to generate simultaneously the full shell of GTOs since (as we shall see later) integrals over the individual components share the same intermediates. Clearly, the calculation of integrals over contracted spherical-harmonic GTOs is a challenging organizational problem, where the number of floating-point operations should be kept as small as possible while keeping memory usage as low as possible.

9.2 Cartesian Gaussians

Cartesian Gaussians and products of two such Gaussians (i.e. the Cartesian overlap distributions) play an important role in the evaluation of molecular integrals. In the present section, we prepare ourselves for the study of integration techniques by examining the analytical properties of single Gaussians and their overlap distributions.

9.2.1 CARTESIAN GAUSSIANS

The primitive Cartesian Gaussians to be considered in this chapter may all be written in the form [1]

$$G_{ijk}(\mathbf{r}, a, \mathbf{A}) = x_A^i y_A^j z_A^k \exp(-ar_A^2) \qquad (9.2.1)$$

An important property of the Cartesian Gaussians is that they may be factorized in the three Cartesian directions

$$G_{ijk}(\mathbf{r}, a, \mathbf{A}) = G_i(x, a, A_x)G_j(y, a, A_y)G_k(z, a, A_z) \qquad (9.2.2)$$

where, for example,

$$G_i(x, a, A_x) = x_A^i \exp(-ax_A^2) \qquad (9.2.3)$$

As we shall see, this factorization simplifies the calculation of integrals over Cartesian Gaussians significantly.

The first few Cartesian Gaussians of unit exponent are illustrated in Figure 9.1. All functions except the spherical Gaussian ($i = 0$) have one node at the origin. The Cartesian Gaussians are either symmetric or antisymmetric with respect to inversion through the origin.

We shall always work with Cartesian GTOs in the unnormalized form (9.2.1). Still, it is of some interest to consider the self-overlap of these fundamental functions. The self-overlap of the

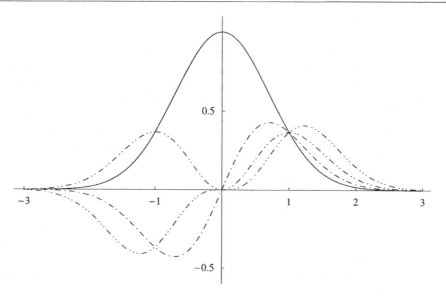

Fig. 9.1. The Cartesian Gaussians $G_i(x, 1, 0)$ for quantum numbers 0, 1, 2 and 3. The number of consecutive dots in the lines corresponds to the quantum number i of the Cartesian Gaussian.

x component of the unnormalized GTO (9.2.3) is given by (see Exercise 6.9)

$$\langle G_i | G_i \rangle = \frac{(2i - 1)!!}{(4a)^i} \sqrt{\frac{\pi}{2a}} \tag{9.2.4}$$

The double factorial is given in (6.5.10).

9.2.2 RECURRENCE RELATIONS FOR CARTESIAN GAUSSIANS

For a number of purposes, we shall need the differentiation property of the Gaussians. The first derivative is given by (omitting arguments for clarity)

$$\frac{\partial G_i}{\partial A_x} = -\frac{\partial G_i}{\partial x} = 2a G_{i+1} - i G_{i-1} \tag{9.2.5}$$

The derivative depends on the exponent and is a linear combination of two undifferentiated Gaussians with incremented and decremented quantum numbers. We may obtain higher derivatives by further differentiation, but recursion is often more useful. From (9.2.5), we obtain

$$\frac{\partial^{q+1} G_i}{\partial A_x^{q+1}} = \left(\frac{\partial}{\partial A_x} \right)^q (2a G_{i+1} - i G_{i-1}) = 2a \frac{\partial^q G_{i+1}}{\partial A_x^q} - i \frac{\partial^q G_{i-1}}{\partial A_x^q} \tag{9.2.6}$$

In the notation

$$G_i^q = \frac{\partial^q G_i}{\partial A_x^q} \tag{9.2.7}$$

we therefore have the following recurrence relation

$$G_i^{q+1} = 2a G_{i+1}^q - i G_{i-1}^q \tag{9.2.8}$$

In this way, we may construct higher derivatives from those of lower orders. This is useful since we normally need all derivatives up to a given order. We also note the trivial recurrence

$$x_A G_i = G_{i+1} \tag{9.2.9}$$

which is used on several occasions later.

9.2.3 THE GAUSSIAN PRODUCT RULE

The product of two spherical Gaussians centred on **A** and **B** may be written in terms of a single spherical Gaussian located on the line segment connecting the two centres. Using (6.2.6)–(6.2.10), we obtain the *Gaussian product rule* [1]

$$\exp(-ax_A^2) \exp(-bx_B^2) = \exp(-\mu X_{AB}^2) \exp(-px_P^2) \tag{9.2.10}$$

Here the *total exponent* p and the *reduced exponent* μ are defined as

$$p = a + b \tag{9.2.11}$$

$$\mu = \frac{ab}{a+b} \tag{9.2.12}$$

while the *centre-of-charge coordinate* P_x and the *relative coordinate* or *Gaussian separation* X_{AB} are given by

$$P_x = \frac{aA_x + bB_x}{p} \tag{9.2.13}$$

$$X_{AB} = A_x - B_x \tag{9.2.14}$$

Note that only the last factor on the right-hand side of (9.2.10) depends on the electronic coordinate x. The first factor

$$K_{ab}^x = \exp(-\mu X_{AB}^2) \tag{9.2.15}$$

is constant and is sometimes referred to as the *pre-exponential factor*. This factor is small when the separation between the centres is large.

The Gaussian product rule greatly simplifies the calculation of integrals. It is illustrated in Figure 9.2 for products of Gaussians of different exponents and separations. We note the dependence on the Gaussian separation (upper plots) and also the tendency of the product Gaussian to resemble most closely the Gaussian with the largest exponent (lower plots).

9.2.4 GAUSSIAN OVERLAP DISTRIBUTIONS

In the following short-hand notation for the Gaussians

$$G_a(\mathbf{r}) = G_{ikm}(\mathbf{r}, a, \mathbf{A}) \tag{9.2.16}$$

$$G_b(\mathbf{r}) = G_{jln}(\mathbf{r}, b, \mathbf{B}) \tag{9.2.17}$$

we define the *Gaussian overlap distribution* of $G_a(\mathbf{r})$ and $G_b(\mathbf{r})$:

$$\Omega_{ab}(\mathbf{r}) = G_a(\mathbf{r})G_b(\mathbf{r}) \tag{9.2.18}$$

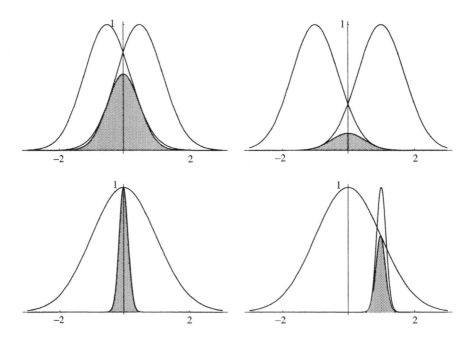

Fig. 9.2. The Gaussian product rule. The shaded areas represent the product of the two individual Gaussians. In the upper plots, the exponents are equal to 1 and the two Gaussians are centred at $\pm\frac{1}{2}$ and ± 1; in the lower plots, the exponents are $\frac{1}{2}$ (for the Gaussian at the origin in both plots) and 25 (for the Gaussian centred at 0 and 1).

Since the two Gaussians factorize into the three Cartesian directions, we may factorize the overlap distribution in the same way

$$\Omega_{ab}(\mathbf{r}) = \Omega_{ij}^x(x, a, b, A_x, B_x)\Omega_{kl}^y(y, a, b, A_y, B_y)\Omega_{mn}^z(z, a, b, A_z, B_z) \tag{9.2.19}$$

where the x component is given by

$$\Omega_{ij}^x(x, a, b, A_x, B_x) = G_i(x, a, A_x)G_j(x, b, B_x) \tag{9.2.20}$$

and likewise for the y and z components.

According to the Gaussian product rule (9.2.10), the overlap distribution (9.2.20) may be written as a single Gaussian positioned at the centre of charge P_x:

$$\Omega_{ij}^x = K_{ab}^x x_A^i x_B^j \exp(-px_P^2) \tag{9.2.21}$$

Noting that we may rewrite x_A and x_B as

$$x_A = x_P + X_{PA} = x_P - \frac{b}{p}X_{AB} \tag{9.2.22}$$

$$x_B = x_P + X_{PB} = x_P + \frac{a}{p}X_{AB} \tag{9.2.23}$$

we find that the overlap distribution (9.2.21) may be expanded in Cartesian Gaussians of orders $0 \le k \le i + j$ centred at **P**

$$\Omega_{ij}^x = K_{ab}^x \sum_{k=0}^{i+j} C_k^{ij} x_P^k \exp(-p x_P^2) \tag{9.2.24}$$

where the expansion coefficients may be obtained by application of the binomial theorem. Using the expansion (9.2.24), we may, for example, reduce all four-centre two-electron integrals over Cartesian Gaussians to a linear combination of two-centre integrals. This approach has indeed been taken in some schemes developed for the evaluation of two-electron integrals [3]. In Section 9.5, we shall take a similar but slightly different approach, expanding the Gaussian overlap distributions in Hermite rather than Cartesian Gaussians. Integrals over Ω_{ij}^x may, however, be obtained also without constructing such expansions explicitly, as will be shown in Section 9.3.

9.2.5 PROPERTIES OF GAUSSIAN OVERLAP DISTRIBUTIONS

The properties of Gaussian overlap distributions are easily inferred from those of the constituent Gaussians. We note in particular the trivial relationships

$$x_A \Omega_{ij}^x = \Omega_{i+1, j}^x \tag{9.2.25}$$

$$x_B \Omega_{ij}^x = \Omega_{i, j+1}^x \tag{9.2.26}$$

which follow from (9.2.9) and the recurrence

$$\Omega_{i, j+1}^x - \Omega_{i+1, j}^x = X_{AB} \Omega_{ij}^x \tag{9.2.27}$$

which follows by subtracting (9.2.25) from (9.2.26). Differentiating the overlap distributions, we obtain the relations

$$\frac{\partial \Omega_{ij}^x}{\partial A_x} = 2a \Omega_{i+1, j}^x - i \Omega_{i-1, j}^x \tag{9.2.28}$$

$$\frac{\partial \Omega_{ij}^x}{\partial B_x} = 2b \Omega_{i, j+1}^x - j \Omega_{i, j-1}^x \tag{9.2.29}$$

which follow directly from the derivative expression (9.2.5).

When working with overlap distributions, it is sometimes more convenient to use the coordinates P_x and X_{AB} rather than A_x and B_x, for example when derivatives are calculated [4]. From (9.2.13) and (9.2.14), we obtain

$$A_x = P_x + \frac{b}{p} X_{AB} \tag{9.2.30}$$

$$B_x = P_x - \frac{a}{p} X_{AB} \tag{9.2.31}$$

which give us the following expressions for the partial derivatives with respect to P_x and X_{AB}:

$$\frac{\partial}{\partial P_x} = \frac{\partial}{\partial A_x} + \frac{\partial}{\partial B_x} \tag{9.2.32}$$

$$\frac{\partial}{\partial X_{AB}} = \frac{b}{p} \frac{\partial}{\partial A_x} - \frac{a}{p} \frac{\partial}{\partial B_x} \tag{9.2.33}$$

The inverse relations are obtained directly from (9.2.13) and (9.2.14) by using the chain rule

$$\frac{\partial}{\partial A_x} = \frac{a}{p}\frac{\partial}{\partial P_x} + \frac{\partial}{\partial X_{AB}} \tag{9.2.34}$$

$$\frac{\partial}{\partial B_x} = \frac{b}{p}\frac{\partial}{\partial P_x} - \frac{\partial}{\partial X_{AB}} \tag{9.2.35}$$

From these expressions, we may establish the following relations

$$\frac{\partial \Omega_{ij}^x}{\partial P_x} = 2a\Omega_{i+1,j}^x + 2b\Omega_{i,j+1}^x - i\Omega_{i-1,j}^x - j\Omega_{i,j-1}^x \tag{9.2.36}$$

$$\frac{\partial \Omega_{ij}^x}{\partial X_{AB}} = 2\mu(\Omega_{i+1,j}^x - \Omega_{i,j+1}^x) + \frac{1}{2p}(2aj\Omega_{i,j-1}^x - 2bi\Omega_{i-1,j}^x) \tag{9.2.37}$$

for differentiation with respect to P_x and X_{AB}.

9.2.6   INTEGRALS OVER SPHERICAL OVERLAP DISTRIBUTIONS

As an introduction to the integration over GTOs, let us consider the simplest of all multicentre molecular integrals – namely, the integral over a spherical Gaussian charge distribution

$$\int_{-\infty}^{\infty} \Omega_{00}^x \, dx = \exp(-\mu X_{AB}^2) \int_{-\infty}^{\infty} \exp(-p x_P^2) \, dx \tag{9.2.38}$$

Using the expression

$$\int_{-\infty}^{\infty} \exp(-p x_P^2) \, dx = \sqrt{\frac{\pi}{p}} \tag{9.2.39}$$

we obtain

$$\int_{-\infty}^{\infty} \Omega_{00}^x \, dx = \sqrt{\frac{\pi}{p}} \exp(-\mu X_{AB}^2) \tag{9.2.40}$$

for the x component. Combing this expression with those for the y and z components, we arrive at the following simple expression for the overlap between two s orbitals separated by a distance R_{AB}:

$$\int \Omega_{ss} \, d\mathbf{r} = \left(\frac{\pi}{p}\right)^{3/2} \exp(-\mu R_{AB}^2) \tag{9.2.41}$$

Note that the overlap integral behaves like a Gaussian with respect to variations in the separation between the Gaussian centres.

9.3   The Obara–Saika scheme for simple integrals

Having discussed the Cartesian Gaussian functions and their overlap distributions, we are now ready to consider the evaluation of the *simple* one-electron integrals. By *simple*, we here mean the standard molecular integrals that do not involve the Coulomb interaction. In the present section, we thus discuss the evaluation of overlap integrals and multipole-moment integrals by the *Obara–Saika scheme* [5], based on the translational invariance of the integrals. We also consider the evaluation of one-electron integrals over differential operators and, in particular, the momentum and kinetic-energy integrals. A different approach to the calculation of integrals (the McMurchie–Davidson scheme), based on the expansion of the overlap distributions in Hermite Gaussians [6], is discussed in Section 9.5.

9.3.1 OVERLAP INTEGRALS

Let us first consider the simple overlap integrals

$$S_{ab} = \langle G_a | G_b \rangle \tag{9.3.1}$$

which, according to (9.2.19), may be factorized in the three Cartesian directions

$$S_{ab} = S_{ij} S_{kl} S_{mn} \tag{9.3.2}$$

where, for example

$$S_{ij} = \int_{-\infty}^{\infty} \Omega_{ij}^x \, dx \tag{9.3.3}$$

We begin by noting that this integral is invariant to an overall translation of the coordinate system along the x axis. Since the overlap integral depends only on the coordinates $\mathbf{A}$ and $\mathbf{B}$ of the two centres A and B, this invariance implies that the overlap integral should be unaffected by identical translations of these two centres in the x direction. The sum of the derivatives of the overlap integral with respect to A_x and B_x must therefore be equal to zero:

$$\frac{\partial S_{ij}}{\partial A_x} + \frac{\partial S_{ij}}{\partial B_x} = 0 \tag{9.3.4}$$

We now insert (9.3.3) and take the differential operators inside the integration sign (which is allowed since the integration limits are independent of A_x and B_x):

$$\int_{-\infty}^{\infty} \frac{\partial \Omega_{ij}^x}{\partial A_x} \, dx + \int_{-\infty}^{\infty} \frac{\partial \Omega_{ij}^x}{\partial B_x} \, dx = 0 \tag{9.3.5}$$

Applying (9.2.28) and (9.2.29), we obtain the *translational recurrence relation*

$$2aS_{i+1,j} - iS_{i-1,j} + 2bS_{i,j+1} - jS_{i,j-1} = 0 \tag{9.3.6}$$

which relates overlap integrals of different Cartesian quantum numbers. The total quantum number is $i + j + 1$ for the first and third terms and $i + j - 1$ for the other two terms. To arrive at a useful recurrence, we must eliminate one of the terms of order $i + j + 1$. For this purpose, we invoke the *horizontal recurrence relation*

$$S_{i,j+1} - S_{i+1,j} = X_{AB} S_{ij} \tag{9.3.7}$$

which follows directly by integration over (9.2.27). Combining the translational and horizontal recurrences to eliminate either $S_{i,j+1}$ or $S_{i+1,j}$, we arrive at the *Obara-Saika recurrence relations* for the Cartesian overlap integrals [5]:

$$S_{i+1,j} = X_{PA} S_{ij} + \frac{1}{2p}(iS_{i-1,j} + jS_{i,j-1}) \tag{9.3.8}$$

$$S_{i,j+1} = X_{PB} S_{ij} + \frac{1}{2p}(iS_{i-1,j} + jS_{i,j-1}) \tag{9.3.9}$$

Starting from the overlap integral for the spherical Gaussians (9.2.40)

$$S_{00} = \sqrt{\frac{\pi}{p}} \exp(-\mu X_{AB}^2) \qquad (9.3.10)$$

we may use the Obara–Saika recurrences to generate the x overlap integrals over primitive Cartesian GTOs for arbitrary quantum numbers. With obvious modifications, the same recurrence relations may be used to generate the y and z overlap integrals. The primitive Cartesian integrals (9.3.2) are then obtained by taking the products of the three factors. To generate the final overlap integrals over contracted spherical-harmonic GTOs, the integrals (9.3.2) must be transformed first to the contracted basis using (9.1.13) and next to the spherical-harmonic representation using (9.1.9).

As with any recursion involving more than one index, the target integrals S_{ij} may be generated from the initial integral S_{00} in many different ways. We may not only apply the Obara–Saika recurrences (9.3.8) and (9.3.9) in different ways, we may also combine these recurrences with the horizontal relation (9.3.7). We may, for example, first generate a set of integrals of the form S_{i0} and then apply the horizontal recurrence to generate the final integrals S_{ij}. Since the generation of the overlap integrals is in any case a trivial matter, we shall not compare and investigate these different schemes in any detail here.

9.3.2 MULTIPOLE-MOMENT INTEGRALS

Having considered the simple overlap integrals in Section 9.3.1, let us now consider the slightly more complicated multipole-moment integrals of the form

$$S_{ab}^{efg} = \langle G_a | x_C^e y_C^f z_C^g | G_b \rangle \qquad (9.3.11)$$

where $\mathbf{C}$ is the origin of the Cartesian multipole moments. Special cases of the multipole integrals are the overlap, dipole and quadrupole integrals. Since the operator and the orbitals in (9.3.11) are separable, the integral may be calculated as

$$S_{ab}^{efg} = S_{ij}^e S_{kl}^f S_{mn}^g \qquad (9.3.12)$$

where, for instance, the x component

$$S_{ij}^e = \langle G_i | x_C^e | G_j \rangle = \int_{-\infty}^{\infty} \Omega_{ij}^x x_C^e \, \mathrm{d}x \qquad (9.3.13)$$

may be treated by a simple extension to the scheme for the overlap integrals.

Like the overlap integrals, the multipole-moment integrals are invariant to an overall translation of the coordinate system. The sum of the derivatives of the integral (9.3.13) with respect to A_x, B_x and C_x must therefore be zero:

$$\frac{\partial S_{ij}^e}{\partial A_x} + \frac{\partial S_{ij}^e}{\partial B_x} + \frac{\partial S_{ij}^e}{\partial C_x} = 0 \qquad (9.3.14)$$

Taking the differential operators inside the integration, we obtain the following translational recurrence relation analogous to (9.3.6):

$$2a S_{i+1,j}^e - i S_{i-1,j}^e + 2b S_{i,j+1}^e - j S_{i,j-1}^e - e S_{ij}^{e-1} = 0 \qquad (9.3.15)$$

Also, in addition to the horizontal recurrence for the orbital quantum numbers (9.3.7), we have the following horizontal recurrences involving the order of the multipole operator:

$$S_{ij}^{e+1} = S_{i+1,j}^e + X_{AC}S_{ij}^e = S_{i,j+1}^e + X_{BC}S_{ij}^e \qquad (9.3.16)$$

as is easily verified. Combining the translational recurrences (9.3.15) with the horizontal recurrences (9.3.7) and (9.3.16), we obtain the Obara–Saika recurrence relations [5]

$$S_{i+1,j}^e = X_{PA}S_{ij}^e + \frac{1}{2p}(iS_{i-1,j}^e + jS_{i,j-1}^e + eS_{ij}^{e-1}) \qquad (9.3.17)$$

$$S_{i,j+1}^e = X_{PB}S_{ij}^e + \frac{1}{2p}(iS_{i-1,j}^e + jS_{i,j-1}^e + eS_{ij}^{e-1}) \qquad (9.3.18)$$

$$S_{ij}^{e+1} = X_{PC}S_{ij}^e + \frac{1}{2p}(iS_{i-1,j}^e + jS_{i,j-1}^e + eS_{ij}^{e-1}) \qquad (9.3.19)$$

from which the full set of Cartesian multipole-moment integrals may be generated. As for the overlap integrals, the Obara–Saika recurrences may be used in conjunction with the horizontal recurrences (9.3.7) and (9.3.16). The final integrals are obtained by multiplying the Cartesian factors according to (9.3.12), followed by a transformation to the contracted spherical-harmonic basis. A transformation of the Cartesian multipole-moment components to spherical-harmonic components may sometimes also be required.

9.3.3 INTEGRALS OVER DIFFERENTIAL OPERATORS

As the next example of the Obara–Saika scheme for simple one-electron integrals, we consider the integrals over the differential operators

$$D_{ab}^{efg} = \langle G_a | \frac{\partial^e}{\partial x^e} \frac{\partial^f}{\partial y^f} \frac{\partial^g}{\partial z^g} | G_b \rangle \qquad (9.3.20)$$

where the superscripts indicate the order of the differentiation with respect to the electronic coordinates. These integrals may again be factorized into three Cartesian components

$$D_{ab}^{efg} = D_{ij}^e D_{kl}^f D_{mn}^g \qquad (9.3.21)$$

where the x component, for example, is now given by

$$D_{ij}^e = \langle G_i | \frac{\partial^e}{\partial x^e} | G_j \rangle \qquad (9.3.22)$$

We first note that this integral may be written in the form of a differentiated overlap integral

$$D_{ij}^e = (-1)^e \frac{\partial^e S_{ij}^0}{\partial B_x^e} = \frac{\partial^e S_{ij}^0}{\partial A_x^e} \qquad (9.3.23)$$

where the first relation comes from the first part of (9.2.5) and the second from translational invariance. To arrive at recurrence relations for these integrals, we simply differentiate the Obara–Saika recurrences for the overlaps (9.3.8) and (9.3.9) with respect to A_x. Noting that the only nonvanishing derivatives of X_{PA} and X_{PB} are given by

$$\frac{\partial X_{PA}}{\partial A_x} = -\frac{b}{p} \qquad (9.3.24)$$

$$\frac{\partial X_{PB}}{\partial A_x} = \frac{a}{p} \qquad (9.3.25)$$

we obtain the following Obara–Saika recurrence relations [5]:

$$D^e_{i+1,j} = X_{PA} D^e_{ij} + \frac{1}{2p}(i D^e_{i-1,j} + j D^e_{i,j-1} - 2be D^{e-1}_{ij}) \qquad (9.3.26)$$

$$D^e_{i,j+1} = X_{PB} D^e_{ij} + \frac{1}{2p}(i D^e_{i-1,j} + j D^e_{i,j-1} + 2ae D^{e-1}_{ij}) \qquad (9.3.27)$$

$$D^{e+1}_{ij} = 2a D^e_{i+1,j} - i D^e_{i-1,j} \qquad (9.3.28)$$

where the derivative recurrence (9.3.28) has been obtain from (9.2.28). Because of the presence of the differential operator, the horizontal recursion is slightly more complicated than for the overlap and multipole-moment integrals. Multiple differentiation of (9.3.7) with respect to A_x gives

$$D^e_{i,j+1} - D^e_{i+1,j} = X_{AB} D^e_{ij} + e D^{e-1}_{ij} \qquad (9.3.29)$$

as the horizontal recurrence relation for the D^e_{ij} integrals. In passing, we note that these integrals may be expressed also directly in terms of the undifferentiated overlap integrals. To the lowest orders, we have

$$D^1_{ij} = 2a S_{i+1,j} - i S_{i-1,j} \qquad (9.3.30)$$

$$D^2_{ij} = 4a^2 S_{i+2,j} - 2a(2i+1) S_{ij} + i(i-1) S_{i-2,j} \qquad (9.3.31)$$

as may easily be verified.

9.3.4 MOMENTUM AND KINETIC-ENERGY INTEGRALS

In the previous two subsections, we have seen how the Obara–Saika scheme may be used to set up simple recurrence relations for the generation of the x, y and z components of the one-electron integrals over multipole-moment and differential operators. These one-dimensional integrals may also be combined to yield other important integrals – namely, the integrals for linear and angular momentum as well as the kinetic-energy integrals:

$$\mathbf{P}_{ab} = -i\langle G_a | \boldsymbol{\nabla} | G_b \rangle \qquad (9.3.32)$$

$$\mathbf{L}_{ab} = -i\langle G_a | \mathbf{r} \times \boldsymbol{\nabla} | G_b \rangle \qquad (9.3.33)$$

$$T_{ab} = -\tfrac{1}{2}\langle G_a | \nabla^2 | G_b \rangle \qquad (9.3.34)$$

Expanding the operators appearing in these integrals and factorizing in the Cartesian directions, we arrive at the following expressions for the z components of the momentum integrals

$$P^z_{ab} = -i S^0_{ij} S^0_{kl} D^1_{mn} \qquad (9.3.35)$$

$$L^z_{ab} = -i(S^1_{ij} D^1_{kl} S^0_{mn} - D^1_{ij} S^1_{kl} S^0_{mn}) \qquad (9.3.36)$$

and for the kinetic-energy integral

$$T_{ab} = -\tfrac{1}{2}(D_{ij}^2 S_{kl}^0 S_{mn}^0 + S_{ij}^0 D_{kl}^2 S_{mn}^0 + S_{ij}^0 S_{kl}^0 D_{mn}^2) \tag{9.3.37}$$

in terms of the basic one-dimensional integrals S_{ij}^e and D_{ij}^e.

Obviously, a large number of integrals may be generated by application of the basic Obara–Saika recurrence relations. Again, with the different integral types, there are often a number of possible approaches. We may thus write the kinetic-energy integrals also in the form

$$T_{ab} = T_{ij} S_{kl} S_{mn} + S_{ij} T_{kl} S_{mn} + S_{ij} S_{kl} T_{mn} \tag{9.3.38}$$

where, for example

$$T_{ij} = -\frac{1}{2} \langle G_i | \frac{\partial^2}{\partial x^2} | G_j \rangle \tag{9.3.39}$$

The Obara–Saika recurrence relations for these one-dimensional kinetic-energy integrals may be obtained from (9.3.26)–(9.3.28) as [5]

$$T_{i+1,j} = X_{\text{PA}} T_{ij} + \frac{1}{2p}(i T_{i-1,j} + j T_{i,j-1}) + \frac{b}{p}(2a S_{i+1,j} - i S_{i-1,j}) \tag{9.3.40}$$

$$T_{i,j+1} = X_{\text{PB}} T_{ij} + \frac{1}{2p}(i T_{i-1,j} + j T_{i,j-1}) + \frac{a}{p}(2b S_{i,j+1} - j S_{i,j-1}) \tag{9.3.41}$$

$$T_{00} = \left[a - 2a^2 \left(X_{\text{PA}}^2 + \frac{1}{2p} \right) \right] S_{00} \tag{9.3.42}$$

In this way, we may generate the kinetic-energy integrals by a simple modification of the recurrences for the overlap integrals.

9.4 Hermite Gaussians

In the previous section, we explored one particular approach to the evaluation of molecular integrals – the Obara–Saika scheme, which makes use of recurrence relations based on the translational invariance of the integrals. In the present section, we introduce a new set of Gaussians, the *Hermite Gaussians*, thereby preparing ourselves for an alternative approach to integral evaluation – the McMurchie–Davidson scheme [6], based on the expansion of Gaussian overlap distributions in one-centre Hermite Gaussians as discussed in Section 9.5.

9.4.1 HERMITE GAUSSIANS

The Hermite Gaussians of exponent p and centred on $\mathbf{P}$ are defined by

$$\Lambda_{tuv}(\mathbf{r}, p, \mathbf{P}) = (\partial/\partial P_x)^t (\partial/\partial P_y)^u (\partial/\partial P_z)^v \exp(-p r_{\mathbf{P}}^2) \tag{9.4.1}$$

where

$$\mathbf{r}_{\mathbf{P}} = \mathbf{r} - \mathbf{P} \tag{9.4.2}$$

Like the Cartesian Gaussians, these functions are separable

$$\Lambda_{tuv}(\mathbf{r}, p, \mathbf{P}) = \Lambda_t(x, p, P_x) \Lambda_u(y, p, P_y) \Lambda_v(z, p, P_z) \tag{9.4.3}$$

where, for example,

$$\Lambda_t(x, p, P_x) = (\partial/\partial P_x)^t \exp(-px_P^2) \qquad (9.4.4)$$

The Hermite Gaussians differ from the Cartesian Gaussians (9.2.3) only in the polynomial factors, which for the Hermite Gaussians are generated by differentiation.

The first four Hermite Gaussians with unit exponents are plotted in Figure 9.3. Like the Cartesian Gaussians, the Hermite Gaussians are symmetric (t even) or antisymmetric (t odd) with respect to inversion through the origin but their nodal structure is quite different. Whereas the number of nodes for the Cartesian Gaussians is one for all $i > 0$, the number of nodes in the Hermite Gaussian Λ_t is t. This follows from their definition by differentiation: Λ_{t+1} has nodes where Λ_t has extrema because the gradient vanishes at the extrema. Since Λ_0 has one extremum only and no nodes, we conclude that Λ_1 has one node and therefore two extrema, that Λ_2 has two nodes and therefore three extrema and so on.

It is possible to consider the use of Hermite Gaussians themselves as basis functions (in place of the Cartesian Gaussians) [7]. Here, we shall consider the use of Hermite Gaussians only as intermediates in the calculation of integrals over Cartesian Gaussians [6]. Their usefulness stems from the fact that they are defined by differentiation, which leads to many simplifications when calculating molecular one- and two-electron integrals, as will become apparent in Section 9.5 (for the one-electron integrals) and, in particular, in Section 9.9 (for the two-electron integrals).

9.4.2 DERIVATIVE AND RECURRENCE RELATIONS FOR HERMITE GAUSSIANS

From the definition of Hermite Gaussians (9.4.4), we obtain the simple differentiation formula

$$\frac{\partial \Lambda_t}{\partial P_x} = -\frac{\partial \Lambda_t}{\partial x} = \Lambda_{t+1} \qquad (9.4.5)$$

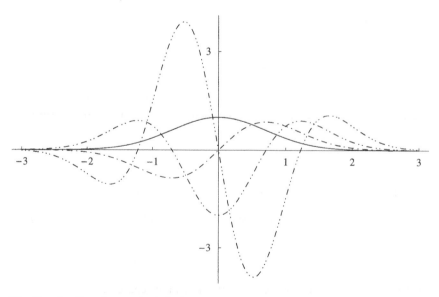

Fig. 9.3. The Hermite Gaussians $\Lambda_t(x, 1, 0)$ for quantum numbers 0, 1, 2 and 3. The number of consecutive dots in the lines corresponds to the quantum number t of the Hermite Gaussian.

which should be compared with the slightly more complicated formula for the Cartesian Gaussians (9.2.5). We also need to know what happens when the Hermite Gaussians are multiplied by x_P. We note that

$$\Lambda_{t+1} = (\partial/\partial P_x)^t \frac{\partial \Lambda_0}{\partial P_x} = 2p(\partial/\partial P_x)^t x_P \Lambda_0 \qquad (9.4.6)$$

To move x_P to the left of the differential operators, we use the commutator

$$\left[(\partial/\partial P_x)^t, x_P\right] = -t(\partial/\partial P_x)^{t-1} \qquad (9.4.7)$$

which may be demonstrated by induction. Inserting (9.4.7) into (9.4.6), we obtain

$$\Lambda_{t+1} = 2p(x_P \Lambda_t - t\Lambda_{t-1}) \qquad (9.4.8)$$

where we put

$$\Lambda_t = 0, \qquad t < 0 \qquad (9.4.9)$$

Equation (9.4.8) gives us the following basic recurrence relation for the Hermite Gaussians

$$x_P \Lambda_t = \frac{1}{2p}\Lambda_{t+1} + t\Lambda_{t-1} \qquad (9.4.10)$$

Note that the corresponding formula for Cartesian Gaussians (9.2.9) is even simpler.

9.4.3 INTEGRALS OVER HERMITE GAUSSIANS

Since the Hermite Gaussians are to be used as one-centre basis functions for the expansion of Cartesian overlap distributions (over which the integration will eventually be carried out), it is important to determine the integrals over these Gaussians. The integral over the x component of the Hermite Gaussians is given by

$$\int_{-\infty}^{+\infty} \Lambda_t(x)\,dx = \int_{-\infty}^{+\infty} (\partial/\partial P_x)^t \exp(-px_P^2)\,dx = (\partial/\partial P_x)^t \int_{-\infty}^{+\infty} \exp(-px_P^2)\,dx \qquad (9.4.11)$$

According to (9.2.39), the integral on the right-hand side is independent of P_x. It therefore vanishes when differentiated and we obtain the following simple result:

$$\int_{-\infty}^{+\infty} \Lambda_t(x)\,dx = \delta_{t0}\sqrt{\frac{\pi}{p}} \qquad (9.4.12)$$

which should be contrasted with the somewhat more complicated expression for the Cartesian Gaussians

$$\int_{-\infty}^{+\infty} G_i(x)\,dx = \begin{cases} (i-1)!!(2a)^{-i/2}\sqrt{\dfrac{\pi}{a}} & \text{for even } i \\ 0 & \text{for odd } i \end{cases} \qquad (9.4.13)$$

obtained using (6E.9.2) and the fact that $G_i(x)$ is antisymmetric for odd values of i.

Equation (9.4.11) illustrates an important technique, which will be used on several occasions. Since the Hermite Gaussians are defined as derivatives, we may take their differential operators outside the integration sign. In this way, we may turn integrals over Hermite Gaussians into differentiated integrals over spherical Gaussians. No such simple scheme is possible for the Cartesian Gaussians, which is the main reason for preferring the Hermite rather than Cartesian Gaussians for expanding the overlap distributions.

9.4.4 HERMITE GAUSSIANS AND HO FUNCTIONS COMPARED

Before we go on to consider the use of Hermite Gaussians as basis functions for overlap distributions, let us establish the relationship between Hermite Gaussians and Hermite polynomials. From Section 6.6.6, we recall that the Hermite polynomials may be generated from the Rodrigues expression (6.6.29)

$$H_n(x) = (-1)^n \exp(x^2) \frac{d^n}{dx^n} \exp(-x^2) \tag{9.4.14}$$

which, upon substitution of $\sqrt{p}\,x_P$ for x in the argument, may be written in the form

$$H_t(\sqrt{p}\,x_P) = p^{-t/2} \exp(px_P^2) \left(\frac{d}{dP_x}\right)^t \exp(-px_P^2) \tag{9.4.15}$$

Inserting the definition of the Hermite Gaussians (9.4.4) on the right-hand side and rearranging, we obtain

$$\Lambda_t(\sqrt{p}\,x_P) = p^{t/2} H_t(\sqrt{p}\,x_P) \exp(-px_P^2) \tag{9.4.16}$$

Hence, apart from a constant, the Hermite Gaussians consist of a Hermite polynomial multiplied by a Gaussian. It is therefore of some interest to compare with the HO functions (6.6.25)

$$\chi_i^{HO}(x) = \left(\frac{2\alpha}{\pi}\right)^{1/4} \frac{1}{\sqrt{2^i i!}} H_i(\sqrt{2\alpha}\,x) \exp(-\alpha x^2) \tag{9.4.17}$$

of HO theory; see Section 6.6.6. Clearly, the Hermite Gaussians and HO functions differ by a factor of $\sqrt{2}$ in the argument of the polynomials. Since the HO functions satisfy the orthonormality relations

$$\int_{-\infty}^{+\infty} \chi_i^{HO}(x)\chi_j^{HO}(x)\,dx = \delta_{ij} \tag{9.4.18}$$

we may conclude that the Hermite Gaussians are not orthogonal.

9.5 The McMurchie–Davidson scheme for simple integrals

Having introduced the Hermite Gaussians in the previous section, we shall now explore the McMurchie–Davidson scheme for the evaluation of molecular integrals over Cartesian Gaussians [6]. In this scheme, we proceed by expanding the Cartesian overlap distributions in one-centre Hermite Gaussians, thereby reducing the integration over two-centre functions to the evaluation over a set of one-centre functions whose analytical properties make them particularly well suited to integration – for the simple one-electron integrals studied in this section and for the more complicated Coulomb integrals to be studied in Section 9.9.

9.5.1 OVERLAP DISTRIBUTIONS EXPANDED IN HERMITE GAUSSIANS

Let us consider the expansion of two-centre Cartesian overlap distributions in Hermite Gaussians. Since the overlap distribution Ω_{ij} is a polynomial of degree $i + j$ in x_P (9.2.24), it may be expanded exactly in the Hermite polynomials of degree $t \leq i + j$. We therefore write

$$\Omega_{ij} = \sum_{t=0}^{i+j} E_t^{ij} \Lambda_t \tag{9.5.1}$$

where the expansion coefficients E_t^{ij} are constant – that is, independent of the electronic coordinates. It is possible to derive explicit expressions for these coefficients, but in practice it is easier to work with recurrence relations. Several such relations may be derived, based on the properties of the Cartesian overlap distributions and the Hermite Gaussians.

We begin by considering the incremented distribution

$$\Omega_{i+1,j} = \sum_{t=0}^{i+j+1} E_t^{i+1,j} \Lambda_t \tag{9.5.2}$$

To relate this expansion to that of Ω_{ij}, we apply first (9.2.25) and then (9.2.22):

$$\Omega_{i+1,j} = x_A \Omega_{ij} = x_P \Omega_{ij} + X_{PA} \Omega_{ij} \tag{9.5.3}$$

We now expand the first term on the right-hand side of (9.5.3) in Hermite Gaussians and eliminate x_P using the Hermite recurrence relation (9.4.10):

$$x_P \Omega_{ij} = \sum_{t=0}^{i+j} E_t^{ij} \left(t \Lambda_{t-1} + \frac{1}{2p} \Lambda_{t+1} \right) = \sum_{t=0}^{i+j+1} \left[(t+1) E_{t+1}^{ij} + \frac{1}{2p} E_{t-1}^{ij} \right] \Lambda_t \tag{9.5.4}$$

where the expansion coefficients are taken to satisfy the relations

$$E_t^{ij} = 0, \qquad t < 0 \text{ or } t > i + j \tag{9.5.5}$$

The second term in (9.5.3) is expanded straightforwardly. Inserting these expansions into (9.5.3) and identifying the terms to same order in t, we arrive at the McMurchie–Davidson recurrence relations for the expansion coefficients [6]

$$E_t^{i+1,j} = \frac{1}{2p} E_{t-1}^{ij} + X_{PA} E_t^{ij} + (t+1) E_{t+1}^{ij} \tag{9.5.6}$$

$$E_t^{i,j+1} = \frac{1}{2p} E_{t-1}^{ij} + X_{PB} E_t^{ij} + (t+1) E_{t+1}^{ij} \tag{9.5.7}$$

where the second expression has been obtained by expanding $\Omega_{i,j+1}$ by analogy with (9.5.2)–(9.5.5). The starting coefficient for the recursion is given by the factor (9.2.15)

$$E_0^{00} = K_{ab}^x \tag{9.5.8}$$

From these expressions, the full set of Hermite-to-Cartesian expansion coefficients may be generated and the overlap distribution may then be expanded in Hermite Gaussians according to (9.5.1).

Combining (9.5.6) and (9.5.7), we arrive at the trivial recurrence relation

$$E_t^{i,j+1} = E_t^{i+1,j} + X_{AB}E_t^{ij} \tag{9.5.9}$$

which also follows directly from the corresponding relation for the overlap distributions (9.2.27).

The Hermite-to-Cartesian coefficients E_t^{ij} are independent of P_x, depending only on the relative separation of the Gaussians X_{AB}:

$$\frac{\partial E_t^{ij}}{\partial P_x} = 0 \tag{9.5.10}$$

The independence of the coefficients of P_x is perhaps not immediately apparent from the recurrence relations (9.5.6) and (9.5.7). However, we note that, according to (9.2.15), the starting coefficients of the recursion (9.5.8) are independent of P_x. Next, we note that, in the recurrences, the coefficients X_{PA} and X_{PB} are both equal to X_{AB} scaled by some constant factor – see (9.2.30) and (9.2.31). The recurrence relations (9.5.6) and (9.5.7) are therefore independent of P_x and we conclude that the coefficients themselves must be independent of P_x (9.5.10).

The expressions (9.5.6) and (9.5.7) are three-term recurrence relations. A simpler set of relations may be obtained from the invariance (9.5.10). Differentiating (9.5.1) with respect to P_x, we obtain [4]

$$\frac{\partial \Omega_{ij}}{\partial P_x} = \sum_{t=0}^{i+j} E_t^{ij} \frac{\partial \Lambda_t}{\partial P_x} \tag{9.5.11}$$

since the coefficients are independent of P_x. Applying (9.2.36) to the left-hand side of this expression and (9.4.5) to the right-hand side, we find

$$2a\Omega_{i+1,j}^x + 2b\Omega_{i,j+1}^x - i\Omega_{i-1,j}^x - j\Omega_{i,j-1}^x = \sum_{t=0}^{i+j} E_t^{ij} \Lambda_{t+1} \tag{9.5.12}$$

Next, inserting the expansion (9.5.1) on the left-hand side and collecting terms to the same order in t, we obtain the following relationship between the coefficients:

$$2aE_t^{i+1,j} + 2bE_t^{i,j+1} - iE_t^{i-1,j} - jE_t^{i,j-1} = E_{t-1}^{ij} \tag{9.5.13}$$

On the left-hand side, we now substitute (9.5.6) and (9.5.7) for $E_t^{i+1,j}$ and $E_t^{i,j+1}$. Using (9.2.11) and (9.2.13) and making the global substitution $t \to t-1$, we arrive at the following simple two-term recurrence relation for the expansion coefficients:

$$2ptE_t^{ij} = iE_{t-1}^{i-1,j} + jE_{t-1}^{i,j-1} \tag{9.5.14}$$

Note that this relation, which is the same for the three Cartesian directions, determines all coefficients with $t > 0$. Restricting the recurrence relations (9.5.6) and (9.5.7) to $t = 0$, we have the following three *two-term recurrences*:

$$E_0^{i+1,j} = X_{PA}E_0^{ij} + E_1^{ij} \tag{9.5.15}$$

$$E_0^{i,j+1} = X_{PB}E_0^{ij} + E_1^{ij} \tag{9.5.16}$$

$$E_t^{ij} = \frac{1}{2pt}(iE_{t-1}^{i-1,j} + jE_{t-1}^{i,j-1}), \qquad t > 0 \tag{9.5.17}$$

from which the full set of coefficients may be generated, starting with (9.5.8). We note the following special cases

$$E_t^{i0} = (2p)^{-t} \binom{i}{t} E_0^{i-t,0} \tag{9.5.18}$$

$$E_t^{0j} = (2p)^{-t} \binom{j}{t} E_0^{0,j-t} \tag{9.5.19}$$

which are easily obtained from (9.5.17) by iteration.

One more set of recurrence relations for the expansion coefficients should be pointed out. Combining (9.5.6) and (9.5.17), we obtain

$$E_t^{i+1,j} = X_{PA} E_t^{ij} + \frac{1}{2p}(iE_t^{i-1,j} + jE_t^{i,j-1} + E_{t-1}^{ij}) \tag{9.5.20}$$

$$E_t^{i,j+1} = X_{PB} E_t^{ij} + \frac{1}{2p}(iE_t^{i-1,j} + jE_t^{i,j-1} + E_{t-1}^{ij}) \tag{9.5.21}$$

where the last equation follows from (9.5.20) and (9.5.9). These recurrence relations, which are reminiscent of the Obara–Saika relations (9.3.8) and (9.3.9), will later prove useful when setting up the Obara–Saika scheme for the Coulomb integrals.

9.5.2 OVERLAP DISTRIBUTIONS FROM HERMITE GAUSSIANS BY RECURSION

In Section 9.5.1, the Cartesian overlap distributions were generated from the one-centre Hermite Gaussians by explicit expansion, using the expansion coefficients E_t^{ij}. These distributions may be generated recursively as well. For this purpose, we introduce the auxiliary distributions

$$\Omega_{ij}^t = K_{ab}^x x_A^i x_B^j \Lambda_t(x_P) \tag{9.5.22}$$

which contain as special cases the Cartesian overlap distributions ($t = 0$) and the Hermite Gaussians ($i = j = 0$). To relate these hybrid, auxiliary functions by recursion, we first multiply the recurrence relation for Hermite Gaussians (9.4.10) by $K_{ab}^x x_A^i x_B^j$ to give

$$x_P \Omega_{ij}^t = \frac{1}{2p} \Omega_{ij}^{t+1} + t\Omega_{ij}^{t-1} \tag{9.5.23}$$

noting from (9.4.9) that

$$\Omega_{ij}^t = 0, \qquad t < 0 \tag{9.5.24}$$

Next, substituting (9.2.22) and (9.2.23) and noting from (9.2.9) that

$$x_A \Omega_{ij}^t = \Omega_{i+1,j}^t \tag{9.5.25}$$

$$x_B \Omega_{ij}^t = \Omega_{i,j+1}^t \tag{9.5.26}$$

we arrive at the McMurchie-Davidson recurrence relations for the auxiliary functions:

$$\Omega_{i+1,j}^t = t\Omega_{ij}^{t-1} + X_{PA} \Omega_{ij}^t + \frac{1}{2p} \Omega_{ij}^{t+1} \tag{9.5.27}$$

$$\Omega_{i,j+1}^t = t\Omega_{ij}^{t-1} + X_{PB} \Omega_{ij}^t + \frac{1}{2p} \Omega_{ij}^{t+1} \tag{9.5.28}$$

Using these relations, we may generate the Cartesian overlap distributions recursively from the Hermite Gaussians.

9.5.3 THE McMURCHIE–DAVIDSON SCHEME FOR MULTIPOLE-MOMENT INTEGRALS

Having investigated the properties of the Hermite Gaussians and having considered the expansion of the Cartesian overlap distributions in such Gaussians, we are now ready to consider the evaluation of one-electron integrals in the McMurchie–Davidson scheme. However, since a rather complete treatment of the more important simple one-electron integrals was given in Section 9.3 in connection with the Obara–Saika scheme, our treatment will be restricted to multipole-moment integrals, bearing in mind that the techniques presented here can be applied also to the other integrals.

Let us consider again the x component of the Cartesian multipole-moment integrals (9.3.11):

$$S_{ij}^e = \langle G_i | x_C^e | G_j \rangle \tag{9.5.29}$$

Inserting the Hermite expansion of the overlap distribution (9.5.1), we obtain

$$S_{ij}^e = \sum_{t=0}^{i+j} E_t^{ij} \int_{-\infty}^{+\infty} x_C^e \Lambda_t \, dx = \sum_{t=0}^{i+j} E_t^{ij} M_t^e \tag{9.5.30}$$

where the Hermite multipole-moment integrals are given by

$$M_t^e = \int_{-\infty}^{+\infty} x_C^e \Lambda_t \, dx \tag{9.5.31}$$

The expansion coefficients E_t^{ij} may be calculated according to the recurrence relations (9.5.6)–(9.5.8) and we here discuss only the evaluation of the Hermite integrals (9.5.31).

First, consider the Hermite integrals with $e = 0$. From (9.4.12), we know that these Hermite integrals reduce to

$$M_t^0 = \delta_{t0} \sqrt{\frac{\pi}{p}} \tag{9.5.32}$$

which vanishes for $t > 0$. The remaining Hermite integrals (with $e > 0$) may be generated by recursion. Incrementing e, we obtain

$$M_t^{e+1} = \int_{-\infty}^{+\infty} x_C^e x_C \Lambda_t \, dx \tag{9.5.33}$$

Making the substitution

$$x_C = x_P + X_{PC} \tag{9.5.34}$$

and applying the basic Hermite recurrence relation (9.4.10) in the usual way, we find

$$x_C \Lambda_t = x_P \Lambda_t + X_{PC} \Lambda_t = \frac{1}{2p} \Lambda_{t+1} + t \Lambda_{t-1} + X_{PC} \Lambda_t \tag{9.5.35}$$

which may be inserted into the Hermite integral (9.5.33) to yield

$$M_t^{e+1} = t M_{t-1}^e + X_{PC} M_t^e + \frac{1}{2p} M_{t+1}^e \tag{9.5.36}$$

Together with (9.5.32), this expression allows us to generate recursively the higher-order multipole moments from those of lower orders, noting that all Hermite integrals with $t > e$ are zero:

$$M_t^e = 0, \qquad t > e \tag{9.5.37}$$

To demonstrate (9.5.37), we apply partial integration to (9.5.31)

$$
M_t^e = \int_{-\infty}^{+\infty} x_C^e \left[\frac{\partial^t \exp(-px_P^2)}{\partial P_x^t} \right] dx = (-1)^t \int_{-\infty}^{+\infty} x_C^e \left[\frac{\partial^t \exp(-px_P^2)}{\partial x^t} \right] dx
$$
$$
= \int_{-\infty}^{+\infty} \left(\frac{\partial^t x_C^e}{\partial x^t} \right) \exp(-px_P^2)\, dx \tag{9.5.38}
$$

and note that the derivative in the integrand vanishes for $t > e$. The final expression for the Cartesian multipole-moment integrals therefore becomes [4]

$$
S_{ij}^e = \sum_{t=0}^{\min(i+j,e)} E_t^{ij} M_t^e \tag{9.5.39}
$$

where the Hermite-to-Cartesian coefficients are calculated recursively from (9.5.6) and (9.5.7), and the Hermite multipole-moment integrals are calculated recursively from (9.5.36).

Let us consider two examples. According to the above discussion, the x component of the overlap integral is simply

$$
S_{ij}^0 = E_0^{ij} \sqrt{\frac{\pi}{p}} \tag{9.5.40}
$$

and the total integral is therefore

$$
S_{ab}^{000} = E_0^{ij} E_0^{kl} E_0^{mn} \left(\frac{\pi}{p} \right)^{3/2} \tag{9.5.41}
$$

If we assume s orbitals, this expression becomes (see (9.5.8))

$$
S_{ab}^{000} = \exp(-\mu R_{AB}^2) \left(\frac{\pi}{p} \right)^{3/2} \tag{9.5.42}
$$

in accordance with (9.2.41). The x component of the dipole integral contains two terms (the first vanishes when both orbitals are of s type):

$$
S_{ij}^1 = (E_1^{ij} + X_{PC} E_0^{ij}) \sqrt{\frac{\pi}{p}} \tag{9.5.43}
$$

as is easily established from the recurrence relation of the Hermite multipole-moment integrals (9.5.36).

9.6 Gaussian quadrature for simple integrals

In the previous sections, we have studied two different approaches to the calculation of the simple one-electron integrals over Cartesian Gaussians – the Obara–Saika and McMurchie–Davidson schemes. Before discussing the evaluation of the more complicated Coulomb integrals by means of these techniques, we shall consider a third method for the evaluation of simple one-electron integrals over Cartesian Gaussians – *Gaussian quadrature* [8,9]. The purpose of this section is not only to present a useful method for the evaluation of the simple one-electron integrals, but also to serve as an introduction to the use of Gaussian quadrature in molecular calculations. As such, the present section provides the necessary mathematical background for understanding the Rys method [10] for the evaluation of Coulomb integrals in Section 9.11.

9.6.1 ORTHOGONAL POLYNOMIALS

Consider a set of polynomials $p_n(x)$ that are *orthogonal* in the sense that

$$\int_a^b p_m(x)p_n(x)W(x)\,\mathrm{d}x = h_n\delta_{mn} \tag{9.6.1}$$

where $W(x)$ is a *weight function* that is positive in the interval $[a, b]$ [8,9]. Standard examples of such *orthogonal polynomials* are the *Legendre polynomials* $P_n(x)$ of Section 6.4.1, the *Laguerre polynomials* $L_n^\alpha(x)$ of Section 6.5.1, the *Hermite polynomials* $H_n(x)$ of Section 6.6.6 and the *Chebyshev polynomials* $T_n(x)$, which fulfil the following *orthogonality relations*:

$$\int_{-1}^1 P_m(x)P_n(x)\,\mathrm{d}x = \frac{2}{2n+1}\delta_{mn} \tag{9.6.2}$$

$$\int_0^\infty L_m^\alpha(x)L_n^\alpha(x)x^\alpha\exp(-x)\,\mathrm{d}x = \frac{\Gamma(n+\alpha+1)}{n!}\delta_{mn} \tag{9.6.3}$$

$$\int_{-\infty}^\infty H_m(x)H_n(x)\exp(-x^2)\,\mathrm{d}x = \sqrt{\pi}2^n n!\delta_{mn} \tag{9.6.4}$$

$$\int_{-1}^1 \frac{T_m(x)T_n(x)}{\sqrt{1-x^2}}\,\mathrm{d}x = \frac{\pi}{2}(1+\delta_{n0})\delta_{mn} \tag{9.6.5}$$

Except for an overall scaling factor, these relations determine the orthogonal polynomials uniquely in the sense that the full set of polynomials to a given degree N may be generated by a Gram–Schmidt orthogonalization of the monomials x^k for $0 \le k \le N$. Alternatively, we may generate the same polynomials using a Rodrigues formula such as those given in Chapter 6 or by means of a *two-term recurrence relation*

$$(n+1)P_{n+1} = (2n+1)xP_n - nP_{n-1} \tag{9.6.6}$$

$$(n+1)L_{n+1}^\alpha = (2n+1+\alpha-x)L_n^\alpha - (n+\alpha)L_{n-1}^\alpha \tag{9.6.7}$$

$$H_{n+1} = 2xH_n - 2nH_{n-1} \tag{9.6.8}$$

$$T_{n+1} = 2xT_n - T_{n-1} \tag{9.6.9}$$

starting in all cases with

$$p_{-1}(x) = 0 \tag{9.6.10}$$

$$p_0(x) = 1 \tag{9.6.11}$$

The reader may wish to derive the recurrence relations for the Legendre, Laguerre and Hermite polynomials from the Rodrigues expressions (6.4.4), (6.5.1) and (6.6.29).

These and other orthogonal polynomials have many properties in common. We shall not attempt to explore in detail their common properties here, but we note (without proof) the nontrivial fact that an orthogonal polynomial $p_n(x)$ of degree n has n distinct *roots* or *zeros*

$$p_n(x_i) = 0, \qquad 1 \le i \le n \tag{9.6.12}$$

in the interval $[a, b]$ interleaving the $n - 1$ roots of the polynomial $p_{n-1}(x)$. As we shall see in the remainder of this section, the roots of orthogonal polynomials play an important role in the theory of Gaussian quadrature.

9.6.2 GAUSSIAN QUADRATURE

Consider a general polynomial $f_k(x)$ of degree k

$$f_k(x) = \sum_{j=0}^{k} c_j x^j \tag{9.6.13}$$

and let x_i be the n distinct roots (9.6.12) of the polynomial $p_n(x)$ orthogonal in the interval $[a, b]$ with weight function $W(x)$; see (9.6.1). If $k < 2n$, then [8,9]

$$\int_a^b f_k(x) W(x)\, dx = \sum_{i=1}^{n} w_i f_k(x_i) \tag{9.6.14}$$

where

$$w_i = \int_a^b W(x) \prod_{\substack{j=1 \\ j \ne i}}^{n} \frac{x - x_j}{x_i - x_j}\, dx \tag{9.6.15}$$

as shown in Section 9.6.3. In the n-point quadrature formula (9.6.14), the x_i are known as the *abscissae* and the w_i as the *weights*. The abscissae and weights are independent of the polynomial $f_k(x)$ but depend on the number of quadrature points n. Depending on what particular weight function and polynomials are used, the resulting quadrature schemes are known as Gauss–Legendre quadrature, Gauss–Laguerre quadrature, Gauss–Hermite quadrature, and so on.

The great advantage of *Gaussian quadrature* over other quadrature schemes is that an n-point integration is exact for all polynomials of degree $k < 2n$. Thus, if the abscissae had been chosen in a different manner – for example, distributed evenly in the interval $[a, b]$ rather than at the nodes of the orthogonal polynomials – then we would have to carry out a $2n$-point quadrature to make integration exact for all polynomials of degree $k < 2n$. In our applications, where the polynomial form in (9.6.13) is of known degree but rather complicated, the use of Gaussian quadrature greatly improves the efficiency of the evaluation, reducing the cost of evaluating the exact integral roughly by a factor of 2.

The application of Gaussian quadrature is not restricted to polynomial functions $f(x)$. Indeed, in most applications of Gaussian quadrature, the functions $f(x)$ are not polynomials but some more complicated function, the analytical integration over which presents difficulties [8]. In such cases, we employ Gaussian quadrature to provide an accurate numerical estimate of the true integral. By carrying out quadrature over a sufficiently large number of points (which is equivalent to approximating the integrand by a polynomial of sufficiently high degree) and by a clever choice of weight function $W(x)$ (which need not be present in the original integrand), we may hope to obtain an accurate numerical estimate of the true integral.

9.6.3 PROOF OF THE GAUSSIAN-QUADRATURE FORMULA

We shall demonstrate that (9.6.14) and (9.6.15) hold for all polynomials $f_k(x)$ of degree $k < 2n$. Such polynomials may be written in the form

$$f_k(x) = Q_q(x)p_n(x) + R_r(x) \qquad (9.6.16)$$

where $p_n(x)$ is the orthogonal polynomial of degree n and $Q_q(x)$ and $R_r(x)$ are polynomials of degrees $q < n$ and $r < n$. In this expression, we may consider $Q_q(x)$ the quotient of $f_k(x)$ and $p_n(x)$, with $R_r(x)$ being the remainder.

With this decomposition of the polynomial (9.6.16), the integral over $f_k(x)$ may be written as

$$\int_a^b f_k(x)W(x)\,\mathrm{d}x = \int_a^b R_r(x)W(x)\,\mathrm{d}x \qquad (9.6.17)$$

since, according to (9.6.1), the polynomial $p_n(x)$ is orthogonal to all polynomials of degree $q < n$:

$$\int_a^b Q_q(x)p_n(x)W(x)\,\mathrm{d}x = 0 \qquad (9.6.18)$$

To evaluate the integral over the remainder in (9.6.17), we note that $R_r(x)$ may be written as a *Lagrange interpolating polynomial*

$$R_r(x) = \sum_{i=1}^{n} R_r(x_i) \prod_{\substack{j=1 \\ j \neq i}}^{n} \frac{x - x_j}{x_i - x_j} \qquad (9.6.19)$$

as may be verified by direct substitution of the n distinct abscissae x_i, using the fact that two polynomials of degree less than n are identical if they have the same values at n points. Inserting this expression in (9.6.17), we obtain

$$\int_a^b f_k(x)W(x)\,\mathrm{d}x = \sum_{i=1}^{n} R_r(x_i) \int_a^b W(x) \prod_{\substack{j=1 \\ j \neq i}}^{n} \frac{x - x_j}{x_i - x_j}\,\mathrm{d}x \qquad (9.6.20)$$

which, combined with (9.6.15), yields the n-point quadrature expression, given as

$$\int_a^b f_k(x)W(x)\,\mathrm{d}x = \sum_{i=1}^{n} w_i R_r(x_i) \qquad (9.6.21)$$

The final quadrature formula (9.6.14) now follows immediately from the fact that

$$f_k(x_i) = Q_q(x_i)p_n(x_i) + R_r(x_i) = R_r(x_i) \qquad (9.6.22)$$

since $p_n(x)$ evaluates to zero at the roots x_i. Whereas the identity (9.6.17) follows from the use of the orthogonal polynomials $p_n(x)$ when rewriting $f_k(x)$ according to (9.6.16), the identity (9.6.22) follows from the use of the roots (9.6.12) as the grid points.

9.6.4 GAUSS–HERMITE QUADRATURE FOR SIMPLE INTEGRALS

Consider the evaluation of the multipole-moment integrals of the type

$$S_{ij}^e = \langle G_i | x_C^e | G_j \rangle \tag{9.6.23}$$

Using the Gaussian product rule (9.2.21), this integral becomes

$$S_{ij}^e = K_{ab}^x \int_{-\infty}^{\infty} x_A^i x_B^j x_C^e \exp(-p x_P^2)\, dx \tag{9.6.24}$$

where K_{ab}^x is given by (9.2.15). Making a simple substitution of variables, this integral may be written in the form

$$S_{ij}^e = \frac{K_{ab}^x}{\sqrt{p}} \int_{-\infty}^{\infty} \left(\frac{x}{\sqrt{p}} + X_{PA} \right)^i \left(\frac{x}{\sqrt{p}} + X_{PB} \right)^j \left(\frac{x}{\sqrt{p}} + X_{PC} \right)^e \exp(-x^2)\, dx \tag{9.6.25}$$

which, according to the theory of Gaussian quadrature, may be evaluated exactly as

$$S_{ij}^e = \frac{K_{ab}^x}{\sqrt{p}} \sum_{\kappa=1}^{\gamma} w_\kappa \left(\frac{x_\kappa}{\sqrt{p}} + X_{PA} \right)^i \left(\frac{x_\kappa}{\sqrt{p}} + X_{PB} \right)^j \left(\frac{x_\kappa}{\sqrt{p}} + X_{PC} \right)^e \tag{9.6.26}$$

where x_κ and w_κ are the γ roots and weights of the Hermite polynomial $H_\gamma(x)$ of degree

$$\gamma = \left[\frac{i+j+e}{2} \right] + 1 \tag{9.6.27}$$

Note that, in (9.6.26), the abscissae and weights are the same for all integrals of the same quantum number $i+j+e$. The Gauss–Hermite scheme for multipole-moment integrals may easily be extended to the evaluation of kinetic-energy integrals.

9.7 Coulomb integrals over spherical Gaussians

Coulomb-interaction integrals cannot be expressed in a closed analytical form. They may, however, be reduced to one-dimensional integrals, the numerical evaluation of which is relatively straight-forward. In the present section, we begin our discussion of Coulomb integrals by considering the electrostatics of simple spherical Gaussian charge distributions. In particular, we shall demonstrate that the potential arising from a spherical Gaussian distribution as well as the Coulomb repulsion between two such distributions may be described by means of a special mathematical function – the Boys function – related to the incomplete gamma function, whose mathematical properties and evaluation are considered in Section 9.8.

9.7.1 SPHERICAL GAUSSIAN CHARGE DISTRIBUTIONS

We consider the electrostatics of two *spherical Gaussian charge distributions* of exponents p and q, centred at $\mathbf{P}$ and $\mathbf{Q}$:

$$\rho_p(\mathbf{r}_P) = \left(\frac{p}{\pi} \right)^{3/2} \exp(-p r_P^2) \tag{9.7.1}$$

$$\rho_q(\mathbf{r}_Q) = \left(\frac{q}{\pi} \right)^{3/2} \exp(-q r_Q^2) \tag{9.7.2}$$

Each of these distributions corresponds to a positive unit charge since from (9.2.39)

$$\int \rho_p(\mathbf{r_P}) \, d\mathbf{r} = \int \rho_q(\mathbf{r_Q}) \, d\mathbf{r} = 1 \tag{9.7.3}$$

In this section, we shall consider the electrostatic potential at $\mathbf{C}$ due to $\rho_p(\mathbf{r_P})$

$$V_p(\mathbf{C}) = \int \frac{\rho_p(\mathbf{r_P})}{r_C} \, d\mathbf{r} \tag{9.7.4}$$

and the energy of repulsion between $\rho_p(\mathbf{r_{1P}})$ and $\rho_q(\mathbf{r_{2Q}})$:

$$V_{pq} = \int\int \frac{\rho_p(\mathbf{r_{1P}})\rho_q(\mathbf{r_{2Q}})}{r_{12}} \, d\mathbf{r_1} \, d\mathbf{r_2} \tag{9.7.5}$$

We expect the electrostatic potential and the repulsion energy to depend on the exponents and on the separations R_{PC} and R_{PQ}. For large separations, they should reduce to the expressions for point charges (i.e. charge distributions with infinite exponents).

9.7.2 THE POTENTIAL FROM A SPHERICAL GAUSSIAN CHARGE DISTRIBUTION

Unlike the simple one-electron integrals discussed in Sections 9.3–9.6, the potential integral (9.7.4) cannot be factorized in the Cartesian directions, the complications arising from the presence of the inverse operator r_C^{-1}. However, using (9.2.39), we may re-express the inverse operator in terms of a one-dimensional integral over a Gaussian function, which is separable in the Cartesian directions [1]:

$$\frac{1}{r_C} = \frac{1}{\sqrt{\pi}} \int_{-\infty}^{+\infty} \exp(-r_C^2 t^2) \, dt \tag{9.7.6}$$

This is the key step in treating integrals involving Coulomb interactions and Gaussian orbitals. We obtain the following expression for the potential $V_p(\mathbf{C})$:

$$V_p(\mathbf{C}) = \frac{p^{3/2}}{\pi^2} \int \exp(-pr_P^2) \left[\int_{-\infty}^{+\infty} \exp(-t^2 r_C^2) \, dt \right] d\mathbf{r} \tag{9.7.7}$$

We now invoke the Gaussian product rule (9.2.10)

$$V_p(\mathbf{C}) = \frac{p^{3/2}}{\pi^2} \int_{-\infty}^{+\infty} \left\{ \int \exp[-(p+t^2)r_S^2] \, d\mathbf{r} \right\} \exp\left(-\frac{pt^2}{p+t^2} R_{PC}^2 \right) dt \tag{9.7.8}$$

where $\mathbf{S}$ is a point on the line connecting $\mathbf{C}$ and $\mathbf{P}$:

$$\mathbf{S} = \frac{p\mathbf{P} + t^2\mathbf{C}}{p+t^2} \tag{9.7.9}$$

Integration over the spatial coordinates now gives

$$V_p(\mathbf{C}) = \frac{2p^{3/2}}{\sqrt{\pi}} \int_0^\infty (p+t^2)^{-3/2} \exp\left(-pR_{PC}^2 \frac{t^2}{p+t^2} \right) dt \tag{9.7.10}$$

where, to reduce the integration interval, we have used the fact that the integrand is a quadratic function in t. The variable substitution

$$u^2 = \frac{t^2}{p+t^2} \tag{9.7.11}$$

gives

$$dt = \sqrt{p}(1 - u^2)^{-3/2} \, du \tag{9.7.12}$$

with integration limits 0 and 1 for u. The integral (9.7.10) now becomes

$$V_p(\mathbf{C}) = \sqrt{\frac{4p}{\pi}} \int_0^1 \exp(-pR_{PC}^2 u^2) \, du \tag{9.7.13}$$

We have thus replaced integration over all space in (9.7.4) by a one-dimensional integration over a finite interval.

The integral appearing on the right-hand side of (9.7.13) is of central importance in the calculation of Coulomb-interaction integrals over Gaussian distributions. It belongs to a class of functions that we shall refer to as the *Boys function*

$$F_n(x) = \int_0^1 \exp(-xt^2)t^{2n} \, dt \tag{9.7.14}$$

The potential from the Gaussian charge distribution of a positive unit charge may therefore be written as

$$V_p(\mathbf{C}) = \sqrt{\frac{4p}{\pi}} F_0(pR_{PC}^2) \tag{9.7.15}$$

in terms of the Boys function. As we shall see, this special function also arises when we calculate the energy of interaction for two Gaussian charge distributions.

9.7.3 THE REPULSION BETWEEN SPHERICAL GAUSSIAN CHARGE DISTRIBUTIONS

The interaction between two charge distributions (9.7.5) may be calculated as the electrostatic energy of the second distribution in the potential due to the first distribution:

$$V_{pq} = \int V_p(\mathbf{r}_2)\rho_q(\mathbf{r}_{2Q}) \, d\mathbf{r}_2 \tag{9.7.16}$$

Since we have expressed the potential from the first charge distribution in terms of the Boys function (9.7.15), this interaction becomes

$$V_{pq} = \sqrt{\frac{4p}{\pi}} \left(\frac{q}{\pi}\right)^{3/2} \int F_0(pr_{2P}^2) \exp(-qr_{2Q}^2) \, d\mathbf{r}_2 \tag{9.7.17}$$

We now insert the definition of the Boys function (9.7.14), use the Gaussian product rule (9.2.10) and integrate over all space. We obtain

$$V_{pq} = \sqrt{\frac{4pq}{\pi}} \int_0^1 \frac{q}{(pt^2 + q)^{3/2}} \exp\left(-\frac{pqt^2 R_{PQ}^2}{pt^2 + q}\right) dt \tag{9.7.18}$$

Making the substitution

$$u^2 = \frac{p+q}{pt^2 + q} t^2 \tag{9.7.19}$$

with

$$dt = \frac{p+q}{(p+q-u^2 p)^{3/2}} \sqrt{q} \, du \tag{9.7.20}$$

we obtain the final result for the repulsion energy [1]

$$V_{pq} = \sqrt{\frac{4\alpha}{\pi}} F_0(\alpha R_{PQ}^2) \tag{9.7.21}$$

where α is the *reduced exponent*

$$\alpha = \frac{pq}{p+q} \tag{9.7.22}$$

The interaction between two spherical Gaussian distributions may therefore also be expressed in terms of the Boys function.

9.7.4 THE ELECTROSTATICS OF SPHERICAL GAUSSIAN DISTRIBUTIONS

To summarize the results from Sections 9.7.2 and 9.7.3, we have found that the following relationships hold for spherical Gaussian distributions of unit charge (9.7.1) and (9.7.2):

$$V_p(\mathbf{C}) = \int \frac{\rho_p(\mathbf{r}_P)}{r_C} \, d\mathbf{r} = \sqrt{\frac{4p}{\pi}} F_0(pR_{PC}^2) \tag{9.7.23}$$

$$V_{pq} = \int\int \frac{\rho_p(\mathbf{r}_{1P})\rho_q(\mathbf{r}_{2Q})}{r_{12}} \, d\mathbf{r}_1 \, d\mathbf{r}_2 = \sqrt{\frac{4\alpha}{\pi}} F_0(\alpha R_{PQ}^2) \tag{9.7.24}$$

These equations replace the usual classical electrostatic expressions R_{PC}^{-1} and R_{PQ}^{-1} for the interactions of unit point charges.

Figure 9.4 illustrates the potentials arising from positive unit Gaussian distributions of different exponents compared with that from a point charge. At large separations, the potential from a

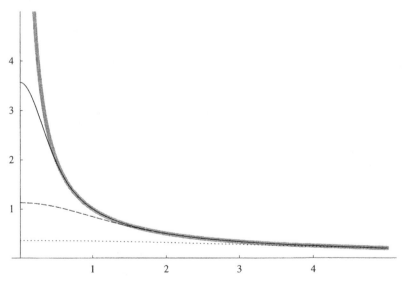

Fig. 9.4. The potentials from spherical Gaussian distributions of unit charge with exponents 10 (full line), 1 (dashed line) and $\frac{1}{10}$ (dotted line) plotted as functions of the distance from the centre of the charge distribution and compared with the corresponding potential from a point charge (thick grey line).

Gaussian distribution approaches closely that of a point-charge potential; at short distances, it is reduced. At the centre of the charge distribution, the potential is finite.

9.8 The Boys function

In Section 9.7, we found that the electrostatics of spherical Gaussian charge distributions are determined by the Boys function $F_n(x)$. Since this function plays such an important role in the evaluation of molecular integrals over Gaussian functions, we shall in this section discuss its properties and evaluation in more detail.

9.8.1 THE BOYS FUNCTION

The *Boys function* of order n is defined by

$$F_n(x) = \int_0^1 \exp(-xt^2)t^{2n}\, dt \tag{9.8.1}$$

for $x \geq 0$ [1,11]. It is a strictly positive function since the integrand is positive:

$$F_n(x) > 0 \tag{9.8.2}$$

Its derivatives are negative

$$\frac{dF_n(x)}{dx} = -F_{n+1}(x) < 0 \tag{9.8.3}$$

and the Boys function is therefore a strictly decreasing function. We also note that

$$F_n(x) - F_{n+1}(x) = \int_0^1 \exp(-xt^2)t^{2n}(1 - t^2)\, dt > 0 \tag{9.8.4}$$

since the integrand is positive within the integration range. We therefore obtain

$$F_n(x) > F_{n+1}(x) \tag{9.8.5}$$

for all n. The values at $x = 0$ may be expressed in closed form

$$F_n(0) = \int_0^1 t^{2n}\, dt = \frac{1}{2n + 1} \tag{9.8.6}$$

which, since the function is strictly decreasing, implies that

$$F_n(x) \leq \frac{1}{2n + 1} \tag{9.8.7}$$

For large values of x, we may determine the Boys function approximately from

$$F_n(x) = \int_0^1 \exp(-xt^2)t^{2n}\, dt \approx \int_0^\infty \exp(-xt^2)t^{2n}\, dt \qquad (x \text{ large}) \tag{9.8.8}$$

The latter integral can be integrated (see (6E.9.2)) to give

$$F_n(x) \approx \frac{(2n - 1)!!}{2^{n+1}}\sqrt{\frac{\pi}{x^{2n+1}}} \qquad (x \text{ large}) \tag{9.8.9}$$

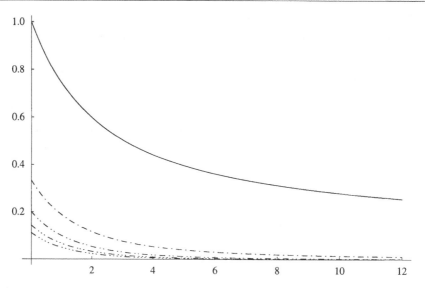

Fig. 9.5. The Boys function $F_n(x)$ for $0 \leq n \leq 4$. The number of consecutive dots in the lines corresponds to the order n of the function.

from which we conclude that the Boys function tends to zero as x tends to infinity. From (9.8.8), we also note that, for all x, the Boys function is bounded as

$$F_n(x) \leq \frac{(2n-1)!!}{2^{n+1}} \sqrt{\frac{\pi}{x^{2n+1}}} \qquad (9.8.10)$$

The properties discussed above are illustrated for $n \leq 4$ in Figure 9.5.

9.8.2 EVALUATION OF THE BOYS FUNCTION

We now consider the evaluation of the Boys function. Since this function is at the centre of integral evaluation, it is important that it is calculated efficiently. Different methods of evaluating the Boys function have been suggested in the literature [11]. Some of these approaches are discussed in the present subsection.

We first note that, for large x, (9.8.9) provides a good estimate. For small x, this approximation breaks down. However, since we know both the function and its derivatives at $x = 0$ in closed form (see (9.8.3) and (9.8.6)), we may easily construct a Taylor expansion for small values:

$$F_n(x) = \sum_{k=0}^{\infty} \frac{(-x)^k}{k!(2n+2k+1)} \qquad (9.8.11)$$

These two approximations – (9.8.9) for large values of x and (9.8.11) truncated at $k = 6$ for small values – are illustrated for $F_0(x)$ in Figure 9.6. Although we have a reasonable approximation to the Boys function by these two methods, they are inadequate for use in integral calculations, in which we would like errors of order 10^{-10} or smaller. Using the above two approximations, this accuracy is attained only in the regions $x < 0.18$ and $x > 19.35$. More accurate methods are therefore needed in practice.

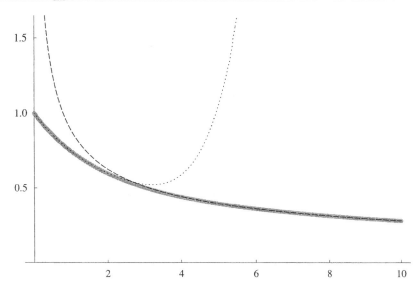

Fig. 9.6. The Boys function $F_0(x)$ (thick grey line) as approximated by the long-range formula (9.8.9) for large x (dashed line) and by a sixth-order Taylor expansion (9.8.11) around $x = 0$ for small x (dotted line).

A practicable alternative is to pretabulate the function at regular intervals x_t for small arguments. During the calculation of the integrals, the Boys function at x is then expanded around the nearest tabulated point $x_t = x - \Delta x$:

$$F_n(x_t + \Delta x) = \sum_{k=0}^{\infty} \frac{F_{n+k}(x_t)(-\Delta x)^k}{k!} \tag{9.8.12}$$

Using intervals of 0.1, convergence to errors smaller than 10^{-14} is obtained after six terms. For large arguments, the asymptotic formula (9.8.9) may be used. Other methods are also used in practice – for large arguments, the asymptotic formula (9.8.9) may be supplemented with more accurate asymptotic expansions, which are valid in a larger region; for small arguments, various polynomial fitting schemes may be applied, based for example on the use of Chebyshev polynomials.

The Boys functions of different orders are related by recursion. Integrating the Boys function by parts, we obtain for upward recursion (see Exercise 9.3)

$$F_{n+1}(x) = \frac{(2n+1)F_n(x) - \exp(-x)}{2x} \tag{9.8.13}$$

and for downward recursion

$$F_n(x) = \frac{2xF_{n+1}(x) + \exp(-x)}{2n+1} \tag{9.8.14}$$

We therefore only need to calculate the Boys function for the highest or the lowest order needed, obtaining the others by downward or upward recursion. For small x, upward recursion is unstable since it involves the difference between two almost equal numbers, making downward recursion the preferred strategy.

9.8.3   THE INCOMPLETE GAMMA FUNCTION

The Boys function is related to the incomplete gamma function, one of the special functions of mathematical physics. From Section 6.5.1, we recall that the *Euler gamma function* is defined as

$$\Gamma(n) = \int_0^\infty \exp(-t)t^{n-1}\,dt \tag{9.8.15}$$

where $n > 0$ is not necessarily an integer. As generalizations of the gamma function, the *incomplete gamma function* $P(n, x)$ and its complement $Q(n, x)$ are given by $(x > 0)$ [8,9]

$$P(n, x) = \frac{\gamma(n, x)}{\Gamma(n)} = \frac{1}{\Gamma(n)} \int_0^x \exp(-t)t^{n-1}\,dt \tag{9.8.16}$$

$$Q(n, x) = \frac{\Gamma(n, x)}{\Gamma(n)} = \frac{1}{\Gamma(n)} \int_x^\infty \exp(-t)t^{n-1}\,dt \tag{9.8.17}$$

and satisfy the relation

$$P(n, x) + Q(n, x) = 1 \tag{9.8.18}$$

For fixed n, the incomplete gamma function $P(n, x)$ increases strictly from 0 to 1 as x tends to infinity (see Figure 9.7).

To establish the relation of the Boys function to the incomplete gamma function, we substitute $u = xt^2$ in $F_{n-1/2}(x)$ and obtain

$$F_{n-1/2}(x) = \frac{1}{2x^n} \int_0^x \exp(-u)u^{n-1}\,du \tag{9.8.19}$$

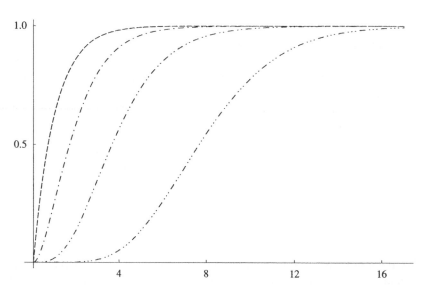

Fig. 9.7.   The incomplete gamma function $P(n, x)$ plotted as a function of x for $n = 1$ (no dots), $n = 2$ (one dot), $n = 4$ (two dots) and $n = 8$ (three dots).

From this expression, we note the following close relationship between the Boys function and the incomplete gamma function:

$$F_n(x) = \frac{\gamma\left(n + \frac{1}{2}, x\right)}{2x^{n+1/2}} = \frac{\Gamma\left(n + \frac{1}{2}\right) P\left(n + \frac{1}{2}, x\right)}{2x^{n+1/2}} \tag{9.8.20}$$

Indeed, in electronic-structure theory, $F_n(x)$ is often loosely referred to as the *incomplete gamma function*. In order to avoid any confusion with the special functions $P(n, x)$ and $\gamma(n, x)$, we have here adopted the term the *Boys function* for $F_n(x)$. From (9.8.20), we may interpret the Boys function as the function $\frac{1}{2}\Gamma(n + \frac{1}{2})x^{-1/2-n}$ weighted by the incomplete gamma function $P(n + \frac{1}{2}, x)$. For large x,

$$F_n(x) = P\left(n + \frac{1}{2}, x\right) \frac{\Gamma\left(n + \frac{1}{2}\right)}{2x^{n+1/2}} \approx \frac{\Gamma\left(n + \frac{1}{2}\right)}{2x^{n+1/2}} \qquad \text{(large } x\text{)} \tag{9.8.21}$$

since $P(n + \frac{1}{2}, x)$ tends to 1 at infinity. Using (6.5.8), this expression can be reduced to (9.8.9).

9.8.4 THE ERROR FUNCTION

Another important special function of mathematical physics is the *error function* [8,9]

$$\text{erf}(x) = \frac{2}{\sqrt{\pi}} \int_0^x \exp(-t^2)\, dt \tag{9.8.22}$$

Using (9.8.16) and (6.5.8) and carrying out a substitution of variables, we find that the error function is closely related to the incomplete gamma function

$$\text{erf}(x) = P\left(\tfrac{1}{2}, x^2\right) \tag{9.8.23}$$

and therefore also to the Boys function:

$$F_0(x) = \sqrt{\frac{\pi}{4x}}\, \text{erf}(\sqrt{x}) \tag{9.8.24}$$

Returning to the expressions (9.7.23) and (9.7.24) of electrostatics in Section 9.7, we note that these may now be cast in the simple form

$$\int \frac{\rho_p(\mathbf{r}_P)}{r_C}\, d\mathbf{r} = \frac{\text{erf}\left(\sqrt{p}R_{PC}\right)}{R_{PC}} \tag{9.8.25}$$

$$\iint \frac{\rho_p(\mathbf{r}_{1P})\rho_q(\mathbf{r}_{2Q})}{r_{12}}\, d\mathbf{r}_1\, d\mathbf{r}_2 = \frac{\text{erf}(\sqrt{\alpha}R_{PQ})}{R_{PQ}} \tag{9.8.26}$$

which makes their relationship to Coulomb's law and electrostatics particularly clear. For example, for a charge distribution of unit exponent, the error function $\text{erf}(R_{PC})$ represents the part of the charge which is inside a sphere centred at $\mathbf{P}$ and which extends to the point $\mathbf{C}$ where the potential is measured. Only this part of the total Gaussian charge distribution contributes to the potential at $\mathbf{C}$. As the exponent p tends to infinity, the error function tends to 1 and we recover the usual expression for a point charge. For an illustration of the error function and its relationship to the Coulomb potential, see Figure 9.8.

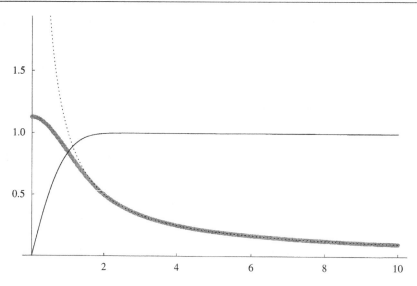

Fig. 9.8. The error function erf(x) (thin black line). Also plotted are the Coulomb potential from a unit point charge $1/x$ (dotted line) and from a Gaussian charge distribution of exponent 1 (thick grey line). The latter potential is equal to the point-charge Coulomb potential multiplied by the error function.

9.8.5   THE COMPLEMENTARY ERROR FUNCTION

The *complementary error function* [8,9] is defined as

$$\text{erfc}(x) = 1 - \text{erf}(x) \tag{9.8.27}$$

and may be used to characterize the deviation of the interaction of unit Gaussian charges from that of point charges:

$$\frac{1}{R_{PC}} - \int \frac{\rho_p(\mathbf{r}_P)}{r_C}\, \mathbf{dr} = \frac{\text{erfc}(\sqrt{p}R_{PC})}{R_{PC}} \tag{9.8.28}$$

$$\frac{1}{R_{PQ}} - \iint \frac{\rho_p(\mathbf{r}_{1P})\rho_q(\mathbf{r}_{2Q})}{r_{12}}\, \mathbf{dr}_1\mathbf{dr}_2 = \frac{\text{erfc}(\sqrt{\alpha}R_{PQ})}{R_{PQ}} \tag{9.8.29}$$

An upper bound to the complementary error function is easily established. From (9.8.22), we note that the complementary error function may be written in the form

$$\text{erfc}(x) = \frac{2}{\sqrt{\pi}} \int_x^\infty \exp(-t^2)\, dt \tag{9.8.30}$$

Rewriting the integrand and integrating by parts, we obtain

$$\text{erfc(x)} = -\frac{1}{\sqrt{\pi}} \int_x^\infty \frac{1}{t}\left[\frac{d}{dt}\exp(-t^2)\right]\, dt = \frac{\exp(-x^2)}{\sqrt{\pi}x} - \int_x^\infty \frac{\exp(-t^2)}{\sqrt{\pi}t^2}\, dt \tag{9.8.31}$$

Since the final integrand is positive, we find that the complementary error function is bounded as

$$\text{erfc}(x) \leq \frac{\exp(-x^2)}{\sqrt{\pi}x} \tag{9.8.32}$$

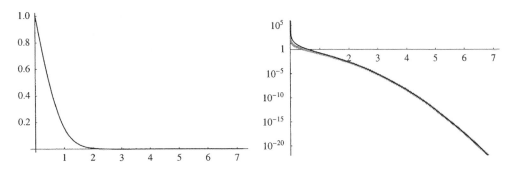

Fig. 9.9. The complementary error function erfc(x). On the left, we have plotted erfc(x) on a linear scale. On the right, we have plotted the function erfc(x)/x of (9.8.28) and (9.8.29) (grey line) and its upper bound $\exp(-x^2)/(\sqrt{\pi}x^2)$ (black line) on a logarithmic scale. The function erfc(x)/x represents the error incurred by calculating interactions between Gaussian charge distributions using point-charge models.

Consequently, we have the following upper bounds to the errors incurred by treating the interactions of unit Gaussians like those of point charges:

$$\frac{1}{R_{PC}} - \int \frac{\rho_p(\mathbf{r}_P)}{r_C}\, d\mathbf{r} \leq \frac{\exp(-pR_{PC}^2)}{\sqrt{\pi p}R_{PC}^2} \tag{9.8.33}$$

$$\frac{1}{R_{PQ}} - \iint \frac{\rho_p(\mathbf{r}_{1P})\rho_q(\mathbf{r}_{2Q})}{r_{12}}\, d\mathbf{r}_1 d\mathbf{r}_2 \leq \frac{\exp(-\alpha R_{PQ}^2)}{\sqrt{\pi\alpha}R_{PQ}^2} \tag{9.8.34}$$

In Figure 9.9, we have illustrated the validity of this approximation, which may be used to decide what integrals may be treated by multipole expansions.

9.8.6 THE CONFLUENT HYPERGEOMETRIC FUNCTION

Many functions of mathematical physics are related to the *Kummer confluent hypergeometric function*, defined as [9]:

$$M(a, b, x) = \sum_{k=0}^{\infty} \frac{(a)_k}{k!(b)_k} x^k \tag{9.8.35}$$

Here $(a)_k$ is the *Pochhammer symbol*

$$(a)_k = \frac{\Gamma(a+k)}{\Gamma(a)} \tag{9.8.36}$$

which for integral k takes on the values

$$(a)_k = \begin{cases} 1 & k = 0 \\ a(a+1)(a+2)\cdots(a+k-1) & k > 0 \\ \dfrac{1}{(a-1)(a-2)\cdots(a-k)} & k < 0 \end{cases} \tag{9.8.37}$$

For negative integers a, $M(a, b, x)$ yields a polynomial provided $b < a < 0$ or $a < 0 < b$; when neither a nor b is a negative integer, $M(a, b, x)$ constitutes an infinite convergent series. Among its many special cases are the Hermite and Laguerre polynomials and the exponential function $\exp(x)$.

We are here interested in the relationship of the Kummer function to the Boys function. From (9.8.35) we obtain, using (9.8.37),

$$M\left(n + \tfrac{1}{2}, n + \tfrac{3}{2}, -x\right) = (2n + 1)\sum_{k=0}^{\infty} \frac{(-x)^k}{k!(2n + 2k + 1)} \tag{9.8.38}$$

Comparing with (9.8.11), we find that the Boys function is a (renormalized) special case of the Kummer function:

$$F_n(x) = \frac{M\left(n + \tfrac{1}{2}, n + \tfrac{3}{2}, -x\right)}{2n + 1} \tag{9.8.39}$$

Therefore, the standard relations that have been established for the Kummer function hold for the Boys function as well. For example, for the Boys function, the general relations [9]

$$M'(a, b, x) = \frac{a}{b}M(a + 1, b + 1, x) \tag{9.8.40}$$

$$M(a + 1, b + 1, x) = \frac{b(x - b + 1)M(a, b, x) + b(b - 1)M(a - 1, b - 1, x)}{ax} \tag{9.8.41}$$

reduce to

$$F'_n(x) = -F_{n+1}(x) \tag{9.8.42}$$

$$F_{n+1}(x) = \frac{(2x + 2n + 1)F_n(x) - (2n - 1)F_{n-1}(x)}{2x} \tag{9.8.43}$$

The first relation is the usual differentiation formula (9.8.3) and the second is equivalent to the recurrence relation (9.8.13), from which it may be obtained by elimination of $\exp(-x)$.

9.9 The McMurchie–Davidson scheme for Coulomb integrals

Having discussed the electrostatics of spherical Gaussian distributions and the Boys function, we are now in a position to consider the evaluation of one- and two-electron Coulomb integrals over nonspherical Gaussians. For this task, several schemes have been developed. In the present chapter, we shall consider three such schemes: the McMurchie–Davidson scheme [6], in which the Cartesian integrals are expanded in an intermediate set of integrals over Hermite Gaussians; the Obara–Saika scheme [5], in which the Cartesian Coulomb integrals are obtained directly from the Boys function by recursion; and the Rys-polynomial scheme [10], in which the evaluation of the Boys function is avoided by the application of a special Gaussian-quadrature scheme. Different modifications and further developments of these schemes exist – in the PRISM scheme, for example, the formation of contracted integrals (for segmented sets) is carried out in a particularly efficient manner [12,13].

In the present section, we consider the evaluation of Cartesian Coulomb integrals by the McMurchie–Davidson scheme. First, we show how Coulomb integrals over Hermite Gaussians may be obtained by differentiating the Boys function, and we derive a set of recurrence relations for obtaining these derivatives. Next, we discuss how Coulomb integrals over Cartesian Gaussians may be obtained from integrals over Hermite Gaussians by expansion or recursion.

9.9.1 HERMITE COULOMB INTEGRALS

In Section 9.7, we discussed the evaluation of Coulomb integrals over spherical Gaussians. We now go one step further and consider nonspherical electron distributions as described by Hermite Gaussians. The one-electron Coulomb integral can then be expressed as

$$V_{tuv}^{efg} = \int \Lambda_{tuv}(\mathbf{r}) \left(\frac{\partial}{\partial C_x} \right)^e \left(\frac{\partial}{\partial C_y} \right)^f \left(\frac{\partial}{\partial C_z} \right)^g r_C^{-1} \, d\mathbf{r} \tag{9.9.1}$$

where, for example, V_{tuv}^{000} is the potential and V_{tuv}^{100}, V_{tuv}^{010} and V_{tuv}^{001} are the three components of the electric field. The two-electron Coulomb-interaction integral is

$$V_{tuv;\tau\upsilon\phi} = \iint \frac{\Lambda_{tuv}(\mathbf{r}_1)\Lambda_{\tau\upsilon\phi}(\mathbf{r}_2)}{r_{12}} \, d\mathbf{r}_1 d\mathbf{r}_2 \tag{9.9.2}$$

where Λ_{tuv} is a Hermite Gaussian of exponent p centred on $\mathbf{P}$ and $\Lambda_{\tau\upsilon\phi}$ a Hermite Gaussian of exponent q centred on $\mathbf{Q}$. Inserting the definition of Hermite Gaussians (9.4.1) and taking the differential operators outside the integration sign, we obtain

$$V_{tuv}^{efg} = \left(\frac{\partial}{\partial P_x} \right)^t \left(\frac{\partial}{\partial P_y} \right)^u \left(\frac{\partial}{\partial P_z} \right)^v \left(\frac{\partial}{\partial C_x} \right)^e \left(\frac{\partial}{\partial C_y} \right)^f \left(\frac{\partial}{\partial C_z} \right)^g \int \frac{\exp(-pr_P^2)}{r_C} \, d\mathbf{r} \tag{9.9.3}$$

and

$$V_{tuv;\tau\upsilon\phi} = \left(\frac{\partial}{\partial P_x} \right)^t \left(\frac{\partial}{\partial P_y} \right)^u \left(\frac{\partial}{\partial P_z} \right)^v \left(\frac{\partial}{\partial Q_x} \right)^\tau \left(\frac{\partial}{\partial Q_y} \right)^\upsilon \left(\frac{\partial}{\partial Q_z} \right)^\phi$$

$$\times \iint \frac{\exp(-pr_{1P}^2)\exp(-qr_{2Q}^2)}{r_{12}} \, d\mathbf{r}_1 \, d\mathbf{r}_2 \tag{9.9.4}$$

The integrals appearing in these expressions are the Coulomb potential and interaction integrals for spherical charge distributions, introduced in Section 9.7.1. Equations (9.7.23) and (9.7.24) may now be used to replace these integrals by the Boys function. Taking the normalization of the charge distributions (9.7.1) and (9.7.2) into account, we obtain

$$V_{tuv}^{efg} = \frac{2\pi}{p} \left(\frac{\partial}{\partial P_x} \right)^t \left(\frac{\partial}{\partial P_y} \right)^u \left(\frac{\partial}{\partial P_z} \right)^v \left(\frac{\partial}{\partial C_x} \right)^e \left(\frac{\partial}{\partial C_y} \right)^f \left(\frac{\partial}{\partial C_z} \right)^g F_0(pR_{PC}^2) \tag{9.9.5}$$

and

$$V_{tuv;\tau\upsilon\phi} = \frac{2\pi^{5/2}}{pq\sqrt{p+q}} \left(\frac{\partial}{\partial P_x} \right)^t \left(\frac{\partial}{\partial P_y} \right)^u \left(\frac{\partial}{\partial P_z} \right)^v \left(\frac{\partial}{\partial Q_x} \right)^\tau \left(\frac{\partial}{\partial Q_y} \right)^\upsilon \left(\frac{\partial}{\partial Q_z} \right)^\phi F_0(\alpha R_{PQ}^2) \tag{9.9.6}$$

The definition of Hermite Gaussians in terms of differentiation has thus enabled us to express integrals over nonspherical distributions as derivatives of integrals over spherical distributions.

These derivatives may be further simplified since the Boys function depends only on the relative separation of the two centres, giving the following major simplifications:

$$V_{tuv}^{efg} = (-1)^{e+f+g} \frac{2\pi}{p} \left(\frac{\partial}{\partial P_x}\right)^{t+e} \left(\frac{\partial}{\partial P_y}\right)^{u+f} \left(\frac{\partial}{\partial P_z}\right)^{v+g} F_0(pR_{PC}^2) \qquad (9.9.7)$$

and

$$V_{tuv;\tau v\phi} = (-1)^{\tau+v+\phi} \frac{2\pi^{5/2}}{pq\sqrt{p+q}} \left(\frac{\partial}{\partial P_x}\right)^{t+\tau} \left(\frac{\partial}{\partial P_y}\right)^{u+v} \left(\frac{\partial}{\partial P_z}\right)^{v+\phi} F_0(\alpha R_{PQ}^2) \qquad (9.9.8)$$

These simplifications occur since we use Hermite (rather than Cartesian) Gaussians to describe the nonspherical distributions. Since the derivatives of the Boys function play such an important role, we introduce the Hermite Coulomb integrals

$$R_{tuv}(p, \mathbf{R}_{PC}) = \left(\frac{\partial}{\partial P_x}\right)^t \left(\frac{\partial}{\partial P_y}\right)^u \left(\frac{\partial}{\partial P_z}\right)^v F_0(pR_{PC}^2) \qquad (9.9.9)$$

The one- and two-electron Hermite Coulomb integrals (9.9.1) and (9.9.2) may now be expressed as [6]

$$V_{tuv}^{efg} = \int \Lambda_{tuv}(\mathbf{r}) \left(\frac{\partial}{\partial C_x}\right)^e \left(\frac{\partial}{\partial C_y}\right)^f \left(\frac{\partial}{\partial C_z}\right)^g r_C^{-1} \, d\mathbf{r}$$

$$= (-1)^{e+f+g} \frac{2\pi}{p} R_{t+e,u+f,v+g}(p, \mathbf{R}_{PC}) \qquad (9.9.10)$$

and

$$V_{tuv;\tau v\phi} = \int\int \frac{\Lambda_{tuv}(\mathbf{r}_1)\Lambda_{\tau v\phi}(\mathbf{r}_2)}{r_{12}} \, d\mathbf{r}_1 \, d\mathbf{r}_2$$

$$= (-1)^{\tau+v+\phi} \frac{2\pi^{5/2}}{pq\sqrt{p+q}} R_{t+\tau,u+v,v+\phi}(\alpha, \mathbf{R}_{PQ}) \qquad (9.9.11)$$

Hence, to calculate Coulomb integrals over Hermite Gaussians, we simply take the derivatives of the Boys function. We also see that field and field-gradient integrals may be calculated in the same way as Coulomb-potential integrals. It remains, however, to develop a method for calculating the derivatives of the Boys function.

9.9.2 THE EVALUATION OF HERMITE COULOMB INTEGRALS

To develop a scheme for evaluating the Hermite Coulomb integrals (9.9.9), we note that the first derivative involves the first-order Boys function

$$R_{100}(p, \mathbf{R}_{PC}) = -2pX_{PC} \int_0^1 \exp(-pR_{PC}^2 t^2) t^2 \, dt = -2pX_{PC} F_1(pR_{PC}^2) \qquad (9.9.12)$$

In general, therefore, higher derivatives are linear combinations of Boys functions of different orders and our task is now to develop a recursive scheme by which the Hermite integrals R_{tuv} for

$t + u + v \leq N$ may be calculated from the Boys functions F_n of order $n \leq N$. To this end, we introduce the auxiliary integrals

$$R_{tuv}^n(p, \mathbf{R}_{PC}) = \left(\frac{\partial}{\partial P_x}\right)^t \left(\frac{\partial}{\partial P_y}\right)^u \left(\frac{\partial}{\partial P_z}\right)^v R_{000}^n(p, \mathbf{R}_{PC}) \tag{9.9.13}$$

where

$$R_{000}^n(p, \mathbf{R}_{PC}) = (-2p)^n F_n(pR_{PC}^2) \tag{9.9.14}$$

Note that the definition (9.9.13) includes the integrals R_{tuv} and F_n as special cases. We must now relate the integrals R_{tuv}^n by recursion. Incrementing t, we obtain

$$R_{t+1,u,v}^n(p, \mathbf{R}_{PC}) = \left(\frac{\partial}{\partial P_x}\right)^t \left(\frac{\partial}{\partial P_y}\right)^u \left(\frac{\partial}{\partial P_z}\right)^v \frac{\partial R_{000}^n(p, \mathbf{R}_{PC})}{\partial P_x} \tag{9.9.15}$$

Using (9.9.14) and differentiating the nth order Boys function, we find

$$R_{t+1,u,v}^n(p, \mathbf{R}_{PC}) = \left(\frac{\partial}{\partial P_x}\right)^t X_{PC} \, R_{0uv}^{n+1}(p, \mathbf{R}_{PC}) \tag{9.9.16}$$

The operator on the right-hand side may be written as

$$\left(\frac{\partial}{\partial P_x}\right)^t X_{PC} = \left[\left(\frac{\partial}{\partial P_x}\right)^t, X_{PC}\right] + X_{PC}\left(\frac{\partial}{\partial P_x}\right)^t = t\left(\frac{\partial}{\partial P_x}\right)^{t-1} + X_{PC}\left(\frac{\partial}{\partial P_x}\right)^t \tag{9.9.17}$$

where we have used (9.4.7). Inserting (9.9.17) in (9.9.16), we obtain the recurrence relation for the Hermite integrals. The formulae for increments in the three indices are [6]

$$R_{t+1,u,v}^n(p, \mathbf{R}_{PC}) = tR_{t-1,u,v}^{n+1}(p, \mathbf{R}_{PC}) + X_{PC} \, R_{tuv}^{n+1}(p, \mathbf{R}_{PC}) \tag{9.9.18}$$

$$R_{t,u+1,v}^n(p, \mathbf{R}_{PC}) = uR_{t,u-1,v}^{n+1}(p, \mathbf{R}_{PC}) + Y_{PC} \, R_{tuv}^{n+1}(p, \mathbf{R}_{PC}) \tag{9.9.19}$$

$$R_{t,u,v+1}^n(p, \mathbf{R}_{PC}) = vR_{t,u,v-1}^{n+1}(p, \mathbf{R}_{PC}) + Z_{PC} \, R_{tuv}^{n+1}(p, \mathbf{R}_{PC}) \tag{9.9.20}$$

In this way, all Hermite Gaussians of order $t + u + v \leq N$ may be calculated from the Boys functions of order $n \leq N$ by recursion. The same recurrences may be used to evaluate the integrals (9.9.11) except that X_{PQ} replaces X_{PC} and so on.

To summarize, we have shown that the Coulomb integrals over Hermite Gaussians (9.9.10) and (9.9.11) may be obtained from the Boys function by three simple recurrence relations (9.9.18)–(9.9.20). The one- and two-electron Coulomb integrals follow essentially the same scheme; field and field-gradient integrals may be calculated in the same way as the one-electron Coulomb integrals. The remaining task is to generate the integrals over Cartesian Gaussians from the Hermite integrals.

9.9.3 CARTESIAN COULOMB INTEGRALS BY HERMITE EXPANSION

Having obtained the Hermite Coulomb integrals, we are now in a position to calculate the Cartesian Coulomb integrals

$$V_{ab}^{efg} = \langle G_a | \left(\frac{\partial}{\partial C_x}\right)^e \left(\frac{\partial}{\partial C_y}\right)^f \left(\frac{\partial}{\partial C_z}\right)^g r_C^{-1} | G_b \rangle \tag{9.9.21}$$

$$g_{abcd} = \langle G_a(\mathbf{r}_1)G_b(\mathbf{r}_1)|\frac{1}{r_{12}}|G_c(\mathbf{r}_2)G_d(\mathbf{r}_2)\rangle \tag{9.9.22}$$

In terms of overlap distributions, these integrals are given by

$$V_{ab}^{efg} = \int \Omega_{ab}(\mathbf{r})\left(\frac{\partial}{\partial C_x}\right)^e\left(\frac{\partial}{\partial C_y}\right)^f\left(\frac{\partial}{\partial C_z}\right)^g r_C^{-1}\mathbf{dr} \tag{9.9.23}$$

$$g_{abcd} = \int\int \frac{\Omega_{ab}(\mathbf{r}_1)\Omega_{cd}(\mathbf{r}_2)}{r_{12}}\mathbf{dr}_1\mathbf{dr}_2 \tag{9.9.24}$$

Inserting the Hermite expansions of the overlap distributions (9.5.1), we obtain

$$V_{ab}^{efg} = \sum_{tuv} E_{tuv}^{ab}\int \Lambda_{tuv}(\mathbf{r})\left(\frac{\partial}{\partial C_x}\right)^e\left(\frac{\partial}{\partial C_y}\right)^f\left(\frac{\partial}{\partial C_z}\right)^g r_C^{-1}\mathbf{dr} \tag{9.9.25}$$

$$g_{abcd} = \sum_{tuv} E_{tuv}^{ab}\sum_{\tau\upsilon\phi} E_{\tau\upsilon\phi}^{cd}\int\int \frac{\Lambda_{tuv}(\mathbf{r}_1)\Lambda_{\tau\upsilon\phi}(\mathbf{r}_2)}{r_{12}}\mathbf{dr}_1\mathbf{dr}_2 \tag{9.9.26}$$

with the notation

$$E_{tuv}^{ab} = E_t^{ij}E_u^{kl}E_v^{mn} \tag{9.9.27}$$

We recall that Λ_{tuv} is a Hermite Gaussian of exponent p centred on $\mathbf{P}$

$$p = a + b \tag{9.9.28}$$

$$\mathbf{P} = \frac{a\mathbf{A} + b\mathbf{B}}{p} \tag{9.9.29}$$

and that $\Lambda_{\tau\upsilon\phi}$ is a Hermite Gaussian of exponent q centred on $\mathbf{Q}$

$$q = c + d \tag{9.9.30}$$

$$\mathbf{Q} = \frac{c\mathbf{C} + d\mathbf{D}}{q} \tag{9.9.31}$$

Inserting the Hermite integrals (9.9.10) and (9.9.11) in the expressions for the Cartesian integrals (9.9.25) and (9.9.26), we obtain the final expressions for Cartesian Coulomb integrals [6]:

$$V_{ab}^{efg} = (-1)^{e+f+g}\frac{2\pi}{p}\sum_{tuv} E_{tuv}^{ab}R_{t+e,u+f,v+g}(p, \mathbf{R}_{PC}) \tag{9.9.32}$$

$$g_{abcd} = \frac{2\pi^{5/2}}{pq\sqrt{p+q}}\sum_{tuv} E_{tuv}^{ab}\sum_{\tau\upsilon\phi}(-1)^{\tau+\upsilon+\phi}E_{\tau\upsilon\phi}^{cd}R_{t+\tau,u+\upsilon,v+\phi}(\alpha, \mathbf{R}_{PQ}) \tag{9.9.33}$$

We see that the Cartesian integrals may be calculated straightforwardly from the Hermite integrals of Section 9.9.2, the basic manipulation being a transformation of an array from one basis (Hermite Gaussian overlap distributions) to another basis (Cartesian Gaussian overlap distributions). In practice, this transformation (9.9.33) is the time-consuming step in the evaluation of two-electron integrals. Note that, once the Hermite Coulomb integrals have been generated, the calculation of field and field-gradient integrals is no more expensive than the calculation of the one-electron Coulomb integrals since the summations are the same.

The nuclear-attraction part of the one-electron Hamiltonian contains one contribution from each nucleus, obtained by multiplying the charge of the nucleus by the potential at its position. Introducing the charges of -1 for the electron and Z_K for the nuclei, we obtain

$$h_{ab}^{\mathrm{NA}} = -\sum_K Z_K V_{ab}^{000}(\mathbf{C}_K) = -\frac{2\pi}{p} \sum_{tuv} E_{tuv}^{ab} \sum_K Z_K R_{tuv}(p, \mathbf{R}_{\mathrm{PC}_K}) \tag{9.9.34}$$

where the summation is over the full set of nuclei. Note that the contributions from all the nuclei may be added together before we transform from Hermite to Cartesian integrals.

9.9.4 CARTESIAN COULOMB INTEGRALS BY HERMITE RECURSION

Having derived formulae for calculating the Coulomb integrals by expansion, we now discuss their calculation by recursion. To avoid complicated notation, we consider only the one-electron potential integrals. One-electron field and field-gradient integrals, as well as two-electron integrals, are obtained by a straightforward modification of the scheme. We introduce the integrals

$$\Phi_{ijklmn}^{tuv} = \int \frac{\Omega_{ij}^t \Omega_{kl}^u \Omega_{mn}^v}{r_{\mathrm{C}}} \, d\mathbf{r} \tag{9.9.35}$$

where we use the mixed Cartesian and Hermite overlap distributions defined in (9.5.22). These integrals contain as special cases the Cartesian and Hermite integrals

$$\Phi_{ijklmn}^{000} = V_{ab}^{000} \tag{9.9.36}$$

$$\Phi_{000000}^{tuv} = K_{ab}^{xyz} \int \frac{\Lambda_{tuv}(\mathbf{r})}{r_{\mathrm{C}}} \, d\mathbf{r} \tag{9.9.37}$$

where K_{ab}^{xyz} is the product of three preexponential factors, one for each Cartesian direction (9.2.15). Using the recurrence relations for the mixed overlap distributions (9.5.27) and (9.5.28), we obtain from (9.9.35) the following recurrences

$$\Phi_{i+1,j,k,l,m,n}^{tuv} = \frac{1}{2p} \Phi_{ijklmn}^{t+1,u,v} + X_{\mathrm{PA}} \Phi_{ijklmn}^{tuv} + t\Phi_{ijklmn}^{t-1,u,v} \tag{9.9.38}$$

$$\Phi_{i,j+1,k,l,m,n}^{tuv} = \frac{1}{2p} \Phi_{ijklmn}^{t+1,u,v} + X_{\mathrm{PB}} \Phi_{ijklmn}^{tuv} + t\Phi_{ijklmn}^{t-1,u,v} \tag{9.9.39}$$

and similarly for increments in the other indices. Starting from the Hermite integrals (9.9.37) with $t \leq N$, $u \leq N$ and $v \leq N$, we may use these relations to generate all Cartesian integrals with $i + j \leq N$, $k + l \leq N$ and $m + n \leq N$.

9.9.5 COMPUTATIONAL CONSIDERATIONS FOR THE ONE-ELECTRON INTEGRALS

The calculation of integrals is a complicated business, where compromises must be made between considerations such as efficiency, memory usage, and programming effort. Moreover, different implementations may be superior in different ways – some for basis sets consisting of highly contracted functions of low angular momentum, others for basis sets with a large number of functions of high angular momentum, and so on. In the following, we shall give some consideration to the different algorithms that may be used in order to calculate integrals efficiently. Although the time-critical task is the calculation of two-electron integrals, we begin by considering in this subsection the one-electron Coulomb integrals within the McMurchie–Davidson scheme, noting that many of the considerations apply also to the two-electron integrals discussed in Section 9.9.6 as well as to the Obara–Saika and Rys schemes discussed in Sections 9.10 and 9.11.

In the McMurchie–Davidson scheme, the one-electron Coulomb integrals may be written in the following manner

$$V^{000}_{ijklmn} = \frac{2\pi}{p} \sum_{tuv} E^{ij}_t E^{kl}_u E^{mn}_v R_{tuv} \tag{9.9.40}$$

In our analysis, we shall assume that both orbitals have angular momentum L and contain p primitive functions. Each shell then contains $(L + 1)(L + 2)/2$ orbitals in the Cartesian basis and $2L + 1$ orbitals in the spherical-harmonic basis.

The obvious way to generate the integrals (9.9.40) is first to calculate the expansion coefficients and Hermite integrals and then to carry out the summation. As the number of Cartesian orbitals in each shell scales as L^2, the number of Cartesian integrals scales as $L^4 p^2$. For each Cartesian integral, we carry out a simultaneous summation over three indices, each of which scales as L, so the total operation count scales as $L^7 p^2$.

We must also consider the cost of generating the expansion coefficients and the Hermite integrals in (9.9.40). The number of Hermite integrals scales as $L^3 p^2$ and their evaluation scales as $L^4 p^2$. The $L^4 p^2$ dependence for the construction arises from the use of the auxiliary four-index integrals R^n_{tuv}, where each index scales as L. The evaluation of the Boys function scales only as $L p^2$. Finally, the number of expansion coefficients E^{ij}_t scales as $L^3 p^2$ and their evaluation also as $L^3 p^2$ since no auxiliary elements are used in the calculation. The evaluation of the primitive Cartesian integrals is therefore dominated by the transformation (9.9.40), with a total cost of $L^7 p^2$.

To produce the final integrals over contracted functions in a spherical-harmonic basis, we must carry out a set of transformations of the primitive Cartesian integrals. Assuming a segmented basis set, the transformation to the contracted basis is a simple data reduction step, with a cost that scales as the number of primitive Cartesian functions – that is, as $L^4 p^2$. The generation of integrals over spherical-harmonic Gaussians requires one transformation for each orbital index. The transformation of the first orbital scales as L^5 and that of the second orbital as L^4. The overall cost of the Cartesian-to-spherical transformation then scales as L^5, assuming that it is carried out after the transformation of primitive to contracted functions. We conclude that the overall cost for the calculation of the one-electron Coulomb integrals (as described here) is $L^7 p^2$.

There is one obvious way to bring this cost down. Rewriting the Cartesian integrals in terms of three partial summations, we obtain

$$V^{000}_{ijklmn} = \frac{2\pi}{p} \sum_t E^{ij}_t \sum_u E^{kl}_u \sum_v E^{mn}_v R_{tuv} \tag{9.9.41}$$

Noting that the upper indices of the expansion coefficients mn scale as $L^2 p^2$, we find that the first partial summation scales as $L^5 p^2$ and produces $L^4 p^2$ intermediates. In the second transformation, we combine these $L^4 p^2$ intermediates with $L^3 p^2$ expansion coefficients, in a process that scales as $L^6 p^2$ and produces $L^5 p^2$ new intermediates. Finally, in the last summation, the upper indices ij are fixed by the requirement that the total angular momentum of each orbital is equal to L. In this step, therefore, we combine $L p^2$ coefficients with $L^5 p^2$ intermediates in a step that scales as $L^5 p^2$ and produces $L^4 p^2$ integrals. The overall scaling of the integration has thus been reduced to $L^6 p^2$.

Table 9.1 Cost and memory requirements of the McMurchie–Davidson scheme for one-electron Coulomb integrals

	Cost	Memory
Boys functions	Lp^2	Lp^2
Hermite integrals	L^4p^2	L^3p^2
Expansion coefficients	L^3p^2	L^3p^2
Cartesian integrals	L^7p^2/L^6p^2	L^4p^2
Primitive contraction	L^4p^2	L^4
Solid harmonics	L^5	L^2

The cost of the two schemes (9.9.40) and (9.9.41) for the evaluation of the one-electron Coulomb integrals is summarized in Table 9.1. At this point, it is worth noting that factors such as L^6p^2 and L^7p^2 are not always good indicators of the overall cost of a given process. For low angular momentum, in particular, lower-order terms may dominate the evaluation, making the cost analysis rather difficult. In practice, except for orbitals of high angular momentum, the three-step transformation (9.9.41) will not be faster than the original one-step transformation (9.9.40).

Let us now consider a somewhat different approach where, in the first *vertical step*, we evaluate integrals of the form

$$V_{i0k0m0}^{000} = \frac{2\pi}{p} \sum_{tuv} E_t^{i0} E_u^{k0} E_v^{m0} R_{tuv} \tag{9.9.42}$$

Next, we transform the integrals to the contracted basis and then, in the *horizontal step*, we apply recurrence relations such as

$$V_{i,j+1,k,l,m,n}^{000} = V_{i+1,j,k,l,m,n}^{000} + X_{AB} V_{ijklmn}^{000} \tag{9.9.43}$$

to transfer angular momentum from the first to the second orbital; see (9.5.9). This step may be carried out after the transformation to the contracted basis since it does not depend on the exponents of the primitive functions. In the final step, the orbitals are transformed to the spherical-harmonic basis.

For this procedure to work, we must in the vertical step (9.9.42) generate integrals not only of angular momentum L, but of all angular momenta from L to $2L$ since these are all needed in the horizontal step (9.9.43). Still, since the functions of nonzero angular momentum are restricted to the first orbital, the number of Cartesian integrals in (9.9.42) scales as L^3p^2 rather than as L^4p^2 for (9.9.40), reducing the cost of this step to L^6p^2 or L^4p^2 (if carried out in a three-step manner). The cost of the various steps of this *vertical–horizontal algorithm* is given in Table 9.2. This algorithm should be contrasted with the original scheme in Table 9.1, with an overall cost of L^7p^2 for the first step (9.9.40) and L^4p^2 for the contraction.

9.9.6 COMPUTATIONAL CONSIDERATIONS FOR THE TWO-ELECTRON INTEGRALS

Having considered the evaluation of the one-electron Coulomb integrals, let us turn our attention to the two-electron repulsion integrals within the McMurchie–Davidson scheme. Clearly, many of the considerations for the one-electron integrals apply also to the two-electron integrals. In addition, some new considerations arise because of the higher complexity of the two-electron integrals.

Table 9.2 Cost and memory requirements of the vertical–horizontal McMurchie–Davidson scheme for one-electron Coulomb integrals

	Cost	Memory
Boys functions	Lp^2	Lp^2
Hermite integrals	L^4p^2	L^3p^2
Expansion coefficients	L^2p^2	L^2p^2
Vertical transformation	L^6p^2/L^4p^2	L^3p^2
Primitive contraction	L^3p^2	L^3
Horizontal transformation	L^6	L^4
Solid harmonics	L^5	L^2

We begin our discussion by noting that the two-electron integrals (9.9.33) may be written in the manner

$$g_{abcd} = \sum_{\substack{tuv \\ \tau\upsilon\phi}} E^{ab}_{tuv} F^{cd}_{\tau\upsilon\phi} \mathfrak{R}_{t+\tau,u+\upsilon,v+\phi}(\alpha, \mathbf{R}_{PQ}) \tag{9.9.44}$$

where we have introduced

$$F^{cd}_{\tau\upsilon\phi} = (-1)^{\tau+\upsilon+\phi} E^{cd}_{\tau\upsilon\phi} \tag{9.9.45}$$

$$\mathfrak{R}_{t+\tau,u+\upsilon,v+\phi}(\alpha, \mathbf{R}_{PQ}) = \frac{2\pi^{5/2}}{pq\sqrt{p+q}} R_{t+\tau,u+\upsilon,v+\phi}(\alpha, \mathbf{R}_{PQ}) \tag{9.9.46}$$

We now observe that the evaluation of the two-electron integrals is best carried out in two steps:

$$g^{cd}_{tuv} = \sum_{\tau\upsilon\phi} F^{cd}_{\tau\upsilon\phi} \mathfrak{R}_{t+\tau,u+\upsilon,v+\phi}(\alpha, \mathbf{R}_{PQ}) \tag{9.9.47}$$

$$g_{abcd} = \sum_{tuv} E^{ab}_{tuv} g^{cd}_{tuv} \tag{9.9.48}$$

The advantage of carrying out this partial summation is that the operation count is brought down from $L^{14}p^4$ for (9.9.44) to $L^{10}p^4$ for (9.9.47) and $L^{11}p^4$ for (9.9.48), clearly a major improvement. Next, we note that, once the first Hermite-to-Cartesian step has been completed, we may carry out the transformation to the contracted spherical-harmonic basis for the first electron before we go on to the second Hermite-to-Cartesian transformation (9.9.48), bringing the number of integrals for the second step (9.9.48) down from L^7p^4 to L^5p^2. The cost is L^7p^4 for the contraction and L^8p^2 for the spherical-harmonic transformation. As a result, the cost of the second step (9.9.48) is reduced from $L^{11}p^4$ to L^9p^2. To conclude our calculation, a contraction and transformation to the spherical-harmonic basis must be carried out for the second electron, at the costs L^6p^2 and L^7, respectively. The cost of the evaluation of the different steps is given in Table 9.3, where, for the Hermite-to-Cartesian steps, we have given also the costs L^8p^4 and L^8p^2 for the alternative partial summations analogous to (9.9.41). For reasons that will become clear shortly, will shall refer to this algorithm as MD4.

As for the one-electron integrals, we may reduce the cost of the integration (at the expense of a somewhat more complicated algorithm) by employing vertical and horizontal transformations for the Hermite-to-Cartesian step. The same arguments apply as for the one-electron integrals in Section 9.9.5, and the costs of the various steps are given in Table 9.4. The overall cost of the evaluation of the two-electron integrals now scales as L^7p^4 (for the p^4 steps) and L^9p^2 (for the p^2 steps). Since we now use the McMurchie–Davidson expansion to generate integrals with nonzero

Table 9.3 Cost and memory requirements of the two-electron MD4 scheme

| | First electron | | Second electron | |
	Cost	Memory	Cost	Memory
Boys functions	Lp^4	Lp^4		
Hermite integrals	L^4p^4	L^3p^4		
Expansion coefficients	L^3p^2	L^3p^2	L^3p^2	L^3p^2
Cartesian integrals	$L^{10}p^4/L^8p^4$	L^7p^4	L^9p^2/L^8p^2	L^6p^2
Primitive contraction	L^7p^4	L^7p^2	L^6p^2	L^6
Solid harmonics	L^8p^2	L^5p^2	L^7	L^4

Table 9.4 Cost and memory requirements of the two-electron MD2 scheme

| | First electron | | Second electron | |
	Cost	Memory	Cost	Memory
Boys functions	Lp^4	Lp^4		
Hermite integrals	L^4p^4	L^3p^4		
Expansion coefficients	L^2p^2	L^2p^2	L^2p^2	L^2p^2
Vertical transformation	L^9p^4/L^7p^4	L^6p^4	L^8p^2/L^6p^2	L^5p^2
Primitive contraction	L^6p^4	L^6p^2	L^5p^2	L^5
Horizontal transformation	L^9p^2	L^7p^2	L^8	L^6
Solid harmonics	L^8p^2	L^5p^2	L^7	L^4

angular momentum only for the first and third indices, we shall refer to this scheme as MD2. In the limit of high angular momentum, the MD2 scheme appears to be a substantial improvement on the MD4 scheme. For integrals of low angular momentum, the simpler MD4 scheme is usually more efficient.

The separate treatment of the two electrons in the McMurchie–Davidson scheme may be very useful in some circumstances. Consider, for example, the calculation of the Coulomb contribution to the Fock matrix, which, in the primitive Cartesian basis, may be written in the manner

$$f_{ab}^{\text{cou}} = \sum_{cd} g_{abcd} D_{cd}^{\text{AO}} \tag{9.9.49}$$

as discussed in Section 10.6.3. Combining these expressions with (9.9.44)–(9.9.48), we find that the contribution to the Fock matrix may be obtained in the following two steps:

$$g_{tuv} = \sum_{cd} g_{tuv}^{cd} D_{cd}^{\text{AO}} \tag{9.9.50}$$

$$f_{ab}^{\text{cou}} = \sum_{tuv} E_{tuv}^{ab} g_{tuv} \tag{9.9.51}$$

reducing the cost for the second electron in direct Hartree–Fock calculations significantly.

9.10 The Obara–Saika scheme for Coulomb integrals

In the McMurchie–Davidson scheme, we begin by calculating the Boys function; next, we generate the Hermite integrals; and, finally, we arrive at the Cartesian integrals by combination of the Hermite integrals. We shall now investigate a different route to the Coulomb integrals – the

Obara–Saika scheme [5] – in which we avoid the intermediate Hermite integrals altogether, obtaining the final Cartesian integrals by recursion directly from the Boys function.

9.10.1   THE OBARA–SAIKA SCHEME FOR ONE-ELECTRON COULOMB INTEGRALS

According to our discussion in Section 9.9, the Coulomb-potential integrals may be written in the form (9.9.32)

$$V_{ijklmn}^{000} = \frac{2\pi}{p} \sum_{tuv} E_t^{ij} E_u^{kl} E_v^{mn} R_{tuv} \tag{9.10.1}$$

To derive the Obara–Saika recurrence relations, we introduce the auxiliary integrals

$$\Theta_{ijklmn}^{N} = \frac{2\pi}{p} (-2p)^{-N} \sum_{tuv} E_t^{ij} E_u^{kl} E_v^{mn} R_{tuv}^{N} \tag{9.10.2}$$

For lower indices equal to 0, these integrals represent the (scaled) Boys function; for 0 upper index, they represent the final Cartesian integrals:

$$\Theta_{000000}^{N} = \frac{2\pi}{p} (-2p)^{-N} K_{ab}^{xyz} R_{000}^{N} = \frac{2\pi}{p} K_{ab}^{xyz} F_N(pR_{PC}^2) \tag{9.10.3}$$

$$\Theta_{ijklmn}^{0} = V_{ijklmn}^{000} \tag{9.10.4}$$

To obtain (9.10.3), we have used (9.5.8) and (9.9.14). Our task is now to develop a set of recurrence relations by means of which we may arrive at the target integrals (9.10.4) starting from the source integrals (9.10.3).

We shall derive the recurrence relations for the Coulomb-potential integrals from the recurrence relations for the expansion coefficients in Section 9.5.1 and for the Hermite integrals in Section 9.9.2. Consider the integral (9.10.2) with i incremented by 1:

$$\Theta_{i+1,j}^{N} = \frac{2\pi}{p} (-2p)^{-N} \sum_{tuv} E_t^{i+1,j} E_u^{kl} E_v^{mn} R_{tuv}^{N} \tag{9.10.5}$$

where, for ease of notation, we have omitted the Cartesian quantum numbers kl and mn for the y and z directions and written $\Theta_{i+1,j}^{N}$ rather than $\Theta_{i+1,j,k,l,m,n}^{N}$. First, we insert the recurrence relation for the expansion coefficients (9.5.20) and replace t by $t+1$ in the summation:

$$\Theta_{i+1,j}^{N} = X_{PA}\Theta_{ij}^{N} + \frac{1}{2p}(i\Theta_{i-1,j}^{N} + j\Theta_{i,j-1}^{N}) - \frac{2\pi}{p}(-2p)^{-N-1} \sum_{tuv} E_t^{ij} E_u^{kl} E_v^{mn} R_{t+1,u,v}^{N} \tag{9.10.6}$$

Next, we invoke the recurrence relation for the Hermite integrals (9.9.18):

$$\Theta_{i+1,j}^{N} = X_{PA}\Theta_{ij}^{N} + \frac{1}{2p}(i\Theta_{i-1,j}^{N} + j\Theta_{i,j-1}^{N}) - \frac{2\pi}{p}(-2p)^{-N-1}X_{PC} \sum_{tuv} E_t^{ij} E_u^{kl} E_v^{mn} R_{tuv}^{N+1}$$

$$+ \frac{2\pi}{p}(-2p)^{-N-2} \sum_{tuv} (2pt E_t^{ij}) E_u^{kl} E_v^{mn} R_{t-1,u,v}^{N+1} \tag{9.10.7}$$

Finally, using the relation (9.5.14) for the expansion coefficients and reintroducing the full set of indices, we arrive at the *Obara–Saika recurrence relations* for the Coulomb-potential integral [5]:

$$\Theta_{i+1,j,k,l,m,n}^{N} = X_{PA}\Theta_{ijklmn}^{N} + \frac{1}{2p}(i\Theta_{i-1,j,k,l,m,n}^{N} + j\Theta_{i,j-1,k,l,m,n}^{N})$$

$$- X_{PC}\Theta_{ijklmn}^{N+1} - \frac{1}{2p}(i\Theta_{i-1,j,k,l,m,n}^{N+1} + j\Theta_{i,j-1,k,l,m,n}^{N+1}) \tag{9.10.8}$$

$$\Theta^N_{i,j+1,k,l,m,n} = X_{PB}\Theta^N_{ijklmn} + \frac{1}{2p}(i\Theta^N_{i-1,j,k,l,m,n} + j\Theta^N_{i,j-1,k,l,m,n})$$

$$- X_{PC}\Theta^{N+1}_{ijklmn} - \frac{1}{2p}(i\Theta^{N+1}_{i-1,j,k,l,m,n} + j\Theta^{N+1}_{i,j-1,k,l,m,n}) \qquad (9.10.9)$$

The recurrence relation (9.10.9) is identical to (9.10.8) except that X_{PA} is replaced by X_{PB} and is most easily obtain from the horizontal recurrence

$$\Theta^N_{i+1,j,k,l,m,n} = \Theta^N_{i,j+1,k,l,m,n} - X_{AB}\Theta^N_{ijklmn} \qquad (9.10.10)$$

Using these recurrences and their analogues for the remaining four Cartesian indices, we may generate the full set of Cartesian integrals starting from the Boys function.

Comparing the recurrence relations (9.10.8) and (9.10.9) with those for the simple one-electron integrals in Section 9.3, we note that the recurrences for the Coulomb-potential integrals are considerably more complicated, containing as many as six terms – three terms with an unmodified upper index N and three terms with an incremented upper index $N + 1$.

9.10.2 THE OBARA–SAIKA SCHEME FOR TWO-ELECTRON COULOMB INTEGRALS

To derive the Obara–Saika recurrence relations for the two-electron Coulomb integral, we begin by setting up the following auxiliary integral

$$\Theta^N_{i_x j_x k_x l_x; i_y j_y k_y l_y; i_z j_z k_z l_z} = \frac{2\pi^{5/2}}{pq\sqrt{p+q}}(-2\alpha)^{-N}$$

$$\times \sum_{\substack{tuv \\ \tau\upsilon\phi}}(-1)^{\tau+\upsilon+\phi}E^{i_x j_x}_t E^{k_x l_x}_\tau E^{i_y j_y}_u E^{k_y l_y}_\upsilon E^{i_z j_z}_v E^{k_z l_z}_\phi R^N_{t+\tau,u+\upsilon,v+\phi} \qquad (9.10.11)$$

From (9.5.8), (9.9.14) and (9.9.33), this integral is seen to contain as special cases the source and target integrals:

$$\Theta^N_{0000;0000;0000} = \frac{2\pi^{5/2}}{pq\sqrt{p+q}}K^{xyz}_{ab}K^{xyz}_{cd}F_N(\alpha R^2_{PQ}) \qquad (9.10.12)$$

$$\Theta^0_{i_x j_x k_x l_x; i_y j_y k_y l_y; i_z j_z k_z l_z} = g_{i_x j_x k_x l_x; i_y j_y k_y l_y; i_z j_z k_z l_z} \qquad (9.10.13)$$

To simplify notation, we shall write the auxiliary integral in the form

$$\Theta^N_{ijkl} = A_N \sum_{\substack{tuv \\ \tau\upsilon\phi}}(-1)^{\tau+\upsilon+\phi}E^{ij}_t E^{kl}_\tau E_{uv\upsilon\phi}R^N_{t+\tau,u+\upsilon,v+\phi} \qquad (9.10.14)$$

where i, j, k, l are the Cartesian quantum numbers for orbitals a, b, c, d in the x *direction*, and where we have introduced the quantities

$$A_N = \frac{2\pi^{5/2}}{pq\sqrt{p+q}}(-2\alpha)^{-N} \qquad (9.10.15)$$

$$E_{uv\upsilon\phi} = E^{i_y j_y}_u E^{k_y l_y}_\upsilon E^{i_z j_z}_v E^{k_z l_z}_\phi \qquad (9.10.16)$$

Note that the Cartesian quantum numbers for the y and z directions are not affected by the recurrence relations to be developed here. Also, the recurrences for the y and z directions may be easily set up by analogy with the recurrences for the x direction.

Our development follows closely that for the one-electron Coulomb integrals in Section 9.10.1. In the incremented integral

$$\Theta^N_{i+1,j,k,l} = A_N \sum_{\substack{tuv \\ \tau\upsilon\phi}} (-1)^{\tau+\upsilon+\phi} E^{i+1,j}_t E^{kl}_\tau E_{uvv\phi} R^N_{t+\tau,u+\upsilon,v+\phi} \tag{9.10.17}$$

we insert (9.5.20) and replace t by $t+1$ to obtain

$$\Theta^N_{i+1,j,k,l} = X_{PA} \Theta^N_{ijkl} + \frac{1}{2p}(i\Theta^N_{i-1,j,k,l} + j\Theta^N_{i,j-1,k,l}) - \frac{1}{2p}Q^N_{ijkl} \tag{9.10.18}$$

where we have introduced

$$Q^N_{ijkl} = -A_N \sum_{\substack{tuv \\ \tau\upsilon\phi}} (-1)^{\tau+\upsilon+\phi} E^{ij}_t E^{kl}_\tau E_{uvv\phi} R^N_{t+\tau+1,u+\upsilon,v+\phi} \tag{9.10.19}$$

which represents the original integral Θ^N_{ijkl} differentiated with respect to Q_x. To evaluate Q^N_{ijkl}, we invoke the Hermite recurrence (9.9.18) in the form

$$R^N_{t+\tau+1,u+\upsilon,v+\phi} = (t+\tau)R^{N+1}_{t+\tau-1,u+\upsilon,v+\phi} + X_{PQ} R^{N+1}_{t+\tau,u+\upsilon,v+\phi} \tag{9.10.20}$$

Inserting this expression in (9.10.19) and rearranging terms, we obtain

$$Q^N_{ijkl} = 2\alpha X_{PQ} A_{N+1} \sum_{\substack{tuv \\ \tau\upsilon\phi}} (-1)^{\tau+\upsilon+\phi} E^{ij}_t E^{kl}_\tau E_{uvv\phi} R^{N+1}_{t+\tau,u+\upsilon,v+\phi}$$

$$+ \frac{\alpha}{p} A_{N+1} \sum_{\substack{tuv \\ \tau\upsilon\phi}} (-1)^{\tau+\upsilon+\phi} (2pt E^{ij}_t) E^{kl}_\tau E_{uvv\phi} R^{N+1}_{t+\tau-1,u+\upsilon,v+\phi}$$

$$+ \frac{\alpha}{q} A_{N+1} \sum_{\substack{tuv \\ \tau\upsilon\phi}} (-1)^{\tau+\upsilon+\phi} E^{ij}_t (2q\tau E^{kl}_\tau) E_{uvv\phi} R^{N+1}_{t+\tau-1,u+\upsilon,v+\phi} \tag{9.10.21}$$

We now invoke (9.5.14) for the expressions $2pt E^{ij}_t$ and $2q\tau E^{kl}_\tau$

$$Q^N_{ijkl} = 2\alpha X_{PQ} \Theta^{N+1}_{ijkl} + i\frac{\alpha}{p} A_{N+1} \sum_{\substack{tuv \\ \tau\upsilon\phi}} (-1)^{\tau+\upsilon+\phi} E^{i-1,j}_{t-1} E^{kl}_\tau E_{uvv\phi} R^{N+1}_{t+\tau-1,u+\upsilon,v+\phi}$$

$$+ j\frac{\alpha}{p} A_{N+1} \sum_{\substack{tuv \\ \tau\upsilon\phi}} (-1)^{\tau+\upsilon+\phi} E^{i,j-1}_{t-1} E^{kl}_\tau E_{uvv\phi} R^{N+1}_{t+\tau-1,u+\upsilon,v+\phi}$$

$$- k\frac{\alpha}{q} A_{N+1} \sum_{\substack{tuv \\ \tau\upsilon\phi}} (-1)^{\tau-1+\upsilon+\phi} E^{ij}_t E^{k-1,l}_{\tau-1} E_{uvv\phi} R^{N+1}_{t+\tau-1,u+\upsilon,v+\phi}$$

$$- l\frac{\alpha}{q} A_{N+1} \sum_{\substack{tuv \\ \tau\upsilon\phi}} (-1)^{\tau-1+\upsilon+\phi} E^{ij}_t E^{k,l-1}_{\tau-1} E_{uvv\phi} R^{N+1}_{t+\tau-1,u+\upsilon,v+\phi} \tag{9.10.22}$$

which yields the expression

$$Q_{ijkl}^{N} = 2\alpha X_{PQ}\Theta_{ijkl}^{N+1} + \frac{\alpha}{p}\left(i\Theta_{i-1,j,k,l}^{N+1} + j\Theta_{i,j-1,k,l}^{N+1}\right) - \frac{\alpha}{q}\left(k\Theta_{i,j,k-1,l}^{N+1} + l\Theta_{i,j,k,l-1}^{N+1}\right) \quad (9.10.23)$$

for the integral Θ_{ijkl}^{N} differentiated with respect to Q_x.

Inserting (9.10.23) in (9.10.18), we obtain the Obara–Saika two-electron recurrence relation [5]

$$\Theta_{i+1,j,k,l}^{N} = X_{PA}\Theta_{ijkl}^{N} - \frac{\alpha}{p}X_{PQ}\Theta_{ijkl}^{N+1} + \frac{i}{2p}\left(\Theta_{i-1,j,k,l}^{N} - \frac{\alpha}{p}\Theta_{i-1,j,k,l}^{N+1}\right)$$
$$+ \frac{j}{2p}\left(\Theta_{i,j-1,k,l}^{N} - \frac{\alpha}{p}\Theta_{i,j-1,k,l}^{N+1}\right) + \frac{k}{2(p+q)}\Theta_{i,j,k-1,l}^{N+1} + \frac{l}{2(p+q)}\Theta_{i,j,k,l-1}^{N+1}$$

$$(9.10.24)$$

Using the horizontal recurrence relation (9.10.10), a similar relation may be written down for increments in j, replacing X_{PA} with X_{PB}. For completeness, we give also the recurrence relation for increments in k for the second electron:

$$\Theta_{i,j,k+1,l}^{N} = X_{QC}\Theta_{ijkl}^{N} + \frac{\alpha}{q}X_{PQ}\Theta_{ijkl}^{N+1} + \frac{k}{2q}\left(\Theta_{i,j,k-1,l}^{N} - \frac{\alpha}{q}\Theta_{i,j,k-1,l}^{N+1}\right)$$
$$+ \frac{l}{2q}\left(\Theta_{i,j,k,l-1}^{N} - \frac{\alpha}{q}\Theta_{i,j,k,l-1}^{N+1}\right) + \frac{i}{2(p+q)}\Theta_{i-1,j,k,l}^{N+1} + \frac{j}{2(p+q)}\Theta_{i,j-1,k,l}^{N+1}$$

$$(9.10.25)$$

Note that (9.10.25) can be obtained from (9.10.24) by index substitution and renaming of variables. Similar recurrence relations are easily set up for the remaining two Cartesian directions. Note that the recurrence (9.10.24) reduces to that for the one-electron Coulomb integrals of Section 9.10.1 – see (9.10.8) – when $k = l = 0$ and when q (i.e. the sum of the exponents for the second electron) tends to infinity.

We have now succeeded in setting up a set of recurrence relations by means of which the two-electron Cartesian integrals may be obtained from the Boys function. The resulting expressions are rather complicated, however, involving as many as eight distinct contributions. Unlike the McMurchie–Davidson scheme, the Obara–Saika scheme does not treat the two electrons separately since the recurrences (9.10.24) and (9.10.25), for example, affect the indices of all four orbitals. In Section 9.10.3, we shall see how the Obara–Saika recurrences may be simplified considerably when used in conjunction with two other types of recurrence relations: the electron-transfer recurrences and the horizontal recurrences.

9.10.3 THE ELECTRON-TRANSFER AND HORIZONTAL RECURRENCE RELATIONS

In the original Obara–Saika scheme, the two-electron integrals are generated from the Boys function by means of rather unwieldy recurrence relations such as (9.10.24). It is possible, however, to break the generation of the two-electron integrals up into several smaller steps, each of which involves a simpler set of recurrences. Consider the following three-step path to the Cartesian integrals.

First, a set of two-electron integrals with $j = k = l = 0$ is generated by means of the following special four-term version of the *Obara–Saika recurrence relation*:

$$\Theta_{i+1,0,0,0}^{N} = X_{PA}\Theta_{i000}^{N} - \frac{\alpha}{p}X_{PQ}\Theta_{i000}^{N+1} + \frac{i}{2p}\left(\Theta_{i-1,0,0,0}^{N} - \frac{\alpha}{p}\Theta_{i-1,0,0,0}^{N+1}\right) \quad (9.10.26)$$

In the next step, we transfer Cartesian powers from the first to the second electron by the *electron-transfer recurrence relation* [14,15]

$$\Theta^N_{i,0,k+1,0} = -\frac{bX_{AB} + dX_{CD}}{q}\Theta^N_{i0k0} + \frac{i}{2q}\Theta^N_{i-1,0,k,0} + \frac{k}{2q}\Theta^N_{i,0,k-1,0} - \frac{p}{q}\Theta^N_{i+1,0,k,0} \qquad (9.10.27)$$

to be proved shortly. In the final step, we transfer Cartesian powers between the orbitals of the same electron, using the simple two-term *horizontal recurrence relations* [16]:

$$\Theta^N_{i,j+1,k,l} = \Theta^N_{i+1,j,k,l} + X_{AB}\Theta^N_{ijkl} \qquad (9.10.28)$$

$$\Theta^N_{i,j,k,l+1} = \Theta^N_{i,j,k+1,l} + X_{CD}\Theta^N_{ijkl} \qquad (9.10.29)$$

In this way, we may build up the full set of Cartesian integrals by the use of three sets of recurrence relations, each of which is considerably simpler than the full Obara–Saika scheme. The horizontal recurrences (9.10.28) and (9.10.29) do not involve the orbital exponents and may therefore be applied *after* the transformation of the integrals to the contracted basis, as will be discussed later. In passing, we note that this procedure may also be used for the one-electron integrals, replacing the first step by the similar recurrence relation

$$\Theta^N_{i+1,0} = X_{PA}\Theta^N_{i0} - X_{PC}\Theta^{N+1}_{i0} + \frac{i}{2p}(\Theta^N_{i-1,0} - \Theta^{N+1}_{i-1,0}) \qquad (9.10.30)$$

omitting the electron-transfer step, and carrying out the horizontal recursion in the final step

$$\Theta^N_{i,j+1} = \Theta^N_{i+1,j} + X_{AB}\Theta^N_{ij} \qquad (9.10.31)$$

using the same notation as for the two-electron integrals. Clearly, in this method, the same set of routines may be used for the one- and two-electron integrals.

We conclude this subsection by deriving the electron-transfer recurrence relation (9.10.27). As in the derivation of the Obara–Saika recurrence relations, we begin by setting up the translational condition on the integrals:

$$\frac{\partial \Theta^N_{i0k0}}{\partial A_x} + \frac{\partial \Theta^N_{i0k0}}{\partial B_x} + \frac{\partial \Theta^N_{i0k0}}{\partial C_x} + \frac{\partial \Theta^N_{i0k0}}{\partial D_x} = 0 \qquad (9.10.32)$$

Differentiating each orbital separately, we obtain

$$2a\Theta^N_{i+1,0,k,0} + 2b\Theta^N_{i1k0} + 2c\Theta^N_{i,0,k+1,0} + 2d\Theta^N_{i0k1} - i\Theta^N_{i-1,0,k,0} - k\Theta^N_{i,0,k-1,0} = 0 \qquad (9.10.33)$$

Applying the horizontal recurrences (9.2.27), we arrive at the expression

$$2p\Theta^N_{i+1,0,k,0} + 2bX_{AB}\Theta^N_{i0k0} + 2q\Theta^N_{i,0,k+1,0} + 2dX_{CD}\Theta^N_{i0k0} - i\Theta^N_{i-1,0,k,0} - k\Theta^N_{i,0,k-1,0} = 0 \qquad (9.10.34)$$

which may be rearranged to yield the transfer relation (9.10.27).

9.10.4 COMPUTATIONAL CONSIDERATIONS FOR THE TWO-ELECTRON INTEGRALS

From our discussion of the Obara–Saika scheme three techniques emerge, differing in the use of the electron-transfer and the horizontal recurrence relations; see Table 9.5. In the OS4 scheme, no use is made of the transfer and horizontal recurrences – the primitive Cartesian integrals Θ^0_{ijkl}

Table 9.5 Cost and memory requirements of the two-electron Obara–Saika scheme

	OS4		OS2		OS1	
	Cost	Memory	Cost	Memory	Cost	Memory
Boys functions	Lp^4	Lp^4	Lp^4	Lp^4	Lp^4	Lp^4
Vertical recursion	$L^{13}p^4$	L^8p^4	L^7p^4	L^6p^4	L^4p^4	L^3p^4
Transfer recursion					L^6p^4	L^6p^4
Primitive contraction	L^8p^4	L^8	L^6p^4	L^6	L^6p^4	L^6
Horizontal recursion 1			L^9	L^7	L^9	L^7
Solid harmonics 1	L^9	L^6	L^8	L^5	L^8	L^5
Horizontal recursion 2			L^8	L^6	L^8	L^6
Solid harmonics 2	L^7	L^4	L^7	L^4	L^7	L^4

are generated directly by the Obara–Saika recurrence relations (9.10.24) and (9.10.25) [5]. The cost of the OS4 scheme is high, scaling as $L^{13}p^4$, where L is the angular momentum and p the number of primitive functions for each contracted orbital. Substantial reductions may be obtained, however, by avoiding unnecessary intermediates in the construction of the integrals.

In the OS2 scheme, we generate the Cartesian integrals of the type Θ^0_{i0k0} from the Obara–Saika recurrence relations (9.10.24) and (9.10.25) and the remaining ones by means of the horizontal relations (9.10.28) and (9.10.29) [16]. The OS2 scheme is a substantial improvement on the OS4 scheme, scaling as L^7p^4 in the vertical part and L^9 in the horizontal part. Finally, in the OS1 scheme, we use the electron-transfer as well as the horizontal recurrence relations (as described in Section 9.10.3), generating in the first step integrals of the type Θ^0_{i000}. The OS1 scheme appears to be the superior one, at least for orbitals of high angular momentum – the cost of the Obara–Saika step has been reduced to L^4p^4, whereas the transfer step scales as L^6p^4 and the horizontal step as L^9.

Comparing the Obara–Saika and McMurchie–Davidson schemes, we find that MD4 scales more favourably than OS4, with a highest cost of $L^{10}p^4$ (or L^8p^4) rather than $L^{13}p^4$. However, a simple order analysis (like that carried out here) is not sufficiently detailed to allow us to distinguish between the two schemes except in the limit of high L. The MD4 scheme involves a rather large number of steps, most of which must be repeated for both electrons. In the OS4 scheme, on the other hand, the key recurrence relations are complicated, involving as many as 15 floating-point operations (9.10.24) compared with 3 in the time-consuming MD4 step (9.9.47). A more satisfactory comparison would require a careful count of all operations.

The MD2 and OS2 schemes are similar, with highest costs of L^9p^4 (or L^7p^4) and L^9p^2 for MD2 and L^7p^4 and L^9 for OS2, but both schemes appear to be slower than OS1 for integrals of high angular momentum. At this point, we note that it is perfectly possible to set up an MD1 scheme, where we first generate – by means of the McMurchie–Davidson algorithm – integrals of the type Θ^0_{i000} and then proceed exactly as for OS1. Thus, for high angular-momentum functions, the efficiency of the calculation appears to be more dependent on the use of electron-transfer and horizontal recurrence relations than on the type of algorithm used for the generation of Cartesian integrals from the Boys function.

9.11 Rys quadrature for Coulomb integrals

In the previous sections, we considered two approaches to the calculation of Coulomb integrals over Gaussian orbitals – the McMurchie–Davidson scheme and the Obara–Saika scheme – both of which employ Boys functions as intermediates. In the present section, a different approach is explored, in which the construction of the Boys function is bypassed altogether and replaced by a Gaussian-quadrature scheme [10], providing an interesting alternative to the McMurchie–Davidson and Obara–Saika schemes.

9.11.1 MOTIVATION FOR THE GAUSSIAN-QUADRATURE SCHEME

In the McMurchie–Davidson and Obara–Saika schemes, the Coulomb integrals are generated as linear combinations of Boys functions $F_n(x)$ of different orders n. Referring back to the discussion of these schemes, we find that the two-electron integrals may be written in the form

$$g_{abcd} = \sum_{n=0}^{L} c_n F_n(\alpha R_{PQ}^2) \qquad (9.11.1)$$

where L is the sum of the total angular momentum of the four orbitals and where the coefficients c_n depend on the coordinates and exponents of these orbitals – see, for example, Section 9.10.2. Since the Boys function (9.8.1) is given by

$$F_n(x) = \int_0^1 t^{2n} \exp(-xt^2)\, dt \qquad (9.11.2)$$

we conclude that the two-electron integrals may be written in the form

$$g_{abcd} = \int_0^1 f_L(t^2) \exp(-\alpha R_{PQ}^2 t^2)\, dt \qquad (9.11.3)$$

where the integrand is an exponential in t^2 times a polynomial of degree L in t^2

$$f_L(t^2) = \sum_{n=0}^{L} c_n t^{2n} \qquad (9.11.4)$$

According to the discussion in Section 9.6, such integrals can be evaluated exactly by *Gaussian quadrature* over $L + 1$ points

$$g_{abcd} = \sum_{\kappa=1}^{L+1} w_\kappa f_L(t_\kappa^2) \qquad (9.11.5)$$

where the abscissae t_κ and weights w_κ are determined by the general rules of Gaussian-quadrature theory. This observation forms the basis for the Rys-polynomial scheme for the evaluation of Coulomb integrals over Gaussian functions [10].

In the following, we shall first show how Gaussian quadrature can be simplified for the special case of even polynomials and weight functions such as those in (9.11.3). The orthogonal polynomials needed for the calculation of Coulomb integrals (9.11.3) are then introduced, and finally we show how a Gaussian-quadrature scheme for the evaluation of Coulomb integrals can be developed based on the McMurchie–Davidson and Obara–Saika schemes.

9.11.2 GAUSSIAN QUADRATURE FOR EVEN POLYNOMIALS AND WEIGHT FUNCTIONS

Consider the Gaussian quadrature for integrals such as

$$I_k = \int_{-a}^{a} f_k(x^2) W(x^2)\, dx \qquad (9.11.6)$$

where $f_k(x^2)$ is a polynomial of degree k in x^2 and where the weight function $W(x^2)$ is symmetric about the origin. According to the general theory of Gaussian quadrature, such integrals can be evaluated exactly in terms of the roots and weights of a polynomial $p_n(x)$ of degree $n > k$, provided the polynomials satisfy the orthonormality relations

$$\int_{-a}^{a} p_n(x) p_m(x) W(x^2)\, dx = \delta_{mn} \tag{9.11.7}$$

In the present subsection, we shall see that the symmetry of the integrand about the origin (9.11.6) reduces the number of quadrature points needed for exact integration by a factor of 2 relative to the requirement $n > k$.

In setting up the quadrature scheme for (9.11.6), the orthogonal polynomials $p_n(x)$ may be generated by a Gram–Schmidt orthogonalization of the monomials x^i using the inner product (9.11.7). In this process, the inner product between an even function and an odd function vanishes trivially and it follows that $p_{2n}(x)$ contains only monomials of even degree in x and that $p_{2n+1}(x)$ contains only monomials of odd degree in x. Moreover, the roots of the orthogonal polynomials are distributed symmetrically about the origin – the even polynomial $p_{2n}(x)$ has n positive and n negative roots, whereas the odd polynomial $p_{2n+1}(x)$ has an additional root at the origin.

The symmetrical distribution of the roots about the origin for $p_n(x)$ leads to identical weights (9.6.15) at the roots x_i and $-x_i$:

$$w_{n;-x_i} = \int_{-a}^{a} W(x^2) \prod_{\substack{j=1 \\ x_j \neq -x_i}}^{n} \frac{x - x_j}{(-x_i) - x_j}\, dx$$

$$= \int_{-a}^{a} W(x^2) \prod_{\substack{j=1 \\ x_j \neq x_i}}^{n} \frac{x - (-x_j)}{(-x_i) - (-x_j)}\, dx = w_{n;x_i} \tag{9.11.8}$$

The quadrature for (9.11.6) can therefore be reduced to a sum over just the positive roots of a polynomial $p_n(x)$ of even degree n. To obtain an exact quadrature for a polynomial $f_k(x^2)$, we must – according to the general theory – use a polynomial $p_n(x)$ with $n > k$. The smallest even integer that satisfies this requirement is $n = 2[k/2] + 2$. The associated polynomial has $[k/2] + 1$ positive roots, and we may consequently obtain the integral exactly as

$$\int_{-a}^{a} f_k(x^2) W(x^2)\, dx = 2 \sum_{i=1}^{[k/2]+1} w_{2[k/2]+2;i} f_k(x_{2[k/2]+2;i}) \tag{9.11.9}$$

in terms of the positive roots and weights of the polynomial $p_{2[k/2]+2}(x)$. We also note that the calculation of the weights can be simplified in the following manner:

$$w_{2[k/2]+2;i} = \int_{-a}^{a} W(x^2) \prod_{\substack{j=1 \\ j \neq i}}^{2[k/2]+2} \frac{x - x_j}{x_i - x_j}\, dx$$

$$= \int_{-a}^{a} W(x^2) \frac{x + x_i}{2x_i} \prod_{\substack{j=1 \\ j \neq i}}^{[k/2]+1} \frac{x^2 - x_j^2}{x_i^2 - x_j^2}\, dx$$

$$= \int_{0}^{a} W(x^2) \prod_{\substack{j=1 \\ j \neq i}}^{[k/2]+1} \frac{x^2 - x_j^2}{x_i^2 - x_j^2}\, dx \tag{9.11.10}$$

where the product is over only the positive roots and the integration over positive x.

Since only the even polynomials $p_{2n}(x)$ are needed for the quadrature and since the integration may be restricted to positive x, we introduce the new polynomials

$$q_n(x) = \sqrt{2} p_{2n}(x) \tag{9.11.11}$$

that satisfy the orthonormality relations

$$\int_0^a q_m(x) q_n(x) W(x^2) \, dx = \delta_{mn} \tag{9.11.12}$$

Using the n roots of $q_n(x)$ in the interval $[0,a]$ to define the weights as in (9.11.10), we now obtain the following simple expression for the Gaussian quadrature of a polynomial of degree k in x^2

$$\int_0^a f_k(x^2) W(x^2) \, dx = \sum_{i=1}^{n} w_{n;i} f_k(x_{n;i}^2) \tag{9.11.13}$$

where the number of quadrature points is given by

$$n = \left[\frac{k}{2} \right] + 1 \tag{9.11.14}$$

This result should be contrasted with the general Gaussian-quadrature scheme, which requires $k + 1$ quadrature points for a general polynomial of degree $2k$.

9.11.3 RYS POLYNOMIALS AND GAUSS–RYS QUADRATURE

We now return to the Gaussian quadrature of the Coulomb integrals such as that given in (9.11.3):

$$g_{abcd} = \int_0^1 f_L(t^2) \exp(-\alpha R_{PQ}^2 t^2) \, dt \tag{9.11.15}$$

In accordance with the discussion of Section 9.11.2, we introduce for this purpose the *J-type Rys polynomials* $J_n^\alpha(x)$ of degree n, orthonormal on the interval $[-1, 1]$ with weight function $\exp(-\alpha x^2)$ [17]:

$$\int_{-1}^1 J_m^\alpha(x) J_n^\alpha(x) \exp(-\alpha x^2) \, dx = \delta_{mn} \tag{9.11.16}$$

Since only the even members of the J-type Rys polynomials are needed, we also introduce the *R-type Rys polynomials* of degree $2n$ [17]

$$R_n^\alpha(x) = \sqrt{2} J_{2n}^\alpha(x) \tag{9.11.17}$$

which are orthogonal on the interval $[0,1]$ with weight function $\exp(-\alpha x^2)$:

$$\int_0^1 R_m^\alpha(x) R_n^\alpha(x) \exp(-\alpha x^2) \, dx = \delta_{mn} \tag{9.11.18}$$

In terms of the roots and weights of these polynomials, we may write the Coulomb integrals as

$$\int_0^1 f_L(x^2) \exp(-\alpha x^2) \, dx = \sum_{i=1}^{n} w_{n;i} f_L(x_{n;i}^2) \tag{9.11.19}$$

where the number of quadrature points is given by

$$n = \left[\frac{L}{2}\right] + 1 \tag{9.11.20}$$

as discussed in Section 9.11.2.

The weight function of the Rys polynomials $J_n^\alpha(x)$ and $R_n^\alpha(x)$ depends on the real parameter $\alpha \geq 0$. These polynomials consequently constitute a manifold of polynomials, one for each α. In Figure 9.10, we have plotted the first four Rys polynomials for several values of the weight parameter α. As α increases, the Rys roots shift towards the origin. At $\alpha = 0$, the weight function becomes equal to unity and the Rys polynomials turn into the scaled Legendre polynomials (9.6.2)

$$J_n^0(x) = \sqrt{\frac{2n+1}{2}} P_n(x) \tag{9.11.21}$$

On the other hand, in the limit as α tends to infinity, we may in (9.11.16) integrate over the full set of real numbers and the Rys polynomials may then be related to the Hermite functions as

$$J_n^\alpha(x) \approx \frac{1}{\sqrt{2^n n!}} \left(\frac{\alpha}{\pi}\right)^{1/4} H_n(\sqrt{\alpha}x) \qquad \text{(large } \alpha) \tag{9.11.22}$$

From these expressions, we conclude that the roots of the Rys polynomials coincide with the roots of the Legendre polynomials for $\alpha = 0$ and with the scaled roots of the Hermite polynomials for large α:

$$x_{n;i}^{\text{J}} = x_{n;i}^{\text{L}} \qquad (\alpha = 0) \tag{9.11.23}$$

$$x_{n;i}^{\text{J}} \approx \alpha^{-1/2} x_{n;i}^{\text{H}} \qquad \text{(large } \alpha) \tag{9.11.24}$$

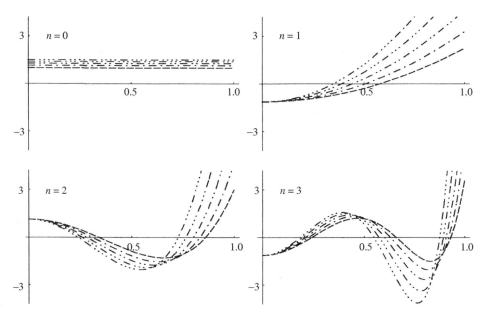

Fig. 9.10. The Rys polynomials $R_n^\alpha(x)$ of degree $2n \leq 6$ for $\alpha = 0, 1, 2, 3, 4$, with the number of consecutive dots in the lines increasing with α.

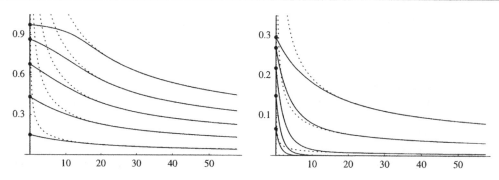

Fig. 9.11. The roots and weights of the Rys polynomials $R_5^\alpha(x)$. On the left, the roots of $R_5^\alpha(x)$ have been plotted as functions α; the dotted lines represent the five positive roots of $H_{10}(\sqrt{\alpha}x)$ and the dots at $\alpha = 0$ the five positive roots of $L_{10}(x)$. On the right, the weights of $R_5^\alpha(x)$ have been plotted as functions α; the dotted lines represent the corresponding five weights of $H_{10}(\sqrt{\alpha}x)$ and the dots at $\alpha = 0$ the five weights of $L_{10}(x)$.

These relationships are illustrated in Figure 9.11, where, in the left-hand plot, we have plotted the five roots of the Rys polynomials $R_5^\alpha(x)$ as a function of α, superimposed on the positive roots of the Hermite polynomial $H_{10}(\sqrt{\alpha}x)$ multiplied by $\alpha^{-1/2}$(dotted lines) and with the positive roots of the Legendre polynomial $L_{10}(x)$ indicated with black dots at $\alpha = 0$.

The weights of the Rys polynomials are related to the weights of the Legendre and Hermite polynomials in the same manner as for the roots (9.11.23) and (9.11.24)

$$w_{n;i}^{\mathrm{J}} = w_{n;i}^{\mathrm{L}} \qquad (\alpha = 0) \qquad (9.11.25)$$

$$w_{n;i}^{\mathrm{J}} \approx \alpha^{-1/2} w_{n;i}^{\mathrm{H}} \qquad (\text{large } \alpha) \qquad (9.11.26)$$

as may be ascertained by considering (9.6.15). On the right in Figure 9.11, we have drawn a plot of the Rys weights of $R_5^\alpha(x)$ analogous to the left-hand plot of the roots. Clearly, for $\alpha = 0$, we may obtain the roots and weights from the Legendre polynomials and, for large α, the roots and the weights are obtained by a simple scaling of the Hermite roots and weights. In the intermediate region, a more elaborate technique must be employed, based, for example, on the use of Taylor expansions or Chebyshev polynomials. Thus, the evaluation of the roots and weights of the Rys polynomials proceeds in a similar manner as for the Boys function in Section 9.8.2, where the short- and long-range forms are obtained in a simple manner and more elaborate techniques must be used in the intermediate region.

9.11.4 THE RYS SCHEME FOR HERMITE COULOMB INTEGRALS

Having introduced and discussed the Rys polynomials, we are now in a position to consider the evaluation of Hermite Coulomb integrals by Gaussian quadrature. In this subsection, we consider the one-electron Hermite Coulomb integrals (9.9.9)

$$R_{tuv}(p, \mathbf{R}_{\mathrm{PC}}) = \left(\frac{\partial}{\partial P_x}\right)^t \left(\frac{\partial}{\partial P_y}\right)^u \left(\frac{\partial}{\partial P_z}\right)^v F_0(pR_{\mathrm{PC}}^2) \qquad (9.11.27)$$

which may be calculated recursively as discussed in Section 9.9.2. The calculation of the corresponding two-electron integrals of the form $R_{tuv}(\alpha, \mathbf{R}_{\mathrm{PQ}})$ may be carried out in exactly the same manner, replacing p by α and $\mathbf{C}$ by $\mathbf{Q}$.

Inserting the definition of the Boys function (9.8.1) in (9.11.27), we may write the Hermite integral in the following manner:

$$R_{tuv}(p, \mathbf{R}_{PC}) = \int_0^1 \left(\frac{\partial}{\partial P_x}\right)^t \left(\frac{\partial}{\partial P_y}\right)^u \left(\frac{\partial}{\partial P_z}\right)^v \exp(-pR_{PC}^2 s^2)\,ds \qquad (9.11.28)$$

We now introduce the auxiliary polynomials

$$\overline{H}_t(s^2) = \exp(pX_{PC}^2 s^2)\left(\frac{\partial}{\partial P_x}\right)^t \exp(-pX_{PC}^2 s^2) \qquad (9.11.29)$$

These polynomials, which are of degree t in s^2, are related to the Hermite polynomials by

$$\overline{H}_t(s^2) = (-\sqrt{p}s)^t H_t(\sqrt{p}X_{PC}s) \qquad (9.11.30)$$

as may be verified using the Rodrigues expression (6.6.29). In terms of the modified Hermite polynomials (9.11.29) and their counterparts for the y and z directions, we may write the Hermite integrals as

$$R_{tuv}(p, \mathbf{R}_{PC}) = \int_0^1 \overline{H}_t(s^2)\overline{H}_u(s^2)\overline{H}_v(s^2)\exp(-pR_{PC}^2 s^2)\,ds \qquad (9.11.31)$$

This expression is in the appropriate form for Gauss–Rys quadrature (9.11.19) and may thus be evaluated exactly as

$$R_{tuv}(p, \mathbf{R}_{PC}) = \sum_{\kappa=1}^{\gamma} w_\kappa \overline{H}_t(s_\kappa^2)\overline{H}_u(s_\kappa^2)\overline{H}_v(s_\kappa^2) \qquad (9.11.32)$$

where the number of quadrature points is given by

$$\gamma = \left[\frac{t+u+v}{2}\right] + 1 \qquad (9.11.33)$$

and where s_κ and w_κ are the abscissae and weights of the Rys polynomial $R_\gamma^\alpha(s)$ with

$$\alpha = pR_{PC}^2 \qquad (9.11.34)$$

To complete the evaluation of the Hermite integrals by Rys quadrature, we must derive a scheme for the evaluation of the modified Hermite polynomials.

To evaluate the modified Hermite polynomials (9.11.29) for a given argument s, we shall employ a recursive scheme. Incrementing t by 1, we obtain

$$\overline{H}_{t+1}(s^2) = \exp(pX_{PC}^2 s^2)\left(\frac{\partial}{\partial P_x}\right)^{t+1}\exp(-pX_{PC}^2 s^2)$$

$$= -2ps^2\exp(pX_{PC}^2 s^2)\left(\frac{\partial}{\partial P_x}\right)^t X_{PC}\exp(-pX_{PC}^2 s^2) \qquad (9.11.35)$$

Using (9.9.17), we arrive at the recurrence relation [18]

$$\overline{H}_{t+1}(s^2) = -2ps^2[X_{PC}\overline{H}_t(s^2) + t\overline{H}_{t-1}(s^2)] \qquad (9.11.36)$$

Starting from the polynomial

$$\overline{H}_0(s^2) = 1 \tag{9.11.37}$$

we may use (9.11.36) to generate $\overline{H}_t(s^2)$ for a given argument s and a given degree t. The recurrences for the y and z directions are similar, replacing X_{PC} by Y_{PC} and Z_{PC}, respectively. Note the similarity with (9.9.18).

In the two-electron case, we may calculate the Hermite integrals in the same manner

$$R_{tuv}(\alpha, \mathbf{R}_{PQ}) = \sum_{\kappa=1}^{\gamma} w_\kappa \overline{H}_t(s_\kappa^2) \overline{H}_u(s_\kappa^2) \overline{H}_v(s_\kappa^2) \tag{9.11.38}$$

using the recurrence relation [18]

$$\overline{H}_{t+1}(s^2) = -2\alpha s^2 [X_{PQ} \overline{H}_t(s^2) + t \overline{H}_{t-1}(s^2)] \tag{9.11.39}$$

and similarly for the other directions.

9.11.5   THE RYS SCHEME FOR CARTESIAN COULOMB INTEGRALS

To calculate the one-electron Coulomb integrals by Rys quadrature, we write the integrals in the standard McMurchie–Davidson form (9.9.32)

$$V_{ijklmn}^{000} = \frac{2\pi}{p} \sum_{tuv} E_t^{ij} E_u^{kl} E_v^{mn} R_{tuv}(p, \mathbf{R}_{PC}) \tag{9.11.40}$$

Inserting the quadrature expression (9.11.32) in (9.11.40), we obtain

$$V_{ijklmn}^{000} = \frac{2\pi}{p} \sum_{\kappa=1}^{\gamma} w_\kappa \sum_{tuv} E_t^{ij} E_u^{kl} E_v^{mn} \overline{H}_t(s_\kappa^2) \overline{H}_u(s_\kappa^2) \overline{H}_v(s_\kappa^2) \tag{9.11.41}$$

This expression may be rearranged to give

$$V_{ijklmn}^{000} = \frac{2\pi}{p} \sum_{\kappa=1}^{\gamma} w_\kappa I_{ij}(s_\kappa^2) I_{kl}(s_\kappa^2) I_{mn}(s_\kappa^2) \tag{9.11.42}$$

where we have introduced the one-dimensional Cartesian integrals

$$I_{ij}(s^2) = \sum_t E_t^{ij} \overline{H}_t(s^2) \tag{9.11.43}$$

$$I_{kl}(s^2) = \sum_u E_u^{kl} \overline{H}_u(s^2) \tag{9.11.44}$$

$$I_{mn}(s^2) = \sum_v E_v^{mn} \overline{H}_v(s^2) \tag{9.11.45}$$

Equation (9.11.42) constitutes the final expression for the one-electron Coulomb integrals in the Rys scheme [10].

We have now developed a complete scheme for the evaluation of one-electron Coulomb integrals by Gaussian quadrature. First, we calculate (by some numerical scheme) the abscissae and weights of the Rys polynomial of degree

$$\gamma = \left[\frac{L}{2}\right] + 1 \tag{9.11.46}$$

where L is the sum of the quantum numbers of the two orbitals (which is the same for all orbitals in the same shell)

$$L = i + j + k + l + m + n \tag{9.11.47}$$

Next, at each abscissa, we calculate the modified Hermite polynomials (9.11.30) using the recurrence relations (9.11.36). The resulting polynomial values are then contracted with the Hermite-to-Cartesian expansion coefficients, yielding the one-dimensional Cartesian integrals (9.11.43)–(9.11.45). The expansion coefficients, which may be obtained from the two-term recurrence relations (9.5.15)–(9.5.17), are the same for all the abscissae. The final Cartesian one-electron Coulomb integral is obtained by carrying out the summation (9.11.42).

The two-electron integrals are obtained in the same manner. Starting from (9.9.33), we obtain the expression [10]

$$g_{abcd} = \frac{2\pi^{5/2}}{pq\sqrt{p+q}} \sum_{\kappa=1}^{\gamma} w_\kappa I_x(s_\kappa^2) I_y(s_\kappa^2) I_z(s_\kappa^2) \tag{9.11.48}$$

The two-dimensional Cartesian integrals are given by

$$I_x(s^2) = \sum_{t\tau} (-1)^\tau E_t^{i_x j_x} E_\tau^{k_x l_x} \overline{H}_{t+\tau}(s^2) \tag{9.11.49}$$

$$I_y(s^2) = \sum_{uv} (-1)^v E_u^{i_y j_y} E_v^{k_y l_y} \overline{H}_{u+v}(s^2) \tag{9.11.50}$$

$$I_z(s^2) = \sum_{v\phi} (-1)^\phi E_v^{i_z j_z} E_\phi^{k_z l_z} \overline{H}_{v+\phi}(s^2) \tag{9.11.51}$$

and are evaluated in terms of the modified Hermite polynomials, obtained by using recurrence relations such as (9.11.39). The number of quadrature points is again obtained from (9.11.46), where L is now the sum of the quantum numbers for all four orbitals. Thus, there is one quadrature point for $(ssss)$ and $(psss)$ integrals, two quadrature points for $(ppss)$ and $(ppps)$ integrals, three for a $(pppp)$ integral, and so on. In general, for an integral with all orbitals of angular momentum l, the number of quadrature points is equal to $2l + 1$; therefore, we must calculate $2l + 1$ abscissae and $2l + 1$ weights. These numbers should be compared with the number of Boys-function evaluations needed for the same integral $(4l + 1)$.

9.11.6 OBARA–SAIKA RECURSION FOR THE TWO-DIMENSIONAL RYS INTEGRALS

In the Rys-quadrature scheme presented above, the one- and two-dimensional integrals were calculated using the McMurchie–Davidson scheme. These integrals may also be obtained from the Obara–Saika scheme, as we shall now discuss.

Let us first consider the one-dimensional integrals needed for the one-electron Coulomb integral. For the x direction, these are given by (9.11.43)

$$I_{ij}(s^2) = \sum_t E_t^{ij} \overline{H}_t(s^2) \tag{9.11.52}$$

We proceed in the same manner as in Section 9.10.1. Incrementing the first index and using the recurrence relation (9.5.20), we obtain

$$I_{i+1,j} = X_{PA}I_{ij} + \frac{1}{2p}(iI_{i-1,j} + jI_{i,j-1}) + \frac{1}{2p}\sum_t E_t^{ij}\overline{H}_{t+1} \tag{9.11.53}$$

where we have omitted the argument s^2 and have substituted t for $t-1$ in the summation. Next, we invoke the recurrence relation (9.11.36)

$$I_{i+1,j} = X_{PA}I_{ij} + \frac{1}{2p}(iI_{i-1,j} + jI_{i,j-1}) - s^2 X_{PC}I_{ij} - \frac{s^2}{2p}\sum_t 2pt E_t^{ij}\overline{H}_{t-1} \tag{9.11.54}$$

and finally the identity (9.5.14). We then arrive at the following recurrence relations for the one-dimensional integrals [19]

$$I_{i+1,j} = (X_{PA} - s^2 X_{PC})I_{ij} + \frac{1}{2p}(1-s^2)(iI_{i-1,j} + jI_{i,j-1}) \tag{9.11.55}$$

$$I_{i,j+1} = (X_{PB} - s^2 X_{PC})I_{ij} + \frac{1}{2p}(1-s^2)(iI_{i-1,j} + jI_{i,j-1}) \tag{9.11.56}$$

where the latter relation is obtained from (9.11.55) by using (9.5.9). Starting from the integral

$$I_{00} = K_{ab}^x \tag{9.11.57}$$

we may then generate the full set of one-dimensional integrals needed for Rys quadrature. The one-dimensional recurrence relations (9.11.55) and (9.11.56) should be compared with their three-dimensional counterparts (9.10.8) and (9.10.9). We note that increments in the upper index for the three-dimensional integrals have been replaced by multiplication by a factor of s^2 for the one-dimensional ones.

The recurrence relations for the two-dimensional two-electron integrals are worked out in the same manner. Consider the integral (9.11.49) for the x direction

$$I_{ijkl} = \sum_{t\tau}(-1)^\tau E_t^{ij} E_\tau^{kl}\overline{H}_{t+\tau} \tag{9.11.58}$$

Incrementing the first index and invoking the recurrence relation (9.5.20), we obtain

$$I_{i+1,j,k,l} = X_{PA}I_{ijkl} + \frac{1}{2p}(iI_{i-1,j,k,l} + jI_{i,j-1,k,l}) + \frac{1}{2p}\sum_{t\tau}(-1)^\tau E_t^{ij} E_\tau^{kl}\overline{H}_{t+\tau+1} \tag{9.11.59}$$

The recurrence relation for the modified Hermite polynomials (9.11.39) in the form

$$\overline{H}_{t+\tau+1} = -2\alpha s^2[X_{PQ}\overline{H}_{t+\tau} + (t+\tau)\overline{H}_{t+\tau-1}] \tag{9.11.60}$$

gives us

$$I_{i+1,j,k,l} = (X_{PA} - \frac{\alpha}{p}s^2 X_{PQ})I_{ijkl} + \frac{1}{2p}(iI_{i-1,j,k,l} + jI_{i,j-1,k,l})$$
$$- \frac{\alpha}{p}s^2\sum_{t\tau}(-1)^\tau t E_t^{ij} E_\tau^{kl}\overline{H}_{t+\tau-1} - \frac{\alpha}{p}s^2\sum_{t\tau}(-1)^\tau E_t^{ij} \tau E_\tau^{kl}\overline{H}_{t+\tau-1} \tag{9.11.61}$$

Finally, for the last two terms, we employ (9.5.17) and obtain the following recurrence relation for increments in the first index [19]:

$$I_{i+1,j,k,l} = \left(X_{PA} - \frac{\alpha}{p}X_{PQ}s^2\right)I_{ijkl} + \frac{1}{2p}\left(1 - \frac{\alpha}{p}s^2\right)(iI_{i-1,j,k,l} + jI_{i,j-1,k,l})$$

$$+ \frac{s^2}{2(p+q)}(kI_{i,j,k-1,l} + lI_{i,j,k,l-1}) \tag{9.11.62}$$

The corresponding recurrence relation for increments in the second index is identical except that X_{PA} is replaced by X_{PB}. The recurrence relation for increments in the third index is obtained from (9.11.62) by interchange of the quantities that refer to particles 1 and 2:

$$I_{i,j,k+1,l} = \left(X_{QC} + \frac{\alpha}{q}X_{PQ}s^2\right)I_{ijkl} + \frac{1}{2q}\left(1 - \frac{\alpha}{q}s^2\right)(kI_{i,j,k-1,l} + lI_{i,j,k,l-1})$$

$$+ \frac{s^2}{2(p+q)}(iI_{i-1,j,k,l} + jI_{i,j-1,k,l}) \tag{9.11.63}$$

and, for the fourth index, we simply replace X_{QC} by X_{QD}. Comparing the recurrence relations (9.11.62) and (9.11.63) with the Obara–Saika recurrences for the six-dimensional integrals (9.10.24) and (9.10.25), we find (as in the one-electron case) that increments in the upper index of the six-dimensional integrals have been replaced by multiplication by the factor s^2 in the two-dimensional integrals.

It is of course possible to simplify the recurrence relations for the two-dimensional two-electron integrals considerably by considering only the integrals I_{i0k0}. We then obtain the following two recurrence relations [19]

$$I_{i+1,0,k,0} = \left(X_{PA} - \frac{\alpha}{p}s^2X_{PQ}\right)I_{i0k0} + \frac{1}{2p}\left(1 - \frac{\alpha}{p}s^2\right)iI_{i-1,0,k,0} + \frac{s^2}{2(p+q)}kI_{i,0,k-1,0} \tag{9.11.64}$$

$$I_{i,0,k+1,0} = \left(X_{QC} + \frac{\alpha}{q}s^2X_{PQ}\right)I_{i0k0} + \frac{1}{2q}\left(1 - \frac{\alpha}{q}s^2\right)kI_{i,0,k-1,0} + \frac{s^2}{2(p+q)}iI_{i-1,0,k,0} \tag{9.11.65}$$

The remaining integrals may then be generated by the standard horizontal recurrence relations. Also, we may specialize (9.11.64) to integrals of the form I_{i000}, resulting in a simple two-term recurrence relation; the final integrals are then obtained using an electron-transfer relation. Finally, we note that there is no need to carry out either the horizontal or the electron-transfer relations until after the six-dimensional integrals have been calculated.

9.11.7 COMPUTATIONAL CONSIDERATIONS FOR THE TWO-ELECTRON INTEGRALS

Just as in the Obara–Saika scheme, we may set up three different algorithms for the evaluation of two-electron integrals by the Rys scheme. In the R4 scheme, we use the four-index recurrence relations such as (9.11.62) and (9.11.63) to generate the two-dimensional integrals (9.11.49)–(9.11.51) at each of the γ abscissae; the final six-dimensional Cartesian integrals are then obtained from the quadrature formula (9.11.48) [10]. In the R2 scheme, we first use the simpler two-index recurrence relations (9.11.64) and (9.11.65) for the two-dimensional integrals; next, we generate by quadrature six-dimensional Cartesian integrals with the angular momentum distributed between two orbitals only; in the final step, we invoke the horizontal recurrence relations to distribute the

Table 9.6 Cost and memory requirements of the two-electron Rys scheme

	R4		R2		R1	
	Cost	Memory	Cost	Memory	Cost	Memory
Rys weights and roots	Lp^4	Lp^4	Lp^4	Lp^4	Lp^4	Lp^4
2-D integrals	L^5p^4	L^5p^4	L^3p^4	L^3p^4	L^2p^4	L^2p^4
6-D integrals	L^9p^4	L^8p^4	L^7p^4	L^6p^4	L^4p^4	L^3p^4
Transfer recursion					L^6p^4	L^6p^4
Primitive contraction	L^8p^4	L^8	L^6p^4	L^6	L^6p^4	L^6
Horizontal recursion 1			L^9	L^7	L^9	L^7
Solid harmonics 1	L^9	L^6	L^8	L^5	L^8	L^5
Horizontal recursion 2			L^8	L^6	L^8	L^6
Solid harmonics 2	L^7	L^4	L^7	L^4	L^7	L^4

angular momentum among all four orbitals [19]. In the R1 scheme, we first generate one-index two-dimensional integrals; next, we generate from these integrals six-dimensional Cartesian integrals with the angular momentum on only one orbital; the final integrals are then obtained by distributing the angular momentum using the electron-transfer recurrence relations and finally the horizontal recurrence relations [15]. The cost of these different schemes is given in Table 9.6.

Comparing with the McMurchie–Davidson scheme in Section 9.9.6 and the Obara–Saika scheme in Section 9.10.4, we note that the cost scales in the same manner for the schemes that employ the electron-transfer relation – the MD1, OS1 and R1 schemes, all of which are dominated by the electron-transfer step (L^6p^4) and by the horizontal step (L^9). For high angular momentum, these appear to be the preferred schemes – in particular the Obara–Saika scheme, which involves a smaller number of steps.

Comparing the vertical–horizontal schemes MD2, OS2 and R2, we note that these perform similarly ($L^9p^4 + L^9p^2$ or $L^7p^4 + L^9p^2$ for MD2, and $L^7p^4 + L^9$ for OS2 and R2) and a more careful analysis is needed to distinguish between them. Finally, for the schemes that do not employ either the transfer or the horizontal recurrences, the cost is similar for MD4 ($L^{10}p^4$ or L^8p^4) and R4 (L^9p^4) and less favourable for OS4 ($L^{13}p^4$).

9.12 Scaling properties of the molecular integrals

Using the techniques developed in the preceding sections, the molecular integrals may be calculated to any desired accuracy. In practice, however, we are interested in calculating the integrals to some fixed absolute accuracy, discarding all integrals whose magnitude is smaller than the allowed error. In large systems, in particular, most integrals are small and it becomes imperative to detect these integrals in advance so that their evaluation can be avoided. In the present section, we shall examine molecular integrals in large systems – in particular, their scaling properties (i.e. the dependence of the number of significant integrals on the size of the system) and techniques for their pre-screening.

9.12.1 LINEAR SCALING OF THE OVERLAP AND KINETIC-ENERGY INTEGRALS

To study the scaling properties of the integrals, we consider a linear model system consisting of n hydrogen atoms, separated from one another by $1a_0$. To each atom, we attach a single primitive Gaussian $1s$ function of unit exponent. In Figure 9.12, we have plotted, for this model system, the number of overlap, kinetic-energy, nuclear-attraction, and electron-repulsion integrals as a function of n. Full lines are used for the total number of integrals and dotted lines for the number

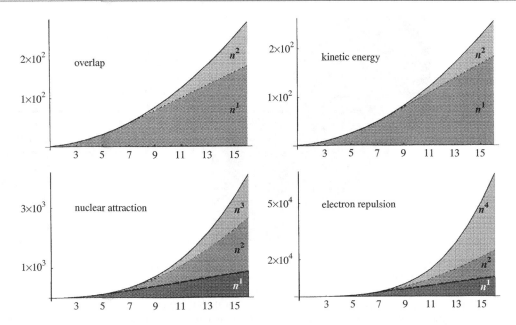

Fig. 9.12. The scaling properties of the molecular integrals. The full line represents the total number of integrals as a function of n, where n is the number of $1s$ AOs of unit exponent in a linear system of n hydrogen atoms, separated from one another by $1a_0$. The dotted line represents the number of significant integrals (i.e. integrals larger than 10^{-10} in magnitude). Finally, for the Coulomb integrals, the dashed line represents the number of integrals that cannot be calculated classically (by multipole expansions) since their charge distributions overlap. Note the differences in the vertical scales (approximately 10^2, 10^2, 10^3 and 10^4), which reflect the different scaling properties of the integrals.

of significant integrals, which we here take to be integrals larger than 10^{-10} in magnitude. For small n, all integrals are significant and must be calculated. However, as the number of atoms increases, many of the integrals become insignificant and may be neglected when the calculations are carried out to a given finite precision.

Let us consider the overlap integrals. For our model system consisting of n atoms, the total number of overlap integrals is equal to n^2. Thus, formally at least, the number of overlap integrals scales quadratically with the size of the system, as indicated by the full line in Figure 9.12. However, as seen from the dotted line in the same figure, the number of *significant* integrals scales only *linearly* with the size of the system. To understand how this linear dependence on the size of the system arises, consider the expression for the overlap between two Gaussian s functions (9.2.41):

$$S_{ab} = \left(\frac{\pi}{a+b}\right)^{3/2} \exp\left(-\frac{ab}{a+b}R_{AB}^2\right) \tag{9.12.1}$$

This integral decreases exponentially with the square of the distance between the functions. To determine the distance d_s where this integral becomes smaller than 10^{-k}, we consider the Gaussian with the smallest exponent a_{min}. Some simple algebra then shows that all s functions separated by more than

$$d_s = \sqrt{a_{min}^{-1} \log\left[\left(\frac{\pi}{2a_{min}}\right)^3 10^{2k}\right]} \tag{9.12.2}$$

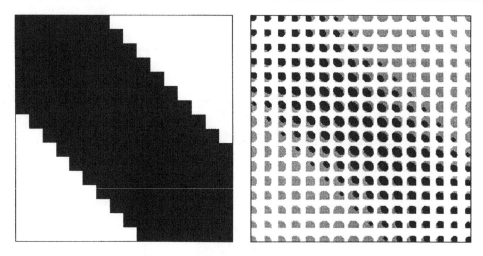

Fig. 9.13. The sparsity of the 16×16 overlap matrix (left) and the 256×256 electron-repulsion matrix (right) for a linear model system of 16 Gaussian $1s$ functions of unit exponent, separated from one another by $1a_0$. In the overlap matrix, the significant integrals (i.e. integrals larger than 10^{-10}) appear in black and the remaining, insignificant integrals in white. In the electron-repulsion matrix, the nonclassical significant integrals appear in black, the classical significant integrals in grey and the insignificant integrals in white.

are smaller than 10^{-k} in magnitude. In particular, in our model system (where the exponents are all equal to 1), all overlaps for functions separated by more than $6.9a_0$ will be smaller than 10^{-10}, giving rise to the banded matrix in Figure 9.13. Clearly, as the length of the chain increases, there is a crossover point, beyond which the number of significant overlap integrals increases only linearly with the size of the system. In our model, this crossover occurs for systems containing about seven atoms – see Figure 9.12.

In a more realistic system, the AOs will be less uniformly distributed in space and there will be a range of orbital exponents. The crossover to linear scaling will then be less sharply defined. Nevertheless, for any molecular system, there will be a characteristic distance d_s beyond which all overlap integrals are negligible. When the size of the system increases beyond d_s, the amount of work required for the evaluation of the overlap integrals will begin to increase linearly with the size of the system – provided, of course, that the evaluation of the insignificant integrals is avoided by the application of some screening mechanism such as testing against d_s.

Since the kinetic-energy integrals are fixed linear combinations of overlap integrals (of higher angular momentum), there will also be a characteristic distance beyond which all kinetic-energy integrals are negligible, leading to linear scaling for these integrals as well – see Figure 9.12. For the kinetic-energy integrals, the crossover to linear scaling appears to occur somewhat later than for the overlap integrals.

9.12.2 QUADRATIC SCALING OF THE COULOMB INTEGRALS

Having considered the scaling properties of the overlap and kinetic-energy integrals, we now turn our attention to the Coulomb integrals and, in particular, to the two-electron integrals. For s orbitals (the only orbitals present in our model system), we may write the two-electron Coulomb integrals in the form

$$g_{abcd} = \sqrt{\frac{4\alpha}{\pi}} S_{ab} S_{cd} F_0(\alpha R_{\mathrm{PQ}}^2) \tag{9.12.3}$$

which is the generalization of (9.7.24) to unnormalized charge distributions, obtained by using the Gaussian product rule (9.2.10) for the overlap distributions and then (9.12.1) and (9.7.24). We recall here that R_{PQ} is the separation between the overlap distributions of the two electrons and that α is the reduced exponent $pq/(p+q)$. To study the scaling properties of the two-electron integrals, we note that the integral (9.12.3) is bounded by

$$g_{abcd} \leq \min\left(\sqrt{\frac{4\alpha}{\pi}} S_{ab} S_{cd}, \frac{S_{ab} S_{cd}}{R_{PQ}}\right) \tag{9.12.4}$$

where the first upper bound represents the short-range limit and the second bound the long-range limit. These bounds for the integrals follow from the corresponding upper bounds to the Boys function (9.8.7) and (9.8.10).

For small systems, all integrals are significant (in the absence of symmetries) and the number of two-electron integrals increases as n^4. However, when the system increases beyond d_s and the number of significant overlap integrals begins to increase as n, the number of significant two-electron integrals will begin to increase as n^2 [20]. This crossover from quartic to *quadratic scaling* is clearly seen in Figure 9.12. In Figure 9.13, we have depicted the sparse structure of the two-electron integral matrix for $n = 16$, with the significant integrals shaded in grey or black, depending on whether they represent classical or nonclassical interactions as explained in Section 9.12.3.

From the long-range upper bound in (9.12.4), we conclude that the number of significant two-electron integrals should eventually increase *linearly* with the size of the system. However, for this to happen, the system must become truly macroscopic in size, with typical separations R_{PQ} greater than 1 m $\approx 10^{10} a_0$. For all practical purposes, therefore, the number of significant two-electron integrals will depend at least quadratically on the size of the system.

9.12.3 LINEAR SCALING OF THE NONCLASSICAL COULOMB INTEGRALS

Although, in large molecular systems, the number of significant two-electron integrals increases quadratically with the size of the system, most of these integrals represent classical interactions between disjoint charge distributions. To discuss the classical and nonclassical interactions, we first rewrite the two-electron integrals (9.12.3) in terms of the error function (9.8.22). Using (9.8.24), we obtain

$$g_{abcd} = \operatorname{erf}(\sqrt{\alpha} R_{PQ}) \frac{S_{ab} S_{cd}}{R_{PQ}} \tag{9.12.5}$$

Next, introducing the complementary error function (9.8.27), we decompose the two-electron integral as

$$g_{abcd} = g_{abcd}^{cls} + g_{abcd}^{non} \tag{9.12.6}$$

where the *classical* and *nonclassical contributions* are given by

$$g_{abcd}^{cls} = \frac{S_{ab} S_{cd}}{R_{PQ}} \tag{9.12.7}$$

$$g_{abcd}^{non} = -\operatorname{erfc}(\sqrt{\alpha} R_{PQ}) \frac{S_{ab} S_{cd}}{R_{PQ}} \tag{9.12.8}$$

For small R_{PQ}, the classical and nonclassical contributions are both large, even when the integral g_{abcd} itself is small. Thus, as R_{PQ} tends to zero, the total integral represents a small difference between a classical term and a nonclassical one, both of which tend to infinity. By contrast, for large R_{PQ}, the complementary error function in (9.12.8) vanishes according to (9.8.32) and the Coulomb interaction is dominated by the classical contribution (9.12.7). In large systems, therefore, most of the two-electron integrals will represent such purely classical interactions.

From the upper bound to the complementary error function (9.8.32), we find that the nonclassical contribution to the two-electron integrals is bounded by

$$|g_{abcd}^{\text{non}}| \le S_{ab}S_{cd}\frac{\exp(-\alpha R_{PQ}^2)}{\sqrt{\pi\alpha}R_{PQ}^2} \tag{9.12.9}$$

and therefore decreases rapidly with increasing R_{PQ}. Indeed, comparing with the expression for the overlap integral (9.12.1), we find that the nonclassical Coulomb contribution decreases more rapidly with R_{PQ} than does the overlap integral with R_{AB}. We thus conclude that, for sufficiently large systems, the number of *nonclassical Coulomb integrals* (i.e. Coulomb integrals for which the nonclassical contribution is significant) should increase *linearly* with the size of the system.

In Figure 9.12, we have plotted the number of nonclassical two-electron integrals using a dashed line. Note that the crossover to linear scaling for the nonclassical Coulomb integrals occurs in the same region as the crossover to linear scaling for the overlap integrals and as the crossover to quadratic scaling for the classical two-electron integrals. The distribution of the significant nonclassical two-electron integrals is also illustrated in Figure 9.13, where these integrals are shaded in black and are seen to be located in a broad band along the matrix diagonal, unlike the classical integrals, which are distributed more evenly over the whole matrix.

To distinguish between the classical and nonclassical Coulomb integrals, it is useful to introduce the concept of the extent of a Gaussian distribution. For a target accuracy of 10^{-k} in the evaluation, we introduce the *extent of a Gaussian distribution* of exponent p as [21]

$$r_p = p^{-1/2}\text{erfc}^{-1}(10^{-k}) \tag{9.12.10}$$

where $\text{erfc}^{-1}(x)$ is the inverse of the complementary error function. We shall now demonstrate that two Gaussians separated by more than the sum of their extents

$$R_{PQ} > r_p + r_q \tag{9.12.11}$$

interact classically. Multiplying (9.12.11) by $\sqrt{pq/(p+q)}$ and inserting the expressions for the Gaussian extents, we obtain

$$\sqrt{\frac{pq}{p+q}}R_{PQ} > \sqrt{\frac{pq}{p+q}}\left(\frac{1}{\sqrt{p}} + \frac{1}{\sqrt{q}}\right)\text{erfc}^{-1}(10^{-k}) = \frac{\sqrt{p}+\sqrt{q}}{\sqrt{p+q}}\text{erfc}^{-1}(10^{-k}) \tag{9.12.12}$$

which, by simple algebraic considerations, gives the inequality

$$\sqrt{\frac{pq}{p+q}}R_{PQ} > \text{erfc}^{-1}(10^{-k}) \tag{9.12.13}$$

Recalling that $\text{erfc}(x)$ is a strictly decreasing function, we obtain

$$\text{erfc}\left(\sqrt{\frac{pq}{p+q}}R_{PQ}\right) < 10^{-k} \tag{9.12.14}$$

Thus, when two Gaussian distributions are separated by more than the sum of their extents, the nonclassical contribution (9.12.8) will be negligible relative to the classical contribution (9.12.7) since

$$\frac{g_{abcd}^{\text{non}}}{g_{abcd}^{\text{cls}}} = -\text{erfc}\left(\sqrt{\frac{pq}{p+q}}R_{\text{PQ}}\right) \tag{9.12.15}$$

and the two-electron integral (9.12.5) may then be calculated classically. In practice, the requirement (9.12.14) is too strict, since it is the overall contribution of the nonclassical term (9.12.8) to, for example, the Hartree–Fock energy rather than its relative contribution to the two-electron integral that matters. Nevertheless, the extent of the Gaussian provides a useful test (9.12.11), easily applied to any pair of Gaussian distributions. We also note that, in our model example, the target accuracy is 10^{-10} and the exponent of all distributions is equal to 2, giving an extent of $3.2a_0$. Therefore, as the system increases beyond twice the Gaussian extent of $6.4a_0$, some of the two-electron integrals should begin to represent purely classical interactions, as seen in Figure 9.12.

The different scaling behaviour of the classical and nonclassical two-electron integrals has important ramifications. Thus, whereas the nonclassical integrals must be evaluated by the standard techniques such the McMurchie–Davidson, Obara–Saika and Rys schemes, the classical integrals may be evaluated, to an accuracy of 10^{-k}, more simply by the multipole method developed in Section 9.13. Moreover, to calculate the total Coulomb contribution to the Fock operator or to the energy in large systems, there is no need to evaluate the individual integrals explicitly. Rather, as discussed in Section 9.14, their contribution may be calculated much more efficiently by the fast multipole method, at a cost that scales only linearly with the size of the system.

To summarize, although the number of significant two-electron integrals scales quadratically with the size of the system, most of these integrals represent classical electrostatic interactions. These classical integrals may either be calculated individually by multipole expansions at a cost that scales quadratically with the size of the system (Section 9.13), or else their total contribution to, for example, the Hartree–Fock energy may be calculated collectively at a cost that scales only linearly (Section 9.14). The significant nonclassical two-electron integrals, by contrast, scale linearly with the size of the system and cannot be calculated by multipole expansions. *Mutatis mutandis*, the same considerations apply to the nuclear-attraction integrals – see Figure 9.12.

9.12.4 THE SCHWARZ INEQUALITY

In Section 9.12.2, we established that, for sufficiently large systems, the number of significant two-electron integrals increases quadratically rather than quartically with the size of the system. For this result to be useful, however, we must develop some method according to which we may avoid the calculation of the small integrals. In the present subsection, we shall establish a useful upper bound to the two-electron integrals based on the Schwarz inequality.

The two-electron integrals can be viewed as a matrix with the electron distributions as row and column labels

$$g_{abcd} = \int\int \frac{\Omega_{ab}(\mathbf{r}_1)\Omega_{cd}(\mathbf{r}_2)}{r_{12}}\,d\mathbf{r}_1\,d\mathbf{r}_2 \tag{9.12.16}$$

Assuming that the orbitals are real, we shall demonstrate that this matrix is positive definite [22]. Let us consider the interaction between two electrons in the same distribution $\rho(\mathbf{r})$:

$$I[\rho] = \int\int \frac{\rho(\mathbf{r}_1)\rho(\mathbf{r}_2)}{r_{12}}\,d\mathbf{r}_1\,d\mathbf{r}_2 \tag{9.12.17}$$

Inserting the Fourier transform of the interaction operator

$$\frac{1}{r_{12}} = \frac{1}{2\pi^2} \int k^{-2} \exp[i\mathbf{k} \cdot (\mathbf{r}_1 - \mathbf{r}_2)] \, d\mathbf{k} \tag{9.12.18}$$

and carrying out the integration over the Cartesian coordinates, we obtain

$$I[\rho] = \frac{1}{2\pi^2} \int k^{-2} |\rho(\mathbf{k})|^2 \, d\mathbf{k} \tag{9.12.19}$$

where we have introduced the distributions

$$\rho(\mathbf{k}) = \int \exp(-i\mathbf{k} \cdot \mathbf{r}) \rho(\mathbf{r}) \, d\mathbf{r} \tag{9.12.20}$$

Since the integrand in (9.12.19) is always positive or zero, we obtain the inequality

$$I[\rho] > 0 \tag{9.12.21}$$

Moreover, expanding the charge distribution $\rho(\mathbf{r})$ in (9.12.17) in a set of one-electron orbital distributions

$$\rho(\mathbf{r}) = \sum_{ab} c_{ab} \Omega_{ab}(\mathbf{r}) \tag{9.12.22}$$

we may write the inequality (9.12.21) in the form

$$\sum_{abcd} c_{ab} g_{abcd} c_{cd} > 0 \tag{9.12.23}$$

where the two-electron integrals are given by (9.12.16). Equation (9.12.23) constitutes the definition of a positive definite matrix. As a corollary, the diagonal elements are positive

$$g_{abab} > 0 \tag{9.12.24}$$

The two-electron integrals therefore satisfy the conditions for inner products, in a metric defined by r_{12}^{-1}. Application of the Schwarz inequality then yields [23]

$$|g_{abcd}| \leq \sqrt{g_{abab}} \sqrt{g_{cdcd}} \tag{9.12.25}$$

Thus, if we calculate the square root of all the diagonal integrals

$$G_{ab} = \sqrt{g_{abab}} \tag{9.12.26}$$

prior to the calculation of the full set of two-electron integrals, we can easily compute upper bounds to the individual integrals as these are generated. Integrals below a given threshold may then be identified in advance and their evaluation avoided. Note that the Schwarz inequality is optimal in the sense that it is exact for all diagonal elements. Also, from the discussion of the evaluation of two-electron integrals in the preceding sections of this chapter, we note that the diagonal integrals (9.12.26) may be calculated in a particularly efficient manner and in a closed form since, for such integrals, the argument to the Boys function αR_{PQ}^2 is always zero.

Let us consider the application of the Schwarz inequality to a two-electron integral of s orbitals (9.12.3). The integrals (9.12.26) are simply obtained from (9.12.3) as

$$G_{ab} = \left(\frac{2p}{\pi}\right)^{1/4} |S_{ab}| \tag{9.12.27}$$

since R_{PQ} is 0 and the Boys function becomes equal to 1. The Schwarz inequality (9.12.25) then gives the following upper bound to the two-electron integral (9.12.3):

$$|g_{abcd}| \leq \sqrt{\frac{2}{\pi}}(pq)^{1/4}|S_{ab}S_{cd}| \tag{9.12.28}$$

Note that this bound is independent of the separation between the distributions R_{PQ}. Since the integral decreases in magnitude with R_{PQ}, the Schwarz inequality provides a best estimate for small R_{PQ}. However, since R_{PQ} is usually considerably smaller than say $100a_0$, this overestimate does not reduce the usefulness of the Schwarz screening severely.

Assuming that R_{PQ} is zero, the Schwarz inequality (9.12.28) for the two-electron integral (9.12.3) may be expressed in the form

$$\sqrt{\frac{4\alpha}{\pi}}|S_{ab}S_{cd}| \leq \sqrt{\frac{2}{\pi}}(pq)^{1/4}|S_{ab}S_{cd}| \tag{9.12.29}$$

or equivalently

$$\frac{pq}{p+q} \leq \frac{1}{2}\sqrt{pq} \tag{9.12.30}$$

Thus, the Schwarz test provides a good upper bound when p and q are similar and a poorer one when they are different. As an illustration of the Schwarz screening, we note that, for the largest system with $n = 16$ in Figure 9.12, 38.8% of the integrals passed the Schwarz screening. Of these integrals, nearly all (37.6%) turned out to be significant. Of course, since all exponents in this system are the same, it is particularly well suited to the Schwarz screening.

9.13 The multipole method for Coulomb integrals

In Section 9.12, we found that, in large molecular systems, most of the significant two-electron integrals represent classical electrostatic interactions between nonoverlapping charge distributions. Although these classical integrals may be evaluated in the same manner as the nonclassical ones, using one of the methods developed in the preceding sections such as the McMurchie–Davidson scheme, the Obara–Saika scheme and the Rys-quadrature scheme, it is usually much more efficient to evaluate these integrals by the *multipole method*, in which the integrals are evaluated by a multipole expansion of the Coulomb interaction [24,25]. In the present section, we discuss the evaluation of two-electron integrals by the multipole method.

9.13.1 THE MULTIPOLE METHOD FOR PRIMITIVE TWO-ELECTRON INTEGRALS

Let us consider the evaluation of the two-electron integral

$$g_{abcd} = \int\int \frac{\Omega_{ab}(\mathbf{r}_1)\Omega_{cd}(\mathbf{r}_2)}{r_{12}}\, d\mathbf{r}_1\, d\mathbf{r}_2 \tag{9.13.1}$$

where the two primitive Gaussian overlap distributions $\Omega_{ab}(\mathbf{r}_1)$ and $\Omega_{cd}(\mathbf{r}_2)$ are centred at $\mathbf{P}$ and $\mathbf{Q}$. To set up a multipole expansion of this integral, we express the positions of the electrons relative to the centres of the overlap distributions as

$$\mathbf{r}_1 = \mathbf{r}_{1P} + \mathbf{P} \tag{9.13.2}$$

$$\mathbf{r}_2 = \mathbf{r}_{2Q} + \mathbf{Q} \tag{9.13.3}$$

and introduce the vectors

$$\Delta \mathbf{r}_{12} = \mathbf{r}_{1P} - \mathbf{r}_{2Q} \tag{9.13.4}$$

$$\mathbf{R}_{QP} = \mathbf{Q} - \mathbf{P} \tag{9.13.5}$$

We shall further assume that, in the regions where the integrand in (9.13.1) is nonnegligible, the following inequality holds:

$$\Delta r_{12} = |\Delta \mathbf{r}_{12}| < R_{QP} = |\mathbf{R}_{QP}| \tag{9.13.6}$$

For the purpose of evaluating the integral (9.13.1), we may then carry out a partial-wave expansion of the inverse interelectronic distance in accordance with (7.4.3):

$$\frac{1}{r_{12}} = \sum_{l=0}^{\infty} \frac{\Delta r_{12}^l}{R_{QP}^{l+1}} P_l(\cos \theta) \tag{9.13.7}$$

Here the $P_l(\cos \theta)$ are the Legendre polynomials in $\cos \theta$ introduced in Section 6.4.1 and θ is the angle between $\Delta \mathbf{r}_{12}$ and $\mathbf{R}_{QP}$:

$$\cos \theta = \frac{\Delta \mathbf{r}_{12} \cdot \mathbf{R}_{QP}}{\Delta r_{12} R_{QP}} \tag{9.13.8}$$

Since, for all θ, the Legendre polynomials satisfy the inequality [9]

$$|P_l(\cos \theta)| \le 1 \tag{9.13.9}$$

the convergence of the partial-wave expansion (9.13.7) follows directly from the assumption (9.13.6). In Section 9.13.2, we shall examine the convergence in more detail.

To develop the multipole expansion of the two-electron integral (9.13.1), we invoke the addition theorem for the spherical harmonics (7.3.3). In Racah's normalization of the spherical harmonics (6.4.10), the addition theorem takes the form

$$P_l(\cos \theta) = \sum_{m=-l}^{l} C_{lm}(\theta_{12}, \varphi_{12}) C_{lm}^*(\theta_{QP}, \varphi_{QP}) \tag{9.13.10}$$

where θ_{12} and φ_{12} are the angular coordinates of $\Delta \mathbf{r}_{12}$, and θ_{QP} and φ_{QP} are the corresponding coordinates of $\mathbf{R}_{QP}$. Inserting this expansion in (9.13.7), we obtain

$$\frac{1}{r_{12}} = \sum_{l=0}^{\infty} \sum_{m=-l}^{l} \Delta r_{12}^l C_{lm}(\theta_{12}, \varphi_{12}) R_{QP}^{-l-1} C_{lm}^*(\theta_{QP}, \varphi_{QP}) \tag{9.13.11}$$

At this point, it would be possible to express this expansion in terms of the solid harmonics (6.4.22)

$$C_{lm}(\mathbf{r}) = r^l C_{lm}(\theta, \varphi) \tag{9.13.12}$$

introduced in Section 6.4.2 to yield [24]

$$\frac{1}{r_{12}} = \sum_{l=0}^{\infty} \sum_{m=-l}^{l} \frac{C_{lm}(\Delta \mathbf{r}_{12}) C_{lm}^*(\mathbf{R}_{QP})}{R_{QP}^{2l+1}} \tag{9.13.13}$$

However, for our purposes, it turns out to be more convenient to work with the *scaled regular and irregular solid harmonics* [25]

$$R_{lm}(\mathbf{r}) = \frac{1}{\sqrt{(l-m)!(l+m)!}} r^l C_{lm}(\theta, \varphi) \tag{9.13.14}$$

$$I_{lm}(\mathbf{r}) = \sqrt{(l-m)!(l+m)!} \, r^{-l-1} C_{lm}(\theta, \varphi) \tag{9.13.15}$$

In terms of these functions, the partial-wave expansion of the inverse electronic separation (9.13.11) may be written compactly as

$$\frac{1}{r_{12}} = \sum_{l=0}^{\infty} \sum_{m=-l}^{l} R_{lm}(\Delta \mathbf{r}_{12}) I_{lm}^*(\mathbf{R}_{QP}) \tag{9.13.16}$$

We shall consider the recursive evaluation of the scaled regular and irregular solid harmonics (9.13.14) and (9.13.15) in Section 9.13.8. At present, we only note that these functions exhibit the same inversion and conjugation symmetries (6.4.17) and (6.4.18) as do the standard solid harmonics.

The next step in the derivation of the multipole expansion of the two-electron integrals is to separate the electronic coordinates according to (9.13.4). Invoking the *addition theorem for the regular solid harmonics* [24]

$$R_{lm}(\mathbf{u} + \mathbf{v}) = \sum_{j=0}^{l} \sum_{k=-j}^{j} R_{l-j,m-k}(\mathbf{u}) R_{jk}(\mathbf{v}) \tag{9.13.17}$$

we may write the expansion (9.13.16) in the following manner:

$$\begin{aligned}
\frac{1}{r_{12}} &= \sum_{l=0}^{\infty} \sum_{m=-l}^{l} R_{lm}(\mathbf{r}_{1P} - \mathbf{r}_{2Q}) I_{lm}^*(\mathbf{R}_{QP}) \\
&= \sum_{l=0}^{\infty} \sum_{m=-l}^{l} \sum_{j=0}^{l} \sum_{k=-j}^{j} R_{l-j,m-k}(\mathbf{r}_{1P}) I_{lm}^*(\mathbf{R}_{QP}) R_{jk}(-\mathbf{r}_{2Q}) \\
&= \sum_{l=0}^{\infty} \sum_{m=-l}^{l} \sum_{j=0}^{\infty} \sum_{k=-j}^{j} (-1)^j R_{lm}(\mathbf{r}_{1P}) I_{l+j,m+k}^*(\mathbf{R}_{QP}) R_{jk}(\mathbf{r}_{2Q}) \tag{9.13.18}
\end{aligned}$$

where, in the last expression, we have rearranged the dummy indices and used the inversion symmetry (6.4.17). For a proof of the addition theorem (9.13.17), we refer to standard textbooks on mathematical physics [26]. We note here that, in a normalization different from that of the scaled regular harmonics (9.13.14), the addition theorem assumes a more complicated form, involving a numerical factor that is different for each term in the expansion (9.13.17). Since the addition theorem of the regular solid harmonics will be used frequently in our discussion, the normalization (9.13.14) is adopted here.

Substituting the partial-wave expansion in the form (9.13.18) into (9.13.1), we arrive at the *bipolar multipole expansion* of the two-electron integral

$$g_{abcd} = \sum_{l=0}^{\infty} \sum_{m=-l}^{l} \sum_{j=0}^{\infty} \sum_{k=-j}^{j} q_{lm}^{ab}(\mathbf{P}) T_{lm,jk}(\mathbf{R}_{QP}) q_{jk}^{cd}(\mathbf{Q}) \tag{9.13.19}$$

where we have introduced the *multipole moments* of the charge distributions

$$q_{lm}^{ab}(\mathbf{P}) = \int \Omega_{ab}(\mathbf{r})R_{lm}(\mathbf{r}_{\mathrm{P}})\,d\mathbf{r} \tag{9.13.20}$$

$$q_{jk}^{cd}(\mathbf{Q}) = \int \Omega_{cd}(\mathbf{r})R_{jk}(\mathbf{r}_{\mathrm{Q}})\,d\mathbf{r} \tag{9.13.21}$$

and the elements of the *interaction matrix*

$$T_{lm,jk}(\mathbf{R}_{\mathrm{QP}}) = (-1)^{j}I_{l+j,m+k}^{*}(\mathbf{R}_{\mathrm{QP}}) \tag{9.13.22}$$

The term *bipolar* is used to describe this expansion since it involves two sets of multipole moments, with different origins. In matrix notation, the multipole expansion (9.13.19) may be written as

$$g_{abcd} = \mathbf{q}^{ab}(\mathbf{P})^{\mathrm{T}}\mathbf{T}(\mathbf{R}_{\mathrm{QP}})\mathbf{q}^{cd}(\mathbf{Q}) \tag{9.13.23}$$

Here $\mathbf{q}^{ab}(\mathbf{P})$ is a column vector of the multipole moments for the overlap distribution Ω_{ab} with origin at $\mathbf{P}$ and $\mathbf{q}^{cd}(\mathbf{Q})$ is interpreted in a similar way. The elements of the multipole vectors $\mathbf{q}$ are ordered as q_{00}, $q_{1,-1}$, $q_{1,0}$, $q_{1,1}$, and so on. Finally, $\mathbf{T}(\mathbf{R}_{\mathrm{QP}})$ is a square matrix containing the interaction moments for the vector $\mathbf{R}_{\mathrm{QP}}$, its columns and rows ordered in the same manner as in the multipole vectors. More general expressions for bipolar multipole expansions can be derived, in which the interacting charge distributions are represented in local coordinate systems [24,27]. Such a formulation is important in the theory of intermolecular forces but unnecessarily complicated for our purposes since we employ a global coordinate system for all charges.

In the multipole expansion of the two-electron integral (9.13.23), the interaction tensor (9.13.22) depends only on the relative position of the origins about the charge distributions – all information about the charge distributions is contained in the multipole moments (9.13.20) and (9.13.21). As an illustration, consider the lowest-order expansion of the two-electron integrals. The moments of the charge distributions are given as

$$q_{00}^{ab}(\mathbf{P}) = \int \Omega_{ab}(\mathbf{r})R_{00}(\mathbf{r}_{\mathrm{P}})\,d\mathbf{r} = \int \Omega_{ab}(\mathbf{r})\,d\mathbf{r} = S_{ab} \tag{9.13.24}$$

$$q_{00}^{cd}(\mathbf{Q}) = \int \Omega_{cd}(\mathbf{r})R_{00}(\mathbf{r}_{\mathrm{Q}})\,d\mathbf{r} = \int \Omega_{cd}(\mathbf{r})\,d\mathbf{r} = S_{cd} \tag{9.13.25}$$

$$T_{00,00}(\mathbf{R}_{\mathrm{QP}}) = R_{\mathrm{QP}}^{-1} \tag{9.13.26}$$

and the integral becomes (see also Exercise 9.4)

$$g_{abcd} \approx \frac{S_{ab}S_{cd}}{R_{\mathrm{PQ}}} \tag{9.13.27}$$

Note that, for spherical overlap distributions centred at $\mathbf{P}$ and $\mathbf{Q}$, (9.13.24) and (9.13.25) are the only nonzero multipole moments and the monopole expansion (9.13.27) then represents an exact expression for the two-electron integral (assuming disjoint charge distributions) – see also the discussion in Section 9.12.3. In the same manner, the multipole expansion (9.13.19) terminates exactly after a finite number of terms whenever the charge distributions of the electrons are one-centre functions, whose centres are chosen as origins of the multipole expansions. In general, however, the bipolar multipole expansion does not terminate and the expansion is then truncated when the remainder is sufficiently small as discussed in Section 9.13.2.

In the multipole expansion of the two-electron integral (9.13.23), the complicated six-dimensional integration of the two-electron integral has been reduced to the much simpler three-dimensional integration of the one-electron multipoles (9.13.20) and (9.13.21). These one-electron integrals may be evaluated by, for example, the Obara–Saika technique of Section 9.3.2, whose recursive structure makes it well suited to the simultaneous generation of the full set of multipole moments needed in the multipole method. We also note that, for each electron, the number of such one-electron multipole-moment integrals to be evaluated is $(L+1)^2$, where L is the order of the multipole expansion. Consequently, the cost of the contraction of the multipole vectors with the interaction matrix to form the final two-electron integral (9.13.19) scales as L^4.

9.13.2 CONVERGENCE OF THE MULTIPOLE EXPANSION

Let us consider the rate of convergence of the multipole expansion. Assuming that the partial-wave expansion in (9.13.7) has been carried out to order L, the error in the inverse interelectronic distance is given as

$$\delta_{12}^L = \sum_{l=L+1}^{\infty} \frac{\Delta r_{12}^l}{R_{QP}^{l+1}} P_l(\cos\theta) \leq \sum_{l=L+1}^{\infty} \frac{\Delta r_{12}^l}{R_{QP}^{l+1}} \tag{9.13.28}$$

Carrying out the summation, we obtain

$$\delta_{12}^L \leq \frac{1}{R_{QP} - \Delta r_{12}} \left(\frac{\Delta r_{12}}{R_{QP}} \right)^{L+1} \tag{9.13.29}$$

Clearly, for rapid convergence, the separation between the origins of the multipole expansions R_{QP} should be large relative to the size of the interacting charge distributions. We also note that, with each increment in the multipole expansion, the error is reduced by a constant factor. Indeed, using (9.13.29), it is possible to control the accuracy of the multipole expansion, although this simple estimate of the error is usually rather conservative. In practice, therefore, the achieved accuracy may be considerably higher than predicted by (9.13.29).

With regard to the convergence of the multipole expansion, it should be noted that two Gaussian distributions will always overlap to some extent, irrespective of their relative separation. The condition for convergence (9.13.6) is therefore, strictly speaking, never satisfied over the whole region of integration. In other words, the multipole expansion constitutes an asymptotic rather than convergent expansion. However, provided the product of the two charge distributions is sufficiently small in the divergent regions, this will not create problems in practice since convergence will be achieved in the important regions before the divergent contributions of the weakly overlapping regions become larger than the allowed error in the integral.

9.13.3 THE MULTIPOLE METHOD FOR CONTRACTED TWO-ELECTRON INTEGRALS

In Section 9.13.1, we considered the evaluation of a primitive two-electron integral by a bipolar multipole expansion, with the origins of $\mathbf{q}^{ab}(\mathbf{P})$ and $\mathbf{q}^{cd}(\mathbf{Q})$ at the centres $\mathbf{P}$ and $\mathbf{Q}$ of the overlap distributions $\Omega_{ab}(\mathbf{r}_1)$ and $\Omega_{cd}(\mathbf{r}_2)$. Let us now consider the evaluation of a two-electron integral over a *contracted* set of overlap distributions, which we write here in the form

$$g_{\alpha\beta\gamma\delta} = \sum_{abcd} d_{ab}^{\alpha\beta} d_{cd}^{\gamma\delta} g_{abcd} \tag{9.13.30}$$

Inserting the multipole expansion (9.13.23) in this expression, we obtain

$$g_{\alpha\beta\gamma\delta} = \sum_{abcd} d_{ab}^{\alpha\beta} d_{cd}^{\gamma\delta} \mathbf{q}^{ab} (\mathbf{P}^{ab})^\mathrm{T} \mathbf{T}(\mathbf{R}_{\mathrm{QP}}^{abcd}) \mathbf{q}^{cd} (\mathbf{Q}^{cd}) \tag{9.13.31}$$

where the notation $\mathbf{P}^{ab}$ is used to remind us that $\mathbf{P}$ is calculated from the positions $\mathbf{A}$ and $\mathbf{B}$ according to (9.2.13) and similarly for $\mathbf{R}_{\mathrm{QP}}^{abcd}$ and $\mathbf{Q}^{cd}$. Since in (9.13.31) the origins of the multipole expansions are different for each pair of primitive overlap distributions, we must carry out a separate expansion for each such pair, making this particular application of the multipole method for the contracted two-electron integral rather expensive.

For contracted integrals, a more attractive approach is to introduce a single, shared origin $\overline{\mathbf{P}}$ for the overlap distributions of the first electron and likewise a common origin $\overline{\mathbf{Q}}$ for the distributions of the second electron. The two-electron integral (9.13.31) then takes the form

$$g_{\alpha\beta\gamma\delta} = \sum_{abcd} d_{ab}^{\alpha\beta} d_{cd}^{\gamma\delta} \mathbf{q}^{ab} (\overline{\mathbf{P}})^\mathrm{T} \mathbf{T}(\mathbf{R}_{\overline{\mathbf{Q}}\overline{\mathbf{P}}}) \mathbf{q}^{cd} (\overline{\mathbf{Q}}) \tag{9.13.32}$$

Carrying out the summations, we now obtain the compact expression

$$g_{\alpha\beta\gamma\delta} = \mathbf{q}^{\alpha\beta} (\overline{\mathbf{P}})^\mathrm{T} \mathbf{T}(\mathbf{R}_{\overline{\mathbf{Q}}\overline{\mathbf{P}}}) \mathbf{q}^{\gamma\delta} (\overline{\mathbf{Q}}) \tag{9.13.33}$$

where the multipole moments of the first electron are given by

$$q_{lm}^{\alpha\beta}(\overline{\mathbf{P}}) = \sum_{ab} d_{ab}^{\alpha\beta} \int \Omega_{ab}(\mathbf{r}) R_{lm}(\mathbf{r}_{\overline{\mathbf{P}}}) \, \mathrm{d}\mathbf{r} \tag{9.13.34}$$

and similarly for the second electron. This procedure is clearly superior to (9.13.31), reducing the quadruple summation over primitive functions to double summations (9.13.34) and requiring the formation of only a single bipolar multipole expansion (9.13.33).

More generally, we may calculate two-electron integrals over arbitrary linear combinations of primitive overlap distributions (representing, for example, the charge distribution of a large part of a molecular system) in the same manner, obtaining the multipole moments by an obvious extension of (9.13.34). We note, however, that this procedure will work only if the multipole expansion for the combined overlap distributions converges – that is, if the resulting combined overlap distributions are sufficiently far removed from each other for the partial-wave expansion to converge.

9.13.4 TRANSLATION OF MULTIPOLE MOMENTS

An obvious strategy for calculating the multipole moments (9.13.34) would be first to set up the contracted charge distribution

$$\Omega_{\alpha\beta}(\mathbf{r}) = \sum_{ab} d_{ab}^{\alpha\beta} \Omega_{ab}(\mathbf{r}) \tag{9.13.35}$$

and then to evaluate the multipole moments

$$q_{lm}^{\alpha\beta}(\overline{\mathbf{P}}) = \int \Omega_{\alpha\beta}(\mathbf{r}) R_{lm}(\mathbf{r}_{\overline{P}}) \, \mathrm{d}\mathbf{r} \tag{9.13.36}$$

by some chosen integration scheme. For distributions such as those present in a contracted two-electron integral, the integration may be carried out exactly, using some modification of the

Obara–Saika or McMurchie–Davidson schemes. For more general charge distributions, representing, for example, a large part of a molecular system, it may be better to carry out the integration numerically, using Gaussian quadrature or some other numerical scheme.

Let us now consider an alternative procedure for the evaluation of the contracted multipole moments (9.13.36). First, we calculate – for each primitive overlap distribution contributing to the two-electron integral – local multipole moments of the type (9.13.20), each with origin at the centre of the primitive overlap distribution:

$$q_{lm}^{ab}(\mathbf{P}) = \int \Omega_{ab}(\mathbf{r}) R_{lm}(\mathbf{r}_\mathrm{P}) \, d\mathbf{r} \tag{9.13.37}$$

Next, we translate the origins $\mathbf{P}$ of these local expansions to the common origin $\overline{\mathbf{P}}$ employed for the evaluation of the two-electron integral (9.13.32). Invoking the addition theorem for the regular solid harmonics (9.13.17), we obtain

$$q_{lm}^{ab}(\overline{\mathbf{P}}) = \int \Omega_{ab}(\mathbf{r}) R_{lm}(\mathbf{r}_\mathrm{P} - \mathbf{R}_{\overline{\mathrm{P}}\mathrm{P}}) \, d\mathbf{r}$$

$$= \sum_{j=0}^{l} \sum_{k=-j}^{j} R_{l-j,m-k}(-\mathbf{R}_{\overline{\mathrm{P}}\mathrm{P}}) \int \Omega_{ab}(\mathbf{r}) R_{jk}(\mathbf{r}_\mathrm{P}) \, d\mathbf{r} \tag{9.13.38}$$

Introducing the notation

$$W_{lm,jk}(\mathbf{R}_{\overline{\mathrm{P}}\mathrm{P}}) = R_{l-j,m-k}(-\mathbf{R}_{\overline{\mathrm{P}}\mathrm{P}}) \tag{9.13.39}$$

we may write the transformation (9.13.38) as

$$q_{lm}^{ab}(\overline{\mathbf{P}}) = \sum_{j=0}^{l} \sum_{k=-j}^{j} W_{lm,jk}(\mathbf{R}_{\overline{\mathrm{P}}\mathrm{P}}) q_{jk}^{ab}(\mathbf{P}) \tag{9.13.40}$$

In matrix notation, the shift of origin from $\mathbf{P}$ to $\overline{\mathbf{P}}$ for the full set of multipole moments may be expressed compactly as a linear transformation

$$\mathbf{q}^{ab}(\overline{\mathbf{P}}) = \mathbf{W}(\mathbf{R}_{\overline{\mathrm{P}}\mathrm{P}}) \mathbf{q}^{ab}(\mathbf{P}) \tag{9.13.41}$$

where $\mathbf{W}(\mathbf{R}_{\overline{\mathrm{P}}\mathrm{P}})$ is the *translation matrix*. Using this expression, we may calculate the integral (9.13.23) from

$$g_{abcd} = \mathbf{q}^{ab}(\mathbf{P})^{\mathrm{T}} \mathbf{W}(\mathbf{R}_{\overline{\mathrm{P}}\mathrm{P}})^{\mathrm{T}} \mathbf{T}(\mathbf{R}_{\overline{\mathrm{Q}}\,\overline{\mathrm{P}}}) \mathbf{W}(\mathbf{R}_{\overline{\mathrm{Q}}\mathrm{Q}}) \mathbf{q}^{cd}(\mathbf{Q}) \tag{9.13.42}$$

Such translations of the origins of the multipole expansions will prove particularly useful in the development of the multipole method for the calculation of Coulomb interactions in large systems, where each multipole moment contributes to a large number of Coulomb interactions with different origins; see Section 9.14.

By inspection of (9.13.39), we see that the translation matrix $\mathbf{W}$ is a lower triangular matrix with unit diagonal elements. This structure implies that the monopole moment at $\mathbf{P}$ makes contributions to all multipoles at $\overline{\mathbf{P}}$, that the dipole moment at $\mathbf{P}$ makes contributions to the dipole and all higher moments at $\overline{\mathbf{P}}$, and so on. In particular, we note that a charge distribution that is represented exactly by a finite number of multipoles centred at $\mathbf{P}$, will require an infinite expansion at a different position $\overline{\mathbf{P}}$ for an exact evaluation.

9.13.5 REAL MULTIPOLE MOMENTS

A disadvantage of the multipole method presented so far is that it is expressed in terms of complex multipole moments and interaction tensors. Since the Coulomb interactions are real, it is more natural (and, it turns out, more efficient) to express the expansions in terms of real moments and interaction tensors. In the remainder of this section, we shall see how the transformation to real expressions may be accomplished.

We begin by decomposing the complex regular and irregular solid harmonics as

$$R_{lm}(\mathbf{r}) = R_{lm}^{c}(\mathbf{r}) + iR_{lm}^{s}(\mathbf{r}) \tag{9.13.43}$$

$$I_{lm}(\mathbf{r}) = I_{lm}^{c}(\mathbf{r}) + iI_{lm}^{s}(\mathbf{r}) \tag{9.13.44}$$

where $R_{lm}^{c}(\mathbf{r})$ and $R_{lm}^{s}(\mathbf{r})$ are the real (cosine) and imaginary (sine) parts of $R_{lm}(\mathbf{r})$:

$$R_{lm}^{c}(\mathbf{r}) = \frac{R_{lm}(\mathbf{r}) + R_{lm}^{*}(\mathbf{r})}{2} \tag{9.13.45}$$

$$R_{lm}^{s}(\mathbf{r}) = \frac{R_{lm}(\mathbf{r}) - R_{lm}^{*}(\mathbf{r})}{2i} \tag{9.13.46}$$

Note that the real and imaginary components are defined for negative as well as positive indices m. Using (9.13.14), (6.4.22) and (6.4.18), we find that the real regular solid harmonics satisfy the following symmetry properties:

$$R_{l,-m}^{c}(\mathbf{r}) = (-1)^{m} R_{lm}^{c}(\mathbf{r}) \tag{9.13.47}$$

$$R_{l,-m}^{s}(\mathbf{r}) = -(-1)^{m} R_{lm}^{s}(\mathbf{r}) \tag{9.13.48}$$

The irregular solid harmonics satisfy analogous relations.

It should be realized that the real regular solid harmonics $R_{lm}^{c}(\mathbf{r})$ and $R_{lm}^{s}(\mathbf{r})$ introduced here differ from the real solid harmonics $S_{lm}(\mathbf{r})$ of Section 6.4.2. Thus, whereas the real functions in Section 6.4.2 were obtained by a unitary transformation of the complex ones (in Racah's normalization), the transformation in (9.13.45) and (9.13.46) is not unitary. It is adopted here since it leads to simpler expressions for the multipole expansions.

It is often convenient to employ a common notation for the cosine and sine components of the solid harmonics. In such cases, we shall indicate real components by the use of Greek letters, adopting the following convention to distinguish between the cosine and sine components

$$R_{l\mu}(\mathbf{r}) = \begin{cases} R_{l\mu}^{c}(\mathbf{r}) & \mu \geq 0 \\ R_{l\mu}^{s}(\mathbf{r}) & \mu < 0 \end{cases} \tag{9.13.49}$$

However, it should be understood that $R_{lm}^{c}(\mathbf{r})$ and $R_{lm}^{s}(\mathbf{r})$ are both defined for positive and negative indices m, obeying the relationships (9.13.47) and (9.13.48).

Having defined the real regular solid harmonics, we now go on to consider the associated multipole moments. The complex multipole moments may be written as

$$q_{lm}(\mathbf{P}) = q_{lm}^{c}(\mathbf{P}) + iq_{lm}^{s}(\mathbf{P}) \tag{9.13.50}$$

where $q_{lm}^{c}(\mathbf{P})$ and $q_{lm}^{s}(\mathbf{P})$ are the real multipole moments, generated by integration over the appropriate charge distribution multiplied by the real regular solid harmonics:

$$q_{lm}^{c}(\mathbf{P}) = \int \rho(\mathbf{r}) R_{lm}^{c}(\mathbf{r}_{\mathrm{P}}) \, d\mathbf{r} \tag{9.13.51}$$

$$q_{lm}^s(\mathbf{P}) = \int \rho(\mathbf{r}) R_{lm}^s(\mathbf{r}_P)\, d\mathbf{r} \tag{9.13.52}$$

As for the real solid harmonics in (9.13.49), we shall use the common notation $q_{l\mu}(\mathbf{P})$ for the cosine and sine components of the complex multipole moments $q_{lm}(\mathbf{P})$.

9.13.6 THE REAL TRANSLATION MATRIX

To obtain an expression for the real translation matrix, we begin by writing (9.13.40) as

$$q_{lm}(\overline{\mathbf{P}}) = \sum_{j=0}^{l}\sum_{k=-j}^{j} W_{lm,jk}(\mathbf{R}_{\overline{P}P})q_{jk}(\mathbf{P}) = \sum_{j=0}^{l}\sum_{k=-j}^{j} R_{l-j,m-k}(-\mathbf{R}_{\overline{P}P})q_{jk}(\mathbf{P}) \tag{9.13.53}$$

Inserting (9.13.43) and (9.13.50) and treating the cosine and sine components separately, we obtain

$$q_{lm}^c(\overline{\mathbf{P}}) = \sum_{jk} R_{l-j,m-k}^c(-\mathbf{R}_{\overline{P}P})q_{jk}^c(\mathbf{P}) - \sum_{jk} R_{l-j,m-k}^s(-\mathbf{R}_{\overline{P}P})q_{jk}^s(\mathbf{P}) \tag{9.13.54}$$

$$q_{lm}^s(\overline{\mathbf{P}}) = \sum_{jk} R_{l-j,m-k}^c(-\mathbf{R}_{\overline{P}P})q_{jk}^s(\mathbf{P}) + \sum_{jk} R_{l-j,m-k}^s(-\mathbf{R}_{\overline{P}P})q_{jk}^c(\mathbf{P}) \tag{9.13.55}$$

where the summation ranges are the same as in (9.13.53). These expressions are now manipulated as illustrated here for one of the contributions to the first term in (9.13.54) (omitting arguments):

$$A_j^{cc} = \sum_k R_{l-j,m-k}^c q_{jk}^c \tag{9.13.56}$$

Considering positive and negative k separately and invoking (9.13.47), we obtain

$$
\begin{aligned}
A_j^{cc} &= \sum_{k\geq 0} R_{l-j,m-k}^c q_{jk}^c + \sum_{k<0} R_{l-j,m-k}^c q_{jk}^c \\
&= \sum_{k\geq 0} R_{l-j,m-k}^c q_{jk}^c + \sum_{k>0} R_{l-j,m+k}^c q_{j,-k}^c \\
&= \sum_{k\geq 0} R_{l-j,m-k}^c q_{jk}^c + \sum_{k>0} (-1)^k R_{l-j,m+k}^c q_{jk}^c \\
&= \sum_{k\geq 0} \left(\tfrac{1}{2}\right)^{\delta_{k0}} [R_{l-j,m-k}^c + (-1)^k R_{l-j,m+k}^c]q_{jk}^c
\end{aligned}
\tag{9.13.57}
$$

Similar manipulations may be carried out for the other terms in (9.13.54) and (9.13.55) as well. Introducing the matrix elements

$$W_{lm,jk}^{cc}(\mathbf{R}_{\overline{P}P}) = \left(\tfrac{1}{2}\right)^{\delta_{k0}} [R_{l-j,m-k}^c(-\mathbf{R}_{\overline{P}P}) + (-1)^k R_{l-j,m+k}^c(-\mathbf{R}_{\overline{P}P})] \tag{9.13.58}$$

$$W_{lm,jk}^{cs}(\mathbf{R}_{\overline{P}P}) = -R_{l-j,m-k}^s(-\mathbf{R}_{\overline{P}P}) + (-1)^k R_{l-j,m+k}^s(-\mathbf{R}_{\overline{P}P}) \tag{9.13.59}$$

$$W_{lm,jk}^{sc}(\mathbf{R}_{\overline{P}P}) = \left(\tfrac{1}{2}\right)^{\delta_{k0}} [R_{l-j,m-k}^s(-\mathbf{R}_{\overline{P}P}) + (-1)^k R_{l-j,m+k}^s(-\mathbf{R}_{\overline{P}P})] \tag{9.13.60}$$

$$W_{lm,jk}^{ss}(\mathbf{R}_{\overline{P}P}) = R_{l-j,m-k}^c(-\mathbf{R}_{\overline{P}P}) - (-1)^k R_{l-j,m+k}^c(-\mathbf{R}_{\overline{P}P}) \tag{9.13.61}$$

we may now write the transformations (9.13.54) and (9.13.55) as

$$q_{lm}^{c}(\bar{\mathbf{P}}) = \sum_{j,k\geq0} W_{lm,jk}^{cc}(\mathbf{R}_{\bar{\mathbf{P}}\mathbf{P}})q_{jk}^{c}(\mathbf{P}) + \sum_{j,k>0} W_{lm,jk}^{cs}(\mathbf{R}_{\bar{\mathbf{P}}\mathbf{P}})q_{jk}^{s}(\mathbf{P}) \tag{9.13.62}$$

$$q_{lm}^{s}(\bar{\mathbf{P}}) = \sum_{j,k\geq0} W_{lm,jk}^{sc}(\mathbf{R}_{\bar{\mathbf{P}}\mathbf{P}})q_{jk}^{c}(\mathbf{P}) + \sum_{j,k>0} W_{lm,jk}^{ss}(\mathbf{R}_{\bar{\mathbf{P}}\mathbf{P}})q_{jk}^{s}(\mathbf{P}) \tag{9.13.63}$$

where the summation ranges have been reduced by one-half.

As is easily verified, the elements (9.13.58)–(9.13.61) have the same symmetries with respect to sign changes of m and k as do the cosine and sine components of the regular solid harmonics (9.13.47) and (9.13.48), for example

$$W_{lm,jk}^{sc} = -(-1)^{m}W_{l\bar{m},jk}^{sc} = (-1)^{k}W_{lm,j\bar{k}}^{sc} \tag{9.13.64}$$

where, to avoid ambiguity, we have used the notation $\bar{m} = -m$. We may therefore carry out each of the summations in (9.13.62) and (9.13.63) over positive or negative indices, with no changes in the expressions. Associating positive and negative indices with the cosine and sine components as in (9.13.49), we may write the transformation of the real multipole moments compactly as

$$q_{l\mu}(\bar{\mathbf{P}}) = \sum_{j=0}^{l}\sum_{\kappa=-j}^{j} W_{l\mu,j\kappa}(\mathbf{R}_{\bar{\mathbf{P}}\mathbf{P}})q_{j\kappa}(\mathbf{P}) \tag{9.13.65}$$

or, in matrix notation,

$$\mathbf{q}(\bar{\mathbf{P}}) = \mathbf{W}(\mathbf{R}_{\bar{\mathbf{P}}\mathbf{P}})\mathbf{q}(\mathbf{P}) \tag{9.13.66}$$

Obviously, this matrix representation has the same form as in the complex case, with identical summation ranges. Since the real multipole moments and translation-matrix elements have one instead of two components as in the complex counterpart to (9.13.66), the real formulation is more compact.

9.13.7 THE REAL INTERACTION MATRIX

Let us finally consider the real expression for the interaction matrix. We proceed in the same manner as for the translation matrix, beginning with (9.13.19), which we write as

$$g_{abcd} = \sum_{lm}\sum_{jk} q_{lm}(\mathbf{P})T_{lm,jk}(\mathbf{R}_{QP})q_{jk}(\mathbf{Q})$$
$$= \sum_{lm}\sum_{jk}(-1)^{j}q_{lm}(\mathbf{P})I_{l+j,m+k}^{*}(\mathbf{R}_{QP})q_{jk}(\mathbf{Q}) \tag{9.13.67}$$

To obtain the real interaction matrix, we substitute in this expression the two-term expansions of the multipole moments (9.13.50) and of the irregular harmonics (9.13.44) in real components. This substitution gives eight distinct terms, one of which is given as (omitting arguments)

$$g_{abcd}^{ccc} = \sum_{lm}\sum_{jk}(-1)^{j}q_{lm}^{c}I_{l+j,m+k}^{c}q_{jk}^{c} \tag{9.13.68}$$

Expanding in positive and negative components, we obtain

$$
\begin{aligned}
g_{abcd}^{ccc} &= \sum_{lm} \sum_{\substack{j \\ k \geq 0}} (-1)^j \left(\tfrac{1}{2}\right)^{\delta_{k0}} q_{lm}^c (I_{l+j,m+k}^c q_{jk}^c + I_{l+j,m-k}^c q_{j,-k}^c) \\
&= \sum_{lm} \sum_{\substack{j \\ k \geq 0}} (-1)^j \left(\tfrac{1}{2}\right)^{\delta_{k0}} q_{lm}^c (I_{l+j,m+k}^c + (-1)^k I_{l+j,m-k}^c) q_{jk}^c \\
&= 2 \sum_{\substack{l \\ m \geq 0}} \sum_{\substack{j \\ k \geq 0}} (-1)^j \left(\tfrac{1}{2}\right)^{\delta_{k0}+\delta_{m0}} q_{lm}^c (I_{l+j,m+k}^c + (-1)^k I_{l+j,m-k}^c) q_{jk}^c \qquad (9.13.69)
\end{aligned}
$$

Treating the remaining seven terms that arise from (9.13.67) in the same manner, we find that the four imaginary terms vanish and that the full expansion (9.13.67) may be written as

$$
\begin{aligned}
g_{abcd} &= \sum_{\substack{l,m \geq 0 \\ j,k \geq 0}} q_{lm}^c(\mathbf{P}) T_{lm,jk}^{cc}(\mathbf{R}_{QP}) q_{jk}^c(\mathbf{Q}) + \sum_{\substack{l,m \geq 0 \\ j,k \geq 0}} q_{lm}^c(\mathbf{P}) T_{lm,jk}^{cs}(\mathbf{R}_{QP}) q_{jk}^s(\mathbf{Q}) \\
&+ \sum_{\substack{l,m \geq 0 \\ j,k \geq 0}} q_{lm}^s(\mathbf{P}) T_{lm,jk}^{sc}(\mathbf{R}_{QP}) q_{jk}^c(\mathbf{Q}) + \sum_{\substack{l,m \geq 0 \\ j,k \geq 0}} q_{lm}^s(\mathbf{P}) T_{lm,jk}^{ss}(\mathbf{R}_{QP}) q_{jk}^s(\mathbf{Q}) \qquad (9.13.70)
\end{aligned}
$$

in terms of the following real components of the interaction matrix

$$
T_{lm,jk}^{cc}(\mathbf{R}_{QP}) = (-1)^j \left(\tfrac{1}{2}\right)^{\delta_{m0}+\delta_{k0}} 2[I_{l+j,m+k}^c(\mathbf{R}_{QP}) + (-1)^k I_{l+j,m-k}^c(\mathbf{R}_{QP})] \qquad (9.13.71)
$$

$$
T_{lm,jk}^{cs}(\mathbf{R}_{QP}) = (-1)^j \left(\tfrac{1}{2}\right)^{\delta_{m0}+\delta_{k0}} 2[I_{l+j,m+k}^s(\mathbf{R}_{QP}) - (-1)^k I_{l+j,m-k}^s(\mathbf{R}_{QP})] \qquad (9.13.72)
$$

$$
T_{lm,jk}^{sc}(\mathbf{R}_{QP}) = (-1)^j \left(\tfrac{1}{2}\right)^{\delta_{m0}+\delta_{k0}} 2[I_{l+j,m+k}^s(\mathbf{R}_{QP}) + (-1)^k I_{l+j,m-k}^s(\mathbf{R}_{QP})] \qquad (9.13.73)
$$

$$
T_{lm,jk}^{ss}(\mathbf{R}_{QP}) = (-1)^j \left(\tfrac{1}{2}\right)^{\delta_{m0}+\delta_{k0}} 2[-I_{l+j,m+k}^c(\mathbf{R}_{QP}) + (-1)^k I_{l+j,m-k}^c(\mathbf{R}_{QP})] \qquad (9.13.74)
$$

The indices m and k of the interaction matrix (9.13.71)–(9.13.74) obey the same symmetries as the indices of the translation matrix; see (9.13.64). Recognizing that q_{l0}^s is zero since q_{l0} is real, we may, by the convention (9.13.49), write the interaction (9.13.70) in exactly the same manner as in the complex case (9.13.67):

$$
g_{abcd} = \sum_{l\mu} \sum_{j\kappa} q_{l\mu}(\mathbf{P}) T_{l\mu,j\kappa}(\mathbf{R}_{QP}) q_{j\kappa}(\mathbf{Q}) \qquad (9.13.75)
$$

The real formulation is more economical since each term involves only real numbers. Also, we shall see that simple recurrence relations may be set up for the evaluation of the regular and irregular solid harmonics, making the generation of the interaction and translation matrices a straightforward matter.

9.13.8 EVALUATION OF THE SCALED SOLID HARMONICS

Let us consider the evaluation of the scaled real solid harmonics defined in (9.13.43) and (9.13.44). We first note that the real regular solid harmonics are related to the standard functions of Section 6.4 as

$$
R_{l,|m|}^c(\mathbf{r}) = \frac{(-1)^m}{\sqrt{(2-\delta_{m0})(l-m)!(l+m)!}} S_{l,|m|}(\mathbf{r}) \qquad (9.13.76)
$$

$$R^s_{l,|m|}(\mathbf{r}) = \frac{(1 - \delta_{m0})(-1)^m}{\sqrt{2(l - m)!(l + m)!}} S_{l,-|m|}(\mathbf{r}) \tag{9.13.77}$$

Using these relations, it is trivial to set up an explicit expression for their evaluation analogous to (6.4.47) for $S_{lm}(\mathbf{r})$. The recurrence relations may be obtained from (6.4.55), (6.4.56) and (6.4.69). Taking into account the different normalizations (9.13.76) and (9.13.77) and also the symmetry relations (9.13.47) and (9.13.48), we obtain after some algebra the following set of recurrence relations for the scaled regular solid harmonics:

$$R^c_{00} = 1 \tag{9.13.78}$$

$$R^s_{00} = 0 \tag{9.13.79}$$

$$R^c_{l+1,l+1} = -\frac{xR^c_{ll} - yR^s_{ll}}{2l + 2} \tag{9.13.80}$$

$$R^s_{l+1,l+1} = -\frac{yR^c_{ll} + xR^s_{ll}}{2l + 2} \tag{9.13.81}$$

$$R^{c/s}_{l+1,m} = \frac{(2l + 1)zR^{c/s}_{lm} - r^2 R^{c/s}_{l-1,m}}{(l + m + 1)(l - m + 1)}, \quad 0 \le m < l \tag{9.13.82}$$

In the last recurrence, the use of $R^{c/s}_{lm}(\mathbf{r})$ indicates that the relation is valid for both $R^c_{lm}(\mathbf{r})$ and $R^s_{lm}(\mathbf{r})$. We note that the recurrence relations for the scaled functions $R_{lm}(\mathbf{r})$ are slightly simpler than those for their unscaled counterparts (6.4.70)–(6.4.73).

Let us now consider the scaled irregular solid harmonics, related to the unscaled real solid harmonics by

$$I^c_{l,|m|}(\mathbf{r}) = (-1)^m \sqrt{\frac{(l - m)!(l + m)!}{2 - \delta_{m0}}} \frac{S_{l,|m|}(\mathbf{r})}{r^{2l+1}} \tag{9.13.83}$$

$$I^s_{l,|m|}(\mathbf{r}) = (-1)^m (1 - \delta_{m0}) \sqrt{\frac{(l - m)!(l + m)!}{2}} \frac{S_{l,-|m|}(\mathbf{r})}{r^{2l+1}} \tag{9.13.84}$$

Again, it is trivial to set up an explicit expression for their evaluation, analogous to (6.4.47) for $S_{lm}(\mathbf{r})$. The recurrence relations are obtained as for the regular harmonics, including r^{-2l-1} in the scaling factor. After some algebra, we arrive at the following recurrence relations for the real irregular solid harmonics:

$$I^c_{00} = \frac{1}{r} \tag{9.13.85}$$

$$I^s_{00} = 0 \tag{9.13.86}$$

$$I^c_{l+1,l+1} = -(2l + 1)\frac{xI^c_{ll} - yI^s_{ll}}{r^2} \tag{9.13.87}$$

$$I^s_{l+1,l+1} = -(2l + 1)\frac{yI^c_{ll} + xI^s_{ll}}{r^2} \tag{9.13.88}$$

$$I^{c/s}_{l+1,m} = \frac{(2l + 1)zI^{c/s}_{lm} - (l^2 - m^2)I^{c/s}_{l-1,m}}{r^2}, \quad 0 \le m < l \tag{9.13.89}$$

Using relations (9.13.78)–(9.13.82) and (9.13.85)–(9.13.89), the solid harmonics required for the evaluation of the bipolar multipole expansions may be easily generated.

9.14 The multipole method for large systems

The multipole method discussed in Section 9.13 may be applied not only to the evaluation of two-electron integrals, but more generally to the evaluation of Coulomb interactions in any system. For example, applied to the evaluation of the Coulomb potential in Hartree–Fock theory, it can be developed to a highly efficient scheme that requires an amount of work that increases only linearly with the size of the molecular system [25].

In the present section, we shall discuss the application of the multipole method to the evaluation of Coulomb interactions in large systems, with special emphasis on aspects related to the linear scaling of the evaluation with the size of the system [25,28,29]. To simplify the presentation, we shall in Sections 9.14.1–9.14.3 assume that the system consists of a set of point charges contained in an equilateral cubic box. The modifications needed to treat Gaussian charge distributions in a molecular system are considered in Section 9.14.4.

9.14.1 THE NAIVE MULTIPOLE METHOD

The direct evaluation of the Coulomb interactions in a system containing point-charge particles involves the explicit calculation of all pairwise interactions according to Coulomb's law:

$$U = \sum_{i>j} \frac{Z_i Z_j}{R_{ij}} \qquad (9.14.1)$$

For a large number N of particles, this procedure becomes unnecessarily expensive since, for a given required accuracy, there is no need to calculate all interactions exactly. A better approach is to apply the multipole method, where the long-range interactions are calculated approximately by multipole expansions, leaving only the short-range interactions to be treated explicitly.

To apply the multipole method, we divide the system into a set of *cubic boxes*, all of the same size. For each box A in the system, we set up a multipole expansion $\mathbf{q}_A$ of its charge distribution, with the origin at the centre of the box. For all pairs of boxes A and B that are sufficiently far away from each other, the interaction between their charge distributions may then be evaluated as a multipole expansion (9.13.75)

$$U_{AB} = \mathbf{q}_A^T \mathbf{T}_{AB} \mathbf{q}_B \qquad (9.14.2)$$

where $\mathbf{T}_{AB}$ is the interaction matrix for the two boxes

$$\mathbf{T}_{AB} = \mathbf{T}(\mathbf{R}_{BA}) \qquad (9.14.3)$$

The remaining interactions are evaluated explicitly according to (9.14.1), summing only over those pairs whose interactions cannot be treated by multipole expansions (9.14.2).

Let us consider which interactions can be calculated by multipole expansions. From the discussion in Section 9.13.1, we recall that the interaction between two charge distributions may be calculated by a multipole expansion if, for all particles in the two systems, the following inequality condition is satisfied

$$|\mathbf{a} - \mathbf{b}| < |\mathbf{R}| \qquad (9.14.4)$$

where $\mathbf{a}$ and $\mathbf{b}$ are the position vectors of the particles relative to the origins of the bipolar multipole expansion and $\mathbf{R}$ the vector connecting these origins – see (9.13.6). This condition is equivalent to the requirement that the charges in the two systems may be enclosed by two spheres that do not intersect – see Figure 9.14. It is noteworthy that, for a multipole expansion to converge, it is not sufficient that the two interacting distributions do not overlap.

Let us consider how this condition can be met in our application of the multipole method to the partitioned system. In the system, two boxes that share a boundary point are said to be *nearest neighbours*, otherwise they are said to be *well separated*. Since nearest neighbours cannot be enclosed by nonintersecting spheres, their interactions cannot be calculated by a multipole expansion. By contrast, a pair of well-separated boxes can always be enclosed by two nonintersecting spheres and their interactions may therefore be obtained from a bipolar expansion with origins at the box centres. In Figure 9.15, the nearest neighbours of box C are shaded light grey and the well-separated boxes dark grey. The sphere around C intersects the nearest neighbours but not the spheres that contain the well-separated boxes.

Let us now consider the evaluation of the interaction of a particle i located in box C with the other particles in the system. There are two distinct contributions to this interaction: the *near-field*

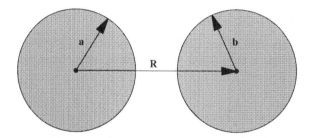

Fig. 9.14. The condition for convergence of the bipolar multipole expansion. The condition $|\mathbf{a} - \mathbf{b}| < |\mathbf{R}|$ implies that the two systems are enclosed in spheres that do not intersect.

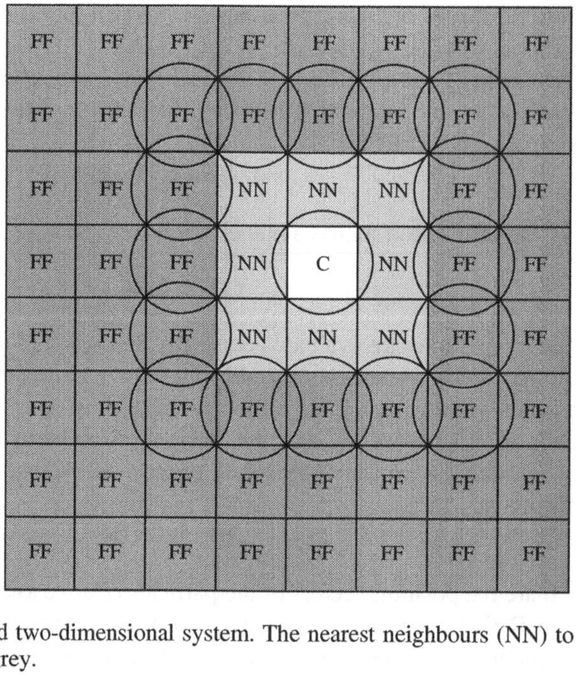

FF	FF	FF	FF	FF	FF	FF	FF
FF	FF	FF	FF	FF	FF	FF	FF
FF	FF	FF	NN	NN	NN	FF	FF
FF	FF	FF	NN	C	NN	FF	FF
FF	FF	FF	NN	NN	NN	FF	FF
FF	FF	FF	FF	FF	FF	FF	FF
FF	FF	FF	FF	FF	FF	FF	FF
FF	FF	FF	FF	FF	FF	FF	FF

Fig. 9.15. A partitioned two-dimensional system. The nearest neighbours (NN) to C are shaded light grey and the FF boxes dark grey.

(NF) contribution from the particles located in C and its nearest neighbours, and the *far-field (FF)* *contribution* from the particles in the boxes that are well separated from C:

$$U_i = U_i^{\mathrm{NF}} + U_i^{\mathrm{FF}} \tag{9.14.5}$$

The NF interactions are obtained directly, by summing over all particles located in the *NF boxes* (defined as C and its nearest neighbours):

$$U_i^{\mathrm{NF}} = \sum_{\substack{j \in \mathrm{NF} \\ j \neq i}} Z_i Z_j R_{ij}^{-1} \tag{9.14.6}$$

The FF interactions, on the other hand, are obtained by calculating multipole expansions with the *FF boxes* (the well-separated boxes) in the following manner:

$$U_i^{\mathrm{FF}} = \sum_{A \in \mathrm{FF}} \mathbf{q}_{iC}^{\mathrm{T}} \mathbf{T}_{\mathrm{CA}} \mathbf{q}_A \tag{9.14.7}$$

Here $\mathbf{q}_{iC}$ contains the multipole moments of particle i with the origin at the centre of box C and $\mathbf{q}_A$ is the multipole expansion of all particles in box A:

$$\mathbf{q}_A = \sum_{i \in A} \mathbf{q}_{iA} \tag{9.14.8}$$

The order of the multipole expansion is adjusted so that the required accuracy is obtained. If a sufficiently large number of particles is contained in each box, this approach is considerably faster than the direct evaluation of all interactions according to (9.14.1) since (9.14.7) involves a summation over boxes whereas (9.14.1) sums over particles.

It is possible to carry out the calculation of the FF interactions more efficiently than indicated by (9.14.7). First, we set up the *FF field vector* at the center of box C:

$$\mathbf{V}_{\mathrm{C}}^{\mathrm{FF}} = \sum_{A \in \mathrm{FF}} \mathbf{T}_{\mathrm{CA}} \mathbf{q}_A \tag{9.14.9}$$

Next, we calculate the interaction with particle i by contracting this field vector with the multipoles of particle i:

$$U_i^{\mathrm{FF}} = \mathbf{q}_{iC}^{\mathrm{T}} \mathbf{V}_{\mathrm{C}}^{\mathrm{FF}} \tag{9.14.10}$$

The advantage of this scheme is that we may use the same field vector $\mathbf{V}_{\mathrm{C}}^{\mathrm{FF}}$ for all particles in box C. Even simpler, the total interaction with all particles in C may be obtained by contracting the field vector (9.14.9) with the multipoles for all particles in C:

$$U_{\mathrm{C}}^{\mathrm{FF}} = \mathbf{q}_{\mathrm{C}}^{\mathrm{T}} \mathbf{V}_{\mathrm{C}}^{FF} \tag{9.14.11}$$

Finally, the interaction among all particles in the system is obtained by summing the interactions (9.14.5) for all particles.

Regarding computational cost of the multipole method, we note that the cost of the FF calculation increases *quadratically* with the size of the system since, for a given box, the number of FF boxes is proportional to the size of the system (assuming that the size of the boxes is kept constant in order to maintain the same accuracy in the calculation). By contrast, the cost of the NF calculation increases only *linearly* with the size of the system since, for a given box, the number of NF boxes is always 27 (9 in a two-dimensional system) – irrespective of the size of the system. In short, the cost of the *naive multipole method* described in this subsection increases quadratically with the size of the system. Still, provided that each box contains a sufficiently large number of particles, this method should perform considerably better than the explicit evaluation of the Coulomb interactions.

9.14.2 THE TWO-LEVEL MULTIPOLE METHOD

From Section 9.13.2, we recall that the multipole expansion may be regarded as an expansion in r/R, where r is the radius of the two spheres containing the interacting boxes and R the distance between the centres of the boxes. For a fixed order of the multipole expansions L, the long-range multipole interactions are therefore calculated to an accuracy higher than the short-range multipole interactions. One way to avoid this waste of accuracy would be to calculate distant interactions using lower-order multipole expansions. However, even though this approach improves the situation somewhat, it does not change it fundamentally – the cost of the calculation still scales quadratically with the size of the system since, for each interacting pair, we must retain at least one term in the expansion. A better solution to this problem is to group the small boxes into larger ones and to calculate the long-range interactions in terms of these larger boxes. By increasing the box size and keeping the ratio r/R approximately constant, we should be able to maintain a uniform accuracy in the evaluation of the short- and long-range interactions.

Let us consider the system of Section 9.14.1 at two levels of subdivision. At level 2, the system has been subdivided twice; at level 3, we have carried out three such subdivisions. These levels of subdivision are illustrated in Figure 9.16 for a two-dimensional system. Our strategy is now to calculate the long-range interactions at the coarse level and the short-range interactions at the fine level. For this purpose, it is useful to introduce some notation.

To describe the relationships between boxes at different levels, we introduce the terms *parent* and *child*. The cells obtained by subdividing a given cell are the *children* of this *parent*. Each parent thus contains eight children (four in a two-dimensional system) and each child has a single

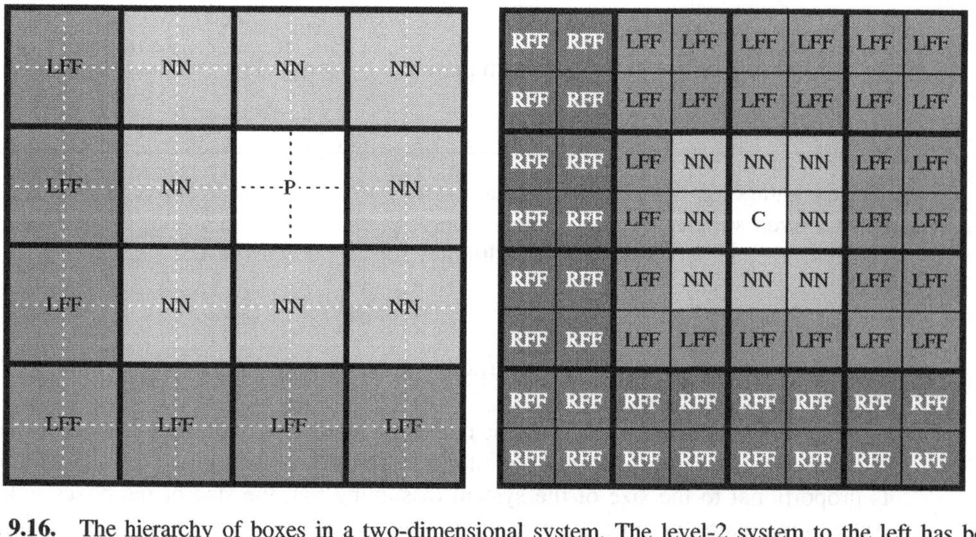

Fig. 9.16. The hierarchy of boxes in a two-dimensional system. The level-2 system to the left has been subdivided twice; the level-3 system to the right has been subdivided three times.

parent. To describe the relationships between boxes at the same level, we retain the terms *nearest neighbour* and *well separated* of Section 9.14.1, as well as the concepts of NF and FF boxes. However, we shall now need to distinguish between two types of FF boxes: The *local FF (LFF) boxes* are those FF boxes that are the children of the nearest neighbours of C's parent P; the remaining FF boxes are called the *remote FF (RFF) boxes*. The RFF boxes are the children of the parent's FF boxes.

Let us now consider the evaluation of the Coulomb interactions of the particles in C with the remaining particles in the system. The interaction of particle i may be calculated as

$$U_i = U_i^{NF} + \sum_{A \in LFF} \mathbf{q}_{iC}^T \mathbf{T}_{CA} \mathbf{q}_A + \sum_{B \in RFF} \mathbf{q}_{iC}^T \mathbf{T}_{CB} \mathbf{q}_B \qquad (9.14.12)$$

where the first term represents the NF interactions (9.14.6) and the second and third terms the LFF and RFF interactions, respectively, expressed as bipolar expansions at the centres of C and its FF boxes. Carrying out partial summations, we first evaluate the FF vectors at the centres of C and its parent P:

$$\mathbf{V}_C^{LFF} = \sum_{A \in LFF} \mathbf{T}_{CA} \mathbf{q}_A \qquad (9.14.13)$$

$$\mathbf{V}_P^{RFF} = \sum_{B \in RFF} \mathbf{T}_{PB} \mathbf{q}_B \qquad (9.14.14)$$

Using the translation relation (9.13.41) in the form

$$\mathbf{q}_{iP} = \mathbf{W}_{PC} \mathbf{q}_{iC} \qquad (9.14.15)$$

the interaction (9.14.12) may be expressed compactly as

$$U_i = U_i^{NF} + \mathbf{q}_{iC}^T (\mathbf{V}_C^{LFF} + \mathbf{W}_{PC}^T \mathbf{V}_P^{RFF}) \qquad (9.14.16)$$

Note that the field vector $\mathbf{V}_C^{LFF}$ is used by all particles located in box C and that $\mathbf{V}_P^{RFF}$ is shared by all children of P, reducing the cost of evaluating the full set of interactions U_i in the system. As we shall see in Section 9.14.3, when generalized to many levels of subdivision, the two-level strategy developed here leads to a scheme according to which the Coulomb interactions may, for a given fixed accuracy, be evaluated at a cost that increases only linearly with the size of the system.

9.14.3 THE FAST MULTIPOLE METHOD

As a generalization of the technique discussed in Section 9.14.2, we now introduce a *hierarchy of boxes* in the interacting system. At the highest level 0, the whole system is contained in a single box. Next, at level 1, the system is subdivided into eight equal-size equilateral cubic boxes. A further subdivision produces a system containing 64 boxes at level 2, and so on. This process is continued until, at the lowest level, each box contains only a small number of particles (10–100).

To calculate the number of levels and boxes in such a hierarchy, let us assume that the total number N of particles is equal to some power of 8 and that the smallest boxes each contain a single particle. At level l, there are 8^l boxes and the deepest level in the hierarchy is given by

$$S = \log_8 N \qquad (9.14.17)$$

Consequently, the total number of boxes in the hierarchy is given as

$$\sum_{l=0}^{\log_8 N} 8^l = \frac{8N-1}{7} \tag{9.14.18}$$

We note that the number of boxes in the hierarchy increases *linearly* with the number of particles and that nearly 90% of the boxes are located at the deepest level. In practice, the number of particles is not always equal to some power of 8, and the smallest boxes contain more than a single particle. Thus, for a system consisting of a million uniformly distributed particles, $S = 5$ gives a system in which about 30 particles are contained in each of the smallest boxes.

Having introduced this hierarchy of boxes, we may now calculate the Coulomb interactions in the system by a clever generalization of the two-level scheme of Section 9.14.2 – *the fast multipole method (FMM)* [28,29]:

1. Evaluate the multipole expansions for all particles $\mathbf{q}_{iA}$ and all boxes $\mathbf{q}_A$ at the deepest level:

$$\mathbf{q}_A = \sum_{i \in A} \mathbf{q}_{iA} = \sum_{i \in A} \mathbf{W}_{Ai} \mathbf{q}_i \tag{9.14.19}$$

   Since each particle belongs to a single box A, the cost of this step increases linearly with the size of the system.

2. Generate multipole expansions for all boxes at higher levels. For each parent box P, there is one contribution from each of its eight children:

$$\mathbf{q}_P = \sum_{C \in P} \mathbf{W}_{PC} \mathbf{q}_C \tag{9.14.20}$$

   Since the number of parent boxes is proportional to N, the cost of this step increases linearly with the size of the system. At this stage, we have generated a multipole expansion for each box B at all levels in the system.

3. For each box B in the system, calculate the field vector from its $6^3 - 3^3 = 189$ LFF boxes:

$$\mathbf{V}_B^{\text{LFF}} = \sum_{F \in \text{LFF}(B)} \mathbf{T}_{BF} \mathbf{q}_F \tag{9.14.21}$$

   When constructing the field vector, we have not included the NF and RFF boxes – the NF interactions are treated at deeper levels and the RFF interactions at higher levels. Since the number of boxes B in (9.14.21) is proportional to N and since there are always 189 terms in the sum, the cost of this step scales linearly with the size of the system.

4. For each box B in the hierarchy, calculate $\mathbf{V}_B^{\text{FF}}$, representing the field generated by *all* FF boxes – the remote ones as well as the local ones. At levels 0 and 1, there are no well separated boxes and therefore no FF. At level 2, each box T has a local FF but no remote FF:

$$\mathbf{V}_T^{\text{FF}} = \mathbf{V}_T^{\text{LFF}} \tag{9.14.22}$$

   At all deeper levels, there is a remote as well as a local contribution to the FF, the RFF of a given box C corresponding (by definition) to the total FF of its parent P:

$$\mathbf{V}_C^{\text{FF}} = \mathbf{V}_C^{\text{LFF}} + \mathbf{V}_C^{\text{RFF}} = \mathbf{V}_C^{\text{LFF}} + \mathbf{W}_{PC}^T \mathbf{V}_P^{\text{FF}} \tag{9.14.23}$$

In short, each box receives an RFF from its parent, which is added to its own LFF; at the next level, the $\mathbf{V}_C^{FF}$ is transferred to its eight children. The process ends when it reaches the deepest level. At this point, we have generated, for each box A at the deepest level, the vector $\mathbf{V}_A^{FF}$, which represents the total field generated by all well-separated boxes in the system. Again, since the amount of work is the same for each box and independent of the size of the system, the cost of this step scales linearly with the size of the system.

5. For each particle in the system, we now calculate its interaction with all well-separated particles:

$$U_i^{FF} = \mathbf{q}_{iA}^T \mathbf{V}_A^{FF} \tag{9.14.24}$$

This step involves a constant amount of work for each particle and the cost is again proportional to the number of particles in the system.

6. At this point, it remains only to add the NF contributions to the interactions, summing over all NF particles (including those contained in box A itself)

$$U_i = \sum_{\substack{j \in NF(A) \\ j \neq i}} Z_i Z_j R_{ij}^{-1} + U_i^{FF} \tag{9.14.25}$$

Since there are 27 such boxes for each particle in the system, the cost scales linearly with the size of the system.

7. If needed, the total Coulomb interaction energy of the system is obtained as

$$U = \tfrac{1}{2} \sum_i U_i \tag{9.14.26}$$

where the factor of one-half arises since each pairwise interaction is included twice.

Following these basic steps, it is possible to calculate the interactions of all particles in a system in a manner that scales linearly with the size of the system.

Clearly, the accuracy and efficiency of the FMM depend on the order of the multipole expansion L and the number of subdivisions S (i.e. the number of particles in the smallest boxes). The number of subdivisions (and therefore the size of the boxes) is usually chosen such that the smallest boxes contain 10–100 particles. The order of the multipole expansion is then adjusted such that the required accuracy is achieved.

However, the performance of the FMM may also be affected by adjusting the size of the NF. Let us introduce the *NF width parameter* Ω, which represents the number of NF boxes that separates a given box in the system from its FF boxes. In our discussion, we have included in the NF only the nearest neighbours – that is, we have used $\Omega = 1$. Often, a better performance is achieved with $\Omega = 2$, including in the NF also the second nearest neighbours, thus reducing the r/R ratio. For a given accuracy, the order of the multipole expansion may then be reduced, improving the overall efficiency of the calculation – provided that the increased cost of the NF interactions does not offset the reduced cost of the FF interactions. In practice, it is usually best to use a small width parameter Ω for low accuracy and large Ω for high accuracy.

9.14.4 THE CONTINUOUS FAST MULTIPOLE METHOD

The FMM described in Section 9.14.3 was developed for point charges. For systems of continuous charge distributions such as Gaussians, the FMM cannot be applied directly since such distributions may interact nonclassically. Also, in a given system, the distributions may vary considerably in size, making the separation into NF and FF interactions more difficult than for point charges. In the present subsection, we describe the *continuous fast multipole method (CFMM)*, which is a generalization of the FMM to systems of continuous charge distributions.

In generalizing the FMM to continuous distributions, we must accomplish two objectives. First, we must ensure that only classical interactions are included in the FF calculations. Second, to reduce as much as possible the number of classical interactions that are evaluated explicitly (i.e. as NF interactions), we must introduce a flexible system for determining the FF region. In the CFMM [21,30], these objectives are achieved by introducing, at each level, a range of NF width parameters

$$\Omega_n = 2n, \qquad n = 1, 2, \ldots \tag{9.14.27}$$

and by associating with each Ω_n a separate *branch* of boxes and distributions. (The reason for restricting the Ω_n to even numbers will become clear shortly.) At a given level in the hierarchy, the branches define the NF regions assigned to their boxes. Thus, the first branch has the smallest width parameter Ω_1 (indicating that each box in each direction is separated from its FF by only two boxes) and is used by the most compact distributions. For more diffuse charge distributions, branches of larger Ω_n are used in order to increase the width of the NF region.

At the deepest level of the hierarchy, each Gaussian distribution is assigned to a branch and a box, based on its extent and position. Let the size of the boxes at level l be given by

$$A_l = \left(\tfrac{1}{2}\right)^l A_0 \tag{9.14.28}$$

where A_0 is the largest box in the hierarchy. A Gaussian distribution of exponent p and extent (9.12.10)

$$r_p = p^{-1/2} \mathrm{erfc}^{-1}(10^{-k}) \tag{9.14.29}$$

is then assigned to the nth branch, whose width parameter is given by

$$\Omega_n = \max\left(\Omega_1, 2\left\lceil \frac{r_p}{A_l} \right\rceil\right) \tag{9.14.30}$$

where $\lceil i \rceil$ is the smallest integer greater than or equal to i. Note that $2\lceil r_p/A_l \rceil$ is the smallest number of boxes needed to ensure classical interactions between two Gaussians of extent r_p. It thus follows that, in a given branch, the distributions inside two boxes interact classically if they are separated by Ω_n or more boxes. Also, two boxes in different branches (at the same level) interact classically when separated by Ω_{mn} or more boxes, where Ω_{mn} is the average of the NF width parameters of the two branches:

$$\Omega_{mn} = \frac{\Omega_m + \Omega_n}{2} \tag{9.14.31}$$

Since Ω_m and Ω_n are even numbers, Ω_{mn} is an integer. Having thus sorted the Gaussian distributions at the deepest level into boxes and branches based on their position and extent, we are now ready to discuss the CFMM evaluation of Coulomb interactions for continuous charge distributions.

In describing the application of the CFMM, we shall refer to the steps of the FMM procedure of Section 9.14.3. Unlike the FMM, where all particles in the same box are treated collectively, the CFMM treats collectively all distributions that belong to the same box and branch. In step 1, we form multipole expansions by integrating over the Gaussian distributions in all boxes in all branches at the deepest level, using the techniques of Section 9.13. This step differs from that of the FMM only in that the summations are replaced by integrations. Next, in step 2, we generate multipole expansions in all boxes at the higher levels. In translating the expansions to the higher levels, we note that two boxes in different branches may have the same parent since, in some cases, the NF width parameter (9.14.30) is reduced by one-half (rounded up to the nearest even integer) as l is reduced by 1. This parent sharing by boxes in different branches improves the efficiency of the CFMM by reducing the number of multipole expansions that must be calculated at higher levels.

Having created the multipole expansions, we proceed to step 3 and generate, for each box in all branches, the LFF field vector $\mathbf{V}_B^{LFF}$ (9.14.21), noting that there are now contributions from all branches at the same level, with the separation into LFF and RFF boxes governed by the Ω_n parameter of each branch. No interactions are counted twice if, for each box, we include in the LFF all boxes not included in its parent's FF. Next, in step 4, we traverse the hierarchy downwards, generating at each level the field vectors $\mathbf{V}_B^{FF}$ by adding the RFF contributions from the parents according to (9.14.23). At the end of this process, we have generated, for each box in each branch at the deepest level, a vector that represents the total field from all FF boxes in the system. Using this vector we calculate, in step 5, for each distribution at the deepest level, its interaction with all FF boxes. Finally, we add the NF contributions for each distribution in step 6, replacing the FMM summation by integration.

It should be understood that the CFMM differs from the FMM in that the work required for the evaluation of the Coulomb interactions depends critically on the extent of the Gaussians. Thus, in the limit where all Gaussians extend over the whole system, there are no FF interactions and the CFMM reduces to the standard nonclassical integration, whose cost scales quartically with the size of the system. Linear scaling of the CFMM is achieved only when the system extends significantly beyond the largest Gaussian extent of the system – in general, the cost of the CFMM is intermediate between that of the explicit Gaussian integration and that of the point-charge FMM.

References

[1]   S. F. Boys, *Proc. Roy. Soc.* **A200**, 542 (1950).
[2]   R. McWeeny, *Nature* **166**, 21 (1950).
[3]   H. Taketa, S. Huzinaga and K. O-ohata, *J. Phys. Soc. Jpn* **21**, 2313 (1966).
[4]   T. Helgaker and P. R. Taylor, in D. R. Yarkony (ed.), *Modern Electronic Structure Theory*, World Scientific, 1995, p. 725.
[5]   S. Obara and A. Saika, *J. Chem. Phys.* **84**, 3963 (1986); *J. Chem. Phys.* **89**, 1540 (1988).
[6]   L. E. McMurchie and E. R. Davidson, *J. Comput. Phys.* **26**, 218 (1978).
[7]   T. Živković and Z. B. Maksić, *J. Chem. Phys.* **49**, 3083 (1968).
[8]   W. H. Press, S. A. Teukolsky, W. T. Vetterling and B. P. Flannery, *Numerical Recipes*, Cambridge University Press, 1986.
[9]   M. Abramowitz and I. A. Stegun, *Handbook of Mathematical Functions*, Dover, 1965.
[10] M. Dupuis, J. Rys and H. F. King, *J. Chem. Phys.* **65**, 111 (1976).
[11] I. Shavitt, *Meth. Comput. Phys.* **2**, 1 (1963).
[12] P. M. W. Gill, *Adv. Quantum Chem.* **25**, 141 (1994).
[13] P. M. W. Gill, M. Head-Gordon and J. A. Pople, *Int. J. Quantum Chem.* **S23**, 269 (1989); P. M. W. Gill, M. Head-Gordon and J. A. Pople, *J. Phys. Chem.* **94**, 5564 (1990); P. M. W. Gill and J. A. Pople, *Int. J. Quantum Chem.* **40**, 753 (1991).
[14] T. P. Hamilton and H. F. Schaefer III, *Chem. Phys.* **150**, 163 (1991).
[15] R. Lindh, U. Ryu and B. Liu, *J. Chem. Phys.* **95**, 5889 (1991).
[16] M. Head-Gordon and J. A. Pople, *J. Chem. Phys.* **89**, 5777 (1988).
[17] H. F. King and M. Dupuis, *J. Comput. Phys.* **21**, 144 (1976).
[18] V. R. Saunders, in G. H. F. Diercksen and S. Wilson (eds.), *Methods in Computational Molecular Physics*, Reidel, 1983, p. 1.
[19] J. Rys, M. Dupuis and H. F. King, *J. Comput. Chem.* **4**, 154 (1983).

[20] D. L. Strout and G. E. Scuseria, *J. Chem. Phys.* **102**, 8448 (1995).
[21] C. A. White, B. G. Johnson, P. M. W. Gill and M. Head-Gordon, *Chem. Phys. Lett.* **230**, 8 (1994).
[22] C. C. J. Roothaan, *Rev. Mod. Phys.* **23**, 69 (1951).
[23] M. Häser and R. Ahlrichs, *J. Comput. Chem.* **10**, 104 (1989).
[24] A. J. Stone, *The Theory of Intermolecular Forces*, Clarendon, 1996.
[25] C. A. White and M. Head-Gordon, *J. Chem. Phys.* **101**, 6593 (1994).
[26] D. M. Brink and G. R. Satchler, *Angular Momentum*, 2nd edn, Clarendon Press, 1968.
[27] C. Hättig and B. A. Hess, *Mol. Phys.* **81**, 813 (1994); C. Hättig, *Chem. Phys. Lett.* **260**, 341 (1996).
[28] L. Greengard and V. I. Rokhlin, *J. Comput. Phys.* **73**, 325 (1987); L. Greengard, *Science* **265**, 909 (1994).
[29] H.-Q. Ding, N. Karasawa and W. A. Goddard III, *Chem. Phys. Lett.* **196**, 6 (1992); *J. Chem. Phys.* **97**, 4309 (1992).
[30] C. A. White, B. G. Johnson, P. M. W. Gill and M. Head-Gordon, *Chem. Phys. Lett.* **253**, 268 (1996).

Further reading

P. M. W. Gill, Molecular integrals over Gaussian basis functions, *Adv. Quantum Chem.* **25**, 141 (1994).
L. Greengard, Fast algorithms for classical physics, *Science* **265**, 909 (1994).
T. Helgaker and P. R. Taylor, Gaussian basis sets and molecular integrals, in D. R. Yarkony (ed.), *Modern Electronic Structure Theory*, World Scientific, 1995, p. 725.
R. Lindh, Integrals of electron repulsion, in P. v. R. Schleyer, N. L. Allinger, T. Clark, J. Gasteiger, P. A. Kollman, H. F. Schaefer III and P. R. Schreiner (eds), *Encyclopedia of Computational Chemistry*, Vol. 2, Wiley, 1998, p. 1337.
V. R. Saunders, An introduction to molecular integral evaluation, in G. H. F. Diercksen, B. T. Sutcliffe and A. Veillard (eds), *Computational Techniques in Quantum Chemistry and Molecular Physics*, Reidel, 1974, p. 347.
V. R. Saunders, Molecular integrals for Gaussian type functions, in G. H. F. Diercksen and S. Wilson (eds), *Methods in Computational Molecular Physics*, Reidel, 1983, p. 1.
I. Shavitt, The history and evolution of Gaussian basis sets, *Isr. J. Chem.* **33**, 357 (1993).
A. J. Stone, *The Theory of Intermolecular Forces*, Clarendon, 1996.
M. C. Strain, G. E. Scuseria and M. J. Frisch, Achieving linear scaling for the electronic quantum Coulomb problem, *Science*, **271**, 51 (1996).

Exercises

EXERCISE 9.1

The self-overlap integral of a Cartesian Gaussian

$$G_{ijk} = x_A^i y_A^j z_A^k \exp(-a r_A^2) \tag{9E.1.1}$$

is given by

$$\langle G_{ijk} | G_{ijk} \rangle = \frac{(2i-1)!!(2j-1)!!(2k-1)!!}{(4a)^{i+j+k}} \left(\frac{\pi}{2a}\right)^{3/2} \tag{9E.1.2}$$

where the double factorial is defined in (6.5.10).

1. Using the McMurchie–Davidson scheme, verify that the general expression (9E.1.2) holds for the special cases $\langle p_x | p_x \rangle$ and $\langle d_{x^2} | d_{x^2} \rangle$.
2. Show that the one-centre even-order multipole s integrals are given by

$$\langle s | x_A^n | s \rangle = (n-1)!!(4a)^{-n/2} \left(\frac{\pi}{2a}\right)^{3/2}, \qquad n = 0, 2, 4, \ldots \tag{9E.1.3}$$

Why do the one-centre multipole integrals vanish for odd n?

3. For $n = 2$, verify the expression

$$\langle s|(-i\partial/\partial x)^n|s\rangle = (n-1)!!a^{n/2}\left(\frac{\pi}{2a}\right)^{3/2}, \qquad n = 0, 2, 4, \ldots \tag{9E.1.4}$$

EXERCISE 9.2

Consider the ground-state hydrogen and helium atoms in a basis of a single Gaussian with exponent a:

$$G_{1s} = \exp(-ar^2) \tag{9E.2.1}$$

1. Show that the overlap and kinetic-energy integrals are given by

$$S_{1s1s}(a) = \left(\frac{\pi}{2a}\right)^{3/2} \tag{9E.2.2}$$

$$T_{1s1s}(a) = \frac{3a}{2}\left(\frac{\pi}{2a}\right)^{3/2} \tag{9E.2.3}$$

and that the Coulomb integrals are given by

$$C_{1s1s}(Z, a) = -ZV_{1s1s}(a) = -Z\frac{\pi}{a} \tag{9E.2.4}$$

$$g_{1s1s1s1s}(a) = \frac{1}{4}\left(\frac{\pi}{a}\right)^{5/2} \tag{9E.2.5}$$

where Z is the nuclear charge.

2. Show that the energy expressions for the hydrogen and helium atoms in the basis of a single Gaussian with exponent a are given by

$$E_H(a) = \frac{3a}{2} - 2\sqrt{\frac{2a}{\pi}} \tag{9E.2.6}$$

$$E_{He}(a) = 3a + (2 - 8\sqrt{2})\sqrt{\frac{a}{\pi}} \tag{9E.2.7}$$

3. Optimize the energies by means of the variation principle and show that we obtain for the hydrogen atom

$$a_H^* = \frac{8}{9\pi} = 0.2829 \tag{9E.2.8}$$

$$E_H(a_H^*) = -\frac{4}{3\pi} = -0.4244 \tag{9E.2.9}$$

and for the helium atom

$$a_{He}^* = \frac{33 - 8\sqrt{2}}{9\pi} = 0.7670 \tag{9E.2.10}$$

$$E_{He}(a_{He}^*) = \frac{8\sqrt{2} - 33}{3\pi} = -2.3010 \tag{9E.2.11}$$

How do these results compare with the exact ground-state energies of the hydrogen and helium atoms?

4. For an atom, the virial theorem of Section 4.2.7 states that the kinetic and total energies are related by

$$E_{\text{kin}} = -E_{\text{tot}} \tag{9E.2.12}$$

Determine the Gaussian exponent for the hydrogen atom from the virial theorem. How does the result obtained in this way compare with that obtained from the variation principle?

EXERCISE 9.3

Prove the recurrence relations for the Boys function (9.8.13) and (9.8.14). Why is the upward recursion numerically unstable for small arguments?

EXERCISE 9.4

Show that, for a sufficiently large separation between the overlap distributions, the two-electron integral (9.9.33) reduces to the simple expression

$$g_{abcd} = \frac{S_{ab}S_{cd}}{R_{PQ}} \tag{9E.4.1}$$

where S_{ab} is the overlap integral. This is the first term in the multipole expansion of the two-electron integral (9.13.27).

EXERCISE 9.5

Show that the expression for the two-electron repulsion integral (9.9.33) reduces to the one-electron potential integral (9.9.32) when the second charge distribution represents a unit point charge.

Solutions

SOLUTION 9.1

1. According to (9E.1.2), the self-overlap integrals are given by

$$\langle p_x | p_x \rangle = \frac{1}{2p} \left(\frac{\pi}{p} \right)^{3/2} \tag{9S.1.1}$$

$$\langle d_{x^2} | d_{x^2} \rangle = \frac{3}{(2p)^2} \left(\frac{\pi}{p} \right)^{3/2} \tag{9S.1.2}$$

where

$$p = 2a \tag{9S.1.3}$$

According to (9.5.41), these integrals may, in the McMurchie–Davidson scheme, be written in the form

$$\langle p_x | p_x \rangle = E_0^{11} E_0^{00} E_0^{00} \left(\frac{\pi}{p} \right)^{3/2} \tag{9S.1.4}$$

$$\langle d_{x^2} | d_{x^2} \rangle = E_0^{22} E_0^{00} E_0^{00} \left(\frac{\pi}{p} \right)^{3/2} \tag{9S.1.5}$$

where, for a one-centre integral, the Hermite-to-Cartesian expansion coefficients for the three Cartesian directions are the same. We also note that, for a one-centre integral, the Hermite recurrence relations (9.5.6)–(9.5.8) may be written in the simplified form

$$E_0^{00} = 1 \tag{9S.1.6}$$

$$E_t^{i+1,j} = E_t^{i,j+1} = \frac{1}{2p}E_{t-1}^{ij} + (t+1)E_{t+1}^{ij} \tag{9S.1.7}$$

since

$$K_{ab}^x = K_{ab}^y = K_{ab}^z = 1 \tag{9S.1.8}$$

$$X_{PA} = Y_{PA} = Z_{PA} = 0 \tag{9S.1.9}$$

We now obtain the following expansion coefficients

$$E_0^{11} = E_1^{01} = \frac{1}{2p}E_0^{00} = \frac{1}{2p} \tag{9S.1.10}$$

$$E_0^{22} = E_1^{12} = \frac{1}{2p}E_0^{11} + 2E_2^{11} = \frac{1}{2p}E_1^{01} + 2\frac{1}{2p}E_1^{01} = \frac{3}{2p}E_1^{01} \tag{9S.1.11}$$

$$= 3\left(\frac{1}{2p}\right)^2 E_0^{00} = 3\left(\frac{1}{2p}\right)^2$$

which, combined with (9S.1.4) and (9S.1.5), yield (9S.1.1) and (9S.1.2) directly.

2. The left-hand side of (9E.1.3) is easily written as

$$\langle s|x_A^n|s\rangle = \langle G_{n/2,0,0}|G_{n/2,0,0}\rangle \tag{9S.1.12}$$

The relation (9E.1.3) now follows directly from (9E.1.2). For odd n, these integrals vanish as the integrand is then antisymmetric about the origin.

3. We use partial integration to obtain

$$\langle s|(-i\partial/\partial x)^2|s\rangle = \left\langle \frac{\partial s}{\partial x}\middle|\frac{\partial s}{\partial x}\right\rangle = 4a^2\langle p_x|p_x\rangle \tag{9S.1.13}$$

from which (9E.1.4) follows for $n = 2$ by application of (9E.1.2).

SOLUTION 9.2

1. The overlap and kinetic-energy integrals (9E.2.2) and (9E.2.3) follow directly from the results in Exercise 9.1, in particular from (9E.1.2) and (9E.1.4). For the evaluation of (9E.2.4) and (9E.2.5), we note that, for s orbitals, (9.9.32) and (9.9.33) reduce to

$$V_{1s1s} = \frac{2\pi}{2a}E_{000}R_{000}(2a, 0) = \frac{\pi}{a}F_0(0) = \frac{\pi}{a} \tag{9S.2.1}$$

$$g_{1s1s1s1s} = \frac{2\pi^{5/2}}{(2a)^2\sqrt{2a+2a}}E_{000}E_{000}R_{000}(a, 0) = \frac{1}{4}\left(\frac{\pi}{a}\right)^{5/2} \tag{9S.2.2}$$

where we have used (9.2.15) and (9.9.27) to obtain $E_{000} = 1$. To simplify the Hermite integrals, we have used (9.9.9) and (9.8.6). Note that, for one-centre functions, the Coulomb integrals may be expressed in a closed analytical form.

2. The energies are given by the expressions

$$E_H(a) = \frac{T_{1s1s} + C_{1s1s}(1)}{S_{1s1s}} = \frac{3a}{2} - 2\sqrt{\frac{2a}{\pi}} \tag{9S.2.3}$$

$$E_{He}(a) = 2\frac{T_{1s1s} + C_{1s1s}(2)}{S_{1s1s}} + \frac{g_{1s1s1s1s}}{S_{1s1s}^2} = 3a + (2 - 8\sqrt{2})\sqrt{\frac{a}{\pi}} \tag{9S.2.4}$$

by substitution of (9E.2.2)–(9E.2.5).

3. Differentiating (9E.2.6) and (9E.2.7) with respect to a and setting the results equal to zero, we obtain

$$\frac{3}{2} - \sqrt{\frac{2}{a_H^*\pi}} = 0 \tag{9S.2.5}$$

$$3 + (1 - 4\sqrt{2})\sqrt{\frac{1}{a_{He}^*\pi}} = 0 \tag{9S.2.6}$$

which have the solutions (9E.2.8) and (9E.2.10). Inserting these solutions into the energy expressions, we obtain (9E.2.9) and (9E.2.11). The calculated hydrogen energy is 15% higher than the true energy of $-\frac{1}{2}$ E_h and the calculated helium energy is 21% higher than the true energy of -2.903724 E_h.

4. From the virial theorem (9E.2.12), we obtain the condition

$$2T_{1s1s} = -C_{1s1s} \tag{9S.2.7}$$

which, upon substitution of (9E.2.3) and (9E.2.4), yields the same solution as the variation principle.

SOLUTION 9.3

The Boys function is defined as

$$F_n(x) = \int_0^1 \exp(-xt^2)t^{2n}\, dt \tag{9S.3.1}$$

Writing the integrand as uv' with

$$u = \exp(-xt^2) \tag{9S.3.2}$$

$$v' = t^{2n} \tag{9S.3.3}$$

we obtain by partial integration

$$F_n(x) = \frac{2x}{2n+1}\int_0^1 \exp(-xt^2)t^{2n+2}\, dt + \left[\exp(-xt^2)\frac{t^{2n+1}}{2n+1}\right]_0^1 \tag{9S.3.4}$$

and consequently

$$F_n(x) = \frac{2x}{2n+1}F_{n+1}(x) + \frac{\exp(-x)}{2n+1} \tag{9S.3.5}$$

which may be rearranged to give the recurrence relations (9.8.13) and (9.8.14). The upward recursion is numerically unstable since, for small x,

$$(2n + 1)F_n(x) \approx \exp(-x) \approx 1 \tag{9S.3.6}$$

where we have used (9.8.6).

SOLUTION 9.4

For large separations

$$R_{PQ} \gg 0 \tag{9S.4.1}$$

we may write the zero-order Boys function as

$$F_0(\alpha R_{PQ}^2) = \sqrt{\frac{\pi}{4\alpha}} \frac{1}{R_{PQ}} \tag{9S.4.2}$$

where we have used (9.8.9). We may now write the Hermite integrals in (9.9.33) in the form

$$R_{t+\tau,u+\upsilon,v+\phi} = \left(\frac{\partial}{\partial P_x}\right)^{t+\tau} \left(\frac{\partial}{\partial P_y}\right)^{u+\upsilon} \left(\frac{\partial}{\partial P_z}\right)^{v+\phi} \sqrt{\frac{\pi}{4\alpha}} \frac{1}{R_{PQ}} \tag{9S.4.3}$$

where we have used (9.9.9) and (9S.4.2). From (9S.4.3), we note that

$$R_{000} \gg R_{t+\tau,u+\upsilon,v+\phi} \tag{9S.4.4}$$

which means that the two-electron integrals (9.9.33) may be written in the simple form

$$g_{abcd} = \frac{2\pi^{5/2}}{pq\sqrt{p+q}} E_{000}^{ab} E_{000}^{cd} \sqrt{\frac{\pi}{4\alpha}} \frac{1}{R_{PQ}} \tag{9S.4.5}$$

where we have neglected higher-order terms. Using (9.7.22), (9S.4.5) may now be rearranged to give

$$g_{abcd} = \left(\frac{\pi}{p}\right)^{3/2} E_{000}^{ab} \left(\frac{\pi}{q}\right)^{3/2} E_{000}^{cd} \frac{1}{R_{PQ}} = \frac{S_{ab}S_{cd}}{R_{PQ}} \tag{9S.4.6}$$

where we have also used (9.5.41) for the overlap integrals.

SOLUTION 9.5

When the second overlap distribution corresponds to a one-centre $1s$ charge distribution, the two-electron integrals in (9.9.33) may be written as

$$g_{abss} = \frac{2\pi^{5/2}}{pq\sqrt{p+q}} \sum_{tuv} E_{tuv}^{ab} R_{tuv}\left(\frac{pq}{p+q}, \mathbf{R}_{PQ}\right) \tag{9S.5.1}$$

Normalizing the second charge contribution to unity, we obtain

$$G_{ab}(q) = \left(\frac{q}{\pi}\right)^{3/2} \frac{2\pi^{5/2}}{pq\sqrt{p+q}} \sum_{tuv} E_{tuv}^{ab} R_{tuv}\left(\frac{pq}{p+q}, \mathbf{R}_{PQ}\right) \tag{9S.5.2}$$

In the limit as q tends to infinity, this integral becomes

$$\lim_{q \to \infty} G_{ab}(q) = \frac{2\pi}{p} \sum_{tuv} E_{tuv}^{ab} R_{tuv}(p, \mathbf{R}_{PQ}) \tag{9S.5.3}$$

which is equivalent to the expression for the Coulomb-potential integral (9.9.32).

10 HARTREE–FOCK THEORY

The Hartree–Fock wave function is the cornerstone of *ab initio* electronic-structure theory. It is obtained by minimizing the expectation value of the electronic energy with respect to the orbitals occupying a single Slater determinant or more generally a single CSF, containing several such determinants in a fixed linear combination. The Hartree–Fock wave function yields total electronic energies that are in error by less than 1% and a wide range of important molecular properties such as dipole moments, electric polarizabilities, electronic excitation energies, magnetizabilities, force constants and nuclear magnetic shieldings are usually reproduced to within 5–10% accuracy. Molecular geometries are particularly well reproduced and are mostly within a few picometres of the true equilibrium structure.

The Hartree–Fock wave function is often used in qualitative studies of molecular systems, particularly larger systems. Indeed, the Hartree–Fock wave function is still the only wave function that can be applied routinely to large systems, and systems containing several hundred atoms have been studied at this level of approximation. For accurate, quantitative studies of molecular systems, the Hartree–Fock wave function is by itself not useful but it constitutes the starting point for more accurate treatments. The nonvariational Møller–Plesset and coupled-cluster approaches, for example, are both based on the premise that the Hartree–Fock function represents a good zero-order approximation to the true wave function and may in fact be viewed as yielding corrections to the Hartree–Fock state.

The purpose of the present chapter is to discuss the structure and construction of restricted Hartree–Fock wave functions. We cover not only the traditional methods of optimization, based on the diagonalization of the Fock matrix, but also second-order methods of optimization, based on an expansion of the Hartree–Fock energy in nonredundant orbital rotations, as well as density-based methods, required for the efficient application of Hartree–Fock theory to large molecular systems. In addition, some important aspects of the Hartree–Fock model are analysed, such as the size-extensivity of the energy, symmetry constraints and symmetry-broken solutions, and the interpretation of orbital energies in the canonical representation.

10.1 Parametrization of the wave function and the energy

In the present section, we discuss the parametrization of the restricted Hartree–Fock wave function and its electronic energy. We begin by reviewing CSFs in Section 10.1.1 and then go on to consider orbital rotations of such CSFs in Section 10.1.2, paying special attention to the concept of redundant and nonredundant orbital rotations. Finally, in Section 10.1.3, we discuss the parametrization of the electronic energy, expanding it to second order in the orbital-rotation parameters.

10.1.1 SINGLET AND TRIPLET CSFs

In *restricted Hartree–Fock (RHF) theory*, the electronic state is represented by a single CSF

$$|\text{CSF}\rangle = \sum_i C_i |i\rangle \tag{10.1.1}$$

where $|i\rangle$ are Slater determinants, with coefficients C_i fixed by the spin symmetry of the wave function. The Slater determinants in (10.1.1) belong to the same orbital configuration – they have identical orbital occupation numbers but different spin-orbital occupation numbers. For a detailed discussion of CSFs and their construction from Slater determinants, see Sections 2.5 and 2.6. We do not consider here the *spatial* symmetry adaptation for degenerate states of non-Abelian point groups since such an adaptation would require the combination of Slater determinants belonging to different orbital configurations.

We shall assume that the Slater determinants in the RHF wave function are constructed from a set of real and orthonormal orbitals. To distinguish orbitals of different occupancies, we use the following conventions: Orbitals that are doubly occupied in all determinants are called *inactive* and are labelled i, j, k, l. Partially occupied orbitals are known as *active* and are distinguished by the labels v, w, x, y, z. For the *virtual* orbitals, which are *unoccupied* in all determinants, we use the indices a, b, c, d, e. Such orbitals are also referred to as *secondary*. Finally, the *mixed* or *general* indices m, n, o, p, q, r, s, t, u are used for orbitals of arbitrary or unspecified occupancies. In the following, we shall assume that the orbitals are sorted in the order inactive, active and virtual.

Let us consider some important examples of CSFs. A closed-shell Hartree–Fock wave function contains only inactive orbitals and the CSF may be written as

$$|\text{CSF}\rangle = \left(\prod_i a_{i\alpha}^\dagger a_{i\beta}^\dagger \right) |\text{vac}\rangle = \hat{A}_c^\dagger |\text{vac}\rangle \tag{10.1.2}$$

where we use the core creation operator or core string

$$\hat{A}_c^\dagger = \prod_i a_{i\alpha}^\dagger a_{i\beta}^\dagger \tag{10.1.3}$$

The CSF of triplet symmetry and single occupancy in two orbitals (with the remaining orbitals doubly occupied) has the three components

$$|\text{CSF}\rangle^{1,1} = \hat{Q}_{vw}^{1,1} \hat{A}_c^\dagger |\text{vac}\rangle = a_{v\alpha}^\dagger a_{w\alpha}^\dagger \left(\prod_i a_{i\alpha}^\dagger a_{i\beta}^\dagger \right) |\text{vac}\rangle \tag{10.1.4}$$

$$|\text{CSF}\rangle^{1,0} = \hat{Q}_{vw}^{1,0} \hat{A}_c^\dagger |\text{vac}\rangle = \frac{1}{\sqrt{2}} (a_{v\alpha}^\dagger a_{w\beta}^\dagger + a_{v\beta}^\dagger a_{w\alpha}^\dagger) \left(\prod_i a_{i\alpha}^\dagger a_{i\beta}^\dagger \right) |\text{vac}\rangle \tag{10.1.5}$$

$$|\text{CSF}\rangle^{1,-1} = \hat{Q}_{vw}^{1,-1} \hat{A}_c^\dagger |\text{vac}\rangle = a_{v\beta}^\dagger a_{w\beta}^\dagger \left(\prod_i a_{i\alpha}^\dagger a_{i\beta}^\dagger \right) |\text{vac}\rangle \tag{10.1.6}$$

where we have used the triplet two-body creation operators (2.3.17)–(2.3.19). By means of the singlet two-body creation operator (2.3.16), the two unpaired electrons may also be coupled to form a singlet CSF:

$$|\text{CSF}\rangle^{0,0} = \hat{Q}_{vw}^{0,0} \hat{A}_c^\dagger |\text{vac}\rangle = \frac{1}{\sqrt{2}} (a_{v\alpha}^\dagger a_{w\beta}^\dagger - a_{v\beta}^\dagger a_{w\alpha}^\dagger) \left(\prod_i a_{i\alpha}^\dagger a_{i\beta}^\dagger \right) |\text{vac}\rangle \tag{10.1.7}$$

We refer to Section 2.6 for a general discussion of the spin adaptation of Slater determinants of a given orbital configuration.

In *unrestricted Hartree–Fock (UHF) theory*, the wave function is represented by a single Slater determinant – no restrictions are enforced on the total spin of the system and the wave function is not required to transform as an irreducible representation of the molecular point group. Since the alpha and beta spin orbitals are separately optimized, they will in general have different spatial forms. The resulting UHF wave function may therefore yield an energy lower than that of RHF theory; see the examples in Section 5.4. Although we are concerned mainly with RHF theory, much of the theory presented in this chapter carries over to unrestricted wave functions with little or no modification. The relationship between RHF and UHF theories is discussed in more detail in Section 10.10.

10.1.2   ORBITAL ROTATIONS

When the orthonormal MOs from which a given CSF is constructed are transformed into a different set of orthonormal MOs, then we obtain a new state, which may be generated from the original CSF by the application of a unitary operator as described in Chapter 3. To ensure that the transformed state remains a spin eigenstate (and thus remains a CSF), the unitary operator must be of singlet symmetry. A real transformation that preserves the orthonormality of the MOs and the spin of the CSF may be written as

$$|\text{CSF}(\boldsymbol{\kappa})\rangle = \exp(-\hat{\kappa})|\text{CSF}\rangle \tag{10.1.8}$$

where $\hat{\kappa}$ is the anti-Hermitian one-electron operator

$$\hat{\kappa} = \sum_{p>q} \kappa_{pq}(E_{pq} - E_{qp}) = \sum_{p>q} \kappa_{pq} E_{pq}^- \tag{10.1.9}$$

For more general orbital-rotation operators, see Section 3.3.3. If we also wish to conserve the spatial symmetry of the CSF, we must retain in the orbital-rotation operator (10.1.9) only those excitation operators that transform as the totally symmetric irreducible representation of the molecular point group. For Abelian point groups, this is accomplished by summing over only those pairs pq where p and q transform as the same irreducible representation.

As it stands, the orbital-rotation operator in (10.1.9) contains parameters that mix all classes of MOs among one another – the inactive orbitals, the active orbitals and the virtual orbitals. These parameters are all necessary in order to carry out a general rotation of the individual MOs. For a general transformation of the *wave function*, however, not all of these parameters are needed; those parameters that are not needed for a general transformation of the wave function are referred to as *redundant*. We shall here use the term *redundant* in a more specialized sense, applying it to those parameters κ_{pq} that are not needed for a general *first-order* transformation of the wave function $-\hat{\kappa}|\text{CSF}\rangle$. A parameter κ_{pq} is thus said to be redundant if the corresponding operator E_{pq}^- in (10.1.9) satisfies the condition

$$E_{pq}^-|\text{CSF}\rangle \equiv 0 \tag{10.1.10}$$

By group-theoretical arguments, it may be shown that, if the set of redundant orbital-rotation operators (10.1.10) constitute a group, then their elimination from the orbital-rotation operator (10.1.9) will not affect our description of the wave function to *any* order in $\hat{\kappa}$ – that is, any state that can be reached with the full set of redundant and nonredundant parameters in (10.1.8) can then also be reached with the set of nonredundant parameters.

To illustrate the identification of redundant parameters, we consider the three CSFs of Section 10.1.1. Since, according to Section 3.3, no diagonal elements κ_{pp} appear in the special orthogonal orbital transformation, we need to examine only the nondiagonal elements. For the closed-shell CSF (10.1.2), we have for rotations among inactive orbitals

$$E_{kl}^- \left(\prod_i a_{i\alpha}^\dagger a_{i\beta}^\dagger \right) |\text{vac}\rangle = 0 \qquad (10.1.11)$$

since, for $k \neq l$, each term gives zero separately:

$$E_{lk} \left(\prod_i a_{i\alpha}^\dagger a_{i\beta}^\dagger \right) |\text{vac}\rangle = E_{kl} \left(\prod_i a_{i\alpha}^\dagger a_{i\beta}^\dagger \right) |\text{vac}\rangle = 0 \qquad (10.1.12)$$

Rotations among inactive orbitals are therefore redundant. For rotations among virtual orbitals, we likewise obtain

$$E_{ab}^- \left(\prod_i a_{i\alpha}^\dagger a_{i\beta}^\dagger \right) |\text{vac}\rangle = 0 \qquad (10.1.13)$$

whereas the occupied–virtual rotations give nonvanishing results:

$$E_{aj}^- \left(\prod_i a_{i\alpha}^\dagger a_{i\beta}^\dagger \right) |\text{vac}\rangle \neq 0 \qquad (10.1.14)$$

For a closed-shell wave function, therefore, the only nonredundant rotations are those that mix inactive and virtual orbitals.

Having considered the closed-shell CSFs, we now turn our attention to the two-electron open-shell CSFs (10.1.4)–(10.1.7). The singlet and triplet CSFs may be written in the general form

$$|\text{CSF}\rangle^{S,M} = \hat{Q}_{vw}^{S,M} \hat{A}_c^\dagger |\text{vac}\rangle \qquad (10.1.15)$$

As for closed shells, it follows that the inactive–inactive and virtual–virtual rotations are redundant. On the other hand, all rotations that mix different classes of orbitals (i.e. the inactive–active, inactive–virtual and active–virtual rotations) are nonredundant. It remains only to investigate the active–active rotations.

Since the active–active excitation operators commute with the core string and annihilate the vacuum state, we may write

$$E_{vw}^- |\text{CSF}\rangle^{S,M} = [E_{vw}^-, \hat{Q}_{vw}^{S,M}] \hat{A}_c^\dagger |\text{vac}\rangle \qquad (10.1.16)$$

and we need examine only the commutator between the excitation operator and the two-body creation operators. According to (2.3.39), this commutator is given by

$$[E_{vw}^-, \hat{Q}_{vw}^{S,M}] = \hat{Q}_{vv}^{S,M} - \hat{Q}_{ww}^{S,M} \qquad (10.1.17)$$

Since the triplet two-body creation operators are antisymmetric in the orbital indices, the commutator vanishes for the triplet state

$$E_{vw}^- |\text{CSF}\rangle^{1,M} = 0 \qquad (10.1.18)$$

but not for the singlet state

$$E_{vw}^- |\text{CSF}\rangle^{0,0} = (\hat{Q}_{vv}^{0,0} - \hat{Q}_{ww}^{0,0}) \hat{A}_c^\dagger |\text{vac}\rangle \qquad (10.1.19)$$

We conclude that the active–active rotations are nonredundant for the singlet state (10.1.7) but redundant for the triplet state (10.1.4)–(10.1.6).

More generally, we conclude that the inactive–inactive and secondary–secondary rotations are always redundant, whereas the inactive–active, inactive–secondary and active–secondary rotations are always nonredundant. The active–active rotations, by contrast, may or may not be redundant depending on the structure of the CSF, and their redundancy must be established for each CSF separately. In Section 12.2.6, we shall see how redundant parameters can be identified for multiconfigurational wave functions and present a simple general prescription for identifying the nonredundant active–active rotations for single-configuration CSF wave functions.

10.1.3 EXPANSION OF THE ENERGY

Let us now consider the electronic energy of the transformed CSF in (10.1.8), expanding the expectation value of the Hamiltonian

$$E(\kappa) = \langle \mathrm{CSF}(\kappa)|\hat{H}|\mathrm{CSF}(\kappa)\rangle \tag{10.1.20}$$

to second order in the orbital-rotation parameters κ around $\kappa = 0$:

$$E(\kappa) = E^{(0)} + \kappa^{\mathrm{T}}\mathbf{E}^{(1)} + \tfrac{1}{2}\kappa^{\mathrm{T}}\mathbf{E}^{(2)}\kappa + \cdots \tag{10.1.21}$$

Here $\mathbf{E}^{(1)}$ is a column vector representing the *electronic gradient* at the expansion point and $\mathbf{E}^{(2)}$ is a matrix representing the *electronic Hessian*. The orbital-rotation parameters κ_{pq} have been arranged as a column vector κ. Occasionally, the same notation κ will be used for these parameters arranged as an (antisymmetric) square matrix of dimension equal to the number of MOs. The correct interpretation of κ will always be clear from the context.

To arrive at explicit expressions for the electronic gradient and Hessian in the Taylor series (10.1.21), we note that the expectation value (10.1.20) may be written in the form

$$E(\kappa) = \langle \mathrm{CSF}| \exp(\hat{\kappa})\hat{H} \exp(-\hat{\kappa})|\mathrm{CSF}\rangle \tag{10.1.22}$$

which may then be expanded in a BCH series (3.1.8):

$$E(\kappa) = \langle \mathrm{CSF}|\hat{H}|\mathrm{CSF}\rangle + \langle \mathrm{CSF}|[\hat{\kappa}, \hat{H}]|\mathrm{CSF}\rangle + \tfrac{1}{2}\langle \mathrm{CSF}|[\hat{\kappa}, [\hat{\kappa}, \hat{H}]]|\mathrm{CSF}\rangle + \cdots \tag{10.1.23}$$

Identifying the terms that occur to the same orders in the Taylor and BCH expansions of the energy, we obtain the following expressions for the electronic energy, the electronic gradient and the electronic Hessian at $\kappa = 0$:

$$E^{(0)} = E(\mathbf{0}) = \langle \mathrm{CSF}|\hat{H}|\mathrm{CSF}\rangle \tag{10.1.24}$$

$$E^{(1)}_{pq} = \left.\frac{\partial E(\kappa)}{\partial \kappa_{pq}}\right|_{\kappa=0} = \langle \mathrm{CSF}|[E_{pq}^{-}, \hat{H}]|\mathrm{CSF}\rangle \tag{10.1.25}$$

$$E^{(2)}_{pqrs} = \left.\frac{\partial^2 E(\kappa)}{\partial \kappa_{pq}\partial \kappa_{rs}}\right|_{\kappa=0} = \frac{1}{2}(1 + P_{pq,rs})\langle \mathrm{CSF}|[E_{pq}^{-}, [E_{rs}^{-}, \hat{H}]]|\mathrm{CSF}\rangle \tag{10.1.26}$$

Here $P_{pq,rs}$ permutes the pair indices pq and rs, making the expression for the Hessian manifestly symmetric. From the antisymmetry of E_{pq}^{-} in (10.1.25), we obtain

$$E^{(1)}_{pq} = -E^{(1)}_{qp} \tag{10.1.27}$$

The elements of the electronic gradient $E^{(1)}_{pq}$ may therefore also be considered the elements of an antisymmetric matrix. Similar relationships exist for the electronic Hessian

$$E^{(2)}_{pqrs} = -E^{(2)}_{qprs} = -E^{(2)}_{pqsr} = E^{(2)}_{qpsr} \tag{10.1.28}$$

in addition to the permutational symmetry between the pair indices pq and rs.

For real wave functions, we may simplify the expressions for the electronic gradient (10.1.25) and Hessian (10.1.26) somewhat:

$$E^{(1)}_{pq} = 2\langle \text{CSF}|[E_{pq}, \hat{H}]|\text{CSF}\rangle \tag{10.1.29}$$

$$E^{(2)}_{pqrs} = (1 + P_{pq,rs})\langle \text{CSF}|[E_{pq}, [E^-_{rs}, \hat{H}]]|\text{CSF}\rangle$$

$$= (1 + P_{pq,rs})\langle \text{CSF}|[E^-_{pq}, [E_{rs}, \hat{H}]]|\text{CSF}\rangle \tag{10.1.30}$$

The expression for the gradient (10.1.29) follows directly from (10.1.25) by expanding the commutator and using the identity

$$\langle \text{CSF}|[E_{qp}, \hat{O}]|\text{CSF}\rangle = -\langle \text{CSF}|[E_{pq}, \hat{O}]|\text{CSF}\rangle \tag{10.1.31}$$

which holds for real wave functions and Hermitian operators $\hat{O}$. For the Hessian, the first expression in (10.1.30) follows in the same way since the operator $[E^-_{rs}, \hat{H}]$ is Hermitian. The second expression in (10.1.30) is obtained in the same manner, using the fact that the expectation value is real.

The expressions for the electronic gradient and Hessian derived in this subsection play an important role in Hartree–Fock theory and will be used repeatedly in the remainder of this chapter. Note carefully, however, that these expressions are valid only at the expansion point $\kappa = 0$. More general expressions, valid for any κ, are considered in Exercise 10.1 but are not needed here. Expressions for higher-order derivatives of the energy may also be derived by the techniques of this subsection but are not considered here since they are not required for the optimization and the characterization of the wave function.

10.2 The Hartree–Fock wave function

Having examined the parametrization of the Hartree–Fock model in Section 10.1, we now turn our attention to the Hartree–Fock wave function itself, obtained by applying the variation principle to the energy expression $E(\kappa)$ in (10.1.20). In the present section, the emphasis is on the structure and characterization of the Hartree–Fock state rather than on its optimization. We shall examine the Hartree–Fock variational conditions, the gradient and Hessian of the optimized wave function, redundant orbital rotations, the Brillouin theorem and size-extensivity. For optimization techniques, see Sections 10.6–10.9.

10.2.1 THE HARTREE–FOCK WAVE FUNCTION

The Hartree–Fock state $|\text{HF}\rangle$ is an electronic wave function of the form $|\text{CSF}(\kappa)\rangle$ for which the expectation value of the Hamiltonian is stationary with respect to unitary variations in the MOs:

$$\delta E(\kappa) = \delta\langle \text{CSF}(\kappa)|\hat{H}|\text{CSF}(\kappa)\rangle = 0 \tag{10.2.1}$$

The variations in the energy are described by means of the orbital-rotation parameters κ of (10.1.9). Once a solution κ^{HF} to the nonlinear equations (10.2.1) has been found, we may generate the corresponding wave function by a unitary transformation:

$$|\text{HF}\rangle = |\text{CSF}(\kappa^{\text{HF}})\rangle = \exp(-\hat{\kappa}^{\text{HF}})|\text{CSF}\rangle \qquad (10.2.2)$$

Although in principle it is possible to work with the optimized state in the representation of the original orbitals (10.2.2), it is much more convenient to express the Hartree–Fock state directly in terms of a set of MOs where $|\text{HF}\rangle$ corresponds to $\kappa = \mathbf{0}$. This is easily achieved by transforming the elementary operators (the MOs) to the Hartree–Fock basis

$$^{\text{HF}}a_P^\dagger = \exp(-\hat{\kappa}^{\text{HF}})a_P^\dagger \exp(\hat{\kappa}^{\text{HF}}) \qquad (10.2.3)$$

In the following, we shall always assume that the optimized Hartree–Fock state is expressed in the transformed basis (10.2.3).

It should be understood that there are many solutions to equations (10.2.1) – for two reasons. First, as for any nonlinear set of equations, there are many independent solutions to the variational conditions, each of which represents a different stationary point (local minimum or saddle point) of the Hartree–Fock energy function $E(\kappa)$. Second, because of the presence of the redundant rotations, each stationary point of $E(\kappa)$ may be represented in infinitely many ways, related to one another by means of unitary transformations, as will be discussed in Section 10.2.2. In the following, we shall (unless otherwise stated) assume that the solution to (10.2.1) corresponds to the global minimum of $E(\kappa)$, as appropriate for an approximation to the electronic ground state.

Let us consider the electronic gradient and Hessian at a stationary point in some detail. In Section 10.1.3, we found that, at $\kappa = \mathbf{0}$, the Hartree–Fock electronic gradient may be written as the expectation value of the commutators of the excitation operators with the Hamiltonian (10.1.29). The variational conditions on the Hartree–Fock energy (10.2.1) may therefore be written in the simple manner

$$E_{pq}^{(1)} = 2\langle\text{HF}|[E_{pq}, \hat{H}]|\text{HF}\rangle = 0 \qquad (10.2.4)$$

recalling, however, that this expression holds only in the Hartree–Fock orbital basis where $\kappa^{\text{HF}} = \mathbf{0}$.

To characterize the stationary point and to distinguish local minima from saddle points, we must consider the second-order variation of the energy $E(\kappa)$. For ground states, in particular, we must require the electronic Hessian $\mathbf{E}^{(2)}$ in (10.1.30) to be positive definite (with respect to the nonredundant orbital rotations). At a stationary point, the expression for this Hessian simplifies somewhat. Invoking the Jacobi identity (1.8.17), we obtain

$$\langle\text{HF}|[E_{rs}^-, [E_{pq}^-, \hat{H}]]|\text{HF}\rangle = \langle\text{HF}|[E_{pq}^-, [E_{rs}^-, \hat{H}]]|\text{HF}\rangle + \langle\text{HF}|[[E_{rs}^-, E_{pq}^-], \hat{H}]|\text{HF}\rangle \qquad (10.2.5)$$

Expanding the commutator in the last term using (2.3.35)

$$[E_{rs}^-, E_{pq}^-] = E_{rq}^-\delta_{ps} - E_{ps}^-\delta_{rq} - E_{rp}^-\delta_{qs} + E_{qs}^-\delta_{rp} \qquad (10.2.6)$$

we note that the last term in (10.2.5) vanishes

$$\langle\text{HF}|[[E_{rs}^-, E_{pq}^-], \hat{H}]|\text{HF}\rangle = 2\langle\text{HF}|[E_{rq}\delta_{ps} - E_{ps}\delta_{rq} - E_{rp}\delta_{qs} + E_{qs}\delta_{rp}, \hat{H}]|\text{HF}\rangle = 0 \qquad (10.2.7)$$

because of the variational conditions (10.2.4). Manipulations like those leading to (10.1.30) then show that the electronic Hessian (10.1.30) at a stationary point and in the Hartree–Fock orbital basis can be written as

$$E^{(2)}_{pqrs} = 2\langle \mathrm{HF}|[E_{pq}, [E^-_{rs}, \hat{H}]]|\mathrm{HF}\rangle = 2\langle \mathrm{HF}|[E^-_{pq}, [E_{rs}, \hat{H}]]|\mathrm{HF}\rangle \qquad (10.2.8)$$

or in the form generated by permuting the pair indices rs and pq. At nonstationary points, the full expression (10.1.30) containing the permutation operator must be used.

10.2.2   REDUNDANT PARAMETERS

In Section 10.1.2, the redundant orbital-rotation parameters κ_{pq} were defined by the following condition on the associated excitation operators:

$$E^-_{pq}|\mathrm{CSF}\rangle \equiv 0 \quad \text{(redundant rotations)} \qquad (10.2.9)$$

It is now easy to see that any optimized CSF is also stationary with respect to the redundant excitation operators. For real wave functions, the electronic gradient in (10.1.25) may be written in the form

$$E^{(1)}_{pq} = 2\langle \mathrm{CSF}|E_{pq}\hat{H} - \hat{H}E_{pq}|\mathrm{CSF}\rangle = -2\langle \mathrm{CSF}|\hat{H}E^-_{pq}|\mathrm{CSF}\rangle \qquad (10.2.10)$$

which is trivially equal to zero for any excitation operator that satisfies the condition (10.2.9):

$$E^{(1)}_{pq} \equiv 0 \quad \text{(redundant rotations)} \qquad (10.2.11)$$

Hence, there is no need to include the redundant orbital-rotation parameters in the optimization – any state that we may arrive at in the space of the nonredundant rotations will remain stationary in the full space of redundant and nonredundant rotations.

In any Newton-based optimization – which, as discussed in Section 10.8, implicitly or explicitly requires the inversion of the Hessian matrix – the inclusion of redundant parameters is not only unnecessary but also undesirable since, at stationary points, these parameters make the electronic Hessian singular. The singularity of the Hessian follows from (10.2.8), which shows that the rows and columns corresponding to redundant rotations vanish at stationary points. Away from the stationary points, however, the Hessian (10.1.30) is nonsingular since the gradient elements that couple the redundant and nonredundant operators in (10.2.5) do not vanish. Still, as the optimization approaches a stationary point, the smallest eigenvalues of the Hessian will tend to zero and may create convergence problems as the stationary point is approached. Therefore, for the optimization of a closed-shell state by a Newton-based method, we should consider only those rotations that mix occupied and virtual orbitals:

$$|\mathrm{CSF}(\boldsymbol{\kappa})\rangle = \exp\left(-\sum_{ai} \kappa_{ai} E^-_{ai}\right)|\mathrm{CSF}\rangle \qquad (10.2.12)$$

For open shells, we must, according to the discussion in Section 10.1.2, also include operators for inactive–active, active–virtual and possibly active–active rotations.

Having established the need to avoid the redundant rotations in Newton-based optimizations of the wave function, let us briefly consider rotations that are purely redundant. According to the discussion in Section 10.1.2, an antisymmetric matrix containing the redundant orbital-rotation parameters has the following block-diagonal structure

$$\boldsymbol{\kappa}^{\mathrm{RD}} = \begin{pmatrix} \boldsymbol{\kappa}_{IJ} & 0 & 0 \\ 0 & \boldsymbol{\kappa}_{VW} & 0 \\ 0 & 0 & \boldsymbol{\kappa}_{AB} \end{pmatrix} \qquad (10.2.13)$$

where the three blocks contain the inactive–inactive elements, the active–active elements and the virtual–virtual elements. Depending on the structure of the CSF, the active–active block may or may not be present in κ^{RD}. The electronic energy is clearly invariant with respect to the redundant orbital-rotation parameters in κ^{RD} since, according to (10.2.9),

$$E(\kappa^{RD}) = \langle \mathrm{CSF}| \exp(\hat{\kappa}^{RD})\hat{H} \exp(-\hat{\kappa}^{RD})|\mathrm{CSF}\rangle = E^{(0)} \tag{10.2.14}$$

The unitary matrix $\exp(-\kappa^{RD})$ transforms the inactive orbitals among themselves, the active orbitals among themselves, and the virtual orbitals among themselves. For the optimization of the wave function, it is convenient to set the redundant parameters equal to zero in each iteration since this choice of κ^{RD} simplifies the orbital-rotation operator $\hat{\kappa}$. In Exercise 10.2, it is shown that, for a closed-shell state, the choice of $\kappa^{RD} = \mathbf{0}$ leads to transformed MOs that (in the least-squares sense) are as similar as possible to the original MOs – that is, to the transformed MOs that have the largest possible overlap with the original ones.

Although we may keep the redundant parameters fixed (equal to zero) during the optimization of the Hartree–Fock state, we are also free to vary them so as to satisfy additional requirements on the solution – that is, requirements that do not follow from the variational conditions. In canonical Hartree–Fock theory (discussed in Section 10.3), the redundant rotations are used to generate a set of orbitals (the canonical orbitals) that diagonalize an effective one-electron Hamiltonian (the Fock operator). This use of the redundant parameters does not in any way affect the final electronic state but leads to a set of MOs with special properties.

10.2.3 THE BRILLOUIN THEOREM

Let us consider the Hartree–Fock variational conditions (10.2.4) in more detail. For a closed-shell electronic state $|\mathrm{cs}\rangle$, the only nonredundant operators are those that mix virtual and inactive orbitals. The nonredundant Hartree–Fock conditions (here divided by a factor of 2)

$$\langle \mathrm{cs}|[E_{ai}, \hat{H}]|\mathrm{cs}\rangle = 0 \tag{10.2.15}$$

may then be written in the form

$$\langle \mathrm{cs}|E_{ai}\hat{H} - \hat{H}E_{ai}|\mathrm{cs}\rangle = -\langle \mathrm{cs}|\hat{H}E_{ai}|\mathrm{cs}\rangle = 0 \tag{10.2.16}$$

The variational conditions therefore imply that the closed-shell Hartree–Fock state does not interact with singly excited states – that is, with any state generated by exciting a single electron from an occupied orbital to an unoccupied orbital:

$$\langle \mathrm{cs}|\hat{H}E_{ai}|\mathrm{cs}\rangle = \langle \mathrm{cs}|\hat{H}|i \to a\rangle = 0 \tag{10.2.17}$$

This result, which is known as *the Brillouin theorem*, expresses the Hartree–Fock conditions for closed-shell electronic states in terms of matrix elements of the Hamiltonian operator.

More generally, we find that the Hartree–Fock variational conditions (10.2.4) may be written in the form

$$\langle \mathrm{HF}|\hat{H}E_{pq}|\mathrm{HF}\rangle = \langle \mathrm{HF}|\hat{H}E_{qp}|\mathrm{HF}\rangle \tag{10.2.18}$$

or equivalently

$$\langle \mathrm{HF}|\hat{H}|q \to p\rangle = \langle \mathrm{HF}|\hat{H}|p \to q\rangle \tag{10.2.19}$$

The Hartree–Fock state is thus characterized by a perfect balance between excitations and deexcitations: for any pair of orbitals p and q, the interaction with the state generated by the excitation of a single electron from p to q is exactly matched by the interaction with the state generated by the opposite excitation. This result is known as *the generalized Brillouin theorem (GBT)* [1]. For closed-shell states, all interactions are trivially equal to zero (due to the structure of the Hartree–Fock state) except those with the singly excited states $|i \rightarrow a\rangle$ and (10.2.19) then reduces to the special condition (10.2.17). For all other states, we may write the GBT (10.2.19) in the more explicit form

$$\langle \mathrm{HF}|\hat{H}|i \rightarrow p\rangle = 0 \tag{10.2.20}$$

$$\langle \mathrm{HF}|\hat{H}|p \rightarrow a\rangle = 0 \tag{10.2.21}$$

$$\langle \mathrm{HF}|\hat{H}|v \rightarrow w\rangle = \langle \mathrm{HF}|\hat{H}|w \rightarrow v\rangle \tag{10.2.22}$$

Hence, according to the GBT, the Hartree–Fock state does not interact with any state generated by the excitation of an electron either from an inactive orbital (10.2.20) or into a virtual orbital (10.2.21). The Hartree–Fock state may interact, however, with states generated by the excitation of an electron between two active orbitals – but always in such a way that the interactions with the states generated by the two opposite excitations are identical (10.2.22).

10.2.4 SIZE-EXTENSIVITY

The Hartree–Fock model is size-extensive. To demonstrate size-extensivity, consider a system composed of two noninteracting closed-shell subsystems. The Hartree–Fock state optimized for system A

$$|\mathrm{HF_A}\rangle = \left(\prod_{i_A} a^\dagger_{i_A\alpha} a^\dagger_{i_A\beta} \right) |\mathrm{vac}\rangle \tag{10.2.23}$$

satisfies the variational conditions for this system

$$\langle \mathrm{HF_A}|[E_{a_A i_A}, \hat{H}_A]| \mathrm{HF_A}\rangle = 0 \tag{10.2.24}$$

and gives the total energy

$$E_A = \langle \mathrm{HF_A}|\hat{H}_A|\mathrm{HF_A}\rangle \tag{10.2.25}$$

and similarly for system B. The orbitals of the two systems A and B do not overlap since the systems do not interact. We first note that the product state

$$|\mathrm{HF_A HF_B}\rangle = \left(\prod_{i_A} a^\dagger_{i_A\alpha} a^\dagger_{i_A\beta} \right) \left(\prod_{i_B} a^\dagger_{i_B\alpha} a^\dagger_{i_B\beta} \right) |\mathrm{vac}\rangle \tag{10.2.26}$$

represents a solution for the supersystem since the stationary condition is fulfilled for this state:

$$\langle \mathrm{HF_B HF_A}|[E_{ai}, \hat{H}_A + \hat{H}_B]|\mathrm{HF_A HF_B}\rangle = 0 \tag{10.2.27}$$

To demonstrate stationarity, we first integrate out the dependence on system B in the term containing $\hat{H}_A$ (10.2.27). This term vanishes if a or i (or both) belong to B. It also vanishes if a and i both belong to A since the orbitals have been optimized for this particular system (10.2.24). For the term in (10.2.27) containing $\hat{H}_B$, we proceed in the same manner. We finally note that the energy of the product state (10.2.26) is equal to the sum of the energies of the separate systems E_A and E_B

$$E_{AB} = \langle HF_B HF_A | \hat{H}_A + \hat{H}_B | HF_A HF_B \rangle = E_A + E_B \tag{10.2.28}$$

The product state thus represents not only a solution for the supersystem (10.2.27) but also a size-extensive solution (10.2.28). Since this argument may be extended to open-shell systems, we conclude that the Hartree–Fock model is size-extensive. In Section 4.3, we demonstrated that exponentially generated wave functions are in general size-extensive. The size-extensivity of the Hartree–Fock model is therefore a direct consequence of the exponential ansatz (10.1.8).

For large systems, there is a noteworthy difference in the behaviour of the error in the Hartree–Fock energy and the error in the Hartree–Fock wave function. Consider m noninteracting subsystems. For monomer I, the exact wave function can be written as

$$|\psi_I^{ex}\rangle = \hat{\psi}_I^{ex} |vac\rangle = C_{HF} \hat{\psi}_I^{HF} |vac\rangle + C_{corr} \hat{\psi}_I^{corr} |vac\rangle \tag{10.2.29}$$

where the states $\hat{\psi}_I^{ex}|vac\rangle$, $\hat{\psi}_I^{HF}|vac\rangle$ and $\hat{\psi}_I^{corr}|vac\rangle$ all are normalized. The wave function for m monomers is

$$|\psi^{ex}(m)\rangle = \left(\prod_I \hat{\psi}_I^{ex}\right) |vac\rangle = C_{HF}^m \left(\prod_I \hat{\psi}_I^{HF}\right) |vac\rangle$$
$$+ C_{corr} C_{HF}^{m-1} \sum_J \hat{\psi}_1^{HF} \cdots \hat{\psi}_{J-1}^{HF} \hat{\psi}_J^{corr} \hat{\psi}_{J+1}^{HF} \cdots \hat{\psi}_m^{HF} |vac\rangle + \cdots \tag{10.2.30}$$

The weight of the Hartree–Fock determinant in the exact wave function is therefore C_{HF}^{2m}, which tends to zero as the number of monomers increases. Nevertheless, the description of each monomer does not deteriorate and the size-extensive Hartree–Fock wave function recovers the same proportion of the exact energy for all values of m. This paradoxical behaviour arises from a difference in the separability of the wave function and of the energy. The wave function is *multiplicatively separable*, depending simultaneously on all the electrons and the errors accumulate in a binomial fashion. The energy, by contrast, is *additively separable* and may be written as a sum of terms, each of which depends only on the wave function for a single monomer:

$$\langle \psi^{ex}(m) | \sum_I \hat{H}_I | \psi^{ex}(m) \rangle = \sum_I \langle \psi_I^{ex} | \hat{H}_I | \psi_I^{ex} \rangle \prod_{J \neq I} \langle \psi_J^{ex} | \psi_J^{ex} \rangle = \sum_I \langle \psi_I^{ex} | \hat{H}_I | \psi_I^{ex} \rangle \tag{10.2.31}$$

Since the Hamiltonian is a sum of local operators, no terms in the energy expression contain products of correlation corrections for several monomers, so the error does not accumulate in the same manner as for the wave function (10.2.30). For example, even though the weight of the Hartree–Fock wave function in the exact wave function of a crystal is close to zero, the description of each monomer may be quite accurate and provide good energies and other properties.

10.3 Canonical Hartree–Fock theory

A closed-shell Hartree–Fock state is represented by a variationally optimized Slater determinant. Such a wave function represents a state where each electron behaves as an independent particle (subject to Fermi correlation as discussed in Section 5.2.8). We should therefore be able to determine this state by solving an *effective one-electron Schrödinger equation*, generating one independent-particle solution (spin orbital) for each electron in the system. The final N-electron state is then obtained as an antisymmetrized product of N independent eigenfunctions of the effective Hamiltonian of the one-electron Schrödinger equation. This is the approach taken in *canonical Hartree–Fock theory*.

As discussed in the present section and also in Sections 10.5 and 10.6, canonical Hartree–Fock theory offers many advantages. However, for open-shell RHF states, this approach is not always possible because of the constraints imposed by the spin- and space-symmetry adaptation of the wave function. In such cases, the wave function must be calculated using the more general methods of Section 10.8; in the present section, we restrict our attention to closed-shell systems. In addition, the application of canonical Hartree–Fock theory is restricted to small and moderately large systems since it requires an amount of work that scales at least cubically with the size of the system; for large systems, the methods of Section 10.7 must be used instead.

10.3.1 THE FOCK OPERATOR

The effective Hamiltonian of the one-electron Schrödinger equation, *the Fock operator*, must (like the exact Hamiltonian $\hat{H}$) be Hermitian and totally symmetric in spin space. It therefore has the form

$$\hat{f} = \sum_{pq} f_{pq} E_{pq} \tag{10.3.1}$$

where the *Fock matrix* $\mathbf{f}$ constitutes a symmetric matrix with elements f_{pq}. In the basis where the Fock matrix is diagonal

$$f_{pq} = \delta_{pq}\varepsilon_p \tag{10.3.2}$$

the one-electron eigenfunctions of the Fock operator

$$\hat{f} a^\dagger_{p\sigma}|\text{vac}\rangle = \varepsilon_p a^\dagger_{p\sigma}|\text{vac}\rangle \tag{10.3.3}$$

are called the *canonical spin orbitals* and the associated eigenvalues are the *orbital energies*. In the absence of higher symmetries, the energy levels of the Fock operator (10.3.1) are doubly degenerate, with one eigenfunction of alpha spin and one of beta spin at each level. Upon solution of the *Hartree–Fock equations* (10.3.3), the N-electron closed-shell Hartree–Fock wave function is constructed as a product of the canonical spin orbitals

$$|\text{cs}\rangle = \left(\prod_i a^\dagger_{i\alpha} a^\dagger_{i\beta}\right)|\text{vac}\rangle \tag{10.3.4}$$

and the total energy is obtained as the expectation value of the Hamiltonian

$$E^{(0)} = \langle \text{cs}|\hat{H}|\text{cs}\rangle \tag{10.3.5}$$

The purpose of the present section is to construct the Fock operator $\hat{f}$ in such a way that the product generated from its eigenvectors (10.3.4) represents a variationally optimized N-electron state. We shall also require that $\hat{f}$ becomes the exact Hamiltonian in the limit where the electrons do not interact. It should be noted that the above requirements do not specify the Fock operator uniquely; the additional requirements needed for a unique specification of this operator will be given in the course of the following discussion.

10.3.2 IDENTIFICATION OF THE ELEMENTS OF THE FOCK OPERATOR

Let us construct the Fock operator (10.3.1) in accordance with the requirements of Section 10.3.1. We recall that, for a closed-shell electronic system, the necessary and sufficient conditions for the optimized state may be written in the form

$$2\langle cs|[E_{ai}, \hat{H}]|cs\rangle = 0 \qquad (10.3.6)$$

The effective operator should therefore be constructed in such a way that a diagonal Fock matrix (10.3.2) ensures the fulfilment of these conditions. Consequently, we require the following identity to hold for all (optimized as well as nonoptimized) choices of orbitals

$$f_{ai} = f_{ia} = N_f \langle cs|[E_{ai}, \hat{H}]|cs\rangle \qquad (10.3.7)$$

where the normalization constant N_f will be fixed later. However, although (10.3.7) is a necessary condition on the matrix elements of the Fock operator, we cannot *in general* identify f_{pq} with $N_f \langle cs|[E_{pq}, \hat{H}]|cs\rangle$ since this would lead to an antisymmetric Fock matrix with vanishing occupied–occupied and virtual–virtual elements, in disagreement with our requirement that the Fock matrix, in the absence of two-electron interactions, should reproduce the true Hamiltonian. We shall therefore rearrange the operator in (10.3.7) in such a way that it can be generalized to all elements f_{pq}:

$$\langle cs|[E_{ai}, \hat{H}]|cs\rangle = \sum_\sigma \langle cs|[a_{a\sigma}^\dagger a_{i\sigma}, \hat{H}]|cs\rangle = \sum_\sigma \langle cs|[a_{a\sigma}^\dagger, \hat{H}]a_{i\sigma}|cs\rangle$$

$$= \sum_\sigma \langle cs|[[a_{a\sigma}^\dagger, \hat{H}], a_{i\sigma}]_+|cs\rangle = -\sum_\sigma \langle cs|[a_{i\sigma}^\dagger, [a_{a\sigma}, \hat{H}]]_+|cs\rangle \qquad (10.3.8)$$

Note that the second and third steps hold for closed-shell states only and that the last step holds for real wave functions only. From the rules for rank reduction in Section 1.8, we also note that the anticommutator $[a_{p\sigma}^\dagger, [a_{q\sigma}, \hat{H}]]_+$ in (10.3.8) is a one-electron operator.

We shall now show that the following identification of the Fock-matrix elements (with the normalization factor N_f equal to $-1/2$)

$$f_{pq} = \tfrac{1}{2} \sum_\sigma \langle cs|[a_{q\sigma}^\dagger, [a_{p\sigma}, \hat{H}]]_+|cs\rangle \qquad (10.3.9)$$

leads to an effective operator $\hat{f}$ in accordance with our stated requirements. First, we note that the virtual–occupied Fock-matrix elements f_{ai} are by construction proportional to the gradient elements $2\langle cs|[E_{ai}, \hat{H}]|cs\rangle$. Next, the permutational symmetry of the matrix elements

$$f_{pq} = \tfrac{1}{2} \sum_\sigma \langle cs|[a_{q\sigma}^\dagger, [a_{p\sigma}, \hat{H}]]_+|cs\rangle = \tfrac{1}{2} \sum_\sigma \langle cs|[a_{p\sigma}^\dagger, [a_{q\sigma}, \hat{H}]]_+|cs\rangle = f_{qp} \qquad (10.3.10)$$

(which, for real wave functions, follows upon expansion of the (anti)commutators and use of the elementary anticommutation relations) shows that $f_{ia} = f_{ai}$. Finally, we consider our requirement that $\hat{f}$ should reproduce the true Hamiltonian in the absence of two-electron interactions. Inserting the one-electron Hamiltonian

$$\hat{h} = \sum_{rs} h_{rs} E_{rs} \qquad (10.3.11)$$

into the Fock operator (10.3.1), we obtain

$$\hat{f}_1 = \frac{1}{2} \sum_{pqrs} h_{rs} \sum_{\sigma} \langle cs|[a_{q\sigma}^{\dagger}, [a_{p\sigma}, E_{rs}]]_+|cs\rangle E_{pq}$$

$$= \frac{1}{2} \sum_{pqrs} h_{rs} \sum_{\sigma} \langle cs|[a_{q\sigma}^{\dagger}, \delta_{pr} a_{s\sigma}]_+|cs\rangle E_{pq}$$

$$= \frac{1}{2} \sum_{pqrs} h_{rs} \sum_{\sigma} \delta_{pr} \delta_{qs} E_{pq} = \sum_{pq} h_{pq} E_{pq} \tag{10.3.12}$$

where we have used the commutator (2.3.37), thereby demonstrating that the identification (10.3.9) leads to a Fock operator that reproduces the true Hamiltonian in the absence of two-electron interactions.

The identification (10.3.9) fixes all the elements of the Fock matrix by a single, simple expression. Nevertheless, it should be emphasized that there is an element of arbitrariness in the construction of the Fock matrix and that our stated requirements on the effective Hamiltonian do *not* fix the Fock matrix uniquely. We may, in particular, add any combination of two-electron terms to the occupied–occupied and virtual–virtual blocks of the Fock matrix without affecting the fulfilment of our requirements. It is therefore important to examine the occupied–occupied and virtual–virtual elements of this matrix in more detail to see if some additional justification for the identification (10.3.9) can be obtained from these elements. Expanding the (anti)commutators, we obtain for the diagonal elements

$$f_{ii} = \langle cs|\hat{H}|cs\rangle - \frac{1}{2} \sum_{\sigma} \langle cs|a_{i\sigma}^{\dagger} \hat{H} a_{i\sigma}|cs\rangle \tag{10.3.13}$$

$$f_{aa} = \frac{1}{2} \sum_{\sigma} \langle cs|a_{a\sigma} \hat{H} a_{a\sigma}^{\dagger}|cs\rangle - \langle cs|\hat{H}|cs\rangle \tag{10.3.14}$$

where we have used the fact that $\langle cs|a_{i\sigma}^{\dagger} a_{i\sigma}$ is equal to $\langle cs|$ since ϕ_i is occupied in $\langle cs|$, and that $a_{a\sigma} a_{a\sigma}^{\dagger}|cs\rangle$ is equal to $|cs\rangle$ since ϕ_a is unoccupied in $|cs\rangle$). With the Fock-matrix elements given by (10.3.9), we may thus identify the diagonal elements of the Fock operator (and in particular its eigenvalues) with the energies associated with the addition and removal of electrons in the closed-shell system – that is, with the effective energies of the electrons in the system. This particular point is discussed in detail in Section 10.5.

Based on the considerations in this subsection, we shall take the following expression as our definition of the Fock operator of a closed-shell system:

$$\hat{f} = \frac{1}{2} \sum_{pq} \sum_{\sigma} \langle cs|[a_{q\sigma}^{\dagger}, [a_{p\sigma}, \hat{H}]]_+|cs\rangle E_{pq} \tag{10.3.15}$$

When constructed from an arbitrary set of orbitals, the Fock matrix is nondiagonal. In the special *canonical representation*, however, the orbitals satisfy the *canonical conditions*

$$f_{pq} = \frac{1}{2} \sum_{\sigma} \langle cs|[a_{q\sigma}^{\dagger}, [a_{p\sigma}, \hat{H}]]_+|cs\rangle = \varepsilon_p \delta_{pq} \tag{10.3.16}$$

and the Fock matrix is diagonal. It should be realized that the canonical conditions (10.3.16) constitute a nonlinear set of equations that cannot be satisfied by a simple diagonalization of the Fock matrix: the canonical orbitals are not simply the eigenvectors of the Fock matrix – they are at the same time the orbitals from which the Fock matrix itself is constructed. The one-electron *Hartree–Fock equations* (10.3.3) are therefore not true eigenvalue equations but are more properly described as pseudo-eigenvalue equations, to be solved in an iterative process where the Fock matrix is repeatedly constructed and diagonalized until a set of orbitals that satisfy the canonical conditions (10.3.16) is obtained – as discussed later in this section and, in particular, in Section 10.6.

10.3.3 THE FOCK MATRIX

In the present subsection, we shall evaluate the elements of the Fock matrix

$$f_{pq} = \tfrac{1}{2} \sum_{\sigma} \langle cs|[a_{q\sigma}^\dagger, [a_{p\sigma}, \hat{H}]]_+|cs\rangle \tag{10.3.17}$$

in terms of molecular integrals and density-matrix elements. We recall that, according to the rules for rank reduction in Section 1.8, $[a_{q\sigma}^\dagger, [a_{p\sigma}, \hat{H}]]_+$ is a one-particle operator. To arrive at an explicit expression for this operator, we consider separately the one- and two-electron parts of the Hamiltonian (2.2.18):

$$\hat{H} = \hat{h} + \hat{g} + h_{\text{nuc}} = \sum_{pq} h_{pq} E_{pq} + \tfrac{1}{2} \sum_{\substack{pqrs \\ \sigma\tau}} g_{pqrs} a_{p\sigma}^\dagger a_{r\tau}^\dagger a_{s\tau} a_{q\sigma} + h_{\text{nuc}} \tag{10.3.18}$$

The constant nuclear term makes no contribution to the Fock matrix. The contribution from the one-electron part follows in the same way as in (10.3.12) and gives

$$\sum_{\sigma}[a_{n\sigma}^\dagger, [a_{m\sigma}, \hat{h}]]_+ = 2h_{mn} \tag{10.3.19}$$

For the two-electron part in (10.3.18), we first evaluate the inner commutator using (1.8.7)

$$
\begin{aligned}
[a_{m\sigma}, \hat{g}] &= \tfrac{1}{2} \sum_{\substack{pqrs \\ \tau\upsilon}} g_{pqrs}[a_{m\sigma}, a_{p\upsilon}^\dagger a_{r\tau}^\dagger a_{s\tau} a_{q\upsilon}] \\
&= \tfrac{1}{2} \sum_{\substack{pqrs \\ \tau\upsilon}} g_{pqrs}(\delta_{mp}\delta_{\sigma\upsilon} a_{r\tau}^\dagger - \delta_{mr}\delta_{\sigma\tau} a_{p\upsilon}^\dagger) a_{s\tau} a_{q\upsilon} \\
&= \sum_{\substack{qrs \\ \tau}} g_{mqrs} a_{r\tau}^\dagger a_{s\tau} a_{q\sigma} \tag{10.3.20}
\end{aligned}
$$

The rank of the two-electron operator has now been reduced to 3/2. Anticommuting this operator with the creation operator $a_{n\sigma}^\dagger$, we finally reduce the rank to 1

$$
\begin{aligned}
\sum_{\sigma}[a_{n\sigma}^\dagger, [a_{m\sigma}, \hat{g}]]_+ &= \sum_{qrs} g_{mqrs} \sum_{\sigma\tau}[a_{n\sigma}^\dagger, a_{r\tau}^\dagger a_{s\tau} a_{q\sigma}]_+ \\
&= \sum_{qrs} g_{mqrs} \sum_{\sigma\tau}(\delta_{nq} a_{r\tau}^\dagger a_{s\tau} - \delta_{ns}\delta_{\sigma\tau} a_{r\tau}^\dagger a_{q\sigma}) \\
&= \sum_{rs}(2g_{mnrs} - g_{msrn})E_{rs} \tag{10.3.21}
\end{aligned}
$$

where we have used (1.8.9). The one-electron operator that appears in the expression for the Fock matrix in (10.3.17) therefore becomes

$$\sum_{\sigma}[a_{q\sigma}^{\dagger}, [a_{p\sigma}, \hat{H}]]_{+} = 2h_{pq} + \sum_{rs}(2g_{pqrs} - g_{psrq})E_{rs} \tag{10.3.22}$$

The evaluation of the elements of the Fock matrix is now trivial since, for a closed-shell state, the one-electron density-matrix elements are given by

$$D_{pq} = \langle \text{cs}|E_{pq}|\text{cs}\rangle = \begin{cases} 2\delta_{pq} & p \text{ occupied} \\ 0 & p \text{ virtual} \end{cases} \tag{10.3.23}$$

Inserting the operator (10.3.22) into (10.3.17) and using this expression for the density-matrix elements, we obtain

$$f_{pq} = h_{pq} + \sum_{i}(2g_{pqii} - g_{piiq}) \tag{10.3.24}$$

as our final expression for the Fock matrix of a closed-shell system.

10.3.4 THE SELF-CONSISTENT FIELD METHOD

The Fock matrix is by design an effective one-electron Hamiltonian whose diagonalization yields a set of MOs, the canonical orbitals, from which a variationally optimized Hartree–Fock wave function may be constructed. However, since the Fock matrix (10.3.24) contains contributions from each occupied canonical orbital, we cannot solve the Hartree–Fock pseudo-eigenvalue equations (10.3.3) in a single step but must instead resort to some iterative scheme.

From an initial set of orthonormal MOs ϕ_i, we calculate the Fock matrix according to (10.3.24). Diagonalization of the Fock matrix in this basis yields

$$\mathbf{fU} = \mathbf{U}\boldsymbol{\varepsilon} \tag{10.3.25}$$

where $\boldsymbol{\varepsilon}$ is a diagonal matrix containing the eigenvalues, and the eigenvectors $\mathbf{U}$ constitute an orthogonal matrix. From $\mathbf{U}$ we obtain a transformed set of MOs (arranged as row vectors)

$$\tilde{\boldsymbol{\phi}} = \boldsymbol{\phi}\mathbf{U} \tag{10.3.26}$$

The Fock matrix is now recalculated from the transformed MOs and an iterative process is thus established, in which the Fock matrix is repeatedly constructed and diagonalized until the MOs from which it is constructed are the same as those generated by its diagonalization. At this stage, we have satisfied the canonical conditions (10.3.16) and the solution is said to be *self-consistent*. In the same parlance, the field generated by the converged Fock potential is said to be self-consistent and the Hartree–Fock function itself is called a *self-consistent field (SCF)* wave function.

From a more general point of view, the SCF iterations are a particular example of fixed-point iterations [2], frequently encountered in numerical analysis, in which we determine a *fixed point* x of a function $f(x)$

$$x = f(x) \tag{10.3.27}$$

from the sequence of iterations of the type

$$x_{n+1} = f(x_n) \tag{10.3.28}$$

Of course, there is no guarantee that the simple-minded fixed-point iterations outlined here will converge at all or that the convergence will be fast. The important questions of convergence and convergence rates of the SCF iterations are discussed in Sections 10.6 and 10.9. At present, we proceed on the assumption that a self-consistent set of solutions has been obtained to the Hartree–Fock equations.

10.3.5 THE VARIATIONAL AND CANONICAL CONDITIONS COMPARED

In calculating the canonical Hartree–Fock orbitals, we carry out a full diagonalization of the Fock matrix. Complete diagonalization is not necessary, however, to ensure that the elements of the electronic gradient vanish. Indeed, any set of orbitals that brings the Fock matrix into block-diagonal form with vanishing occupied–virtual elements

$$f_{ai} = f_{ia} = 0 \tag{10.3.29}$$

constitutes a bona fide set of MOs in terms of which the Hartree–Fock wave function may be constructed. In a sense, the only thing that matters for the construction of the Hartree–Fock wave function is the correct separation of the orbital space into an occupied subspace and a virtual subspace. The detailed form of the occupied orbitals does not matter; we may rotate the occupied orbitals among themselves in any way we like without affecting the total wave function. Indeed, this fact was established in Section 10.1.2 with the identification of the occupied–occupied rotations as redundant for closed-shell electronic systems.

At this point, it is instructive to compare the two sets of conditions we have set up for the closed-shell wave function – the variational conditions and the canonical conditions. As discussed in Section 3.2, we may, from a given set of spin orbitals, generate any other set of spin orbitals according to the expressions

$$\tilde{a}_{p\sigma}^{\dagger} = \exp(-\hat{\kappa}) a_{p\sigma}^{\dagger} \exp(\hat{\kappa}) \tag{10.3.30}$$

Substituting these operators and their adjoints into the variational conditions (10.2.15) and into the diagonal conditions (10.3.16) and noting that the Hamiltonian is kept fixed (constructed from the original operators), we obtain the following expressions for the variational conditions

$$\langle cs|[E_{ai}, \exp(\hat{\kappa}^{var})\hat{H}\exp(-\hat{\kappa}^{var})]|cs\rangle = 0 \tag{10.3.31}$$

$$\hat{\kappa}^{var} = \sum_{ai} \kappa_{ai} E_{ai}^{-} \tag{10.3.32}$$

and for the canonical conditions

$$\frac{1}{2}\sum_{\sigma}\langle cs|[a_{q\sigma}^{\dagger}, [a_{p\sigma}, \exp(\hat{\kappa}^{can})\hat{H}\exp(-\hat{\kappa}^{can})]]_+|cs\rangle = \delta_{pq}\varepsilon_p \tag{10.3.33}$$

$$\hat{\kappa}^{can} = \sum_{p>q} \kappa_{pq} E_{pq}^{-} \tag{10.3.34}$$

In either case, the number of orbital-rotation parameters is equal to the number of conditions. In setting up the variational conditions (10.3.31), the invariances of the wave function have been exploited to minimize the number of orbital-rotation parameters and there are only OV parameters present in $\hat{\kappa}^{var}$ (where O and V are the numbers of occupied and virtual orbitals, respectively). In setting up the canonical conditions, on the other hand, the invariances of the wave function are exploited to make the orbitals canonical and the full set of $n(n-1)/2$ orbital parameters is present in $\hat{\kappa}^{\text{can}}$ (where n is the total number of orbitals). Both sets of conditions (10.3.31) and (10.3.33) constitute a nonlinear set of equations, which must be solved iteratively. We note that, for large systems, it becomes important to use the variational conditions rather than the more involved canonical conditions for the efficient optimization of the wave function.

10.4 The RHF total energy and orbital energies

In the preceding section, we found that the Hartree–Fock wave function of a closed-shell electronic system may be calculated as an antisymmetrized product of canonical spin orbitals, each of which is an eigenfunction of the Fock operator with the orbital energy as the eigenvalue. In the present section, we review the results of the preceding section, comparing the effective one-electron Fock operator with the true Hamiltonian and the orbital energies with the total energy of the electronic system.

10.4.1 THE HAMILTONIAN AND THE FOCK OPERATOR

The Hamiltonian operator may be written in the form

$$\hat{H} = \hat{h} + \hat{g} + h_{\text{nuc}} = \sum_{pq} h_{pq} E_{pq} + \tfrac{1}{2} \sum_{pqrs} g_{pqrs} e_{pqrs} + h_{\text{nuc}} \tag{10.4.1}$$

and contains the usual one-electron and two-electron contributions in addition to a constant term that represents the nuclear–nuclear repulsions. The Hamiltonian contains the full set of electronic interactions in a given basis and is independent of the electronic state studied. The Fock operator, by contrast, is an effective one-electron operator, designed with a particular electronic state in mind:

$$\hat{f} = \tfrac{1}{2} \sum_{pq} \sum_{\sigma} \langle \text{HF} | [a_{q\sigma}^{\dagger}, [a_{p\sigma}, \hat{H}]]_{+} | \text{HF} \rangle E_{pq} \tag{10.4.2}$$

The Fock operator may be written in the form

$$\hat{f} = \hat{h} + \hat{V} \tag{10.4.3}$$

where the two-electron part of the true Hamiltonian $\hat{g}$ is replaced by an effective one-electron potential, known as the *Fock potential*:

$$\hat{V} = \sum_{pq} V_{pq} E_{pq} = \sum_{pq} \sum_{i} (2g_{pqii} - g_{piiq}) E_{pq} \tag{10.4.4}$$

There are two different contributions to the Fock potential, representing distinct physical interactions among the electrons, as will be discussed shortly.

10.4.2 THE CANONICAL REPRESENTATION AND ORBITAL ENERGIES

The Fock operator is in general nondiagonal. However, in the special canonical representation of the optimized Hartree–Fock wave function, it becomes diagonal:

$$\hat{f} = \sum_{p} \varepsilon_p E_{pp} \tag{10.4.5}$$

The canonical spin orbitals are thus the eigenfunctions of the Fock operator

$$\hat{f} a_{p\sigma}^\dagger |\text{vac}\rangle = \varepsilon_p a_{p\sigma}^\dagger |\text{vac}\rangle \tag{10.4.6}$$

with the orbital energies given by the expression

$$\varepsilon_p = h_{pp} + \sum_i (2g_{ppii} - g_{piip}) \tag{10.4.7}$$

The first term represents the kinetic energy of the electron and its attractive interaction with the stationary nuclei. The second term represents the repulsive interaction with the remaining electrons in the system. In a more explicit representation, we may write the orbital energies in the form

$$\varepsilon_p = \int \phi_p(\mathbf{r}_1) \left(-\frac{1}{2}\nabla^2 - \sum_K \frac{Z_K}{r_{1K}} \right) \phi_p(\mathbf{r}_1)\, d\mathbf{r}_1 + 2\sum_i \int\int \frac{\phi_p(\mathbf{r}_1)\phi_p(\mathbf{r}_1)\phi_i(\mathbf{r}_2)\phi_i(\mathbf{r}_2)}{r_{12}}\, d\mathbf{r}_1\, d\mathbf{r}_2$$
$$- \sum_i \int\int \frac{\phi_p(\mathbf{r}_1)\phi_i(\mathbf{r}_1)\phi_i(\mathbf{r}_2)\phi_p(\mathbf{r}_2)}{r_{12}}\, d\mathbf{r}_1\, d\mathbf{r}_2 \tag{10.4.8}$$

assuming real orbitals. An electron in ϕ_p thus experiences a classical electrostatic *Coulomb potential*, generated by the nuclear framework and by the charge distribution of the remaining electrons, as well as a nonclassical *exchange potential*, which corrects the classical repulsion for Fermi correlation. Note that the self-interaction term that occurs in the Coulomb part of (10.4.8) is cancelled by a similar exchange contribution.

For the occupied and virtual orbitals, the expression for the orbital energies (10.4.7) may be written as

$$\varepsilon_i = h_{ii} + \sum_{j\neq i}(2g_{iijj} - g_{ijji}) + g_{iiii} \tag{10.4.9}$$

$$\varepsilon_a = h_{aa} + \sum_j (2g_{aajj} - g_{ajja}) \tag{10.4.10}$$

The orbital energy of an *occupied* MO thus represents the interaction between an electron in orbital ϕ_i and the remaining $N - 1$ electrons in the system. The orbital energy of a *virtual* MO, on the other hand, represents the interaction of an electron in orbital ϕ_a with the electrons in the neutral N-electron system. We shall in Section 10.5 see that, in accordance with these observations, we may interpret $-\varepsilon_i$ as an ionization potential and $-\varepsilon_a$ as an electron affinity.

The special diagonal property of the canonical orbitals is exploited in the optimization of the Hartree–Fock state. The canonical orbitals of closed-shell states are obtained in an iterative process, where the Fock matrix in each iteration is rebuilt and diagonalized until self-consistency is achieved as described in Section 10.3.4 and in more detail in Section 10.6. The final electronic state may be obtained as an antisymmetrized product of the canonical spin orbitals

$$|\text{HF}\rangle = \left(\prod_i a_{i\alpha}^\dagger a_{i\beta}^\dagger \right) |\text{vac}\rangle \tag{10.4.11}$$

Using the commutator relation

$$[\hat{f}, a_{p\sigma}^\dagger] = \varepsilon_p a_{p\sigma}^\dagger \tag{10.4.12}$$

we find that the Hartree–Fock state is an eigenfunction of the Fock operator with an eigenvalue equal to the sum of the orbital energies of the occupied canonical spin orbitals

$$\hat{f}|\text{HF}\rangle = 2\sum_i \varepsilon_i|\text{HF}\rangle \qquad (10.4.13)$$

in accordance with our interpretation of the Hartree–Fock wave function as an independent-particle model.

10.4.3 THE HARTREE–FOCK ENERGY

The Hartree–Fock energy is the expectation value of the true Hamiltonian operator (10.4.1)

$$E^{(0)} = \langle\text{HF}|\hat{H}|\text{HF}\rangle \qquad (10.4.14)$$

and may be calculated in the usual way

$$E^{(0)} = \sum_{pq} D_{pq}h_{pq} + \frac{1}{2}\sum_{pqrs} d_{pqrs}g_{pqrs} + h_{\text{nuc}} \qquad (10.4.15)$$

from the one- and two-electron density-matrix elements

$$D_{pq} = \langle\text{HF}|E_{pq}|\text{HF}\rangle \qquad (10.4.16)$$

$$d_{pqrs} = \langle\text{HF}|e_{pqrs}|\text{HF}\rangle \qquad (10.4.17)$$

For a closed-shell electronic Hartree–Fock state, the only nonzero density-matrix elements are those with all indices inactive

$$D_{ij} = 2\delta_{ij} \qquad (10.4.18)$$

$$d_{ijkl} = D_{ij}D_{kl} - \frac{1}{2}D_{il}D_{kj} = 4\delta_{ij}\delta_{kl} - 2\delta_{il}\delta_{kj} \qquad (10.4.19)$$

Using these elements, the *total Hartree–Fock energy* becomes

$$E^{(0)} = 2\sum_i h_{ii} + \sum_{ij}(2g_{iijj} - g_{ijji}) + h_{\text{nuc}} \qquad (10.4.20)$$

The first term represents the kinetic energy of the electrons and their attractive interactions with the stationary nuclei. The second term represents the averaged repulsive interactions of the electrons in the system. Like the orbital energies (10.4.7), this term contains Coulomb as well as exchange contributions, but we note the different relative weights of the one-electron and two-electron contributions in (10.4.7) and (10.4.20).

10.4.4 HUND'S RULE FOR SINGLET AND TRIPLET STATES

The total energies of the singlet and triplet states where one electron of a closed-shell system has been promoted from the occupied orbital ϕ_i to the virtual orbital ϕ_a may, according to Exercise 10.3, be written as

$$^1E_{ai} = \langle\text{HF}|\hat{S}_{ia}^{0,0}\hat{H}\hat{S}_{ai}^{0,0}|\text{HF}\rangle = \langle\text{HF}|\hat{H}|\text{HF}\rangle + \varepsilon_a - \varepsilon_i + 2g_{aiia} - g_{aaii} \qquad (10.4.21)$$

$$^3E_{ai} = \langle\text{HF}|\hat{T}_{ia}^{0,0}\hat{H}\hat{T}_{ai}^{0,0}|\text{HF}\rangle = \langle\text{HF}|\hat{H}|\text{HF}\rangle + \varepsilon_a - \varepsilon_i - g_{aaii} \qquad (10.4.22)$$

where the excitation operators are given in (2.3.21) and (2.3.23). The difference between the energies of the singlet and triplet states then becomes

$$^{1}E_{ai} - {}^{3}E_{ai} = 2g_{aiia} > 0 \qquad (10.4.23)$$

since, according to (9.12.24), the two-electron integrals constitute a positive definite matrix. For closed-shell Hartree–Fock states, therefore, the energy of a singlet excited state is higher than the energy of the corresponding triplet state – assuming that the singlet can be properly represented by a single electronic configuration. Inequality (10.4.23) is a particular case of *Hund's rule*, which states that, for a given electronic configuration, the state of highest spin multiplicity is lowest in energy.

10.4.5 THE FLUCTUATION POTENTIAL

It is important to realize that, although the Hartree–Fock wave function is an eigenfunction of the Fock operator with an eigenvalue equal to the sum of the orbital energies, this eigenvalue is not the same as the Hartree–Fock energy, which is the expectation value of the true Hamiltonian. Let us introduce the *fluctuation potential* as the difference between the two-electron part of the true Hamiltonian and the Fock potential

$$\hat{\Phi} = \hat{g} - \hat{V} \qquad (10.4.24)$$

The true Hamiltonian may now be written as the sum of the Fock operator, the fluctuation potential and the nuclear–nuclear repulsion operator

$$\hat{H} = \hat{f} + \hat{\Phi} + h_{\text{nuc}} \qquad (10.4.25)$$

With this decomposition of the Hamiltonian, we obtain the following expression for the Hartree–Fock energy

$$E^{(0)} = 2\sum_{i} \varepsilon_{i} + \langle \text{HF}|\hat{\Phi}|\text{HF}\rangle + h_{\text{nuc}} \qquad (10.4.26)$$

The Hartree–Fock energy may be viewed as the sum of the orbital energies corrected to first order in perturbation theory with the fluctuation potential as the perturbation operator. Going to higher orders in perturbation theory, we obtain corrections to the Hartree–Fock energy which take into account the effects of electron correlation; see Section 5.8 and, for a full discussion, Chapter 14.

In qualitative MO theory, the total energy of the system is often identified with the sum of the orbital energies (10.4.7)

$$2\sum_{i} \varepsilon_{i} = 2\sum_{i} h_{ii} + 2\sum_{ij}(2g_{iijj} - g_{ijji}) = \langle \text{HF}|\hat{h} + 2\hat{g}|\text{HF}\rangle \qquad (10.4.27)$$

Compared with the true Hartree–Fock energy (10.4.20), the electronic repulsions are counted twice whereas the nuclear repulsions are neglected. Although this may be a useful approximation, it should be understood that the orbital energies measure the binding of the individual electrons to the system and do not constitute a partitioning of the total energy among the occupied orbitals.

From (10.4.26) and (10.4.27), we obtain the following simple expression for the expectation value of the fluctuation potential:

$$\langle \text{HF}|\hat{\Phi}|\text{HF}\rangle = -\langle \text{HF}|\hat{g}|\text{HF}\rangle \qquad (10.4.28)$$

The expectation value of the fluctuation potential is therefore not a small term, being equal to the negative of the total electron-repulsion energy.

10.5　Koopmans' theorem

In Sections 10.3 and 10.4, we introduced the Fock operator as an effective Hamiltonian for the calculation of Hartree–Fock orbitals (the canonical orbitals) and orbital energies by the repeated diagonalization of the Fock matrix. In the present section, we consider in more detail the properties of the canonical orbitals and the associated orbitals energies – the eigenfunctions and eigenvalues, respectively, of the Fock operator. In particular, we shall introduce Koopmans' theorem and identify the ionization potentials and electron affinities of a closed-shell system with the negative energies of the canonical orbitals.

10.5.1　KOOPMANS' THEOREM FOR IONIZATION POTENTIALS

Consider the removal of an electron from the Hartree–Fock state, not necessarily constructed from the canonical orbitals. The (presumably) neutral N-electron closed-shell system is represented by the optimized Hartree–Fock state $|\mathrm{HF}\rangle$ of energy

$$E_{\mathrm{HF}}^N = \langle \mathrm{HF}|\hat{H}|\mathrm{HF}\rangle \tag{10.5.1}$$

Let us first assume that the ionized system is described by the single-configuration state $a_{i\sigma}|\mathrm{HF}\rangle$. The *ionization potential (IP)* is then given by

$$
\begin{aligned}
\mathrm{IP}_i &= \langle \mathrm{HF}|a_{i\sigma}^\dagger \hat{H} a_{i\sigma}|\mathrm{HF}\rangle - E_{\mathrm{HF}}^N \\
&= \langle \mathrm{HF}|a_{i\sigma}^\dagger a_{i\sigma}\hat{H}|\mathrm{HF}\rangle - \langle \mathrm{HF}|[a_{i\sigma}^\dagger, [a_{i\sigma}, \hat{H}]]_+|\mathrm{HF}\rangle - E_{\mathrm{HF}}^N \\
&= -f_{ii}
\end{aligned}
\tag{10.5.2}
$$

where we have used the definition (10.3.17) for the closed-shell Fock matrix. According to (10.5.2), we may identify the negative diagonal element of the Fock marix $-f_{ii}$ with the IP for the state $a_{i\sigma}|\mathrm{HF}\rangle$. In the special case of the canonical orbitals, we may identify the negative orbital energies with the IPs.

The identification of IPs with the negative diagonal elements of the Fock matrix (10.5.2) does not tell us which MOs are best suited to the description of the ionization processes. To answer this question, we expand the cationic wave function in the configurations generated from the Hartree–Fock state by application of the annihilation operators of the occupied spin orbitals:

$$|N-1\rangle = \sum_i C_i a_{i\sigma}|\mathrm{HF}\rangle \tag{10.5.3}$$

To determine the independent cationic states, we invoke the variation principle and solve the following standard CI eigenvalue problem (see Section 4.2.3) for the states in (10.5.3):

$$\mathbf{H}^{N-1}\mathbf{C} = E^{N-1}\mathbf{C} \tag{10.5.4}$$

Here E^{N-1} is the energy of the cationic system and the matrix elements of the Hamiltonian operator are given by

$$H_{ij}^{N-1} = \langle \mathrm{HF}|a_{i\sigma}^\dagger \hat{H} a_{j\sigma}|\mathrm{HF}\rangle \tag{10.5.5}$$

Using (10.3.17), we may rewrite these matrix elements in the following way:

$$
\begin{aligned}
H_{ij}^{N-1} &= \langle \mathrm{HF}|a_{i\sigma}^\dagger [\hat{H}, a_{j\sigma}]|\mathrm{HF}\rangle + \langle \mathrm{HF}|a_{i\sigma}^\dagger a_{j\sigma}\hat{H}|\mathrm{HF}\rangle \\
&= -\langle \mathrm{HF}|[a_{i\sigma}^\dagger, [a_{j\sigma}, \hat{H}]]_+|\mathrm{HF}\rangle + \delta_{ij}\langle \mathrm{HF}|\hat{H}|\mathrm{HF}\rangle \\
&= -f_{ji} + \delta_{ij}E_{\mathrm{HF}}^N
\end{aligned}
\tag{10.5.6}
$$

The cationic eigenvalue problem (10.5.4) may now be written in the form

$$\mathbf{f}^{occ}\mathbf{C} = -(E^{N-1} - E^N_{HF})\mathbf{C} \tag{10.5.7}$$

where $\mathbf{f}^{occ}$ is the occupied–occupied block of the Fock matrix. The solution of (10.5.7) yields a set of orthonormal cationic states, the eigenvalues of which represent the negative IPs.

The cationic eigenvalue equations can be set up and solved in any basis that satisfies the variational conditions for the neutral system. In the canonical representation, the solution is particularly simple, however, since the Fock matrix in (10.5.7) is then already diagonal. The cationic energy is therefore given by

$$E^{N-1}_i = E^N_{HF} - \varepsilon_i \tag{10.5.8}$$

and the ionized state (10.5.3) reduces to a single term, generated from the Hartree–Fock state by the application of the associated annihilation operator:

$$|N - 1\rangle_{i\sigma} = a_{i\sigma}|HF\rangle \tag{10.5.9}$$

We conclude that the occupied canonical orbitals are the MOs best suited to describe ionization processes in the sense that they are the solutions to the small CI problem (10.5.7). Furthermore, we identify the orbital energies with the negative IPs according to (10.5.8). This result is known as *Koopmans' theorem*.

10.5.2 KOOPMANS' THEOREM FOR ELECTRON AFFINITIES

In the same way as for IPs in Section 10.5.1, we can study the addition of an electron by introducing the CI wave function

$$|N + 1\rangle = \sum_a C_a a^\dagger_{a\sigma}|HF\rangle \tag{10.5.10}$$

and solving the corresponding eigenvalue equations. The Hamiltonian matrix elements now become

$$H^{N+1}_{ab} = f_{ab} + \delta_{ab}E^N_{HF} \tag{10.5.11}$$

In the canonical representation, the solutions to the anionic CI equations are given by

$$E^{N+1}_a = E^N_{HF} + \varepsilon_a \tag{10.5.12}$$

$$|N + 1\rangle_{a\sigma} = a^\dagger_{a\sigma}|HF\rangle \tag{10.5.13}$$

which should be compared with the cationic solutions (10.5.8) and (10.5.9). Thus, within the one-particle model, the virtual canonical orbitals represent the best MOs for *adding* electrons to the N-electron system and their negative orbital energies may be identified with *electron affinities (EAs)*. Note carefully, however, that the virtual canonical orbitals are not the best orbitals for describing *excitations* from the occupied MOs since excitation processes represent the motion in the field of an $N - 1$ electron system.

To summarize, we have found that the IPs and EAs of a closed-shell molecular system may be identified with the negative orbital energies of the canonical orbitals:

$$IP^{KT}_i = E^{N-1}_i - E^N_{HF} = -\varepsilon_i \tag{10.5.14}$$

$$EA^{KT}_a = E^N_{HF} - E^{N+1}_a = -\varepsilon_a \tag{10.5.15}$$

These identifications are based on Koopmans' theorem, which states that the canonical orbitals are optimal for the description of ionized states, yielding the same results as a CI calculation in the configurations generated by the application of the full set of annihilation or creation operators to the neutral state.

10.5.3 IONIZATION POTENTIALS OF H_2O AND N_2

According to Koopmans' theorem, we may identify the negative orbital energies with IPs:

$$IP_i^{KT} = -\varepsilon_i \qquad (10.5.16)$$

This identification is based on a simple model for the open-shell ionized system, in which the ionic Hartree–Fock wave function is not allowed to relax upon ionization but is instead constructed from the frozen MOs of the neutral system. A more satisfactory procedure would be to relax the MOs of the ionized system – that is, to calculate the ionic wave function by the standard (open-shell) Hartree–Fock procedure. This approach is known as the ΔSCF *method* since the IPs are obtained as the difference between SCF (Hartree–Fock) energies of the neutral and ionic systems:

$$IP_i^{\Delta SCF} = E_{HF,i}^{N-1} - E_{HF}^{N} \qquad (10.5.17)$$

Since the ΔSCF method neglects electron correlation, we do not expect the energies (10.5.17) to be close to the exact solution. We would expect, however, the more balanced ΔSCF model (where the orbital relaxation is considered in both states) to give more accurate results than the simpler Koopmans' approximation (10.5.16).

In the present subsection, we calculate IPs of the water molecule and the nitrogen molecule by Koopmans' method and by the ΔSCF method, comparing the results with each other and also with results from FCI calculations and from experiment. In Table 10.1, we have listed the lowest IPs for the water molecule as calculated in the cc-pVDZ basis using Koopmans' method, the ΔSCF method and the FCI method at the reference geometry of Section 5.3.3. Comparing the ΔSCF and FCI methods, we find that the correlation corrections constitute 6–11% of the total IPs (i.e. 40–56 mE_h) and are positive in all three cases. Correlation corrections to the IPs are indeed often positive (but by no means always so) since there are fewer electrons – in particular, fewer electron pairs – to correlate in the cation than in the neutral system.

Turning to the IPs calculated by Koopmans' method, we note that these are always larger than those obtained by the ΔSCF method. This must be so since a relaxation of the MOs of the ionized system will always lower its energy and hence the IPs. From Table 10.1, we see that the relaxation effect is of the order of 0.1 E_h. Since relaxation and correlation corrections often are in opposite directions, there is considerable scope for cancellation of errors in Koopmans' method. Indeed,

Table 10.1 The lowest IPs of H_2O (in E_h) calculated in the cc-pVDZ basis by Koopmans' method, by the ΔSCF method and by the FCI method at the reference geometry of Section 5.3.3

	$1b_2$	$3a_1$	$1b_1$
Koopmans' method	0.7002	0.5510	0.4895
ΔSCF method	0.6433	0.4464	0.3945
FCI	0.6836	0.5027	0.4350

from an inspection of Table 10.1, we note that the IPs calculated by Koopmans' method are at least as accurate as those obtained by the ΔSCF method, being in error by only 2–13% relative to FCI, compared with errors in the range 6–11% for the ΔSCF method.

Having compared the valence IPs in a small basis, let us now consider the core and valence IPs in a sequence of basis sets. In Tables 10.2 and 10.3, we have listed the core and valence IPs of H_2O and N_2 calculated using Koopmans' method and the ΔSCF method. For comparison, we have also listed the vertical experimental IPs. The core–valence cc-pCVXZ basis sets have been used in order to describe the relaxation of the core orbitals accurately in the ΔSCF method. All calculations have been carried out at the optimized geometry of the neutral molecule in the given basis.

For both methods, the calculated IPs converge rapidly towards the basis-set limit and it appears that, at the cc-pCVQZ level, this limit has been reached to within a few millihartrees. The variations with the basis are smaller than 0.04 E_h, which should be compared with the much larger deviations from experiment (0.73 E_h and smaller).

The effects of the ΔSCF orbital relaxation are rather different for the core and valence IPs. Thus, the relaxation energies for the core IPs of N_2 and H_2O are 0.27 and 0.74 E_h, respectively, whereas they are less than 0.1 E_h for the valence IPs. However, since the valence IPs are lower than the core IPs, the relative effects are larger for the valence IPs (24% of the ΔSCF result for the water $1b_1$ ionization).

The correlation contributions to the IPs (as estimated by comparing the ΔSCF results with experiment) vary in sign as well as in magnitude. For H_2O, the correction is positive except for the $2a_1$ ionization; for N_2, the correction is negative except for the $1\pi_u$ ionization. For the core

Table 10.2 The vertical IPs of H_2O (in E_h) calculated using Koopmans' method and the ΔSCF method at the optimized geometries of the neutral molecule

		$1a_1$	$2a_1$	$1b_2$	$3a_1$	$1b_1$
Koopmans' method	cc-pCVDZ	20.547	1.342	0.706	0.569	0.494
	cc-pCVTZ	20.550	1.352	0.722	0.578	0.505
	cc-pCVQZ	20.556	1.357	0.728	0.582	0.509
ΔSCF method	cc-pCVDZ	19.854	1.260	0.648	0.485	0.401
	cc-pCVTZ	19.815	1.265	0.656	0.489	0.409
	cc-pCVQZ	19.813	1.266	0.658	0.489	0.410
Experiment[a]		19.83	1.18	0.676	0.540	0.463

[a]K. Siegbahn, C. Nordling, C. Johansson, J. Hedman, P. F. Heden, K. Hamrin, U. Gelius, T. Bergmark, L. O. Werme, R. Manne and Y. Baer, *ESCA Applied to Free Molecules*, North–Holland, 1969.

Table 10.3 The vertical IPs of N_2 (in E_h) calculated using Koopmans' method and the ΔSCF method at the optimized geometries of the neutral molecule

		$1\sigma_g$	$1\sigma_u$	$2\sigma_g$	$2\sigma_u$	$1\pi_u$	$3\sigma_g$
Koopmans' method	cc-pCVDZ	15.679	15.675	1.487	0.767	0.618	0.629
	cc-pCVTZ	15.669	15.665	1.493	0.767	0.626	0.635
	cc-pCVQZ	15.670	15.666	1.494	0.768	0.627	0.637
ΔSCF method	cc-pCVDZ	15.431	15.426	1.404	0.737	0.572	0.586
	cc-pCVTZ	15.401	15.395	1.401	0.733	0.575	0.588
	cc-pCVQZ	15.400	15.394	1.400	0.732	0.575	0.588
Experiment[a]		15.06	15.06	1.37	0.684	0.617	0.570

[a]See reference given in Table 10.2.

IPs, the correlation contribution is small for H_2O (0.02 E_h) but it reduces the IPs by as much as 0.34 E_h for N_2. For the valence IPs, the corrections are all smaller than 0.1 E_h.

Comparing with experiment, we find that, for the core IPs, the ΔSCF method is more accurate than Koopmans' approximation. For the valence IPs, the two methods give similar accuracy for H_2O (mostly bracketing the experimental value). For N_2, by contrast, the ΔSCF method is closer to experiment except for the $1\pi_u$ ionization. For H_2O, a correct assignment of the experimental IPs is obtained both in Koopmans' approximation and from the ΔSCF method. For N_2, however, the assignment of the $3\sigma_g$ and $1\pi_u$ ionizations would be incorrect if based on the results obtained by either method.

10.6 The Roothaan–Hall self-consistent field equations

In this section, we present the classical Roothaan–Hall formulation of Hartree–Fock theory, in which the MOs are expanded in a set of AOs whose expansion coefficients are used as the variational parameters. As demonstrated in Section 10.6.1, the variational conditions on the Hartree–Fock energy then give a set of Fock eigenvalue equations similar to those in Section 10.3 but expressed in the AO basis, thereby avoiding the expensive transformation of the two-electron integrals to the MO basis. Next, in Section 10.6.2, we describe how the convergence of the Roothaan–Hall scheme may be accelerated and made more robust by using the direct inversion in the iterative subspace (DIIS) algorithm. Finally, in Section 10.6.3, we discuss the direct Hartree–Fock method, a particular implementation of the Roothaan–Hall theory that enables the study of large molecular systems since it avoids not only the transformation of the two-electron integrals to the MO basis, but also their storage on disk.

10.6.1 THE ROOTHAAN–HALL EQUATIONS

In most applications of Hartree–Fock theory to molecular systems, the MOs ϕ_p are expanded in a set of AOs χ_μ, usually a set of contracted Gaussian functions:

$$\phi_p = \sum_\mu \chi_\mu C_{\mu p} \tag{10.6.1}$$

The MO coefficients $C_{\mu p}$ of this expansion may be used as the variational parameters for the optimization of the Hartree–Fock energy. For a closed-shell state, the Hartree–Fock energy (10.4.20) is given as

$$E(\mathbf{C}) = 2 \sum_i h_{ii} + \sum_{ij} (2g_{iijj} - g_{ijji}) + h_{\text{nuc}} \tag{10.6.2}$$

Since the energy must be optimized subject to the orthonormality conditions on the MOs

$$\langle \phi_i | \phi_j \rangle = \delta_{ij} \tag{10.6.3}$$

we introduce the Hartree–Fock Lagrangian

$$L(\mathbf{C}) = E(\mathbf{C}) - 2 \sum_{ij} \lambda_{ij} (\langle \phi_i | \phi_j \rangle - \delta_{ij}) \tag{10.6.4}$$

For real orbitals, the multipliers λ_{ij} satisfy the symmetry conditions

$$\lambda_{ij} = \lambda_{ji} \tag{10.6.5}$$

and therefore constitute a real symmetric matrix $\boldsymbol{\lambda}$.

The variational conditions on the Hartree–Fock energy (10.6.2) may now be expressed in the following unconstrained form:

$$\frac{\partial L(\mathbf{C})}{\partial C_{\mu k}} = 0 \tag{10.6.6}$$

The first derivatives of the Lagrangian (10.6.4) with respect to the MOs are given as:

$$\frac{\partial L(\mathbf{C})}{\partial C_{\mu k}} = 4h_{\mu k} + 4\sum_j (2g_{\mu kjj} - g_{\mu jjk}) - 4\sum_j S_{\mu j}\lambda_{jk} \tag{10.6.7}$$

Note that the integrals on the right-hand side of this equation are in a mixed representation, with one index in the AO basis and the remaining indices in the MO basis. Substituting this expression into the variational conditions (10.6.6), we arrive at the following conditions for the optimized Hartree–Fock state:

$$f_{\mu k} = \sum_j S_{\mu j}\lambda_{jk} \tag{10.6.8}$$

Multiplication from the left by a set of orthonormal virtual orbitals that are orthogonal to the occupied MOs and by the set of occupied MOs gives

$$f_{ak} = 0 \tag{10.6.9}$$

$$f_{ik} = \lambda_{ik} \tag{10.6.10}$$

The first set of equations (10.6.9) represent the optimization condition, f_{ak} being proportional to the Hartree–Fock gradient (10.1.25). These equations are sufficient to determine the Hartree–Fock state. The Lagrange multipliers may be determined from the second set of equations (10.6.10).

Since the multiplier matrix $\boldsymbol{\lambda}$ is symmetric (10.6.5), it may be diagonalized by an orthogonal transformation among the occupied orbitals:

$$\boldsymbol{\lambda} = \mathbf{U}\boldsymbol{\varepsilon}\mathbf{U}^T \tag{10.6.11}$$

The Hartree–Fock energy is invariant to such transformations. The off-diagonal Lagrange multipliers may therefore be eliminated by an orthogonal transformation to a set of occupied MOs that satisfy the canonical conditions

$$f_{ik} = \delta_{ik}\varepsilon_i \tag{10.6.12}$$

In the canonical basis, we may express the variational conditions (10.6.8) in the form

$$\sum_v f_{\mu v}^{AO} C_{vk} = \varepsilon_k \sum_v S_{\mu v} C_{vk} \tag{10.6.13}$$

where the elements of the AO Fock matrix are given as

$$f_{\mu v}^{AO} = h_{\mu v} + \sum_i (2g_{\mu vii} - g_{\mu iiv}) \tag{10.6.14}$$

Since $\mathbf{f}^{AO}$ is symmetric, we may extend (10.6.13) to give also a set of orthonormal virtual MOs that are orthogonal to the occupied MOs. These virtual orbitals satisfy the same canonical conditions as the occupied MOs (10.6.12):

$$f_{ab} = \delta_{ab}\varepsilon_a \tag{10.6.15}$$

In matrix form, the Hartree–Fock variational conditions (10.6.13) may now be written as

$$\mathbf{f}^{AO}\mathbf{C} = \mathbf{SC\varepsilon} \tag{10.6.16}$$

where ε is a diagonal matrix containing the orbital energies (i.e. the Lagrange multipliers in the canonical representation). Equations (10.6.16) are the equivalent of the Hartree–Fock equations in the MO basis (10.3.25).

It should be noted that the AO Fock matrix (10.6.14) may be evaluated entirely in the AO basis according to the expression

$$f_{\mu\nu}^{AO} = h_{\mu\nu} + \sum_{\rho\sigma} D_{\rho\sigma}^{AO} \left(g_{\mu\nu\rho\sigma} - \tfrac{1}{2} g_{\mu\sigma\rho\nu} \right) \tag{10.6.17}$$

where the elements of the symmetric matrix $\mathbf{D}^{AO}$ are given by

$$D_{\rho\sigma}^{AO} = 2 \sum_{i} C_{\rho i} C_{\sigma i} \tag{10.6.18}$$

As is easily verified, $\mathbf{D}^{AO}$ is the AO representation of the one-electron density matrix $\mathbf{D}$ for a closed-shell system:

$$\mathbf{D}^{AO} = \mathbf{CDC}^{T} \tag{10.6.19}$$

This result follows since the density matrix $\mathbf{D}$ in (10.4.16) is a diagonal matrix with nonzero elements 2 for the occupied MOs (10.4.18).

We have now succeeded in expressing the Hartree–Fock equations in the AO basis, avoiding any transformation to the MO basis. The pseudo-eigenvalue equations (10.6.16) are called *the Roothaan–Hall equations* [3,4]. In Exercise 10.4, the Roothaan–Hall SCF procedure is used to calculate the Hartree–Fock wave function for HeH$^+$ in the STO-3G basis.

10.6.2 DIIS CONVERGENCE ACCELERATION

In our discussion so far, we have assumed that the SCF fixed-point iterations converge. In practice, a naive implementation of the SCF method, in which the Fock matrix in each iteration is constructed from the eigenvectors of the Fock matrix generated in the previous iteration, may converge slowly or may diverge. There is thus a need to develop methods that improve and accelerate the convergence of the SCF method. Several such schemes have been proposed; the most successful of these is discussed in the present subsection.

Let us begin by outlining a naive implementation of the SCF method. The steps that constitute the nth iteration are the following:

1. Construct the density matrix $\mathbf{D}_n^{AO}$ from the orbitals $\mathbf{C}_n$ according to (10.6.18).
2. Generate the AO Fock matrix $\mathbf{f}_n^{AO}$ from $\mathbf{D}_n^{AO}$ according to (10.6.17).
3. Transform $\mathbf{f}_n^{AO}$ to the MO basis $\mathbf{f}_n^{MO}$ and generate the error vector $\mathbf{e}_n$ containing the elements $[\mathbf{f}_n^{MO}]_{ia}$ of the occupied–virtual block of $\mathbf{f}_n^{MO}$.

   (a) If $\|\mathbf{e}_n\| < \tau$, where τ is some pre-set threshold, terminate the iterations.
   (b) Otherwise, generate a new set of orbitals $\mathbf{C}_{n+1}$ by diagonalizing the Fock matrix $\mathbf{f}_n^{AO}$ according to (10.6.16) and return to step 1 for iteration $n + 1$.

Although this sequence of steps may lead to convergence, it may also fail to converge or it may converge very slowly.

To improve upon the convergence of the SCF method, we note that the SCF iterations represent an iterative search for an effective one-electron Hamiltonian $\mathbf{f}^{AO}$. After n iterations, we have generated n effective Hamiltonians $\mathbf{f}_i^{AO}$ and error vectors $\mathbf{e}_i$ with $i \leq n$. At this point, rather than generating the next set of orbitals $\mathbf{C}_{n+1}$ from the last Hamiltonian $\mathbf{f}_n^{AO}$, we may use the information from the preceding iterations to generate the orbitals from an *averaged effective Hamiltonian*

$$\bar{\mathbf{f}}_n^{AO} = \sum_{i=1}^{n} w_i \mathbf{f}_i^{AO} \tag{10.6.20}$$

To ensure that the one-electron part of the true Hamiltonian is recovered exactly in the averaged operator, the weights are chosen such that

$$\sum_{i=1}^{n} w_i = 1 \tag{10.6.21}$$

Referring back to (10.6.17), we see that this procedure is equivalent to the calculation of the AO Fock matrix from an averaged one-electron density matrix

$$\bar{\mathbf{D}}_n^{AO} = \sum_{i=1}^{n} w_i \mathbf{D}_i^{AO} \tag{10.6.22}$$

In other words, we shall attempt to generate an improved Fock operator (10.6.20) by averaging over the previous (and current) effective two-electron potentials, thereby hoping to stabilize the iterations. This approach should be especially effective in situations where the naive SCF iterations oscillate between two potentials by allowing the orbitals to adjust to an averaged potential.

Obviously, if this procedure is to work, we must select the weights in such a way that an improved effective potential is generated. The weights are sometimes selected by simple rules of thumb. However, if we assume a near-linear relationship between the error vector

$$[\mathbf{e}_n]_{ia} = [\mathbf{f}_n^{MO}]_{ia} \tag{10.6.23}$$

and the effective Hamiltonian $\mathbf{f}_n^{AO}$, a better strategy is to combine the potentials in such a manner that the norm of the averaged error vector

$$\bar{\mathbf{e}}_n = \sum_{i=1}^{n} w_i \mathbf{e}_i \tag{10.6.24}$$

becomes as small as possible. In this manner, we utilize the error vectors $\mathbf{e}_i$ not only to test for convergence as in the naive SCF method but also to determine the optimal weights w_i for setting up an improved effective Hamiltonian for the generation of the orbitals $\mathbf{C}_{n+1}$. The resulting method, which differs from the original SCF procedure only in the replacement of $\mathbf{f}_n^{AO}$ by $\bar{\mathbf{f}}_n^{AO}$ in step 3(b), is known as the method of *direct inversion in the iterative subspace (DIIS)* [5]. It should be emphasized that the underlying assumption of the DIIS acceleration method is the linearity between the effective potentials and the error vectors. However, since the orbitals in terms of which the error vectors are calculated change from iteration to iteration, this requirement is not strictly fulfilled.

Let us now consider how we may calculate the DIIS weights $\mathbf{w}$ for the nth iteration. We wish to minimize the norm of the averaged error vector

$$\|\bar{\mathbf{e}}_n\|^2 = \sum_{i,j=1}^{n} w_i B_{ij} w_j \tag{10.6.25}$$

where

$$B_{ij} = \langle \mathbf{e}_i | \mathbf{e}_j \rangle \tag{10.6.26}$$

subject to the constraint (10.6.21). For this purpose, we construct the Lagrangian

$$L = \sum_{i,j=1}^{n} w_i B_{ij} w_j - 2\lambda \left(\sum_{i=1}^{n} w_i - 1 \right) \tag{10.6.27}$$

where λ is the undetermined multiplier and the factor 2 is introduced for convenience. Minimization of L with respect to the coefficients and the multiplier gives the linear equations

$$\begin{pmatrix} B_{11} & B_{12} & \dots & B_{1n} & -1 \\ B_{21} & B_{22} & \dots & B_{2n} & -1 \\ \vdots & \vdots & \ddots & \vdots & \vdots \\ B_{n1} & B_{n2} & \dots & B_{nn} & -1 \\ -1 & -1 & \dots & -1 & 0 \end{pmatrix} \begin{pmatrix} w_1 \\ w_2 \\ \vdots \\ w_n \\ \lambda \end{pmatrix} = \begin{pmatrix} 0 \\ 0 \\ \vdots \\ 0 \\ -1 \end{pmatrix} \tag{10.6.28}$$

from which the weights $\mathbf{w}$ may be obtained.

A disadvantage of the DIIS method over the standard SCF method is the need to store previous AO Fock matrices and error vectors. In practice, only the matrices and vectors from a few iterations are stored – not only to save disk space and memory but also since the oldest matrices usually carry little information and their inclusion in the averaging may in fact degrade the performance.

In Table 10.4, we have listed, for each iteration of the standard SCF method and for the DIIS method, the expectation value of the energy relative to the converged energy for the water molecule in the cc-pVQZ basis. (We have also listed the results obtained by the Newton trust-region method, to be discussed in Sections 10.8.1 and 10.9.5.) The first set of calculations have been carried out close to the equilibrium bond distance (i.e. at the reference geometry of Section 5.3.3) where the exact wave function is dominated by the Hartree–Fock configuration. The second set of calculations have been performed with the bond distances stretched by a factor of 2 and represent a situation where several configurations are nearly degenerate with the Hartree–Fock configuration. In all calculations, the starting orbitals are obtained by diagonalizing the one-electron Hamiltonian.

The convergence patterns are different for the different methods and for the two geometries. At the equilibrium geometry, the naive SCF method exhibits a typical linear behaviour with a fairly constant reduction in the error in each iteration. Thus, from iteration 11 to convergence, the error – that is, the difference between the actual and the converged energy – is reduced by a constant factor of about 0.4 in each iteration. The DIIS method converges significantly faster but with a somewhat unpredictable behaviour of the error term – in some of the iterations, the error term is reduced by a factor similar to that of the SCF method; in other iterations, the error is reduced by an order of magnitude or more. At the stretched geometry, the standard SCF method exhibits a striking oscillatory behaviour and does not converge. The DIIS method, on the other hand, converges in 15 iterations with a very smooth behaviour after the first five iterations, where some slight oscillation is present. The efficacy of the DIIS scheme is clearly demonstrated by the fact that the DIIS iterations at $2R_{\text{ref}}$ converge more rapidly than the SCF iterations at R_{ref}. We finally note that, in terms of iterations (but not computer time), the SCF and the DIIS methods are both inferior to Newton's method, to be discussed in Section 10.8. A comparison of the SCF and Newton methods is given in Section 10.9.

Table 10.4 Convergence of Hartree–Fock calculations on the water molecule in the cc-pVQZ basis using the standard SCF method, the DIIS method and the Newton trust-region method. For each iteration, we have listed the energy relative to the converged energy in E_h. The calculations have been carried out at the geometries R_{ref} and $2R_{ref}$ of Section 5.3.3

	R_{ref}			$2R_{ref}$		
	SCF	DIIS	Newton	SCF	DIIS	Newton
1	16.00647	16.00647	16.00647	16.49326	16.49326	16.49326
2	9.58984	9.58984	8.88965	12.73424	12.73427	7.88928
3	4.61743	4.64733	4.96737	6.52430	1.07245	4.56122
4	5.07608	0.31858	3.44710	12.31403	2.75895	3.37801
5	2.72273	0.02346	1.00855	6.40243	0.12441	2.44947
6	2.95707	0.01010	0.11676	12.28473	0.26336	1.99168
7	1.60597	0.00005	0.00078	6.39597	0.00887	0.61365
8	1.10266	0.00000	0.00000	12.28278	0.00202	0.07316
9	0.53622			6.34555	0.00035	0.01495
10	0.27230			12.28263	0.00018	0.00140
11	0.12028			6.39553	0.00014	0.00005
12	0.05426			12.28264	0.00010	0.00000
13	0.02324			6.39552	0.00005	
14	0.01010			12.28264	0.00001	
15	0.00431			6.39552	0.00000	
16	0.00185			12.28264		
17	0.00079			6.39552		
18	0.00034			12.28264		
19	0.00015			6.39552		
20	0.00006			12.28264		
21	0.00003					
22	0.00001					
23	0.00000					

10.6.3 INTEGRAL-DIRECT HARTREE–FOCK THEORY

The time-critical step of a Roothaan–Hall SCF iteration is usually the evaluation of the two-electron contribution to the Fock matrix. In each iteration, a large number of two-electron integrals are needed to construct the Fock matrix. Since the same AO integrals are used in each iteration, these integrals may be written to disk and read in once in each iteration. In this way, the time needed to generate the two-electron integrals is reduced significantly.

In large calculations, this approach becomes impractical since, for more than a few hundred basis functions, it becomes difficult to store all integrals on disk. In such cases, therefore, the integrals are instead recomputed in each iteration, reducing disk requirements at the expense of increased computer time. As soon as a batch of two-electron integrals has been produced, it is multiplied by the appropriate densities and their contributions to the Fock matrix are calculated. This procedure is known as the *direct SCF or direct Hartree–Fock method*, since the integrals are calculated on the fly and used directly rather than stored on disk and read in as required [6]. It should be noted that integral-direct methods can be applied to any method in which the two-electron integrals are needed only in the AO basis. It may also be applied to the second-order optimization schemes discussed in Section 10.8.

In direct SCF calculations, each two-electron integral is used a few times and then discarded. For example, in the standard SCF scheme, each integral makes a few contributions to the Fock matrix according to (10.6.17). Clearly, if an integral is very small or if it is to be combined with small density-matrix elements, its contribution to the Fock matrix may be neglected. Therefore, before a batch of integrals is evaluated, it is important to make an estimate (preferably an upper bound as described in Section 9.12.4) of the magnitude of the integrals. Comparing this estimate with the one-electron densities with which they will be combined according to (10.6.17), we can decide if the integral should be calculated and, if so, to what precision. It turns out that, in extended systems, most of the two-electron integrals may be discarded altogether since their contribution to the Fock matrix and to the energy is negligible. The efficient implementation of integral screening can therefore reduce the cost of direct calculations significantly. We note that, in the conventional (nondirect) SCF scheme, the magnitudes of the density elements are unknown when the integrals are calculated and we must therefore assume that each integral makes a nonnegligible contribution to the Fock matrices. Pre-screening therefore becomes less effective in such calculations. In fact, for larger systems, the conventional SCF method (if it could be carried out) would be slower than the direct SCF method since – even though each integral is calculated only once – many integrals are calculated which are never needed in the calculations at all.

The SCF method is formally a fourth-order method in the sense that the calculation of the Fock matrix scales as n^4, where n is the number of AOs. For extended systems, where very few of the AOs overlap with one another, it is advantageous to partition the AO Fock matrix as [7]

$$f_{\mu\nu} = h_{\mu\nu} + f_{\mu\nu}^{\text{cou}} + f_{\mu\nu}^{\text{exc}} \qquad (10.6.29)$$

where the Coulomb and exchange contributions are given by

$$f_{\mu\nu}^{\text{cou}} = \sum_{\rho\sigma} g_{\mu\nu\rho\sigma} D_{\rho\sigma}^{\text{AO}}$$

$$= \int \chi_\mu(\mathbf{r}_1)\chi_\nu(\mathbf{r}_1) \left[\int \frac{1}{r_{12}} \sum_{\rho\sigma} D_{\rho\sigma}^{\text{AO}} \chi_\rho(\mathbf{r}_2)\chi_\sigma(\mathbf{r}_2)\,d\mathbf{r}_2 \right] d\mathbf{r}_1 \qquad (10.6.30)$$

$$f_{\mu\nu}^{\text{exc}} = -\frac{1}{2} \sum_{\rho\sigma} g_{\mu\sigma\rho\nu} D_{\rho\sigma}^{\text{AO}}$$

$$= -\frac{1}{2} \sum_\sigma \int \chi_\mu(\mathbf{r}_1)\chi_\sigma(\mathbf{r}_1) \left[\int \frac{1}{r_{12}} \sum_\rho D_{\rho\sigma}^{\text{AO}} \chi_\rho(\mathbf{r}_2)\chi_\nu(\mathbf{r}_2)\,d\mathbf{r}_2 \right] d\mathbf{r}_1 \qquad (10.6.31)$$

The Coulomb term is the classical electrostatic interaction between the two charge distributions $\chi_\mu\chi_\nu$ and $\sum_{\rho\sigma} D_{\rho\sigma}^{\text{AO}} \chi_\rho\chi_\sigma$; the exchange term represents a nonclassical correction arising from the Fermi correlation of the electrons.

Since the contribution to $f_{\mu\nu}^{\text{cou}}$ from $\chi_\rho\chi_\sigma$ falls off only as the inverse of the separation between the regions where the charge distributions $\chi_\mu\chi_\nu$ and $\chi_\rho\chi_\sigma$ are located, the Coulomb term (10.6.30) makes a long-range contribution to the Fock matrix. The long-range contributions to $f_{\mu\nu}^{\text{cou}}$, arising from the nonoverlapping distributions $\chi_\mu\chi_\nu$ and $\chi_\rho\chi_\sigma$, can be evaluated by multipole expansions of $\chi_\mu\chi_\nu$ and $\chi_\rho\chi_\sigma$ as explained in Section 9.13, leaving only the relatively small number of overlapping distributions to be evaluated explicitly in terms of two-electron integrals. The exchange correction, on the other hand, is short-range and $f_{\mu\nu}^{\text{exc}}$ will have contributions only from orbitals χ_σ and χ_ρ that overlap significantly with χ_μ and χ_ν, respectively.

By combining the separation into Coulomb and exchange terms with the multipole technique of Section 9.14, it is possible to arrive at an operation count that scales linearly with the size of the system, allowing us to carry out direct Hartree–Fock calculations for very large systems. However, to achieve such a linear scaling of the cost, it is necessary to formulate the Hartree–Fock theory differently, in a manner that does not require the diagonalization of the Fock operator and the calculation of MOs. In Section 10.7, we develop such a scheme, based on the direct optimization of the one-electron density matrix in the AO basis.

10.7 Density-based Hartree–Fock theory

For small and moderately large molecular systems, the classical Roothaan–Hall method discussed in Section 10.6 represents a highly successful implementation of Hartree–Fock theory, in particular when combined with the DIIS acceleration scheme and the direct evaluation of the Fock matrix. For very large systems, however, the Roothaan–Hall method becomes less useful since the cost of the diagonalization of the Fock matrix scales cubically with the size of the system. For such systems, it is better to employ a scheme that avoids the diagonalization step altogether, replacing it by a direct optimization of the one-electron density matrix in the AO basis. Such a scheme requires no operations other than the multiplication and addition of AO matrices, whose sparsity ensures a linear scaling of the computational cost for large systems. Combined with the FMM for the calculation of the Coulomb interactions described in Section 9.14, the density-based optimization scheme developed in the present section leads to a highly efficient formulation of Hartree–Fock theory, whose computational cost increases linearly with the size of the system, making it possible to study truly large systems at the Hartree–Fock level.

10.7.1 DENSITY-MATRIX FORMULATION OF HARTREE–FOCK THEORY

For a closed-shell molecular electronic system, the Hartree–Fock energy (10.4.15) may, according to (10.4.19), be written as

$$E = \sum_{pq} D_{pq} h_{pq} + \frac{1}{2} \sum_{pqrs} \left(D_{pq} D_{rs} - \frac{1}{2} D_{ps} D_{rq} \right) g_{pqrs} + h_{\text{nuc}} \tag{10.7.1}$$

where the summations are over the full set of n occupied and virtual MOs, which satisfy the usual orthonormality relations

$$\mathbf{C}^{\text{T}} \mathbf{S} \mathbf{C} = \mathbf{1}_n \tag{10.7.2}$$

and where the one-electron density matrix $\mathbf{D}$ is a diagonal one, whose elements are 2 for the $N/2$ occupied MOs and 0 for the virtual MOs:

$$\mathbf{D} = 2 \begin{pmatrix} \mathbf{1}_{N/2} & \mathbf{0} \\ \mathbf{0} & \mathbf{0} \end{pmatrix} \tag{10.7.3}$$

Inserting (10.7.3) in (10.7.1), we arrive at the standard expression for the Hartree–Fock energy

$$E = 2 \sum_{i} h_{ii} + \sum_{ij} (2g_{iijj} - g_{ijji}) + h_{\text{nuc}} \tag{10.7.4}$$

However, since this expression involves the transformation of the two-electron integrals to the MO basis, it is not well suited to calculations on large molecules. A more useful expression is obtained by introducing in (10.7.1) the one-electron AO density matrix

$$\mathbf{D}^{\text{AO}} = \mathbf{C} \mathbf{D} \mathbf{C}^{\text{T}} \tag{10.7.5}$$

The Hartree–Fock energy may now be written in the form

$$E(\mathbf{D}^{AO}) = \text{Tr } \mathbf{D}^{AO}\mathbf{h}^{AO} + \tfrac{1}{4}\text{Tr } \mathbf{D}^{AO}\mathbf{G}^{AO}(\mathbf{D}^{AO}) + h_{\text{nuc}} \quad (10.7.6)$$

where $\mathbf{h}^{AO}$ is the one-electron Hamiltonian matrix in the AO basis and the matrix $\mathbf{G}^{AO}$ represents the two-electron interactions:

$$G_{\mu\nu}^{AO}(\mathbf{A}) = \sum_{\rho\sigma}(2g_{\mu\nu\rho\sigma} - g_{\mu\sigma\rho\nu})A_{\rho\sigma} \quad (10.7.7)$$

In (10.7.6), we have indicated that the Hartree–Fock energy $E(\mathbf{D}^{AO})$ is now to be regarded as a function of the one-electron density matrix $\mathbf{D}^{AO}$ [8,9].

However, not all one-electron matrices constitute a valid density matrix of a single-determinant wave function – that is, not all one-electron matrices may be expressed in the form (10.7.5), where $\mathbf{C}$ satisfies (10.7.2) and $\mathbf{D}$ is given by (10.7.3). Thus, we cannot simply consider variations of the form

$$\delta E = E(\delta\mathbf{D}^{AO}) \quad (10.7.8)$$

since the variations in the density matrix $\delta\mathbf{D}^{AO}$ may then be inconsistent with the requirements on the density matrix. Thus, before we can consider the calculation of the Hartree–Fock energy by optimization of the one-electron density matrix $\mathbf{D}^{AO}$, we must examine the properties of this matrix and establish the conditions that it must satisfy for the energy $E(\mathbf{D}^{AO})$ to represent the expectation value of a single-determinant wave function.

10.7.2 PROPERTIES OF THE MO DENSITY MATRIX

Consider a closed-shell electronic system described by a single determinant $|0\rangle$ constructed from a set of real orthonormal MOs. In the basis of these MOs, the one-electron density matrix

$$D_{pq} = \langle 0|E_{pq}|0\rangle \quad (10.7.9)$$

has the diagonal structure (10.7.3), with the diagonal elements equal to 2 for the occupied orbitals and 0 for the virtual ones. All other closed-shell determinants may be generated from this reference state by a unitary transformation of the orbitals. In the original MO basis, the density matrix for such a rotated state is given by

$$D_{pq}(\boldsymbol{\kappa}) = \langle 0|\exp(\hat{\kappa})E_{pq}\exp(-\hat{\kappa})|0\rangle \quad (10.7.10)$$

where $\boldsymbol{\kappa}$ is an antisymmetric matrix containing the parameters of the orbital-rotation operator

$$\hat{\kappa} = \sum_{p>q}\kappa_{pq}E_{pq}^{-} \quad (10.7.11)$$

To calculate the density matrix of the transformed state $\exp(-\hat{\kappa})|0\rangle$ in the original MO basis, we must evaluate the operator $\exp(\hat{\kappa})E_{pq}\exp(-\hat{\kappa})$. Combining (3.2.18) and (3.2.24) and using the fact that $\exp(\boldsymbol{\kappa})$ is orthogonal, we obtain

$$\exp(\hat{\kappa})a_{p\sigma}^{\dagger}\exp(-\hat{\kappa}) = \sum_{r}a_{r\sigma}^{\dagger}[\exp(\boldsymbol{\kappa})]_{rp} = \sum_{r}a_{r\sigma}^{\dagger}[\exp(-\boldsymbol{\kappa})]_{pr} \quad (10.7.12)$$

$$\exp(\hat{\kappa})a_{q\sigma}\exp(-\hat{\kappa}) = \sum_{s}a_{s\sigma}[\exp(\boldsymbol{\kappa})]_{sq} \quad (10.7.13)$$

which gives the following expression for the elements of the density matrix of the transformed state in the basis of the original MOs:

$$D_{pq}(\kappa) = \sum_{rs}[\exp(-\kappa)]_{pr}D_{rs}[\exp(\kappa)]_{sq} \tag{10.7.14}$$

In the MO basis, therefore, all density matrices representing a single determinant containing $N/2$ doubly occupied orbitals may be written as a symmetric matrix with $N/2$ eigenvalues equal to 2 and the remaining eigenvalues equal to 0:

$$\mathbf{D}(\kappa) = 2\exp(-\kappa)\begin{pmatrix} \mathbf{1}_{N/2} & \mathbf{0} \\ \mathbf{0} & \mathbf{0} \end{pmatrix}\exp(\kappa) \tag{10.7.15}$$

Conversely, all such matrices constitute the density matrix of some determinant containing $N/2$ doubly occupied orbitals.

For large systems, we cannot diagonalize the density matrix to find out whether it represents a closed-shell determinant. We shall therefore express the conditions on the density matrix in an alternative form, which does not require its diagonalization. Introducing

$$\rho = \tfrac{1}{2}\mathbf{D} \tag{10.7.16}$$

we find that the conditions on the one-electron density matrix of a closed-shell determinant of $N/2$ doubly occupied orbitals can be expressed succinctly as [8]

$$\rho = \rho^{\mathrm{T}} \tag{10.7.17}$$

$$\mathrm{Tr}\,\rho = \tfrac{1}{2}N \tag{10.7.18}$$

$$\rho^2 = \rho \tag{10.7.19}$$

Whereas the *symmetry condition* (10.7.17) and the *trace condition* (10.7.18) are shared by all N-electron wave functions, the *idempotency condition* (10.7.19) is a special property of the scaled density matrix of a closed-shell determinant of doubly occupied orbitals, arising since the eigenvalues of such matrices do not change when they are squared. In the next subsection, we shall consider the corresponding conditions on the AO density matrix, in terms of which the optimization of the Hartree–Fock state will be carried out.

10.7.3 PROPERTIES OF THE AO DENSITY MATRIX

In this subsection, we examine the properties of the scaled AO density matrix of a single-determinant wave function:

$$\mathbf{R} = \mathbf{C}\rho\mathbf{C}^{\mathrm{T}} \tag{10.7.20}$$

It is a general result of linear algebra that the eigenvalues of a symmetric matrix $\mathbf{A}$ are unaffected by similarity transformations $\mathbf{TAT}^{-1}$ [10]. From the orthonormality relation (10.7.2), we find that the inverse of $\mathbf{C}$ is given as

$$\mathbf{C}^{-1} = \mathbf{C}^{\mathrm{T}}\mathbf{S} \tag{10.7.21}$$

from which it follows that

$$\mathbf{RS} = \mathbf{C}\rho\mathbf{C}^{\mathrm{T}}\mathbf{S} = \mathbf{C}\rho\mathbf{C}^{-1} \tag{10.7.22}$$

constitutes a similarity transformation of the density matrix ρ. Consequently, $\mathbf{RS}$ must satisfy the same trace and idempotency conditions as ρ, as is easily verified:

$$\text{Tr } \mathbf{RS} = \text{Tr } \mathbf{C}\rho\mathbf{C}^{\mathrm{T}}\mathbf{S} = \text{Tr } \rho\mathbf{C}^{\mathrm{T}}\mathbf{SC} = \text{Tr } \rho \tag{10.7.23}$$

$$\mathbf{RSRS} = \mathbf{C}\rho\mathbf{C}^{\mathrm{T}}\mathbf{SC}\rho\mathbf{C}^{\mathrm{T}}\mathbf{S} = \mathbf{C}\rho\rho\mathbf{C}^{\mathrm{T}}\mathbf{S} = \mathbf{C}\rho\mathbf{C}^{\mathrm{T}}\mathbf{S} = \mathbf{RS} \tag{10.7.24}$$

The scaled AO density matrix of a closed-shell determinant $\mathbf{R}$ therefore satisfies the following three conditions

$$\mathbf{R}^{\mathrm{T}} = \mathbf{R} \tag{10.7.25}$$

$$\text{Tr } \mathbf{RS} = \tfrac{1}{2}N \tag{10.7.26}$$

$$\mathbf{RSR} = \mathbf{R} \tag{10.7.27}$$

where the last relation has been obtained from (10.7.24) by multiplying it from the right by $\mathbf{S}^{-1}$. These conditions are suitable for the study of large molecular systems, involving only elementary algebraic manipulations (i.e. multiplications and additions) of the (sparse) density and overlap matrices in the AO basis. In Section 10.7.6, we shall use these conditions to establish a scheme according to which we may generate, from a matrix that does not satisfy these three conditions, a new matrix that satisfies the conditions.

10.7.4 EXPONENTIAL PARAMETRIZATION OF THE AO DENSITY MATRIX

Having established the conditions for AO density matrices, let us now consider the transformation of one AO density matrix into another. As discussed in Section 10.7.2, the MO density matrices are related by

$$\rho(\boldsymbol{\kappa}) = \exp(-\boldsymbol{\kappa})\rho \exp(\boldsymbol{\kappa}) \tag{10.7.28}$$

where $\boldsymbol{\kappa}$ is antisymmetric. Transforming this matrix to the AO basis according to (10.7.20) and introducing the orthonormality relation (10.7.2), we obtain

$$\mathbf{R}(\boldsymbol{\kappa}) = \mathbf{C}\rho(\boldsymbol{\kappa})\mathbf{C}^{\mathrm{T}} = \mathbf{C} \exp(-\boldsymbol{\kappa})\mathbf{C}^{\mathrm{T}}\mathbf{SC}\rho\mathbf{C}^{\mathrm{T}}\mathbf{SC} \exp(\boldsymbol{\kappa})\mathbf{C}^{\mathrm{T}} \tag{10.7.29}$$

Recalling that $\mathbf{C}^{\mathrm{T}}\mathbf{S}$ is the inverse of $\mathbf{C}$ and using (3.1.5), this expression may be written as

$$\mathbf{R}(\boldsymbol{\kappa}) = \exp(-\mathbf{C}\boldsymbol{\kappa}\mathbf{C}^{\mathrm{T}}\mathbf{S})\mathbf{R} \exp(\mathbf{SC}\boldsymbol{\kappa}\mathbf{C}^{\mathrm{T}}) \tag{10.7.30}$$

Next, introducing the antisymmetric matrix

$$\mathbf{X} = \mathbf{C}\boldsymbol{\kappa}\mathbf{C}^{\mathrm{T}} \tag{10.7.31}$$

we arrive at our final expression for the transformed AO density matrix

$$\mathbf{R}(\mathbf{X}) = \exp(-\mathbf{XS})\mathbf{R} \exp(\mathbf{SX}) \tag{10.7.32}$$

Using this relation, we may generate, from a given single-determinant AO density matrix $\mathbf{R}$, the AO density of any other determinant in terms of the antisymmetric matrix $\mathbf{X}$. Note that, because of the presence of $\mathbf{S}$, (10.7.32) does not constitute an orthogonal transformation – see Section 3.1.6, where transformations such as (10.7.32) are discussed. Clearly, if $\mathbf{S}$ becomes the unit matrix, then the nonorthogonal transformation (10.7.32) reduces to the orthogonal one in (10.7.28).

Since (10.7.32) constitutes a relation between AO density matrices, it follows that it must satisfy conditions (10.7.25)–(10.7.27). Still, it is instructive to verify these conditions explicitly, assuming that $\mathbf{R}$ represents the density matrix of a single determinant. First, the symmetry condition (10.7.25) follows trivially from the antisymmetry of $\mathbf{X}$:

$$\mathbf{R(X)}^\mathrm{T} = [\exp(-\mathbf{XS})\mathbf{R}\exp(\mathbf{SX})]^\mathrm{T} = \exp(-\mathbf{XS})\mathbf{R}^\mathrm{T}\exp(\mathbf{SX}) = \mathbf{R(X)} \qquad (10.7.33)$$

Next, the trace (10.7.26) is verified by a cyclic permutation of the matrices:

$$\mathrm{Tr}\,\mathbf{R(X)S} = \mathrm{Tr}\exp(-\mathbf{XS})\mathbf{R}\exp(\mathbf{SX})\mathbf{S} = \mathrm{Tr}\,\mathbf{R}\exp(\mathbf{SX})\mathbf{S}\exp(-\mathbf{XS}) = \mathrm{Tr}\,\mathbf{RS} = \tfrac{1}{2}N \qquad (10.7.34)$$

We have here used the relation

$$\exp(\mathbf{SX})\mathbf{S} = \mathbf{S}\exp(\mathbf{XS}) \qquad (10.7.35)$$

which is easily seen to hold by expansion of the exponentials or from (3.1.40). The idempotency (10.7.27) is demonstrated as follows:

$$\mathbf{R(X)SR(X)} = \exp(-\mathbf{XS})\mathbf{R}\exp(\mathbf{SX})\mathbf{S}\exp(-\mathbf{XS})\mathbf{R}\exp(\mathbf{SX})$$
$$= \exp(-\mathbf{XS})\mathbf{RSR}\exp(\mathbf{SX}) = \exp(-\mathbf{XS})\mathbf{R}\exp(\mathbf{SX}) = \mathbf{R(X)} \qquad (10.7.36)$$

We have thus verified that the unconstrained exponential parametrization (10.7.32) indeed satisfies the symmetry, trace and idempotency constraints on the AO density matrix.

The transformation (10.7.32) appears to be a rather complicated one, requiring the evaluation of matrix exponentials. However, as discussed in Section 3.1.6, we may write the transformed density matrix as an *asymmetric BCH expansion*

$$\mathbf{R(X)} = \mathbf{R} + [\mathbf{R},\mathbf{X}]_\mathrm{S} + \tfrac{1}{2}[[\mathbf{R},\mathbf{X}]_\mathrm{S},\mathbf{X}]_\mathrm{S} + \cdots \qquad (10.7.37)$$

where the *S commutators* are defined as

$$[\mathbf{A},\mathbf{X}]_\mathrm{S} = \mathbf{ASX} - \mathbf{XSA} \qquad (10.7.38)$$

To low orders, the expansion (10.7.37) is easily verified. For a given antisymmetric matrix $\mathbf{X}$, we may then form the transformed density (10.7.32) by carrying out the expansion (10.7.37) to a sufficiently high order, thereby reducing the density transformation to a sequence of matrix multiplications and additions. Usually, only a few terms are used. However, since this truncation of the BCH expansion violates the trace and idempotency conditions, the resulting matrix must be purified as discussed in Section 10.7.6.

10.7.5 THE REDUNDANCY OF THE EXPONENTIAL PARAMETRIZATION

The exponential parametrization of the AO density matrix is redundant in the sense that, for certain nonzero $\mathbf{X}$, the density matrix does not change $\mathbf{R(X)} = \mathbf{R}$. To identify the redundancies, we first note that, since $\mathbf{RS}$ has eigenvalues 1 and 0, it constitutes a projector. Thus, partitioning the MO coefficient matrix $\mathbf{C}$ into two rectangular blocks containing the occupied and virtual MOs

$$\mathbf{C} = (\mathbf{C}_\mathrm{occ}\ \ \mathbf{C}_\mathrm{vrt}) \qquad (10.7.39)$$

we find that the matrices

$$\mathbf{P} = \mathbf{RS} \qquad (10.7.40)$$

$$\mathbf{Q} = \mathbf{1}_n - \mathbf{RS} \qquad (10.7.41)$$

project onto the occupied and virtual MO spaces, respectively:

$$\mathbf{PC} = \mathbf{C}\rho\mathbf{C}^{\mathrm{T}}\mathbf{SC} = \mathbf{C}\rho = (\mathbf{C}_{\mathrm{occ}} \quad \mathbf{0}) \tag{10.7.42}$$

$$\mathbf{QC} = (\mathbf{0} \quad \mathbf{C}_{\mathrm{vrt}}) \tag{10.7.43}$$

Next, using the idempotency of $\mathbf{P}$ and the orthogonality relation

$$\mathbf{PQ} = \mathbf{0} \tag{10.7.44}$$

we note that $\mathbf{R}$ commutes with $\mathbf{PXP}^{\mathrm{T}}$ and $\mathbf{QXQ}^{\mathrm{T}}$ in the sense that their S commutators vanish

$$[\mathbf{R}, \mathbf{PXP}^{\mathrm{T}}]_{\mathrm{S}} = \mathbf{RSPXP}^{\mathrm{T}} - \mathbf{PXP}^{\mathrm{T}}\mathbf{SR} = \mathbf{0} \tag{10.7.45}$$

$$[\mathbf{R}, \mathbf{QXQ}^{\mathrm{T}}]_{\mathrm{S}} = \mathbf{RSQXQ}^{\mathrm{T}} - \mathbf{QXQ}^{\mathrm{T}}\mathbf{SR} = \mathbf{0} \tag{10.7.46}$$

The asymmetric BCH expansions of $\mathbf{R(PXP)}$ and $\mathbf{R(QXQ)}$ therefore terminate after one term:

$$\mathbf{R(PXP}^{\mathrm{T}}) = \exp(-\mathbf{PXP}^{\mathrm{T}}\mathbf{S})\mathbf{R}\exp(\mathbf{SPXP}^{\mathrm{T}}) = \mathbf{R(0)} \tag{10.7.47}$$

$$\mathbf{R(QXQ}^{\mathrm{T}}) = \exp(-\mathbf{QXQ}^{\mathrm{T}}\mathbf{S})\mathbf{R}\exp(\mathbf{SQXQ}^{\mathrm{T}}) = \mathbf{R(0)} \tag{10.7.48}$$

The redundancies of the exponential parametrization are thus related to rotations among the occupied orbitals and among the virtual ones, which do not change the density matrix of the determinant, in agreement with our identification of redundant orbital rotations in Section 10.1.2.

By group-theoretical arguments similar to those invoked for (3.1.22), it can be shown that, for any antisymmetric matrix $\mathbf{X}$, the matrix $\exp(\mathbf{SX})$ may be written in the form

$$\exp(\mathbf{SX}) = \exp[\mathbf{S}(\mathbf{PX}_1\mathbf{P}^{\mathrm{T}} + \mathbf{QX}_1\mathbf{Q}^{\mathrm{T}})]\exp[\mathbf{S}(\mathbf{PX}_2\mathbf{Q}^{\mathrm{T}} + \mathbf{QX}_2\mathbf{P}^{\mathrm{T}})] \tag{10.7.49}$$

where the antisymmetric matrices $\mathbf{X}_1$ and $\mathbf{X}_2$ are in general different from $\mathbf{X}$ and where the matrices $\mathbf{SPX}_1\mathbf{P}^{\mathrm{T}}$ and $\mathbf{SQX}_1\mathbf{Q}^{\mathrm{T}}$ commute. Substituting (10.7.49) in (10.7.32) and using (10.7.47) and (10.7.48), we find that a general transformation of the density matrix may be written as

$$\mathbf{R(X)} = \exp[-(\mathbf{PXQ}^{\mathrm{T}} + \mathbf{QXP}^{\mathrm{T}})\mathbf{S}]\mathbf{R}\exp[\mathbf{S}(\mathbf{PXQ}^{\mathrm{T}} + \mathbf{QXP}^{\mathrm{T}})] \tag{10.7.50}$$

Alternatively, we may write the AO density matrix in the form

$$\mathbf{R(X}_{\mathrm{ov}}) = \exp(-\mathbf{X}_{\mathrm{ov}}\mathbf{S})\mathbf{R}\exp(\mathbf{SX}_{\mathrm{ov}}) \tag{10.7.51}$$

where the antisymmetric matrix $\mathbf{X}_{\mathrm{ov}}$ satisfies the projection relation

$$\mathbf{X}_{\mathrm{ov}} = \mathbf{PX}_{\mathrm{ov}}\mathbf{Q}^{\mathrm{T}} + \mathbf{QX}_{\mathrm{ov}}\mathbf{P}^{\mathrm{T}} \tag{10.7.52}$$

This parametrization may readily be specialized to the orthonormal basis of MOs, where the nonredundant matrix $\mathbf{X}_{\mathrm{ov}}$ has nonzero elements only in the occupied–virtual blocks.

10.7.6 PURIFICATION OF THE DENSITY MATRIX

The exponential parametrization developed in Sections 10.7.4 and 10.7.5 enables us to transform between different 'pure' density matrices, all of which satisfy conditions (10.7.25)–(10.7.27) exactly. Often, however, we are presented not with a pure density matrix but with a 'contaminated' one, which satisfies these conditions only approximately. We shall here consider how to purify such a contaminated density matrix, so that it becomes a valid density matrix for a single-determinant wave function.

The symmetry condition (10.7.25) is trivially satisfied by a simple averaging of $\mathbf{R}$ and requires no further attention. We also note that, provided we can come up with a purification scheme for the idempotency of $\mathbf{RS}$ (10.7.27), we need not worry about its trace (10.7.26) since the idempotency ensures that it has an integral trace. In short, for the purposes of purifying the density matrix, we need only concern ourselves with the idempotency.

We consider a matrix $\mathbf{P} = \mathbf{RS}$ in (10.7.40) that is contaminated in the sense that it almost satisfies the idempotency condition

$$\mathbf{P}^2 - \mathbf{P} \approx \mathbf{0} \tag{10.7.53}$$

and we wish to purify this matrix such that it becomes idempotent:

$$(\mathbf{P} + \delta\mathbf{P})(\mathbf{P} + \delta\mathbf{P}) - (\mathbf{P} + \delta\mathbf{P}) = \mathbf{0} \tag{10.7.54}$$

The matrix equation (10.7.54) is satisfied when the sum of the squares of all elements on the left-hand side is equal to zero:

$$\mathrm{Tr}[(\mathbf{P} + \delta\mathbf{P})^2 - (\mathbf{P} + \delta\mathbf{P})]^2 = 0 \tag{10.7.55}$$

Solving this equation to first order in $\delta\mathbf{P}$, we obtain

$$\mathrm{Tr}\delta\mathbf{P}(4\mathbf{P}^3 - 6\mathbf{P}^2 + 2\mathbf{P}) = 0 \tag{10.7.56}$$

which is equivalent to the condition that the matrix multiplying $\delta\mathbf{P}$ should vanish:

$$4\mathbf{P}^3 - 6\mathbf{P}^2 + 2\mathbf{P} = \mathbf{0} \tag{10.7.57}$$

This equation forms the basis for the following fixed-point problem for $\mathbf{P}$

$$\mathbf{P}_{n+1} = 3\mathbf{P}_n^2 - 2\mathbf{P}_n^3 \tag{10.7.58}$$

or equivalently, in terms of $\mathbf{R}$,

$$\mathbf{R}_{n+1} = 3\mathbf{R}_n\mathbf{S}\mathbf{R}_n - 2\mathbf{R}_n\mathbf{S}\mathbf{R}_n\mathbf{S}\mathbf{R}_n \tag{10.7.59}$$

This fixed-point iteration provides a highly efficient scheme for the purification of density matrices [8]. As shown in Section 10.7.7, it converges quadratically in the local region and has robust global convergence characteristics.

10.7.7 CONVERGENCE OF THE PURIFICATION SCHEME

In this subsection, we shall first consider the local and then the global convergence properties of the fixed-point iteration (10.7.59). Let us assume that, in the nth iteration, we have generated the density matrix

$$\mathbf{R}_n = \mathbf{R}_\infty + \tau \tag{10.7.60}$$

where $\mathbf{R}_\infty$ is idempotent in the sense of (10.7.27) and where τ is a symmetric matrix that represents its contamination. Before the nth iteration is carried out, the deviation from idempotency is linear in τ:

$$\mathbf{R}_n\mathbf{S}\mathbf{R}_n - \mathbf{R}_n = (\mathbf{R}_\infty + \tau)\mathbf{S}(\mathbf{R}_\infty + \tau) - (\mathbf{R}_\infty + \tau)$$
$$= \mathbf{R}_\infty\mathbf{S}\tau + \tau\mathbf{S}\mathbf{R}_\infty - \tau + O(\tau^2) \tag{10.7.61}$$

After the nth iteration, the partially purified density matrix may be written as

$$\mathbf{R}_{n+1} = \mathbf{R}_\infty + \mathbf{R}_\infty \mathbf{S}\tau + \tau \mathbf{S}\mathbf{R}_\infty - 2\mathbf{R}_\infty \mathbf{S}\tau \mathbf{S}\mathbf{R}_\infty + O(\tau^2) \tag{10.7.62}$$

Inserting this expression for $\mathbf{R}_{n+1}$ in the idempotency condition (10.7.27), we find that all terms vanish to first order in τ:

$$\mathbf{R}_{n+1}\mathbf{S}\mathbf{R}_{n+1} - \mathbf{R}_{n+1} = O(\tau^2) \tag{10.7.63}$$

The fixed-point iteration thus converges *quadratically* towards idempotency. Therefore, for any density matrix sufficiently close to idempotency, the fixed-point iteration (10.7.59) converges rapidly to an idempotent matrix.

We now turn our attention to the global convergence properties of the purification scheme (10.7.59). To simplify the discussion, consider the scalar fixed-point iteration

$$x_{n+1} = f(x_n) \tag{10.7.64}$$

where the function $f(x)$ is given by

$$f(x) = 3x^2 - 2x^3 \tag{10.7.65}$$

In Figure 10.1, we have plotted this function, superimposed on the straight line $g(x) = \frac{1}{2}$ with the points of intersections marked. We note that $f(x)$ has stationary points at 0 and 1 (also marked), corresponding to the eigenvalues of the purified density matrix. From an inspection of the figure, we find that any fixed-point iteration (10.7.64) that begins in the interval

$$\frac{1 - \sqrt{3}}{2} < x_0 < \frac{1}{2} \tag{10.7.66}$$

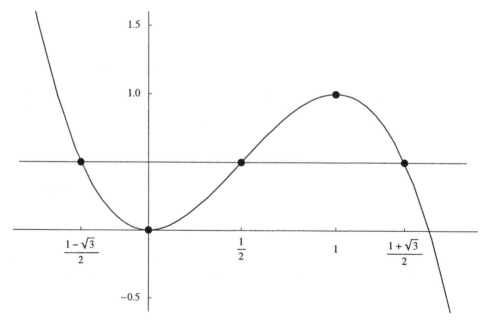

Fig. 10.1. The function $f(x) = 3x^2 - 2x^3$ superimposed on the line $g(x) = \frac{1}{2}$.

will proceed in the interval $0 < x_n < \frac{1}{2}$ and converge towards 0 from above. Similarly, for an initial x_0 in the interval

$$\frac{1}{2} < x_0 < \frac{1 + \sqrt{3}}{2} \tag{10.7.67}$$

the iterations will proceed within $\frac{1}{2} < x_n < 1$ and converge to 1 from below. More generally, the fixed-point iteration (10.7.64) may be shown to converge to 0 or 1 for initial values in the larger range

$$\frac{1 - \sqrt{5}}{2} < x_0 < \frac{1 + \sqrt{5}}{2} \tag{10.7.68}$$

although the convergence will not always be to the stationary point closest to x_0 if x_0 is outside the range defined by (10.7.66) and (10.7.67). We conclude that the purification scheme (10.7.59) is robust and rapidly convergent for most of the contaminated matrices that occur in practice.

10.7.8 THE HARTREE–FOCK ENERGY AND THE VARIATIONAL CONDITIONS

According to the discussions in the preceding subsections, we may write the Hartree–Fock energy (10.7.6) in the form

$$E(\mathbf{X}) = 2\mathrm{Tr}\,\mathbf{R}(\mathbf{X})\mathbf{h}^{\mathrm{AO}} + \mathrm{Tr}\,\mathbf{R}(\mathbf{X})\mathbf{G}^{\mathrm{AO}}(\mathbf{R}(\mathbf{X})) + h_{\mathrm{nuc}} \tag{10.7.69}$$

where $\mathbf{R}(\mathbf{X})$ is given by (10.7.32) and where we have used the notation (10.7.7). We recall that the elements of the antisymmetric matrix $\mathbf{X}$ may be chosen freely but to avoid redundancies (which is often important for optimizations), it should comply with the projection relation

$$\mathbf{X} = \mathbf{P}\mathbf{X}\mathbf{Q}^{\mathrm{T}} + \mathbf{Q}\mathbf{X}\mathbf{P}^{\mathrm{T}} \tag{10.7.70}$$

The independent parameters of $\mathbf{X}$ are the elements $X_{\mu\nu}$ with $\mu > \nu$. Let us introduce the *elementary matrices* or *matrix units* $\mathbf{E}_{\mu\nu}$ with elements

$$[\mathbf{E}_{\mu\nu}]_{\rho\sigma} = \delta_{\mu\rho}\delta_{\nu\sigma} \tag{10.7.71}$$

and their antisymmetric combinations

$$\mathbf{E}_{\mu\nu}^{-} = \mathbf{E}_{\mu\nu} - \mathbf{E}_{\nu\mu} \tag{10.7.72}$$

We may now write $\mathbf{X}$ in terms of the independent parameters as

$$\mathbf{X} = \sum_{\mu > \nu} X_{\mu\nu}(\mathbf{E}_{\mu\nu} - \mathbf{E}_{\nu\mu}) = \sum_{\mu > \nu} X_{\mu\nu}\mathbf{E}_{\mu\nu}^{-} \tag{10.7.73}$$

by analogy with the orbital-rotation operator of second quantization (3.3.22). In passing, we note that the excitation operators E_{pq} and the matrices $\mathbf{E}_{\mu\nu}$ satisfy the same commutation relations.

For the optimization of the energy (10.7.69), we may in principle apply any scheme developed for the unconstrained minimization of multivariate functions – for example, some globally convergent modification of the Newton method or some quasi-Newton scheme. Expanding the energy to second order by analogy with (10.1.21), we obtain

$$E(\mathbf{x}) = E^{(0)} + \mathbf{x}^{\mathrm{T}}\mathbf{E}^{(1)} + \tfrac{1}{2}\mathbf{x}^{\mathrm{T}}\mathbf{E}^{(2)}\mathbf{x} + \cdots \tag{10.7.74}$$

where the vector $\mathbf{x}$ contains the independent elements of the antisymmetric matrix $\mathbf{X}$. To obtain the expressions for the electronic gradient and the Hessian, we expand the energy (10.7.69) to second order:

$$
\begin{aligned}
E(\mathbf{X}) = E^{(0)} &+ \mathrm{Tr}(2\mathbf{h}^{\mathrm{AO}} + \mathbf{G}^{\mathrm{AO}}(\mathbf{R}))[\mathbf{R}, \mathbf{X}]_{\mathrm{S}} + \mathrm{Tr}\, \mathbf{RG}^{\mathrm{AO}}([\mathbf{R}, \mathbf{X}]_{\mathrm{S}}) \\
&+ \tfrac{1}{2}\mathrm{Tr}(2\mathbf{h}^{\mathrm{AO}} + \mathbf{G}^{\mathrm{AO}}(\mathbf{R}))[[\mathbf{R}, \mathbf{X}]_{\mathrm{S}}, \mathbf{X}]_{\mathrm{S}} + \tfrac{1}{2}\mathrm{Tr}\, \mathbf{RG}^{\mathrm{AO}}([[\mathbf{R}, \mathbf{X}]_{\mathrm{S}}, \mathbf{X}]_{\mathrm{S}}) \\
&+ \mathrm{Tr}\, \mathbf{G}^{\mathrm{AO}}([\mathbf{R}, \mathbf{X}]_{\mathrm{S}})[\mathbf{R}, \mathbf{X}]_{\mathrm{S}} + O(\mathbf{X}^3)
\end{aligned}
\tag{10.7.75}
$$

We note that, for two symmetric matrices $\mathbf{A}$ and $\mathbf{B}$, the following identity holds

$$
\mathrm{Tr}\, \mathbf{A}\mathbf{G}^{\mathrm{AO}}(\mathbf{B}) = \mathrm{Tr}\, \mathbf{B}\mathbf{G}^{\mathrm{AO}}(\mathbf{A}) \quad (\mathbf{A} \text{ and } \mathbf{B} \text{ symmetric})
\tag{10.7.76}
$$

which may be used to simplify the second-order expansion of the energy:

$$
\begin{aligned}
E(\mathbf{X}) = E^{(0)} &+ 2\mathrm{Tr}\, \mathbf{f}^{\mathrm{AO}}[\mathbf{R}, \mathbf{X}]_{\mathrm{S}} + \mathrm{Tr}\, \mathbf{f}^{\mathrm{AO}}[[\mathbf{R}, \mathbf{X}]_{\mathrm{S}}, \mathbf{X}]_{\mathrm{S}} \\
&+ \mathrm{Tr}\, \mathbf{G}^{\mathrm{AO}}([\mathbf{R}, \mathbf{X}]_{\mathrm{S}})[\mathbf{R}, \mathbf{X}]_{\mathrm{S}} + O(\mathbf{X}^3)
\end{aligned}
\tag{10.7.77}
$$

Taking the first derivative at the point of expansion, we obtain the electronic gradient:

$$
E^{(1)}_{\mu\nu} = 2\mathrm{Tr}\, \mathbf{f}^{\mathrm{AO}}[\mathbf{R}, \mathbf{E}^-_{\mu\nu}]_{\mathrm{S}} = 2\mathrm{Tr}\, \mathbf{E}^-_{\mu\nu}(\mathbf{f}^{\mathrm{AO}}\mathbf{R}\mathbf{S} - \mathbf{S}\mathbf{R}\mathbf{f}^{\mathrm{AO}})
\tag{10.7.78}
$$

This result may be simplified by using the identities

$$
\mathrm{Tr}\, \mathbf{E}^-_{\mu\nu}\mathbf{A} = 2\mathrm{Tr}\, \mathbf{E}_{\mu\nu}\mathbf{A} \quad (\mathbf{A} \text{ antisymmetric})
\tag{10.7.79}
$$

$$
\mathrm{Tr}\, \mathbf{E}_{\mu\nu}\mathbf{M} = M_{\nu\mu} \quad (\text{all } \mathbf{M})
\tag{10.7.80}
$$

and we arrive at the following expression for the *Hartree–Fock electronic gradient*:

$$
E^{(1)}_{\mu\nu} = 4(\mathbf{S}\mathbf{R}\mathbf{f}^{\mathrm{AO}} - \mathbf{f}^{\mathrm{AO}}\mathbf{R}\mathbf{S})_{\mu\nu}
\tag{10.7.81}
$$

Differentiating the second-order energy (10.7.77) twice, we may identify the elements of the *electronic Hessian* as

$$
\begin{aligned}
E^{(2)}_{\mu\nu\rho\sigma} &= (1 + P_{\mu\nu,\rho\sigma})\, \mathrm{Tr}\, \mathbf{f}^{\mathrm{AO}}[[\mathbf{R}, \mathbf{E}^-_{\mu\nu}]_{\mathrm{S}}, \mathbf{E}^-_{\rho\sigma}]_{\mathrm{S}} \\
&+ (1 + P_{\mu\nu,\rho\sigma})\, \mathrm{Tr}\, \mathbf{G}^{\mathrm{AO}}([\mathbf{R}, \mathbf{E}^-_{\mu\nu}]_{\mathrm{S}})[\mathbf{R}, \mathbf{E}^-_{\rho\sigma}]_{\mathrm{S}}
\end{aligned}
\tag{10.7.82}
$$

where the operator $P_{\mu\nu,\rho\sigma}$ permutes the pair indices $\mu\nu$ and $\rho\sigma$. As we have no need for the explicit form of the Hessian, we shall not develop (10.7.82) further. In Section 10.8.2, we shall instead evaluate products of the Hessian matrix with vectors, without constructing the Hessian elements (10.7.82) directly.

At convergence, the elements of the Hartree–Fock gradient (10.7.81) are all zero and we obtain for $\mu > \nu$:

$$
E^{(1)}_{\mu\nu} = 0
\tag{10.7.83}
$$

Extending the definition of the gradient (10.7.81) to a full antisymmetric matrix (noting that the diagonal elements are zero by symmetry), we find that we may write the *Hartree–Fock variational conditions* in the form

$$
\mathbf{f}^{\mathrm{AO}}\mathbf{R}\mathbf{S} = \mathbf{S}\mathbf{R}\mathbf{f}^{\mathrm{AO}}
\tag{10.7.84}
$$

The corresponding MO conditions are obtained by multiplying (10.7.84) from the left by $\mathbf{C}^T$ and from the right by $\mathbf{C}$. Using (10.7.20) and then the orthonormality relation (10.7.2), we obtain

$$\mathbf{f}\rho = \rho\mathbf{f} \tag{10.7.85}$$

In the MO basis, therefore, the Fock matrix commutes with the density matrix at convergence.

For a first-order minimization of the Hartree–Fock energy, we need to calculate explicitly in each iteration only the energy and the gradient. For a given pure density $\mathbf{R}_n$, the energy and the gradient may be obtained from the one-electron Hamiltonian, the Fock matrix and the overlap matrix according to the expressions

$$E_n^{(0)} = \text{Tr } \mathbf{R}_n\mathbf{h}^{AO} + \text{Tr } \mathbf{R}_n\mathbf{f}_n^{AO} + h_{\text{nuc}} \tag{10.7.86}$$

$$\mathbf{E}_n^{(1)} = -4\mathbf{f}_n^{AO}\mathbf{R}_n\mathbf{S} + 4\mathbf{S}\mathbf{R}_n\mathbf{f}_n^{AO} \tag{10.7.87}$$

where $\mathbf{f}_n^{AO}$ is the Fock matrix calculated from $\mathbf{R}_n$. For a second-order minimization of the energy, we also need the Hessian; see Section 10.8.2. In either case, we generate in each iteration – by some suitable scheme – an antisymmetric matrix $\mathbf{X}_n$ in terms of which we obtain the new density matrix

$$\mathbf{R}_{n+1} = \exp(-\mathbf{X}_n\mathbf{S})\mathbf{R}_n\exp(\mathbf{S}\mathbf{X}_n) \tag{10.7.88}$$

The updated density matrix is then obtained by truncating the BCH expansion at some low order, followed by the purification as described in Section 10.7.6.

10.7.9 THE DENSITY-BASED SCF METHOD

In Section 10.7.8, we outlined the optimization of the Hartree–Fock energy by a direct minimization of the energy function (10.7.69) with respect to the density matrix, using a standard first- or second-order scheme. This approach has many advantages – for example, the quadratic convergence of Newton's method close to the minimum. However, for the optimization of the Hartree–Fock energy, the SCF method has in practice proved to be more useful – in particular, when combined with the DIIS acceleration scheme. In this subsection, we shall therefore present a density-matrix formulation of the Roothaan–Hall DIIS approach, avoiding the diagonalization of the Fock matrix.

In the orbital-based SCF method as formulated in Section 10.6.1, the key step in each iteration is the solution of the Roothaan–Hall equations (10.6.16)

$$\mathbf{f}_n^{AO}\mathbf{C} = \mathbf{S}\mathbf{C}\varepsilon \tag{10.7.89}$$

followed by the construction of the AO density matrix in the form

$$\mathbf{R}_{n+1} = \mathbf{C}^T\rho\mathbf{C} \tag{10.7.90}$$

From this density, a new Fock operator is calculated and this process is repeated until convergence – that is, until the density obtained from (10.7.90) becomes identical to the density from which the Fock matrix was calculated. Since this approach involves the solution of a generalized eigenvalue problem (10.7.89), whose cost scales cubically with the size of the system, it becomes prohibitively expensive for large systems. In the *density-based SCF method*, we avoid the diagonalization step altogether, replacing it by a direct optimization of the AO density matrix.

We begin by noting that the solution of the Roothaan–Hall equations is equivalent to the optimization of the sum of the orbital energies of the occupied MOs

$$E_\varepsilon = 2\sum_i \varepsilon_i = 2\sum_i \langle \phi_i | \hat{f} | \phi_i \rangle \tag{10.7.91}$$

subject to the orthonormality constraints on the occupied MOs:

$$\langle \phi_i | \phi_j \rangle = \delta_{ij} \tag{10.7.92}$$

The optimization is here with respect to the MO coefficients, with the density matrix in the Fock operator $\hat{f}$ kept fixed. To see the equivalence, we introduce the Lagrangian

$$L = E_\varepsilon - 2\sum_{ij} \lambda_{ij}(\langle \phi_i | \phi_j \rangle - \delta_{ij}) \tag{10.7.93}$$

and find that the constrained variational conditions on (10.7.91) may be expressed as

$$f_{\mu k} = \sum_j S_{\mu j}\lambda_{jk} \tag{10.7.94}$$

These are the same equations as (10.6.8) and the optimization of (10.7.91) subject to (10.7.92) is therefore equivalent to the solution of the Roothaan–Hall equations. Since the AO density matrix is invariant to rotations among the occupied orbitals, the solution to (10.7.94) produces the same density matrix as do the canonical orbitals.

We now consider the density-based optimization of E_ε. In the AO basis, we may express (10.7.91) in terms of the density matrix $\mathbf{R}$ and the Fock matrix $\mathbf{f}_n^{AO}$, without reference to the MOs

$$E_\varepsilon = 2\text{Tr}\rho\varepsilon = 2\text{Tr}\rho\mathbf{C}^T\mathbf{f}_n^{AO}\mathbf{C} = 2\text{Tr}\,\mathbf{R}\mathbf{f}_n^{AO} \tag{10.7.95}$$

where, in the second step, we have used the relation obtained by left-multiplying (10.7.89) by $\mathbf{C}^T$ and invoking the orthonormality relation (10.7.2). Introducing the usual exponential parametrization of the AO density matrix

$$\mathbf{R} = \exp(-\mathbf{XS})\mathbf{R}_n \exp(\mathbf{SX}) \tag{10.7.96}$$

we find that the solution of the Roothaan–Hall eigenvalue problem implies the conditions

$$\frac{\partial E_\varepsilon(\mathbf{X})}{\partial X_{\mu\nu}} = 0 \tag{10.7.97}$$

on the function

$$E_\varepsilon(\mathbf{X}) = 2\text{Tr}\exp(-\mathbf{XS})\mathbf{R}_n \exp(\mathbf{SX})\mathbf{f}_n^{AO} \tag{10.7.98}$$

For the electronic ground state, we thus minimize $E_\varepsilon(\mathbf{X})$ with respect to $\mathbf{X}$ and calculate the new density matrix according to (10.7.96) from the solution $\mathbf{X}_n$:

$$\mathbf{R}_{n+1} = \exp(-\mathbf{X}_n\mathbf{S})\mathbf{R}_n \exp(\mathbf{SX}_n) \tag{10.7.99}$$

The minimization of $E_\varepsilon(\mathbf{X})$ is discussed in Section 10.7.10. Since the density matrix $\mathbf{R}_{n+1}$ obtained in this manner is the same as that obtained from (10.7.90), the resulting SCF iterations, generated in the density-based optimization of E_ε, are the same as those of the standard Roothaan–Hall scheme.

10.7.10 OPTIMIZATION OF THE SCF ORBITAL-ENERGY FUNCTION

In each SCF iteration, the Hartree–Fock orbital-energy function (10.7.98) may be minimized using any of the standard iterative methods of unconstrained optimization. Moreover, since the optimization of $E_\varepsilon(\mathbf{X})$ requires no re-evaluation of the Fock matrix, each step is relatively inexpensive.

To set up a second-order scheme for the minimization of (10.7.98), let $\mathbf{R}_{ni}$ be our current approximation to the next density matrix $\mathbf{R}_{n+1}$. We wish to determine an antisymmetric matrix $\mathbf{X}$ from which an improved approximation to $\mathbf{R}_{n+1}$ is obtained as:

$$\mathbf{R}_{n,i+1} = \exp(-\mathbf{XS})\mathbf{R}_{ni}\exp(\mathbf{SX}) \tag{10.7.100}$$

The corresponding energy function now becomes

$$E_\varepsilon(\mathbf{X})_{ni} = 2\mathrm{Tr}\exp(-\mathbf{XS})\mathbf{R}_{ni}\exp(\mathbf{SX})\mathbf{f}_n^{AO} \tag{10.7.101}$$

Setting the derivative of the second-order expansion of (10.7.101) with respect to $X_{\mu\nu}$ equal to zero, we obtain an equation for $X_{\mu\nu}$ with $\mu > \nu$:

$$4\mathrm{Tr}\,\mathbf{f}_n^{AO}[\mathbf{R}_{ni}, \mathbf{E}_{\mu\nu}]_S = -2\mathrm{Tr}\,\mathbf{f}_n^{AO}[[\mathbf{R}_{ni}, \mathbf{E}_{\mu\nu}]_S, \mathbf{X}]_S - 2\mathrm{Tr}\,\mathbf{f}_n^{AO}[[\mathbf{R}_{ni}, \mathbf{X}]_S, \mathbf{E}_{\mu\nu}]_S \tag{10.7.102}$$

Using (10.7.80), we then arrive at a set of linear equations for $\mathbf{X}$ as the upper triangle of the matrix equation

$$\mathbf{f}_n^{AO}\mathbf{R}_{ni}\mathbf{S} - \mathbf{S}\mathbf{R}_{ni}\mathbf{f}_n^{AO} = \mathbf{f}_n^{AO}\mathbf{XS}\mathbf{R}_{ni}\mathbf{S} + \mathbf{S}\mathbf{R}_{ni}\mathbf{SX}\mathbf{f}_n^{AO} - \tfrac{1}{2}\mathbf{SX}(\mathbf{f}_n^{AO}\mathbf{R}_{ni}\mathbf{S} + \mathbf{S}\mathbf{R}_{ni}\mathbf{f}_n^{AO})$$
$$- \tfrac{1}{2}(\mathbf{f}_n^{AO}\mathbf{R}_{ni}\mathbf{S} + \mathbf{S}\mathbf{R}_{ni}\mathbf{f}_n^{AO})\mathbf{XS} \tag{10.7.103}$$

Next, introducing the following notation for the symmetric and antisymmetric parts of a matrix

$$[\mathbf{M}]^S = \tfrac{1}{2}(\mathbf{M} + \mathbf{M}^T) \tag{10.7.104}$$

$$[\mathbf{M}]^A = \tfrac{1}{2}(\mathbf{M} - \mathbf{M}^T) \tag{10.7.105}$$

we may write (10.7.103) in the more compact form

$$[\mathbf{f}_n^{AO}\mathbf{R}_{ni}\mathbf{S}]^A = [\mathbf{f}_n^{AO}\mathbf{XS}\mathbf{R}_{ni}\mathbf{S}]^A - [[\mathbf{f}_n^{AO}\mathbf{R}_{ni}\mathbf{S}]^S\mathbf{XS}]^A \tag{10.7.106}$$

The solution to these equations yields $\mathbf{X}_i$, which may be used to generate a new $\mathbf{R}_{n,i+1}$ according to (10.7.100). This process is repeated until, at convergence, we have obtained the density matrix $\mathbf{R}_{n+1}$ from which the new Fock operator is constructed.

As an alternative to determining the stationary value of E_ε by the second-order method just described, we may employ a first-order scheme, which requires only the gradient

$$\mathbf{E}^{(1)} = 8[\mathbf{S}\mathbf{R}_{ni}\mathbf{f}_n^{AO}]^A \tag{10.7.107}$$

in order to generate the new step. Such a first-order method may, for example, be the conjugate-gradient method.

10.7.11 LINEAR SCALING OF THE DENSITY-BASED SCF SCHEME

The density-based optimization of the Hartree–Fock energy is well suited to applications to large systems [9,11–15]. For such systems, we note the following important advantages of the density-based scheme over the Roothaan–Hall scheme: it involves only one-electron matrices calculated in the AO basis (i.e. the one-electron density matrix, the one-electron Hamiltonian, the Fock operator and the overlap matrix); it requires no matrix manipulations except the multiplication and addition of sparse matrices. The computations may then be carried out at a cost that scales linearly with the size of the system, for the following reasons.

First, as discussed for the integrals in Section 9.12, for large systems, the AO matrices entering the calculations become sparse and the number of nonzero elements increases only linearly with the size of the system. The same is true also for the density matrix, which, for large systems, becomes sparse with negligible elements between well-separated AOs.

Second, the evaluation of the nonzero elements of the AO matrices may be carried out at a cost that scales linearly with the size of the system. In particular, the long-range Coulomb contributions to the Fock matrix may be evaluated efficiently using the FMM of Section 9.14. Moreover, because of the sparsity of the AO density matrix, the number of nonnegligible contributions to the exchange interaction scales linearly as well.

Third, all manipulations required for the density-based optimization of the Hartree–Fock energy may be carried out as matrix additions and multiplications. Owing to the sparsity of the matrices, the cost of these manipulations scales linearly with the size of the system.

10.8   Second-order optimization

In the preceding sections, we discussed the optimization of the Hartree–Fock wave function by the iterative SCF procedure, using an approach based on the diagonalization of the Fock matrix as well as an approach based on the optimization of the AO density matrix. The SCF method, in particular when accelerated with the DIIS scheme, is simple to implement and works well in many cases, especially for closed-shell systems close to the equilibrium geometry. Nevertheless, in some cases, convergence may be difficult to achieve even with the DIIS method. In such cases, it may be better to use one of the standard methods of numerical analysis.

In this section, we shall apply the second-order Newton method to the optimization of the Hartree–Fock energy, considering both a method that works with the AO density matrix and is applicable to large systems, and a method that carries out rotations among the MOs and is applicable to general single-configuration states [16–18]. Our discussion is here restricted to minimizations – in the discussion of MCSCF theory in Chapter 12, we shall consider also the second-order localization of saddle points.

10.8.1   NEWTON'S METHOD

In the nth iteration of *Newton's method*, we first generate a local model of the energy function $E(\mathbf{z})$ by truncating the Taylor expansion of the energy at second order in $\mathbf{z}$:

$$Q_n(\mathbf{z}) = E_n^{(0)} + \Delta \mathbf{z}^{\mathrm{T}} \mathbf{E}_n^{(1)} + \tfrac{1}{2} \Delta \mathbf{z}^{\mathrm{T}} \mathbf{E}_n^{(2)} \Delta \mathbf{z} \tag{10.8.1}$$

In this expression, $\mathbf{z}$ may refer either to the MO parameters $\boldsymbol{\kappa}$ of the expansion (10.1.21) or to the AO parameters $\mathbf{x}$ of the expansion (10.7.74). The coefficients $E_n^{(0)}$, $\mathbf{E}_n^{(1)}$ and $\mathbf{E}_n^{(2)}$ in (10.8.1) represent the energy, gradient and Hessian constructed at the expansion point $\mathbf{z} = \mathbf{0}$. The Newton step $\Delta \mathbf{z}_n$ is generated by optimizing this quadratic function with respect to $\mathbf{z}$. Thus, in the nth iteration, we solve the linear Newton equations

$$\mathbf{E}_n^{(2)} \Delta \mathbf{z}_n = -\mathbf{E}_n^{(1)} \tag{10.8.2}$$

and the parameters $\mathbf{z}_{n+1}$ that define the expansion point of the next iteration become

$$\mathbf{z}_{n+1} = \mathbf{z}_n + \Delta\mathbf{z}_n \tag{10.8.3}$$

The Newton sequence of iterations is thus established. The iterations are continued until the norm of the gradient $\mathbf{E}_n^{(1)}$ is smaller than a given threshold or until some other convergence criterion is fulfilled. The most important characteristic of Newton's method is the *second-order local convergence* – that is, sufficiently close to the minimizer, the error in iteration $n+1$ is quadratic in the error in iteration n.

The simple Newton algorithm outlined above does not converge unconditionally. Indeed, unless we are sufficiently close to the minimizer of $E(\mathbf{z})$, there is no guarantee that a step to the stationary point of the quadratic model $Q_n(\mathbf{z})$ will bring us closer to the minimizer of the true function $E(\mathbf{z})$. To understand this, we note that the second-order model $Q_n(\mathbf{z})$ gives a reasonable representation of the true energy function $E(\mathbf{z})$ only in some restricted region around the expansion point – the *trust region*. In practice, we define the trust region as a hypersphere of radius h_n around the expansion point, $\|\mathbf{z}\| \leq h_n$. Only if there is a minimizer of $Q_n(\mathbf{z})$ inside the trust region can we expect that a step to this point will bring us closer to the minimizer of the true function $E(\mathbf{z})$.

In the *trust-region* or *restricted-step method* [19], the Newton step (10.8.2) is taken only if it is smaller than or equal to the *trust radius* h_n:

$$\|\Delta\mathbf{z}_n\| \leq h_n \tag{10.8.4}$$

If the Newton step (10.8.2) is larger than the trust radius, it is rejected and a new step is generated to the boundary of the trust region. This step is calculated by minimizing the Lagrangian

$$L_n(\mathbf{z}) = E_n^{(0)} + \Delta\mathbf{z}^{\mathrm{T}}\mathbf{E}_n^{(1)} + \frac{1}{2}\Delta\mathbf{z}^{\mathrm{T}}\mathbf{E}_n^{(2)}\Delta\mathbf{z} - \frac{\mu}{2}(\Delta\mathbf{z}^{\mathrm{T}}\Delta\mathbf{z} - h_n^2) \tag{10.8.5}$$

where we have introduced an undetermined multiplier μ for the constraint that the step should be to the boundary of the trust region. Optimizing this function with respect to $\Delta\mathbf{z}$, we obtain the modified Newton equations

$$(\mathbf{E}_n^{(2)} - \mu\mathbf{1})\Delta\mathbf{z}_n = -\mathbf{E}_n^{(1)} \tag{10.8.6}$$

where the multiplier or *level-shift parameter* μ is adjusted so that the step is to the boundary of the trust region

$$\|\Delta\mathbf{z}_n(\mu)\| = h_n \tag{10.8.7}$$

A number of solutions satisfy equations (10.8.6) and (10.8.7). In minimizations, the solution of the lowest energy is selected. The trust radius h_n is updated from iteration to iteration by a feedback mechanism described in more detail in Section 12.3.

The trust-region method constitutes a powerful and robust method for the minimization or more generally for the optimization of nonlinear functions. In the global region of the optimization – where the trust region does not encompass the true minimizer and the Newton step must be rejected – the trust-region method gives a reliable step in the right direction. In the local region of the search, the Newton step is inside the trust region and Newton's method will then usually converge to the stationary point in a few iterations, with the error decreasing quadratically in each iteration. The local convergence rates of the Newton method and other methods used in Hartree–Fock theory are compared in Section 10.9.5.

In the Newton trust-region method, we must in each iteration solve a linear set of equations (10.8.2) or (10.8.6). The number of nonredundant parameters is usually so large that these equations are best solved by an *iterative method*, avoiding the explicit inversion or decomposition of the Hessian matrix. In iterative methods, a number of linear transformations of *trial vectors* $\mathbf{z}$

$$\boldsymbol{\sigma} = \mathbf{E}^{(2)}\mathbf{z} \tag{10.8.8}$$

are carried out in each iteration. In principle, such matrix–vector products may be calculated by first constructing the Hessian $\mathbf{E}^{(2)}$ explicitly and then multiplying it by the trial vectors $\mathbf{z}$. In practice, however, the dimension of the Hessian is often so large that this approach is not possible. For example, in calculations involving 100 occupied and 400 virtual orbitals, the number of nonredundant rotations is 40 000 and the Hessian contains about 1.6 billion elements. Clearly, it is not practical to calculate or store such a matrix in memory. In the present section, we shall therefore describe how the linear transformations (10.8.8) can be carried out without calculating the Hessian explicitly, making it possible to treat cases of large dimensions.

To illustrate the convergence of the Newton trust-region method, we have listed in Table 10.4, for each iteration, the errors in the energies in the optimization of the Hartree–Fock wave function at the cc-pVQZ level. The convergence is smooth – both at the equilibrium geometry and at the stretched geometry – and the convergence is quadratic close to the minimum. In terms of iterations, the performance of Newton's method is seen to be better than that of the SCF method and of the DIIS method. The smooth convergence of Newton's method and the fact that it can be applied to any Hartree–Fock wave function combine to make it a powerful algorithm indeed. We shall see, however, that each Newton step is an expensive one – more expensive than the SCF step or the DIIS step. Newton's method is therefore usually not used in situations where the standard SCF or DIIS algorithms work.

The remainder of Section 10.8 consists of two parts. In Section 10.8.2, we discuss the implementation of Newton's method within the framework of density-based Hartree–Fock theory, in particular the transformation of trial vectors (10.8.8). Next, in Sections 10.8.3–10.8.8, we consider Newton's method in orbital-based Hartree–Fock theory. The orbital-based treatment is more general than the density-based one in that it enables us to treat more general Hartree–Fock states, thus preparing us for the multiconfigurational states of MCSCF theory in Chapter 12.

10.8.2 DENSITY-BASED FORMULATION OF NEWTON'S METHOD

In density-based Hartree–Fock theory, the Newton iterations are established by minimizing the second-order energy (10.8.1) obtained by truncating (10.7.77) at second order in $\mathbf{X}$:

$$Q(\mathbf{X}) = E^{(0)} + 2\mathrm{Tr}\,\mathbf{f}^{\mathrm{AO}}[\mathbf{R}, \mathbf{X}]_\mathrm{S} + \mathrm{Tr}\,\mathbf{f}^{\mathrm{AO}}[[\mathbf{R}, \mathbf{X}]_\mathrm{S}, \mathbf{X}]_\mathrm{S} + \mathrm{Tr}\,\mathbf{G}^{\mathrm{AO}}([\mathbf{R}, \mathbf{X}]_\mathrm{S})[\mathbf{R}, \mathbf{X}]_\mathrm{S} \tag{10.8.9}$$

Taking the derivatives of $Q(\mathbf{X})$ with respect to $X_{\mu\nu}$ and setting the result equal to zero, we obtain the Newton equations

$$2\mathrm{Tr}\,\mathbf{f}^{\mathrm{AO}}[\mathbf{R}, \mathbf{E}_{\mu\nu}^-]_\mathrm{S} = -\mathrm{Tr}\,\mathbf{f}^{\mathrm{AO}}[[\mathbf{R}, \mathbf{X}]_\mathrm{S}, \mathbf{E}_{\mu\nu}^-]_\mathrm{S} - \mathrm{Tr}\,\mathbf{f}^{\mathrm{AO}}[[\mathbf{R}, \mathbf{E}_{\mu\nu}^-]_\mathrm{S}, \mathbf{X}]_\mathrm{S}$$
$$- 2\mathrm{Tr}\,\mathbf{G}^{\mathrm{AO}}([\mathbf{R}, \mathbf{X}]_\mathrm{S})[\mathbf{R}, \mathbf{E}_{\mu\nu}^-]_\mathrm{S} \tag{10.8.10}$$

where we have used (10.7.76). Invoking the Jacobi identity generalized to S commutators

$$[\mathbf{A}, [\mathbf{B}, \mathbf{C}]_\mathrm{S}]_\mathrm{S} + [\mathbf{B}, [\mathbf{C}, \mathbf{A}]_\mathrm{S}]_\mathrm{S} + [\mathbf{C}, [\mathbf{A}, \mathbf{B}]_\mathrm{S}]_\mathrm{S} = \mathbf{0} \tag{10.8.11}$$

and the identity

$$\mathrm{Tr}\,\mathbf{A}[\mathbf{B}, \mathbf{E}_{\mu\nu}]_\mathrm{S} = -\mathrm{Tr}\,\mathbf{A}[\mathbf{B}, \mathbf{E}_{\nu\mu}]_\mathrm{S} \quad (\mathbf{A} \text{ and } \mathbf{B} \text{ symmetric}) \tag{10.8.12}$$

for symmetric matrices, we may rewrite (10.8.10) as

$$4\text{Tr } \mathbf{f}^{\text{AO}}[\mathbf{R}, \mathbf{E}_{\mu\nu}]_{\text{S}} = -4\text{Tr } \mathbf{f}^{\text{AO}}[[\mathbf{R}, \mathbf{X}]_{\text{S}}, \mathbf{E}_{\mu\nu}]_{\text{S}} - 2\text{Tr } \mathbf{f}^{\text{AO}}[[\mathbf{X}, \mathbf{E}_{\mu\nu}]_{\text{S}}, \mathbf{R}]_{\text{S}}$$
$$- 4\text{Tr } \mathbf{G}^{\text{AO}}([\mathbf{R}, \mathbf{X}]_{\text{S}})[\mathbf{R}, \mathbf{E}_{\mu\nu}]_{\text{S}} \tag{10.8.13}$$

Expanding the S commutators and using the notation (10.7.104) and (10.7.105) for the symmetric and antisymmetric parts of a matrix, we arrive at a set of linear equations for $\mathbf{X}$, given as the upper triangle of the matrix equation

$$[2\mathbf{f}^{\text{AO}}[\mathbf{RSX}]^{\text{S}}\mathbf{S} + 2\mathbf{G}^{\text{AO}}([\mathbf{RSX}]^{\text{S}})\mathbf{RS} + \mathbf{SX}[\mathbf{f}^{\text{AO}}\mathbf{RS}]^{\text{A}}]^{\text{A}} = -[\mathbf{f}^{\text{AO}}\mathbf{RS}]^{\text{A}} \tag{10.8.14}$$

This equation is equivalent to (10.8.2), with the right-hand side proportional to the negative gradient and the left-hand side to the Hessian multiplied by $\mathbf{X}$. Its solution $\mathbf{X}$ may be used to generate an updated density matrix (10.7.88) and the Newton sequence is then iterated until convergence. At the stationary point, the last term on the left-hand side vanishes since it is proportional to the gradient.

For large systems, the solution to the Newton equations must be obtained by an iterative, direct algorithm. From (10.8.14), we note that the required linear transformation of an antisymmetric trial matrix $\mathbf{b}$ may be expressed as the antisymmetric part of the matrix

$$\sigma = 2\mathbf{f}^{\text{AO}}[\mathbf{RSb}]^{\text{S}}\mathbf{S} + 2\mathbf{G}^{\text{AO}}([\mathbf{RSb}]^{\text{S}})\mathbf{RS} + \mathbf{Sb}[\mathbf{f}^{\text{AO}}\mathbf{RS}]^{\text{A}} \tag{10.8.15}$$

In addition to matrix multiplications and additions, the linear transformation requires only the construction of $\mathbf{G}^{\text{AO}}([\mathbf{RSb}]^{\text{S}})$. For sufficiently large systems, the cost of these operations scales linearly with the size of the system, implying that Newton's method may be applied to large systems at a cost that increases only linearly with the size of the system.

10.8.3 THE ELECTRONIC GRADIENT IN ORBITAL-BASED HARTREE–FOCK THEORY

Having considered in Section 10.8.2 Newton's method in the density-based formulation of Hartree–Fock theory, we now turn our attention to the implementation of Newton's method in the orbital-based formulation. After a discussion of the gradient in this subsection, the generalized Fock matrix is discussed in Sections 10.8.4 and 10.8.5, the Hessian in Section 10.8.6, and the linear transformation of trial vectors in Sections 10.8.7 and 10.8.8.

For a real CSF, the electronic gradient (10.1.29) may be written in the form

$$E_{mn}^{(1)} = 2\langle\text{CSF}|[E_{mn}, \hat{H}]|\text{CSF}\rangle$$
$$= 2\sum_{\sigma}[\langle\text{CSF}|a_{m\sigma}^{\dagger}[a_{n\sigma}, \hat{H}]|\text{CSF}\rangle + \langle\text{CSF}|[a_{m\sigma}^{\dagger}, \hat{H}]a_{n\sigma}|\text{CSF}\rangle]$$
$$= 2\sum_{\sigma}[\langle\text{CSF}|a_{m\sigma}^{\dagger}[a_{n\sigma}, \hat{H}]|\text{CSF}\rangle - \langle\text{CSF}|a_{n\sigma}^{\dagger}[a_{m\sigma}, \hat{H}]|\text{CSF}\rangle] \tag{10.8.16}$$

Introducing the *generalized Fock matrix* by the expression

$$F_{mn} = \sum_{\sigma}\langle\text{CSF}|a_{m\sigma}^{\dagger}[a_{n\sigma}, \hat{H}]|\text{CSF}\rangle \tag{10.8.17}$$

we may now write the electronic gradient as [20]

$$E_{mn}^{(1)} = 2(F_{mn} - F_{nm}) \tag{10.8.18}$$

The gradient is thus proportional to the antisymmetric part of the generalized Fock matrix. The evaluation of the gradient is trivial once we have obtained an expression for the generalized Fock matrix. Note that a symmetric generalized Fock matrix is a sufficient condition for a stationary wave function.

To arrive at an explicit expression for the generalized Fock matrix (10.8.17) in terms of integrals and density-matrix elements, we may proceed as for the Fock matrix in Section 10.3.3. Writing the Hamiltonian as

$$\hat{H} = \sum_{pq} h_{pq} E_{pq} + \tfrac{1}{2} \sum_{pqrs} g_{pqrs} e_{pqrs} + h_{\text{nuc}} \tag{10.8.19}$$

the operator entering the generalized Fock matrix (10.8.17) becomes

$$\sum_{\sigma} a^{\dagger}_{m\sigma}[a_{n\sigma}, \hat{H}] = \sum_{q} h_{nq} E_{mq} + \sum_{qrs} g_{nqrs} e_{mqrs} \tag{10.8.20}$$

where we have used (10.3.20). Since this is a two-electron operator, the generalized Fock matrix must (unlike the Fock matrix in Section 10.3.3) be evaluated using both the one- and two-electron density-matrix elements

$$D_{pq} = \langle \text{CSF}|E_{pq}|\text{CSF}\rangle \tag{10.8.21}$$

$$d_{pqrs} = \langle \text{CSF}|e_{pqrs}|\text{CSF}\rangle = \langle \text{CSF}|E_{pq}E_{rs} - \delta_{rq}E_{ps}|\text{CSF}\rangle \tag{10.8.22}$$

Inserting (10.8.20) into (10.8.17) and using the expressions (10.8.21) and (10.8.22), we arrive at the following expression for the generalized Fock matrix in terms of integrals and densities:

$$F_{mn} = \sum_{q} D_{mq} h_{nq} + \sum_{qrs} d_{mqrs} g_{nqrs} \tag{10.8.23}$$

This expression should be compared with that for the total electronic energy (10.4.15)

$$E^{(0)} = \sum_{pq} D_{pq} h_{pq} + \tfrac{1}{2} \sum_{pqrs} d_{pqrs} g_{pqrs} + h_{\text{nuc}}$$

$$= \tfrac{1}{2} \sum_{pq} (D_{pq} h_{pq} + \delta_{pq} F_{pq}) + h_{\text{nuc}} \tag{10.8.24}$$

Note that, because of the different weights on the one- and two-electron parts, the electronic energy cannot be calculated as the trace of the generalized Fock matrix.

10.8.4   THE INACTIVE AND ACTIVE FOCK MATRICES

The expression for the nonsymmetric generalized Fock matrix (10.8.23) given in the previous subsection is completely general and valid for any kind of wave function. In many cases, however, the evaluation of the generalized Fock matrix may be simplified considerably by taking into account the different occupancies of the inactive, active and virtual orbitals [20].

We first note that the elements of the density matrices (10.8.21) and (10.8.22) vanish if one or more of the indices are virtual. If one index is inactive, their evaluation may be simplified as follows:

$$D_{iq} = 2\delta_{iq} \tag{10.8.25}$$

$$d_{pqis} = \langle \text{CSF}|E_{pq}E_{is} - \delta_{iq}E_{ps}|\text{CSF}\rangle = 2\delta_{is}D_{pq} - \delta_{iq}D_{ps} \tag{10.8.26}$$

Using these expressions, we may rewrite the expression for the generalized Fock matrix in (10.8.23) such that only density-matrix elements with all indices active are referenced. Keeping in mind that the generalized Fock matrix is not symmetric, we begin by calculating the elements of the generalized Fock matrix where the first index is inactive and the second is unspecified

$$F_{in} = \sum_q D_{iq} h_{nq} + \sum_{qrs} d_{rsiq} g_{rsnq}$$

$$= 2h_{ni} + \sum_{qrs} (2\delta_{iq} D_{rs} - \delta_{si} D_{rq}) g_{rsnq}$$

$$= 2h_{ni} + \sum_{rs} D_{rs} (2g_{rsni} - g_{rins})$$

$$= 2(^{I}F_{ni} + {}^{A}F_{ni}) \tag{10.8.27}$$

We have here introduced *the inactive and active Fock matrices*

$$^{I}F_{mn} = h_{mn} + \sum_i (2g_{mnii} - g_{miin}) \tag{10.8.28}$$

$$^{A}F_{mn} = \sum_{vw} D_{vw} \left(g_{mnvw} - \tfrac{1}{2} g_{mwvn} \right) \tag{10.8.29}$$

Next, when the first index is active, we obtain

$$F_{vn} = \sum_q D_{vq} h_{nq} + \sum_{qrs} d_{vqrs} g_{nqrs}$$

$$= \sum_w D_{vw} h_{nw} + \sum_{iqs} d_{vqis} g_{isnq} + \sum_{wqs} d_{vqws} g_{wsnq}$$

$$= \sum_w D_{vw} h_{nw} + \sum_{wi} D_{vw} (2g_{iinw} - g_{iwni}) + \sum_{wxy} d_{vxwy} g_{wynx}$$

$$= \sum_w {}^{I}F_{nw} D_{vw} + Q_{vn} \tag{10.8.30}$$

where we have introduced the auxiliary **Q** *matrix* given by

$$Q_{vm} = \sum_{wxy} d_{vwxy} g_{mwxy} \tag{10.8.31}$$

Finally, when the first index is virtual, the element of the generalized Fock matrix vanishes:

$$F_{an} = 0 \tag{10.8.32}$$

When the generalized Fock matrix is evaluated from (10.8.27), (10.8.30) and (10.8.32), only two-electron density-matrix elements with all four indices active are used. If we had used (10.8.23) directly, we would instead have had to reference density elements with indices labelling the full set of occupied MOs. From the definition of the inactive and active Fock matrices (10.8.28) and (10.8.29), we see that these matrices are symmetric – unlike the generalized Fock matrix (10.8.23), which is symmetric only at stationary points. In Section 10.8.5, we shall consider the cost of the evaluation of the Fock matrices in more detail.

In closed-shell electronic systems, there are no active orbitals and the inactive Fock matrix (10.8.28) is identical to the Fock matrix (10.3.24) of canonical Hartree–Fock theory. Since there are no contributions from the active Fock matrix and from the **Q** matrix, the generalized Fock matrix then becomes

$$\mathbf{F} = 2 \begin{pmatrix} {}^{\mathrm{I}}\mathbf{F}_{IJ} & \vdots & {}^{\mathrm{I}}\mathbf{F}_{IA} \\ \cdots & \vdots & \cdots \\ 0 & \vdots & 0 \end{pmatrix} \quad \text{(closed shells)} \tag{10.8.33}$$

The electronic gradient for the nonredundant occupied–virtual orbital rotations is therefore

$$E_{ai}^{(1)} = 2(F_{ai} - F_{ia}) = -4{}^{\mathrm{I}}F_{ai} \quad \text{(closed shells)} \tag{10.8.34}$$

where we have used (10.8.18) and (10.8.27). At a stationary point, the occupied–virtual block of the inactive Fock matrix vanishes and the generalized Fock matrix becomes symmetric. For the CSFs of singlet or triplet symmetries and single occupancy in two orbitals discussed in Section 10.1.1, the **Q** matrix is determined in Exercise 10.5.

Important simplifications occur also for high-spin CSFs – that is, for CSFs where all singly occupied orbitals have the same (alpha or beta) spin. The high-spin active density matrices have the simple form

$$D_{vw}^{\text{high-spin}} = \delta_{vw} \tag{10.8.35}$$

$$d_{vwxy}^{\text{high-spin}} = \delta_{vw}\delta_{xy} - \delta_{vy}\delta_{wx} \tag{10.8.36}$$

simplifying the construction of the Fock matrices considerably. In particular, we find that the high-spin **Q** matrix may be calculated as

$$Q_{vm}^{\text{high-spin}} = \sum_{w} (g_{mvww} - g_{mwwv}) \tag{10.8.37}$$

which now has the same structure as the inactive and active Fock matrices (10.8.28) and (10.8.29), but with equal weights on the Coulomb and exchange contributions. Note that the **Q** matrix vanishes for high-spin doublets, where only a single active orbital is present.

We finally note from (10.3.22) that the sum of the inactive and active Fock matrices can be written in the form

$$ {}^{\mathrm{I}}F_{mn} + {}^{\mathrm{A}}F_{mn} = \tfrac{1}{2} \sum_{\sigma} \langle \text{CSF} | [a_{n\sigma}^{\dagger}, [a_{m\sigma}, \hat{H}]]_{+} | \text{CSF} \rangle \tag{10.8.38}$$

which is a generalization of (10.3.17) to CSFs with both inactive and active orbitals. For one inactive index, (10.8.38) becomes equal to half the Fock-matrix element F_{in} in (10.8.27).

10.8.5 COMPUTATIONAL COST FOR THE CALCULATION OF THE FOCK MATRIX

An important characteristic of any computational task is the number of *floating-point operations* that it requires. Consider the calculation of the generalized Fock matrix (10.8.23). There are nO nonzero elements in this matrix, where n is the total number of MOs and O the number of occupied MOs. For each element F_{mn}, we must carry out a series of multiplications and additions according to (10.8.23) – more precisely, O multiplications and additions for the one-electron part and O^3 multiplications and additions for the two-electron part. The total number of floating-point operations required for the construction of the generalized Fock matrix is therefore $2nO^2 + 2nO^4$. It is customary to count only the multiplications, however, giving us an *operation count* of $nO^2 + nO^4$. For all but the smallest basis sets, the last term dominates, and we say that the construction of the generalized Fock matrix scales as nO^4.

The operation count of nO^4 tells us that the computational cost for the construction of the generalized Fock matrix increases by a factor of $2^5 = 32$ if the size of the system increases by a factor of 2 (for example, by carrying out a calculation on two molecules instead of one). The construction of the generalized Fock matrix (10.8.23) is therefore said to be a fifth-order process. However, if we keep the number of electrons constant but increase the number of orbitals by a factor of 2 (for example, if we wish to carry out a more accurate calculation on the same system), then the cost for the calculation of the generalized Fock matrix increases by a factor of only 2. When quoting operation counts, therefore, it is important to keep track of the dependency on the different orbital types. To distinguish between the different orbital types, we shall use I and A for the total numbers of inactive and active orbitals, O and V for the total numbers of occupied and virtual orbitals, and n for the total number of orbitals.

We have seen that the evaluation of the generalized Fock matrix according to (10.8.23) is an expensive process, with an operation count of nO^4. Let us now consider the construction of the generalized Fock matrix from the inactive and active Fock matrices and the **Q** matrix as discussed in Section 10.8.4. The construction of the inactive Fock matrix (10.8.28) is a third-order process which scales as n^2I since, for each of the n^2 elements of the inactive Fock matrix, we must sum over I terms in the two-electron part. Further reductions arise since the inactive Fock matrix is symmetric, leaving us with only $n^2I/2$ operations. However, constant factors such as 1/2 are usually less important and more difficult to keep track of than the overall scaling of the operation count and are often disregarded in discussions of computational cost.

From (10.8.29) and (10.8.31), we note that the construction of the active Fock matrix and of the **Q** matrix requires n^2A^2 and nA^4 operations, respectively. The construction of the generalized Fock matrix from the inactive and active Fock matrices and from the **Q** matrix thus scales as $\max(n^2I, n^2A^2, nA^4)$, which should be compared with nO^4 for (10.8.23). The use of inactive and active Fock matrices is clearly advantageous for systems with a large number of inactive orbitals and a small number of active orbitals. If all the occupied orbitals are active, then the two procedures become equivalent. Conversely, for closed-shell systems, the construction of the generalized Fock matrix via the inactive Fock matrix requires only n^2I operations. If only those elements that are needed for the gradient are calculated (the inactive–virtual elements), then the operation count reduces to I^2V.

Our discussion has been concerned with the calculation of the generalized Fock matrix (10.8.23) from a set of integrals in the MO basis. Two important observations should be made in this respect. First, the calculation of the MO integrals from the AO integrals is in itself an expensive undertaking, with a cost that scales as n^5 but which may be reduced by factors such as O/n depending on the exact range of integrals needed. Second, for closed-shell systems, where only the inactive Fock matrix is required, the AO Fock matrix is most easily calculated directly from the AO integrals according to (10.6.17) in a process that scales as n^4 or lower (in larger systems as discussed in Section 10.6.3). The transformation of the AO Fock matrix to the MO basis is a third-order process.

10.8.6 THE ELECTRONIC HESSIAN IN ORBITAL-BASED HARTREE–FOCK THEORY

The Hartree–Fock electronic Hessian is given in (10.1.30). Since the commutator that appears in this expression for the Hessian is a rank-2 operator, it may be expressed in terms of one- and two-electron density matrices. The construction of the electronic Hessian may be simplified in the same way as for the gradient in Section 10.8.3. We first note that $[E_{rs}^-, \hat{H}]$ is a Hermitian operator. As for the electronic gradient, we may therefore write

$$\langle \text{CSF}|[E_{pq}, [E_{rs}^-, \hat{H}]]|\text{CSF}\rangle = G_{pqrs} - G_{qprs} = (1 - P_{pq})G_{pqrs} \qquad (10.8.39)$$

where the element G_{pqrs} takes the form of a generalized Fock matrix (10.8.17):

$$G_{pqrs} = \sum_\sigma \langle \text{CSF}|a_{p\sigma}^\dagger[a_{q\sigma}, [E_{rs}^-, \hat{H}]]|\text{CSF}\rangle \qquad (10.8.40)$$

To evaluate this expression, we first use the Jacobi identity (1.8.17) and obtain

$$G_{pqrs} = G_1 + G_2 + G_3 \qquad (10.8.41)$$

where

$$G_1 = \sum_\sigma \langle \text{CSF}|a_{p\sigma}^\dagger[[a_{q\sigma}, E_{rs}^-], \hat{H}]|\text{CSF}\rangle \qquad (10.8.42)$$

$$G_2 = \sum_\sigma \langle \text{CSF}|a_{p\sigma}^\dagger[E_{rs}^-, [a_{q\sigma}, \hat{h}]]|\text{CSF}\rangle \qquad (10.8.43)$$

$$G_3 = \sum_\sigma \langle \text{CSF}|a_{p\sigma}^\dagger[E_{rs}^-, [a_{q\sigma}, \hat{g}]]|\text{CSF}\rangle \qquad (10.8.44)$$

We have partitioned the Hamiltonian here in the usual way (10.3.18). Using the commutator in (2.3.37), the first term is rewritten in terms of the generalized Fock matrix:

$$G_1 = (1 - P_{rs})\sum_\sigma \langle \text{CSF}|a_{p\sigma}^\dagger[\delta_{qr}a_{s\sigma}, \hat{H}]|\text{CSF}\rangle = -(1 - P_{rs})\delta_{qs}F_{pr} \qquad (10.8.45)$$

Applying the same commutator twice, the second term becomes

$$G_2 = (1 - P_{rs})\sum_{\substack{mn \\ \sigma}} h_{mn}\langle \text{CSF}|a_{p\sigma}^\dagger[E_{rs}, [a_{q\sigma}, E_{mn}]]|\text{CSF}\rangle$$

$$= (1 - P_{rs})\sum_{\substack{mn \\ \sigma}} h_{mn}\delta_{mq}\langle \text{CSF}|a_{p\sigma}^\dagger[E_{rs}, a_{n\sigma}]|\text{CSF}\rangle$$

$$= -(1 - P_{rs})\sum_{\substack{mn \\ \sigma}} h_{mn}\delta_{mq}\delta_{nr}\langle \text{CSF}|a_{p\sigma}^\dagger a_{s\sigma}|\text{CSF}\rangle$$

$$= -(1 - P_{rs})D_{ps}h_{qr} = (1 - P_{rs})D_{pr}h_{qs} \qquad (10.8.46)$$

The third term may be written in the form

$$G_3 = (1 - P_{rs})\sum_\sigma \langle \text{CSF}|a_{p\sigma}^\dagger[E_{rs}, [a_{q\sigma}, \hat{g}]]|\text{CSF}\rangle$$

$$= (1 - P_{rs})\sum_{\substack{tmn \\ \sigma}} g_{qtmn}\langle \text{CSF}|a_{p\sigma}^\dagger[E_{rs}, E_{mn}a_{t\sigma}]|\text{CSF}\rangle \qquad (10.8.47)$$

where we have used (10.3.20). Further expansion of the commutator

$$[E_{rs}, E_{mn}a_{t\sigma}] = -E_{mn}a_{s\sigma}\delta_{tr} + E_{rn}a_{t\sigma}\delta_{ms} - E_{ms}a_{t\sigma}\delta_{rn} \qquad (10.8.48)$$

yields three double sums:

$$G_3 = (1 - P_{rs})\left(-\sum_{mn} g_{qrmn}d_{psmn} + \sum_{tn} g_{qtsn}d_{ptrn} - \sum_{tm} g_{qtmr}d_{ptms}\right)$$

$$= (1 - P_{rs})\sum_{mn}(g_{qsmn}d_{prmn} + g_{qmns}d_{pmrn} + g_{qmns}d_{pmnr})$$

$$= (1 - P_{rs})Y_{pqrs} \tag{10.8.49}$$

where we have introduced the matrix

$$Y_{pqrs} = \sum_{mn}[(d_{pmrn} + d_{pmnr})g_{qmns} + d_{prmn}g_{qsmn}] \tag{10.8.50}$$

with the permutational symmetry

$$Y_{pqrs} = Y_{rspq} \tag{10.8.51}$$

This gives us the following expression for the elements of (10.8.41):

$$G_{pqrs} = (1 - P_{rs})(D_{pr}h_{qs} - F_{pr}\delta_{qs} + Y_{pqrs}) \tag{10.8.52}$$

Combining (10.8.39) and (10.1.30), we arrive at the following final expression for the electronic Hessian:

$$E^{(2)}_{pqrs} = (1 + P_{pq,rs})(1 - P_{pq})(1 - P_{rs})(D_{pr}h_{qs} - F_{pr}\delta_{qs} + Y_{pqrs})$$

$$= (1 - P_{pq})(1 - P_{rs})[2D_{pr}h_{qs} - (F_{pr} + F_{rp})\delta_{qs} + 2Y_{pqrs}] \tag{10.8.53}$$

The construction of the Hessian matrix thus requires the symmetric part of the generalized Fock matrix and the matrix $\mathbf{Y}$. Since the pair indices pq and rs correspond to the nonredundant excitation operators, only n^2O^2 elements are needed. In accordance with the discussion in Section 10.1.3, the Hessian (10.8.53) is symmetric with respect to permutations of pq and rs, but antisymmetric with respect to permutations of p and q and permutations of r and s.

The construction of the Hartree–Fock electronic Hessian (10.8.53) is dominated by the evaluation of the matrix $\mathbf{Y}$ (10.8.50). The two-electron densities vanish if any of the indices are virtual and the operation count for the matrix $\mathbf{Y}$ is thus at most n^2O^4. The construction of the full Hessian is therefore a sixth-order process, although this cost may be reduced by taking into account the simplifications of the density matrices in (10.8.25) and (10.8.26). The construction of the Hessian should be compared with the calculation of the gradient, which is a fifth-order process, and with the calculation of the energy (10.8.24), which is a fourth-order process – assuming that the two-electron integrals are available in the MO basis.

Using (10.8.25) and (10.8.26), we may simplify the evaluation of the Hessian in the same way as we simplified the construction of the gradient in Section 10.8.4. As an illustration, we evaluate the Hessian for a closed-shell system, where the only nonredundant rotations are those that mix the occupied and virtual MOs. The density-matrix elements vanish if any index is virtual and $\mathbf{Y}$ vanishes if the first or the third index is virtual, giving us the following simplifications:

$$E^{(2)}_{aibj} = (1 - P_{ai})(1 - P_{bj})[2D_{ab}h_{ij} - (F_{ab} + F_{ba})\delta_{ij} + 2Y_{aibj}]$$

$$= 2D_{ij}h_{ab} - (F_{ij} + F_{ji})\delta_{ab} + 2Y_{iajb} \tag{10.8.54}$$

Inserting the definition of **Y** in (10.8.54) and using (10.4.18) and (10.4.19) for the density matrices and (10.8.28) and (10.8.33) for the generalized Fock matrix, we obtain

$$E^{(2)}_{aibj} = 4(\delta_{ij}{}^{\mathrm{I}}F_{ab} - \delta_{ab}{}^{\mathrm{I}}F_{ij} + 4g_{aibj} - g_{abij} - g_{ajib}) \tag{10.8.55}$$

which constitutes the working equation for the electronic Hessian of a closed-shell system. In the canonical basis (10.3.16), the electronic Hessian becomes

$$E^{(2)}_{aibj} = 4[\delta_{ab}\delta_{ij}(\varepsilon_a - \varepsilon_i) + 4g_{aibj} - g_{abij} - g_{ajib}] \tag{10.8.56}$$

In this representation, the contributions from the effective one-electron part of the Hamiltonian (10.4.3) are restricted to the diagonal elements of the Hessian and consist of differences between the orbital energies of the virtual and occupied orbitals. These differences constitute the zero-order energies of Møller–Plesset perturbation theory and are usually considerably larger than the two-electron integrals, which appear both off and on the diagonal. In the canonical representation, therefore, the electronic Hessian is usually diagonally dominant. From a computational point of view, the diagonal dominance is important, ensuring, for example, rapid convergence of iterative optimization algorithms.

The evaluation of the Hessian elements in (10.8.55) requires only I^2V^2 operations and thus scales as the fourth power of the size of the system (provided the two-electron MO integrals are available). This cost may be compared with the calculation of the corresponding inactive–virtual elements of the inactive Fock matrix (needed for the gradient of closed-shell states), which requires I^2V operations.

10.8.7 LINEAR TRANSFORMATIONS IN THE MO BASIS

In the present subsection, we describe how linear transformations of trial vectors by the Hessian (10.8.8) in MO-based Hartree–Fock theory may be carried out without constructing the Hessian explicitly. From the definition of the Hessian (10.1.26) and from (10.2.5), we may write the pq component of the linearly transformed vector (10.8.8) in the form

$$\sigma_{pq} = \sum_{r>s} \langle \mathrm{CSF}|[E^-_{pq}, [E^-_{rs}, \hat{H}]] - \tfrac{1}{2}[[E^-_{pq}, E^-_{rs}], \hat{H}]|\mathrm{CSF}\rangle \kappa_{rs}$$

$$= 2\langle \mathrm{CSF}| \left[E_{pq}, \left[\sum_{rs}\kappa_{rs}E_{rs}, \hat{H} \right] \right] |\mathrm{CSF}\rangle - \langle \mathrm{CSF}| \left[\left[E_{pq}, \sum_{rs}\kappa_{rs}E_{rs} \right], \hat{H} \right] |\mathrm{CSF}\rangle \tag{10.8.57}$$

where we consider κ_{pq} the elements of an antisymmetric square matrix κ. The commutator of $\hat{H}$ with the anti-Hermitian operator $\hat{\kappa}$ is Hermitian and may be treated as an effective Hamiltonian, constructed from a set of one-index transformed integrals (see Exercise 2.4) [21]

$$\hat{H}_\kappa = \left[\sum_{pq}\kappa_{pq}E_{pq}, \hat{H} \right] = \sum_{pq}h^\kappa_{pq}E_{pq} + \tfrac{1}{2}\sum_{pqrs}g^\kappa_{pqrs}e_{pqrs} \tag{10.8.58}$$

The *one-index transformed integrals* are defined as

$$h^\kappa_{pq} = \sum_m (\kappa_{pm}h_{mq} + \kappa_{qm}h_{pm}) \tag{10.8.59}$$

$$g^\kappa_{pqrs} = \sum_m (\kappa_{pm}g_{mqrs} + \kappa_{qm}g_{pmrs} + \kappa_{rm}g_{pqms} + \kappa_{sm}g_{pqrm}) \tag{10.8.60}$$

and have the same symmetries as the ordinary Hamiltonian integrals – twofold permutational symmetry for the one-electron integrals and eightfold for the two-electron integrals. Substituting (10.8.58) in (10.8.57), we now obtain

$$\sigma_{pq} = 2\langle \text{CSF}|[E_{pq}, \hat{H}_\kappa]|\text{CSF}\rangle + \sum_r (\kappa_{rp}\langle \text{CSF}|[E_{rq}, \hat{H}]|\text{CSF}\rangle - \kappa_{qr}\langle \text{CSF}|[E_{pr}, \hat{H}]|\text{CSF}\rangle)$$

(10.8.61)

The linearly transformed vector σ may thus be evaluated in terms of gradient elements obtained from the standard integrals (the second term) and gradient elements constructed from one-index transformed integrals (the first term). In Section 10.8.4, we demonstrated that electronic gradients are often best calculated from the inactive Fock matrix, the active Fock matrix and the $\mathbf{Q}$ matrix. Gradients involving one-index transformed integrals may in the same way be obtained from the inactive Fock matrix $^{\text{I}}\mathbf{F}^\kappa$, the active Fock matrix $^{\text{A}}\mathbf{F}^\kappa$ and the $\mathbf{Q}^\kappa$ matrix, all constructed from one-index transformed integrals.

The operation count for the one-index transformation in (10.8.60) is n^2O^3 since the transformed integrals needed for the electronic gradient have one general and three occupied indices. The cost of the evaluation of the gradient from the one-index transformed integrals is nO^4, the same as for the gradient in Section 10.8.5. Thus, if (10.8.61) is used, then the operation count for the linear transformation in (10.8.8) is n^2O^3 – an order of magnitude less than that for the explicit construction of the Hessian (10.8.53), which requires n^2O^4 operations.

10.8.8   LINEAR TRANSFORMATIONS IN THE AO BASIS

In Section 10.8.7, we demonstrated how the key step in the orbital-based second-order optimization of the Hartree–Fock state – namely, the linear transformation of trial vectors with the electronic Hessian (needed in the iterative solution of the Newton equations) – may be carried out at a cost (n^2O^3) considerably smaller than that of the construction of the electronic Hessian itself (n^2O^4). Unfortunately, the method developed in Section 10.8.7 requires the transformation of the full set of AO integrals to the MO basis, a computationally expensive task. A more efficient scheme is obtained by recasting the equations in terms of AO integrals, as done in the present subsection [18].

In the MO basis, the transformed trial vectors are calculated in terms of gradients constructed from standard MO integrals as well as gradients constructed from one-index transformed MO integrals; see (10.8.61). Therefore, if we can calculate the generalized Fock matrix (with and without one-index transformed integrals) directly in the AO basis, then we would also be able to generate the transformed trial vectors in this basis. Since the generalized Fock matrix may be obtained from the inactive and active Fock matrices and from the $\mathbf{Q}$ matrix, our task reduces to the evaluation of these matrices (constructed from standard integrals and from one-index transformed integrals) directly in the AO basis. This task is easily accomplished for closed-shell and high-spin states – the $\mathbf{Q}$ matrix then either vanishes or reduces to a simple linear combination of Coulomb and exchange integrals, which may be evaluated in the same manner as the two-electron part of the inactive and active Fock matrices.

Let us first consider the matrices constructed from the standard integrals – that is, from integrals that have not been one-index transformed. The calculation of the inactive Fock matrix in the AO basis is trivial; see Section 10.6.1. The active Fock matrix and $\mathbf{Q}^{\text{high-spin}}$ may be evaluated in the same manner; see Exercise 10.6. It remains to consider the matrices constructed from one-index transformed integrals. Using (10.8.28), (10.8.59) and (10.8.60), the inactive Fock matrix containing one-index transformed integrals can be expressed as

$$
{}^{I}F_{mn}^{\kappa} = h_{mn}^{\kappa} + \sum_{i}(2g_{mnii}^{\kappa} - g_{miin}^{\kappa})
$$

$$
= \sum_{p}({}^{I}F_{mp}\kappa_{np} + {}^{I}F_{pn}\kappa_{mp}) + \sum_{ip}(4g_{mnip} - g_{mpin} - g_{npim})\kappa_{ip} \qquad (10.8.62)
$$

The first term is obtained from $^{I}\mathbf{F}$ and can be calculated in the AO basis as described in Section 10.6.1. Introducing the modified density matrix

$$
p_{\rho\sigma} = 4\sum_{iq}\kappa_{iq}C_{\rho i}C_{\sigma q} \qquad (10.8.63)
$$

we may rewrite the last term in the form

$$
\sum_{ip}(4g_{mnip} - g_{mpin} - g_{npim})\kappa_{ip} = \sum_{\rho\sigma}\left(g_{mn\rho\sigma} - \tfrac{1}{4}g_{m\sigma\rho n} - \tfrac{1}{4}g_{n\sigma\rho m}\right)p_{\rho\sigma} \qquad (10.8.64)
$$

which may be calculated in the AO basis from the two-index quantity

$$
M_{\mu\nu} = \sum_{\rho\sigma}\left(g_{\mu\nu\rho\sigma} - \tfrac{1}{4}g_{\mu\sigma\rho\nu} - \tfrac{1}{4}g_{\nu\sigma\rho\mu}\right)p_{\rho\sigma} \qquad (10.8.65)
$$

In Exercise 10.6, this technique is extended to the calculation of the active Fock matrix and the matrix $\mathbf{Q}^{\text{high-spin}}$.

The operation count for setting up the inactive Fock matrices (10.6.17) and (10.8.62) in the AO basis is n^4. A closed-shell Newton iteration thus scales as n^4 when carried out in the AO basis. It is considerably more expensive to carry out the Newton iterations in the MO basis, since the MO transformation of the two-electron integrals scales as n^5. We also note that, even though the cost of a single AO trial-vector transformation (10.8.8) is the same as for a single Roothaan–Hall SCF iteration (n^4), the Newton iterations are still more expensive than the SCF iterations since a number of trial vectors are transformed in each Newton iteration.

We conclude that, for closed-shell and high-spin states, second-order optimizations can be carried out in the AO basis at a cost of n^4 for each trial-vector transformation (10.8.8). For other open-shell CSF states, it is more difficult to simplify the construction of the $\mathbf{Q}$ matrix in order to carry out a second-order optimization in the AO basis. However, with the possible exception of the two-electron open-shell singlet state (10.1.7), Hartree–Fock wave functions for other than high-spin states are of little interest except for systems of high spatial symmetry. In Exercise 10.7, an STO-3G Hartree–Fock wave function for HeH^{+} is calculated using Newton's method.

10.9   The SCF method as an approximate second-order method

In the present chapter, two methods have been introduced for the optimization of Hartree–Fock wave functions: the SCF method of Sections 10.3–10.6 and Newton's method of Section 10.8. We shall now develop the SCF method from a perspective different from that of the previous sections – namely, as an approximate second-order method. In this way, we shall be able to establish a stronger connection between the SCF method and Newton's method. Our analysis will also make it possible to examine under what conditions the SCF method can be applied to open-shell systems and we shall be able to analyse the local convergence characteristics of the SCF method.

10.9.1   THE GBT VECTOR

We begin this section by considering a slight variation on Newton's method as described in Section 10.8. For a given CSF, we define the *GBT vector* as

$$W_{mn}(\kappa) = 2\langle \text{CSF}(\kappa)|[E_{mn}(\kappa), \hat{H}]|\text{CSF}(\kappa)\rangle = 2\langle \text{CSF}|[E_{mn}, \exp(\hat{\kappa})\hat{H}\exp(-\hat{\kappa})]|\text{CSF}\rangle \quad (10.9.1)$$

where the excitation operators in the transformed basis are given by the expression

$$E_{mn}(\kappa) = \exp(-\hat{\kappa})E_{mn}\exp(\hat{\kappa}) \quad (10.9.2)$$

Expanding the GBT vector around the point $\kappa = \mathbf{0}$, we obtain

$$\mathbf{W}(\kappa) = \mathbf{W}^{(0)} + \mathbf{W}^{(1)}\kappa + \cdots \quad (10.9.3)$$

where the zero- and first-order terms are given by

$$W_{mn}^{(0)} = 2\langle \text{CSF}|[E_{mn}, \hat{H}]|\text{CSF}\rangle \quad (10.9.4)$$

$$W_{mnpq}^{(1)} = 2\langle \text{CSF}|[E_{mn}, [E_{pq}^-, \hat{H}]]|\text{CSF}\rangle \quad (10.9.5)$$

At $\kappa = \mathbf{0}$, the zero-order GBT vector $\mathbf{W}^{(0)}$ is identical to the electronic gradient $\mathbf{E}^{(1)}$ (10.1.29), whereas the *Jacobian* (i.e. the first-derivative matrix) $\mathbf{W}^{(1)}$ differs from the electronic Hessian $\mathbf{E}^{(2)}$ (10.1.30) in that the nested commutator is not symmetrized. From Section 10.2.1, we conclude that $\mathbf{W}^{(1)}$ and $\mathbf{E}^{(2)}$ are identical at stationary points; at nonstationary points, they differ in terms that are proportional to the GBT vector.

A stationary point of the electronic energy occurs when $\mathbf{W}(\kappa) = \mathbf{0}$. Applying Newton's method for the solution of nonlinear equations, we determine the Hartree–Fock state by truncating the GBT expansion (10.9.3) at first order and setting the resulting vector equal to zero. In the nth iteration, we solve the nonsymmetric linear Newton equations

$$\mathbf{W}_n^{(1)}\Delta\kappa_n = -\mathbf{W}_n^{(0)} \quad (10.9.6)$$

in the same manner as for the energy-based Newton optimization in Section 10.8.1. The step generated by (10.9.6) differs from that generated by the linear equations (10.8.2) in terms that are quadratic in the gradient. Since the GBT vector is identical to the electronic gradient only at $\kappa = \mathbf{0}$, we must in each iteration transform the orbitals to the basis of the new expansion point $\kappa_n + \Delta\kappa_n$ (as done in the energy-based Newton optimization). At convergence, the zero-order GBT vector (10.9.4) vanishes and we have thus satisfied the variational conditions (10.2.4) or, equivalently, the GBT of Section 10.2.3.

10.9.2   THE FOCK OPERATOR

Our interest in the GBT-based optimization of Section 10.9.1 is not so much that it offers an alternative to the energy-based Newton's method of Section 10.8 but rather that it provides us with a strategy for re-examining the SCF method. In particular, we shall investigate whether it is possible to carry out a GBT-based optimization of the Hartree–Fock wave function by replacing the exact Hamiltonian of the GBT vector (10.9.1) by an effective one-electron Hermitian operator in the form

$$\hat{f} = \sum_{pq} f_{pq}E_{pq} \quad (10.9.7)$$

We shall see that, although the process of identification differs considerably from that of Section 10.3, it leads us to the same effective one-electron Hamiltonian (i.e. the Fock operator). Moreover, the identification presented here allows us to compare the SCF and Newton methods of optimization and to investigate in more detail under what conditions an effective Fock operator may be set up for open-shell RHF wave functions.

If we are to succeed in our task of setting up an effective one-electron Hamiltonian, then the elements of the Fock matrix must be chosen in such a way that the approximate GBT vector

$$w_{mn}(\kappa) = 2\langle \mathrm{CSF}(\kappa)|[E_{mn}(\kappa), \hat{f}]|\mathrm{CSF}(\kappa)\rangle = 2\langle \mathrm{CSF}|[E_{mn}, \exp(\hat{\kappa})\hat{f}\exp(-\hat{\kappa})]|\mathrm{CSF}\rangle \qquad (10.9.8)$$

reproduces the true GBT vector (10.9.1) *exactly* at the expansion point – that is, we must require the following conditions to be satisfied for the full set of nonredundant excitation operators E_{mn}:

$$\langle \mathrm{CSF}|[E_{mn}, \hat{f}]|\mathrm{CSF}\rangle = \langle \mathrm{CSF}|[E_{mn}, \hat{H}]|\mathrm{CSF}\rangle \qquad (10.9.9)$$

Since the number of nonredundant rotations is always smaller than the number of Fock-matrix elements, these equations are not sufficient to determine $\hat{f}$ completely. The remaining elements of the Fock matrix are chosen so that the approximate Jacobian

$$w^{(1)}_{mnpq} = 2\langle \mathrm{CSF}|[E_{mn}, [E^-_{pq}, \hat{f}]]|\mathrm{CSF}\rangle \qquad (10.9.10)$$

resembles the true Jacobian (10.9.5) as closely as possible, thus improving the local behaviour of the approximate GBT vector (and presumably the convergence of the iterations).

Let us now assume that we have succeeded in replacing the full two-electron Hamiltonian by some useful Fock operator that satisfies the requirements (10.9.9). In each iteration, we may then solve the variational conditions

$$\mathbf{w}(\kappa) = \mathbf{0} \qquad (10.9.11)$$

To solve (10.9.11) for a given operator $\hat{f}$, we note that from (3.2.18) and (3.2.24) we have

$$\exp(\hat{\kappa})\hat{f}\exp(-\hat{\kappa}) = \sum_{\substack{mn\\\sigma}} f_{mn}\exp(\hat{\kappa})a^{\dagger}_{m\sigma}\exp(-\hat{\kappa})\exp(\hat{\kappa})a_{n\sigma}\exp(-\hat{\kappa})$$

$$= \sum_{\substack{mn\\\sigma}}[\exp(\kappa)\mathbf{f}\exp(-\kappa)]_{mn}a^{\dagger}_{m\sigma}a_{n\sigma} = \sum_{mn}\tilde{f}_{mn}E_{mn} \qquad (10.9.12)$$

Inserting (10.9.12) into (10.9.8), we obtain, after some simple algebra,

$$w_{mn}(\kappa) = 2\sum_{p}(D_{mp}\tilde{f}_{np} - \tilde{f}_{pm}D_{pn}) \qquad (10.9.13)$$

Assuming that the one-electron density matrix is diagonal (which holds for all CSFs generated by the spin adaptation of a single orbital configuration), (10.9.11) becomes

$$(D_{mm} - D_{nn})\tilde{f}_{nm} = 0 \qquad (10.9.14)$$

This equation may be satisfied by choosing

$$\mathbf{U} = \exp(-\kappa) \qquad (10.9.15)$$

so that the matrix $\tilde{\mathbf{f}}$ is diagonal

$$\tilde{\mathbf{f}} = \mathbf{U}^{\mathrm{T}}\mathbf{f}\mathbf{U} = \boldsymbol{\varepsilon} \qquad (10.9.16)$$

The above procedure must be repeated for self-consistency since the Fock operator itself depends on the optimized wave function. In this way, we have established a fixed-point sequence of iterations (where each iteration involves the diagonalization of the Fock matrix) which – if it converges – will lead us to the optimized Hartree–Fock wave function. Convergence of the iterations is not guaranteed, however, but will depend on how we choose those elements of the Fock matrix that are not determined by the zero-order conditions (10.9.9).

10.9.3 IDENTIFICATION FROM THE GRADIENT

We now return to the identification of Fock-matrix elements from the zero-order conditions (10.9.9). Since the one-electron density matrix is assumed to be diagonal, we obtain for the left-hand side of (10.9.9)

$$2\langle \text{CSF}|[E_{mn}, \hat{f}]|\text{CSF}\rangle = 2(D_{mm} - D_{nn})f_{nm} \tag{10.9.17}$$

whereas the right-hand side of (10.9.9) is given by

$$2\langle \text{CSF}|[E_{mn}, \hat{H}]|\text{CSF}\rangle = E_{mn}^{(1)} \tag{10.9.18}$$

Equating (10.9.17) and (10.9.18), we arrive at the following expression for those elements of the Fock matrix that correspond to the nonredundant orbital rotations:

$$f_{nm} = \frac{E_{mn}^{(1)}}{2(D_{mm} - D_{nn})} \tag{10.9.19}$$

This expression is symmetric with respect to permutation of m and n since the electronic gradient is antisymmetric in m and n (10.1.27). We note that only those parameters f_{nm} that correspond to orbitals with different occupations can be determined from the condition (10.9.9).

For the nonredundant inactive–active, inactive–virtual and active–virtual rotations of a Hartree–Fock state, the occupations are different and (10.9.19) can be used to fix the Fock-matrix elements. For closed-shell systems, for example, the only nonredundant rotations are those that mix occupied and virtual orbitals. The corresponding elements of the Fock matrix may be identified uniquely as

$$f_{ai} = f_{ia} = \frac{E_{ai}^{(1)}}{2(D_{aa} - D_{ii})} = \frac{-4^{\text{I}}F_{ia}}{2(0 - 2)} = {}^{\text{I}}F_{ia} \tag{10.9.20}$$

where we have used (10.8.34) for the electronic gradient and (10.8.25) for the density. Since the inactive Fock matrix is given by (10.8.28), this result agrees with the expression for the Fock matrix (10.3.24).

For the nonredundant active–active rotations of a spatially nondegenerate CSF, the occupations of the active orbitals are all equal to 1 and the conditions (10.9.19) then do not fix the Fock-matrix elements – in such cases, the Fock method cannot be used to determine the wave function. As an illustration, consider the two-electron open-shell system of Section 10.1.1, from which a singlet state and a triplet state can be constructed. For the triplet configuration, the rotations among the singly occupied active orbitals are redundant and the full set of Fock-matrix elements needed to reproduce the GBT vector can therefore be deduced from (10.9.19). For the singlet configuration, by contrast, the rotations between the two singly occupied active orbitals are nonredundant and it becomes impossible to construct a Fock matrix that reproduces the GBT vector.

10.9.4   IDENTIFICATION FROM THE HESSIAN

The conditions on the zero-order approximate GBT vector determine those elements that correspond to nonredundant orbital rotations. To fix the remaining elements of the Fock matrix, we consider the Jacobian. By matching as closely as possible the elements of the exact and approximate Jacobians (10.9.5) and (10.9.10), we hope to improve the local description of the GBT vector and thus the convergence of the optimization. In general, however, we cannot reproduce the Jacobian exactly by this procedure. This is not too critical since the form of the Jacobian does not affect the identification of the stationary point, only the path that is taken towards this point.

Let us first consider closed-shell states. For such states, the elements of the approximate Jacobian (10.9.10)

$$w^{(1)}_{aibj} = 4(\delta_{ij} f_{ab} - \delta_{ab} f_{ij}) \tag{10.9.21}$$

contain exactly those parameters of the effective Hamiltonian that are not fixed by the zero-order conditions (10.9.19). These elements may therefore be chosen freely so as to match the exact Jacobian as closely as possible. At the stationary point, the exact Jacobian is identical to the electronic Hessian, which for closed-shell states takes the form (10.8.55)

$$E^{(2)}_{aibj} = 4(\delta_{ij}{}^{\mathrm{I}}F_{ab} - \delta_{ab}{}^{\mathrm{I}}F_{ij} + 4g_{aibj} - g_{abij} - g_{ajib}) \tag{10.9.22}$$

A comparison of (10.9.21) and (10.9.22) suggests the following identifications for the occupied–occupied and virtual–virtual elements of the Fock matrix:

$$f_{ij} = {}^{\mathrm{I}}F_{ij} \tag{10.9.23}$$

$$f_{ab} = {}^{\mathrm{I}}F_{ab} \tag{10.9.24}$$

We have now fixed all the parameters of the effective one-electron Hamiltonian for a closed-shell state

$$\hat{f} = \sum_{pq} {}^{\mathrm{I}}F_{pq} E_{pq} \tag{10.9.25}$$

The resulting Fock operator is identical to the one determined in Sections 10.3.2 and 10.3.3.

For open shells, the identification of the Fock-matrix elements is more difficult. For high-spin states, it is straightforward to identify the elements that correspond to the nonredundant rotations since these rotations mix orbitals of different occupations. The identification of the elements that correspond to redundant rotations is less straightforward. In some cases, such as for the two-electron singlet state, an unambiguous identification cannot be made since rotations between singly occupied orbitals are not redundant.

10.9.5   CONVERGENCE RATES

In the local region, the norm of the gradient may be used as a measure of the distance to the stationary point. We may therefore characterize the local convergence rate of an algorithm by comparing the gradients at two consecutive iterations. Let us consider Newton's method first. Expanding the gradient in iteration $n + 1$ in terms of the derivatives of iteration n, we obtain

$$\mathbf{E}^{(1)}_{n+1} = \mathbf{E}^{(1)}_n + \mathbf{E}^{(2)}_n \Delta\kappa_n + O(\Delta\kappa_n^{(2)}) \tag{10.9.26}$$

where the step $\Delta\kappa_n$ has been obtained from the Newton equations (10.8.2)

$$\Delta\kappa_n = -(\mathbf{E}^{(2)}_n)^{-1}\mathbf{E}^{(1)}_n \tag{10.9.27}$$

Substituting (10.9.27) in (10.9.26), we find that the gradient in iteration $n + 1$ is proportional to the square of the gradient in iteration n:

$$\mathbf{E}_{n+1}^{(1)} = \mathbf{E}_n^{(1)} - \mathbf{E}_n^{(2)}(\mathbf{E}_n^{(2)})^{-1}\mathbf{E}_n^{(1)} + O(\Delta\kappa_n^2) = O(\Delta\kappa_n^2) = O[(\mathbf{E}_n^{(1)})^2] \quad (10.9.28)$$

Newton's method is thus quadratically convergent. A detailed analysis of the convergence rate of Newton's method in terms of error vectors is given in Section 11.5.2.

Having established the local convergence rate of Newton's method, let us turn our attention to the SCF method. Before considering the local convergence, it should be noted that, strictly speaking, the SCF method is not equivalent to the application of Newton's method with an approximate Jacobian since, in each SCF step, we determine the stationary point of a *nonlinear* approximation to the true GBT vector (by diagonalization of the Fock matrix). Nevertheless, close to the stationary point, the SCF steps become so small that they are dominated by the linear term in the expansion of the approximate GBT vector of (10.9.8). For a closed-shell system, the resulting SCF step therefore becomes

$$\Delta\kappa_n = -(\mathbf{w}_n^{(1)})^{-1}\mathbf{E}_n^{(1)} \quad (10.9.29)$$

where $\mathbf{w}^{(1)}$ is given by (10.9.10). Inserting (10.9.29) in (10.9.26), we obtain

$$\mathbf{E}_{n+1}^{(1)} = [\mathbf{1} - \mathbf{E}_n^{(2)}(\mathbf{w}_n^{(1)})^{-1}]\mathbf{E}_n^{(1)} + O(\Delta\kappa_n^2) \quad (10.9.30)$$

If the matrix $\mathbf{1} - \mathbf{E}_n^{(2)}(\mathbf{w}_n^{(1)})^{-1}$ has eigenvalues larger than 1, then the gradient in iteration $n + 1$ will have elements greater than those of the previous iteration and the SCF iterations may diverge. If, on the other hand, the eigenvalues are all smaller than 1, then the error in iteration $n + 1$ will be smaller than that in the previous iteration and the SCF iterations will converge (if we are sufficiently close to the minimum).

Let us be more specific and consider the special case of a closed-shell system. The approximate Jacobian and the exact Hessian are then given by the expressions (10.9.21) and (10.9.22):

$$w_{aibj}^{(1)} = 4(^{\mathrm{I}}F_{ab}\delta_{ij} - {}^{\mathrm{I}}F_{ij}\delta_{ab}) \quad (10.9.31)$$

$$E_{aibj}^{(2)} = 4(\delta_{ij}{}^{\mathrm{I}}F_{ab} - \delta_{ab}{}^{\mathrm{I}}F_{ij} + 4g_{aibj} - g_{abij} - g_{ajib}) \quad (10.9.32)$$

For closed-shell systems, we may therefore write the Hessian matrix as

$$\mathbf{E}^{(2)} = \mathbf{w}^{(1)} + \mathbf{M} \quad (10.9.33)$$

where $\mathbf{M}$ contains the two-electron contributions

$$M_{aibj} = 4(4g_{aibj} - g_{abij} - g_{ajib}) \quad (10.9.34)$$

Inserting (10.9.33) into the general expression (10.9.30), we obtain

$$\mathbf{E}_{n+1}^{(1)} = -\mathbf{M}_n(\mathbf{w}_n^{(1)})^{-1}\mathbf{E}_n^{(1)} + O(\Delta\kappa_n^2) \quad (10.9.35)$$

which shows that the local criterion for convergence in the SCF method is that the eigenvalues of $-\mathbf{M}_n(\mathbf{w}_n^{(1)})^{-1}$ must be numerically smaller than 1, otherwise the iterations may diverge.

For a comparison of Newton's method and the SCF method, we return to Table 10.4. For each iteration, we have here listed the errors in the energy obtained using the second-order trust-region method, the naive SCF method and the DIIS method for the water molecule in the cc-pVQZ basis. In the first few iterations, the trust-region method does not converge faster than the SCF and DIIS methods – indeed, the DIIS method is fastest in this region. In the local region, however, the trust-region method reduces to Newton's method and exhibits quadratic convergence, with the number of zeros in the error doubling in each iteration. The trust-region method is now faster than the SCF and DIIS methods, which converge only linearly, with the error in each iteration decreasing by a constant factor in accordance with (10.9.35). The superiority of Newton's method in the local region is particularly evident at the stretched geometry.

When comparing the different methods, we should consider not only the number of iterations but also their relative costs. The costs of the SCF and DIIS steps are similar. Each trust-region step is much more expensive, however, requiring from three to six linear transformations of the type in (10.8.8) – each at the cost of a single SCF iteration. For the cases in Table 10.4, therefore, the DIIS procedure is the most cost-efficient. In general, the best strategy may be to begin with the DIIS method but to switch to the second-order trust-region method if DIIS experiences problems.

10.9.6 THE SCF AND NEWTON METHODS COMPARED

Newton's method is based on a local quadratic model of the energy surface. The orbital rotations generated by this method therefore behave incorrectly for large rotations. In particular, the Newton equations are not periodic in the orbital-rotation parameters as we would expect from a consideration of the global behaviour of the energy function (10.1.22). The one-electron approximation of the SCF method, by contrast, is correct only to zero order in the rotations, but – provided the effective Hamiltonian (i.e. the Fock operator) is a reasonable one – it exhibits the correct global behaviour. In particular, it is periodic in the orbital rotations.

The Newton step is obtained by solving a set of linear equations involving the Hessian matrix; the SCF step is generated by diagonalizing the (significantly smaller) matrix representation of the Fock operator. The application of either model leads to an iterative scheme, which is repeated until the stationary point is reached and the gradient vanishes. For quadratic energy functions, Newton's method reaches the stationary point in a single iteration, whereas the SCF method requires a sequence of iterations. For one-electron systems or for noninteracting electrons, the SCF approach reaches the stationary point in a single iteration, whereas a sequence of iterations is required in Newton's method.

The SCF method is based on the existence of an effective one-electron Fock operator. For closed-shell systems and for some classes of open-shell systems, the RHF Fock operator is easily set up. For other open-shell systems, it may be impossible to set up a Fock operator that reproduces the exact gradient at the expansion point or it may be difficult to choose a reasonable set of parameters for those parts of the Fock operator that are not defined by the zero-order variational conditions. By contrast, the second-order trust-region method discussed in Section 10.8 is robust and can be applied straightforwardly to any open-shell system.

10.10 Singlet and triplet instabilities in RHF theory

We have so far considered RHF theory, where a well-defined spin multiplicity and spatial symmetry is imposed on the solution. In UHF theory, these restrictions are relaxed and the solution may then have an energy lower than that of the corresponding RHF wave function. In the present section, we examine the relationship between the RHF and UHF solutions in greater detail.

10.10.1 ORBITAL-ROTATION OPERATORS IN RHF AND UHF THEORIES

In RHF theory, the wave function is taken as a spin eigenfunction that transforms as an irreducible representation of the molecular point group – for example, as $^1A_1$ in the C_{2v} point group. The RHF wave function corresponds to a variationally optimized CSF, parametrized as

$$|\text{CSF}(\boldsymbol{\kappa}^{\text{RHF}})\rangle = \exp(-\hat{\kappa}^{\text{RHF}})|\text{CSF}\rangle \tag{10.10.1}$$

where $\hat{\kappa}^{\text{RHF}}$ contains only the totally symmetric real singlet part of the orbital-rotation operator (3.3.19) of Section 3.3.3:

$$\hat{\kappa}^{\text{RHF}} = \sum_{p>q} {}^{\text{R}}\kappa_{pq}^{0,0}(E_{pq} - E_{qp}) \tag{10.10.2}$$

In Abelian point groups, the requirement that $\hat{\kappa}^{\text{RHF}}$ is totally symmetric is easily satisfied by summing only pairs of MOs p and q that belong to the same irreducible representation of the point group.

In UHF theory, the requirement that the wave function transforms as an irreducible representation of a given spin multiplicity is lifted, but the wave function is still an eigenfunction of $\hat{S}_z$ since pure alpha and beta spin orbitals are used. The UHF wave function is written as a single Slater determinant

$$|\text{SD}(\boldsymbol{\kappa}^{\text{UHF}})\rangle = \exp(-\hat{\kappa}^{\text{UHF}})|\text{SD}\rangle \tag{10.10.3}$$

where $\hat{\kappa}^{\text{UHF}}$ also contains the triplet excitation operators of (3.3.19)

$$\hat{\kappa}^{\text{UHF}} = \sum_{p>q} {}^{\text{R}}\kappa_{pq}^{0,0}(E_{pq} - E_{qp}) + \sum_{p>q} {}^{\text{R}}\kappa_{pq}^{1,0}(\hat{T}_{pq}^{1,0} - \hat{T}_{qp}^{1,0}) \tag{10.10.4}$$

with no restrictions on the symmetries of the orbitals p and q, relaxing the requirement that the orbitals and thus the Slater determinant transform as irreducible representations of the molecular point group. The operators $\hat{T}_{pq}^{1,1}$ and $\hat{T}_{pq}^{1,-1}$ are not included in (10.10.4) since the resulting wave function would then not remain an eigenfunction of $\hat{S}_z$. Disregarding the differences that result from point-group symmetry adaptation, we note that the UHF model (10.10.3) contains twice as many variational parameters as the RHF model (10.10.1). Those rotations that are present in $\hat{\kappa}^{\text{UHF}}$ but not in $\hat{\kappa}^{\text{RHF}}$ are called *symmetry-breaking rotations*. It is the presence of the symmetry-breaking rotations that may lead to an energy lowering of the UHF energy relative to that of the RHF energy.

The form of $\hat{\kappa}^{\text{UHF}}$ in (10.10.4) is useful for discussing the differences between RHF and UHF theory and, in particular, for examining what happens when an already optimized RHF state is reoptimized in the full set of symmetry-breaking variational parameters. For the direct optimization of the UHF wave function (10.10.3) itself, it is more convenient to work with excitation operators that are not spin tensor operators. Decomposing the singlet and triplet excitation operators in alpha and beta parts (see Section 2.3.4)

$$E_{pq} = E_{pq}^{\alpha} + E_{pq}^{\beta} \tag{10.10.5}$$

$$\hat{T}_{pq}^{1,0} = \frac{1}{\sqrt{2}}(E_{pq}^{\alpha} - E_{pq}^{\beta}) \tag{10.10.6}$$

where

$$E_{pq}^{\alpha} = a_{p\alpha}^{\dagger} a_{q\alpha} \tag{10.10.7}$$

$$E_{pq}^{\beta} = a_{p\beta}^{\dagger} a_{q\beta} \tag{10.10.8}$$

we note that the UHF orbital-rotation operator may be written in the simpler form

$$\hat{\kappa}^{UHF} = \sum_{p>q} {}^{R}\kappa_{pq}^{\alpha}(E_{pq}^{\alpha} - E_{qp}^{\alpha}) + \sum_{p>q} {}^{R}\kappa_{pq}^{\beta}(E_{pq}^{\beta} - E_{qp}^{\beta}) \tag{10.10.9}$$

There is here one set of parameters for the alpha spin orbitals and one set for the beta spin orbitals. Indeed, these are the parameters used in UHF theory for carrying out an unrestricted optimization of the wave function, generating separate sets of alpha and beta spin orbitals.

10.10.2 RHF INSTABILITIES FOR NONDEGENERATE ELECTRONIC STATES

Let us assume that we have carried out a ground-state RHF calculation on a molecular system. The RHF energy is then stationary and stable (i.e. a local minimum) with respect to all nonredundant totally symmetric singlet rotations. An important question is now whether the energy is also stationary with respect to the symmetry-breaking rotations. Furthermore, if the RHF energy is indeed found to be a stationary point with respect to such rotations, is it a minimum or is it a saddle point? Of course, since we have made no attempt to minimize the energy with respect to the symmetry-breaking rotations, we should not expect the energy to be stable (i.e. a local minimum) with respect to these rotations. Nevertheless, it turns out that the RHF energy often is stable – not only with respect to the totally symmetric singlet rotations but also with respect to symmetry-breaking rotations.

When discussing the stationarity and stability of the RHF energy with respect to symmetry-breaking rotations, it is useful to treat nondegenerate and degenerate electronic states separately. Nondegenerate electronic states are examined in the present subsection; the stationarity of degenerate states is considered in Section 10.10.3. By simple group-theoretical arguments, it follows that any nondegenerate RHF state must be *stationary* with respect to symmetry-breaking rotations. A singlet electronic state, for example, must be stationary with respect to triplet rotations since the direct product of the irreducible representations of the singlet bra state, the singlet Hamiltonian, the triplet rotation operator and the singlet ket state does not transform as the singlet irreducible representation:

$$\langle {}^{1}HF|[\hat{T}_{pq}^{1,M}, \hat{H}]|{}^{1}HF\rangle = 0 \tag{10.10.10}$$

In the same manner, it follows that any spatially nondegenerate state (including all closed-shell states) must be stationary with respect to rotations that mix orbitals of different symmetries. These arguments cannot be applied to degenerate states since the product of a nontotally symmetric operator with the wave function may contain a component of the same symmetry as the wave function, making it possible for symmetry-breaking rotations to make nonzero contributions to the gradient.

Although we have established that all nondegenerate RHF states are *stationary* with respect to the symmetry-breaking rotations, it does not follow that these states are *stable* with respect to such rotations. To establish stability of a given RHF state, we cannot resort to group-theoretical arguments but must examine the electronic Hessian to determine whether the eigenvalues with respect to the symmetry-breaking rotations are positive. We note that, since the eigenvalues change as we change the geometry of the system (or otherwise perturb the system), stability must be established for each geometry separately.

For nondegenerate states, the Hessian separates into blocks of pure singlet and triplet symmetries. RHF instabilities are classified as *singlet and triplet instabilities*, depending on whether the negative eigenvalues occur in the singlet or the triplet block of the electronic Hessian [22,23]. Triplet instabilities occur more often than singlet instabilities and indicate that a spin-contaminated UHF wave function of lower energy may be found by relaxing the spin-symmetry restrictions. Triplet instabilities are frequently encountered as molecular bonds are stretched – in such situations, the RHF wave function makes a poor approximation to the true wave function as the electron pairs are disrupted and the electrons are recoupled – see the discussion in Section 5.4. Close to the molecular equilibrium configuration, however, the RHF wave function constitutes a better approximation to the true wave function and triplet instabilities are less frequently encountered. Examples of triplet instabilities of closed-shell electronic states are given in Sections 10.10.4 and 10.10.5.

Singlet instabilities are less common than triplet instabilities and occur when the negative eigenvalue of the Hessian is associated with a singlet rotation that does not transform as the totally symmetric representation of the molecular point group – that is, rotations among orbitals of different irreducible representations. (For nondegenerate states, the singlet and triplet parts of the electronic Hessian separate further into blocks belonging to the different irreducible representations of the molecular point group.) When the restrictions are lifted, singlet instabilities lead to spatially symmetry-broken solutions (i.e. to electronic states that cannot be classified according to the irreducible representations of the molecular point group). Singlet instabilities indicate the presence of several electronic configurations close in energy. If some geometrical distortion belongs to the same irreducible representation as the negative Hessian eigenvalue, then an optimization of the molecular geometry in this direction may lead to an equilibrium structure of a lower point-group symmetry. An example of a singlet instability is given in Section 10.10.6.

Singlet and triplet instabilities often manifest themselves when second- and higher-order molecular properties are calculated from the wave function. At the onset of the instability, in particular, the electronic Hessian becomes singular and the response of the wave function becomes infinite to perturbations of the same symmetry as the instability. Triplet instabilities affect, for example, the calculation of indirect nuclear spin–spin coupling constants (which are primarily mediated by the triplet Fermi-contact operator of Exercise 2.1), giving anomalous coupling constants that bear no relationship to the true couplings of the paramagnetic nuclei. Singlet instabilities are, for example, responsible for anomalies in the molecular Hessian, producing force constants that tend to infinity at the onset of the instability. In the following, we shall consider two examples of triplet instabilities (in the hydrogen and water molecules) and one example of singlet instabilities (in the allyl radical).

10.10.3 RHF ENERGIES OF DEGENERATE ELECTRONIC STATES

Degenerate electronic states differ from nondegenerate states in that the elements of the electronic gradient associated with symmetry-breaking rotations are no longer zero by symmetry. For triplet states, for example, we obtain

$$\langle {}^3\text{HF}|[\hat{T}_{pq}^{1,M}, \hat{H}]|{}^3\text{HF}\rangle \neq 0 \qquad (10.10.11)$$

since the direct product of the irreducible representations of the triplet state and the triplet orbital-rotation operator contains a component of triplet spin. In the absence of any higher symmetries, therefore, the RHF gradient will be nonzero and there will consequently always be a spin-contaminated UHF state of lower energy. Note that spin-contaminated triplet states occur even when the RHF Hessian is positive definite with respect to triplet rotations.

Likewise, for spatially degenerate states, orbital rotations may exist that transform as a nontotally symmetric irreducible representation of the molecular point group and which will then have nonzero gradient elements. Consider the $|^3P\rangle$ state of the carbon atom, for which $1s^2 2s^2 2p_x\alpha 2p_y\alpha$ is one of three spatially degenerate components with spin projection $M = 1$. Rotations between s and d orbitals will have nonzero gradient elements

$$\langle ^3P|[E_{ds}, \hat{H}]|^3P\rangle \neq 0 \qquad (10.10.12)$$

since $E_{ds}|^3P\rangle$ contains a component of the same symmetry as $|^3P\rangle$. The use of such operators will therefore lead to a symmetry-broken Hartree–Fock state with the s and d orbitals slightly mixed (irrespective of the structure of the Hessian), leading to a small deviation from the correct atomic symmetry.

10.10.4 TRIPLET INSTABILITIES IN H_2

To illustrate the connection between spin-contaminated UHF solutions and triplet instabilities in RHF theory, we consider the RHF and UHF energies of the hydrogen molecule in the cc-pVQZ basis. In Figure 10.2, we have plotted the RHF and UHF potential-energy curves of H_2 as well as the lowest RHF Hessian eigenvalues for singlet and triplet orbital rotations of σ_u symmetry. For bond distances shorter than $2.298a_0$, the RHF state is stable with respect to singlet and triplet rotations and the RHF and UHF solutions coincide. At $2.298a_0$, the triplet Hessian becomes singular and the RHF and UHF energies separate. Beyond this point, the RHF solution is still stable with respect to singlet rotations but unstable with respect to triplet rotations, indicating the presence of a spin-contaminated UHF solution of lower energy. In Exercise 10.8, the electronic Hessian for triplet rotations is derived for a closed-shell system and triplet instabilities are found to occur in an STO-3G basis for internuclear distances longer than $2.65a_0$.

10.10.5 TRIPLET INSTABILITIES IN H_2O

The connection between RHF triplet instabilities and UHF solutions of lower energy is also seen in the C_{2v} dissociation of the water molecule. As in Section 5.4.9, the basis set is cc-pVDZ and the HOH bond angle is fixed at $110.565°$. The four plots in Figure 10.3 illustrate the behaviour

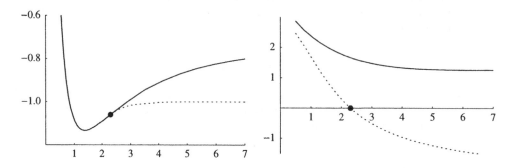

Fig. 10.2. Triplet instability in H_2. On the left, we have plotted the ground-state potential-energy curves (in E_h) for the RHF wave function (full line) and the UHF wave function (dotted line) in the cc-pVQZ basis as a function of the internuclear distance (in a_0). On the right, we have plotted the lowest RHF Hessian eigenvalues (in E_h) of singlet symmetry (full line) and triplet symmetry (dotted line) as a function of the internuclear distance. The onset of the triplet instability at $2.298a_0$ has been indicated by a dot.

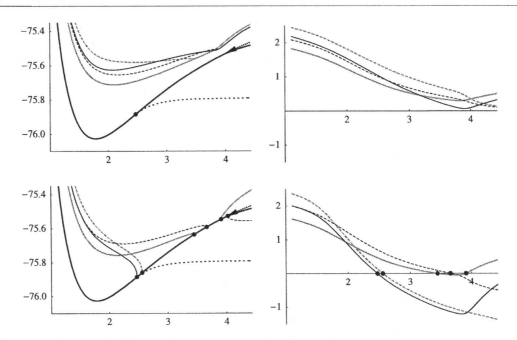

Fig. 10.3. Triplet instabilities in the C_{2v} water molecule. On the left, we have plotted the potential-energy curves for the ground-state RHF wave function (full line) and the ground-state UHF wave function (dotted line) in the cc-pVDZ basis as a function of the internuclear OH distance. The presence of a second RHF solution of $^1A_1$ symmetry is indicated with a dotted line. Superimposed on these ground-state curves are the potential-energy curves for the lowest excited states of the symmetry species A_1 (full line), A_2 (dashed line), B_1 (full grey line) and B_2 (dashed grey line), calculated using the RPA method. The upper plot contains the singlet excited states; the lower plot contains the triplet excited states. On the right, we have plotted the lowest RHF Hessian eigenvalues of singlet symmetry (upper plot) and of triplet symmetry (lower plot) of the different irreducible representations. Atomic units are used.

of the ground-state energies, the excitation energies and the Hessian eigenvalues as the bonds are stretched.

The plots on the left in Figure 10.3 contain the potential-energy curves for the ground-state RHF and UHF wave functions. Superimposed on these curves, we have plotted the energy of the lowest excited state of each irreducible representation. These energies have been obtained by adding to the Hartree–Fock ground-state energy the excitation energies from Hartree–Fock *linear response theory*, also known as the *random-phase approximation (RPA)* [24,25]. The upper plot contains the singlet excited states; the lower plot the triplet states. On the right, we have plotted the lowest Hessian eigenvalues of each representation, with the singlet eigenvalues at the top and the triplet eigenvalues at the bottom.

The lowest singlet and triplet Hessian eigenvalues of all irreducible representations approach zero as the bonds are stretched. The singlet and triplet eigenvalues behave quite differently in that the singlet Hessian eigenvalues remain positive at all plotted distances whereas the triplet eigenvalues become negative when the bonds are stretched (although the B_1 eigenvalue becomes positive again and the same may later happen at least for the A_1 eigenvalue). Recall, however, that only the totally symmetric A_1 singlet Hessian eigenvalue is positive by construction (with a minimum of 0.06 E_h at $3.9a_0$); the remaining eigenvalues may or may not become negative as the OH bonds are stretched.

The minimum in the A_1 singlet Hessian eigenvalue at $3.9a_0$ arises because of an avoided crossing with a second Hartree–Fock state as indicated by the dotted line terminated by an arrow in Figure 10.3. This second RHF solution represents an excited (albeit unphysical) state of $^1A_1$ symmetry. (For distances shorter than $4.0a_0$, the calculation of the excited RHF state becomes difficult and no plot has been attempted in this region.) Multiple RHF solutions close in energy are often found in regions where the Hartree–Fock wave function provides an inadequate description of the electronic system and where it is necessary to go beyond the Hartree–Fock model for a proper description of the electronic system.

Before considering the RPA excitation energies, we note that these have been calculated based on information in the electronic Hessian and that their behaviour should reflect closely that of the eigenvalues. We thus find that the singlet excitation energies behave smoothly, much like the corresponding eigenvalues. The triplet excitation energies, by contrast, show a more complicated behaviour. At a bond distance of $2.64a_0$, the $^3A_1$ excitation energy becomes zero, signalling the onset of the triplet instability. Beyond $2.64a_0$, the triplet Hessian is indefinite and the excitation energies can no longer be calculated by RPA theory. In the same region, the spin-contaminated UHF energy falls below the RHF energy. The abnormal behaviour of the triplet excitation energies for stretched bond distances reflects the breakdown of the RHF approximation in this region. All triplet excitation energies disappear when the corresponding eigenvalues become negative. At longer bond distances, two of the excited states reappear – one above and the other below the RHF ground-state energy curve. The vanishing of the excitation energies at the onset of a triplet instability is an example of the anomalies encountered in second- and higher-order molecular properties at such points.

10.10.6 SINGLET INSTABILITIES IN THE ALLYL RADICAL

To illustrate the problems that may arise in connection with singlet instabilities, we consider the allyl radical C_3H_5, a planar molecular species of C_{2v} symmetry. When the geometry of this system is optimized at the Hartree–Fock level with no spatial symmetry restrictions applied, a minimum of C_s symmetry is obtained – in disagreement with experiment and also with more elaborate calculations. In the following, we shall study the allyl radical at the RHF and UHF levels of theory in order to understand this behaviour [26]. The allyl radical is shown in Figure 10.4, which also illustrates the coordinates used in this study.

When the molecular structure of the allyl radical is optimized at the cc-pVDZ level in the C_{2v} point group, the RHF equilibrium geometry has a CC bond distance $R_e = 2.5980a_0$ and a CCC bond angle $\Theta_e = 124.49°$. The symmetry of the ground-state RHF wave function is $^2A_2$ and the orbital configuration is $6a_1^2 4b_2^2 1b_1^2 1a_2^1$, with the unpaired electron occupying the nonbonding π orbital. If the RHF energy is plotted along the symmetric CC stretching coordinate δ (with $\delta = 0$ corresponding to R_e) with all other internal coordinates fixed at their equilibrium values, we obtain the C_{2v} potential-energy curve displayed in Figure 10.5 (full line) with a minimum at $\delta = 0$.

Fig. 10.4. The structure of the allyl radical. On the left, the symmetric stretching coordinate δ for the carbon–carbon bond distance; on the right, the corresponding antisymmetric stretching coordinate Δ.

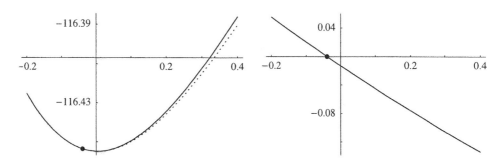

Fig. 10.5. On the left, the RHF (full line) and UHF (dotted line) total energies of the allyl radical plotted as functions of the symmetric stretching coordinate δ in Figure 10.4. On the right, the lowest singlet Hessian eigenvalue of symmetry B_2 of the RHF wave function. Atomic units are used.

At face value, the RHF wave function appears to give a satisfactory description of the molecular system. However, if we also plot the lowest singlet RHF Hessian eigenvalue, we obtain a monotonically decreasing curve, which goes through zero at $\delta = -0.0395a_0$ (i.e. at a CC bond distance of $2.5585a_0$), indicating that a solution of lower energy would be obtained by relaxing the restriction that the wave function should transform as an irreducible representation of the C_{2v} point group. The C_{2v} potential-energy curve of the corresponding UHF wave function (which does not transform as any of the irreducible representations of the C_{2v} point group) is also plotted in Figure 10.5 (dotted line), superimposed on the RHF curve (full line). The difference between the RHF and UHF curves is exceedingly small – in particular at the C_{2v} equilibrium structure of the RHF wave function, where the UHF energy is only about 0.01 mE$_\text{h}$ lower than the RHF energy. When both CC bonds are stretched by $0.4a_0$ from the minimum, the difference between the two curves is still only 4.2 mE$_\text{h}$. The C_{2v}-adapted delocalized MOs of the RHF wave function are slightly distorted in the UHF wave function – the nonbonding A_2 MO develops a weakly bonding character between the central carbon atom and one of the terminal carbon atoms, and a weakly antibonding character between the central atom and the other terminal carbon atom.

The negative Hessian eigenvalue of the RHF wave function in Figure 10.5 belongs to the B_2 representation of the C_{2v} point group, the same representation as the antisymmetric stretching of the CC bonds. We should therefore be able to generate an allyl system of lower energy by distorting the molecule into C_s symmetry, contracting one of the CC bonds and stretching the other one by the same amount. In Figure 10.6, we have plotted the resulting C_s potential-energy curve, obtained by calculating the energy along the antisymmetric CC stretching coordinate Δ in Figure 10.4 with δ fixed at $0.2a_0$ and with the remaining internal coordinates fixed at their C_{2v} equilibrium values.

In Figure 10.6, there are two independent solutions of C_{2v} symmetry: the uppermost dot represents the singlet-unstable $^2A_2$ RHF solution; the lowermost dot represents the corresponding symmetry-broken UHF solution. The UHF solution does not transform as any of the four irreducible representations of the C_{2v} molecular point group, but may instead be classified as a $^2A''$ electronic state in the C_s subgroup. At nonzero displacements, the symmetry-broken UHF wave function turns into a *restricted* wave function of $^2A''$ symmetry in the C_s molecular point group (dotted lines). The double-well structure arises since the molecule may be distorted in two equivalent ways, depending on which bond is stretched and which bond is contracted.

The uppermost potential-energy curve in Figure 10.6 (full line) represents an RHF wave function of $^2A''$ symmetry with one negative Hessian eigenvalue of A'' symmetry. At $\Delta = 0$, this wave

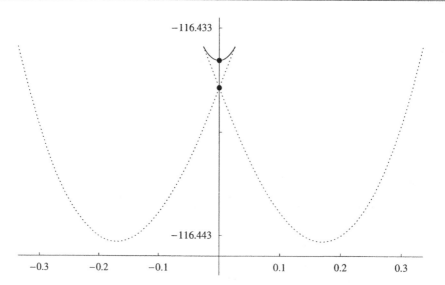

Fig. 10.6. The Hartree–Fock energies of the allyl radical as functions of the antisymmetric stretching coordinate Δ in Figure 10.4 (atomic units). The uppermost dot, which represents the symmetry-restricted C_{2v} ground-state RHF energy, lies on a C_s potential-energy curve (full line) with one negative singlet Hessian eigenvalue. The lowermost dot, which represents the corresponding symmetry-broken UHF solution, lies on two equivalent C_s potential-energy curves (dotted lines) with a positive definite Hessian.

function correlates with the symmetry-restricted RHF solution of $^2A_2$ symmetry and one negative Hessian eigenvalue of B_2 symmetry. The potential-energy curves that contain the RHF and UHF solutions at $\Delta = 0$ become degenerate at $|\Delta| \approx 0.03a_0$, where the lowest Hessian eigenvalue of A'' symmetry becomes zero for both states.

It should be emphasized that both the RHF solution and the UHF solution in Figure 10.6 are approximations to the same wave function. The $^2A_2$ RHF wave function transforms in the same manner as the true wave function but is higher in energy than the symmetry-broken UHF solution. The difference in energy is very small, however, compared with the true excitation energies of the system. Neither wave function may be classified as 'best' and, for a satisfactory resolution of this symmetry dilemma of Hartree–Fock theory, a more advanced wave function must be used, as described in Section 12.8.

10.11 Multiple solutions in Hartree–Fock theory

Close to the molecular equilibrium geometry, where the wave function is normally dominated by a single electronic configuration, the calculation of the Hartree–Fock ground state is usually unproblematic, with convergence to the same stationary point irrespective of what method is used and what orbitals are used in the first iteration. In other cases – in particular, far away from the equilibrium geometry, where several configurations may be important – convergence is more difficult to achieve. In such regions, several stationary points (local minima) may be close in energy and all of these may be reached – depending on what orbitals are used in the first iteration and what algorithm is used to converge the orbitals. In such cases, physical insight is needed in order to decide which minimum represents the desired solution. To illustrate the existence of multiple minima, we consider RHF/cc-pVDZ calculations on the water molecule in C_{2v} symmetry, with a bond angle of $110.565°$ and a bond length of $8R_e$. Although this example is special in that there is essentially no interaction left among the atoms, it illustrates some of the considerations that must be made in cases where several minima may be obtained or are expected.

Let us first consider the orbital configurations of the fragment system. Figure 5.10 shows that the ground-state H_2O electron configuration $1a_1^2 2a_1^2 1b_2^2 3a_1^2 1b_1^2$ becomes, in the limit of no interatomic interactions, $O_{1s}^2 O_{2s}^2 (c_1 O_{2p_y} + c_2 H_{b_2})^2 (c_3 O_{2p_z} + c_4 H_{a_1})^2 O_{2p_x}^2$, where H_{a_1} and H_{b_2} are respectively the symmetric and antisymmetric combinations of the $1s$ orbitals on the two hydrogens. Since the fragments do not interact, the coefficients c_1, c_2, c_3 and c_4 must be equal to 1 or 0. In this way, we obtain for the fragmentation products the four configurations in Table 10.5.

The first solution in Table 10.5 represents the closed-shell anionic oxygen atom O^{2-} in the ^1S ground state with the inert-gas configuration $1s^2 2s^2 2p^6$. The second solution corresponds to a cationic oxygen atom O^{2+} and two hydrogen anions H^-. The oxygen atom is a mixture of the states ^1S and ^1D obtained by coupling of the two electrons in the p_x orbital; the ^3P state does not appear because of the double occupancy of all orbitals. For hydrogen, the double occupancy of the symmetric and antisymmetric $1s$ orbitals gives the total energy of two H^- fragments in the ^1S ground state. The two remaining configurations in Table 10.5 both give rise to a neutral oxygen atom of configuration $1s^2 2s^2 2p^4$ with contributions from the ^1S and ^1D states, again with no contribution from the ^3P state. The symmetric hydrogen orbital of the third configuration and the antisymmetric hydrogen orbital of the fourth configuration both dissociate into a mixture of an ionic and a covalent states as discussed in Section 5.2.9.

All four configurations in Table 10.5 represent true local minima of the Hartree–Fock energy function. It is noteworthy, however, that none of these minima represent the correct ground-state products of the dissociation with an oxygen atom in the ^3P ground state and two hydrogen atoms in the ^2S ground state, and that the true wave function representing these products has *zero* overlap with the Hartree–Fock states in Table 10.5.

One might argue that these peculiar features of the Hartree–Fock solutions in the dissociation limit arise since we have applied the RHF model in a region where it should not be applied. Still, it is not always clear when such regions are encountered. At an OH separation of $3R_e$, for example, several solutions exist, all of which correspond to the dissociation products O + 2H. The two lowest solutions are almost degenerate and describe nearly nonbonding situations, corresponding to the third and fourth configurations in Table 10.5. The third solution at $3R_e$ has higher energy and represents the desired solution, in which each of the orbitals $3a_1$ and $1b_2$ has significant contributions from both oxygen and hydrogen.

Table 10.5 The Hartree–Fock fragmentation products of H_2O in C_{2v} symmetry

Configuration	Energy (E_h)	Fragments	Atomic configurations	Atomic terms
$O_{1s}^2 O_{2s}^2 O_{2p_y}^2 O_{2p_z}^2 O_{2p_x}^2$	-74.410876	O^{2-}	$1s^2 2s^2 2p^6$	^1S
$O_{1s}^2 O_{2s}^2 H_{b_2}^2 H_{a_1}^2 O_{2p_x}^2$	-74.050033	O^{2+}	$1s^2 2s^2 2p^2$	$^1S + {}^1D$
		H^-	$1s^2$	^1S
$O_{1s}^2 O_{2s}^2 O_{2p_y}^2 H_{a_1}^2 O_{2p_x}^2$	-75.393278	O	$1s^2 2s^2 2p^4$	$^1S + {}^1D$
		H^-	$1s^2$	^1S
		H	$1s^1$	^2S
$O_{1s}^2 O_{2s}^2 H_{b_2}^2 O_{2p_z}^2 O_{2p_x}^2$	-75.393278	O	$1s^2 2s^2 2p^4$	$^1S + {}^1D$
		H^-	$1s^2$	^1S
		H	$1s^1$	^2S

References

[1] B. Levy and G. Berthier, *Int. J. Quantum Chem.* **2**, 307 (1968).
[2] R. L. Burden and J. D. Faires, *Numerical Analysis*, 5th edn, PWS Publishing Company, 1993.
[3] C. C. J. Roothaan, *Rev. Mod. Phys.* **23**, 69 (1951).
[4] G. G. Hall, *Proc. Roy. Soc. Lond.* **A205**, 541 (1951).
[5] P. Pulay, *Chem. Phys. Lett.* **73**, 393 (1980); *J. Comput. Chem.* **3**, 556 (1982).
[6] J. Almlöf, K. Fægri, Jr and K. Korsell, *J. Comput. Chem.* **3**, 385 (1982).
[7] J. Almlöf, in B. O. Roos (ed.), *Lecture Notes in Quantum Chemistry II*, Lecture Notes in Chemistry Vol. 64, Springer-Verlag, 1994, p. 1.
[8] R. McWeeny, *Rev. Mod. Phys.* **32**, 335 (1960).
[9] R. W. Nunes and D. Vanderbilt, *Phys. Rev.* **B50**, 17611 (1994).
[10] W. H. Press, S. A. Teukolsky, W. T. Vetterling and B. P. Flannery, *Numerical Recipes*, Cambridge University Press, 1986.
[11] X.-P. Li, R. W. Nunes and D. Vanderbilt, *Phys. Rev.* **B47**, 10891 (1993).
[12] C. Ochsenfeld and M. Head-Gordon, *Chem. Phys. Lett.* **270**, 399 (1997).
[13] D. L. Strout and G. E. Scuseria, *J. Chem. Phys.* **102**, 8448 (1995).
[14] M. C. Strain, G. E. Scuseria and M. J. Frisch, *Science* **271**, 51 (1996).
[15] J. M. Millam and G. E. Scuseria, *J. Chem. Phys.* **106**, 5569 (1997).
[16] L. G. Yaffe and W. A. Goddard III, *Phys. Rev.* **A13**, 1682 (1976).
[17] E. Dalgaard and P. Jørgensen, *J. Chem. Phys.* **69**, 3833 (1978).
[18] G. B. Bacskay, *Chem. Phys.* **61**, 385 (1981).
[19] R. Fletcher, *Practical Methods of Optimization*, 2nd edn, Wiley, 1987.
[20] P. E. M. Siegbahn, J. Almlöf, A. Heiberg and B. O. Roos, *J. Chem. Phys.* **74**, 2384 (1981).
[21] J. Olsen, D. L. Yeager and P. Jørgensen, *Adv. Chem. Phys.* **54**, 1 (1983).
[22] D. J. Thouless, *The Quantum Mechanics of Many-Body Systems*, Academic Press, 1961.
[23] J. Čížek and J. Paldus, *J. Chem. Phys.* **47**, 3976 (1967).
[24] T. H. Dunning and V. McKoy, *J. Chem. Phys.* **47**, 1735 (1967).
[25] J. Oddershede, P. Jørgensen and D. L. Yeager, *Comp. Phys. Rep.* **2**, 33 (1984).
[26] J. Paldus and A. Veillard, *Mol. Phys.* **35**, 445 (1978).

Further reading

J. Almlöf, Notes on Hartree–Fock theory and related topics, in B. O. Roos (ed.), *Lecture Notes in Quantum Chemistry II*, Lecture Notes in Chemistry Vol. 64, Springer-Verlag, 1994, p. 1.

J. Almlöf, Direct methods in electronic structure theory, in D. R. Yarkony (ed.), *Modern Electronic Structure Theory*, World Scientific, 1995, p. 110.

Exercises

EXERCISE 10.1

In this exercise, we shall determine the gradient of the closed-shell Hartree–Fock energy

$$E_{PQ}^{(1)}(\kappa) = \frac{\partial \langle \mathrm{cs}| \exp(\hat{\kappa})\hat{H}\exp(-\hat{\kappa})|\mathrm{cs}\rangle}{\partial \kappa_{PQ}} \tag{10E.1.1}$$

for a nonvanishing $\hat{\kappa}$ that contains only real parameters κ_{PQ}. In Exercise 3.5, the derivative of an exponential operator was obtained as

$$\frac{\partial \exp(-\hat{\kappa})}{\partial \kappa_{PQ}} = -\sum_{RS} \frac{\exp[\mathrm{i}(\lambda_R - \lambda_S)] - 1}{\mathrm{i}(\lambda_R - \lambda_S)}(U_{PR}^* U_{QS} - U_{QR}^* U_{PS})\bar{a}_R^\dagger \bar{a}_S \exp(-\hat{\kappa}) \tag{10E.1.2}$$

where U_{PQ} are the elements of the unitary matrix $\mathbf{U}$ that diagonalizes κ and λ_P are the real eigenvalues of $i\kappa$:

$$i\kappa = \mathbf{U}\lambda\mathbf{U}^\dagger \tag{10E.1.3}$$

The $\bar{a}_P^\dagger$ are the creation operators in the eigenvector basis of κ:

$$\bar{a}_P^\dagger = \sum_R a_R^\dagger U_{RP} \tag{10E.1.4}$$

1. Show that the elements of the gradient (10E.1.1) are given by

$$E_{PQ}^{(1)}(\kappa) = \sum_{RS} \frac{\exp[i(\lambda_R - \lambda_S)] - 1}{i(\lambda_R - \lambda_S)} (U_{PR}^* U_{QS} - U_{QR}^* U_{PS})$$

$$\times \langle cs| \exp(\hat{\kappa})[\bar{a}_R^\dagger \bar{a}_S, \hat{H}] \exp(-\hat{\kappa})|cs\rangle \tag{10E.1.5}$$

2. Specialize (10E.1.5) to the case $\kappa = \mathbf{0}$.

EXERCISE 10.2

In the optimization of a closed-shell electronic system, the nonredundant orbital-rotation parameters

$$\kappa = \begin{pmatrix} \mathbf{0} & \kappa_{AI}^{\mathrm{T}} \\ -\kappa_{AI} & \mathbf{0} \end{pmatrix} \tag{10E.2.1}$$

determine the transformation from the original MOs $\phi_p(\mathbf{0})$ to a set of improved MOs $\phi_p(\kappa)$:

$$\phi_p(\kappa) = \sum_q \phi_q(\mathbf{0}) \exp(-\kappa)_{qp} \tag{10E.2.2}$$

A further rotation of $\phi_p(\kappa)$ involving the redundant parameters

$$\rho = \kappa^{\mathrm{RD}} = \begin{pmatrix} \kappa_{IJ} & \mathbf{0} \\ \mathbf{0} & \kappa_{AB} \end{pmatrix} \tag{10E.2.3}$$

gives a new MO basis

$$\psi_p(\rho) = \sum_q \phi_q(\kappa) \exp(-\rho)_{qp} \tag{10E.2.4}$$

which is equivalent to (10E.2.2) in the sense that it gives the same energy – that is, the total energy is invariant to the transformation (10E.2.4). In this exercise, we shall show that the choice of redundant parameters $\rho = \mathbf{0}$ minimizes the function

$$D(\rho) = \sum_p \|\phi_p(\mathbf{0}) - \psi_p(\rho)\|^2 \tag{10E.2.5}$$

which represents the difference between the original and optimized MOs.

1. Show that the difference measure $D(\rho)$ may be expressed in the matrix form

$$D(\rho) = 2\mathrm{Tr}[\mathbf{1} - \exp(-\kappa)\exp(-\rho)] \tag{10E.2.6}$$

2. Show that the first-order variation of $D(\rho)$ at $\rho = \mathbf{0}$ becomes

$$\delta D(\rho) = 2\mathrm{Tr}\,\mathbf{W}\delta\rho \tag{10E.2.7}$$

where

$$\mathbf{W} = \begin{pmatrix} \mathbf{W}_{IJ} & \mathbf{0} \\ \mathbf{0} & \mathbf{W}_{AB} \end{pmatrix} = \begin{pmatrix} \exp(-\boldsymbol{\kappa})_{IJ} & \mathbf{0} \\ \mathbf{0} & \exp(-\boldsymbol{\kappa})_{AB} \end{pmatrix} \tag{10E.2.8}$$

3. Show that the first-order variation (10E.2.7) vanishes if

$$\mathbf{W} = \mathbf{W}^{\mathrm{T}} \tag{10E.2.9}$$

4. Show that the exponential of (10E.2.1) may be written in the form

$$\exp(-\boldsymbol{\kappa}) = \begin{pmatrix} \cos \mathbf{A} & -\boldsymbol{\kappa}_{AI}^{\mathrm{T}}\mathbf{B}^{-1}\sin \mathbf{B} \\ \boldsymbol{\kappa}_{AI}\mathbf{A}^{-1}\sin \mathbf{A} & \cos \mathbf{B} \end{pmatrix} \tag{10E.2.10}$$

where

$$\mathbf{A} = (\boldsymbol{\kappa}_{AI}^{\mathrm{T}}\boldsymbol{\kappa}_{AI})^{1/2} \tag{10E.2.11}$$

$$\mathbf{B} = (\boldsymbol{\kappa}_{AI}\boldsymbol{\kappa}_{AI}^{\mathrm{T}})^{1/2} \tag{10E.2.12}$$

5. Use (10E.2.10) to show that $\boldsymbol{\rho} = \mathbf{0}$ represents a stationary point.
6. Show that, at the stationary point $\boldsymbol{\rho} = \mathbf{0}$, the second-order variation of $D(\boldsymbol{\rho})$ becomes

$$\delta^2 D = -\mathrm{Tr}\,\mathbf{W}\delta\boldsymbol{\rho}\delta\boldsymbol{\rho} \tag{10E.2.13}$$

7. Show that $\mathbf{W}$ is positive definite for small $\boldsymbol{\kappa}$ and use this to show that (10E.2.13) is positive, implying that the stationary point $\boldsymbol{\rho} = \mathbf{0}$ is a minimum. Thus, for closed-shell wave functions, $\boldsymbol{\rho} = \mathbf{0}$ gives the largest overlap between the rotated and original MOs.

EXERCISE 10.3

In this exercise, we consider the total energies of the singlet and triplet states obtained by promoting one electron from the occupied canonical orbital ϕ_i to the virtual canonical orbital ϕ_a of a closed-shell Hartree–Fock state. Using (2.3.21) and (2.3.23), the total energies of the singlet and triplet states can be written as

$$^1E_{ai} = \langle \mathrm{HF} | \hat{S}_{ia}^{0,0} \hat{H} \hat{S}_{ai}^{0,0} | \mathrm{HF} \rangle \tag{10E.3.1}$$

$$^3E_{ai} = \langle \mathrm{HF} | \hat{T}_{ia}^{1,0} \hat{H} \hat{T}_{ai}^{1,0} | \mathrm{HF} \rangle \tag{10E.3.2}$$

1. Show that the singlet total energy can be expressed as

$$^1E_{ai} = \langle \mathrm{HF} | \hat{H} | \mathrm{HF} \rangle + \tfrac{1}{2}\langle \mathrm{HF} | E_{ia}[\hat{H}, E_{ai}] | \mathrm{HF} \rangle \tag{10E.3.3}$$

2. Use (13B.2.2) in Box 13.2 to show that the singlet energy may be written in the form

$$^1E_{ai} = \langle \mathrm{HF} | \hat{H} | \mathrm{HF} \rangle + \varepsilon_a - \varepsilon_i + 2g_{aiia} - g_{aaii} \tag{10E.3.4}$$

3. Show that the triplet and singlet energies are related as

$$^3E_{ai} = {}^1E_{ai} - 2\langle \mathrm{HF} | E_{ia}^{\beta} \hat{H} E_{ai}^{\alpha} | \mathrm{HF} \rangle \tag{10E.3.5}$$

where

$$E_{ai}^{\alpha} = a_{a\alpha}^{\dagger} a_{i\alpha} \tag{10E.3.6}$$

$$E_{ai}^{\beta} = a_{a\beta}^{\dagger} a_{i\beta} \tag{10E.3.7}$$

4. Show that the energy of the triplet state becomes

$$^3E_{ai} = \langle \text{HF}|\hat{H}|\text{HF}\rangle + \varepsilon_a - \varepsilon_i - g_{aaii} \qquad (10\text{E}.3.8)$$

EXERCISE 10.4

In this exercise, we carry out an STO-3G Roothaan–Hall optimization of the $1\sigma^2$ Hartree–Fock wave function of HeH$^+$ at the internuclear distance of $1.463796a_0$. The HeH$^+$ system is chosen rather than H$_2$ since the HeH$^+$ orbitals are not determined by symmetry. The AO integrals are given in Table 10E.4.1, the optimized MOs in Table 10E.4.2 and the corresponding MO integrals in Table 10E.4.3. Atomic units are used throughout this exercise.

1. We use s_He as the first guess of the occupied MO. Determine the corresponding normalized virtual MO.
2. For this occupied MO, determine the density matrix (10.6.18) and the Fock matrix (10.6.17) in the AO basis.
3. Determine the gradient.
4. Carrying out a Roothaan–Hall iteration, determine the improved occupied MO.
5. For the second iteration, answer Exercises 10.4.1–10.4.3 for the improved occupied MO from the first iteration.
6. The iterative procedure may be continued until convergence – that is, until the norm of the gradient is smaller than 10^{-5} E$_\text{h}$. In Table 10E.4.4, we have listed the energy and the gradient of each iteration. For all iterations, determine the ratio $\|\mathbf{E}_{n+1}^{(1)}\|/\|\mathbf{E}_n^{(1)}\|$.
7. For small steps, the ratio $\|\mathbf{E}_{n+1}^{(1)}\|/\|\mathbf{E}_n^{(1)}\|$ may be predicted from (10.9.35). Using the MO integrals in Table 10E.4.3, compare this predicted ratio with those obtained in Exercise 10.4.6.

Table 10E.4.1 The STO-3G AO integrals (in atomic units) of HeH$^+$ at the bond distance of $1.463796a_0$. The indices A and B refer to the s orbital on He (s_He) and to the s orbital on H (s_H), respectively

S_AA	S_BA	S_BB	h_AA	h_BA	h_BB
1.00000000	0.53659972	1.00000000	−2.59804029	−1.43111581	−1.73143519
g_AAAA	g_BAAA	g_BABA	g_BBAA	g_BBBA	g_BBBB
1.05571322	0.44374250	0.22411632	0.59066195	0.36721343	0.77460615

Table 10E.4.2 The optimized MOs and orbital energies of HeH$^+$

MO	AO expansion	Orbital energy (E$_\text{h}$)
1σ	$0.87664738\ s_\text{He} + 0.20247591\ s_\text{H}$	−1.63255846
2σ	$0.79741078\ s_\text{He} − 1.16763782\ s_\text{H}$	−0.17271555

Table 10E.4.3 The MO integrals of HeH$^+$ (atomic units). The indices 1 and 2 refer to the 1σ and 2σ MOs, respectively

S_{11}	S_{21}	S_{22}	h_{11}	h_{21}	h_{22}
1.00000000	0.00000000	1.00000000	−2.57565040	−0.17297495	−1.34761434
g_{1111}	g_{2111}	g_{2121}	g_{2211}	g_{2221}	g_{2222}
0.94309194	0.17297495	0.14536406	0.66013143	−0.03732934	0.75251097

Table 10E.4.4 The energy and gradient of each iteration of the Roothaan–Hall optimization of the HeH^+ Hartree–Fock wave function (E_h)

n	E_n	$E_n^{(1)}$
1	−2.77405675	−0.75730764
2	−2.83940081	−0.15531885
3	−2.84183410	−0.02513981
4	−2.84189672	−0.00389095
5	−2.84189822	−0.00059792
6	−2.84189825	−0.00009178
7	−2.84189825	−0.00001409
8	−2.84189825	−0.00000216

8. In Exercise 4.1, it is shown that the error in the energy can be estimated from the gradient and Hessian as

$$E_n - E_{cnv} \approx \tfrac{1}{2} \mathbf{E}_n^{(1)^{\mathrm{T}}} (\mathbf{E}_n^{(2)})^{-1} \mathbf{E}_n^{(1)} \tag{10E.4.1}$$

Calculate the Hessian from the integrals in Table 10E.4.3. From this Hessian and from the gradients in Table 10E.4.4, estimate the error in the energy in each iteration.

EXERCISE 10.5

Consider the triplet state in (10.1.4)–(10.1.6) and the singlet state in (10.1.7), which we write in the form

$$|CSF\rangle^{S,M} = \hat{Q}_{vw}^{S,M} |cs\rangle \tag{10E.5.1}$$

where $|cs\rangle$ is the closed-shell part of the wave function.

1. Show that, for the states (10E.5.1), the elements of the active one-electron density matrix are given by

$$D_{xy} = \delta_{vxy} + \delta_{wxy} \tag{10E.5.2}$$

where the generalized Kronecker delta δ_{vxy} vanishes except when all indices are identical. *Hint*: Use the commutation relation (2.3.39).

2. Show that, for the states (10E.5.1), the elements of the active two-electron density matrix are given by

$$d_{xyzu} = \delta_{vzu}\delta_{wxy} + \delta_{vxy}\delta_{wzu} + (-1)^S(\delta_{vxu}\delta_{wzy} + \delta_{vzy}\delta_{wxu}) \tag{10E.5.3}$$

where we have supplemented the usual indices for active orbitals v, w, x, y and z with the index u.

3. Show that the elements of the $\mathbf{Q}$ matrix in (10.8.31) for the states (10E.5.1) are given by

$$Q_{vm} = g_{mvww} + (-1)^S g_{mwwv} \tag{10E.5.4}$$

EXERCISE 10.6

1. Set up a scheme for calculating the active Fock matrix in (10.8.29) in the AO basis.
2. Show how the active Fock matrix, when constructed from the one-index transformed integrals in (10.8.59) and (10.8.60), can be set up in the AO basis.

3. Show how the $\mathbf{Q}^{\text{high-spin}}$ matrix can be set up in the AO basis. Consider the cases where this matrix is calculated from the standard integrals as in (10.8.37) and from the one-index transformed integrals (10.8.59) and (10.8.60).

EXERCISE 10.7

In this exercise, we carry out a Newton optimization of the Hartree–Fock wave function for HeH^+ at the bond distance $1.463796a_0$ in the STO-3G basis. The AO integrals are listed in Table 10E.4.1. All numbers quoted are in atomic units.

1. In the first iteration, we use the occupied and unoccupied MOs

$$\mathbf{C} = (\mathbf{C}_{1\sigma}\,\mathbf{C}_{2\sigma}) = \begin{pmatrix} 1.00000000 & 0.63590453 \\ 0.00000000 & -1.18506311 \end{pmatrix} \tag{10E.7.1}$$

where the first column contains the coefficients of the helium $1s$ function and the second column the corresponding orthonormalized virtual MO, see Exercise 10.4.1. Verify that, in this basis, the inactive Fock matrix becomes

$$^{\mathrm{I}}\mathbf{F} = \begin{pmatrix} -1.54232707 & 0.18932691 \\ 0.18932691 & -0.22284004 \end{pmatrix} \tag{10E.7.2}$$

Determine the gradient for the Newton step.

2. Determine the Hessian for the Newton step. *Hint*: The only element of the electronic Hessian is given by

$$E_{2121}^{(2)} = 2\langle \mathrm{CSF}|[E_{21}, [E_{21}^-, \hat{H}]]|\mathrm{CSF}\rangle \tag{10E.7.3}$$

It may be calculated directly from (10.8.55) but more efficiently as the electronic gradient with a one-index transformed Hamiltonian

$$E_{2121}^{(2)} = 2\langle \mathrm{CSF}|[E_{21}, [\hat{\kappa}, \hat{H}]]|\mathrm{CSF}\rangle = -4\,^{\mathrm{I}}F_{21}^{\kappa} \tag{10E.7.4}$$

where the only nonzero elements of κ are $\kappa_{21} = -\kappa_{12} = 1$. For the evaluation of Fock matrices from one-index transformed Hamiltonians, see Section 10.8.8.

3. Determine the orbital-rotation parameter by solving the Newton equations.
4. Determine the orthogonal matrix $\exp(-\kappa)$ corresponding to this orbital-rotation parameter.
5. Determine the improved MOs.

Table 10E.7.1 The energy and gradient in each iteration of the Newton optimization of the Hartree–Fock wave function of HeH^+ (E_h)

n	E_n	$E_n^{(1)}$
1	-2.77405675	-0.75730764
2	-2.84001158	0.13778209
3	-2.84189797	0.00168681
4	-2.84189825	0.00000029

6. If the Newton iterations are continued, convergence to a gradient norm smaller than 10^{-5} E_h is reached in four iterations – see Table 10E.7.1. By comparing gradients, verify that the convergence is quadratic.

EXERCISE 10.8

In this exercise, we discuss the changes in the energy that arise from triplet rotations of a closed-shell Hartree–Fock wave function. In particular, we shall examine triplet instabilities of such systems.

1. In the notation

$$\hat{T}_{ai}^- = \hat{T}_{ai} - \hat{T}_{ia} \tag{10E.8.1}$$

$$\hat{T}_{ai} = E_{ai}^\alpha - E_{ai}^\beta \tag{10E.8.2}$$

where the alpha and beta excitation operators are defined in (10.10.7) and (10.10.8), we parametrize the triplet variations of the closed-shell Hartree–Fock wave function as

$$|\mathrm{HF}(\boldsymbol{\kappa})\rangle = \exp\left(-\sum_{ai} \kappa_{ai} \hat{T}_{ai}^-\right) |\mathrm{HF}\rangle \tag{10E.8.3}$$

Show that the triplet gradient vanishes and that the triplet Hessian becomes

$$G_{aibj} = \tfrac{1}{2}(1 + P_{ai,bj})\langle\mathrm{HF}|[\hat{T}_{ai}^-, [\hat{T}_{bj}^-, \hat{H}]]|\mathrm{HF}\rangle \tag{10E.8.4}$$

2. Verify that, for an optimized RHF state, the triplet Hessian (10E.8.4) reduces to

$$G_{aibj} = 2\langle\mathrm{HF}|[\hat{T}_{ai}^-, [\hat{T}_{bj}^-, \hat{H}]]|\mathrm{HF}\rangle \tag{10E.8.5}$$

3. Show that the triplet Hessian in (10E.8.5) is related to the corresponding singlet Hessian (10.2.8) in the following manner:

$$G_{aibj} = E_{aibj}^{(2)} - 8\langle\mathrm{HF}|[E_{ai}^\alpha, [E_{bj}^\beta - E_{jb}^\beta, \hat{H}]]|\mathrm{HF}\rangle \tag{10E.8.6}$$

Does this expression hold only for an optimized RHF state?

4. Evaluate the matrix element in (10E.8.6) and show that the triplet and singlet Hessians are related by

$$G_{aibj} = E_{aibj}^{(2)} - 16g_{aibj} \tag{10E.8.7}$$

5. Show that, in the canonical basis, the triplet Hessian (10E.8.7) becomes

$$G_{aibj} = 4[\delta_{ab}\delta_{ij}(\varepsilon_a - \varepsilon_i) - g_{abij} - g_{ajib}] \tag{10E.8.8}$$

6. From the RHF/STO-3G total energy, orbital energies and two-electron integrals of the H_2 molecule listed in Table 10E.8.1, determine by linear interpolation the bond distance at which the RHF solution becomes unstable.

Table 10E.8.1 The RHF/STO-3G total energy, orbital energies and nonzero two-electron integrals of the H_2 molecule (E_h)

	$2.5a_0$	$2.6a_0$	$2.7a_0$	$2.8a_0$
E_{HF}	−0.922197	−0.916982	−0.911210	−0.905020
ε_g	−0.443869	−0.435455	−0.427370	−0.419617
ε_u	0.071097	0.063575	0.056249	0.049121
g_{gggg}	0.434460	0.430686	0.426840	0.422932
g_{uugg}	0.388960	0.387118	0.385248	0.383353
g_{ugug}	0.102943	0.104359	0.105821	0.107327
g_{uuuu}	0.385084	0.383654	0.382192	0.380699

Solutions

SOLUTION 10.1

1. Carrying out the differentiation in (10E.1.1), we obtain

$$E_{PQ}^{(1)} = \langle \mathrm{cs}| \frac{\partial \exp(\hat{\kappa})}{\partial \kappa_{PQ}} \hat{H} \exp(-\hat{\kappa}) + \exp(\hat{\kappa}) \hat{H} \frac{\partial \exp(-\hat{\kappa})}{\partial \kappa_{PQ}} |\mathrm{cs}\rangle \qquad (10S.1.1)$$

The derivative of $\exp(-\hat{\kappa})$ is given in (10E.1.2). Taking its adjoint, we obtain the derivative of $\exp(\hat{\kappa})$:

$$\begin{aligned}
\frac{\partial \exp(\hat{\kappa})}{\partial \kappa_{PQ}} &= \left[\frac{\partial \exp(-\hat{\kappa})}{\partial \kappa_{PQ}} \right]^{\dagger} \\
&= \sum_{RS} \frac{\exp[i(\lambda_S - \lambda_R)] - 1}{i(\lambda_R - \lambda_S)} (U_{PR} U_{QS}^* - U_{QR} U_{PS}^*) \exp(\hat{\kappa}) \bar{a}_S^{\dagger} \bar{a}_R \\
&= \sum_{RS} \frac{\exp[i(\lambda_R - \lambda_S)] - 1}{i(\lambda_R - \lambda_S)} (U_{PR}^* U_{QS} - U_{QR}^* U_{PS}) \exp(\hat{\kappa}) \bar{a}_R^{\dagger} \bar{a}_S \qquad (10S.1.2)
\end{aligned}$$

We now arrive at (10E.1.5) by inserting (10E.1.2) and (10S.1.2) in (10S.1.1).

2. When $\kappa = \mathbf{0}$, the matrix $i\kappa$ is already diagonal. The unitary matrix of eigenvectors $\mathbf{U}$ then reduces to the identity matrix and the eigenvalues become zero:

$$U_{PQ} = \delta_{PQ} \qquad (10S.1.3)$$

$$\lambda_P = 0 \qquad (10S.1.4)$$

Inserting (10S.1.3) into (10E.1.5) and taking into account the limit

$$\lim_{a \to 0} \frac{\exp(ia) - 1}{ia} = 1 \qquad (10S.1.5)$$

we obtain

$$E_{PQ}^{(1)}(\mathbf{0}) = \sum_{RS} (\delta_{PR} \delta_{QS} - \delta_{QR} \delta_{PS}) \langle \mathrm{cs}| [a_R^{\dagger} a_S, \hat{H}] |\mathrm{cs}\rangle = 2\langle \mathrm{cs}| [a_P^{\dagger} a_Q, \hat{H}] |\mathrm{cs}\rangle \qquad (10S.1.6)$$

in agreement with (10.1.29).

SOLUTION 10.2

1. Expression (10E.2.6) is obtained by expanding (10E.2.5) as

$$D(\boldsymbol{\rho}) = \sum_p \langle \phi_p(\mathbf{0}) - \psi_p(\boldsymbol{\rho}) | \phi_p(\mathbf{0}) - \psi_p(\boldsymbol{\rho}) \rangle$$

$$= 2\sum_p \left[1 - \langle \phi_p(\mathbf{0}) | \psi_p(\boldsymbol{\rho}) \rangle \right]$$

$$= 2\sum_p \left[1 - \sum_q \langle \phi_p(\mathbf{0}) | \phi_q(\boldsymbol{\kappa}) \rangle \exp(-\boldsymbol{\rho})_{qp} \right]$$

$$= 2\sum_p \left[1 - \sum_{qr} \langle \phi_p(\mathbf{0}) | \phi_r(\mathbf{0}) \rangle \exp(-\boldsymbol{\kappa})_{rq} \exp(-\boldsymbol{\rho})_{qp} \right]$$

$$= 2\sum_p \left[1 - \sum_q \exp(-\boldsymbol{\kappa})_{pq} \exp(-\boldsymbol{\rho})_{qp} \right] \tag{10S.2.1}$$

where we have used (10E.2.4) and then (10E.2.2).

2. Evaluating the variation in (10E.2.6) around $\boldsymbol{\rho} = \mathbf{0}$, we obtain:

$$\delta D(\boldsymbol{\rho}) = -2\mathrm{Tr}\exp(-\boldsymbol{\kappa})\delta\exp(-\boldsymbol{\rho}) = 2\mathrm{Tr}\exp(-\boldsymbol{\kappa})\delta\boldsymbol{\rho} \tag{10S.2.2}$$

Since $\boldsymbol{\rho}$ is block diagonal, the off-diagonal blocks of $\exp(-\boldsymbol{\kappa})$ will not contribute to the variation and we may write (10S.2.2) as (10E.2.7), where $\mathbf{W}$ contains only the occupied and virtual blocks of $\exp(-\boldsymbol{\kappa})$ in (10E.2.8).

3. Expansion of the trace in (10E.2.7) and use of the antisymmetry of $\delta\boldsymbol{\rho}$ give

$$\delta D(\boldsymbol{\rho}) = 2\sum_{pq} W_{pq}\,\delta\rho_{qp} = 2\sum_{p>q}(W_{pq} - W_{qp})\delta\rho_{qp} \tag{10S.2.3}$$

Setting $\delta D(\boldsymbol{\rho})$ equal to zero and noting that $\delta\boldsymbol{\rho}$ is nonzero only if p and q are either both occupied or both virtual, we obtain the variational conditions

$$W_{ij} = W_{ji} \tag{10S.2.4}$$

$$W_{ab} = W_{ba} \tag{10S.2.5}$$

Because of the block-diagonal structure of $\mathbf{W}$, these conditions are equivalent to the matrix equation (10E.2.9).

4. We first note that the square of $\boldsymbol{\kappa}$ may be written in the form

$$\begin{pmatrix} \mathbf{0} & \boldsymbol{\kappa}_{AI}^{\mathrm{T}} \\ -\boldsymbol{\kappa}_{AI} & \mathbf{0} \end{pmatrix} \begin{pmatrix} \mathbf{0} & \boldsymbol{\kappa}_{AI}^{\mathrm{T}} \\ -\boldsymbol{\kappa}_{AI} & \mathbf{0} \end{pmatrix} = -\begin{pmatrix} \mathbf{A} & \mathbf{0} \\ \mathbf{0} & \mathbf{B} \end{pmatrix}^2 \tag{10S.2.6}$$

where $\mathbf{A}$ and $\mathbf{B}$ are given by (10E.2.11) and (10E.2.12). Expansion of the exponential now gives

$$\exp(-\boldsymbol{\kappa}) = \sum_{n=0}^{\infty} \frac{(-1)^n}{(2n)!} \begin{pmatrix} \mathbf{A} & \mathbf{0} \\ \mathbf{0} & \mathbf{B} \end{pmatrix}^{2n} - \boldsymbol{\kappa} \sum_{n=0}^{\infty} \frac{(-1)^n}{(2n+1)!} \begin{pmatrix} \mathbf{A} & \mathbf{0} \\ \mathbf{0} & \mathbf{B} \end{pmatrix}^{2n}$$

$$= \begin{pmatrix} \cos\mathbf{A} & \mathbf{0} \\ \mathbf{0} & \cos\mathbf{B} \end{pmatrix} - \boldsymbol{\kappa} \begin{pmatrix} \mathbf{A} & \mathbf{0} \\ \mathbf{0} & \mathbf{B} \end{pmatrix}^{-1} \begin{pmatrix} \sin\mathbf{A} & \mathbf{0} \\ \mathbf{0} & \sin\mathbf{B} \end{pmatrix} \tag{10S.2.7}$$

Inserting (10E.2.1) and carrying out some simple block-matrix multiplications, we arrive at (10E.2.10).

5. According to (10E.2.9), we must show that $\mathbf{W}$ in (10E.2.8) is symmetric. The nonzero elements of $\mathbf{W}$ are restricted to the diagonal blocks $\mathbf{W}_{IJ}$ and $\mathbf{W}_{AB}$, which according to (10E.2.10) are given by $\cos\mathbf{A}$ and $\cos\mathbf{B}$, both of which are symmetric.

6. With $\boldsymbol{\rho} = \mathbf{0}$ as the expansion point, the second-order variation of $\exp(-\boldsymbol{\rho})$ is $\frac{1}{2}\delta\rho^2$. Using (10E.2.6) and (10E.2.8), we see that the second-order variation of $D(\boldsymbol{\rho})$ gives (10E.2.13).

7. In (10E.2.11) and (10E.2.12), $\mathbf{A}$ and $\mathbf{B}$ are both positive semidefinite symmetric matrices that can be expressed in the form

$$\mathbf{A} = \mathbf{X}^A \boldsymbol{\lambda}^A (\mathbf{X}^A)^T \tag{10S.2.8}$$

$$\mathbf{B} = \mathbf{X}^B \boldsymbol{\lambda}^B (\mathbf{X}^B)^T \tag{10S.2.9}$$

where the elements of the diagonal matrices $\boldsymbol{\lambda}^A$ and $\boldsymbol{\lambda}^B$ are nonnegative. The matrix

$$\mathbf{W} = \begin{pmatrix} \cos\mathbf{A} & \mathbf{0} \\ \mathbf{0} & \cos\mathbf{B} \end{pmatrix} = \begin{pmatrix} \mathbf{X}^A \cos\boldsymbol{\lambda}^A (\mathbf{X}^A)^T & \mathbf{0} \\ \mathbf{0} & \mathbf{X}^B \cos\boldsymbol{\lambda}^B (\mathbf{X}^B)^T \end{pmatrix} \tag{10S.2.10}$$

is therefore positive definite provided the eigenvalues are smaller than $\pi/2$ in magnitude, which is true for small κ. Since the second-order variation

$$\delta^2 D = \text{Tr}\, \delta\boldsymbol{\rho}^T \mathbf{W} \delta\boldsymbol{\rho} \tag{10S.2.11}$$

is a sum of positive quantities, the stationary point is a minimum.

SOLUTION 10.3

1. We use (2.3.29) to express $\hat{S}^{0,0}_{pq}$ in terms of E_{pq} and arrive at (10E.3.3) after some simple algebra:

$$\begin{aligned} {}^1E_{ai} &= \tfrac{1}{2}\langle\text{HF}|E_{ia}\hat{H}E_{ai}|\text{HF}\rangle \\ &= \tfrac{1}{2}\langle\text{HF}|E_{ia}[\hat{H}, E_{ai}]|\text{HF}\rangle + \tfrac{1}{2}\langle\text{HF}|E_{ia}E_{ai}\hat{H}|\text{HF}\rangle \\ &= \tfrac{1}{2}\langle\text{HF}|E_{ia}[\hat{H}, E_{ai}]|\text{HF}\rangle + \langle\text{HF}|\hat{H}|\text{HF}\rangle \end{aligned} \tag{10S.3.1}$$

2. Inserting (13B.2.2) and making use of the relation

$$\langle\text{HF}|E_{ia}E_{bj}|\text{HF}\rangle = 2\delta_{ij}\delta_{ab} \tag{10S.3.2}$$

and the fact that only the single-excitation part of (13B.2.2) contributes, we obtain for the singlet total energy (10E.3.3):

$$\begin{aligned} {}^1E_{ai} &= \langle\text{HF}|\hat{H}|\text{HF}\rangle + \tfrac{1}{2}\sum_b f_{ba}\langle\text{HF}|E_{ia}E_{bi}|\text{HF}\rangle \\ &\quad - \tfrac{1}{2}\sum_j f_{ij}\langle\text{HF}|E_{ia}E_{aj}|\text{HF}\rangle + \tfrac{1}{2}\sum_{bj}(2g_{bjia} - g_{baij})\langle\text{HF}|E_{ia}E_{bj}|\text{HF}\rangle \\ &= \langle\text{HF}|\hat{H}|\text{HF}\rangle + \varepsilon_a - \varepsilon_i + 2g_{aiia} - g_{aaii} \end{aligned} \tag{10S.3.3}$$

In using (13B.2.2), note that $L_{pqrs} = 2g_{pqrs} - g_{psrq}$ and that, for closed-shell states, ${}^1F_{pq} = f_{pq}$.

3. Expanding the triplet excitation operator in (10E.3.2) in accordance with (2.3.23) and using (2.3.21) for the singlet excitation operator, we obtain (10E.3.5)

$$
\begin{aligned}
{}^3E_{ai} &= \tfrac{1}{2}\langle \mathrm{HF}|(E_{ia}^\alpha - E_{ia}^\beta)\hat{H}(E_{ai}^\alpha - E_{ai}^\beta)|\mathrm{HF}\rangle \\
&= \tfrac{1}{2}\langle \mathrm{HF}|(E_{ia}^\alpha + E_{ia}^\beta)\hat{H}(E_{ai}^\alpha + E_{ai}^\beta)|\mathrm{HF}\rangle - 2\langle \mathrm{HF}|E_{ia}^\alpha \hat{H} E_{ai}^\beta|\mathrm{HF}\rangle \\
&= \langle \mathrm{HF}|\hat{S}_{ia}^{0,0}\hat{H}\hat{S}_{ai}^{0,0}|\mathrm{HF}\rangle - 2\langle \mathrm{HF}|E_{ia}^\beta \hat{H} E_{ai}^\alpha|\mathrm{HF}\rangle
\end{aligned}
\tag{10S.3.4}
$$

4. Having evaluated the matrix element

$$
\langle \mathrm{HF}|E_{ia}^\beta \hat{H} E_{ai}^\alpha|\mathrm{HF}\rangle = g_{aiia} \tag{10S.3.5}
$$

we obtain (10E.3.8) from (10E.3.5) and (10E.3.4).

SOLUTION 10.4

1. Orthogonalizing against the occupied MO, we obtain the following virtual MO for the first iteration:

$$
\mathbf{C}_{2\sigma}^{\mathrm{T}} = (0.63590453,\ -1.18506311) \tag{10S.4.1}
$$

2. The AO density and Fock matrices of the first iteration are given as

$$
\mathbf{D}^{\mathrm{AO}} = \begin{pmatrix} 2.00000000 & 0.00000000 \\ 0.00000000 & 0.00000000 \end{pmatrix} \tag{10S.4.2}
$$

$$
\mathbf{f}^{\mathrm{AO}} = \begin{pmatrix} -1.54232707 & -0.98737331 \\ -0.98737331 & -0.77422761 \end{pmatrix} \tag{10S.4.3}
$$

3. The electronic gradient of the first iteration is given by

$$
E_{2\sigma 1\sigma}^{(1)} = -4f_{2\sigma 1\sigma} = -4\mathbf{C}_{2\sigma}^{\mathrm{T}}\mathbf{f}^{\mathrm{AO}}\mathbf{C}_{1\sigma} = -0.75730764 \tag{10S.4.4}
$$

4. Solving the Roothaan–Hall generalized eigenvalue problem, we obtain the following occupied MO for the second iteration:

$$
\mathbf{C}_{1\sigma}^{\mathrm{T}} = (0.90168730,\ 0.16505094) \tag{10S.4.5}
$$

5. Proceeding as in the first iteration, we obtain the following virtual MO, density matrix, Fock matrix and gradient for the second iteration:

$$
\mathbf{C}_{2\sigma}^{\mathrm{T}} = (0.76898282,\ -1.17351300) \tag{10S.4.6}
$$

$$
\mathbf{D}^{\mathrm{AO}} = \begin{pmatrix} 1.62607999 & 0.29764867 \\ 0.29764867 & 0.05448363 \end{pmatrix} \tag{10S.4.7}
$$

$$
\mathbf{f}^{\mathrm{AO}} = \begin{pmatrix} -1.58154779 & -1.04817484 \\ -1.04817484 & -0.82278488 \end{pmatrix} \tag{10S.4.8}
$$

$$
E_{2\sigma 1\sigma}^{(1)} = -0.15531885 \tag{10S.4.9}
$$

Table 10S.4.1 The ratio between the gradients in two consecutive Roothaan–Hall iterations, the error in the energy in each iteration, and the estimated error in the energy (E_h)

n	$\|\mathbf{E}_{n+1}^{(1)}\|/\|\mathbf{E}_n^{(1)}\|$	$E_n - E_{cnv}$	$\frac{1}{2}\mathbf{E}_n^{(1)\mathrm{T}}(\mathbf{E}_{cnv}^{(2)})^{-1}\mathbf{E}_n^{(1)}$
1	0.2051	0.0678415010	0.0580103131
2	0.1619	0.0024974469	0.0024401071
3	0.1548	0.0000641517	0.0000639271
4	0.1537	0.0000015322	0.0000015313
5	0.1535	0.0000000362	0.0000000362
6	0.1535	0.0000000009	0.0000000009
7	0.1535	0.0000000000	0.0000000000

6. The ratios are listed in Table 10S.4.1.
7. We first calculate the $\mathbf{M}$ and $\mathbf{w}^{(1)}$ contributions to the electronic Hessian (10.9.33):

$$M_{2121} = 12g_{2121} - 4g_{2211} = -0.89615696 \tag{10S.4.10}$$

$$w_{2121}^{(1)} = 4\varepsilon_2 - 4\varepsilon_1 = 5.83937164 \tag{10S.4.11}$$

The predicted ratio is then obtained from (10.9.35)

$$-\frac{M_{2121}}{w_{2121}^{(1)}} = 0.15346805 \tag{10S.4.12}$$

For small steps, this ratio agrees with those in Table 10S.4.1.

8. Calculating the Hessian from (10.8.56), we obtain

$$E_{2121}^{(2)} = 4(\varepsilon_2 - \varepsilon_1 + 3g_{2121} - g_{2211}) = 4.94321469 \tag{10S.4.13}$$

In Table 10S.4.1, we have listed the true errors of each iteration as well as the estimated errors as calculated from (10E.4.1), using the gradient in Table 10E.4.4 and the Hessian (10S.4.13). For the first iterations, a better agreement is obtained if the estimates are made from $\mathbf{E}_n^{(2)}$ rather than $\mathbf{E}_{cnv}^{(2)}$.

SOLUTION 10.5

In evaluating the density-matrix elements for the two-electron singlet and triplet states, we shall make frequent use of the relation

$$\langle cs|(\hat{Q}_{vw}^{S,M})^\dagger \hat{Q}_{xy}^{S,M}|cs\rangle = \delta_{vx}\delta_{wy} + (-1)^S \delta_{vy}\delta_{wx} \tag{10S.5.1}$$

noting that only one of the terms on the right-hand side will be nonzero since v and w are different. To establish (10S.5.1), use the relation

$$\hat{Q}_{pq}^{S,M} = (-1)^S \hat{Q}_{qp}^{S,M} \tag{10S.5.2}$$

discussed in Section 2.3.3. The general strategy we shall follow is to reduce the density-matrix elements to the form (10S.5.1), making use of the commutator relation (2.3.39).

1. Using (2.3.39) and then (10S.5.1), we obtain for the one-electron density matrix

$$D_{xy} = \langle cs|(\hat{Q}_{vw}^{S,M})^\dagger E_{xy}\hat{Q}_{vw}^{S,M}|cs\rangle = \langle cs|(\hat{Q}_{vw}^{S,M})^\dagger(\hat{Q}_{xw}^{S,M}\delta_{vy} + \hat{Q}_{vx}^{S,M}\delta_{yw})|cs\rangle$$

$$= \delta_{xv}\delta_{vy} + \delta_{xw}\delta_{yw} = \delta_{vxy} + \delta_{wxy} \tag{10S.5.3}$$

2. To evaluate the two-electron density matrix

$$d_{xyzu} = \langle cs | (\hat{Q}_{vw}^{S,M})^\dagger (E_{xy}E_{zu} - \delta_{zy}E_{xu}) \hat{Q}_{vw}^{S,M} | cs \rangle \tag{10S.5.4}$$

we first use (2.3.39) several times:

$$(E_{xy}E_{zu} - \delta_{zy}E_{xu}) \hat{Q}_{vw}^{S,M} | cs \rangle = \delta_{vu}(\hat{Q}_{xw}^{S,M}\delta_{zy} + \hat{Q}_{zx}^{S,M}\delta_{wy}) | cs \rangle + \delta_{wu}(\hat{Q}_{xz}^{S,M}\delta_{vy} + \hat{Q}_{vx}^{S,M}\delta_{zy}) | cs \rangle$$
$$- \delta_{zy}(\hat{Q}_{xw}^{S,M}\delta_{vu} + \hat{Q}_{vx}^{S,M}\delta_{wu}) | cs \rangle \tag{10S.5.5}$$

The density-matrix element (10S.5.4) may now be evaluated as

$$d_{xyzu} = \delta_{vux}\delta_{zy} + \delta_{vuz}\delta_{wxy} + (-1)^S \delta_{vxu}\delta_{wzy} + \delta_{vxy}\delta_{wzu}$$
$$+ (-1)^S \delta_{vzy}\delta_{wxu} + \delta_{wxu}\delta_{zy} - \delta_{vxu}\delta_{zy} - \delta_{wxu}\delta_{zy}$$
$$= \delta_{vzu}\delta_{wxy} + \delta_{vxy}\delta_{wzu} + (-1)^S \delta_{vxu}\delta_{wzy} + (-1)^S \delta_{vzy}\delta_{wxu} \tag{10S.5.6}$$

3. Using (10S.5.6) and recalling that v and w are different (so that all Kronecker deltas where both appear vanish), we obtain

$$Q_{vm} = \sum_{yzu} d_{vyzu}g_{myzu} = \sum_{yzu} [\delta_{vy}\delta_{wzu}g_{myzu} + (-1)^S \delta_{vu}\delta_{wzy}g_{myzu}] = g_{mvww} + (-1)^S g_{mwwv} \tag{10S.5.7}$$

SOLUTION 10.6

1. In the AO basis, the active Fock matrix (10.8.29) is given by

$$^A F_{\mu\nu} = \sum_{vw} D_{vw} \left(g_{\mu v \nu w} - \tfrac{1}{2} g_{\mu w \nu v} \right) \tag{10S.6.1}$$

Expanding the MOs in AOs, we obtain

$$^A F_{\mu\nu} = \sum_{vw} D_{vw} \left(g_{\mu\nu\rho\sigma} - \tfrac{1}{2} g_{\mu\sigma\rho\nu} \right) C_{\sigma w} C_{\rho v} = \sum_{\rho\sigma} D_{\rho\sigma}^{AO} \left(g_{\mu\nu\rho\sigma} - \tfrac{1}{2} g_{\mu\sigma\rho\nu} \right) \tag{10S.6.2}$$

in terms of the density-matrix elements

$$D_{\rho\sigma}^{AO} = \sum_{vw} D_{vw} C_{\sigma w} C_{\rho v} \tag{10S.6.3}$$

2. Introducing one-index transformed integrals in (10.8.29), the active Fock matrix becomes

$$^A F_{mn}^\kappa = \sum_{vw} D_{vw} \left(g_{mnvw}^\kappa - \tfrac{1}{2} g_{mwvn}^\kappa \right)$$
$$= \sum_p (^A F_{mp}\kappa_{np} + {}^A F_{pn}\kappa_{mp}) + \sum_{vwp} D_{vw} \left(2g_{mnvp} - \tfrac{1}{2} g_{mpvn} - \tfrac{1}{2} g_{mvpn} \right) \kappa_{wp} \tag{10S.6.4}$$

The first two terms in (10S.6.4) are constructed from the usual active Fock matrix, which in Exercise 10.6.1 was set up in the AO basis. The last term may be rewritten as

$$\sum_{vwp} D_{vw} \left(2g_{mnvp} - \tfrac{1}{2} g_{mpvn} - \tfrac{1}{2} g_{mvpn} \right) \kappa_{wp} = \sum_{\rho\sigma} K_{\rho\sigma} \left(2g_{mn\rho\sigma} - \tfrac{1}{2} g_{m\sigma\rho n} - \tfrac{1}{2} g_{m\rho\sigma n} \right) \tag{10S.6.5}$$

where we have introduced the one-electron AO matrix

$$K_{\rho\sigma} = \sum_{vwp} D_{vw} \kappa_{wp} C_{\sigma p} C_{\rho v} \tag{10S.6.6}$$

All terms in (10S.6.4) may therefore be calculated in the AO basis, avoiding the transformation of the two-electron integrals to the MO basis.

3. Comparing $\mathbf{Q}^{\text{high-spin}}$ in (10.8.37) and the two-electron part of $^{\text{I}}\mathbf{F}$ in (10.8.28), we note that they are identical except for different weights on the Coulomb and exchange contributions. With small modifications, $\mathbf{Q}^{\text{high-spin}}$ may therefore be calculated in the same manner as the two-electron part of $^{\text{I}}\mathbf{F}$ – see Sections 10.6.1 and 10.8.8.

SOLUTION 10.7

1. The inactive Fock matrix is best calculated by first setting it up in the AO basis as done in Exercise 10.4.2 and then transforming to the MO basis according to

$$^{\text{I}}\mathbf{F} = \mathbf{C}^{\text{T}} \, ^{\text{I}}\mathbf{F}^{\text{AO}} \mathbf{C} \tag{10S.7.1}$$

From this matrix, we may easily extract the gradient:

$$E_{21}^{(1)} = -4^{\text{I}}F_{21} = -0.75730764 \tag{10S.7.2}$$

2. We shall evaluate the Hessian according to (10E.7.4), using the rotation matrix

$$\kappa = \begin{pmatrix} 0 & -1 \\ 1 & 0 \end{pmatrix} \tag{10S.7.3}$$

Substituting (10S.7.3) into (10.8.62), we obtain

$$E_{2121}^{(2)} = 4(^{\text{I}}F_{22} - \, ^{\text{I}}F_{11} - \mathbf{C}_{2\sigma}^{\text{T}} \mathbf{M} \mathbf{C}_{1\sigma}) \tag{10S.7.4}$$

where the elements of $\mathbf{M}$ are given in (10.8.65). To generate $\mathbf{M}$, we first calculate, using (10E.7.1) and (10S.7.3), the modified density matrix in (10.8.63):

$$\mathbf{p} = \begin{pmatrix} -2.54361813 & 4.74025244 \\ 0.00000000 & 0.00000000 \end{pmatrix} \tag{10S.7.5}$$

Next, combining the elements of $\mathbf{p}$ with the two-electron AO integrals according to (10.8.65), we obtain

$$\mathbf{M} = \begin{pmatrix} -0.29093991 & -0.46755147 \\ -0.46755147 & -0.34704310 \end{pmatrix} \tag{10S.7.6}$$

Table 10S.7.1 The relationship between the gradients of two consecutive iterations in the second-order Newton optimization of the RHF/STO-3G HeH$^+$ system (E_{h})

n	$E_{n+1}^{(1)}$	$(E_n^{(1)})^2$	$E_{n+1}^{(1)}/(E_{n+1}^{(1)})^2$
1	0.13778209	0.57351486	0.2402
2	0.00168681	0.01898390	0.0889
3	0.00000029	0.00000285	0.1031

which gives us the following MO matrix element:

$$\mathbf{C}_{2\sigma}^{\mathrm{T}}\mathbf{M}\mathbf{C}_{1\sigma} = 0.36906799 \tag{10S.7.7}$$

Substituting (10S.7.7) and the diagonal elements of the inactive Fock matrix (10E.7.2) into (10S.7.4), we obtain the following Hessian element:

$$E_{2121}^{(2)} = 4(-0.22284004 + 1.54232707 - 0.36906799) = 3.80167616 \tag{10S.7.8}$$

3. We obtain, using the gradient (10S.7.2) and the Hessian (10S.7.8):

$$\kappa_{21} = \frac{-E_{21}^{(1)}}{E_{2121}^{(2)}} = \frac{0.75730764}{3.80167616} = 0.19920362 \tag{10S.7.9}$$

4. To evaluate the exponential matrix $\exp(-\kappa)$, we use the closed-form expression derived in Exercise 3.3.3:

$$\exp(-\kappa) = \begin{pmatrix} \cos\kappa_{21} & \sin\kappa_{21} \\ -\sin\kappa_{21} & \cos\kappa_{21} \end{pmatrix} = \begin{pmatrix} 0.98022448 & 0.19788876 \\ -0.19788876 & 0.98022448 \end{pmatrix} \tag{10S.7.10}$$

5. The improved set of orbitals is now obtained by a unitary transformation:

$$\mathbf{C}_{\mathrm{new}} = \mathbf{C}\exp(-\kappa) = \begin{pmatrix} 0.85438612 & 0.82121795 \\ 0.23451067 & -1.16162788 \end{pmatrix} \tag{10S.7.11}$$

6. See Table 10S.7.1. The error constant is thus about 0.1.

SOLUTION 10.8

1. Following the same procedure as in Section 10.1.3, we find that the triplet gradient is given by

$$G_{ai} = \langle\mathrm{HF}|[\hat{T}_{ai}^{-}, \hat{H}]|\mathrm{HF}\rangle \tag{10S.8.1}$$

and the triplet Hessian by (10E.8.4). Since $\hat{T}_{ai}^{-}$ is a triplet spin tensor operator, the gradient (10S.8.1) must vanish for a singlet state such as the closed-shell RHF state:

$$G_{ai} = 0 \tag{10S.8.2}$$

The vanishing of the triplet gradient may also be demonstrated explicitly, without invoking group theory. Using (10E.8.1) and (10E.8.2), we first write the gradient (10S.8.1) in the form

$$G_{ai} = 2\langle\mathrm{HF}|[E_{ai}^{\alpha}, \hat{H}]|\mathrm{HF}\rangle - 2\langle\mathrm{HF}|[E_{ai}^{\beta}, \hat{H}]|\mathrm{HF}\rangle \tag{10S.8.3}$$

The two terms on the right-hand side cancel, as may be seen by a relabelling of the spins or more formally by introducing the unitary spin-flip operator $\hat{U}_x$ defined in (3E.4.9) of Exercise 3.4:

$$\begin{aligned}
\langle\mathrm{HF}|[E_{ai}^{\alpha}, \hat{H}]|\mathrm{HF}\rangle &= \langle\mathrm{HF}|\hat{U}_x^{\dagger}\hat{U}_x[E_{ai}^{\alpha}, \hat{H}]\hat{U}_x^{\dagger}\hat{U}_x|\mathrm{HF}\rangle \\
&= \langle\mathrm{HF}|\hat{U}_x^{\dagger}[\hat{U}_x E_{ai}^{\alpha}\hat{U}_x^{\dagger}, \hat{U}_x\hat{H}\hat{U}_x^{\dagger}]\hat{U}_x|\mathrm{HF}\rangle \\
&= \langle\mathrm{HF}|[E_{ai}^{\beta}, \hat{H}]|\mathrm{HF}\rangle
\end{aligned} \tag{10S.8.4}$$

The last equality follows since the Hamiltonian is unaffected by the spin-flip operator and since the sign change introduced by applying the spin-flip operator to the closed-shell state $|\mathrm{HF}\rangle$

$$\hat{U}_x|\mathrm{HF}\rangle = (-1)^{N/2}|\mathrm{HF}\rangle \tag{10S.8.5}$$

is cancelled by the same change in $\langle\mathrm{HF}|\hat{U}_x^{\dagger}$.

2. The arguments are the same as for the singlet Hessian in Section 10.2.1. First, invoking the Jacobi identity (1.8.17), we obtain

$$\langle \text{HF}|[\hat{T}^-_{ai}, [\hat{T}^-_{bj}, \hat{H}]]|\text{HF}\rangle = \langle \text{HF}|[\hat{T}^-_{bj}, [\hat{T}^-_{ai}, \hat{H}]]|\text{HF}\rangle + \langle \text{HF}|[[\hat{T}^-_{ai}, \hat{T}^-_{bj}], \hat{H}]|\text{HF}\rangle \quad (10S.8.6)$$

Next, since all commutators of the type $[E^\alpha_{pq}, E^\beta_{rs}]$ vanish, we note the identity

$$[\hat{T}_{pq}, \hat{T}_{rs}] = [\hat{E}_{pq}, \hat{E}_{rs}] \quad (10S.8.7)$$

The last term in (10S.8.6) is therefore the same as in the singlet Hessian – that is, a linear combination of gradient elements – and vanishes for an optimized RHF state due to the Brillouin theorem; see Section 10.2.1. The triplet Hessian (10E.8.4) may thus be expressed in the more compact form

$$G_{aibj} = \langle \text{HF}|[\hat{T}^-_{ai}, [\hat{T}^-_{bj}, \hat{H}]]|\text{HF}\rangle = 2\langle \text{HF}|[\hat{T}_{ai}, [\hat{T}^-_{bj}, \hat{H}]]|\text{HF}\rangle \quad (10S.8.8)$$

To obtain the last expression, we have used the relation

$$\hat{T}^\dagger_{ia} = \hat{T}_{ai} \quad (10S.8.9)$$

and the fact that the matrix element in (10S.8.8) is real.

3. With the notation

$$E^{\sigma-}_{pq} = E^\sigma_{pq} - E^\sigma_{qp} \quad (10S.8.10)$$

we may write the singlet and tripet Hessians (10.2.8) and (10E.8.5) of an optimized RHF state in the form

$$E^{(2)}_{aibj} = 2\langle \text{HF}|[E^\alpha_{ai} + E^\beta_{ai}, [E^{\alpha-}_{bj} + E^{\beta-}_{bj}, \hat{H}]]|\text{HF}\rangle \quad (10S.8.11)$$

$$G_{aibj} = 2\langle \text{HF}|[E^\alpha_{ai} - E^\beta_{ai}, [E^{\alpha-}_{bj} - E^{\beta-}_{bj}, \hat{H}]]|\text{HF}\rangle \quad (10S.8.12)$$

The relationship between these elements is therefore given by

$$G_{aibj} = E^{(2)}_{aibj} - 4\langle \text{HF}|[E^\alpha_{ai}, [E^{\beta-}_{bj}, \hat{H}]]|\text{HF}\rangle - 4\langle \text{HF}|[E^\beta_{ai}, [E^{\alpha-}_{bj}, \hat{H}]]|\text{HF}\rangle$$

$$= E^{(2)}_{aibj} - 8\langle \text{HF}|[E^\alpha_{ai}, [E^{\beta-}_{bj}, \hat{H}]]|\text{HF}\rangle \quad (10S.8.13)$$

where the last expression follows by application of the unitary spin-flip operator as in (10S.8.4). This relationship holds also for an unoptimized RHF state since the same linear combination of gradient elements appears in G_{aibj} and $E^{(2)}_{aibj}$.

4. Expanding the commutator in (10E.8.6), we find that only two terms are nonzero:

$$\langle \text{HF}|[E^\alpha_{ai}, [E^{\beta-}_{bj}, \hat{H}]]|\text{HF}\rangle = \langle \text{HF}|\hat{H}E^\beta_{bj}E^\alpha_{ai}|\text{HF}\rangle + \langle \text{HF}|E^\beta_{jb}\hat{H}E^\alpha_{ai}|\text{HF}\rangle \quad (10S.8.14)$$

Table 10S.8.1 The RHF/STO-3G singlet and triplet Hessians at different internuclear separations of the H_2 molecule (E_h)

	$2.5a_0$	$2.6a_0$	$2.7a_0$	$2.8a_0$
$E^{(2)}_{ugug}$	1.739340	1.699956	1.663336	1.629464
G_{ugug}	0.092252	0.030212	−0.029800	−0.087768

Inserting the usual expression for the Hamiltonian, we note that there are contributions only from the two-electron double excitations:

$$
\langle \text{HF}|[E_{ai}^{\alpha}, [E_{bj}^{\beta-}, \hat{H}]]|\text{HF}\rangle = \frac{1}{2} \sum_{pqrs} g_{pqrs} \langle \text{HF}|E_{pq}E_{rs}E_{bj}^{\beta}E_{ai}^{\alpha}|\text{HF}\rangle
$$

$$
+ \frac{1}{2} \sum_{pqrs} g_{pqrs} \langle \text{HF}|E_{jb}^{\beta}E_{pq}E_{rs}E_{ai}^{\alpha}|\text{HF}\rangle
$$

$$
= 2g_{aibj} \tag{10S.8.15}
$$

We now obtain (10E.8.7) by substituting this expression in (10E.8.6).

5. Inserting (10.8.56) into (10E.8.7), we arrive at (10E.8.8).

6. For H_2 in the STO-3G basis, the only nonredundant rotation is the one that mixes the gerade and ungerade orbitals. The singlet and triplet Hessians thus contain only one element:

$$
E_{ugug}^{(2)} = 4(\varepsilon_u - \varepsilon_g - g_{gguu} + 3g_{gugu}) \tag{10S.8.16}
$$

$$
G_{ugug} = 4(\varepsilon_u - \varepsilon_g - g_{gguu} - g_{gugu}) \tag{10S.8.17}
$$

The calculated Hessian elements are listed in Table 10S.8.1. Linear interpolation shows that the RHF solution becomes unstable at $2.65a_0$, which should be compared with $2.30a_0$ obtained in the cc-pVQZ basis in Section 10.10.4.

11 CONFIGURATION-INTERACTION THEORY

The *configuration-interaction (CI) wave function* consists of a linear combination of Slater determinants, the expansion coefficients of which are variationally determined. Owing to the simple structure of the wave function, the CI method has been extensively and successfully applied in quantum chemistry. The method is flexible and can give highly accurate wave functions for small closed- and open-shell molecular systems with varying importance of static and dynamical correlation; it may be used to describe complex electronic-structure problems such as bond breakings and excited states. The principal shortcoming of the CI method is that it does not provide a compact description of electron correlation and that it is difficult to apply to large molecules because of the rapid growth in the number of configurations needed to recover a substantial part of the correlation energy for large systems. Furthermore, when attempts are made at compact descriptions of correlation by truncating the CI expansion, the resulting CI wave function invariably suffers from lack of size-extensivity.

We begin this chapter with a short general discussion of CI expansions and configuration selection in Section 11.1. Next, in Section 11.2, we consider the problems associated with designing size-extensive configuration spaces. Section 11.3 presents a simple model calculation to illustrate the size-extensivity problem of truncated CI wave functions. We then turn to the more technical aspects of CI theory. Having considered the parametrization of the CI model in Section 11.4, we discuss iterative techniques for solving the CI eigenvalue problem for selected roots in Section 11.5. Next, in Section 11.6, we analyse the structure of Slater determinants in more detail, introducing the idea of representing the determinants as products of alpha and beta strings. Having examined the sparseness of the determinantal representation of the Hamiltonian in Section 11.7, we then develop in Section 11.8 various direct techniques for solving the CI eigenvalue problem, avoiding the explicit construction of the Hamiltonian operator in the determinantal representation. Finally, after describing the transformation of the CI wave function between different orbital representations in Section 11.9, we conclude this chapter by considering symmetry-broken CI solutions in Section 11.10.

11.1 The CI model

The underlying ideas of CI theory were introduced in Chapter 5 – see, in particular, the discussions of FCI wave functions in Section 5.3 and of truncated CI wave functions in Section 5.6. We refer also to Chapter 4, where the variation principle and the concept of size-extensivity are discussed. In the present section, we summarize the main ideas of CI theory and, in particular, the design of configuration spaces.

11.1.1   THE CI MODEL

In the CI method, the electronic wave function is constructed as a linear combination of Slater determinants

$$|\mathbf{C}\rangle = \sum_i C_i |i\rangle \tag{11.1.1}$$

with the linear coefficients C_i determined by a variational optimization of the expectation value of the electronic energy:

$$\frac{\partial}{\partial C_i} \frac{\langle \mathbf{C}|\hat{H}|\mathbf{C}\rangle}{\langle \mathbf{C}|\mathbf{C}\rangle} = 0 \tag{11.1.2}$$

As described in Section 4.2, these conditions are equivalent to an eigenvalue problem for the coefficients and the energy

$$\mathbf{HC} = E\mathbf{C} \tag{11.1.3}$$

where $\mathbf{H}$ is the Hamiltonian matrix with the elements

$$H_{ij} = \langle i|\hat{H}|j\rangle \tag{11.1.4}$$

and $\mathbf{C}$ is a vector containing the expansion coefficients C_i. Equation (11.1.3) corresponds to a standard Hermitian eigenvalue problem of linear algebra. The construction of the CI wave function may therefore be accomplished by diagonalizing the Hamiltonian matrix in the usual manner, but more often special iterative techniques are employed for extracting selected eigenvalues and eigenvectors, as discussed in Section 11.5.

In the CI expansion (11.1.1), the basis functions $|i\rangle$ are Slater determinants. As discussed in Section 2.5, a more compact representation is obtained by expanding the wave function in the spin-symmetrized CSFs rather than in Slater determinants. Still, in the present chapter, we shall always assume that the CI wave function is expanded in the simpler basis of Slater determinants. As discussed in Section 2.6.6, the linear transformation between these two alternative representations of the CI wave function can be carried out rather easily. We may therefore adapt the determinantal techniques of the present chapter to CSF expansions simply by expanding the CSFs in determinants just before any calculation or manipulation is to be carried out on the CI wave function and transforming back again immediately afterwards. Alternatively, we may employ any of the many special techniques that have been developed for performing the computational tasks of CI theory directly in the CSF basis – without recourse to the basis of Slater determinants. However, for high efficiency and generality, the simpler determinantal basis is more useful.

11.1.2   FULL CI WAVE FUNCTIONS

The CI method is completely general with respect to the choice of configurations. A rigorous but impractical approach is to include in the expansion (11.1.1) the full set of determinants (of the appropriate spin and space symmetry) generated by distributing all electrons among all orbitals. Such expansions are called *full CI (FCI) expansions*.

For FCI wave functions, the number of Slater determinants increases very rapidly with the number of electrons and with the number of orbitals. This behaviour is illustrated in Table 11.1,

Table 11.1 The number of Slater determinants N_{det} with spin projection zero obtained by distributing $2k$ electrons among $2k$ orbitals

$2k$	N_{det}
2	4
4	36
6	400
8	4 900
10	63 504
12	853 776
14	11 778 624
16	165 636 900
18	2 363 904 400
20	34 134 779 536

where we have listed the number of determinants with spin projection zero obtained by distributing an even number of $2k$ electrons among $2k$ orbitals for $1 \leq k \leq 10$.

In Table 11.1, the number of Slater determinants N_{det} in the FCI wave function appears to grow by a constant factor of about 16 with each increment in k. To see how this exponential growth comes about, we note that N_{det} is the product of the number of determinants for the alpha and beta electrons separately – see Section 11.6 for details. Thus, for a system with n orbitals containing k alpha electrons and k beta electrons, the number of determinants is given by

$$N_{det} = \binom{n}{k}^2 \tag{11.1.5}$$

For $n = 2k$, we obtain the simpler expression

$$N_{det}(n = 2k) \approx \frac{16^k}{k\pi} \quad \text{(large } k\text{)} \tag{11.1.6}$$

where we have used the *Stirling approximation* for factorials [1]

$$k! \approx k^{k+1/2}\sqrt{2\pi}e^{-k} \quad \text{(large } k\text{)} \tag{11.1.7}$$

For large k, the number of determinants thus increases by a factor of 16 for each new pair of electrons and orbitals.

As a second illustration, consider the number of determinants in the limit of a large number of orbitals

$$N_{det} \approx \frac{1}{2k\pi}\left(\frac{n}{k}\right)^{2k} \quad \text{(large } k, n/k\text{)} \tag{11.1.8}$$

as appropriate for highly correlated calculations. For a fixed number of electrons $2k$, the number of determinants N_{det} increases as n^{2k} with the number of orbitals n. These two examples should suffice to illustrate the intractability of the FCI expansion for any but the smallest electronic systems. Indeed, the usefulness of the FCI model is mostly that, for small systems, it may provide benchmarks for other determinantal wave-function models, as illustrated for the hydrogen molecule and the water molecule in Chapter 5.

11.1.3 TRUNCATED CI WAVE FUNCTIONS: CAS AND RAS EXPANSIONS

Although the FCI wave function itself quickly becomes intractable, we may obtain useful approximations by *truncating* the FCI expansion. In the early days of CI calculations, truncated CI wave functions were commonly constructed by selecting *individual configurations*. This selection was sometimes based on physical intuition but more often on perturbative estimates. Today, such methods have largely been abandoned since it has been realized that the relatively small number of configurations that can be selected individually cannot accurately recover the important dynamical correlation effects in molecules – see, for example, the discussion of the helium atom in Chapter 7. Also, the selection of individual configurations becomes particularly problematic in studies of, for example, potential-energy surfaces, where the relative importance of the configurations changes across the surface. Modern truncation techniques are therefore based on the selection of whole *classes of configurations*, as discussed in the remainder of this subsection.

In the design of configuration spaces smaller than that of FCI, it is important to distinguish between static and dynamical correlation. *Static correlation* is treated by retaining, in the truncated expansion, the dominant configurations of the FCI expansion as well as those configurations that are nearly degenerate with the dominant configurations. These configurations – selected for an adequate description of static correlation – are often referred to as the *reference configurations* of the CI wave function, and the configuration space spanned by the reference configurations is called the *reference space*. *Dynamical correlation* is subsequently treated by adding to the wave function configurations generated by excitations out of the reference space – for example, all single and double excitations, all single, double and triple excitations, and so on.

In the treatment of static correlation, the concept of *active orbitals* plays an important role. Thus, in *complete active space (CAS) wave functions*, no restrictions are placed on the occupations of the active orbitals but the remaining orbitals are either always doubly occupied (the *inactive orbitals*) or always unoccupied (the *secondary orbitals*) in the wave function [2]. Typically, the core orbitals of a system are treated as inactive and the valence orbitals as active – the CAS reference space then consists of all configurations obtained by distributing the valence electrons in all possible ways among the active orbitals, keeping the core orbitals doubly occupied in all configurations. In the optimization of the CASSCF wave function, the orbitals and CAS configurations are simultaneously optimized, yielding a highly flexible although usually not very accurate wave function since no account is taken of dynamical correlation. For discussion and illustration of CAS wave functions, see Section 5.5.

A CAS calculation provides a set of orbitals and a reference space for treating static correlation. To recover the dynamical correlation energy, we include in the CI expansion all configurations generated from the reference space by carrying out excitations up to a given excitation level. In the singles-and-doubles model, for example, we include configurations generated by carrying out single and double excitations from the full set of reference configurations. Such wave functions are called CISD (when the reference space contains a single configuration) or MRSDCI [3] (when the reference space contains more than one configuration). The orbitals of the CISD and MRSDCI wave functions are usually obtained from preceding Hartree–Fock or MCSCF calculations and no further optimization of the orbitals is attempted in the construction of the final CI wave function.

A unified framework for treating static and dynamical correlation is the concept of a *restricted active space (RAS)* [4]. As for CAS wave functions, the orbitals of RAS wave functions are divided into three classes – inactive, active and secondary orbitals. In addition, there is a further subdivision of the active orbitals into three categories – the RAS1, RAS2 and RAS3 orbitals. In the construction of the RAS wave function, a lower limit is placed on the allowed number of electrons in RAS1 and an upper limit on the allowed number of electrons in RAS3; no constraints are placed on the occupations in RAS2. As an example, an MRSDCI wave function with a CAS reference space may be realized within the RAS framework: RAS1 corresponds to the inactive orbitals, RAS2 to the active CAS orbitals and RAS3 to the secondary orbitals. RAS1 must have at most two holes and RAS3 at most two electrons; in RAS2, no constraints are imposed except that the total number of electrons should be equal to N. This unified approach to the construction of configuration expansions, which will be used throughout this chapter and also in Chapter 12, illustrates the strategy that is usually taken to the selection of configurations: the orbitals are divided into different classes and all configurations obeying certain rules with respect to the occupancies of these classes are included in the CI expansion. Such a procedure is a powerful one in that it simplifies the design and optimization of the wave function considerably.

Concerning the orbitals in CI theory, the following points should be noted. For the treatment of *static* correlation, a simultaneous optimization of the orbitals and the reference configurations is necessary since several configurations may make similar contributions to the final state, making it necessary to treat all the important reference configurations in the same manner. Also, across a potential-energy surface, the relative importance of the various configurations may change, making it necessary to include the full set of reference configurations in the orbital optimization so as to avoid a bias in the description. The static correlation energy is therefore usually recovered by MCSCF rather than CI wave functions. By contrast, for the treatment of *dynamical* correlation, any reoptimization of the Hartree–Fock or MCSCF orbitals is unnecessary since the quality of the description depends not so much on the detailed form of the orbitals as on their nodal structure and overall shape. Moreover, to recover a substantial part of the dynamical correlation energy, a large number of excited configurations are needed, making the simultaneous optimization of orbitals and configuration coefficients difficult, if not impossible.

Although the approach of truncating the FCI wave function by selecting classes of configurations has met with considerable success, major problems persist. A particularly severe problem is the lack of size-extensivity, to which we now turn our attention.

11.2 Size-extensivity and the CI model

From the discussion in the preceding section, it is clear that FCI wave functions are not generally applicable and that, in practice, we must resort to truncated CI expansions. In Section 4.3, it was demonstrated that the CI method is size-extensive only if the CI variational space of a compound system corresponds to the direct product of the CI spaces of the individual systems. In truncated CI expansions, obtained by retaining no higher than, for example, double excitations with respect to the Hartree–Fock configuration, this condition is not fulfilled and the description is consequently not size-extensive.

In the present section, we examine size-extensivity in more detail in relation to the truncated CI wave function. We begin by formulating the size-extensivity of FCI wave functions, separating the energy and the wave function into Hartree–Fock and correlation contributions. We then turn to truncated CI wave functions and recapitulate how truncation of the FCI expansion destroys the size-extensivity of the wave function. Next, we introduce the Davidson correction – a nonrigorous but sometimes useful correction that can be applied to compensate for the lack of size-extensivity of truncated CI wave functions. Some numerical examples are then given to illustrate the performance of the truncated CI wave function (with and without the Davidson correction applied) for systems of different sizes.

11.2.1 FCI WAVE FUNCTIONS

Assume that the FCI wave functions are known for the two systems A and B. In the intermediate normalization, the FCI wave function for system A may be written as

$$|\psi_A^{FCI}\rangle = \hat{\psi}_A^{FCI}|vac\rangle = (\hat{\psi}_A^{HF} + \hat{\psi}_A^{corr})|vac\rangle \tag{11.2.1}$$

where $\hat{\psi}_A^{HF}$ is an operator string that generates the normalized Hartree–Fock reference state and $\hat{\psi}_A^{corr}$ is a sum of strings that generates the correlation part of the wave function. The FCI wave function satisfies the equation

$$\hat{H}_A|\psi_A^{FCI}\rangle = E_A^{FCI}|\psi_A^{FCI}\rangle \tag{11.2.2}$$

where $\hat{H}_A$ is the Hamiltonian of system A. The FCI energy of the subsystem E_A^{FCI} can be separated into the Hartree–Fock energy E_A^{HF} and the correlation energy E_A^{corr}

$$E_A^{FCI} = E_A^{HF} + E_A^{corr} \tag{11.2.3}$$

$$E_A^{HF} = \langle\psi_A^{HF}|\hat{H}_A|\psi_A^{HF}\rangle \tag{11.2.4}$$

$$E_A^{corr} = \frac{\langle\psi_A^{FCI}|\hat{H}_A - E_A^{HF}|\psi_A^{FCI}\rangle}{\langle\psi_A^{FCI}|\psi_A^{FCI}\rangle} \tag{11.2.5}$$

Similar equations hold for system B.

Consider now the supersystem containing A and B placed infinitely far apart so that they do not interact. From Section 4.3, the product wave function

$$|\psi_{AB}^{FCI}\rangle = \hat{\psi}_A^{FCI}\hat{\psi}_B^{FCI}|vac\rangle \tag{11.2.6}$$

is known to represent a size-extensive solution for the compound system:

$$(\hat{H}_A + \hat{H}_B)|\psi_{AB}^{FCI}\rangle = (E_A^{FCI} + E_B^{FCI})|\psi_{AB}^{FCI}\rangle \tag{11.2.7}$$

Note that, because of the algebraic properties of the second-quantization strings $\hat{\psi}_A^{FCI}$ and $\hat{\psi}_B^{FCI}$, the Pauli antisymmetry principle is satisfied for the product wave function $\hat{\psi}_A^{FCI}\hat{\psi}_B^{FCI}|vac\rangle$. The FCI wave function and energy of the compound system can now be expanded as

$$|\psi_{AB}^{FCI}\rangle = (\hat{\psi}_A^{HF}\hat{\psi}_B^{HF} + \hat{\psi}_A^{HF}\hat{\psi}_B^{corr} + \hat{\psi}_A^{corr}\hat{\psi}_B^{HF} + \hat{\psi}_A^{corr}\hat{\psi}_B^{corr})|vac\rangle \tag{11.2.8}$$

$$E_{AB}^{FCI} = E_A^{HF} + E_B^{HF} + E_A^{corr} + E_B^{corr} \tag{11.2.9}$$

Since the Hartree–Fock energy $E_A^{HF} + E_B^{HF}$ is size-extensive (Section 10.2.4), the FCI correlation energy $E_A^{corr} + E_B^{corr}$ must also be size-extensive.

From the above discussion, it follows that FCI expectation values of additively separable operators are size-extensive. For example, the expectation value of the additively separable operator

$$\hat{O}_{AB} = \hat{O}_A + \hat{O}_B \tag{11.2.10}$$

separates into the expectation values for the noninteracting subsystems

$$\frac{\langle\psi_{AB}^{FCI}|\hat{O}_{AB}|\psi_{AB}^{FCI}\rangle}{\langle\psi_{AB}^{FCI}|\psi_{AB}^{FCI}\rangle} = \frac{\langle\psi_A^{FCI}|\hat{O}_A|\psi_A^{FCI}\rangle}{\langle\psi_A^{FCI}|\psi_A^{FCI}\rangle} + \frac{\langle\psi_B^{FCI}|\hat{O}_B|\psi_B^{FCI}\rangle}{\langle\psi_B^{FCI}|\psi_B^{FCI}\rangle} \tag{11.2.11}$$

If we assume that $\hat{O}_A$ and $\hat{O}_B$ are one-electron operators such as

$$\hat{O}_A = \sum_{pq} [O_A]_{p_A q_A} E_{p_A q_A} \qquad (11.2.12)$$

then the expectation value of $\hat{O}_{AB}$ may be written in terms of density matrices as

$$\frac{\langle \psi_{AB}^{FCI} | \hat{O}_{AB} | \psi_{AB}^{FCI} \rangle}{\langle \psi_{AB}^{FCI} | \psi_{AB}^{FCI} \rangle} = \sum_{pq} \frac{\langle \psi_A^{FCI} | E_{p_A q_A} | \psi_A^{FCI} \rangle}{\langle \psi_A^{FCI} | \psi_A^{FCI} \rangle} [O_A]_{p_A q_A} + \sum_{pq} \frac{\langle \psi_B^{FCI} | E_{p_B q_B} | \psi_B^{FCI} \rangle}{\langle \psi_B^{FCI} | \psi_B^{FCI} \rangle} [O_B]_{p_B q_B}$$

$$(11.2.13)$$

The error in the expectation value of $\hat{O}_{AB}$ therefore does not depend on the error in the product wave function $|\psi_{AB}^{FCI}\rangle$ but rather on the errors in the density matrices of the subsystems. For the wave function of a system containing many monomers, the accumulation of small errors in the subsystems will inevitably lead to a product wave function with a large error. No such accumulation of errors occurs for expectation values of additively separable operators, which are as accurate as those of the subsystems.

11.2.2 TRUNCATED CI WAVE FUNCTIONS

In the FCI model, the wave function for the compound system contains the product $\hat{\psi}_A^{corr} \hat{\psi}_B^{corr}$. For truncated CI wave functions, this term is only partly included. As we shall soon see, it is the failure to incorporate the full product term $\hat{\psi}_A^{corr} \hat{\psi}_B^{corr}$ in the compound wave function that leads to a lack of size-extensivity in the truncated CI model.

Consider the truncated *CI-doubles (CID)* expansion, which contains the Hartree–Fock determinant together with all double excitations out of this determinant, and assume that, because of local symmetry, all single excitations vanish. The wave functions for the individual systems $\hat{\psi}_A^{CID}$ and $\hat{\psi}_B^{CID}$ may again be written in the form (11.2.1). The product of these two states therefore contains the term $\hat{\psi}_A^{corr} \hat{\psi}_B^{corr}$. However, since this term contains only quadruple excitations, it is not included in the wave function for the compound system (where only double excitations are allowed). We may therefore write the truncated CID function for the compound system in the form

$$|\psi_{AB}^{CID}\rangle = (\hat{\psi}_A^{HF} \hat{\psi}_B^{HF} + \hat{\psi}_A^{HF} \hat{\psi}_B^{corr} + \hat{\psi}_A^{corr} \hat{\psi}_B^{HF})|\text{vac}\rangle \qquad (11.2.14)$$

where $\hat{\psi}_A^{corr}$ and $\hat{\psi}_B^{corr}$ are the correlation terms for the fragments. Note that, in this expression, $|\psi_{AB}^{CID}\rangle$ does not correspond to the result of a CID calculation on the compound system but is instead constructed from the wave functions of the subsystems A and B. The total energy may again be partitioned into a size-extensive Hartree–Fock contribution

$$E_{AB}^{HF} = E_A^{HF} + E_B^{HF} \qquad (11.2.15)$$

and a correlation contribution

$$E_{AB}^{corr} = \frac{\langle \psi_{AB}^{CID} | \hat{H}_A + \hat{H}_B - E_{AB}^{HF} | \psi_{AB}^{CID} \rangle}{\langle \psi_{AB}^{CID} | \psi_{AB}^{CID} \rangle} \qquad (11.2.16)$$

Expansion of the numerator in this expression for the correlation energy now gives

$$\langle \psi_{AB}^{CID} | \hat{H}_A + \hat{H}_B - E_{AB}^{HF} | \psi_{AB}^{CID} \rangle = \langle \psi_{AB}^{CID} | \hat{H}_A - E_A^{HF} | \psi_{AB}^{CID} \rangle + \langle \psi_{AB}^{CID} | \hat{H}_B - E_B^{HF} | \psi_{AB}^{CID} \rangle$$

$$= \langle \psi_A^{CID} | \hat{H}_A - E_A^{HF} | \psi_A^{CID} \rangle + \langle \psi_B^{CID} | \hat{H}_B - E_B^{HF} | \psi_B^{CID} \rangle \qquad (11.2.17)$$

where we have used the steps (insert (11.2.14))

$$\langle \psi_{AB}^{CID} | \hat{H}_A - E_A^{HF} | \psi_{AB}^{CID} \rangle = \langle \psi_A^{corr} | \hat{H}_A - E_A^{HF} | \psi_A^{corr} \rangle + \langle \psi_A^{corr} | \hat{H}_A | \psi_A^{HF} \rangle + \langle \psi_A^{HF} | \hat{H}_A | \psi_A^{corr} \rangle$$

$$= \langle \psi_A^{CID} | \hat{H}_A - E_A^{HF} | \psi_A^{CID} \rangle \tag{11.2.18}$$

and similarly for system B. Next, expanding the denominator in the expression for the correlation energy, we obtain

$$\langle \psi_{AB}^{CID} | \psi_{AB}^{CID} \rangle = 1 + \langle \psi_A^{corr} | \psi_A^{corr} \rangle + \langle \psi_B^{corr} | \psi_B^{corr} \rangle$$

$$= \langle \psi_A^{CID} | \psi_A^{CID} \rangle + \langle \psi_B^{CID} | \psi_B^{CID} \rangle - 1 \tag{11.2.19}$$

Inserting the expanded numerator and denominator in (11.2.16), we arrive at the following expression for the correlation energy of the truncated wave function:

$$E_{AB}^{corr} = \frac{\langle \psi_A^{CID} | \hat{H}_A - E_A^{HF} | \psi_A^{CID} \rangle + \langle \psi_B^{CID} | \hat{H}_B - E_B^{HF} | \psi_B^{CID} \rangle}{\langle \psi_A^{CID} | \psi_A^{CID} \rangle + \langle \psi_B^{CID} | \psi_B^{CID} \rangle - 1} \tag{11.2.20}$$

We conclude that the correlation energy of the truncated CID wave function (11.2.14) is not size-extensive. By contrast, for a product wave function such as the FCI wave function

$$| \psi_{AB}^{FCI} \rangle = \hat{\psi}_A^{FCI} \hat{\psi}_B^{FCI} | vac \rangle \tag{11.2.21}$$

expansion of the numerator and denominator in the correlation energy (11.2.16) leads to a separation of the total energy as required by size-extensivity.

Our discussion has shown that, in calculations on two noninteracting systems, the use of truncated CI expansions – obtained, for example, by retaining no higher than double excitations with respect to the Hartree–Fock configuration – introduces an unphysical coupling of the two systems since, for a given determinant, the excitations allowed in one system depend on the excitations present in the other system. In CISD calculations, for example, a single excitation in one system may be coupled to at most single excitations in the other system rather than to the full CISD wave function of that system, as we would expect from a physical consideration of the system.

11.2.3 THE DAVIDSON CORRECTION

For truncated CI wave functions, the correlation energy is not size-extensive. However, since the numerator in the expression for the correlation energy (11.2.20) is size-extensive, we should obtain a size-extensive expression for the correlation energy by replacing the denominator by a constant number. Assuming that the correlation contribution to the wave function is small, we replace the denominator by 1. Equation (11.2.20) now reduces to the following size-extensive expression for the correlation energy:

$$E_{AB}^{corr^*} = \langle \psi_{AB}^{CID} | \hat{H}_A + \hat{H}_B - E_{AB}^{HF} | \psi_{AB}^{CID} \rangle \tag{11.2.22}$$

Let us now, for system A, consider the difference between the variational correlation energy (11.2.16) and the size-extensive correlation energy (11.2.22):

$$E_A^{corr^*} - E_A^{corr} = \frac{\langle \psi_A^{CID} | \hat{H}_A - E_A^{HF} | \psi_A^{CID} \rangle}{1 + \langle \psi_A^{corr} | \psi_A^{corr} \rangle} \langle \psi_A^{corr} | \psi_A^{corr} \rangle \tag{11.2.23}$$

The intermediately normalized CID wave function in (11.2.23) is proportional to the normalized state

$$|\psi_A^{CID}\rangle_{norm} = C_0(|\psi_A^{HF}\rangle + |\psi_A^{corr}\rangle) \tag{11.2.24}$$

where

$$C_0 = \frac{1}{\sqrt{1 + \langle \psi_A^{corr}|\psi_A^{corr}\rangle}} \tag{11.2.25}$$

which allows us to express (11.2.23) as

$$E_A^{corr*} - E_A^{corr} = E_A^{corr}\frac{1 - C_0^2}{C_0^2} \tag{11.2.26}$$

Since C_0^2 is assumed to be close to 1, we may replace $(1 - C_0^2)/C_0^2$ by $1 - C_0^2$ and (11.2.26) now reduces to

$$E_A^Q = E_A^{corr}(1 - C_0^2) \tag{11.2.27}$$

which is the *Davidson correction* [5], sometimes added to a variationally determined, truncated CI energy such as the CISD energy to correct for the lack of size-extensivity. Equivalently, this correction may be considered to correct for the lack of quadruple or higher excitations, as indicated by the use of the superscript Q in (11.2.27). Obviously, since the correlation terms $\hat{\psi}_A^{corr}$ and $\hat{\psi}_B^{corr}$ obtained in the calculation on the compound system (11.2.14) differ from those obtained in calculations on the individual systems, the Davidson correction does not lead to a rigorously size-extensive energy.

It should be realized that our discussion does not provide a rigorous justification for the use of the expectation value (11.2.22) or the Davidson correction (11.2.27) as approximations to size-extensive correlation energies. So far, our justification for the Davidson correction is only that it has been obtained by a (presumably) small modification of the variational correlation energy (11.2.16) and that, for truncated wave functions of the particular form (11.2.14), the expression (11.2.22) for the correlation energy is size-extensive. Indeed, the final justification for the use of the Davidson correction is an empirical one, based on a large number of calculations.

11.2.4 A NUMERICAL STUDY OF SIZE-EXTENSIVITY

For few-electron systems dominated by the Hartree–Fock configuration, the most important contributions to the correlation energy originate from the single and double excitations out of the Hartree–Fock configuration. However, as the number of electrons increases, the contributions from higher-order excitations become more important. As an illustration, we here examine, at the CISD level, molecular systems containing from one to eight noninteracting water monomers. The cc-pVDZ basis is employed to allow for comparisons with the FCI results of Tables 5.5 and 5.9.

In Table 11.2, we have listed the correlation energies per monomer obtained from CISD calculations with and without the Davidson correction applied. Two series of calculations are listed. In the first series, the geometry (with a bond distance of R_{ref}) is close to the equilibrium structure and represents an electronic structure dominated by dynamical correlation. The second series (with a bond distance of $2R_{ref}$) represents a different situation, characterized by large static correlation effects.

Consider first the calculations at equilibrium. For the monomer, the CISD wave function is dominated by the Hartree–Fock determinant and recovers as much as 94.5% of the correlation

Table 11.2 The CISD monomer correlation energy (in mE$_h$) for one to eight noninteracting water molecules, calculated in the cc-pVDZ basis with and without the Davidson correction applied. The weight of the Hartree–Fock configuration in the CISD wave function W_{HF} is also listed. The calculations have been carried out for the C_{2v} water molecule with $R = R_{ref}$ and $R = 2R_{ref}$ where $R_{ref} = 1.84345a_0$. The HOH bond angle is fixed at 110.565°. At R_{ref} and $2R_{ref}$, the Hartree–Fock energies are -76.0240 and -75.5877 E$_h$, respectively, and the FCI correlation energies are -217.8 and -364.0 mE$_h$, respectively

	$R = R_{ref}$			$R = 2R_{ref}$		
	CISD	CISD + Q	W_{HF}	CISD	CISD + Q	W_{HF}
1	−205.8	−216.1	0.9499	−291.9	−353.2	0.7901
2	−196.5	−213.4	0.9140	−251.6	−307.3	0.7786
3	−188.7	−210.2	0.8863	−229.5	−283.1	0.7663
4	−182.1	−206.8	0.8641	−214.2	−266.6	0.7556
5	−176.3	−203.5	0.8456	−202.6	−254.1	0.7463
6	−171.1	−200.2	0.8299	−193.3	−244.0	0.7380
7	−166.5	−197.1	0.8164	−185.6	−235.6	0.7306
8	−162.4	−194.1	0.8046	−179.0	−228.3	0.7241

energy. The Davidson correction reduces the error by a factor of more than 5, giving a total correlation energy within 1% of the FCI result. As the number of monomers increases, however, the proportion of the correlation energy recovered by the CISD expansion decreases steadily: with four monomers included, the CISD wave function recovers less than 84% of the FCI correlation energy and the Davidson correction gives an energy no more accurate than the uncorrected monomer energy; with eight monomers, the CISD wave function recovers only about 75% of the FCI correlation energy and the Davidson correction is no longer useful. The systematic degradation of the CISD description is illustrated in Figure 11.1, where the corrected and uncorrected CISD monomer correlation energies are plotted as functions of the number of noninteracting water monomers.

These results for the water molecule are typical of systems with small static correlation effects. For up to ten electrons, the CISD energy with the Davidson correction added is in good agreement

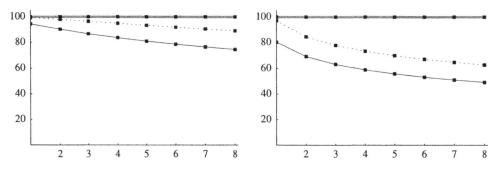

Fig. 11.1. The cc-pVDZ correlation energy per water monomer (as a percentage of the FCI correlation energy) as a function of the number of noninteracting water molecules. The grey line represents the FCI energy, the full black line the CISD energy, and the dotted line the CISD energy with the Davidson correction applied. The calculations have been carried out for the C_{2v} water molecule at the OH separations $R = R_{ref}$ (left) and $R = 2R_{ref}$ (right); see Table 11.2.

with the FCI energy. As the number of electrons increases, the CISD description deteriorates until, for large systems, only a fraction of the correlation energy is recovered. It is important to realize that the systematic degradation of the CISD description with the number of electrons is not related to the presence of static correlation. For the water monomer, for example, the simpler and less expensive MP2 method recovers 94% of the correlation energy; see Table 5.12. Moreover, since the MP2 energy is size-extensive, the same proportion of the correlation energy is recovered for any number of noninteracting monomers. In short, for small systems, the CISD and MP2 models give results of similar quality; for large systems, the MP2 model is superior.

Let us now examine the CISD description at the stretched geometry. For the monomer, the weight of the Hartree–Fock configuration in the FCI and CISD wave functions is 0.589 and 0.790, respectively, indicating the presence of large static correlation effects. Accordingly, the CISD wave function recovers only 80.2% of the FCI correlation energy; with the Davidson correction added, the error is reduced to 3%. As the number of water monomers increases, the degradation in the performance is even more severe than for the equilibrium geometry – see Figure 11.1. For eight monomers, for example, the CISD wave function recovers only about one-half of the FCI correlation energy and the Davidson correction does not improve this result significantly. It is noteworthy that, as the number of monomers increases, the Davidson correction remains more or less constant since, in (11.2.27), the increase in $1 - C_0^2$ is balanced by the reduction in the magnitude of $E_{\text{corr}}^{\text{CISD}}$.

Having discussed the energy of the noninteracting water molecules, we turn our attention to the wave function itself. In Figure 11.2, we have plotted the weight of the Hartree–Fock determinant in the FCI and CISD wave functions as a function of the number of noninteracting monomers. For the CISD wave function, the weights have been obtained from the calculations listed in Table 11.2; for the multiplicatively separable FCI wave function, the weights for m monomers have been calculated as 0.9411^m and 0.5897^m, where 0.9411 and 0.5897 are the Hartree–Fock weights for a single monomer – see Table 5.9.

The most notable feature of Figure 11.2 is the decreasing weight of the Hartree–Fock determinant with the number of monomers – in particular, for the FCI wave function. At equilibrium (which is dominated by dynamical correlation), the Hartree–Fock weight in the FCI wave function is 0.9411 for a single monomer and 0.6153 for eight monomers; for the stretched system (which contains significant static correlation), the Hartree–Fock weight is only 0.5897 for a single

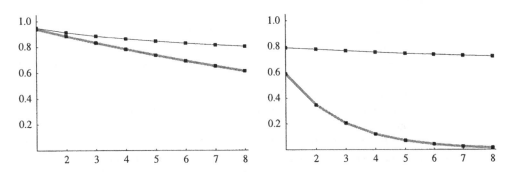

Fig. 11.2. The weight of the Hartree–Fock determinant in the FCI wave function (grey line) and in the CISD wave function (black line) as a function of the number of noninteracting water monomers in the cc-pVDZ basis. The plot on the left corresponds to the molecular equilibrium geometry; the plot on the right represents a situation where the OH bonds have been stretched to twice the equilibrium bond distance. For details on the calculations, see Table 11.2.

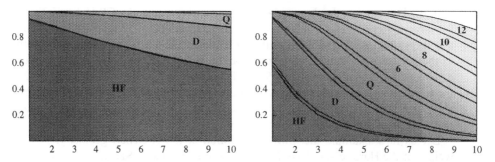

Fig. 11.3. The weights of the excitation levels in the FCI wave function for a system containing one to ten noninteracting water monomers. The plot on the left corresponds to the molecular equilibrium geometry in the cc-pVDZ basis; the plot on the right represents a situation where the OH bonds have been stretched to twice the equilibrium bond distance. For details on the calculations, see Table 11.2.

monomer and as little as 0.0146 for eight monomers. It should be emphasized, however, that the decreasing weight of the Hartree–Fock determinant in the FCI wave function occurs because of the cumulative collective weight of a large number of independent excitations in the noninteracting monomers rather than because of some inherent deficiency of the Hartree–Fock description.

Comparing the two plots in Figure 11.2, we find that the Hartree–Fock configuration plays a more dominant role at the CISD level than at the FCI level, in particular for the stretched molecules. This behaviour occurs because the CISD wave function lacks the ability to describe properly the independent excitations in the noninteracting monomers.

More insight into the shortcomings of truncated CI wave functions can be obtained from an inspection of the weights of the various excitation levels in the FCI wave function for different numbers of monomers. In Figure 11.3, we have plotted the FCI/cc-pVDZ weights of the excitations up to order 12 for one to ten noninteracting water monomers at the R_{ref} and $2R_{ref}$ reference geometries. The weights for m monomers have been obtained from those for a single monomer as described in Exercise 11.1. Although the two plots represent the situation for noninteracting monomers, the distribution of the excitation levels should be representative of interacting systems as well.

Close to the equilibrium geometry, the FCI wave function for a single monomer is dominated by the Hartree–Fock configuration with a weight of 94.1%. The doubles contribute 5.7% and the remaining excitations only 0.2%. As the number of monomers increases, the Hartree–Fock dominance is reduced. For ten monomers, the Hartree–Fock configuration contributes only 54.5% to the wave function, which now has large contributions from the doubles, quadruples and sextuples (32.7%, 9.6% and 1.9%, respectively).

As the OH bonds are stretched, the higher-order excitations become even more important. For a single monomer, the FCI wave function is still dominated by the Hartree–Fock configuration (59.0%), but the doubles now contribute as much as 33.3% and the remaining excitations 7.7%. For ten stretched monomers, the Hartree–Fock configuration contributes only 0.5% to the wave function, which now contains important contributions from a number of high-order excitations, with the largest single contribution of 15.9% coming from the eightfold excitations. Note that, as the number of monomers increases, the odd-order excitations become more important.

From Figure 11.3, it is clear that, as the system increases, higher-order excitations become more important and cannot be neglected for an accurate description of the electronic system. In Section 11.3, we shall see that, in the FCI wave function, the average excitation level increases *linearly* with the size of the system. Any approach such as CISD, based on the truncation of the FCI wave function at a given excitation level, will therefore necessarily provide a nonuniform description of systems of different sizes. Thus, although the CISD wave function gives a reasonable description for a single water molecule at equilibrium, it does not improve on the Hartree–Fock description for large systems since the important excitation levels are much higher than those accounted for at the CISD level.

11.3 A CI model system for noninteracting hydrogen molecules

In the present section, we present a simple model system of m noninteracting hydrogen molecules, analysing and comparing the behaviour of the FCI and truncated CI models for a varying number of hydrogen molecules. The purpose of this simple exercise is to illustrate and explain some of the concepts introduced in Section 11.2.

11.3.1 THE CID WAVE FUNCTION AND ENERGY

The Hamiltonian for the supersystem of m noninteracting hydrogen molecules may be written as

$$\hat{H} = \sum_{I=1}^{m} \hat{H}_I \tag{11.3.1}$$

where $\hat{H}_I$ is the Hamiltonian for the Ith molecule (disregarding the pure nuclear contributions):

$$\hat{H}_I = \sum_{p_I q_I} h_{p_I q_I} E_{p_I q_I} + \frac{1}{2} \sum_{p_I q_I r_I s_I} g_{p_I q_I r_I s_I} e_{p_I q_I r_I s_I} \tag{11.3.2}$$

The summations are over the two gerade and ungerade MOs that together constitute the minimal basis of the Ith molecule.

Our CID ground-state trial function for the supersystem contains the Hartree–Fock wave function together with all configurations that are doubly excited with respect to this state. Quadruply and more highly excited configurations may be included as well; see Exercise 11.3. We may disregard the simultaneous single excitations in different molecules since these configurations of ungerade local symmetry cannot contribute to the total wave function. Also, we need not take into account intermolecular excitations where electrons are transferred between different molecules since such configurations do not interact with the CID wave function. In short, we need consider only the m doubly excited intramolecular states $|I\rangle = \hat{D}_I|\text{HF}\rangle$, where $\hat{D}_I$ creates a doubly excited state for system I – for example, $\hat{D}_1 = a_{u_1\alpha}^\dagger a_{u_1\beta}^\dagger a_{g_1\beta} a_{g_1\alpha}$. In this basis, the CI wave function may be written as

$$|\text{CID}\rangle = |\text{HF}\rangle + \sum_{I=1}^{m} c_I|I\rangle \tag{11.3.3}$$

The coefficients in this expansion are equal by symmetry. We may therefore write our CID wave function as

$$|\text{CID}\rangle = |\text{HF}\rangle + C_D(m)|D\rangle_m \tag{11.3.4}$$

where $|D\rangle_m$ is the normalized combination of doubly excited determinants

$$|D\rangle_m = \frac{1}{\sqrt{m}} \sum_{I=1}^{m} |I\rangle \tag{11.3.5}$$

In the basis of $|\text{HF}\rangle$ and $|D\rangle_m$, the CI eigenvalue problem becomes

$$\begin{pmatrix} 0 & \sqrt{m}\,K \\ \sqrt{m}\,K & 2\Delta \end{pmatrix} \begin{pmatrix} 1 \\ C_D(m) \end{pmatrix} = E_{\text{corr}}^{\text{CID}}(m) \begin{pmatrix} 1 \\ C_D(m) \end{pmatrix} \tag{11.3.6}$$

where

$$K = g_{gugu} \tag{11.3.7}$$

$$2\Delta = E_{\text{HF}}^* - E_{\text{HF}} \tag{11.3.8}$$

$$E_{\text{HF}} = 2h_{gg} + g_{gggg} \tag{11.3.9}$$

$$E_{\text{HF}}^* = 2h_{uu} + g_{uuuu} \tag{11.3.10}$$

In (11.3.6), we have subtracted mE_{HF} from the diagonal elements to obtain the variational correlation energy directly for the supersystem $E_{\text{corr}}^{\text{CID}}(m)$. Carrying out the multiplications in (11.3.6), we find

$$C_D(m) = \frac{\sqrt{m}\,K}{E_{\text{corr}}^{\text{CID}}(m) - 2\Delta} \tag{11.3.11}$$

$$E_{\text{corr}}^{\text{CID}}(m) = \sqrt{m}\,K C_D(m) \tag{11.3.12}$$

Solving these equations for $E_{\text{corr}}^{\text{CID}}(m)$ and $C_D(m)$, we obtain

$$E_{\text{corr}}^{\text{CID}}(m) = \Delta - \sqrt{\Delta^2 + mK^2} \tag{11.3.13}$$

$$C_D(m) = \frac{\Delta - \sqrt{\Delta^2 + mK^2}}{\sqrt{m}\,K} \tag{11.3.14}$$

for the CID energy and the wave function of m noninteracting hydrogen molecules.

For $m = 1$, the energy (11.3.13) corresponds to the FCI solution since, in a minimal-basis H_2 calculation, the Hartree–Fock state and the doubly excited state are the only states of Σ_g^+ symmetry. The exact FCI minimal-basis correlation energy of m noninteracting H_2 molecules is therefore

$$E_{\text{corr}}^{\text{FCI}}(m) = mE_{\text{corr}}^{\text{CID}}(1) = m\left(\Delta - \sqrt{\Delta^2 + K^2}\right) \tag{11.3.15}$$

which should be compared with the CID expression (11.3.13). In particular, we note that, whereas the FCI expression depends linearly on the number of molecules, the CID energy depends in a complicated, unphysical manner on the size of the system. In the limit of an infinite system, we may write the CID energy (11.3.13) as

$$E_{\text{corr}}^{\text{CID}}(m) \approx -\sqrt{m}\,K \quad \text{(large } m\text{)} \tag{11.3.16}$$

which demonstrates that, for an infinitely large system, the CID correlation energy per molecule tends to zero – see Figure 11.1.

Let us now compare the CID and exact correlation energies for small K/Δ and small m. Expanding the CID and FCI energies (11.3.13) and (11.3.15) around $K = 0$, we obtain the following molecular correlation energies:

$$\frac{E_{\text{corr}}^{\text{FCI}}(m)}{m} = -\frac{K^2}{2\Delta}\left[1 - \frac{1}{4}\left(\frac{K}{\Delta}\right)^2 + \frac{1}{8}\left(\frac{K}{\Delta}\right)^4 + \cdots\right] \tag{11.3.17}$$

$$\frac{E_{\text{corr}}^{\text{CID}}(m)}{m} = -\frac{K^2}{2\Delta}\left[1 - \frac{1}{4}m\left(\frac{K}{\Delta}\right)^2 + \frac{1}{8}m^2\left(\frac{K}{\Delta}\right)^4 + \cdots\right] \tag{11.3.18}$$

Clearly, the leading terms are identical in the two expressions but, to higher orders in K/Δ, the CID expression depends nonlinearly on the number of particles.

11.3.2 THE DAVIDSON CORRECTION

It is of some interest to see how the CID expansion (11.3.18) is modified by the Davidson correction (11.2.27):

$$E_{\text{Q}}^{\text{CID}}(m) = E_{\text{corr}}^{\text{CID}}(m)(1 - C_0^2) = E_{\text{corr}}^{\text{CID}}(m)\frac{C_{\text{D}}^2(m)}{1 + C_{\text{D}}^2(m)} \tag{11.3.19}$$

where we have used (11.2.25) to rewrite the correction in terms of $C_{\text{D}}(m)$. Expansion of (11.3.14) gives

$$C_{\text{D}}^2(m) = \frac{1}{4}m\left(\frac{K}{\Delta}\right)^2 - \frac{1}{8}m^2\left(\frac{K}{\Delta}\right)^4 + \cdots \tag{11.3.20}$$

Using (11.3.18)–(11.3.20), we obtain

$$\frac{E_{\text{corr}}^{\text{CID}}(m) + E_{\text{Q}}^{\text{CID}}(m)}{m} = \frac{E_{\text{corr}}^{\text{CID}}(m)}{m}[1 + C_{\text{D}}^2(m) - C_{\text{D}}^4(m) + \cdots]$$

$$= -\frac{K^2}{2\Delta}\left[1 - \frac{1}{8}m^2\left(\frac{K}{\Delta}\right)^4 + \cdots\right] \tag{11.3.21}$$

The Davidson-corrected CID energy is size-extensive to fifth order in K and to first order in m – for higher orders in the correlation correction, the unphysical dependence on the number of particles persists. Note that the Davidson correction works by eliminating completely the term that depends linearly on m in the energy expression (11.3.18). The modified CID energy (11.3.21) is therefore not correct for a single molecule. These considerations are consistent with the general observation that the Davidson correction should not be applied when only a few (two or three) electron pairs are correlated.

11.3.3 THE CID ONE-ELECTRON DENSITY MATRIX

Let us examine the one-electron density-matrix elements of the CID wave function:

$$D_{p_I q_I}^{\text{CID}}(m) = \langle\text{CID}|E_{p_I q_I}|\text{CID}\rangle \tag{11.3.22}$$

For m monomers, the normalized CID wave function may be written in the form

$$|\text{CID}\rangle = \frac{1}{\sqrt{1 + C_D^2(m)}} \left[|\text{HF}\rangle + m^{-1/2} C_D(m) \sum_{I=1}^{m} |I\rangle \right] \qquad (11.3.23)$$

Inserting this expansion in (11.3.22) and carrying out the integrations, we obtain the following nonzero density-matrix elements:

$$D_{g_I g_I}^{\text{CID}}(m) = \frac{2 + 2(1 - m^{-1})C_D^2(m)}{1 + C_D^2(m)} \qquad (11.3.24)$$

$$D_{u_I u_I}^{\text{CID}}(m) = \frac{2m^{-1} C_D^2(m)}{1 + C_D^2(m)} \qquad (11.3.25)$$

Thus, the probability of locating the electrons in the antibonding orbital of subsystem I depends on the number of systems in the calculation, being inversely proportional to the number of noninteracting monomers. For $m = 1$, we recover the FCI one-electron density matrix

$$D_{g_I g_I}^{\text{FCI}}(1) = \frac{2}{1 + C_D^2(1)} \qquad (11.3.26)$$

$$D_{u_I u_I}^{\text{FCI}}(1) = \frac{2C_D^2(1)}{1 + C_D^2(1)} \qquad (11.3.27)$$

but, in the limit of large m, where $C_D(m)$ tends to -1, the density matrix and hence the description of the electronic system approaches that of the Hartree–Fock wave function

$$D_{g_I g_I}^{\text{HF}}(m) = 2 \qquad (11.3.28)$$

$$D_{u_I u_I}^{\text{HF}}(m) = 0 \qquad (11.3.29)$$

Hence, whereas the CID density matrix is identical to the FCI density matrix for a single system, it approaches the Hartree–Fock density matrix for a large number of systems.

11.3.4 THE FCI DISTRIBUTION OF EXCITATION LEVELS

The normalized FCI wave function for the m noninteracting hydrogen molecules may be written in the form

$$|\text{FCI}\rangle = [1 + C_D^2(1)]^{-m/2} \left[\prod_I (1 + C_D(1)\hat{D}_I) \right] |\text{HF}\rangle \qquad (11.3.30)$$

Let us introduce the weights of the Hartree–Fock configuration and of the doubly excited configuration in a single molecule

$$W_{\text{HF}} = \frac{1}{1 + C_D^2(1)} \qquad (11.3.31)$$

$$W_D = \frac{C_D^2(1)}{1 + C_D^2(1)} \qquad (11.3.32)$$

Expanding the FCI wave function (11.3.30) and collecting terms to order $2k$ in the excitation levels, we note that there are $\binom{m}{k}$ such contributions, each with a weight $W_{\mathrm{HF}}^{m-k}W_{\mathrm{D}}^{k}$. The total weight of the $2k$-fold excitations is therefore

$$\mathrm{W}_{2k} = \binom{m}{k} W_{\mathrm{D}}^{k} W_{\mathrm{HF}}^{m-k} \qquad (11.3.33)$$

Hence, the excitation levels of the FCI wave function are populated according to the *binomial distribution* [6].

In statistics, the binomial distribution describes the number of *successes* that occur in m *independent trials*, when the probability of success in each trial is the same. In our case, the m independent trials correspond to the m noninteracting systems; success of a trial corresponds to the double excitation of the electrons in a system, each of which occurs with the probability W_{D}. According to the theory of binomial distributions, the *average number* of double excitations (i.e. the average number of successes in m trials) and the *standard deviation* in the number of double excitations are given as

$$\overline{W} = mW_{\mathrm{D}} \qquad (11.3.34)$$

$$\sigma_W = \sqrt{mW_{\mathrm{HF}}W_{\mathrm{D}}} \qquad (11.3.35)$$

We note that the average excitation level increases linearly with the number of systems m. For large m, the binomial distribution approaches the *Gaussian (normal) distribution*

$$W_{2k} = \frac{1}{\sqrt{2\pi\sigma_W^2}} \exp\left[-\frac{1}{2}\left(\frac{k-\overline{W}}{\sigma_W}\right)^2\right] \qquad (11.3.36)$$

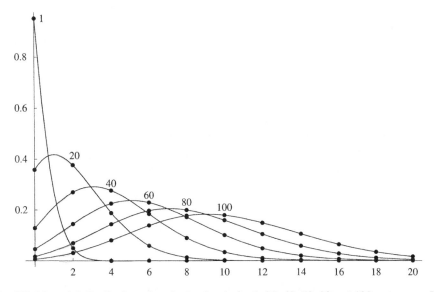

Fig. 11.4. The binomial distribution of excitation levels for 1, 20, 40, 60, 80 and 100 monomers. In a single monomer, the weight of the doubles is 0.05. The plotted points correspond to the proportion of systems found at a given excitation level as calculated from the expression (11.3.33).

In Figure 11.4, we have plotted the binomial distribution of excitation levels for 1, 20, 40, 60, 80 and 100 noninteracting monomers, assuming $W_D = 0.05$. According to (11.3.34), the mean numbers of double excitations in these supersystems are 0.05, 1, 2, 3, 4 and 5, corresponding to mean excitation levels of 0.1, 2, 4, 6, 8 and 10, respectively. As the number of monomers increases, the distribution approaches the Gaussian (11.3.36). Clearly, any approximation to the FCI wave function based on the truncation of the expansion at a fixed excitation level can describe accurately only systems in which the typical excitation level is lower than the truncation level. The use of CI expansions for systems containing many electrons is therefore, at best, laborious.

11.4 Parametrization of the CI model

Before we can discuss the optimization of the CI state, we must establish a useful parametrization of the wave function and the energy. Also, to apply the standard methods of numerical analysis to the optimization of the CI wave function, we must determine the first and second derivatives of the CI energy with respect to the variational parameters of the wave function. In the present section, we accomplish these tasks, laying the foundation for the discussion of optimization techniques in Section 11.5.

11.4.1 THE CI EXPANSION

In the CI model, the wave function is represented as a linear combination of Slater determinants:

$$|\mathbf{C}\rangle = \sum_i C_i |i\rangle \tag{11.4.1}$$

Because of the simplicity of this expression, we might be tempted to employ the linear coefficients of this expansion as the variational parameters of our wave function. However, because of the constraint imposed by the normalization condition (assuming real coefficients)

$$\sum_i C_i^2 = 1 \tag{11.4.2}$$

this particular parametrization is an inconvenient one. Thus, if the configuration space contains M determinants, then $|\mathbf{C}\rangle$ contains M expansion coefficients, while the set of M normalized states is represented by only $M - 1$ independent parameters. The linear parametrization (11.4.1) therefore contains a *redundancy* related to the normalization. The purpose of the present subsection is to develop a more convenient, unconstrained parametrization of the wave function, in which this redundancy is easily identified and controlled.

The normalization constraint (11.4.2) is easily eliminated by introducing the unconstrained parametrization

$$|\mathbf{C}\rangle = \frac{\sum_i C_i |i\rangle}{\sqrt{\mathbf{C}^T \mathbf{C}}} \tag{11.4.3}$$

which is normalized for all $\mathbf{C}$ (where $\mathbf{C}$ is the vector containing the elements C_i). In most applications, it is convenient to describe the variations in the wave function with respect to some normalized *reference state*

$$|0\rangle = \sum_i C_i^{(0)}|i\rangle \tag{11.4.4}$$

$$\sum_i (C_i^{(0)})^2 = 1 \tag{11.4.5}$$

For this purpose, we rewrite the wave function (11.4.3) as

$$|\mathbf{C}\rangle = \frac{\sum_i (C_i^{(0)} + c_i)|i\rangle}{\sqrt{(\mathbf{C}^{(0)} + \mathbf{c})^{\mathrm{T}}(\mathbf{C}^{(0)} + \mathbf{c})}} \tag{11.4.6}$$

in terms of a new set of variational parameters $\mathbf{c}$. This parametrization of the CI state is uncon-strained and reduces to the reference state when $\mathbf{c}$ vanishes – a useful property since the variations in the wave function are often small. Furthermore, all CI states $|\mathbf{C}_\alpha\rangle$ generated from (11.4.6) by the coefficients

$$\mathbf{c}_\alpha = \mathbf{c} + \alpha(\mathbf{C}^{(0)} + \mathbf{c}) \tag{11.4.7}$$

are independent of the numerical parameter α:

$$|\mathbf{C}_\alpha\rangle = \frac{(1 + \alpha)\sum_i (C_i^{(0)} + c_i)|i\rangle}{\sqrt{(1 + \alpha)(\mathbf{C}^{(0)} + \mathbf{c})^{\mathrm{T}}(1 + \alpha)(\mathbf{C}^{(0)} + \mathbf{c})}} = |\mathbf{C}\rangle \tag{11.4.8}$$

The vector $\mathbf{C}^{(0)} + \mathbf{c}$ therefore represents the redundant component of this parametrization. Since this redundancy depends on the value of the variational parameters $\mathbf{c}$, it is difficult to control – a shortcoming of the present parametrization of the CI wave function.

To improve upon our representation of the CI state, we now modify our parametrization (11.4.6) so that the redundancy becomes parallel to the reference state and is therefore easily controlled. Thus, introducing the projection operator

$$\hat{P} = 1 - |0\rangle\langle 0| \tag{11.4.9}$$

we write the CI state as

$$|\mathbf{C}\rangle = \frac{|0\rangle + \hat{P}|\mathbf{c}\rangle}{\sqrt{1 + \langle \mathbf{c}|\hat{P}|\mathbf{c}\rangle}} \tag{11.4.10}$$

where $|\mathbf{c}\rangle$ is given by

$$|\mathbf{c}\rangle = \sum_i c_i|i\rangle \tag{11.4.11}$$

In simplifying the denominator in (11.4.10), we have used the idempotency and Hermiticity of the projection operator:

$$\hat{P}^2 = \hat{P} \tag{11.4.12}$$

$$\hat{P} = \hat{P}^\dagger \tag{11.4.13}$$

To identify the redundancy of the new parametrization (11.4.10), we note that, because of the presence of the projection operator, all coefficients of the form

$$\mathbf{c}_\alpha = \mathbf{c} + \alpha\mathbf{C}^{(0)} \tag{11.4.14}$$

represent the same electronic state

$$|\mathbf{C}_\alpha\rangle = |\mathbf{C}\rangle \tag{11.4.15}$$

The redundant component is therefore $\mathbf{C}^{(0)}$, which is independent of $\mathbf{c}$. This 'global' redundancy is easier to control than the 'local' redundancy in (11.4.7). In particular, all states may now be reduced to a *canonical form* in a simple manner. For a given state $|\mathbf{C}\rangle$, the canonical reduction is accomplished by applying the projection operator

$$\mathbf{C}_{\text{can}} = \hat{P}|\mathbf{C}\rangle = |\mathbf{C}\rangle - \langle 0|\mathbf{C}\rangle|0\rangle \tag{11.4.16}$$

In the canonical form, we have eliminated the arbitrariness associated with the redundancy (11.4.14), making it easy to compare two different states. In the next subsection, this parametrization is used to derive expressions for the derivatives of the CI energy.

11.4.2 THE CI ENERGY

Having established a useful parametrization of the wave function in Section 11.4.1, we now express the CI energy in terms of this parametrization. A simple substitution of the wave function (11.4.10) in the expectation value for the energy gives us

$$E(\mathbf{c}) = \langle \mathbf{C}|\hat{H}|\mathbf{C}\rangle = \frac{\langle 0|\hat{H}|0\rangle + 2\langle \mathbf{c}|\hat{P}\hat{H}|0\rangle + \langle \mathbf{c}|\hat{P}\hat{H}\hat{P}|\mathbf{c}\rangle}{1 + \langle \mathbf{c}|\hat{P}|\mathbf{c}\rangle} \tag{11.4.17}$$

In matrix representation, this expression may be written as

$$E(\mathbf{c}) = \frac{E^{(0)} + 2\mathbf{c}^{\mathrm{T}}\mathbf{PHC}^{(0)} + \mathbf{c}^{\mathrm{T}}\mathbf{PHPc}}{1 + \mathbf{c}^{\mathrm{T}}\mathbf{Pc}} \tag{11.4.18}$$

where $E^{(0)}$ is the expectation value of the reference state

$$E^{(0)} = \langle 0|\hat{H}|0\rangle \tag{11.4.19}$$

and $\mathbf{P}$ is the matrix representation

$$P_{\mathrm{ij}} = \langle i|\hat{P}|j\rangle \tag{11.4.20}$$

of the projection operator (11.4.9).

To be able to use the energy expression (11.4.18) in optimizations, we must take its derivatives with respect to the variational parameters $\mathbf{c}$ at $\mathbf{c} = \mathbf{0}$:

$$E^{(n)}_{i_1 i_2 \cdots i_n} = \left. \frac{\partial^n E}{\partial c_{i_1} \partial c_{i_2} \cdots \partial c_{i_n}} \right|_{\mathbf{c}=\mathbf{0}} \tag{11.4.21}$$

These derivatives are most easily obtained by rewriting (11.4.18) in the form

$$(1 + \mathbf{c}^{\mathrm{T}}\mathbf{Pc})E(\mathbf{c}) = E^{(0)} + 2\mathbf{c}^{\mathrm{T}}\mathbf{PHC}^{(0)} + \mathbf{c}^{\mathrm{T}}\mathbf{PHPc} \tag{11.4.22}$$

At $\mathbf{c} = \mathbf{0}$, the *electronic gradient* and the *electronic Hessian* of the CI energy are given by the expressions

$$\mathbf{E}^{(1)} = 2\mathbf{PHC}^{(0)} = 2(\mathbf{H} - E^{(0)}\mathbf{1})\mathbf{C}^{(0)} \tag{11.4.23}$$

$$\mathbf{E}^{(2)} = 2\mathbf{P}(\mathbf{H} - E^{(0)}\mathbf{1})\mathbf{P} \tag{11.4.24}$$

For the third derivatives, we obtain the following expression at $\mathbf{c} = \mathbf{0}$:

$$E^{(3)}_{ijk} = -2P_{ij}E^{(1)}_k - 2P_{ik}E^{(1)}_j - 2P_{jk}E^{(1)}_i \tag{11.4.25}$$

More generally, we find that the higher derivatives of the energy are related to those of lower orders by

$$E^{(n)}_{i_1 i_2 \cdots i_n} = -2 \sum_{k<l} P_{i_k i_l} E^{(n-2)}_{i_1 \cdots i_{k-1} i_{k+1} \cdots i_{l-1} i_{l+1} \cdots i_n} \tag{11.4.26}$$

(see Exercise 11.4). In particular, we note that, if the gradient (11.4.23) vanishes at the point of expansion, then so do all odd-order derivatives:

$$\mathbf{E}^{(1)} = \mathbf{0} \Rightarrow \mathbf{E}^{(2n+1)} = \mathbf{0} \quad (n \text{ integer}) \tag{11.4.27}$$

We emphasize, however, that the above expressions for the derivatives (11.4.23)–(11.4.27) are valid only at $\mathbf{c} = \mathbf{0}$.

The derivatives obtained in the present subsection are valid for the expectation value of any Hermitian operator, not just the Hamiltonian. For the special case of the Hamiltonian, we obtain the CI eigenvalue equation by setting the gradient (11.4.23) equal to zero

$$\mathbf{E}^{(1)} = \mathbf{0} \quad \Leftrightarrow \quad \mathbf{H}\mathbf{C}^{(0)} = E^{(0)}\mathbf{C}^{(0)} \tag{11.4.28}$$

as discussed in Section 11.1.1. In the next section, we shall employ the expressions for the gradient and Hessian derived in the present section to set up efficient schemes for the optimization of the CI wave function.

11.5 Optimization of the CI wave function

The CI wave functions are solutions to the eigenvalue problem (11.1.3). For large CI expansions, containing millions of determinants, it is impossible to construct and store the Hamiltonian matrix $\mathbf{H}$. Moreover, even if such a matrix could be set up, it would not be possible to diagonalize it by the standard techniques of numerical analysis such as the Givens and Jacobi methods [7,8]. Usually, however, a complete diagonalization of the Hamiltonian matrix is not required since we are typically interested in only a few of its lowest eigenvalues. Selected eigenvalues of the Hamiltonian can be determined by *iterative methods*. In these methods, the individual elements of the Hamiltonian matrix are not needed – the eigenvectors and eigenvalues are instead generated by a sequence of linear transformations or contractions of the form

$$\sigma = \mathbf{H}\mathbf{C} \tag{11.5.1}$$

where $\mathbf{C}$ is some trial vector and σ the linearly transformed trial vector. As we shall see in Section 11.8, such transformations may be carried out at a cost significantly lower than that of the construction of the Hamiltonian matrix. In the present section, we shall consider the optimization of the CI wave function by such iterative methods, assuming that linear transformations of the type (11.5.1) can be carried out efficiently.

Iterative methods can be developed either for solving the CI optimization problem (11.1.2) or for solving the CI eigenvalue problem (11.1.3). Departing from the standard practice, we shall here pursue the former course and consider the optimization of the CI energy by Newton's method. We begin by deriving, in Section 11.5.1, an expression for the CI Newton step. Next, in Section 11.5.2, we discuss the convergence rate of Newton's method as applied to CI energies, illustrating the theory with calculations carried out on the water molecule. In Section 11.5.3, we introduce some important approximations to the Newton step, in particular the popular method due to Davidson. Numerical examples of these methods are given in Section 11.5.4.

11.5.1 THE NEWTON STEP

To apply *Newton's method* to the optimization of the CI energy, we proceed in the usual manner and expand the energy to second order in the variational parameters:

$$E(\mathbf{c}) = E^{(0)} + \mathbf{c}^\mathrm{T}\mathbf{E}^{(1)} + \tfrac{1}{2}\mathbf{c}^\mathrm{T}\mathbf{E}^{(2)}\mathbf{c} + \cdots \tag{11.5.2}$$

Truncating the expansion at second order and setting the gradient to zero, we obtain the *Newton step*:

$$\mathbf{E}^{(2)}\mathbf{c} = -\mathbf{E}^{(1)} \tag{11.5.3}$$

Because of the redundancy along the reference vector $\mathbf{C}^{(0)}$ (which here corresponds to our present approximation to the solution), the gradient (11.4.23) is zero in the direction of the reference vector $\mathbf{C}^{(0)}$

$$\mathbf{C}^{(0)^\mathrm{T}}\mathbf{E}^{(1)} = 0 \tag{11.5.4}$$

and the Hessian (11.4.24) is singular

$$\mathbf{E}^{(2)}\mathbf{C}^{(0)} = \mathbf{0} \tag{11.5.5}$$

There are therefore infinitely many solutions to the linear equations (11.5.3). However, from the parametrization of the wave function (11.4.10), it is clear that we may restrict ourselves to the solution that is orthogonal to the redundancy $\mathbf{C}^{(0)}$:

$$\mathbf{C}^{(0)^\mathrm{T}}\mathbf{c} = 0 \tag{11.5.6}$$

This *canonical* solution is unaffected if a multiple of $\mathbf{C}^{(0)}\mathbf{C}^{(0)^\mathrm{T}}$ is added to the Hessian in (11.5.3) and we may therefore obtain our solution from the following equations

$$(\mathbf{E}^{(2)} + 2\alpha\mathbf{C}^{(0)}\mathbf{C}^{(0)^\mathrm{T}})\mathbf{c} = -\mathbf{E}^{(1)} \tag{11.5.7}$$

where α is some numerical constant different from zero. The modified Hessian matrix

$$\mathbf{G}^{(2)} = \mathbf{E}^{(2)} + 2\alpha\mathbf{C}^{(0)}\mathbf{C}^{(0)^\mathrm{T}} = 2\mathbf{P}(\mathbf{H} - E^{(0)}\mathbf{1})\mathbf{P} + 2\alpha\mathbf{C}^{(0)}\mathbf{C}^{(0)^\mathrm{T}} \tag{11.5.8}$$

is nonsingular and the Newton step may now be written in the form

$$\mathbf{c} = -(\mathbf{G}^{(2)})^{-1}\mathbf{E}^{(1)} \tag{11.5.9}$$

in terms of the Hessian (11.5.8) and the gradient (11.4.23).

To obtain an explicit expression for the inverted Hessian, we rewrite (11.5.8) as

$$\mathbf{G}^{(2)} = 2(\mathbf{H} - E^{(0)}\mathbf{1}) + 2(\mathbf{C}^{(0)}, (\mathbf{H} - E^{(0)}\mathbf{1})\mathbf{C}^{(0)}) \begin{pmatrix} \alpha & -1 \\ -1 & 0 \end{pmatrix} \begin{pmatrix} \mathbf{C}^{(0)^\mathrm{T}} \\ \mathbf{C}^{(0)^\mathrm{T}}(\mathbf{H} - E^{(0)}\mathbf{1}) \end{pmatrix} \tag{11.5.10}$$

This matrix may be inverted by applying the *Sherman–Morrison formula* [7] in the form (see Exercise 11.5)

$$(\mathbf{A} + \mathbf{u}\mathbf{S}\mathbf{u}^{\mathrm{T}})^{-1} = \mathbf{A}^{-1} - \mathbf{A}^{-1}\mathbf{u}(\mathbf{S}^{-1} + \mathbf{u}^{\mathrm{T}}\mathbf{A}^{-1}\mathbf{u})^{-1}\mathbf{u}^{\mathrm{T}}\mathbf{A}^{-1} \tag{11.5.11}$$

with

$$\mathbf{A} = 2(\mathbf{H} - E^{(0)}\mathbf{1}) \tag{11.5.12}$$

$$\mathbf{S} = 2\begin{pmatrix} \alpha & -1 \\ -1 & 0 \end{pmatrix}, \qquad \mathbf{S}^{-1} = \frac{1}{2}\begin{pmatrix} 0 & -1 \\ -1 & -\alpha \end{pmatrix} \tag{11.5.13}$$

$$\mathbf{u} = (\mathbf{C}^{(0)}, (\mathbf{H} - E^{(0)}\mathbf{1})\mathbf{C}^{(0)}) \tag{11.5.14}$$

Noting that

$$\mathbf{u}^{\mathrm{T}}\mathbf{A}^{-1}\mathbf{u} = \frac{1}{2}\begin{pmatrix} \mathbf{C}^{(0)\mathrm{T}}(\mathbf{H} - E^{(0)}\mathbf{1})^{-1}\mathbf{C}^{(0)} & 1 \\ 1 & 0 \end{pmatrix} \tag{11.5.15}$$

we arrive at the following expression for the inverse modified Hessian matrix:

$$(\mathbf{G}^{(2)})^{-1} = \frac{1}{2}(\mathbf{H} - E^{(0)}\mathbf{1})^{-1} + \frac{1}{2\alpha}\mathbf{C}^{(0)}\mathbf{C}^{(0)\mathrm{T}}$$
$$- \frac{1}{2}\frac{(\mathbf{H} - E^{(0)}\mathbf{1})^{-1}\mathbf{C}^{(0)}\mathbf{C}^{(0)\mathrm{T}}(\mathbf{H} - E^{(0)}\mathbf{1})^{-1}}{\mathbf{C}^{(0)\mathrm{T}}(\mathbf{H} - E^{(0)}\mathbf{1})^{-1}\mathbf{C}^{(0)}} \tag{11.5.16}$$

Upon multiplication from the right by the negative electronic gradient (11.4.23), the Newton step (11.5.7) becomes

$$\mathbf{c} = -\mathbf{C}^{(0)} + \frac{(\mathbf{H} - E^{(0)}\mathbf{1})^{-1}\mathbf{C}^{(0)}}{\mathbf{C}^{(0)\mathrm{T}}(\mathbf{H} - E^{(0)}\mathbf{1})^{-1}\mathbf{C}^{(0)}} \tag{11.5.17}$$

which is independent of α and orthogonal to $\mathbf{C}^{(0)}$. The contribution from the second term in (11.5.16) vanishes because of the orthogonality (11.5.4). The new normalized eigenvector is then obtained from (11.4.10) as

$$\mathbf{C} = \frac{(\mathbf{H} - E^{(0)}\mathbf{1})^{-1}\mathbf{C}^{(0)}}{\mathbf{C}^{(0)\mathrm{T}}(\mathbf{H} - E^{(0)}\mathbf{1})^{-1}\mathbf{C}^{(0)}\sqrt{1 + \mathbf{c}^{\mathrm{T}}\mathbf{P}\mathbf{c}}} \tag{11.5.18}$$

Avoiding the explicit evaluation of the normalization constant in (11.5.18), we may write the normalized, improved CI vector in the form

$$\mathbf{C} = \frac{(\mathbf{H} - E^{(0)}\mathbf{1})^{-1}\mathbf{C}^{(0)}}{\|(\mathbf{H} - E^{(0)}\mathbf{1})^{-1}\mathbf{C}^{(0)}\|} \tag{11.5.19}$$

The method established by iterating (11.5.19) is known as the *inverse-iteration method with the Rayleigh quotient* or simply as the *Rayleigh method* [7,9]. As should be clear from our discussion, the Rayleigh method is just Newton's method applied to the eigenvalue problem (11.4.28).

11.5.2 CONVERGENCE RATE OF NEWTON'S METHOD FOR THE CI ENERGY

Sufficiently close to the solution, Newton's method exhibits *quadratic convergence*. To see how this convergence arises, we expand the electronic gradient and Hessian calculated at $\mathbf{c} = \mathbf{0}$ about the exact solution at $\mathbf{c} = \mathbf{c}_{\mathrm{ex}}$ (in a pseudo matrix notation):

$$\mathbf{E}^{(1)}(\mathbf{0}) = \mathbf{E}^{(2)}_{\text{ex}}(\mathbf{0} - \mathbf{c}_{\text{ex}}) + \tfrac{1}{2}\mathbf{E}^{(3)}_{\text{ex}}(\mathbf{0} - \mathbf{c}_{\text{ex}})^2 + O(\mathbf{c}^3_{\text{ex}}) \tag{11.5.20}$$

$$\mathbf{E}^{(2)}(\mathbf{0}) = \mathbf{E}^{(2)}_{\text{ex}} + \mathbf{E}^{(3)}_{\text{ex}}(\mathbf{0} - \mathbf{c}_{\text{ex}}) + O(\mathbf{c}^2_{\text{ex}}) \tag{11.5.21}$$

In these expressions, $\mathbf{E}^{(2)}_{\text{ex}}$ and $\mathbf{E}^{(3)}_{\text{ex}}$ are the derivatives of the energy function (11.4.18) at $\mathbf{c} = \mathbf{c}_{\text{ex}}$ in a coordinate system with the origin at $\mathbf{C}^{(0)}$. Note that $\mathbf{E}^{(3)}_{\text{ex}}$ is nonzero since the derivatives are not taken at the origin and consequently that (11.4.27) does not apply. After the Newton step has been taken, the error becomes

$$
\begin{aligned}
\mathbf{c} - \mathbf{c}_{\text{ex}} &= -\mathbf{E}^{(2)^{-1}}\mathbf{E}^{(1)} - \mathbf{c}_{\text{ex}} \\
&= -[\mathbf{E}^{(2)}_{\text{ex}} + \mathbf{E}^{(3)}_{\text{ex}}(\mathbf{0} - \mathbf{c}_{\text{ex}}) + O(\mathbf{c}^2_{\text{ex}})]^{-1}[\mathbf{E}^{(2)}_{\text{ex}}(\mathbf{0} - \mathbf{c}_{\text{ex}}) + \tfrac{1}{2}\mathbf{E}^{(3)}_{\text{ex}}(\mathbf{0} - \mathbf{c}_{\text{ex}})^2 + O(\mathbf{c}^3_{\text{ex}})] - \mathbf{c}_{\text{ex}} \\
&= \tfrac{1}{2}\mathbf{E}^{(2)^{-1}}_{\text{ex}}\mathbf{E}^{(3)}_{\text{ex}}(\mathbf{0} - \mathbf{c}_{\text{ex}})^2 + O(\mathbf{c}^3_{\text{ex}})
\end{aligned}
\tag{11.5.22}
$$

The new error is quadratic in the old one, with a leading term $\mathbf{E}^{(2)^{-1}}_{\text{ex}}\mathbf{E}^{(3)}_{\text{ex}}(\mathbf{0} - \mathbf{c}_{\text{ex}})^2/2$, thereby establishing the well-known quadratic convergence of Newton's method.

Owing to the special structure of the CI energy function, Newton's method converges even faster than quadratically for CI energies. According to Exercise 11.4.3, the third derivative at $\mathbf{c} = \mathbf{c}_{\text{ex}}$ with $\mathbf{C}^{(0)}$ at the origin becomes

$$E^{(3)}_{ijk} = 2\frac{[\mathbf{Pc}_{\text{ex}}]_i E^{(2)}_{jk} + [\mathbf{Pc}_{\text{ex}}]_j E^{(2)}_{ik} + [\mathbf{Pc}_{\text{ex}}]_k E^{(2)}_{ij}}{1 + \mathbf{c}^{\text{T}}_{\text{ex}}\mathbf{Pc}_{\text{ex}}} = O(\mathbf{c}_{\text{ex}}) \tag{11.5.23}$$

since $\mathbf{E}^{(1)}(\mathbf{c}_{\text{ex}})$ is zero. The third derivative of the CI energy is therefore linear in $\mathbf{c}_{\text{ex}}$, implying that the error after the Newton step (11.5.22) is *cubic* in the old error. This remarkable local rate of convergence of the CI iterations arises because of the special properties of the odd-order derivatives of the CI wave function. Moreover, because of the simple structure of the energy function and because usually we have a good initial estimate of the solution (since the Hamiltonian matrix is diagonally dominant), global convergence problems seldom occur in CI calculations – typically, the Newton iterations start close to the solution and only a few iterations (three to five) are needed for convergence.

In Table 11.3, the convergence of the inverse iterations (i.e. Newton's method) is illustrated by calculations on the water molecule at the CISD/cc-pVDZ level, carried out at the geometries of Table 11.2. At both geometries, the Hartree–Fock wave function is used as the initial guess. For each iteration, the error in the energy is listed.

Table 11.3 Convergence of the inverse-iteration method (i.e. Newton's method) for the CI energy. The calculations are carried out for water at the CISD/cc-pVDZ level at the geometries of Table 11.2. For each iteration, the difference between the calculated and converged energies (in E_h) is listed

Iteration	$R = R_{\text{ref}}$	$R = 2R_{\text{ref}}$
1	0.2057981191	0.2919278031
2	0.1031691961	0.2888220715
3	0.0000739193	0.0557665473
4	0.0000000000	0.0010645471
5		0.0000000064
6		0.0000000000

Close to the equilibrium geometry (i.e. with OH bond distances R_{ref}), the Hartree–Fock wave function constitutes a good initial guess and convergence to 10^{-10} E_h in the energy is obtained in three iterations, each of which requires the solution of a set of linear equations. The first iteration gives a rather modest improvement, reducing the error by about 50%. In the second iteration, however, a dramatic improvement is observed – in fact, the reduction is in this particular case even better than cubic. The last iteration displays cubic convergence, with the error reduced from 7×10^{-5} to 2×10^{-13} E_h.

In the second set of iterations (with OH bond distances $2R_{ref}$), the Hartree–Fock wave function represents a poorer starting point. The initial convergence is therefore slow but cubic convergence is established after three to four iterations. The convergence patterns of Table 11.3 are typical of Newton's method applied to the CI energy – after a few iterations of slow convergence, rapid cubic convergence is observed as the local region is entered.

The small number of Newton iterations required for convergence of the CI energy is quite remarkable. However, in judging the performance of Newton's method, we should keep in mind that each Newton iteration requires the solution of a set of linear equations. For each such set of equations, a number of linear transformations (11.5.1) must be carried out. In the next subsection, we shall see that it is possible to develop less expensive schemes, which do not require the solution of linear equations in each iteration. Nevertheless, Newton's method offers certain advantages over other schemes that, in some cases, may offset its high computational cost.

First, by carrying out the iterations (11.5.19) with $E^{(0)}$ fixed at some preselected value (not necessarily an expectation value), we automatically converge to the eigenvalue closest to $E^{(0)}$. Thus, by scanning over a range of values for $E^{(0)}$, the complete eigenvalue spectrum in a given interval may be recovered. Second, the convergence rate of Newton's method is the same for all states – it converges as rapidly to highly excited states as it does to the ground state. This behaviour is in contrast to that of the more commonly used (and less expensive) methods discussed in Section 11.5.3, which, for the calculation of a given excited state, often require at least some states of lower energy to be calculated as well. Newton's method is therefore particularly well suited to excited states.

11.5.3 APPROXIMATE NEWTON SCHEMES

In the previous subsections, we developed Newton's method for the optimization of CI wave functions and energies. Newton's method requires only a few *macro (outer) iterations*, but, in each macro iteration (i.e. in each Newton step), a relatively large number of *micro (inner) iterations* are needed for solving the linear equations (11.5.3). Each micro iteration requires the multiplication of the Hamiltonian matrix by a trial vector (11.5.1). The total number of micro iterations needed for convergence may therefore become quite large with Newton's method.

It is natural to enquire whether the expensive Newton step can be replaced by a less expensive *quasi-Newton step* (in which the exact Hessian is replaced by an approximate one), thereby increasing the number of macro iterations but reducing the number of micro iterations. We shall therefore investigate whether, in the Newton step, it is advantageous to replace $\mathbf{H}$ in $(\mathbf{H} - E^{(0)}\mathbf{1})^{-1}$ by some approximate Hamiltonian $\mathbf{H}_0$ so that $(\mathbf{H}_0 - E^{(0)}\mathbf{1})^{-1}$ is easily obtained. Replacing $(\mathbf{H} - E^{(0)}\mathbf{1})^{-1}$ by $(\mathbf{H}_0 - E^{(0)}\mathbf{1})^{-1}$ in (11.5.16) and multiplying by the negative gradient (11.4.23) from the right, we obtain the following *quasi-Newton step* (see Exercise 11.6 for an alternative derivation) [10]

$$\mathbf{c} = -(\mathbf{H}_0 - E^{(0)}\mathbf{1})^{-1} \left[(\mathbf{H} - E^{(0)}\mathbf{1})\mathbf{C}^{(0)} - \frac{\mathbf{C}^{(0)\mathrm{T}}(\mathbf{H}_0 - E^{(0)}\mathbf{1})^{-1}(\mathbf{H} - E^{(0)}\mathbf{1})\mathbf{C}^{(0)}}{\mathbf{C}^{(0)\mathrm{T}}(\mathbf{H}_0 - E^{(0)}\mathbf{1})^{-1}\mathbf{C}^{(0)}}\mathbf{C}^{(0)} \right] \quad (11.5.24)$$

which, like the exact Newton step, is orthogonal to $\mathbf{C}^{(0)}$. Each macro iteration now requires only one evaluation that involves the full Hamiltonian matrix – namely, the calculation of $\mathbf{HC}^{(0)}$. The number of micro iterations therefore becomes the same as the number of macro iterations. Obviously, the usefulness of this approach depends on the availability of good approximations to $\mathbf{H}$ that can be easily inverted.

It is customary to neglect the second term in (11.5.24), in which case the quasi-Newton step reduces to the *Davidson step* [11]:

$$\mathbf{c}^{\text{dav}} = -(\mathbf{H}_0 - E^{(0)}\mathbf{1})^{-1}(\mathbf{H} - E^{(0)}\mathbf{1})\mathbf{C}^{(0)} \qquad (11.5.25)$$

The *Davidson method* has proved highly successful in electronic-structure theory. Still, this method should be applied with some caution – in particular, if attempts are made at improving the approximate Hamiltonian $\mathbf{H}_0$. Thus, from (11.5.25), we note that the Davidson step is not orthogonal to $\mathbf{C}^{(0)}$. In fact, in the limit when $\mathbf{H}_0$ becomes the full matrix $\mathbf{H}$, the Davidson step becomes parallel to $\mathbf{C}^{(0)}$

$$\mathbf{c}^{\text{dav}}(\mathbf{H}_0 \rightarrow \mathbf{H}) = -\mathbf{C}^{(0)} \qquad (11.5.26)$$

and no new direction is generated. In general, therefore, the convergence of the Davidson method slows down as the agreement between $\mathbf{H}_0$ and $\mathbf{H}$ is improved. By contrast, the full quasi-Newton step (11.5.24) turns into the exact, cubically convergent Newton step when $\mathbf{H}_0$ is replaced by $\mathbf{H}$, as we would expect of a robust algorithm.

Once a correction vector $\mathbf{c}$ has been obtained from (11.5.24) or (11.5.25), an improved eigenvector may be generated according to (11.4.10). Usually, however, it is better to regard $\mathbf{C}^{(0)}$ and $\mathbf{c}$ as vectors spanning a two-dimensional *trial-vector subspace* and to calculate the optimal linear combination of these vectors from the eigenvector of the Hamiltonian projected onto this subspace. This approach is possible since the elements of the projected Hamiltonian are easily obtained from $\mathbf{HC}^{(0)}$, which was required to generate $\mathbf{c}$ in the first place. More generally, we may, in the course of the optimization, save the initial approximation to the eigenvector together with all the correction vectors generated by (11.5.24) or (11.5.25), spanning a larger and larger subspace within which we may obtain an increasingly more accurate approximation to the solution. In effect, we solve the CI eigenvalue problem in the subspace spanned by vectors generated by the quasi-Newton step or the Davidson step. Since this approach requires all generated vectors and their linear transformations to be stored, it may become impractical for large CI expansions. In such cases, we may keep the dimension of the projected space at some fixed number, discarding old vectors as new ones are generated.

The approximate Hamiltonian matrix $\mathbf{H}_0$ should be chosen so that it is a good approximation to the exact Hamiltonian $\mathbf{H}$ and so that $\mathbf{H}_0 - E^{(0)}\mathbf{1}$ is easily inverted. Usually, $\mathbf{H}_0$ is taken to be a diagonal matrix with elements equal to those of $\mathbf{H}$. This approximation is a reasonable one since the exact Hamiltonian is often dominated by the diagonal elements. If necessary, the diagonal approximation can be refined by calculating the Hamiltonian exactly in a small subspace that corresponds to the dominant part of the wave function.

11.5.4 CONVERGENCE RATE OF QUASI-NEWTON SCHEMES FOR THE CI ENERGY

To illustrate the performance of the quasi-Newton method for the optimization of the CI energy, we have in Table 11.4 listed the energy error in each iteration for a series of calculations on the water molecule at the CISD/cc-pVDZ level. The calculations have been carried out at the two

Table 11.4 Convergence of the quasi-Newton scheme for calculations on water at the CISD/cc-pVDZ level using reduced subspaces of different dimensions. For each iteration, the error in the energy (in E_h) is listed. For geometries, see Table 11.2

Iteration	$R = R_{ref}$				$R = 2R_{ref}$			
	1	2	3	All	1	2	3	All
1	0.20580	0.20580	0.20580	0.20580	0.29194	0.29194	0.29194	0.29194
2	0.02373	0.00660	0.00660	0.00660	0.25524	0.05129	0.05129	0.05129
3	0.00641	0.00044	0.00026	0.00026	0.59421	0.00896	0.00576	0.00576
4	0.00142	0.00005	0.00002	0.00002	0.41099	0.00267	0.00149	0.00149
5	0.00038	0.00001	0.00000	0.00000	0.50127	0.00115	0.00057	0.00057
6	0.00010	0.00000			0.61162	0.00062	0.00013	0.00013
7	0.00003				0.61961	0.00035	0.00002	0.00002
8	0.00001				0.79931	0.00021	0.00001	0.00001
9	0.00000				0.96208	0.00012	0.00000	0.00000
10					0.96048	0.00007		
11					0.10609	0.00005		
12					1.14219	0.00003		
13					1.12792	0.00002		
14					1.35859	0.00001		
15					1.43131	0.00001		

standard geometries with bond distances R_{ref} and $2R_{ref}$. In each calculation, the Hartree–Fock wave function provides the initial guess and H_0 corresponds to the diagonal part of H.

At both geometries, four series of iterations are listed. In the first series, the improved eigenvector is generated simply by adding the correction vector c to the reference vector $C^{(0)}$ according to (11.4.10). In the second series, the improved vector is generated by solving an eigenvalue problem in the two-dimensional linear subspace spanned by c and the previous approximation to the eigenvector. In the third series, we obtain the improved vector from a three-dimensional eigenvalue problem in the subspace of c and the previous two approximations to the eigenvector. In the last series, the reduced subspace contains all previous eigenvector approximations as well as the correction vector c.

At R_{ref}, all four schemes converge linearly. With the simple update scheme (11.4.10), the convergence is rather slow, with the error reduced by a factor of only 3–4 in each iteration. With two vectors in the reduced eigenvector subspace, a significant improvement is observed, in particular in the first few iterations. The three-dimensional subspace performs even better, with a reduction factor greater than 10 in all iterations and convergence to an error of 10^{-5} E_h in five iterations. Recalling that four Newton iterations are needed for the same convergence (see Table 11.3), this performance is quite satisfactory indeed. Also, since convergence has already been reached after four iterations, little is gained by retaining in the trial subspace all the generated approximate eigenvectors. Indeed, within the reported number of digits, there is no difference between the three-dimensional and full-dimensional subspaces.

At the distorted geometry, the convergence is somewhat slower than at equilibrium since the initial approximation (i.e. the Hartree–Fock determinant) has a lower weight in the final wave function and since the diagonal approximation to the exact Hamiltonian is less accurate. Most important, we note that the simple update scheme (11.4.10) leads to divergence and that, for satisfactory performance, at least a three-dimensional reduced subspace is needed. Again, within the reported number of digits, there is no difference in performance between the three-dimensional and full-dimensional reduced subspaces, although, for convergence to 10^{-10} E_h, 15 iterations are needed with the three-dimensional subspace and 13 with the full-dimensional space.

When only the Davidson part of the quasi-Newton step is used, we observe nearly identical convergence if all vectors are retained in the eigenvector subspace. With truncation of the subspace, the convergence of the Davidson method degrades somewhat relative to the quasi-Newton method. Thus, at the equilibrium geometry, the Davidson method converges to 10^{-10} E_h in 15 iterations with two vectors in the subspace, compared with the 12 iterations for the quasi-Newton method.

Comparing Tables 11.3 and 11.4, we find that, in terms of macro iterations, the Newton method converges faster than the quasi-Newton method. The quasi-Newton method works better than the Newton method in the first few iterations, but the local cubic convergence of the Newton method then takes over and ensures that this method gives the smallest number of macro iterations. However, since each Newton iteration is an order of magnitude more expensive than each quasi-Newton or Davidson iteration, the quasi-Newton and Davidson methods are far more cost-effective.

11.6 Slater determinants as products of alpha and beta strings

Nowadays, FCI wave functions containing more than a billion Slater determinants can be calculated in a straightforward fashion. For CI expansions of this size, it is necessary to develop a compact way of addressing and storing information about the occupations of the Slater determinants. Furthermore, when truncated expansions are employed, we must have a simple way of identifying and addressing the determinants that belong to the expansion. In the present section, we shall see how the separation of Slater determinants into two parts – one for the alpha electrons and one for the beta electrons – provides a compact way of representing the determinants belonging to a particular expansion [12]. This separation of alpha and beta electrons constitutes an essential ingredient of the direct CI algorithms described in Section 11.8.

The nonrelativistic Hamiltonian (2.2.18) is a spin-free operator – that is, a spin tensor operator of zero rank; see Section 2.3. Determinants of different spin projections therefore give vanishing Hamiltonian matrix elements and we may restrict the determinants of the CI expansion to have the same spin projection. If the total number of electrons is N and the spin projection is M, the numbers of electrons with alpha and beta spins are given by

$$N_\alpha = \tfrac{1}{2}N + M \tag{11.6.1}$$

$$N_\beta = \tfrac{1}{2}N - M \tag{11.6.2}$$

The determinants in the CI expansion therefore have N_α electrons of alpha spin and N_β electrons of beta spin. With the alpha spin orbitals preceding the beta spin orbitals, each Slater determinant may be written as

$$|I_\alpha I_\beta\rangle = \hat{\alpha}_{I_\alpha} \hat{\beta}_{I_\beta} |\text{vac}\rangle \tag{11.6.3}$$

where $\hat{\alpha}_{I_\alpha}$ and $\hat{\beta}_{I_\beta}$ are the *alpha and beta strings* – that is, ordered products of alpha and beta spin orbitals, respectively [12]. In a given *spin string* (i.e. alpha or beta string), the orbitals are written in ascending (canonical) order.

In setting up an FCI wave function, we first generate all possible alpha (beta) strings that may be formed by distributing $N_\alpha(N_\beta)$ electrons among n orbitals and next include in the FCI wave function all possible products of such alpha and beta strings. We may then regard the Slater determinants of the FCI expansion as constituting a matrix, where the row and column indices point to the alpha and beta strings, respectively. The CI coefficients have the same structure and may be represented by a matrix $\mathbf{C}$ with elements $C_{I_\alpha I_\beta}$ and dimensions equal to the numbers of alpha and beta strings. For singlet states, the coefficient matrix is symmetric; see Exercise 11.7.

For N_α alpha electrons distributed among n orbitals, the number of alpha strings (i.e. the number of rows in $\mathbf{C}$) is given by the binomial coefficient

$$N_{\text{str}}^\alpha = \binom{n}{N_\alpha} = \frac{n!}{(n - N_\alpha)! N_\alpha!} \tag{11.6.4}$$

since each orbital is at most singly occupied. A similar expression gives the number of beta strings (i.e. the number of columns in $\mathbf{C}$). The total number of Slater determinants is therefore given by the expression

$$N_{\text{det}} = \binom{n}{N_\alpha} \binom{n}{N_\beta} \tag{11.6.5}$$

Usually, the number of alpha and beta strings is much smaller than the number of Slater determinants. In FCI calculations with $N_\alpha = N_\beta = N/2$, the number of alpha and beta strings is the same and the total number of determinants is given by $\binom{n}{N/2}^2$, which is the expression used to generate Table 11.1.

As an example, consider a system with three alpha electrons and three beta electrons distributed among five orbitals. The $\binom{5}{3} = 10$ alpha and beta strings of this system are listed in Table 11.5. There are $10 \times 10 = 100$ Slater determinants in the wave function and the CI coefficients may be arranged as a 10×10 matrix.

In RAS theory, the constraints on the orbital occupations usually eliminate some of the alpha and beta strings as well as a large number of the product strings (determinants). Consider the RAS expansion where orbitals 1 and 2 in Table 11.5 belong to RAS1 (which should contain at least three electrons), orbital 3 belongs to RAS2, and orbitals 4 and 5 to RAS3 (which should contain at most two electrons). All strings in Table 11.5 are then allowed except for $\hat{\alpha}_{10}$ and $\hat{\beta}_{10}$, which are forbidden since they cannot be combined with another string to yield a determinant with three or more electrons in RAS1. Similarly, a large number of string products are forbidden. For example, since $\hat{\alpha}_8$ has two electrons in RAS3, it can be combined only with $\hat{\beta}_1$, the only beta string with no electrons in RAS3.

The allowed combinations of strings can be identified in a convenient way by introducing classes of strings based on the numbers of electrons present in RAS1 and RAS3. For the RAS expansion

Table 11.5 The alpha and beta strings generated by distributing three electrons among five orbitals

	Alpha strings	Beta strings
1	$a_{1\alpha}^\dagger a_{2\alpha}^\dagger a_{3\alpha}^\dagger$	$a_{1\beta}^\dagger a_{2\beta}^\dagger a_{3\beta}^\dagger$
2	$a_{1\alpha}^\dagger a_{2\alpha}^\dagger a_{4\alpha}^\dagger$	$a_{1\beta}^\dagger a_{2\beta}^\dagger a_{4\beta}^\dagger$
3	$a_{1\alpha}^\dagger a_{3\alpha}^\dagger a_{4\alpha}^\dagger$	$a_{1\beta}^\dagger a_{3\beta}^\dagger a_{4\beta}^\dagger$
4	$a_{2\alpha}^\dagger a_{3\alpha}^\dagger a_{4\alpha}^\dagger$	$a_{2\beta}^\dagger a_{3\beta}^\dagger a_{4\beta}^\dagger$
5	$a_{1\alpha}^\dagger a_{2\alpha}^\dagger a_{5\alpha}^\dagger$	$a_{1\beta}^\dagger a_{2\beta}^\dagger a_{5\beta}^\dagger$
6	$a_{1\alpha}^\dagger a_{3\alpha}^\dagger a_{5\alpha}^\dagger$	$a_{1\beta}^\dagger a_{3\beta}^\dagger a_{5\beta}^\dagger$
7	$a_{2\alpha}^\dagger a_{3\alpha}^\dagger a_{5\alpha}^\dagger$	$a_{2\beta}^\dagger a_{3\beta}^\dagger a_{5\beta}^\dagger$
8	$a_{1\alpha}^\dagger a_{4\alpha}^\dagger a_{5\alpha}^\dagger$	$a_{1\beta}^\dagger a_{4\beta}^\dagger a_{5\beta}^\dagger$
9	$a_{2\alpha}^\dagger a_{4\alpha}^\dagger a_{5\alpha}^\dagger$	$a_{2\beta}^\dagger a_{4\beta}^\dagger a_{5\beta}^\dagger$
10	$a_{3\alpha}^\dagger a_{4\alpha}^\dagger a_{5\alpha}^\dagger$	$a_{3\beta}^\dagger a_{4\beta}^\dagger a_{5\beta}^\dagger$

Table 11.6 RAS string classes for the alpha strings of Table 11.5. The restrictions are at least three electrons in RAS1 (orbitals 1 and 2) and at most two in RAS3 (orbitals 4 and 5). The beta strings are classified in the same manner

String class	Strings	Electrons in RAS1	Electrons in RAS3
T_{I}^{α}	$\hat{\alpha}_1$	2	0
T_{II}^{α}	$\hat{\alpha}_2, \hat{\alpha}_5$	2	1
$T_{\mathrm{III}}^{\alpha}$	$\hat{\alpha}_3, \hat{\alpha}_4, \hat{\alpha}_6, \hat{\alpha}_7$	1	1
T_{IV}^{α}	$\hat{\alpha}_8, \hat{\alpha}_9$	1	2

Table 11.7 The allowed combinations of the string classes of Table 11.6. For the allowed combinations, the number of string products (determinants) is given

	T_{I}^{β}	T_{II}^{β}	T_{III}^{β}	T_{IV}^{β}
T_{I}^{α}	$1 \times 1 = 1$	$1 \times 2 = 2$	$1 \times 4 = 4$	$1 \times 2 = 2$
T_{II}^{α}	$2 \times 1 = 2$	$2 \times 2 = 4$	$2 \times 4 = 8$	forbidden
$T_{\mathrm{III}}^{\alpha}$	$4 \times 1 = 4$	$4 \times 2 = 8$	forbidden	forbidden
T_{IV}^{α}	$2 \times 1 = 2$	forbidden	forbidden	forbidden

discussed above, we obtain the four *string classes* in Table 11.6 (and similarly for the beta strings). The RAS expansion now consists of a set of matrices, each containing alpha and beta strings that belong to a given pair of classes. The allowed class combinations are listed in Table 11.7. The total number of product strings that satisfy the RAS conditions is obtained by summing the products in Table 11.7, yielding a total of 37 Slater determinants in the RAS wave function (compared with 100 in the FCI wave function). In this way, by organizing the strings into classes, the RAS CI matrix reduces to a set of smaller matrices – one for each allowed combination of string classes.

As we shall see in Section 11.8, the separation of determinants into alpha and beta strings is a powerful technique, which greatly simplifies the calculation of CI wave functions. However, as a first application of string theory, we shall in Section 11.7 calculate the number of nonzero matrix elements in the determinantal representation of the Hamiltonian operator.

11.7 The determinantal representation of the Hamiltonian operator

In second quantization, the singlet one- and two-electron spin tensor operators have the following representations:

$$\hat{h} = \sum_{pq} h_{pq} E_{pq} \tag{11.7.1}$$

$$\hat{g} = \frac{1}{2} \sum_{pqrs} g_{pqrs} (E_{pq} E_{rs} - \delta_{rq} E_{ps}) \tag{11.7.2}$$

Disregarding permutational symmetries, the one-electron operator is thus represented by n^2 integrals h_{pq} and the two-electron operator by n^4 integrals g_{pqrs}. Alternatively, for a given CI expansion, the same operators can be represented as

$$h_{I_{\alpha} I_{\beta}, J_{\alpha} J_{\beta}}^{\mathrm{SD}} = \langle I_{\alpha} I_{\beta} | \hat{h} | J_{\alpha} J_{\beta} \rangle \tag{11.7.3}$$

$$g_{I_{\alpha} I_{\beta}, J_{\alpha} J_{\beta}}^{\mathrm{SD}} = \langle I_{\alpha} I_{\beta} | \hat{g} | J_{\alpha} J_{\beta} \rangle \tag{11.7.4}$$

where we have written the Slater determinants in terms of spin strings. Indeed, this determinantal representation of the Hamiltonian would appear to be a natural one for carrying out the linear transformations $\mathbf{HC}$ in (11.5.1). However, since $\hat{h}$ and $\hat{g}$ couple only determinants that differ in at most one and two occupations, respectively, the matrices $\mathbf{h}^{SD}$ and $\mathbf{g}^{SD}$ of this representation are very sparse.

In the present section, we shall determine the number of nonzero matrix elements in the determinantal representation of the one- and two-electron operators (11.7.3) and (11.7.4). Our purpose is twofold. First, the large number of zero elements in $\mathbf{g}^{SD}$ will be used to argue in favour of methods where the products $\mathbf{h}^{SD}\mathbf{C}$ and $\mathbf{g}^{SD}\mathbf{C}$ are constructed directly from the integrals h_{pq} and g_{pqrs}, avoiding the explicit calculation of the matrix representations $\mathbf{h}^{SD}$ and $\mathbf{g}^{SD}$. Second, the number of nonzero elements in $\mathbf{h}^{SD}$ and $\mathbf{g}^{SD}$ is used to establish lower bounds to the operation counts for the direct construction of the products $\mathbf{h}^{SD}\mathbf{C}$ and $\mathbf{g}^{SD}\mathbf{C}$, assuming that, in order to carry out these matrix–vector multiplications, at least one multiplication and one addition are required for each nonzero matrix element. We shall restrict our analysis to FCI expansions, the generalization to RAS expansions being straightforward.

Consider first the matrix representation of the one-electron operator $\hat{h}$. For a given determinant $|J_\alpha J_\beta\rangle$, $h^{SD}_{I_\alpha I_\beta, J_\alpha J_\beta}$ is nonzero only for the following choices of $|I_\alpha I_\beta\rangle$:

$$|I_\alpha I_\beta\rangle = |J_\alpha J_\beta\rangle \tag{11.7.5}$$

$$|I_\alpha I_\beta\rangle = E^\alpha_{pq}|J_\alpha J_\beta\rangle \tag{11.7.6}$$

$$|I_\alpha I_\beta\rangle = E^\beta_{pq}|J_\alpha J_\beta\rangle \tag{11.7.7}$$

where

$$E^\alpha_{pq} = a^\dagger_{p\alpha}a_{q\alpha} \tag{11.7.8}$$

$$E^\beta_{pq} = a^\dagger_{p\beta}a_{q\beta} \tag{11.7.9}$$

are the alpha and beta parts of the singlet one-electron excitation operator

$$E_{pq} = E^\alpha_{pq} + E^\beta_{pq} \tag{11.7.10}$$

The number of determinants $E^\alpha_{pq}|J_\alpha J_\beta\rangle$ different from $|J_\alpha J_\beta\rangle$ is equal to the number of occupied alpha orbitals multiplied by the number of unoccupied alpha orbitals. Using the same argument for $E^\beta_{pq}|J_\alpha J_\beta\rangle$, we find that the total number of nonvanishing elements in $\mathbf{h}^{SD}$ is

$$L_1 = [1 + N_\alpha(n - N_\alpha) + N_\beta(n - N_\beta)]N_{det} \tag{11.7.11}$$

This number is usually very small compared with the number of matrix elements N^2_{det}.

For the two-electron operator $\hat{g}$, the matrix elements $g^{SD}_{I_\alpha I_\beta, J_\alpha J_\beta}$ are nonzero either if one or more of the relations (11.7.5)–(11.7.7) hold or if one of the following relations is satisfied:

$$|I_\alpha I_\beta\rangle = E^\alpha_{pq}E^\beta_{rs}|J_\alpha J_\beta\rangle \tag{11.7.12}$$

$$|I_\alpha I_\beta\rangle = E^\alpha_{pq}E^\alpha_{rs}|J_\alpha J_\beta\rangle \tag{11.7.13}$$

$$|I_\alpha I_\beta\rangle = E^\beta_{pq}E^\beta_{rs}|J_\alpha J_\beta\rangle \tag{11.7.14}$$

The number of single excitations (11.7.5)–(11.7.7) was calculated above and the number of double excitations arising from $E^\alpha_{pq}E^\beta_{rs}$ in (11.7.12) may be obtained in the same manner. In (11.7.13), we note that only true double excitations $E^\alpha_{pq}E^\alpha_{rs}|J_\alpha J_\beta\rangle$ (i.e. excitations with all four indices different) should be counted; with these restrictions, the operators $E_{pq}^{\alpha}E_{rs}^{\alpha}$, $E_{rs}^{\alpha}E_{pq}^{\alpha}$, $E_{ps}^{\alpha}E_{rq}^{\alpha}$ and $E_{rq}^{\alpha}E_{ps}^{\alpha}$ all produce the same excited determinant. Using the same argument for $E_{pq}^{\beta}E_{rs}^{\beta}|J_{\alpha}J_{\beta}\rangle$ in (11.7.14), we obtain the total number of nonvanishing matrix elements as

$$L_2 = N_{\text{det}} \left[1 + N_{\alpha}(n - N_{\alpha}) + N_{\beta}(n - N_{\beta}) + N_{\alpha}N_{\beta}(n - N_{\alpha})(n - N_{\beta}) \right.$$
$$\left. + \tfrac{1}{4}N_{\alpha}(N_{\alpha} - 1)(n - N_{\alpha})(n - N_{\alpha} - 1) + \tfrac{1}{4}N_{\beta}(N_{\beta} - 1)(n - N_{\beta})(n - N_{\beta} - 1) \right] \quad (11.7.15)$$

Retaining only the leading terms (i.e. only the terms quartic in n_{α}, n_{β} and N), we arrive at the expression

$$L_2 \approx \left[\tfrac{1}{4}N_{\alpha}^2(n - N_{\alpha})^2 + \tfrac{1}{4}N_{\beta}^2(n - N_{\beta})^2 + N_{\alpha}N_{\beta}(n - N_{\alpha})(n - N_{\beta}) \right] N_{\text{det}} \quad (11.7.16)$$

for the number of nonzero elements of the two-electron operator in the determinantal representation.

For a closed-shell system with 12 electrons distributed among 12 orbitals, we have $N_{\alpha} = N_{\beta} = n/2 = 6$. There are $853\,776$ Slater determinants (see Table 11.1) and therefore a total of 7.29×10^{11} zero and nonzero elements in the FCI determinantal representation of any (one- or two-electron) operator. The numbers of nonzero one- and two-electron elements are $L_1 \approx 6.23 \times 10^7$ and $L_2 \approx 1.66 \times 10^9$. Thus, in the determinantal representation of the Hamiltonian operator, there are almost 500 zero matrix elements for each nonzero element. Conversely, in the orbital representation, the one-electron operator (11.7.1) is represented by only 144 elements and the two-electron operator (11.7.2) by $20\,736$ elements, disregarding permutational symmetries. The determinantal matrix representation is clearly a sparse one, unsuitable for storing the one- and two-electron operators.

11.8 Direct CI methods

In Section 11.7, we found that the determinantal matrix representation of a two-electron operator in an FCI calculation with 12 electrons distributed among 12 orbitals has a total of 7.29×10^{11} elements, of which 1.66×10^9 are different from zero. Even if only the nonzero elements are considered, it is impractical to store this matrix. However, in Section 11.5, we found that solutions to the CI eigenvalue problem may be arrived at by carrying out a relatively small number of Hamiltonian contractions of the type $\mathbf{HC}$, bypassing the explicit calculation of the Hamiltonian matrix $\mathbf{H}$. Of course, for this scheme to be practical, these contractions must be carried out efficiently. In the present section, we develop methods for carrying out vector contractions of operators in the determinantal representation directly from the second-quantization representation of the operator – that is, directly from the integrals h_{pq} and g_{pqrs}. This is the idea of *direct CI* introduced by Roos [13].

11.8.1 GENERAL CONSIDERATIONS

Our task here is to develop efficient methods for carrying out calculations of the type

$$\sigma_{I_{\alpha}I_{\beta}} = \sum_{J_{\alpha}J_{\beta}} \langle I_{\alpha}I_{\beta}|\hat{H}|J_{\alpha}J_{\beta}\rangle C_{J_{\alpha}J_{\beta}} \quad (11.8.1)$$

where I_{α} and I_{β} are the alpha and beta strings introduced in Section 11.6 and the summation is over all pairs of such strings. Naively, we might first fall back on the Slater–Condon rules for matrix elements and imagine the calculation of the contributions from the product $\langle I_\alpha I_\beta | \hat{H} | J_\alpha J_\beta \rangle C_{J_\alpha J_\beta}$ as proceeding in the following four steps:

1. Align matching orbitals in $|I_\alpha I_\beta\rangle$ and $|J_\alpha J_\beta\rangle$ and determine the phase factor required for their alignment.

2. Determine the number of pairs of unmatched orbitals in $|I_\alpha I_\beta\rangle$ and $|J_\alpha J_\beta\rangle$. If the determinants differ in more than two pairs, the matrix element is zero.

3. If the matrix element does not vanish, use the occupations of the determinants to identify the excitations that connect the two determinants and retrieve the corresponding integrals. Multiply the integrals by phase factors and add the contributions to give the matrix element $\langle I_\alpha I_\beta | \hat{H} | J_\alpha J_\beta \rangle$.

4. Multiply the resulting matrix element by $C_{J_\alpha J_\beta}$ and add the product to the appropriate element of $\sigma_{I_\alpha I_\beta}$.

Although conceptually simple, this scheme is inefficient since each Hamiltonian matrix element is examined individually and all nonzero elements are calculated explicitly. This procedure therefore scales quadratically in the number of determinants.

A more efficient procedure is obtained by determining, for each Slater determinant, all determinants to which it is connected by single and double excitations, thereby identifying the nonzero elements of the Hamiltonian matrix directly. This procedure scales only linearly in the number of determinants and is therefore used in all modern CI programs. For this strategy to work, however, an efficient technique must be developed for generating all single and double excitations relative to a given Slater determinant. This is accomplished by writing the determinants as products of strings and by using graphs to index and order these strings as described in Section 11.8.2.

Modern computers are especially effective at elementary linear algebraic operations such as matrix–vector and matrix–matrix multiplications, which provide a simple and predictable flow of data. The key computational steps of a direct CI algorithm should therefore consist of such operations. At the same time, the CI algorithm should require as few operations as possible. The operation count for the contraction **HC** cannot be smaller than the number of nonzero elements in **H**, which therefore represents a lower bound to the operation count. Unfortunately, the requirement of a low operation count is often in conflict with the requirement of a simple data flow. In the present section, two alternative direct CI algorithms are developed – one that maximizes the use of matrix–matrix multiplications (Section 11.8.3) and one that minimizes the operation count (Section 11.8.4). Obviously, the enormous number of nonzero Hamiltonian matrix elements makes it necessary to perform rather lengthy calculations – no matter what scheme is adopted. Still, the algorithms presented in the following subsections have made possible the calculation of CI wave functions containing several billions of determinants.

11.8.2 ORDERING AND ADDRESSING OF SPIN STRINGS

When the Hamiltonian is applied to a Slater determinant $|J_\alpha J_\beta\rangle$, one or two excitation operators E^σ_{pq} will act on the strings $|J_\alpha\rangle$ and $|J_\beta\rangle$ and produce a new set of strings $E^\sigma_{pq}|J_\sigma\rangle$. It is then important to develop a good strategy for resolving the addresses of these new strings. A simple strategy would be first to generate the strings in some order, storing their occupations in an intermediate array as they are generated. Subsequently, the absolute addresses of the generated strings may be obtained by comparing their occupations with those of the strings in the canonical order. Although simple, this scheme is expensive since the resolution of the addresses requires an expensive search. What is needed is a simple connection between the occupation of a string and its address. Such a connection is presented in this subsection.

The basic idea is to represent the strings as paths in a diagram and to introduce an ordering of these paths [14]. Let us consider the general case, with N electrons in n orbitals. Each string can be represented as a vector of n ordered pairs of integers (k, m), where k is the orbital index and m the number of electrons in the orbitals up to k. This set of ordered pairs defines a *path* in an $n \times N$ diagram, obtained by drawing lines or *arcs* between the *vertices* (k, m). All paths start at the *head* $(0,0)$ and end at the *tail* (n, N). Each path contains two types of arcs – diagonal arcs and vertical arcs. The *diagonal arcs* connect vertices (k, m) and $(k + 1, m + 1)$, indicating that orbital $k + 1$ is occupied. The *vertical arcs* connect vertices (k, m) and $(k + 1, m)$, indicating that orbital $k + 1$ is unoccupied. A vertex is called *allowed* if it is visited by at least one path.

As an illustration, we consider the case with three electrons in five orbitals, with at least one electron in RAS1 (orbitals 1 and 2) and at most two electrons in RAS3 (orbitals 4 and 5). The latter constraint does not affect the number of strings generated but becomes important when we combine alpha and beta strings to comply with the RAS constraints. The allowed strings are the first nine listed in Table 11.5. In Figure 11.5, the corresponding paths have been drawn in the same

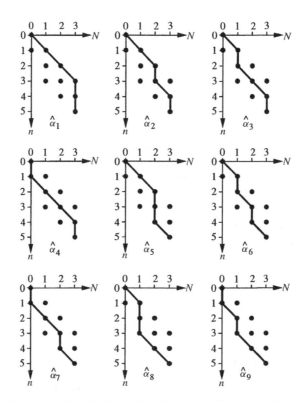

Fig. 11.5. The nine paths representing the allowed RAS strings in Table 11.5. Each path contains three diagonal arcs (representing the occupied orbitals of the string) and two vertical arcs (representing the unoccupied orbitals).

order. The first and last paths thus represent the strings $\hat{\alpha}_1$ and $\hat{\alpha}_9$ – the remaining strings give rise to paths located inside the polygon delimited by these 'extreme' paths. The allowed vertices are located either inside this polygon or on its boundary and are marked by dots.

The purpose of the graphical representation is to develop a simple scheme for resolving the address of a given string from its occupations. Before such a scheme can be devised, we must establish a canonical ordering of the strings. We shall here use *reverse lexical ordering*. In this ordering scheme, string A comes before string B if, in the last occupation where they differ, A has the lower orbital number. Thus, in Table 11.5, $a_1^\dagger a_2^\dagger a_4^\dagger$ comes before $a_1^\dagger a_3^\dagger a_5^\dagger$ since 4 is smaller than 5; likewise, $a_1^\dagger a_2^\dagger a_4^\dagger$ comes before $a_1^\dagger a_3^\dagger a_4^\dagger$ since 2 is smaller than 3. The strings and paths in Table 11.5 and in Figure 11.5 are listed in the reverse lexical order. To generate the strings in this order, we start with the string that has the lowest possible orbital indices in ascending order – in our case, $a_1^\dagger a_2^\dagger a_3^\dagger$. To generate the next string, we identify the lowest index that may be raised without affecting the ascending order or the RAS constraints. This index is then raised and all indices to its left are set to their lowest possible values. In our example, the second string becomes $a_1^\dagger a_2^\dagger a_4^\dagger$. Repeating this procedure, we obtain first $a_1^\dagger a_3^\dagger a_4^\dagger$ and next the remaining strings in the reverse lexical order.

Let us now see how the ordering number of a given string (its address) can be obtained from its occupations – that is, from the positions of the diagonal arcs in its path. We begin by associating with each allowed vertex (k, m) a *vertex weight* $W_{k,m}$, equal to the number of different paths from $(0,0)$ to (k, m). Since all such paths must come from either $(k - 1, m - 1)$ or $(k - 1, m)$, we have the following recurrence relation between the vertex weights

$$W_{k,m} = W_{k-1,m} + W_{k-1,m-1} \qquad (11.8.2)$$

where the weights of the (forbidden) vertices outside the polygon are set equal to zero. Starting from the head vertex with weight $W_{0,0} = 1$, the weights of all allowed vertices are readily obtained from this relation. In Figure 11.6, a string graph containing the vertex weights of the allowed vertices for the RAS strings in Table 11.5 is given.

Next, we introduce *arc weights* in such a manner that the ordering number of a path (the *path weight*) can be obtained by summing the weights of each arc in the path:

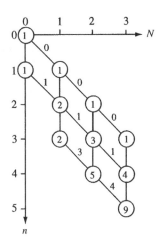

Fig. 11.6. String graph with vertex and arc weights of the graphs for the RAS strings in Table 11.5.

$$I_{m_1,m_2,\dots m_n} = 1 + \sum_{k=1}^{n} Y_{k,m_k}^{m_k-m_{k-1}} \tag{11.8.3}$$

Here $Y_{k,m}^0$ is the weight of the vertical arc ending at (k, m) and $Y_{k,m}^1$ the weight of the diagonal arc ending at (k, m). Since the unoccupied orbitals do not contribute to the ordering of the strings, we set the weights of the vertical arcs equal to zero:

$$Y_{k,m}^0 = 0 \tag{11.8.4}$$

To establish the weights of the diagonal arcs, we note that the subgraph with head (0,0) and tail (k, m) contains $W_{k,m}$ paths with weights from 1 to $W_{k,m}$. Likewise, in the extended subgraph from (0,0) to $(k+1, m+1)$, there are $W_{k+1,m+1}$ paths with weights from 1 to $W_{k+1,m+1}$. In our convention – that is, in the reverse lexical ordering scheme – the paths with the highest weights are those with the highest associated occupied orbital indices. Of the paths in the extended subgraph, the one with the highest weight $W_{k+1,m+1}$ must therefore correspond to the path that has orbital $k + 1$ occupied and which is connected to the path with the highest weight $W_{k,m}$ in the smaller subgraph. We thus have the relation

$$W_{k+1,m+1} = W_{k,m} + Y_{k+1,m+1}^1 \tag{11.8.5}$$

so that

$$Y_{k+1,m+1}^1 = W_{k+1,m+1} - W_{k,m} = W_{k,m+1} \tag{11.8.6}$$

where we have used (11.8.2). In Figure 11.6, we have given the weights of the diagonal arcs for the RAS example discussed above. Note that the weight of each arc is equal to the weight of the vertex located diagonally to the right and above the arc.

Once the arc weights have been determined, the path weights – that is, the ordering numbers of the strings – are obtained from the arc weights according to (11.8.3). We have thus arrived at a simple addressing scheme, in which the address of a given string is calculated as a simple sum of arc weights, one for each occupied orbital. Each arc weight depends only on the orbital number and its position in the string (see Figure 11.6) and is easily retrieved from a small two-dimensional array. Using this scheme, the reverse lexical ordering of Table 11.5 may be readily verified.

It is sometimes necessary to depart from the reverse lexical order. In RAS calculations, for example, we may wish to sort the strings so as to reflect the class structure. The reverse lexical ordering then serves as the first step in the indexing of the strings, in terms of which any further reordering may be accomplished by setting up pointers.

11.8.3 THE N-RESOLUTION METHOD

For the purpose of developing direct CI algorithms, it is convenient to rewrite the spin-free Hamiltonian operator (2.2.18) slightly as

$$\hat{H} = \sum_{pq} h_{pq} E_{pq} + \tfrac{1}{2} \sum_{pqrs} g_{pqrs}(E_{pq}E_{rs} - \delta_{rq}E_{ps})$$

$$= \sum_{pq} k_{pq} E_{pq} + \tfrac{1}{2} \sum_{pqrs} g_{pqrs} E_{pq} E_{rs} \tag{11.8.7}$$

where we have introduced the effective one-electron integrals

$$k_{pq} = h_{pq} - \tfrac{1}{2} \sum_{r} g_{prrq} \tag{11.8.8}$$

and disregarded the nuclear–nuclear contributions. The sigma vector (11.8.1) may now be written as

$$\sigma_{I_\alpha I_\beta} = \sigma_{I_\alpha I_\beta}^{(1)} + \sigma_{I_\alpha I_\beta}^{(2)} \tag{11.8.9}$$

where

$$\sigma_{I_\alpha I_\beta}^{(1)} = \sum_{pq} \sum_{J_\alpha J_\beta} k_{pq} \langle I_\alpha I_\beta | E_{pq} | J_\alpha J_\beta \rangle C_{J_\alpha J_\beta} \tag{11.8.10}$$

$$\sigma_{I_\alpha I_\beta}^{(2)} = \frac{1}{2} \sum_{pqrs} \sum_{J_\alpha J_\beta} g_{pqrs} \langle I_\alpha I_\beta | E_{pq} E_{rs} | J_\alpha J_\beta \rangle C_{J_\alpha J_\beta} \tag{11.8.11}$$

We shall here concentrate on the calculation of the two-electron contribution to the sigma vector (11.8.11) since the simpler one-electron contribution (11.8.10) is easily obtained as a by-product of the two-electron calculation. We begin by inserting the resolution of the identity between the excitation operators in (11.8.11). We then obtain the following expression for the two-electron contribution

$$\sigma_{I_\alpha I_\beta}^{(2)} = \frac{1}{2} \sum_{\substack{K_\alpha K_\beta J_\alpha J_\beta \\ pqrs}} \langle I_\alpha I_\beta | E_{pq} | K_\alpha K_\beta \rangle g_{pqrs} \langle K_\alpha K_\beta | E_{rs} | J_\alpha J_\beta \rangle C_{J_\alpha J_\beta} \tag{11.8.12}$$

where K_α and K_β run over all alpha and beta strings. Since the resolution of the identity has been applied at the N-electron level, the resulting algorithm is known as the *N-resolution method* [12].

The N-resolution method is now developed by rewriting (11.8.12) as a sequence of three partial summations:

$$D_{rs,K_\alpha K_\beta} = \sum_{J_\alpha J_\beta} \langle K_\alpha K_\beta | E_{rs} | J_\alpha J_\beta \rangle C_{J_\alpha J_\beta} \tag{11.8.13}$$

$$G_{pq,K_\alpha K_\beta} = \frac{1}{2} \sum_{rs} g_{pqrs} D_{rs,K_\alpha K_\beta} \tag{11.8.14}$$

$$\sigma_{I_\alpha I_\beta}^{(2)} = \sum_{pq,K_\alpha K_\beta} \langle I_\alpha I_\beta | E_{pq} | K_\alpha K_\beta \rangle G_{pq,K_\alpha K_\beta} \tag{11.8.15}$$

In the first of these three steps, the matrix **D** is generated. Separating the excitation operator E_{rs} into alpha and beta parts (11.7.10), we may write this matrix as a sum of two distinct contributions:

$$D_{rs,K_\alpha K_\beta} = \sum_{J_\beta} \langle K_\beta | E_{rs}^\beta | J_\beta \rangle C_{K_\alpha J_\beta} + \sum_{J_\alpha} \langle K_\alpha | E_{rs}^\alpha | J_\alpha \rangle C_{J_\alpha K_\beta} \tag{11.8.16}$$

To evaluate the first contribution, the addressing rules of Section 11.8.2 are employed to identify the nonzero matrix elements $\langle K_\beta | E_{rs}^\beta | J_\beta \rangle$. For each nonzero element, the elements of $C_{K_\alpha J_\beta}$ with a fixed index J_β constitute a vector, the elements of which are addressed by K_α. This vector is then scaled by the nonzero element $\langle K_\beta | E_{rs}^\beta | J_\beta \rangle$ and added to the corresponding vector obtained by fixing the indices rs and K_β in $D_{rs,K_\alpha K_\beta}$. The same method is used for the second contribution in (11.8.16). This procedure, which uses elementary vector operations, is well suited to modern computers. The number of operations required to construct the matrix **D** is equal to the number of nonzero matrix elements $\langle K_\alpha K_\beta | E_{rs} | J_\alpha J_\beta \rangle$, which is given by L_1 of (11.7.11).

In the second, time-consuming step of the N-resolution method (11.8.14), the matrix **D** from the first step (11.8.13) is multiplied by the two-electron integrals in a matrix-multiplication step that involves as many as $n^4 N_{\text{det}}$ operations. Although $\mathbf{D}$ is a sparse matrix, this sparsity cannot easily be exploited to reduce the cost of the matrix-multiplication step. However, exploiting the permutational symmetry of the real two-electron integrals, we may rewrite (11.8.14) as

$$G_{pq,K_\alpha K_\beta} = \tfrac{1}{2} \sum_{r \geq s} g_{pqrs}(D_{rs,K_\alpha K_\beta} + D_{sr,K_\alpha K_\beta})(1 + \delta_{sr})^{-1} \qquad (11.8.17)$$

which reduces the number of operations by a factor of 2. The permutational symmetry of $G_{pq,K_\alpha K_\beta}$ in pq may be used to reduce the overall cost of this step by a further factor of 2, yielding a total of $n^4 N_{\text{det}}/4$ operations.

In the third and final step of the N-resolution method, the matrix $\mathbf{G}$ from the second step is contracted with the matrix elements of the E_{pq} operator (11.8.15). This step is similar to the first step and requires L_1 operations. The calculation of the two-electron contribution to the sigma vector therefore proceeds in three steps, two of which require L_1 operations and one that requires $n^4 N_{\text{det}}/4$ operations. We note that the one-electron contribution to the sigma vector may be written in the form

$$\sigma^{(1)}_{I_\alpha I_\beta} = \sum_{pq} k_{pq} D_{pq,I_\alpha I_\beta} \qquad (11.8.18)$$

and may thus be obtained by a multiplication of the matrix (11.8.13) by the effective one-electron integrals (11.8.8) at little extra cost ($n^2 N_{\text{det}}$).

From the discussion in this subsection, we conclude that the total cost of calculating the sigma vector by the N-resolution method is

$$M_N = \tfrac{1}{4} n^4 N_{\text{det}} \qquad (11.8.19)$$

neglecting all terms of lower orders in n. It is of interest to compare this cost with the theoretical lower bound of Section 11.7. For closed shells, the number of nonzero elements in the Hamiltonian matrix is given by $\tfrac{3}{8} N^2 (n - N/2)^2 N_{\text{det}}$, where $N/2 = N_\alpha = N_\beta$; see (11.7.16). For $N = n$, the N-resolution method thus requires two to three times as many operations as the theoretical limit. When the number of electrons N is much smaller or much larger than the number of orbitals n, then M_N becomes even larger relative to the theoretical minimum. The poor performance of the N-resolution method arises since the sparseness of the matrix $D_{rs,K_\alpha K_\beta}$ is not exploited in the time-consuming second step (11.8.14). Moreover, in the same step, we generate many elements $G_{pq,K_\alpha K_\beta}$ that are not needed in the third step (11.8.15) since, in a given $|K_\alpha K_\beta\rangle$, p must be unoccupied and q occupied to give nonzero contributions to the sigma vector.

11.8.4 THE MINIMAL OPERATION-COUNT METHOD

In the time-consuming step of the N-resolution method, we did not exploit the separation of excitation operators into alpha and beta spin parts as in (11.7.10). We shall now consider a different method that more fully exploits the separation into the alpha and beta spin spaces. This algorithm is known as the *minimal operation-count (MOC) method* [4] since it yields an operation count that, to leading orders in the numbers of electrons and orbitals, is identical to the theoretical minimum (11.7.16).

Introducing the spin-separated excitation operators (11.7.10) in the one-electron part of the sigma vector (11.8.10), we obtain

$$\sigma^{(1)}_{I_\alpha I_\beta} = \sigma^\alpha_{I_\alpha I_\beta} + \sigma^\beta_{I_\alpha I_\beta} \qquad (11.8.20)$$

where, for example,

$$\sigma^\alpha_{I_\alpha I_\beta} = \sum_{J_\alpha J_\beta} \sum_{pq} k_{pq} \langle I_\alpha I_\beta | E^\alpha_{pq} | J_\alpha J_\beta \rangle C_{J_\alpha J_\beta} = \sum_{J_\alpha J_\beta} \sum_{pq} k_{pq} \langle I_\alpha | E^\alpha_{pq} | J_\alpha \rangle \delta_{I_\beta J_\beta} C_{J_\alpha J_\beta} \tag{11.8.21}$$

The one-electron alpha and beta contributions to the sigma vector may then be written as

$$\sigma^\alpha_{I_\alpha I_\beta} = \sum_{J_\alpha} \sum_{pq} k_{pq} \langle I_\alpha | E^\alpha_{pq} | J_\alpha \rangle C_{J_\alpha I_\beta} \tag{11.8.22}$$

$$\sigma^\beta_{I_\alpha I_\beta} = \sum_{J_\beta} \sum_{pq} k_{pq} \langle I_\beta | E^\beta_{pq} | J_\beta \rangle C_{I_\alpha J_\beta} \tag{11.8.23}$$

and we may treat the alpha and beta excitations separately.

We now introduce the matrix representations of the one-electron operator in the alpha and beta spaces

$$k^\alpha_{I_\alpha J_\alpha} = \sum_{pq} k_{pq} \langle I_\alpha | E^\alpha_{pq} | J_\alpha \rangle \tag{11.8.24}$$

$$k^\beta_{I_\beta J_\beta} = \sum_{pq} k_{pq} \langle I_\beta | E^\beta_{pq} | J_\beta \rangle \tag{11.8.25}$$

where the elements of the excitation operators are again easily obtained within the small alpha and beta spaces. These matrices are sparse – for example, each row and column in (11.8.24) contains only $1 + N_\alpha(n - N_\alpha)$ nonzero elements. In terms of these matrices, the one-electron contributions to the sigma vector may be written in the form

$$\sigma^\alpha_{I_\alpha I_\beta} = \sum_{J_\alpha} k^\alpha_{I_\alpha J_\alpha} C_{J_\alpha I_\beta} \tag{11.8.26}$$

$$\sigma^\beta_{I_\alpha I_\beta} = \sum_{J_\beta} k^\beta_{I_\beta J_\beta} C_{I_\alpha J_\beta} \tag{11.8.27}$$

The sparsity of the matrices (11.8.24) and (11.8.25) may be exploited in the following manner. To generate, for example, all the elements of row I_α of σ^α in (11.8.26), we identify in turn each nonzero element (I_α, J_α) of $\mathbf{k}^\alpha$ and calculate its contributions to σ^α by multiplying it by the elements of row J_α of $\mathbf{C}$. This process requires $1 + N_\alpha(n - N_\alpha)$ operations for each of the N_{det} elements of the sigma vector. Adding also the operation count for σ^β, we obtain

$$M^{(1)}_{\text{MOC}} = [N_\alpha(n - N_\alpha) + N_\beta(n - N_\beta)]N_{\text{det}} \tag{11.8.28}$$

where we have retained leading terms only. The operation count for the one-electron contribution (11.8.20) is, to leading order, identical to the theoretical lower limit L_1 in (11.7.11). This low count has been attained by carrying out the alpha and beta transformations (11.8.26) and (11.8.27) separately in the small alpha and beta spaces, where the nonzero elements are easily identified. Since each nonzero matrix element $k^\alpha_{I_\alpha J_\alpha}$ is used for all beta strings, the construction of $k^\alpha_{I_\alpha J_\alpha}$ constitutes only a small part of the computation.

Turning to the more time-critical two-electron part of the transformation, we note that the sigma vector may be written in terms of three contributions:

$$\sigma^{(2)}_{I_\alpha I_\beta} = \sigma^{\alpha\alpha}_{I_\alpha I_\beta} + \sigma^{\beta\beta}_{I_\alpha I_\beta} + \sigma^{\alpha\beta}_{I_\alpha I_\beta} \tag{11.8.29}$$

where

$$\sigma_{I_\alpha I_\beta}^{\alpha\alpha} = \frac{1}{2} \sum_{J_\alpha} \sum_{pqrs} g_{pqrs} \langle I_\alpha | E_{pq}^\alpha E_{rs}^\alpha | J_\alpha \rangle C_{J_\alpha I_\beta} \tag{11.8.30}$$

$$\sigma_{I_\alpha I_\beta}^{\beta\beta} = \frac{1}{2} \sum_{J_\beta} \sum_{pqrs} g_{pqrs} \langle I_\beta | E_{pq}^\beta E_{rs}^\beta | J_\beta \rangle C_{I_\alpha J_\beta} \tag{11.8.31}$$

$$\sigma_{I_\alpha I_\beta}^{\alpha\beta} = \sum_{J_\alpha J_\beta} \sum_{pqrs} g_{pqrs} \langle I_\alpha | E_{pq}^\alpha | J_\alpha \rangle \langle I_\beta | E_{rs}^\beta | J_\beta \rangle C_{J_\alpha J_\beta} \tag{11.8.32}$$

The alpha–alpha and beta–beta contributions may be calculated in much the same way as the corresponding one-electron contributions, but complications arise in the alpha–beta contribution, where the alpha and beta excitations are coupled.

Concentrating first on the alpha–alpha and beta–beta contributions (11.8.30) and (11.8.31), we follow the same procedure as in the one-electron case and introduce the matrix representations of the two-electron operator in the alpha and beta spaces:

$$G_{I_\alpha J_\alpha}^\alpha = \frac{1}{2} \sum_{pqrs} g_{pqrs} \langle I_\alpha | E_{pq}^\alpha E_{rs}^\alpha | J_\alpha \rangle \tag{11.8.33}$$

$$G_{I_\beta J_\beta}^\beta = \frac{1}{2} \sum_{pqrs} g_{pqrs} \langle I_\beta | E_{pq}^\beta E_{rs}^\beta | J_\beta \rangle \tag{11.8.34}$$

Like their one-electron counterparts, these matrices are sparse. From the discussion in Section 11.7, their rows and columns contain, to leading orders, $N_\sigma^2 (n - N_\sigma)^2/4$ nonzero elements. Since these elements are easily identified, each transformation

$$\sigma_{I_\alpha I_\beta}^{\alpha\alpha} = \sum_{J_\alpha} G_{I_\alpha J_\alpha}^\alpha C_{J_\alpha I_\beta} \tag{11.8.35}$$

$$\sigma_{I_\alpha I_\beta}^{\beta\beta} = \sum_{J_\beta} G_{I_\beta J_\beta}^\beta C_{I_\alpha J_\beta} \tag{11.8.36}$$

may be carried out with costs

$$M_{\text{MOC}}^{\alpha\alpha} = \frac{1}{4} N_\alpha^2 (n - N_\alpha)^2 N_{\text{det}} \tag{11.8.37}$$

$$M_{\text{MOC}}^{\beta\beta} = \frac{1}{4} N_\beta^2 (n - N_\beta)^2 N_{\text{det}} \tag{11.8.38}$$

where again we have retained only the leading terms.

The mixed alpha–beta term (11.8.32) is more complicated than the other two terms and is most easily treated by rewriting it in the form

$$\sigma_{I_\alpha I_\beta}^{\alpha\beta} = \sum_{pq} \sigma_{I_\alpha I_\beta}^{\alpha\beta}[pq] \tag{11.8.39}$$

where

$$\sigma_{I_\alpha I_\beta}^{\alpha\beta}[pq] = \sum_{J_\beta} \left[\sum_{J_\alpha} \langle I_\alpha | E_{pq}^\alpha | J_\alpha \rangle C_{J_\alpha J_\beta} \right] \left[\sum_{rs} g_{pqrs} \langle I_\beta | E_{rs}^\beta | J_\beta \rangle \right] \tag{11.8.40}$$

Introducing the intermediate quantities

$$D_{I_\alpha J_\beta}^\alpha[pq] = \sum_{J_\alpha} \langle I_\alpha | E_{pq}^\alpha | J_\alpha \rangle C_{J_\alpha J_\beta} \tag{11.8.41}$$

$$G_{I_\beta J_\beta}^\beta[pq] = \sum_{rs} g_{pqrs} \langle I_\beta | E_{rs}^\beta | J_\beta \rangle \tag{11.8.42}$$

we obtain the following expression for the individual contributions to (11.8.39):

$$\sigma^{\alpha\beta}_{I_\alpha I_\beta}[pq] = \sum_{J_\beta} D^\alpha_{I_\alpha J_\beta}[pq] G^\beta_{I_\beta J_\beta}[pq] \tag{11.8.43}$$

The elements $D^\alpha_{I_\alpha J_\beta}[pq]$ vanish if the row index I_α corresponds to a string with orbital p unoccupied or orbital q occupied and different from p. To obtain an algorithm with a minimal operation count, we must avoid operations on the vanishing rows, which is easily achieved by restricting the index I_α in $\sigma^{\alpha\beta}_{I_\alpha I_\beta}[pq]$ and $D^\alpha_{I_\alpha J_\beta}[pq]$ to run over only those strings for which $a^\dagger_{q\alpha} a_{p\alpha}|I_\alpha\rangle$ is nonzero. When the pq contribution (11.8.43) is added to the final alpha–beta sigma vector (11.8.39), its rows must be expanded (scattered) from the restricted list of nonzero alpha strings to the complete list of alpha strings.

To determine the cost of evaluating the alpha–beta contribution to the sigma vector (11.8.39), we note that, for a given I_α, the number of index pairs pq with nonzero contributions is $1 + N_\alpha(n - N_\alpha)$. Retaining only the leading terms, the number of nonzero elements $\sigma^{\alpha\beta}_{I_\alpha I_\beta}[pq]$ is then $N_\alpha(n - N_\alpha)N_{\text{det}}$. For each element $\sigma^{\alpha\beta}_{I_\alpha I_\beta}[pq]$, there are $1 + N_\beta(n - N_\beta)$ terms in the summation over J_β in (11.8.43). The cost for all alpha–beta contributions (11.8.43) is therefore

$$M^{\alpha\beta}_{\text{MOC}} = N_\alpha N_\beta(n - N_\alpha)(n - N_\beta)N_{\text{det}} \tag{11.8.44}$$

The cost of evaluating the intermediates (11.8.41) and (11.8.42) is smaller, as is easily verified. Combining the operations counts for (11.8.37), (11.8.38) and (11.8.44), we find that the overall cost of evaluating the two-electron contribution to the sigma vector is

$$M^{(2)}_{\text{MOC}} = \left[\tfrac{1}{4}N^2_\alpha(n - N_\alpha)^2 + \tfrac{1}{4}N^2_\beta(n - N_\beta)^2 + N_\alpha N_\beta(n - N_\alpha)(n - N_\beta)\right]N_{\text{det}} \tag{11.8.45}$$

which is identical to the theoretical limit (11.7.16). We have thus set up an algorithm that, in the leading terms, has the lowest possible operation count – that is, one multiplication and one addition for each nonzero element in the determinantal representation of the Hamiltonian.

The key to the success of the MOC algorithm is the use of string excitation matrix elements of the form $\langle I_\sigma|E^\sigma_{pq}|J_\sigma\rangle$ and $\langle I_\sigma|E^\sigma_{pq}E^\sigma_{rs}|J_\sigma\rangle$, which ensures that redundant operations are avoided. For a given string $|I_\sigma\rangle$ and for a given excitation operator E^σ_{pq}, the occupations of $|J_\sigma\rangle = E^\sigma_{pq}|I_\sigma\rangle$ can straightforwardly be obtained, and the graphical representation and ordering of the strings can then be used to resolve its address. This approach allows for an efficient evaluation of the nonzero matrix elements $\langle I_\sigma|E^\sigma_{pq}|J_\sigma\rangle$ and, by a straightforward extension, $\langle I_\sigma|E^\sigma_{pq}E^\sigma_{rs}|J_\sigma\rangle$. Moreover, since the evaluated matrix elements are reused many times, their evaluation takes only a fraction of the time required for a direct CI iteration.

The MOC algorithm performs poorly when the number of electrons of alpha spin is small (one or two). With a single alpha electron, the only string for which $a^\dagger_{q\alpha} a_{p\alpha}|I_\alpha\rangle$ is nonzero is $I_\alpha = a^\dagger_{p\alpha}|\text{vac}\rangle$. The operation count for (11.8.42) then becomes comparable with the operation count for (11.8.43). Moreover, each element $G^\beta_{I_\beta J_\beta}[pq]$ is only used once in (11.8.43) and the efficiency of the MOC algorithm is greatly reduced. A similar degradation in performance occurs when the number of alpha electrons is close to the total number of orbitals. In [16], an efficient algorithm is presented for the case where the number of electrons is significantly smaller than the number of orbitals.

In Table 11.8, we compare the operation count of the MOC method with that of the N-resolution method discussed in Section 11.8.3. In all cases, the MOC method is the more efficient one in terms of floating-point operations, requiring (to leading order) no more than the theoretical limit. The

Table 11.8 Cost of the N-resolution method relative to that of the MOC method for different numbers of alpha and beta electrons relative to the number of orbitals n

$N_\alpha/n = N_\beta/n$	Cost of N-resolution method/ cost of MOC method
1/10	21
1/4	4.7
1/2	2.7
3/4	4.7
9/10	21

performance of the N-resolution method is particularly poor for few or many electrons relative to the number of orbitals. It should be noted, however, that it is often difficult to judge the efficiency of an algorithm from the operation count alone. An algorithm that requires few operations may not always be the fastest, in particular when the operation counts differ by a factor of 2 or less. Issues related to computational overhead, vectorization and parallelization may offset the advantages of fewer operations. In general, however, MOC is the method of choice for FCI determinant calculations.

11.8.5 DIRECT CI ALGORITHMS FOR RAS CALCULATIONS

In the present subsection, we discuss how the FCI algorithms developed in Section 11.8.4 can be modified for RAS expansions, where restrictions are imposed on the occupations of the active orbitals. As an example, we shall consider RAS expansions that contain all single and double excitations from a closed-shell Hartree–Fock state (CISD). The RAS active orbitals are then divided into occupied and virtual orbitals, with the restriction that no more than two electrons may occupy the virtual orbitals.

As discussed in Section 11.6, RAS strings are sorted in classes, each characterized by the number of electrons present in the different orbital spaces. In CISD theory, there are three such classes: class 0, with no electrons in the virtual space; class 1, with one electron in the virtual space; and class 2, with two electrons in the virtual space. The total number of alpha or beta strings is then

$$N_{\text{str}} = 1 + OV + \tfrac{1}{4}O(O-1)V(V-1) \tag{11.8.46}$$

where O is the number of occupied orbitals and V the number of virtual orbitals. The RAS CI coefficients are likewise organized in matrices or blocks with elements $C^{st}_{I^s_\alpha, I^t_\beta}$, where s and t label the alpha and beta classes and I^s_α and I^t_β represent the strings of these classes. In the CISD case, there are six blocks ($\mathbf{C}^{00}$, $\mathbf{C}^{01}$, $\mathbf{C}^{10}$, $\mathbf{C}^{11}$, $\mathbf{C}^{20}$ and $\mathbf{C}^{02}$) and the total number of determinants is

$$N_{\text{det}} = 1 + 2OV + O^2V^2 + \tfrac{1}{2}O(O-1)V(V-1) \tag{11.8.47}$$

The key step of a direct CI iteration, the contraction of the Hamiltonian with a trial vector, is separated into contributions from a given block of $\mathbf{C}$ to a given block of $\boldsymbol{\sigma}$

$$\sigma^{st}_{I^s_\alpha I^t_\beta} = \sum_{\substack{mn \\ J^u_\alpha J^v_\beta}} \langle I^s_\alpha I^t_\beta | E_{mn} | J^u_\alpha J^v_\beta \rangle C^{uv}_{J^u_\alpha J^v_\beta} h_{mn} + \frac{1}{2}\sum_{\substack{mnpq \\ J^u_\alpha J^v_\beta}} \langle I^s_\alpha I^t_\beta | E_{mn} E_{pq} - \delta_{np} E_{mq} | J^u_\alpha J^v_\beta \rangle C^{uv}_{J^u_\alpha J^v_\beta} g_{mnpq} \tag{11.8.48}$$

In this expression, only those excitations that change the classes of $\mathbf{C}^{uv}$ to those of $\boldsymbol{\sigma}^{st}$ will contribute. Thus, a given block $\mathbf{C}^{uv}$ makes no contributions to $\boldsymbol{\sigma}^{st}$ if the classes differ by more than a double excitation. In the CISD case, for example, σ^{20} receives no contribution from $\mathbf{C}^{02}$ since a fourfold excitation is needed to match the string classes of these blocks. Another consequence of the class structure is that, for a given pair of sigma and coefficient blocks, only certain integrals may contribute. Thus, for CISD wave functions, the only excitations that connect the strings of $\mathbf{C}^{00}$ and σ^{02} are of the type $E_{ai}E_{ai}$, so only the integrals g_{aiai} may contribute.

To develop efficient RAS algorithms, several new problems must be solved. Compare first the number of strings in FCI and RAS calculations. In an FCI expansion with $M = 0$, the number of strings is equal to the square root of the number of Slater determinants. From a small number of strings, we can thus obtain an efficient representation of a much larger number of determinants. In RAS calculations, the ratio between the number of strings and the number of determinants is often less favourable. In CISD theory, for example, the number of strings (11.8.46) is $O^2V^2/4$ and the number of determinants (11.8.47) is $3O^2V^2/2$, so their numbers are comparable. Moreover, in RAS theory, the number of strings can become so large that the strings cannot be stored and a more compact representation of the information in the strings is needed. We shall return to this problem shortly.

Another complication in RAS theory is the presence of blocks such as $\mathbf{C}^{02}$ that contain a large number of beta strings but few alpha strings – in our case one. For such blocks, the MOC algorithm of Section 11.8.4 becomes inefficient. Consider the contribution to σ^{02} from $\mathbf{C}^{02}$ that involves two-electron integrals with four virtual indices

$$^{4v}\sigma^{02}_{I^0_\alpha I^2_\beta} = \frac{1}{2}\sum_{\substack{abcd \\ J^2_\beta}} \langle I^2_\beta | a^\dagger_{a\beta} a^\dagger_{c\beta} a_{d\beta} a_{b\beta} | J^2_\beta \rangle g_{abcd} C^{02}_{I^0_\alpha J^2_\beta} \tag{11.8.49}$$

For convenience, the two-electron excitation operator has been expressed in terms of creation and annihilation operators (2.2.16). In the MOC algorithm, this contribution is evaluated by first constructing the intermediate matrix

$$G_{I^2_\beta J^2_\beta} = \frac{1}{2}\sum_{abcd} \langle I^2_\beta | a^\dagger_{a\beta} a^\dagger_{c\beta} a_{d\beta} a_{b\beta} | J^2_\beta \rangle g_{abcd} \tag{11.8.50}$$

which is subsequently contracted with the coefficients

$$^{4v}\sigma^{02}_{I^0_\alpha I^2_\beta} = \sum_{J^2_\beta} G_{I^2_\beta J^2_\beta} C^{02}_{I^0_\alpha J^2_\beta} \tag{11.8.51}$$

For this procedure to be efficient, each element $G_{I^2_\beta J^2_\beta}$ must be used with many alpha strings, reducing the evaluation of (11.8.50) to a fraction of the total work. With only one alpha string as above, the evaluation of $\mathbf{G}$ becomes the time-consuming step.

Several algorithms may be devised for an efficient processing of such terms. The one to be described here eliminates the need for constructing and storing the large number of strings that may occur in RAS calculations. Recall that, so far, strings have been used to specify the occupations of the full set of orbitals – in CISD theory, for example, a single string specifies the occupations of both the occupied and the virtual orbitals. Instead, we may introduce separate strings for the different orbital spaces. In the CISD case, we thus introduce one set of strings for the occupied orbitals and another set of strings for the virtual orbitals. The $O(O - 1)V(V - 1)/4$ strings (with two holes in the occupied space and two electrons in the virtual space) are then replaced by $O(O - 1)/2$ strings for the occupied space and $V(V - 1)/2$ strings for the virtual space. The number of strings that must be generated and stored is thereby significantly reduced, becoming again much smaller than the number of Slater determinants. This separation of strings into *substrings for orbital subspaces* is a generalization of the idea of specifying determinants as products of spin strings – just like the spin strings were introduced for a compact representation of determinants, the substrings are introduced for a compact representation of the strings in the full orbital space.

The separation of strings into substrings can be utilized to make the CI algorithms more efficient. We first note that each block of the coefficient matrix now becomes multidimensional. For example, the $\mathbf{C}^{02}$ block has the elements $C^{02}_{I^{o(0)}_\alpha I^{v(0)}_\alpha I^{o2}_\beta I^{v(2)}_\beta}$, where $I^{o(n)}_\sigma$ are the strings in the occupied space with n holes and $I^{v(n)}_\sigma$ the strings in the virtual space with n electrons. Next, consider again the evaluation of (11.8.49). Since the excitation operators in the virtual space commute with strings containing occupied orbitals, we obtain

$$^{4v}\sigma^{02}_{I^{o(0)}_\alpha I^{v(0)}_\alpha I^{o2}_\beta I^{v(2)}_\beta} = \frac{1}{2} \sum_{\substack{abcd \\ J^{v(2)}_\beta}} \langle I^{v(2)}_\beta | a^\dagger_{a\beta} a^\dagger_{c\beta} a_{d\beta} a_{b\beta} | J^{v(2)}_\beta \rangle C^{02}_{I^{o(0)}_\alpha I^{v(0)}_\alpha I^{o2}_\beta J^{v(2)}_\beta} g_{abcd} \qquad (11.8.52)$$

This contribution to σ^{02} can be obtained by first constructing the excitations for the virtual strings

$$G_{I^{v(2)}_\beta J^{v(2)}_\beta} = \frac{1}{2} \sum_{abcd} \langle I^{v(2)}_\beta | a^\dagger_{a\beta} a^\dagger_{c\beta} a_{d\beta} a_{b\beta} | J^{v(2)}_\beta \rangle g_{abcd} \qquad (11.8.53)$$

and then contracting with the coefficient matrix

$$^{4v}\sigma^{02}_{I^{o(0)}_\alpha I^{v(0)}_\alpha I^{o2}_\beta I^{v(2)}_\beta} = \sum_{J^{v(2)}_\beta} G_{I^{v(2)}_\beta J^{v(2)}_\beta} C^{02}_{I^{o(0)}_\alpha I^{v(0)}_\alpha I^{o2}_\beta J^{v(2)}_\beta} \qquad (11.8.54)$$

Note the similarity with the separation of determinants into spin strings: the excitation operators affect only one of the two subspace beta strings and the unaffected string is removed from the second-quantization matrix element. Having split the strings into two classes, we need to construct $\mathbf{G}$ only in the subspace of $V(V-1)/2$ virtual strings, whereas the original algorithm required its construction in the full space of $O(O-1)V(V-1)/4$ doubly excited strings. Moreover, a given $G_{I^{v(2)}_\beta J^{v(2)}_\beta}$ is now reused for $O(O-1)/2$ different strings, whereas, in the original algorithm, a given element $G_{I^2_\beta J^2_\beta}$ was used for a single string.

For strings containing only a few electrons, the $\mathbf{G}$ matrices have a particularly simple form, which can be used to further simplify the computations. Consider, for example, the elements $G_{I^{v(2)}_\beta J^{v(2)}_\beta}$. The strings with two electrons in the virtual space are the following

$$|I^{v(2)}_\beta\rangle = a^\dagger_{e\beta} a^\dagger_{f\beta} |\text{vac}\rangle, \quad e < f \qquad (11.8.55)$$

$$|J^{v(2)}_\beta\rangle = a^\dagger_{g\beta} a^\dagger_{h\beta} |\text{vac}\rangle, \quad g < h \qquad (11.8.56)$$

and the elements of $\mathbf{G}$ become

$$G_{efgh} = \frac{1}{2} \sum_{abcd} \langle \text{vac} | a_{f\beta} a_{e\beta} a^\dagger_{a\beta} a^\dagger_{c\beta} a_{d\beta} a_{b\beta} a^\dagger_{g\beta} a^\dagger_{h\beta} |\text{vac}\rangle g_{abcd} = g_{egfh} - g_{ehfg} \qquad (11.8.57)$$

The contribution to σ^{02} can then be written directly in terms of the CI coefficients and the integrals as

$$^{4v}\sigma^{02}_{I^{o(0)}_\alpha I^{v(0)}_\alpha I^{o2}_\beta, ef} = \sum_{g<h} C^{02}_{I^{o(0)}_\alpha I^{v(0)}_\alpha I^{o2}_\beta, gh} (g_{egfh} - g_{ehfg}) \qquad (11.8.58)$$

These examples should suffice to illustrate how the separation of strings into substrings and the use of explicit expressions for matrix elements between strings containing only one or two electrons allow us to develop efficient RAS algorithms.

11.8.6 SIMPLIFICATIONS FOR WAVE FUNCTIONS OF ZERO PROJECTED SPIN

For states with an even number of electrons, the CI calculation may be carried out on the component with zero spin projection M. In the present subsection, we show that this component of the wave function has an additional symmetry that may be used to reduce the number of independent parameters and the cost of a direct CI iteration.

We begin by introducing the unitary operator [15]

$$\hat{U}_{\alpha\beta} = \exp[i\pi(\hat{S}_x - \tfrac{1}{2}\hat{N})] \tag{11.8.59}$$

which interchanges alpha and beta spins (see Exercise 3.4)

$$\hat{U}_{\alpha\beta} a^\dagger_{p\alpha} \hat{U}^\dagger_{\alpha\beta} = a^\dagger_{p\beta} \tag{11.8.60}$$

$$\hat{U}_{\alpha\beta} a^\dagger_{p\beta} \hat{U}^\dagger_{\alpha\beta} = a^\dagger_{p\alpha} \tag{11.8.61}$$

In Exercise 11.7, it is shown that the effect of $\hat{U}_{\alpha\beta}$ on an N-electron wave function of total spin S and zero spin projection is given by

$$\hat{U}_{\alpha\beta}|0\rangle = (-1)^{N/2-S}|0\rangle \tag{11.8.62}$$

This symmetry arises from the invariance of the electronic wave function to time reversal. The effect of $\hat{U}_{\alpha\beta}$ on a Slater determinant with zero spin projection is readily obtained as

$$\hat{U}_{\alpha\beta}|I_\alpha I_\beta\rangle = \hat{\beta}_{I_\alpha}\hat{\alpha}_{I_\beta}|\text{vac}\rangle = (-1)^{N/2}|I_\beta I_\alpha\rangle \tag{11.8.63}$$

Expanding the CI wave function as

$$|0\rangle = \sum_{I_\alpha I_\beta} C_{I_\alpha I_\beta}|I_\alpha I_\beta\rangle \tag{11.8.64}$$

and applying (11.8.62) and (11.8.63), we obtain

$$C_{I_\alpha I_\beta} = (-1)^S C_{I_\beta I_\alpha} \tag{11.8.65}$$

Introducing the normalized spin combinations

$$|\overline{IJ}\rangle = \begin{cases} |I_\alpha J_\beta\rangle & I = J \\ \dfrac{1}{\sqrt{2}}[|I_\alpha J_\beta\rangle + (-1)^S|J_\alpha I_\beta\rangle] & I > J \end{cases} \tag{11.8.66}$$

we may now expand the CI wave function in the *spin-combined basis* as

$$|0\rangle = \sum_I C_{II}|\overline{II}\rangle + \sqrt{2}\sum_{I>J} C_{IJ}|\overline{IJ}\rangle \tag{11.8.67}$$

effectively reducing the number of independent CI parameters by a factor of 2.

The time-reversal symmetry may be used also to reduce the cost of generating the sigma vector in direct CI. Consider the following element of the sigma vector in the spin-combined basis:

$$\langle\overline{IJ}|\hat{H}|0\rangle = \frac{1}{\sqrt{2}}[\langle I_\alpha J_\beta|\hat{H}|0\rangle + (-1)^S\langle J_\alpha I_\beta|\hat{H}|0\rangle] \tag{11.8.68}$$

Applying (11.8.62) and (11.8.63), we obtain

$$(-1)^S \langle J_\alpha I_\beta | \hat{H} | 0 \rangle = \langle I_\alpha J_\beta | \hat{U}_{\alpha\beta} \hat{H} \hat{U}_{\alpha\beta}^\dagger | 0 \rangle \tag{11.8.69}$$

Since the Hamiltonian $\hat{H}$ is a number-conserving singlet operator, it commutes with both $\hat{N}$ and $\hat{S}_x$ and therefore with $\hat{U}_{\alpha\beta}$. Equation (11.8.69) may then be written as

$$(-1)^S \langle J_\alpha I_\beta | \hat{H} | 0 \rangle = \langle I_\alpha J_\beta | \hat{H} | 0 \rangle \tag{11.8.70}$$

and the elements of the sigma vector in the spin-combined basis becomes

$$\langle \overline{IJ} | \hat{H} | 0 \rangle = \sqrt{2} \; \langle I_\alpha J_\beta | \hat{H} | 0 \rangle \tag{11.8.71}$$

For wave functions with $M = 0$ and a well-defined total spin S, the number of independent variational parameters and the computational cost can thus be reduced by a factor of 2.

11.8.7 DENSITY MATRICES

The one- and two-electron density matrices allow for an efficient evaluation of the expectation values of one- and two-electron operators. Algorithms for the calculation of CI density matrices can be developed by analogy with the direct CI algorithms discussed in Sections 11.8.3–11.8.5. We here restrict our discussion to the calculation of the one-electron density matrix $\mathbf{D}$ for FCI wave functions.

The one-electron density matrix can be written as a sum of two spin components

$$\mathbf{D} = \mathbf{D}^\alpha + \mathbf{D}^\beta \tag{11.8.72}$$

where the beta component is given as

$$D_{pq}^\beta = \langle \mathbf{C} | E_{pq}^\beta | \mathbf{C} \rangle \tag{11.8.73}$$

and similarly for the alpha component. Expanding the CI state $|\mathbf{C}\rangle$ in Slater determinants, we obtain for the beta component

$$D_{pq}^\beta = \sum_{\substack{I_\alpha I_\beta \\ J_\alpha J_\beta}} C_{I_\alpha I_\beta} \langle I_\alpha I_\beta | E_{pq}^\beta | J_\alpha J_\beta \rangle C_{J_\alpha J_\beta} = \sum_{\substack{I_\alpha I_\beta \\ J_\beta}} C_{I_\alpha I_\beta} \langle I_\beta | E_{pq}^\beta | J_\beta \rangle C_{I_\alpha J_\beta} \tag{11.8.74}$$

A given string J_β contributes to the density matrix only if $E_{pq}^\beta | J_\beta \rangle$ is nonzero. The nonzero excitations $E_{pq}^\beta | J_\beta \rangle$ and the resulting beta strings I_β and phase factors $\langle I_\beta | E_{pq}^\beta | J_\beta \rangle$ can be generated using the graphical techniques of Section 11.8.2. The corresponding element of the density matrix is then updated with

$$^{J_\beta} D_{pq}^\beta = \langle I_\beta | E_{pq}^\beta | J_\beta \rangle \sum_{I_\alpha} C_{I_\alpha I_\beta} C_{I_\alpha J_\beta} \tag{11.8.75}$$

The alpha part of the density matrix is obtained in the same manner. The operation count for calculating the one-electron density matrix (11.8.72) is identical to the operation count of one direct CI iteration for a one-electron operator. Likewise, a well-designed algorithm for the construction of the two-electron density matrix will have an operation count identical to the count of a direct CI iteration for a two-electron operator.

11.9 CI orbital transformations

A wave function can often be expressed in different orbital bases. This flexibility may be exploited to express the optimized CI wave function in an orbital basis with particularly attractive features for the analysis of the wave function. Thus, it is often useful to express the wave function in the natural-orbital basis. The transformation to this basis can of course be achieved by repeating the whole calculation in the new basis, but this procedure is rather cumbersome. We here present an alternative, inexpensive scheme for expressing the CI wave function in the new basis, avoiding the reoptimization of the wave function [17].

Assume that we have a set of orbitals $\mathbf{a}^{(0)\dagger}$ that we wish to transform to a different, possibly nonorthogonal set $\mathbf{a}^{(n)\dagger}$:

$$a_{p\sigma}^{(n)\dagger} = \sum_q a_{q\sigma}^{(0)\dagger} D_{qp} \tag{11.9.1}$$

To allow for a general (nonorthogonal) transformation, we do not impose the unitary condition on $\mathbf{D}$ and we shall use the machinery of biorthogonal orbitals developed in Section 1.9. The Slater determinants in the $\mathbf{a}^{(0)\dagger}$ basis are denoted by $|i^{(0)}\rangle$ and the Slater determinants in the final basis by $|i^{(n)}\rangle$. Our goal is then to determine a vector $\mathbf{C}^{(n)}$ such that

$$\sum_i C_i^{(n)}|i^{(n)}\rangle = \sum_i C_i^{(0)}|i^{(0)}\rangle \tag{11.9.2}$$

For a general transformation matrix $\mathbf{D}$, the determinants $|i^{(n)}\rangle$ must span the full configuration space for this identity to be satisfied. However, we shall see that, for certain restricted forms of $\mathbf{D}$, it is possible to satisfy (11.9.2) when $|i^{(0)}\rangle$ and $|i^{(n)}\rangle$ fulfil the same RAS constraints.

The method to be presented proceeds in n steps, where n is the number of orbitals. In each step, only one orbital is changed, and the remaining ones are fixed. In the first step, we replace the first orbital of the initial set by the first orbital of the final set; in the second step, the second orbital is replaced; and so on. Thus, after the kth step of this scheme, the set of orbitals is given by

$$\mathbf{a}^{(k)\dagger} = \begin{pmatrix} a_{1\alpha}^{(n)\dagger} \\ a_{2\alpha}^{(n)\dagger} \\ \vdots \\ a_{k\alpha}^{(n)\dagger} \\ a_{k+1,\alpha}^{(0)\dagger} \\ \vdots \\ a_{n\alpha}^{(0)\dagger} \end{pmatrix} \tag{11.9.3}$$

For simplicity, we here transform only the orbitals that have alpha spin. The orbitals with beta spin may be included in the vector (11.9.3) with no complications except of notation. In each step, we generate a new set of determinants and determine a new set of CI coefficients such that the identity

$$\sum_i C_i^{(k)}|i^{(k)}\rangle = \sum_i C_i^{(k-1)}|i^{(k-1)}\rangle \tag{11.9.4}$$

is fulfilled. For this procedure to work, we must be prepared to work with nonorthogonal orbitals. Even when both the initial and the final sets are orthonormal, the intermediate sets generated in the course of our transformation are nonorthogonal since they contain orbitals from both the initial and final sets.

Let us examine the kth step in more detail. We note that there is a very simple relationship between the orbital sets of steps $k - 1$ and k since all orbitals except orbital k are identical:

$$a_{p\alpha}^{(k-1)\dagger} = a_{p\alpha}^{(k)\dagger}, \quad p \neq k \tag{11.9.5}$$

$$a_{k\alpha}^{(k-1)\dagger} \neq a_{k\alpha}^{(k)\dagger} \tag{11.9.6}$$

It is therefore easy to set up the relationships between the determinants of steps $k - 1$ and k:

$$|i^{(k-1)}\rangle = \begin{cases} |i^{(k)}\rangle & |i\rangle \text{ does not contain orbital } k \\ a_{k\alpha}^{(k-1)\dagger}\bar{a}_{k\alpha}^{(k)}|i^{(k)}\rangle & |i\rangle \text{ contains orbital } k \end{cases} \tag{11.9.7}$$

Since the orbitals are nonorthogonal, we have introduced annihilation operators $\bar{a}_{p\alpha}^{(m)}$ that are biorthogonal to the creation operators in the sense that the anticommutation relations

$$[a_{p\alpha}^{(m)\dagger}, \bar{a}_{q\alpha}^{(m)}]_+ = \delta_{pq} \tag{11.9.8}$$

are satisfied – see Section 1.9. In (11.9.7), all orbitals occupied in $\bar{a}_{k\alpha}^{(k)}|i^{(k)}\rangle$ belong to the orbitals of step $k - 1$, and the corresponding determinant $|i^{(k-1)}\rangle$ of step $k - 1$ is therefore generated simply by applying the creation operator $a_{k\alpha}^{(k-1)\dagger}$.

It is possible to write (11.9.7) as a single equation of the form

$$|i^{(k-1)}\rangle = (1 - a_{k\alpha}^{(k)\dagger}\bar{a}_{k\alpha}^{(k)})|i^{(k)}\rangle + a_{k\alpha}^{(k-1)\dagger}\bar{a}_{k\alpha}^{(k)}|i^{(k)}\rangle \tag{11.9.9}$$

If orbital k is not occupied in the determinant, then only the first term survives and gives the same result as in (11.9.7). Conversely, if orbital k is occupied, then only the second term survives and gives the result of (11.9.7). Expression (11.9.9) therefore holds for all determinants. To make it more useful, we should, on the right-hand side, refer only to quantities of the kth step, avoiding the creation operator belonging to the orbitals of step $k - 1$. To achieve this, we note that we can expand any orbital of step $k - 1$ in terms of the orbitals of step k:

$$a_{k\alpha}^{(k-1)\dagger} = \sum_p a_{p\alpha}^{(k)\dagger} t_{pk} \tag{11.9.10}$$

The calculation of the transformation coefficients t_{pk} will be discussed later. We only note here that, in each step k, we need one set of n transformation coefficients $\{t_{pk}, p = 1, n\}$, the numerical values of which must in some manner depend on the transformation matrix $\mathbf{D}$ in (11.9.1).

Combining (11.9.9) and (11.9.10), we obtain the following relationship between the determinants of steps $k - 1$ and k:

$$|i^{(k-1)}\rangle = \left[1 + \sum_p (t_{pk} - \delta_{pk})a_{p\alpha}^{(k)\dagger}\bar{a}_{k\alpha}^{(k)}\right]|i^{(k)}\rangle \tag{11.9.11}$$

All quantities on the left refer to step $k - 1$ and all quantities on the right to step k. Inserting this expression in the right-hand side of (11.9.4) and rearranging, we obtain an equation for the CI coefficients of step k in terms of the CI coefficients of step $k - 1$ and the transformation coefficients t_{pk}:

$$\sum_i \Delta C_i^{(k)} |i^{(k)}\rangle = \left[\sum_p (t_{pk} - \delta_{pk}) a_{p\alpha}^{(k)\dagger} \bar{a}_{k\alpha}^{(k)} \right] \sum_i C_i^{(k-1)} |i^{(k)}\rangle \tag{11.9.12}$$

$$\Delta C_i^{(k)} = C_i^{(k)} - C_i^{(k-1)} \tag{11.9.13}$$

Multiplying (11.9.12) from the left by a determinant in the biorthonormal basis $\langle \bar{j}^{(k)}|$, we arrive at the following equation for the CI corrections:

$$\Delta C_j^{(k)} = \sum_i \langle \bar{j}^{(k)} | \sum_p (t_{pk} - \delta_{pk}) a_{p\alpha}^{(k)\dagger} \bar{a}_{k\alpha}^{(k)} |i^{(k)}\rangle C_i^{(k-1)} \tag{11.9.14}$$

The corrections to the expansion coefficients (11.9.14) arising from the replacement of orbital k as described by (11.9.10) may thus be obtained by carrying out a direct CI iteration (i.e. contraction) involving a one-electron operator with one fixed orbital index. It should be noted that, although (11.9.14) involves the use of a nonorthogonal basis, it can be treated using the standard direct CI techniques of Section 11.8 since, in a biorthogonal basis, the anticommutation relations (11.9.8) are the same as in an orthonormal basis. The cost of one CI contraction of the type (11.9.14) is less than that of a standard one-electron CI contraction since one index is fixed. On the other hand, to complete the transformation from one orbital basis to another, we must carry out a sequence of n direct CI contractions, one for each orbital in the basis set. The combined cost of these partial contractions is equivalent to that of a single full one-electron contraction. Clearly, this scheme represents a major improvement on the alternative approach – namely, a full reoptimization of the CI wave function in the new orbital basis.

To complete our prescription for obtaining the transformed CI coefficients, we must discuss how to obtain the coefficients t_{pk} in (11.9.14) from the orbital transformation coefficients D_{pq} in (11.9.1). Returning to the transformation of step k (11.9.10), we note that the creation operator on the left-hand side is equal to that of the initial set:

$$a_{k\alpha}^{(k-1)\dagger} = a_{k\alpha}^{(0)\dagger} \tag{11.9.15}$$

On the right-hand side of (11.9.10), the first k operators have been replaced and are the same as in the final set, whereas the remaining operators correspond to those of the initial set:

$$a_{p\alpha}^{(k)\dagger} = \begin{cases} a_{p\alpha}^{(n)\dagger} & p \leq k \\ a_{p\alpha}^{(0)\dagger} & p > k \end{cases} \tag{11.9.16}$$

We may therefore rewrite (11.9.10) in the following form

$$a_{k\alpha}^{(0)\dagger} = \sum_{p \leq k} a_{p\alpha}^{(n)\dagger} t_{pk} + \sum_{p > k} a_{p\alpha}^{(0)\dagger} t_{pk} \tag{11.9.17}$$

where the provenance of each operator in step k (i.e. the initial or final set) has been made explicit.

Equation (11.9.17) must hold in each step. For further manipulation, it is convenient to partition the matrix $\mathbf{t}$ into an upper triangular matrix $\mathbf{t}^U$ with elements

$$t_{pk}^U = \begin{cases} t_{pk} & p \leq k \\ 0 & p > k \end{cases} \tag{11.9.18}$$

and a strictly lower triangular matrix $\mathbf{t}^L$ with elements

$$t_{pk}^L = \begin{cases} 0 & p \leq k \\ t_{pk} & p > k \end{cases} \tag{11.9.19}$$

This partitioning allows us to write (11.9.17) in the simpler form

$$a_{k\alpha}^{(0)\dagger} = \sum_p a_{p\alpha}^{(n)\dagger} t_{pk}^U + \sum_p a_{p\alpha}^{(0)\dagger} t_{pk}^L \tag{11.9.20}$$

We now introduce the orbital transformation matrix $\mathbf{D}$ of (11.9.1) by expanding the creation operators that belong to the final set in terms of the operators of the initial set:

$$a_{k\alpha}^{(0)\dagger} = \sum_p a_{p\alpha}^{(0)\dagger} [\mathbf{D}\mathbf{t}^U + \mathbf{t}^L]_{pk} \tag{11.9.21}$$

This expression gives us the following matrix equation relating $\mathbf{D}$ and $\mathbf{t}$:

$$\mathbf{D}\mathbf{t}^U + \mathbf{t}^L = \mathbf{1} \tag{11.9.22}$$

Introducing the LU decomposition $\mathbf{D} = \mathbf{L}\mathbf{U}$, where $\mathbf{L}$ is a unit lower triangular matrix (i.e. a lower triangular matrix with diagonal elements equal to 1) and $\mathbf{U}$ is an upper triangular matrix [7], (11.9.22) may be written in the form

$$\mathbf{L}\mathbf{U}\mathbf{t}^U + \mathbf{t}^L = \mathbf{1} \tag{11.9.23}$$

Since the inverse of an upper triangular matrix is another upper triangular matrix, this equation has the solution

$$\mathbf{t}^U = \mathbf{U}^{-1} \tag{11.9.24}$$

$$\mathbf{t}^L = \mathbf{1} - \mathbf{L} \tag{11.9.25}$$

We may therefore obtain $\mathbf{t}$ by a LU decomposition of the orbital-transformation matrix $\mathbf{D}$ followed by an inversion of the upper triangular matrix $\mathbf{U}$.

We have thus succeeded in developing an algorithm for carrying out transformations of the CI coefficients in response to orthogonal or nonorthogonal transformations of the MOs. We have only considered the transformation where all orbitals have alpha spin – the same procedure may of course be used also for beta electrons. The total cost of transforming all orbitals is no higher than that of one direct CI contraction for a one-electron operator.

It should be realized that, for a general orbital transformation, the full CI space $|i^{(n)}\rangle$ is usually required to satisfy the identity (11.9.2) – even for wave functions that, in the original basis, have been expanded in a restricted configuration space. For RAS wave functions, we must therefore

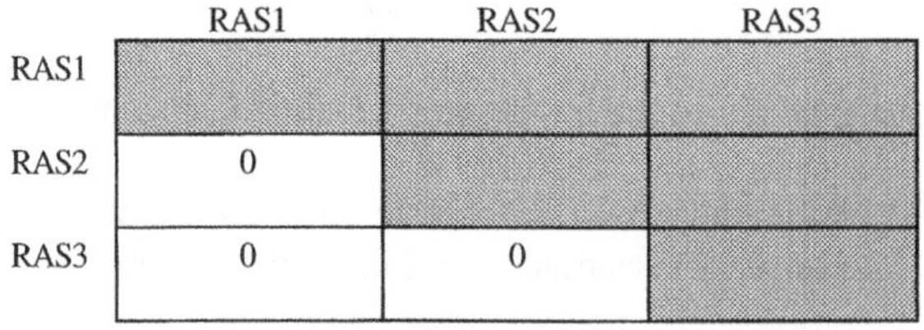

	RAS1	RAS2	RAS3
RAS1			
RAS2	0		
RAS3	0	0	

Fig. 11.7. The structure of the orbital-transformation matrices $\mathbf{t}$ and $\mathbf{D}$ that preserves the RAS constraints.

impose certain restrictions on the orbital-rotation matrix in order to ensure that $|i^{(n)}\rangle$ is spanned by the same space as $|i^{(0)}\rangle$.

Let us assume that the RAS orbitals are sorted in the order RAS1, RAS2, RAS3. Because of the RAS constraints, the original CI coefficients $C_i^{(0)}$ are zero for determinants with RAS1 occupations smaller than some limit or with RAS3 occupations larger than some limit. To ensure that the corresponding transformed coefficients $C_i^{(n)}$ are zero, we must impose restrictions on t_{pk} in (11.9.14). Note from (11.9.14) that the index p refers to a creation operator and the index k to an annihilation operator. If there are nonzero elements t_{pk} with p belonging to RAS3 and k to RAS1 or RAS2, then there will be nonzero coefficients $C_i^{(n)}$ for determinants $|i^{(n)}\rangle$ that violate the RAS constraints. Similarly, nonzero elements t_{pk} for p belonging to RAS2 and k to RAS1 will give rise to configurations that violate the RAS conditions. To ensure that the transformed wave function satisfies the original RAS constraints, we must insist that $\mathbf{t}$ has the block structure depicted in Figure 11.7.

To infer from the allowed form of $\mathbf{t}$ to the allowed form of $\mathbf{D}$, we note that it can be shown that, if $\mathbf{A}$ has the upper triangular block structure of Figure 11.7, then $\mathbf{L}$ in the LU decomposition $\mathbf{A} = \mathbf{LU}$ will have the same structure. Hence, if we require $\mathbf{D}$ to have the structure of Figure 11.7, then $\mathbf{t}$ in (11.9.25) will have the same structure. We have thus identified the structure of $\mathbf{D}$ that obeys the RAS constraints. Note that the matrix in Figure 11.7 is not sufficiently flexible for a general transformation – for example, for transforming the RAS wave function to the natural-orbital basis.

11.10 Symmetry-broken CI solutions

In Section 10.10.6, we examined the allyl radical at the Hartree–Fock level and obtained a spatially symmetry-broken UHF solution of a slightly lower energy than the corresponding RHF solution. Figure 10.6 displays the Hartree–Fock potential-energy curves as functions of the antisymmetric stretching coordinate Δ of Figure 10.4. From the discussion in Section 10.10.6, we recall that, strictly speaking, it is only for the symmetric system with $\Delta = 0$ that a symmetry-broken UHF solution exists. For nonzero displacements, all solutions in Figure 10.6 are, technically speaking, different RHF solutions. For convenience, we shall here refer to the energy curves in their entirety as either unrestricted or restricted, depending on which solution they represent at $\Delta = 0$.

Let us now consider the treatment of the allyl radical at the CI level. Obviously, we may base our CI calculation on the RHF or UHF orbitals. At the FCI level, the choice of orbitals does not matter since, in a complete expansion, the same solution is recovered in either case. At the truncated level, however, the RHF- and UHF-based CI wave functions will differ. In Figure 11.8, we have plotted the CISD/cc-pVDZ potential-energy curves obtained using the RHF orbitals (full line) and the UHF orbitals (dotted lines). These curves should be compared with the corresponding Hartree–Fock curves in Figure 10.6.

The structures of the Hartree–Fock potential-energy curves in Figure 10.6 and the CISD curves in Figure 11.8 are quite similar – in each case, the restricted curve has a symmetric minimum and the unrestricted curves an unsymmetric minimum, although, for the CISD wave function, the unphysical, symmetry-broken minima are shallower. Somewhat surprisingly, we find that the RHF-based CISD curve is below the corresponding UHF-based curve. A similar situation was encountered in Chapter 5, where, in Figures 5.18 and 5.19, we compared the water dissociation curves at the spin-restricted and spin-unrestricted coupled-cluster and Møller–Plesset levels, respectively. For some OH bond distances, the spin-restricted energies were found to be lower than the unrestricted energies at the correlated levels.

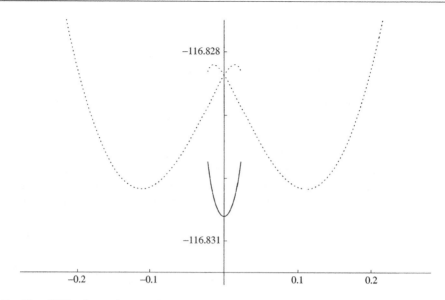

Fig. 11.8. The CISD electronic energies (in E_h) of the allyl radical as functions of the antisymmetric stretching coordinate Δ (a_0) in Figure 10.4. The full line corresponds to the CISD energy based on the RHF solution; the dotted lines correspond to CISD energies based on the UHF solution.

In conclusion, the CISD wave function is unable to mend the deficiencies of the Hartree–Fock solutions for the allyl radical. Of course, we may further reduce the discrepancies between the UHF- and RHF-based solutions by including in the CI expansion excitations higher than doubles. Quite apart from being very expensive, such a solution is not very attractive since the symmetry-broken solutions would probably persist for rather high excitation levels. A more attractive and fundamental solution to the symmetry problem would be to select the most important configurations for the description of the allyl radical and to carry out an MCSCF calculation where the orbitals and configurations are simultaneously optimized. As discussed in the next chapter, such a procedure yields a wave function that exhibits none of the unphysical behaviour characterizing the Hartree–Fock-based solutions.

References

[1]   G. Arfken, *Mathematical Methods for Physicists*, Academic Press, 1970.
[2]   B. O. Roos, P. R. Taylor and P. E. M. Siegbahn, *Chem. Phys.* **48**, 157 (1980).
[3]   P. E. M. Siegbahn, in G. H. F. Diercksen and S. Wilson (eds), *Methods in Computational Molecular Physics*, Reidel, 1983, p. 189.
[4]   J. Olsen, B. O. Roos, P. Jørgensen and H. J. Aa. Jensen, *J. Chem. Phys.* **89**, 2185 (1988).
[5]   S. R. Langhoff and E. R. Davidson, *Int. J. Quantum Chem.* **8**, 61 (1974).
[6]   W. Feller, *An Introduction to Probability Theory and its Applications*, Wiley, 1957.
[7]   W. H. Press, S. A. Teukolsky, W. T. Vetterling and B. P. Flannery, *Numerical Recipes*, Cambridge University Press, 1986.
[8]   G. H. Golub and C. F. Van Loan, *Matrix Computations*, 2nd edn, Johns Hopkins University Press, 1989.
[9]   B. N. Parlett, *The Symmetric Eigenvalue Problem*, Prentice Hall, 1980.
[10] J. Olsen, P. Jørgensen and J. Simons, *Chem. Phys. Lett.* **169**, 463 (1990).
[11] E. R. Davidson, *J. Comput. Phys.* **17**, 87 (1975).
[12] P. J. Knowles and N. C. Handy, *Chem. Phys. Lett.* **111**, 315 (1984).

[13] B. Roos, *Chem. Phys. Lett.* **15**, 153 (1972).

[14] W. Duch, *GRMS or Graphical Representation of Model Spaces I: Basics*, Lecture Notes in Chemistry Vol. 42, Springer-Verlag, 1986.

[15] J. Linderberg and Y. Öhrn, *Propagators in Quantum Chemistry*, Academic Press, 1973.

[16] R. J. Harrison and S. Zarrabian, *Chem. Phys. Lett.* **158**, 393 (1989).

[17] P.-Å. Malmqvist, *Int. J. Quantum Chem.* **30**, 479 (1986).

Further reading

J. Hinze (ed.), *The Unitary Group for the Evaluation of Electronic Energy Matrix Elements*, Lecture Notes in Chemistry Vol. 22, Springer-Verlag, 1981.

Exercises

EXERCISE 11.1

Assume that, for a single molecule, the normalized FCI wave function can be written as

$$|\text{FCI}(1)\rangle = \hat{\psi}|\text{vac}\rangle = \sum_{i \geq 0} C_i \hat{\psi}_i |\text{vac}\rangle \tag{11E.1.1}$$

where $C_i \hat{\psi}_i |\text{vac}\rangle$ is the part of the wave function that is ith-fold excited with respect to the reference determinant $\hat{\psi}_0 |\text{vac}\rangle$ and where $\hat{\psi}_i |\text{vac}\rangle$ is normalized so that the weight of the ith-fold excitation level is given by

$$W_i = C_i^2 \tag{11E.1.2}$$

We shall in this exercise consider m noninteracting molecules represented by the wave function

$$|\text{FCI}(m)\rangle = \left[\prod_{I=1}^{m} \hat{\psi}_I \right] |\text{vac}\rangle = \left[\prod_{I=1}^{m} \left(\sum_{i \geq 0} C_i \hat{\psi}_{iI} \right) \right] |\text{vac}\rangle \tag{11E.1.3}$$

and, in particular, determine the weights $W_i(m)$ of the excitation levels i for the supersystem from the weights of a single monomer.

1. Show that the weights of the lowest excitation levels are given by

$$W_0(m) = W_0^m \tag{11E.1.4}$$

$$W_1(m) = m W_0^{m-1} W_1 \tag{11E.1.5}$$

$$W_2(m) = m W_0^{m-1} W_2 + \frac{m(m-1)}{2} W_0^{m-2} W_1^2 \tag{11E.1.6}$$

2. For higher excitations, the weights become rather complicated but can be simplified by assuming that only the even-order weights W_{2k} are nonzero. Show that the weights of the lowest even-order excitation levels are then given as

$$W_2(m) = m W_0^{m-1} W_2 \tag{11E.1.7}$$

$$W_4(m) = m W_0^{m-1} W_4 + \frac{m(m-1)}{2} W_0^{m-2} W_2^2 \tag{11E.1.8}$$

$$W_6(m) = m W_0^{m-1} W_6 + m(m-1) W_0^{m-2} W_2 W_4 + \frac{m(m-1)(m-2)}{6} W_0^{m-3} W_2^3 \tag{11E.1.9}$$

Table 11E.1.1 The weights W_n of different excitation levels for the ground-state water molecule in the cc-pVDZ basis at the geometries R_{ref} and $2R_{ref}$

	W_0	W_2	W_4	W_6
R_{ref}	0.9410	0.0565	0.0014	0.000015
$2R_{ref}$	0.5897	0.3330	0.0397	0.00044

3. In the cc-pVDZ basis, we obtain for the ground-state water molecule at the reference geometry R_{ref} and at the stretched geometry $2R_{ref}$ the excitation weights listed in Table 11E.1.1. Using (11E.1.4) and (11E.1.7)–(11E.1.9), determine the corresponding weights for $m = 2, 4, 8, 16$. For each system, determine also the weight of the part of the wave function with excitation levels higher than 6.

EXERCISE 11.2

Consider a dimer of two noninteracting H_2 molecules each described in a minimal basis.

1. Show that the ground-state FCI wave function can be parametrized as

$$|\text{FCI}\rangle = (1 + c_1\hat{D}_1 + c_2\hat{D}_2 + c_{12}\hat{D}_1\hat{D}_2)|\text{HF}\rangle \tag{11E.2.1}$$

where the operator

$$\hat{D}_I = a_{u_I\alpha}^\dagger a_{u_I\beta}^\dagger a_{g_I\beta} a_{g_I\alpha} \tag{11E.2.2}$$

creates a double excitation from the Hartree–Fock state for molecule I.

2. Show that the FCI eigenvalue problem becomes

$$\begin{pmatrix} 0 & K & K & 0 \\ K & 2\Delta & 0 & K \\ K & 0 & 2\Delta & K \\ 0 & K & K & 4\Delta \end{pmatrix} \begin{pmatrix} c_0 \\ c_1 \\ c_2 \\ c_{12} \end{pmatrix} = E_{\text{corr}}^{\text{FCI}}(2) \begin{pmatrix} c_0 \\ c_1 \\ c_2 \\ c_{12} \end{pmatrix} \tag{11E.2.3}$$

where $E_{\text{corr}}^{\text{FCI}}(2)$ is the FCI correlation energy of the H_2 dimer and

$$K = g_{gugu} \tag{11E.2.4}$$

$$2\Delta = E_{\text{HF}}^* - E_{\text{HF}} \tag{11E.2.5}$$

$$E_{\text{HF}} = 2h_{gg} + g_{gggg} \tag{11E.2.6}$$

$$E_{\text{HF}}^* = 2h_{uu} + g_{uuuu} \tag{11E.2.7}$$

3. Assume now that we are interested only in those solutions to (11E.2.3) that have both a nonzero Hartree–Fock component and the same coefficients for the two doubly excited determinants. Using intermediate normalization

$$c_0 = 1 \tag{11E.2.8}$$

show that these solutions satisfy the relation

$$c_{12} = c^2 \tag{11E.2.9}$$

where

$$c_1 = c_2 = c \tag{11E.2.10}$$

4. Show that the ground-state FCI solution becomes

$$E_{\text{corr}}^{\text{FCI}}(2) = 2\Delta - 2\sqrt{\Delta^2 + K^2} \tag{11E.2.11}$$

$$c = \frac{\Delta - \sqrt{\Delta^2 + K^2}}{K} \tag{11E.2.12}$$

5. Write the FCI wave function obtained in Exercise 11.2.4 in a separable form.
6. Show that this FCI wave function can be written as a coupled-cluster wave function:

$$|\text{FCI}\rangle = \exp(c\hat{D}_1 + c\hat{D}_2)|\text{HF}\rangle \tag{11E.2.13}$$

EXERCISE 11.3

Consider m noninteracting hydrogen molecules, each described in a minimal basis. The Hartree–Fock state can then be written as

$$|\text{HF}\rangle = \left(\prod_{I=1}^{m} \hat{\psi}_I^{\text{HF}}\right)|\text{vac}\rangle \tag{11E.3.1}$$

where

$$\hat{\psi}_I^{\text{HF}} = a_{g_I\alpha}^\dagger a_{g_I\beta}^\dagger \tag{11E.3.2}$$

For each molecule, only a double excitation may occur, with the same weight in all molecules. The wave function for the m noninteracting systems can be written as

$$|0\rangle = |\text{HF}\rangle + c_2|2\rangle + c_4|4\rangle + c_6|6\rangle + c_8|8\rangle + \cdots \tag{11E.3.3}$$

where $|2\rangle$, $|4\rangle$, $|6\rangle$ and $|8\rangle$ are normalized states containing the double, quadruple, sextuple and octuple excitations relative to $|\text{HF}\rangle$.

1. Show that the double excitations in (11E.3.3) are given as

$$|2\rangle = \frac{1}{\sqrt{m}} \sum_{I=1}^{m} \hat{D}_I|\text{HF}\rangle \tag{11E.3.4}$$

where

$$\hat{D}_I = a_{u_I\alpha}^\dagger a_{u_I\beta}^\dagger a_{g_I\beta} a_{g_I\alpha} \tag{11E.3.5}$$

2. Show that the quadruple excitations in (11E.3.3) can be described by the following normalized state:

$$|4\rangle = \sqrt{\frac{2}{m(m-1)}} \sum_{I>J} \hat{D}_I\hat{D}_J|\text{HF}\rangle \tag{11E.3.6}$$

Determine the normalized states for the sextuple $|6\rangle$ and octuple $|8\rangle$ excitations.

3. Show that, up to octuple excitations, the Hamiltonian matrix becomes

$$
\mathbf{H} = \begin{pmatrix}
mE_{HF} & \sqrt{m}K & 0 & 0 & 0 \\
\sqrt{m}K & mE_{HF} + 2\Delta & \sqrt{2(m-1)}K & 0 & 0 \\
0 & \sqrt{2(m-1)}K & mE_{HF} + 4\Delta & \sqrt{3(m-2)}K & 0 \\
0 & 0 & \sqrt{3(m-2)}K & mE_{HF} + 6\Delta & \sqrt{4(m-3)}K \\
0 & 0 & 0 & \sqrt{4(m-3)}K & mE_{HF} + 8\Delta
\end{pmatrix}
\tag{11E.3.7}
$$

where the notation is that of (11E.2.4)–(11E.2.7).

4. The CI eigenvalue problem can be truncated at different levels:

$$
|D\rangle = |HF\rangle + C_2|2\rangle
\tag{11E.3.8}
$$

$$
|DQ\rangle = |HF\rangle + C_2|2\rangle + C_4|4\rangle
\tag{11E.3.9}
$$

$$
|DQ6\rangle = |HF\rangle + C_2|2\rangle + C_4|4\rangle + C_6|6\rangle
\tag{11E.3.10}
$$

$$
|DQ68\rangle = |HF\rangle + C_2|2\rangle + C_4|4\rangle + C_6|6\rangle + C_8|8\rangle
\tag{11E.3.11}
$$

For each CI wave function, determine the largest number of molecules m for which we recover more than 50%, 90% and 99% of the FCI correlation energy. The integrals needed to evaluate the Hamiltonian (11E.3.7) are listed in Table 5.1, where the first orbital is gerade and the second ungerade.

EXERCISE 11.4

The CI energy in (11.4.18) may be written as

$$
E(\mathbf{c}) = \frac{\varepsilon(\mathbf{c})}{S(\mathbf{c})}
\tag{11E.4.1}
$$

where

$$
\varepsilon(\mathbf{c}) = E^{(0)} + 2\mathbf{c}^T\mathbf{PHC}^{(0)} + \mathbf{c}^T\mathbf{PHPc}
\tag{11E.4.2}
$$

$$
S(\mathbf{c}) = 1 + \mathbf{c}^T\mathbf{Pc}
\tag{11E.4.3}
$$

1. Show that the first, second and third derivatives become

$$
E_i^{(1)}(\mathbf{c}) = 2\frac{[\mathbf{PHC}^{(0)}]_i + [\mathbf{P}(\mathbf{H} - E(\mathbf{c})\mathbf{1})\mathbf{Pc}]_i}{S(\mathbf{c})}
\tag{11E.4.4}
$$

$$
E_{ij}^{(2)}(\mathbf{c}) = 2\frac{[\mathbf{P}(\mathbf{H} - E(\mathbf{c})\mathbf{1})\mathbf{P}]_{ij} - [\mathbf{Pc}]_i E_j^{(1)}(\mathbf{c}) - [\mathbf{Pc}]_j E_i^{(1)}(\mathbf{c})}{S(\mathbf{c})}
\tag{11E.4.5}
$$

$$
E_{ijk}^{(3)}(\mathbf{c}) = -2\frac{P_{ij}E_k^{(1)}(\mathbf{c}) + P_{ik}E_j^{(1)}(\mathbf{c}) + P_{jk}E_i^{(1)}(\mathbf{c})}{S(\mathbf{c})}
$$

$$
- 2\frac{[\mathbf{Pc}]_i E_{jk}^{(2)}(\mathbf{c}) + [\mathbf{Pc}]_j E_{ik}^{(2)}(\mathbf{c}) + [\mathbf{Pc}]_k E_{ij}^{(2)}(\mathbf{c})}{S(\mathbf{c})}
\tag{11E.4.6}
$$

2. Show that, for $n \geq 3$, a general expression for the nth derivative is given by

$$E_{i_1 \cdots i_n}^{(n)}(\mathbf{c}) = -\frac{2}{S(\mathbf{c})} \sum_{k=1}^{n} [\mathbf{Pc}]_{i_k} E_{i_1 \cdots i_{k-1} i_{k+1} \cdots i_n}^{(n-1)}(\mathbf{c}) - \frac{2}{S(\mathbf{c})} \sum_{k<m}^{n} P_{i_k i_m} E_{i_1 \cdots i_{k-1} i_{k+1} \cdots i_{m-1} i_{m+1} \cdots i_n}^{(n-2)}(\mathbf{c})$$

(11E.4.7)

For $\mathbf{c} = \mathbf{0}$, we therefore obtain

$$E_{i_1 \cdots i_n}^{(n)}(\mathbf{0}) = -2 \sum_{k<m}^{n} P_{km} E_{i_1 \cdots i_{k-1} i_{k+1} \cdots i_{m-1} i_{m+1} \cdots i_n}^{(n-2)}(\mathbf{0})$$

(11E.4.8)

3. Assume that $\mathbf{E}^{(1)}(\mathbf{c}) = \mathbf{0}$ and simplify the expressions for $\mathbf{E}^{(2)}$ and $\mathbf{E}^{(3)}$.

EXERCISE 11.5

1. Let an elementary matrix – that is, the unit matrix plus a rank-1 matrix – be given by

$$\mathbf{E} = \mathbf{1} + \mathbf{U}\mathbf{V}^{\mathrm{T}}$$

(11E.5.1)

where $\mathbf{U}$ and $\mathbf{V}$ are column vectors. Assume that its inverse is another elementary matrix

$$\mathbf{E}^{-1} = \mathbf{1} + \mathbf{Z}\mathbf{Y}^{\mathrm{T}}$$

(11E.5.2)

where $\mathbf{Z}$ and $\mathbf{Y}$ are column vectors. Determine $\mathbf{Z}$ and $\mathbf{Y}$ and show that

$$\mathbf{E}^{-1} = \mathbf{1} - \frac{\mathbf{U}\mathbf{V}^{\mathrm{T}}}{1 + \mathbf{V}^{\mathrm{T}}\mathbf{U}}$$

(11E.5.3)

2. Use (11E.5.3) to show that, for a nonsingular matrix $\mathbf{A}$, the following relation holds for all column vectors $\mathbf{U}$ and $\mathbf{V}$:

$$(\mathbf{A} + \mathbf{U}\mathbf{V}^{\mathrm{T}})^{-1} = \mathbf{A}^{-1} - \frac{(\mathbf{A}^{-1}\mathbf{U})(\mathbf{V}^{\mathrm{T}}\mathbf{A}^{-1})}{1 + \mathbf{V}^{\mathrm{T}}\mathbf{A}^{-1}\mathbf{U}}$$

(11E.5.4)

3. Consider now $\mathbf{A}$ with a rank-2 matrix added in the form

$$\mathbf{B} = \mathbf{A} + (\mathbf{U}\mathbf{V})\mathbf{S}\begin{pmatrix} \mathbf{U}^{\mathrm{T}} \\ \mathbf{V}^{\mathrm{T}} \end{pmatrix}$$

(11E.5.5)

where $\mathbf{S}$ is a 2×2 matrix. A natural ansatz for the inverse of $\mathbf{B}$ is given by

$$\mathbf{B}^{-1} = \mathbf{A}^{-1} + \mathbf{A}^{-1}(\mathbf{U}\mathbf{V})\mathbf{M}\begin{pmatrix} \mathbf{U}^{\mathrm{T}} \\ \mathbf{V}^{\mathrm{T}} \end{pmatrix}\mathbf{A}^{-1}$$

(11E.5.6)

Show that the inverse of $\mathbf{B}$ is obtained by using the 2×2 matrix

$$\mathbf{M} = -\left[\mathbf{S}^{-1} + \begin{pmatrix} \mathbf{U}^{\mathrm{T}} \\ \mathbf{V}^{\mathrm{T}} \end{pmatrix}\mathbf{A}^{-1}(\mathbf{U}\mathbf{V})\right]^{-1}$$

(11E.5.7)

EXERCISE 11.6

In the following, we shall consider the solution of the eigenvalue problem

$$\mathbf{H}\mathbf{C} = E\mathbf{C}$$

(11E.6.1)

We assume that the Hamiltonian can be decomposed into a dominant part and a correction term

$$\mathbf{H} = \mathbf{H}_0 + \mathbf{H}_1 \tag{11E.6.2}$$

and we express the true eigenvector and the eigenvalue in terms of a zero-order part and a correction term:

$$\mathbf{C} = \mathbf{C}^{(0)} + \mathbf{C}^{(1)} \tag{11E.6.3}$$

$$E = E^{(0)} + E^{(1)} \tag{11E.6.4}$$

It is assumed that the approximate eigenvector $\mathbf{C}^{(0)}$ is not an eigenvector of $\mathbf{H}_0$ and that the approximate eigenvalue $E^{(0)}$ is given by

$$E^{(0)} = \frac{\mathbf{C}^{(0)\mathrm{T}} \mathbf{H} \mathbf{C}^{(0)}}{\mathbf{C}^{(0)\mathrm{T}} \mathbf{C}^{(0)}} \tag{11E.6.5}$$

Show that the correction vector can be determined from (11E.6.1) as

$$\mathbf{C}^{(1)} = -(\mathbf{H}_0 - E^{(0)}\mathbf{1})^{-1} \left[(\mathbf{H} - E^{(0)}\mathbf{1})\mathbf{C}^{(0)} - \frac{\mathbf{C}^{(0)\mathrm{T}} (\mathbf{H}_0 - E^{(0)}\mathbf{1})^{-1} (\mathbf{H} - E^{(0)}\mathbf{1})\mathbf{C}^{(0)}}{\mathbf{C}^{(0)\mathrm{T}} (\mathbf{H}_0 - E^{(0)}\mathbf{1})^{-1} \mathbf{C}^{(0)}} \mathbf{C}^{(0)} \right] \tag{11E.6.6}$$

by neglecting the quadratic correction terms and requiring the correction vector to be orthogonal to the zero-order solution:

$$\mathbf{C}^{(1)\mathrm{T}} \mathbf{C}^{(0)} = 0 \tag{11E.6.7}$$

Note that (11E.6.6) is identical to the correction vector in (11.5.24).

EXERCISE 11.7

In this exercise, we consider a symmetry relation that exists between the expansion coefficients of the determinants in a CI wave function $|0\rangle$ with an even number of electrons $2N$, total spin S, and spin projection $M = 0$.

1. In a configuration containing $2m$ singly occupied orbitals, the determinants occur in pairs $|\mathbf{p}\rangle^{\mathrm{d}}$ and $|-\mathbf{p}\rangle^{\mathrm{d}}$ – for example, $|+, +, -, -\rangle^{\mathrm{d}}$ and $|-, -, +, +\rangle^{\mathrm{d}}$ in the relative notation of Section 2.6.1. In addition, $|\mathbf{p}\rangle^{\mathrm{d}}$ may contain a set of doubly occupied orbitals. Show that, in the genealogical expansion of a given CSF, the contributions from the paired determinants are related as

$$^{\mathrm{d}}\langle\mathbf{p}|\mathbf{t}\rangle^{\mathrm{c}} = (-1)^{N_t^-} \, ^{\mathrm{d}}\langle-\mathbf{p}|\mathbf{t}\rangle^{\mathrm{c}} \tag{11E.7.1}$$

where N_t^- is the number of spin-down couplings in the CSF – that is, the number of minus signs in the relative notation of $|\mathbf{t}\rangle^{\mathrm{c}}$.

2. Use (11E.7.1) to show that the corresponding CSFs satisfy the relation

$$\hat{U}_{\alpha\beta}|\mathrm{CSF}\rangle = (-1)^{N-S}|\mathrm{CSF}\rangle \tag{11E.7.2}$$

where $\hat{U}_{\alpha\beta}$ of (11.8.59) changes the spin projection of all the electrons in $|\mathrm{CSF}\rangle$, including those in the closed-shell orbitals – see (11.8.60) and (11.8.61).

3. A wave function $|0\rangle$ expanded in CSFs satisfies the same symmetry relation as the individual CSFs (11E.7.2):

$$\hat{U}_{\alpha\beta}|0\rangle = (-1)^{N-S}|0\rangle \tag{11E.7.3}$$

Let us now consider the expansion of $|0\rangle$ in determinants:

$$|0\rangle = \sum_{I_\alpha I_\beta} C_{I_\alpha I_\beta} \hat{\alpha}_{I_\alpha} \hat{\beta}_{I_\beta} |\text{vac}\rangle \tag{11E.7.4}$$

Show that the coefficients of this expansion satisfy the relation

$$C_{I_\alpha I_\beta} = (-1)^S C_{I_\beta I_\alpha} \tag{11E.7.5}$$

EXERCISE 11.8

Consider an FCI expansion with eight electrons distributed among six orbitals and $M = 0$.

1. Calculate the number of strings and the number of determinants in this expansion.
2. For this system, draw the string graph and determine the vertex and arc weights in the reverse lexical order.
3. Draw the string paths in the reverse lexical order.

EXERCISE 11.9

Consider the CI expansion with six electrons in six orbitals and $M = 0$; at most two electrons are allowed in orbitals 4, 5 and 6.

1. Draw the string graph and determine the vertex and arc weights in the reverse lexical order.
2. Draw the string paths in the reverse lexical order.
3. Group the strings according to the total number of electrons in orbitals 4, 5 and 6. Determine the number of strings in each group.
4. Find the allowed combinations of alpha and beta strings and determine the total number of determinants in the CI expansion.

EXERCISE 11.10

Consider a closed-shell Hartree–Fock state of 10 doubly occupied orbitals and assume that the virtual space contains 100 orbitals. Formulate the RAS constraints and determine the number of strings and determinants for CI expansions containing:

1. the Hartree–Fock state and all single and double excitations out of this state;
2. the Hartree–Fock state, all single and double excitations into the 25 first virtual orbitals and all single excitations into the remaining 75 virtual orbitals.

EXERCISE 11.11

We shall in this exercise perform an FCI/STO-3G calculation of the totally symmetric singlet ground state of HeH^+ at a bond distance of $1.463796a_0$. The calculation is carried out in the basis of the Hartree–Fock orbitals of Exercise 10.4. The MO integrals are listed in Table 10E.4.3. All numbers quoted are in atomic units.

1. Determine the Slater determinants with $M = 0$ and set up the Hamiltonian matrix in the basis of these determinants.

2. Determine the CSFs with $M = 0$ and transform the Hamiltonian matrix to this basis.

3. We now solve the CI eigenvalue problem using the Davidson method in a two-dimensional subspace. Each iteration consists of the following steps:

   (a) For the current approximate energy E_n and wave function $\mathbf{C}_n$, calculate the residual vector

$$\mathbf{R}_n = (\mathbf{H} - E_n \mathbf{1})\mathbf{C}_n \tag{11E.11.1}$$

   (b) Generate a new direction $\mathbf{d}$ by orthogonalizing the generated vector

$$\mathbf{P} = (\mathbf{H}_0 - E_n \mathbf{1})^{-1} \mathbf{R}_n \tag{11E.11.2}$$

   against $\mathbf{C}_n$ and normalizing.

   (c) Set up and diagonalize the 2×2 Hamiltonian matrix in the basis of $\mathbf{C}_n$ and $\mathbf{d}$:

$$\mathbf{H}_{\text{sub}} = \begin{pmatrix} E_n & \mathbf{C}_n^{\text{T}} \mathbf{H} \mathbf{d} \\ \mathbf{d}^{\text{T}} \mathbf{H} \mathbf{C}_n & \mathbf{d}^{\text{T}} \mathbf{H} \mathbf{d} \end{pmatrix} \tag{11E.11.3}$$

   The lowest eigenvalue represents the improved energy E_{n+1} and the improved CI wave function is obtained from the normalized associated eigenvector $(x_1, x_2)^{\text{T}}$:

$$\mathbf{C}_{n+1} = x_1 \mathbf{C}_n + x_2 \mathbf{d} \tag{11E.11.4}$$

   Carry out the first Davidson iteration, using $E_1 = -2.84189825$ E$_h$ and $\mathbf{C}_1^{\text{T}} = (1, 0, 0)$. In the optimization, let $\mathbf{H}_0$ be equal to the diagonal part of $\mathbf{H}$.

4. The total energy, wave function and norm of the residual of the first four iterations are listed in Table 11E.11.1, along with the converged energy E_{cnv} and eigenvector $\mathbf{C}_{\text{cnv}}$. For each iteration, determine $E_n - E_{\text{cnv}}$ and $\|\mathbf{C}_n - \mathbf{C}_{\text{cnv}}\|$ and compare with the norm of the residual.

5. The lowest eigenvalue of the Hamiltonian matrix of Exercise 11.11.2 may also be determined by Newton's method. Perform one Newton iteration with $\mathbf{C}_1^{\text{T}} = (1, 0, 0)$ as the starting vector.

6. Compare $E_n - E_{\text{cnv}}$, $\|\mathbf{R}_n\|$ and $\|\mathbf{C}_n - \mathbf{C}_{\text{cnv}}\|$ for $n = 1$ and $n = 2$ in the Newton optimization of Exercise 11.11.5. Do the errors exhibit the expected cubic convergence? Compare with the errors of the Davidson method.

Table 11E.11.1 The total energy (E$_h$), wave function and norm of the residual for the first four Davidson iterations of an FCI/STO-3G calculation on the ground state of HeH$^+$ at the bond distance of 1.463796a$_0$

Iteration	E_n	$\mathbf{C}_n$			$\|\mathbf{R}_n\|$
1	−2.84189825	1.00000000	0.00000000	0.00000000	0.14536406
2	−2.85118738	0.99796447	0.00000000	−0.06377239	0.01896688
3	−2.85151440	0.99781617	−0.01723918	−0.06376291	0.00511676
4	−2.85152616	0.99766162	−0.01754262	−0.06605714	0.00035141
convergence	−2.85152628	0.99765449	−0.01786503	−0.06607849	0.00000000

Solutions

SOLUTION 11.1

1. For $W_0(m)$, none of the monomers is excited. The weight of this part of the wave function is therefore

$$W_0(m) = (C_0^m)^2 = W_0^m \tag{11S.1.1}$$

There are m contributions to $W_1(m)$, each of which is singly excited in one monomer and otherwise unexcited. The combined weight of the single excitations thus becomes

$$W_1(m) = m(C_0^{m-1} C_1)^2 = m W_0^{m-1} W_1 \tag{11S.1.2}$$

Two kinds of terms contribute to $W_2(m)$: m terms that are doubly excited in one monomer and $m(m-1)/2$ terms that are singly excited in two monomers. The total weight of the doubly excited configurations therefore becomes

$$W_2(m) = m W_0^{m-1} W_2 + \tfrac{1}{2} m(m-1) W_0^{m-2} W_1^2 \tag{11S.1.3}$$

2. For $W_4(m)$, two kinds of terms are generated – m terms that are quadruply excited in one monomer and $m(m-1)/2$ terms that are doubly excited in two monomers:

$$W_4(m) = m W_0^{m-1} W_4 + \tfrac{1}{2} m(m-1) W_0^{m-2} W_2^2 \tag{11S.1.4}$$

For $W_6(m)$, three kinds of terms contribute: m terms that are sextuply excited in one monomer, $m(m-1)$ terms that are doubly excited in one monomer and quadruply excited in another, and finally $m(m-1)(m-2)/6$ terms that are doubly excited in three monomers. The total weight becomes:

$$W_6(m) = m W_0^{m-1} W_6 + m(m-1) W_0^{m-2} W_2 W_4 + \tfrac{1}{6} m(m-1)(m-2) W_0^{m-3} W_2^3 \tag{11S.1.5}$$

3. The calculated weights are listed in Table 11S.1.1.

SOLUTION 11.2

1. For each of the two molecules, there is a gerade and an ungerade orbital and the Hartree–Fock state may be written as $a_{g_I \alpha}^\dagger a_{g_I \beta}^\dagger |\text{vac}\rangle$. Single and triple excitations create states of local $^1\Sigma_u^+$ symmetry that do not contribute to the total wave function. The only states of $^1\Sigma_g^+$ symmetry are the Hartree–Fock state and the doubly excited states on either one or both molecules.

Table 11S.1.1 The weights of the even-order excitation levels for m ground-state water molecules, calculated in the cc-pVDZ basis at the reference geometry R_{ref} and at the stretched geometry $2R_{\text{ref}}$

	m	$W_0(m)$	$W_2(m)$	$W_4(m)$	$W_6(m)$	$W_{\text{rest}}(m)$
R_{ref}	2	0.8855	0.1063	0.0058	0.0002	0.0022
	4	0.7841	0.1883	0.0216	0.0016	0.0044
	8	0.6148	0.2953	0.0694	0.0106	0.0099
	16	0.3779	0.3631	0.1725	0.0540	0.0324
$2R_{\text{ref}}$	2	0.3477	0.3927	0.1577	0.0270	0.0748
	4	0.1209	0.2731	0.2639	0.1426	0.1994
	8	0.0146	0.0661	0.1384	0.1787	0.6022
	16	0.0002	0.0019	0.0084	0.0235	0.9659

2. The FCI eigenvalue problem (11E.2.3) is obtained by setting up the standard eigenvalue problem and subtracting from both sides the ground-state Hartree–Fock energy. The matrix elements may be evaluated by the standard methods of second quantization.

3. Expanding (11E.2.3) and using (11E.2.8), we obtain the following three equations:

$$2Kc = E_{\text{corr}}^{\text{FCI}}(2) \tag{11S.2.1}$$

$$K + 2\Delta c + Kc_{12} = E_{\text{corr}}^{\text{FCI}}(2)c \tag{11S.2.2}$$

$$2Kc + 4\Delta c_{12} = E_{\text{corr}}^{\text{FCI}}(2)c_{12} \tag{11S.2.3}$$

Solving (11S.2.3) for c_{12} and using (11S.2.1), we obtain

$$c_{12} = \frac{2Kc}{E_{\text{corr}}^{\text{FCI}}(2) - 4\Delta} = \frac{E_{\text{corr}}^{\text{FCI}}(2)}{E_{\text{corr}}^{\text{FCI}}(2) - 4\Delta} \tag{11S.2.4}$$

Next, substituting this result into (11S.2.2) and solving for c, we obtain

$$[E_{\text{corr}}^{\text{FCI}}(2) - 2\Delta]c = K + \frac{KE_{\text{corr}}^{\text{FCI}}(2)}{E_{\text{corr}}^{FCI}(2) - 4\Delta} = \frac{2K[E_{\text{corr}}^{\text{FCI}}(2) - 2\Delta]}{E_{\text{corr}}^{\text{FCI}}(2) - 4\Delta} \tag{11S.2.5}$$

which yields

$$c = \frac{2K}{E_{\text{corr}}^{\text{FCI}}(2) - 4\Delta} \tag{11S.2.6}$$

Comparing (11S.2.6) with (11S.2.4), we obtain (11E.2.9).

4. From (11S.2.2) and (11E.2.9), we obtain

$$K + 2\Delta c + Kc^2 = E_{\text{corr}}^{\text{FCI}}(2)c \tag{11S.2.7}$$

Using (11S.2.1), we may now eliminate c from (11S.2.7) to give

$$[E_{\text{corr}}^{\text{FCI}}(2)]^2 - 4\Delta E_{\text{corr}}^{\text{FCI}}(2) - 4K^2 = 0 \tag{11S.2.8}$$

from which we obtain the two solutions

$$E_{\text{corr}}^{\text{FCI}}(2) = 2(\Delta \pm \sqrt{\Delta^2 + K^2}) \tag{11S.2.9}$$

where the negative sign gives the lowest energy (11E.2.11). Inserting (11S.2.9) into (11S.2.1), we obtain (11E.2.12).

5. Using (11E.2.9) and (11E.2.10), we may separate the FCI wave function (11E.2.1) as

$$|\text{FCI}\rangle = (1 + c\hat{D}_1 + c\hat{D}_2 + c^2\hat{D}_1\hat{D}_2)|\text{HF}\rangle = (1 + c\hat{D}_1)(1 + c\hat{D}_2)|\text{HF}\rangle \tag{11S.2.10}$$

6. Expansion of the exponential in (11E.2.13) gives (11S.2.10) directly for the FCI wave function.

SOLUTION 11.3

1. The double excitations can be written as

$$|\text{D}\rangle = \sum_{I=1}^{m} C_I\hat{D}_I|\text{HF}\rangle \tag{11S.3.1}$$

where the C_I are all identical. The double excitations in (11E.3.3) can therefore be described in terms of the state

$$|2\rangle = \frac{1}{\sqrt{m}} \sum_{I=1}^{m} \hat{D}_I |\mathrm{HF}\rangle \qquad (11\mathrm{S}.3.2)$$

which is normalized, being a sum of m orthogonal and normalized vectors.

2. The quadruple excitations are given by

$$|Q\rangle = \sum_{I>J} C_{IJ} \hat{D}_I \hat{D}_J |\mathrm{HF}\rangle \qquad (11\mathrm{S}.3.3)$$

where the $m(m-1)/2$ coefficients C_{IJ} must be identical. The quadruple excitations can therefore be described in terms of the normalized state

$$|4\rangle = \sqrt{\frac{2}{m(m-1)}} \sum_{I>J} \hat{D}_I \hat{D}_J |\mathrm{HF}\rangle \qquad (11\mathrm{S}.3.4)$$

The sextuple and octuple excitations can likewise be described by the states

$$|6\rangle = \sqrt{\frac{6}{m(m-1)(m-2)}} \sum_{I>J>K} \hat{D}_I \hat{D}_J \hat{D}_K |\mathrm{HF}\rangle \qquad (11\mathrm{S}.3.5)$$

$$|8\rangle = \sqrt{\frac{24}{m(m-1)(m-2)(m-3)}} \sum_{I>J>K>L} \hat{D}_I \hat{D}_J \hat{D}_K \hat{D}_L |\mathrm{HF}\rangle \qquad (11\mathrm{S}.3.6)$$

3. The zeros in the Hamiltonian matrix occur for states differing by more than a double excitation. In the notation of (11E.2.4)–(11.E.2.7), we obtain for the first diagonal elements

$$\langle \mathrm{HF}|\hat{H}|\mathrm{HF}\rangle = \langle \mathrm{HF}_1 \cdots \mathrm{HF}_m| \sum_I \hat{H}_I |\mathrm{HF}_1 \cdots \mathrm{HF}_m\rangle = \sum_I \langle \mathrm{HF}_I|\hat{H}_I|\mathrm{HF}_I\rangle = mE_{\mathrm{HF}} \quad (11\mathrm{S}.3.7)$$

$$\langle 2|\hat{H}|2\rangle = \frac{1}{m} \sum_{IJK} \langle \hat{D}_I|\hat{H}_J|\hat{D}_K\rangle = \frac{1}{m} \sum_{IJ} \langle \hat{D}_I|\hat{H}_J|\hat{D}_I\rangle$$

$$= \frac{1}{m} \sum_I [(m-1)E_{\mathrm{HF}} + E_{\mathrm{HF}}^*] = mE_{\mathrm{HF}} + 2\Delta \qquad (11\mathrm{S}.3.8)$$

and for the first off-diagonal elements

$$\langle \mathrm{HF}|\hat{H}|2\rangle = \frac{1}{\sqrt{m}} \sum_{IJ} \langle \mathrm{HF}|\hat{H}_I|\hat{D}_J\rangle = \sqrt{m}K \qquad (11\mathrm{S}.3.9)$$

$$\langle 4|\hat{H}|2\rangle = \sqrt{\frac{2}{m(m-1)}} \sqrt{\frac{1}{m}} \sum_{\substack{KL \\ I>J}} \langle \hat{D}_I \hat{D}_J|\hat{H}_K|\hat{D}_L\rangle$$

$$= \frac{1}{m} \sqrt{\frac{2}{m-1}} \sum_{I>J} [\langle \hat{D}_I \hat{D}_J|\hat{H}_I|\hat{D}_J\rangle + \langle \hat{D}_I \hat{D}_J|\hat{H}_J|\hat{D}_I\rangle]$$

$$= \frac{1}{m} \sqrt{\frac{2}{m-1}} \left(\frac{m(m-1)}{2}K + \frac{m(m-1)}{2}K \right) = \sqrt{2(m-1)}K \qquad (11\mathrm{S}.3.10)$$

Table 11S.3.1 The largest number of H_2 molecules that, for a given truncated CI wave function, give at least 50%, 90% and 99% of the exact correlation energy

Excitation level	50%	90%	99%
D	163	11	1
DQ	578	61	14
DQ6	1112	138	41
DQ68	1715	232	79

In general, we may write the Hamiltonian matrix elements as

$$\langle 2M|\hat{H}|2M\rangle = mE_{\mathrm{HF}} + 2M\Delta \tag{11S.3.11}$$

$$\langle 2M|\hat{H}|2M-2\rangle = \sqrt{M(m-M+1)}K \tag{11S.3.12}$$

where M is the excitation level, giving rise to the Hamiltonian matrix in (11E.3.7).

4. The numbers in Table 11S.3.1 have been obtained by numerical diagonalization of the Hamiltonian (11E.3.7) for different m, followed by comparison with the exact correlation energy of -0.010954 E_h per molecule as obtained from (11.3.15).

SOLUTION 11.4

1. To calculate the derivatives of the energy, it is advantageous to rewrite (11E.4.1) in the form

$$E(\mathbf{c})S(\mathbf{c}) = \varepsilon(\mathbf{c}) \tag{11S.4.1}$$

Taking the derivatives on both sides of this expression with respect to c_i, c_j and c_k, we obtain the three equations

$$E_i^{(1)}(\mathbf{c})S(\mathbf{c}) + 2E(\mathbf{c})[\mathbf{Pc}]_i = 2[\mathbf{PHC}^{(0)}]_i + 2[\mathbf{PHPc}]_i \tag{11S.4.2}$$

$$E_{ij}^{(2)}(\mathbf{c})S(\mathbf{c}) + 2E_i^{(1)}(\mathbf{c})[\mathbf{Pc}]_j + 2E_j^{(1)}(\mathbf{c})[\mathbf{Pc}]_i + 2E(\mathbf{c})P_{ij} = 2[\mathbf{PHP}]_{ij} \tag{11S.4.3}$$

$$E_{ijk}^{(3)}(\mathbf{c})S(\mathbf{c}) + 2E_{ij}^{(2)}(\mathbf{c})[\mathbf{Pc}]_k + 2E_{ik}^{(2)}(\mathbf{c})[\mathbf{Pc}]_j + 2E_{jk}^{(2)}(\mathbf{c})[\mathbf{Pc}]_i$$
$$+ 2E_i^{(1)}(\mathbf{c})P_{jk} + 2E_j^{(1)}(\mathbf{c})P_{ik} + 2E_k^{(1)}(\mathbf{c})P_{ij} = 0 \tag{11S.4.4}$$

which may be rewritten to yield (11E.4.4)–(11E.4.6).

2. Assume that (11E.4.7) holds for a given n:

$$S(\mathbf{c})E_{i_1\cdots i_n}^{(n)}(\mathbf{c}) = -2\sum_{k=1}^n [\mathbf{Pc}]_{i_k} E_{i_1\cdots i_{k-1}i_{k+1}\cdots i_n}^{(n-1)}(\mathbf{c}) - 2\sum_{k<m} P_{i_k i_m} E_{i_1\cdots i_{k-1}i_{k+1}\cdots i_{m-1}i_{m+1}\cdots i_n}^{(n-2)}(\mathbf{c}) \tag{11S.4.5}$$

Differentiating this equation with respect to $c_{i_{n+1}}$, we obtain

$$2[\mathbf{Pc}]_{i_{n+1}} E_{i_1\cdots i_n}^{(n)} + S(\mathbf{c})E_{i_1\cdots i_{n+1}}^{(n+1)}(\mathbf{c}) = -2\sum_{k=1}^n [\mathbf{Pc}]_{i_k} E_{i_1\cdots i_{k-1}i_{k+1}\cdots i_{n+1}}^{(n)}(\mathbf{c}) - 2\sum_{k=1}^n P_{i_k i_{n+1}} E_{i_1\cdots i_{k-1}i_{k+1}\cdots i_n}^{(n-1)}(\mathbf{c})$$

$$-2\sum_{k<m} P_{i_k i_m} E_{i_1\cdots i_{k-1}i_{k+1}\cdots i_{m-1}i_{m+1}\cdots i_{n+1}}^{(n-1)}(\mathbf{c}) \tag{11S.4.6}$$

or equivalently

$$E^{(n+1)}_{i_1\cdots i_{n+1}}(\mathbf{c}) = -\frac{2}{S(\mathbf{c})}\sum_{k=1}^{n+1}[\mathbf{Pc}]_{i_k}E^{(n+1-1)}_{i_1\cdots i_{k-1}i_{k+1}\cdots i_{n+1}}(\mathbf{c}) - \frac{2}{S(\mathbf{c})}\sum_{k<m}^{n+1}P_{i_k i_m}E^{(n+1-2)}_{i_1\cdots i_{k-1}i_{k+1}\cdots i_{m-1}i_{m+1}\cdots i_{n+1}}(\mathbf{c})$$

$$(11S.4.7)$$

Thus, if (11E.4.7) holds for n, then (11S.4.7) shows that it holds for $n + 1$ as well. Since it holds for $n = 3$, the induction is complete.

3. If $\mathbf{E}^{(1)}(\mathbf{c})$ is zero, then the second and third derivatives (11E.4.5) and (11E.4.6) simplify to

$$E^{(2)}_{ij}(\mathbf{c}) = \frac{2}{S(\mathbf{c})}[\mathbf{P}[\mathbf{H} - E(\mathbf{c})\mathbf{1}]\mathbf{P}]_{ij} \qquad (11S.4.8)$$

$$E^{(3)}_{ijk}(\mathbf{c}) = -\frac{2}{S(\mathbf{c})}([\mathbf{Pc}]_i E^{(2)}_{jk}(\mathbf{c}) + [\mathbf{Pc}]_j E^{(2)}_{ik}(\mathbf{c}) + [\mathbf{Pc}]_k E^{(2)}_{ij}(\mathbf{c})) \qquad (11S.4.9)$$

SOLUTION 11.5

1. From the requirement

$$(\mathbf{1} + \mathbf{UV}^\mathrm{T})(\mathbf{1} + \mathbf{ZY}^\mathrm{T}) = \mathbf{1} \qquad (11S.5.1)$$

it follows that

$$[\mathbf{U}(\mathbf{V}^\mathrm{T}\mathbf{Z}) + \mathbf{Z}]\mathbf{Y}^\mathrm{T} = -\mathbf{UV}^\mathrm{T} \qquad (11S.5.2)$$

This equation may be satisfied if we write

$$\mathbf{Z} = \alpha\mathbf{U} \qquad (11S.5.3)$$

$$\mathbf{Y} = \beta\mathbf{V} \qquad (11S.5.4)$$

where α and β are numbers. Inserting (11S.5.3) and (11S.5.4) in (11S.5.2), we obtain the condition

$$\alpha\beta = -\frac{1}{\mathbf{V}^\mathrm{T}\mathbf{U} + 1} \qquad (11S.5.5)$$

The inverse (11E.5.3) is now obtained from (11S.5.3)–(11S.5.5).

2. Writing

$$\mathbf{A} + \mathbf{UV}^\mathrm{T} = \mathbf{A}(\mathbf{1} + \mathbf{A}^{-1}\mathbf{UV}^\mathrm{T}) \qquad (11S.5.6)$$

we may apply (11E.5.3) to obtain

$$(\mathbf{A} + \mathbf{UV}^\mathrm{T})^{-1} = [\mathbf{1} + (\mathbf{A}^{-1}\mathbf{U})\mathbf{V}^\mathrm{T}]^{-1}\mathbf{A}^{-1} = \left[\mathbf{1} - \frac{(\mathbf{A}^{-1}\mathbf{U})\mathbf{V}^\mathrm{T}}{1 + \mathbf{V}^\mathrm{T}\mathbf{A}^{-1}\mathbf{U}}\right]\mathbf{A}^{-1} \qquad (11S.5.7)$$

which upon expansion gives (11E.5.4).

3. Our task is to determine $\mathbf{M}$ such that the following equation is satisfied:

$$\mathbf{BB}^{-1} = \left[\mathbf{A} + (\mathbf{UV})\mathbf{S}\begin{pmatrix}\mathbf{U}^\mathrm{T}\\\mathbf{V}^\mathrm{T}\end{pmatrix}\right]\left[\mathbf{A}^{-1} + \mathbf{A}^{-1}(\mathbf{UV})\mathbf{M}\begin{pmatrix}\mathbf{U}^\mathrm{T}\\\mathbf{V}^\mathrm{T}\end{pmatrix}\mathbf{A}^{-1}\right] = \mathbf{1} \qquad (11S.5.8)$$

Expanding the product and carrying out some simple manipulations, we arrive at the following equivalent equation for $\mathbf{M}$:

$$\mathbf{M} + \mathbf{S} + \mathbf{S}\begin{pmatrix}\mathbf{U}^\mathrm{T}\\\mathbf{V}^\mathrm{T}\end{pmatrix}\mathbf{A}^{-1}(\mathbf{UV})\mathbf{M} = \mathbf{0} \qquad (11S.5.9)$$

Solving this equation for $\mathbf{M}$, we obtain

$$\mathbf{M} = -\left[\mathbf{1} + \mathbf{S}\begin{pmatrix}\mathbf{U}^{\mathrm{T}}\\\mathbf{V}^{\mathrm{T}}\end{pmatrix}\mathbf{A}^{-1}(\mathbf{UV})\right]^{-1}\mathbf{S} \tag{11S.5.10}$$

which is equivalent to (11E.5.7).

SOLUTION 11.6

Substituting (11E.6.2), (11E.6.3) and (11E.6.4) into the eigenvalue equation (11E.6.1) and neglecting all terms quadratic in the corrections, we obtain

$$(\mathbf{H}_0 - E^{(0)}\mathbf{1})\mathbf{C}^{(1)} = -[(\mathbf{H} - E^{(0)}\mathbf{1})\mathbf{C}^{(0)} - E^{(1)}\mathbf{C}^{(0)}] \tag{11S.6.1}$$

which may be written in the form

$$\mathbf{C}^{(1)} = -(\mathbf{H}_0 - E^{(0)}\mathbf{1})^{-1}[(\mathbf{H} - E^{(0)}\mathbf{1})\mathbf{C}^{(0)} - E^{(1)}\mathbf{C}^{(0)}] \tag{11S.6.2}$$

Multiplication by $\mathbf{C}^{(0)\mathrm{T}}$ from the left and use of relation (11E.6.7) give an expression for the energy correction:

$$E^{(1)} = \frac{\mathbf{C}^{(0)\mathrm{T}}(\mathbf{H}_0 - E^{(0)}\mathbf{1})^{-1}(\mathbf{H} - E^{(0)}\mathbf{1})\mathbf{C}^{(0)}}{\mathbf{C}^{(0)\mathrm{T}}(\mathbf{H}_0 - E^{(0)}\mathbf{1})^{-1}\mathbf{C}^{(0)}} \tag{11S.6.3}$$

Inserting this expression into (11S.6.2), we arrive at (11E.6.6).

SOLUTION 11.7

1. From (2.6.10), we obtain the following CSF expansion coefficients for the two determinants:

$$^{\mathrm{d}}\langle\mathbf{p}|\mathbf{t}\rangle^{\mathrm{c}} = \prod_{k=1}^{2m} C_{t_k,p_k}^{T_k,P_k} \tag{11S.7.1}$$

$$^{\mathrm{d}}\langle-\mathbf{p}|\mathbf{t}\rangle^{\mathrm{c}} = \prod_{k=1}^{2m} C_{t_k,-p_k}^{T_k,-P_k} \tag{11S.7.2}$$

Using (2.6.5) and (2.6.6), we find that the coupling coefficients are related as follows:

$$C_{1/2,1/2}^{S,M} = C_{1/2,-1/2}^{S,-M} = \sqrt{\frac{S+M}{2S}} \tag{11S.7.3}$$

$$C_{-1/2,1/2}^{S,M} = -C_{-1/2,-1/2}^{S,-M} = -\sqrt{\frac{S+1-M}{2(S+1)}} \tag{11S.7.4}$$

which allows us to establish the following relation between (11S.7.1) and (11S.7.2):

$$^{\mathrm{d}}\langle-\mathbf{p}|\mathbf{t}\rangle^{\mathrm{c}} = (-1)^{N_t^-}\prod_{k=1}^{2m} C_{t_k,p_k}^{T_k,P_k} = (-1)^{N_t^-}\,{}^{\mathrm{d}}\langle\mathbf{p}|\mathbf{t}\rangle^{\mathrm{c}} \tag{11S.7.5}$$

where N_t^- is the number of spin-down couplings in the CSF.

2. The operator $\hat{U}_{\alpha\beta}$ can be written in the product form

$$\hat{U}_{\alpha\beta} = \hat{U}^{os}_{\alpha\beta}\hat{U}^{cs}_{\alpha\beta} \tag{11S.7.6}$$

where $\hat{U}^{os}_{\alpha\beta}$ operates on the singly occupied open-shell orbitals and $\hat{U}^{cs}_{\alpha\beta}$ on the doubly occupied closed-shell orbitals. Applying $\hat{U}^{os}_{\alpha\beta}$ to the CSF, we obtain

$$\hat{U}^{os}_{\alpha\beta}|\text{CSF}\rangle = \hat{U}^{os}_{\alpha\beta}\sum_i |^i\mathbf{p}\rangle^{\text{d}} \ ^{\text{d}}\langle ^i\mathbf{p}|\mathbf{t}\rangle^{\text{c}} = \sum_i |-^ip\rangle^{\text{d}} \ ^{\text{d}}\langle ^i\mathbf{p}|\mathbf{t}\rangle^{\text{c}}$$

$$= (-1)^{N^-_t}\sum_i |-^i\mathbf{p}\rangle^{\text{d}} \ ^{\text{d}}\langle -^i\mathbf{p}|\mathbf{t}\rangle^{\text{c}} = (-1)^{N^-_t}|\text{CSF}\rangle \tag{11S.7.7}$$

The numbers of spin-down and spin-up couplings N^-_t and N^+_t are related to the total spin S and the number of open-shell electrons $2m$ by

$$N^+_t - N^-_t = 2S \tag{11S.7.8}$$

$$N^+_t + N^-_t = 2m \tag{11S.7.9}$$

which gives us

$$N^-_t = m - S \tag{11S.7.10}$$

The permutation of the unpaired electrons in the CSF can thus be written as

$$\hat{U}^{os}_{\alpha\beta}|\text{CSF}\rangle = (-1)^{m-S}|\text{CSF}\rangle \tag{11S.7.11}$$

To establish the corresponding relation for the permutation of the closed-shell electrons, we note that the $2N-2m$ closed-shell electrons occupy $N-m$ orbitals. Assuming as usual that the spin orbitals are occupied in the order $|\ldots, i\alpha, i\beta, j\alpha, j\beta, \ldots|$, we obtain

$$\hat{U}^{cs}_{\alpha\beta}|\text{CSF}\rangle = (-1)^{N-m}|\text{CSF}\rangle \tag{11S.7.12}$$

Combining (11S.7.11) and (11S.7.12), we obtain the following result for the application of the spin-flip operator to a CSF with $2m$ singly occupied orbitals and total spin projection zero:

$$\hat{U}_{\alpha\beta}|\text{CSF}\rangle = (-1)^{N-m}(-1)^{m-S}|\text{CSF}\rangle = (-1)^{N-S}|\text{CSF}\rangle \tag{11S.7.13}$$

3. Applying the spin-flip operator to the wave function $|0\rangle$, we obtain

$$\hat{U}_{\alpha\beta}|0\rangle = \sum_{I_\alpha I_\beta} C_{I_\alpha I_\beta}\hat{U}_{\alpha\beta}|\alpha(I_\alpha)\beta(I_\beta)\rangle = \sum_{I_\alpha I_\beta} C_{I_\alpha I_\beta}|\beta(I_\alpha)\alpha(I_\beta)\rangle$$

$$= \sum_{I_\alpha I_\beta} C_{I_\beta I_\alpha}|\beta(I_\beta)\alpha(I_\alpha)\rangle = (-1)^N\sum_{I_\alpha I_\beta} C_{I_\beta I_\alpha}|\alpha(I_\alpha)\beta(I_\beta)\rangle \tag{11S.7.14}$$

where the sign factor $(-1)^N$ arises from moving the N beta electrons past the N alpha electrons. On the other hand, using (11S.7.13), we find

$$\hat{U}_{\alpha\beta}|0\rangle = (-1)^{N-S}\sum_{I_\alpha I_\beta} C_{I_\alpha I_\beta}|\alpha(I_\alpha)\beta(I_\beta)\rangle \tag{11S.7.15}$$

Comparing (11S.7.14) and (11S.7.15), we obtain

$$(-1)^N C_{I_\beta I_\alpha} = (-1)^{N-S} C_{I_\alpha I_\beta} \tag{11S.7.16}$$

from which we arrive at (11E.7.5).

SOLUTION 11.8

1. According to (11.6.5), the total number of determinants is given by

$$N_{det} = \binom{n}{N_\alpha}\binom{n}{N_\beta} = \binom{6}{4}^2 = 15^2 = 225 \tag{11S.8.1}$$

2. See Figure 11S.8.1.
3. See Figure 11S.8.2.

SOLUTION 11.9

1. Since at least one of the first three orbitals must be occupied, the extreme string paths are $a_1^\dagger a_2^\dagger a_3^\dagger$ and $a_3^\dagger a_5^\dagger a_6^\dagger$. The string graph with vertex and arc weights is given in Figure 11S.9.1.
2. See Figure 11S.9.2.
3. The string classes are listed in Table 11S.9.1.
4. The allowed combinations of string classes are listed in Table 11S.9.2. The total number of determinants is 118.

SOLUTION 11.10

1. In this system, RAS1 contains at least 18 electrons in the 10 occupied orbitals, RAS2 is empty, and RAS3 contains at most 2 electrons in the 100 virtual orbitals. These restrictions give rise

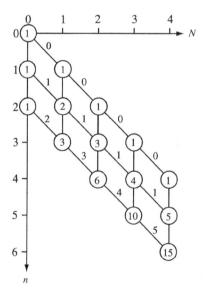

Fig. 11S.8.1. String graph with vertex and arc weights for four electrons distributed among six orbitals without restrictions on the orbital occupations.

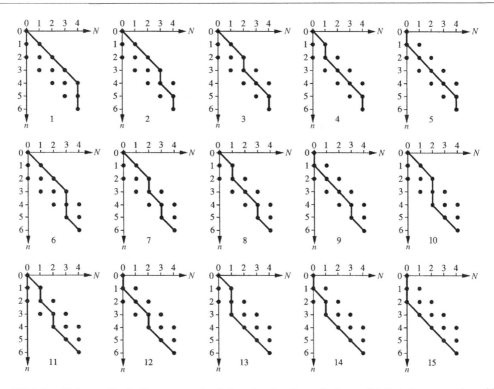

Fig. 11S.8.2. String paths in the reverse lexical order for four electrons distributed among six orbitals without restrictions on the orbital occupations.

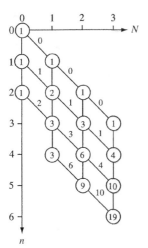

Fig. 11S.9.1. String graph with vertex and arc weights for four electrons distributed among six orbitals with the restriction that at most two electrons are allowed in the last three orbitals.

to the three string classes in Table 11S.10.1. For each class, we have calculated the number of (alpha or beta) strings according to the expression

$$N_{\text{str}}^{\sigma} = \binom{10}{N_1}\binom{100}{N_3} \qquad (11S.10.1)$$

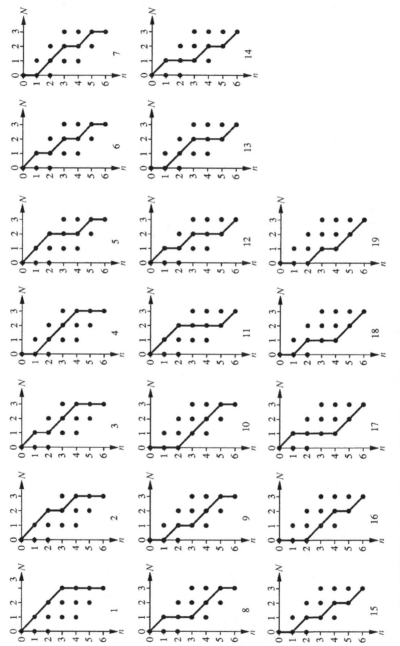

Fig. 11S.9.2.   String paths for four electrons in six orbitals with at most two electrons in the last three orbitals.

Table 11S.9.1   String classes for a system of four electrons distributed among six orbitals with at most two electrons in the last three orbitals

String class	Strings	Electrons in 4, 5 and 6
T_{I}^{α}	$\hat{\alpha}_1$	0
T_{II}^{α}	$\hat{\alpha}_2, \hat{\alpha}_3, \hat{\alpha}_4, \hat{\alpha}_5, \hat{\alpha}_6, \hat{\alpha}_7, \hat{\alpha}_{11}, \hat{\alpha}_{12}, \hat{\alpha}_{13}$	1
$T_{\mathrm{III}}^{\alpha}$	$\hat{\alpha}_8, \hat{\alpha}_9, \hat{\alpha}_{10}, \hat{\alpha}_{14}, \hat{\alpha}_{15}, \hat{\alpha}_{16}, \hat{\alpha}_{17}, \hat{\alpha}_{18}, \hat{\alpha}_{19}$	2

Table 11S.9.2   The allowed combinations of the string classes of Table 11S.9.1. For each combination, the number of string products (determinants) is given

	T_{I}^{β}	T_{II}^{β}	T_{III}^{β}
T_{I}^{α}	1	9	9
T_{II}^{α}	9	81	forbidden
$T_{\mathrm{III}}^{\alpha}$	9	forbidden	forbidden

where N_1 and N_3 are the numbers of electrons in RAS1 and RAS3, respectively. The number of determinants in each allowed combination of classes is listed in Table 11S.10.2. The total number of determinants is 1 447 501.

2. In this system, RAS1 contains at least 18 electrons in 10 orbitals, RAS2 contains 25 orbitals with no constraints on the occupations, and RAS3 contains at most one electron in 75 orbitals. These restrictions give rise to the five classes in Table 11S.10.3, where we have calculated the number of strings of each class from the expression

$$N_{\mathrm{str}}^{\sigma} = \binom{10}{N_1}\binom{25}{N_2}\binom{75}{N_3} \qquad (11S.10.2)$$

in which N_1 is the number of electrons in RAS1, N_2 the number of electrons in RAS2, and N_3 the number of electrons in RAS3. In Table 11S.10.4, we have listed the number of determinants for each allowed combination of classes. The total number of determinants becomes 635 251.

Table 11S.10.1   String classes and the number of strings in each class for Exercise 11.10.1

	N_1	N_2	N_3	Number of strings
T_{I}^{σ}	10	0	0	1
T_{II}^{σ}	9	0	1	1 000
$T_{\mathrm{III}}^{\sigma}$	8	0	2	222 750

Table 11S.10.2   String combinations and the number of determinants of each combination for Exercise 11.10.1

	T_{I}^{β}	T_{II}^{β}	T_{III}^{β}
T_{I}^{α}	1	1 000	222 750
T_{II}^{α}	1 000	1 000 000	forbidden
$T_{\mathrm{III}}^{\alpha}$	222 750	forbidden	forbidden

Table 11S.10.3 String classes and the number of strings in each class for Exercise 11.10.2

	N_1	N_2	N_3	Number of strings
T_I^σ	10	0	0	1
T_{II}^σ	9	1	0	250
T_{III}^σ	9	0	1	750
T_{IV}^σ	8	2	0	13 500
T_V^σ	8	1	1	84 375

Table 11S.10.4 String combinations and the number of determinants of each combination for Exercise 11.10.2

	T_I^β	T_{II}^β	T_{III}^β	T_{IV}^β	T_V^β
T_I^α	1	250	750	13 500	84 375
T_{II}^α	250	62 500	187 500	forbidden	forbidden
T_{III}^α	750	187 500	forbidden	forbidden	forbidden
T_{IV}^α	13 500	forbidden	forbidden	forbidden	forbidden
T_V^α	84 375	forbidden	forbidden	forbidden	forbidden

SOLUTION 11.11

1. There are four Slater determinants with $M = 0$:

$$|1\rangle^d = a_{1\alpha}^\dagger a_{1\beta}^\dagger |\text{vac}\rangle \tag{11S.11.1}$$

$$|2\rangle^d = a_{2\alpha}^\dagger a_{1\beta}^\dagger |\text{vac}\rangle \tag{11S.11.2}$$

$$|3\rangle^d = a_{1\alpha}^\dagger a_{2\beta}^\dagger |\text{vac}\rangle \tag{11S.11.3}$$

$$|4\rangle^d = a_{2\alpha}^\dagger a_{2\beta}^\dagger |\text{vac}\rangle \tag{11S.11.4}$$

Evaluating the Hamiltonian in the basis of these determinants, we obtain the elements listed in Table 11S.11.1. The zeros occur because of the Brillouin theorem.

Table 11S.11.1 The elements of the Hamiltonian (E_h) for the HeH$^+$ system in the determinantal basis (11S.11.1)–(11S.11.4) at the internuclear distance of $1.463796a_0$

	Matrix element	STO-3G
11	$2h_{11} + g_{1111} + h_{\text{nuc}}$	-2.84189825
21	$h_{21} + g_{2111}$	0.00000000
22	$h_{11} + h_{22} + g_{2211} + h_{\text{nuc}}$	-1.89682271
31	$h_{21} + g_{2111}$	0.00000000
32	g_{2121}	0.14536406
33	$h_{11} + h_{22} + g_{2211} + h_{\text{nuc}}$	-1.89682271
41	g_{2121}	0.14536406
42	$h_{21} + g_{2221}$	-0.21030429
43	$h_{21} + g_{2221}$	-0.21030429
44	$2h_{22} + g_{2222} + h_{\text{nuc}}$	-0.57640710

2. From the determinants in Exercise 11.11.1, we may construct three singlet CSFs:

$$|1\rangle^c = |1\rangle^d \tag{11S.11.5}$$

$$|2\rangle^c = \frac{1}{\sqrt{2}}(|2\rangle^d + |3\rangle^d) \tag{11S.11.6}$$

$$|3\rangle^c = |4\rangle^d \tag{11S.11.7}$$

The only matrix elements modified by the CSF transformation are

$$H_{21} = \frac{1}{\sqrt{2}}H_{31}^{dd} + \frac{1}{\sqrt{2}}H_{21}^{dd} = 0 \tag{11S.11.8}$$

$$H_{22} = \frac{1}{2}H_{33}^{dd} + \frac{1}{2}H_{22}^{dd} + H_{32}^{dd} = -1.75145864 \tag{11S.11.9}$$

$$H_{32} = \frac{1}{\sqrt{2}}H_{43}^{dd} + \frac{1}{\sqrt{2}}H_{42}^{dd} = -0.29741518 \tag{11S.11.10}$$

giving rise to the following Hamiltonian matrix:

$$\mathbf{H} = \begin{pmatrix} -2.84189825 & 0.00000000 & 0.14536406 \\ 0.00000000 & -1.75145864 & -0.29741518 \\ 0.14536406 & -0.29741518 & -0.57640710 \end{pmatrix} \tag{11S.11.11}$$

3. From (11E.11.1), we obtain the residual vector

$$\mathbf{R}_1 = \begin{pmatrix} 0.00000000 \\ 0.00000000 \\ 0.14536406 \end{pmatrix} \tag{11S.11.12}$$

Next, we generate

$$\mathbf{P} = \begin{pmatrix} 0.00000000 \\ 0.00000000 \\ 0.06416448 \end{pmatrix} \tag{11S.11.13}$$

using (11E.11.2) but taking care to avoid the singularity in $(\mathbf{H}_0 - E_1\mathbf{1})^{-1}$, which is balanced by a zero in $\mathbf{R}_1$. Normalizing (11S.11.13), we obtain the vector that represents the new direction:

$$\mathbf{d} = \begin{pmatrix} 0 \\ 0 \\ 1 \end{pmatrix} \tag{11S.11.14}$$

Diagonalization of the Hamiltonian matrix constructed in the basis of $\mathbf{C}_1$ and $\mathbf{d}$

$$\mathbf{H}_{sub} = \begin{pmatrix} -2.84189825 & 0.14536406 \\ 0.14536406 & -0.57640710 \end{pmatrix} \tag{11S.11.15}$$

gives the following lowest eigenvalue and associated eigenvector:

$$E_2 = -2.85118738 \tag{11S.11.16}$$

$$\begin{pmatrix} x_1 \\ x_2 \end{pmatrix} = \begin{pmatrix} 0.99796447 \\ -0.06377239 \end{pmatrix} \tag{11S.11.17}$$

Finally, we arrive at the improved approximate CI wave function by using (11E.11.4):

$$\mathbf{C}_2 = \begin{pmatrix} 0.99796447 \\ 0.00000000 \\ -0.06377239 \end{pmatrix} \tag{11S.11.18}$$

Using $\mathbf{C}_2$ and E_2 in place of $\mathbf{C}_1$ and E_1, we may begin a new CI iteration.

4. As seen from Table 11S.11.2, the energy converges significantly faster than the wave function. Whereas the norm of the residual vector is roughly linear in the error in the wave function, the error in the energy is quadratic in the error of the wave function.

5. The gradient and Hessian are obtained from (11.4.23) and (11.4.24) as

$$\mathbf{E}^{(1)} = 2(\mathbf{H} - E_1\mathbf{1})\mathbf{C}_1 = \begin{pmatrix} 0.00000000 \\ 0.00000000 \\ 0.29072812 \end{pmatrix} \tag{11S.11.19}$$

$$\mathbf{E}^{(2)} = 2\mathbf{P}_1(\mathbf{H} - E_1\mathbf{1})\mathbf{P}_1 = \begin{pmatrix} 0.00000000 & 0.00000000 & 0.00000000 \\ 0.00000000 & 2.18087922 & -0.59483037 \\ 0.00000000 & -0.59483037 & 4.53098231 \end{pmatrix} \tag{11S.11.20}$$

where $\mathbf{P}_1$ is the projection matrix

$$\mathbf{P}_1 = \mathbf{1} - \mathbf{C}_1\mathbf{C}_1^{\mathrm{T}} \tag{11S.11.21}$$

The Hessian (11S.11.20) is clearly singular. As discussed in Section 11.5.1, we may instead solve the Newton equations by using the modified, nonsingular Hessian (11.5.8):

$$\mathbf{G}^{(2)} = \mathbf{E}^{(2)} + \mathbf{C}_1\mathbf{C}_1^{\mathrm{T}} = \begin{pmatrix} 1.00000000 & 0.00000000 & 0.00000000 \\ 0.00000000 & 2.18087922 & -0.59483037 \\ 0.00000000 & -0.59483037 & 4.53098231 \end{pmatrix} \tag{11S.11.22}$$

From the modified Newton equations

$$\mathbf{G}^{(2)}\mathbf{c}_1 = -\mathbf{E}^{(1)} \tag{11S.11.23}$$

we then obtain the solution vector

$$\mathbf{c}_1 = \begin{pmatrix} 0.00000000 \\ -0.01815064 \\ -0.06654731 \end{pmatrix} \tag{11S.11.24}$$

from which the following improved CI vector is generated:

$$\mathbf{C}_2 = \frac{\mathbf{C}_1 + \mathbf{c}_1}{\|\mathbf{C}_1 + \mathbf{c}_1\|} = \begin{pmatrix} 0.99762946 \\ -0.01810762 \\ -0.06638956 \end{pmatrix} \tag{11S.11.25}$$

The energy calculated as the expectation value of $\mathbf{C}_2$ is -2.85152603 E_{h}.

Table 11S.11.2 Convergence of the Davidson iterations for the HeH$^+$ system (E_{h})

Iteration	$E_n - E_{\mathrm{cnv}}$	$\|\mathbf{C}_n - \mathbf{C}_{\mathrm{cnv}}\|$	$\|\mathbf{R}_n\|$
1	0.00962802	0.06849108	0.14536406
2	0.00033890	0.01801593	0.01896688
3	0.00001188	0.00240411	0.00511676
4	0.00000011	0.00032320	0.00035141

Table 11S.11.3 Convergence of the Newton iterations for the HeH$^+$ system (E_h)

Iteration	$E_n - E_{cnv}$	$\|\mathbf{C}_n - \mathbf{C}_{cnv}\|$	$\|\mathbf{R}_n\|$
1	0.00962802	0.06849108	0.14536406
2	0.00000024	0.00039526	0.00066411

6. All three error measures in Table 11S.11.3 show cubic convergence. Comparing with the Davidson method in Table 11S.11.2, we find that, for this particular system, one Newton iteration corresponds to three Davidson iterations.

12 MULTICONFIGURATIONAL SELF-CONSISTENT FIELD THEORY

Electronic wave functions are often dominated by more than one electronic configuration. Examples of multiconfigurational wave functions are found in the electronic ground state of the ozone molecule, in the dissociation of the nitrogen molecule and in symmetry-forbidden reactions. The presence of several important configurations poses a difficult challenge for *ab initio* electronic-structure theory. The single-configuration Hartree–Fock approximation, by its very construction, is incapable of representing systems dominated by several configurations. By the same argument, methods designed to improve on the Hartree–Fock description by taking into account the effects of dynamical correlation, such as the coupled-cluster and Møller–Plesset methods, are also not suitable for such systems. Furthermore, to carry out a CI calculation 'on top of the Hartree–Fock calculation' is problematic as well since the Hartree–Fock model is inappropriate as an orbital generator: the orbitals generated self-consistently in the field of a single electronic configuration may have little or no relevance to a multiconfigurational system.

An obvious solution to the multiconfigurational problem is to carry out a CI calculation where the orbitals are variationally optimized simultaneously with the coefficients of the electronic configurations, thereby ensuring that the orbitals employed in the wave function are optimal for the problem at hand and do not introduce a bias (towards a particular configuration) in the calculations. This approach is referred to as the *multiconfigurational self-consistent field (MCSCF) method*. As discussed in Section 5.5, the MCSCF method represents a flexible solution to the multiconfigurational problem in quantum chemistry – it may be used either as a wave function in its own right (for a qualitative description of the electronic system) or as an orbital generator for more elaborate treatments of the electronic structure. In the present chapter, we discuss the MCSCF model with emphasis on the technical aspects related to its construction and optimization.

12.1 The MCSCF model

In MCSCF theory, the wave function is written as a linear combination of Slater determinants or CSFs

$$|\kappa, \mathbf{C}\rangle = \exp(-\hat{\kappa}) \sum_i C_i |i\rangle \tag{12.1.1}$$

and the wave function is constructed by variationally optimizing the expectation value of the energy

$$E = \min_{\kappa, \mathbf{C}} \frac{\langle \kappa, \mathbf{C} | \hat{H} | \kappa, \mathbf{C} \rangle}{\langle \kappa, \mathbf{C} | \kappa, \mathbf{C} \rangle} \tag{12.1.2}$$

with respect to the parameters of the orbital-rotation operator $\exp(-\hat{\kappa})$ (which carries out unitary transformations of the orbitals as discussed in Section 3.2) and the CI coefficients C_i. Except for the presence of $\exp(-\hat{\kappa})$, the parametrization of the MCSCF model (12.1.1) is identical to that of the CI model (11.1.1). The orbital-rotation operator allows us to tailor the orbitals to the CI expansion, which is particularly important for systems with several important electronic configurations and large static correlation effects; see the discussion in Section 5.5. At convergence, the MCSCF wave function is usually expanded in an orbital basis where $\hat{\kappa}$ is zero; the MCSCF wave function (12.1.1) then takes the form of a standard CI wave function (11.1.1), the only difference being that the MCSCF orbitals are variationally optimal for the wave function.

In practice, the simultaneous optimization of orbitals and CI coefficients is a difficult nonlinear problem, which severely restricts the length of MCSCF expansions relative to those of CI wave functions. By itself, the MCSCF model is therefore not suited to the treatment of dynamical correlation (for which large basis sets and long configuration expansions are needed) but it may be used in conjunction with a subsequent correlation treatment by a more extensive multireference CI wave function, providing the reference configurations and orbitals needed for such treatments.

Often, the most difficult question arising in an MCSCF calculation is the selection of the configuration space. In early applications of MCSCF theory to chemical problems, the configurations were selected individually, based on physical insight and intuition. Nowadays, the selection of the configurations proceeds in a more systematic manner, with whole classes of configurations selected simultaneously, as discussed for truncated CI wave functions in Section 11.1.3. The underlying idea of such selection schemes is the partitioning of the orbital space into subspaces, characterized by certain restrictions with respect to the occupations of the configurations entering the MCSCF wave function.

In *complete active space (CAS) calculations*, the total orbital space is partitioned into a set of inactive orbitals, a set of active orbitals and a set of secondary orbitals [1]. In the *CASSCF wave function*, the *inactive orbitals* are doubly occupied in all configurations, the *active orbitals* are subject to no restrictions on their occupations and the *secondary orbitals* are unoccupied in all configurations. In *restricted active space (RAS) calculations*, the active orbital space is further divided into three subspaces [2–4]: *RAS1*, with an upper limit on the allowed number of holes in each configuration; *RAS2*, with no restrictions enforced on the occupations; and *RAS3*, with an upper limit on the allowed number of electrons in each configuration. In CASSCF and RASSCF wave functions, the CI coefficients and the orbitals are variationally optimized, using the configurations generated within the CAS and RAS restrictions. For examples of CAS and RAS expansions, see Sections 5.5 and 11.1.

Nowadays, MCSCF calculations can be carried out routinely for several hundred orbitals, of which 50–100 are inactive and 10–20 active. CASSCF wave functions may contain several million determinants, generated by distributing 12–14 electrons among the same number of active orbitals. RASSCF wave functions, which are more difficult to optimize, usually contain fewer configurations. However, because of the additional restrictions on the occupations of the active orbital space in such wave functions, significantly larger active spaces can be treated in RAS theory than in CAS theory, allowing us to recover the main effects of dynamical correlation in small molecular systems using RASSCF wave functions.

The optimization of MCSCF wave functions represents a difficult computational problem. The MCSCF energy function is highly nonlinear, with a potentially large number of local solutions and sometimes strong couplings among the different degrees of freedom. Indeed, the problems that are sometimes encountered in connection with the optimization of Hartree–Fock wave functions are often strongly compounded in MCSCF theory, where we aim to optimize simultaneously the

MOs and the configuration expansion. To be successful, we must choose the parametrization of the MCSCF wave function with care and apply an algorithm for the optimization that is robust as well as efficient. The first attempts at developing MCSCF optimization schemes, which borrowed heavily from the standard first-order methods of single-configuration Hartree–Fock theory, were not successful. With the introduction of second-order methods and the exponential parametrization of the orbital space, the calculation of MCSCF wave functions became routine. Still, even with the application of second-order methods, the optimization of MCSCF wave functions can be difficult – more difficult than for the other wave functions treated in this book. A large part of the present chapter is therefore devoted to the discussion of MCSCF optimization techniques.

The MCSCF method is not a black-box method and cannot easily be applied by nonspecialists. The optimized MCSCF state should always be inspected carefully to ensure that it properly represents the desired electronic state. Nevertheless, the MCSCF model is a highly flexible one and it still constitutes the only approach that, in a balanced manner, can describe bond breakings and molecular dissociation processes, where an unbiased treatment of several electronic configurations is required. The small number of active orbitals that can be treated in MCSCF theory usually makes it impossible to treat dynamical correlation and only a qualitatively correct description of the electronic system can therefore be expected.

12.2 The MCSCF energy and wave function

In the present section, we introduce the tools needed for optimization and manipulation of the MCSCF wave function. Having introduced the parametrization of the wave function in Section 12.2.1, we consider the expansion of the MCSCF energy in Section 12.2.2 and derive expressions for the electronic gradient and Hessian in Section 12.2.3. Some special properties of the MCSCF gradient and Hessian are then considered in Section 12.2.4 and 12.2.5. After a discussion of the important question of redundant orbital rotations in Section 12.2.6, we conclude this section with a discussion of the structure of MCSCF stationary points in Sections 12.2.7 and 12.2.8.

12.2.1 THE PARAMETRIZATION OF THE MCSCF STATE

Our parametrization of the MCSCF state is a straightforward combination of the parametrizations of the Hartree–Fock state in Chapter 10 and of the CI state in Chapter 11. The orbital variations are carried out by means of an exponential unitary operator and the variations in the configuration space are expressed by means of a configuration vector added orthogonally to the reference state:

$$|\mathbf{C}\rangle = \exp(-\hat{\kappa})\frac{|0\rangle + \hat{P}|\mathbf{c}\rangle}{\sqrt{1 + \langle\mathbf{c}|\hat{P}|\mathbf{c}\rangle}} \tag{12.2.1}$$

The *reference state* $|0\rangle$ represents our current approximation to the electronic state

$$|0\rangle = \sum_i C_i^{(0)}|i\rangle \tag{12.2.2}$$

and is normalized to unity

$$\langle 0|0\rangle = \sum_i (C_i^{(0)})^2 = 1 \tag{12.2.3}$$

All variations in the MCSCF configuration space will be carried out relative to this reference state. In (12.2.1), the state $|\mathbf{c}\rangle$ contains the free parameters

$$|\mathbf{c}\rangle = \sum_i c_i |i\rangle \qquad (12.2.4)$$

and $\hat{P}$ is the projection operator

$$\hat{P} = 1 - \hat{O} = 1 - |0\rangle\langle 0| \qquad (12.2.5)$$

which projects out the component of the reference state $|0\rangle$ from $|\mathbf{c}\rangle$.

The advantage of the present parametrization of the MCSCF wave function (12.2.1) over the somewhat simpler parametrization (12.1.1) is that the former is normalized and allows for a simpler treatment of the redundancy in the configuration space – see the discussion in Section 11.4.1. In general, redundancies are undesirable as they may interfere with the optimization of the wave function. In (12.2.1), however, the redundancy of the CI space is parallel to the reference vector since $|\mathbf{c}\rangle = \alpha|0\rangle$ does not change $|\mathbf{C}\rangle$ from $|0\rangle$. The redundancy is therefore easily controlled and will not create problems in our development of MCSCF optimization algorithms. On the contrary, we shall be able to turn this particular redundancy to our advantage, using it in Section 12.4 to develop a scheme that allows us to converge more easily to the desired stationary point. An alternative, nonredundant parametrization will be presented in Section 12.6.

Let us briefly comment on the parametrization of the MCSCF orbitals. The unitary parametrization of the orbital space in (12.2.1) was developed in Chapter 3. We shall here be concerned only with real rotations that preserve spin. Following the discussion in Section 3.3, we may then write $\hat{\kappa}$ in the form

$$\hat{\kappa} = \sum_{p>q} \kappa_{pq} E_{pq}^- \qquad (12.2.6)$$

where the operators E_{pq}^- are antisymmetric combinations of excitation operators

$$E_{pq}^- = E_{pq} - E_{qp} \qquad (12.2.7)$$

The elements κ_{pq} with $p > q$ constitute a set of real parameters that may be regarded as the elements of a vector $\boldsymbol{\kappa}$. Occasionally, we shall use the same notation $\boldsymbol{\kappa}$ for an antisymmetric matrix containing the elements κ_{pq}, with $\kappa_{qp} = -\kappa_{pq}$ for $p > q$. This convention should not create problems, the meaning always being clear from the context.

The summation in (12.2.6) should be interpreted as involving only the nonredundant orbital rotations. A discussion of Hartree–Fock orbital redundancies was given in Section 10.1.2. In MCSCF theory, the question of orbital redundancies is more difficult because of the coupling to the configuration space. The MCSCF redundancies are therefore considered in more detail in Section 12.2.6, where we develop a set of simple conditions that allow us to identify the redundant operators in (12.2.6).

12.2.2 THE TAYLOR EXPANSION OF THE MCSCF ENERGY

Let us now consider the Taylor expansion of the MCSCF energy around the reference state. The energy of the MCSCF state (12.2.1) may be written in the form

$$E(\mathbf{c}, \boldsymbol{\kappa}) = \frac{W(\mathbf{c}, \boldsymbol{\kappa})}{S(\mathbf{c}, \boldsymbol{\kappa})} \qquad (12.2.8)$$

Here the numerator is given by the expression

$$W(\mathbf{c}, \kappa) = ((\langle \mathbf{c}|\hat{P} + \langle 0|) \exp(\hat{\kappa})\hat{H} \exp(-\hat{\kappa})(|0\rangle + \hat{P}|\mathbf{c}\rangle)) \tag{12.2.9}$$

and the denominator contains the overlap

$$S(\mathbf{c}, \kappa) = ((\langle \mathbf{c}|\hat{P} + \langle 0|) \exp(\hat{\kappa}) \exp(-\hat{\kappa})(|0\rangle + \hat{P}|\mathbf{c}\rangle)) = \langle 0|0\rangle + \langle \mathbf{c}|\hat{P}|\mathbf{c}\rangle \tag{12.2.10}$$

Since $S(\mathbf{c}, \kappa)$ is independent of κ, we shall in the following omit κ from the argument and write the overlap as $S(\mathbf{c})$. In the wave function and the energy, a general MCSCF parameter point is represented by the vector

$$\lambda = \begin{pmatrix} \mathbf{c} \\ \kappa \end{pmatrix} \tag{12.2.11}$$

where $\mathbf{c}$ and κ contain the configuration and orbital parameters, respectively.

In our development of MCSCF optimization theory in Section 12.3, the reference state will in each iteration be represented by the zero vector:

$$\lambda^{(0)} = \begin{pmatrix} \mathbf{c}^{(0)} \\ \kappa^{(0)} \end{pmatrix} = \begin{pmatrix} \mathbf{0} \\ \mathbf{0} \end{pmatrix} \tag{12.2.12}$$

We therefore begin each MCSCF iteration by a transformation of the Hamiltonian and the wave function in accordance with the step generated in the previous iteration. In practice, this means that we must transform the Hamiltonian one- and two-electron integrals from the AO basis to the MO basis in each iteration.

For an implementation of second-order optimization methods and for a characterization of the optimized state, we expand the MCSCF energy (12.2.8) to second order in the variational parameters. In terms of the parameters λ, the second-order MCSCF energy $Q(\lambda)$ is given by

$$Q(\lambda) = E^{(0)} + \mathbf{E}^{(1)\mathrm{T}}\lambda + \tfrac{1}{2}\lambda^{\mathrm{T}}\mathbf{E}^{(2)}\lambda \tag{12.2.13}$$

Here $E^{(0)}$ is the MCSCF energy at the expansion point

$$E^{(0)} = \langle 0|\hat{H}|0\rangle \tag{12.2.14}$$

and $\mathbf{E}^{(1)}$ and $\mathbf{E}^{(2)}$ are the electronic gradient and Hessian, respectively, at this point. The gradient vector and the Hessian matrix may be written in the following block form

$$\mathbf{E}^{(1)} = \begin{pmatrix} {}^{\mathrm{c}}\mathbf{E}^{(1)} \\ {}^{\mathrm{o}}\mathbf{E}^{(1)} \end{pmatrix} \tag{12.2.15}$$

$$\mathbf{E}^{(2)} = \begin{pmatrix} {}^{\mathrm{cc}}\mathbf{E}^{(2)} & {}^{\mathrm{co}}\mathbf{E}^{(2)} \\ {}^{\mathrm{oc}}\mathbf{E}^{(2)} & {}^{\mathrm{oo}}\mathbf{E}^{(2)} \end{pmatrix} \tag{12.2.16}$$

where the left superscripts c and o indicate differentiation with respect to configuration and orbital parameters, respectively:

$$^{\mathrm{c}}E_i^{(1)} = \left.\frac{\partial E}{\partial c_i}\right|_{\lambda=0} \tag{12.2.17}$$

$$^{\mathrm{o}}E_{pq}^{(1)} = \left.\frac{\partial E}{\partial \kappa_{pq}}\right|_{\lambda=0} \tag{12.2.18}$$

$$^{cc}E_{i,j}^{(2)} = \frac{\partial^2 E}{\partial c_i \partial c_j}\Bigg|_{\lambda=0} \tag{12.2.19}$$

$$^{oc}E_{pq,i}^{(2)} = {}^{co}E_{i,pq}^{(2)} = \frac{\partial^2 E}{\partial \kappa_{pq} \partial c_i}\Bigg|_{\lambda=0} \tag{12.2.20}$$

$$^{oo}E_{pq,rs}^{(2)} = \frac{\partial^2 E}{\partial \kappa_{pq} \partial \kappa_{rs}}\Bigg|_{\lambda=0} \tag{12.2.21}$$

Note that all derivatives are taken at the expansion point $\lambda = \mathbf{0}$. At this point, the derivatives take a particularly simple form, especially those with respect to κ_{pq} (see Exercise 10.1).

12.2.3 THE MCSCF ELECTRONIC GRADIENT AND HESSIAN

The elements of the MCSCF gradient and Hessian may be obtained by direct differentiation of the energy expression (12.2.8). In practice, it is easier to rewrite the energy in the form

$$S(\mathbf{c})E(\mathbf{c}, \boldsymbol{\kappa}) = W(\mathbf{c}, \boldsymbol{\kappa}) \tag{12.2.22}$$

Differentiation of this expression yields (in an obvious notation)

$$\mathbf{E}^{(1)} = \mathbf{W}^{(1)} - E^{(0)}\mathbf{S}^{(1)} \tag{12.2.23}$$

$$\mathbf{E}^{(2)} = \mathbf{W}^{(2)} - \mathbf{E}^{(1)}\mathbf{S}^{(1)\mathrm{T}} - \mathbf{S}^{(1)}\mathbf{E}^{(1)\mathrm{T}} - E^{(0)}\mathbf{S}^{(2)} \tag{12.2.24}$$

where we have assumed that the orbitals are real and where we have used the fact that

$$S^{(0)} = 1 \tag{12.2.25}$$

Further simplifications occur since the gradient $\mathbf{S}^{(1)}$ vanishes and since $^{cc}\mathbf{S}^{(2)}$ is the only nonzero block of $\mathbf{S}^{(2)}$; see (12.2.10). Carrying out the differentiations in (12.2.23) and (12.2.24), we arrive at the following compact expressions for the MCSCF electronic gradient:

$$^cE_i^{(1)} = 2\langle i|\hat{P}\hat{H}|0\rangle = 2\langle i|\hat{H}|0\rangle - 2C_i^{(0)}E^{(0)} \tag{12.2.26}$$

$$^oE_{pq}^{(1)} = \langle 0|[E_{pq}^-, \hat{H}]|0\rangle \tag{12.2.27}$$

As expected, the configuration part of the MCSCF gradient is similar to that of the CI wave function (11.4.23) and its orbital part is similar to that of the Hartree–Fock wave function (10.1.25). The elements of the MCSCF electronic Hessian are given by

$$^{cc}E_{i,j}^{(2)} = 2\langle i|\hat{P}(\hat{H} - E^{(0)})\hat{P}|j\rangle = 2\langle i|\hat{H} - E^{(0)}|j\rangle - C_i^{(0)}E_j^{(1)} - C_j^{(0)}E_i^{(1)} \tag{12.2.28}$$

$$^{oc}E_{pq,i}^{(2)} = 2\langle i|\hat{P}[E_{pq}^-, \hat{H}]|0\rangle = 2\langle i|[E_{pq}^-, \hat{H}]|0\rangle - 2C_i^{(0)}E_{pq}^{(1)} \tag{12.2.29}$$

$$^{oo}E_{pq,rs}^{(2)} = \tfrac{1}{2}(1 + P_{pq,rs})\langle 0|[E_{pq}^-, [E_{rs}^-, \hat{H}]]|0\rangle \tag{12.2.30}$$

where $P_{pq,rs}$ permutes the pair indices pq and rs. Again, we may compare the configuration–configuration part of the MCSCF Hessian with the CI electronic Hessian (11.4.24) and the orbital–orbital part with the Hartree–Fock electronic Hessian (10.1.26).

12.2.4 INVARIANCE OF THE SECOND-ORDER MCSCF ENERGY

As noted in Section 12.2.1, the parametrization of the wave function (12.2.1) is redundant in the sense that the MCSCF energy is unaffected by the addition of $|c\rangle = \alpha|0\rangle$ to the wave function. Since the MCSCF energy must be invariant to each order separately, we obtain for the second-order energy (12.2.13):

$$Q(\lambda + \alpha \zeta^{(0)}) = Q(\lambda) \tag{12.2.31}$$

$$\zeta^{(0)} = \begin{pmatrix} \mathbf{C}^{(0)} \\ \mathbf{0} \end{pmatrix} \tag{12.2.32}$$

This invariance may be verified by calculating the products of the MCSCF gradient and Hessian with (12.2.32). Using the expressions that contain the projection operator in (12.2.26), (12.2.28) and (12.2.29), we obtain trivially

$$\mathbf{E}^{(1)^{\mathrm{T}}} \zeta^{(0)} = 0 \tag{12.2.33}$$

$$\mathbf{E}^{(2)} \zeta^{(0)} = \mathbf{0} \tag{12.2.34}$$

The gradient and the Hessian contain no components along the reference CI vector and the second-order energy is thus invariant as indicated in (12.2.31). Note that (12.2.34) represents an eigenvalue problem, demonstrating that the electronic Hessian in the parametrization (12.2.1) is singular, containing at least one zero eigenvalue.

In our discussion of MCSCF theory, we shall find it convenient to work with the matrix representations of the projectors $\hat{P}$ and $\hat{O}$ of (12.2.5). These Hermitian matrices are given by

$$\mathbf{P} = \mathbf{1} - \mathbf{O} = \mathbf{1} - \zeta^{(0)} \zeta^{(0)^{\mathrm{T}}} \tag{12.2.35}$$

For these projectors, we note the following expressions involving the MCSCF electronic gradient and Hessian

$$\mathbf{PE}^{(1)} = \mathbf{E}^{(1)} \tag{12.2.36}$$

$$\mathbf{PE}^{(2)}\mathbf{P} = \mathbf{E}^{(2)} \tag{12.2.37}$$

and

$$\mathbf{OE}^{(1)} = \mathbf{0} \tag{12.2.38}$$

$$\mathbf{OE}^{(2)} = \mathbf{E}^{(2)}\mathbf{O} = \mathbf{0} \tag{12.2.39}$$

These simple relations are direct consequences of the redundancy of the CI parametrization (12.2.1), as manifested in (12.2.33) and (12.2.34).

12.2.5 RANK-1 CONTRIBUTIONS TO THE MCSCF ELECTRONIC HESSIAN

The elements of the MCSCF Hessian matrix in (12.2.28)–(12.2.30) contain contributions that depend on the electronic gradient and vanish at the stationary point. For future reference and manipulation, it is convenient to separate these gradient-containing terms from the remaining contributions to the electronic Hessian. Thus, introducing the notation

$$\mathbf{K}^{(2)} = \begin{pmatrix} {}^{cc}\mathbf{K}^{(2)} & {}^{co}\mathbf{K}^{(2)} \\ {}^{oc}\mathbf{K}^{(2)} & {}^{oo}\mathbf{K}^{(2)} \end{pmatrix} \tag{12.2.40}$$

where

$$^{cc}K_{i,j}^{(2)} = 2\langle i|\hat{H} - E^{(0)}|j\rangle \tag{12.2.41}$$

$$^{oc}K_{pq,i}^{(2)} = 2\langle i|[E_{pq}^-, \hat{H}]|0\rangle \tag{12.2.42}$$

$$^{oo}K_{pq,rs}^{(2)} = \tfrac{1}{2}(1 + P_{pq,rs})\langle 0|[E_{pq}^-, [E_{rs}^-, \hat{H}]]|0\rangle \tag{12.2.43}$$

we may write the MCSCF electronic Hessian in the form

$$\mathbf{E}^{(2)} = \mathbf{K}^{(2)} - \overline{\mathbf{E}}^{(1)}\boldsymbol{\zeta}^{(0)\mathrm{T}} - \boldsymbol{\zeta}^{(0)}\overline{\mathbf{E}}^{(1)\mathrm{T}} \tag{12.2.44}$$

where

$$\overline{\mathbf{E}}^{(1)} = \begin{pmatrix} ^{c}\mathbf{E}^{(1)} \\ 2^{o}\mathbf{E}^{(1)} \end{pmatrix} \tag{12.2.45}$$

Note the presence of the factor 2 in the orbital part of $\overline{\mathbf{E}}^{(1)}$. No significance should be attached to the form (12.2.44) except that it is convenient in many situations, displaying separately the rank-1 contributions.

We may separate the MCSCF electronic gradient in the same manner. Thus, introducing the vector

$$\mathbf{K}^{(1)} = \begin{pmatrix} ^{c}\mathbf{K}^{(1)} \\ ^{o}\mathbf{K}^{(1)} \end{pmatrix} \tag{12.2.46}$$

where

$$^{c}K_i^{(1)} = 2\langle i|\hat{H}|0\rangle \tag{12.2.47}$$

$$^{o}K_{pq}^{(1)} = \langle 0|[E_{pq}^-, \hat{H}]|0\rangle \tag{12.2.48}$$

we may write the MCSCF gradient (12.2.26) and (12.2.27) in the form

$$\mathbf{E}^{(1)} = \mathbf{K}^{(1)} - 2E^{(0)}\boldsymbol{\zeta}^{(0)} \tag{12.2.49}$$

by analogy with the expression for the Hessian (12.2.44).

12.2.6 REDUNDANT ORBITAL ROTATIONS

The MCSCF wave function in the form (12.2.1) is parametrized in terms of a set of configuration coefficients $\mathbf{c}$ and a set of orbital-rotation parameters $\boldsymbol{\kappa}$. For the purpose of optimizing the wave function, it is important to ensure that this set of parameters contains no redundancies except for that in the configuration space parallel to reference vector; see Section 12.2.4. Redundancies are undesirable since they may interfere with the optimization, introducing singularities in the Hessian and creating large rotations that may make it difficult or even impossible to achieve convergence. Moreover, redundancies may introduce complications when the wave function is subsequently employed to calculate molecular properties by the standard techniques of response theory.

We have already pointed out that the parametrization (12.2.1) contains a redundancy in the configuration part of the wave function since the addition of a vector parallel to the reference state does not change the energy. We shall not eliminate this redundancy – it has been identified and will be accommodated for in our design of the MCSCF optimization algorithm. The remaining parameters in $\mathbf{c}$ are nonredundant (at least in the absence of special symmetries) and need not concern us further here. Instead, we shall concentrate on identifying the redundancies in the orbital-rotation operator $\hat{\kappa}$.

Orbital redundancies were discussed in connection with Hartree–Fock theory in Sections 10.1.2 and 10.2.2. We recall that, in Hartree–Fock theory, all rotations that mix the inactive orbitals among one another (i.e. the inactive–inactive rotations) were identified as redundant, as were the virtual–virtual rotations. On the other hand, the inactive–virtual, inactive–active and active–virtual rotations were all classified as nonredundant.

In Hartree–Fock theory, complications arise only for open-shell systems, where the active–active rotations are in some cases redundant, in other cases nonredundant. For instance, for open-shell states constructed by distributing two electrons between two orbitals, we found in Section 10.1.2 that the active–active rotations are redundant for the triplet state but nonredundant for the singlet state. We can easily imagine that the situation becomes even more complicated in the MCSCF case, where the wave function is generated by optimizing simultaneously the orbital-rotation parameters and a (potentially) large number of CI coefficients. Fortunately, for the more common MCSCF models such as those based on the CAS and RAS concepts, the question of redundancies is simple and unexpected redundancies will only rarely arise.

A complete analysis of the redundancy problem for a nonlinearly parametrized wave function such as the MCSCF wave function is complicated and will not be attempted here. Instead, we shall be content with obtaining a set of simple conditions that are sufficient for identifying most (if not all) of the redundancies that occur in practice. As a starting point for our discussion, we classify a set of parameters A as redundant if there exists a smaller set B such that all states that can be described by A can also be described by B. For MCSCF wave functions in the form (12.2.1), this definition of redundancy means that, for any choice of $\mathbf{c}^A$ and $\hat{\kappa}^A$ of a redundant parameter set A, we may find a set of parameters $\hat{\kappa}^B$ and $\mathbf{c}^B$ belonging to a smaller set B such that

$$\frac{\exp(-\hat{\kappa}^B)(|0\rangle + \hat{P}|\mathbf{c}^B\rangle)}{\sqrt{1 + \langle \mathbf{c}^B|\hat{P}|\mathbf{c}^B\rangle}} = \frac{\exp(-\hat{\kappa}^A)(|0\rangle + \hat{P}|\mathbf{c}^A\rangle)}{\sqrt{1 + \langle \mathbf{c}^A|\hat{P}|\mathbf{c}^A\rangle}} \tag{12.2.50}$$

However, because of the nonlinear parametrization, this criterion is difficult to work with. We therefore adopt a more pragmatic approach and consider only those redundancies that may be identified by expanding the MCSCF wave function to first order. A parametrization A is then said to be redundant if there exists a smaller set B such that, for any value of $\mathbf{c}^A$ and $\hat{\kappa}^A$, we may find a set of parameters $\hat{\kappa}^B$ and $\mathbf{c}^B$ such that

$$-\hat{\kappa}^B|0\rangle + \hat{P}|\mathbf{c}^B\rangle = -\hat{\kappa}^A|0\rangle + \hat{P}|\mathbf{c}^A\rangle \tag{12.2.51}$$

This equation is obtained from (12.2.50) by retaining only the linear terms in the expansion of the wave functions. Assuming further that B is contained in A, we find that A is redundant provided there is a nontrivial (nonzero) solution to the linear equations (omitting the superscript A)

$$\hat{\kappa}|0\rangle = \hat{P}|\mathbf{c}\rangle \tag{12.2.52}$$

Clearly, our MCSCF parametrization is redundant in the sense that (12.2.52) is satisfied for $\mathbf{c} = \mathbf{C}^{(0)}$ and $\kappa = \mathbf{0}$. However, as noted above, we are not interested in this redundancy and shall instead concentrate on identifying redundancies among the orbital-rotation parameters.

According to (12.2.52), the orbital rotation induced by E_{pq}^- is redundant if its effect on $|0\rangle$ may be spanned by the first-order variations in the configuration space and in the remaining orbital-rotation parameters:

$$E_{pq}^-|0\rangle = \sum_i c_i \hat{P}|i\rangle + \sum_{rs \neq pq} \kappa_{rs} E_{rs}^-|0\rangle \tag{12.2.53}$$

Since $E_{pq}^-|0\rangle$ is orthogonal to the reference state

$$\langle 0|E_{pq}^-|0\rangle = \langle 0|E_{pq}|0\rangle - \langle 0|E_{qp}|0\rangle = 0 \tag{12.2.54}$$

we may write (12.2.53) in the form

$$E_{pq}^-|0\rangle = \sum_i c_i|i\rangle + \sum_{rs \neq pq} \kappa_{rs} E_{rs}^-|0\rangle \tag{12.2.55}$$

without the projection operator. This expression, which constitutes a sufficient rather than necessary condition for orbital redundancies, is much simpler to work with than the general condition (12.2.50). In practice, the even simpler condition

$$E_{pq}^-|0\rangle = \sum_i c_i|i\rangle \tag{12.2.56}$$

is usually sufficient to identify the redundancies that arise in MCSCF theory. Any search for redundant rotations should therefore begin with (12.2.56), which, for Hartree–Fock wave functions, reduces to the simple condition (see Section 10.1.2)

$$E_{pq}^-|\text{CSF}\rangle = 0 \tag{12.2.57}$$

If problems persist and additional redundancies are suspected in the MCSCF wave function, these redundancies may be detected from (12.2.55) or by solution of the homogeneous linear equations (12.2.52).

Let us now be more specific and consider redundancies in CASSCF theory using (12.2.56). We first note that all inactive–inactive and secondary–secondary orbital rotations are redundant; their redundancy may be demonstrated using the same arguments as for the Hartree–Fock function in Section 10.1.2, establishing that $E_{ij}^-|0\rangle$ and $E_{ab}^-|0\rangle$ are zero so that (12.2.56) is satisfied for $\mathbf{c} = \mathbf{0}$. Second, we observe that, for the active–active rotations, $E_{vw}^-|0\rangle$ is a linear combination of configurations with all inactive orbitals doubly occupied and all secondary orbitals unoccupied. Since all such configurations are represented in the set of CAS configurations $|i\rangle$, the redundancy condition (12.2.56) is satisfied. In CAS theory, therefore, all active–active rotations are redundant. Analysing the RASSCF wave function in the same manner, we may readily establish that all RAS1–RAS1, RAS2–RAS2 and RAS3–RAS3 rotations are redundant.

We may summarize our findings by stating that, in CASSCF and RASSCF theory, all *intraspace rotations* are redundant. By contrast, all *interspace rotations*, such as the inactive–active rotations and the RAS1–RAS2 rotations, are (presumably) nonredundant. Still, no proof has here been given that all interspace rotations are nonredundant; we have only used (12.2.56) to identify certain rotations as redundant – some of the remaining rotations may therefore yet turn out to be redundant. In practice, redundant interspace rotations are very seldom encountered in CAS and RAS calculations and then only owing to the presence of special symmetries; one such rare example is given in Exercise 12.1. By contrast, if the MCSCF wave function is generated in a less systematic fashion by selecting the configurations individually, then the redundancies may become more difficult to identify and we may have to search for these using the more general condition (12.2.55).

With the present discussion in mind, it is instructive to reconsider the active–active rotations of the singlet and triplet open-shell RHF wave functions of Section 10.1.2. We first note that, even though the triplet RHF wave function in (10.1.5) is a single-configuration wave function, it may be regarded as a CAS wave function generated by distributing two electrons in all possible ways between two orbitals consistent with triplet symmetry. The active–active rotation is therefore redundant in the triplet RHF wave function. By contrast, there are three different ways of distributing two electrons between two orbitals consistent with singlet symmetry; the singlet CAS wave function thus contains three configurations. The singlet open-shell RHF wave function (10.1.7) is therefore not a CAS wave function and we cannot conclude that the active–active rotation is redundant. On the contrary, the active–active rotation is needed to generate the best single-configuration RHF approximation.

12.2.7 THE MCSCF ELECTRONIC GRADIENT AT STATIONARY POINTS

We are now ready to consider the requirements for an optimized MCSCF state. In the present subsection, we examine the electronic gradient at stationary points; in Section 12.2.8, we consider the electronic Hessian. According to (12.2.26) and (12.2.27), the stationary condition for an MCSCF wave function may be written as

$$^{c}E_i^{(0)} = 2\langle i|\hat{P}\hat{H}|0\rangle = 0 \tag{12.2.58}$$

$$^{o}E_{pq}^{(1)} = \langle 0|[E_{pq}^-, \hat{H}]|0\rangle = 0 \tag{12.2.59}$$

sometimes referred to as the *generalized Brillouin theorem (GBT)* – see the discussion of Hartree–Fock theory in Section 10.2.3. Obviously, the MCSCF variational conditions are satisfied for all rotations that have been included in the optimization of the wave function, in particular those that are totally symmetric in spin and ordinary space. They may, however, be satisfied for other rotations as well. First, the MCSCF variational conditions are automatically satisfied for the redundant rotations discussed in Section 12.2.6. Second, the variational conditions are automatically satisfied for certain classes of rotations that are not totally symmetric in spin or ordinary space.

Let us first demonstrate explicitly the automatic fulfilment of the stationary conditions for the redundant rotations. Inserting the expansion (12.2.53) in the expression for the orbital gradient (12.2.59), we find that the gradient with respect to a redundant orbital rotation may be expressed as a linear combination of the gradients with respect to the nonredundant orbital and configuration parameters:

$$
\begin{aligned}
^{o}E_{pq}^{(1)} &= \langle 0|[E_{pq}^-, \hat{H}]|0\rangle = -2\langle 0|\hat{H}E_{pq}^-|0\rangle \\
&= -2\sum_i c_i\langle 0|\hat{H}\hat{P}|i\rangle - 2\sum_{rs \neq pq} \kappa_{rs}\langle 0|\hat{H}E_{rs}^-|0\rangle \\
&= -\sum_i c_i\,^{c}E_i^{(1)} + \sum_{rs \neq pq} \kappa_{rs}\,^{o}E_{rs}^{(1)} = 0
\end{aligned}
\tag{12.2.60}
$$

Of course, this result is just what we would expect for a redundant variational parameter at the stationary point. We now pass on to a less trivial point – namely, the gradient with respect to symmetry-breaking variational parameters at the stationary point.

In the optimization of an MCSCF state, the imposition of symmetry restrictions on the wave function may lead to an energy which is higher than that obtained with no such restrictions imposed. The situation is here similar to that for the RHF wave function. From the discussion in Section 10.10, we recall that the Hartree–Fock orbital gradient is usually but not always zero for the symmetry-breaking orbital rotations. The same is true for the MCSCF orbital gradient. More precisely, if the MCSCF state is *nondegenerate* – belonging, for instance, to the totally symmetric singlet irreducible representation – then the orbital gradient must be zero for all symmetry-breaking orbital rotations since the direct product of the irreducible representations of the bra state, the ket state and E_{pq}^- in (12.2.59) has the symmetry of the (nontotally symmetric) irreducible representation of E_{pq}^-. If, on the other hand, the MCSCF wave function belongs to a *degenerate* representation, then the orbital gradient may be *nonzero* for the symmetry-breaking rotations. Nonzero elements may arise since the direct product of the degenerate irreducible representations of the bra and ket states in (12.2.59) with the irreducible representation of E_{pq}^- now may contain a component of the totally symmetric representation. However, to avoid symmetry breaking of the wave function, these orbital rotations must be excluded from the optimization – an awkward feature of MCSCF theory, which is particularly bothersome in the context of molecular properties.

For the configuration part of the MCSCF gradient (12.2.58), the situation with respect to symmetry-breaking variations is simpler: the gradient associated with a configuration $\langle a|$ belonging to a symmetry different from that of the reference state $|0\rangle$ is always zero, for degenerate as well as nondegenerate reference states. This result follows from the observation that the gradient

$$^{\mathrm{c}}E_a^{(0)} = 2\langle a|\hat{H}|0\rangle - 2E^{(0)}\langle a|0\rangle \tag{12.2.61}$$

is just a linear combination of inner products of vectors that are orthogonal if they belong to different irreducible representations.

12.2.8 THE MCSCF ELECTRONIC HESSIAN AT STATIONARY POINTS

In our discussion of exact and approximate wave functions in Section 4.2.3, it was pointed out that, in the limit of the exact solution, the wave function representing the Kth electronic state should be stationary with respect to all variational parameters and that the electronic Hessian should have $K - 1$ negative eigenvalues, corresponding to the $K - 1$ negative excitation energies associated with this state. Accordingly, we require the variationally optimized MCSCF state not only to have a zero electronic gradient $\mathbf{E}^{(1)}$ for the full set of variational parameters but also to have an electronic Hessian $\mathbf{E}^{(2)}$ with $K - 1$ negative eigenvalues.

In the optimization of the MCSCF wave function, the imposition of symmetry restrictions in the optimization allows us to enforce the correct number of negative eigenvalues on only that part of the Hessian that is associated with the totally symmetric rotations; no control may be exercised over the nontotally symmetric part of the Hessian. Occasionally, therefore, the imposition of symmetry restrictions may lead to an MCSCF wave function that has the correct number of negative eigenvalues associated with the totally symmetric variations, but in addition some negative eigenvalues associated with symmetry-breaking rotations. Such instabilities occur whenever the unrestricted MCSCF wave function has a lower energy than the restricted wave function – for degenerate as well as for nondegenerate states and also in cases where the symmetry-breaking elements in the gradient are all zero. The situation is once again similar to that of RHF theory and the reader is referred to Section 10.10, where such instabilities are discussed in more detail.

Assuming now that the electronic Hessian has the correct number of negative eigenvalues, let us consider the 'location' of these eigenvalues in the Hessian. The electronic Hessian (12.2.16) contains four blocks – the diagonal blocks $^{\mathrm{cc}}\mathbf{E}^{(2)}$ and $^{\mathrm{oo}}\mathbf{E}^{(2)}$ and the off-diagonal blocks $^{\mathrm{co}}\mathbf{E}^{(2)}$ and $^{\mathrm{oc}}\mathbf{E}^{(2)}$, which couple the configuration and orbital spaces. Usually, the negative eigenvalues of the Hessian are located in the configuration–configuration block ${}^{cc}\mathbf{E}^{(2)}$ in the sense that a diagonalization of ${}^{cc}\mathbf{E}^{(2)}$ gives the same number of negative eigenvalues as does the diagonalization of $\mathbf{E}^{(2)}$. Such a situation is desirable since it means that the CI energy (and perforce the MCSCF energy) is an upper bound to the exact energy. Occasionally, however, the number of negative eigenvalues may be different in ${}^{cc}\mathbf{E}^{(2)}$ and $\mathbf{E}^{(2)}$. This situation is sometimes referred to as *root flipping*.

The term *root flipping* suggests that the negative eigenvalues in the MCSCF Hessian rightfully belong to the configuration–configuration block of the Hessian. However, this expression should not be taken too literally. Even though root flipping does imply that the MCSCF energy is not necessarily an upper bound to the exact energy, the MCSCF state may still give a reasonable description of the desired state. Indeed, root flipping may sometimes be unavoidable – for example, when the excited state has a character very different from that of the lower-lying states. Still, in cases where root flipping occurs, it may often be desirable to carry out a new MCSCF calculation in a larger configuration space, capable of representing the excited state of interest.

12.3 The MCSCF Newton trust-region method

In the present chapter, two different but related methods for the optimization of the MCSCF wave function are treated. The first method, which we discuss in this section, is a straightforward application of Newton's method for the optimization of nonlinear functions, modified so as to ensure global convergence. The second method, discussed in Section 12.4, is based on the solution of an eigenvalue equation which gives steps similar to those in the present section and which reduces to the standard eigenvalue problem of CI theory when no orbital optimization is included.

12.3.1 THE NEWTON STEP

In Newton's method, we determine the stationary point of the MCSCF energy by constructing a sequence of local quadratic models of the energy function. Each such model is generated by expanding the true MCSCF energy to second order around the optimizer of the previous iteration according to (12.2.13). Thus, in each iteration, we construct the following model surface of the MCSCF energy function

$$Q(\boldsymbol{\lambda}) = E^{(0)} + \mathbf{E}^{(1)^{\mathrm{T}}}\boldsymbol{\lambda} + \tfrac{1}{2}\boldsymbol{\lambda}^{\mathrm{T}}\mathbf{E}^{(2)}\boldsymbol{\lambda} \tag{12.3.1}$$

and determine its stationary point by differentiating (12.3.1) with respect to $\boldsymbol{\lambda}$ and setting the result equal to zero:

$$\mathbf{E}^{(2)}\boldsymbol{\lambda} = -\mathbf{E}^{(1)} \tag{12.3.2}$$

According to the discussion in Section 12.2.4, there are an infinite number of solutions to (12.3.2) related to the redundancy in the MCSCF wave function

$$\boldsymbol{\lambda} = \boldsymbol{\lambda}^* + \alpha\boldsymbol{\zeta}^{(0)} \tag{12.3.3}$$

Here $\boldsymbol{\lambda}^*$ is a particular solution to (12.3.2), α is some numerical parameter, and the reference vector $\boldsymbol{\zeta}^{(0)}$ is defined in (12.2.32). We resolve the redundancy problem by projecting away the component of the solution along $\boldsymbol{\zeta}^{(0)}$, thus requiring our solution to satisfy the equation

$$\boldsymbol{\lambda} = \mathbf{P}\boldsymbol{\lambda} \tag{12.3.4}$$

where $\mathbf{P}$ is the projection matrix (12.2.35). In the following, we shall always assume that the step vector satisfies (12.3.4); we may therefore freely switch between formulae with and without the projection matrix included.

In the vicinity of the true stationary point, the fixed-point iterations based on (12.3.2) will converge rapidly to the true optimizer with the characteristic second-order convergence rate of Newton's method discussed in Section 11.5.2. Further away from the optimizer, the Newton step may not necessarily lead us towards the true optimizer of the function since the stationary point of the local surface may no longer resemble the true optimizer of the function. In such cases, we should not apply the Newton step (12.3.2) directly but instead determine our step based on some other strategy. Such a strategy is presented in Section 12.3.2.

12.3.2 THE LEVEL-SHIFTED NEWTON STEP

In Newton's method, we determine in each iteration the global optimizer of the second-order model. In the global region of the search, such a strategy may sometimes give very long steps, steps that cannot be justified based on the second-order expansion of the energy function. In the *trust-region method* [5], we resolve this problem by optimizing the local quadratic function (12.3.1) within a restricted region around the expansion point – the *trust region*:

$$\|\mathbf{P}\lambda\| \leq h \tag{12.3.5}$$

In this expression, h is the *trust radius* – a positive numerical parameter that defines the region in which the second-order surface is assumed to be a good approximation to the exact surface. In calculating the length of the step (12.3.5), we have introduced the projection matrix since the trust radius applies only to the component of the step orthogonal to $\zeta^{(0)}$.

For harmonic surfaces, the second-order expansion (12.3.1) reproduces the true surface exactly at all points and h may be set equal to infinity; for highly anharmonic surfaces, the second-order expansion is a good model only in some small region around the expansion point and h is chosen correspondingly small. In practice, of course, there is no a priori way of selecting h uniquely. Instead, we proceed by choosing some reasonable value for h in the first iteration; in the subsequent iterations, h is updated based on our experience with the function, as discussed in Section 12.3.6. For the moment, we shall assume that some reasonable value of h has been selected.

How do we select the best step consistent with the inequality constraint (12.3.5)? Obviously, if the Newton step (12.3.2) is shorter than the trust radius, then this step is an acceptable one and we may proceed simply by taking the Newton step. In each iteration, therefore, we should always try the Newton step first so as to benefit from its rapid rate of convergence in the vicinity of the optimizer. If it turns out that the Newton step is too long (i.e. longer than h), then the second-order model has no stationary point inside the trust region and we must instead proceed by determining the best step to the boundary of this region. For minimizations, an obvious strategy would be to go to the minimizer on the boundary of the trust region. For excited states, the strategy is less obvious but it turns out that a useful approach is to go to a saddle point on the boundary of the trust region.

Thus, for all optimizations, we must be able to locate the stationary points on the boundary of the trust region. To determine the stationary points on the boundary of a trust region of radius h, we optimize the second-order surface (12.3.1) subject to the equality condition

$$\|\mathbf{P}\lambda\| = h \tag{12.3.6}$$

We thus set up the Lagrangian

$$L(\boldsymbol{\lambda}, \mu) = E^{(0)} + \mathbf{E}^{(1)^{\mathrm{T}}}\boldsymbol{\lambda} + \tfrac{1}{2}\boldsymbol{\lambda}^{\mathrm{T}}\mathbf{E}^{(2)}\boldsymbol{\lambda} - \tfrac{1}{2}\mu(\boldsymbol{\lambda}^{\mathrm{T}}\mathbf{P}\boldsymbol{\lambda} - h^2) \qquad (12.3.7)$$

where μ is some (as yet) undetermined Lagrange multiplier to be chosen such that (12.3.6) is satisfied. The stationary points of the Lagrangian are obtained from the equation

$$\frac{\partial L(\boldsymbol{\lambda}, \mu)}{\partial \boldsymbol{\lambda}} = \mathbf{E}^{(1)} + \mathbf{E}^{(2)}\boldsymbol{\lambda} - \mu\mathbf{P}\boldsymbol{\lambda} = \mathbf{0} \qquad (12.3.8)$$

This equation determines a *level-shifted Newton step*

$$(\mathbf{E}^{(2)} - \mu\mathbf{1})\mathbf{P}\boldsymbol{\lambda} = -\mathbf{E}^{(1)} \qquad (12.3.9)$$

which reduces to the Newton step (12.3.2) for $\mu = 0$. Assuming that the *level-shift parameter* μ does not coincide with any of the eigenvalues of $\mathbf{E}^{(2)}$ (including the zero eigenvalue), we may write the level-shifted Newton step in the form

$$\boldsymbol{\lambda}(\mu) = -(\mathbf{E}^{(2)} - \mu\mathbf{1})^{-1}\mathbf{E}^{(1)} \qquad (12.3.10)$$

On the left-hand side of this equation, we have, in accordance with (12.2.36) and (12.2.37), dispensed with the projection matrix $\mathbf{P}$ and we have indicated that the step $\boldsymbol{\lambda}(\mu)$ depends on the level-shift parameter μ.

The level-shifted Newton step (12.3.10) becomes infinite whenever μ approaches one of the eigenvalues of the Hessian since the inverted matrix then becomes singular. An exception occurs when μ approaches the zero eigenvalue since the singularity at $\mu = 0$ is orthogonal to the gradient. In this case, we recover the Newton step (12.3.2). We also note that the step (12.3.10) approaches zero when the level-shift parameter tends to plus or minus infinity. More generally, we may interpret the level-shifted Newton step (12.3.10) as defining a path or trajectory $\boldsymbol{\lambda}(\mu)$ through the space of vectors $\boldsymbol{\lambda}$, with each value of μ corresponding to a particular point on this path. As we shall see in Section 12.3.3, the level-shifted Newton equation gives several such paths, each of which is known as a *Levenberg–Marquardt trajectory* [5].

Our task is now to choose μ so that the step has the correct length (12.3.6) and so that the step is in the 'right' direction. Before we proceed, however, we note that the linear equations (12.3.9) are seldom solved by inverting explicitly the level-shifted Hessian as indicated in (12.3.10). Instead, the MCSCF calculations are carried out using iterative methods to solve the linear set of equations (12.3.9). The numerical instabilities as μ approaches zero therefore does not matter in practice. In Section 12.5, we shall see how these linear equations are solved for large MCSCF expansions.

12.3.3 THE LEVEL-SHIFT PARAMETER

To analyse the dependence of the level-shifted Newton step (12.3.10) on the numerical parameter μ, it is convenient to transform the equations to the diagonal representation of the Hessian. We emphasize that this transformation is carried out only for the purpose of the analysis – in practice, no diagonalization is ever attempted. Diagonalizing the MCSCF electronic Hessian, we obtain

$$\boldsymbol{\varepsilon} = \mathbf{U}^{\mathrm{T}}\mathbf{E}^{(2)}\mathbf{U} \qquad (12.3.11)$$

where $\boldsymbol{\varepsilon}$ is a diagonal matrix and $\mathbf{U}$ an orthogonal matrix. Equation (12.3.10) may now be written in the form

$$\boldsymbol{\chi}(\mu) = -(\boldsymbol{\varepsilon} - \mu\mathbf{1})^{-1}\boldsymbol{\varphi} \qquad (12.3.12)$$

where we have introduced the following notation for the step and the gradient in the Hessian eigenvector basis

$$\chi = \mathbf{U}^T \lambda \tag{12.3.13}$$

$$\varphi = \mathbf{U}^T \mathbf{E}^{(1)} \tag{12.3.14}$$

Likewise, the step length may be written in the simple form

$$\|\lambda(\mu)\| = \sqrt{\sum_i \frac{\varphi_i^2}{(\varepsilon_i - \mu)^2}} \tag{12.3.15}$$

where we assume that the eigenvalues have been sorted in ascending order $\varepsilon_i \leq \varepsilon_{i+1}$ and that the zero eigenvalue associated with the redundancy is omitted.

To illustrate the selection of the level-shift parameter in (12.3.12), let us consider the simple two-dimensional function

$$f(x, y) = x^2 + 10y^2 \tag{12.3.16}$$

plotted on the left in Figure 12.1. This function, which has a global minimum at the origin, is assumed to be a quadratic approximation to some more complicated function that we wish to minimize. We further assume that our current estimate of the minimum corresponds to the point $(-15, 10)$ and consider three trust regions, with trust radius h equal to 8, 16.13 and 24. In Figure 12.1, the boundaries of these trust regions have been superimposed as dashed concentric circles on the contour plot. Also shown in this plot are the Levenberg–Marquardt trajectories, generated by the level-shifted Newton step as we let the level-shift parameter μ take on all possible real numbers.

On the right in Figure 12.1, we have plotted the step length (12.3.15) as a function of μ for this model surface. The step-length function consists of $k + 1$ branches, where k is the number

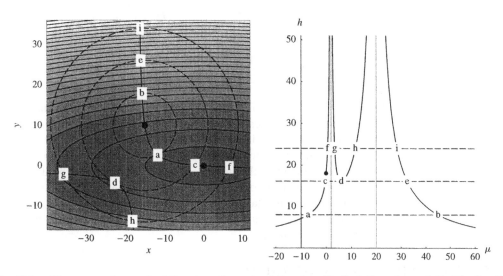

Fig. 12.1. The trust-region method for a two-dimensional function. On the left, we have plotted the function $x^2 + 10y^2$ with superimposed Levenberg–Marquardt trajectories and concentric circles representing three different trust regions centred at $(-15, 10)$. On the right, we have plotted the corresponding step-length function (12.3.15) with superimposed horizontal lines representing the radii of the three trust regions.

of eigenvalues (in this case two). The step length is positive for all values of μ, tends to zero at plus and minus infinity, and has asymptotes at the eigenvalues ε_i (which for the model function are 2 and 20). The width of the peaks separating the branches is related to the magnitude of the associated gradient φ_i. For the model function, the gradient in the eigenvector representation is $(-30, 200)^T$ at the expansion point.

Superimposed on the step-length function in Figure 12.1, we have plotted three dashed horizontal lines, which represent the different values of the trust radius h. The intersections of these lines with the step-length function (12.3.15) correspond to the values of the level-shift parameter μ that give steps to the stationary points on the boundary of the trust region. In the contour plot, these stationary points represent intersections of the trust regions with the Levenberg–Marquardt trajectories, labelled by the same letters as the intersections of the step-length function. For $h = 24$, the Newton step is inside the trust region.

In general, there can be at most $2k$ stationary points on the boundary of a trust region, where k is the number of distinct eigenvalues: there is always one solution in the leftmost branch $\mu < \varepsilon_1$ and one solution in the rightmost branch $\mu > \varepsilon_k$ (e.g. for $h = 8$ in the plot); in addition, there may be one (for $h = 16.13$) or two solutions (for $h = 24$) in each of the branches $\varepsilon_i < \mu < \varepsilon_{i+1}$. Note that, if the Newton step is too large, then there is always a solution $\mu < 0$. If ε_1 is negative, this conclusion follows trivially from the existence of a solution in the branch $\mu < \varepsilon_1$. If ε_1 is positive, then it follows from the fact that the Newton step – which is too big – corresponds to $\mu = 0$ and a smaller step in this branch can be generated by reducing the level-shift parameter below zero.

12.3.4 STEP CONTROL FOR GROUND STATES

To identify the appropriate value of the level-shift parameter in an optimization, consider the first- and second-order changes in the energy function [6]:

$$\Delta E_1 = \mathbf{E}^{(1)T}\boldsymbol{\lambda} = \sum_i (\mu - \varepsilon_i)\left(\frac{\varphi_i}{\varepsilon_i - \mu}\right)^2 \tag{12.3.17}$$

$$\Delta E_2 = \mathbf{E}^{(1)T}\boldsymbol{\lambda} + \frac{1}{2}\boldsymbol{\lambda}^T\mathbf{E}^{(2)}\boldsymbol{\lambda} = \sum_i \left(\mu - \frac{1}{2}\varepsilon_i\right)\left(\frac{\varphi_i}{\varepsilon_i - \mu}\right)^2 \tag{12.3.18}$$

In a ground-state optimization, we shall require the individual energy contributions from all modes (eigenvalues) to be negative. If the Hessian is positive definite, this requirement may be fulfilled by selecting the unique solution $\mu \leq 0$. If the Hessian has negative eigenvalues (which may happen in the early stages of a minimization), we may select the unique solution $\mu < \varepsilon_1$. Combining these observations, we conclude that the step in a minimization is found by determining the level-shift parameter that fulfils the condition $\mu \leq \min(\varepsilon_1, 0)$. There is always a unique solution in this region, and the search for a ground-state energy is thus unambiguous. It should be noted, however, that the solution in the branch $\mu \leq \min(\varepsilon_1, 0)$ does not necessarily represent the *global* minimum on the boundary of the trust region.

12.3.5 STEP CONTROL FOR EXCITED STATES

In calculations of excited states, the selection of the level-shift parameter is less clear-cut than for minimizations. For example, in an optimization of the first excited state, we would like the final electronic Hessian to have one and only one negative eigenvalue. The search is therefore for a first-order saddle point. Thus, if the current Hessian is positive definite, we would like to move in a direction where a negative eigenvalue is most likely to appear. Intuitively, this goal can best be achieved by increasing the energy in the direction associated with the lowest eigenvalue and by reducing it in the remaining directions. From a consideration of the energy contributions (12.3.17) and (12.3.18), we select a level shift that satisfies the first-order conditions $\varepsilon_1 < \mu < \varepsilon_2$ and also the second-order conditions $\varepsilon_1 < 2\mu < \varepsilon_2$. The first-order condition determines the branch and the second-order condition selects one of the two intersections in that branch; see Figure 12.1. It should be emphasized, however, that it is not always possible to satisfy these conditions. For example, for $h = 8$ in Figure 12.1, there is no solution in the branch dictated by the first-order condition $\varepsilon_1 < \mu < \varepsilon_2$. In such cases, we may select a level shift that corresponds to the solution associated with a greater trust radius and scale the resulting step so as to agree with the original trust radius. We also note that, even in cases where both conditions are satisfied, it may not always be clear which of the two intersections in a given branch should be chosen.

In general, the optimization of an excited MCSCF state is more difficult than the optimization of the ground state. For minimizations, the trust-region procedure is robust and globally convergent, although convergence to the global minimum is not guaranteed. For excited states, the trust-region procedure may not always converge or it may converge to a state different from that desired – for example, to a spurious state with an erroneous (unphysical) location of the negative eigenvalues. It is always necessary to analyse the converged wave function to ascertain that it has the desired structure and that it represents the electronic state under study.

12.3.6 TRUST-RADIUS UPDATE SCHEMES

To conclude the discussion of the trust-region method, we describe how the trust radius h is updated. In iteration n, we generate a vector $\boldsymbol{\lambda}_n$, from which we obtain the reference wave function for the next iteration. However, we may also use this vector to predict the energy in the next iteration $n + 1$ according to the expression

$$Q_n(\boldsymbol{\lambda}_n) = E_n^{(0)} + \mathbf{E}_n^{(1)^{\mathrm{T}}}\boldsymbol{\lambda}_n + \tfrac{1}{2}\boldsymbol{\lambda}_n^{\mathrm{T}}\mathbf{E}_n^{(2)}\boldsymbol{\lambda}_n \qquad (12.3.19)$$

In the next iteration, the true energy differs from the second-order energy in terms that are of third and higher orders in the step:

$$E_{n+1}^{(0)} = Q_n(\boldsymbol{\lambda}_n) + O(\boldsymbol{\lambda}_n^3) \qquad (12.3.20)$$

Let us consider the ratio between the observed and predicted changes in the energy:

$$r = \frac{E_{n+1}^{(0)} - E_n^{(0)}}{Q_n(\boldsymbol{\lambda}_n) - E_n^{(0)}} = 1 + \frac{O(\boldsymbol{\lambda}_n^3)}{Q_n(\boldsymbol{\lambda}_n) - E_n^{(0)}} \qquad (12.3.21)$$

The closer this ratio is to unity, the more quadratic is the surface. Assuming a smooth surface, we may update the trust radius h in each iteration based on the closeness of r to unity. The following scheme works reasonably well for minimizations:

1. If $r > 0.75$, increase the trust radius $h_{n+1} = 1.2h_n$
2. If $0.25 < r < 0.75$, do not change the trust radius $h_{n+1} = h_n$
3. If $0 < r < 0.25$, reduce the trust radius $h_{n+1} = 0.7h_n$
4. If $r < 0$, reject the step $\boldsymbol{\lambda}_n$ and generate a new step of trust radius $h_n \to 0.7h_n$

Note that the step $\boldsymbol{\lambda}_n$ is rejected if the energy does not decrease in the iteration. We then go back to the previous iteration and generate a new step $\boldsymbol{\lambda}_n$ with a reduced trust radius. The trust-region method is not particularly sensitive to the details of the updating scheme for h. The important point is that h is allowed to adjust itself to the circumstances in the course of the optimization.

12.4 The Newton eigenvector method

In the second-order method of Section 12.3, we obtain the MCSCF step from the modified Newton equations (12.3.9). The search is controlled by a level-shift parameter, chosen so as to locate a stationary point of the appropriate structure, representing either the electronic ground state or some excited state. For this method to work, we must in each iteration first determine the appropriate interval for the level-shift parameter and then solve the linear equations (12.3.9). In view of the large number of parameters in an MCSCF calculation, this two-step procedure is cumbersome. In the present section, we discuss a different realization of the second-order trust-region method, replacing the two-step level-shift procedure of Section 12.3 by a single eigenvalue problem. As we shall see, this *Newton eigenvector method* provides a natural search control in the sense that the correction to the Kth electronic state is represented by the Kth eigenvector of a symmetric matrix, allowing us to generate the correction to the MCSCF wave function in a single step.

12.4.1 THE MCSCF EIGENVALUE PROBLEM

Before we develop the MCSCF eigenvector method, it is instructive to note that, provided the eigenvector is normalized, we may write the CI energy as a quadratic form. Thus, differentiating the Lagrangian in the form

$$L(\zeta, E) = \zeta^T \mathbf{H} \zeta - E(\zeta^T \zeta - 1) \tag{12.4.1}$$

with respect to the CI vector ζ, we obtain the standard CI eigenvalue problem

$$\mathbf{H}\zeta = E\zeta \tag{12.4.2}$$

Note that the Lagrange multiplier E cannot take on any value – only those that correspond to the eigenvalues of the Hamiltonian $\mathbf{H}$ are allowed.

We now proceed to set up a similar Lagrangian for the optimization of the MCSCF energy. Writing the parameters of the MCSCF state in the form

$$\zeta = \zeta^{(0)} + \mathbf{P}\lambda \tag{12.4.3}$$

in terms of the vector $\zeta^{(0)}$ in (12.2.32) and introducing the modified Hessian matrix (often referred to as the *augmented Hessian*)

$$\mathbf{G} = \mathbf{E}^{(2)} + \zeta^{(0)} \mathbf{E}^{(1)^T} + \mathbf{E}^{(1)} \zeta^{(0)^T} \tag{12.4.4}$$

we find that the second-order MCSCF energy (12.3.1) may be written as

$$Q(\zeta) = E^{(0)} + \tfrac{1}{2}\zeta^T \mathbf{G}\zeta \tag{12.4.5}$$

This quadratic form has been made possible because we use the intermediate normalization (12.4.3) of ζ, which allows us to incorporate the MCSCF gradient terms as rank-1 corrections to the Hessian (12.4.4). This function is now to be optimized subject to the constraints

$$\zeta^T \mathbf{O}\zeta = 1 \tag{12.4.6}$$

$$\zeta^T \mathbf{P}\zeta = h^2 \tag{12.4.7}$$

The first constraint ensures that the new vector has unit overlap with the reference state $\boldsymbol{\zeta}^{(0)}$ (intermediate normalization) and the second condition that the step is to the boundary of the trust region. Introducing the Lagrange multipliers ν and μ, we may write the energy function (12.4.5) as a quadratic form

$$L(\boldsymbol{\zeta}, \mu, \nu) = \tfrac{1}{2}\boldsymbol{\zeta}^{\mathrm{T}}\mathbf{G}\boldsymbol{\zeta} - \tfrac{1}{2}\nu(\boldsymbol{\zeta}^{\mathrm{T}}\mathbf{O}\boldsymbol{\zeta} - 1) - \tfrac{1}{2}\mu(\boldsymbol{\zeta}^{\mathrm{T}}\mathbf{P}\boldsymbol{\zeta} - h^2) \qquad (12.4.8)$$

This Lagrangian (which is equivalent to the MCSCF energy function (12.3.1) except for a constant contribution $E^{(0)}$) is the MCSCF analogue to the CI Lagrangian (12.4.1). It forms the basis for our optimization of the MCSCF wave function. Note that, whereas the CI Lagrangian represents the *exact* energy subject to the normalization constraint, the MCSCF Lagrangian represents the *second-order* energy (12.4.5) subject to the constraints (12.4.6) and (12.4.7).

Differentiating the MCSCF Lagrangian (12.4.8) with respect to $\boldsymbol{\zeta}$ and setting the result equal to zero, we obtain the equations

$$\mathbf{G}\boldsymbol{\zeta} = \nu\mathbf{O}\boldsymbol{\zeta} + \mu\mathbf{P}\boldsymbol{\zeta} \qquad (12.4.9)$$

For a fixed value of the multiplier μ, these equations have the form of a generalized eigenvalue problem with eigenvector $\boldsymbol{\zeta}$ and eigenvalue ν. To bring out more clearly the eigensystem structure, we introduce the parameter α and the matrix $\mathbf{T}(\alpha)$ given by

$$\mu = \nu\alpha^2 \qquad (12.4.10)$$

$$\mathbf{T}(\alpha) = \mathbf{O} + \alpha^2\mathbf{P} \qquad (12.4.11)$$

We may now express (12.4.9) in the form of a generalized eigenvalue problem

$$\mathbf{G}\boldsymbol{\zeta} = \nu\mathbf{T}(\alpha)\boldsymbol{\zeta} \qquad (12.4.12)$$

analogous to the CI equations (12.4.2). For a given α, the solution to (12.4.12) is an eigenpair where the eigenvector corresponds to the new MCSCF vector $\boldsymbol{\zeta}$ and the eigenvalue to the multiplier ν. The solution to (12.4.12) is discussed in Section 12.4.2. Assuming that $\boldsymbol{\zeta}$ has been scaled in accordance with (12.4.6), we note that the projection of (12.4.12) from the left by $\mathbf{P}$ leads to the standard level-shifted Newton equations

$$(\mathbf{E}^{(2)} - \nu\alpha^2\mathbf{1})\mathbf{P}\boldsymbol{\zeta} = -\mathbf{E}^{(1)} \qquad (12.4.13)$$

After the substitution (12.4.10), these equations become identical to (12.3.9) – the fundamental equations of Section 12.3.

12.4.2 THE NEWTON EIGENVECTOR METHOD

In the present subsection, we shall explore the solution of the generalized eigenvalue problem (12.4.12) for the optimization of the MCSCF wave function. The strategy that emerges from a comparison with (12.4.13) is that the magnitude of the step is to be controlled by α and the direction of the step by the eigenvector $\boldsymbol{\zeta}$ and hence by the eigenvalue ν. We note that the eigenvectors of (12.4.12) are not necessarily orthogonal since they are obtained from a generalized rather than standard symmetric eigenvalue problem. For α restricted to real positive numbers, the matrix $\mathbf{T}(\alpha)$ is positive definite since the inequality

$$\mathbf{V}^{\mathrm{T}}\mathbf{T}(\alpha)\mathbf{V} = \mathbf{V}^{\mathrm{T}}\mathbf{O}\mathbf{V} + \alpha^2\mathbf{V}^{\mathrm{T}}\mathbf{P}\mathbf{V} = \|\mathbf{O}\mathbf{V}\|^2 + \alpha^2\|\mathbf{P}\mathbf{V}\|^2 > 0 \qquad (12.4.14)$$

holds for any vector $\mathbf{V}$. The equation (12.4.12) is then a symmetric generalized eigenvalue problem with a positive definite metric $\mathbf{T}(\alpha)$ and may therefore be recast as a standard symmetric eigenvalue problem. We first note that, for the allowed values of α, we may take $\mathbf{T}(\alpha)$ to any rational power (not necessarily positive)

$$\mathbf{T}^{a}(\alpha) = \mathbf{O} + \alpha^{2a}\mathbf{P} \tag{12.4.15}$$

For integral powers n, this expression follows by expansion of $\mathbf{T}^{n}(\alpha)$. The result may then be verified for rational powers by using the relation

$$[\mathbf{T}^{1/n}(\alpha)]^{n} = \mathbf{T}(\alpha) \tag{12.4.16}$$

Multiplying the generalized eigenvalue equations (12.4.12) from the left by the inverse square root of $\mathbf{T}(\alpha)$, we arrive at a standard eigenvalue problem

$$[\mathbf{T}^{-1/2}(\alpha)\mathbf{G}\mathbf{T}^{-1/2}(\alpha)]\mathbf{T}^{1/2}(\alpha)\boldsymbol{\zeta} = \nu\mathbf{T}^{1/2}(\alpha)\boldsymbol{\zeta} \tag{12.4.17}$$

We next introduce the 'gradient-scaled' augmented Hessian

$$\mathbf{G}(\alpha) = \alpha^{2}\mathbf{T}^{-1/2}(\alpha)\mathbf{G}\mathbf{T}^{-1/2}(\alpha) = \mathbf{E}^{(2)} + \alpha\boldsymbol{\zeta}^{(0)}\mathbf{E}^{(1)^{\mathrm{T}}} + \alpha\mathbf{E}^{(1)}\boldsymbol{\zeta}^{(0)^{\mathrm{T}}} \tag{12.4.18}$$

where the last expression has been obtained by using the relationships (12.2.36)–(12.2.39) and (12.4.15). The eigenvalue equations (12.4.17) may now be written as

$$\mathbf{G}(\alpha)\boldsymbol{\xi}(\alpha) = \mu\boldsymbol{\xi}(\alpha) \tag{12.4.19}$$

where

$$\boldsymbol{\xi}(\alpha) = \mathbf{T}^{1/2}(\alpha)\boldsymbol{\zeta} \tag{12.4.20}$$

The matrix $\mathbf{G}(\alpha)$ is symmetric since α is real. The eigenvalue problem (12.4.19) therefore has real eigenvalues and orthogonal eigenvectors.

By itself, the eigenvalue problem (12.4.19) does not determine the normalization of the eigenvectors. However, if we multiply (12.4.20) from the left by the transpose of $\boldsymbol{\zeta}^{(0)}$, we obtain

$$\boldsymbol{\zeta}^{(0)^{\mathrm{T}}}\boldsymbol{\zeta}(\alpha) = \boldsymbol{\zeta}^{(0)^{\mathrm{T}}}(\mathbf{O} + \alpha\mathbf{P})\boldsymbol{\zeta} = \boldsymbol{\zeta}^{(0)^{\mathrm{T}}}\boldsymbol{\zeta} = 1 \tag{12.4.21}$$

which shows that the requirement of intermediate normalization on $\boldsymbol{\zeta}$ means that the eigenvectors $\boldsymbol{\xi}(\alpha)$ must be intermediately normalized as well. In this eigenvector normalization, the MCSCF step $\boldsymbol{\zeta}$ is obtained from (12.4.20) as

$$\boldsymbol{\zeta}(\alpha) = \mathbf{T}^{-1/2}(\alpha)\boldsymbol{\xi}(\alpha) = (\mathbf{O} + \alpha^{-1}\mathbf{P})\boldsymbol{\xi}(\alpha) = \boldsymbol{\zeta}^{(0)} + \alpha^{-1}\mathbf{P}\boldsymbol{\xi}(\alpha) \tag{12.4.22}$$

In short, to obtain the MCSCF step, we first solve the eigenvalue problem (12.4.19) in the intermediate normalization (12.4.21) and next generate the step from (12.4.22).

It is instructive to verify that, upon substitution in (12.4.22), our solution to the eigenvalue problem (12.4.19) represents a solution to the level-shifted Newton equations (12.3.9) as well. Projecting (12.4.19) from the left by $\mathbf{P}$ and using (12.4.18) and (12.4.20), we find

$$(\mathbf{E}^{(2)} + \alpha\mathbf{E}^{(1)}\boldsymbol{\zeta}^{(0)^{\mathrm{T}}})\mathbf{T}^{1/2}(\alpha)\boldsymbol{\zeta} = \mu\mathbf{P}\mathbf{T}^{1/2}(\alpha)\boldsymbol{\zeta} \tag{12.4.23}$$

Inserting (12.4.15) for $\mathbf{T}^{1/2}(\alpha)$ and rearranging, we obtain

$$(\mathbf{E}^{(2)} - \mu\mathbf{1})\mathbf{P}\boldsymbol{\zeta} = -\mathbf{E}^{(1)} \tag{12.4.24}$$

which is equivalent to (12.3.9) in the intermediate normalization (12.4.6). The step vector obtained from an eigenvector of $\mathbf{G}(\alpha)$ with eigenvalue μ is therefore identical to the Newton step with a level shift μ (12.3.9).

Let us summarize the results so far. We have demonstrated that the variational condition on the MCSCF Lagrangian (12.4.8) may be cast in the form of a symmetric eigenvalue problem (12.4.19). From its eigenvectors, we may extract the solution to the variational problem according to (12.4.22). The solution contains two parts. The first part is parallel to the reference state and ensures that the condition of intermediate normalization (12.4.6) is satisfied. The second part is orthogonal to the reference state; its magnitude depends on the numerical value of α, which should be adjusted so that the step is to the boundary of the trust region (12.4.7). For each new value of α, the eigenvalue problem (12.4.19) must be solved again since the augmented Hessian $\mathbf{G}(\alpha)$ is a function of α. In practice, however, this is not a problem since the eigenvalue equations are solved iteratively in a small subspace of trial vectors; see Section 12.5. Any adjustment of α (to ensure a step of correct magnitude) may therefore readily be corrected for by solving the eigenvalue problem in the small trial subspace.

12.4.3 NORM-EXTENDED OPTIMIZATION

The optimization of MCSCF wave functions by the solution of the Newton eigenvalue problem (12.4.19) using α for step control is known as *norm-extended optimization (NEO)* [7]. Before we can discuss step-size control in the NEO method, we must consider the choice of eigenvector in (12.4.22). For this purpose, it is instructive to compare the eigenvalues μ_i of the augmented Hessian $\mathbf{G}(\alpha)$ with the eigenvalues ε_i of the true Hessian $\mathbf{E}^{(2)}$. We note the following relationship between the augmented and true Hessians:

$$\mathbf{P}\mathbf{G}(\alpha)\mathbf{P} = \mathbf{G}(0) = \mathbf{E}^{(2)} \tag{12.4.25}$$

The application of the projection matrix effectively reduces the dimension of the augmented Hessian, projecting out the component along the reference vector. The eigenvalue associated with this projection is exactly zero. The remaining eigenvalues of the projected augmented Hessian (which are identical to the eigenvalues of $\mathbf{E}^{(2)}$) are related to the eigenvalues of $\mathbf{G}(\alpha)$ according to the Hylleraas–Undheim theorem of Section 4.2.4. Thus, arranging the eigenvalues in ascending order, we obtain

$$\mu_1 \leq \varepsilon_1 \leq \mu_2 \leq \cdots \leq \mu_n \leq \varepsilon_n \leq \mu_{n+1} \tag{12.4.26}$$

where the zero eigenvalue of $\mathbf{E}^{(2)}$ (associated with the reference vector) has been omitted from ε_i.

From the relationship (12.4.26), it is clear that, by selecting the eigenvector associated with the Kth eigenvalue of the augmented Hessian, we obtain a step (12.4.22) consistent with the first-order conditions on the energy for an optimization of the Kth electronic state – see Sections 12.3.4 and 12.3.5. Thus, for ground-state calculations, we choose the first eigenvector of the augmented Hessian; for calculations of the first excited state, we choose the second eigenvector, and so on. In this way, the Newton eigenvector method allows for a simple control over the direction of the step vector. There is no need first to calculate the eigenvalues to identify the appropriate range for μ and then to solve the linear set of equations (12.3.9) as done in the level-shifted Newton method of Section 12.3. Instead, we obtain a step in the right direction by selecting the appropriate eigenvector of the augmented Hessian.

Having selected the eigenvector, we must consider the magnitude of the step vector as determined by the step-control parameter α. The parameter α is used for control of the eigenvector step

(12.4.22) in much the same way as the level-shift parameter μ is used for control of the Newton step (12.3.10). We recall that α has been restricted to positive values.

Since the Newton eigenvector method is equivalent to the level-shifted Newton method of Section 12.3, it yields the same set of solutions. In our discussion of the level-shifted Newton method, we noted that, for excited states, it is not always possible to obtain a solution consistent with the first-order condition on the energy and the condition on the length of the step. The same situation may arise in the NEO method as well: for excited states, it may not always be possible to generate a step (12.4.22) to the boundary of the trust region. In such situations, we may instead adjust α so as to yield the shortest possible step and then scale the resulting step down to the desired length.

The variation of the step length with α is rather complicated and will not be analysed in detail here. Some results can be stated briefly, however. First, consider the limit when α tends to zero. In this limit, the eigenvalues of the augmented Hessian $\mathbf{G}(0)$ coincide with the eigenvalues of the true Hessian $\mathbf{E}^{(2)}$. As the stationary point is approached and the gradient becomes zero, the eigenvector associated with the zero eigenvalue of $\mathbf{E}^{(2)}$ becomes equal to the Newton step plus the normalized reference vector. The eigenvector method therefore becomes numerically unstable in the region where the Newton step is very small. In the local region, we should therefore instead apply the Newton step (12.3.2) directly. For excited states, the Newton step (12.3.2) is always preferable in this region since it requires only the solution of a single set of equations. The Newton eigenvector method, by contrast, requires the simultaneous calculation of the full set of eigenvectors associated with the lower states.

Let us now consider the limit when α tends to infinity. From the structure of the augmented Hessian (12.4.18), we note that exactly two eigenvalues (of opposite signs) become infinite in this region since

$$(\boldsymbol{\zeta}^{(0)}\mathbf{E}^{(1)^{\mathrm{T}}} + \mathbf{E}^{(1)}\boldsymbol{\zeta}^{(0)^{\mathrm{T}}})(\boldsymbol{\zeta}^{(0)} \pm g\mathbf{E}^{(1)}) = \pm\frac{1}{g}(\boldsymbol{\zeta}^{(0)} \pm g\mathbf{E}^{(1)}) \tag{12.4.27}$$

$$g = (\mathbf{E}^{(1)^{\mathrm{T}}}\mathbf{E}^{(1)})^{-1/2} \tag{12.4.28}$$

The associated steps become infinitely small and oppositely directed along the gradient, as may be verified by applying (12.4.22). These solutions are thus infinitely small steepest-descent and steepest-ascent steps and correspond to the steps generated with infinitely large (positive and negative) μ in the level-shifted Newton method.

12.4.4 THE AUGMENTED-HESSIAN METHOD

The Newton eigenvector method is often applied with a fixed value of $\alpha = 1$, which is not adjusted in the course of the optimization. Step-size control is instead accomplished by an overall scaling of the intermediately normalized step generated by (12.4.19). This scheme is known as the *augmented-Hessian method* [8]. For $\alpha = 1$, the configuration–configuration block of the augmented Hessian is related to the CI Hamiltonian matrix in the following simple manner:

$$^{\mathrm{cc}}G_{ij}(1) = {}^{\mathrm{cc}}K_{ij}^{(2)} = 2\langle i|\hat{H} - E^{(0)}|j\rangle \tag{12.4.29}$$

To obtain this expression, we have applied equations (12.4.18), (12.2.28) and (12.2.41). Therefore, for an MCSCF wave function with no orbital-rotation parameters (i.e. for a CI wave function), the augmented-Hessian eigenvalue problem (12.4.19) with $\alpha = 1$ is equivalent to the standard CI eigenvalue problem (12.4.2):

$$\mathbf{H}\boldsymbol{\zeta} = \left(E^{(0)} + \tfrac{1}{2}\mu\right)\boldsymbol{\zeta} \tag{12.4.30}$$

The augmented-Hessian method may therefore be regarded as a generalization of the CI eigenvalue method to the MCSCF wave function.

12.5 Computational considerations

Having outlined the general strategy for optimizing MCSCF wave functions, we now turn our attention to the more practical aspects of the optimization. The number of parameters in an MCSCF wave function is usually so large (up to several million parameters) that the explicit calculation of the electronic Hessian and its subsequent inversion or diagonalization is out of the question. Even the storage of the Hessian matrix is usually not possible. Instead, we solve the level-shifted Newton equations (12.3.9)

$$(\mathbf{E}^{(2)} - \mu\mathbf{1})\mathbf{P}\boldsymbol{\lambda} = -\mathbf{E}^{(1)} \tag{12.5.1}$$

or alternatively the Newton eigenvalue equations (12.4.19)

$$(\mathbf{E}^{(2)} + \alpha\boldsymbol{\zeta}^{(0)}\mathbf{E}^{(1)^{\mathrm{T}}} + \alpha\mathbf{E}^{(1)}\boldsymbol{\zeta}^{(0)^{\mathrm{T}}})\boldsymbol{\xi} = \mu\boldsymbol{\xi} \tag{12.5.2}$$

by some iterative technique, the key computational steps of which are the repeated multiplication of the level-shifted Hessian or the augmented Hessian by trial vectors $\mathbf{v}$. Referring to the expression (12.2.44) for the electronic Hessian

$$\mathbf{E}^{(2)} = \mathbf{K}^{(2)} - \overline{\mathbf{E}}^{(1)}\boldsymbol{\zeta}^{(0)^{\mathrm{T}}} - \boldsymbol{\zeta}^{(0)}\overline{\mathbf{E}}^{(1)^{\mathrm{T}}} \tag{12.5.3}$$

where $\mathbf{K}^{(2)}$ is defined in (12.2.40)–(12.2.43) and the scaled gradient $\overline{\mathbf{E}}^{(1)}$ in (12.2.45), we find that the computationally intensive part of the solution of the linear equations (12.5.1) and of the eigenvalue equations (12.5.2) is the repeated calculation of matrix–vector products of the form

$$\boldsymbol{\sigma} = \mathbf{K}^{(2)}\mathbf{v} \tag{12.5.4}$$

where $\mathbf{v}$ is some trial vector and $\boldsymbol{\sigma}$ the transformed vector. The multiplications that involve the rank-1 contributions to the level-shifted Hessian in (12.5.1) or to the augmented Hessian in (12.5.2) are trivial and require no further comment.

The numerical techniques employed for the solution of the MCSCF equations (12.5.1) and (12.5.2) are similar to those for CI wave functions. The standard approach is to build up a small basis of vectors spanning the solution to the desired accuracy; see the discussion of iterative techniques for the optimization of CI wave functions in Section 11.5. An additional complication in MCSCF calculations is the incorporation of the step-size parameters μ and α, the values of which must be updated in the course of the optimization to generate steps in the desired direction and of the specified length. In practice, these parameters may be incorporated and adjusted at little or no extra cost, in particular for ground-state calculations.

In the remainder of the present section, we first consider the calculation of the electronic gradient $\mathbf{E}^{(1)}$ in Section 12.5.1 and the evaluation of the matrix–vector products (12.5.4) in Section 12.5.2, building on the results for Hartree–Fock wave functions in Chapter 10 and for CI wave functions in Chapter 11. Next, some aspects particular to the optimization of MCSCF wave functions are discussed in Section 12.5.3. After a discussion of the structure of the MCSCF Hessian (based on a numerical example) in Section 12.5.4, we consider three examples of MCSCF optimizations in Section 12.5.5.

12.5.1 THE MCSCF ELECTRONIC GRADIENT

The structure of the MCSCF orbital gradient (12.2.27) is identical to that of the Hartree–Fock gradient (10.1.25). Indeed, we may calculate the MCSCF orbital gradient in the same way as we calculate the gradient for an open-shell Hartree–Fock wave function, recognizing that open-shell RHF wave functions have the same three classes of orbitals (inactive, active and secondary) as MCSCF wave functions. There is thus no need to rederive the expressions for the MCSCF orbital gradient here – we need only recapitulate the expressions obtained in Chapter 10.

Following the discussion in Section 10.8.3, the most general expression for the MCSCF orbital gradient (12.2.27) is

$$^{0}E_{mn}^{(1)} = 2(F_{mn} - F_{nm}) \tag{12.5.5}$$

where $\mathbf{F}$ is *the generalized Fock matrix* [9] given by

$$F_{mn} = \sum_{q} D_{mq}h_{nq} + \sum_{qrs} d_{mqrs}g_{nqrs} \tag{12.5.6}$$

here calculated from the MCSCF density-matrix elements

$$D_{pq} = \langle 0|E_{pq}|0\rangle \tag{12.5.7}$$

$$d_{pqrs} = \langle 0|e_{pqrs}|0\rangle \tag{12.5.8}$$

The calculation of the generalized Fock matrix according to (12.5.6) is quite expensive, however, scaling as n^5 (where n is the total number of orbitals) and requiring the construction of the two-electron density matrix in the full MO basis.

When the MO space is divided into inactive, active and secondary orbitals, the construction of the generalized Fock matrix may be simplified considerably. Following the procedure in Section 10.8.4, we then arrive at the following useful expressions for the generalized Fock matrix:

$$F_{in} = 2\,^{\mathrm{I}}F_{ni} + 2\,^{\mathrm{A}}F_{ni} \tag{12.5.9}$$

$$F_{vn} = \sum_{w} \,^{\mathrm{I}}F_{nw}D_{vw} + Q_{vn} \tag{12.5.10}$$

$$F_{an} = 0 \tag{12.5.11}$$

where we have introduced the *inactive* and *active Fock matrices*

$$^{\mathrm{I}}F_{mn} = h_{mn} + \sum_{i}(2g_{mnii} - g_{miin}) \tag{12.5.12}$$

$$^{\mathrm{A}}F_{mn} = \sum_{vw} D_{vw}\left(g_{mnvw} - \tfrac{1}{2}g_{mwvn}\right) \tag{12.5.13}$$

and also the auxiliary $\mathbf{Q}$ *matrix*

$$Q_{vm} = \sum_{wxy} d_{vwxy}g_{mwxy} \tag{12.5.14}$$

The only two-electron density-matrix elements now needed are those with all four indices active. Likewise, a full MO transformation of the two-electron integrals is no longer necessary. Since the active orbital space is quite small, the construction of the MCSCF orbital gradient from (12.5.9)–(12.5.14) is usually considerably less expensive than from (12.5.6).

Turning our attention to the configuration part of the electronic gradient (12.2.26), we note that this part is identical to the CI gradient derived in Section 11.4.2. It may therefore be calculated using the direct CI techniques of Section 11.8. As for the orbital gradient, it is possible to reduce the computational cost by taking into account the special structure of the MCSCF orbital space, recalling that the expansion usually involves only a small number of active orbitals. In Exercise 12.2, it is shown that the configuration part of the electronic gradient may be written in the form

$$^{c}E_{i}^{(1)} = 2 \sum_{vw} \langle i|E_{vw}|0\rangle {}^{\mathrm{I}}F_{vw} + \sum_{vwxy} \langle i|e_{vwxy}|0\rangle g_{vwxy} - 2C_{i}^{(0)}(E^{(0)} - {}^{\mathrm{I}}E) \qquad (12.5.15)$$

where we have introduced the *inactive energy*

$$^{\mathrm{I}}E = \sum_{i} (h_{ii} + {}^{\mathrm{I}}F_{ii}) \qquad (12.5.16)$$

Thus, provided the inactive Fock matrix (12.5.12) and the inactive energy (12.5.16) are available (the inactive Fock matrix is also needed for the orbital gradient), we may evaluate the configuration part of the MCSCF gradient (12.5.15) using the direct CI techniques of Section 11.8, referencing only active orbitals. The cost of constructing the MCSCF configuration gradient is equivalent to that of one direct CI iteration in the small set of active orbitals.

In conclusion, the MCSCF orbital and configuration gradients may be calculated using the techniques previously developed for the Hartree–Fock and CI wave functions, taking advantage of the fact that, in an MCSCF wave function, only a small number of active orbitals are variably occupied.

12.5.2 MCSCF HESSIAN TRANSFORMATIONS

As noted in the introduction to this section, the electronic Hessian is never explicitly set up in large-scale MCSCF calculations – the number of variational parameters would make this task impossible. Solving instead the MCSCF equations iteratively, the Hessian is needed for only a fairly small number of linear transformations (compared with the number of variational parameters). Such transformations can be carried out quite efficiently using the techniques described in this subsection [10].

Disregarding the trivial contributions from the rank-1 parts of the level-shifted or augmented Hessian, we concentrate here on the linear transformation (12.5.4) involving the nonseparable matrix $\mathbf{K}^{(2)}$. Partitioning this matrix in the usual manner, we may write the linear transformation in the form

$$\begin{pmatrix} ^{c}\boldsymbol{\sigma} \\ ^{o}\boldsymbol{\sigma} \end{pmatrix} = \begin{pmatrix} ^{cc}\mathbf{K}^{(2)} & ^{co}\mathbf{K}^{(2)} \\ ^{oc}\mathbf{K}^{(2)} & ^{oo}\mathbf{K}^{(2)} \end{pmatrix} \begin{pmatrix} \mathbf{c} \\ \boldsymbol{\kappa} \end{pmatrix} = \begin{pmatrix} ^{cc}\mathbf{K}^{(2)}\mathbf{c} + {}^{co}\mathbf{K}^{(2)}\boldsymbol{\kappa} \\ ^{oc}\mathbf{K}^{(2)}\mathbf{c} + {}^{oo}\mathbf{K}^{(2)}\boldsymbol{\kappa} \end{pmatrix} \qquad (12.5.17)$$

where the elements of $\mathbf{K}^{(2)}$ are defined in (12.2.41)–(12.2.43). Inserting these expressions in (12.5.17) and carrying out some simple algebra, we obtain

$$[^{cc}\mathbf{K}^{(2)}\mathbf{c}]_{i} = 2\langle i|\hat{H} - E^{(0)}|\mathbf{c}\rangle \qquad (12.5.18)$$

$$[^{co}\mathbf{K}^{(2)}\boldsymbol{\kappa}]_{i} = 2\langle i|\hat{H}_{\kappa}|0\rangle \qquad (12.5.19)$$

$$[^{oc}\mathbf{K}^{(2)}\mathbf{c}]_{pq} = \langle 0|[E_{pq}^{-}, \hat{H}]|\mathbf{c}\rangle + \langle \mathbf{c}|[E_{pq}^{-}, \hat{H}]|0\rangle \qquad (12.5.20)$$

$$[^{oo}\mathbf{K}^{(2)}\boldsymbol{\kappa}]_{pq} = \langle 0|[E_{pq}^{-}, \hat{H}_{\kappa}]|0\rangle + ([^{o}\mathbf{E}^{(1)}, \boldsymbol{\kappa}])_{pq} \qquad (12.5.21)$$

Here the state $|\mathbf{c}\rangle$ is given by (12.2.4) and the effective operator

$$\hat{H}_\kappa = [\hat{\kappa}, \hat{H}] = \sum_{pq} \kappa_{pq}[E_{pq}^-, \hat{H}] \tag{12.5.22}$$

is a bona fide Hamiltonian operator

$$\hat{H}_\kappa = \sum_{pq} h_{pq}^\kappa E_{pq} + \tfrac{1}{2} \sum_{pqrs} g_{pqrs}^\kappa e_{pqrs} \tag{12.5.23}$$

with one- and two-electron effective integrals of the form

$$h_{pq}^\kappa = \sum_o (\kappa_{po} h_{oq} + \kappa_{qo} h_{po}) \tag{12.5.24}$$

$$g_{pqrs}^\kappa = \sum_o (\kappa_{po} g_{oqrs} + \kappa_{qo} g_{pors} + \kappa_{ro} g_{pqos} + \kappa_{so} g_{pqro}) \tag{12.5.25}$$

(see Exercise 2.4). To arrive at (12.5.21), we have used the Jacobi identity (1.8.17) and rewritten one of the resulting terms using (10.2.6) as

$$\langle 0|[[\hat{\kappa}, E_{pq}^-], \hat{H}]|0\rangle = ([^o\mathbf{E}^{(1)}, \boldsymbol{\kappa}])_{pq} \tag{12.5.26}$$

The right-hand side of this equation is a matrix element of the commutator between an antisymmetric matrix with elements $^oE_{pq}^{(1)}$ and an antisymmetric matrix with elements κ_{pq}.

An inspection of expressions (12.5.18)–(12.5.21) for the Hessian reveals that the transformed vector (12.5.17) may be obtained from contributions whose evaluation has been discussed elsewhere. The configuration–configuration contribution (12.5.18) and the configuration–orbital contribution (12.5.19) may be calculated straightforwardly using the direct CI techniques developed in Section 11.8. In the configuration–orbital contribution, the usual Hamiltonian $\hat{H}$ has been replaced by a one-index transformed Hamiltonian $\hat{H}_\kappa$ with elements obtained by a sequence of one-index transformations (12.5.24) and (12.5.25) of the usual Hamiltonian integrals. Again, to reduce the computational cost, we may incorporate the contributions from the inactive orbitals in the same manner as for the MCSCF configuration gradient in Section 12.5.1.

Proceeding to the orbital–configuration (12.5.20) and orbital–orbital (12.5.21) contributions to the linear transformations, we note that these consist mainly of gradient-like terms. Thus, we calculate (12.5.20) as an MCSCF electronic gradient, replacing the usual MCSCF densities by the transition densities

$$D_{pq}^{\mathbf{c}} = \langle 0|E_{pq}|\mathbf{c}\rangle + \langle \mathbf{c}|E_{pq}|0\rangle \tag{12.5.27}$$

$$d_{pqrs}^{\mathbf{c}} = \langle 0|e_{pqrs}|\mathbf{c}\rangle + \langle \mathbf{c}|e_{pqrs}|0\rangle \tag{12.5.28}$$

The orbital–orbital contribution (12.5.21) may be calculated as an orbital gradient as well, using the one-index transformed Hamiltonian in (12.5.23). In addition, a small correction term, involving the orbital gradient and the orbital-rotation parameters, must be calculated.

Although the MCSCF electronic Hessian is only very rarely constructed explicitly (except perhaps the orbital–orbital block), we shall briefly comment on its calculation. Again, we concentrate on the nonseparable $\mathbf{K}^{(2)}$ part of the Hessian (12.5.3). The orbital–orbital block (12.2.43) may be calculated in the same way as the Hartree–Fock Hessian in Section 10.8.6. The configuration–configuration block (12.2.41) is identical to the CI Hessian and may be obtained using the techniques of Section 11.8. Finally, the coupling block (12.2.42) may be rewritten as

$$^{\text{oc}}K^{(2)}_{pq,i} = \langle 0|[E^-_{pq}, \hat{H}]|i\rangle + \langle i|[E^-_{pq}, \hat{H}]|0\rangle \tag{12.5.29}$$

and calculated as an MCSCF orbital gradient, employing one set of the transition densities

$$D^i_{pq} = \langle 0|E_{pq}|i\rangle + \langle i|E_{pq}|0\rangle \tag{12.5.30}$$

$$d^i_{pqrs} = \langle 0|e_{pqrs}|i\rangle + \langle i|e_{pqrs}|0\rangle \tag{12.5.31}$$

for each configuration $|i\rangle$. Although many simplifications are possible because of the special structure of for example CAS wave functions, the explicit calculation of the MCSCF Hessian is still an expensive undertaking. Nevertheless, the relatively small $^{\text{oc}}\mathbf{K}^{(2)}$ block is sometimes calculated explicitly – in particular, in connection with approximate second-order optimization techniques.

12.5.3 INNER AND OUTER ITERATIONS

The optimization an MCSCF wave function is carried out as a sequence of inner and outer iterations. The *outer iterations* are those discussed in Sections 12.3 and 12.4 – the individual iterations of the second-order methods carried out to converge the wave function. The *inner iterations* are those carried out to solve the linear equations (12.5.1) or the eigenvalue equations (12.5.2). The inner and outer iterations are also referred to as *micro and macro iterations*.

In problems involving inner and outer iterations (such as the optimization of an MCSCF wave function), we must make sure that the inner iterations are carried out in such a manner that the error in the solution vector does not impede the progress of the outer iterations. In MCSCF calculations, we must solve the linear equations (12.5.1) or alternatively the eigenvalue equations (12.5.2) to such an accuracy that the progress of the outer iterations towards the stationary point is not adversely affected. On the other hand, for reasons of efficiency, we must make sure also that the inner iterations are not carried out to an accuracy higher than that dictated by the outer iterations. In the following, we shall briefly discuss the *target accuracy* to which the solution vectors of the inner iterations should be converged to ensure a steady progress of the outer iterations.

The target accuracy of the inner iterations depends on whether the calculation takes place in the local or the global part of the optimization. In each outer iteration in the local region, we solve the Newton equations (12.3.2) by means of a series of inner iterations. The resulting step vector λ_{app} is not an exact solution to the linear equations (12.3.2), but contains an error described by the residual vector $\mathbf{R}$:

$$\mathbf{R} = \mathbf{E}^{(1)} + \mathbf{E}^{(2)}\lambda_{\text{app}} \tag{12.5.32}$$

To estimate the maximum error $\|\mathbf{R}\|$ that will not impede the convergence of the outer iterations, we recall that the second-order convergence of the exact solution λ is characterized by the relationship:

$$\|\mathbf{E}^{(1)}(\lambda)\| \approx a\|\mathbf{E}^{(1)}\|^2 \tag{12.5.33}$$

Here $\mathbf{E}^{(1)}(\lambda)$ is the gradient at the next point λ and a is a parameter characteristic of the optimization; see the discussion in Section 10.9.5. The numerical value of a remains fairly constant from iteration to iteration and may be estimated by comparing the gradients of two consecutive iterations. Expanding the gradient at λ_{app} around $\mathbf{0}$, we obtain

$$\mathbf{E}^{(1)}(\lambda_{\text{app}}) = \mathbf{E}^{(1)} + \mathbf{E}^{(2)}\lambda_{\text{app}} + O(\lambda^2_{\text{app}}) = \mathbf{R} + O(\lambda^2_{\text{app}}) \tag{12.5.34}$$

using (12.5.32). Next, using (12.5.34) in the second-order conditions (12.5.33), we find that the requirement

$$\|\mathbf{R}\| \ll a \|\mathbf{E}^{(1)}\|^2 \tag{12.5.35}$$

is sufficient to ensure that the second-order convergence (12.5.33) of Newton's method in the local region is not spoilt. In the global region, the requirements on the residual vector are less stringent. In practice, a residual norm of about one-tenth of the gradient norm is sufficient in this region.

In MCSCF calculations, the orbitals are usually considerably more difficult to converge than the configuration coefficients. The trial vectors of the inner iterations may therefore be divided into two classes of vectors – one class containing pure *configuration trial vectors* with zero orbital rotations and another class containing pure *orbital trial vectors* with zero configuration coefficients. This strategy allows us to concentrate on that part of the variational space that is most difficult to converge. In each inner iteration, we generate either a configuration trial vector or an orbital trial vector, depending on what part of the gradient is larger. Usually, the number of orbital trial vectors exceeds the number of configuration vectors significantly. For this strategy to be useful, the multiplication of the electronic Hessian by an orbital vector or by a configuration vector must be less expensive than the multiplication by a full vector with no zero elements. Indeed, from (12.5.18)–(12.5.21), we see that the transformed configuration and orbital vectors can be calculated independently without duplication of work, and that the combined cost of one configuration vector and one orbital vector is equal to the cost for the combined trial vector.

12.5.4 THE STRUCTURE OF THE MCSCF ELECTRONIC HESSIAN

Although in large-scale calculations the MCSCF electronic Hessian matrix $\mathbf{E}^{(2)}$ in (12.2.16) is never constructed explicitly, some knowledge of the numerical structure of this matrix is helpful in enabling us to design optimization algorithms that converge in as few iterations as possible. In the present subsection, we examine the numerical structure of the electronic Hessian in more detail, basing our discussion on a simple example – an electronic Hessian of the $^1\Sigma_g^+$ ground state of the H_2 molecule.

Table 12.1 contains the electronic Hessian of an optimized valence CASSCF wave function of the H_2 molecule at the experimental bond distance of $1.40a_0$. The wave function is a linear combination of the $|1\sigma_g^2\rangle$ and $|1\sigma_u^2\rangle$ configurations, the orbitals of which have been variationally optimized in the cc-pVDZ basis:

$$|\text{CAS}\rangle = C_1 |1\sigma_g^2\rangle + C_2 |1\sigma_u^2\rangle \tag{12.5.36}$$

Table 12.1 The electronic Hessian of an optimized CASSCF/cc-pVDZ ground-state wave function of the H_2 molecule at the equilibrium bond distance of $1.40a_0$ (in E_h). Also listed are the Hessian eigenvalues

	$1\sigma_g^2$	$1\sigma_u^2$	$1\sigma_g \rightarrow 2\sigma_g$	$1\sigma_g \rightarrow 3\sigma_g$	$1\sigma_u \rightarrow 2\sigma_u$	$1\sigma_u \rightarrow 3\sigma_u$
$1\sigma_g^2$	0.037	0.337	−0.026	−0.050	−0.015	0.013
$1\sigma_u^2$	0.337	3.075	−0.240	−0.455	−0.136	0.119
$1\sigma_g \rightarrow 2\sigma_g$	−0.026	−0.240	3.489	−0.164	0.044	−0.037
$1\sigma_g \rightarrow 3\sigma_g$	−0.050	−0.455	−0.164	8.556	0.017	0.066
$1\sigma_u \rightarrow 2\sigma_u$	−0.015	−0.136	0.044	0.017	0.072	0.007
$1\sigma_u \rightarrow 3\sigma_u$	0.013	0.119	−0.037	0.066	0.007	0.194
Eigenvalues	0.000	2.960	3.611	8.599	0.064	0.189

The active space contains the $1\sigma_g$ and $1\sigma_u$ orbitals; the secondary space comprises the σ orbitals $2\sigma_g, 2\sigma_u, 3\sigma_g, 3\sigma_u$ and the π orbitals $1\pi_{gx}, 1\pi_{ux}, 1\pi_{gy}, 1\pi_{uy}$. There are no inactive orbitals. The variational space comprises a total of six parameters: two CI parameters and four orbital-rotation parameters. As indicated in Table 12.1, where the electronic Hessian is listed, the nonredundant totally symmetric orbital rotations are those that mix the active $1\sigma_g$ orbital with the secondary $2\sigma_g$ and $3\sigma_g$ orbitals and those that mix the active $1\sigma_u$ orbital with the secondary $2\sigma_u$ and $3\sigma_u$ orbitals. The final wave function has a total energy of $-1.146908\ E_h$, with orbital occupations of 1.976 and 0.024 for the gerade and ungerade orbitals, respectively. This energy may be compared with the STO minimal-basis energy of Section 5.2 ($-1.107\ E_h$) and the FCI/cc-pVDZ energy of Section 5.3 ($-1.174\ E_h$).

In Table 12.1 we have also listed the eigenvalues of the MCSCF Hessian. Since the Hessian is (mostly) diagonally dominant, each eigenvector has one dominant element and the eigenvalues are similar to the diagonal Hessian elements. In the table we have therefore listed the Hessian eigenvalues in the columns of the corresponding dominant contributions to the eigenvectors. As expected, there is one zero eigenvalue, related to the redundancy of the parametrization of the CI expansion.

Before examining the MCSCF Hessian in Table 12.1, we note that the electronic Hessian is not uniquely determined by the wave function. Thus, in CASSCF theory, we may carry out orbital rotations among the inactive–inactive, secondary–secondary and active–active orbitals without affecting the energy. This freedom may be exploited to increase the diagonal dominance of the Hessian and thus improve the convergence of the optimization. From the expressions for the RHF Hessian (10.8.55) and (10.8.56) in Chapter 10, we note that, for Hartree–Fock wave functions, the diagonal dominance is more pronounced in the canonical basis, where the Fock matrix is diagonal. In practice, it has been found that, for MCSCF wave functions, the use of orbitals that diagonalize the diagonal blocks of the matrix $^{\mathrm{I}}\mathbf{F} + {^{\mathrm{A}}\mathbf{F}}$ usually increases the diagonal dominance of Hessian. The Hessian in Table 12.1 has been calculated from such orbitals.

Let us now examine the $^{cc}\mathbf{E}^{(2)}$ block in Table 12.1. In the expression (12.2.28) for the Hessian elements, the contributions from the projection operator to $^{cc}\mathbf{E}^{(2)}$ are proportional to the CI gradient – these contributions vanish at convergence and are usually small otherwise. The CI part of the Hessian is therefore dominated by the shifted and scaled Hamiltonian matrix with elements $2\langle i|\hat{H} - E^{(0)}|j\rangle$. This matrix is usually diagonally dominant and has eigenvalues that are approximations to twice the excitation energies. As discussed in Section 11.5, iterative methods involving this matrix usually converge in a rather small number of iterations.

The diagonal elements of the CI Hessian are given by $2\langle i|\hat{H}|i\rangle - 2E^{(0)}$. For states dominated by a single configuration, there will therefore be at least one diagonal element close to zero. For example, the small diagonal element of $0.037\ E_h$ is $^{cc}E_{11}^{(0)}$ since $\langle 1\sigma_g^2|\hat{H}|1\sigma_g^2\rangle$ is close to the optimized MCSCF energy. From a slightly different point of view, we note that a small variation in the dominant coefficient will not change the wave function much, as indicated by the smallness of the first diagonal element in Table 12.1. Also, in the limit of a single configuration, variations in the dominant coefficient become redundant and give rise to a zero Hessian matrix element.

The numerical value of the largest diagonal element in the configuration block of the Hessian in Table 12.1 is $3.08\ E_h$. Moreover, a diagonalization of $^{cc}\mathbf{E}^{(0)}$ gives the eigenvalue $3.11\ E_h$ as well as a zero eigenvalue. The $|1\sigma_g^2\rangle \rightarrow |1\sigma_u^2\rangle$ excitation energy should therefore be about $1.55\ E_h$. Referring back to the two-configuration STO minimal-basis calculation in Table 5.2, we find that the corresponding excitation energy is $1.28\ E_h$. In view of the differences between these two calculations, the agreement is reasonable.

Turning our attention to the orbital–orbital Hessian, we note that the MCSCF orbital rotations in Table 12.1 fall into two broad groups. Rotations that mix orbitals with very different occupations change the wave function significantly and give rise to large diagonal Hessian elements. Conversely, rotations that mix orbitals with similar occupations change the wave function only slightly and give rise to small diagonal Hessian elements. As limiting cases, the rotations among the doubly occupied inactive orbitals and among the unoccupied secondary orbitals of a CAS wave function are redundant and have zero diagonal elements in the Hessian (if calculated).

This grouping of the orbital rotations is seen in the $^{oo}\mathbf{E}^{(2)}$ block of Table 12.1. The rotations of the strongly occupied $1\sigma_g$ orbital with the unoccupied $2\sigma_g$ and $3\sigma_g$ orbitals have large diagonal Hessian elements; the rotations of the weakly occupied $1\sigma_u$ orbital with the unoccupied $2\sigma_u$ and $3\sigma_u$ orbitals have small Hessian elements. We emphasize that the smallness of these elements arises since the corresponding rotations do not change the wave function much and should not be taken as an indication of nearby excited states.

Let us finally consider the coupling blocks $^{oc}\mathbf{E}^{(2)}$ and $^{co}\mathbf{E}^{(2)}$. The magnitude of the elements in these blocks is small and similar to that of the near-redundant orbital rotations in the $^{oo}\mathbf{E}^{(2)}$ block of the Hessian. Since the CI parameters couple with the near-redundant orbital rotations in a rather unpredictable manner, the neglect of this coupling may lead to slow convergence.

12.5.5   EXAMPLES OF MCSCF OPTIMIZATIONS

In the present subsection, we consider some numerical examples of the convergence of valence CASSCF calculations on H_2 and H_2O using the NEO method discussed in Section 12.4. The H_2 calculations in Tables 12.2 and 12.3 were carried out in the cc-pVDZ basis at the experimental bond distance of $1.40a_0$; the H_2O calculation in Table 12.4 was carried out in the cc-pVQZ basis at the reference geometry of Section 5.3.3. In all calculations, the initial set of CI coefficients was generated by carrying out three direct CI iterations in the CAS space of an MO basis obtained either from a canonical Hartree–Fock calculation or by the diagonalization of the MP2 one-electron

Table 12.2 The convergence of a valence CASSCF/cc-pVDZ calculation on H_2 starting with the MP2 natural orbitals (atomic units). The converged MCSCF energy is -1.14690814 E_h

Iteration	Energy error	Gradient norm	Ratio (12.3.21)	Step length
1	0.00011261	0.03308166	–	0.02916
2	0.00000004	0.00030047	0.99156	0.00082
3	0.00000000	0.00000037	1.00083	–

Table 12.3 The convergence of a valence CASSCF/cc-pVDZ calculation on H_2 starting with canonical Hartree–Fock orbitals (atomic units). The converged MCSCF energy is -1.14690814 E_h

Iteration	Energy error	Gradient norm	Ratio (12.3.21)	Step length
1	0.01546096	0.01829828	–	0.69524
2	0.00335950	0.10475454	0.48312	0.15511
3	0.00013770	0.01380173	1.04146	0.05410
4	0.00000046	0.00141876	1.00860	0.00209
5	0.00000000	0.00000174	1.00048	–

Table 12.4 The convergence of a valence CASSCF/cc-pVDZ calculation on H_2O starting with the MP2 natural orbitals (atomic units). The converged MCSCF energy is -76.11653159 E_h. For each iteration, the number of inner iterations n_{inner} required for the solution of the Newton eigenvalue problem is listed

Iteration	n_{inner}	Energy error	Gradient norm	Ratio (12.3.21)	Step length
1	8	0.00399760	0.16030687	–	0.71005
2^a	0	0.01025851	0.27385050	−0.72849	no step
1^b	0	0.00399760	0.16030687	–	0.47573
2	5	0.00223496	0.12773140	0.26827	0.11159
3	9	0.00010050	0.01744743	1.04250	0.07486
4	9	0.00000630	0.00473616	0.89828	0.01570
5	9	0.00000005	0.00022996	1.04575	0.00171
6	0	0.00000000	0.00000689	1.00574	–

^{a}Rejected point; no step attempted.
^{b}Step from iteration 1 scaled by a factor of $\frac{2}{3}$ (backstep).

density matrix (the MP2 natural orbitals). In each outer iteration, the Newton eigenvalue problem was solved by carrying out a sequence of inner iterations, using the Davidson method.

Let us first consider the calculations on the H_2 system. In Table 12.2, the MP2 natural orbitals are used in the first iteration; in Table 12.3, we use the canonical Hartree–Fock orbitals. Because of the different choices of orbitals, the optimizations proceed rather differently. For the optimization based on the MP2 natural orbitals, the optimization begins in the local region; each step corresponds to a Newton step with no step-length restrictions. Quadratic convergence is therefore observed in all outer iterations – see the reduction in the gradient and step norms in Table 12.2. The ratio parameter r (12.3.21), which probes the quadratic dominance of the energy function, is close to 1 in all iterations.

By contrast, the canonical Hartree–Fock orbitals constitute a less favourable starting point. In the first iteration of Table 12.3, the Hessian has one negative eigenvalue and the step-size control is activated with an initial trust radius of 0.7. From the second iteration, however, the optimization proceeds in the local region and quadratic convergence is observed. Still, a total of five outer iterations are needed for convergence, compared with only three iterations for the optimization in Table 12.2. These calculations illustrate the importance of using a good initial set of orbitals for the efficient optimization of MCSCF wave functions. Usually, the orbitals obtained by diagonalizing the MP2 density matrix constitute a better initial set of orbitals than do the canonical Hartree–Fock orbitals, where the virtual orbitals with the lowest orbital energies are too diffuse to describe electron correlation.

Another important aspect of MCSCF optimizations is illustrated by the calculation in Table 12.3. In the first iteration, the structure of the Hessian is incorrect (with one negative eigenvalue). Still, by selecting the lowest eigenvector of the augmented Hessian, the search is guided into the local region (with the correct Hessian structure), and the optimization then proceeds to the correct electronic state. By contrast, if instead a set of unrestricted Newton iterations is carried out starting with the canonical orbitals, then the optimization leads to a stationary point of energy -1.129128 E_h (i.e. 18 mE_h above the minimum) with occupation numbers 1.9998 and 0.0002, respectively, for the $1\sigma_g$ and $1\sigma_u$ orbitals. This stationary point has one negative Hessian eigenvalue and does not represent a satisfactory wave function, either for the ground state or for the first excited state.

We now turn our attention to the convergence of the MCSCF wave function for H_2O in Table 12.4. In the first iteration, a step of length 0.71 is taken to the boundary of the trust region.

At the new point, however, the energy increases rather than decreases and the ratio parameter r, which tests for the harmonicity of the energy surface, is negative. The first step is therefore rejected as described in Section 12.3 and we resume the optimization with a reduced trust radius of 0.47. With this smaller step, the Hessian is positive definite and local steps (unrestricted Newton steps) are taken inside the trust region in all subsequent iterations. Nevertheless, the energy function still has a notable anharmonic character: the ratio parameter r fluctuates strongly and the reduction in the gradient is less than that expected from quadratic convergence. Still, convergence to a gradient norm less than 10^{-5} is obtained in six iterations, which is fairly typical of a CAS calculation with a reasonable set of initial orbitals. However, usually there are no backsteps: MCSCF optimizations typically consist of two or three global iterations followed by a sequence of quadratically convergent Newton iterations.

12.6 Exponential parametrization of the configuration space

So far in our discussion of CI and MCSCF theory, we have found it convenient to employ a redundant parametrization of the configuration space – see Section 11.4.1 for the CI wave function and Section 12.2.1 for the MCSCF wave function. This parametrization has the advantage of simplicity, which for many purposes such as the optimization of the wave function offsets the disadvantages arising from the presence of the redundancy. Nevertheless, in some situations, it is best to remove this redundancy and to use a parametrization where the electronic states are described by a set of independent, nonredundant variables. One such situation arises in the state-averaged MCSCF scheme discussed in Section 12.7, where several MCSCF states are generated simultaneously in the same orbital basis.

In this section, we introduce an exponential unitary parametrization of the configuration space. In this parametrization, transformations among orthogonal states in the configuration space are generated by means of an exponential unitary operator in much the same way as unitary orbital rotations are carried out by means the exponential operator $\exp(-\hat{\kappa})$ of Chapter 3. Just as the transformations in the orbital space are usually directed towards determining a set of occupied orbitals, the transformations in the configuration space are concerned with determining a single reference state or a small set of reference states. However, the large dimension of the configuration space poses special problems that need to be addressed.

In Section 12.6.1, we introduce general unitary rotations in the CI space. Next, we develop in Section 12.6.2 a machinery for rotating a single reference state and consider in Section 12.6.3 the related problem of setting up an orthonormal basis for the orthogonal complement space. Finally, in Section 12.6.4, we extend the scheme of Sections 12.6.2 and 12.6.3 to cases involving several reference states.

12.6.1 GENERAL EXPONENTIAL PARAMETRIZATION OF THE CONFIGURATION SPACE

We assume that we have a set of states $|P\rangle$ that constitutes an orthonormal basis in the configuration space of dimension m, spanned, for example, by a set of m Slater determinants. From the states $|P\rangle$, we may generate a new set of orthonormal states $|\tilde{P}\rangle$ by a unitary transformation

$$|\tilde{P}\rangle = \sum_Q |Q\rangle U_{QP} \tag{12.6.1}$$

where the coefficients are the elements of an m-dimensional unitary matrix $\mathbf{U}$. According to the discussion in Section 3.1, we may write the unitary matrix $\mathbf{U}$ as

$$\mathbf{U} = \exp(-\mathbf{S}) \tag{12.6.2}$$

where $\mathbf{S}$ is an anti-Hermitian matrix, the elements of which are independent except for the simple requirement

$$S_{QR} = -S_{RQ}^* \tag{12.6.3}$$

The minus sign in the exponential (12.6.2) is conventional. A nonredundant parametrization of the configuration space may therefore be written as

$$|\tilde{P}\rangle = \sum_Q |Q\rangle [\exp(-\mathbf{S})]_{QP} \tag{12.6.4}$$

Like the orbital transformation in Section 3.2, this unitary transformation can be given an operator representation

$$|\tilde{P}\rangle = \exp(-\hat{S})|P\rangle \tag{12.6.5}$$

where the anti-Hermitian operator $\hat{S}$ is given by

$$\hat{S} = \sum_{QR} S_{QR}|Q\rangle\langle R| \tag{12.6.6}$$

The equivalence of the matrix and operator representations (12.6.4) and (12.6.5) is readily established by expanding the exponential operator in (12.6.5):

$$
\begin{aligned}
|\hat{P}\rangle &= \left(1 - \hat{S} + \tfrac{1}{2}\hat{S}^2 + \cdots\right)|P\rangle \\
&= \sum_Q |Q\rangle \left[\delta_{QP} - S_{QP} + \tfrac{1}{2}(\mathbf{S}^2)_{QP} + \cdots\right] \\
&= \sum_Q |Q\rangle [\exp(-\mathbf{S})]_{QP}
\end{aligned}
\tag{12.6.7}
$$

Often, it is convenient to write $\hat{S}$ in such a way that its anti-Hermiticity is explicitly indicated. Writing the diagonal elements of $\mathbf{S}$ in the form $\mathrm{i}^{\mathrm{I}}S_{PP}$, we obtain

$$\hat{S} = \sum_{Q>R}(S_{QR}|Q\rangle\langle R| - S_{QR}^*|R\rangle\langle Q|) + \mathrm{i}\sum_Q {}^{\mathrm{I}}S_{QQ}|Q\rangle\langle Q| \tag{12.6.8}$$

A *special unitary transformation* $\exp(-\hat{S})$ is obtained by omitting the diagonal elements of this expression; see Section 3.1.3. Moreover, an *orthogonal transformation* of the configuration states is obtained by employing in $\exp(-\hat{S})$ a real antisymmetric matrix $\mathbf{S}$.

As already noted, there are many similarities between the exponential unitary transformations in configuration space and in orbital space. Comparing with the results for orbital transformations in Chapter 3, we note that the operators $\hat{S}$ and $\hat{\kappa}$ are both anti-Hermitian, as are the matrices $\mathbf{S}$ and $\boldsymbol{\kappa}$. Moreover, whereas $\exp(-\hat{S})|P\rangle$ represents a unitarily transformed configuration state, $\exp(-\hat{\kappa})|P\rangle$ represents a state where the spin orbitals have been unitarily transformed as $\exp(-\hat{\kappa})a_P^\dagger \exp(\hat{\kappa})$.

12.6.2 EXPONENTIAL PARAMETRIZATION FOR A SINGLE REFERENCE STATE

In the N-electron configuration space, we are interested mostly in transformations of a single electronic state or perhaps of a small number of such states. In particular, we are not interested in transformations within the orthogonal-complement space, only in obtaining a basis for this space that can be generated and manipulated easily. The purpose of the present and the next subsection is to modify the general unitary operator of Section 12.6.1 for the efficient transformation of a *single reference state* [11]. In this subsection, we consider the parametrization of the reference state; in Section 12.6.3, we generate an orthonormal basis for the orthogonal complement.

Let us denote our reference state by $|0\rangle$ and let the $m - 1$ states $|K\rangle$, with $K \neq 0$, be a basis for the orthogonal complement to $|0\rangle$. All m states are expanded in a set of m orthonormal configurations $|i\rangle$ (Slater determinants or CSFs). As we shall shortly demonstrate, the exponential unitary operator

$$\exp(-\hat{R}) = \exp\left[-\sum_{K \neq 0}(R_K|K\rangle\langle 0| - R_K^*|0\rangle\langle K|)\right] \tag{12.6.9}$$

may, when applied to the reference state

$$|\tilde{0}\rangle = \exp(-\hat{R})|0\rangle \tag{12.6.10}$$

generate an arbitrary normalized state

$$|\tilde{0}\rangle = \tilde{C}_0|0\rangle + \sum_{K \neq 0}\tilde{C}_K|K\rangle \tag{12.6.11}$$

$$\sum_i |\tilde{C}_i|^2 = 1 \tag{12.6.12}$$

with the phase of $|\tilde{0}\rangle$ chosen such that $\tilde{C}_0$ is real and positive. The number of parameters in the anti-Hermitian operator $\hat{R}$ is $m - 1$, one less than the dimension of the configuration space since the reference state $|0\rangle$ is omitted from the summation in (12.6.9). For a general unitary transformation, the parameters R_K are complex:

$$R_K = {}^R R_K + i\,{}^I R_K \tag{12.6.13}$$

For real states, we may replace R_K and R_K^* in (12.6.9) by ${}^R R_K$ and carry out orthogonal rather than unitary transformations.

In Exercise 12.3, it is shown that (12.6.10) may be written as

$$\exp(-\hat{R})|0\rangle = \cos d\,|0\rangle - \frac{\sin d}{d}\sum_{K \neq 0} R_K|K\rangle \tag{12.6.14}$$

where

$$d = \sqrt{\sum_{K \neq 0}|R_K|^2} \tag{12.6.15}$$

The equivalence with (12.6.11) is now established by inserting in (12.6.14) the coefficients

$$R_K = -\frac{\arccos \tilde{C}_0}{\sqrt{1 - \tilde{C}_0^2}}\tilde{C}_K \tag{12.6.16}$$

It then follows that $d = \arccos \tilde{C}_0$ is a positive number $0 \le d \le \pi$, and we may make the identifications

$$\cos d = \tilde{C}_0 \tag{12.6.17}$$

$$\sin d = \sqrt{1 - \tilde{C}_0^2} \tag{12.6.18}$$

where $\sin d$ is always positive but $\cos d$ may be positive or negative.

12.6.3 A BASIS FOR THE ORTHOGONAL COMPLEMENT TO THE REFERENCE STATE

The unitary operator $\exp(-\hat{R})$ of Section 12.6.2 may be used not only to generate an orthogonal transformation of the reference state $|0\rangle$, but also to generate an orthonormal basis for the orthogonal complement to the transformed state $|\tilde{0}\rangle$:

$$|\tilde{M}\rangle = \exp(-\hat{R})|M\rangle \tag{12.6.19}$$

In Exercise 12.3, it is shown that the transformed functions are given by

$$|\tilde{M}\rangle = |M\rangle + \frac{\sin d}{d} R_M^* |0\rangle + \frac{\cos d - 1}{d^2} R_M^* \sum_{K \neq 0} R_K |K\rangle$$

$$= |M\rangle + \frac{\sin d}{d} R_M^* |0\rangle - \frac{1}{2} \left[\frac{\sin(d/2)}{d/2} \right]^2 R_M^* \sum_{K \neq 0} R_K |K\rangle \tag{12.6.20}$$

In principle, therefore, we may first use $\exp(-\hat{R})$ to generate the desired reference state $|\tilde{0}\rangle$ from $|0\rangle$ according to (12.6.14) and next perform the necessary adjustments of the basis $|M\rangle$ for the orthogonal complement according to (12.6.20). For large configuration spaces, however, this procedure becomes impractical since each function in (12.6.20) is a complicated linear combination of the full set of m configurations $|i\rangle$, precluding its construction and storage for all but the smallest expansions.

Fortunately, there is a simple way to generate an orthonormal basis for the orthogonal complement to $|0\rangle$. The strategy is to regard the reference state and the orthogonal-complement basis as generated by the application of an exponential unitary operator to the configuration basis $|i\rangle$. We begin the construction of our basis by singling out a particular *reference configuration* $|o\rangle$ in the reference state – for example, the configuration with the largest coefficient:

$$|0\rangle = C_o |o\rangle + \sum_{i \neq o} C_i |i\rangle \tag{12.6.21}$$

We assume that the phase of $|o\rangle$ is chosen so that C_o is real and positive. Next, we regard the reference state as generated by a unitary transformation of the reference configuration $|o\rangle$:

$$|0\rangle = \exp(-\hat{\tau})|o\rangle \tag{12.6.22}$$

From the preceding discussion, we find that the associated exponential operator is given by

$$\exp(-\hat{\tau}) = \exp\left[-\sum_{i \neq o} (\tau_i |i\rangle \langle o| - \tau_i^* |o\rangle \langle i|) \right] \tag{12.6.23}$$

where

$$\tau_i = -\frac{d}{\sin d} C_i \tag{12.6.24}$$

$$d = \arccos C_o \tag{12.6.25}$$

The orthogonal-complement basis may now be generated by applying the exponential operator $\exp(-\hat{\tau})$ to each basis vector $|i\rangle$ except $|o\rangle$. From the expansion (12.6.21) and from the identities (12.6.24) and (12.6.25), we obtain

$$|M\rangle = |m\rangle + \frac{\sin d}{d} \tau_m^* |o\rangle + \frac{\cos d - 1}{d^2} \tau_m^* \sum_{i \neq o} \tau_i |i\rangle$$

$$= |m\rangle - C_m^* |o\rangle - \frac{C_m^*}{1 + C_o} \sum_{i \neq o} C_i |i\rangle \tag{12.6.26}$$

which may be rearranged to give

$$|M\rangle = |m\rangle - \frac{C_m^*}{1 + C_o}(|0\rangle + |o\rangle) \tag{12.6.27}$$

Note that the parametrization of the orthogonal complement (12.6.27) depends on the choice of reference configuration in (12.6.21). It is a simple exercise to verify that $|M\rangle$ is indeed normalized and that it is orthogonal to the reference state $|0\rangle$ and to the remaining basis functions of the orthogonal complement. We have thus set up an orthonormal basis $|M\rangle$ for the orthogonal complement to the reference state $|0\rangle$, where each basis function is easily constructed from the expansion coefficients of the reference state.

12.6.4 EXPONENTIAL PARAMETRIZATION FOR SEVERAL REFERENCE STATES

In Section 12.6.2, we saw how any given normalized electronic state in configuration space may be generated from a single reference state by an exponential mapping. Next, in Section 12.6.3, we set up a simple orthonormal basis for the orthogonal complement to the reference state. In the present subsection, we shall generalize this scheme to the generation of a set of n arbitrary orthonormal states $|\tilde{0}_i\rangle$ from n orthonormal reference states $|0_i\rangle$ by the application of a unitary operator

$$|\tilde{0}_i\rangle = \hat{U}|0_i\rangle, \qquad i \leq n \tag{12.6.28}$$

We shall also introduce a compact and simple representation for an orthonormal basis $|K\rangle$ for the orthogonal complement to the reference states. This development is of interest in connection with the simultaneous optimization of several states discussed in Section 12.7, where the reference states $|0_i\rangle$ are approximations to a set of n electronic states and we improve our description by carrying out simultaneous unitary transformations among these states.

Assume that an orthonormal basis is given such that the first n elements correspond to the reference states $|0_i\rangle$ and the remaining $m - n$ states constitute a basis $|K\rangle$ for the orthogonal complement. The unitary operator

$$\hat{U} = \exp(-\hat{R}) \tag{12.6.29}$$

where

$$\hat{R} = \sum_{k>l}(R_{kl}|0_k\rangle\langle 0_l| - R_{kl}^*|0_l\rangle\langle 0_k|) + \sum_{Kl}(R_{Kl}|K\rangle\langle 0_l| - R_{Kl}^*|0_l\rangle\langle K|) \tag{12.6.30}$$

represents an obvious generalization of the unitary transformation of a single state (12.6.9). In (12.6.30), the first part rotates the reference states among themselves and the second part mixes the reference states with the orthogonal complement. By a suitable choice of parameters in (12.6.30), we may transform the set of reference states into any desired set of n orthonormal states, which will then constitute the new reference set:

$$|\tilde{0}_k\rangle = \exp(-\hat{R})|0_k\rangle \tag{12.6.31}$$

Although no proof is given here that this parametrization is unique and general, we note that the number of independent parameters (real and imaginary) in $\hat{R}$ is $2n(m-n) + n^2 - n$. This number is equal to the number of independent parameters for n orthonormal states in an m-dimensional space, excluding a phase factor for each of the states.

Although we are interested primarily in the reference states and their transformation, we must also be able to generate, for any given set of reference states $|0_k\rangle$, a convenient basis for the orthogonal complement to the reference states. The detailed form of the orthogonal-complement basis is unimportant – what matters is only that it is easy to generate. This generation of a basis for the orthogonal complement is discussed in the remainder of the present subsection. For convenience, we restrict the treatment to real expansions. Our strategy is to construct a unitary operator as a product of n elementary unitary operators $\hat{U}_i$, which, when applied to the m configurations $|i\rangle$, generates a set of m orthonormal states, of which the first n correspond to the reference set. In setting up this scheme, we shall make extensive use of rank-1 operators of the form

$$\hat{U}_{ab} = 1 - \frac{(|a\rangle - |b\rangle)(\langle a| - \langle b|)}{1 - \langle a|b\rangle} \tag{12.6.32}$$

where $|a\rangle$ and $|b\rangle$ are real states that are normalized but not necessarily orthogonal to each other. This operator is unitary and Hermitian

$$\hat{U}_{ab}\hat{U}_{ab}^{\dagger} = \hat{U}_{ab}^{\dagger}\hat{U}_{ab} = 1 \tag{12.6.33}$$

$$\hat{U}_{ab} = \hat{U}_{ab}^{\dagger} \tag{12.6.34}$$

as is easily verified. Moreover, the operator $\hat{U}_{ab}$ interchanges $|a\rangle$ and $|b\rangle$

$$\hat{U}_{ab}|a\rangle = |b\rangle \tag{12.6.35}$$

$$\hat{U}_{ab}|b\rangle = |a\rangle \tag{12.6.36}$$

but leaves all vectors orthogonal to $|a\rangle$ and $|b\rangle$ unchanged.

Let us now consider the problem of unitarily transforming the set of m configurations $|i\rangle$ to a new orthonormal set, the first n elements of which are the reference states $|0_i\rangle$. From our discussion of $\hat{U}_{ab}$, we know that the rank-1 unitary operator

$$\hat{U}_1 = 1 - \frac{(|0_1\rangle - |1\rangle)(\langle 0_1| - \langle 1|)}{1 - \langle 0_1|1\rangle} \tag{12.6.37}$$

transforms the configuration $|1\rangle$ into the first reference state $|0_1\rangle$ and that, applied to the remaining $m-1$ configurations $|i\rangle$ in the configuration space, it generates an orthonormal basis for the orthogonal complement to $|0_1\rangle$. However, none of the remaining $m-1$ states correspond to any of the states of the reference space.

To recover the second reference state $|0_2\rangle$, we introduce a unitary operator $\hat{U}_2$ with the properties that $\hat{U}_1\hat{U}_2|1\rangle$ remains equal to $|0_1\rangle$, whereas $\hat{U}_1\hat{U}_2|2\rangle$ becomes $|0_2\rangle$. The operator

$$\hat{U}_2 = 1 - \frac{(\hat{U}_1|0_2\rangle - |2\rangle)(\langle 0_2|\hat{U}_1 - \langle 2|)}{1 - \langle 0_2|\hat{U}_1|2\rangle} \tag{12.6.38}$$

accomplishes this task as is readily verified:

$$\hat{U}_1\hat{U}_2|1\rangle = \hat{U}_1 \left[|1\rangle - \frac{(\hat{U}_1|0_2\rangle - |2\rangle)(\langle 0_2|0_1\rangle - \langle 2|1\rangle)}{1 - \langle 0_2|\hat{U}_1|2\rangle} \right] = \hat{U}_1|1\rangle = |0_1\rangle \tag{12.6.39}$$

$$\hat{U}_1\hat{U}_2|2\rangle = \hat{U}_1 \left[|2\rangle - \frac{(\hat{U}_1|0_2\rangle - |2\rangle)(\langle 0_2|\hat{U}_1|2\rangle - 1)}{1 - \langle 0_2|\hat{U}_1|2\rangle} \right] = \hat{U}_1\hat{U}_1|0_2\rangle = |0_2\rangle \tag{12.6.40}$$

Therefore, applying the unitary product operator $\hat{U}_1\hat{U}_2$ to the m states $|i\rangle$, we generate $|0_1\rangle$ and $|0_2\rangle$ and a set of $m - 2$ orthonormal states spanning the orthogonal complement to $|0_1\rangle$ and $|0_2\rangle$.

We now invoke induction to set up the general unitary transformation. Assume that, in step $k - 1 < n$, we have generated the operators

$$\hat{U}_i = 1 - \frac{(\hat{U}_{i-1}\cdots\hat{U}_1|0_i\rangle - |i\rangle)(\langle 0_i|\hat{U}_1\cdots\hat{U}_{i-1} - \langle i|)}{1 - \langle 0_i|\hat{U}_1\cdots\hat{U}_{i-1}|i\rangle} \tag{12.6.41}$$

with $i \leq k - 1$ and where the product $\hat{U}_1\hat{U}_2\cdots\hat{U}_{k-1}$ transforms the first $k - 1$ configurations $|i\rangle$ into the first $k - 1$ reference states $|0_i\rangle$:

$$\hat{U}_1\hat{U}_2\cdots\hat{U}_{k-1}|i\rangle = |0_i\rangle, \qquad i \leq k - 1 \tag{12.6.42}$$

Next, we introduce the operator $\hat{U}_k$ according to (12.6.41) and obtain, for $i \leq k - 1$,

$$\hat{U}_1\cdots\hat{U}_{k-1}\hat{U}_k|i\rangle = \hat{U}_1\cdots\hat{U}_{k-1} \left[|i\rangle - \frac{(\hat{U}_{k-1}\cdots\hat{U}_1|0_k\rangle - |k\rangle)(\langle 0_k|0_i\rangle - \langle k|i\rangle)}{1 - \langle 0_k|\hat{U}_1\cdots\hat{U}_{k-1}|k\rangle} \right]$$
$$= \hat{U}_1\cdots\hat{U}_{k-1}|i\rangle = |0_i\rangle \tag{12.6.43}$$

and

$$\hat{U}_1\cdots\hat{U}_{k-1}\hat{U}_k|k\rangle = \hat{U}_1\cdots\hat{U}_{k-1} \left[|k\rangle - \frac{(\hat{U}_{k-1}\cdots\hat{U}_1|0_k\rangle - |k\rangle)(\langle 0_k|\hat{U}_1\cdots\hat{U}_{k-1}|k\rangle - 1)}{1 - \langle 0_k|\hat{U}_1\cdots\hat{U}_{k-1}|k\rangle} \right]$$
$$= \hat{U}_1\cdots\hat{U}_{k-1}\hat{U}_{k-1}\cdots\hat{U}_1|0_k\rangle = |0_k\rangle \tag{12.6.44}$$

We have thus obtained n elementary operators $\hat{U}_i$ which, when applied to the first n configurations, generate the reference vectors

$$|0_i\rangle = \hat{U}_1\cdots\hat{U}_n|i\rangle, \qquad i \leq n \tag{12.6.45}$$

and, when applied to the remaining configurations, generate a basis for the orthogonal complement to the reference space

$$|K_i\rangle = \hat{U}_1\cdots\hat{U}_n|i\rangle, \qquad i > n \tag{12.6.46}$$

Each of the n unitary operators $\hat{U}_i$ is a simple rank-1 operator, easily constructed from the (usually small set of) n reference functions $|0_i\rangle$ and the first n configurations $|i\rangle$.

In practice, we need never explicitly construct the matrix representation of the elementary operators $\hat{U}_i$. As an example, assume that the matrix elements

$$\langle i|\hat{F}|0\rangle = F_i \tag{12.6.47}$$

of the operator $\hat{F}$ are known for the m configurations and that we wish to obtain the representation of $\hat{F}$ in the (usually) much smaller space of the n reference states. We may then calculate this quantity in the following manner:

$$\langle 0_i|\hat{F}|0\rangle = \langle i|\hat{U}_n \cdots \hat{U}_1\hat{F}|0\rangle = (\mathbf{U}_n \cdots \mathbf{U}_1\mathbf{F})_i \tag{12.6.48}$$

where $\mathbf{U}_i$ and $\mathbf{F}$ are the matrix representations of the operators $\hat{U}_i$ and $\hat{F}$. Since the elementary matrices $\mathbf{U}_i$, with $i \leq n$, have a simple rank-1 structure, we may calculate $\mathbf{U}_n \cdots \mathbf{U}_1\mathbf{F}$ as a sequence of simple vector–matrix manipulations. Consider, for example, the calculation of $\mathbf{U}_1\mathbf{F}$. If we let $\mathbf{C}_1$ represent $|0_1\rangle$, the matrix representation of $|0_1\rangle - |1\rangle$ becomes

$$[\mathbf{d}_1]_i = [\mathbf{C}_1]_i - \delta_{1i} \tag{12.6.49}$$

Using (12.6.37), we obtain

$$\mathbf{U}_1\mathbf{F} = \left(1 - \frac{\mathbf{d}_1\mathbf{d}_1^{\mathrm{T}}}{1 - [\mathbf{C}_1]_1}\right)\mathbf{F} = \mathbf{F} - \frac{\mathbf{d}_1^{\mathrm{T}}\mathbf{F}}{1 - [\mathbf{C}_1]_1}\mathbf{d}_1 \tag{12.6.50}$$

Repeating this step n times, we obtain (12.6.48) without ever constructing the matrices $\mathbf{U}_i$ explicitly.

12.7 MCSCF theory for several electronic states

An important area of application of MCSCF theory is the simultaneous study of several electronic states. Such studies may be carried out in different ways, two of which we outline here.

12.7.1 SEPARATE OPTIMIZATION OF THE INDIVIDUAL STATES

One approach is first to calculate wave functions separately for each state of interest. The wave functions obtained in this way are not orthogonal since they have been generated independently in separate calculations. To arrive at a more convenient description of the electronic system in terms of a set of orthogonal and noninteracting states, we may subsequently carry out a small CI calculation in the basis of the MCSCF wave functions. For this procedure to work, we must be able to calculate matrix elements between electronic states expanded in different MOs – the MCSCF orbitals of the electronic states are different since they have been optimized for each state individually.

For CAS and RAS expansions, the matrix elements between nonorthogonal states are conveniently obtained using the procedure of Section 11.9. For each pair of states, we determine a biorthonormal basis – see Section 1.9.3. The overlap and Hamiltonian matrix elements of this basis are obtained by transforming the integrals and CI vectors as described in Sections 1.9.3 and 11.9, respectively. A subsequent diagonalization of the resulting small Hamiltonian matrix (suitably weighted by the overlap matrix) then produces a set of orthonormal, noninteracting electronic states [12]. This procedure represents a rather straightforward extension of single-state MCSCF theory and yields a description comparable to that of single-state wave functions. Since this popular approach does not involve any new computational techniques, we shall not pursue it further here.

12.7.2 STATE-AVERAGED MCSCF THEORY

A different approach to the study of several electronic states involves the simultaneous MCSCF optimization of all electronic states in a common orbital basis. In this approach, the orbitals are not optimized separately for each state but are instead determined to minimize an averaged energy of all states [13]. This is accomplished by introducing an energy function to which each electronic state contributes with a weight factor. This state-averaged energy is minimized in the usual manner. Obviously, in this *state-averaged MCSCF approach*, the description of each electronic state suffers from the fact that the orbitals are not optimal for each state separately. On the other hand, the method provides directly a set of orthonormal, noninteracting electronic states. Moreover, the use of the state-averaged orbitals simplifies the subsequent treatment of the electronic states in a CI calculation, where several orthogonal eigenstates are simultaneously generated. Also, the state-averaged MCSCF approach is well suited to the subsequent treatment of the wave functions by response-theory methods for the calculation of various coupling-matrix elements.

Let the configuration space be spanned by a set of m configurations $|i\rangle$ and let us assume that we wish the orbitals to be optimal in an average sense for n electronic states

$$|I\rangle = \sum_i C_{Ii}|i\rangle \tag{12.7.1}$$

that are orthonormal and noninteracting

$$\langle I|J\rangle = \delta_{IJ} \tag{12.7.2}$$

$$\langle I|\hat{H}|J\rangle = E_I\delta_{IJ} \tag{12.7.3}$$

To achieve this goal, we optimize a state-averaged energy function consisting of a linear combination of the energies of the individual states $|I\rangle$:

$$E = \sum_I w_I E_I = \sum_I w_I \langle I|\hat{H}|I\rangle \tag{12.7.4}$$

The weight coefficients w_I are fixed numerical parameters. They may be chosen based on some physical considerations, but are usually the same for all electronic states $|I\rangle$.

The parametrization of the state-averaged MCSCF energy (12.7.4) adopted in the present subsection differs from the single-state parametrization introduced in Section 12.2.1 Since we require the states to be orthogonal (12.7.2), it is no longer convenient to employ, as our primary variational parameters, the individual CI coefficients of the expansions (12.7.1). Instead, we base our description on the exponential parametrization of the configuration space of Section 12.6 and write the MCSCF wave functions in the form

$$|\tilde{I}\rangle = \exp(-\hat{\kappa})\exp(-\hat{R})|I\rangle \tag{12.7.5}$$

where $\hat{\kappa}$ is the usual orbital-rotation operator

$$\hat{\kappa} = \sum_{p>q} \kappa_{pq} E_{pq}^- \tag{12.7.6}$$

and $\hat{R}$ is a linear combination of *state-transfer operators*:

$$\hat{R} = \sum_{\substack{K>J \\ J\le n}} R_{KJ}(|K\rangle\langle J| - |J\rangle\langle K|) \tag{12.7.7}$$

The summation in (12.7.7) is over pairs of states $|J\rangle$ and $|K\rangle$; J refers to any of the n states (12.7.1) for which the MCSCF energy (12.7.4) is optimized and K refers to these states and also to the orthogonal complement. The states $|K\rangle$ are ordered such that the n states $|J\rangle$ precede the $m - n$ states of the orthogonal complement. For a discussion of this parametrization of the configuration space, see Section 12.6.4. Note that, in the present subsection, we deviate somewhat from the notation of Section 12.6.4, including in the set $|K\rangle$ the reference states as well as the orthogonal-complement states.

The state-averaged energy of the transformed MCSCF states may now be written as

$$E(\mathbf{R}, \kappa) = \sum_I w_I \langle \tilde{I}|\hat{H}|\tilde{I}\rangle = \sum_I w_I \langle I| \exp(\hat{R}) \exp(\hat{\kappa})\hat{H} \exp(-\hat{\kappa}) \exp(-\hat{R})|I\rangle \tag{12.7.8}$$

The state-averaged electronic gradient and Hessian are obtained from a BCH expansion (3.1.8) of the energy (12.7.8). In an obvious notation, we obtain for the gradient

$$^cE_{KJ}^{(1)} = \sum_I w_I \langle I|[|K\rangle\langle J| - |J\rangle\langle K|, \hat{H}]|I\rangle = -2\left(w_J - \sum_I \delta_{KI}w_I\right)\langle J|\hat{H}|K\rangle \tag{12.7.9}$$

$$^oE_{pq}^{(1)} = \sum_I w_I \langle I|[E_{pq}^-, \hat{H}]|I\rangle \tag{12.7.10}$$

and for the Hessian

$$^{cc}E_{KJ,K'J'}^{(2)} = \sum_I w_I \langle I|[|K\rangle\langle J| - |J\rangle\langle K|, [|K'\rangle\langle J'| - |J'\rangle\langle K'|, \hat{H}]]|I\rangle \tag{12.7.11}$$

$$^{co}E_{KJ,pq}^{(2)} = \sum_I w_I \langle I|[|K\rangle\langle J| - |J\rangle\langle K|, [E_{pq}^-, \hat{H}]]|I\rangle \tag{12.7.12}$$

$$^{oo}E_{pq,rs}^{(2)} = \tfrac{1}{2}(1 + P_{pq,rs}) \sum_I w_I \langle I|[E_{pq}^-, [E_{rs}^-, \hat{H}]]|I\rangle \tag{12.7.13}$$

The gradient and Hessian may be expressed in the configuration basis, using the transformations described in Section 12.6.4. Second-order procedures may be carried out as described in Sections 12.3 and 12.4 for conventional single-state MCSCF wave functions, modified for the simultaneous optimization of several states. After the parameters $\mathbf{R}$ and κ have been determined, the new orbitals may be obtained as described in Chapter 3 and the new CI expansions as described in Section 12.6. At convergence, we see from the expression for the configuration gradient (12.7.9) that the reference CI states do not interact with states belonging to the orthogonal complement. However, the configuration-gradient element $^cE_{KJ}^{(1)}$ vanishes trivially when K belongs to the reference states and the weight factors are equal. It is then convenient to diagonalize the reference-space Hamiltonian in each iteration to ensure that (12.7.3) is satisfied.

12.8   Removal of RHF instabilities in MCSCF theory

In Chapter 10, we found that triplet and singlet instabilities occur when RHF wave functions are used to describe the breaking of chemical bonds. Such instabilities arise since bond breakings cannot be described adequately by a single electronic configuration. In MCSCF theory, several configurations are allowed and bond breakings can therefore be described with the appropriate choice of configuration space. For this reason, MCSCF calculations are less affected by instabilities than are RHF calculations. As an illustration, we here discuss MCSCF calculations where the OH bonds of the water molecule and the CC bonds of the allyl radical are being broken. The corresponding RHF calculations were discussed in Section 10.10.

12.8.1   BOND BREAKING IN H_2O

In a valence CASSCF description of the C_{2v}-conserving bond breaking in H_2O, the active space contains the oxygen $2s$ and $2p$ orbitals and the hydrogen $1s$ orbitals. This choice of orbitals gives a (3,0,1,2) active space, where the numbers of active orbitals in the four irreducible representations are listed in the order (A_1, A_2, B_1, B_2). As in the RHF calculation in Section 10.10.5, we have used the cc-pVDZ basis and the HOH angle is fixed at $110.565°$.

Figure 12.2 illustrates the behaviour of the ground-state energy, the lowest excitation energy of each symmetry and the lowest Hessian eigenvalues of the CASSCF wave function as the OH

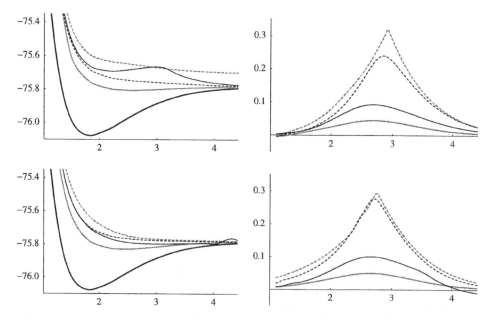

Fig. 12.2. The CASSCF dissociation of the C_{2v} water molecule. On the left, we have plotted the potential-energy curve of the ground-state valence CASSCF wave function (full line) in the cc-pVDZ basis as a function of the internuclear OH distance. Superimposed on the ground-state curve are the potential-energy curves of the lowest excited states of the symmetry species A_1 (full line), A_2 (dashed line), B_1 (full grey line) and B_2 (dashed grey line), calculated using MCSCF linear response theory. The upper plot contains the singlet excited states; the lower plot contains the triplet excited states. On the right, we have plotted the lowest MCSCF Hessian eigenvalues of singlet symmetry (upper plot) and of triplet symmetry (lower plot) of the different irreducible representations. Atomic units are used.

bonds are broken. Similar plots are given for the RHF wave function in Figure 10.3. On the left, we have plotted the potential-energy curve of the ground state with the curves of the lowest excited states of each irreducible representation superimposed. The excited-state energies have been obtained from the ground-state energy by adding the excitation energies from MCSCF linear response theory, obtained by solving a generalized eigenvalue problem involving the electronic Hessian [14,15]. These excitation energies may be taken as probes of the structure and quality of the MCSCF Hessian. The upper plot contains the singlet excited curves, the lower plot the triplet excited curves. On the right, we have plotted the lowest Hessian eigenvalues of each symmetry, with the singlet eigenvalues at the top and the triplet eigenvalues at the bottom.

In contrast to the RHF wave function in Figure 10.3, the valence CASSCF wave function gives a proper description of the bond breaking and no triplet instabilities occur. Also, the excited energy curves obtained from MCSCF linear response calculations show no anomalies, in contrast to those obtained at RHF level. In the equilibrium region, the RHF and valence CAS curves are similar; in the bond-breaking region, only the CASSCF wave function gives an adequate description of the electronic system. Note the avoided crossing for the lowest $^1A_1$ excited state at a separation of about $3a_0$.

The MCSCF singlet and triplet Hessian eigenvalues are similar, with a maximum in the bond-breaking region. For short internuclear separations, the MCSCF Hessian eigenvalues are an order of magnitude smaller than those of the RHF wave function in Figure 10.3. The lowest MCSCF eigenvalues arise from orbital rotations between orbitals of similar occupation numbers – between weakly occupied active orbitals and secondary orbitals for short internuclear separations; between the inactive orbitals and a nearly doubly occupied orbital for large separations. In the RHF wave function, no such rotations are present. For large separations, different spin couplings of the fragments give rise to several close-lying states of triplet symmetry and the Hessian eigenvalues may become negative – as happens for the lowest Hessian eigenvalue of $^3A_1$ symmetry at a bond distance of about $4a_0$.

12.8.2 THE GROUND STATE OF THE ALLYL RADICAL

In our study of the allyl radical C_3H_5 in Section 10.10.6, we found that the ground-state $^2A_2$ RHF wave function contains a singlet instability at the C_{2v} molecular equilibrium geometry – see Figure 10.5, where the RHF and UHF energies are plotted as functions of the symmetric CC stretching coordinate δ. When the symmetry restrictions are lifted at the UHF level, the molecule becomes distorted (relative to the C_{2v} symmetry) and assumes a $^2A''$ RHF minimum structure of C_s symmetry with two different CC bond lengths – see the potential-energy curves for the antisymmetric CC stretching coordinate Δ in Figure 10.6. In Section 11.10, the allyl radical was investigated at the more elaborate CISD level (where dynamical correlation is recovered); at this level, the singlet instability is not removed.

A simple π-electron MO description of the allyl radical gives a doubly occupied bonding π orbital ($1b_1$), a singly occupied nonbonding π orbital ($1a_2$) and an unoccupied antibonding π orbital ($2b_1$). The ground-state $^2A_2$ RHF wave function has the electronic configuration $(\sigma)1b_1^2 1a_2^1$ where (σ) represents the σ part of the configuration (with six doubly occupied σ orbitals of A_1 symmetry and four doubly occupied σ orbitals of B_2 symmetry). The instability occurs as it is energetically favourable for the nonbonding $1a_2$ π orbital to distort and develop a weakly bonding character between the central carbon atom and one of the terminal carbon atoms, and a weakly antibonding character with the other terminal carbon atom. At the UHF level, this situation leads to an unphysical symmetry-broken solution, where one of the CC bonds is stronger than the other bond.

Since the symmetry-broken Hartree–Fock solution arises from competition between bonding and antibonding effects in the π space, we may be able to remove the singlet instability at the CASSCF level by using an active orbital space containing all three π orbitals. In addition to the $^2A_2$ RHF configuration $(\sigma)1b_1^2 1a_2^1$, the CASSCF wave function then contains the two configurations $(\sigma)1b_1^1 1a_2^1 2b_1^1$ and $(\sigma)1a_2^1 2b_1^2$ of the same symmetry. In other words, the MCSCF expansion contains all configurations obtained by distributing two electrons between the bonding and antibonding π orbitals. In Figure 12.3, we have plotted the MCSCF potential-energy curve and the lowest Hessian eigenvalue using the same symmetric stretching coordinate δ as for the RHF wave function in Figure 10.5. The lowest Hessian eigenvalue, which has B_2 symmetry, is now positive for all displacements. The $^2A_2$ potential-energy curve represents a true minimum and the singlet instability of the RHF wave function has been removed. At the RHF equilibrium (i.e. for zero displacement in Figure 12.3), the π occupation numbers are 1.907 for the bonding orbital, 1 for the nonbonding orbital and 0.093 for the antibonding orbital.

On the left in Figure 12.4, we have plotted the CASSCF potential-energy curve for the same antisymmetric stretching coordinate Δ as for the single-configuration RHF and UHF wave functions in Figure 10.6. The CASSCF potential-energy curve has a minimum of C_{2v} symmetry, confirming that the three-configuration CASSCF wave function represents a qualitatively correct

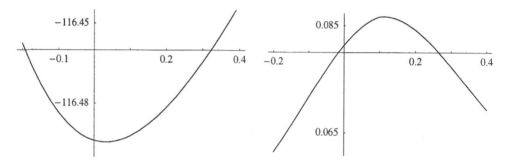

Fig. 12.3. On the left, the CASSCF energy of the allyl radical as a function of the symmetric stretching coordinate δ in Figure 10.4. On the right, the lowest singlet Hessian eigenvalue of symmetry B_2 for the CASSCF wave function. Atomic units are used.

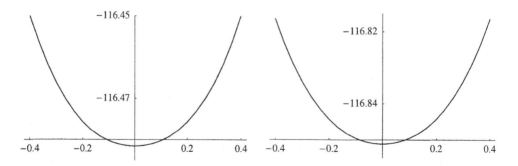

Fig. 12.4. The CAS (left) and MRSDCI (right) energies as functions of the antisymmetric stretching coordinate Δ in the allyl radical (atomic units).

description of the electronic structure in the allyl radical. For high accuracy, we must include also the effects of dynamical correlation. If we perform an MRSDCI calculation using the above CAS space as the reference space, we obtain the potential-energy curve plotted on the right in Figure 12.4. The MRSDCI energy curve is similar to the CASSCF curve, but is lower in energy by about 370 mE$_h$. The MRSDCI potential-energy curve should be contrasted with the CISD curves in Figure 11.8, which are similar to the Hartree–Fock curves in Figure 10.6.

References

[1]   B. O. Roos, P. R. Taylor and P. E. M. Siegbahn, *Chem. Phys.* **48**, 157 (1980).
[2]   J. Olsen, B. O. Roos, P. Jørgensen and H. J. Aa. Jensen, *J. Chem. Phys.* **89**, 2185 (1988).
[3]   H. J. Aa. Jensen, P. Jørgensen, T. Helgaker and J. Olsen, *Chem. Phys. Lett.* **162**, 355 (1989).
[4]   P.-Å. Malmqvist, A. Rendell and B. O. Roos, *J. Phys. Chem.* **94**, 5477 (1990).
[5]   R. Fletcher, *Practical Methods of Optimization*, 2nd edn, Wiley, 1987.
[6]   J. Simons, P. Jørgensen, H. Taylor and J. Ozment, *J. Phys. Chem.* **87**, 2745 (1983).
[7]   H. J. Aa. Jensen and P. Jørgensen, *J. Chem. Phys.* **80**, 1204 (1984).
[8]   B. H. Lengsfield III, *J. Chem. Phys.* **73**, 382 (1980).
[9]   P. E. M. Siegbahn, J. Almlöf, A. Heiberg and B. O. Roos, *J. Chem. Phys.* **74**, 2384 (1981).
[10] J. Olsen, D. L. Yeager and P. Jørgensen, *Adv. Chem. Phys.* **54**, 1 (1983).
[11] E. Dalgaard, *Chem. Phys. Lett.* **65**, 559 (1979).
[12] P.-Å. Malmqvist and B. O. Roos, *Chem. Phys. Lett.* **155**, 189 (1989).
[13] H.-J. Werner and W. Meyer, *J. Chem. Phys.* **74**, 5794 (1981).
[14] D. L. Yeager and P. Jørgensen, *Chem. Phys. Lett.* **65**, 77 (1979).
[15] E. Dalgaard, *J. Chem. Phys.* **72**, 816 (1980).

Further reading

B. O. Roos, The multiconfigurational (MC) self-consistent field (SCF) theory, in B. O. Roos (ed.), *Lecture Notes in Quantum Chemistry*, Lecture Notes in Chemistry Vol. 58, Springer-Verlag, 1992, p. 177.
B. O. Roos, The complete active space self-consistent field method and its applications in electronic structure calculations, *Adv. Chem. Phys.* **69**, 399 (1987).
R. Shepard, The multiconfiguration self-consistent field method, *Adv. Chem. Phys.* **69**, 63 (1987).
H.-J. Werner, Matrix-formulated direct multiconfiguration self-consistent field and multiconfiguration reference configuration-interaction methods, *Adv. Chem. Phys.* **69**, 1 (1987).

Exercises

EXERCISE 12.1

Consider the normalized two-configuration wave function for helium in the $1s2p_x(^1P)$ state:

$$|0\rangle = \frac{c_1}{\sqrt{2}}(a^\dagger_{1s\alpha}a^\dagger_{2p_x\beta} - a^\dagger_{1s\beta}a^\dagger_{2p_x\alpha})|\text{vac}\rangle + \frac{c_2}{\sqrt{2}}(a^\dagger_{2s\alpha}a^\dagger_{2p_x\beta} - a^\dagger_{2s\beta}a^\dagger_{2p_x\alpha})|\text{vac}\rangle \qquad (12E.1.1)$$

1. Show that the excitation operator

$$E^-_{2s1s} = \sum_\sigma (a^\dagger_{2s\sigma}a_{1s\sigma} - a^\dagger_{1s\sigma}a_{2s\sigma}) \qquad (12E.1.2)$$

satisfies the relation

$$E_{2s1s}^-|0\rangle = -\frac{c_2}{\sqrt{2}}(a_{1s\alpha}^\dagger a_{2p_x\beta}^\dagger - a_{1s\beta}^\dagger a_{2p_x\alpha}^\dagger)|\text{vac}\rangle + \frac{c_1}{\sqrt{2}}(a_{2s\alpha}^\dagger a_{2p_x\beta}^\dagger - a_{2s\beta}^\dagger a_{2p_x\alpha}^\dagger)|\text{vac}\rangle \quad (12E.1.3)$$

and is thus redundant in the sense of (12.2.56).

2. Assume next that we have a set of V virtual s orbitals $a_{ns\sigma}^\dagger$ with $n > 2$, and that we wish to optimize the $1s$ and $2s$ orbitals of $|0\rangle$ in the parametrization (12E.1.1). Show that the operators

$$E_{ns2s}^- = \sum_\sigma (a_{n\sigma}^\dagger a_{2s\sigma} - a_{2s\sigma}^\dagger a_{n\sigma}) \quad (12E.1.4)$$

with $n > 2$ are redundant in the sense of (12.2.55). *Hint*: Show the relation

$$E_{ns2s}^-|0\rangle = \frac{c_2}{c_1}E_{ns1s}^-|0\rangle \quad (12E.1.5)$$

3. Determine the number of nonredundant parameters for the optimization of the wave function (12E.1.1).

4. Show that a normalized orbital $1s'$ can be defined such that the state (12E.1.1) may be expressed in terms of a single configuration:

$$|0\rangle = \frac{1}{\sqrt{2}}(a_{1s'\alpha}^\dagger a_{2p_x\beta}^\dagger - a_{1s'\beta}^\dagger a_{2p_x\alpha}^\dagger)|\text{vac}\rangle \quad (12E.1.6)$$

Equations (12E.1.1) and (12E.1.6) thus represent two different parametrizations of the same state.

5. For the optimization of (12E.1.6), the virtual space consists of the $V + 1$ s orbitals $a_{ns\sigma}^\dagger$ with $n > 1$. Show that the number of nonredundant parameters is the same as for the optimization of the wave function in the form (12E.1.1).

EXERCISE 12.2

Assume that we are carrying out an MCSCF calculation of the electronic state

$$|0\rangle = \sum_i C_i^{(0)}|i\rangle \quad (12E.2.1)$$

where the determinants share a common set of inactive orbitals. Show that, for this state, the Hamiltonian matrix elements $\langle i|\hat{H}|0\rangle$ may be written in the form

$$\langle i|\hat{H}|0\rangle = \sum_{vw}{}^I F_{vw}\langle i|E_{vw}|0\rangle + \frac{1}{2}\sum_{vwxy} g_{vwxy}\langle i|e_{vwxy}|0\rangle + C_i^{(0)}\,{}^I E \quad (12E.2.2)$$

where ${}^I\mathbf{F}$ is the inactive Fock matrix in (12.5.12) and ${}^I E$ the inactive energy in (12.5.16). Combining this result with (12.2.26), we arrive at the expression for the gradient (12.5.15).

EXERCISE 12.3

Let $|K\rangle$ be an orthonormal basis for the orthogonal complement to the reference state $|0\rangle$ and introduce the exponential operator

$$\exp(-\hat{R}) = \exp\left[-\sum_{K\neq 0}(R_K|K\rangle\langle 0| - R_K^*|0\rangle\langle K|)\right] \quad (12E.3.1)$$

1. Show that, when this operator is applied to reference state $|0\rangle$, we obtain

$$\exp(-\hat{R})|0\rangle = \cos d\,|0\rangle - \frac{\sin d}{d}\sum_{K\neq 0} R_K|K\rangle \tag{12E.3.2}$$

where

$$d = \sqrt{\sum_{K\neq 0}|R_K|^2} \tag{12E.3.3}$$

2. Show that, when the operator (12E.3.1) is applied to a basis vector $|M\rangle$ of the orthogonal complement, we obtain

$$\exp(-\hat{R})|M\rangle = |M\rangle + \frac{\sin d}{d}R_M^*|0\rangle + \frac{\cos d - 1}{d^2}R_M^*\sum_{K\neq 0}R_K|K\rangle \tag{12E.3.4}$$

Solutions

SOLUTION 12.1

1. Applying (12E.1.2) to (12E.1.1) and using the elementary properties of the creation and annihilation operators, we obtain

$$E_{2s1s}^-|0\rangle = \frac{c_1}{\sqrt{2}}E_{2s1s}(a_{1s\alpha}^\dagger a_{2p_x\beta}^\dagger - a_{1s\beta}^\dagger a_{2p_x\alpha}^\dagger)|\text{vac}\rangle - \frac{c_2}{\sqrt{2}}E_{1s2s}(a_{2s\alpha}^\dagger a_{2p_x\beta}^\dagger - a_{2s\beta}^\dagger a_{2p_x\alpha}^\dagger)|\text{vac}\rangle$$

$$= \frac{c_1}{\sqrt{2}}(a_{2s\alpha}^\dagger a_{2p_x\beta}^\dagger - a_{2s\beta}^\dagger a_{2p_x\alpha}^\dagger)|\text{vac}\rangle - \frac{c_2}{\sqrt{2}}(a_{1s\alpha}^\dagger a_{2p_x\beta}^\dagger - a_{1s\beta}^\dagger a_{2p_x\alpha}^\dagger)|\text{vac}\rangle \tag{12S.1.1}$$

which is equivalent to (12E.1.3).
2. Applying the excitation operator (12E.1.4) to (12E.1.1), we obtain

$$E_{ns2s}^-|0\rangle = \frac{c_2}{\sqrt{2}}(a_{ns\alpha}^\dagger a_{2p_x\beta}^\dagger - a_{ns\beta}^\dagger a_{2p_x\alpha}^\dagger)|\text{vac}\rangle = \frac{c_2}{c_1}E_{ns1s}^-|0\rangle \tag{12S.1.2}$$

3. The number of nonredundant orbital rotations E_{ns1s}^- for $n > 2$ is V and there is one nonredundant CI parameter. The total number of nonredundant parameters is therefore $V + 1$.
4. Introducing the normalized $1s'$ orbital

$$a_{1s'\sigma}^\dagger = c_1 a_{1s\sigma}^\dagger + c_2 a_{2s\sigma}^\dagger \tag{12S.1.3}$$

we may rewrite the two-configuration state (12.E.1.1) as a single-configuration state (12E.1.6).
5. The number of nonredundant orbital rotations E_{ns1s}^- for $n > 1$ is $V + 1$ and there are no CI parameters. The total number of nonredundant parameters is thus again $V + 1$.

SOLUTION 12.2

Expanding the Hamiltonian in the usual manner, we obtain

$$\langle i|\hat{H}|0\rangle = \sum_{pq} h_{pq}\langle i|E_{pq}|0\rangle + \frac{1}{2}\sum_{pqrs} g_{pqrs}\langle i|e_{pqrs}|0\rangle \tag{12S.2.1}$$

We first evaluate the one-electron contribution, decomposing it into two parts:

$$\sum_{pq} h_{pq} \langle i|E_{pq}|0\rangle = \sum_{vq} h_{vq} \langle i|E_{vq}|0\rangle + \sum_{jq} h_{jq} \langle i|E_{jq}|0\rangle$$

$$= \sum_{vw} h_{vw} \langle i|E_{vw}|0\rangle + 2\sum_j h_{jj} \langle i|0\rangle \qquad (12S.2.2)$$

(Recall that the density-matrix elements vanish if any of the indices are secondary.) To evaluate the two-electron contribution, we first decompose it as follows:

$$\frac{1}{2}\sum_{pqrs} g_{pqrs} \langle i|e_{pqrs}|0\rangle = \frac{1}{2}\sum_{pqjs} g_{pqjs} \langle i|e_{pqjs}|0\rangle + \frac{1}{2}\sum_{jqvs} g_{jqvs} \langle i|e_{jqvs}|0\rangle$$

$$+ \frac{1}{2}\sum_{wqvs} g_{wqvs} \langle i|e_{wqvs}|0\rangle \qquad (12S.2.3)$$

Next, using the relation

$$\langle i|e_{jspq}|0\rangle = \langle i|e_{pqjs}|0\rangle = 2\delta_{js}\langle i|E_{pq}|0\rangle - \delta_{qj}\langle i|E_{ps}|0\rangle \qquad (12S.2.4)$$

we may rewrite (12S.2.3) as

$$\frac{1}{2}\sum_{pqrs} g_{pqrs} \langle i|e_{pqrs}|0\rangle = \sum_{pqj} g_{pqjj} \langle i|E_{pq}|0\rangle - \frac{1}{2}\sum_{pqj} g_{pjjq} \langle i|E_{pq}|0\rangle$$

$$+ \sum_{jvq} g_{jjvq} \langle i|E_{vq}|0\rangle - \frac{1}{2}\sum_{jqv} g_{jqvj} \langle i|E_{vq}|0\rangle$$

$$+ \frac{1}{2}\sum_{vwxy} g_{vwxy} \langle i|e_{vwxy}|0\rangle \qquad (12S.2.5)$$

The first two terms give nonzero contributions for $p = q = k$ and for $p = w$ and $q = v$, the latter being identical to those from the third and fourth terms. Collecting terms, we then obtain the following expression for the two-electron part:

$$\frac{1}{2}\sum_{pqrs} g_{pqrs} \langle i|e_{pqrs}|0\rangle = 2\sum_{kj} \left(g_{kkjj} - \tfrac{1}{2}g_{kjjk}\right) \langle i|0\rangle + 2\sum_{jvw} \left(g_{jjvw} - \tfrac{1}{2}g_{jwvj}\right) \langle i|E_{vw}|0\rangle$$

$$+ \frac{1}{2}\sum_{vwxy} g_{vwxy} \langle i|e_{vwxy}|0\rangle \qquad (12S.2.6)$$

Substituting (12S.2.2) and (12S.2.6) into (12S.2.1) and rearranging, we obtain

$$\langle i|\hat{H}|0\rangle = \sum_{vw} \left[h_{vw} + \sum_j (2g_{vwjj} - g_{vjjw}) \right] \langle i|E_{vw}|0\rangle + \frac{1}{2}\sum_{vwxy} g_{vwxy} \langle i|e_{vwxy}|0\rangle$$

$$+ \sum_j \left[2h_{jj} + \sum_k (2g_{jjkk} - g_{jkkj}) \right] \langle i|0\rangle \qquad (12S.2.7)$$

Recalling the definition of the inactive Fock matrix (12.5.12) and the inactive energy (12.5.16), we arrive at the desired expression (12E.2.2).

SOLUTION 12.3

1. Expanding the exponential in (12E.3.1), we obtain

$$\exp(-\hat{R})|0\rangle = \sum_{n=0}^{\infty} \frac{1}{(2n)!} \hat{R}^{2n}|0\rangle - \sum_{n=0}^{\infty} \frac{1}{(2n+1)!} \hat{R}^{2n+1}|0\rangle \tag{12S.3.1}$$

To simplify the expansion, we note the relations

$$\hat{R}|0\rangle = \sum_{K\neq 0} R_K |K\rangle \tag{12S.3.2}$$

$$\hat{R}^2|0\rangle = -d^2|0\rangle \tag{12S.3.3}$$

where we have used (12E.3.3). Using these relations in (12S.3.1) and rearranging slightly, we arrive at (12E.3.2):

$$\exp(-\hat{R})|0\rangle = \sum_{n=0}^{\infty} \frac{(-1)^n}{(2n)!} d^{2n}|0\rangle - \frac{1}{d} \sum_{n=0}^{\infty} \frac{(-1)^n}{(2n+1)!} d^{2n+1} \hat{R}|0\rangle$$

$$= \cos d |0\rangle - \frac{\sin d}{d} \sum_{K\neq 0} R_K |K\rangle \tag{12S.3.4}$$

2. Expanding the exponential, we first obtain

$$\exp(-\hat{R})|M\rangle = |M\rangle + \sum_{n=1}^{\infty} \frac{1}{(2n)!} \hat{R}^{2n}|M\rangle - \sum_{n=0}^{\infty} \frac{1}{(2n+1)!} \hat{R}^{2n+1}|M\rangle \tag{12S.3.5}$$

To simplify this expansion, we first note the two relations

$$\hat{R}|M\rangle = -R_M^*|0\rangle \tag{12S.3.6}$$

$$\hat{R}^{2n+1}|M\rangle = -(-1)^n d^{2n} R_M^*|0\rangle \tag{12S.3.7}$$

where the latter follows from (12S.3.6) and (12S.3.3). Using these relations, we obtain

$$\exp(-\hat{R})|M\rangle = |M\rangle + \frac{R_M^*}{d^2} \sum_{n=1}^{\infty} \frac{(-1)^n d^{2n}}{(2n)!} \hat{R}|0\rangle + \frac{R_M^*}{d} \sum_{n=0}^{\infty} \frac{(-1)^n d^{2n+1}}{(2n+1)!} |0\rangle \tag{12S.3.8}$$

which gives us (12E.3.4).

13 COUPLED-CLUSTER THEORY

The coupled-cluster method represents the most successful approach to accurate many-electron molecular wave functions. It can be applied to relatively large systems and is capable of recovering a large part of the correlation energy. It is size-extensive and presents few if any problems with respect to optimization. It does, however, require the existence of a reasonably accurate single-determinant wave function and cannot – at least in its more common formulation – be applied to systems with degenerate or nearly degenerate electronic configurations. In practice, therefore, the application of the coupled-cluster method is restricted to systems that are dominated by a single electronic configuration and the coupled-cluster wave function is best regarded as providing an accurate correction to the Hartree–Fock description.

The present chapter contains a general discussion of the coupled-cluster method and also a detailed exposition of the coupled-cluster singles-and-doubles (CCSD) model. We begin our presentation of coupled-cluster theory in Section 13.1, where we introduce the underlying physical model of coupled-cluster theory and the concept of clusters. In Section 13.2, we introduce the important exponential ansatz of coupled-cluster theory and employ this ansatz to study in more detail the structure of the coupled-cluster wave function and its optimization. Following these introductory sections, we go on to discuss various aspects of coupled-cluster theory such as size-extensivity in Section 13.3 and optimization techniques in Section 13.4. The variational Lagrangian formulation of coupled-cluster theory and the Hellmann–Feynman theorem are then discussed in Section 13.5, followed by a treatment of the calculation of excited states and excitation energies in Section 13.6. In Section 13.7, we turn our attention to an important special case of coupled-cluster theory – the CCSD model, for which all expressions needed for the calculation of the energy and the optimization of the wave function are explicitly derived. In Section 13.8, we consider some important treatments of the correlation problem that are modifications of the standard coupled-cluster theory (i.e. the Brueckner and quadratic CI models). We conclude this chapter with a discussion of open-shell systems in Section 13.9. In this chapter, multireference systems are not discussed since at present no consensus regarding the treatment of such systems in the context of coupled-cluster theory has been established.

13.1 The coupled-cluster model

The purpose of the present section is to introduce the coupled-cluster model. First, in Section 13.1.1, we consider the description of virtual excitation processes and correlated electronic states by means of pair clusters. Next, in Section 13.1.2, we introduce the coupled-cluster model as a generalization of the concept of pair clusters. After a discussion of connected and disconnected clusters in Section 13.1.3, we consider the conditions for the optimized coupled-cluster state in Section 13.1.4.

13.1.1 PAIR CLUSTERS

In the independent-particle model, the wave function, which corresponds to a product of creation operators working on the vacuum state, describes an uncorrelated motion of the electrons. The variationally optimized electronic system is represented by a set of occupied spin orbitals from which no virtual excitations ever occur since the electrons do not interact. As a refinement to this model, we note that, within the orbital picture, the correlated motion of interacting electrons manifests itself in virtual excitations of electrons from occupied to unoccupied spin orbitals. With each such excitation, we may associate an amplitude, representing the probability that this particular excitation will occur as a result of interactions among the electrons.

To a first approximation, we may restrict ourselves to a pairwise-correlated treatment of the electrons. Consider two electrons that, in the independent-particle model, occupy the spin orbitals I and J. As a result of the instantaneous interaction between the two electrons, their motion is disturbed and the electrons are excited to a different set of spin orbitals A and B, initially unoccupied. With each excitation process, we associate an *amplitude* t_{IJ}^{AB}. Our description of the motion of the two electrons is thus improved in the following manner

$$a_I^\dagger a_J^\dagger \to a_I^\dagger a_J^\dagger + \sum_{A>B} t_{IJ}^{AB} a_A^\dagger a_B^\dagger \tag{13.1.1}$$

and this expansion is known as a *two-electron cluster* or a *pair cluster*. We now introduce an operator that describes this 'correlation process' of two electrons initially found in the spin orbitals I and J of the reference state. Introducing the notation

$$\hat{\tau}_{IJ}^{AB} = a_A^\dagger a_I a_B^\dagger a_J \tag{13.1.2}$$

and using the relation

$$\hat{\tau}_{IJ}^{AB} \hat{\tau}_{IJ}^{CD} = 0 \tag{13.1.3}$$

we may write the pair cluster IJ in the following way

$$\left[\prod_{A>B} (1 + t_{IJ}^{AB} \hat{\tau}_{IJ}^{AB}) \right] a_I^\dagger a_J^\dagger | \rangle = a_I^\dagger a_J^\dagger | \rangle + \sum_{A>B} t_{IJ}^{AB} a_A^\dagger a_B^\dagger | \rangle \tag{13.1.4}$$

assuming that I and J are unoccupied in $|\rangle$. Each operator of the form $1 + t_{IJ}^{AB} \hat{\tau}_{IJ}^{AB}$ provides an improved correlated description of the electronic system, generating a superposition of the original state with a new state that represents the outcome of the excitation process.

Assuming a Hartree–Fock reference state and introducing cluster expansions of type (13.1.4) for all pairs of occupied spin orbitals, we arrive at the following expression for our pair-correlated electronic state [1]:

$$|\text{CCD}\rangle = \left[\prod_{A>B, I>J} (1 + t_{IJ}^{AB} \hat{\tau}_{IJ}^{AB}) \right] |\text{HF}\rangle \tag{13.1.5}$$

Since the excitation operators (13.1.2) commute among one another, there are no problems with the order of the operators in this expression. The resulting wave function (13.1.5) corresponds to a particularly simple realization of the *coupled-cluster model*, in which only double excitations are allowed: the *coupled-cluster doubles (CCD) wave function*. For a complete specification of this model, we must also describe the method by which the amplitudes are determined. We shall return to this point in Section 13.1.4, following the introduction and discussion of the general coupled-cluster model in Sections 13.1.2 and 13.1.3.

13.1.2 THE COUPLED-CLUSTER WAVE FUNCTION

Excitations within the pair clusters of Section 13.1.1 provide the dominant contributions to the description of the complicated correlated motion of interacting electrons. The pair clusters dominate since the correlated motion is especially important for electrons that are close to each other and since at most two electrons (with opposite spins) may coincide in space. However, for an accurate treatment of the correlated motion of the electrons (within the orbital model), we must consider clusters of all sizes. For the three-electron clusters, we thus introduce amplitudes t_{IJK}^{ABC} that represent the simultaneous interaction of three electrons, resulting in the excitation of three electrons from three occupied spin orbitals to three unoccupied ones. Furthermore, we must allow excitations to occur also within 'clusters' containing a single electron. Such one-electron processes represent a relaxation of the spin orbitals and occur since the Hartree–Fock mean field experienced by each electron before the excitations were introduced is modified by the many-electron excitation processes occurring within the remaining clusters.

For a general discussion of the coupled-cluster method, we introduce the generic notation $\hat{\tau}_\mu$ for an excitation operator of unspecified excitation level (single, double, etc.) and t_μ for the associated amplitude. In our model, the effect of a given excitation μ on any state $|\ \rangle$ is to change this state as follows

$$|\ \rangle \to (1 + t_\mu \hat{\tau}_\mu)|\ \rangle \tag{13.1.6}$$

and the result of such *correlating operators* working on the Hartree–Fock state is the *coupled-cluster wave function*

$$|\mathrm{CC}\rangle = \left[\prod_\mu (1 + t_\mu \hat{\tau}_\mu)\right] |\mathrm{HF}\rangle \tag{13.1.7}$$

Several points should be noted about the form of this wave function. First, the coupled-cluster state is manifestly in a product form, leading to a size-extensive treatment of the electronic system as discussed in Sections 4.3 and 13.3. In this respect, the coupled-cluster model differs fundamentally from the linear CI model, for which the corresponding wave function is not in a product form:

$$|\mathrm{CI}\rangle = \left(1 + \sum_\mu C_\mu \hat{\tau}_\mu\right) |\mathrm{HF}\rangle \tag{13.1.8}$$

Second, the order of the operators in (13.1.7) is unimportant since the excitation operators $\hat{\tau}_\mu$ commute:

$$[\hat{\tau}_\mu, \hat{\tau}_\nu] = 0 \tag{13.1.9}$$

Commutation occurs since the excitation operators always excite from the set of occupied Hartree–Fock spin orbitals to the virtual ones – see (13.1.2) for the double-excitation operators. The creation and annihilation operators of the excitation operators therefore anticommute.

13.1.3 CONNECTED AND DISCONNECTED CLUSTERS

Let us now analyse the coupled-cluster state in terms of determinants

$$|\mu\rangle = \hat{\tau}_\mu |\mathrm{HF}\rangle \tag{13.1.10}$$

Upon expansion of the product operator in (13.1.7), we obtain the expression [1]

$$|\text{CC}\rangle = \left(1 + \sum_\mu t_\mu \hat{\tau}_\mu + \sum_{\mu > \nu} t_\mu t_\nu \hat{\tau}_\mu \hat{\tau}_\nu + \cdots \right) |\text{HF}\rangle$$

$$= |\text{HF}\rangle + \sum_\mu t_\mu |\mu\rangle + \sum_{\mu > \nu} t_\mu t_\nu |\mu\nu\rangle + \cdots \qquad (13.1.11)$$

where, in the summation, the restriction $\mu > \nu$ applies since each excitation operator appears only once in the product (13.1.7). Clearly, we cannot associate each excitation operator and its amplitude with a single unique determinant – the operator $\hat{\tau}_\mu$ not only generates the determinant $|\mu\rangle$ but also gives rise to a large number of determinants in collaboration with the other excitation operators, for example

$$|\mu\nu\rangle = \hat{\tau}_\mu \hat{\tau}_\nu |\text{HF}\rangle = \hat{\tau}_\nu \hat{\tau}_\mu |\text{HF}\rangle \qquad (13.1.12)$$

As seen from the expansion (13.1.11), the amplitude of such a composite excitation is obtained by taking the product of the amplitudes of the contributing excitations.

Owing to the presence of product excitations in the coupled-cluster state (13.1.7), each determinant may be reached in several distinct ways. For example, the introduction of a composite excitation operator equivalent to the product of two separate excitations

$$\hat{\tau}_{\mu\nu} = \hat{\tau}_\mu \hat{\tau}_\nu \qquad (13.1.13)$$

means that the determinant $|\mu\nu\rangle$ may be reached by at least two distinct processes, with an overall amplitude equal to the sum of the individual amplitudes:

$$t_{\mu\nu}^{\text{total}} = t_{\mu\nu} + t_\mu t_\nu + \cdots \qquad (13.1.14)$$

With respect to the determinant $|\mu\nu\rangle$, the amplitude $t_{\mu\nu}$ is referred to as a *connected cluster amplitude* and $t_\mu t_\nu$ as a *disconnected cluster amplitude*. In general, high-order excitations can be reached by a large number of processes or mechanisms, each contributing to the total amplitude with a weight equal to the product of the amplitudes of the individual excitations.

From (13.1.11) it is apparent that a coupled-cluster wave function – generated, for example, by all possible single- and double-excitation operators $\hat{\tau}_\mu$ – contains contributions from *all* determinants entering the FCI wave function (although the number of free parameters is usually much smaller). In practice, therefore, we cannot work with the coupled-cluster state in the expanded form (13.1.11) but we must instead retain the wave function in the more compact form (13.1.7), avoiding references to the individual determinants.

13.1.4 THE COUPLED-CLUSTER SCHRÖDINGER EQUATION

Given the product ansatz for the coupled-cluster wave function (13.1.7), let us consider its optimization. We recall that, in CI theory, the wave function (13.1.8) is determined by minimizing the expectation value of the Hamiltonian with respect to the linear expansion coefficients:

$$E_{\text{CI}} = \min_{C_\mu} \frac{\langle \text{CI}|\hat{H}|\text{CI}\rangle}{\langle \text{CI}|\text{CI}\rangle} \qquad (13.1.15)$$

By analogy with CI theory, therefore, we might attempt to determine the coupled-cluster state by minimizing the expectation value of the Hamiltonian with respect to the amplitudes:

$$E_{\min} = \min_{t_\mu} \frac{\langle \mathrm{CC} | \hat{H} | \mathrm{CC} \rangle}{\langle \mathrm{CC} | \mathrm{CC} \rangle} \tag{13.1.16}$$

We recall that the derivatives of the CI wave function with respect to the variational parameters may be written in a particularly simple form

$$\frac{\partial}{\partial C_\mu} | \mathrm{CI} \rangle = | \mu \rangle \tag{13.1.17}$$

giving rise to a standard eigenvalue problem for the CI coefficients (11.1.3)

$$\langle \mu | \hat{H} | \mathrm{CI} \rangle = E_{\mathrm{CI}} \langle \mu | \mathrm{CI} \rangle \tag{13.1.18}$$

By contrast, the nonlinear parametrization of the coupled-cluster model (13.1.7) means that the derivatives of the coupled-cluster state become complicated functions of the amplitudes

$$\frac{\partial}{\partial t_\mu} | \mathrm{CC} \rangle = \left[\prod_\nu (1 + t_\nu \hat{\tau}_\nu) \right] | \mu \rangle \tag{13.1.19}$$

The variational conditions on the amplitudes (13.1.16) therefore give rise to an intractable set of nonlinear equations for the amplitudes

$$\langle \mu | \left[\prod_\nu (1 + t_\nu \hat{\tau}_\nu^\dagger) \right] \hat{H} | \mathrm{CC} \rangle = E_{\min} \langle \mu | \left[\prod_\nu (1 + t_\nu \hat{\tau}_\nu^\dagger) \right] | \mathrm{CC} \rangle \tag{13.1.20}$$

which involves the full set of FCI determinants and high-order products of the amplitudes. The variational minimization of the coupled-cluster energy is thus a complicated undertaking, which can be carried out only for small molecular systems. We shall therefore abandon the variation principle as a basis for the optimization of the coupled-cluster wave function.

To establish a different principle for the optimization of the coupled-cluster wave function, we note that, for the linear CI wave function, the variational minimization of the energy is entirely equivalent to the solution of the *projected Schrödinger equation* (13.1.18). By contrast, for nonlinearly parametrized wave functions, the solution of the projected Schrödinger equation is, in general, not equivalent to the minimization of the energy. For such models, therefore, we may regard the solution of the projected Schrödinger equation as an alternative to the minimization of the energy. In particular, applied to the coupled-cluster model, projection of the Schrödinger equation against those determinants that enter the coupled-cluster state (13.1.11) with connected amplitudes

$$\langle \mu | = \langle \mathrm{HF} | \hat{\tau}_\mu^\dagger \tag{13.1.21}$$

gives us the projected coupled-cluster equations

$$\langle \mu | \hat{H} | \mathrm{CC} \rangle = E \langle \mu | \mathrm{CC} \rangle \tag{13.1.22}$$

where the coupled-cluster energy is obtained by projection against the Hartree–Fock state

$$E = \langle \mathrm{HF} | \hat{H} | \mathrm{CC} \rangle \tag{13.1.23}$$

since

$$\langle HF|CC \rangle = 1 \qquad (13.1.24)$$

Like the variational coupled-cluster conditions (13.1.20), the projected equations (13.1.22) are nonlinear in the amplitudes. However, unlike the variational conditions, the expansion of the wave function in (13.1.22) and (13.1.23) terminates after a few terms since the Hamiltonian operator couples determinants that differ by no higher than double excitations, making the solution of the projected equations and the calculation of the energy tractable. Of course, the calculated coupled-cluster energy no longer represents an upper bound to the FCI energy. In practice, the deviation from the variational energy turns out to be small and of little practical consequence.

The nonvariational character of the coupled-cluster energy may be observed in different situations. For example, in the course of the optimization of the wave function, we may sometimes find that the partially optimized energy is below the final, fully optimized coupled-cluster energy – see Table 13.2. Similarly, the coupled-cluster energy calculated in a given basis and at a given truncation level may sometimes be below the FCI energy – see Table 5.11, where the coupled-cluster energies of the stretched water molecule are lower than the FCI energy at some levels of truncation. In general, this somewhat unsystematic behaviour of the coupled-cluster energy (compared with a variationally determined energy) is irrelevant since the calculated energies are nevertheless rather accurate.

The small difference between the energy calculated as an expectation value and by projection is illustrated in Figure 13.1, where we have plotted both the standard CCSD energy (13.1.23) and the energy obtained from the expression

$$E_{\text{ave}} = \frac{\langle CC|\hat{H}|CC \rangle}{\langle CC|CC \rangle} \qquad (13.1.25)$$

for the water molecule in the cc-pVDZ basis. The left-hand plot shows that the two energies are essentially the same, the difference being scarcely detectable. The right-hand plot, which contains the errors relative to FCI, reveals that, for this system, the expectation value is always lower than the standard CCSD energy, even though they have been obtained from the same amplitudes. Obviously, a variational minimization of (13.1.25) would lead to a further (presumably small) lowering of the energy. All things considered, we conclude that the more complicated variational expression (13.1.25) does not improve the description significantly.

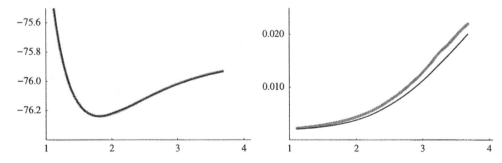

Fig. 13.1. The CCSD energy (in E_h) calculated in the standard manner (grey line) and as an expectation value (black line) for the water molecule in the cc-pVDZ basis, at a fixed bond angle of $110.565°$. On the left, we have plotted the potential-energy curve as a function of the bond distance (in a_0); on the right, we have plotted the difference relative to the FCI energy.

13.2   The coupled-cluster exponential ansatz

In Section 13.1, the coupled-cluster wave function was expressed as a product of correlating operators working on the Hartree–Fock state (13.1.7). This expression for the coupled-cluster wave function is useful for displaying the relation to CI theory and for exhibiting the fundamental role played by the excitation operators in coupled-cluster theory. In general, however, the coupled-cluster wave function is expressed as the exponential of an operator applied to the Hartree–Fock determinant. In the present section, we introduce and explore this exponential ansatz for the coupled-cluster wave function.

13.2.1   THE EXPONENTIAL ANSATZ

In the product form of the coupled-cluster wave function (13.1.7), the spin-orbital excitation operators satisfy the relation

$$\hat{\tau}_\mu^2 = 0 \tag{13.2.1}$$

and the correlating operators may therefore be written as exponentials of the excitation operators:

$$1 + t_\mu \hat{\tau}_\mu = \exp(t_\mu \hat{\tau}_\mu) \tag{13.2.2}$$

Using the commutators (13.1.9), we may then write the wave function in the form

$$|\text{CC}\rangle = \exp(\hat{T})|\text{HF}\rangle \tag{13.2.3}$$

known as the *exponential ansatz* for the coupled-cluster wave function [2–4], where the *cluster operator*

$$\hat{T} = \sum_\mu t_\mu \hat{\tau}_\mu \tag{13.2.4}$$

is a linear combination of excitation operators multiplied by the associated (connected) cluster amplitudes. In the spin-orbital basis, expansion of the exponential operator in (13.2.3) leads to the same expression as does the expansion of the product function (13.1.7), producing contributions from all possible configurations that may be obtained by excitations from the Hartree–Fock state – as appropriate for an electronic-structure model that emphasizes the physical process of excitations rather than the resulting configurations.

13.2.2   THE COUPLED-CLUSTER HIERARCHY OF EXCITATION LEVELS

In coupled-cluster theory, the wave function is written as an exponential of a cluster operator (13.2.4) acting on a single-determinant reference state (13.2.3). In constructing the wave function, the excitations included in the cluster operator are not selected individually. Rather, a *hierarchy of approximations* is established by partitioning the cluster operator into classes comprising all *single (one-electron) excitations*, all *double (two-electron) excitations*, all *triple (three-electron) excitations*, and so on. We may then write the cluster operator in the form

$$\hat{T} = \hat{T}_1 + \hat{T}_2 + \cdots + \hat{T}_N \tag{13.2.5}$$

where, for example, the one- and two-electron parts are given by

$$\hat{T}_1 = \sum_{AI} t_I^A a_A^\dagger a_I = \sum_{AI} t_I^A \hat{\tau}_I^A \tag{13.2.6}$$

$$\hat{T}_2 = \sum_{\substack{A>B \\ I>J}} t_{IJ}^{AB} a_A^\dagger a_I a_B^\dagger a_J = \tfrac{1}{4} \sum_{AIBJ} t_{IJ}^{AB} a_A^\dagger a_I a_B^\dagger a_J = \tfrac{1}{4} \sum_{AIBJ} t_{IJ}^{AB} \hat{\tau}_{IJ}^{AB} \qquad (13.2.7)$$

As before, the indices I and J are used for the occupied Hartree–Fock spin orbitals and the indices A and B for the unoccupied (virtual) spin orbitals. The cluster amplitudes t_{IJ}^{AB} are antisymmetric with respect to permutations of A and B and permutations of I and J.

Each excitation operator in (13.2.5) excites at least one electron from an occupied to a virtual spin orbital. Since there are N electrons in the system, the expansion (13.2.5) terminates after $\hat{T}_N$. We further note that the excitation operators given by (13.2.6) and (13.2.7) satisfy the commutation relation (13.1.9) and that the following relationship holds

$$\hat{\tau}_\mu^\dagger |\text{HF}\rangle = 0 \qquad (13.2.8)$$

since it is impossible to excite an electron from an unoccupied orbital. We also note that

$$[\hat{T}_i, \hat{T}_j] = 0 \qquad (13.2.9)$$

as a simple consequence of the commutation relation for the individual excitation operators.

To compare the excitation-based coupled-cluster model with the configuration-based CI model, we expand the exponential operator in (13.2.3) and collect terms to the same order in the excitation level:

$$\exp(\hat{T})|\text{HF}\rangle = \sum_{i=0}^{N} \hat{C}_i |\text{HF}\rangle \qquad (13.2.10)$$

The lowest-order 'configuration' operators $\hat{C}_i$ are given by

$$\hat{C}_0 = 1 \qquad (13.2.11)$$

$$\hat{C}_1 = \hat{T}_1 \qquad (13.2.12)$$

$$\hat{C}_2 = \hat{T}_2 + \tfrac{1}{2}\hat{T}_1^2 \qquad (13.2.13)$$

$$\hat{C}_3 = \hat{T}_3 + \hat{T}_1\hat{T}_2 + \tfrac{1}{6}\hat{T}_1^3 \qquad (13.2.14)$$

$$\hat{C}_4 = \hat{T}_4 + \hat{T}_1\hat{T}_3 + \tfrac{1}{2}\hat{T}_2^2 + \tfrac{1}{2}\hat{T}_1^2\hat{T}_2 + \tfrac{1}{24}\hat{T}_1^4 \qquad (13.2.15)$$

These equations show which excitation processes contribute at each excitation level. Thus, the quadruply excited configurations are generated by five distinct mechanisms, where, for instance, the disconnected $\hat{T}_2^2$ term represents the independent interactions within two distinct pairs of electrons and the connected $\hat{T}_4$ term describes the simultaneous interaction of four electrons. The disconnected terms represent interactions of product clusters within disjoint sets of electrons and vanish whenever two or more spin-orbital indices are identical.

Without truncation, the FCI and full coupled-cluster functions contain the same number of parameters since there is then one connected cluster amplitude for each determinant. In this special case, the CI and coupled-cluster models provide linear and nonlinear parametrizations of the same state and there is then no obvious advantage in employing the more complicated exponential parametrization. The advantages of the cluster parametrization become apparent only upon truncation and are related to the fact that, even at the truncated level, the coupled-cluster state contains contributions from all determinants in the FCI wave function, with weights obtained from the different excitation processes leading to the determinants.

The most common approximation in coupled-cluster theory is to truncate the cluster operator at the doubles level, yielding the *coupled-cluster singles-and-doubles (CCSD) model* [5]. In this model, the $\hat{T}_2$ operator describes the important electron-pair interactions and $\hat{T}_1$ carries out the orbital relaxations induced by the field set up by the pair interactions. The CCSD wave function contains contributions from all determinants of the FCI wave function, although the highly excited determinants, generated by disconnected clusters, are in general less accurately described than those that also contain connected contributions. However, the disconnected contributions may also in many situations be dominant. For example, even though the $\hat{T}_4$ and $\hat{T}_3\hat{T}_1$ contributions to the quadruples (13.2.15) are neglected at the CCSD level, these determinants are still quite accurately described since the $\hat{T}_2^2$ contributions, which represent the simultaneous but independent interactions within disjoint pairs of electrons, usually constitute the dominant contributions to the $\hat{C}_4$ amplitudes.

To investigate the importance of the disconnected clusters, we have in Table 13.1 listed the weights of the various excitation levels in the (normalized) CCSD and FCI wave functions of the water molecule in the cc-pVDZ basis. At the reference geometry R_{ref}, the Hartree–Fock determinant is a good zero-order approximation and the singles and doubles weights are similar at the CCSD and FCI levels, differing by only about 5%. Turning our attention to the higher excitations (which in CCSD theory are represented by disconnected clusters), we note that the CCSD wave function recovers as much as 84% of the FCI quadruples weight. The quadruples are therefore well described in CCSD theory, mainly by means of $\hat{T}_2^2$. On the other hand, the CCSD wave function is incapable of describing the triple excitations, recovering as little as 4% of their total weight in the FCI wave function – apparently, the triples are not well represented by disconnected clusters.

At the stretched geometry $2R_{ref}$, the single-determinant reference state is no longer a good zero-order approximation. The CCSD model is consequently a poor one, underestimating the doubles by as much as 14%. Again the triples are poorly described but we note that the disconnected quadruples represent as much as 81% of the total FCI quadruples weight. In conclusion, the size-extensive CCSD model appears to recover most of the effects of the quadruples – at least for systems without large contributions from static correlation. The CCSD model is discussed in detail in Section 13.7.

For high accuracy, we must take into account also the connected triple excitations. Truncating the cluster expansion (13.2.5) at the $\hat{T}_3$ level, we arrive at the *coupled-cluster singles-doubles-and-triples (CCSDT) model* [6,7]. Although highly accurate for the description of dynamical correlation, the CCSDT model is computationally very demanding and can be applied only to small systems.

Table 13.1 The weights of the excitation levels in the normalized CCSD and FCI wave functions for the water molecule in the cc-pVDZ basis

Excitation level	R_{ref}		$2R_{ref}$	
	CCSD	FCI	CCSD	FCI
0	0.94410	0.94100	0.65114	0.58966
1	0.00056	0.00053	0.02494	0.02680
2	0.05413	0.05650	0.28762	0.33300
3	0.00002	0.00055	0.00371	0.01040
4	0.00115	0.00137	0.03225	0.03970
5	0.00000	0.00002	0.00006	0.00080
6	0.00001	0.00002	0.00028	0.00044

Unlike the CCSD model, the CCSDT and higher models – such as the *coupled-cluster singles-doubles-triples-and-quadruples (CCSDTQ) model* [8] – are therefore not treated in detail in this chapter. Fortunately, accurate coupled-cluster models have been developed that include the effects of the connected triples in an approximate fashion. We defer the discussion of such approximate CCSDT treatments to Chapter 14.

13.2.3 THE PROJECTED COUPLED-CLUSTER EQUATIONS

In a given orbital basis, the full coupled-cluster wave function satisfies the Schrödinger equation

$$\hat{H}\exp(\hat{T})|\text{HF}\rangle = E\exp(\hat{T})|\text{HF}\rangle \tag{13.2.16}$$

Truncated coupled-cluster wave functions cannot satisfy this equation exactly and, as discussed in Section 13.1.4, we use projection to determine the wave function. The optimized coupled-cluster wave function then satisfies the Schrödinger equation (13.2.16) projected onto the Hartree–Fock state and onto the excited projection manifold

$$\langle\mu| = \langle\text{HF}|\hat{\tau}_\mu^\dagger \tag{13.2.17}$$

The resulting projected coupled-cluster equations may be written as

$$\langle\text{HF}|\hat{H}\exp(\hat{T})|\text{HF}\rangle = E \tag{13.2.18}$$

$$\langle\mu|\hat{H}\exp(\hat{T})|\text{HF}\rangle = E\langle\mu|\exp(\hat{T})|\text{HF}\rangle \tag{13.2.19}$$

In the CCSD model, for example, the excited projection manifold comprises the full set of all singly and doubly excited determinants, giving rise to one equation (13.2.19) for each connected amplitude. For the full coupled-cluster wave function, the number of equations is equal to the number of determinants and the solution of the projected equations recovers the FCI wave function. The nonlinear equations (13.2.19) must be solved iteratively, substituting in each iteration the coupled-cluster energy as calculated from (13.2.18).

For many purposes, it is convenient to express the projected coupled-cluster equations in a slightly different form. First, we multiply the Schrödinger equation (13.2.16) from the left by the operator $\exp(-\hat{T})$ to obtain

$$\exp(-\hat{T})\hat{H}\exp(\hat{T})|\text{HF}\rangle = E|\text{HF}\rangle \tag{13.2.20}$$

which may be regarded as a Schrödinger equation with an effective, non-Hermitian *similarity-transformed Hamiltonian*:

$$\hat{H}^{\text{T}} = \exp(-\hat{T})\hat{H}\exp(\hat{T}) \tag{13.2.21}$$

Projecting the similarity-transformed Schrödinger equation (13.2.20) against the same determinants as in (13.2.18) and (13.2.19), we arrive at the following set of equations for the coupled-cluster amplitudes and energy:

$$\langle\text{HF}|\exp(-\hat{T})\hat{H}\exp(\hat{T})|\text{HF}\rangle = E \tag{13.2.22}$$

$$\langle\mu|\exp(-\hat{T})\hat{H}\exp(\hat{T})|\text{HF}\rangle = 0 \tag{13.2.23}$$

As will be shown shortly, these similarity-transformed equations are equivalent to (13.2.18) and (13.2.19), yielding the same amplitudes and the same energy upon solution. In the following, we shall refer to (13.2.23) as the *linked coupled-cluster equations* and to (13.2.19) as the *unlinked coupled-cluster equations*. The reason for the use of the terms 'linked' and 'unlinked' is that, in diagrammatic coupled-cluster theory, the energy-independent (similarity-transformed) equations (13.2.23) give rise to only linked diagrams, whereas the energy-dependent equations (13.2.19) give rise to unlinked as well as to linked diagrams [9]. As we shall see in Section 13.3, even though the linked and unlinked coupled-cluster equations are equivalent and both provide a size-extensive treatment of the electronic system, the treatment of size-extensivity is rather different in the two formulations.

For the full coupled-cluster wave function, the equivalence of the equations (13.2.16) and (13.2.20) is trivial; for truncated cluster expansions, on the other hand, the equivalence of the linked and unlinked forms of the amplitude equations is less obvious and requires special attention. First, the equivalence of the energy expressions (13.2.18) and (13.2.22) is easily established since for any choice of amplitudes

$$\langle \mathrm{HF}| \exp(-\hat{T}) = \langle \mathrm{HF}| \tag{13.2.24}$$

Next, to demonstrate the equivalence of the amplitude equations (13.2.19) and (13.2.23), we introduce the unsymmetric matrices $\mathbf{T}^+$ and $\mathbf{T}^-$ with elements given by

$$T_{\mu\nu}^{\pm} = \langle \mu| \exp(\pm\hat{T})|\nu\rangle = \langle \mu|\nu\rangle \pm \langle \mu|\hat{T}|\nu\rangle + \tfrac{1}{2}\langle \mu|\hat{T}^2|\nu\rangle \pm \cdots \tag{13.2.25}$$

The lower triangular structure of these two matrices, illustrated in Figure 13.2, follows from the fact that $\hat{T}$ contains only excitation operators and it is also retained if the definition of $\mathbf{T}^{\pm}$ is extended to include the Hartree–Fock state. Since the upper-triangular elements are 0 and since the diagonal elements are equal to 1, the matrices $\mathbf{T}^{\pm}$ are nonsingular for any projection manifold, irrespective of whether or not we omit, for example, excited determinants higher than doubles. We note that the diagonal blocks with μ and ν belonging to the same excitation level – both singles or both doubles, for instance – are equal to the identity matrices.

	HF	S	D	⋯	N
HF	1	0	0	⋯	0
S		1	0	⋯	0
D			1	⋯	0
⋮	⋮	⋮	⋮	⋱	0
N					1

Fig. 13.2. The lower triangular structure of the matrices $\mathbf{T}^{\pm}$ in (13.2.25). The diagonal blocks contain identity matrices, the lower triangular blocks are nonzero and the upper triangular blocks zero.

Let us now assume that the similarity-transformed amplitude equations (13.2.23) are satisfied. Since the matrix $\mathbf{T}^+$ is nonsingular, the conditions (13.2.23) are equivalent to the conditions

$$A_\mu = \sum_\nu T^+_{\mu\nu} \langle \nu | \exp(-\hat{T}) \hat{H} \exp(\hat{T}) | \text{HF} \rangle$$

$$= \sum_\nu \langle \mu | \exp(\hat{T}) | \nu \rangle \langle \nu | \exp(-\hat{T}) \hat{H} \exp(\hat{T}) | \text{HF} \rangle = 0 \qquad (13.2.26)$$

where μ and ν belong to the excitation manifold of (13.2.23). From the structure of $\mathbf{T}^+$, we note that $\langle \mu | \exp(\hat{T})$ contains determinants of excitation levels lower than or equal to that of $\langle \mu |$. Assuming that the projection manifold is *closed under de-excitation* (i.e. if $\langle \mu |$ is a member of the projection manifold, then so is $\langle \mu | \hat{T} \rangle$), we may invoke the resolution of the identity

$$\langle \mu | \exp(\hat{T}) = \sum_\nu \langle \mu | \exp(\hat{T}) | \nu \rangle \langle \nu | + \langle \mu | \exp(\hat{T}) | \text{HF} \rangle \langle \text{HF} | \qquad (13.2.27)$$

where the summation is again restricted to the excitation manifold. Equation (13.2.26) can now be written as

$$A_\mu = \langle \mu | \exp(\hat{T}) \exp(-\hat{T}) \hat{H} \exp(\hat{T}) | \text{HF} \rangle - \langle \mu | \exp(\hat{T}) | \text{HF} \rangle \langle \text{HF} | \exp(-\hat{T}) \hat{H} \exp(\hat{T}) | \text{HF} \rangle$$

$$= \langle \mu | \hat{H} \exp(\hat{T}) | \text{HF} \rangle - E \langle \mu | \exp(\hat{T}) | \text{HF} \rangle \qquad (13.2.28)$$

where we have used (13.2.27) and (13.2.22). We conclude that the conditions (13.2.26) are equivalent to the unlinked equations (13.2.19) if the projection manifold is closed under de-excitation. We have thus established the equivalence of the linked and unlinked coupled-cluster conditions (13.2.23) and (13.2.19) for the standard models CCSD, CCSDT, and so on. Equivalence also holds for all models containing only even-order excitations such as CCD. Finally, the projection manifold is closed and equivalence is maintained if, at the highest excitation level, only selected excitations are retained – for example, if selected triple excitations are included in addition to all singles and doubles.

Let us now compare the coupled-cluster equations in the linked and unlinked forms. We begin by reiterating that these two forms of the coupled-cluster equations are equivalent for the standard models in the sense that they have the same solutions. Moreover, applied at the important CCSD level of theory, neither form is superior to the other, requiring about the same number of floating-point operations. The energy-dependent unlinked form (13.2.19) exhibits more closely the relationship with CI theory, where the projected equations may be written in a similar form (13.1.18). On the other hand, the linked form (13.2.23) has some important advantages over the unlinked one (13.2.19), making it the preferred form in most situations.

First, when the similarity-transformed Hamiltonian is expressed as a BCH expansion (see Section 13.2.5), the coupled-cluster equations may be shown to be no higher than quartic in the cluster amplitudes – for any truncation of the cluster operator. As a bonus, the nested commutators of the BCH expansion reduce the rank of the operators, further simplifying the algebra. Second, although the linked and unlinked equations yield the same, size-extensive wave function, the linked equations have the useful additional property of being size-extensive term by term (see Section 13.3), allowing for a simple control of size-extensivity upon modification of the coupled-cluster equations and making this particular form a useful starting point for the development of perturbation theory as discussed in Section 14.3. Third, the similarity-transformed linked equations also constitute the starting point for the development of a coupled-cluster approach for the calculation of excited states in Section 13.6. Fourth, by carrying out an explicit similarity transformation of the Hamiltonian using the singles amplitudes, the algebra of the coupled-cluster equations is significantly simplified, as discussed for the CCSD wave function in Section 13.7. For these reasons, we shall, in the remainder of this chapter, mostly employ the coupled-cluster equations in the linked form.

13.2.4 THE COUPLED-CLUSTER ENERGY

In coupled-cluster theory, the electronic energy is obtained from (13.2.22):

$$E = \langle \mathrm{HF}| \exp(-\hat{T})\hat{H} \exp(\hat{T})|\mathrm{HF}\rangle = \langle \mathrm{HF}|\hat{H} \exp(\hat{T})|\mathrm{HF}\rangle \tag{13.2.29}$$

Expanding the cluster amplitudes, we obtain

$$E = \langle \mathrm{HF}|\hat{H} \left(1 + \hat{T} + \tfrac{1}{2}\hat{T}^2 + \cdots \right) |\mathrm{HF}\rangle = \langle \mathrm{HF}|\hat{H} \left(1 + \hat{T}_2 + \tfrac{1}{2}\hat{T}_1^2 \right) |\mathrm{HF}\rangle \tag{13.2.30}$$

Cluster operators higher than doubles do not contribute to the energy since $\hat{H}$ is a two-particle operator. Because of the Brillouin theorem, the one-particle operators contribute only to second order:

$$\langle \mathrm{HF}|\hat{H}\hat{T}_1|\mathrm{HF}\rangle = 0 \tag{13.2.31}$$

As a result, only singles and doubles amplitudes contribute directly to the coupled-cluster energy – irrespective of the truncation level in the cluster operator. Of course, the higher-order excitations contribute indirectly since all amplitudes are coupled by the projected equations (13.2.23).

13.2.5 THE COUPLED-CLUSTER AMPLITUDE EQUATIONS

Having examined the coupled-cluster energy and seen that it is no higher than quadratic in the cluster amplitudes, let us now turn our attention to the structure of the linked projected coupled-cluster equations (13.2.23):

$$\langle \mu | \exp(-\hat{T})\hat{H} \exp(\hat{T})|\mathrm{HF}\rangle = 0 \tag{13.2.32}$$

Since $\hat{T}$ is not anti-Hermitian, it gives rise to a nonunitary transformation and the similarity-transformed Hamiltonian operator is therefore non-Hermitian. Naively, we would expect the BCH expansion (3.1.7) of the similarity-transformed Hamiltonian to yield an infinite sequence of nested commutators. Nevertheless, we shall see that the expansion terminates after five terms.

Let the *up rank* s_A^+ and the *down rank* s_A^- of a string of elementary operators $\hat{A}$ be given by

$$s_A^+ = \tfrac{1}{2}(n_v^c + n_o^a) \tag{13.2.33}$$

$$s_A^- = \tfrac{1}{2}(n_o^c + n_v^a) \tag{13.2.34}$$

where n_v^c and n_o^c are the numbers of creation operators in $\hat{A}$ for the virtual and occupied spin orbitals, respectively; likewise, n_o^a and n_v^a are the numbers of occupied and virtual annihilation operators. We also introduce the *excitation rank* s_A as the difference between the up and down ranks

$$s_A = s_A^+ - s_A^- \tag{13.2.35}$$

and note that the sum of the up and down ranks is equal to the particle rank of $\hat{A}$. In Section 13.2.8, we prove the following *cluster-commutation condition* for the vanishing of k nested commutators of $\hat{A}$ with cluster operators:

$$k > 2s_A^- \Rightarrow [[\ldots [[\hat{A}, \hat{T}_{n_1}], \hat{T}_{n_2}], \ldots], \hat{T}_{n_k}] = 0 \tag{13.2.36}$$

Since the highest down rank of the Hamiltonian is 2, it follows that the BCH expansion of the similarity-transformed Hamiltonian is no higher than quartic in the amplitudes:

$$\exp(-\hat{T})\hat{H}\exp(\hat{T}) = \hat{H} + [\hat{H}, \hat{T}] + \tfrac{1}{2}[[\hat{H}, \hat{T}], \hat{T}] + \tfrac{1}{6}[[[\hat{H}, \hat{T}], \hat{T}], \hat{T}] + \tfrac{1}{24}[[[[\hat{H}, \hat{T}], \hat{T}], \hat{T}], \hat{T}]$$
$$\tag{13.2.37}$$

The projected coupled-cluster Schrödinger equation (13.2.32) therefore yields at most *quartic equations* in the cluster amplitudes – even for the full cluster expansion. The BCH expansion terminates because of the special structure of the cluster operators, which are linear combinations of commuting excitation operators of the form (13.2.6) and (13.2.7).

Although the similarity-transformed Hamiltonian is quartic in the cluster amplitudes, the equations for the cluster amplitudes (13.2.32) need not contain all the amplitudes to this order. In Section 13.2.8, we use the cluster-commutation condition (13.2.36) to show that, for a general operator $\hat{O}$ of particle rank m_O, the state

$$|n_1 n_2 \ldots n_k\rangle = [[\ldots [[\hat{O}, \hat{T}_{n_1}], \hat{T}_{n_2}], \ldots], \hat{T}_{n_k}]|\text{HF}\rangle \tag{13.2.38}$$

is a linear combination of determinants with excitation ranks s in the range

$$\sum_{i=1}^{k} n_i - m_O \leq s \leq \sum_{i=1}^{k} n_i + m_O - k \tag{13.2.39}$$

where n_i is the excitation rank of $\hat{T}_{n_i}$. Using these conditions, we may set up the following expressions for the CCSD amplitude equations:

$$\langle \mu_1 | \hat{H} | \text{HF} \rangle + \langle \mu_1 | [\hat{H}, \hat{T}_1] | \text{HF} \rangle + \langle \mu_1 | [\hat{H}, \hat{T}_2] | \text{HF} \rangle + \tfrac{1}{2} \langle \mu_1 | [[\hat{H}, \hat{T}_1], \hat{T}_1] | \text{HF} \rangle$$
$$+ \langle \mu_1 | [[\hat{H}, \hat{T}_1], \hat{T}_2] | \text{HF} \rangle + \tfrac{1}{6} \langle \mu_1 | [[[\hat{H}, \hat{T}_1], \hat{T}_1], \hat{T}_1] | \text{HF} \rangle = 0 \tag{13.2.40}$$

$$\langle \mu_2 | \hat{H} | \text{HF} \rangle + \langle \mu_2 | [\hat{H}, \hat{T}_1] | \text{HF} \rangle + \langle \mu_2 | [\hat{H}, \hat{T}_2] | \text{HF} \rangle + \tfrac{1}{2} \langle \mu_2 | [[\hat{H}, \hat{T}_1], \hat{T}_1] | \text{HF} \rangle$$
$$+ \langle \mu_2 | [[\hat{H}, \hat{T}_1], \hat{T}_2] | \text{HF} \rangle + \tfrac{1}{2} \langle \mu_2 | [[\hat{H}, \hat{T}_2], \hat{T}_2] | \text{HF} \rangle$$
$$+ \tfrac{1}{6} \langle \mu_2 | [[[\hat{H}, \hat{T}_1], \hat{T}_1], \hat{T}_1] | \text{HF} \rangle + \tfrac{1}{2} \langle \mu_2 | [[[\hat{H}, \hat{T}_1], \hat{T}_1], \hat{T}_2] | \text{HF} \rangle$$
$$+ \tfrac{1}{24} \langle \mu_2 | [[[[\hat{H}, \hat{T}_1], \hat{T}_1], \hat{T}_1], \hat{T}_1] | \text{HF} \rangle = 0 \tag{13.2.41}$$

Whereas the singles occur to fourth order, the doubles appear only *quadratically* in these expressions. Moreover, from (13.2.39), it is easily verified that, except for the singles and the doubles, the amplitudes of the highest excitation level always occur only *linearly* in the coupled-cluster equations. These additional simplifications in the coupled-cluster equations – beyond what is dictated by the termination of the BCH expansion (13.2.37) – occur because of the restrictions on the excitation levels in the projection space $\langle \mu_i |$. If instead we had calculated the energy from the variation principle, the expansion of the exponentials would not terminate (except, of course, for the fact that we have only a finite number of electrons to excite from the occupied spin orbitals). The evaluation of the energy and the amplitudes in such a model is therefore considerably more difficult than in standard coupled-cluster theory.

13.2.6 COUPLED-CLUSTER THEORY IN THE CANONICAL REPRESENTATION

In coupled-cluster theory, it is often convenient to work in the canonical representation of the spin orbitals. The Hamiltonian operator is then partitioned into the Fock operator $\hat{f}$, the fluctuation potential $\hat{\Phi}$ and the nuclear–nuclear contribution h_{nuc} as discussed in Section 10.4.5:

$$\hat{H} = \hat{f} + \hat{\Phi} + h_{\text{nuc}} \qquad (13.2.42)$$

In the canonical representation, the Fock operator may be written in terms of the orbital energies ε_P as

$$\hat{f} = \sum_P \varepsilon_P a_P^\dagger a_P \qquad (13.2.43)$$

The commutator of the Fock operator with the cluster operator is found to be

$$[\hat{f}, \hat{T}] = \sum_\mu \varepsilon_\mu t_\mu \hat{\tau}_\mu \qquad (13.2.44)$$

where ε_μ is the sum of all unoccupied orbital energies minus the sum of all occupied orbital energies of the spin orbitals in $\hat{\tau}_\mu$, for example,

$$\varepsilon_{AI} = \varepsilon_A - \varepsilon_I \qquad (13.2.45)$$

$$\varepsilon_{AIBJ} = \varepsilon_A - \varepsilon_I + \varepsilon_B - \varepsilon_J \qquad (13.2.46)$$

Inserting the Hamiltonian (13.2.42) into the expressions for the energy (13.2.22) and the nonlinear equations (13.2.23), we obtain the following expressions for the coupled-cluster energy and the amplitude equations:

$$E = E_{\text{HF}} + \langle \text{HF} | \hat{\Phi} \left(\hat{T}_2 + \tfrac{1}{2} \hat{T}_1^2 \right) | \text{HF} \rangle \qquad (13.2.47)$$

$$\varepsilon_\mu t_\mu = -\langle \mu | \exp(-\hat{T}) \hat{\Phi} \exp(\hat{T}) | \text{HF} \rangle \qquad (13.2.48)$$

where we have used the commutator (13.2.44) and the fact that all higher commutators of the Fock operator vanish since the down rank of the Fock operator is $\tfrac{1}{2}$. As we shall see in Section 14.3, (13.2.47) and (13.2.48) form a convenient starting point for the development of a perturbation theory with the Fock operator as the zero-order Hamiltonian. Moreover, in Section 13.4, we shall use (13.2.48) to set up an efficient iterative scheme for the optimization of the coupled-cluster wave function.

13.2.7 COMPARISON OF THE CI AND COUPLED-CLUSTER HIERARCHIES

Owing to the presence of the disconnected clusters, coupled-cluster wave functions truncated at a given excitation level also contain contributions from determinants corresponding to higher-order excitations. The terms that are missing relative to FCI represent higher-order connected clusters and the associated disconnected clusters. By contrast, CI wave functions truncated at the same level contain contributions from determinants only up to this level. Since the disconnected contributions to the energy are significant (and dominant for extended systems), the accuracy

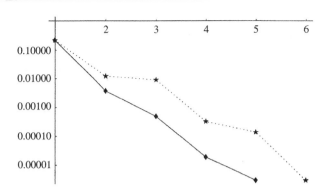

Fig. 13.3. The error (with respect to FCI) in the total energy (E_h) of coupled-cluster wave functions (full line) and CI wave functions (dotted line) at different excitation levels for the water molecule at the equilibrium geometry in the cc-pVDZ basis.

achieved in a coupled-cluster calculation is expected to be higher than that of the corresponding CI calculation, in particular for large systems. In Figure 13.3, we have plotted the errors in the coupled-cluster and CI energies of the water molecule at the reference geometry R_{ref} as a function of the excitation level [10]. The errors are relative to the FCI limit in the cc-pVDZ basis.

In the coupled-cluster calculations, the error decreases by about the same factor at each excitation level. Also, at a given truncation level, the error is significantly smaller than that of the CI energy, which converges in a less systematic and satisfactory manner because of the neglect of higher-order disconnected clusters. As an example, the CCSD error falls between the CISDT and CISDTQ errors since, in CI theory, the disconnected term $\hat{T}_2^2$ is accounted for only at the SDTQ and higher levels. Finally, it should be emphasized that Figure 13.3 is typical of small electronic systems. For large systems, the quality of the CI description (at a given truncation level) deteriorates dramatically, as previously discussed in Sections 11.2 and 11.3. The coupled-cluster description, by contrast, is unaffected by the number of electrons in the system.

13.2.8 CLUSTER-COMMUTATION CONDITIONS AND OPERATOR RANKS

In Section 13.2.5, a set of inequalities was used to identify the nonvanishing commutators of the Hamiltonian and the cluster operators. In the present subsection, these inequalities are derived. Let us first consider a simple string of elementary creation and annihilation operators $\hat{A}$. The *up and down ranks* of $\hat{A}$, introduced in Section 13.2.5, are defined as

$$s_A^+ = \tfrac{1}{2}(n_v^c + n_o^a) \tag{13.2.49}$$

$$s_A^- = \tfrac{1}{2}(n_o^c + n_v^a) \tag{13.2.50}$$

where n_o^c and n_v^c are the numbers of occupied and virtual creation operators in $\hat{A}$, respectively, and n_o^a and n_v^a are the corresponding numbers of annihilation operators. Note that the up and down ranks are always defined relative to the reference determinant. The *particle rank* m_A and *excitation rank* s_A of $\hat{A}$ are then given as

$$m_A = s_A^+ + s_A^- \tag{13.2.51}$$

$$s_A = s_A^+ - s_A^- \tag{13.2.52}$$

If s_A is positive, then $\hat{A}|\text{HF}\rangle$ represents an excited determinant of *excitation rank (level)* s_A relative to the Hartree–Fock state. Conversely, if s_A is negative, then $\hat{A}^\dagger|\text{HF}\rangle$ represents an excited determinant of excitation rank $|s_A|$.

We now examine the nested commutator of the string $\hat{A}$ with k cluster operators $\hat{T}_{n_i}$ such as those in (13.2.6) and (13.2.7):

$$\hat{\Omega} = [[\dots[[\hat{A}, \hat{T}_{n_1}], \hat{T}_{n_2}], \dots], \hat{T}_{n_k}] \tag{13.2.53}$$

For each cluster operator $\hat{T}_{n_i}$, the particle rank and the excitation rank are both equal to n_i. If the commutator does not vanish, then its particle and excitation ranks are given by

$$m_\Omega = m_A + n_T - k \tag{13.2.54}$$

$$s_\Omega = s_A + n_T \tag{13.2.55}$$

where n_T is the total excitation rank of all cluster operators:

$$n_T = \sum_{i=1}^{k} n_i \tag{13.2.56}$$

In calculating the particle rank m_Ω, we have added the particle ranks of all the operators and subtracted k since each commutator in (13.2.53) reduces the rank by 1, as discussed in Section 1.8. In calculating the excitation level s_Ω, we have added the excitation ranks of all operators, noting that commutation does not change the excitation rank of the operators.

Subtracting (13.2.55) from (13.2.54), we obtain

$$m_\Omega - s_\Omega = m_A - s_A - k \tag{13.2.57}$$

Inserting (13.2.51) and (13.2.52) on both sides, we find that the down ranks of the commutator $\hat{\Omega}$ and the operator $\hat{A}$ are related by

$$2s_\Omega^- = 2s_A^- - k \tag{13.2.58}$$

In other words, *each commutator reduces the down rank of the operator by one-half and increases the up rank by the same amount*. Since the down rank of $\hat{\Omega}$ cannot be negative, we obtain the following condition on the down rank of $\hat{A}$

$$2s_A^- \geq k \tag{13.2.59}$$

for the commutator $\hat{\Omega}$ not to vanish.

Let us now consider the nested commutator of a general number-conserving m-particle operator $\hat{O}$:

$$\hat{\Omega} = [[\dots[[\hat{O}, \hat{T}_{n_1}], \hat{T}_{n_2}], \dots], \hat{T}_{n_k}] \tag{13.2.60}$$

The excitation rank s_O of a given term in the operator $\hat{O}$ satisfies the condition

$$-m_O \leq s_O \leq m_O \tag{13.2.61}$$

Moreover, for this particular term to give a nonzero contribution to the commutator, its down rank s_O^- must, according to (13.2.59), satisfy the condition

$$2s_O^- \geq k \tag{13.2.62}$$

Subtracting (13.2.52) from (13.2.51), we may write this condition as

$$m_O - s_O \geq k \tag{13.2.63}$$

Combining the conditions (13.2.61) and (13.2.63), we arrive at the following condition, which a given term in $\hat{O}$ must satisfy to give a nonzero contribution to the commutator:

$$-m_O \leq s_O \leq m_O - k \tag{13.2.64}$$

Since the total excitation rank of a nonzero commutator is equal to the sum of the excitation ranks of the individual operators, we have established the following range of excitation ranks of the commutator (13.2.60):

$$n_T - m_O \leq s_\Omega \leq n_T + m_O - k \tag{13.2.65}$$

This condition has solutions only for

$$k \leq 2m_O \tag{13.2.66}$$

For $k = 2m_O$, the commutator has a sharply defined excitation rank $n_T - m_O$; in all other cases, the commutator possesses a range of excitation ranks (13.2.65).

13.3 Size-extensivity in coupled-cluster theory

As discussed in Section 4.3, a computational method is said to be size-extensive if a calculation on the compound system AB consisting of two noninteracting systems A and B yields a total energy equal to the sum of the energies obtained in separate calculations on the two subsystems. This property of the coupled-cluster model is demonstrated for the linked formulation in Section 13.3.1, leading to the concept of termwise size-extensivity in Section 13.3.2. In Section 13.3.3, we consider size-extensivity in the unlinked formulation of coupled-cluster theory, demonstrating how size-extensivity in this case arises from a cancellation of terms that individually violate size-extensivity.

13.3.1 SIZE-EXTENSIVITY IN LINKED COUPLED-CLUSTER THEORY

We assume that we have calculated the coupled-cluster wave functions for the individual systems A and B

$$|CC_A\rangle = \exp(\hat{T}_A)|HF_A\rangle \tag{13.3.1}$$

$$|CC_B\rangle = \exp(\hat{T}_B)|HF_B\rangle \tag{13.3.2}$$

The energies of these systems are given by

$$E_A = \langle HF_A|\hat{H}_A^T|HF_A\rangle \tag{13.3.3}$$

$$E_B = \langle HF_B|\hat{H}_B^T|HF_B\rangle \tag{13.3.4}$$

and the amplitudes fulfil the projected equations

$$\langle \mu_A|\hat{H}_A^T|HF_A\rangle = 0 \tag{13.3.5}$$

$$\langle \mu_B|\hat{H}_B^T|HF_B\rangle = 0 \tag{13.3.6}$$

in terms of the effective, similarity-transformed Hamiltonians

$$\hat{H}_A^T = \exp(-\hat{T}_A)\hat{H}_A \exp(\hat{T}_A) \tag{13.3.7}$$

$$\hat{H}_B^T = \exp(-\hat{T}_B)\hat{H}_B \exp(\hat{T}_B) \tag{13.3.8}$$

The Hamiltonian for the compound system of the noninteracting subsystems is given by

$$\hat{H}_{AB} = \hat{H}_A + \hat{H}_B \tag{13.3.9}$$

We wish to demonstrate that the product wave function

$$|CC_{AB}\rangle = \exp(\hat{T}_{AB})|HF_{AB}\rangle \tag{13.3.10}$$

where

$$|HF_{AB}\rangle = |HF_A HF_B\rangle \tag{13.3.11}$$

$$\hat{T}_{AB} = \hat{T}_A + \hat{T}_B \tag{13.3.12}$$

represents a size-extensive solution for the combined system.

The key point in establishing the size-extensivity of coupled-cluster theory in the linked formulation is to note that the similarity-transformed Hamiltonian of the compound system

$$\hat{H}_{AB}^T = \exp(-\hat{T}_{AB})\hat{H}_{AB} \exp(\hat{T}_{AB}) \tag{13.3.13}$$

separates into the corresponding Hamiltonians of the subsystems

$$\hat{H}_{AB}^T = \hat{H}_A^T + \hat{H}_B^T \tag{13.3.14}$$

This separation is obtained by using (13.2.9), (13.3.9) and (13.3.12) together with the commutators

$$[\hat{H}_A, \hat{T}_B] = 0 \tag{13.3.15}$$

$$[\hat{H}_B, \hat{T}_A] = 0 \tag{13.3.16}$$

The energy of the compound system may therefore be written as

$$\begin{aligned}
E_{AB} &= \langle HF_{AB}|\hat{H}_{AB}^T|HF_{AB}\rangle \\
&= \langle HF_A HF_B|\hat{H}_A^T + \hat{H}_B^T|HF_A HF_B\rangle \\
&= \langle HF_A|\hat{H}_A^T|HF_A\rangle + \langle HF_B|\hat{H}_B^T|HF_B\rangle \\
&= E_A + E_B
\end{aligned} \tag{13.3.17}$$

and is therefore equal to the sum of the energies of the subsystems.

It remains to establish that the product wave function (13.3.10) indeed represents a solution to the Schrödinger equation of the compound system:

$$\langle \mu_{AB}|\hat{H}_{AB}^T|HF_{AB}\rangle = 0 \tag{13.3.18}$$

There are three cases to be considered, depending on whether the bra state in (13.3.18) represents an excitation within system A, an excitation within system B or an excitation involving both systems. For these three cases, the left-hand side in (13.3.18) may be written as

$$\langle \mu_{A} HF_{B} | \hat{H}_{A}^{T} + \hat{H}_{B}^{T} | HF_{A} HF_{B} \rangle = \langle \mu_{A} | \hat{H}_{A}^{T} | HF_{A} \rangle \qquad (13.3.19)$$

$$\langle HF_{A} \mu_{B} | \hat{H}_{A}^{T} + \hat{H}_{B}^{T} | HF_{A} HF_{B} \rangle = \langle \mu_{B} | \hat{H}_{B}^{T} | HF_{B} \rangle \qquad (13.3.20)$$

$$\langle \mu_{A} \mu_{B} | \hat{H}_{A}^{T} + \hat{H}_{B}^{T} | HF_{A} HF_{B} \rangle = 0 \qquad (13.3.21)$$

where we have used the orthogonality of the Hartree–Fock state and the excited determinants to simplify the equations. For example, the first term on the left-hand side of (13.3.21) gives

$$\langle \mu_{A} \mu_{B} | \hat{H}_{A}^{T} | HF_{A} HF_{B} \rangle = \langle \mu_{A} | \hat{H}_{A}^{T} | HF_{A} \rangle \langle \mu_{B} | HF_{B} \rangle \qquad (13.3.22)$$

which vanishes because of the orthogonality of $|\mu_{B}\rangle$ and $|HF_{B}\rangle$. It should be realized that, for truncated cluster expansions such as CCSD, some of the product states $\langle \mu_{A}\mu_{B}|$ in (13.3.21) may not be contained in the projection manifold $\langle \mu_{AB}|$ for the compound system. However, this does not matter as the projections are zero anyway.

Since the right-hand sides of the first two equations (13.3.19) and (13.3.20) are identical to the left-hand sides of the coupled-cluster equations of the subsystems (13.3.5) and (13.3.6), we conclude that expressions (13.3.19)–(13.3.21) vanish and hence that the coupled-cluster equations of the compound system are satisfied by the product wave function $|CC_{AB}\rangle$ in (13.3.10). In keeping with the terminology of Section 4.3.1, the cluster operator $\hat{T}_{AB}$ in (13.3.12) is said to be additively separable. It gives rise to a multiplicatively separable wave operator

$$\exp(\hat{T}_{AB}) = \exp(\hat{T}_{A} + \hat{T}_{B}) = \exp(\hat{T}_{A}) \exp(\hat{T}_{B}) \qquad (13.3.23)$$

and thus produces a multiplicatively separable wave function $|CC_{AB}\rangle$ when applied to the Hartree–Fock state. Note that, although the state $|CC_{AB}\rangle$ contains simultaneous excitations in A and B, no such excitations are present in the cluster operator $\hat{T}_{AB}$.

In the present subsection, we have established that the coupled-cluster model is size-extensive – at least in the similarity-transformed, linked formulation of the theory. Note that nothing has been assumed about the nature of the cluster operators for the subsystems – size-extensivity occurs irrespective of how we truncate these operators and what excitation operators are left out. In Section 13.3.3, we shall see that the coupled-cluster state is also size-extensive in the unlinked formulation of the coupled-cluster equations.

13.3.2 TERMWISE SIZE-EXTENSIVITY

The use of a similarity-transformed Hamiltonian in linked coupled-cluster theory means that the energy and amplitude equations contain terms that consist either of the Hamiltonian itself or of nested commutators of the Hamiltonian with cluster operators. For a system containing two noninteracting subsystems A and B, these nested commutators separate additively into nested commutators, each involving a single subsystem, for example,

$$[[\hat{H}_{A} + \hat{H}_{B}, \hat{T}_{1A} + \hat{T}_{1B}], \hat{T}_{1A} + \hat{T}_{1B}] = [[\hat{H}_{A}, \hat{T}_{1A}], \hat{T}_{1A}] + [[\hat{H}_{B}, \hat{T}_{1B}], \hat{T}_{1B}] \qquad (13.3.24)$$

The additive separability of each commutator leads to a formulation of coupled-cluster theory where each term (i.e. each expectation-value expression) in the energy or in the amplitude equations is separately size-extensive. The linked equations are therefore said to be *termwise size-extensive*. No terms that violate the size-extensivity arise and no cancellation of such terms ever occurs.

Termwise size-extensivity makes the linked formulation of coupled-cluster theory particularly convenient for developing approximate but rigorously size-extensive models: from the energy or amplitude equations, we may omit any commutator contribution without destroying size-extensivity. We shall see examples of this technique in our discussion of quadratic CI theory in Section 13.8.2 and in the development of perturbation theory in Section 14.6.

13.3.3 SIZE-EXTENSIVITY IN UNLINKED COUPLED-CLUSTER THEORY

In Section 13.3.1, size-extensivity was demonstrated for the linked formulation of coupled-cluster theory. In this subsection, we consider size-extensivity in the alternative, unlinked formulation of the theory, thereby establishing size-extensivity also in those (rare) cases where the linked and unlinked formulations differ. More important, the discussion in the present subsection illustrates that, in unlinked coupled-cluster theory, size-extensivity arises from a cancellation of contributions that individually violate size-extensivity. This behaviour is in contrast to that of linked coupled-cluster theory, where size-extensivity occurs separately for all commutators that contribute to the equations as discussed in Section 13.3.2.

In the unlinked formulation of coupled-cluster theory, the expression for the energy (13.2.18) is identical to that of the linked formulation (13.2.22) and size-extensivity follows simply:

$$E_{AB} = \langle HF_A HF_B | (\hat{H}_A + \hat{H}_B) \exp(\hat{T}_A + \hat{T}_B) | HF_A HF_B \rangle$$
$$= \langle HF_A | \hat{H}_A \exp(\hat{T}_A) | HF_A \rangle + \langle HF_B | \hat{H}_B \exp(\hat{T}_B) | HF_B \rangle$$
$$= E_A + E_B \tag{13.3.25}$$

Turning our attention to the amplitude equations, these are given as follows for two separate, noninteracting systems

$$\langle \mu_A | \hat{H}_A \exp(\hat{T}_A) | HF_A \rangle = E_A \langle \mu_A | \exp(\hat{T}_A) | HF_A \rangle \tag{13.3.26}$$

$$\langle \mu_B | \hat{H}_B \exp(\hat{T}_B) | HF_B \rangle = E_B \langle \mu_B | \exp(\hat{T}_B) | HF_B \rangle \tag{13.3.27}$$

For the compound system, we have three cases to consider:

$$\langle \mu_A HF_B | (\hat{H}_A + \hat{H}_B) \exp(\hat{T}_A + \hat{T}_B) | HF_A HF_B \rangle = (E_A + E_B) \langle \mu_A HF_B | \exp(\hat{T}_A + \hat{T}_B) | HF_A HF_B \rangle \tag{13.3.28}$$

$$\langle HF_A \mu_B | (\hat{H}_A + \hat{H}_B) \exp(\hat{T}_A + \hat{T}_B) | HF_A HF_B \rangle = (E_A + E_B) \langle HF_A \mu_B | \exp(\hat{T}_A + \hat{T}_B) | HF_A HF_B \rangle \tag{13.3.29}$$

$$\langle \mu_A \mu_B | (\hat{H}_A + \hat{H}_B) \exp(\hat{T}_A + \hat{T}_B) | HF_A HF_B \rangle = (E_A + E_B) \langle \mu_A \mu_B | \exp(\hat{T}_A + \hat{T}_B) | HF_A HF_B \rangle \tag{13.3.30}$$

We note that these equations contain terms that involve the systems A and B simultaneously so that their left- and right-hand sides are not separately size-extensive. Nevertheless, by a suitable rearrangement of the equations, we may show that the terms that violate size-extensivity cancel, making the equations rigorously size-extensive. For example, expanding the two sides of (13.3.28), we obtain

$$\langle \mu_A | \hat{H}_A \exp(\hat{T}_A) | HF_A \rangle + \langle \mu_A | \exp(\hat{T}_A) | HF_A \rangle \langle HF_B | \hat{H}_B \exp(\hat{T}_B) | HF_B \rangle$$
$$= (E_A + E_B) \langle \mu_A | \exp(\hat{T}_A) | HF_A \rangle \tag{13.3.31}$$

Upon inspection, we find that the term that violates size-extensivity on the left-hand side cancels a similar term on the right-hand side, reducing the equations to those for system A in (13.3.26). In the same manner, (13.3.29) reduces to (13.3.27) for system B. Finally, the equations for simultaneous excitations in the two systems (13.3.30) reduce to a linear combination of conditions (13.3.26) and (13.3.27):

$$\langle \mu_A | (\hat{H}_A - E_A) \exp(\hat{T}_A) | HF_A \rangle \langle \mu_B | \exp(\hat{T}_B) | HF_B \rangle$$
$$+ \langle \mu_B | (\hat{H}_B - E_B) \exp(\hat{T}_B) | HF_B \rangle \langle \mu_A | \exp(\hat{T}_A) | HF_A \rangle = 0 \qquad (13.3.32)$$

Clearly, the fulfilment of conditions (13.3.26) and (13.3.27) for the noninteracting systems A and B implies the fulfilment of conditions (13.3.28)–(13.3.30) for the compound system – as expected for a size-extensive computational model. Note that, in our demonstration of size-extensivity in the unlinked formulation, no assumptions were made about the projection space. Thus, although the linked and unlinked coupled-cluster equations may give different solutions when the projection space is not closed under de-excitations, both solutions are size-extensive.

In conclusion, in the unlinked formulation of coupled-cluster theory, the amplitude equations yield solutions that are size-extensive but not termwise so. In general, therefore, it is not trivial to make size-extensive approximations to the unlinked coupled-cluster equations and the linked equations are to be preferred for such purposes.

13.3.4 A NUMERICAL STUDY OF SIZE-EXTENSIVITY

To illustrate the concepts of additive and multiplicative separability, let us examine the cc-pVDZ energies and wave functions of the CCSD and FCI models for a set of noninteracting water monomers. In the first series of calculations, the water monomers are at the reference geometry R_{ref}; in the second series, their bond distances are stretched to $2R_{ref}$. Similar calculations were described for the CISD model in Section 11.2.4.

At the equilibrium geometry, the CCSD monomer correlation energy of -214.1 mE_h represents as much as 98% of the FCI correlation energy of -217.8 mE_h. At the stretched geometry, there are large static correlation effects and the CCSD energy of -341.9 mE_h represents only 94% of the FCI energy of -364.0 mE_h. Since the CCSD model is size-extensive, it recovers the same proportion of the total correlation energy for any number of noninteracting monomers. By contrast, the CISD model deteriorates markedly with the number of monomers, as discussed in Section 11.2.4. Thus, for eight monomers at R_{ref}, the CISD model recovers just -162.4 mE_h per monomer – much less than -205.8 mE_h, obtained for a single monomer.

In Figure 13.4 we have plotted, for the normalized CCSD and FCI wave functions, the weights of the Hartree–Fock determinant and of the six lowest excitation levels as a function of the number of monomers. Considering first the even-order excitations close to the equilibrium geometry (upper left-hand plot), we note that the CCSD model reproduces the FCI weights well. While it overestimates the Hartree–Fock weight slightly, it underestimates the weights of the doubles, quadruples and sextuples. Since the CCSD wave function is multiplicatively separable, the overlap with the FCI wave function is S^m, where m is the number of monomers and S the overlap for a single monomer. The error in the wave function therefore increases with the number of monomers.

According to Table 13.1, the CCSD wave function does not produce an accurate weight for the sextuples in a single monomer. On the other hand, from Figure 13.4, we note that the sextuples are accurately reproduced for several monomers, even though the CCSD model contains parameters only for the single and double excitations. This behaviour occurs since, for a system of several

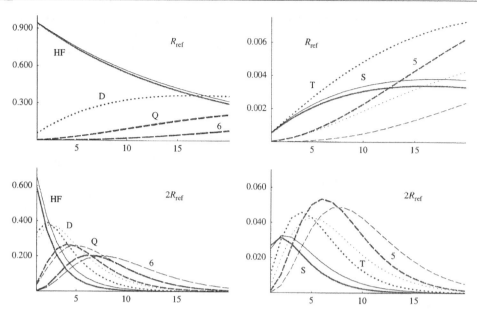

Fig. 13.4. The weights of the Hartree–Fock determinant and of the six lowest excitation levels in the normalized CCSD (thin black lines) and FCI (thick grey lines) wave functions, plotted as functions of the number of noninteracting water monomers in the cc-pVDZ basis.

monomers, the sextuples are dominated by disconnected contributions from double excitations that occur independently within three separate monomers, each of which is well described at the CCSD level. Thus, the fact that the higher even-order excitations are described more accurately for several monomers than for a single monomer should not be taken to imply that the CCSD wave function becomes more accurate as the system increases – it merely reflects the increasing dominance of disconnected excitations in large systems.

Regarding the odd-order excitations at the reference geometry (upper right-hand plot in Figure 13.4), only for the singles is there some sort of agreement between the CCSD and FCI curves. The triples, in particular, are strongly underestimated by the CCSD model, indicating the need to include connected triples for high accuracy. Nevertheless, since the weights of these odd-order excitations are much smaller than those of the most important even-order excitations, these deficiencies do not invalidate the CCSD model.

At the stretched geometry (the lower plots in Figure 13.4), the CCSD wave function differs significantly from the FCI wave function – for all numbers of monomers. The behaviour of the FCI curves for the single excitations and for all even-order excitations is only qualitatively reproduced. Whereas the Hartree–Fock weight is overestimated for all numbers of monomers, the weights of the higher-order excitations are underestimated for few monomers but become overestimated as the number of monomers increases. For an accurate description of the electronic system, connected triples are needed at this geometry.

13.4 Coupled-cluster optimization techniques

For the solution of the coupled-cluster amplitude equations, a number of techniques have been proposed. In the present section, some of these methods are discussed. We begin in Section 13.4.1

by presenting the standard Newton method for the solution of the nonlinear equations (13.2.23). In Section 13.4.2, we introduce an important modification of Newton's method for the optimization of coupled-cluster wave functions, based on the Møller–Plesset partitioning of the Hamiltonian. A particularly efficient implementation of this method, which employs the DIIS acceleration scheme, is presented in Section 13.4.3. In Section 13.4.4, we conclude our discussion of optimization techniques with some illustrations of the optimization of coupled-cluster wave functions.

13.4.1 NEWTON'S METHOD

Consider the left-hand side of the coupled-cluster amplitude equations (13.2.23). Its elements constitute a vector function

$$\Omega_\mu(\mathbf{t}) = \langle\mu|\exp(-\hat{T})\hat{H}\exp(\hat{T})|\text{HF}\rangle \tag{13.4.1}$$

that can be expanded around the set of amplitudes of the current iteration $\mathbf{t}^{(n)}$:

$$\Omega(\mathbf{t}^{(n)} + \Delta\mathbf{t}) = \Omega^{(0)}(\mathbf{t}^{(n)}) + \Omega^{(1)}(\mathbf{t}^{(n)})\Delta\mathbf{t} + \cdots \tag{13.4.2}$$

Here $\Omega^{(0)}(\mathbf{t}^{(n)})$ is the *vector function* and $\Omega^{(1)}(\mathbf{t}^{(n)})$ the *Jacobian matrix* calculated from the amplitudes $\mathbf{t}^{(n)}$:

$$\Omega_\mu^{(0)}(\mathbf{t}^{(n)}) = \langle\mu|\exp(-\hat{T}^{(n)})\hat{H}\exp(\hat{T}^{(n)})|\text{HF}\rangle \tag{13.4.3}$$

$$\Omega_{\mu\nu}^{(1)}(\mathbf{t}^{(n)}) = \langle\mu|\exp(-\hat{T}^{(n)})[\hat{H}, \hat{\tau}_\nu]\exp(\hat{T}^{(n)})|\text{HF}\rangle \tag{13.4.4}$$

In these expressions, the cluster operator $\hat{T}^{(n)}$ is constructed from the amplitudes $\mathbf{t}^{(n)}$. In Newton's method, an iterative scheme is established by setting the expanded vector function equal to zero and neglecting terms that are nonlinear in $\Delta\mathbf{t}$:

$$\Omega^{(1)}(\mathbf{t}^{(n)})\Delta\mathbf{t}^{(n)} = -\Omega^{(0)}(\mathbf{t}^{(n)}) \tag{13.4.5}$$

The improved estimate of the amplitudes now becomes

$$\mathbf{t}^{(n+1)} = \mathbf{t}^{(n)} + \Delta\mathbf{t}^{(n)} \tag{13.4.6}$$

Equations (13.4.5) and (13.4.6) are iterated until convergence in the amplitudes.

Newton's method requires the solution of the linear set of equations (13.4.5) in each iteration. Since the number of amplitudes is usually very large, an iterative technique must be invoked for solving these equations. We must then be prepared to carry out the linear transformations

$$\sigma_\mu = \sum_\nu \Omega_{\mu\nu}^{(1)} t_\nu \tag{13.4.7}$$

a large number of times, making the direct application of Newton's method expensive for coupled-cluster wave functions. Since more efficient techniques, which do not rely on the solution of the linear equations (13.4.5), have been developed for the optimization of coupled-cluster wave functions, the solution of these equations is not discussed further here. We note, however, that the linear transformations (13.4.7) cannot be avoided in response theory and that they also appear in coupled-cluster calculations of excited states, as discussed in Section 13.6.

13.4.2   THE PERTURBATION-BASED QUASI-NEWTON METHOD

We shall now set up a coupled-cluster optimization scheme that avoids the solution of the linear equations (13.4.5) in each iteration. In the canonical representation, we may invoke (13.2.44) to write the coupled-cluster vector function (13.4.3) and its Jacobian (13.4.4) of the nth iteration in the form

$$\Omega_\mu^{(0)}(\mathbf{t}^{(n)}) = \varepsilon_\mu t_\mu^{(n)} + \langle\mu|\exp(-\hat{T}^{(n)})\hat{\Phi}\exp(\hat{T}^{(n)})|\text{HF}\rangle \tag{13.4.8}$$

$$\Omega_{\mu\nu}^{(1)}(\mathbf{t}^{(n)}) = \varepsilon_\mu\delta_{\mu\nu} + \langle\mu|\exp(-\hat{T}^{(n)})[\hat{\Phi},\hat{\tau}_\nu]\exp(\hat{T}^{(n)})|\text{HF}\rangle \tag{13.4.9}$$

The Jacobian consists of a purely diagonal part, which involves differences of the orbital energies such as (13.2.45) and (13.2.46), and a nondiagonal part, which contains the fluctuation potential. In Møller–Plesset theory (see Sections 5.8 and 14.2), the orbital energies appear to zero order and the fluctuation potential to first order in the perturbation, suggesting that the Jacobian should be diagonally dominant in the canonical representation and providing some justification for ignoring the contributions from the fluctuation potential to the Jacobian. We then obtain from (13.4.5) the following *quasi-Newton equations* for the optimization of the coupled-cluster wave function [11]:

$$\Delta t_\mu^{(n)} = -\varepsilon_\mu^{-1}\Omega_\mu^{(0)}(\mathbf{t}^{(n)}) \tag{13.4.10}$$

Because of its relationship to Møller–Plesset theory (as discussed in Section 14.2), we shall refer to (13.4.10) as the *perturbation correction* to the coupled-cluster amplitudes. The first correction to the wave function $\Delta t_\mu^{(0)}$ is obtained from (13.4.10) using zero cluster amplitudes, and higher corrections are generated by repeating this procedure.

The perturbation-based quasi-Newton scheme (13.4.10) is fairly robust, usually converging to six decimal places in the energy in 10–20 iterations, although as many as 50 iterations may be needed in difficult cases. As for Hartree–Fock theory in Section 10.6.2, the convergence may be improved significantly by application of the DIIS acceleration scheme.

13.4.3   DIIS ACCELERATION OF THE QUASI-NEWTON METHOD

Within the DIIS framework [12], the amplitudes $\mathbf{t}^{(n+1)}$ generated in the nth iteration are not calculated directly from $\Delta\mathbf{t}^{(n)}$ according to (13.4.6). Instead, the new amplitudes $\mathbf{t}^{(n+1)}$ are obtained by a linear interpolation among the previous estimates of the amplitudes (13.4.6):

$$\mathbf{t}^{(n+1)} = \sum_{k=1}^{n} w_k(\mathbf{t}^{(k)} + \Delta\mathbf{t}^{(k)}) \tag{13.4.11}$$

where $\Delta\mathbf{t}^{(k)}$ is obtained from (13.4.10) and where the interpolation weights sum to unity:

$$\sum_{k=1}^{n} w_k = 1 \tag{13.4.12}$$

To determine the DIIS weights, we associate an error vector with each set of amplitudes $\mathbf{t}^{(k)}$. A natural choice would be to use the vector function $\mathbf{\Omega}^{(0)}(\mathbf{t}^{(k)})$ since this is the quantity to be minimized in our optimization. Alternatively, we may use the scaled vector function $\Delta\mathbf{t}^{(k)}$ of (13.4.10) as error vector and determine the interpolation coefficients by minimizing the norm of the averaged error vector

$$\Delta \mathbf{t}^{\mathrm{ave}} = \sum_{k=1}^{n} w_k \Delta \mathbf{t}^{(k)} \tag{13.4.13}$$

subject to condition (13.4.12) [13]. As discussed in Section 10.6.2, the interpolated amplitudes are obtained by the solution of a simple, linear set of equations of dimension $n + 1$. The new amplitudes $\mathbf{t}^{(n+1)}$ are then determined from (13.4.11).

It should be realized that DIIS by itself does not constitute an optimization algorithm. On its own, DIIS does not produce an amplitude vector that cannot be written as a linear combination of those already generated. Rather, DIIS provides an improved mechanism for utilizing the information contained in the quasi-Newton corrections (13.4.10). As we shall see in Section 13.4.4, the acceleration achieved by the DIIS procedure is often quite dramatic, significantly reducing the number of iterations needed for convergence.

13.4.4 EXAMPLES OF COUPLED-CLUSTER OPTIMIZATIONS

To illustrate the convergence rates of the perturbation-based quasi-Newton scheme with and without DIIS acceleration, we have in Table 13.2 listed the errors in the energy in each iteration of CCSD/cc-pVDZ calculations on the water molecule at the bond distances of R_{ref} and $2R_{\mathrm{ref}}$. At R_{ref} and $2R_{\mathrm{ref}}$, the converged CCSD energies are -76.238116 and -75.929633 E_{h}, respectively (see Table 5.11).

Table 13.2 Errors in the energy (E_{h}) in each iteration of CCSD calculations on the water molecule in the cc-pVDZ basis obtained using the perturbation-based quasi-Newton (QN) optimization algorithm with and without the DIIS acceleration scheme

	R_{ref}		$2R_{\mathrm{ref}}$	
Iteration	QN	QN with DIIS	QN	QN with DIIS
1	0.009387	0.009387	0.032698	0.032698
2	0.004536	0.004536	0.063687	0.063687
3	0.001233	0.000495	0.019485	0.036548
4	0.000488	0.000015	0.022514	0.008349
5	0.000188	0.000006	0.010350	0.006523
6	0.000078	−0.000004	0.009594	0.001866
7	0.000032	−0.000001	0.005557	0.001044
8	0.000014	−0.000001	0.004722	0.000706
9	0.000006	0.000000	0.003146	−0.000022
10	0.000003		0.002590	0.000073
11	0.000001		0.001871	0.000022
12	0.000000		0.001523	0.000003
13			0.001151	0.000000
14			0.000932	
15			0.000721	
16			0.000582	
⋮			⋮	
47			0.000001	
48			0.000001	
49			0.000000	

At the equilibrium distance, the Hartree–Fock determinant has an FCI weight of 0.941 and dominates the coupled-cluster wave function. The Jacobian is diagonally dominant and quite accurately represented by the orbital-energy differences in the quasi-Newton step (13.4.10). The quasi-Newton method converges rapidly, displaying a characteristic linear behaviour. The DIIS acceleration scheme improves the convergence moderately, reducing the number of iterations by about one-third. Since the DIIS scheme is not applied until after the second iteration, the first two iterations are identical in the two schemes.

At the stretched geometry with a bond distance of $2R_{\text{ref}}$, the situation is rather different. The Hartree–Fock determinant is no longer dominant, carrying a weight of only 0.589 in the FCI wave function. The Jacobian is not as diagonally dominant as at equilibrium and is no longer well represented by the zero-order approximation. The unaccelerated perturbation scheme converges slowly and requires 49 iterations for an accuracy of six decimal digits. The DIIS scheme improves the convergence significantly, requiring only 13 iterations for the same accuracy. In general, the DIIS-accelerated quasi-Newton method represents the most efficient and also the simplest procedure for obtaining the coupled-cluster amplitudes. However, in difficult cases, it may still be necessary to invoke the full Newton method to arrive at a converged set of amplitudes.

13.5 The coupled-cluster variational Lagrangian

For reasons of computational practicality and efficiency, molecular electronic coupled-cluster energies are determined using a nonvariational projection technique. The chief deficiency of this approach is not so much the loss of boundedness (since the coupled-cluster energy is nevertheless rather accurate), but the difficulties that it creates for the calculation of properties as the conditions for the Hellmann–Feynman theorem are not satisfied – even in the limit of a complete one-electron basis. Fortunately, as discussed in Section 4.2.8, this situation may be remedied by the construction of a variational Lagrangian [14]. In this formulation, the conditions of the Hellmann–Feynman theorem are fulfilled and molecular properties may be calculated by a procedure that is essentially the same as for variational wave functions. The Lagrangian formulation of the energy is also related to a variational treatment of coupled-cluster theory applicable to excited states, as discussed in Section 13.6.

13.5.1 THE COUPLED-CLUSTER LAGRANGIAN

We consider the variational reformulation of the coupled-cluster energy in the presence of a perturbation $\alpha\hat{V}$. The coupled-cluster energy is obtained from the standard expression

$$E(\alpha, \mathbf{t}) = \langle \text{HF}|(\hat{H} + \alpha\hat{V})\exp(\hat{T})\text{HF}\rangle \tag{13.5.1}$$

where $\mathbf{t}$ is a vector containing the connected cluster amplitudes, which satisfy the conditions

$$\langle \mu|\exp(-\hat{T})(\hat{H} + \alpha\hat{V})\exp(\hat{T})|\text{HF}\rangle = 0 \tag{13.5.2}$$

Following the discussion in Section 4.2.8, the total energy can be calculated from the Lagrangian [15]

$$L(\alpha, \mathbf{t}, \bar{\mathbf{t}}) = \langle \text{HF}|(\hat{H} + \alpha\hat{V})\exp(\hat{T})|\text{HF}\rangle + \sum_{\nu} \bar{t}_{\nu}\langle \nu|\exp(-\hat{T})(\hat{H} + \alpha\hat{V})\exp(\hat{T})|\text{HF}\rangle \tag{13.5.3}$$

where the cluster amplitudes and their multipliers are determined variationally:

$$\frac{\partial L}{\partial \bar{t}_\mu} = \langle \mu | \exp(-\hat{T})(\hat{H} + \alpha \hat{V}) \exp(\hat{T}) | \mathrm{HF} \rangle = 0 \tag{13.5.4}$$

$$\frac{\partial L}{\partial t_\mu} = \langle \mathrm{HF} | (\hat{H} + \alpha \hat{V}) \hat{\tau}_\mu \exp(\hat{T}) | \mathrm{HF} \rangle + \sum_\nu \bar{t}_\nu \langle \nu | \exp(-\hat{T}) [\hat{H} + \alpha \hat{V}, \hat{\tau}_\mu] \exp(\hat{T}) | \mathrm{HF} \rangle = 0 \tag{13.5.5}$$

Note that the variational conditions for the multipliers (13.5.4) are equivalent to the original equations for the amplitudes, and that, in the absence of the perturbation, these conditions reduce to the standard projected equations for the cluster amplitudes (13.2.23). At the same time, the variational conditions for the amplitudes (13.5.5) reduce to the equations

$$\langle \mathrm{HF} | \hat{H} \hat{\tau}_\mu \exp(\hat{T}^{(0)}) | \mathrm{HF} \rangle + \sum_\nu \bar{t}_\nu^{(0)} \langle \nu | \exp(-\hat{T}^{(0)}) [\hat{H}, \hat{\tau}_\mu] \exp(\hat{T}^{(0)}) | \mathrm{HF} \rangle = 0 \tag{13.5.6}$$

where $\hat{T}^{(0)}$ is the cluster operator and $\bar{t}_\mu^{(0)}$ the multipliers in the absence of the perturbation. This set of equations determines the multipliers.

The calculation of the zero-order multipliers and Lagrangian requires the solution of one set of linear equations. The resulting expression for the coupled-cluster energy is then variational with respect to the amplitudes as well as their multipliers. Note that the linear equations for the multipliers (13.5.6) are similar in structure to the Newton equations for the amplitudes (13.4.5), both containing the coupled-cluster Jacobian (13.4.4).

13.5.2 THE HELLMANN–FEYNMAN THEOREM

Having arrived at a variational expression for the coupled-cluster energy (13.5.3), we are now in a position to consider the evaluation of the gradient with respect to the perturbational parameter α according to the Hellmann–Feynman theorem. According to the discussion in Section 4.2.8, the total derivative of the energy with respect to α corresponds to the partial derivative of the Lagrangian with respect the same parameter. The first-order change in the energy due to the perturbation therefore becomes

$$\frac{\mathrm{d}E}{\mathrm{d}\alpha}\bigg|_{\alpha=0} = \frac{\partial L}{\partial \alpha}\bigg|_{\alpha=0} = \langle \mathrm{HF} | \hat{V} \exp(\hat{T}^{(0)}) | \mathrm{HF} \rangle + \sum_\nu \bar{t}_\nu^{(0)} \langle \nu | \exp(-\hat{T}^{(0)}) \hat{V} \exp(\hat{T}^{(0)}) | \mathrm{HF} \rangle \tag{13.5.7}$$

The reader may verify this expression by taking the total (not partial) derivative of the Lagrangian and eliminating all terms that involve derivatives of the amplitudes, using the conditions (13.5.4) and (13.5.5) at $\alpha = 0$.

The usefulness of the Lagrangian expression (13.5.7) stems from the fact that, like the usual Hellmann–Feynman expression, it does not contain the perturbed amplitudes. To make the connection with the Hellmann–Feynman theorem more transparent, we write (13.5.7) as a transition expectation value

$$\frac{\mathrm{d}E}{\mathrm{d}\alpha}\bigg|_{\alpha=0} = \langle \Lambda | \hat{V} | \mathrm{CC} \rangle \tag{13.5.8}$$

where the bra state is given by

$$\langle \Lambda | = \langle \mathrm{HF} | + \sum_\nu \bar{t}_\nu^{(0)} \langle \nu | \exp(-\hat{T}^{(0)}) \tag{13.5.9}$$

Equation (13.5.8), which represents the generalization of the Hellmann–Feynman theorem to coupled-cluster wave functions, is shown in Exercise 13.1 to give size-extensive first-order properties. For variational wave functions, the Hellmann–Feynman theorem contains the real average value of the operator $\hat{V}$ (4.2.51) rather than a transition expectation value as in (13.5.8). Likewise, to ensure real properties, we may in coupled-cluster theory work in terms of the manifestly real expression

$$\left.\frac{dE}{d\alpha}\right|_{\alpha=0} = \frac{1}{2}(\langle\Lambda|\hat{V}|\text{CC}\rangle + \langle\Lambda|\hat{V}|\text{CC}\rangle^*) \tag{13.5.10}$$

rather than in terms of (13.5.8). In the following, however, we shall assume that all quantities are real and continue to work with (13.5.8). We further note that, in the absence of the perturbation, the coupled-cluster energy may also be expressed in terms of the bra state (13.5.9):

$$E_{\text{var}} = \langle\Lambda|\hat{H}|\text{CC}\rangle \tag{13.5.11}$$

An immediate computational advantage of this expression over the standard form

$$E = \langle\text{HF}|\hat{H}|\text{CC}\rangle \tag{13.5.12}$$

is that (13.5.11) is variational and therefore less sensitive to numerical errors in the amplitudes than (13.5.12). Moreover, the form of the Lagrangian (13.5.11) suggests that the coupled-cluster energy may be obtained as a solution to a variational problem. In Section 13.6, we explore this idea further and demonstrate that it leads to an eigenvalue problem, from which we may extract size-extensive energies for excited states.

13.5.3 LAGRANGIAN DENSITY MATRICES

Let us assume that, in the spin-orbital basis, the Hamiltonian operator may be written in the form

$$\hat{H} = \sum_{PQ} h_{PQ} a_P^\dagger a_Q + \frac{1}{2} \sum_{PQRS} g_{PQRS} a_P^\dagger a_R^\dagger a_S a_Q + h_{\text{nuc}} \tag{13.5.13}$$

Calculating the energy according to (13.5.12), we obtain

$$E = \sum_{PQ} \overline{D}_{PQ} h_{PQ} + \frac{1}{2} \sum_{PQRS} \overline{d}_{PQRS} g_{PQRS} + h_{\text{nuc}} \tag{13.5.14}$$

where we have introduced the one- and two-electron *coupled-cluster density matrices*

$$\overline{D}_{PQ} = \langle\text{HF}|a_P^\dagger a_Q|\text{CC}\rangle \tag{13.5.15}$$

$$\overline{d}_{PQRS} = \langle\text{HF}|a_P^\dagger a_R^\dagger a_S a_Q|\text{CC}\rangle \tag{13.5.16}$$

Unlike the density matrices discussed in Section 1.7, the coupled-cluster density matrices are not Hermitian and may give complex eigenvalues upon diagonalization. For the calculation of the energy and in general the expectation value of any Hermitian operator, it is sufficient to consider the real symmetric part of the coupled-cluster density matrix, the eigenvalues of which are always real.

As discussed in Section 13.5.2, for the calculation of first-order properties, it is in general more useful to consider the Lagrangian formulation of the coupled-cluster energy. Following (13.5.11), we may then calculate the energy according to the expression

$$E_{\text{var}} = \sum_{PQ} \overline{D}_{PQ}^\Lambda h_{PQ} + \tfrac{1}{2} \sum_{PQRS} \overline{d}_{PQRS}^\Lambda g_{PQRS} + h_{\text{nuc}} \tag{13.5.17}$$

where we have introduced the elements of the *coupled-cluster Lagrangian density matrices* [16]:

$$\overline{D}_{PQ}^\Lambda = \langle \Lambda | a_P^\dagger a_Q | \text{CC} \rangle \tag{13.5.18}$$

$$\overline{d}_{PQRS}^\Lambda = \langle \Lambda | a_P^\dagger a_R^\dagger a_S a_Q | \text{CC} \rangle \tag{13.5.19}$$

In terms of the Lagrangian densities, we may calculate coupled-cluster first-order properties in the same way as for variational wave functions, contracting the density-matrix elements with the molecular integrals [17]. The Lagrangian density matrices are also known as the *variational* or *relaxed density matrices*. For a closed-shell CCSD wave function, an expression for the one-electron variational density matrix is derived in Exercise 13.5.

13.6 The equation-of-motion coupled-cluster method

In coupled-cluster theory, we arrive at a correlated description of the electronic ground state by applying an exponential operator to an uncorrelated wave function, usually the Hartree–Fock state:

$$|\text{CC}\rangle = \exp(\hat{T})|\text{HF}\rangle \tag{13.6.1}$$

Assuming that a truncated coupled-cluster operator $\hat{T}$ has been chosen and optimized for the electronic ground state as described in the previous sections of this chapter, we now turn our attention to *excited states*, asking how such states can be investigated within the framework of coupled-cluster theory. In particular, we would like to set up a theory according to which size-extensive excited-state energies can be calculated that are of the same quality as the ground state. Ideally, we would like the excited states and the ground state to constitute an orthonormal set, thus making the identification with the true electronic states unambiguous.

13.6.1 THE EQUATION-OF-MOTION COUPLED-CLUSTER MODEL

A somewhat naive approach to excited-state coupled-cluster theory would be to carry out a separate, independent calculation for each state of interest, using in each case some appropriately chosen zero-order reference determinant. Although this approach has the merit of treating ground and excited states in a uniform manner, there are several problems associated with it. First, it is usually difficult and often impossible to determine an adequate zero-order reference determinant for an excited state. Second, the individual calculation of separate states is expensive – each state requires at least the same computational effort as does the ground state. Third, the states generated in this way are not orthogonal, making it difficult to identify the excited states.

The problems associated with nonorthogonality may be solved by resorting to a linear, CI-type parametrization of the excited states. Obviously, the linear variational space must now contain, as a special case, the coupled-cluster ground state $|\text{CC}\rangle$. We thus explore the possibility of calculating the excited states by a linear expansion in the space spanned by all states of the form

$$|\mathbf{c}\rangle = \sum_{\mu} c_{\mu}\hat{\tau}_{\mu}|\text{CC}\rangle = \exp(\hat{T})\sum_{\mu} c_{\mu}\hat{\tau}_{\mu}|\text{HF}\rangle \tag{13.6.2}$$

where the summation is over the identity operator $\hat{\tau}_0$ as well as over the excitation operators present in the cluster operator $\hat{T}$. Together with the recipe for determining the expansion coefficients discussed in Section 13.6.2, the expansion (13.6.2) constitutes the basis for the *equation-of-motion coupled-cluster (EOM-CC) model* [18]. Note that the EOM-CC excited states (13.6.2) may be regarded as generated from a conventional CI state

$$|\mathbf{C}\rangle = \sum_{\mu} C_{\mu}\hat{\tau}_{\mu}|\text{HF}\rangle \tag{13.6.3}$$

by the application of an exponential operator containing the ground-state cluster amplitudes, in the same way as the coupled-cluster ground state (for which all coefficients except c_0 are zero) is obtained from the Hartree–Fock state by the application of the exponential operator.

Comparing the EOM-CC and CI representations of electronic states, we first note that, whereas the EOM-CC ground state is represented by a single term, the CI ground state contains a large number of terms. Also, whereas the EOM-CC ground state is size-extensive and contains contributions from all determinants of the FCI wave function, the CI ground state is usually obtained from a truncated CI expansion and is not size-extensive. Regarding excited states, the CI model suffers from the same deficiencies as for the ground state – the description is neither compact nor size-extensive. The CI coefficients have the dual purpose of describing the zero-order character of the excited state and dynamical correlation. In the EOM-CC model, on the other hand, the linear variational parameters reflect directly the differences between the ground state and the excited states, while the exponential describes the correlation effects common to all states.

Of course, the distinction between the description of electron correlation and the character of the excitation is not a sharp one, even in EOM-CC theory. In particular, since the cluster operator in (13.6.2) has been optimized for the ground state (13.6.1), it does not provide an optimal description of correlation in the excited states. Some adjustment for the description of correlation must therefore be made by the expansion coefficients. Still, since the EOM-CC ground-state expansion contains a single term, we may interpret the remaining coefficients as reflecting the character of the excitation. Accordingly, we shall refer to excited states as dominated by a single or a double replacement, depending on whether the expansion coefficients in (13.6.2) are dominated by the singles or doubles coefficients.

Obviously, the EOM-CC model will be useful only if we are able to develop a practical method for calculating electronic states that are, in some sense, orthogonal to one another as well as size-extensive. In the remainder of this section, we shall see that it is indeed possible to satisfy these requirements by the application of the variation principle. We begin by introducing the EOM-CC eigenvalue problem in Section 13.6.2. In Section 13.6.3, we investigate the structure of the EOM-CC Hamiltonian matrix; next, in Section 13.6.4, we go on to consider the eigenvalues and eigenvectors. Finally, in Section 13.6.5, we demonstrate that the EOM-CC energies are size-extensive.

13.6.2 THE EOM-CC EIGENVALUE PROBLEM

As noted in Section 13.6.1, we would like the EOM-CC excited states (13.6.2) to be orthonormal. Orthonormality in the usual sense presents problems of the sort discussed in Section 13.1.4. Indeed, even the calculation of the norm of the ground state $\langle CC|CC\rangle$ is cumbersome since the excitation operators in $\hat{T}$ do not commute with the de-excitation operators in $\hat{T}^\dagger$. We solve this problem by resorting to biorthogonality, expanding the bra states in a set of configurations that, together with the ket states (13.6.2), constitute a biorthonormal set. Adopting the notation

$$|A) = \exp(\hat{T})|A\rangle \tag{13.6.4}$$

$$(A| = \langle A|\exp(-\hat{T}) \tag{13.6.5}$$

the EOM-CC basis vectors may be written as

$$|\mu) = \exp(\hat{T})|\mu\rangle = \exp(\hat{T})\hat{\tau}_\mu|\text{HF}\rangle \tag{13.6.6}$$

$$(\mu| = \langle\mu|\exp(-\hat{T}) = \langle\text{HF}|\hat{\tau}_\mu^\dagger\exp(-\hat{T}) \tag{13.6.7}$$

Note that, in addition to the excited projection manifold, the basis includes the states

$$|0) = |\text{HF}) = |\text{CC}\rangle \tag{13.6.8}$$

$$(0| = (\text{HF}| = \langle\text{HF}| \tag{13.6.9}$$

in accordance with our interpretation of $\hat{\tau}_0$ as the identity operator. In the spin-orbital basis, biorthonormality follows from the orthonormality of the determinants $|\mu\rangle$:

$$(\mu|\nu) = \langle\mu|\nu\rangle = \delta_{\mu\nu} \tag{13.6.10}$$

In spin-adapted theory, it is also convenient to employ a biorthogonal basis for $\langle\mu|$ and $|\mu\rangle$; see Section 13.7.5. In such cases, condition (13.6.10) still holds and no extra complications are introduced in the theory as presented here.

In the biorthonormal basis (13.6.6) and (13.6.7), we may set up CI-like wave functions of the form

$$|\mathbf{c}) = \sum_\mu c_\mu|\mu) \tag{13.6.11}$$

$$(\bar{\mathbf{c}}| = \sum_\mu \bar{c}_\mu(\mu| \tag{13.6.12}$$

and express the energy as a pseudo-expectation value

$$E(\mathbf{c}, \bar{\mathbf{c}}) = \frac{(\bar{\mathbf{c}}|\hat{H}|\mathbf{c})}{(\bar{\mathbf{c}}|\mathbf{c})} \tag{13.6.13}$$

Overbars are used for the bra coefficients since they are numerically different from the ket coefficients. The EOM-CC states are now determined by applying the variation principle. Differentiation with respect to the ket and bra coefficients yields the unsymmetric eigenvalue problems (assuming real expansion coefficients)

$$\mathbf{Hc} = E\mathbf{c} \tag{13.6.14}$$

$$\bar{\mathbf{c}}^\text{T}\mathbf{H} = \bar{\mathbf{c}}^\text{T}E \tag{13.6.15}$$

where $\mathbf{c}$ and $\bar{\mathbf{c}}$ are column vectors containing the expansion coefficients, and the elements of the unsymmetric real matrix $\mathbf{H}$ are given by

$$H_{\mu\nu} = (\mu|\hat{H}|\nu) \tag{13.6.16}$$

The eigenvectors of the Hamiltonian can be chosen such that

$$\bar{\mathbf{c}}_i^{\mathrm{T}}\mathbf{c}_j = \delta_{ij} \tag{13.6.17}$$

We have thus reduced the problem of calculating the EOM-CC states to the problem of diagonalizing an unsymmetric matrix. As is usual in CI theory, no attempt is made at a complete diagonalization of the Hamiltonian since only a few of the lowest energies and states are of interest. We shall return to the EOM-CC eigenvalue problem in Section 13.6.4, following a discussion of the structure of the Hamiltonian matrix in Section 13.6.3. We here note that, since (13.6.16) is not the matrix representation of a Hermitian Hamiltonian in a linear variational space, the eigenvalues of (13.6.14) and (13.6.15) are not upper bounds to the exact energies and may therefore become complex. In the limit of a full expansion, however, the FCI eigenvalues are recovered. In practice, if $|\mathbf{c}\rangle$ provides a good representation of the excited state, no imaginary components are encountered in the solution.

From the structure of the ket and bra configurations (13.6.6) and (13.6.7), we find that the EOM-CC expectation value (13.6.13) may be written in the equivalent form

$$E(\mathbf{c}, \bar{\mathbf{c}}) = \frac{\langle\bar{\mathbf{c}}|\hat{H}^{\mathrm{T}}|\mathbf{c}\rangle}{\langle\bar{\mathbf{c}}|\mathbf{c}\rangle} = \frac{\langle\bar{\mathbf{c}}|\exp(-\hat{T})\hat{H}\exp(\hat{T})|\mathbf{c}\rangle}{\langle\bar{\mathbf{c}}|\mathbf{c}\rangle} \tag{13.6.18}$$

where $|\mathbf{c}\rangle$ and $\langle\bar{\mathbf{c}}|$ are conventional CI expansions

$$|\mathbf{c}\rangle = \sum_{\mu} c_{\mu}|\mu\rangle \tag{13.6.19}$$

$$\langle\bar{\mathbf{c}}| = \sum_{\mu} \bar{c}_{\mu}\langle\mu| \tag{13.6.20}$$

We may thus regard EOM-CC theory as conventional CI theory with a similarity-transformed Hamiltonian: The similarity-transformed Hamiltonian carries the information about electron correlation and the configuration expansions carry the information about the excitation structure of the electronic states.

13.6.3 THE SIMILARITY-TRANSFORMED HAMILTONIAN AND THE JACOBIAN

For an optimized coupled-cluster ground state, the matrix representation of the similarity-transformed Hamiltonian has a special structure. Since its first column represents the left-hand side of the ground-state coupled-cluster equations (13.2.22) and (13.2.23), we obtain

$$(\mu|\hat{H}|\mathrm{HF}) = \begin{cases} E_0 & \mu = 0 \\ 0 & \mu > 0 \end{cases} \tag{13.6.21}$$

where E_0 is the ground-state energy. Restricting ourselves to the block of the Hamiltonian matrix that refers to the excited projection manifold, we obtain

$$H_{\mu\nu} = (\mu|\hat{H}|\nu) = \langle\mathrm{HF}|\hat{\tau}_{\mu}^{\dagger}\hat{H}^{\mathrm{T}}\hat{\tau}_{\nu}|\mathrm{HF}\rangle = \langle\mathrm{HF}|\hat{\tau}_{\mu}^{\dagger}[\hat{H}^{\mathrm{T}}, \hat{\tau}_{\nu}]|\mathrm{HF}\rangle + \langle\mathrm{HF}|\hat{\tau}_{\mu}^{\dagger}\hat{\tau}_{\nu}\hat{H}^{\mathrm{T}}|\mathrm{HF}\rangle \tag{13.6.22}$$

The first term may be written in the form

$$\langle \text{HF}|\hat{\tau}_\mu^\dagger[\hat{H}^{\text{T}}, \hat{\tau}_v]|\text{HF}\rangle = (\mu|[\hat{H}, \hat{\tau}_v]|\text{HF}) \tag{13.6.23}$$

and for the second term we invoke the resolution of the identity:

$$\langle \text{HF}|\hat{\tau}_\mu^\dagger\hat{\tau}_v\hat{H}^{\text{T}}|\text{HF}\rangle = \sum_\lambda \langle \text{HF}|\hat{\tau}_\mu^\dagger\hat{\tau}_v|\lambda\rangle\langle\lambda|\hat{H}^{\text{T}}|\text{HF}\rangle = \langle \text{HF}|\hat{\tau}_\mu^\dagger\hat{\tau}_v|\text{HF}\rangle\langle \text{HF}|\hat{H}^{\text{T}}|\text{HF}\rangle = \delta_{\mu v}E_0$$
$$\tag{13.6.24}$$

In deriving (13.6.24), we have first assumed that the projection manifold is closed under de-excitations and then used (13.6.21) to restrict the summation to the Hartree–Fock state. The matrix elements of the similarity-transformed Hamiltonian (13.6.22) that correspond to the excited-state manifold may therefore be written in the form

$$H_{\mu v} = (\mu|[\hat{H}, \hat{\tau}_v]|\text{HF}) + \delta_{\mu v}E_0 \tag{13.6.25}$$

For an optimized coupled-cluster state, we may thus write the EOM-CC Hamiltonian matrix in the partitioned form

$$\mathbf{H} = \begin{pmatrix} 0 & \boldsymbol{\eta}^{\text{T}} \\ \mathbf{0} & \mathbf{A} \end{pmatrix} + E_0\mathbf{1} \tag{13.6.26}$$

where the elements of the column vector $\boldsymbol{\eta}$ are given by

$$\eta_\mu = (\text{HF}|\hat{H}|\mu) \tag{13.6.27}$$

and the elements of the *coupled-cluster Jacobian matrix* $\mathbf{A}$ are given by

$$A_{\mu v} = (\mu|[\hat{H}, \hat{\tau}_v]|\text{HF}) \tag{13.6.28}$$

The unsymmetric Jacobian matrix has previously appeared in the optimization of the coupled-cluster wave function (13.4.4) and in the calculation of the coupled-cluster Lagrange multipliers (13.5.6).

13.6.4 SOLUTION OF THE EOM-CC EIGENVALUE PROBLEM

We now turn our attention to the EOM-CC eigenvalue problems (13.6.14) and (13.6.15). With each eigenvalue, there is an associated pair of right and left eigenvectors, which we write in the following way

$$\mathbf{c} = \begin{pmatrix} s \\ \mathbf{t} \end{pmatrix} \tag{13.6.29}$$

$$\bar{\mathbf{c}} = \begin{pmatrix} \bar{s} \\ \bar{\mathbf{t}} \end{pmatrix} \tag{13.6.30}$$

where s and $\bar{s}$ are the coefficients associated with the reference state $\mu = 0$ and $\mathbf{t}$ and $\bar{\mathbf{t}}$ contain the coefficients associated with the excited configurations $\mu > 0$. For convenience, we shall consider the level-shifted equations

$$\Delta\mathbf{Hc} = \Delta E\mathbf{c} \tag{13.6.31}$$

$$\bar{\mathbf{c}}^{\text{T}}\Delta\mathbf{H} = \bar{\mathbf{c}}^{\text{T}}\Delta E \tag{13.6.32}$$

where the Hamiltonian and the eigenvalues are given by

$$\Delta \mathbf{H} = \mathbf{H} - E_0 \mathbf{1} = \begin{pmatrix} 0 & \boldsymbol{\eta}^{\mathrm{T}} \\ \mathbf{0} & \mathbf{A} \end{pmatrix} \tag{13.6.33}$$

$$\Delta E = E - E_0 \tag{13.6.34}$$

The eigenvectors of $\Delta \mathbf{H}$ are thus the same as for $\mathbf{H}$ and its nonzero eigenvalues correspond to the excitation energies from the ground state.

Consider first the eigenvectors $\mathbf{c}_0$ and $\bar{\mathbf{c}}_0$ associated with the electronic ground state. From the structure of the effective Hamiltonian (13.6.33), we see immediately that the vector

$$\mathbf{c}_0 = \begin{pmatrix} 1 \\ \mathbf{0} \end{pmatrix} \tag{13.6.35}$$

represents a right eigenvector with eigenvalue $\Delta E_0 = 0$. Clearly, this solution represents the electronic ground state. To ensure normalization (13.6.17), the associated left eigenvector is written in the form

$$\bar{\mathbf{c}}_0 = \begin{pmatrix} 1 \\ \bar{\mathbf{t}}_0 \end{pmatrix} \tag{13.6.36}$$

with the first element $\bar{s}_0 = 1$. Inserting this expression in the eigenvalue equations (13.6.32) with $\Delta E_0 = 0$, we obtain the partitioned equations

$$(1 \ \ \bar{\mathbf{t}}_0^{\mathrm{T}}) \begin{pmatrix} 0 & \boldsymbol{\eta}^{\mathrm{T}} \\ \mathbf{0} & \mathbf{A} \end{pmatrix} = (0 \ \ \mathbf{0}) \tag{13.6.37}$$

which upon expansion yields the linear set of equations

$$\bar{\mathbf{t}}_0^{\mathrm{T}} \mathbf{A} = -\boldsymbol{\eta}^{\mathrm{T}} \tag{13.6.38}$$

for the undetermined coefficients in the left eigenvector (13.6.36). These equations are identical to the multiplier equations for the variational Lagrangian (13.5.6). The construction of the variational coupled-cluster Lagrangian (13.5.11) is thus equivalent to calculating the expectation value of the electronic ground state in the EOM-CC formalism, with the Lagrangian state $\langle \Lambda |$ corresponding to the EOM-CC left eigenvector for the ground state.

We next consider the excited-state solutions to the EOM-CC eigenvalue equations, which for the Kth state may be written in the form

$$\begin{pmatrix} 0 & \boldsymbol{\eta}^{\mathrm{T}} \\ \mathbf{0} & \mathbf{A} \end{pmatrix} \begin{pmatrix} s_K \\ \mathbf{t}_K \end{pmatrix} = \Delta E_K \begin{pmatrix} s_K \\ \mathbf{t}_K \end{pmatrix} \tag{13.6.39}$$

$$(\bar{s}_K \ \ \bar{\mathbf{t}}_K^{\mathrm{T}}) \begin{pmatrix} 0 & \boldsymbol{\eta}^{\mathrm{T}} \\ \mathbf{0} & \mathbf{A} \end{pmatrix} = (\bar{s}_K \ \ \bar{\mathbf{t}}_K^{\mathrm{T}}) \Delta E_K \tag{13.6.40}$$

Expanding, we find that the excitation energies are obtained from the unsymmetric eigenvalue problems

$$\mathbf{A} \mathbf{t}_K = \Delta E_K \mathbf{t}_K \tag{13.6.41}$$

$$\bar{\mathbf{t}}_K^{\mathrm{T}} \mathbf{A} = \bar{\mathbf{t}}_K^{\mathrm{T}} \Delta E_K \tag{13.6.42}$$

and that the reference-state coefficients are given by

$$s_K = \Delta E_K^{-1} \boldsymbol{\eta}^T \mathbf{t}_K \tag{13.6.43}$$

$$\bar{s}_K = 0 \tag{13.6.44}$$

The EOM-CC excitation energies therefore correspond to the eigenvalues of the coupled-cluster Jacobian matrix $\mathbf{A}$. Since the Jacobian is unsymmetric, there is no mathematical guarantee that the calculated eigenvalues are real. In practice, however, this is not a problem and the calculated excitation energies are real for any reasonably accurate ground-state wave function.

The key assumption underlying the calculation of excitation energies as eigenvalues of the Jacobian (13.6.41) and (13.6.42) is that the first column vector of the EOM-CC Hamiltonian (13.6.33) vanishes. This assumption is satisfied for the standard coupled-cluster models such as CCSD and CCSDT, for which the first column contains the coupled-cluster vector function (13.2.23). By contrast, in other models, where certain terms are omitted as in quadratic CI (discussed in Section 13.8.2), the first column does not vanish and EOM-CC theory can no longer be applied.

13.6.5 SIZE-EXTENSIVITY OF THE EOM-CC ENERGIES

Having developed EOM-CC theory, we now turn to the crucial question of size-extensivity. The ground-state energy corresponds to the usual coupled-cluster energy and is necessarily size-extensive. We therefore consider only the excited-state energies, for which it is sufficient to analyse the eigenvalues of the Jacobian matrix (13.6.28) [19].

Consider the compound system AB consisting of the two noninteracting subsystems A and B. In a basis where $\{\hat{\tau}_A^\dagger, \hat{\tau}_B^\dagger, \hat{\tau}_{AB}^\dagger\}$ and $\{\hat{\tau}_A, \hat{\tau}_B, \hat{\tau}_{AB}\}$ constitute the row and column labels, respectively, the Jacobian may be written in the partitioned form

$$\mathbf{A} = \begin{pmatrix} \mathbf{A}_{A,A} & \mathbf{A}_{A,B} & \mathbf{A}_{A,AB} \\ \mathbf{A}_{B,A} & \mathbf{A}_{B,B} & \mathbf{A}_{B,AB} \\ \mathbf{A}_{AB,A} & \mathbf{A}_{AB,B} & \mathbf{A}_{AB,AB} \end{pmatrix} \tag{13.6.45}$$

It is straightforward to show that the blocks $\mathbf{A}_{A,B}$, $\mathbf{A}_{B,A}$, $\mathbf{A}_{AB,A}$ and $\mathbf{A}_{AB,B}$ vanish for noninteracting systems. For example, the elements of $\mathbf{A}_{A,B}$ are zero since

$$A_{A,B} = \langle \mathrm{HF} | \hat{\tau}_A^\dagger \exp(-\hat{T}_{AB}) [\hat{H}_A + \hat{H}_B, \hat{\tau}_B] \exp(\hat{T}_{AB}) | \mathrm{HF} \rangle$$

$$= \langle \mathrm{HF} | \hat{\tau}_A^\dagger \exp(-\hat{T}_B) [\hat{H}_B, \hat{\tau}_B] \exp(\hat{T}_B) | \mathrm{HF} \rangle = 0 \tag{13.6.46}$$

In general, therefore, the Jacobian has the following structure for noninteracting systems:

$$\mathbf{A} = \begin{pmatrix} \mathbf{A}_{A,A} & \mathbf{0} & \mathbf{A}_{A,AB} \\ \mathbf{0} & \mathbf{A}_{B,B} & \mathbf{A}_{B,AB} \\ \mathbf{0} & \mathbf{0} & \mathbf{A}_{AB,AB} \end{pmatrix} \tag{13.6.47}$$

Because of the special block structure of the Jacobian, its eigenvalues are identical to the eigenvalues of $\mathbf{A}_{A,A}$, $\mathbf{A}_{B,B}$ and $\mathbf{A}_{AB,AB}$, representing excitations within system A, excitations within system B and excitations that refer to both systems. Adding to the ground-state energy the

excitation energies that refer to a single system, we obtain size-extensive excited-state energies, associated with physical processes that occur within a single system. The EOM-CC theory
thus represents an interesting illustration of the fact that, through a careful selection of the
configuration basis, size-extensive energies can be generated within a truncated, linear variational
approach.

Let us now consider the excitation energies of $\mathbf{A}_{AB,AB}$, the elements of which may be written as

$$
\begin{aligned}
A_{A_iB_j,A_kB_l} &= \langle HF|\hat{\tau}^\dagger_{A_iB_j} \exp(-\hat{T}_{AB})[\hat{H}_A + \hat{H}_B, \hat{\tau}_{A_kB_l}] \exp(\hat{T}_{AB})|HF\rangle \\
&= \langle HF|\hat{\tau}^\dagger_{A_iB_j} \exp(-\hat{T}_A)[\hat{H}_A, \hat{\tau}_{A_kB_l}] \exp(\hat{T}_A)|HF\rangle \\
&\quad + \langle HF|\hat{\tau}^\dagger_{A_iB_j} \exp(-\hat{T}_B)[\hat{H}_B, \hat{\tau}_{A_kB_l}] \exp(\hat{T}_B)|HF\rangle \\
&= [\mathbf{A}_{A,A}]_{ik}\delta_{B_jB_l} + \delta_{A_iA_k}[\mathbf{A}_{B,B}]_{jl}
\end{aligned}
\tag{13.6.48}
$$

For a full EOM-CC expansion, where the A_iB_j correspond to a direct-product basis, we obtain

$$
\mathbf{A}_{AB,AB} = \mathbf{A}_{A,A} \otimes \mathbf{1}_{B,B} + \mathbf{1}_{A,A} \otimes \mathbf{A}_{B,B}
\tag{13.6.49}
$$

with excitation energies (eigenvalues) that are equal to the sum of the excitation energies in
the systems A and B. In this limit, therefore, size-extensivity is retained and we obtain the same
excitation energies as in FCI. For truncated EOM-CC expansions, however, $\mathbf{A}_{AB,AB}$ does not reduce
to a direct-product matrix and its eigenvalues will no longer represent size-extensive energies.

Our analysis has shown that, for a truncated expansion, the coupled-cluster Jacobian of two
noninteracting systems contains all the excitation energies of the individual systems but in addition
a set of solutions that are not size-extensive, representing simultaneous excitations in the two
systems. For those eigenvalues of $\mathbf{A}_{A,A}$ and $\mathbf{A}_{B,B}$ that are well separated from those of $\mathbf{A}_{AB,AB}$,
the size-extensive excitations of the noninteracting systems will remain nearly decoupled also
when interactions are introduced and all blocks of $\mathbf{A}$ become nonzero. However, if there are near
degeneracies among the eigenvalues of $\mathbf{A}_{AB,AB}$ and those of $\mathbf{A}_{A,A}$ or $\mathbf{A}_{B,B}$, then these will mix even
for small interactions, thus blurring the distinction between the solutions that are size-extensive
and those that are not. If only the lowest excitation energies are studied, the presence of solutions
that are not size-extensive usually presents no problems. Nevertheless, for an extended interacting
system, the presence of such solutions may affect also the lower excitation energies, making the
calculations of such excitation energies unreliable.

13.6.6 FINAL COMMENTS

For the study of excited states and molecular properties, the EOM-CC model constitutes a
conceptually simple approach closely related to the CI model, with the emphasis shifted away
from excitations from determinants towards excitations from a more general state. However,
the EOM-CC approach is somewhat restricted in the sense that it can be applied only to the
standard coupled-cluster models such as CCSD and CCSDT. For the calculation of molecular
properties and excitation energies, the response-function approach – based on the solution of the
time-dependent Schrödinger equation – constitutes a more general framework, also encompassing,
for example, the quadratic CI model of Section 13.8.2 and the iterative hybrid CC2 and CC3
models of Section 14.6.2 [20,21].

A discussion of the response-function model is beyond the scope of the present monograph. We
here note that, for the standard coupled-cluster models such as CCSD, the same size-extensive

excitation energies are obtained in the response-function and EOM-CC approaches, but that the molecular properties obtained in the two approaches differ. In short, the close relationship to the CI model leads to certain deficiencies in the EOM-CC model, resulting, for example, in a lack of size-extensivity in the calculation of molecular properties such as the polarizability.

13.7 The closed-shell CCSD model

We now consider an important special case of coupled-cluster theory – the CCSD model for a closed-shell system [5]. The cluster operator is restricted to contain only singles and doubles operators of singlet symmetry. The importance of this model can be seen from the fact that, for any coupled-cluster wave function, the singles and doubles amplitudes are the only amplitudes that contribute directly to the coupled-cluster energy. Of the two types of amplitudes, the doubles are more important for the total energy, as can be understood from a consideration of the Brillouin theorem. However, although the singles may have little effect on the total energy, they are important for the description of even simple molecular properties such as the dipole moment, for which it is important to treat singles and doubles at the same level. This is also true when the CCSD wave function is further refined by a perturbational correction or when the CCSD wave function is used for the calculation of excitation energies. In such cases, the singles play an important role in describing the dominant excitation processes as well as the relaxation of the orbitals and the wave function.

13.7.1 PARAMETRIZATION OF THE CCSD CLUSTER OPERATOR

We wish to generate a singlet CCSD state by the application of the exponential cluster operator to a closed-shell Hartree–Fock state:

$$|\text{CC}\rangle = \exp(\hat{T}_1 + \hat{T}_2)|\text{HF}\rangle \qquad (13.7.1)$$

Therefore, only those terms in $\hat{T}_1$ and $\hat{T}_2$ that transform as singlet operators should be retained in the cluster operator. From the discussion of tensor operators in Section 2.3, it follows that the singlet cluster operators should satisfy (2.3.1) and (2.3.2):

$$[\hat{S}_\pm, \hat{T}_i] = 0 \qquad (13.7.2)$$

$$[\hat{S}_z, \hat{T}_i] = 0 \qquad (13.7.3)$$

These equations impose constraints on the singles and doubles cluster amplitudes in the spin-orbital basis.

We first consider the singles operator $\hat{T}_1$. Inserting this operator in (13.7.3), we obtain the requirement

$$[\hat{S}_z, \hat{T}_1] = \sum_{a\sigma i\tau} t_{i\tau}^{a\sigma}[\hat{S}_z, a_{a\sigma}^\dagger a_{i\tau}] = \sum_{a\sigma i\tau} t_{i\tau}^{a\sigma}([\hat{S}_z, a_{a\sigma}^\dagger]a_{i\tau} + a_{a\sigma}^\dagger[\hat{S}_z, a_{i\tau}])$$

$$= \sum_{a\sigma i\tau} (\sigma - \tau) t_{i\tau}^{a\sigma} a_{a\sigma}^\dagger a_{i\tau} = 0 \qquad (13.7.4)$$

where σ and τ run over $-\frac{1}{2}$ and $+\frac{1}{2}$. The requirement (13.7.3) is thus satisfied if $\sigma = \tau$. With this restriction, the first set of requirements (13.7.2) may be written in the form

$$[\hat{S}_+, \hat{T}_1] = \sum_{ai}(t_{i\beta}^{a\beta} - t_{i\alpha}^{a\alpha})a_{a\alpha}^{\dagger}a_{i\beta} = 0 \tag{13.7.5}$$

$$[\hat{S}_-, \hat{T}_1] = \sum_{ai}(t_{i\alpha}^{a\alpha} - t_{i\beta}^{a\beta})a_{a\beta}^{\dagger}a_{i\alpha} = 0 \tag{13.7.6}$$

both of which can be satisfied by requiring the alpha and beta amplitudes to be identical. The final form of the singles cluster operator is therefore

$$\hat{T}_1 = \sum_{ai} t_i^a E_{ai} \tag{13.7.7}$$

where E_{ai} is the singlet one-electron excitation operator discussed in Section 2.3.4. This form of $\hat{T}_1$ can be obtained also directly from the discussion of spin-free operators in Section 2.2.1. Expression (13.7.7) thus contains all linearly independent combinations of the singlet one-electron excitation operators.

The singlet doubles operator $\hat{T}_2$ may be obtained in a similar way, using the same requirements as for $\hat{T}_1$ in (13.7.4)–(13.7.6). This procedure gives

$$\hat{T}_2 = \frac{1}{2}\sum_{abij} t_{ij}^{ab} E_{ai}E_{bj} \tag{13.7.8}$$

where the cluster amplitudes satisfy the symmetry relation

$$t_{ij}^{ab} = t_{ji}^{ba} \tag{13.7.9}$$

In Exercise 13.2, the $\hat{T}_2$ operator is obtained in different manner, by explicit spin coupling of the creation and annihilation operators. The singlet spin requirements are thus easily satisfied, reducing the number of independent parameters significantly – by a factor of 4 for the singles and by a factor of about 8 for the doubles.

13.7.2 THE CCSD ENERGY EXPRESSION

As discussed in Section 13.2.4, the CCSD energy may be calculated according to (13.2.30), which we write in the form

$$E_{\text{CCSD}} = E_{\text{HF}} + \frac{1}{2}\langle\text{HF}|[[\hat{H}, \hat{T}_1], \hat{T}_1]|\text{HF}\rangle + \langle\text{HF}|[\hat{H}, \hat{T}_2]|\text{HF}\rangle \tag{13.7.10}$$

where the first term is the Hartree–Fock energy

$$E_{\text{HF}} = \langle\text{HF}|\hat{H}|\text{HF}\rangle \tag{13.7.11}$$

and the remaining two terms are the correlation corrections from the singles and doubles amplitudes. Inserting the CCSD operators (13.7.7) and (13.7.8) in the expression for the energy (13.7.10), we obtain

$$E_{\text{CCSD}} = E_{\text{HF}} + \frac{1}{2}\sum_{aibj} t_i^a t_j^b\langle\text{HF}|[[\hat{H}, E_{ai}], E_{bj}]|\text{HF}\rangle + \frac{1}{2}\sum_{aibj} t_{ij}^{ab}\langle\text{HF}|[\hat{H}, E_{ai}E_{bj}]|\text{HF}\rangle \tag{13.7.12}$$

The commutators may be evaluated explicitly by the techniques of Section 1.8. However, we shall encounter these commutators so frequently in our discussion of the CCSD model, in particular in Sections 13.7.7 and 13.7.8, that it is a good idea to tabulate them here once and for all.

In Box 13.1, we have listed all nonzero commutators between the Hamiltonian operator and the single-excitation operators. Usually, we are not so much interested in the commutators themselves as in their application to the Hartree–Fock state; see Box 13.2. In these boxes [22], the operators $P_{ijk...}^{abc...}$ carry out permutations of the pair indices in the following manner:

$$P_{ij}^{ab} A_{ij}^{ab} = A_{ij}^{ab} + A_{ji}^{ba} \tag{13.7.13}$$

$$P_{ijk}^{abc} A_{ijk}^{abc} = A_{ijk}^{abc} + A_{ikj}^{acb} + A_{jik}^{bac} + A_{jki}^{bca} + A_{kij}^{cab} + A_{kji}^{cba} \tag{13.7.14}$$

In Box 13.2, we have also introduced the notation

$$L_{pqrs} = 2g_{pqrs} - g_{psrq} \tag{13.7.15}$$

In Exercise 13.3, (13B.1.1) of Box 13.1 and (13B.2.2) of Box 13.2 are derived.

In Boxes 13.1 and 13.2, we have listed only those commutators that contain the single-excitation operators. Commutators involving double and higher excitations are easily expressed in terms of the single excitations, for instance:

$$\langle \mathrm{HF} | [\hat{H}, E_{ai} E_{bj}] | \mathrm{HF} \rangle = \langle \mathrm{HF} | [[\hat{H}, E_{ai}], E_{bj}] | \mathrm{HF} \rangle \tag{13.7.16}$$

This expression follows since, upon expansion of the commutators, all terms that contain excitation operators to the left of the Hamiltonian vanish. We may therefore write the CCSD energy (13.7.12) in the form

$$E_{\mathrm{CCSD}} = E_{\mathrm{HF}} + \frac{1}{2} \sum_{aibj} (t_{ij}^{ab} + t_i^a t_j^b) \langle \mathrm{HF} | [[\hat{H}, E_{ai}], E_{bj}] | \mathrm{HF} \rangle \tag{13.7.17}$$

Using the commutator (13B.2.3) from Box 13.2, we obtain the expression

$$E_{\mathrm{CCSD}} = E_{\mathrm{HF}} + \sum_{aibj} (t_{ij}^{ab} + t_i^a t_j^b) L_{iajb} \tag{13.7.18}$$

for the energy of a closed-shell CCSD state.

13.7.3 THE T1-TRANSFORMED HAMILTONIAN

In the CCSD approximation, the similarity-transformed Hamiltonian may be written as

$$\exp(-\hat{T}_2 - \hat{T}_1) \hat{H} \exp(\hat{T}_1 + \hat{T}_2) = \exp(-\hat{T}_2) \tilde{\hat{H}} \exp(\hat{T}_2) \tag{13.7.19}$$

where we have introduced the *T1-transformed Hamiltonian* [23]

$$\tilde{\hat{H}} = \exp(-\hat{T}_1) \hat{H} \exp(\hat{T}_1) \tag{13.7.20}$$

Box 13.1   The nonzero commutators between the Hamiltonian and the single-excitation operators

$$[\hat{H}, E_{ai}] = \sum_p h_{pa} E_{pi} - \sum_p h_{ip} E_{ap} + \sum_{pqr} g_{pqra} e_{pqri} - \sum_{pqr} g_{pqir} e_{pqar} \qquad (13\text{B}.1.1)$$

$$[[\hat{H}, E_{ai}], E_{bj}] = -P_{ij}^{ab} h_{ib} E_{aj} - P_{ij}^{ab} \sum_{pq} (g_{pbiq} e_{pjaq} + g_{pqja} e_{pqbi})$$

$$\qquad\qquad\qquad + \sum_{pq} (g_{pbqa} e_{pjqi} + g_{jpiq} e_{bpaq}) \qquad (13\text{B}.1.2)$$

$$[[[\hat{H}, E_{ai}], E_{bj}], E_{ck}] = P_{ijk}^{abc} \sum_p (g_{kbip} e_{cjap} - g_{pbic} e_{pjak}) \qquad (13\text{B}.1.3)$$

$$[[[[\hat{H}, E_{ai}], E_{bj}], E_{ck}], E_{dl}] = \tfrac{1}{2} P_{ijkl}^{abcd} g_{ibkd} e_{ajcl} \qquad (13\text{B}.1.4)$$

Box 13.2   The Hamiltonian and its nonzero commutators with the single-excitation operators applied to the Hartree–Fock state. The commutators are listed in Box 13.1

$$\hat{H}|\text{HF}\rangle = \sum_i (h_{ii} + {}^{\mathrm{I}}F_{ii})|\text{HF}\rangle + \sum_{ai} {}^{\mathrm{I}}F_{ai} E_{ai}|\text{HF}\rangle + \tfrac{1}{2} \sum_{aibj} g_{aibj} E_{ai} E_{bj}|\text{HF}\rangle \qquad (13\text{B}.2.1)$$

$$[\hat{H}, E_{ai}]|\text{HF}\rangle = 2{}^{\mathrm{I}}F_{ia}|\text{HF}\rangle + \left(\sum_b {}^{\mathrm{I}}F_{ba} E_{bi} - \sum_j {}^{\mathrm{I}}F_{ij} E_{aj} + \sum_{bj} L_{bjia} E_{bj} \right)|\text{HF}\rangle$$

$$\qquad\qquad + \left(\sum_{bjc} g_{bjca} E_{bj} E_{ci} - \sum_{bjk} g_{bjik} E_{bj} E_{ak} \right)|\text{HF}\rangle \qquad (13\text{B}.2.2)$$

$$[[\hat{H}, E_{ai}], E_{bj}]|\text{HF}\rangle = 2L_{iajb}|\text{HF}\rangle - P_{ij}^{ab} \left({}^{\mathrm{I}}F_{ib} E_{aj} + \sum_k L_{ikjb} E_{ak} - \sum_c L_{cajb} E_{ci} \right)|\text{HF}\rangle$$

$$\qquad\qquad - P_{ij}^{ab} \left(\sum_{ck} g_{ibck} E_{aj} E_{ck} + \sum_{ck} g_{ikcb} E_{ak} E_{cj} \right)|\text{HF}\rangle$$

$$\qquad\qquad + \left(\sum_{kl} g_{ikjl} E_{ak} E_{bl} + \sum_{cd} g_{cadb} E_{ci} E_{dj} \right)|\text{HF}\rangle \qquad (13\text{B}.2.3)$$

$$[[[\hat{H}, E_{ai}], E_{bj}], E_{ck}]|\text{HF}\rangle = -P_{ijk}^{abc} L_{jbic} E_{ak}|\text{HF}\rangle$$

$$\qquad\qquad + P_{ijk}^{abc} \left(\sum_l g_{iljc} E_{al} E_{bk} - \sum_d g_{ibdc} E_{aj} E_{dk} \right)|\text{HF}\rangle \qquad (13\text{B}.2.4)$$

$$[[[[\hat{H}, E_{ai}], E_{bj}], E_{ck}], E_{dl}]|\text{HF}\rangle = P_{ijk}^{abc} (g_{kbid} E_{cj} E_{al} + g_{lbic} E_{dj} E_{ak})|\text{HF}\rangle$$

$$\qquad\qquad = \tfrac{1}{2} P_{ijkl}^{abcd} g_{idjc} E_{al} E_{bk}|\text{HF}\rangle \qquad (13\text{B}.2.5)$$

The T1 transformation does not affect the particle rank of the Hamiltonian. Indeed, the only complication that occurs upon the transformation (13.7.20) is a loss of symmetry in the one- and two-electron integrals, with only the particle-permutation symmetry of the two-electron integrals retained as discussed in Section 13.7.4. Therefore, if we are prepared to work with integrals of reduced symmetry, the T1 transformation of the Hamiltonian (13.7.20) will simplify the subsequent manipulation of the coupled-cluster equations considerably, effectively reducing the complexity of the CCSD equations to that of the CCD equations.

To determine the form of the T1-transformed Hamiltonian (13.7.20), we must consider the transformed creation and annihilation operators. The creation operators transform as

$$\tilde{a}^\dagger_{p\sigma} = \exp(-\hat{T}_1)a^\dagger_{p\sigma}\exp(\hat{T}_1) = a^\dagger_{p\sigma} + [a^\dagger_{p\sigma}, \hat{T}_1] \tag{13.7.21}$$

The higher-order, nested commutators vanish as discussed in Section 13.2.8. Expanding $\hat{T}_1$ in (13.7.21) and evaluating the resulting commutators using (2.3.36), we may express the transformed creation operator as a linear combination of untransformed operators:

$$\tilde{a}^\dagger_{p\sigma} = a^\dagger_{p\sigma} - \sum_{ai}\delta_{pi}t^a_i a^\dagger_{a\sigma} \tag{13.7.22}$$

Similarly, the transformed annihilation operators may be written as linear combinations of untransformed annihilation operators:

$$\tilde{a}_{p\sigma} = a_{p\sigma} + \sum_{ai}\delta_{pa}t^a_i a_{i\sigma} \tag{13.7.23}$$

It should be noted that, according to (13.7.22) and (13.7.23),

$$\tilde{a}^\dagger_{a\sigma} = a^\dagger_{a\sigma} \tag{13.7.24}$$

$$\tilde{a}_{i\sigma} = a_{i\sigma} \tag{13.7.25}$$

The occupied creation operators and unoccupied annihilation operators are therefore the only elementary operators affected by the T1 transformation. We may then write the transformed creation and annihilation operators in the form

$$\tilde{a}^\dagger_{p\sigma} = \sum_q a^\dagger_{q\sigma}x_{qp} \tag{13.7.26}$$

$$\tilde{a}_{p\sigma} = \sum_q a_{q\sigma}y_{qp} \tag{13.7.27}$$

Here the transformation coefficients x_{qp} and y_{qp} are elements of the matrices

$$\mathbf{x} = \mathbf{1} - \mathbf{t_1} \tag{13.7.28}$$

$$\mathbf{y} = \mathbf{1} + \mathbf{t_1^T} \tag{13.7.29}$$

where we have introduced the square matrix $\mathbf{t_1}$ with elements

$$[\mathbf{t_1}]_{pq} = \begin{cases} t^p_q & p \text{ virtual, } q \text{ occupied} \\ 0 & \text{otherwise} \end{cases} \tag{13.7.30}$$

With the occupied orbitals preceding the virtual ones, $\mathbf{t}_1$ may be partitioned as

$$\mathbf{t}_1 = \begin{pmatrix} 0 & 0 \\ t_i^a & 0 \end{pmatrix} \tag{13.7.31}$$

where only the lower off-diagonal block is nonzero.

13.7.4 THE T1-TRANSFORMED INTEGRALS

Using the expressions for the transformed creation and annihilation operators (13.7.26) and (13.7.27), the T1-transformed Hamiltonian (13.7.20) may be written as

$$\tilde{H} = \sum_{pq} \tilde{h}_{pq} E_{pq} + \frac{1}{2} \sum_{pqrs} \tilde{g}_{pqrs} e_{pqrs} + h_{\text{nuc}} \tag{13.7.32}$$

in terms of the modified molecular integrals

$$\tilde{h}_{pq} = \sum_{rs} x_{pr} y_{qs} h_{rs} \tag{13.7.33}$$

$$\tilde{g}_{pqrs} = \sum_{\substack{tu \\ mn}} x_{pt} y_{qu} x_{rm} y_{sn} g_{tumn} \tag{13.7.34}$$

The T1-transformed integrals are not symmetric with respect to permutations of the orbital indices

$$\tilde{h}_{pq} \neq \tilde{h}_{qp} \tag{13.7.35}$$

$$\tilde{g}_{pqrs} \neq \tilde{g}_{qprs} \neq \tilde{g}_{pqsr} \neq \tilde{g}_{qpsr} \tag{13.7.36}$$

except that the particle-permutation symmetry of the two-electron integrals is retained

$$\tilde{g}_{pqrs} = \tilde{g}_{rspq} \tag{13.7.37}$$

as is readily verified from (13.7.34). The only symmetry of the integrals assumed in Boxes 13.1 and 13.2 is the particle-permutation symmetry (13.7.37). The expressions in the boxes are therefore valid also for the T1-transformed Hamiltonian (13.7.32). We also note that the T1-transformed one-electron integrals can be written as

$$\tilde{h}_{pq} = h_{pq} - \sum_{r} (\mathbf{t}_1)_{pr} h_{rq} + \sum_{s} h_{ps} (\mathbf{t}_1^{\mathrm{T}})_{qs} - \sum_{rs} (\mathbf{t}_1)_{pr} h_{rs} (\mathbf{t}_1^{\mathrm{T}})_{qs} \tag{13.7.38}$$

From the structure of $\mathbf{t}_1$ in (13.7.31), we find that the second and fourth terms vanish if p corresponds to an occupied orbital and that the third and fourth terms vanish if q refers to a virtual orbital. A similar result holds for the two-electron integrals and we obtain

$$\tilde{h}_{ia} = h_{ia} \tag{13.7.39}$$

$$\tilde{g}_{iajb} = g_{iajb} \tag{13.7.40}$$

simplifying the calculation of the projected coupled-cluster equations in certain cases.

Let us now consider the evaluation of T1-transformed integrals from the AO integrals. One strategy would be first to generate the standard integrals in the MO basis

$$\phi_p = \sum_\mu C_{\mu p} \chi_\mu \tag{13.7.41}$$

by transforming the AO integrals using the MO coefficients as transformation coefficients:

$$h_{pq} = \sum_{\mu\nu} C_{\mu p} C_{vq} h_{\mu\nu} \tag{13.7.42}$$

$$g_{pqrs} = \sum_{\mu\nu\rho\sigma} C_{\mu p} C_{vq} C_{\rho r} C_{\sigma s} g_{\mu\nu\rho\sigma} \tag{13.7.43}$$

The final T1-transformed integrals are then generated from the MO integrals (13.7.42) and (13.7.43) in a second transformation, using the elements of $\mathbf{x}$ and $\mathbf{y}$ as transformation coefficients.

However, a better strategy for evaluating the T1-transformed integrals is first to combine $\mathbf{x}$ and $\mathbf{y}$ with the matrix of MO coefficients $\mathbf{C}$ according to the expressions

$$\mathbf{X} = \mathbf{Cx}^{\mathrm{T}} = \mathbf{C}(\mathbf{1} - \mathbf{t}_1^{\mathrm{T}}) \tag{13.7.44}$$

$$\mathbf{Y} = \mathbf{Cy}^{\mathrm{T}} = \mathbf{C}(\mathbf{1} + \mathbf{t}_1) \tag{13.7.45}$$

We may now obtain the T1-transformed integrals directly from the AO integrals in the same manner as we calculate the standard MO integrals

$$\tilde{h}_{pq} = \sum_{\mu\nu} X_{\mu p} Y_{vq} h_{\mu\nu} \tag{13.7.46}$$

$$\tilde{g}_{pqrs} = \sum_{\mu\nu\rho\sigma} X_{\mu p} Y_{vq} X_{\rho r} Y_{\sigma s} g_{\mu\nu\rho\sigma} \tag{13.7.47}$$

the only differences being the use of two transformation matrices $\mathbf{X}$ and $\mathbf{Y}$ and the resulting lower symmetry of the MO integrals.

13.7.5 REPRESENTATION OF THE CCSD PROJECTION MANIFOLD

To solve the projected coupled-cluster equations (13.2.23) for the CCSD wave function, we need a representation of the projected space for the singles and doubles manifolds. For the ket states, we choose the simple basis

$$\left| \begin{matrix} a \\ i \end{matrix} \right\rangle = E_{ai} |\mathrm{HF}\rangle \tag{13.7.48}$$

$$\left| \begin{matrix} ab \\ ij \end{matrix} \right\rangle = E_{ai} E_{bj} |\mathrm{HF}\rangle \tag{13.7.49}$$

with $ai \geq bj$. The adjoints of these states

$$\left\langle \begin{matrix} a \\ i \end{matrix} \right| = \langle \mathrm{HF}| E_{ia} \tag{13.7.50}$$

$$\left\langle \begin{matrix} ab \\ ij \end{matrix} \right| = \langle \mathrm{HF}| E_{jb} E_{ia} \tag{13.7.51}$$

have the following complicated overlaps with the kets (see Exercise 13.4):

$$\left\langle \begin{array}{c} a \\ i \end{array} \middle| \begin{array}{c} c \\ k \end{array} \right\rangle = 2\delta_{ai,ck} \tag{13.7.52}$$

$$\left\langle \begin{array}{c} ab \\ ij \end{array} \middle| \begin{array}{c} cd \\ kl \end{array} \right\rangle = 4\delta_{ac}\delta_{bd}\delta_{jl}\delta_{ik} + 4\delta_{bc}\delta_{ad}\delta_{jk}\delta_{il} - 2\delta_{ac}\delta_{bd}\delta_{il}\delta_{jk} - 2\delta_{bc}\delta_{ad}\delta_{jl}\delta_{ik}$$

$$= 2P_{ij}^{ab}(2\delta_{aibj,ckdl} - \delta_{ajbi,ckdl}) = 2P_{kl}^{cd}(2\delta_{aibj,ckdl} - \delta_{ajbi,ckdl}) \tag{13.7.53}$$

Because of the nonorthogonality, it is inconvenient to use (13.7.50) and (13.7.51) as a basis for the bra states and the projection space. Clearly, it would be better to use a biorthogonal basis. For the sake of future development, it proves convenient to use for the new basis the normalization

$$\left\langle \begin{array}{c} \bar{a} \\ i \end{array} \middle| \begin{array}{c} c \\ k \end{array} \right\rangle = \delta_{ai,ck} \tag{13.7.54}$$

$$\left\langle \begin{array}{c} \overline{ab} \\ ij \end{array} \middle| \begin{array}{c} cd \\ kl \end{array} \right\rangle = P_{kl}^{cd}\delta_{aibj,ckdl} = P_{ij}^{ab}\delta_{aibj,ckdl} \tag{13.7.55}$$

From the overlaps (13.7.52) and (13.7.53), we see that such a basis must satisfy the equations

$$\left\langle \begin{array}{c} a \\ i \end{array} \middle| = 2 \left\langle \begin{array}{c} \bar{a} \\ i \end{array} \middle| \tag{13.7.56}$$

$$\left\langle \begin{array}{c} ab \\ ij \end{array} \middle| = 4 \left\langle \begin{array}{c} \overline{ab} \\ ij \end{array} \middle| - 2 \left\langle \begin{array}{c} \overline{ab} \\ ji \end{array} \middle| \tag{13.7.57}$$

and the bra basis may therefore be obtained from the adjoints of the kets as follows [24]:

$$\left\langle \begin{array}{c} \bar{a} \\ i \end{array} \middle| = \frac{1}{2} \left\langle \begin{array}{c} a \\ i \end{array} \middle| \tag{13.7.58}$$

$$\left\langle \begin{array}{c} \overline{ab} \\ ij \end{array} \middle| = \frac{1}{3} \left\langle \begin{array}{c} ab \\ ij \end{array} \middle| + \frac{1}{6} \left\langle \begin{array}{c} ab \\ ji \end{array} \middle| \tag{13.7.59}$$

For projection, the important point is the existence of a biorthonormal basis that satisfies (13.7.54) and (13.7.55); its explicit representation is not required to carry out the projection.

The bra states are easily normalized so as to form a biorthonormal basis by a simple scaling of the doubly excited states with $a = b$ and $i = j$:

$$\left\langle \begin{array}{c} \widetilde{ab} \\ ij \end{array} \middle| = \frac{1}{1 + \delta_{ai,bj}} \left\langle \begin{array}{c} \overline{ab} \\ ij \end{array} \middle| \tag{13.7.60}$$

However, since the bra states are used only for projection of the coupled-cluster equations (which is zero for the optimized wave function), normalization is unimportant. In fact, use of the biorthogonal basis (13.7.59) rather than the biorthonormal basis (13.7.60) simplifies our algebraic manipulations considerably.

13.7.6 THE NORM OF THE CCSD WAVE FUNCTION

To illustrate the use of the biorthogonal basis, let us consider the norm of the CCSD state. Expanding in a full biorthonormal set of basis functions, we find that the squared norm of the

CCSD state is given by

$$\|\text{CCSD}\|^2 = \langle \text{CCSD}|\text{CCSD}\rangle = \sum_\mu \langle \text{CCSD}|\mu\rangle\langle\tilde{\mu}|\text{CCSD}\rangle \tag{13.7.61}$$

To calculate this norm, we would need a basis for the full set of configurations – that is, all configurations entering the corresponding FCI wave function. Restricting ourselves instead to the norm of the CCSD wave function projected onto the singles and doubles spaces, we obtain

$$\|\text{CCSD}\|^2_{\text{HFSD}} = 1 + \sum_{ai}\left\langle \text{CCSD}\left|\begin{matrix}a\\i\end{matrix}\right.\right\rangle\left\langle\left.\begin{matrix}\bar{a}\\i\end{matrix}\right|\text{CCSD}\right\rangle + \sum_{ai\geq bj}\left\langle \text{CCSD}\left|\begin{matrix}ab\\ij\end{matrix}\right.\right\rangle\left\langle\left.\begin{matrix}\widetilde{ab}\\ij\end{matrix}\right|\text{CCSD}\right\rangle \tag{13.7.62}$$

For the singles, we get

$$\left\langle\left.\begin{matrix}\bar{a}\\i\end{matrix}\right|\text{CCSD}\right\rangle = \left\langle\left.\begin{matrix}\bar{a}\\i\end{matrix}\right|\hat{T}_1\right|\text{HF}\right\rangle = \sum_{ck} t_k^c \left\langle\left.\begin{matrix}\bar{a}\\i\end{matrix}\right|\begin{matrix}c\\k\end{matrix}\right\rangle = \sum_{ck} t_k^c \delta_{ai,ck} = t_i^a \tag{13.7.63}$$

and (13.7.56) then gives

$$\left\langle\text{CCSD}\left|\begin{matrix}a\\i\end{matrix}\right.\right\rangle = 2t_i^a \tag{13.7.64}$$

For the unnormalized doubles, we obtain

$$\left\langle\left.\begin{matrix}\overline{ab}\\ij\end{matrix}\right|\text{CCSD}\right\rangle = \left\langle\left.\begin{matrix}\overline{ab}\\ij\end{matrix}\right|\hat{T}_2 + \frac{1}{2}\hat{T}_1^2\right|\text{HF}\right\rangle = \frac{1}{2}\sum_{cdkl}(t_{kl}^{cd} + t_k^c t_l^d)\left\langle\left.\begin{matrix}\overline{ab}\\ij\end{matrix}\right|E_{ck}E_{dl}\right|\text{HF}\right\rangle$$

$$= \frac{1}{2}P_{ij}^{ab}\sum_{cdkl}(t_{kl}^{cd} + t_k^c t_l^d)\delta_{aibj,ckdl} = t_{ij}^{ab} + t_i^a t_j^b = T_{ij}^{ab} \tag{13.7.65}$$

and therefore

$$\left\langle\left.\begin{matrix}\widetilde{ab}\\ij\end{matrix}\right|\text{CCSD}\right\rangle = (1 + \delta_{ai,bj})^{-1}T_{ij}^{ab} \tag{13.7.66}$$

$$\left\langle\text{CCSD}\left|\begin{matrix}ab\\ij\end{matrix}\right.\right\rangle = 2(2T_{ij}^{ab} - T_{ji}^{ab}) \tag{13.7.67}$$

where, for the last expression, we have used (13.7.57). Inserting the projections (13.7.63), (13.7.64), (13.7.66) and (13.7.67) in the norm expansion (13.7.62), we obtain the final expression as a linear combination of squared amplitudes

$$\|\text{CCSD}\|^2_{\text{HFSD}} = 1 + 2\sum_{ai}(t_i^a)^2 + \sum_{aibj}(2T_{ij}^{ab} - T_{ji}^{ab})T_{ij}^{ab} \tag{13.7.68}$$

It is possible to derive expressions for higher excitations as well but, for an exact evaluation of the norm, we would need to sum over all configurations in the FCI wave function.

13.7.7 THE CCSD SINGLES PROJECTION

Let us now consider the evaluation of the projected coupled-cluster equations [25,26]. We begin with the singles projection here, discussing the doubles projection in Section 13.7.8. The equation for projection against the singles space is given by:

$$\Omega_{ai} = \left\langle\left.\begin{matrix}\bar{a}\\i\end{matrix}\right|\exp(-\hat{T}_2)\tilde{\hat{H}}\exp(\hat{T}_2)\right|\text{HF}\right\rangle \tag{13.7.69}$$

Our strategy is first to expand in powers of $\hat{T}_2$ and then to write the resulting expressions in terms of nested commutators between the Hamiltonian $\tilde{\hat{H}}$ and single-excitation operators such as E_{ck}. Using Box 13.2, we may then determine the effect of this commutator expression on the Hartree–Fock state. The final projection is obtained by using the overlap expressions (13.7.54) and (13.7.55). Following this strategy, we first expand (13.7.69) in $\hat{T}_2$ to obtain

$$\Omega_{ai} = \left\langle \begin{matrix} \bar{a} \\ i \end{matrix} \middle| \tilde{\hat{H}} + [\tilde{\hat{H}}, \hat{T}_2] \middle| \text{HF} \right\rangle \tag{13.7.70}$$

Higher-order terms do not contribute since the T1-transformed Hamiltonian is a one- and two-electron operator. The first term gives the following contribution

$$\left\langle \begin{matrix} \bar{a} \\ i \end{matrix} \middle| \tilde{\hat{H}} \middle| \text{HF} \right\rangle = \sum_{ck} {}^{\text{I}}\tilde{F}_{ck} \left\langle \begin{matrix} \bar{a} \\ i \end{matrix} \middle| E_{ck} \middle| \text{HF} \right\rangle = \sum_{ck} {}^{\text{I}}\tilde{F}_{ck} \delta_{ai,ck} = {}^{\text{I}}\tilde{F}_{ai} \tag{13.7.71}$$

where the first equality follows after substituting (13B.2.1) and the second after using (13.7.54). The matrix ${}^{\text{I}}\tilde{\mathbf{F}}$ is the standard inactive Fock matrix (10.8.28) calculated from the T1-transformed integrals (13.7.46) and (13.7.47).

To evaluate the second term in (13.7.70), we rewrite the expression in terms of commutators working on the Hartree–Fock state:

$$\left\langle \begin{matrix} \bar{a} \\ i \end{matrix} \middle| [\tilde{\hat{H}}, \hat{T}_2] \middle| \text{HF} \right\rangle = P_1 + P_2 \tag{13.7.72}$$

$$P_1 = \sum_{cdkl} t_{kl}^{cd} \left\langle \begin{matrix} \bar{a} \\ i \end{matrix} \middle| E_{ck}[\tilde{\hat{H}}, E_{dl}] \middle| \text{HF} \right\rangle \tag{13.7.73}$$

$$P_2 = \frac{1}{2} \sum_{cdkl} t_{kl}^{cd} \left\langle \begin{matrix} \bar{a} \\ i \end{matrix} \middle| [[\tilde{\hat{H}}, E_{ck}], E_{dl}] \middle| \text{HF} \right\rangle \tag{13.7.74}$$

Inserting (13B.2.2) and (13B.2.3) in P_1 and P_2 and retaining only the nonzero contributions (from rank considerations), we obtain

$$P_1 = 2 \sum_{cdkl} t_{kl}^{cd} \left\langle \begin{matrix} \bar{a} \\ i \end{matrix} \middle| {}^{\text{I}}\tilde{F}_{ld} E_{ck} \middle| \text{HF} \right\rangle \tag{13.7.75}$$

$$P_2 = -\sum_{cdkl} t_{kl}^{cd} \left\langle \begin{matrix} \bar{a} \\ i \end{matrix} \middle| {}^{\text{I}}\tilde{F}_{kd} E_{cl} + \sum_{j} \tilde{L}_{kjld} E_{cj} - \sum_{b} \tilde{L}_{bcld} E_{bk} \middle| \text{HF} \right\rangle \tag{13.7.76}$$

which reduce to the expressions

$$P_1 = 2 \sum_{dl} t_{il}^{ad} {}^{\text{I}}\tilde{F}_{ld} \tag{13.7.77}$$

$$P_2 = -\sum_{dk} t_{ki}^{ad} {}^{\text{I}}\tilde{F}_{kd} - \sum_{dkl} t_{kl}^{ad} \tilde{L}_{kild} + \sum_{cdk} t_{ki}^{cd} \tilde{L}_{adkc} \tag{13.7.78}$$

Expressions (13.7.71), (13.7.77) and (13.7.78) may now be combined to give the final projection

$$\Omega_{ai} = \Omega_{ai}^{\text{A1}} + \Omega_{ai}^{\text{B1}} + \Omega_{ai}^{\text{C1}} + \Omega_{ai}^{\text{D1}} \tag{13.7.79}$$

where

$$\Omega_{ai}^{A1} = \sum_{ckd} u_{ki}^{cd} \tilde{g}_{adkc} \tag{13.7.80}$$

$$\Omega_{ai}^{B1} = -\sum_{ckl} u_{kl}^{ac} \tilde{g}_{kilc} \tag{13.7.81}$$

$$\Omega_{ai}^{C1} = \sum_{ck} u_{ik}^{ac}\, {}^{I}\tilde{F}_{kc} \tag{13.7.82}$$

$$\Omega_{ai}^{D1} = {}^{I}\tilde{F}_{ai} \tag{13.7.83}$$

in terms of the modified amplitudes

$$u_{ij}^{ab} = 2t_{ij}^{ab} - t_{ji}^{ab} \tag{13.7.84}$$

The computational implementation and the cost of evaluation of these four contributions to the projected coupled-cluster equations are discussed in Section 13.7.9, after we have considered the doubles projection in Section 13.7.8.

13.7.8 THE CCSD DOUBLES PROJECTION

Having discussed the CCSD singles projection, we now consider the evaluation of the doubles projection

$$\Omega_{aibj} = \left\langle \overline{\substack{ab \\ ij}} \middle| \exp(-\hat{T}_2)\tilde{\hat{H}}\exp(\hat{T}_2) \middle| \mathrm{HF} \right\rangle \tag{13.7.85}$$

Using the same technique as for the singles in Section 13.7.7, we expand (13.7.85) in powers of $\hat{T}_2$ to obtain

$$\Omega_{aibj} = \left\langle \overline{\substack{ab \\ ij}} \middle| \tilde{\hat{H}} + [\tilde{\hat{H}}, \hat{T}_2] + \frac{1}{2}[[\tilde{\hat{H}}, \hat{T}_2], \hat{T}_2] \middle| \mathrm{HF} \right\rangle \tag{13.7.86}$$

Evaluation of the first term yields

$$\left\langle \overline{\substack{ab \\ ij}} \middle| \tilde{\hat{H}} \middle| \mathrm{HF} \right\rangle = \tilde{g}_{aibj} \tag{13.7.87}$$

directly. The second term in (13.7.86) is evaluated in the same way as the second contribution to the singles projection (13.7.72):

$$\left\langle \overline{\substack{ab \\ ij}} \middle| [\tilde{\hat{H}}, \hat{T}_2] \middle| \mathrm{HF} \right\rangle = Q_1 + Q_2 \tag{13.7.88}$$

$$Q_1 = \sum_{cdkl} t_{kl}^{cd} \left\langle \overline{\substack{ab \\ ij}} \middle| E_{ck}[\tilde{\hat{H}}, E_{dl}] \middle| \mathrm{HF} \right\rangle \tag{13.7.89}$$

$$Q_2 = \frac{1}{2} \sum_{cdkl} t_{kl}^{cd} \left\langle \overline{\substack{ab \\ ij}} \middle| [[\tilde{\hat{H}}, E_{ck}], E_{dl}] \middle| \mathrm{HF} \right\rangle \tag{13.7.90}$$

Proceeding as for the P_1 and P_2 terms in (13.7.72), we obtain

$$Q_1 = P_{ij}^{ab}\left(\sum_c t_{ij}^{ac}\,^{\mathrm{I}}\tilde{F}_{bc} - \sum_k t_{ik}^{ab}\,^{\mathrm{I}}\tilde{F}_{kj} + \sum_{ck} t_{ik}^{ac}\tilde{L}_{bjkc}\right) \qquad (13.7.91)$$

$$Q_2 = -P_{ij}^{ab}\sum_{ck}(t_{kj}^{ac}\tilde{g}_{bcki} + t_{ki}^{ac}\tilde{g}_{bjkc}) + \sum_{cd}t_{ij}^{cd}\tilde{g}_{acbd} + \sum_{kl}t_{kl}^{ab}\tilde{g}_{kilj} \qquad (13.7.92)$$

Finally, we consider the third term:

$$\frac{1}{2}\left\langle\overline{\begin{matrix}ab\\ij\end{matrix}}\,\middle|\,[[\hat{\tilde{H}},\hat{T}_2],\hat{T}_2]\middle|\mathrm{HF}\right\rangle = R_1 + R_2 + R_3 \qquad (13.7.93)$$

$$R_1 = \frac{1}{2}\sum_{\substack{cdef\\klmn}}t_{kl}^{cd}t_{mn}^{ef}\left\langle\overline{\begin{matrix}ab\\ij\end{matrix}}\,\middle|\,E_{ck}E_{em}[[\hat{\tilde{H}},E_{dl}],E_{fn}]\middle|\mathrm{HF}\right\rangle \qquad (13.7.94)$$

$$R_2 = \frac{1}{2}\sum_{\substack{cdef\\klmn}}t_{kl}^{cd}t_{mn}^{ef}\left\langle\overline{\begin{matrix}ab\\ij\end{matrix}}\,\middle|\,E_{ck}[[[\hat{\tilde{H}},E_{dl}],E_{em}],E_{fn}]\middle|\mathrm{HF}\right\rangle \qquad (13.7.95)$$

$$R_3 = \frac{1}{8}\sum_{\substack{cdef\\klmn}}t_{kl}^{cd}t_{mn}^{ef}\left\langle\overline{\begin{matrix}ab\\ij\end{matrix}}\,\middle|\,[[[[\hat{\tilde{H}},E_{ck}],E_{dl}],E_{em}],E_{fn}]\middle|\mathrm{HF}\right\rangle \qquad (13.7.96)$$

Using the same techniques as above, we obtain

$$R_1 = P_{ij}^{ab}\sum_{cdkl}t_{ik}^{ac}t_{jl}^{bd}\tilde{L}_{kcld} \qquad (13.7.97)$$

$$R_2 = -P_{ij}^{ab}\sum_{cdkl}(t_{ik}^{ab}t_{lj}^{cd}\tilde{L}_{lckd} + t_{ik}^{ac}t_{lj}^{bd}\tilde{L}_{kcld} + t_{ij}^{ac}t_{kl}^{bd}\tilde{L}_{kcld}) \qquad (13.7.98)$$

$$R_3 = \tfrac{1}{2}P_{ij}^{ab}\sum_{cdkl}(t_{kl}^{ab}t_{ij}^{cd} + t_{ki}^{ac}t_{lj}^{bd} + t_{kj}^{ad}t_{li}^{bc})\tilde{g}_{kcld} \qquad (13.7.99)$$

In the evaluation of (13.7.93), we might as well have used the untransformed Hamiltonian $\hat{H}$ since, upon expansion of $\hat{\tilde{H}}$, all terms containing $\hat{T}_1$ vanish by rank considerations. The two-electron integrals in (13.7.97)–(13.7.99) therefore all have occupied–virtual–occupied–virtual indices and the untransformed and T1-transformed integrals are identical, see (13.7.39) and (13.7.40).

We have now obtained all contributions to Ω_{aibj} and collect them in the following way:

$$\Omega_{aibj} = \Omega_{aibj}^{\mathrm{A2}} + \Omega_{aibj}^{\mathrm{B2}} + P_{ij}^{ab}(\Omega_{aibj}^{\mathrm{C2}} + \Omega_{aibj}^{\mathrm{D2}} + \Omega_{aibj}^{\mathrm{E2}}) \qquad (13.7.100)$$

The individual terms are given by

$$\Omega_{aibj}^{\mathrm{A2}} = \tilde{g}_{aibj} + \sum_{cd}t_{ij}^{cd}\tilde{g}_{acbd} \qquad (13.7.101)$$

$$\Omega_{aibj}^{\mathrm{B2}} = \sum_{kl}t_{kl}^{ab}\left(\tilde{g}_{kilj} + \sum_{cd}t_{ij}^{cd}\tilde{g}_{kcld}\right) \qquad (13.7.102)$$

$$\Omega_{aibj}^{\mathrm{C2}} = -\frac{1}{2}\sum_{ck}t_{kj}^{bc}\left(\tilde{g}_{kiac} - \frac{1}{2}\sum_{dl}t_{li}^{ad}\tilde{g}_{kdlc}\right) - \sum_{ck}t_{ki}^{bc}\left(\tilde{g}_{kjac} - \frac{1}{2}\sum_{dl}t_{lj}^{ad}\tilde{g}_{kdlc}\right) \qquad (13.7.103)$$

$$\Omega_{aibj}^{D2} = \frac{1}{2} \sum_{ck} u_{jk}^{bc} \left(\tilde{L}_{aikc} + \frac{1}{2} \sum_{dl} u_{il}^{ad} \tilde{L}_{ldkc} \right)$$ (13.7.104)

$$\Omega_{aibj}^{E2} = \sum_c t_{ij}^{ac} \left({}^{I}\tilde{F}_{bc} - \sum_{dkl} u_{kl}^{bd} \tilde{g}_{ldkc} \right) - \sum_k t_{ik}^{ab} \left({}^{I}\tilde{F}_{kj} + \sum_{cdl} u_{lj}^{cd} \tilde{g}_{kdlc} \right)$$ (13.7.105)

The collection of the terms to obtain (13.7.100) requires a rewriting of some of the terms in (13.7.87), (13.7.91), (13.7.92) and (13.7.97)–(13.7.99). In particular, some of the integrals $\tilde{L}_{pqrs}$ are expanded according to (13.7.15) to obtain the modified amplitudes (13.7.84).

13.7.9 COMPUTATIONAL CONSIDERATIONS

Together with the four contributions to the singles projection Ω_{ai} in (13.7.79), the doubles projection Ω_{aibj} in (13.7.100) contains all the terms needed for the optimization of the CCSD cluster amplitudes. Clearly, the evaluation of the singles and doubles projections is complicated, involving a number of contributions in which amplitudes and integrals are combined. It is therefore not surprising that a variety of schemes have been developed for the evaluation of Ω_{ai} and Ω_{aibj}. We shall consider some general aspects of their evaluation here but do not give a detailed discussion of each term.

In the expressions for the singles and doubles projections (13.7.79) and (13.7.100), the integrals appear in the T1-transformed MO basis, related to the standard MO integrals according to (13.7.33) and (13.7.34). Since the AO integrals are relatively expensive to evaluate and since the same MO integrals are used in each CCSD iteration, an obvious strategy would be to begin the coupled-cluster calculation by evaluating the AO integrals, which are next transformed to the MO basis once and for all. In each iteration, we then first calculate the T1-transformed MO integrals and next combine these with the amplitudes according to the expressions in Section 13.7.7 and 13.7.8. In this *MO-based CCSD algorithm*, the evaluation of the MO integrals (an n^5 task) makes a negligible contribution to the total computational cost, which is dominated by the time it takes to solve the amplitude equations.

From an order analysis of the contributions to the CCSD equations, it is seen that the most expensive contribution is the A2 term (13.7.101), which scales as $V^4 O^2$. The B2, C2 and D2 terms are sixth-order terms as well but less expensive, involving lower powers in the number of virtual orbitals ($V^2 O^4$, $V^3 O^3$ and $V^3 O^3$, respectively). The remaining terms are of lower orders in the number of orbitals. The total cost therefore scales as $V^4 O^2$. This cost should be compared, for example, with the cost of the Hartree–Fock method, which, for small systems, scales as n^4.

Although the computational cost of the CCSD model is high, on modern computers the bottle-neck is often not the operation count but disk storage. Consider a CCSD calculation with 400 basis functions, 40 electrons and no point group symmetry. The storage requirements for the two-electron integrals are 26 Gb (disregarding extra requirements for the transformation) and the memory requirements for the amplitudes about 230 Mb. Assuming an overall cost for each iteration of $V^4 O^2$ and a computer performance of about 500 Mflops, each iteration requires four to five hours of computer time. On many computer systems, therefore, the limiting factor is the storage requirements rather than the number of floating-point operations. In such situations, an algorithm that calculates the projections directly from the AO integrals, eliminating the need for storing the full set of integrals on disk, would be useful.

We do not go into the details of such an integral-direct CCSD algorithm, indicating only how the most expensive term A2 can be obtained directly from the AO integrals – the remaining terms in (13.7.79) and (13.7.100) can be obtained from the AO integrals in a similar manner [23,27,28]. Expanding the two-electron integrals according to (13.7.47), we obtain for the A2 term:

$$\Omega_{aibj} = \sum_{\mu\lambda} X_{\mu a} X_{\lambda b} \left[\sum_{\nu\sigma} \left(Y_{\nu i} Y_{\sigma j} + \sum_{cd} t_{ij}^{cd} Y_{\nu c} Y_{\sigma d} \right) g_{\mu\nu\lambda\sigma} \right] \tag{13.7.106}$$

Introducing the auxiliary matrix

$$T_{ij}^{\nu\sigma} = Y_{\nu i} Y_{\sigma j} + \sum_{cd} t_{ij}^{cd} Y_{\nu c} Y_{\sigma d} \tag{13.7.107}$$

we may calculate Ω_{aibj} in two steps

$$A_{ij}^{\mu\lambda} = \sum_{\nu\sigma} T_{ij}^{\nu\sigma} g_{\mu\nu\lambda\sigma} \tag{13.7.108}$$

$$\Omega_{aibj} = \sum_{\mu\lambda} X_{\mu a} X_{\lambda b} A_{ij}^{\mu\lambda} \tag{13.7.109}$$

The steps (13.7.107) and (13.7.109) may be carried out as fifth-order processes, but (13.7.108) requires $n^4 O^2$ operations. Disregarding the cost of re-evaluating the AO integrals in each iteration, the step (13.7.108) is the most expensive one in *this AO-based CCSD algorithm* and the evaluation of the CCSD projection therefore now requires slightly more operations than in the MO-based algorithm discussed above. The storage requirements are, however, significantly reduced and the AO-based algorithm may be applied to larger systems.

13.8 Special treatments of coupled-cluster theory

In the previous sections of this chapter, we presented the conventional treatment of the correlation problem in coupled-cluster theory. In the present section, we turn our attention briefly to three special correlation treatments within the general framework of coupled-cluster theory: orbital-optimized and Brueckner coupled-cluster theories in Section 13.8.1 and quadratic CI theory in Section 13.8.2.

13.8.1 ORBITAL-OPTIMIZED AND BRUECKNER COUPLED-CLUSTER THEORIES

In standard coupled-cluster theory, the single-excitation manifold is parametrized in terms of the exponential operator

$$\exp(\hat{T}_1) = \exp\left(\sum_{AI} t_I^A a_A^\dagger a_I \right) \tag{13.8.1}$$

An alternative parametrization of the singles manifold is obtained by employing the orthogonal orbital-rotation operator

$$\exp(-\hat{\kappa}) = \exp\left[-\sum_{AI} \kappa_{AI} (a_A^\dagger a_I - a_I^\dagger a_A) \right] \tag{13.8.2}$$

which carries out rotations of the occupied and virtual spin orbitals

$$\tilde{a}_P^\dagger = \exp(-\hat{\kappa})a_P^\dagger \exp(\hat{\kappa}) \tag{13.8.3}$$

using only those parameters κ_{AI} that are nonredundant with respect to transformations of a single-determinant state; see Section 10.1.2. Unlike the orthogonal orbital-rotation operator, the orbital transformation generated by (13.8.1) affects only the occupied spin orbitals, producing a nonorthogonal set of spin orbitals:

$$\bar{a}_I^\dagger = \exp(\hat{T}_1)a_I^\dagger \exp(-\hat{T}_1) \tag{13.8.4}$$

$$\bar{a}_A^\dagger = \exp(\hat{T}_1)a_A^\dagger \exp(-\hat{T}_1) = a_A^\dagger \tag{13.8.5}$$

We also note that, when (13.8.1) and (13.8.2) are applied to the same single-determinant reference state

$$\exp(\hat{T}_1)|\mathrm{RF}\rangle = \left[1 + \sum_{AI} t_I^A a_A^\dagger a_I + O(t^2)\right]|\mathrm{RF}\rangle \tag{13.8.6}$$

$$\exp(-\hat{\kappa})|\mathrm{RF}\rangle = \left[1 - \sum_{AI} \kappa_{AI}(a_A^\dagger a_I - a_I^\dagger a_A) + O(\kappa^2)\right]|\mathrm{RF}\rangle$$

$$= \left[1 - \sum_{AI} \kappa_{AI} a_A^\dagger a_I + O(\kappa^2)\right]|\mathrm{RF}\rangle \tag{13.8.7}$$

the same electronic state is produced to first order.

In standard coupled-cluster theory, we use the Hartree–Fock orbitals and then determine a set of nonzero single-excitation amplitudes together with higher-excitation amplitudes. Alternatively, we may use $\exp(\hat{T}_1)$ to generate an orbital transformation to a basis in which the single-excitation amplitudes vanish. Since (13.8.1) and (13.8.2) generate the same state to first order, we may use $\exp(-\hat{\kappa})$ rather than $\exp(\hat{T}_1)$ to generate this orbital transformation, as done in *orbital-optimized coupled-cluster (OCC) theory* [29,30]. The OCC ansatz for the wave function is

$$|\mathrm{OCC}\rangle = \exp(-\hat{\kappa}) \exp(\hat{T}_\mathrm{O})|\mathrm{RF}\rangle \tag{13.8.8}$$

where the cluster operator contains no singles amplitudes

$$\hat{T}_\mathrm{O} = \hat{T}_2 + \hat{T}_3 + \cdots + \hat{T}_N \tag{13.8.9}$$

and the orbitals are determined from a set of variational conditions. Since the orbital-rotation operator and the cluster operator do not commute, the order of the operators in (13.8.8) is significant and chosen so as to simplify the algebra.

Our goal in OCC theory is to solve the OCC Schrödinger equation

$$\exp(-\hat{T}_\mathrm{O}) \exp(\hat{\kappa})\hat{H} \exp(-\hat{\kappa}) \exp(\hat{T}_\mathrm{O})|\mathrm{RF}\rangle = E_\mathrm{OCC}|\mathrm{RF}\rangle \tag{13.8.10}$$

where the OCC energy is obtained in the usual manner, by projecting (13.8.10) against the reference state

$$E_\mathrm{OCC} = \langle\mathrm{RF}| \exp(-\hat{T}_\mathrm{O}) \exp(\hat{\kappa})\hat{H} \exp(-\hat{\kappa}) \exp(\hat{T}_\mathrm{O})|\mathrm{RF}\rangle \tag{13.8.11}$$

whereas the cluster amplitudes are determined by projection against the manifold $|\mu_O\rangle$ spanned by $\hat{T}_O$

$$\langle \mu_O | \exp(-\tilde{T}_O) \exp(\hat{\kappa}) \hat{H} \exp(-\hat{\kappa}) \exp(\hat{T}_O) | \text{RF} \rangle = 0 \tag{13.8.12}$$

Note that $\langle \mu_O |$ contains no singly excited determinants. The orbital-rotation parameters are determined variationally by optimizing the energy E_{OCC} subject to the requirement that (13.8.12) remains satisfied. This is achieved by optimizing the Lagrangian

$$L(\kappa, \mathbf{t}_O, \bar{\mathbf{t}}_O) = \langle \text{RF} | \exp(-\hat{T}_O) \exp(\hat{\kappa}) \hat{H} \exp(-\hat{\kappa}) \exp(\hat{T}_O) | \text{RF} \rangle$$
$$+ \sum_{\mu_O} \bar{t}_{\mu_O} \langle \mu_O | \exp(-\hat{T}_O) \exp(\hat{\kappa}) \hat{H} \exp(-\hat{\kappa}) \exp(\hat{T}_O) | \text{RF} \rangle \tag{13.8.13}$$

where the $\bar{t}_{\mu_O}$ are the Lagrange multipliers. As in MCSCF theory, we express the OCC state and its current approximation in a basis where the orbital-rotation parameters vanish at the point of expansion. In this basis, the first derivatives of the Lagrangian become

$$\frac{\partial L}{\partial \bar{t}_{\mu_O}} = \langle \mu_O | \exp(-\hat{T}_O) \hat{H} \exp(\hat{T}_O) | \text{RF} \rangle \tag{13.8.14}$$

$$\frac{\partial L}{\partial t_{\mu_O}} = \langle \text{RF} | \exp(-\hat{T}_O) \hat{H} \hat{\tau}_{\mu_O} \exp(\hat{T}_O) | \text{RF} \rangle + \sum_{\nu_O} \bar{t}_{\nu_O} \langle \nu_O | \exp(-\hat{T}_O) [\hat{H}, \hat{\tau}_{\mu_O}] \exp(\hat{T}_O) | \text{RF} \rangle \tag{13.8.15}$$

$$\frac{\partial L}{\partial \kappa_{AI}} = \langle \text{RF} | \exp(-\hat{T}_O) [\hat{H}, a_I^\dagger a_A - a_A^\dagger a_I] \exp(\hat{T}_O) | \text{RF} \rangle$$
$$+ \sum_{\mu_O} \bar{t}_{\mu_O} \langle \mu_O | \exp(-\hat{T}_O) [\hat{H}, a_I^\dagger a_A - a_A^\dagger a_I] \exp(\hat{T}_O) | \text{RF} \rangle \tag{13.8.16}$$

The variational condition on the Lagrangian then requires (13.8.14)–(13.8.16) to vanish for the optimized state. To express the OCC variational conditions in a compact manner, we introduce the OCC states

$$|\text{OCC}\rangle = \exp(\hat{T}_O) | \text{RF} \rangle \tag{13.8.17}$$

$$\langle \Lambda | = \langle \text{RF} | \exp(-\hat{T}_O) + \sum_{\mu_O} \bar{t}_{\mu_O} \langle \mu_O | \exp(-\hat{T}_O) \tag{13.8.18}$$

which satisfy the normalization condition

$$\langle \Lambda | \text{OCC} \rangle = 1 \tag{13.8.19}$$

We may now express the vanishing of the first derivatives (13.8.14)–(13.8.16) in the form

$$\langle \mu_O | \exp(-\hat{T}_O) \hat{H} | \text{OCC} \rangle = 0 \tag{13.8.20}$$

$$\langle \Lambda | [\hat{H}, \hat{\tau}_{\mu_O}] | \text{OCC} \rangle = 0 \tag{13.8.21}$$

$$\langle \Lambda | [\hat{H}, a_I^\dagger a_A - a_A^\dagger a_I] | \text{OCC} \rangle = 0 \tag{13.8.22}$$

These conditions may be viewed as the *generalized Brillouin theorem (GBT)* for the OCC state and should be compared with the MCSCF conditions (12.2.58) and (12.2.59). Whereas (13.8.21) and

(13.8.22) depend explicitly on the Lagrange multipliers, the cluster-amplitude conditions (13.8.20) and the optimized energy depend only implicitly on the multipliers.

The variational conditions require the GBT to be satisfied for the occupied–virtual orbital-rotation operators (13.8.22). For an optimized OCC state, these conditions are trivially satisfied also for the occupied–occupied and virtual–virtual orbital-rotation operators. Consider the occupied–occupied operator $a_I^\dagger a_J - a_J^\dagger a_I$. Invoking the BCH expansion, this operator may be rewritten as

$$a_I^\dagger a_J - a_J^\dagger a_I = [a_I^\dagger a_J - a_J^\dagger a_I, \hat{T}_O] + \exp(\hat{T}_O)(a_I^\dagger a_J - a_J^\dagger a_I)\exp(-\hat{T}_O) \tag{13.8.23}$$

and we obtain the following expression for the Brillouin matrix element:

$$\langle\Lambda|[\hat{H}, a_I^\dagger a_J - a_J^\dagger a_I]|\text{OCC}\rangle = \langle\Lambda|[\hat{H}, [a_I^\dagger a_J - a_J^\dagger a_I, \hat{T}_O]]|\text{OCC}\rangle$$
$$+ \langle\Lambda|[\hat{H}, \exp(\hat{T}_O)(a_I^\dagger a_J - a_J^\dagger a_I)\exp(-\hat{T}_O)]|\text{OCC}\rangle \tag{13.8.24}$$

For the optimized OCC state, the first term on the right-hand side vanishes due to (13.8.21) since $[a_I^\dagger a_J - a_J^\dagger a_I, \hat{T}_O]$ can be expressed in terms of the excitation operators $\hat{\tau}_{\mu_O}$. The second term may be rewritten as

$$\langle\Lambda|[\hat{H}, \exp(\hat{T}_O)(a_I^\dagger a_J - a_J^\dagger a_I)\exp(-\hat{T}_O)]|\text{OCC}\rangle$$
$$= -\sum_{\mu_O}\bar{t}_{\mu_O}\langle\mu_O|(a_I^\dagger a_J - a_J^\dagger a_I)\exp(-\hat{T}_O)\hat{H}\exp(\hat{T}_O)|\text{RF}\rangle \tag{13.8.25}$$

Since the state $\langle\mu_O|(a_I^\dagger a_J - a_J^\dagger a_I)$ is spanned by the projection manifold $\langle\mu_O|$, this term vanishes because of (13.8.20). We thus obtain for the optimized OCC state:

$$\langle\Lambda|[\hat{H}, a_I^\dagger a_J - a_J^\dagger a_I]|\text{OCC}\rangle = 0 \tag{13.8.26}$$

In short, for the OCC optimization, the same orbital-rotation operators are redundant as for the optimization of the single-determinant reference state $|\text{RF}\rangle$.

An algorithm for the optimization of the OCC state may be obtained as in standard coupled-cluster theory by setting up a vector function where the orbitals, cluster amplitudes and multipliers are the variables with (13.8.14)–(13.8.16) constituting the vector function at the expansion point. Expanding to first order and making a diagonal approximation to the Jacobian, we arrive at an iterative algorithm similar to (13.4.10) in standard theory. DIIS may be applied for convergence acceleration.

To illustrate the difference between the energy calculated using standard coupled-cluster theory and OCC theory, we have in Figure 13.5 for the water molecule plotted the CCSD and *OCC doubles (OCCD)* energies in the cc-pVDZ basis as functions of the bond distance [22]. Somewhat surprisingly, perhaps, no significant advantages are obtained by using optimized orbitals for the calculation of the energy. Note that the weight of the Hartree–Fock determinant in the FCI wave function varies significantly with the bond distance, from about 0.94 at equilibrium to about 0.60 for the largest plotted bond distance.

At this point, we mention that the orbital-rotation parameters may also be determined by extending the projection manifold to the single excitations, replacing the orbital conditions (13.8.22) by the amplitude equations (13.8.20) for the singles. This approach is called *Brueckner coupled-cluster (BCC) theory* [5,31,32]. In BCC theory, neither the energy nor the amplitude equations depend on the multipliers and no multipliers must be set up to obtain the BCC wave function.

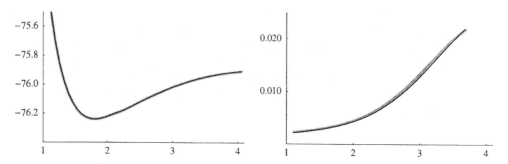

Fig. 13.5. The CCSD energy (grey line) and the OCCD energy (black line) for the water molecule in the cc-pVDZ basis at a fixed bond angle of $110.565°$. On the left, we have plotted the potential-energy curve as a function of the bond distance (a_0); on the right, we have plotted the difference relative to the FCI energy. All energies are in E_h.

In OCC and BCC theories, the wave function has the form of a standard coupled-cluster wave function with vanishing singles amplitudes. However, unlike standard coupled-cluster theory, where the orbitals are determined in a separate optimization of the reference state |HF⟩, the OCC and BCC orbitals are determined simultaneously with the optimization of the cluster amplitudes, making them more suitable for the description of correlation, in a manner reminiscent of MCSCF theory. In practice, the differences between the standard coupled-cluster wave functions and the BCC and OCC wave functions are small, except in systems characterized by Hartree–Fock singlet instabilities such as the allyl radical in Section 10.10.6. In such cases, the Hartree–Fock instability makes the standard approach unsuitable – the BCC and OCC models, by contrast, suffer from no such instabilities.

13.8.2 QUADRATIC CONFIGURATION-INTERACTION THEORY

One popular modification of the standard coupled-cluster model is the *quadratic configuration-interaction (QCI) model*, originally introduced as a size-extensive amendment of the CI model [33]. We here discuss the *QCI singles-and-doubles (QCISD) model* within the framework of similarity-transformed (linked) coupled-cluster theory, from which it is obtained by omitting certain commutators in the CCSD equations. Expanding the remaining commutators, we then go on to express the QCISD equations in a form that illustrates its historical connection to CISD theory.

The starting point for our discussion is equation (13.2.30) for the CCSD energy and equations (13.2.40) and (13.2.41) for the CCSD amplitudes. Omitting from these equations all terms containing commutators that are quadratic or higher in $\hat{T}_1$ and from (13.2.41) also the term containing $[[\hat{H}, \hat{T}_1], \hat{T}_2]$, we arrive at the equations that define the QCISD model:

$$E_{\text{QCI}} = \langle \text{HF}|\hat{H}(1 + \hat{T}_2)|\text{HF}\rangle \tag{13.8.27}$$

$$\langle \mu_1|\hat{H} + [\hat{H}, \hat{T}_1] + [\hat{H}, \hat{T}_2] + [[\hat{H}, \hat{T}_1], \hat{T}_2]|\text{HF}\rangle = 0 \tag{13.8.28}$$

$$\langle \mu_2|\hat{H} + [\hat{H}, \hat{T}_1] + [\hat{H}, \hat{T}_2] + \tfrac{1}{2}[[\hat{H}, \hat{T}_2], \hat{T}_2]|\text{HF}\rangle = 0 \tag{13.8.29}$$

The omissions made in QCISD theory relative to CCSD theory may be justified by the observation that the singles amplitudes are usually smaller than the doubles amplitudes. For further justification, we may invoke perturbation theory, noting that, in Møller–Plesset theory, the doubles amplitudes appear to first order and the singles to second order in the fluctuation potential. Carrying out an order analysis of the terms in the CCSD equations, we find that the QCISD equations are obtained by omitting all terms of orders higher than 4 in the energy and in the singles equations, and all terms of orders higher than 3 in the doubles equations.

Although these arguments provide some justification for QCISD theory *vis-à-vis* CCSD theory, it should be realized that the QCISD model was designed as a revision of CISD theory rather than as a simplification of CCSD theory. For a comparison with CISD, we express the similarity-transformed linked QCISD equations in the corresponding energy-dependent unlinked form. Expanding the commutators in (13.8.28) and (13.8.29), we obtain

$$\langle\mu_1|\hat{H}(1 + \hat{T}_1 + \hat{T}_2 + \hat{T}_1\hat{T}_2)|\text{HF}\rangle = \langle\mu_1|\hat{T}_1\hat{H}(1 + \hat{T}_2)|\text{HF}\rangle \tag{13.8.30}$$

$$\langle\mu_2|\hat{H}\left(1 + \hat{T}_1 + \hat{T}_2 + \tfrac{1}{2}\hat{T}_2^2\right)|\text{HF}\rangle = \langle\mu_2|\hat{T}_2\hat{H}(1 + \hat{T}_2)|\text{HF}\rangle + \langle\mu_2|\hat{T}_1\hat{H}|\text{HF}\rangle \tag{13.8.31}$$

Next, invoking the relations

$$\langle\mu_1|\hat{T}_1 = t_{\mu_1}\langle\text{HF}| \tag{13.8.32}$$

$$\langle\mu_2|\hat{T}_2 = t_{\mu_2}\langle\text{HF}| \tag{13.8.33}$$

the energy expression (13.8.27), and the Brillouin condition to eliminate the last term in (13.8.31), we may write the QCISD equations (13.8.27)–(13.8.29) in the form

$$\langle\text{HF}|\hat{H}(1 + \hat{T}_1 + \hat{T}_2)|\text{HF}\rangle = E_{\text{QCI}} \tag{13.8.34}$$

$$\langle\mu_1|\hat{H}(1 + \hat{T}_1 + \hat{T}_2 + \hat{T}_1\hat{T}_2)|\text{HF}\rangle = E_{\text{QCI}}t_{\mu_1} \tag{13.8.35}$$

$$\langle\mu_2|\hat{H}\left(1 + \hat{T}_1 + \hat{T}_2 + \tfrac{1}{2}\hat{T}_2^2\right)|\text{HF}\rangle = E_{\text{QCI}}t_{\mu_2} \tag{13.8.36}$$

reminiscent of an eigenvalue problem. Indeed, neglecting the terms that are *quadratic* in the cluster amplitudes, we obtain the standard CISD equations in the intermediate normalization:

$$\langle\text{HF}|\hat{H}(1 + \hat{T}_1 + \hat{T}_2)|\text{HF}\rangle = E_{\text{CI}} \tag{13.8.37}$$

$$\langle\mu_1|\hat{H}(1 + \hat{T}_1 + \hat{T}_2)|\text{HF}\rangle = E_{\text{CI}}t_{\mu_1} \tag{13.8.38}$$

$$\langle\mu_2|\hat{H}(1 + \hat{T}_1 + \hat{T}_2)|\text{HF}\rangle = E_{\text{CI}}t_{\mu_2} \tag{13.8.39}$$

However, in doing so, we have sacrificed size-extensivity since we have not consistently eliminated complete commutator expressions in the energy-independent linked equations. For example, in going from (13.8.36) to (13.8.39), we have omitted only the first of the two terms arising from the last commutator in (13.8.29):

$$\tfrac{1}{2}\langle\mu_2|[[\hat{H}, \hat{T}_2], \hat{T}_2]|\text{HF}\rangle = \tfrac{1}{2}\langle\mu_2|\hat{H}\hat{T}_2^2|\text{HF}\rangle - \langle\mu_2|\hat{T}_2\hat{H}\hat{T}_2|\text{HF}\rangle \tag{13.8.40}$$

The second term on the right-hand side of this equation appears as a contribution to the right-hand side of (13.8.36). Comparing the CISD and QCISD equations, we now see that the term 'quadratic' in 'quadratic CI' refers to the quadratic terms introduced in (13.8.35) and (13.8.36) to ensure size-extensivity.

To summarize, the QCISD model constitutes a size-extensive revision of the CISD model closely related to the CCSD model, from which it may be obtained by omitting some of the higher-order terms in the linked equations. The terms omitted in the QCISD equations are not computationally demanding and the cost of the QCISD model is the same as for CCSD. For the calculation of total electronic energies, the performance of the two models is very similar – if anything, the CCSD model performs slightly better than the QCISD model for systems with large singles amplitudes. Moreover, for calculations of molecular properties and excitation energies, the CCSD model is superior. There appear to be few compelling reasons to preserve QCISD as a separate model in quantum chemistry.

13.9 High-spin open-shell coupled-cluster theory

Except for the closed-shell CCSD theory of Section 13.7, the theory presented in this chapter has been that of spin-unrestricted coupled-cluster theory. Spin-unrestricted coupled-cluster theory has the advantage of conceptual simplicity and general applicability and is widely used for open-shell systems. Still, there are considerable disadvantages associated with the spin-unrestricted approach, making it worthwhile to look for an alternative approach for open-shell systems. First, spin-unrestricted coupled-cluster theory suffers from spin contamination, which may adversely affect the calculation of excitation processes and spin-dependent (magnetic) properties. Second, spin-unrestricted theory is expensive since, in the spin-orbital basis, we work with separate sets of orbitals for the alpha and beta spins.

In the closed-shell theory of Section 13.7, we developed a spin-restricted theory in which both of these problems are solved. Thus, for closed-shell systems, we may calculate a singlet coupled-cluster wave function at a fraction of the cost of the corresponding spin-unrestricted wave function. Unfortunately, for open-shell systems, the problems are more complicated and it is no longer obvious how we should best satisfy the Schrödinger and spin equations in coupled-cluster theory.

Closed-shell coupled-cluster theory may be viewed as a special case of spin-orbital coupled-cluster theory, where a closed-shell single-determinant Hartree–Fock state is chosen as the reference state and where the excitation manifold is generated from this state. High-spin open-shell states may likewise be treated in a spin-orbital coupled-cluster framework since such states can be represented by a single-determinant reference state and the associated excitation manifold. In the present section, we discuss generalizations of the spin-orbital formalism, designed so as to reduce the spin contamination of the unrestricted coupled-cluster wave function for high-spin states. In particular, we describe spin-restricted coupled-cluster theory [34], where we use a spin-restricted Hartree–Fock reference state and supplement the coupled-cluster Schrödinger equation with the coupled-cluster spin equations, thereby ensuring that the average value of $\hat{S}^2$ is $S(S+1)$.

13.9.1 SPIN-RESTRICTED COUPLED-CLUSTER THEORY

Let us consider the simultaneous solution of the Schrödinger and spin equations for high-spin open-shell states. The coupled-cluster wave function may be expressed in terms of a single-determinant reference function with the excitation operators chosen to yield excitations from this reference state. The full coupled-cluster wave function, being equivalent to the FCI wave function in the same orbital basis, is a simultaneous eigenfunction of the Schrödinger equation and the spin equations:

$$\hat{H} \exp(\hat{T})|\text{HF}\rangle = E \exp(\hat{T})|\text{HF}\rangle \tag{13.9.1}$$

$$\hat{S}^2 \exp(\hat{T})|\text{HF}\rangle = S(S+1) \exp(\hat{T})|\text{HF}\rangle \tag{13.9.2}$$

$$\hat{S}_z \exp(\hat{T})|\text{HF}\rangle = S \exp(\hat{T})|\text{HF}\rangle \tag{13.9.3}$$

To satisfy these equations, we need only solve the Schrödinger equation (13.9.1) explicitly. If its solution is nondegenerate, the solution of the spin equations (13.9.2) and (13.9.3) follows trivially from the commutation of the Hamiltonian operator with the spin operators.

Let us now consider the solution of the coupled-cluster equations for the energy and the spin in a truncated rather than full space, assuming that the single-determinant high-spin reference state has been determined in an RHF calculation

$$|\text{RHF}\rangle = \left(\prod_v a^\dagger_{v\alpha} \right) \left(\prod_i a^\dagger_{i\alpha} a^\dagger_{i\beta} \right) |\text{vac}\rangle \tag{13.9.4}$$

and therefore satisfies the spin equations

$$\hat{S}^2 |\text{RHF}\rangle = S(S+1)|\text{RHF}\rangle \tag{13.9.5}$$

$$\hat{S}_z |\text{RHF}\rangle = S|\text{RHF}\rangle \tag{13.9.6}$$

The cluster operator $\hat{T}$ of the truncated wave function contains only a subset of the full set of excitation operators. For example, in a CCSD calculation, we include in the cluster operator only the single and double excitations out of the Hartree–Fock reference state. The projection manifold $|\mu\rangle$ is initially chosen in the usual manner – that is, such that it spans all the states that may be created when the cluster operator works on the RHF reference state. The total number of amplitudes (and states in the projection manifold) is denoted by N_D.

In the linked form, the truncated coupled-cluster equations corresponding to the full equations (13.9.1)–(13.9.3) may now be written as

$$\langle \text{RHF}|\hat{H}^\text{T}|\text{RHF}\rangle = E \tag{13.9.7}$$

$$\langle \text{RHF}|[\hat{S}^2]^\text{T}|\text{RHF}\rangle = S(S+1) \tag{13.9.8}$$

$$\langle \text{RHF}|[\hat{S}_z]^\text{T}|\text{RHF}\rangle = S \tag{13.9.9}$$

where the amplitudes satisfy the equations

$$\langle \mu|\hat{H}^\text{T}|\text{RHF}\rangle = 0 \tag{13.9.10}$$

$$\langle \mu|[\hat{S}^2]^\text{T}|\text{RHF}\rangle = 0 \tag{13.9.11}$$

$$\langle \mu|[\hat{S}_z]^\text{T}|\text{RHF}\rangle = 0 \tag{13.9.12}$$

According to (13.9.5) and (13.9.6) and the relation (13.2.24), the spin equations (13.9.8) and (13.9.9) are trivially satisfied and need not concern us further. To obtain the coupled-cluster energy (13.9.7), we first determine the amplitudes that satisfy the projected equations (13.9.10)–(13.9.12) and then substitute these into (13.9.7). However, there are $3N_\text{D}$ projected equations to be solved but only N_D amplitudes in the wave function. In general, therefore, we can satisfy only a subset of the projected equations (13.9.10)–(13.9.12). In spin-unrestricted theory, we solve the N_D equations (13.9.10) and ignore the spin equations (and consequently do not require the reference state to be a spin eigenfunction). In spin-restricted theory, we adopt a different strategy: we first solve the spin equations (13.9.11) and (13.9.12). These equations do not fix the amplitudes uniquely. However, by combining these equations with an appropriately chosen subset of the projected equations (13.9.10), the amplitudes may be determined uniquely, yielding the spin-restricted coupled-cluster wave function.

Let us consider the solution to the projected spin equations. Since (13.9.8) and (13.9.9) are trivially satisfied for the reference state, we may write the projected spin equations (13.9.11) and (13.9.12) in the equivalent unlinked form

$$\langle \mu | \hat{S}^2 \exp(\hat{T}) | \text{RHF} \rangle = S(S+1) \langle \mu | \exp(\hat{T}) | \text{RHF} \rangle \tag{13.9.13}$$

$$\langle \mu | \hat{S}_z \exp(\hat{T}) | \text{RHF} \rangle = S \langle \mu | \exp(\hat{T}) | \text{RHF} \rangle \tag{13.9.14}$$

as discussed for the projected Schrödinger equation in Section 13.2.3. To identify which of these equations are trivially satisfied, we shall consider the spin properties of the cluster operator and of the projection manifold.

First, the cluster operator may be chosen to include only those excitation operators that conserve the spin projection of the RHF state so as to satisfy the commutation relation

$$[\hat{S}_z, \hat{T}] = 0 \tag{13.9.15}$$

This condition gives, for example, the singles cluster operator

$$\hat{T}_1 = \sum_{ai\sigma} t_{i\sigma}^{a\sigma} a_{a\sigma}^\dagger a_{i\sigma} + \sum_{vi} t_{i\beta}^{v\beta} a_{v\beta}^\dagger a_{i\beta} + \sum_{av} t_{v\alpha}^{a\alpha} a_{a\alpha}^\dagger a_{v\alpha} \tag{13.9.16}$$

Note that this operator contains only those excitation operators that do not annihilate $|\text{RHF}\rangle$, so operators such as $a_{a\beta}^\dagger a_{v\beta}$ and $a_{v\alpha}^\dagger a_{i\alpha}$ are not included. This choice of excitation operators ensures that the spin-projection equation is trivially satisfied for the associated projection manifold:

$$\hat{S}_z | \mu \rangle = S | \mu \rangle \tag{13.9.17}$$

We shall not consider conditions (13.9.14) further here, assuming always that the excitation operators have been chosen such that (13.9.15) and therefore (13.9.14) are satisfied.

Next, by diagonalization of $\hat{S}^2$ in the basis of the states $|\mu\rangle$, the projection manifold may be partitioned into two manifolds. The first manifold contains all CSFs $|^S\mu\rangle$ of total spin $S(S+1)$; the second manifold contains the remaining states $|^Q\mu\rangle$:

$$\hat{S}^2 |^S\mu\rangle = S(S+1) |^S\mu\rangle \tag{13.9.18}$$

$$\hat{S}^2 |^Q\mu\rangle \neq S(S+1) |^Q\mu\rangle \tag{13.9.19}$$

As discussed in more detail in Section 13.9.3, the states $|^Q\mu\rangle$ are not necessarily spin eigenstates (CSFs). We denote the number of $|^S\mu\rangle$ states by N_S and the number of $|^Q\mu\rangle$ states by N_Q:

$$N_D = N_S + N_Q \tag{13.9.20}$$

Having partitioned the projection space in this manner, we note that the projected spin equations

$$\langle ^S\mu | \hat{S}^2 \exp(\hat{T}) | \text{RHF} \rangle = S(S+1) \langle ^S\mu | \exp(\hat{T}) | \text{RHF} \rangle \tag{13.9.21}$$

are trivially satisfied. In short, the only spin equations that are not trivially satisfied in spin-restricted coupled-cluster theory are the N_Q equations

$$\langle ^Q\mu | \hat{S}^2 \exp(\hat{T}) | \text{RHF} \rangle = S(S+1) \langle ^Q\mu | \exp(\hat{T}) | \text{RHF} \rangle \tag{13.9.22}$$

Thus, by solving these N_Q equations, we ensure that the N_D spin equations (13.9.11) are satisfied. To determine the coupled-cluster amplitudes uniquely, we then solve the N_S projected equations (13.9.10) in the subset of the states $|^S\mu\rangle$. In spin-restricted theory, we thus satisfy only the N_S projections of the Schrödinger equation against $|^S\mu\rangle$, ignoring the N_Q projections against $|^Q\mu\rangle$.

Let us summarize the *spin-restricted coupled-cluster ansatz* for high-spin open-shell systems [34]. The reference wave function is an RHF state and the projection manifold consists of determinants of the same spin projection as the reference state – for example, all states obtained from the reference state by single and double spin-orbital replacements. The spin-restricted coupled-cluster energy is calculated in the usual manner

$$E = \langle \mathrm{RHF}|\hat{H}^\mathrm{T}|\mathrm{RHF}\rangle \tag{13.9.23}$$

from a coupled-cluster wave function that satisfies the spin equations in the full projection manifold

$$\langle\mu|[\hat{S}^2]^\mathrm{T}|\mathrm{RHF}\rangle = 0 \tag{13.9.24}$$

$$\langle\mu|[\hat{S}_z]^\mathrm{T}|\mathrm{RHF}\rangle = 0 \tag{13.9.25}$$

but where the Schrödinger equation is solved only in the manifold spanned by the CSFs of total spin $S(S+1)$:

$$\langle^S\mu|\hat{H}^\mathrm{T}|\mathrm{RHF}\rangle = 0 \tag{13.9.26}$$

$$\langle^Q\mu|\hat{H}^\mathrm{T}|\mathrm{RHF}\rangle \neq 0 \tag{13.9.27}$$

In practice, the spin-projection equations (13.9.25) are satisfied by including in the cluster operator only those excitation operators that preserve the spin projection, and the total-spin equations (13.9.24) are satisfied by solving equations (13.9.22).

The spin-restricted coupled-cluster theory presented here is given in the spin-orbital basis. As such, there are no immediate computational savings relative to the spin-unrestricted theory. However, the spin equations (13.9.22) may be used to reduce the number of parameters in the projected equations (13.9.26) and thereby the computational cost of the evaluation. However, we do not here go into the practical details of such an approach to the calculation of spin-restricted coupled-cluster wave functions.

13.9.2 TOTAL SPIN OF THE SPIN-RESTRICTED COUPLED-CLUSTER WAVE FUNCTION

Spin-restricted coupled-cluster theory does not provide a wave function that is an eigenfunction of the total spin, just as it does not provide an eigenfunction of the Hamiltonian. Rather, spin-restricted theory gives a wave function for which the spin equations are solved in the truncated projection manifold and the Schrödinger equation in the subspace of spin-adapted CSFs $|^S\mu\rangle$. In a sense, spin-restricted coupled-cluster theory may thus be said to provide a balanced treatment of energy and spin. Neither the spin equations nor the energy equations are satisfied in the orthogonal complement $|^\perp\mu\rangle$ to the projection manifold:

$$\langle^\perp\mu|\hat{H}^\mathrm{T}|\mathrm{RHF}\rangle \neq 0 \tag{13.9.28}$$

$$\langle^\perp\mu|[\hat{S}^2]^\mathrm{T}|\mathrm{RHF}\rangle \neq 0 \tag{13.9.29}$$

However, although spin-restricted coupled-cluster theory does not provide a spin-adapted wave function, it does afford a description where the expectation value of the spin operator is exact.

To show this, consider the evaluation of the average value of the total spin according to the Hellmann–Feynman theorem as discussed in Section 13.5. We first set up a variational Lagrangian for the spin-restricted coupled-cluster energy with the spin operator added as a perturbation, introducing a set of Lagrange multipliers for the projected Schrödinger equations (13.9.26) and for the nontrivial spin equations (13.9.22):

$$L(\alpha) = \langle \mathrm{RHF} | (\hat{H} + \alpha \hat{S}^2) \exp(\hat{T}) | \mathrm{RHF} \rangle + \sum_\mu {}^S\bar{t}_\mu \langle {}^S\mu | \exp(-\hat{T})(\hat{H} + \alpha \hat{S}^2) \exp(\hat{T}) | \mathrm{RHF} \rangle$$

$$+ \sum_\mu {}^Q\bar{t}_\mu \langle {}^Q\mu | [\hat{S}^2 - S(S+1)] \exp(\hat{T}) | \mathrm{RHF} \rangle \tag{13.9.30}$$

The spin-projection constraints (13.9.14) are not included as these are trivially satisfied by the choice of excitation operators. Differentiating this Lagrangian with respect to the perturbational parameter α at $\alpha = 0$, we obtain

$$\left. \frac{dL}{d\alpha} \right|_{\alpha=0} = \langle \mathrm{RHF} | \hat{S}^2 \exp(\hat{T}) | \mathrm{RHF} \rangle + \sum_\mu {}^S\bar{t}_\mu \langle {}^S\mu | \exp(-\hat{T}) \hat{S}^2 \exp(\hat{T}) | \mathrm{RHF} \rangle \tag{13.9.31}$$

The first term may be simplified by using (13.9.8) and the last term vanishes because of (13.9.24):

$$\left. \frac{dL}{d\alpha} \right|_{\alpha=0} = S(S+1) \tag{13.9.32}$$

The average value of the total spin calculated as a derivative of the spin-restricted Lagrangian is therefore equal to $S(S+1)$, in agreement with the exact case.

13.9.3   THE PROJECTION MANIFOLD IN SPIN-RESTRICTED THEORY

In spin-restricted coupled-cluster theory, the projection manifold is partitioned into two spaces: a space containing CSFs of the same spin symmetry as the reference state and a space containing the remaining states. As an illustration, consider a spin-restricted CCSD calculation for a doublet reference state

$$|\mathrm{RHF}\rangle = a_{v\alpha}^\dagger a_{i\alpha}^\dagger a_{i\beta}^\dagger |cs\rangle \tag{13.9.33}$$

where $|cs\rangle$ contains all the doubly occupied orbitals except for the orbital ϕ_i. A double excitation from ϕ_i into the virtual orbitals ϕ_a and ϕ_b gives the two determinants

$$|1\rangle = a_{a\alpha}^\dagger a_{b\beta}^\dagger a_{v\alpha}^\dagger |cs\rangle \tag{13.9.34}$$

$$|2\rangle = a_{a\beta}^\dagger a_{b\alpha}^\dagger a_{v\alpha}^\dagger |cs\rangle \tag{13.9.35}$$

Diagonalizing $\hat{S}^2$ in the basis of these determinants, we obtain the two states:

$$|{}^S1\rangle = \frac{1}{\sqrt{2}} |1\rangle - \frac{1}{\sqrt{2}} |2\rangle \tag{13.9.36}$$

$$|{}^Q1\rangle = \frac{1}{\sqrt{2}} |1\rangle + \frac{1}{\sqrt{2}} |2\rangle \tag{13.9.37}$$

The first state is a pure doublet CSF (see the following two paragraphs) and belongs to the $|{}^S\mu\rangle$ manifold; the second state, which is a mixture of a doublet and a quartet, belongs to $|{}^Q\mu\rangle$.

To see that the state $|^Q1\rangle$ is not a pure CSF, we note that the determinants (13.9.34) and (13.9.35) represent three electrons in three singly occupied orbitals. There are three independent ways of distributing these electrons, giving rise to two doublets and one quartet CSF; see Figure 2.1. In the CCSD model, only one of these CSFs is obtained since the third determinant needed to generate the full set of CSFs is triply excited relative to $|\text{RHF}\rangle$:

$$|3\rangle = a_{a\alpha}^\dagger a_{b\alpha}^\dagger a_{v\beta}^\dagger |\text{cs}\rangle \tag{13.9.38}$$

From this determinant and from the two doubly excited determinants (13.9.34) and (13.9.35), we may construct the following three CSFs, as described in Section 2.6.4:

$$|^21\rangle = \frac{1}{\sqrt{2}}|1\rangle - \frac{1}{\sqrt{2}}|2\rangle \tag{13.9.39}$$

$$|^22\rangle = \sqrt{\frac{2}{3}}|3\rangle - \frac{1}{\sqrt{6}}|1\rangle - \frac{1}{\sqrt{6}}|2\rangle \tag{13.9.40}$$

$$|^41\rangle = \frac{1}{\sqrt{3}}|3\rangle + \frac{1}{\sqrt{3}}|1\rangle + \frac{1}{\sqrt{3}}|2\rangle \tag{13.9.41}$$

In particular, we note that $|^Q1\rangle$ in (13.9.37) may be expressed as

$$|^Q1\rangle = \sqrt{\frac{2}{3}}|^41\rangle - \frac{1}{\sqrt{3}}|^22\rangle \tag{13.9.42}$$

in terms of a quartet and a doublet.

From the simple example in this subsection, we see that the spin-adapted CSFs cannot be characterized in terms of a definite excitation level for spin orbitals relative to the reference state. Thus, although the first doublet $|^21\rangle$ obtained by distributing three electrons among three orbitals is a pure doubly excited state, the second doublet $|^22\rangle$ and the quartet $|^41\rangle$ are mixed doubly and triply excited states.

In calculations of molecular properties other than the energy, it may be better to select excitations based on orbital rather than spin-orbital excitations. In an orbital-based spin-restricted CCSD wave function, for example, we would include the excitation operators that correspond to at most two orbital occupancies in the virtual space, at most two holes in the inactive space and an arbitrary occupancy in the active space. Excitation spaces selected in this manner were discussed in connection with the RAS concept of CI theory; see Section 11.1.3. The orbital-based scheme ensures that $|^Q\mu\rangle$ contains no components of CSFs of total spin S. In an orbital-based spin-restricted CCSD approach, for example, both $|^21\rangle$ and $|^22\rangle$ would be contained in the projection space $|^S\mu\rangle$.

13.9.4 SPIN-ADAPTED CCSD THEORY

The open-shell coupled-cluster states discussed so far are not pure spin states. We shall now consider the generation of *spin-adapted coupled-cluster states* – that is, coupled-cluster wave functions that satisfy the spin equations exactly [35,36]. Again, we consider a state generated by the application of a cluster operator to an open-shell RHF state of the correct spin symmetry:

$$|\text{CC}\rangle = \exp(\hat{T})|\text{RHF}\rangle \tag{13.9.43}$$

To ensure that the coupled-cluster state has the same symmetry as the reference state for any choice of amplitudes, we must require the excitation operators in the cluster operator to be singlet operators, satisfying (2.3.1) and (2.3.2):

$$[\hat{S}_{\pm}, \hat{\tau}_{\mu}] = 0 \tag{13.9.44}$$

$$[\hat{S}_{z}, \hat{\tau}_{\mu}] = 0 \tag{13.9.45}$$

Singlet excitation operators may be generated by the genealogical coupling of doublets of creation operators $\{a_{p\alpha}^{\dagger}, a_{p\beta}^{\dagger}\}$ and annihilation operators $\{-a_{p\beta}, a_{p\alpha}\}$ as described in Section 2.6.7. In the spin-orbital basis, the excitation operators $\hat{\tau}_{\mu}$ contain the inactive annihilation doublets $\{-a_{i\beta}, a_{i\alpha}\}$ and the virtual creation doublets $\{a_{a\alpha}^{\dagger}, a_{a\beta}^{\dagger}\}$. To obtain excitation operators of singlet symmetry, we must, for the active orbitals, include the pairs $\{a_{v\alpha}^{\dagger}, a_{v\beta}^{\dagger}\}$ and $\{-a_{v\beta}, a_{v\alpha}\}$. However, for the high-spin states, the spin-orbital excitation operators contain only $a_{v\beta}^{\dagger}$ and $a_{v\alpha}$ (i.e. the $m_{s} = -\frac{1}{2}$ components). Singlet excitation operators compatible with the specification of the excitation manifold therefore cannot be formed for high-spin states. Consequently, the singlet commutation relations (13.9.44) are incompatible with the cluster commutation relations

$$[\hat{\tau}_{\mu}, \hat{\tau}_{\nu}] = 0 \tag{13.9.46}$$

A coupled-cluster approach based on spin adaptation of the excitation operators may be formulated, but the concomitant loss of commutation among the operators results in a vastly increased algebraic complexity of the theory. For instance, the BCH expansion of the similarity-transformed Hamiltonian no longer terminates after four commutators but only after eight. In passing, we note that the components of the singlet operators not included in the spin-orbital formulation vanish when the singlet operator works directly on the reference state. The difference between spin-restricted and spin-adapted coupled-cluster theories therefore arises from terms that are nonlinear in the cluster amplitudes.

To illustrate the above problems, we consider again the doublet state (13.9.33)

$$|\text{RHF}\rangle = a_{v\alpha}^{\dagger} a_{i\alpha}^{\dagger} a_{i\beta}^{\dagger} |\text{cs}\rangle \tag{13.9.47}$$

and examine the single excitations among the doubly occupied orbital ϕ_{i}, the singly occupied orbital ϕ_{v} and the unoccupied orbital ϕ_{a}. In the spin-orbital basis, we obtain the following excitation operators

$$E_{av}^{\alpha} = a_{a\alpha}^{\dagger} a_{v\alpha} \tag{13.9.48}$$

$$E_{vi}^{\beta} = a_{v\beta}^{\dagger} a_{i\beta} \tag{13.9.49}$$

$$E_{ai}^{\alpha} = a_{a\alpha}^{\dagger} a_{i\alpha} \tag{13.9.50}$$

$$E_{ai}^{\beta} = a_{a\beta}^{\dagger} a_{i\beta} \tag{13.9.51}$$

which commute with one another and with $\hat{S}_{z}$ but not with $\hat{S}_{\pm}$. To satisfy the commutation relations (13.9.44), we must introduce the excitation operators

$$E_{av}^{\beta} = a_{a\beta}^{\dagger} a_{v\beta} \tag{13.9.52}$$

$$E_{vi}^{\alpha} = a_{v\alpha}^{\dagger} a_{i\alpha} \tag{13.9.53}$$

and symmetrization then yields the following singlet excitation operators:

$$E_{av} = E_{av}^{\alpha} + E_{av}^{\beta} \tag{13.9.54}$$

$$E_{vi} = E_{vi}^{\alpha} + E_{vi}^{\beta} \tag{13.9.55}$$

$$E_{ai} = E_{ai}^{\alpha} + E_{ai}^{\beta} \tag{13.9.56}$$

However, these operators do not commute among themselves since

$$[E_{av}, E_{vi}] = E_{ai} \tag{13.9.57}$$

The operators E_{av}^{β} and E_{vi}^{α} are incompatible with a spin-orbital description of the excitation manifold and vanish when applied to the reference state.

To avoid the complicated algebra of a fully spin-adapted coupled-cluster theory, a *partially spin-adapted coupled-cluster theory* has been developed, in which we require the linearized coupled-cluster wave function to be a spin eigenfunction [37,38]:

$$\hat{S}^2 (1 + \hat{T})|\text{RHF}\rangle = S(S + 1)(1 + \hat{T})|\text{RHF}\rangle \tag{13.9.58}$$

In the partially spin-adapted approach, there is no need to introduce, for example, the excitation operator (13.9.52) since the state

$$E_{av}^{\alpha}|\text{RHF}\rangle = a_{a\alpha}^{\dagger} a_{i\alpha}^{\dagger} a_{i\beta}^{\dagger}|\text{cs}\rangle \tag{13.9.59}$$

is still a doublet, just like $|\text{RHF}\rangle$. We also note that the partially spin-adapted approach may be obtained from the spin-restricted coupled-cluster approach by neglecting the nonlinear terms in (13.9.11).

References

[1] O. Sinanoğlu, *J. Chem. Phys.* **36**, 706 (1962).
[2] J. Čížek, *J. Chem. Phys.* **45**, 4256 (1966); *Adv. Chem. Phys.* **14**, 35 (1969).
[3] J. Paldus and J. Čížek, in D. W. Smith and W. B. McRae (eds), *Energy, Structure and Reactivity*, Wiley, 1973, p. 198.
[4] F. E. Harris, *Int. J. Quantum Chem.* **S11**, 403 (1977).
[5] G. D. Purvis III and R. J. Bartlett, *J. Chem. Phys.* **76**, 1910 (1982).
[6] J. Noga and R. J. Bartlett, *J. Chem. Phys.* **86**, 7041 (1987); *J. Chem. Phys.* **89**, 3401 (1988) (erratum).
[7] G. E. Scuseria and H. F. Schaefer III, *Chem. Phys. Lett.* **152**, 382 (1988).
[8] S. A. Kucharski and R. J. Bartlett, *J. Chem. Phys.* **97**, 4282 (1992).
[9] R. J. Bartlett, in D. R. Yarkony (ed.), *Modern Electronic Structure Theory*, World Scientific, 1995, p. 1047.
[10] J. Olsen, P. Jørgensen, H. Koch, A. Balkova and R. J. Bartlett, *J. Chem. Phys.* **104**, 8007 (1996).
[11] R. J. Bartlett and G. D. Purvis, *Int. J. Quantum Chem.* **14**, 561 (1978).
[12] P. Pulay, *Chem. Phys. Lett.* **73**, 393 (1980); *J. Comput. Chem.* **3**, 556 (1982).
[13] G. E. Scuseria, T. J. Lee and H. F. Schaefer III, *Chem. Phys. Lett.* **130**, 236 (1986).
[14] T. Helgaker and P. Jørgensen, *Theor. Chim. Acta* **75**, 111 (1989).
[15] H. Koch, H. J. Aa. Jensen, P. Jørgensen, T. Helgaker, G. E. Scuseria and H. F. Schaefer III, *J. Chem. Phys.* **92**, 4924 (1990).
[16] G. Fitzgerald, R. J. Harrison and R. J. Bartlett, *J. Chem. Phys.* **85**, 5143 (1986).
[17] J. E. Rice and R. D. Amos, *Chem. Phys. Lett.* **122**, 585 (1985).
[18] J. F. Stanton and R. J. Bartlett, *J. Chem. Phys.* **98**, 7029 (1993).

[19] H. Koch, H. J. Aa. Jensen, P. Jørgensen and T. Helgaker, *J. Chem. Phys.* **93**, 3345 (1990).

[20] J. Olsen and P. Jørgensen, *J. Chem. Phys.* **82**, 3235 (1985).

[21] O. Christiansen, P. Jørgensen and C. Hättig, *Int. J. Quantum Chem.* **68**, 1 (1998).

[22] H. Koch, private communication.

[23] H. Koch, O. Christiansen, R. Kobayashi, P. Jørgensen and T. Helgaker, *Chem. Phys. Lett.* **228**, 233 (1994).

[24] P. Pulay, S. Sæbø, and W. Meyer, *J. Chem. Phys.* **81**, 1901 (1984).

[25] T. J. Lee and J. E. Rice, *Chem. Phys. Lett.* **150**, 406 (1988).

[26] G. E. Scuseria, C. L. Janssen and H. F. Schaefer, *J. Chem. Phys.* **89**, 7382 (1988).

[27] H. Koch, A. Sánchez de Merás, T. Helgaker and O. Christiansen, *J. Chem. Phys.* **104**, 4157 (1996).

[28] C. Hampel, K. A. Peterson and H.-J. Werner, *Chem. Phys. Lett.* **190**, 1 (1992).

[29] C. D. Sherrill, A. I. Krylov, E. F. C. Byrd and M. Head-Gordon, *J. Chem. Phys.* **109**, 4171 (1998).

[30] T. B. Pedersen, H. Koch and C. Hättig, *J. Chem. Phys.* **110**, 8318 (1999).

[31] R. A. Chiles and C. E. Dykstra, *J. Chem. Phys.* **74**, 4544 (1981).

[32] N. C. Handy, J. A. Pople, M. Head-Gordon, K. Raghavachari and G. W. Trucks, *Chem. Phys. Lett.* **164**, 185 (1989).

[33] J. A. Pople, M. Head-Gordon and K. Raghavachari, *J. Chem. Phys.* **87**, 5968 (1987).

[34] P. G. Szalay and J. Gauss, *J. Chem. Phys.* **107**, 9028 (1997).

[35] C. L. Janssen and H. F. Schaefer III, *Theor. Chim. Acta* **79**, 1 (1991).

[36] X. Li and J. Paldus, *J. Chem. Phys.* **101**, 8812 (1994).

[37] P. J. Knowles, C. Hampel and H.-J. Werner, *J. Chem. Phys.* **99**, 5219 (1993); *J. Chem. Phys.* **112**, 3106 (2000) (erratum).

[38] P. Neogrády, M. Urban and I. Hubač, *J. Chem. Phys.* **100**, 3706 (1994).

Further reading

R. J. Bartlett, Coupled-cluster theory: an overview of recent developments, in D. R. Yarkony (ed.), *Modern Electronic Structure Theory*, World Scientific, 1995, p. 1047.

T. J. Lee and G. E. Scuseria, Achieving chemical accuracy with coupled-cluster theory, in S. R. Langhoff (ed.), *Quantum-Mechanical Electronic Structure Calculations with Chemical Accuracy*, Kluwer, 1995, p.47.

P. R. Taylor, Coupled-cluster methods in quantum chemistry, in B. O. Roos (ed.), *Lecture Notes in Quantum Chemistry II*, Lecture Notes in Chemistry Vol. 64, Springer-Verlag, 1994, p. 125.

Exercises

EXERCISE 13.1

According to (13.5.8), first-order molecular properties may in coupled-cluster theory be calculated as

$$\left. \frac{dE}{d\alpha} \right|_{\alpha=0} = \langle \Lambda | \hat{V} | CC \rangle \tag{13E.1.1}$$

where $\hat{V}$ is the perturbation operator and where the bra state is defined in terms of the zero-order amplitudes $\hat{T}^{(0)}$ and the zero-order Lagrange multipliers $\bar{t}_\nu^{(0)}$ as follows:

$$\langle \Lambda | = \langle HF | + \sum_\nu \bar{t}_\nu^{(0)} \langle \nu | \exp\left(-\hat{T}^{(0)}\right) \tag{13E.1.2}$$

The zero-order amplitudes satisfy the usual coupled-cluster equations

$$\langle \mu | \exp(-\hat{T}^{(0)}) \hat{H} \exp(\hat{T}^{(0)}) | HF \rangle = 0 \tag{13E.1.3}$$

while the zero-order multipliers are obtained from the linear equations (13.5.6), which we write in the form

$$\eta_\mu + \sum_\nu \bar{t}_\nu^{(0)} A_{\nu\mu} = 0 \tag{13E.1.4}$$

$$\eta_\mu = \langle \text{HF} | \hat{H} \hat{\tau}_\mu | \text{CC} \rangle \tag{13E.1.5}$$

$$A_{\nu\mu} = \langle \nu | \exp(-\hat{T}^{(0)}) [\hat{H}, \hat{\tau}_\mu] | \text{CC} \rangle \tag{13E.1.6}$$

For two noninteracting systems A and B, show that (13E.1.1) gives size-extensive first-order molecular properties.

EXERCISE 13.2

The purpose of this exercise is to develop the singlet doubles operator (13.7.8) by spin coupling of the creation and annihilation operators.

1. In Exercise 2.9, we used the genealogical scheme to determine spin-adapted singlet two-electron excitation operators, valid when all four indices are different:

$$\hat{T}_{pqrs}^\pm = E_{pq} E_{rs} \pm E_{ps} E_{rq} \tag{13E.2.1}$$

   Determine the permutational symmetries of $\hat{T}_{pqrs}^\pm$ for distinct p, q, r and s.
2. Consider next the cases when some indices are identical: $p = r$ and $q \neq s$; $p \neq r$ and $q = s$; and $p = r$ and $q = s$. For these cases, determine the number of linearly independent singlet two-electron excitation operators and express these operators in terms of $\hat{T}_{pqrs}^\pm$.
3. Consider the singlet double excitations for a closed-shell state. Show that the singlet two-electron excitation space is spanned by the operators

$$\hat{T}_{aibj}^+ = E_{ai} E_{bj} + E_{aj} E_{bi}, \quad a \geq b, i \geq j \tag{13E.2.2}$$

$$\hat{T}_{aibj}^- = E_{ai} E_{bj} - E_{aj} E_{bi}, \quad a > b, i > j \tag{13E.2.3}$$

4. According to (13E.2.2) and (13E.2.3), the singlet two-electron excitation space can be parametrized as

$$\hat{T}_2 = \sum_{\substack{a \geq b \\ i \geq j}} c_{ij}^{ab} \hat{T}_{aibj}^+ + \sum_{\substack{a > b \\ i > j}} d_{ij}^{ab} \hat{T}_{aibj}^- \tag{13E.2.4}$$

   Show that this excitation space can also be parametrized as

$$\hat{T}_2 = \frac{1}{2} \sum_{aibj} t_{ij}^{ab} E_{ai} E_{bj} \tag{13E.2.5}$$

   where the excitation amplitudes satisfy the permutation relation

$$t_{ij}^{ab} = t_{ji}^{ba} \tag{13E.2.6}$$

EXERCISE 13.3

1. Verify (13B.1.1) of Box 13.1:

$$[\hat{H}, E_{ai}] = \sum_p h_{pa} E_{pi} - \sum_p h_{ip} E_{ap} + \sum_{pqr} g_{pqra} e_{pqri} - \sum_{pqr} g_{pqir} e_{pqar} \qquad (13E.3.1)$$

2. Verify (13B.2.2) of Box 13.2:

$$[\hat{H}, E_{ai}]|\text{HF}\rangle = 2{}^{\text{I}}F_{ia}|\text{HF}\rangle + \left(\sum_b {}^{\text{I}}F_{ba} E_{bi} - \sum_j {}^{\text{I}}F_{ij} E_{aj} + \sum_{bj} L_{bjia} E_{bj} \right) |\text{HF}\rangle$$

$$+ \left(\sum_{bjc} g_{bjca} E_{bj} E_{ci} - \sum_{bjk} g_{bjik} E_{bj} E_{ak} \right) |\text{HF}\rangle \qquad (13E.3.2)$$

EXERCISE 13.4

Verify that the overlap between the doubly excited states (13.7.49) and (13.7.51) is given by the expression

$$\left\langle \begin{array}{c|c} ab & cd \\ ij & kl \end{array} \right\rangle = 4\delta_{ac}\delta_{bd}\delta_{ik}\delta_{jl} + 4\delta_{ad}\delta_{bc}\delta_{il}\delta_{jk} - 2\delta_{ac}\delta_{bd}\delta_{il}\delta_{jk} - 2\delta_{ad}\delta_{bc}\delta_{ik}\delta_{jl}$$

$$= 2P_{ij}^{ab}(2\delta_{aibj,ckdl} - \delta_{ajbi,ckdl}) = 2P_{kl}^{cd}(2\delta_{aibj,ckdl} - \delta_{ajbi,ckdl}) \qquad (13E.4.1)$$

EXERCISE 13.5

In coupled-cluster theory, first-order one-electron properties may be calculated as the (quasi) expectation value

$$\langle \Lambda|\hat{V}|\text{CC}\rangle = \sum_{pq} D_{pq}^{\Lambda} V_{pq} \qquad (13E.5.1)$$

from the elements of the variational density matrix

$$D_{pq}^{\Lambda} = \langle \Lambda|E_{pq}|\text{CC}\rangle \qquad (13E.5.2)$$

where

$$|\text{CC}\rangle = \exp(\hat{T})|\text{HF}\rangle \qquad (13E.5.3)$$

$$\langle \Lambda| = \left(\langle \text{HF}| + \sum_{\nu} \bar{t}_{\nu} \langle \nu| \right) \exp(-\hat{T}) \qquad (13E.5.4)$$

In this exercise, we consider the evaluation of the variational density matrix (13E.5.2) for a CCSD wave function, for which the cluster operator is written in the usual way

$$\hat{T} = \hat{T}_1 + \hat{T}_2 = \sum_{ai} t_i^a E_{ai} + \frac{1}{2} \sum_{aibj} t_{ij}^{ab} E_{ai} E_{bj} \qquad (13E.5.5)$$

in terms of the singles and doubles amplitudes.

1. Show that, for a closed-shell CCSD wave function, the variational one-electron density matrix (13E.5.2) may be decomposed as

$$D_{pq}^{\Lambda} = D_{pq}^{0} + D_{pq}^{1} + D_{pq}^{2} \qquad (13E.5.6)$$

where

$$D_{pq}^{0} = \langle \mathrm{HF}| E_{pq} \exp(\hat{T})|\mathrm{HF}\rangle \qquad (13E.5.7)$$

$$D_{pq}^{1} = \sum_{ai} \bar{t}_{i}^{a} \left\langle \overline{\begin{matrix} a \\ i \end{matrix}} \,\middle|\, \exp(-\hat{T}) E_{pq} \exp(\hat{T}) \,\middle|\, \mathrm{HF} \right\rangle \qquad (13E.5.8)$$

$$D_{pq}^{2} = \sum_{ai \geq bj} \bar{t}_{ij}^{ab} \left\langle \overline{\begin{matrix} ab \\ ij \end{matrix}} \,\middle|\, \exp(-\hat{T}) E_{pq} \exp(\hat{T}) \,\middle|\, \mathrm{HF} \right\rangle \qquad (13E.5.9)$$

2. Show that the first contribution to the variational density matrix (13E.5.7) may be expressed as

$$D_{pq}^{0} = \delta_{pq} \langle \mathrm{HF}|E_{pp}|\mathrm{HF}\rangle + 2t_{p}^{q} \qquad (13E.5.10)$$

where it is understood that the amplitudes vanish whenever the lower index is virtual or the upper index occupied.

3. Show that the second contribution to the variational density matrix (13E.5.8) may be decomposed as

$$D_{pq}^{1} = D_{pq}^{11} + D_{pq}^{12} \qquad (13E.5.11)$$

where

$$D_{pq}^{11} = \sum_{ai} \bar{t}_{i}^{a} \left\langle \overline{\begin{matrix} a \\ i \end{matrix}} \,\middle|\, E_{pq} + [E_{pq}, \hat{T}_{1}] \,\middle|\, \mathrm{HF} \right\rangle \qquad (13E.5.12)$$

$$D_{pq}^{12} = \sum_{ai} \bar{t}_{i}^{a} \left\langle \overline{\begin{matrix} a \\ i \end{matrix}} \,\middle|\, [E_{pq}, \hat{T}_{2}] + \frac{1}{2}[[E_{pq}, \hat{T}_{1}], \hat{T}_{1}] \,\middle|\, \mathrm{HF} \right\rangle \qquad (13E.5.13)$$

4. Show that the first contribution to (13E.5.11) may be evaluated as

$$D_{pq}^{11} = \bar{t}_{q}^{p} + \sum_{i} \bar{t}_{i}^{p} t_{i}^{q} - \sum_{a} \bar{t}_{q}^{a} t_{p}^{a} \qquad (13E.5.14)$$

5. Demonstrate that the second contribution to (13E.5.11) becomes

$$D_{pq}^{12} = \sum_{ai} \bar{t}_{i}^{a} (2t_{ip}^{aq} - t_{ip}^{qa} - t_{i}^{q} t_{p}^{a}) \qquad (13E.5.15)$$

Hint: Verify and use the identity

$$[E_{pq}, \hat{T}_{2}] + \tfrac{1}{2}[[E_{pq}, \hat{T}_{1}], \hat{T}_{1}]|\mathrm{HF}\rangle$$

$$= \sum_{ckdl} t_{kl}^{cd} E_{ck}[E_{pq}, E_{dl}]|\mathrm{HF}\rangle + \tfrac{1}{2} \sum_{ckdl} (t_{kl}^{cd} + t_{k}^{c} t_{l}^{d})[[E_{pq}, E_{ck}], E_{dl}]|\mathrm{HF}\rangle \qquad (13E.5.16)$$

6. We now begin the evaluation of the last contribution to the variational density matrix (13E.5.9). Show that it consists of the following two parts

$$D_{pq}^{2} = D_{pq}^{21} + D_{pq}^{22} \qquad (13E.5.17)$$

where

$$D_{pq}^{21} = \frac{1}{2} \sum_{aibj} \tilde{t}_{ij}^{ab} \left\langle \overline{\begin{matrix} ab \\ ij \end{matrix}} \middle| [E_{pq}, \hat{T}_2] \middle| HF \right\rangle \tag{13E.5.18}$$

$$D_{pq}^{22} = \frac{1}{2} \sum_{aibj} \tilde{t}_{ij}^{ab} \left\langle \overline{\begin{matrix} ab \\ ij \end{matrix}} \middle| [[E_{pq}, \hat{T}_1], \hat{T}_2] \middle| HF \right\rangle \tag{13E.5.19}$$

In these expressions, we have introduced the modified multipliers

$$\tilde{t}_{ij}^{ab} = \tilde{t}_{ji}^{ba} = (1 + \delta_{ai,bj})\bar{t}_{ij}^{ab}, \quad ai \geq bj \tag{13E.5.20}$$

which are symmetric with respect to permutation of the pair indices ai and bj.

7. Show that the first contribution to (13E.5.17) becomes

$$D_{pq}^{21} = \sum_{aij} \tilde{t}_{ij}^{ap} t_{ij}^{aq} - \sum_{aib} \tilde{t}_{iq}^{ab} t_{ip}^{ab} \tag{13E.5.21}$$

8. Show that the second contribution to (13E.5.17) becomes

$$D_{pq}^{22} = -\sum_{aibj} \tilde{t}_{ij}^{ab} (t_{ip}^{ab} t_j^q + t_{ij}^{aq} t_p^b) \tag{13E.5.22}$$

Hint: Verify and use the relationship

$$[[E_{pq}, \hat{T}_1], \hat{T}_2]|HF\rangle = \sum_{ckdl} t_{kl}^{cd} E_{ck}[[E_{pq}, \hat{T}_1], E_{dl}]|HF\rangle + \frac{1}{2} \sum_{ckdl} t_{kl}^{cd}[[[E_{pq}, \hat{T}_1], E_{ck}], E_{dl}]|HF\rangle \tag{13E.5.23}$$

Summing the various contributions obtained in this exercise, we arrive at the following expression for the variational CCSD density matrix:

$$D_{pq}^\Lambda = D_{pq}^0 + D_{pq}^{11} + D_{pq}^{12} + D_{pq}^{21} + D_{pq}^{22}$$

$$= \delta_{pq}\langle HF|E_{pp}|HF\rangle + 2t_p^q + \bar{t}_q^p + \sum_i \bar{t}_i^p t_i^q - \sum_a \bar{t}_q^a t_p^a + \sum_{ai} \bar{t}_i^a (2t_{ip}^{aq} - t_{ip}^{qa} - t_i^q t_p^a)$$

$$+ \sum_{ija} \tilde{t}_{ij}^{ap} t_{ij}^{aq} - \sum_{abi} \tilde{t}_{iq}^{ab} t_{ip}^{ab} - \sum_{aibj} \tilde{t}_{ij}^{ab} (t_{ip}^{ab} t_j^q + t_{ij}^{aq} t_p^b) \tag{13E.5.24}$$

EXERCISE 13.6

In this exercise, we compare the linked and unlinked coupled-cluster equations with the FCI eigenvalue problem for the H_2 molecule in a minimal basis, containing the gerade (g) and ungerade (u) MOs.

1. Show that the cluster operator becomes

$$\hat{T} = \hat{T}_2 = \frac{1}{2} t_{gg}^{uu} E_{ug} E_{ug} \tag{13E.6.1}$$

and that the biorthogonal basis (13.7.59) can be expressed as

$$\langle D| = \left\langle \overline{\begin{matrix} uu \\ gg \end{matrix}} \middle| = \langle vac|a_{u\beta} a_{u\alpha} \tag{13E.6.2}$$

2. Show that the unlinked coupled-cluster equations can be expressed as

$$\langle \text{HF}|\hat{H}(1+\hat{T}_2)|\text{HF}\rangle = E \tag{13E.6.3}$$

$$\langle \text{D}|\hat{H}(1+\hat{T}_2)|\text{HF}\rangle = E\langle \text{D}|\hat{T}_2|\text{HF}\rangle \tag{13E.6.4}$$

3. Show that the linked coupled-cluster equations can be expressed as

$$\langle \text{HF}|\hat{H}(1+\hat{T}_2)|\text{HF}\rangle = E \tag{13E.6.5}$$

$$\langle \text{D}|(1-\hat{T}_2)\hat{H}(1+\hat{T}_2)|\text{HF}\rangle = 0 \tag{13E.6.6}$$

4. Show that the linked and unlinked coupled-cluster equations are equivalent.
5. Show that (13E.6.3) and (13E.6.4) give the FCI eigenvalue problem in the intermediate normalization.

Solutions

SOLUTION 13.1

For two noninteracting systems, the perturbed Hamiltonian of Section 13.5 separates into two Hamiltonians, one for each system:

$$\hat{H}_{\text{AB}} + \alpha\hat{V}_{\text{AB}} = \hat{H}_{\text{A}} + \hat{H}_{\text{B}} + \alpha\hat{V}_{\text{A}} + \alpha\hat{V}_{\text{B}} \tag{13S.1.1}$$

As discussed in Section 13.3.1, the separability of the unperturbed Hamiltonian in (13S.1.1) ensures that the equation for the amplitudes (13E.1.3) separates into two equations, one for each system. This gives rise to a set of separated unperturbed amplitudes

$$\hat{T}_{\text{AB}}^{(0)} = \hat{T}_{\text{A}}^{(0)} + \hat{T}_{\text{B}}^{(0)} \tag{13S.1.2}$$

ensuring that the similarity-transformed Hamiltonian separates as well

$$\hat{H}_{\text{AB}}^{\text{T}} + \alpha\hat{V}_{\text{AB}}^{\text{T}} = \hat{H}_{\text{A}}^{\text{T}} + \hat{H}_{\text{B}}^{\text{T}} + \alpha\hat{V}_{\text{A}}^{\text{T}} + \alpha\hat{V}_{\text{B}}^{\text{T}} \tag{13S.1.3}$$

where, for example,

$$\hat{H}_{\text{AB}}^{\text{T}} = \exp(-\hat{T}_{\text{AB}}^{(0)})\hat{H}_{\text{AB}}\exp(\hat{T}_{\text{AB}}^{(0)}) \tag{13S.1.4}$$

$$\hat{H}_{\text{A}}^{\text{T}} = \exp(-\hat{T}_{\text{A}}^{(0)})\hat{H}_{\text{A}}\exp(\hat{T}_{\text{A}}^{(0)}) \tag{13S.1.5}$$

$$\hat{V}_{\text{A}}^{\text{T}} = \exp(-\hat{T}_{\text{A}}^{(0)})\hat{V}_{\text{A}}\exp(\hat{T}_{\text{A}}^{(0)}) \tag{13S.1.6}$$

and similarly for the other operators in (13S.1.3). With this separation of the operators, we find that the first-order property may be calculated as

$$\langle \Lambda_{\text{AB}}|\hat{V}_{\text{AB}}|\text{CC}_{\text{AB}}\rangle = \langle \text{HF}_{\text{A}}|\hat{V}_{\text{A}}^{\text{T}}|\text{HF}_{\text{A}}\rangle + \sum_{\nu_{\text{A}}}\bar{t}_{\nu_{\text{A}}}^{(0)}\langle \nu_{\text{A}}|\hat{V}_{\text{A}}^{\text{T}}|\text{HF}_{\text{A}}\rangle + \langle \text{HF}_{\text{B}}|\hat{V}_{\text{B}}^{\text{T}}|\text{HF}_{\text{B}}\rangle$$

$$+ \sum_{\nu_{\text{B}}}\bar{t}_{\nu_{\text{B}}}^{(0)}\langle \nu_{\text{B}}|\hat{V}_{\text{B}}^{\text{T}}|\text{HF}_{\text{B}}\rangle + \sum_{\nu_{\text{A}}\nu_{\text{B}}}\bar{t}_{\nu_{\text{A}}\nu_{\text{B}}}^{(0)}\langle \nu_{\text{A}}\nu_{\text{B}}|\hat{V}_{\text{A}}^{\text{T}} + \hat{V}_{\text{B}}^{\text{T}}|\text{HF}_{\text{A}}\text{HF}_{\text{B}}\rangle \tag{13S.1.7}$$

where the last term vanishes since the matrix element is always zero. To establish full separability of the first-order property, it remains to demonstrate that the multipliers in (13S.1.7) are identical to those obtained when the subsystems are studied separately.

Let us consider the linear multiplier equations (13E.1.4) for the two noninteracting systems. In the notation of Section 13.6.5, where $\{\hat{\tau}_A^\dagger, \hat{\tau}_B^\dagger, \hat{\tau}_{AB}^\dagger\}$ and $\{\hat{\tau}_A, \hat{\tau}_B, \hat{\tau}_{AB}\}$ constitute the row and column labels, respectively, the multiplier equations may be written in the form

$$(\eta_A, \eta_B, \mathbf{0}) + (\bar{\mathbf{t}}_A^{(0)}, \bar{\mathbf{t}}_B^{(0)}, \bar{\mathbf{t}}_{AB}^{(0)}) \begin{pmatrix} \mathbf{A}_{A,A} & \mathbf{0} & \mathbf{A}_{A,AB} \\ \mathbf{0} & \mathbf{A}_{B,B} & \mathbf{A}_{B,AB} \\ \mathbf{0} & \mathbf{0} & \mathbf{A}_{AB,AB} \end{pmatrix} = \mathbf{0} \tag{13S.1.8}$$

In setting up this equation, we have made use of the fact that η in (13E.1.5) contains only components that refer to the individual subsystems. For details about the structure of the Jacobian matrix $\mathbf{A}$, we refer to Section 13.6.5 and in particular to (13.6.47). The first two components of (13S.1.8) may be written in expanded form

$$\eta_A + \bar{\mathbf{t}}_A^{(0)} \mathbf{A}_{AA} = \mathbf{0} \tag{13S.1.9}$$

$$\eta_B + \bar{\mathbf{t}}_B^{(0)} \mathbf{A}_{BB} = \mathbf{0} \tag{13S.1.10}$$

and the multipliers $\bar{\mathbf{t}}_A^{(0)}$ and $\bar{\mathbf{t}}_B^{(0)}$ are therefore the same as those obtained when the two systems are studied separately. The component $\bar{\mathbf{t}}_{AB}^{(0)}$ is nonzero but does not contribute in (13S.1.7) and will not be determined. We may now write (13S.1.7) in the separated form

$$\langle \Lambda_{AB}|\hat{V}_{AB}|CC_{AB}\rangle = \langle HF_A|\hat{V}_A^T|HF_A\rangle + \langle HF_B|\hat{V}_B^T|HF_B\rangle$$
$$+ \sum_{\nu_A} \bar{t}_{\nu_A}^{(0)}\langle \nu_A|\hat{V}_A^T|HF_A\rangle + \sum_{\nu_B} \bar{t}_{\nu_B}^{(0)}\langle \nu_B|\hat{V}_B^T|HF_B\rangle \tag{13S.1.11}$$

which we recognize as the sum of the properties of the two subsystems

$$\langle \Lambda_{AB}|\hat{V}_{AB}|CC_{AB}\rangle = \langle \Lambda_A|\hat{V}_A|CC_A\rangle + \langle \Lambda_B|\hat{V}_B|CC_B\rangle \tag{13S.1.12}$$

thereby establishing size-extensivity of first-order coupled-cluster properties.

SOLUTION 13.2

1. The following symmetries follow directly from the definition (13E.2.1):

$$\hat{T}_{pqrs}^\pm = \pm\hat{T}_{rqps}^\pm = \pm\hat{T}_{psrq}^\pm = \hat{T}_{rspq}^\pm \tag{13S.2.1}$$

2. Consider first the case when $p = r$ and $q \neq s$. The creation operators $a_{p\alpha}^\dagger$ and $a_{p\beta}^\dagger$ can then be coupled only to the singlet $\hat{Q}_{pp}^{0,0}$ – see Section 2.3.3. The annihilation operators must therefore be coupled to a singlet as well – that is, to $(\hat{Q}_{qs}^{0,0})^\dagger$. Thus, only one singlet two-electron excitation operator can be obtained in this case. Similar arguments give a single singlet operator in the other two cases as well. Whereas $\hat{T}_{pqrs}^-$ vanishes in all three cases, $\hat{T}_{pqrs}^+$ does not vanish and can be used to represent the singlet operator.

3. Linear independence is ensured by requiring $a \geq b$ and $i \geq j$. The equality is included only for $\hat{T}_{pqrs}^+$ and not for $\hat{T}_{pqrs}^-$, which vanishes for identical indices – in accordance with the fact that, for two identical indices, spin coupling gives only one singlet excitation operator.

4. To rewrite the restricted summations in (13E.2.4) as free summations, we first introduce the amplitudes

$$C_{ij}^{ab} = (1 + \delta_{ab})(1 + \delta_{ij})c_{ij}^{ab} \tag{13S.2.2}$$

$$D_{ij}^{ab} = (1 - \delta_{ab})(1 - \delta_{ij})d_{ij}^{ab} \tag{13S.2.3}$$

which are required to satisfy the symmetry relations

$$C_{ij}^{ab} = C_{ij}^{ba} = C_{ji}^{ab} = C_{ji}^{ba} \tag{13S.2.4}$$

$$D_{ij}^{ab} = -D_{ij}^{ba} = -D_{ji}^{ab} = D_{ji}^{ba} \tag{13S.2.5}$$

We now arrive at (13E.2.5) in the following manner:

$$\hat{T}_2 = \sum_{\substack{a \geq b \\ i \geq j}} c_{ij}^{ab} \hat{T}_{aibj}^{+} + \sum_{\substack{a > b \\ i > j}} d_{ij}^{ab} \hat{T}_{aibj}^{-} = \frac{1}{4} \sum_{aibj} C_{ij}^{ab} \hat{T}_{aibj}^{+} + \frac{1}{4} \sum_{aibj} D_{ij}^{ab} \hat{T}_{aibj}^{-}$$

$$= \frac{1}{2} \sum_{aibj} C_{ij}^{ab} E_{ai} E_{bj} + \frac{1}{2} \sum_{aibj} D_{ij}^{ab} E_{ai} E_{bj} = \frac{1}{2} \sum_{aibj} t_{ij}^{ab} E_{ai} E_{bj} \tag{13S.2.6}$$

where we have introduced the amplitudes

$$t_{ij}^{ab} = C_{ij}^{ab} + D_{ij}^{ab} \tag{13S.2.7}$$

which satisfy the symmetry relation

$$t_{ij}^{ab} = t_{ji}^{ba} \tag{13S.2.8}$$

Note that the permutational symmetry of these amplitudes is lower than that of the original amplitudes (13S.2.4) and (13S.2.5), enabling us to replace the two quadruple summations in (13S.2.6) by a single quadruple summation.

SOLUTION 13.3

1. The contributions from the one-electron Hamiltonian are obtained straightforwardly. For the two-electron Hamiltonian, we use the commutator (2.3.38):

$$\frac{1}{2} \sum_{pqrs} g_{pqrs} [e_{pqrs}, E_{ai}] = -\frac{1}{2} \sum_{qrs} g_{iqrs} e_{aqrs} + \frac{1}{2} \sum_{prs} g_{pars} e_{pirs} - \frac{1}{2} \sum_{pqs} g_{pqis} e_{pqas} + \frac{1}{2} \sum_{pqr} g_{pqra} e_{pqri}$$

$$= \sum_{pqr} g_{pqra} e_{pqri} - \sum_{pqr} g_{pqir} e_{pqar} \tag{13S.3.1}$$

2. Applying (13E.3.1) to the Hartree–Fock state, we obtain one contribution from each of the four summations on the right-hand side. From a consideration of the effects of the different excitation operators on the Hartree–Fock state, we obtain for the two one-electron contributions:

$$\sum_{p} h_{pa} E_{pi} |\mathrm{HF}\rangle = \left(2h_{ia} + \sum_{b} h_{ba} E_{bi} \right) |\mathrm{HF}\rangle \tag{13S.3.2}$$

$$-\sum_{p} h_{ip} E_{ap} |\mathrm{HF}\rangle = -\sum_{j} h_{ij} E_{aj} |\mathrm{HF}\rangle \tag{13S.3.3}$$

The two-electron contributions are more involved but are obtained in essentially the same manner, expanding the two-electron excitation operator in one-electron operators and considering the possible contributions from occupied and virtual orbitals. Thus, using the notation (13.7.15) for the two-electron integrals, we obtain

$$
\sum_{pqr} g_{pqra} e_{ripq} |\text{HF}\rangle = \sum_{pjr} g_{pjra}(E_{ri}E_{pj} - \delta_{ip}E_{rj})|\text{HF}\rangle
$$

$$
= \left[2\sum_{jr} g_{jjra}E_{ri} + \sum_{bjr} g_{bjra}(E_{bj}E_{ri} - E_{bi}\delta_{rj}) - 2\sum_{j} g_{ijja} - \sum_{jb} g_{ijba}E_{bj} \right]|\text{HF}\rangle
$$

$$
= \left[2\sum_{j} L_{iajj} + \sum_{jb} L_{bajj}E_{bi} + \sum_{bjc} g_{bjca}E_{bj}E_{ci} + \sum_{bj} L_{bjia}E_{bj} \right]|\text{HF}\rangle \quad (13S.3.4)
$$

and

$$
-\sum_{pqr} g_{pqir} e_{arpq} |\text{HF}\rangle = -\sum_{pjr} g_{pjir}(E_{ar}E_{pj} - \delta_{pr}E_{aj})\text{HF}\rangle
$$

$$
= -\left(2\sum_{kj} g_{jjik}E_{ak} + \sum_{bkj} g_{bjik}E_{ak}E_{bj} - \sum_{kj} g_{kjik}E_{aj} \right)|\text{HF}\rangle
$$

$$
= -\left(\sum_{kj} L_{ijkk}E_{aj} + \sum_{bjk} g_{bjik}E_{ak}E_{bj} \right)|\text{HF}\rangle \quad (13S.3.5)
$$

Combining (13S.3.2)–(13S.3.5) and introducing the inactive Fock matrix, we arrive at (13E.3.2).

SOLUTION 13.4

The general strategy for evaluating such an overlap is to move the excitation operators away from the ket towards the bra. Thus, moving E_{ck} to the left, we obtain

$$
\langle\text{HF}|E_{jb}E_{ia}E_{ck}E_{dl}|\text{HF}\rangle = \langle\text{HF}|E_{jb}E_{ck}E_{ia}E_{dl}|\text{HF}\rangle + \langle\text{HF}|E_{jb}[E_{ia}, E_{ck}]E_{dl}|\text{HF}\rangle \quad (13S.4.1)
$$

Whereas the first overlap on the right-hand side is simplified by introducing commutators, the second overlap is evaluated by moving E_{dl} towards the bra:

$$
\langle\text{HF}|E_{jb}E_{ia}E_{ck}E_{dl}|\text{HF}\rangle = \langle\text{HF}|[E_{jb}, E_{ck}][E_{ia}, E_{dl}]|\text{HF}\rangle + \langle\text{HF}|E_{jb}E_{dl}[E_{ia}, E_{ck}]|\text{HF}\rangle
$$
$$
+ \langle\text{HF}|E_{jb}[[E_{ia}, E_{ck}], E_{dl}]|\text{HF}\rangle \quad (13S.4.2)
$$

Finally, we introduce commutators in all three terms:

$$
\langle\text{HF}|E_{jb}E_{ia}E_{ck}E_{dl}|\text{HF}\rangle = \langle\text{HF}|[E_{jb}, E_{ck}][E_{ia}, E_{dl}]|\text{HF}\rangle + \langle\text{HF}|[E_{jb}, E_{dl}][E_{ia}, E_{ck}]|\text{HF}\rangle
$$
$$
+ \langle\text{HF}|[E_{jb}, [[E_{ia}, E_{ck}], E_{dl}]]|\text{HF}\rangle \quad (13S.4.3)
$$

Evaluating each term, we obtain:

$$
\langle\text{HF}|[E_{jb}, E_{ck}][E_{ia}, E_{dl}]|\text{HF}\rangle = \delta_{ad}\delta_{bc}\langle\text{HF}|E_{jk}E_{il}|\text{HF}\rangle = 4\delta_{ad}\delta_{bc}\delta_{il}\delta_{jk} \quad (13S.4.4)
$$

$$\langle \text{HF}|[E_{jb}, E_{dl}][E_{ia}, E_{ck}]|\text{HF}\rangle = \delta_{ac}\delta_{bd}\langle \text{HF}|E_{jl}E_{ik}|\text{HF}\rangle = 4\delta_{ac}\delta_{bd}\delta_{ik}\delta_{jl} \tag{13S.4.5}$$

$$\langle \text{HF}|[E_{jb}, [[E_{ia}, E_{ck}], E_{dl}]]|\text{HF}\rangle = -\delta_{ac}\delta_{il}\langle \text{HF}|[E_{jb}, E_{dk}]|\text{HF}\rangle - \delta_{ik}\delta_{ad}\langle \text{HF}|[E_{jb}, E_{cl}]|\text{HF}\rangle$$

$$= -2\delta_{ac}\delta_{bd}\delta_{il}\delta_{jk} - 2\delta_{ad}\delta_{bc}\delta_{ik}\delta_{jl} \tag{13S.4.6}$$

We arrive at (13E.4.1) by inserting these expressions into (13S.4.3).

SOLUTION 13.5

1. Expression (13E.5.6) follows directly by substituting (13E.5.3) and (13E.5.4) into (13E.5.2), recalling that only the singles and doubles amplitudes and multipliers contribute to the CCSD wave function.

2. Expanding the exponential in (13E.5.7), we obtain the contributions to the density matrix:

$$D^0_{pq} = \langle \text{HF}|E_{pq}|\text{HF}\rangle + \langle \text{HF}|E_{pq}\hat{T}_1|\text{HF}\rangle \tag{13S.5.1}$$

From this expression, the density element (13E.5.10) follows by substitution of the one-electron part of the cluster operator (13E.5.5) and evaluation of a simple matrix element.

3. Carrying out a BCH expansion of the similarity-transformed excitation operator in (13E.5.8) and retaining only those terms that do not vanish upon evaluation, we arrive at (13E.5.11).

4. Expression (13E.5.14) follows upon substitution of the one-electron part of the cluster operator (13E.5.5) into (13E.5.12) and use of (13.7.54) for the biorthogonal basis.

5. To verify (13E.5.16), insert the two-electron part of the cluster operator (13E.5.5) and expand the resulting commutator:

$$[E_{pq}, \hat{T}_2]|\text{HF}\rangle = \tfrac{1}{2}\sum_{ckdl} t^{cd}_{kl}(E_{ck}[E_{pq}, E_{dl}] + [E_{pq}, E_{ck}]E_{dl})|\text{HF}\rangle$$

$$= \sum_{ckdl} t^{cd}_{kl}(E_{ck}[E_{pq}, E_{dl}] + \tfrac{1}{2}[[E_{pq}, E_{ck}], E_{dl}])|\text{HF}\rangle \tag{13S.5.2}$$

We have here also used the permutational symmetry of the amplitudes. From this expression, (13E.5.16) follows directly. Next, substituting (13E.5.16) into (13E.5.13) and using the commutator expression

$$[[E_{pq}, E_{ck}], E_{dl}] = -\delta_{qc}\delta_{pl}E_{dk} - \delta_{pk}\delta_{qd}E_{cl} \tag{13S.5.3}$$

we obtain the desired relation for (13E.5.13)

$$D^{12}_{pq} = \sum_{ai} \bar{t}^a_i \sum_{ckdl} t^{cd}_{kl} \left\langle \left. \begin{matrix} \bar{a} \\ i \end{matrix} \right| E_{ck}(\delta_{qd}E_{pl} - \delta_{pl}E_{dq}) \right| \text{HF}\right\rangle$$

$$- \frac{1}{2}\sum_{ai} \bar{t}^a_i \sum_{ckdl} (t^{cd}_{kl} + t^c_k t^d_l) \left\langle \left. \begin{matrix} \bar{a} \\ i \end{matrix} \right| (\delta_{qc}\delta_{pl}E_{dk} + \delta_{pk}\delta_{qd}E_{cl}) \right| \text{HF}\right\rangle$$

$$= \sum_{ai} \bar{t}^a_i (2t^{aq}_{ip} - t^{qa}_{ip} - t^q_i t^a_p) \tag{13S.5.4}$$

where we have first used the fact that $E_{ck}(\delta_{qd}E_{pl} - \delta_{pl}E_{dq})|\text{HF}\rangle$ has only one single-excitation component $2\delta_{qd}\delta_{pl}E_{ck}|\text{HF}\rangle$ and next used the expression for the biorthogonal overlap (13.7.54).

6. Introducing a free summation over the multipliers

$$\sum_{ai \geq bj} \bar{t}_{ij}^{ab} \left\langle \overline{\begin{matrix} ab \\ ij \end{matrix}} \right| = \frac{1}{2} \sum_{aibj} \tilde{t}_{ij}^{ab} \left\langle \overline{\begin{matrix} ab \\ ij \end{matrix}} \right| \tag{13S.5.5}$$

where the $\tilde{t}_{ij}^{ab}$ are defined in (13E.5.20) and carrying out a BCH expansion of the similarity-transformed excitation operator in (13E.5.9), we arrive at (13E.5.17), retaining only those terms that give a nonzero contribution upon evaluation.

7. Recognizing that only the first term in (13S.5.2) has a doubles component, we obtain from the biorthogonality relation (13.7.55)

$$\begin{aligned}
D_{pq}^{21} &= \frac{1}{2} \sum_{\substack{aibj \\ ckdl}} \tilde{t}_{ij}^{ab} t_{kl}^{cd} \left\langle \overline{\begin{matrix} ab \\ ij \end{matrix}} \right| E_{ck}[E_{pq}, E_{dl}] \Big| \text{HF} \right\rangle \\
&= \frac{1}{2} \sum_{\substack{aibj \\ ckdl}} \tilde{t}_{ij}^{ab} t_{kl}^{cd} \left\langle \overline{\begin{matrix} ab \\ ij \end{matrix}} \right| E_{ck}(E_{pl}\delta_{qd} - E_{dq}\delta_{pl}) \Big| \text{HF} \right\rangle \\
&= \sum_{\substack{aibj \\ ckdl}} \tilde{t}_{ij}^{ab} t_{kl}^{cd} (\delta_{ca}\delta_{pb}\delta_{ki}\delta_{lj}\delta_{qd} - \delta_{ca}\delta_{db}\delta_{ki}\delta_{qj}\delta_{pl}) \\
&= \sum_{aij} \tilde{t}_{ij}^{ap} t_{ij}^{aq} - \sum_{aib} \tilde{t}_{iq}^{ab} t_{ip}^{ab}
\end{aligned} \tag{13S.5.6}$$

8. We first verify (13E.5.23). Carrying out some simple algebra, where we use the symmetry of the doubles amplitudes, we obtain

$$\begin{aligned}
[[E_{pq}, \hat{T}_1], \hat{T}_2]|\text{HF}\rangle &= \frac{1}{2} \sum_{ckdl} t_{kl}^{cd}[[E_{pq}, \hat{T}_1], E_{ck}]E_{dl}|\text{HF}\rangle + \frac{1}{2} \sum_{ckdl} t_{kl}^{cd} E_{dl}[[E_{pq}, \hat{T}_1], E_{ck}]|\text{HF}\rangle \\
&= \sum_{ckdl} t_{kl}^{cd} E_{ck}[[E_{pq}, \hat{T}_1], E_{dl}]|\text{HF}\rangle + \frac{1}{2} \sum_{ckdl} t_{kl}^{cd}[[[E_{pq}, \hat{T}_1], E_{ck}], E_{dl}]|\text{HF}\rangle
\end{aligned} \tag{13S.5.7}$$

Only the first term has a doubles component and will contribute to the density matrix (13E.5.19):

$$\sum_{ckdl} t_{kl}^{cd} E_{ck}[[E_{pq}, \hat{T}_1], E_{dl}]|\text{HF}\rangle = -\sum_{ckdl}(t_{kp}^{cd}t_l^q + t_{kl}^{cq}t_p^d)E_{ck}E_{dl}|\text{HF}\rangle \tag{13S.5.8}$$

Inserting this expression into (13E.5.19) and making use of the biorthogonality relation (13.7.55), we arrive at (13E.5.22):

$$D_{pq}^{22} = -\frac{1}{2} \sum_{\substack{aibj \\ ckdl}} \tilde{t}_{ij}^{ab}(t_{kp}^{cd}t_l^q + t_{kl}^{cq}t_p^d) \left\langle \overline{\begin{matrix} ab \\ ij \end{matrix}} \right| E_{ck}E_{dl} \Big| \text{HF} \right\rangle = -\sum_{aibj} \tilde{t}_{ij}^{ab}(t_{ip}^{ab}t_j^q + t_{ij}^{aq}t_p^b) \tag{13S.5.9}$$

SOLUTION 13.6

1. Since the minimal basis contains no singly excited states of $^1\Sigma_g^+$ symmetry, there is no contribution from $\hat{T}_1$ to $\hat{T}$. The $\hat{T}_2$ operator, on the other hand, contains one excitation of $^1\Sigma_g^+$ symmetry:

$$\hat{T}_2 = \frac{1}{2} t_{gg}^{uu} E_{ug} E_{ug} \tag{13S.6.1}$$

From the expression for the adjoint (13.7.59), we first obtain

$$\left\langle \begin{matrix} \overline{uu} \\ gg \end{matrix} \right| = \frac{1}{2} \left\langle \begin{matrix} uu \\ gg \end{matrix} \right|$$ (13S.6.2)

According to (13.7.49), the state on the right-hand side is given by

$$\left| \begin{matrix} uu \\ gg \end{matrix} \right\rangle = E_{ug} E_{ug} |\mathrm{HF}\rangle = 2 a_{u\alpha}^\dagger a_{u\beta}^\dagger |\mathrm{vac}\rangle$$ (13S.6.3)

Substituting (13S.6.3) into (13S.6.2), we obtain (13E.6.2).

2. The unlinked coupled-cluster Schrödinger equation is given by

$$\hat{H} \exp(\hat{T}_2) |\mathrm{HF}\rangle = E \exp(\hat{T}_2) |\mathrm{HF}\rangle$$ (13S.6.4)

Upon expansion, the exponential terminates after the linear term:

$$\hat{H}(1 + \hat{T}_2)|\mathrm{HF}\rangle = E(1 + \hat{T}_2)|\mathrm{HF}\rangle$$ (13S.6.5)

Projecting against $\langle \mathrm{HF}|$ and $\langle \mathrm{D}|$, we obtain (13E.6.3) and (13E.6.4).

3. The linked coupled-cluster Schrödinger equation is given by

$$\exp(-\hat{T}_2)\hat{H}\exp(\hat{T}_2)|\mathrm{HF}\rangle = E|\mathrm{HF}\rangle$$ (13S.6.6)

Since the exponential terminates after the linear term, we obtain

$$(1 - \hat{T}_2)\hat{H}(1 + \hat{T}_2)|\mathrm{HF}\rangle = E|\mathrm{HF}\rangle$$ (13S.6.7)

Projecting against $\langle \mathrm{HF}|$ and $\langle \mathrm{D}|$, we arrive at (13E.6.5) and (13E.6.6).

4. The unlinked and linked energy expressions (13E.6.3) and (13E.6.5) are identical. It therefore only remains to show that the linked equations (13E.6.6) are identical to the unlinked ones (13E.6.4). Rearranging (13E.6.6) and inserting the resolution of the identity, we obtain

$$\langle \mathrm{D}|\hat{H}(1 + \hat{T}_2)|\mathrm{HF}\rangle = \langle \mathrm{D}|\hat{T}_2\hat{H}(1 + \hat{T}_2)|\mathrm{HF}\rangle = \langle \mathrm{D}|\hat{T}_2|\mathrm{HF}\rangle\langle \mathrm{HF}|\hat{H}(1 + \hat{T}_2)|\mathrm{HF}\rangle$$ (13S.6.8)

Recognizing that the right-hand side contains the energy (13E.6.5), we find that this equation is identical to the unlinked equation (13E.6.4).

5. Invoking the resolution of the identity, we may rewrite the unlinked coupled-cluster equations (13E.6.3) and (13E.6.4) in the form

$$\langle \mathrm{HF}|\hat{H}|\mathrm{HF}\rangle + \langle \mathrm{HF}|\hat{H}|\mathrm{D}\rangle\langle \mathrm{D}|\hat{T}_2|\mathrm{HF}\rangle = E$$ (13S.6.9)

$$\langle \mathrm{D}|\hat{H}|\mathrm{HF}\rangle + \langle \mathrm{D}|\hat{H}|\mathrm{D}\rangle\langle \mathrm{D}|\hat{T}_2|\mathrm{HF}\rangle = E\langle \mathrm{D}|\hat{T}_2|\mathrm{HF}\rangle$$ (13S.6.10)

where $|\mathrm{D}\rangle = a_{u\alpha}^\dagger a_{u\beta}^\dagger |\mathrm{vac}\rangle$. Using (13.7.55), we obtain for the matrix element of the doubles operator:

$$\langle \mathrm{D}|\hat{T}_2|\mathrm{HF}\rangle = \frac{1}{2} t_{gg}^{uu} \left\langle \begin{matrix} \overline{uu} \\ gg \end{matrix} \middle| \begin{matrix} uu \\ gg \end{matrix} \right\rangle = t_{gg}^{uu}$$ (13S.6.11)

Equations (13S.6.9) and (13S.6.10) may now be written in the form

$$\begin{pmatrix} \langle \mathrm{HF}|\hat{H}|\mathrm{HF}\rangle & \langle \mathrm{HF}|\hat{H}|\mathrm{D}\rangle \\ \langle \mathrm{D}|\hat{H}|\mathrm{HF}\rangle & \langle \mathrm{D}|\hat{H}|\mathrm{D}\rangle \end{pmatrix} \begin{pmatrix} 1 \\ t_{gg}^{uu} \end{pmatrix} = E \begin{pmatrix} 1 \\ t_{gg}^{uu} \end{pmatrix}$$ (13S.6.12)

In the intermediate normalization of the eigenvector, these equations represent the FCI eigenvalue problem.

14 PERTURBATION THEORY

In many situations, it is relatively easy (in terms of computer time and resources) to arrive at a reasonably accurate description of the exact wave function, useful for qualitative or preliminary investigations. We may, for example, apply the variation principle to a single-determinant wave function and obtain the Hartree–Fock state. This solution usually recovers more than 99% of the total electronic energy, and other molecular properties are often calculated within 5–10% of the exact values. Although this accuracy may be adequate for some purposes, in many cases higher accuracy is desired and corrections must be applied to the Hartree–Fock solution. In Chapter 13, we explored one possibility, the coupled-cluster approach, which is capable of providing highly accurate solutions to the Schrödinger equation, sufficiently close to the FCI solution to yield results of chemical accuracy. However, the calculation of coupled-cluster wave functions is expensive and it may be profitable to explore an alternative approach to accurate calculations: *perturbation theory*. Perturbation theory provides a different route to the solution of the Schrödinger equation, allowing us to approach the exact solution in a systematic fashion based on an order-by-order expansion of the wave function and the energy.

This chapter, which presents nondegenerate perturbation theory in the context of molecular electronic-structure theory, is divided into seven sections. In Section 14.1, we present the fundamentals of Rayleigh–Schrödinger perturbation theory (RSPT), deriving formulae for the energy and the wave function to arbitrary orders and discussing topics such as Wigner's $2n + 1$ rule, the Hylleraas functional and size-extensivity. Next, in Section 14.2, we give an exposition of Møller–Plesset perturbation theory (MPPT), the most successful application of RSPT in quantum chemistry, where the Fock operator and the Slater determinants in the canonical representation constitute the zero-order system. To low orders, the resulting perturbation expansion gives an efficient scheme for the calculation of electronic energies.

Although the Møller–Plesset energy corrections are size-extensive to each order in the perturbation, they are not termwise size-extensive – that is, the individual terms that contribute to each order in the perturbation are not size-extensive. This problem, common to all methods based on the standard machinery of Rayleigh–Schrödinger theory, is solved by the introduction of coupled-cluster perturbation theory (CCPT) in Section 14.3, where the perturbational energy corrections are derived within the framework of the similarity-transformed formulation of coupled-cluster theory. The resulting expressions are, to each order in the perturbation, equivalent to those of MPPT but contain only size-extensive contributions, with no cancellation of unphysical terms as in MPPT. Next, in Section 14.4, we consider the explicit evaluation of Møller–Plesset corrections to fourth order in the perturbation, based on the termwise size-extensive CCPT formulation of the theory.

Although perturbation theory constitutes a practical tool in quantum chemistry, it is worth keeping in mind that the perturbation expansion in many cases does not converge, as illustrated in Section 5.8.4. It is therefore appropriate to include in this chapter a discussion of the convergence properties of the perturbation expansions in electronic-structure theory. Such a discussion is given in Section 14.5.

In Section 14.6, we examine the relationship between coupled-cluster theory and perturbation theory. After a perturbation analysis of the coupled-cluster models, we consider various hybrid methods, in which perturbational corrections are applied (iteratively and noniteratively) within the framework of coupled-cluster theory. In particular, we apply a triples correction to the CCSD energy, arriving at the highly successful CCSD(T) approximation to the FCI electronic energy.

Most of the theory in the present chapter is concerned with perturbational corrections to states that are dominated by a single electronic configuration, usually represented by a Hartree–Fock wave function. However, in Section 14.7, we consider multiconfigurational generalizations of Møller–Plesset perturbation theory, in particular the second-order perturbation theory developed within the framework of CASSCF theory.

14.1 Rayleigh–Schrödinger perturbation theory

The basic idea of quantum-mechanical perturbation theory is to partition the Hamiltonian operator into two parts

$$\hat{H} = \hat{H}_0 + \hat{U} \tag{14.1.1}$$

where $\hat{H}_0$ is some zero-order Hamiltonian and

$$\hat{U} = \hat{H} - \hat{H}_0 \tag{14.1.2}$$

is the perturbation. The orthonormal eigensolutions to the zero-order Hamiltonian

$$\hat{H}_0 |i^{(0)}\rangle = E_i^{(0)} |i^{(0)}\rangle \tag{14.1.3}$$

are then used to expand the eigenfunction $|0\rangle$ of the exact Hamiltonian (14.1.1):

$$\hat{H} |0\rangle = E |0\rangle \tag{14.1.4}$$

For this purpose, we select an appropriate zero-order state $|0^{(0)}\rangle$ and expand the differences $|0\rangle - |0^{(0)}\rangle$ and $E - E_0^{(0)}$ in orders of $\hat{U}$. Postponing a detailed discussion of convergence of the perturbation expansion to Section 14.5, we note here only that such an order-by-order procedure may converge provided $\hat{H}_0$ incorporates the main features of $\hat{H}$ and provided $\hat{U}$ is in some sense significantly smaller than $\hat{H}_0$.

In the present section, we do not specify the form of the Hamiltonian operator. The results obtained here are therefore valid for any choice of zero-order operator and for any choice of perturbation operator. In the remaining sections of this chapter, we shall be less general and specify in detail the unperturbed and perturbed Hamiltonian operators. In particular, we shall concern ourselves with the application of perturbation theory to the calculation of the electron correlation energy, but it should be understood that the techniques presented here are valid also in other situations. For example, RSPT as presented here provides a convenient theoretical framework for the systematic study of time-independent molecular properties.

We begin this section by deriving the general expressions for the RSPT energy and wave function in Section 14.1.1. For the calculation of the energy to order n, the expressions derived in Section 14.1.1 require the wave function to order $n - 1$. In Section 14.1.2, we develop an alternative scheme, in which the wave function to order n determines the energy to order $2n + 1$, reducing the cost of the calculation in many cases. Next, in Section 14.1.3, we consider the Hylleraas functional, from which the even-order RSPT energies may be obtained by a variational minimization. Finally, in Section 14.1.4, we consider size-extensivity in RSPT.

14.1.1 RSPT ENERGIES AND WAVE FUNCTIONS

We assume that we have partitioned the Hamiltonian operator as in (14.1.1) and that we have at our disposal the exact solutions to the zero-order Hamiltonian (14.1.3). The exact wave function and energy of the full Hamiltonian (14.1.4) are to zero order approximated by $|0^{(0)}\rangle$ and $E^{(0)}$ and may, for higher accuracy, be expanded in orders of the perturbation (14.1.2):

$$|0\rangle = \sum_{k=0}^{\infty} |0^{(k)}\rangle \tag{14.1.5}$$

$$E = \sum_{k=0}^{\infty} E^{(k)} \tag{14.1.6}$$

The zero-order terms in these expansions are known. To determine the higher-order terms (the corrections), we substitute the expansions into the Schrödinger equation (14.1.4)

$$(\hat{H}_0 + \hat{U}) \sum_{k=0}^{\infty} |0^{(k)}\rangle = \left(\sum_{k=0}^{\infty} E^{(k)} \right) \sum_{k=0}^{\infty} |0^{(k)}\rangle \tag{14.1.7}$$

Collecting terms to order n in the perturbation, we obtain

$$(\hat{H}_0 - E^{(0)})|0^{(n)}\rangle = -\hat{U}|0^{(n-1)}\rangle + \sum_{k=1}^{n} E^{(k)}|0^{(n-k)}\rangle \tag{14.1.8}$$

where we have expressed the residual of the zero-order Schrödinger equation with respect to the nth-order wave function $|0^{(n)}\rangle$ (the left-hand side of the equation) in terms of all lower-order wave functions (the right-hand side). We would like to apply this equation recursively, generating higher and higher corrections to the wave function from those of lower orders. We note, however, that (14.1.8) does not determine the component of $|0^{(0)}\rangle$ in $|0^{(n)}\rangle$ since, for any scalar a,

$$(\hat{H}_0 - E^{(0)})(|0^{(n)}\rangle + a|0^{(0)}\rangle) = (\hat{H}_0 - E^{(0)})|0^{(n)}\rangle \tag{14.1.9}$$

We remove this arbitrariness by requiring the corrections to be orthogonal to the zero-order wave function:

$$\langle 0^{(0)}|0^{(k)}\rangle = 0, \qquad k > 0 \tag{14.1.10}$$

This condition is equivalent to the requirement that $|0\rangle$ is intermediately normalized:

$$\langle 0^{(0)}|0\rangle = \sum_{k=0}^{\infty} \langle 0^{(0)}|0^{(k)}\rangle = \langle 0^{(0)}|0^{(0)}\rangle = 1 \tag{14.1.11}$$

With this condition imposed, the master equation (14.1.8) determines the nth-order correction uniquely in terms of the lower-order corrections and the zero-order wave function.

In the intermediate normalization, the exact energy and its corrections may be cast in a particularly simple form. Thus, we extract the exact energy by multiplying the Schrödinger equation (14.1.4) from the left by the zero-order wave function:

$$E = \langle 0^{(0)} | \hat{H} | 0 \rangle \tag{14.1.12}$$

In the same way, the zero-order energy is extracted from the zero-order equation (14.1.3) by multiplying it by the zero-order wave function:

$$E^{(0)} = \langle 0^{(0)} | \hat{H}_0 | 0^{(0)} \rangle \tag{14.1.13}$$

Finally, all corrections to the zero-order energy are obtained from the master equation (14.1.8), multiplying it from the left by the zero-order wave function:

$$E^{(n)} = \langle 0^{(0)} | \hat{U} | 0^{(n-1)} \rangle, \qquad n > 0 \tag{14.1.14}$$

Note that, in this formulation, the wave function to order $n - 1$ determines the energy to order n. The exact energy (14.1.12) is recovered by summing all energy corrections in accordance with (14.1.6).

To obtain closed-form expressions for the wave-function corrections, we multiply (14.1.8) from the left by the inverse of $\hat{H}_0 - E^{(0)}$. This operator has a singularity that appears when it is applied to a wave function containing a component of $|0^{(0)}\rangle$. However, neither side of (14.1.8) contains a component of the zero-order wave function. In the left-hand side, it is absent because of intermediate normalization (14.1.10). For the right-hand side, we find using (14.1.14):

$$\langle 0^{(0)} | \left(-\hat{U} | 0^{(n-1)} \rangle + \sum_{k=1}^{n} E^{(k)} | 0^{(n-k)} \rangle \right) = -E^{(n)} + E^{(n)} = 0 \tag{14.1.15}$$

The inverse of $\hat{H}_0 - E^{(0)}$ is therefore well defined when applied to either side of (14.1.8) and we may thus write

$$|0^{(n)}\rangle = -(\hat{H}_0 - E^{(0)})^{-1} \left(\hat{U} | 0^{(n-1)} \rangle - \sum_{k=1}^{n} E^{(k)} | 0^{(n-k)} \rangle \right) \tag{14.1.16}$$

It is possible to put this equation in a more convenient form by introducing the projection operator

$$\hat{P} = 1 - |0^{(0)}\rangle \langle 0^{(0)}| \tag{14.1.17}$$

Recalling that neither $|0^{(n)}\rangle$ nor the right-hand side of (14.1.8) contains any component of $|0^{(0)}\rangle$, we immediately obtain

$$|0^{(n)}\rangle = -\hat{P}(\hat{H}_0 - E^{(0)})^{-1} \hat{P} \left(\hat{U} | 0^{(n-1)} \rangle - \sum_{k=1}^{n} E^{(k)} | 0^{(n-k)} \rangle \right) \tag{14.1.18}$$

where we have 'padded' the inverse operator from both sides by the projection operator. Since $\hat{P}$ annihilates the component of the zero-order wave function from the last term in (14.1.18), we may write (14.1.16) in the form

$$|0^{(n)}\rangle = -\hat{P}(\hat{H}_0 - E^{(0)})^{-1} \hat{P} \left(\hat{U} | 0^{(n-1)} \rangle - \sum_{k=1}^{n-1} E^{(k)} | 0^{(n-k)} \rangle \right) \tag{14.1.19}$$

Thus, provided $|0^{(k)}\rangle$ and $E^{(k)}$ are known for $k < n$, the relations (14.1.14) and (14.1.19) determine $E^{(n)}$ and $|0^{(n)}\rangle$ uniquely.

We have now derived the basic equations of *Rayleigh–Schrödinger perturbation theory (RSPT)*. To illustrate the Rayleigh–Schrödinger hierarchy, let us write out explicitly the lowest-order corrections to the energy

$$E^{(1)} = \langle 0^{(0)}|\hat{U}|0^{(0)}\rangle \tag{14.1.20}$$

$$E^{(2)} = \langle 0^{(0)}|\hat{U}|0^{(1)}\rangle \tag{14.1.21}$$

$$E^{(3)} = \langle 0^{(0)}|\hat{U}|0^{(2)}\rangle \tag{14.1.22}$$

and to the wave function

$$|0^{(1)}\rangle = -\hat{P}(\hat{H}_0 - E^{(0)})^{-1}\hat{P}\hat{U}|0^{(0)}\rangle \tag{14.1.23}$$

$$|0^{(2)}\rangle = -\hat{P}(\hat{H}_0 - E^{(0)})^{-1}\hat{P}(\hat{U} - E^{(1)})|0^{(1)}\rangle \tag{14.1.24}$$

$$|0^{(3)}\rangle = -\hat{P}(\hat{H}_0 - E^{(0)})^{-1}\hat{P}[(\hat{U} - E^{(1)})|0^{(2)}\rangle - E^{(2)}|0^{(1)}\rangle] \tag{14.1.25}$$

Clearly, we have at our disposal a systematic procedure for improving our description of the electronic system. As demonstrated in Section 5.8, this technique can be very powerful, providing, for the calculation of electron correlation energies, an attractive alternative to the CI and coupled-cluster techniques of Chapters 11 and 13.

Other forms of perturbation theory have been suggested. For example, in *Brillouin–Wigner perturbation theory (BWPT)*, the energy is obtained by an iterative procedure as discussed in Exercise 14.1. However, BWPT is not size-extensive and has proved less useful than RSPT.

14.1.2 WIGNER'S $2n + 1$ RULE

According to the perturbation theory presented in Section 14.1.1, the energy to order n is obtained from the wave function to order $n - 1$. For example, according to (14.1.22), we obtain the third-order energy from the second-order wave function. However, this expression is not necessarily the best one from a computational point of view. A more convenient expression for the third-order energy is arrived at by substituting the second-order wave function (14.1.24) in (14.1.22) and rearranging the resulting expression by means of (14.1.23):

$$\begin{aligned} E^{(3)} &= \langle 0^{(0)}|\hat{U}|0^{(2)}\rangle \\ &= \langle 0^{(0)}|\hat{U}\hat{P}(\hat{H}_0 - E^{(0)})^{-1}\hat{P}(E^{(1)} - \hat{U})|0^{(1)}\rangle \\ &= \langle 0^{(1)}|\hat{U} - E^{(1)}|0^{(1)}\rangle \end{aligned} \tag{14.1.26}$$

It has thus been demonstrated that the third-order energy correction can be calculated directly from the first-order wave function and energy. In situations where the calculation of the second-order wave function is difficult, (14.1.26) can lead to significant computational savings compared with the original expression (14.1.22).

The question now arises as to what savings can be made by making similar rearrangements of the energy to orders higher than 3. More specifically, we may ask what orders of the wave function are needed to calculate the energy to order n. The answer is that the energy to order $2n + 1$ may be calculated from the wave function to orders n and lower; see Exercise 14.2. This result, known as *Wigner's* $2n + 1$ *rule* [1], may lead to substantial computational savings and it is therefore of particular significance for the efficient implementation of perturbation theory as a practical tool in quantum chemistry. We shall therefore here present an alternative derivation of Rayleigh–Schrödinger theory, which – unlike the approach of Section 14.1.1 – directly yields expressions for the energy in agreement with Wigner's $2n + 1$ rule. The procedure is based on the variational Lagrangian technique [2] described in Section 4.2.8.

The material presented in this subsection is self-contained and independent of the results in Section 14.1.1. The notation is also different from that of the preceding subsection, being based on a vector–matrix representation of the electronic states and operators. Thus, we expand the exact and perturbed wave functions in some discrete orthonormal basis $|i\rangle$:

$$|0\rangle = \sum_i C_i |i\rangle \tag{14.1.27}$$

$$|0^{(n)}\rangle = \sum_i C_i^{(n)} |i\rangle \tag{14.1.28}$$

In standard matrix notation, the Schrödinger equation for the exact state may be written as

$$\mathbf{H}(\alpha)\mathbf{C}(\alpha) = E(\alpha)\mathbf{C}(\alpha) \tag{14.1.29}$$

where α is a parameter that represents the strength of the perturbation. The Hamiltonian matrix in (14.1.29) is now partitioned into the zero-order matrix $\mathbf{H}_0$ and the perturbation $\mathbf{U}$

$$\mathbf{H}(\alpha) = \mathbf{H}_0 + \alpha\mathbf{U} \tag{14.1.30}$$

We wish to determine the dependence of the wave function $\mathbf{C}(\alpha)$ and the energy $E(\alpha)$ on α and introduce for this purpose power expansions:

$$E(\alpha) = E^{(0)} + \alpha E^{(1)} + \alpha^2 E^{(2)} + \cdots \tag{14.1.31}$$

$$\mathbf{C}(\alpha) = \mathbf{C}^{(0)} + \alpha\mathbf{C}^{(1)} + \alpha^2\mathbf{C}^{(2)} + \cdots \tag{14.1.32}$$

Our task is thus to derive useful formulae for the calculation of $E^{(n)}$ and $\mathbf{C}^{(n)}$ to a given order n in the perturbation. From these expansions, we may extract an approximation to the exact energy and to the exact wave function in the presence of the perturbation by setting $\alpha = 1$. Clearly, the quality of the calculated energy and wave function depends on the level of truncation in the expansions – provided, of course, that the series converges.

We assume that the zero-order energy and the zero-order wave function at $\alpha = 0$ are known:

$$\mathbf{H}_0\mathbf{C}^{(0)} = E^{(0)}\mathbf{C}^{(0)} \tag{14.1.33}$$

In seeking an order-by-order solution to the Schrödinger equation (14.1.29), it is convenient to use intermediate normalization

$$\mathbf{C}^{(0)^\mathrm{T}}\mathbf{C}(\alpha) = 1 \tag{14.1.34}$$

Introducing the projection operator onto the orthogonal complement to the unperturbed reference state

$$\mathbf{P} = \mathbf{1} - \mathbf{C}^{(0)}\mathbf{C}^{(0)^\mathrm{T}} \tag{14.1.35}$$

we may write the solution vector $\mathbf{C}(\alpha)$ in the form

$$\mathbf{C}(\alpha) = \mathbf{C}^{(0)} + \mathbf{PC}(\alpha) \tag{14.1.36}$$

which makes the intermediate normalization manifest.

We solve the Schrödinger equation (14.1.29) by a projection technique, in a manner reminiscent of that employed for the coupled-cluster energy and wave function in Section 13.2. Thus, the original equation (14.1.29) is equivalent to the following two equations, obtained by projection against the zero-order state and its orthogonal complement:

$$\mathbf{C}^{(0)^T} [\mathbf{H}(\alpha) - E(\alpha)\mathbf{1}]\mathbf{C}(\alpha) = 0 \tag{14.1.37}$$

$$\mathbf{P}[\mathbf{H}(\alpha) - E(\alpha)\mathbf{1}]\mathbf{C}(\alpha) = \mathbf{0} \tag{14.1.38}$$

From the first projection we extract the energy

$$E(\alpha) = \mathbf{C}^{(0)^T} \mathbf{H}(\alpha)\mathbf{C}(\alpha) = \mathbf{C}^{(0)^T} \mathbf{H}(\alpha)\mathbf{C}^{(0)} + \alpha \mathbf{C}^{(0)^T} \mathbf{UPC}(\alpha) \tag{14.1.39}$$

whereas the second projection determines the electronic state $\mathbf{C}(\alpha)$

$$\mathbf{P}[\mathbf{H}(\alpha) - E(\alpha)\mathbf{1}]\mathbf{PC}(\alpha) + \alpha \mathbf{PUC}^{(0)} = \mathbf{0} \tag{14.1.40}$$

To obtain (14.1.39) and (14.1.40), we have inserted (14.1.36) for $\mathbf{C}(\alpha)$ and made use of the zero-order equation (14.1.33) for $\mathbf{C}^{(0)}$.

In Section 14.1.1, we obtained the perturbed energies and wave functions by expanding (14.1.39) and (14.1.40) in orders of the perturbation and by solving the resulting equations to each order separately. We shall here proceed in a different manner but the reader may first wish to derive, from (14.1.39) and (14.1.40), the energy and wave-function corrections in the form

$$E^{(n)} = \mathbf{C}^{(0)^T} \mathbf{UC}^{(n-1)} \tag{14.1.41}$$

$$\mathbf{C}^{(n)} = -\mathbf{P}(\mathbf{H}_0 - E^{(0)}\mathbf{1})^{-1}\mathbf{P}\left(\mathbf{UC}^{(n-1)} - \sum_{k=1}^{n-1} E^{(k)}\mathbf{C}^{(n-k)} \right) \tag{14.1.42}$$

and recognize that these are the matrix representations of (14.1.14) and (14.1.19).

To derive perturbational energy expressions that comply with Wigner's $2n + 1$ rule, we introduce the variational Lagrangian for the Rayleigh–Schrödinger energy. Thus, we regard the energy $E(\alpha)$ in (14.1.39) as calculated variationally subject to the constraints (14.1.40). To this end, we multiply the constraints (14.1.40) by a set of Lagrange multipliers $\overline{\mathbf{C}}(\alpha)$ and add the result to the energy (14.1.39). After some slight rearrangement, the resulting *RSPT Lagrangian* may be written in the form

$$L[\alpha, \mathbf{C}(\alpha), \overline{\mathbf{C}}(\alpha)] = \mathbf{C}^{(0)^T} \mathbf{H}(\alpha)\mathbf{C}^{(0)} + \alpha[\mathbf{C}(\alpha) + \overline{\mathbf{C}}(\alpha)]^T \mathbf{PUC}^{(0)}$$
$$+ \overline{\mathbf{C}}^T(\alpha)\mathbf{P}[\mathbf{H}(\alpha) - E(\alpha)\mathbf{1}]\mathbf{PC}(\alpha) \tag{14.1.43}$$

where $\mathbf{C}(\alpha)$ and the associated multiplier vector $\overline{\mathbf{C}}(\alpha)$ appear in a symmetric fashion. We also note that, due to the presence of the projection operator, we may add any multiple of $\mathbf{C}^{(0)}$ to the multipliers in the Lagrangian without changing its value

$$L[\alpha, \mathbf{C}(\alpha), \overline{\mathbf{C}}(\alpha) + a\mathbf{C}^{(0)}] = L[\alpha, \mathbf{C}(\alpha), \overline{\mathbf{C}}(\alpha)] \tag{14.1.44}$$

We eliminate this arbitrariness by insisting on the intermediate normalization

$$\mathbf{C}^{(0)^{\mathrm{T}}}\overline{\mathbf{C}}(\alpha) = 1 \tag{14.1.45}$$

or equivalently

$$\overline{\mathbf{C}}(\alpha) = \mathbf{C}^{(0)} + \mathbf{P}\overline{\mathbf{C}}(\alpha) \tag{14.1.46}$$

in the same manner as for the original set of parameters $\mathbf{C}(\alpha)$; see (14.1.34) and (14.1.36).

We now consider the variational conditions on the RSPT Lagrangian. Differentiating the Lagrangian (14.1.43), we obtain

$$\frac{\partial L(\alpha)}{\partial \overline{\mathbf{C}}(\alpha)} = \mathbf{P}[\mathbf{H}(\alpha) - E(\alpha)\mathbf{1}]\mathbf{P}\mathbf{C}(\alpha) + \alpha\mathbf{P}\mathbf{U}\mathbf{C}^{(0)} = \mathbf{0} \tag{14.1.47}$$

$$\frac{\partial L(\alpha)}{\partial \mathbf{C}(\alpha)} = \mathbf{P}[\mathbf{H}(\alpha) - E(\alpha)\mathbf{1}]\mathbf{P}\overline{\mathbf{C}}(\alpha) + \alpha\mathbf{P}\mathbf{U}\mathbf{C}^{(0)} = \mathbf{0} \tag{14.1.48}$$

The variational conditions are clearly the same for the wave-function parameters $\mathbf{C}(\alpha)$ and for their multipliers $\overline{\mathbf{C}}(\alpha)$. Since also the zero-order parameters in (14.1.36) and the zero-order multipliers in (14.1.46) are identical, we conclude that $\mathbf{C}(\alpha)$ and $\overline{\mathbf{C}}(\alpha)$ must be identical to all orders

$$\overline{\mathbf{C}}^{(n)} = \mathbf{C}^{(n)} \tag{14.1.49}$$

where

$$\overline{\mathbf{C}}(\alpha) = \mathbf{C}^{(0)} + \alpha\overline{\mathbf{C}}^{(1)} + \alpha^2\overline{\mathbf{C}}^{(2)} + \cdots \tag{14.1.50}$$

As we shall see, this equivalence will simplify the order expansion of the Lagrangian considerably, yielding practical expressions for calculating the perturbed energies in accordance with the $2n + 1$ rule.

Let us now consider the functional form of the RSPT Lagrangian with respect to the perturbation strength α. A power series expansion in α yields the expression

$$L(\alpha) = L^{(0)} + \alpha L^{(1)} + \alpha^2 L^{(2)} + \cdots \tag{14.1.51}$$

where the zero-order term is given by

$$L^{(0)} = \mathbf{C}^{(0)^{\mathrm{T}}}\mathbf{H}_0\mathbf{C}^{(0)} = E^{(0)} \tag{14.1.52}$$

and for higher orders we have

$$L^{(n)} = \mathbf{C}^{(0)^{\mathrm{T}}}\mathbf{U}\mathbf{C}^{(n-1)} + \sum_{k=1}^{n-1} \overline{\mathbf{C}}^{(k)^{\mathrm{T}}}(\mathbf{H}_0 - E^{(0)}\mathbf{1})\mathbf{C}^{(n-k)}$$

$$+ \sum_{k=1}^{n-1} \overline{\mathbf{C}}^{(k)^{\mathrm{T}}}\mathbf{U}\mathbf{C}^{(n-k-1)} - \sum_{k=1}^{n-2}\sum_{m=1}^{n-k-1} E^{(k)}\overline{\mathbf{C}}^{(m)^{\mathrm{T}}}\mathbf{C}^{(n-k-m)} \tag{14.1.53}$$

These expressions are obtained directly from the functional form of the Lagrangian as given in (14.1.43), by collecting terms of order n. In this process, we have made repeated use of the fact that the projection operator $\mathbf{P}$ projects away the zero-order parameters $\mathbf{C}^{(0)}$ and multipliers $\overline{\mathbf{C}}^{(0)}$.

We now observe that, for the optimal choices of $\mathbf{C}(\alpha)$ and $\overline{\mathbf{C}}(\alpha)$, the Lagrangian (14.1.43) and the original energy expression (14.1.39) coincide for all α since the multiplier terms in the

Lagrangian vanish in these cases. The Lagrangian and the energy expression must therefore also be identical to each order in n

$$E_{\text{opt}}^{(n)} = L_{\text{opt}}^{(n)} \tag{14.1.54}$$

and we may hence calculate the perturbed energies from (14.1.53) rather than from (14.1.41). At first sight, this approach does not appear to be a particularly useful one since the functional form of the perturbed Lagrangians (14.1.53) is considerably more complicated than that of the original perturbed energies (14.1.41). However, the crucial difference between these two expressions is the fact that the Lagrangian (14.1.53) is variational in the perturbed wave functions $\mathbf{C}^{(k)}$ and their associated multipliers $\overline{\mathbf{C}}^{(k)}$

$$\frac{\partial L^{(n)}}{\partial \mathbf{C}^{(k)}} = \mathbf{0} \tag{14.1.55}$$

$$\frac{\partial L^{(n)}}{\partial \overline{\mathbf{C}}^{(k)}} = \mathbf{0} \tag{14.1.56}$$

These variational relations are easily obtained from (14.1.47) and (14.1.48). Consider, for example, the first conditions (14.1.55). Using the chain rule, we may write the variational conditions (14.1.48) in the form

$$\frac{\partial L(\alpha)}{\partial C_i^{(k)}} = \sum_m \frac{\partial L(\alpha)}{\partial C_m(\alpha)} \frac{\partial C_m(\alpha)}{\partial C_i^{(k)}} = 0 \tag{14.1.57}$$

Since (14.1.57) holds for arbitrary values of α, all derivatives with respect to α must vanish separately:

$$\frac{\partial^n}{\partial \alpha^n} \frac{\partial L(\alpha)}{\partial \mathbf{C}^{(k)}} = \frac{\partial L^{(n)}}{\partial \mathbf{C}^{(k)}} = \mathbf{0} \tag{14.1.58}$$

The variational conditions (14.1.55) and (14.1.56), which are not shared by the simpler energy expression (14.1.41), can be used to generate the wave-function and multiplier corrections. Here, we shall use these conditions only to simplify the perturbed energies (14.1.53).

We first make the general observation that any function that is linear as well as variational in some parameter must be independent of this parameter since the variational requirement forces the term multiplying this parameter in the function to be zero. Therefore, we may use the variational conditions on the parameters $\mathbf{C}^{(k)}$ and $\overline{\mathbf{C}}^{(k)}$ in the Lagrangian $L^{(n)}$ to eliminate all parameters that occur only linearly in the Lagrangian. By elimination of a given parameter, we here mean that all terms containing this parameter are simultaneously discarded. For example, in the Lagrangian $L^{(n)}$ of (14.1.53), all multipliers occur linearly. As such, they are all candidates for elimination although this would only lead us back to the original energy expression (14.1.41). Obviously, nothing would be gained by this strategy, in particular since, for each vector $\overline{\mathbf{C}}^{(k)}$ that is eliminated, there is one remaining equivalent vector $\mathbf{C}^{(k)}$ – see (14.1.49).

Clearly, we must be cleverer in our choice of vectors for elimination. We first note that, if two different vectors appear in the same term in the Lagrangian, only one of them can be eliminated even though they may both appear linearly in the Lagrangian. The reason for this 'principle of exclusion' is that the removal of the first vector from the expression alters the dependency of $L^{(n)}$ on the second vector since, by dropping all terms containing the first vector, we drop also at least one term containing the second vector. Hence, upon elimination of the first vector, the Lagrangian is no longer variational in the second vector, which must therefore be retained. Guided by this principle, we examine all pairs of vectors $\overline{\mathbf{C}}^{(k)}$ and $\mathbf{C}^{(l)}$ that occur in the Lagrangian (14.1.53) with a view to eliminate those of higher orders. In particular, we shall attempt to eliminate simultaneously $\overline{\mathbf{C}}^{(k)}$ and $\mathbf{C}^{(k)}$ since these are numerically identical and the elimination of only one term would not yield any computational savings.

For RSPT Lagrangians of even order $2n$, an inspection of (14.1.53) reveals that vectors $\overline{\mathbf{C}}^{(k)}$ with $k > n - 1$ and $\mathbf{C}^{(k)}$ with $k > n$ never occur in the same term. They therefore do not interfere with each other and may be simultaneously eliminated, giving the expression

$$L^{(2n)} = \overline{\mathbf{C}}^{(n-1)^{\mathrm{T}}}\mathbf{U}\mathbf{C}^{(n)} - \sum_{k=1}^{2n-2} \sum_{m=\max(1,n-k)}^{\min(n-1,2n-k-1)} E^{(k)}\overline{\mathbf{C}}^{(m)^{\mathrm{T}}}\mathbf{C}^{(2n-m-k)} \tag{14.1.59}$$

Thus, Lagrangians of order $2n$ may be calculated from multipliers of orders $k \leq n - 1$ and wave-function parameters of orders $k \leq n$. For odd-order perturbations, a similar analysis reveals that vectors $\overline{\mathbf{C}}^{(k)}$ with $k > n$ and $\mathbf{C}^{(k)}$ with $k > n$ never appear in the same term in Lagrangians of order $2n + 1$. Elimination of these vectors from (14.1.53) then yields the following expression, containing only vectors of orders $k \leq n$:

$$L^{(2n+1)} = \overline{\mathbf{C}}^{(n)^{\mathrm{T}}}\mathbf{U}\mathbf{C}^{(n)} - \sum_{k=1}^{2n-1} \sum_{m=\max(1,n-k+1)}^{\min(n,2n-k)} E^{(k)}\overline{\mathbf{C}}^{(m)^{\mathrm{T}}}\mathbf{C}^{(2n+1-m-k)} \tag{14.1.60}$$

Combining the results for perturbations of even and odd orders, we conclude that the wave-function parameters to order n determine the RSPT Lagrangian to order $2n + 1$ (*Wigner's $2n + 1$ rule*) [1] and that the associated multipliers to order n determine the Lagrangian to order $2n + 2$ (the $2n + 2$ *rule*) [2].

Since, for the optimal values of the wave-function parameters and the multipliers, the Lagrangian (14.1.53) coincides with the energy (14.1.41) to each order, we arrive at the following expression for the perturbed energies

$$E^{(2n+i)} = \mathbf{C}^{(n+i-1)^{\mathrm{T}}}\mathbf{U}\mathbf{C}^{(n)} - \sum_{k=1}^{2n+i-2} \sum_{m=\max(1,n+i-k)}^{\min(n+i-1,2n+i-k-1)} E^{(k)}\mathbf{C}^{(m)^{\mathrm{T}}}\mathbf{C}^{(2n+i-m-k)} \tag{14.1.61}$$

Here we have combined the expressions for even and odd orders by introducing an integer i, where $i = 0$ for even-order energies and $i = 1$ for odd-order energies. Also, we have taken the opportunity to write the energy entirely in terms of the wave-function parameters $\mathbf{C}^{(k)}$ since, according to (14.1.49), the vectors $\overline{\mathbf{C}}^{(k)}$ and $\mathbf{C}^{(k)}$ are identical to all orders.

Equation (14.1.61) is our final expression for the perturbed energies. It expresses the energy to order $2n + 1$ in terms of wave-function corrections of orders n and lower, in accordance with Wigner's $2n + 1$ rule. As such, this expression is superior to the original one (14.1.41), which requires wave-function corrections of orders $n - 1$ and lower to evaluate the energy to order n. From (14.1.61), it is easy to generate energy corrections to any order in accordance with Wigner's $2n + 1$ rule. For easy reference, we here give the expressions to orders 5 and lower:

$$E^{(1)} = \mathbf{C}^{(0)^{\mathrm{T}}}\mathbf{U}\mathbf{C}^{(0)} \tag{14.1.62}$$

$$E^{(2)} = \mathbf{C}^{(0)^{\mathrm{T}}}\mathbf{U}\mathbf{C}^{(1)} \tag{14.1.63}$$

$$E^{(3)} = \mathbf{C}^{(1)^{\mathrm{T}}}(\mathbf{U} - E^{(1)}\mathbf{1})\mathbf{C}^{(1)} \tag{14.1.64}$$

$$E^{(4)} = \mathbf{C}^{(1)^{\mathrm{T}}}(\mathbf{U} - E^{(1)}\mathbf{1})\mathbf{C}^{(2)} - E^{(2)}\mathbf{C}^{(1)^{\mathrm{T}}}\mathbf{C}^{(1)} \tag{14.1.65}$$

$$E^{(5)} = \mathbf{C}^{(2)^{\mathrm{T}}}(\mathbf{U} - E^{(1)}\mathbf{1})\mathbf{C}^{(2)} - 2E^{(2)}\mathbf{C}^{(1)^{\mathrm{T}}}\mathbf{C}^{(2)} - E^{(3)}\mathbf{C}^{(1)^{\mathrm{T}}}\mathbf{C}^{(1)} \tag{14.1.66}$$

Only the first two energies are identical to those obtained from the original RSPT expression (14.1.41) – all others contain contributions from lower-order energy corrections. The wave-function corrections are obtained from (14.1.42).

A few comments should be made before leaving this topic. First, although the energy expressions that obey the $2n + 1$ rule as given by (14.1.61) have been arrived at by means of a variational Lagrangian, it should be realized that these final expressions are not themselves variational. Indeed, the variational property of the Lagrangian was lost during the process of elimination. Also, the variational property that made this elimination possible has nothing to do with upper bounds in the sense of, for example, Hartree–Fock and CI energies – it is related only to the stationary requirements on the variational parameters. Therefore, although the Lagrangian (14.1.53) is variational, it does not represent an upper bound to the calculated energy.

In our discussion of perturbational energies, we have found that the wave-function parameters of order n determine the energy to order $2n + 1$, but that the Lagrange multipliers obey an even stronger $2n + 2$ rule. The $2n + 2$ rule is a general result for the multipliers of any variational Lagrangian and occurs since the multipliers enter the Lagrangian in a linear fashion [2]. The $2n + 2$ rule is of considerable computational significance in, for example, the calculation of energy derivatives. For the first five energy corrections, we have in Table 14.1 listed the required wave-function and multiplier corrections. The present case is rather special, however, in that, order by order, the multipliers are numerically identical to the wave-function parameters. Therefore, the benefits of the $2n + 2$ rule are of no significance for the calculation of RSPT energies.

14.1.3 THE HYLLERAAS FUNCTIONAL

A hierarchy of functionals exists from which the Rayleigh–Schrödinger energies of even orders and wave functions of general orders may be obtained by a variational procedure. In the *Hylleraas variation method*, the energy of order $2n$ is calculated by a minimization of the *Hylleraas functional* of order $2n$ [3]:

$$E^{(2n)} = \min_{\boldsymbol{\zeta}} J_{\mathrm{H}}^{(2n)}(\boldsymbol{\zeta}) \tag{14.1.67}$$

At the minimum of the Hylleraas functional (to be constructed below), the variational parameters $\boldsymbol{\zeta}$ determine the wave function to order n. In some situations, the Hylleraas method represents a useful alternative to standard perturbation theory, as we shall see in the present subsection.

The Hylleraas functional is easily related to the variational Lagrangian (14.1.53) of Rayleigh–Schrödinger theory. We recall that the RSPT Lagrangian $L^{(2n)}(\mathbf{C}^{(k)}, \overline{\mathbf{C}}^{(k)})$ is symmetric with respect to the wave-function parameters $\mathbf{C}^{(k)}$ and the multipliers $\overline{\mathbf{C}}^{(k)}$, and that the optimal values of $\mathbf{C}^{(k)}$ and $\overline{\mathbf{C}}^{(k)}$ are identical. We now introduce a reduced Lagrangian according to the prescription

$$J^{(2n)}(\mathbf{C}^{(k)}) = L^{(2n)}(\mathbf{C}^{(k)}, \mathbf{C}^{(k)}) \tag{14.1.68}$$

Table 14.1 Wigner's $2n + 1$ rule for the wave-function parameters and the $2n + 2$ rule for the Lagrange multipliers

	$E^{(0)}$	$E^{(1)}$	$E^{(2)}$	$E^{(3)}$	$E^{(4)}$	$E^{(5)}$
$\mathbf{C}^{(k)}$	0	0	1	1	2	2
$\overline{\mathbf{C}}^{(k)}$	0	0	0	1	1	2

Clearly, the stationary points of the reduced Lagrangian coincide with the stationary points of the full Lagrangian. Moreover, the reduced Lagrangian is linear in $\mathbf{C}^{(k)}$ for $k > n$, just like the full Lagrangian. We may therefore manipulate the reduced Lagrangian (14.1.68) in the same manner as we manipulated the Lagrangian (14.1.53) in Section 14.1.2. Thus, we generate the Rayleigh–Schrödinger wave-function corrections from the conditions

$$\frac{\partial J^{(2n)}}{\partial \mathbf{C}^{(k)}} = 2\frac{\partial L^{(2n)}}{\partial \mathbf{C}^{(k)}} = \mathbf{0}, \quad k > n \tag{14.1.69}$$

and calculate the even-order energies from the stationary point of the reduced Lagrangian:

$$E_{\text{opt}}^{(2n)} = J_{\text{opt}}^{(2n)} \tag{14.1.70}$$

At the stationary point defined by the variational conditions (14.1.69), we simplify the reduced Lagrangian according to the $2n + 1$ rule, arriving at an expression of the form

$$J_{\text{opt}}^{(2n)} = \mathbf{C}^{(n)\text{T}}(\mathbf{H}_0 - E^{(0)}\mathbf{1})\mathbf{C}^{(n)} + 2\mathbf{C}^{(n)\text{T}}\left(\mathbf{U}\mathbf{C}^{(n-1)} - \sum_{k=1}^{n-1}E^{(k)}\mathbf{C}^{(n-k)}\right)$$
$$- \sum_{k=1}^{2n-2}\sum_{m=\max(1,n-k+1)}^{\min(n-1,2n-k-1)}E^{(k)}\mathbf{C}^{(m)\text{T}}\mathbf{C}^{(2n-k-m)} \tag{14.1.71}$$

In the process leading to this expression, we have abstained from modifying or eliminating any of the terms that contain $\mathbf{C}^{(n)}$. The reduced Lagrangian (14.1.71) therefore remains variational with respect to $\mathbf{C}^{(n)}$. Based on this observation, we introduce the *Hylleraas functional* as

$$J_{\text{H}}^{(2n)}(\boldsymbol{\zeta}) = \boldsymbol{\zeta}^{\text{T}}(\mathbf{H}_0 - E^{(0)}\mathbf{1})\boldsymbol{\zeta} + 2\boldsymbol{\zeta}^{\text{T}}\left(\mathbf{U}\mathbf{C}^{(n-1)} - \sum_{k=1}^{n-1}E^{(k)}\mathbf{C}^{(n-k)}\right)$$
$$- \sum_{k=1}^{2n-2}\sum_{m=\max(1,n-k+1)}^{\min(n-1,2n-k-1)}E^{(k)}\mathbf{C}^{(m)\text{T}}\mathbf{C}^{(2n-k-m)} \tag{14.1.72}$$

with $\boldsymbol{\zeta}$ required to be orthogonal to $\mathbf{C}^{(0)}$. Note that the Hylleraas functional is a function of $\boldsymbol{\zeta}$ only. The remaining parameters $\mathbf{C}^{(k)}$ with $k < n$ are fixed and must be known before the functional is constructed. Differentiating (14.1.72) with respect to $\boldsymbol{\zeta}$ and comparing with the Rayleigh–Schrödinger corrections (14.1.42), the reader may easily verify that the stationary point of the Hylleraas functional occurs at $\boldsymbol{\zeta} = \mathbf{C}^{(n)}$. Moreover, the stationary point is a minimum (for ground-state calculations) since the Hessian matrix of the Hylleraas functional

$$\frac{\mathrm{d}^2 J_{\text{H}}^{(2n)}}{\mathrm{d}\boldsymbol{\zeta}^2} = 2(\mathbf{H}_0 - E^{(0)}\mathbf{1}) \tag{14.1.73}$$

is positive semidefinite. Finally, according to (14.1.70), the value at the minimum gives the energy to order $2n$.

In conclusion, we may calculate the nth-order Rayleigh–Schrödinger correction to the ground-state wave function by minimizing the Hylleraas functional $J_{\text{H}}^{(2n)}$. The minimum of the functional corresponds to the energy $E^{(2n)}$. This scheme constitutes the *Hylleraas variation method*. By repeated application of the Hylleraas variation method, we may generate the wave function to any order in the perturbation.

In some cases, it is better to calculate $E^{(2n)}$ by the Hylleraas variation method than from the standard expression of perturbation theory. Assume that, instead of the exact wave function $\mathbf{C}^{(n)}$, we have at our disposal only an approximation:

$$\tilde{\mathbf{C}}^{(n)} = \mathbf{C}^{(n)} + \boldsymbol{\delta} \tag{14.1.74}$$

Inserting the approximate vector $\tilde{\mathbf{C}}^{(n)}$ into the Hylleraas functional, we obtain

$$J_{\mathrm{H}}^{(2n)}(\tilde{\mathbf{C}}^{(n)}) = E^{(2n)} + \boldsymbol{\delta}^{\mathrm{T}}(\mathbf{H}_0 - E^{(0)}\mathbf{1})\boldsymbol{\delta} \tag{14.1.75}$$

Thus, if the error in $\mathbf{C}^{(n)}$ is of order $\|\boldsymbol{\delta}\|$, then the error in $E^{(2n)}$ will be positive and proportional to $\|\boldsymbol{\delta}\|^2$ provided the energy has been obtained from the Hylleraas functional. If instead we had used the standard Lagrangian, simplified to comply with the $2n + 1$ and $2n + 2$ rules, an error of magnitude $\|\boldsymbol{\delta}\|$ in the wave function $\mathbf{C}^{(n)}$ would give rise to an error (positive or negative) proportional to $\|\boldsymbol{\delta}\|$ in the energy.

The Hylleraas variation method has the advantage that we can calculate the wave-function corrections $\mathbf{C}^{(n)}$ from variations in a nonlinear set of parameters that define the electronic state. We are thus not restricted to a linear variational space. As a bonus, the error in $E^{(2n)}$ is quadratic in the error in the nth-order wave function. The lower-order corrections ($\mathbf{C}^{(k)}$ with $k < n$) must, however, be accurately calculated.

14.1.4 SIZE-EXTENSIVITY IN RSPT

The question of size-extensivity in RSPT is a difficult one. Although it is easy to verify that size-extensive energies are obtained to low orders [4], the general proof of size-extensivity for a separable zero-order Hamiltonian in RSPT is unknown to the authors. We shall here examine the expressions for the RSPT energy to third order, verifying that these are size-extensive. At the same time, we shall see that the RSPT energies are not termwise size-extensive. In short, in Rayleigh–Schrödinger theory, size-extensivity occurs order by order rather than term by term. In Section 14.3 we shall demonstrate that, in CCPT, size-extensivity occurs term by term as well as order by order, making this particular formulation of perturbation theory especially well suited to molecular calculations.

We consider a supersystem consisting of two noninteracting subsystems A and B. The Hamiltonian is separable

$$\hat{H}_{\mathrm{AB}} = \hat{H}_{\mathrm{A}} + \hat{H}_{\mathrm{B}} \tag{14.1.76}$$

and we assume that the unperturbed Hamiltonian (and hence the perturbation) is separable as well:

$$\hat{H}_{0,\mathrm{AB}} = \hat{H}_{0,\mathrm{A}} + \hat{H}_{0,\mathrm{B}} \tag{14.1.77}$$

$$\hat{U}_{\mathrm{AB}} = \hat{U}_{\mathrm{A}} + \hat{U}_{\mathrm{B}} \tag{14.1.78}$$

Since the zero-order Hamiltonian is additively separable, the zero-order wave function (which is an eigenfunction of the zero-order Hamiltonian) is multiplicatively separable:

$$|0_{\mathrm{AB}}^{(0)}\rangle = |0_{\mathrm{A}}^{(0)}0_{\mathrm{B}}^{(0)}\rangle \tag{14.1.79}$$

We shall now examine the size-extensivity of the RSPT energies to third order as given by (14.1.62)–(14.1.64).

The first-order Rayleigh–Schrödinger energy (14.1.62) may be calculated from the unperturbed compound state and is given by

$$
\begin{aligned}
E_{AB}^{(1)} &= \langle 0_A^{(0)} 0_B^{(0)} | \hat{U}_A + \hat{U}_B | 0_A^{(0)} 0_B^{(0)} \rangle \\
&= \langle 0_A^{(0)} | \hat{U}_A | 0_A^{(0)} \rangle \langle 0_B^{(0)} | 0_B^{(0)} \rangle + \langle 0_A^{(0)} | 0_A^{(0)} \rangle \langle 0_B^{(0)} | \hat{U}_B | 0_B^{(0)} \rangle \\
&= E_A^{(1)} + E_B^{(1)}
\end{aligned}
\tag{14.1.80}
$$

The first-order energy is thus size-extensive, reducing to the sum of the energies obtained by applying RSPT to the two subsystems separately.

For the second- and third-order Rayleigh–Schrödinger energies, we need the first-order wave function (14.1.16). For the compound system, it may be written in the form

$$
|0_{AB}^{(1)}\rangle = -(\hat{K}_A + \hat{K}_B)^{-1}(\hat{V}_A + \hat{V}_B)|0_{AB}^{(0)}\rangle
\tag{14.1.81}
$$

where we have introduced the notation

$$
\hat{K}_A = \hat{H}_{0,A} - E_A^{(0)}
\tag{14.1.82}
$$

$$
\hat{V}_A = \hat{U}_A - E_A^{(1)}
\tag{14.1.83}
$$

and likewise for system B. The wave function (14.1.81) consists of two contributions. Invoking the operator identity

$$
(\hat{A} + \hat{B})^{-1} = \hat{A}^{-1} - (\hat{A} + \hat{B})^{-1}\hat{B}\hat{A}^{-1}
\tag{14.1.84}
$$

the first contribution may be rewritten as

$$
\begin{aligned}
-(\hat{K}_A + \hat{K}_B)^{-1}\hat{V}_A|0_{AB}^{(0)}\rangle &= -\hat{K}_A^{-1}\hat{V}_A|0_A^{(0)} 0_B^{(0)}\rangle + (\hat{K}_A + \hat{K}_B)^{-1}\hat{K}_B\hat{K}_A^{-1}\hat{V}_A|0_A^{(0)} 0_B^{(0)}\rangle \\
&= -\hat{K}_A^{-1}\hat{V}_A|0_A^{(0)} 0_B^{(0)}\rangle = |0_A^{(1)} 0_B^{(0)}\rangle
\end{aligned}
\tag{14.1.85}
$$

where we have commuted $\hat{K}_B$ with $\hat{K}_A^{-1}\hat{V}_A$ and used the relation

$$
\hat{K}_B|0_B^{(0)}\rangle = 0
\tag{14.1.86}
$$

The second contribution to (14.1.81) may be rewritten in the same way and we obtain the simple expression

$$
|0_{AB}^{(1)}\rangle = |0_A^{(1)} 0_B^{(0)}\rangle + |0_A^{(0)} 0_B^{(1)}\rangle
\tag{14.1.87}
$$

for the first-order perturbed compound wave function. Thus, the perturbed wave function may be written as a superposition of two states, each being a product of the unperturbed and perturbed states of the subsystems. It is important to realize that the perturbed wave function is *not* in product form – that is, it is not multiplicatively separable. Rather, it consists of those terms in the product wave function that are correct to first order. The form (14.1.87) is, however, sufficient to ensure size-extensivity of the second- and third-order energies, to which we now turn our attention.

Size-extensivity of the second-order RSPT energy is easily established. Thus, inserting the perturbed wave function (14.1.87) in the energy expression (14.1.63), we obtain the size-extensive correction

$$
\begin{aligned}
E_{\mathrm{AB}}^{(2)} &= \langle 0_{\mathrm{AB}}^{(0)} | \hat{U}_{\mathrm{A}} + \hat{U}_{\mathrm{B}} | 0_{\mathrm{AB}}^{(1)} \rangle \\
&= \langle 0_{\mathrm{A}}^{(0)} 0_{\mathrm{B}}^{(0)} | \hat{U}_{\mathrm{A}} | 0_{\mathrm{A}}^{(1)} 0_{\mathrm{B}}^{(0)} \rangle + \langle 0_{\mathrm{A}}^{(0)} 0_{\mathrm{B}}^{(0)} | \hat{U}_{\mathrm{B}} | 0_{\mathrm{A}}^{(0)} 0_{\mathrm{B}}^{(1)} \rangle \\
&= E_{\mathrm{A}}^{(2)} + E_{\mathrm{B}}^{(2)}
\end{aligned}
\tag{14.1.88}
$$

Finally, we consider the third-order energy (14.1.64), which in the notation

$$
\hat{V}_{\mathrm{AB}} = \hat{V}_{\mathrm{A}} + \hat{V}_{\mathrm{B}}
\tag{14.1.89}
$$

may be written as

$$
\begin{aligned}
E_{\mathrm{AB}}^{(3)} &= \langle 0_{\mathrm{AB}}^{(1)} | \hat{V}_{\mathrm{AB}} | 0_{\mathrm{AB}}^{(1)} \rangle \\
&= \langle 0_{\mathrm{A}}^{(1)} 0_{\mathrm{B}}^{(0)} | \hat{V}_{\mathrm{AB}} | 0_{\mathrm{A}}^{(1)} 0_{\mathrm{B}}^{(0)} \rangle + \langle 0_{\mathrm{A}}^{(1)} 0_{\mathrm{B}}^{(0)} | \hat{V}_{\mathrm{AB}} | 0_{\mathrm{A}}^{(0)} 0_{\mathrm{B}}^{(1)} \rangle \\
&\quad + \langle 0_{\mathrm{A}}^{(0)} 0_{\mathrm{B}}^{(1)} | \hat{V}_{\mathrm{AB}} | 0_{\mathrm{A}}^{(1)} 0_{\mathrm{B}}^{(0)} \rangle + \langle 0_{\mathrm{A}}^{(0)} 0_{\mathrm{B}}^{(1)} | \hat{V}_{\mathrm{AB}} | 0_{\mathrm{A}}^{(0)} 0_{\mathrm{B}}^{(1)} \rangle
\end{aligned}
\tag{14.1.90}
$$

For the first contribution to (14.1.90), we obtain

$$
\langle 0_{\mathrm{A}}^{(1)} 0_{\mathrm{B}}^{(0)} | \hat{V}_{\mathrm{AB}} | 0_{\mathrm{A}}^{(1)} 0_{\mathrm{B}}^{(0)} \rangle = \langle 0_{\mathrm{A}}^{(1)} | \hat{V}_{\mathrm{A}} | 0_{\mathrm{A}}^{(1)} \rangle \langle 0_{\mathrm{B}}^{(0)} | 0_{\mathrm{B}}^{(0)} \rangle + \langle 0_{\mathrm{A}}^{(1)} | 0_{\mathrm{A}}^{(1)} \rangle \langle 0_{\mathrm{B}}^{(0)} | \hat{V}_{\mathrm{B}} | 0_{\mathrm{B}}^{(0)} \rangle
\tag{14.1.91}
$$

where the last term vanishes since

$$
\langle 0_{\mathrm{B}}^{(0)} | \hat{V}_{\mathrm{B}} | 0_{\mathrm{B}}^{(0)} \rangle = \langle 0_{\mathrm{B}}^{(0)} | \hat{U}_{\mathrm{B}} | 0_{\mathrm{B}}^{(0)} \rangle - E_{\mathrm{B}}^{(1)} = 0
\tag{14.1.92}
$$

It should be realized that, when multiplied by $\langle 0_{\mathrm{A}}^{(1)} | 0_{\mathrm{A}}^{(1)} \rangle$, each of the two terms in (14.1.92) gives rise to an energy contribution in (14.1.90) that is not size-extensive – such a cancellation of terms that are not size-extensive is a general feature of RSPT and occurs exactly to each order in the perturbation. The RSPT energy corrections are thus not termwise size-extensive.

Combining (14.1.91) and (14.1.92), we obtain for the first contribution to the third-order energy (14.1.90):

$$
\langle 0_{\mathrm{A}}^{(1)} 0_{\mathrm{B}}^{(0)} | \hat{V}_{\mathrm{AB}} | 0_{\mathrm{A}}^{(1)} 0_{\mathrm{B}}^{(0)} \rangle = E_{\mathrm{A}}^{(3)}
\tag{14.1.93}
$$

In the same manner, the fourth term in (14.1.90) yields

$$
\langle 0_{\mathrm{A}}^{(0)} 0_{\mathrm{B}}^{(1)} | \hat{V}_{\mathrm{AB}} | 0_{\mathrm{A}}^{(0)} 0_{\mathrm{B}}^{(1)} \rangle = E_{\mathrm{B}}^{(3)}
\tag{14.1.94}
$$

and the remaining two terms vanish from the orthogonality of states of different subsystems. We have thus found that the third-order RSPT energy is size-extensive

$$
E_{\mathrm{AB}}^{(3)} = \langle 0_{\mathrm{AB}}^{(1)} | \hat{V}_{\mathrm{AB}} | 0_{\mathrm{AB}}^{(1)} \rangle = E_{\mathrm{A}}^{(3)} + E_{\mathrm{B}}^{(3)}
\tag{14.1.95}
$$

but that this size-extensivity occurs only after a cancellation of terms in $\langle 0_{\mathrm{AB}}^{(1)} | \hat{V}_{\mathrm{AB}} | 0_{\mathrm{AB}}^{(1)} \rangle$ that individually are not size-extensive. More generally, the presence of terms that are not size-extensive may be anticipated from an inspection of the expression for the RSPT energies in the $2n + 1$ form (14.1.61). The presence of such terms is computationally inconvenient as well as theoretically unsatisfactory. In particular, when making approximations or modifications to the RSPT energies, we should be careful since the omission of individual terms may vitiate size-extensivity.

Returning now to the RSPT energies in the $n + 1$ form (14.1.41), we note that this form does not contradict the requirement of termwise size-extensivity. Indeed, if the RSPT energies are calculated from (14.1.41) (which is not advisable as it disobeys the $2n + 1$ rule), then size-extensivity is guaranteed if we assume that the perturbed wave function may be written in the form

$$|0_{AB}^{(n)}\rangle = \sum_{k=0}^{n} |0_A^{(n-k)}0_B^{(k)}\rangle \tag{14.1.96}$$

This form is best understood as a product wave function from which we have extracted all terms correct to order n in the perturbation. To see how size-extensivity arises from the form (14.1.96), we note that

$$E_{AB}^{(n)} = \langle 0_{AB}^{(0)}|\hat{U}_A + \hat{U}_B|0_{AB}^{(n-1)}\rangle = \sum_{k=0}^{n-1}\langle 0_A^{(0)}0_B^{(0)}|\hat{U}_A + \hat{U}_B|0_A^{(n-k-1)}0_B^{(k)}\rangle$$

$$= \langle 0_A^{(0)}|\hat{U}_A|0_A^{(n-1)}\rangle + \langle 0_B^{(0)}|\hat{U}_B|0_B^{(n-1)}\rangle = E_A^{(n)} + E_B^{(n)} \tag{14.1.97}$$

demonstrating that the energy to order n is size-extensive. It is therefore natural to look for a formalism where the perturbed wave function appears in the form (14.1.96), thereby guaranteeing size-extensivity. We shall return to this problem in Section 14.3.

14.2 Møller–Plesset perturbation theory

In Section 14.1, we discussed perturbation theory in general terms, without specifying the zero-order Hamiltonian. The success of perturbation theory depends critically on our ability to provide a suitable zero-order operator. In electronic-structure theory, the most common zero-order Hamiltonian is the Fock operator, which in the canonical spin-orbital representation may be written in terms of orbital energies as

$$\hat{H}_0^{MP} = \sum_{p\sigma} \varepsilon_{p\sigma} a_{p\sigma}^\dagger a_{p\sigma} \tag{14.2.1}$$

The zero-order states are the Hartree–Fock determinant and the determinants excited with respect to this state. The resulting theory is known as *Møller–Plesset perturbation theory (MPPT)* [5]. For systems with small static correlation contributions, the Hartree–Fock wave function provides an adequate zero-order approximation to the FCI wave function. In such situations, the Møller–Plesset partitioning of the Hamiltonian is both appealing and well motivated: the averaged electron–electron interactions are incorporated in the zero-order operator and the perturbation operator (the fluctuation potential) represents the difference between the averaged and instantaneous interactions.

A related perturbation scheme is based on the Epstein–Nesbet partitioning of the Hamiltonian, where the zero-order operator contains those parts of the Hamiltonian that conserve the spin-orbital occupations

$$\hat{H}_0^{EN} = \sum_p h_{pp} \sum_\sigma a_{p\sigma}^\dagger a_{p\sigma} + \frac{1}{2}\sum_{pq} g_{ppqq} \sum_{\sigma\tau} a_{p\sigma}^\dagger a_{q\tau}^\dagger a_{q\tau} a_{p\sigma} + \frac{1}{2}\sum_{pq} g_{pqqp} \sum_\sigma a_{p\sigma}^\dagger a_{q\sigma}^\dagger a_{p\sigma} a_{q\sigma} \tag{14.2.2}$$

As for the Fock operator, the eigenstates of the zero-order *Epstein–Nesbet Hamiltonian* are determinants. Although occasionally used, the Epstein–Nesbet approach has been considerably less successful than that of MPPT.

If the wave function contains several determinants with large weights because of large static correlation contributions, multireference-based techniques should be used rather than single-reference methods such as MPPT. The perturbational treatment of such systems is more complicated than the treatment of single-reference systems. We discuss multireference perturbation theory in Section 14.7.

14.2.1 THE ZERO-ORDER MPPT SYSTEM

In this section, we consider *spin-unrestricted MPPT*, taking as our unperturbed state the spin-unrestricted Hartree–Fock wave function and as our zero-order Hamiltonian the Fock operator. A spin-restricted treatment suitable for closed-shell states is given in Section 14.4, following the discussion of CCPT in Section 14.3.

In the canonical representation, the Fock matrix of Section 10.4 is diagonal and the *zero-order Møller–Plesset Hamiltonian* may be written as

$$\hat{H}_0^{\mathrm{MP}} = \hat{f} = \sum_P \varepsilon_P a_P^\dagger a_P \tag{14.2.3}$$

where the ε_P are the orbital energies and the summation is over all spin orbitals. We follow the same convention as before: the indices P, Q, R and S are used for unspecified (occupied or virtual) spin orbitals; I, J, K and L for the occupied spin orbitals; and A, B, C and D for the virtual spin orbitals. In this notation, the electronic Hamiltonian becomes

$$\hat{H} = \sum_{PQ} h_{PQ} a_P^\dagger a_Q + \tfrac{1}{2} \sum_{PQRS} g_{PQRS} a_P^\dagger a_R^\dagger a_S a_Q + h_{\mathrm{nuc}} \tag{14.2.4}$$

The perturbation operator

$$\hat{\Phi} = \hat{H} - \hat{f} - h_{\mathrm{nuc}} \tag{14.2.5}$$

is know as the *fluctuation potential* since it creates fluctuations of the electrons with respect to the mean-field description of Hartree–Fock theory. The Hartree–Fock reference state satisfies the zero-order Schrödinger equation

$$\hat{f}|\mathrm{HF}\rangle = \sum_I \varepsilon_I |\mathrm{HF}\rangle = E_{\mathrm{MP}}^{(0)}|\mathrm{HF}\rangle \tag{14.2.6}$$

and its orthogonal complement is spanned by all the excited determinants. In keeping with the notation of Section 13.7.5, we write the singly, doubly and triply excited determinants in the following way

$$\left|\begin{matrix} A \\ I \end{matrix}\right\rangle = \hat{\tau}_I^A |\mathrm{HF}\rangle = a_A^\dagger a_I |\mathrm{HF}\rangle \tag{14.2.7}$$

$$\left|\begin{matrix} AB \\ IJ \end{matrix}\right\rangle = \hat{\tau}_{IJ}^{AB} |\mathrm{HF}\rangle = a_A^\dagger a_I a_B^\dagger a_J |\mathrm{HF}\rangle, \quad \begin{cases} A > B \\ I > J \end{cases} \tag{14.2.8}$$

$$\left|\begin{matrix} ABC \\ IJK \end{matrix}\right\rangle = \hat{\tau}_{IJK}^{ABC} |\mathrm{HF}\rangle = a_A^\dagger a_I a_B^\dagger a_J a_C^\dagger a_K |\mathrm{HF}\rangle, \quad \begin{cases} A > B > C \\ I > J > K \end{cases} \tag{14.2.9}$$

Higher excitations may be generated by an obvious extension of this scheme, avoiding duplicates and fixing the phase factors by applying suitable restrictions on the spin-orbital indices. An unspecified determinant may be written as

$$|\mu\rangle = \hat{\tau}_\mu |\mathrm{HF}\rangle \tag{14.2.10}$$

where $\hat{\tau}_0$ is interpreted as the identity operator, generating the Hartree–Fock ground state. The Hartree–Fock and excited determinants constitute an orthonormal set of states:

$$\langle\mu|v\rangle = \delta_{\mu v} \tag{14.2.11}$$

In the closed-shell Møller–Plesset theory of Section 14.4, the construction of an orthonormal basis is more difficult and we shall then instead use a biorthogonal representation.

Like the Hartree–Fock determinant, the zero-order excited states (14.2.10) are eigenfunctions of the Fock operator

$$\hat{f}|\mu\rangle = E_\mu^{(0)}|\mu\rangle = (E_{\mathrm{MP}}^{(0)} + \varepsilon_\mu)|\mu\rangle \tag{14.2.12}$$

where, for example,

$$\varepsilon_{ABIJ} = \varepsilon_A + \varepsilon_B - \varepsilon_I - \varepsilon_J \tag{14.2.13}$$

We also note the following simple commutation relation

$$[\hat{f}, \hat{\tau}_\mu] = \varepsilon_\mu \hat{\tau}_\mu \tag{14.2.14}$$

which holds for all excitation operators, including the identity operator $\hat{\tau}_0$ since ε_0 is zero.

According to the Brillouin theorem of Section 10.2.3, the Hamiltonian does not couple the Hartree–Fock wave function with singly excited determinants:

$$\left\langle {A \atop I}\middle|\hat{H}\middle|\mathrm{HF}\right\rangle = \langle\mathrm{HF}|\hat{\tau}_I^{A\dagger}\hat{H}|\mathrm{HF}\rangle = 0 \tag{14.2.15}$$

Also, since the Hamiltonian is a two-electron operator, it does not couple determinants that differ by more than two levels of excitation. Therefore, only doubly excited configurations (14.2.8) interact directly with the Hartree–Fock state through the Hamiltonian operator, simplifying the expressions generated by applying RSPT to the Hartree–Fock wave function.

14.2.2 THE MP1 WAVE FUNCTION

According to the general expression of Rayleigh–Schrödinger theory (14.1.23), the first-order Møller–Plesset wave function is given by

$$|\mathrm{MP1}\rangle = -\hat{P}(\hat{f} - E_{\mathrm{MP}}^{(0)})^{-1}\hat{P}\hat{\Phi}|\mathrm{HF}\rangle \tag{14.2.16}$$

Invoking the resolution of the identity, we obtain

$$|\mathrm{MP1}\rangle = -\sum_{\mu>0}\hat{P}(\hat{f} - E_{\mathrm{MP}}^{(0)})^{-1}|\mu\rangle\langle\mu|\hat{\Phi}|\mathrm{HF}\rangle = -\sum_{\mu>0}(E_\mu^{(0)} - E_{\mathrm{MP}}^{(0)})^{-1}|\mu\rangle\langle\mu|\hat{\Phi}|\mathrm{HF}\rangle \tag{14.2.17}$$

where the summation is over only the excited determinants. Noting that

$$\langle\mu|\hat{\Phi}|\mathrm{HF}\rangle = \langle\mu|\hat{H}|\mathrm{HF}\rangle, \qquad \mu > 0 \tag{14.2.18}$$

and using (14.2.12) to simplify the energy denominator, we obtain the following expression for the MP1 wave function:

$$|\text{MP1}\rangle = -\sum_{\mu>0} |\mu\rangle \varepsilon_\mu^{-1} \langle \mu | \hat{H} | \text{HF} \rangle \tag{14.2.19}$$

Although the summation is over the full set of excited determinants, only doubles contribute, as discussed in Section 14.2.1. We may therefore write the MP1 correction in the form

$$|\text{MP1}\rangle = -\sum_{\mu_2} |\mu_2\rangle \varepsilon_{\mu_2}^{-1} \langle \mu_2 | \hat{H} | \text{HF} \rangle \tag{14.2.20}$$

where μ_2 is a generic notation for a doubly excited determinant.

It is useful to regard the MP1 correction as generated by the application of an operator to the Hartree–Fock state:

$$|\text{MP1}\rangle = \hat{T}_2^{(1)} |\text{HF}\rangle \tag{14.2.21}$$

$$\hat{T}_2^{(1)} = \sum_{\mu_2} t_{\mu_2}^{(1)} \hat{\tau}_{\mu_2} = \sum_{\substack{A>B \\ I>J}} t_{IJ}^{AB(1)} a_A^\dagger a_I a_B^\dagger a_J \tag{14.2.22}$$

The perturbation operator $\hat{T}_2^{(1)}$ is reminiscent of the doubles cluster operator (13.2.7) of coupled-cluster theory. It is a linear combination of excitation operators, each multiplied by a first-order amplitude of the form

$$t_{IJ}^{AB(1)} = -\frac{\langle \text{HF} | [a_J^\dagger a_B a_I^\dagger a_A, \hat{H}] | \text{HF} \rangle}{\varepsilon_A + \varepsilon_B - \varepsilon_I - \varepsilon_J} \tag{14.2.23}$$

In obtaining (14.2.23), we have used the fact that the cluster operator (14.2.22) gives zero when applied to the bra state $\langle \text{HF} |$. The connection with coupled-cluster theory is a close one, which we shall explore in Section 14.3. At this stage, the reader may wish to compare the expression for the Møller–Plesset amplitudes (14.2.23) with the expression (13.4.10) used in the perturbation-based optimization of coupled-cluster wave functions.

14.2.3 THE MP2 WAVE FUNCTION

Having established the form of the MP1 wave-function correction, we now turn our attention to the more complicated MP2 correction. According to the general expression (14.1.24), we may write the MP2 wave-function correction in the form

$$|\text{MP2}\rangle = -\hat{P}(\hat{f} - E_{\text{MP}}^{(0)})^{-1} \hat{P}(\hat{\Phi} - E_{\text{MP}}^{(1)}) |\text{MP1}\rangle \tag{14.2.24}$$

and, invoking the resolution of the identity, we obtain

$$|\text{MP2}\rangle = -\sum_{\mu>0} |\mu\rangle \varepsilon_\mu^{-1} \langle \mu | \hat{\Phi} - E_{\text{MP}}^{(1)} | \text{MP1} \rangle \tag{14.2.25}$$

The MP1 correction (14.2.21) is a linear combination of doubly excited Slater determinants (14.2.22). Hence, by inspection of (14.2.25), we conclude that the second-order correction involves single, double, triple and quadruple excitations from the Hartree–Fock state:

$$|\text{MP2}\rangle = (\hat{T}_1^{(2)} + \hat{T}_2^{(2)} + \hat{T}_3^{(2)} + \hat{T}_4^{(2)}) |\text{HF}\rangle \tag{14.2.26}$$

Each contribution to the MP2 correction may be written in the form

$$\hat{T}_n^{(2)}|\text{HF}\rangle = -\sum_{\mu_n}|\mu_n\rangle\varepsilon_{\mu_n}^{-1}\langle\mu_n|\hat{\Phi} - E_{\text{MP}}^{(1)}|\text{MP1}\rangle \tag{14.2.27}$$

where $n \leq 4$. The MP1 wave function that appears in this expression is given by (14.2.21) but an expression for the first-order energy $E_{\text{MP}}^{(1)}$ has not yet been derived.

The Møller–Plesset energy corrections are discussed in Section 14.2.4. Here we need only the first-order correction, which, according to the general Rayleigh–Schrödinger expression (14.1.62), may be calculated as the Hartree–Fock expectation value of the fluctuation potential:

$$E_{\text{MP}}^{(1)} = \langle\text{HF}|\hat{\Phi}|\text{HF}\rangle \tag{14.2.28}$$

Using this expression for the MP1 energy and (14.2.21) for the MP1 wave function in (14.2.27), we find

$$\langle\mu_n|\hat{\Phi} - E_{\text{MP}}^{(1)}|\text{MP1}\rangle = \langle\mu_n|\hat{\Phi}\hat{T}_2^{(1)}|\text{HF}\rangle - \langle\mu_n|\hat{T}_2^{(1)}|\text{HF}\rangle\langle\text{HF}|\hat{\Phi}|\text{HF}\rangle \tag{14.2.29}$$

Use of the resolution of the identity in the second term gives

$$\langle\mu_n|\hat{\Phi} - E_{\text{MP}}^{(1)}|\text{MP1}\rangle = \langle\mu_n|[\hat{\Phi}, \hat{T}_2^{(1)}]|\text{HF}\rangle + \sum_{\mu_2}\langle\mu_n|\hat{T}_2^{(1)}|\mu_2\rangle\langle\mu_2|\hat{H}|\text{HF}\rangle \tag{14.2.30}$$

where the identity contribution has been used to introduce a commutator in the first term and the summation is restricted to μ_2 because of the Brillouin theorem. Whereas the first term vanishes for $n = 4$ since the commutator is a rank-3 operator, the last term vanishes for $n \neq 4$ since $\hat{T}_2^{(1)}|\mu_2\rangle$ is a linear combination of quadruply excited determinants. For quadruples, the last term may be written in a more compact form by invoking the resolution of the identity once more:

$$\sum_{\mu_2}\langle\mu_4|\hat{T}_2^{(1)}|\mu_2\rangle\langle\mu_2|\hat{H}|\text{HF}\rangle = \langle\mu_4|\hat{T}_2^{(1)}\hat{H}|\text{HF}\rangle \tag{14.2.31}$$

Substituting (14.2.30) in (14.2.27), we obtain

$$\hat{T}_n^{(2)}|\text{HF}\rangle = \begin{cases} -\sum_{\mu_n}|\mu_n\rangle\varepsilon_{\mu_n}^{-1}\langle\mu_n|[\hat{\Phi}, \hat{T}_2^{(1)}]|\text{HF}\rangle & n = 1, 2, 3 \\ -\sum_{\mu_n}|\mu_n\rangle\varepsilon_{\mu_n}^{-1}\langle\mu_n|\hat{T}_2^{(1)}\hat{H}|\text{HF}\rangle & n = 4 \end{cases} \tag{14.2.32}$$

for the four contributions to the MP2 wave function. For $n = 1$ and $n = 3$, we may substitute $\hat{H}$ for $\hat{\Phi}$ in (14.2.32); for $n = 2$, this substitution is not possible.

A more compact expression may be derived for the quadruples in (14.2.32). Expanding $\hat{T}_2^{(1)}$ in amplitudes and excitation operators, we obtain for the quadruples

$$\hat{T}_4^{(2)}|\text{HF}\rangle = -\sum_{\mu_2}\sum_{\mu_4}|\mu_4\rangle\varepsilon_{\mu_4}^{-1}t_{\mu_2}^{(1)}\langle\mu_4|\hat{\tau}_{\mu_2}\hat{H}|\text{HF}\rangle \tag{14.2.33}$$

To simplify this expression, we first note that it may be written in the more explicit form

$$\hat{T}_4^{(2)}|\text{HF}\rangle = -\frac{1}{(4!)^2}\sum_{\substack{ABCD \\ IJKL}} a_A^\dagger a_I a_B^\dagger a_J a_C^\dagger a_K a_D^\dagger a_L|\text{HF}\rangle$$

$$\times \frac{1}{(2!)^2}\sum_{\substack{A'B' \\ I'J'}} t_{I'J'}^{A'B'(1)} \frac{\langle\text{HF}|a_L^\dagger a_D a_K^\dagger a_C a_J^\dagger a_B a_I^\dagger a_A a_{A'}^\dagger a_{I'} a_{B'}^\dagger a_{J'}\hat{H}|\text{HF}\rangle}{\varepsilon_A + \varepsilon_B + \varepsilon_C + \varepsilon_D - \varepsilon_I - \varepsilon_J - \varepsilon_K - \varepsilon_L} \tag{14.2.34}$$

The factorials appear since the summations are free rather than restricted as in (14.2.8) and (14.2.9). We may now contract the operator string in the denominator in the usual fashion, noting that the two virtual indices A' and B' may match the indices A, B, C and D in $4 \times 3 = 12$ different ways and similarly for the occupied indices. We then obtain

$$\hat{T}_4^{(2)}|\mathrm{HF}\rangle = -\frac{(4 \times 3)^2}{(2!)^2(4!)^2} \sum_{\substack{ABCD \\ IJKL}} a_A^\dagger a_I a_B^\dagger a_J a_C^\dagger a_K a_D^\dagger a_L |\mathrm{HF}\rangle \frac{t_{IJ}^{AB^{(1)}} \langle \mathrm{HF}|a_L^\dagger a_D a_K^\dagger a_C \hat{H}|\mathrm{HF}\rangle}{\varepsilon_A + \varepsilon_B + \varepsilon_C + \varepsilon_D - \varepsilon_I - \varepsilon_J - \varepsilon_K - \varepsilon_L} \tag{14.2.35}$$

which is equivalent to the expression

$$\hat{T}_4^{(2)}|\mathrm{HF}\rangle = -\sum_{\mu_2 \nu_2} |\mu_2 \nu_2\rangle (\varepsilon_{\mu_2} + \varepsilon_{\nu_2})^{-1} t_{\mu_2}^{(1)} \langle \nu_2 | \hat{H} | \mathrm{HF} \rangle \tag{14.2.36}$$

Note that the energy denominator for the product state is just the sum of the denominators for the individual states. According to (14.2.23), we may write

$$\langle \nu_2 | \hat{H} | \mathrm{HF} \rangle = -t_{\nu_2}^{(1)} \varepsilon_{\nu_2} \tag{14.2.37}$$

and (14.2.36) therefore reduces to

$$\hat{T}_4^{(2)}|\mathrm{HF}\rangle = \sum_{\mu_2 \nu_2} |\mu_2 \nu_2\rangle (\varepsilon_{\mu_2} + \varepsilon_{\nu_2})^{-1} \varepsilon_{\nu_2} t_{\mu_2}^{(1)} t_{\nu_2}^{(1)} \tag{14.2.38}$$

Symmetrizing the indices, we arrive at the following simple expression for the quadruples contribution to the MP2 state

$$\hat{T}_4^{(2)}|\mathrm{HF}\rangle = \frac{1}{2} \sum_{\mu_2 \nu_2} t_{\mu_2}^{(1)} t_{\nu_2}^{(1)} |\mu_2 \nu_2\rangle = \frac{1}{2} \hat{T}_2^{(1)^2} |\mathrm{HF}\rangle \tag{14.2.39}$$

and we are finally able to write the MP2 correction as

$$|\mathrm{MP2}\rangle = (\hat{T}_1^{(2)} + \hat{T}_2^{(2)} + \hat{T}_3^{(2)} + \tfrac{1}{2}\hat{T}_2^{(1)^2})|\mathrm{HF}\rangle \tag{14.2.40}$$

Thus, the second-order quadruple excitations decouple into products of two first-order double excitations. The remaining second-order corrections do not decouple in this fashion. The decoupling of the quadruple excitations is a special feature of Møller–Plesset theory.

At this stage, it is interesting to compare the Møller–Plesset corrections (14.2.21) and (14.2.40) with the coupled-cluster wave function as analysed in Section 13.2.2 – see, in particular (13.2.12)– (13.2.15). The MP1 correction contains only the connected first-order doubles – no disconnected terms appear at this level. To second order, the MP2 correction contains connected contributions from the second-order singles, doubles and triples amplitudes as well as a disconnected contribution from the first-order doubles – there are no contributions from the connected quadruples to the MP2 wave function.

From a technical point of view, we note that the identification of the MP2 quadruples as disconnected products of doubles was not easy, requiring a fair amount of tedious algebra. Clearly, it would be convenient if the identification of connected and disconnected terms in the Møller–Plesset wave functions could be made directly, without having to go through extensive algebraic manipulations in each case – this will be achieved in the CCPT of Section 14.3.

14.2.4 THE MØLLER–PLESSET ENERGIES

Having discussed the Møller–Plesset corrections to the wave function, we now turn our attention to the energies. From the MP2 wave function derived in the previous subsection, we may in principle calculate energies to fifth order in the fluctuation potential. We shall here be somewhat less ambitious and restrict ourselves to energies that are correct to third order in the perturbation. According to the $2n + 1$ rule, we may calculate these corrections from the first-order wave function, derived in Section 14.2.2.

Before we begin, it is worth recalling that the zero-order Møller–Plesset energy does not correspond to the Hartree–Fock energy but rather to the sum of the energies of the occupied Hartree–Fock spin orbitals – see (14.2.6). To arrive at the Hartree–Fock energy, we must also calculate the first-order energy correction (14.1.20). From the definition of the fluctuation potential (14.2.5), we obtain

$$E_{MP}^{(1)} = \langle HF|\hat{\Phi}|HF\rangle = \langle HF|\hat{H}|HF\rangle - \langle HF|\hat{f}|HF\rangle - h_{nuc} \tag{14.2.41}$$

The Hartree–Fock energy is therefore equal to the sum of the zero- and first-order Møller–Plesset energies (with the nuclear-repulsion energy added):

$$E_{HF} = E_{MP}^{(0)} + E_{MP}^{(1)} + h_{nuc} \tag{14.2.42}$$

Thus, to first order in the fluctuation, the electron interactions are treated in the mean-field approximation; the correlation corrections arise to second and higher orders.

For a first estimate of the electron correlation energy, we consider the MP2 energy, which according to RSPT theory (14.1.21) is given by

$$E_{MP}^{(2)} = \langle HF|\hat{\Phi}|MP1\rangle = \langle HF|\hat{H}\hat{T}_2^{(1)}|HF\rangle \tag{14.2.43}$$

Here we have replaced the fluctuation potential (14.2.5) by the full Hamiltonian and introduced the first-order wave function in the form (14.2.21). As a final rearrangement of the second-order energy expression, we introduce a commutator between the Hamiltonian and the cluster operator:

$$E_{MP}^{(2)} = \langle HF|[\hat{H}, \hat{T}_2^{(1)}]|HF\rangle \tag{14.2.44}$$

We shall later see that any Møller–Plesset energy correction may be written as the Hartree–Fock expectation value of such commutators. As discussed in Section 13.3.2, these commutators are important for establishing termwise size-extensivity.

Let us now apply the same strategy to the third-order Møller–Plesset energy. We begin with the standard third-order RSPT expression (14.1.64):

$$E_{MP}^{(3)} = \langle MP1|\hat{\Phi} - E_{MP}^{(1)}|MP1\rangle \tag{14.2.45}$$

Introducing the cluster operator (14.2.21) and inserting the expression for the first-order perturbed energy (14.2.41), we get

$$E_{MP}^{(3)} = \langle HF|\hat{T}_2^{(1)\dagger}\hat{\Phi}\hat{T}_2^{(1)}|HF\rangle - \langle HF|\hat{T}_2^{(1)\dagger}\hat{T}_2^{(1)}|HF\rangle\langle HF|\hat{\Phi}|HF\rangle \tag{14.2.46}$$

Next, invoking the resolution of the identity, we obtain

$$\langle HF|\hat{T}_2^{(1)\dagger}\hat{T}_2^{(1)}\hat{\Phi}|HF\rangle = \langle HF|\hat{T}_2^{(1)\dagger}\hat{T}_2^{(1)}|HF\rangle\langle HF|\hat{\Phi}|HF\rangle \tag{14.2.47}$$

since only the Hartree–Fock state has a nonzero overlap with $\langle \mathrm{HF} | \hat{T}_2^{(1)\dagger} \hat{T}_2^{(1)}$. We may therefore write the third-order energy correction in the form

$$E_{\mathrm{MP}}^{(3)} = \langle \mathrm{HF} | \hat{T}_2^{(1)\dagger} \hat{\Phi} \hat{T}_2^{(1)} - \hat{T}_2^{(1)\dagger} \hat{T}_2^{(1)} \hat{\Phi} | \mathrm{HF} \rangle \tag{14.2.48}$$

which is easily expressed in terms of commutators

$$E_{\mathrm{MP}}^{(3)} = \langle \mathrm{HF} | [\hat{T}_2^{(1)\dagger}, [\hat{\Phi}, \hat{T}_2^{(1)}]] | \mathrm{HF} \rangle \tag{14.2.49}$$

At present, this is our final form of the third-order energy correction – we have succeeded in expressing the MP3 energy as the Hartree–Fock expectation value of a commutator that contains the fluctuation potential.

14.2.5 EXPLICIT EXPRESSIONS FOR MPPT WAVE FUNCTIONS AND ENERGIES

Explicit expressions for Møller–Plesset energies in terms of integrals and orbital energies will be derived in Section 14.4 (for closed-shell systems) and are not our main concern in the present section. However, to develop some appreciation of the contents of the expressions obtained so far, we shall pause to consider in greater detail the lowest-order Møller–Plesset wave-function and energy corrections, using the simple spin-unrestricted representation.

The MP1 wave function was discussed in Section 14.2.2 and is given by (14.2.21)–(14.2.23). Only the two-electron part of the Hamiltonian contributes to (14.2.23). Inserting this part of the Hamiltonian and carrying out some straightforward algebra, we arrive at the following expression for the amplitudes

$$t_{IJ}^{AB(1)} = -\frac{1}{2} \sum_{PQRS} \frac{g_{PQRS} \langle \mathrm{HF} | [a_J^\dagger a_B a_I^\dagger a_A, a_P^\dagger a_R^\dagger a_S a_Q] | \mathrm{HF} \rangle}{\varepsilon_A + \varepsilon_B - \varepsilon_I - \varepsilon_J} = -\frac{g_{AIBJ}^{\mathrm{a}}}{\varepsilon_A + \varepsilon_B - \varepsilon_I - \varepsilon_J} \tag{14.2.50}$$

where we have introduced the antisymmetrized two-electron integrals

$$g_{AIBJ}^{\mathrm{a}} = g_{AIBJ} - g_{AJBI} \tag{14.2.51}$$

Inserting the amplitudes (14.2.50) in the wave function (14.2.21), we obtain the final expression for the MP1 wave-function correction:

$$|\mathrm{MP1}\rangle = - \sum_{\substack{A>B \\ I>J}} \frac{g_{AIBJ}^{\mathrm{a}}}{\varepsilon_A + \varepsilon_B - \varepsilon_I - \varepsilon_J} \left| \begin{matrix} AB \\ IJ \end{matrix} \right\rangle \tag{14.2.52}$$

Each doubly excited determinant gives a contribution to the MP1 wave function that is proportional to the antisymmetrized two-electron integral weighted by a denominator containing the associated orbital energies.

We now consider the second-order energy correction, discussed in Section 14.2.4. Substituting in (14.2.44) the expression for the two-electron part of the Hamiltonian and the expression for the perturbed doubles operator $\hat{T}_2^{(1)}$, we obtain, after some algebra, the following simple expression for the energy correction:

$$E_{\mathrm{MP}}^{(2)} = - \sum_{\substack{A>B \\ I>J}} \frac{|g_{AIBJ}^{\mathrm{a}}|^2}{\varepsilon_A + \varepsilon_B - \varepsilon_I - \varepsilon_J} \tag{14.2.53}$$

As always in second-order RSPT, the energy correction is negative. There is one contribution from each doubly excited determinant, obtained by weighting the squared antisymmetrized two-electron integral by the associated orbital energy differences. The calculation of the MP2 energy scales as the fourth power of the size of the system, although the initial transformation of the two-electron integrals from the AO to the MO basis is more expensive, scaling as the fifth power. We shall derive a similar expression, more suitable for closed-shell calculations, in Section 14.4.

At this point, a remark on terminology is in order. By the term 'the MP2 energy', we usually mean the total electronic energy, obtained by summing all corrections to second order:

$$E_{\text{MP2}} = E_{\text{MP}}^{(0)} + E_{\text{MP}}^{(1)} + E_{\text{MP}}^{(2)} + h_{\text{nuc}} = E_{\text{HF}} + E_{\text{MP}}^{(2)} \qquad (14.2.54)$$

We should accordingly refer to the individual contributions to (14.2.54) as 'energy corrections' and speak for example of $E_{\text{MP}}^{(2)}$ as 'the MP2 energy correction' rather than as 'the MP2 energy'.

14.2.6 SIZE-EXTENSIVITY IN MØLLER–PLESSET THEORY

It was suggested in Section 14.2.4 that it is advantageous to express the Møller–Plesset energy corrections in terms of nested commutators. We shall in this subsection see that such commutator expressions are convenient since they make the energy corrections termwise size-extensive, just like the similarity-transformed formulation of coupled-cluster theory makes the coupled-cluster equations termwise size-extensive, as shown in Section 13.3.2. We first consider the perturbed wave function for two noninteracting systems and then go on to consider the separability of the Møller–Plesset energies.

For two noninteracting systems A and B, the first-order Møller–Plesset wave function may be written in the form

$$|\text{MP1}_{\text{AB}}\rangle = \hat{T}_{2\text{AB}}^{(1)}|\text{HF}_{\text{AB}}\rangle = \sum_{\mu_{2\text{AB}}} t_{\mu_{2\text{AB}}}^{(1)} |\mu_{2\text{AB}}\rangle \qquad (14.2.55)$$

with the amplitudes given by

$$t_{\mu_{2\text{AB}}}^{(1)} = -\varepsilon_{\mu_{2\text{AB}}}^{-1} \langle \text{HF}_{\text{AB}}|[\hat{\tau}_{\mu_{2\text{AB}}}^{\dagger}, \hat{H}_{\text{AB}}]|\text{HF}_{\text{AB}}\rangle \qquad (14.2.56)$$

The de-excitation operator in (14.2.56) may be written as the product of two operators – one that affects system A and one that affects system B:

$$\hat{\tau}_{\mu_{2\text{AB}}}^{\dagger} = \hat{\tau}_{\text{A}}^{\dagger} \hat{\tau}_{\text{B}}^{\dagger} \qquad (14.2.57)$$

The operator $\hat{\tau}_{\text{A}}^{\dagger}$ may be the identity operator (in which case it does not affect the Hartree–Fock ket) or it may be a string of creation and annihilation operators for the occupied and virtual spin orbitals of system A, respectively (in which case it annihilates the Hartree–Fock ket). The same considerations apply to $\hat{\tau}_{\text{B}}^{\dagger}$. Since operators referring to different noninteracting subsystems commute, we may write the matrix element in (14.2.56) as

$$\langle \text{HF}_{\text{AB}}|[\hat{\tau}_{\mu_{2\text{AB}}}^{\dagger}, \hat{H}_{\text{AB}}]|\text{HF}_{\text{AB}}\rangle = \langle \text{HF}_{\text{AB}}|[\hat{\tau}_{\text{A}}^{\dagger}, \hat{H}_{\text{A}}]\hat{\tau}_{\text{B}}^{\dagger}|\text{HF}_{\text{AB}}\rangle + \langle \text{HF}_{\text{AB}}|[\hat{\tau}_{\text{B}}^{\dagger}, \hat{H}_{\text{B}}]\hat{\tau}_{\text{A}}^{\dagger}|\text{HF}_{\text{AB}}\rangle \quad (14.2.58)$$

which vanishes for all excitation operators (14.2.57) that affect both systems. We therefore conclude that the first-order amplitudes in (14.2.55) vanish for all determinants that represent simultaneous excitations in the two subsystems. Furthermore, the remaining intrasystem amplitudes become identical to those obtained in calculations on the separate subsystems. We may therefore write the first-order doubles amplitudes as

$$\hat{T}_{2\text{AB}}^{(1)} = \hat{T}_{2\text{A}}^{(1)} + \hat{T}_{2\text{B}}^{(1)} \qquad (14.2.59)$$

in accordance with the general formula (14.1.96) of RSPT. We see how the separability of the amplitudes follows directly from the commutator expressions for the corrections and from the additivity of the Hamiltonian.

Turning to second-order perturbation theory, we would like to verify that the MP2 wave function for noninteracting systems agrees with the general RSPT expression (14.1.96). As discussed in Section 14.2.3, the MP2 wave function may be written in the form (14.2.40), containing disconnected quadruples contributions as well as connected singles, doubles and triples contributions. The connected contributions may be written in the commutator form (14.2.32)

$$\hat{T}_{nAB}^{(2)}|HF_{AB}\rangle = - \sum_{\mu_{nAB}} |\mu_{nAB}\rangle \varepsilon_{\mu_{nAB}}^{-1} \langle HF_{AB}|[\hat{\tau}_{nAB}^\dagger, [\hat{\Phi}_{AB}, \hat{T}_{2AB}^{(1)}]]|HF_{AB}\rangle, \qquad n = 1, 2, 3$$

(14.2.60)

Expanding the inner commutator, we find that all terms that refer to both systems vanish since operators associated with two noninteracting systems commute:

$$[\hat{\Phi}_{AB}, \hat{T}_{2AB}^{(1)}] = [\hat{\Phi}_A + \hat{\Phi}_B, \hat{T}_{2A}^{(1)} + \hat{T}_{2B}^{(1)}] = [\hat{\Phi}_A, \hat{T}_{2A}^{(1)}] + [\hat{\Phi}_B, \hat{T}_{2B}^{(1)}]$$

(14.2.61)

The additivity of the contributions (14.2.60) now follows by the same argument that led to the additivity of the first-order amplitudes (14.2.59). The connected contributions to the MP2 wave function may therefore be written in the same way as the first-order correction:

$$\hat{T}_{nAB}^{(2)} = \hat{T}_{nA}^{(2)} + \hat{T}_{nB}^{(2)}, \qquad n = 1, 2, 3$$

(14.2.62)

By contrast, the disconnected quadruples do not separate as in (14.2.62). Inserting (14.2.59) in the expression for the quadruples (14.2.39), we obtain

$$\hat{T}_{4AB}^{(2)} = \tfrac{1}{2}(\hat{T}_{2AB}^{(1)})^2 = \hat{T}_{4A}^{(2)} + \hat{T}_{4B}^{(2)} + \hat{T}_{2A}^{(1)}\hat{T}_{2B}^{(1)}$$

(14.2.63)

The product $\hat{T}_{2A}^{(1)}\hat{T}_{2B}^{(1)}$ is needed to obtain a wave function consistent with (14.1.96)

$$\begin{aligned}
|MP2_{AB}\rangle &= (\hat{T}_{1AB}^{(2)} + \hat{T}_{2AB}^{(2)} + \hat{T}_{3AB}^{(2)} + \hat{T}_{4AB}^{(2)})|HF_{AB}\rangle \\
&= (\hat{T}_{1A}^{(2)} + \hat{T}_{2A}^{(2)} + \hat{T}_{3A}^{(2)} + \hat{T}_{4A}^{(2)})|HF_{AB}\rangle + (\hat{T}_{1B}^{(2)} + \hat{T}_{2B}^{(2)} + \hat{T}_{3B}^{(2)} + \hat{T}_{4B}^{(2)})|HF_{AB}\rangle \\
&\quad + \hat{T}_{2A}^{(1)}\hat{T}_{2B}^{(1)}|HF_{AB}\rangle
\end{aligned}$$

(14.2.64)

and the MP2 wave operator may accordingly be written in the form

$$\hat{T}_{AB}^{(2)} = \hat{T}_A^{(2)} + \hat{T}_B^{(2)} + \hat{T}_A^{(1)}\hat{T}_B^{(1)}$$

(14.2.65)

as anticipated from (14.1.96). The MP2 wave-function correction is thus generated from operators for the individual systems and from products of such operators.

Having considered the MP1 and MP2 wave-function corrections, we conclude this subsection by considering the energy corrections. The first-order correction separates trivially from the additivity of the fluctuation potential:

$$E_{MPAB}^{(1)} = \langle HF_{AB}|\hat{\Phi}_A + \hat{\Phi}_B|HF_{AB}\rangle = E_{MPA}^{(1)} + E_{MPB}^{(1)}$$

(14.2.66)

In the same way, the additivity of the second-order correction follows from the additivity of the commutator (14.2.61):

$$\begin{aligned}
E_{MPAB}^{(2)} &= \langle HF_{AB}|[\hat{\Phi}_{AB}, \hat{T}_{2AB}^{(1)}]|HF_{AB}\rangle \\
&= \langle HF_{AB}|[\hat{\Phi}_A, \hat{T}_{2A}^{(1)}] + [\hat{\Phi}_B, \hat{T}_{2B}^{(1)}]|HF_{AB}\rangle \\
&= E_{MPA}^{(2)} + E_{MPB}^{(2)}
\end{aligned}$$

(14.2.67)

The reader may verify that the intersystem terms are similarly absent from the third-order energy correction (14.2.49). Thus, separating the fluctuation potential and the amplitudes into terms pertaining to each subsystem and expanding the nested commutators, we obtain

$$E_{\text{MPAB}}^{(3)} = \langle \text{HF}_{\text{AB}} | [\hat{T}_{2\text{AB}}^{(1)\dagger}, [\hat{\Phi}_{\text{AB}}, \hat{T}_{2\text{AB}}^{(1)}]] | \text{HF}_{\text{AB}} \rangle = E_{\text{MPA}}^{(3)} + E_{\text{MPB}}^{(3)} \qquad (14.2.68)$$

Again, we find that the energy of the compound system separates directly into the energies of the subsystems.

The advantage of the commutator form of the energy corrections should now be evident. The commutators ensure that the additivity of the energy follows directly from the separability of the Hartree–Fock wave function and from the additivity of the fluctuation potential and the amplitudes. The energy corrections are *termwise separable* – no terms that violate the size-extensivity arise and no cancellation ever occurs. We sometimes express this by stating that the commutators provide a *linked* form of the energy corrections.

In conclusion, it is possible to restructure the Møller–Plesset energy and wave-function corrections such that their separability and size-extensivity become apparent. However, the separability of the corrections is not obvious in the original formulation of Møller–Plesset theory but becomes transparent only when commutators are introduced. In Section 14.3, we shall develop coupled-cluster perturbation theory, where the connected (termwise size-extensive) commutator form arises naturally, without the need to restructure the expressions by hand.

14.3 Coupled-cluster perturbation theory

We have seen that RSPT provides energy corrections that are size-extensive to each order in the perturbation. The size-extensivity of RSPT is remarkable in the sense that the energy corrections are generated within the framework of a linear parametrization of the wave function, which, as discussed in Section 4.3.2, does not automatically give rise to size-extensive energies. Still, RSPT is unsatisfactory in the sense that size-extensivity occurs as a result of the cancellation of unphysical terms in the energy expressions; see Section 14.1.4. In Møller–Plesset theory, we were able to eliminate these terms by expressing the energies in a commutator form. The resulting commutator expressions are size-extensive not only order by order, but also term by term.

In this section, we present an alternative formulation of MPPT, designed to give corrections that are termwise size-extensive. In Section 13.3, termwise size-extensivity was seen to arise in the similarity-transformed representation of the coupled-cluster equations. Developing perturbation theory in the same representation, we shall find that some of the results arrived at laboriously in MPPT – for example, termwise size-extensivity and the factorization of the quadruples contribution to the second-order wave function – are now obtained effortlessly.

14.3.1 THE SIMILARITY-TRANSFORMED EXPONENTIAL ANSATZ OF CCPT

In our discussion of perturbation theory so far, we have assumed a linear parametrization of the wave function. We now take a different approach, adopting for the perturbed wave function an exponential ansatz

$$|0\rangle = \exp(\hat{T})|\text{HF}\rangle \qquad (14.3.1)$$

where $\hat{T}$ is the cluster operator in the spin-orbital representation; see Section 13.2. Thus, we write in the usual notation

$$\hat{T} = \sum_{\mu} t_{\mu} \hat{\tau}_{\mu} \tag{14.3.2}$$

where t_{μ} are the (real) cluster amplitudes and $\hat{\tau}_{\mu}$ the excitation operators with respect to the Hartree–Fock state:

$$|\mu\rangle = \hat{\tau}_{\mu} |\text{HF}\rangle \tag{14.3.3}$$

The excitation operators, which were introduced in Section 14.2.1, commute among themselves but not with their adjoints. In the spin-orbital basis, the determinants generated by the excitation operators (14.3.3) are orthonormal.

As discussed in Section 13.2.3, the full (untruncated) coupled-cluster energy may be calculated from the expression

$$E = \langle \text{HF}| \exp(-\hat{T}) \hat{H} \exp(\hat{T}) |\text{HF}\rangle \tag{14.3.4}$$

where the amplitudes satisfy the projected Schrödinger equation

$$\langle \mu| \exp(-\hat{T}) \hat{H} \exp(\hat{T}) |\text{HF}\rangle = 0 \tag{14.3.5}$$

Supplemented with a specification of the unperturbed system, the expressions for the wave function (14.3.1), for the electronic energy (14.3.4) and for the Schrödinger equation (14.3.5) constitute the starting point for perturbation theory in this section. As for MPPT in Section 14.2.1, we partition the Hamiltonian into the zero-order Fock operator $\hat{f}$, the first-order fluctuation potential $\hat{\Phi}$ and the nuclear–nuclear repulsion:

$$\hat{H} = \hat{f} + \hat{\Phi} + h_{\text{nuc}} \tag{14.3.6}$$

The zero-order wave functions are the Hartree–Fock state and the excited determinants (14.3.3). In *coupled-cluster perturbation theory (CCPT)*, we expand the full coupled-cluster wave function (14.3.1), the coupled-cluster electronic energy (14.3.4) and the similarity-transformed Schrödinger equation (14.3.5) in orders of the fluctuation potential, generating in this way expressions that are equivalent to those of MPPT but more convenient from a theoretical as well as a practical point of view.

Before we study the CCPT expansion in any detail, it is convenient to rewrite the electronic energy and the Schrödinger equation in terms of the Fock operator and the fluctuation potential. From the commutator (14.2.14) between the Fock operator and the excitation operators, we obtain the following commutators between the Fock and cluster operators

$$[\hat{f}, \hat{T}] = \sum_{\mu} \varepsilon_{\mu} t_{\mu} \hat{\tau}_{\mu} \tag{14.3.7}$$

$$[[\hat{f}, \hat{T}], \hat{T}] = [[[\hat{f}, \hat{T}], \hat{T}], \hat{T}] = \cdots = 0 \tag{14.3.8}$$

Invoking these relations, we may write the similarity-transformed Fock operator in the form

$$\hat{f}^{\text{T}} = \hat{f} + \sum_{\mu} \varepsilon_{\mu} t_{\mu} \hat{\tau}_{\mu} \tag{14.3.9}$$

using the notation

$$\hat{A}^{\text{T}} = \exp(-\hat{T}) \hat{A} \exp(\hat{T}) \tag{14.3.10}$$

introduced in Section 13.2.3. Note that the similarity-transformed operators are not Hermitian.

Equation (14.3.9) gives the following simple expressions for matrix elements involving the similarity-transformed Fock operator:

$$\langle \mathrm{HF} | \hat{f}^{\mathrm{T}} | \mathrm{HF} \rangle = \sum_I \varepsilon_I = E_0 \tag{14.3.11}$$

$$\langle \mu | \hat{f}^{\mathrm{T}} | \mathrm{HF} \rangle = \varepsilon_\mu t_\mu \tag{14.3.12}$$

The coupled-cluster energy (14.3.4) and the similarity-transformed coupled-cluster equations (14.3.5) may now be written as

$$E = E_0 + \langle \mathrm{HF} | \hat{\Phi}^{\mathrm{T}} | \mathrm{HF} \rangle + h_{\mathrm{nuc}} \tag{14.3.13}$$

$$\varepsilon_\mu t_\mu = -\langle \mu | \hat{\Phi}^{\mathrm{T}} | \mathrm{HF} \rangle \tag{14.3.14}$$

These equations, which express the energy and amplitudes in terms of the similarity-transformed fluctuation potential, will be used extensively in our discussion of CCPT.

14.3.2 THE CCPT AMPLITUDE EQUATIONS

The CCPT wave functions and energies are parametrized in terms of the cluster operator $\hat{T}$. Therefore, before considering the expansions of the wave function and the energy, we must determine the expansion of the cluster operator in orders of the fluctuation potential

$$\hat{T} = \hat{T}^{(0)} + \hat{T}^{(1)} + \hat{T}^{(2)} + \cdots \tag{14.3.15}$$

and hence the expansion of the cluster amplitudes

$$t_\mu = t_\mu^{(0)} + t_\mu^{(1)} + t_\mu^{(2)} + \cdots \tag{14.3.16}$$

Substituting these amplitudes in the Schrödinger equation (14.3.14), we obtain the amplitude equations:

$$\varepsilon_\mu t_\mu^{(n)} = -\langle \mu | [\hat{\Phi}^{\mathrm{T}}]^{(n)} | \mathrm{HF} \rangle \tag{14.3.17}$$

where $[\hat{\Phi}^{\mathrm{T}}]^{(n)}$ contains the nth-order part of the similarity-transformed fluctuation potential. The zero-order amplitudes vanish since the fluctuation potential contains no terms to zero order:

$$t_\mu^{(0)} = 0 \tag{14.3.18}$$

To third order in the perturbation, the perturbed equations (14.3.17) become:

$$\varepsilon_\mu t_\mu^{(1)} = -\langle \mu | \hat{\Phi} | \mathrm{HF} \rangle \tag{14.3.19}$$

$$\varepsilon_\mu t_\mu^{(2)} = -\langle \mu | [\hat{\Phi}, \hat{T}^{(1)}] | \mathrm{HF} \rangle \tag{14.3.20}$$

$$\varepsilon_\mu t_\mu^{(3)} = -\langle \mu | [\hat{\Phi}, \hat{T}^{(2)}] | \mathrm{HF} \rangle - \tfrac{1}{2} \langle \mu | [[\hat{\Phi}, \hat{T}^{(1)}], \hat{T}^{(1)}] | \mathrm{HF} \rangle \tag{14.3.21}$$

These equations may be solved in succession, yielding higher and higher corrections to the amplitudes and the cluster operator.

At this point, it is appropriate to comment on the relation between the excitation levels and the perturbation order. To first order in the fluctuation potential (14.3.19), only the double excitations contribute – the single excitations do not contribute because of the Brillouin theorem and the higher-order excitations cannot be coupled to the Hartree–Fock state by a two-electron operator. Further corrections to the doubles are generated by the higher-order equations.

The singles and triples make their first appearance in the second-order equations (14.3.20) and are modified by higher-order corrections to the amplitudes. The quadruples do not enter to second order since the commutator of $\hat{\Phi}$ and $\hat{T}^{(1)}$ is a three-electron operator – see the discussion of excitation ranks and commutators in Section 13.2.8. In general, the nth-order excitations enter first to order $n-1$ since the particle ranks of the commutators in the equations of order $n-1$ are at most n. The only exception to this rule are the singles, which – because of the Brillouin theorem – enter the equations to second order. These results are summarized in Table 14.2.

14.3.3 THE CCPT WAVE FUNCTIONS

Expanding the coupled-cluster wave function in orders of the perturbation, we obtain

$$|0\rangle = |0^{(0)}\rangle + |0^{(1)}\rangle + |0^{(2)}\rangle + \cdots \tag{14.3.22}$$

From a knowledge of the perturbed cluster amplitudes discussed in Section 14.3.2, we may calculate the wave function (14.3.22) to any order according to the expression

$$|0^{(n)}\rangle = [\exp(\hat{T})]^{(n)}|\text{HF}\rangle \tag{14.3.23}$$

Expanding the exponential (14.3.23) in $\hat{T}$ as given by (14.3.15) and collecting terms to the same order in the fluctuation potential, we arrive at the following expressions for the perturbed wave functions:

$$|0^{(0)}\rangle = |\text{HF}\rangle \tag{14.3.24}$$

$$|0^{(1)}\rangle = \hat{T}^{(1)}|\text{HF}\rangle \tag{14.3.25}$$

$$|0^{(2)}\rangle = \left(\hat{T}^{(2)} + \tfrac{1}{2}\hat{T}^{(1)}\hat{T}^{(1)}\right)|\text{HF}\rangle \tag{14.3.26}$$

$$|0^{(3)}\rangle = \left(\hat{T}^{(3)} + \hat{T}^{(2)}\hat{T}^{(1)} + \tfrac{1}{6}\hat{T}^{(1)}\hat{T}^{(1)}\hat{T}^{(1)}\right)|\text{HF}\rangle \tag{14.3.27}$$

In setting up these expressions, we have used the fact that $\hat{T}^{(0)}$ vanishes (14.3.18) and that the cluster operators commute. Expanding the terms in excitation levels according to Table 14.2, we

Table 14.2 The CCPT nonzero perturbational corrections to the cluster operators

	0	1	2	3
$\hat{T}_1$	0	0^{a}	$\hat{T}_1^{(2)}$	$\hat{T}_1^{(3)}$
$\hat{T}_2$	0	$\hat{T}_2^{(1)}$	$\hat{T}_2^{(2)}$	$\hat{T}_2^{(3)}$
$\hat{T}_3$	0	0	$\hat{T}_3^{(2)}$	$\hat{T}_3^{(3)}$
$\hat{T}_4$	0	0	0	$\hat{T}_4^{(3)}$

[a]No first-order contribution due to the Brillouin theorem.

obtain

$$|0^{(1)}\rangle = \hat{T}_2^{(1)}|\text{HF}\rangle \tag{14.3.28}$$

$$|0^{(2)}\rangle = \hat{T}_1^{(2)}|\text{HF}\rangle + \hat{T}_2^{(2)}|\text{HF}\rangle + \hat{T}_3^{(2)}|\text{HF}\rangle + \tfrac{1}{2}\hat{T}_2^{(1)}\hat{T}_2^{(1)}|\text{HF}\rangle \tag{14.3.29}$$

$$|0^{(3)}\rangle = \hat{T}_1^{(3)}|\text{HF}\rangle + \hat{T}_2^{(3)}|\text{HF}\rangle + (\hat{T}_3^{(3)} + \hat{T}_1^{(2)}\hat{T}_2^{(1)})|\text{HF}\rangle$$
$$+ (\hat{T}_4^{(3)} + \hat{T}_2^{(2)}\hat{T}_2^{(1)})|\text{HF}\rangle + \hat{T}_3^{(2)}\hat{T}_2^{(1)}|\text{HF}\rangle + \tfrac{1}{6}\hat{T}_2^{(1)}\hat{T}_2^{(1)}\hat{T}_2^{(1)}|\text{HF}\rangle \tag{14.3.30}$$

In agreement with the MP1 expression (14.2.21), the first-order wave function (14.3.28) contains contributions only from the connected doubles. To second order (14.3.29), there are contributions from the disconnected quadruples as well as from the connected singles, doubles and triples – in agreement with the MP2 expression (14.2.40). However, whereas the MPPT expression was obtained after extensive algebraic manipulations, (14.3.29) was obtained in a simple manner from the general expressions of CCPT. To high orders, a large number of disconnected cluster amplitudes appear in the wave-function corrections – see for example (14.3.30).

14.3.4 THE CCPT ENERGIES

We now consider the evaluation of the coupled-cluster energy corrections:

$$E = E^{(0)} + E^{(1)} + E^{(2)} + \cdots \tag{14.3.31}$$

According to the discussion in Section 14.1.1, we may calculate these energies from the equation

$$E^{(n)} = \langle 0^{(0)}|\hat{\Phi}|0^{(n-1)}\rangle \tag{14.3.32}$$

by substituting the expression for the perturbed wave function (14.3.23). It would appear that the calculation of the energy to a given order is a rather complicated business, requiring the combination of a large number of disconnected amplitude contributions – see, for example, the expression for the third-order wave function (14.3.30). However, most of the terms vanish from a consideration of particle rank.

Rather than calculating the energy from the general RSPT expression (14.3.32), we shall here use the equivalent form obtained from the coupled-cluster energy (14.3.13) (see the discussion at the end of this subsection), writing the energy correction as

$$E^{(n)} = \langle \text{HF}|[\hat{\Phi}^{\text{T}}]^{(n)}|\text{HF}\rangle, \qquad n > 0 \tag{14.3.33}$$

We now expand the similarity-transformed fluctuation potential in a BCH series as in (13.2.37). From a consideration of particle ranks (see the discussion in Section 13.2.8), only single and double commutators contribute to the perturbed energy, which may be written in the form

$$E^{(n)} = \langle \text{HF}|\hat{\Phi}^{(n)}|\text{HF}\rangle + \langle \text{HF}|[\hat{\Phi}, \hat{T}_2]^{(n)}|\text{HF}\rangle + \tfrac{1}{2}\langle \text{HF}|[[\hat{\Phi}, \hat{T}_1], \hat{T}_1]^{(n)}|\text{HF}\rangle, \qquad n > 0 \tag{14.3.34}$$

where $\hat{\Phi}^{(n)}$ vanishes except to first order and only the singles and doubles amplitudes contribute to the energy. In deriving (14.3.34), we have also used the Brillouin theorem.

In accordance with the $n + 1$ rule (which states that the nth-order wave function determines the energy to order $n + 1$), the CCPT energies to fourth order are now given by

$$E^{(0)} = E_0 \tag{14.3.35}$$

$$E^{(1)} = \langle \text{HF}|\hat{\Phi}|\text{HF}\rangle \tag{14.3.36}$$

$$E^{(2)} = \langle \mathrm{HF}|[\hat{\Phi}, \hat{T}_2^{(1)}]|\mathrm{HF}\rangle \qquad (14.3.37)$$

$$E^{(3)} = \langle \mathrm{HF}|[\hat{\Phi}, \hat{T}_2^{(2)}]|\mathrm{HF}\rangle \qquad (14.3.38)$$

$$E^{(4)} = \langle \mathrm{HF}|[\hat{\Phi}, \hat{T}_2^{(3)}]|\mathrm{HF}\rangle \qquad (14.3.39)$$

where we have used $\hat{T}_1^{(1)} = 0$ to simplify the expressions. To fourth order in the perturbation, no disconnected terms appear in the energies. In fact, when the energies are calculated according to the $n + 1$ rule, the first disconnected contribution appears to fifth order in the perturbation:

$$E^{(5)} = \langle \mathrm{HF}|[\hat{\Phi}, \hat{T}_2^{(4)}]|\mathrm{HF}\rangle + \tfrac{1}{2}\langle \mathrm{HF}|[[\hat{\Phi}, \hat{T}_1^{(2)}], \hat{T}_1^{(2)}]|\mathrm{HF}\rangle \qquad (14.3.40)$$

In this sense, the calculation of the CCPT energies is simpler than the calculation of the CCPT wave functions; see (14.3.28)–(14.3.30). As for the coupled-cluster energy in Section 13.2.4, only the singles and doubles amplitudes contribute directly to the energy. However, although the triples and higher amplitudes do not contribute directly, they still make an indirect contribution; see Section 14.3.2. We shall in Section 14.3.8 see that, in the CCPT energy expressions that comply with the $2n + 1$ rule rather than with the $n + 1$ rule, disconnected wave-function contributions appear to lower orders.

The CCPT energies derived in this subsection are all expressed in terms of commutators. To clarify the relation to the RSPT expression (14.3.32), we may expand the commutators in (14.3.34), retaining only those terms where all cluster operators appear to the right of $\hat{\Phi}$. The remaining terms may be discarded since $\langle \mathrm{HF}|\hat{T}^{(k)}$ vanishes for all k. The resulting expression agrees with the corresponding RSPT expression (14.3.32) to all orders, as expected from the basic equivalence of the two approaches to the calculation of perturbed energies.

14.3.5 SIZE-EXTENSIVITY IN CCPT

The arguments for size-extensivity in CCPT follow closely those for linked coupled-cluster theory in Section 13.3. Thus, whereas the order-by-order size-extensivity of the energy follows from the exponential ansatz for the wave function, the termwise size-extensivity follows from the use of a similarity-transformed Hamiltonian, assuming that both the zero-order Fock operator and the first-order fluctuation operator separate for noninteracting systems.

Let us consider the size-extensivity of the CCPT expressions in more detail. In Sections 14.3.2 and 14.3.4, we found that the CCPT cluster amplitudes and energies are, to each order, expressed in terms of commutators between the fluctuation operator and the cluster operators. For noninteracting systems, the cluster operators separate. This is seen by induction, assuming that the cluster operators separate to order n and using the separability of the commutators

$$[\hat{\Phi}_{\mathrm{AB}}, \hat{T}_{\mathrm{AB}}^{(n)}] = [\hat{\Phi}_\mathrm{A} + \hat{\Phi}_\mathrm{B}, \hat{T}_\mathrm{A}^{(n)} + \hat{T}_\mathrm{B}^{(n)}] = [\hat{\Phi}_\mathrm{A}, \hat{T}_\mathrm{A}^{(n)}] + [\hat{\Phi}_\mathrm{B}, \hat{T}_\mathrm{B}^{(n)}] \qquad (14.3.41)$$

to show that the cluster operators also separate to order $n + 1$. From these observations, we conclude that CCPT expressions involving two noninteracting systems contain no terms pertaining to both systems, thus giving energy corrections that are termwise size-extensive.

Let us also consider the CCPT wave function of two noninteracting systems. Since the cluster operator $\hat{T}_{\mathrm{AB}}$ for such systems separates, we may write the exponential operator in the following way:

$$[\exp(\hat{T}_{\mathrm{AB}})]^{(n)} = [\exp(\hat{T}_\mathrm{A})\exp(\hat{T}_\mathrm{B})]^{(n)} = \sum_{k=0}^{n} [\exp(\hat{T}_\mathrm{A})]^{(n-k)}[\exp(\hat{T}_\mathrm{B})]^{(k)} \qquad (14.3.42)$$

The nth-order CCPT wave function thus reduces to a linear combination of all possible product states of combined order n

$$|0_{AB}^{(n)}\rangle = \sum_{k=0}^{n} |0_A^{(n-k)} 0_B^{(k)}\rangle \tag{14.3.43}$$

as anticipated for size-extensive treatments of noninteracting systems (14.1.96).

14.3.6 THE CCPT LAGRANGIAN

So far in our discussion of CCPT, we have not obtained expressions that satisfy Wigner's $2n + 1$ rule. The energy corrections presented in Section 14.3.4 obey the $n + 1$ rule rather than the $2n + 1$ rule since the wave function to order n is required to calculate the energy to order $n + 1$. To derive energy expressions that obey the $2n + 1$ rule while retaining the termwise size-extensivity of CCPT, we proceed as in Section 14.1.2 and set up a variational Lagrangian for the CCPT energies.

The CCPT energies are calculated from the energy expression (14.3.13) subject to the constraints (14.3.14). Introducing one undetermined Lagrange multiplier $\bar{t}_\mu$ for each constraint (i.e. one multiplier for each amplitude t_μ), we arrive at the CCPT Lagrangian

$$L(\mathbf{t}, \bar{\mathbf{t}}) = E_0 + \sum_\mu \varepsilon_\mu t_\mu \bar{t}_\mu + \langle \mathrm{HF}|\hat{\Phi}^\mathrm{T}|\mathrm{HF}\rangle + \langle \bar{t}|\hat{\Phi}^\mathrm{T}|\mathrm{HF}\rangle \tag{14.3.44}$$

where

$$\langle \bar{t}| = \sum_\mu \bar{t}_\mu \langle \mu| \tag{14.3.45}$$

The variational conditions on the CCPT Lagrangian are given by

$$L_\mu = \frac{\partial L}{\partial \bar{t}_\mu} = 0 \tag{14.3.46}$$

$$\bar{L}_\mu = \frac{\partial L}{\partial t_\mu} = 0 \tag{14.3.47}$$

which hold for all perturbational strengths. Noting that

$$\frac{\partial \hat{\Phi}^\mathrm{T}}{\partial t_\mu} = [\hat{\Phi}^\mathrm{T}, \hat{\tau}_\mu] \tag{14.3.48}$$

we obtain the following expressions for the differentiated Lagrangian:

$$L_\mu = \varepsilon_\mu t_\mu + \langle \mu|\hat{\Phi}^\mathrm{T}|\mathrm{HF}\rangle \tag{14.3.49}$$

$$\bar{L}_\mu = \varepsilon_\mu \bar{t}_\mu + \langle \mathrm{HF}|\hat{\Phi}^\mathrm{T}|\mu\rangle + \langle \bar{t}|[\hat{\Phi}^\mathrm{T}, \hat{\tau}_\mu]|\mathrm{HF}\rangle \tag{14.3.50}$$

Whereas the variational conditions (14.3.46) are identical to the original equations for the amplitudes (14.3.14), the variational conditions (14.3.47) determine the Lagrange multipliers. Note that, since $\hat{\Phi}^\mathrm{T}$ is not Hermitian, the second term in (14.3.49) is different from the second term in (14.3.50), even for real spin orbitals.

To arrive at the CCPT energies, we now expand the Lagrangian in orders of the perturbation:

$$L = L^{(0)} + L^{(1)} + L^{(2)} + \cdots \tag{14.3.51}$$

Next, we determine the perturbed amplitudes and multipliers by expanding the variational conditions

$$L_\mu = L_\mu^{(0)} + L_\mu^{(1)} + L_\mu^{(2)} + \cdots \tag{14.3.52}$$

$$\bar{L}_\mu = \bar{L}_\mu^{(0)} + \bar{L}_\mu^{(1)} + \bar{L}_\mu^{(2)} + \cdots \tag{14.3.53}$$

and solving these to each order separately:

$$L_\mu^{(0)} = L_\mu^{(1)} = L_\mu^{(2)} = \cdots = 0 \tag{14.3.54}$$

$$\bar{L}_\mu^{(0)} = \bar{L}_\mu^{(1)} = \bar{L}_\mu^{(2)} = \cdots = 0 \tag{14.3.55}$$

The solution of the amplitude equations (carried out in Section 14.3.7) yields the perturbed amplitudes and multipliers:

$$t_\mu = t_\mu^{(0)} + t_\mu^{(1)} + t_\mu^{(2)} + \cdots \tag{14.3.56}$$

$$\bar{t}_\mu = \bar{t}_\mu^{(0)} + \bar{t}_\mu^{(1)} + \bar{t}_\mu^{(2)} + \cdots \tag{14.3.57}$$

To a given order n, the energy is obtained by inserting the perturbed amplitudes and multipliers into the expression for the Lagrangian of order n – the energy $E^{(n)}$ and Lagrangian $L^{(n)}$ calculated from the optimized amplitudes and multipliers are identical since the constraints are satisfied for the optimized amplitudes and multipliers.

To simplify the Lagrangian expressions for the energy, we shall exploit (in Section 14.3.8) the variational property of the perturbed Lagrangians to arrive at expressions that obey the $2n + 1$ rule: even-order Lagrangians $L^{(2n)}$ and odd-order Lagrangians $L^{(2n+1)}$ are linear in all parameters of order greater than n; all terms containing such parameters may then be discarded because of the variational conditions. For even-order Lagrangians, an additional simplification is possible since the elimination of amplitudes and multipliers from $L^{(2n)}$ in agreement with the $2n + 1$ rule does not affect the terms that contain parameters of order n. The Lagrangian is therefore still variational with respect to these parameters. Since the multipliers $\bar{\mathbf{t}}^{(n)}$ appear linearly in this Lagrangian, we may eliminate from $L^{(2n)}$ all terms containing $\bar{\mathbf{t}}^{(n)}$ in agreement with the $2n + 2$ rule for the multipliers. For the parameters $\mathbf{t}^{(n)}$, by contrast, such a simplification is impossible since $\mathbf{t}^{(n)}$ appears quadratically in the Lagrangian – see Section 14.1.2 for a more detailed discussion of the $2n + 1$ and $2n + 2$ rules.

14.3.7 THE CCPT VARIATIONAL EQUATIONS

The amplitudes and multipliers needed to calculate the CCPT energy to fifth order in the fluctuation potential are obtained by solving the amplitude equations (14.3.46) and (14.3.47) to second order. To first order in the perturbation, we obtain

$$\varepsilon_\mu t_\mu^{(0)} = 0 \tag{14.3.58}$$

$$\varepsilon_\mu \bar{t}_\mu^{(0)} = 0 \tag{14.3.59}$$

$$\varepsilon_\mu t_\mu^{(1)} = -\langle \mu | \hat{\Phi} | \mathrm{HF} \rangle \tag{14.3.60}$$

$$\varepsilon_\mu \bar{t}_\mu^{(1)} = -\langle \mathrm{HF} | \hat{\Phi} | \mu \rangle \tag{14.3.61}$$

Assuming real spin orbitals, the amplitudes and multipliers are identical to zero and first orders:

$$t_\mu^{(0)} = \bar{t}_\mu^{(0)} = 0 \tag{14.3.62}$$

$$\bar{t}_\mu^{(1)} = t_\mu^{(1)} \tag{14.3.63}$$

The zero-order parameters vanish as indicated in (14.3.62). From the Brillouin condition, we conclude that the first-order amplitudes (14.3.60) and multipliers (14.3.61) contain double excitations only. From the first-order amplitudes and multipliers, we may calculate the energy to third order in the fluctuation potential.

To calculate the energy to fourth and fifth orders, we also need the second-order amplitudes and multipliers. Collecting the terms to second order in the variational conditions, we obtain

$$\varepsilon_\mu t_\mu^{(2)} = -\langle \mu | [\hat{\Phi}, \hat{T}^{(1)}] | HF \rangle \tag{14.3.64}$$

$$\varepsilon_\mu \bar{t}_\mu^{(2)} = -\langle HF | [\hat{\Phi}, \hat{T}^{(1)}] | \mu \rangle - \langle \bar{t}^{(1)} | [\hat{\Phi}, \hat{\tau}_\mu] | HF \rangle \tag{14.3.65}$$

Since $[\hat{\Phi}, \hat{T}^{(1)}]$ is a rank-3 operator, the right-hand side of (14.3.64) vanishes for excitations higher than triples. The second-order amplitudes $t_\mu^{(2)}$ therefore contain only singles, doubles and triples – unlike for MPPT in (14.2.32), the second-order quadruples amplitudes are zero in CCPT.

With regard to the perturbed multipliers in (14.3.65), we note that the first term on the right-hand side vanishes for all excitations. For real spin orbitals, the multiplier equations may therefore be written in the form

$$\varepsilon_\mu \bar{t}_\mu^{(2)} = -\langle HF | [[\hat{\tau}_\mu^\dagger, \hat{\Phi}], \hat{T}^{(1)}] | HF \rangle \tag{14.3.66}$$

Invoking the Jacobi identity (1.8.17), this expression may be further rearranged and combined with (14.3.64) to yield

$$\varepsilon_\mu \bar{t}_\mu^{(2)} = \varepsilon_\mu t_\mu^{(2)} - \langle HF | [\hat{\Phi}, [\hat{T}^{(1)}, \hat{\tau}_\mu^\dagger]] | HF \rangle$$
$$= \varepsilon_\mu t_\mu^{(2)} + \langle HF | \hat{\Phi} \hat{\tau}_\mu^\dagger \hat{T}^{(1)} - \hat{\tau}_\mu^\dagger \hat{T}^{(1)} \hat{\Phi} | HF \rangle \tag{14.3.67}$$

Because of the Brillouin theorem, the last term vanishes unless μ corresponds to a quadruple excitation. Moreover, we have already established that the second-order amplitudes are zero except for the single, double and triple excitations. From these observations, we conclude that the second-order amplitudes and multipliers are identical for the single, double and triple excitations. For the quadruple excitations, the amplitudes vanish and the multipliers are obtained from (14.3.67). For quintuple and higher excitations, the amplitudes and multipliers both vanish. Equation (14.3.67) may thus be written as

$$\bar{t}_{\mu_n}^{(2)} = \begin{cases} t_{\mu_n}^{(2)} & n = 1, \ 2, \ 3 \\ -\varepsilon_{\mu_n}^{-1} \langle \mu_n | \hat{T}^{(1)} \hat{H} | HF \rangle & n = 4 \end{cases} \tag{14.3.68}$$

Comparing with (14.2.32), we find that the second-order multipliers of CCPT are identical to the MPPT amplitudes – for the quadruple as well as the lower-order excitations.

To summarize, to second order in CCPT, the amplitudes and multipliers are identical except for the second-order quadruples, which vanish for the amplitudes. Compared with MPPT, the amplitudes are the same except for the second-order quadruples, which in MPPT are nonzero and identical to the CCPT multipliers. Still, the second-order wave functions are identical in MPPT and CCPT: in CCPT, the quadruples appear as disconnected amplitudes; in MPPT, they are generated from the second-order equations without exploiting the disconnected nature of these amplitudes, as discussed in Section 14.2.3.

14.3.8 CCPT ENERGIES THAT OBEY THE $2n + 1$ RULE

Having derived the equations that determine the amplitudes and multipliers to second order, we now turn to the energies. In Section 14.3.4, we gave expressions for the CCPT energies that agree with the $n + 1$ rule. To second order, these expressions agree also with the $2n + 1$ rule – see (14.3.35)–(14.3.37). For completeness, we here list the first- and second-order energies:

$$E^{(1)} = \langle \mathrm{HF} | \hat{\Phi} | \mathrm{HF} \rangle \tag{14.3.69}$$

$$E^{(2)} = \langle \mathrm{HF} | [\hat{\Phi}, \hat{T}^{(1)}] | \mathrm{HF} \rangle \tag{14.3.70}$$

The reader may easily verify that these expressions are obtained from the variational Lagrangians $L^{(1)}$ and $L^{(2)}$ after dropping all terms that do not comply with the $2n + 1$ and $2n + 2$ rules.

To calculate the third-order CCPT energy, we collect all terms to third order in the Lagrangian (14.3.44):

$$L^{(3)} = \sum_{\mu} \varepsilon_{\mu} t_{\mu}^{(1)} \bar{t}_{\mu}^{(2)} + \sum_{\mu} \varepsilon_{\mu} t_{\mu}^{(2)} \bar{t}_{\mu}^{(1)} + \langle \mathrm{HF} | [\hat{\Phi}, \hat{T}^{(2)}] | \mathrm{HF} \rangle + \tfrac{1}{2} \langle \mathrm{HF} | [[\hat{\Phi}, \hat{T}^{(1)}], \hat{T}^{(1)}] | \mathrm{HF} \rangle$$

$$+ \langle \bar{t}^{(1)} | [\hat{\Phi}, \hat{T}^{(1)}] | \mathrm{HF} \rangle + \langle \bar{t}^{(2)} | \hat{\Phi} | \mathrm{HF} \rangle \tag{14.3.71}$$

Since this expression is variational, we may omit all terms containing the linear parameters $\mathbf{t}^{(2)}$ or $\bar{\mathbf{t}}^{(2)}$ in agreement with the $2n + 1$ rule. In addition, the fourth term in (14.3.71) vanishes since the excitation rank of the commutator is at least 2. This leaves us with a single term:

$$E^{(3)} = \langle \bar{t}^{(1)} | [\hat{\Phi}, \hat{T}^{(1)}] | \mathrm{HF} \rangle \tag{14.3.72}$$

Comparing with the $n + 1$ expression (14.3.38), we note that we cannot derive (14.3.72) simply by omitting the term that contains $\mathbf{t}^{(2)}$ in (14.3.38) – we must first set up the Lagrangian (14.3.71) and only then omit the terms that are not required by the $2n + 1$ rule.

Proceeding in the same manner for the fourth-order energy, we first set up the variational Lagrangian

$$L^{(4)} = \sum_{\mu} \varepsilon_{\mu} t_{\mu}^{(1)} \bar{t}_{\mu}^{(3)} + \sum_{\mu} \varepsilon_{\mu} t_{\mu}^{(2)} \bar{t}_{\mu}^{(2)} + \sum_{\mu} \varepsilon_{\mu} t_{\mu}^{(3)} \bar{t}_{\mu}^{(1)} + \langle \mathrm{HF} | [\hat{\Phi}, \hat{T}^{(3)}] | \mathrm{HF} \rangle$$

$$+ \langle \mathrm{HF} | [[\hat{\Phi}, \hat{T}^{(2)}], \hat{T}^{(1)}] | \mathrm{HF} \rangle + \tfrac{1}{6} \langle \mathrm{HF} | [[[\hat{\Phi}, \hat{T}^{(1)}], \hat{T}^{(1)}], \hat{T}^{(1)}] | \mathrm{HF} \rangle + \langle \bar{t}^{(1)} | [\hat{\Phi}, \hat{T}^{(2)}] | \mathrm{HF} \rangle$$

$$+ \tfrac{1}{2} \langle \bar{t}^{(1)} | [[\hat{\Phi}, \hat{T}^{(1)}], \hat{T}^{(1)}] | \mathrm{HF} \rangle + \langle \bar{t}^{(2)} | [\hat{\Phi}, \hat{T}^{(1)}] | \mathrm{HF} \rangle + \langle \bar{t}^{(3)} | \hat{\Phi} | \mathrm{HF} \rangle \tag{14.3.73}$$

Next, we drop all terms containing $\mathbf{t}^{(3)}$ in agreement with the $2n + 1$ rule and all terms containing $\bar{\mathbf{t}}^{(2)}$ and $\bar{\mathbf{t}}^{(3)}$ in agreement with the $2n + 2$ rule. In addition, we observe that the fifth and sixth terms in (14.3.73) vanish since the commutators involve too high excitation ranks. We are then left with just two terms:

$$E^{(4)} = \langle \bar{t}^{(1)} | [\hat{\Phi}, \hat{T}^{(2)}] | \mathrm{HF} \rangle + \tfrac{1}{2} \langle \bar{t}^{(1)} | [[\hat{\Phi}, \hat{T}^{(1)}], \hat{T}^{(1)}] | \mathrm{HF} \rangle \tag{14.3.74}$$

This expression is clearly an improvement on the original $n + 1$ expression (14.3.39). For later reference, we also derive the fifth-order expression. Straightforward expansion of the Lagrangian

Table 14.3 Contributions from connected amplitudes and multipliers to the CCPT energies

Energy	$E^{(1)}$	$E^{(2)}$	$E^{(3)}$	$E^{(4)}$	$E^{(5)}$
Excitations	HF	D	D	SDT	SDT*

*Contains contributions from disconnected quadruples in the multipliers.

and elimination of terms in accordance with the $2n + 1$ rule give the expression

$$E^{(5)} = \tfrac{1}{2}\langle\mathrm{HF}|[[\hat{\Phi}, \hat{T}^{(2)}], \hat{T}^{(2)}]|\mathrm{HF}\rangle + \langle\bar{t}^{(1)}|[[\hat{\Phi}, \hat{T}^{(2)}], \hat{T}^{(1)}]|\mathrm{HF}\rangle$$
$$+ \langle\bar{t}^{(2)}|[\hat{\Phi}, \hat{T}^{(2)}]|\mathrm{HF}\rangle + \tfrac{1}{2}\langle\bar{t}^{(2)}|[[\hat{\Phi}, \hat{T}^{(1)}], \hat{T}^{(1)}]|\mathrm{HF}\rangle \qquad (14.3.75)$$

where the amplitudes and multipliers appear only to second order in the perturbation.

We have thus arrived at a compact set of expressions for the perturbed energies that agree with Wigner's $2n + 1$ rule and which are size-extensive term by term as well as order by order. To summarize, we give in Table 14.3 the contributions to the CCPT energies from the various excitation levels. To second and third orders, there are contributions from the connected doubles only. Singles and triples make their first appearance to fourth order and quadruples have disconnected contributions to the fifth-order energy.

14.3.9 SIZE-EXTENSIVITY OF THE CCPT LAGRANGIAN

In our discussion of size-extensivity in Section 14.3.5, we found that the CCPT energies that comply with the $n + 1$ rule are size-extensive term by term as well as order by order. Since the Lagrange multipliers may not separate for noninteracting systems, we must likewise make sure that the energies derived from the variational Lagrangian (14.3.44) are size-extensive in the same manner.

From our previous discussions, we note that, for two noninteracting systems A and B, the Lagrangian may be written in the form

$$L_{\mathrm{AB}} = E_{0\mathrm{A}} + E_{0\mathrm{B}} + \sum_{\mu\mathrm{A}} \varepsilon_{\mu\mathrm{A}} t_{\mu\mathrm{A}} \bar{t}_{\mu\mathrm{A}} + \sum_{\mu\mathrm{B}} \varepsilon_{\mu\mathrm{B}} t_{\mu\mathrm{B}} \bar{t}_{\mu\mathrm{B}}$$
$$+ \langle\mathrm{HF}|\hat{\Phi}_{\mathrm{A}}^{\mathrm{T}}|\mathrm{HF}\rangle + \langle\mathrm{HF}|\hat{\Phi}_{\mathrm{B}}^{\mathrm{T}}|\mathrm{HF}\rangle + \langle\bar{t}_{\mathrm{AB}}|\hat{\Phi}_{\mathrm{B}}^{\mathrm{T}}|\mathrm{HF}\rangle + \langle\bar{t}_{\mathrm{AB}}|\hat{\Phi}_{\mathrm{A}}^{\mathrm{T}}|\mathrm{HF}\rangle \qquad (14.3.76)$$

Clearly, only multipliers that refer to a single subsystem (A or B) contribute to the total Lagrangian, ensuring that the energies obtained from the Lagrangian are termwise size-extensive.

14.4 Møller–Plesset theory for closed-shell systems

Having discussed Møller–Plesset theory in general terms, we now consider its implementation for closed-shell electronic systems. We begin by carrying out some preparations related to the parametrization of the closed-shell wave function in Section 14.4.1 and setting up the Lagrangian in Section 14.4.2. In Section 14.4.3, we consider the perturbed amplitudes needed for the calculation of the energies and finally, in Section 14.4.4, we derive the Møller–Plesset energy corrections for closed-shell systems, expressed in terms of MO integrals and orbital energies.

14.4.1 THE CLOSED-SHELL ZERO-ORDER SYSTEM

As in the spin-orbital basis, we separate the Hamiltonian of a closed-shell system into the zero-order Fock operator, the first-order fluctuation potential and a nuclear contribution:

$$\hat{H} = \hat{f} + \hat{\Phi} + h_{\text{nuc}} \tag{14.4.1}$$

In the canonical orbital basis, the Fock operator is given by

$$\hat{f} = \sum_p \varepsilon_p E_{pp} \tag{14.4.2}$$

where the orbital energies are the eigenvalues of the inactive Fock matrix:

$$\varepsilon_p = {}^{\text{I}}F_{pq}\delta_{pq} \tag{14.4.3}$$

For a discussion of the closed-shell Fock operator and the fluctuation potential, see Section 10.4.

For closed-shell Møller–Plesset perturbation theory, we shall use the exponential CCPT parametrization of Section 14.3. To calculate the energy to fourth order, we must determine the singlet wave function to second order, constructing the cluster operator $\hat{T}$ from singles, doubles and triples operators of singlet spin symmetry. We shall employ operators of the form

$$\hat{T}_1 = \sum_{ai} t_i^a E_{ai} \tag{14.4.4}$$

$$\hat{T}_2 = \frac{1}{2} \sum_{\substack{ab \\ ij}} t_{ij}^{ab} E_{ai} E_{bj} \tag{14.4.5}$$

$$\hat{T}_3 = \frac{1}{6} \sum_{\substack{abc \\ ijk}} t_{ijk}^{abc} E_{ai} E_{bj} E_{ck} \tag{14.4.6}$$

where the summations over the occupied and virtual orbital indices are free, and the following symmetry relations are understood:

$$t_{ij}^{ab} = t_{ji}^{ba} \tag{14.4.7}$$

$$t_{ijk}^{abc} = t_{ikj}^{acb} = t_{jik}^{bac} = t_{jki}^{bca} = t_{kij}^{cab} = t_{kji}^{cba} \tag{14.4.8}$$

Unlike for the spin-orbital basis, there are no symmetries for permutations separately among the occupied indices or among the virtual indices. When applied to the Hartree–Fock state, the excitation operators in (14.4.4)–(14.4.6) generate singlet combinations of excited CSFs, for which we use the notation

$$\left| \begin{matrix} a \\ i \end{matrix} \right\rangle = E_{ai}|\text{HF}\rangle \tag{14.4.9}$$

$$\left| \begin{matrix} ab \\ ij \end{matrix} \right\rangle = E_{ai}E_{bj}|\text{HF}\rangle \tag{14.4.10}$$

$$\left| \begin{matrix} abc \\ ijk \end{matrix} \right\rangle = E_{ai}E_{bj}E_{ck}|\text{HF}\rangle \tag{14.4.11}$$

We shall also use the following generic notation for the cluster operator and for the excited configurations

$$\hat{T} = \sum_\mu g_\mu t_\mu \hat{\tau}_\mu \tag{14.4.12}$$

$$|\mu\rangle = \hat{\tau}_\mu |\text{HF}\rangle \qquad (14.4.13)$$

where the summation over the spin-adapted excitation operators is unrestricted in accordance with (14.4.4)–(14.4.6). The degeneracy factors g_μ in (14.4.12) contain the constants present in the definition of the cluster operators – that is, 1, $\frac{1}{2}$ and $\frac{1}{6}$ for the single, double and triple excitations, respectively.

For the singles and doubles, the singlet parametrization of the cluster operator was discussed in Section 13.7.1. We shall not carry out a similar, rigorous derivation of the singlet operator for the triples (14.4.6) but note that the given parametrization of the triples space is redundant. Thus, for a set of three different virtual orbitals a, b and c and three different occupied orbitals i, j and k, there are six amplitudes t^{abc}_{ijk}, t^{abc}_{ikj}, t^{abc}_{jik}, t^{abc}_{jki}, t^{abc}_{kij} and t^{abc}_{kji} not related by the permutational symmetries (14.4.8). But from our discussion of the genealogical coupling scheme in Section 2.6, we know that it is only possible to construct five independent singlet states by distributing six electrons among six orbitals – see the branching diagram in Figure 2.1. The six amplitudes not related by the permutational symmetries (14.4.8) therefore cannot be independent and our parametrization of the triples (14.4.6) is redundant. Indeed, in Exercise 14.4, it is shown that the following linear combination of triple excitation operators is equal to zero:

$$E_{ai}E_{bj}E_{ck} + E_{ai}E_{bk}E_{cj} + E_{aj}E_{bi}E_{ck} + E_{aj}E_{bk}E_{ci} + E_{ak}E_{bi}E_{cj} + E_{ak}E_{bj}E_{ci} = 0 \qquad (14.4.14)$$

The doubles operator (14.4.5), by contrast, is nonredundant, as may be verified by inspection of the branching diagram, which shows that it is possible to distribute four electrons among four orbitals so as to yield two singlet configurations.

It is possible to set up a parametrization of the triples space that is nonredundant. The resulting parametrization would be more complicated, however, involving linear combinations of excitation operators. In practice, it is easier to work with the redundant parametrization of the triples space in (14.4.6). The redundancy (14.4.14) will not interfere with our solution of the Schrödinger equation – we shall arrive at the correct perturbed energies provided we are able to satisfy the CCPT equations. The redundancy does imply, however, that there are many different *amplitudes* that satisfy the CCPT equations, but these amplitudes all yield the same energies.

14.4.2 THE CLOSED-SHELL VARIATIONAL LAGRANGIAN

Let us set up the CCPT Lagrangian for a closed-shell state parametrized by the cluster operators (14.4.4)–(14.4.6). We shall follow the approach of Section 14.3, indicating only the necessary modifications. First, we derive the equivalent of (14.3.7) for the singlet Fock operator (14.4.2) and the singlet cluster operators (14.4.4)–(14.4.6). We readily obtain the following relationships

$$[\hat{f}, \hat{T}_1] = \sum_{ai} \varepsilon^a_i t^a_i E_{ai} \qquad (14.4.15)$$

$$[\hat{f}, \hat{T}_2] = \frac{1}{2} \sum_{aibj} \varepsilon^{ab}_{ij} t^{ab}_{ij} E_{ai}E_{bj} \qquad (14.4.16)$$

$$[\hat{f}, \hat{T}_3] = \frac{1}{6} \sum_{aibjck} \varepsilon^{abc}_{ijk} t^{abc}_{ijk} E_{ai}E_{bj}E_{ck} \qquad (14.4.17)$$

where, for example,

$$\varepsilon^{abc}_{ijk} = \varepsilon_a + \varepsilon_b + \varepsilon_c - \varepsilon_i - \varepsilon_j - \varepsilon_k \qquad (14.4.18)$$

We may write these commutators in the generic form

$$[\hat{f}, \hat{T}] = \sum_{\mu} g_{\mu} \varepsilon_{\mu} t_{\mu} \hat{\tau}_{\mu} \tag{14.4.19}$$

and obtain for the similarity-transformed closed-shell Fock operator

$$\hat{f}^{\mathrm{T}} = \hat{f} + \sum_{\mu} g_{\mu} \varepsilon_{\mu} t_{\mu} \hat{\tau}_{\mu} \tag{14.4.20}$$

by analogy with the spin-orbital representation (14.3.9).

The modifications necessary in the orbital representation become apparent when we consider the projected coupled-cluster equations. For the singles and doubles, we shall still assume that $\langle \overline{\mu}_1 |$ and $\langle \overline{\mu}_2 |$ are related to $|\mu_1\rangle$ and $|\mu_2\rangle$ in the biorthogonal fashion (13.7.54) and (13.7.55)

$$\left\langle \begin{array}{c|c} \overline{a} & c \\ i & k \end{array} \right\rangle = \delta_{ai,ck} \tag{14.4.21}$$

$$\left\langle \begin{array}{c|c} \overline{ab} & cd \\ ij & kl \end{array} \right\rangle = P_{ij}^{ab} \delta_{aibj,ckdl} = P_{kl}^{cd} \delta_{aibj,ckdl} \tag{14.4.22}$$

where

$$P_{ij}^{ab} A_{ij}^{ab} = A_{ij}^{ab} + A_{ji}^{ba} \tag{14.4.23}$$

Since the triples operator (14.4.11) is redundant, we cannot set up a projection basis $\langle \overline{\mu} |$ that is biorthogonal to the linear combination of CSFs $|\mu\rangle$ in (14.4.13). We shall simply assume that the $\langle \overline{\mu}_3 |$ constitute a linearly independent basis for the space spanned by the linearly dependent vectors $\langle \mu_3 |$ but we shall not specify their detailed form.

The lack of a biorthogonal representation of the triples space means that the projected equations

$$\langle \overline{\mu} | \hat{H}^{\mathrm{T}} | \mathrm{HF} \rangle = 0 \tag{14.4.24}$$

become more complicated than in the spin-orbital representation. Inserting the partitioned Hamiltonian (14.4.1), we obtain as usual

$$\langle \overline{\mu} | \hat{f}^{\mathrm{T}} | \mathrm{HF} \rangle = -\langle \overline{\mu} | \hat{\Phi}^{\mathrm{T}} | \mathrm{HF} \rangle \tag{14.4.25}$$

Next, substituting (14.4.20) for the similarity-transformed Fock operator, we obtain

$$\sum_{\nu} g_{\nu} \varepsilon_{\nu} t_{\nu} \langle \overline{\mu} | \nu \rangle = -\langle \overline{\mu} | \hat{\Phi}^{\mathrm{T}} | \mathrm{HF} \rangle \tag{14.4.26}$$

which differs from (14.3.14) by the presence of the sum on the left-hand side. For projections against the singles and doubles spaces, we may, as usual, invoke the biorthogonality (14.4.21) and (14.4.22) to obtain

$$\varepsilon_{\mu_i} t_{\mu_i} = -\langle \overline{\mu}_i | \hat{\Phi}^{\mathrm{T}} | \mathrm{HF} \rangle, \qquad i = 1, 2 \tag{14.4.27}$$

but this simplification is impossible for the triples in the redundant representation (14.4.6).

To derive closed-shell Møller–Plesset energies that comply with Wigner's $2n + 1$ rule, we now set up the variational Lagrangian. Introducing one Lagrange multiplier for each variational condition (i.e. one multiplier for each basis vector in the projection space), we obtain

$$L(\mathbf{t}, \overline{\mathbf{t}}) = E_0 + \sum_{\overline{\mu}\nu} g_{\nu} \varepsilon_{\nu} t_{\nu} \overline{t}_{\overline{\mu}} \langle \overline{\mu} | \nu \rangle + \langle \mathrm{HF} | \hat{\Phi}^{\mathrm{T}} | \mathrm{HF} \rangle + \langle \overline{t} | \hat{\Phi}^{\mathrm{T}} | \mathrm{HF} \rangle \tag{14.4.28}$$

We must carefully distinguish the summation ranges for μ and ν. Whereas the summation over the amplitudes t_{ν} is unrestricted, the summation over the multipliers $\overline{t}_{\overline{\mu}}$ is restricted, avoiding the linearly dependent excitations. Thus, for the doubles, the summation over $\bar{t}_{\bar{\mu}}$ involves only the excitations $E_{ai}E_{bj}$ with $ai \geq bj$. Likewise, for the triples, the summation is over a set of linearly independent excitations with multipliers $\bar{t}_{\bar{\mu}}$, but their precise specification is not needed for our development.

Because of the redundant parametrization of the wave function, there are fewer multipliers than amplitudes in the Lagrangian. The variational conditions on the Lagrangian are obtained by differentiating (14.4.28) with respect to the multipliers and the amplitudes and setting the resulting expressions equal to zero. Whereas differentiation with respect to the multipliers yields the projected equations in the form (14.4.26), differentiation with respect to the amplitudes yields a linear set of equations for the multipliers

$$\sum_{\bar{v}} \varepsilon_\mu \bar{t}_{\bar{v}} \langle \bar{v} | \mu \rangle = -\langle \mathrm{HF} | \hat{\Phi}^\mathrm{T} | \mu \rangle - \langle \bar{t} | [\hat{\Phi}^\mathrm{T}, \hat{\tau}_\mu] | \mathrm{HF} \rangle \tag{14.4.29}$$

where we have used the fact that each independent cluster amplitude t_μ occurs $1/g_\mu$ times; see (14.4.7) and (14.4.8). Equation (14.4.29) should be compared with its spin-unrestricted counterpart (14.3.47) and (14.3.50).

14.4.3 THE CLOSED-SHELL WAVE-FUNCTION CORRECTIONS

To evaluate the energy corrections to fourth order, we need $t_{ij}^{ab^{(1)}}$, $\bar{t}_{ij}^{ab^{(1)}}$, $t_i^{a^{(2)}}$, $t_{ij}^{ab^{(2)}}$ and $t_{ijk}^{abc^{(2)}}$. Note that $t_{ij}^{ab^{(1)}}$ and $\bar{t}_{ij}^{ab^{(1)}}$ differ since we use a biorthogonal basis. We must therefore set up the first- and second-order equations for the amplitudes and the first-order equations for the multipliers. Whereas the first-order equations take the form

$$\sum_v g_v \varepsilon_v t_v^{(1)} \langle \bar{\mu} | v \rangle = -\langle \bar{\mu} | \hat{\Phi} | \mathrm{HF} \rangle \tag{14.4.30}$$

$$\sum_{\bar{v}} \varepsilon_\mu \bar{t}_{\bar{v}}^{(1)} \langle \bar{v} | \mu \rangle = -\langle \mathrm{HF} | \hat{\Phi} | \mu \rangle \tag{14.4.31}$$

the second-order equations for the amplitudes become

$$\sum_v g_v \varepsilon_v t_v^{(2)} \langle \bar{\mu} | v \rangle = -\langle \bar{\mu} | [\hat{\Phi}, \hat{T}^{(1)}] | \mathrm{HF} \rangle \tag{14.4.32}$$

As before, only the doubles contribute to the first-order equations, which may therefore be written in the form

$$\varepsilon_{ij}^{ab} t_{ij}^{ab^{(1)}} = -\left\langle \begin{array}{c} \overline{ab} \\ ij \end{array} \middle| \hat{\Phi} \middle| \mathrm{HF} \right\rangle \tag{14.4.33}$$

$$\varepsilon_{ij}^{ab} \bar{t}_{ij}^{ab^{(1)}} = -\left\langle \mathrm{HF} \middle| \hat{\Phi} \middle| \begin{array}{c} ab \\ ij \end{array} \right\rangle \tag{14.4.34}$$

These equations are not identical since the bra and ket states are different, being the basis functions of a biorthogonal representation of the doubles space.

To second order in the fluctuation potential, there are contributions from the singles, doubles and triples. For the singles and doubles, we may invoke biorthogonality to simplify the equations, but for the triples no such simplification is possible. Thus, for the singles and doubles, we may write (14.4.32) in the form

$$\varepsilon_i^a t_i^{a^{(2)}} = -\left\langle \begin{array}{c} \overline{a} \\ i \end{array} \middle| [\hat{\Phi}, \hat{T}_2^{(1)}] \middle| \mathrm{HF} \right\rangle \tag{14.4.35}$$

$$\varepsilon_{ij}^{ab} t_{ij}^{ab(2)} = \left\langle \frac{ab}{ij} \middle| [\hat{\Phi}, \hat{T}_2^{(1)}] \middle| \mathrm{HF} \right\rangle \tag{14.4.36}$$

but the triples equations are left in the unresolved form

$$\frac{1}{6} \sum_{\substack{abc \\ ijk}} \varepsilon_{ijk}^{abc} t_{ijk}^{abc(2)} \left\langle \overline{\mu}_3 \middle| \frac{abc}{ijk} \right\rangle = -\langle \overline{\mu}_3 | [\hat{\Phi}, \hat{T}_2^{(1)}] | \mathrm{HF} \rangle \tag{14.4.37}$$

It only remains to evaluate the right-hand sides of (14.4.33)–(14.4.37). For this purpose, we shall make extensive use of the results from CCSD theory in Section 13.7 and, in particular, Boxes 13.1 and 13.2.

Consider first the right-hand side of the first-order amplitude equations (14.4.33). Replacing the fluctuation potential with the full Hamiltonian and using (13B.2.1) of Box 13.2, we immediately obtain

$$\left\langle \frac{ab}{ij} \middle| \hat{\Phi} \middle| \mathrm{HF} \right\rangle = \left\langle \frac{ab}{ij} \middle| \hat{H} \middle| \mathrm{HF} \right\rangle = g_{aibj} \tag{14.4.38}$$

Making the same replacement on the right-hand side of the multiplier equations (14.4.34) and using (14.4.10), we obtain

$$\left\langle \mathrm{HF} \middle| \hat{\Phi} \middle| \frac{ab}{ij} \right\rangle = \langle \mathrm{HF} | [[\hat{H}, E_{ai}], E_{bj}] | \mathrm{HF} \rangle = 2L_{iajb} \tag{14.4.39}$$

In the final step, we have used (13B.2.3) with the integrals L_{iajb} defined as in (13.7.15):

$$L_{pqrs} = 2g_{pqrs} - g_{psrq} \tag{14.4.40}$$

Combining (14.4.38) and (14.4.39) with (14.4.33) and (14.4.34), we obtain

$$\varepsilon_{ij}^{ab} t_{ij}^{ab(1)} = -g_{aibj} \tag{14.4.41}$$

$$\varepsilon_{ij}^{ab} \overline{t}_{ij}^{ab(1)} = -2L_{iajb} \tag{14.4.42}$$

which are the final expressions for the first-order doubles amplitudes and multipliers in the orbital basis. Note the difference between the amplitudes and their multipliers. The evaluation of the first-order amplitudes and multipliers from the MO integrals scales as $O^2 V^2$.

Turning our attention to the second-order equations, we first note that

$$\langle \overline{\mu} | [\hat{\Phi}, \hat{T}_2^{(1)}] | \mathrm{HF} \rangle = \langle \overline{\mu} | [\hat{H}, \hat{T}_2^{(1)}] | \mathrm{HF} \rangle - \langle \overline{\mu} | [\hat{f}, \hat{T}_2^{(1)}] | \mathrm{HF} \rangle \tag{14.4.43}$$

Expressing the first term on the right-hand side in terms of the commutators of Box 13.2 and evaluating the commutator of the last term using (14.4.16), we obtain

$$\langle \overline{\mu} | [\hat{\Phi}, \hat{T}_2^{(1)}] | \mathrm{HF} \rangle = \frac{1}{2} \sum_{abij} t_{ij}^{ab(1)} \langle \overline{\mu} | [[\hat{H}, E_{ai}], E_{bj}] | \mathrm{HF} \rangle + \sum_{abij} t_{ij}^{ab(1)} \langle \overline{\mu} | E_{ai} [\hat{H}, E_{bj}] | \mathrm{HF} \rangle$$

$$- \frac{1}{2} \sum_{abij} \varepsilon_{ij}^{ab} t_{ij}^{ab(1)} \langle \overline{\mu} | E_{ai} E_{bj} | \mathrm{HF} \rangle \tag{14.4.44}$$

From this equation, we may obtain explicit expressions for the singles, doubles and triples right-hand side by referring to Box 13.2. After some rearrangement and use of the biorthogonal relations

(14.4.21) and (14.4.22), we obtain for the singles and doubles

$$\left\langle \begin{matrix} \overline{a} \\ i \end{matrix} \middle| [\hat{\Phi}, \hat{T}_2^{(1)}] \middle| \mathrm{HF} \right\rangle = -\sum_{dkl} t_{kl}^{ad^{(1)}} L_{kild} + \sum_{cdk} t_{ki}^{cd^{(1)}} L_{adkc} \tag{14.4.45}$$

$$\left\langle \begin{matrix} \overline{ab} \\ ij \end{matrix} \middle| [\hat{\Phi}, \hat{T}_2^{(1)}] \middle| \mathrm{HF} \right\rangle = \sum_{cd} t_{ij}^{cd^{(1)}} g_{acbd} + \sum_{kl} t_{kl}^{ab^{(1)}} g_{kilj}$$

$$+ P_{ij}^{ab} \sum_{ck} (t_{ik}^{ac^{(1)}} L_{bjkc} - t_{kj}^{ac^{(1)}} g_{bcki} - t_{ki}^{ac^{(1)}} g_{bjkc}) \tag{14.4.46}$$

To obtain (14.4.45) and (14.4.46), we have used the fact that the inactive Fock matrix is diagonal. These expressions may be obtained also from the corresponding expressions for the projected CCSD equations (13.7.72) and (13.7.88) in Section 13.7.

For the triples, inspection of particle ranks shows that only the middle term on the right-hand side of (14.4.44) gives a nonzero contribution. From (13B.2.2), we now obtain the following expression for the right-hand side:

$$\langle \overline{\mu}_3 | [\hat{\Phi}, \hat{T}_2^{(1)}] | \mathrm{HF} \rangle = \frac{1}{6} \sum_{\substack{abc \\ ijk}} \left\langle \overline{\mu}_3 \middle| \begin{matrix} abc \\ ijk \end{matrix} \right\rangle P_{ijk}^{abc} \left(\sum_d t_{ij}^{ad^{(1)}} g_{ckbd} - \sum_l t_{il}^{ab^{(1)}} g_{cklj} \right) \tag{14.4.47}$$

where we have introduced the symmetrization operator

$$P_{ijk}^{abc} A_{ijk}^{abc} = A_{ijk}^{abc} + A_{ikj}^{acb} + A_{jik}^{bac} + A_{jki}^{bca} + A_{kij}^{cab} + A_{kji}^{cba} \tag{14.4.48}$$

by analogy with the two-particle symmetrizer (14.4.23).

Having determined the right-hand sides of the second-order amplitude equations, we may go on to calculate the amplitudes from (14.4.35)–(14.4.37). For the singles and doubles, the solutions are unique and are obtained by combining (14.4.35) with (14.4.45) for the singles and (14.4.36) with (14.4.46) for the doubles. Proceeding in the same manner for the triples, we obtain by combining the left-hand side (14.4.37) with the right-hand side (14.4.47):

$$\sum_{\substack{abc \\ ijk}} \left[\varepsilon_{ijk}^{abc} t_{ijk}^{abc^{(2)}} + P_{ijk}^{abc} \left(\sum_d t_{ij}^{ad^{(1)}} g_{ckbd} - \sum_l t_{il}^{ab^{(1)}} g_{cklj} \right) \right] \left\langle \overline{\mu}_3 \middle| \begin{matrix} abc \\ ijk \end{matrix} \right\rangle = 0 \tag{14.4.49}$$

These linear equations may be satisfied in many different ways since there are more unknown amplitudes $t_{ijk}^{abc^{(2)}}$ than there are equations (one equation for each basis vector $\langle \overline{\mu}_3 |$). Selecting the solution for which each term in the outer summation in (14.4.49) is separately zero, we obtain the following final expressions for the second-order perturbed amplitudes:

$$\varepsilon_i^a t_i^{a^{(2)}} = \sum_{dkl} t_{kl}^{ad^{(1)}} L_{kild} - \sum_{cdk} t_{ki}^{cd^{(1)}} L_{adkc} \tag{14.4.50}$$

$$\varepsilon_{ij}^{ab} t_{ij}^{ab^{(2)}} = -\sum_{cd} t_{ij}^{cd^{(1)}} g_{acbd} - \sum_{kl} t_{kl}^{ab^{(1)}} g_{kilj} - P_{ij}^{ab} \sum_{ck} \left(t_{ik}^{ac^{(1)}} L_{bjkc} - t_{kj}^{ac^{(1)}} g_{bcki} - t_{ki}^{ac^{(1)}} g_{bjkc} \right)$$

$$\tag{14.4.51}$$

$$\varepsilon_{ijk}^{abc} t_{ijk}^{abc^{(2)}} = -P_{ijk}^{abc} \left(\sum_d t_{ij}^{ad^{(1)}} g_{ckbd} - \sum_l t_{il}^{ab^{(1)}} g_{cklj} \right) \tag{14.4.52}$$

Together with the first-order expressions for the amplitudes (14.4.41) and (14.4.42), these equations determine the closed-shell spin-restricted CCPT Lagrangian. Whereas the evaluation of the first-order amplitudes scales as O^2V^2, the evaluation of the second-order amplitudes scales as O^2V^3 for the singles, O^2V^4 for the doubles and O^3V^4 for the triples.

14.4.4 THE CLOSED-SHELL ENERGY CORRECTIONS

In this subsection, we give the closed-shell Møller–Plesset energy corrections to fourth order. As noted in Section 14.2.4, the sum of the zero- and first-order Møller–Plesset energies is equal to the Hartree–Fock energy. We therefore proceed directly to the higher-order corrections, treating in turn the closed-shell MP2, MP3 and MP4 energies.

The second-order Møller–Plesset energy correction is given in (14.3.70):

$$E^{(2)} = \langle \mathrm{HF}|[\hat{\Phi}, \hat{T}^{(1)}]|\mathrm{HF}\rangle \tag{14.4.53}$$

Replacing the fluctuation potential by the Hamiltonian and inserting (14.4.5) for the perturbed doubles amplitudes, we obtain

$$E^{(2)} = \tfrac{1}{2} \sum_{aibj} t_{ij}^{ab^{(1)}} \langle \mathrm{HF}|[[\hat{H}, E_{ai}], E_{bj}]|\mathrm{HF}\rangle \tag{14.4.54}$$

According to (13B.2.3), this expression may be written as

$$E^{(2)} = \sum_{aibj} t_{ij}^{ab^{(1)}} L_{iajb} \tag{14.4.55}$$

which is similar to the connected doubles contribution to the CCSD energy in (13.7.18). We may eliminate the amplitudes from this expression by using (14.4.41), obtaining [6,7]:

$$E^{(2)} = -\sum_{aibj} \frac{g_{aibj} L_{iajb}}{\varepsilon_a + \varepsilon_b - \varepsilon_i - \varepsilon_j} \tag{14.4.56}$$

The second-order Møller–Plesset energy can thus be calculated in a simple manner from the MO integrals and the orbital energies, in a process that scales as O^2V^2. It should be recalled, however, that the transformation of the two-electron integrals from the AO to the MO basis is a fifth-order process, scaling as On^4, where n is the number of AOs. It is instructive to compare the closed-shell MP2 expression in the orbital basis (14.4.56) with the corresponding energy in the spin-orbital basis (14.2.53). Indeed, we may arrive at (14.4.56) by introducing in (14.2.53) a spin-orbital basis of doubly occupied orbitals and carrying out the spin integrations.

We now proceed to consider the third-order Møller–Plesset energy (14.3.72):

$$E^{(3)} = \langle \bar{t}^{(1)}|[\hat{\Phi}, \hat{T}^{(1)}]|\mathrm{HF}\rangle \tag{14.4.57}$$

Introducing the multipliers (14.4.42), we may write this expression in the form

$$E^{(3)} = \sum_{ai \geq bj} \bar{t}_{ij}^{ab^{(1)}} \left\langle \left. \frac{\overline{ab}}{ij} \right| [\hat{\Phi}, \hat{T}_2^{(1)}] \right| \mathrm{HF} \right\rangle = \frac{1}{2} \sum_{aibj} \tilde{t}_{ij}^{ab^{(1)}} \left\langle \left. \frac{\overline{ab}}{ij} \right| [\hat{\Phi}, \hat{T}_2^{(1)}] \right| \mathrm{HF} \right\rangle \tag{14.4.58}$$

To avoid the restricted summation, we have introduced weighted multipliers that are symmetric with respect to the simultaneous permutation of the occupied and virtual indices:

$$\tilde{t}_{ij}^{ab^{(1)}} = (1 + \delta_{ai,bj}) \bar{t}_{ij}^{ab^{(1)}} \tag{14.4.59}$$

$$\tilde{t}_{ij}^{ab(1)} = \tilde{t}_{ji}^{ba(1)} \tag{14.4.60}$$

To evaluate (14.4.58), we use (14.4.46) and obtain the following expression for the third-order Møller–Plesset energy [8]:

$$E^{(3)} = \sum_{aibj} \tilde{t}_{ij}^{ab(1)} X_{ij}^{ab} \tag{14.4.61}$$

$$X_{ij}^{ab} = \tfrac{1}{2} \sum_{cd} t_{ij}^{cd(1)} g_{acbd} + \tfrac{1}{2} \sum_{kl} t_{kl}^{ab(1)} g_{kilj} + \sum_{ck} (t_{ik}^{ac(1)} L_{bjkc} - t_{kj}^{ac(1)} g_{bcki} - t_{ki}^{ac(1)} g_{bjkc}) \tag{14.4.62}$$

The MP3 energy thus consists of several contributions, each containing one first-order doubles amplitude (14.4.41), one first-order doubles multiplier (14.4.42) and one integral. The MP3 energy is thus more complicated than the MP2 energy (14.4.55) but involves the same building blocks (MO integrals and orbital energies). The most time-consuming step is the evaluation of the first term in (14.4.62), which scales as $O^2 V^4$. The corresponding term in the CCSD amplitude equations (13.7.101) is also the most expensive in CCSD calculations.

Let us finally consider the fourth-order Møller–Plesset energy, which, according to (14.3.74), may be written in the following form:

$$E^{(4)} = \langle \bar{t}^{(1)} | [\hat{\Phi}, \hat{T}^{(2)}] | \text{HF} \rangle + \tfrac{1}{2} \langle \bar{t}^{(1)} | [[\hat{\Phi}, \hat{T}^{(1)}], \hat{T}^{(1)}] | \text{HF} \rangle \tag{14.4.63}$$

There are four distinct contributions to the MP4 energy:

$$E^{(4)} = S + D + T + Q \tag{14.4.64}$$

where

$$S = \langle \bar{t}^{(1)} | [\hat{\Phi}, \hat{T}_1^{(2)}] | \text{HF} \rangle \tag{14.4.65}$$

$$D = \langle \bar{t}^{(1)} | [\hat{\Phi}, \hat{T}_2^{(2)}] | \text{HF} \rangle \tag{14.4.66}$$

$$T = \langle \bar{t}^{(1)} | [\hat{\Phi}, \hat{T}_3^{(2)}] | \text{HF} \rangle \tag{14.4.67}$$

$$Q = \tfrac{1}{2} \langle \bar{t}^{(1)} | [[\hat{\Phi}, \hat{T}_2^{(1)}], \hat{T}_2^{(1)}] | \text{HF} \rangle \tag{14.4.68}$$

Whereas the first three terms in (14.4.64) contain the second-order singles, doubles and triples amplitudes, respectively, the last term contains only products of the first-order doubles amplitudes (i.e. disconnected quadruples). We shall consider each of these terms in turn.

The singles part of the MP4 energy (14.4.65) may be written in the form

$$S = \frac{1}{2} \sum_{\substack{abc \\ ijk}} \tilde{t}_{ij}^{ab(1)} t_k^{c(2)} \left\langle \overline{\begin{matrix} ab \\ ij \end{matrix}} \,\middle|\, [\hat{H}, E_{ck}] \middle| \text{HF} \right\rangle \tag{14.4.69}$$

where we have replaced the fluctuation potential by the Hamiltonian. Using (13B.2.2), we find, after some simple algebra, that the singles part may be written as

$$S = \sum_{\substack{ab \\ ij}} \tilde{t}_{ij}^{ab(1)} \left(\sum_c t_j^{c(2)} g_{aibc} - \sum_k t_k^{b(2)} g_{aikj} \right) \tag{14.4.70}$$

which requires an amount of work scaling as $O^2 V^3$.

The doubles part of the MP4 energy (14.4.66) is identical to the MP3 energy (14.4.57) except that the second-order doubles amplitudes are used rather than the first-order amplitudes. We may therefore use the MP3 expressions (14.4.61) and (14.4.62) for the doubles part of MP4, replacing the first-order amplitudes in (14.4.62) by the second-order ones. This term scales as $O^2 V^4$.

We now come to the more complicated triples part of the MP4 energy. To evaluate this term, we replace the fluctuation potential by the Hamiltonian, insert (14.4.6) for the triples cluster operator and obtain after some rearrangement

$$T = \langle \bar{t}^{(1)} | [\hat{H}, \hat{T}_3^{(2)}] | \mathrm{HF} \rangle = M_1 + M_2 + M_3 \tag{14.4.71}$$

where

$$M_1 = \frac{1}{2} \sum_{\substack{abc \\ ijk}} t_{ijk}^{abc(2)} \langle \bar{t}^{(1)} | E_{ai} E_{bj} [\hat{H}, E_{ck}] | \mathrm{HF} \rangle \tag{14.4.72}$$

$$M_2 = \frac{1}{2} \sum_{\substack{abc \\ ijk}} t_{ijk}^{abc(2)} \langle \bar{t}^{(1)} | E_{ai} [[\hat{H}, E_{bj}], E_{ck}] | \mathrm{HF} \rangle \tag{14.4.73}$$

$$M_3 = \frac{1}{6} \sum_{\substack{abc \\ ijk}} t_{ijk}^{abc(2)} \langle \bar{t}^{(1)} | [[[\hat{H}, E_{ai}], E_{bj}], E_{ck}] | \mathrm{HF} \rangle \tag{14.4.74}$$

The evaluation of each of these terms follows the usual procedure, using the expressions in Box 13.2. M_1 is zero, which follows from (13B.2.2) since the inactive Fock matrix is diagonal in the canonical representation. For M_2, we use (13B.2.3). After some simple algebra, we are led to the following expression:

$$M_2 = \sum_{\substack{ab \\ ij}} \tilde{t}_{ij}^{ab(1)} \left(\sum_{cdk} t_{ijk}^{acd(2)} L_{bckd} - \sum_{ckl} t_{ikl}^{abc(2)} L_{kjlc} \right) \tag{14.4.75}$$

Completing our derivation of the triples part of the MP4 energy, we obtain from (13B.2.4)

$$M_3 = -\sum_{\substack{ab \\ ij}} \tilde{t}_{ij}^{ab(1)} \left(\sum_{cdk} t_{kji}^{acd(2)} g_{kdbc} - \sum_{ckl} t_{lki}^{abc(2)} g_{kjlc} \right) \tag{14.4.76}$$

The evaluation of M_2 and M_3 scales as $O^3 V^4$. The calculation of the triples contribution to the MP4 energy thus scales as the seventh power of the size of the system.

We finally come to the quadruples part of the MP4 energy correction. Introducing the explicit expression for the multipliers, we obtain

$$Q = \frac{1}{4} \sum_{\substack{ab \\ ij}} \tilde{t}_{ij}^{ab(1)} \left\langle \frac{\overline{ab}}{ij} \middle| [[\hat{H}, \hat{T}_2^{(1)}], \hat{T}_2^{(1)}] \middle| \mathrm{HF} \right\rangle \tag{14.4.77}$$

A similar matrix element is needed for the evaluation of the CCSD amplitudes equations; see (13.7.93). An explicit expression for (14.4.77) is obtained by inserting (13.7.97)–(13.7.99) in (13.7.93), omitting the tilde on the integrals. We do not give the resulting expression since the substitution is trivial and since it is contained in the complete expression for the MP4 energy correction given below. The most expensive steps in the evaluation of Q scale as $O^3 V^3$.

Collecting all contributions, we arrive at the following expression for the closed-shell MP4 energy [9–12]:

$$E^{(4)} = \sum_{abij} \tilde{t}_{ij}^{ab(1)} (S_{ij}^{ab} + D_{ij}^{ab} + T_{ij}^{ab} + Q_{ij}^{ab}) \tag{14.4.78}$$

where

$$S_{ij}^{ab} = \sum_c t_j^{c^{(2)}} g_{aibc} - \sum_k t_k^{b^{(2)}} g_{aikj} \tag{14.4.79}$$

$$D_{ij}^{ab} = \frac{1}{2} \sum_{cd} t_{ij}^{cd^{(2)}} g_{acbd} + \frac{1}{2} \sum_{kl} t_{kl}^{ab^{(2)}} g_{kilj} + \sum_{ck} (t_{ik}^{ac^{(2)}} L_{bjkc} - t_{kj}^{ac^{(2)}} g_{bcki} - t_{ki}^{ac^{(2)}} g_{bjkc}) \tag{14.4.80}$$

$$T_{ij}^{ab} = \sum_{cdk} \left(t_{ijk}^{acd^{(2)}} L_{bckd} - t_{kji}^{acd^{(2)}} g_{kdbc} \right) - \sum_{ckl} \left(t_{ikl}^{abc^{(2)}} L_{kjlc} - t_{lki}^{abc^{(2)}} g_{kjlc} \right) \tag{14.4.81}$$

$$Q_{ij}^{ab} = \frac{1}{2} \sum_{kl} t_{kl}^{ab^{(1)}} \sum_{cd} t_{ij}^{cd^{(1)}} g_{kcld} + \sum_{ck} t_{ik}^{ac^{(1)}} \sum_{dl} (t_{jl}^{bd^{(1)}} - t_{lj}^{bd^{(1)}}) L_{kcld}$$

$$+ \frac{1}{2} \sum_{ck} t_{ki}^{ac^{(1)}} \sum_{dl} t_{lj}^{bd^{(1)}} g_{kcld} + \frac{1}{2} \sum_{dk} t_{kj}^{ad^{(1)}} \sum_{cl} t_{li}^{bc^{(1)}} g_{kcld}$$

$$- \sum_k t_{ik}^{ab^{(1)}} \sum_{cdl} t_{lj}^{cd^{(1)}} L_{lckd} - \sum_c t_{ij}^{ac^{(1)}} \sum_{dkl} t_{kl}^{bd^{(1)}} L_{kcld} \tag{14.4.82}$$

Clearly, a large number of distinct terms are needed for the evaluation of the MP4 energy. The computational costs of these terms vary considerably. The evaluation of the singles part scales as O^2V^3, the doubles part as O^2V^4, the triples part as O^3V^4 and the quadruples part as O^3V^3. In calculating these costs, we have assumed that partial summations are carried out as indicated by the multiple summation signs in the expressions. As already noted, the evaluation of the amplitudes scales as O^3V^4, which therefore is the overall cost of the MP4 energy.

In the calculation of the MP4 energy, the evaluation of the contributions from the triples amplitudes is the most time-consuming step. To reduce the computational cost, the *MP4(SDQ) approximation*, in which the triples part (14.4.81) is omitted, is sometimes used [11]

$$E_{\text{SDQ}}^{(4)} = \sum_{abij} \tilde{t}_{ij}^{ab^{(1)}} (S_{ij}^{ab} + D_{ij}^{ab} + Q_{ij}^{ab}) \tag{14.4.83}$$

The motivation for introducing the MP4(SDQ) approximation – other than the computational savings relative to full the MP4 correction – is that it represents the fourth-order perturbational estimate of the CCSD energy.

14.5 Convergence in perturbation theory

In this section, we examine the important question of convergence in perturbation theory and discuss, in particular, the identification of divergent expansions. We begin in Section 14.5.1 by discussing perturbation theory for a model system containing two states, obtaining a simple condition for the convergence of the perturbation expansion for this system. Next, in Section 14.5.2, we summarize the general conditions for convergence of perturbation expansions in finite-dimensional spaces. The divergence of perturbation expansions is here traced to the occurrence of intruder states, which are studied in detail in Section 14.5.3 for the gap-shifted two-state model. Typical examples of intruders are then examined in Section 14.5.4 for the model system and in Section 14.5.5 for Møller–Plesset theory applied to the HF molecule, illustrating that divergences may easily arise in applications of Møller–Plesset perturbation theory. It is sometimes possible to turn a divergent expansion into a convergent one, obtaining in this manner an accurate approximation to the exact energy. In Section 14.5.6, we present one such technique: analytic continuation.

14.5.1 A TWO-STATE MODEL

If we restrict the Fock space to two states, the Hamiltonian may be written as

$$\mathbf{H} = \begin{pmatrix} \alpha & \delta \\ \delta & \beta \end{pmatrix} \tag{14.5.1}$$

where we shall assume that $\beta > \alpha$. Using the diagonal part as the zero-order Hamiltonian, we obtain the partitioning

$$\mathbf{H} = \mathbf{H}_0 + \mathbf{U} \tag{14.5.2}$$

where

$$\mathbf{H}_0 = \begin{pmatrix} \alpha & 0 \\ 0 & \beta \end{pmatrix} \tag{14.5.3}$$

$$\mathbf{U} = \begin{pmatrix} 0 & \delta \\ \delta & 0 \end{pmatrix} \tag{14.5.4}$$

We note that this partitioning of the Hamiltonian corresponds to that of Epstein–Nesbet theory; the Møller–Plesset partitioning has nonzero diagonal elements also in $\mathbf{U}$.

Let us now calculate the Rayleigh–Schrödinger energy corrections and investigate whether the perturbation expansion converges. The energy corrections can be obtained by the general procedure of (14.1.14) and (14.1.16) and (because of the simple structure of $\mathbf{H}_0$ and $\mathbf{U}$) analytic expressions are obtained. However, in this simple case, we may obtain a closed-form expression for the energy (see Exercise 14.5):

$$E = \frac{\alpha + \beta}{2} \pm \frac{\sqrt{(\beta - \alpha)^2 + 4\delta^2}}{2} \tag{14.5.5}$$

Expanding this expression in orders of δ, we obtain the Rayleigh–Schrödinger energy corrections. The equivalence of this expansion to that of standard perturbation theory follows from the uniqueness of Taylor expansions.

Restricting ourselves to the ground state, we may write the energy in the following form, recalling that $\beta > \alpha$:

$$E = \frac{\alpha + \beta}{2} - \frac{\beta - \alpha}{2} \sqrt{1 + \frac{4\delta^2}{(\beta - \alpha)^2}} \tag{14.5.6}$$

Expanding the square root of this expression in a Taylor series, we obtain

$$E = \sum_{n=0}^{\infty} E^{(2n)} \tag{14.5.7}$$

where only even-order terms contribute:

$$E^{(0)} = \alpha \tag{14.5.8}$$

$$E^{(2n)} = (-1)^n \frac{(2n - 3)!!2^{n-1}}{n!} \frac{\delta^{2n}}{(\beta - \alpha)^{2n-1}} \tag{14.5.9}$$

To establish a criterion for convergence of this series, we note that (14.5.7) represents an expansion of the square root in (14.5.6). Since the expansion of $\sqrt{1 + x}$ in x converges if $|x| < 1$, we obtain the convergence condition

$$\frac{4\delta^2}{(\beta - \alpha)^2} < 1 \tag{14.5.10}$$

or equivalently

$$|\delta| < \frac{\beta - \alpha}{2} \tag{14.5.11}$$

This simple result quantifies our notion that the coupling must be small for the two-state problem to converge. Thus, for problems that can be reduced to a two-state system with the partitioning (14.5.2)–(14.5.4), a simple comparison of the zero-order energy gap $\beta - \alpha$ with the coupling δ allows us to predict whether or not the perturbation expansion converges and also to establish the rate of convergence, which is rapid for $|\delta| \ll \beta - \alpha$ but slow for $|\delta| \approx \beta - \alpha$.

14.5.2 CONDITIONS FOR CONVERGENCE

For perturbation expansions in a general finite-dimensional space, it is impossible to derive analytic expressions for the eigenvalues and the strategy of the preceding subsection cannot be used. To establish whether a given expansion converges, we may instead use the criteria developed by Kato [13]. We here give a simplified discussion of the theory [14], referring to the monograph of Kato for a detailed account.

Consider the partitioned Hamiltonian

$$\mathbf{H}(z) = \mathbf{H}_0 + z\mathbf{U} \tag{14.5.12}$$

for a complex *strength parameter* z. The eigenvalue problem (14.1.29) for the kth state

$$\mathbf{H}(z)\mathbf{C}_k(z) = E_k(z)\mathbf{C}_k(z) \tag{14.5.13}$$

defines the energy function $E_k(z)$ where $z = 0$ represents the zero-order problem and $z = 1$ the physical problem. In general, the expansion of $E_k(z)$ in z

$$E_k(z) = \sum_{n=0}^{\infty} E_k^{(n)} z^n \tag{14.5.14}$$

has a finite *radius of convergence* R such that the expansion converges for $|z| < R$ and diverges for $|z| > R$. Our perturbation expansion (with $z = 1$) thus converges for $R > 1$ and diverges for $R < 1$.

We now define a *point of degeneracy* of $E_k(z)$ as a point ζ where the state k is degenerate with another state l:

$$E_k(\zeta) = E_l(\zeta) = E_{kl} \tag{14.5.15}$$

We note that, for real and symmetric matrices $\mathbf{H}_0$ and $\mathbf{U}$, such points always occur in conjugate pairs (ζ, ζ^*), as follows by complex conjugation of the equations

$$(\mathbf{H}_0 + \zeta\mathbf{U})\mathbf{C}_k(\zeta) = E_{kl}\mathbf{C}_k(\zeta) \tag{14.5.16}$$

$$(\mathbf{H}_0 + \zeta\mathbf{U})\mathbf{C}_l(\zeta) = E_{kl}\mathbf{C}_l(\zeta) \tag{14.5.17}$$

The points of degeneracy in the complex plane are of interest since it may be shown that the radius of convergence R of the expansion (14.5.14) is the radius of the largest circle around $z = 0$ that contains no points of degeneracy of $E_k(z)$. The convergence of the physical expansion ($z = 1$) thus depends on the behaviour of $E_k(z)$ inside the complex unit circle $|z| \leq 1$ – for example, a

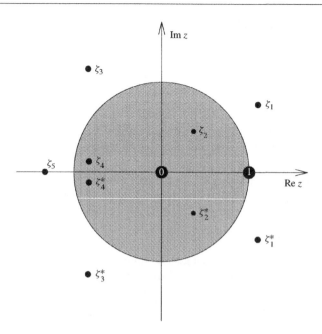

Fig. 14.1. Points of degeneracy in the complex plane. The conjugate points (ζ_2, ζ_2^*) correspond to a front-door intruder and the points (ζ_4, ζ_4^*) to a back-door intruder.

degeneracy close to -1 will destroy the convergence if it is located inside the unit circle. In short, the convergence does not depend so much on the agreement of $\mathbf{H}_0$ and $\mathbf{H}$ (in terms of some matrix norm) as on our ability to select a zero-order operator $\mathbf{H}_0$ such that the eigenvalues of $\mathbf{H}_0 + z\mathbf{U}$ are nondegenerate for any complex strength parameter z inside the unit circle.

A state that becomes degenerate with the reference state at a point of degeneracy ζ inside the unit circle $|\zeta| \leq 1$ is called an *intruder state*. In this terminology, the requirement for convergence is simply the absence of intruders. In Figure 14.1, we have illustrated the occurrence of points of degeneracy and intruders. Intruders are conveniently classified according to the sign of the real part of the point of degeneracy ζ. An intruder with Re $\zeta > 0$ is called a *front-door intruder*; conversely, a *back-door intruder* has Re $\zeta < 0$.

Being the eigenvectors of a nonphysical Hamiltonian $\mathbf{H}(\zeta)$ with $\zeta \neq 1$, the intruders have no physical meaning. Still, front-door intruders may sometimes be given a physical interpretation, in particular if Im ζ is small. Since these intruders are located in the region between the true physical state ($z = 1$) and the zero-order state ($z = 0$), they appear when the ordering of the physical states differs from that of the zero-order states, indicating an essential deficiency of the zero-order Hamiltonian. Back-door intruders, which correspond to a negative strength parameter z in $\mathbf{H}(z)$, are more difficult to interpret.

14.5.3 INTRUDERS IN THE GENERAL TWO-STATE MODEL

Let us consider a generalization of the two-state model system examined in Section 14.5.1, where we have introduced a *gap-shift parameter* γ in the partitioned Hamiltonian (14.5.2):

$$\mathbf{H}_0 = \begin{pmatrix} \alpha & 0 \\ 0 & \beta + \gamma \end{pmatrix} \tag{14.5.18}$$

$$\mathbf{U} = \begin{pmatrix} 0 & \delta \\ \delta & -\gamma \end{pmatrix} \tag{14.5.19}$$

Note that γ does not appear in the physical Hamiltonian – that is, in $\mathbf{H}(z) = \mathbf{H}_0 + z\mathbf{U}$ for $z = 1$. Rather, it changes the zero-order energy gap in $\mathbf{H}_0$, introducing at the same time a compensating, diagonal element in $\mathbf{U}$. A more general partitioning would be obtained by also adding a common shift to the energy of the zero-order states. However, as explained in Exercise 14.6, such a common shift does not modify the perturbation expansion except that it changes the zero- and first-order energy corrections such that their sum is constant.

Solving the secular equations using $\mathbf{H}(z)$, we obtain the two solutions

$$E_\pm(z) = \frac{\alpha + \beta + (1 - z)\gamma}{2} \pm \frac{\sqrt{[(\beta - \alpha) + (1 - z)\gamma]^2 + 4\delta^2 z^2}}{2} \tag{14.5.20}$$

which depend on the gap shift only for $z \neq 1$ and reduce to the physical energies (14.5.5) for $z = 1$. To locate the points of degeneracy, we set the expression under the square-root sign equal to zero and obtain

$$(4\delta^2 + \gamma^2)z^2 - 2\gamma(\beta - \alpha + \gamma)z + (\beta - \alpha + \gamma)^2 = 0 \tag{14.5.21}$$

with the two conjugate solutions

$$\zeta_\pm = \frac{\beta - \alpha + \gamma}{4\delta^2 + \gamma^2}(\gamma \pm 2\delta i) \tag{14.5.22}$$

Note that, whereas the unshifted problem has pure imaginary points of degeneracy

$$\zeta_\pm(\gamma = 0) = \pm\frac{\beta - \alpha}{2\delta}i \tag{14.5.23}$$

the shifted problem has complex points of degeneracy. In the two-dimensional case, these points may never become real but, for large gap shifts and small couplings, they may come arbitrarily close to the real axis. For $|\zeta_\pm| < 1$, the points of degeneracy become intruders. Equation (14.5.22) shows that back-door intruders may occur for gap shifts in the interval $-(\beta - \alpha) < \gamma < 0$ and that other shifts may give rise to front-door intruders.

Let us now investigate, for the gap-shifted two-state model, the condition for convergence $|\zeta_\pm| > 1$, which may be written in the form

$$\frac{(\beta - \alpha + \gamma)^2}{4\delta^2 + \gamma^2} > 1 \tag{14.5.24}$$

Solving this equation for γ, we obtain

$$\gamma > \gamma_c = \frac{4\delta^2 - (\beta - \alpha)^2}{2(\beta - \alpha)} \tag{14.5.25}$$

Thus, for a fixed energy gap $\beta - \alpha$ and a fixed coupling δ, we can always find a gap shift $\gamma > \gamma_c$ for which the expansion converges. Conversely, solving (14.5.24) for δ, we obtain

$$|\delta| < \frac{\beta - \alpha}{2}\sqrt{1 + \frac{2\gamma}{\beta - \alpha}} \tag{14.5.26}$$

which should be compared with (14.5.11) in unshifted perturbation theory.

In Figure 14.2, we have illustrated the positions of the points of degeneracy for the gap-shifted two-state model as a function of γ with $\beta - \alpha = 1$ and $\delta = \frac{3}{4}$. As is easily verified, the points of degeneracy are located on two circles with centres

$$c_\zeta = \frac{1}{2} \pm \frac{\beta - \alpha}{4\delta} \mathrm{i} \tag{14.5.27}$$

and radius

$$r_\zeta = \frac{1}{2} \sqrt{1 + \left(\frac{\beta - \alpha}{2\delta} \right)^2} \tag{14.5.28}$$

For large (positive or negative) gap shifts, the points of degeneracy in Figure 14.2 approach $z = 1$, with a vanishing imaginary component. Conversely, for zero gap shifts, the degeneracies are located diametrically opposite to $z = 1$ on the two circles, with a vanishing real component. A second pair of oppositely located degeneracies is also noted in Figure 14.2. Thus, for the gap shift

$$\gamma_{\mathrm{s}} = -(\beta - \alpha) \tag{14.5.29}$$

the zero-order Hamiltonian becomes degenerate. Conversely, for the gap shift

$$\gamma_{\mathrm{m}} = \frac{4\delta^2}{\beta - \alpha} \tag{14.5.30}$$

the points of degeneracy are located as far away from the zero-order system as possible, presumably leading to the most rapidly convergent series.

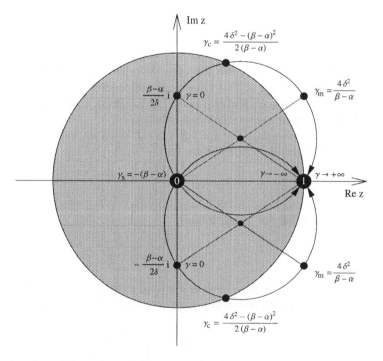

Fig. 14.2. The location of the points of degeneracy for different gap-shift parameters, illustrated for a system with $\beta - \alpha = 1$ and $\delta = 3/4$.

We conclude this subsection by illustrating numerically the convergence for different gap shifts. In Figure 14.3, we have plotted, on a logarithmic scale, the magnitudes of the errors in the perturbation expansions for different values of γ, using the following Hamiltonian:

$$\mathbf{H}(z) = \begin{pmatrix} 1 & 0 \\ 0 & 2+\gamma \end{pmatrix} + z \begin{pmatrix} 0 & \frac{3}{4} \\ \frac{3}{4} & -\gamma \end{pmatrix} \qquad (14.5.31)$$

The energies have been calculated from the expansion

$$E = \sum_{n=0}^{\infty} E^{(n)} \qquad (14.5.32)$$

$$E^{(0)} = \min(\alpha, \beta+\gamma) \qquad (14.5.33)$$

$$E^{(1)} = \begin{cases} 0, & \alpha \le \beta+\gamma \\ -\gamma, & \alpha > \beta+\gamma \end{cases} \qquad (14.5.34)$$

$$E^{(n)} = (n-2)! \frac{|\beta-\alpha+\gamma|}{(\beta-\alpha+\gamma)^n} \sum_{i=1}^{[n/2]} \frac{(-1)^i}{(n-2i)!i!(i-1)!} \gamma^{n-2i}\delta^{2i}, \qquad n > 1 \quad (14.5.35)$$

which, unlike the unshifted expansion (14.5.7)–(14.5.9), contain terms of odd as well as even orders in the perturbation.

As is easily verified from (14.5.11), the unshifted perturbation series generated by (14.5.31) is divergent but from (14.5.25) we note that the series should be convergent for gap shifts greater than $\gamma_c = \frac{5}{8}$. These results are illustrated in Figure 14.3, which also shows that a gap shift of $\gamma_m = \frac{9}{4}$

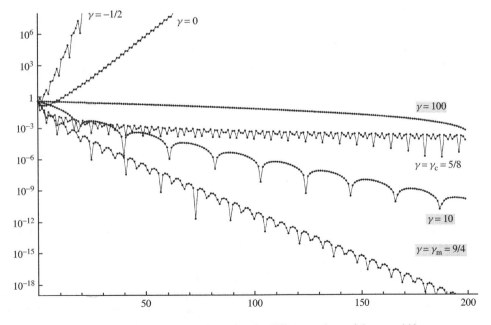

Fig. 14.3. Convergence of the perturbation expansion for different values of the gap-shift parameter γ using $\alpha = 1$ and $\beta = 2$ in the zero-order Hamiltonian (14.5.18) and $\delta = \frac{3}{4}$ in the perturbation (14.5.19). For each order of the expansion, the error has been plotted on a logarithmic scale.

yields a fairly rapidly convergent series and that a gap shift of 100 yields slow but smooth convergence, with several hundred terms needed to improve on the nonconvergent series with $\gamma_c = \frac{5}{8}$.

14.5.4 PROTOTYPICAL INTRUDERS

Two types of intruders are frequently encountered in Møller–Plesset perturbation theory: backdoor intruders generated by high-order excitations (from the Hartree–Fock state) and front-door intruders generated by low-order excitations. Although these types by no means exhaust all possibilities, they are sufficiently common to warrant special attention in the two-state model. In the present subsection, we examine these intruders in the two-dimensional case, preparing ourselves for the discussion of the more complicated Møller–Plesset expansions in Section 14.5.5.

In the two-state model, the degeneracies are trivially identified in the complex plane. In more complicated cases (such as in practical applications of Møller–Plesset theory), the identification of intruders is more difficult. In practice, a full search within the unit circle cannot be undertaken and we must be content with a more restricted search. In particular, much useful information may be gained by investigating the behaviour of the energy for z on the real axis, attempting to identify avoided crossings, whose presence reveals the existence of degeneracies further out in the complex plane. Of course, the identification of an avoided crossing on the real axis is not sufficient to establish divergence since the degeneracy need not be associated with an intruder. Still, a strong avoided crossing may be taken as the tell-tale sign of a divergence, prompting us to treat the perturbation corrections with caution. Moreover, as we shall see in the present and subsequent subsections, by projecting the system onto a two-dimensional space, we may estimate the location of the degeneracy from (14.5.22) and establish divergence by use of (14.5.26).

To determine the position of the avoided crossing in the two-state model, we minimize the difference between the two energies in (14.5.20) with respect to z:

$$\Delta E(z) = E_+(z) - E_-(z) = \sqrt{[(\beta - \alpha) + (1 - z)\gamma]^2 + 4\delta^2 z^2} \tag{14.5.36}$$

Elementary calculus then shows that the avoided crossing occurs at the position of the real part of the points of degeneracy (14.5.22)

$$z_{\min} = \operatorname{Re} \zeta_\pm = \frac{\beta - \alpha + \gamma}{4\delta^2 + \gamma^2}\gamma \tag{14.5.37}$$

and that the corresponding energy gap is given by the expression

$$\Delta E(z_{\min}) = 2\frac{|(\beta - \alpha + \gamma)\delta|}{\sqrt{4\delta^2 + \gamma^2}} \tag{14.5.38}$$

If the coupling δ between the zero-order states is small compared with γ, then the avoided crossing becomes pronounced, with the two curves approaching closely at $z_{\min}$. On the other hand, if δ and $\beta - \alpha + \gamma$ are numerically large, then the two states are well separated at $z_{\min}$, indicating that the points of degeneracy are located far away from the real axis.

In Figure 14.4, we have illustrated the behaviour of the two-state model for two sets of parameters, which represent a back-door intruder dominated by high-order excitations (to the left) and a front-door intruder dominated by low-order excitations (to the right). The numerical values of the parameters were obtained from the Møller–Plesset calculations discussed in Section 14.5.5. For the *high-excitation back-door intruder*, the parameters are $\beta - \alpha = 12.32$, $\delta = -0.00034$ and

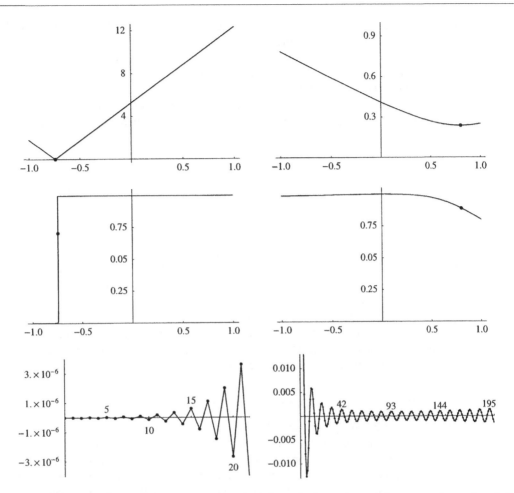

Fig. 14.4. Intruders in two-dimensional model calculations. In the two upper figures, we have plotted the energy difference $\Delta E(z)$ (14.5.36) on the real axis and marked the avoided crossings (14.5.38) with dots; in the middle figures, we have plotted the coefficient of the lowest zero-order state in the ground state of $\mathbf{H}(z)$; in the lower figures, we have plotted the error in the energy of the corresponding perturbation series. In the left-hand plots, the numerical values of α, β, δ and γ are 0, 12.32, -0.00034 and -7.07, respectively; in the right-hand plots, we have used for these parameters 0, 0.07, 0.12 and 0.34, respectively.

$\gamma = -7.07$ – that is, we have a large zero-order gap, a small coupling and a gap shift satisfying $-(\beta - \alpha) < \gamma < 0$. The corresponding point of degeneracy is obtained from (14.5.22) as $\zeta_\pm = -0.743 \pm 0.00007i$. For the *low-excitation front-door intruder*, we have used $\beta - \alpha = 0.07$, $\delta = 0.12$ and $\gamma = 0.34$ – that is, we have a small zero-order gap, a large coupling and a positive gap shift. The points of degeneracy become $\zeta_\pm = 0.800 \pm 0.579i$ – just inside the unit circle. The upper curves in Figure 14.4 are plots of the energy difference (14.5.36) along the real axis; the middle curve gives the coefficient of the lowest zero-order state in the lowest eigenfunction of $\mathbf{H}(z)$; the lower plots are of the errors in the energy calculated from the corrections (14.5.35).

In the plot corresponding to the back-door intruder, an avoided crossing is observed at $z = -0.74$. As expected from the smallness of the coupling, the energy difference (14.5.36) is a piecewise linear curve and the total energies approach each other closely, to within 0.0005, at the minimum (14.5.37). From the plotted coefficient in Figure 14.4, we note that the lowest state changes its composition abruptly at the avoided crossing. Besides the divergence of the energy series evident in the lowest plot in Figure 14.4, the most prominent feature of the expansion is the alternating sign of the corrections. This behaviour of high-excitation back-door intruders may be understood from (14.5.35). For a given n and for a small δ, the sum over i is dominated by the term of lowest order in δ. The energy correction may therefore be written as

$$E^{(n)} \approx -\frac{\gamma^{n-2}\delta^2}{(\beta - \alpha + \gamma)^{n-1}} \qquad (14.5.39)$$

which alternates for a back-door intruder – that is, provided γ is negative and does not change the ordering of the zero-order states $\beta - \alpha + \gamma > 0$.

In the plots corresponding to the low-excitation front-door intruder, the large coupling leads to a weak avoided crossing at $z = 0.800$, with the two energy curves approaching no closer than 0.24. In practice, avoided crossings of this type may be difficult to detect. Thus, in the middle plot of Figure 14.4, we note that the coefficient of the zero-order state changes only slightly in the region of the avoided crossing. Since the intruder is located close to the unit circle, the divergence of the expansion is exceedingly slow and noticeable only for orders higher than 150. The energy corrections do not alternate but oscillate slowly, with a period extending over 10 corrections. The slow oscillations in the sign of the corrections occur as a result of a complicated interplay between δ and γ in the sum in (14.5.35), which is no longer dominated by a single term but consists of several equally important alternating terms.

14.5.5 CONVERGENCE OF THE MØLLER–PLESSET SERIES

In high-dimensional spaces, an exhaustive search for intruder states is too cumbersome to be practicable – in particular, since, for each value of z, we must calculate the energy of two different states. As discussed in Section 14.5.4, we may instead search for avoided crossings on the real axis, associating these with degeneracies and possible intruders in the complex plane.

In Figure 14.5, we have plotted three such scans on the real axis, representing calculations on the HF molecule with the Møller–Plesset partitioning of the Hamiltonian. On the left, the molecule is at its equilibrium geometry with $R_e = 91.694$ pm and the cc-pVDZ basis is used; in the middle scan, diffuse functions have been added at the aug-cc-pVDZ level but without the diffuse d fluorine functions and the diffuse p hydrogen functions; on the right, the calculations have been carried out with a stretched bond distance of $2.5R_e$ in the cc-pVDZ basis. For each set of calculations, four plots are given, depicting: (1) the energy of the lowest state (the ground state at $z = 1$); (2) the energy difference between the first two states; (3) the coefficient of the ground-state Hartree–Fock determinant in the lowest state; and (4) the expectation value of x^2, representing the spatial extent of the ground state. Avoided crossings (indicative of a degeneracy in the complex plane) are identified as local minima in the difference plot, accompanied by a change in the composition of the lowest state.

In the cc-pVDZ calculations at the equilibrium geometry, no avoided crossings are observed in the interval $[-1, 1]$; there is no minimum in the difference curve and the lowest state is dominated by the Hartree–Fock determinant. Consequently, we expect the Møller–Plesset expansion to converge. This prediction is supported by the plotted expansion in Figure 14.6. This figure contains, in addition to the Møller–Plesset energies (relative to the corresponding FCI energy) for the three sets of calculations represented in Figure 14.5, the results for a fourth set of calculations,

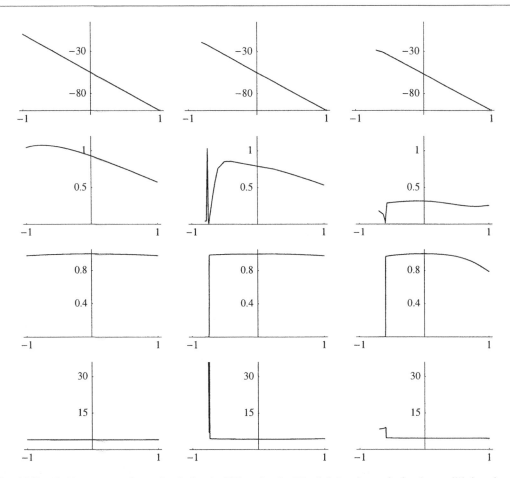

Fig. 14.5. $E_k(z)$ scans on the real axis for the HF molecule. The left-hand scan is for the equilibrium bond distance of R_e in the cc-pVDZ basis; the middle scan for the same bond distance in the aug-cc-pVDZ basis; the right-hand scan for a bond distance of $2.5R_e$ in the cc-pVDZ basis. The upper plot is of the energy of the lowest state; the second plot is of the energy difference between the first two states; the third plot is of the coefficient of the ground-state Hartree–Fock determinant in the lowest state; the fourth plot is of the expectation value of x^2 (the spatial extent of the system). Atomic units are used.

carried out in the aug-cc-pVDZ basis at the stretched geometry of HF. Note the different scales in the plots in Figure 14.6.

In the aug-cc-pVDZ plots in Figure 14.5, an avoided crossing occurs at $z = -0.743$. From the shape of the energy difference curve, we conclude that the degeneracy is located close to the real axis. The Møller–Plesset expansion should therefore diverge for this system, as confirmed by Figure 14.6. To order 14, the corrections decrease in magnitude, oscillating about the FCI energy; to higher orders, the corrections increase, leading to divergence. It is remarkable that, for such a simple system (the ground-state HF molecule at equilibrium) the Møller–Plesset series diverges in this small basis.

The oscillatory behaviour of the expansion may be understood by examining the subspace spanned by the two lowest eigenstates at the avoided crossing. In the basis that diagonalizes $\mathbf{H}_0$

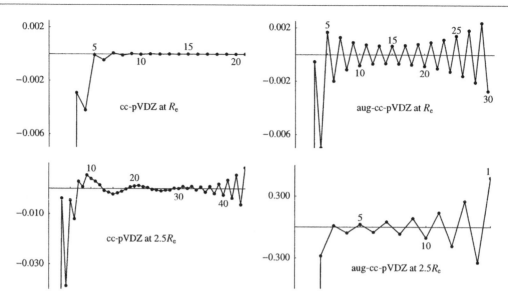

Fig. 14.6. Møller–Plesset expansions for the HF molecule. For each order in the perturbation series, we have plotted the difference between the Møller–Plesset energy and the corresponding FCI energy (in E_h).

in this subspace, the partitioned Hamiltonian becomes

$$\mathbf{H}_0 = \begin{pmatrix} -54.65980 & 0.00000 \\ 0.00000 & -49.41060 \end{pmatrix} \tag{14.5.40}$$

$$\mathbf{U} = \begin{pmatrix} -44.91005 & -0.00034 \\ -0.00034 & -37.84247 \end{pmatrix} \tag{14.5.41}$$

which corresponds to $\beta - \alpha = 12.31678$, $\gamma = -7.06758$ and $\delta = -0.00034$ (see Exercise 14.6 regarding the large diagonal elements of $\mathbf{U}$). Clearly, the divergence of this series is caused by a back-door intruder with a very small coupling, giving rise to alternating corrections such as those discussed in Section 14.5.4 – indeed, the parameters in Section 14.5.4 were obtained from this particular calculation. From the sharpness of the avoided crossing and the closeness of the energies, we conclude that the degeneracy is located close to the real axis. Indeed, using (14.5.22), we obtain $\zeta_\pm = -0.743 \pm 0.00007i$, with the expected small imaginary part.

To gain further insight into the nature of the intruder and the source of the divergence, we note that the physical Hamiltonian $\mathbf{H}_0 + \mathbf{U}$ in the two-dimensional space has two eigenstates of energies -99.56985 and -87.25307 E_h. Whereas the first is the physical ground state, the second state is a highly excited one, dominated by quadruple and higher excitations. For $z > -0.7$ the two states hardly interact, but for $z \approx -0.743$ the large negative gap shift makes the two states nearly degenerate, leading to the abrupt changes of the states, with the lowest state becoming the highly excited one. This change of character is revealed also by the change in the diffuseness of the lowest state in Figure 14.5.

We now go on to consider the third set of calculations in Figure 14.5. When the HF bond is stretched to $2.5R_e$ in the cc-pVDZ basis, the plots reveal an avoided crossing at $z = -0.598$. At this point, there is a sharp minimum in the difference plot and the coefficient of the Hartree–Fock determinant changes abruptly from about 1 to nearly 0. In addition, there is a weak avoided crossing at $z = 0.8$. Based on these observations, we expect the perturbation series to diverge, as illustrated in Figure 14.6. Note, however, that the behaviour of the corrections is rather different from (and more complicated than) that observed in the aug-cc-pVDZ calculation at equilibrium. Following a few irregular initial corrections, the corrections 10–30 exhibit a slowly undulating pattern, after which the corrections begin to alternate and diverge, in much the same manner as for the aug-cc-pVDZ equilibrium calculation.

To understand this behaviour, let us examine the avoided crossing at $z = 0.8$, which arises from an interaction between the two lowest states of the physical Hamiltonian (a possible low-excitation front-door intruder). In the subspace spanned by these states, we obtain the following partitioned Hamiltonian in the diagonal basis of the zero-order Hamiltonian:

$$\mathbf{H}_0 = \begin{pmatrix} -56.28832 & 0.00000 \\ 0.00000 & -55.87498 \end{pmatrix} \tag{14.5.42}$$

$$\mathbf{U} = \begin{pmatrix} -43.65655 & 0.12252 \\ 0.12252 & -43.99542 \end{pmatrix} \tag{14.5.43}$$

This Hamiltonian gives the parameters for the low-excitation front-door intruder discussed in the preceding subsection. The points of degeneracy are $\zeta_\pm = 0.801 \pm 0.579i$. The strong coupling gives the points of degeneracy large imaginary components, making the avoided crossing caused by this intruder barely noticeable on the real axis. Indeed, since these points are so close to the unit circle, it is not clear whether or not the corresponding points of degeneracy of the full cc-pVDZ system represent an intruder. By contrast, the avoided crossing at $z = -0.598$ is undoubtedly a weakly coupled back-door intruder, similar to the one observed in the aug-cc-pVDZ calculation at the equilibrium geometry.

In short, the Møller–Plesset expansion at $2.5R_e$ consists of several distinct parts. After the initial irregular corrections, the corrections up to orders 30–35 are dictated by a (potential) low-excitation front-door intruder strongly coupled to the ground state, with periodic energy corrections similar to those in the two-state problem; to higher orders, a high-excitation back-door intruder dominates the expansion, giving rise to an alternating, divergent expansion. Thus, the ultimate asymptotic behaviour is here determined by the intruder with the smallest radius of convergence.

Divergence in Møller–Plesset theory is typically caused by diffuse, multiply excited states whose energy is high relative to the ground state both for the physical Hamiltonian $\mathbf{H}(1)$ and for the zero-order Hamiltonian $\mathbf{H}(0)$ [15,16]. How can such high-energy states become intruders? From the top plots in Figure 14.5, we note that the energy $E(z)$ of the lowest state is nearly linear in z, with a large negative slope. The first-order term therefore dominates the behaviour of $E(z)$. In Møller–Plesset theory, the first-order energy is the expectation value of the fluctuation potential. This energy is equal to the negative electron-repulsion energy (10.4.28) and may be one or two orders of magnitude larger than the higher-order corrections. Therefore, for molecules with a strong electron repulsion, $E(z)$ may become higher than the energy of the continuum or ionized states as z approaches -1.

This behaviour of $E(z)$ for the ground state should be contrasted with that for states with many electrons excited to diffuse orbitals. Although such states have high $E(z)$ for both the physical and zero-order systems, their electron repulsion is weaker since the electrons are well separated. Consequently, the slope of $E(z)$ is smaller than in the ground state so the ground- and excited-state curves may cross for negative $z > -1$. As a result, the Møller–Plesset expansion often diverges for states with a large electron-repulsion energy, provided the basis contains the functions needed to represent diffuse, multiply excited states with weaker electronic interactions.

As an illustration, consider the equilibrium HF molecule in the diffuse aug-cc-pVDZ basis. From (14.5.40) and (14.5.41), we obtain in the two-state model $E_1^{(0)} = -54.65980$ E$_h$ and $E_1^{(1)} = -44.91005$ E$_h$ for the ground electronic state, and $E_2^{(0)} = -49.41060$ E$_h$ and $E_2^{(1)} = -37.84247$ E$_h$ for the intruder. Neglecting higher-order terms, we may identify the point of degeneracy ζ from the equation

$$E_1^{(0)} + \zeta E_1^{(1)} = E_2^{(0)} + \zeta E_2^{(1)} \tag{14.5.44}$$

which yields a value of ζ in agreement with the observed point of degeneracy at -0.743.

In conclusion, the Møller–Plesset expansion is a rather unpredictable one, which may diverge even for simple systems. Because of the special form of the zero-order operator, the Møller–Plesset series will either alternate or otherwise oscillate with long periods. Nevertheless, the low-order corrections may be quite useful in many situations, as demonstrated in Chapter 15.

14.5.6 ANALYTIC CONTINUATION

Over the years, many techniques have been developed to improve the convergence rate of convergent sequences and to turn divergent sequences into convergent ones. We shall here consider one of the simpler approaches, *analytic continuation* [17], which enables us to remove the divergence caused by back-door intruders.

We assume that the only intruder within the unit circle is a back-door intruder ζ_B. The perturbation expansion therefore converges for $|z| < |\zeta_B|$. In the method of analytic continuation, we choose a new point of expansion z' closer to the physical state $z = 1$. We here take z' to be a real

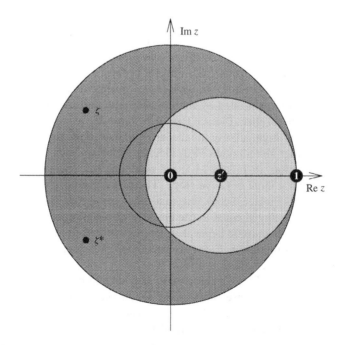

Fig. 14.7. Analytic continuation. Because of the presence of the intruder at ζ and ζ^*, the expansion at $z = 0$ is divergent. However, the expansion around z', which lies inside the region of convergence for expansions around $z = 0$, is convergent.

Table 14.4 Analytic continuation. The error in the Møller–Plesset energy (in E_h) relative to the FCI energy of $-100.24664\ E_h$ for expansions around $z = 0.0$ and $z' = 0.4$

Order	$z = 0.0$	$z' = 0.4$
0	45.45833	31.86676
1	0.21599	0.10883
2	0.00718	0.00456
3	0.00666	0.00173
4	-0.00034	0.00020
5	0.00133	0.00010
6	-0.00066	0.00000
7	0.00063	0.00001
8	-0.00048	0.00000
9	0.00043	0.00000
10	-0.00038	0.00000
11	0.00035	0.00000
12	-0.00033	0.00000

positive number $0 < z' < |\zeta_B|$. At the new point, the energy is re-expanded as

$$E_k(z - z') = \sum_{n=0}^{\infty} E'^{(n)}_k (z - z')^n \tag{14.5.45}$$

Since $z' < |\zeta_B|$, the original expansion (14.5.14) converges at z' and we may therefore determine the coefficients $E'^{(n)}_k$ by requiring the derivatives of (14.5.45) at $z - z'$ to be identical to those of the original expansion:

$$E'^{(n)}_k = \sum_{m=n}^{\infty} \frac{m!}{n!} E^{(m)}_k (z - z')^{m-n} \tag{14.5.46}$$

The region of convergence for $E_k(z - z')$ is now a disc about z' with a radius R' determined by the intruder closest to z'. If $z = 1$ is inside this circle, we can calculate our physical energy from (14.5.45). Otherwise, we introduce a new point of expansion z'' closer to $z = 1$ than z' and re-expand about z''. This process is continued until $z = 1$ is inside the convergence circle. The process of analytic continuation is illustrated in Figure 14.7.

As an example of analytic continuation, consider the Møller–Plesset calculation on the HF molecule at the equilibrium geometry in the aug-cc-pVDZ basis. As discussed in Section 14.5.4, this expansion is divergent with a back-door intruder at about $z = -0.7$. The expansion around $z' = 0.4$ should be convergent for $z = 1$ since $|1 - 0.4| < |-0.7 - 0.4|$. In Table 14.4, we have listed the convergent expansion for $z = 1$ around $z' = 0.4$ as well as the divergent expansion around $z' = 0.0$.

14.6 Perturbative treatments of coupled-cluster wave functions

We now have at our disposal two different approaches to the evaluation of electronic wave functions and energies, both developed within the framework of the exponential cluster ansatz: the coupled-cluster hierarchy of Chapter 13 and the Møller–Plesset hierarchy presented in this chapter. In coupled-cluster theory, the wave functions are classified according to the level of truncation in the cluster operator (13.2.5). For example, in the CCSD approximation, we disregard all contributions from connected triples and higher excitations and determine the singles and doubles amplitudes by solving the associated projected Schrödinger equation (13.2.23). At the next level (CCSDT), we include the triple excitations and determine the wave function by solving the projected equation in the singles, doubles and triples space. By including higher and higher excitations in the cluster operator and solving the associated projected Schrödinger equation, we generate more and more accurate approximations to the FCI wave function.

In the Møller–Plesset hierarchy of approximations, the level of theory is classified according to the order in the fluctuation potential to which the wave function and the energy are correct. Thus, at the MP2 level the energy is correct to second order in the fluctuation potential, at the MP3 level the energy is correct to third order, and so on. Clearly, the truncation and projection approach of coupled-cluster theory and the perturbation approach of Møller–Plesset theory should yield the same solution as we go to sufficiently high levels in the hierarchies (provided, of course, that the perturbation series converges).

In this section, we study the relationship between coupled-cluster and Møller–Plesset theories in greater detail. We begin by carrying out a perturbation analysis of the coupled-cluster wave functions and energies in Section 14.6.1. We then go on to consider two sets of hybrid methods, where the coupled-cluster approximations are improved upon by means of perturbation theory. In Section 14.6.2, we consider a set of hybrid coupled-cluster wave functions, obtained by simplifying the projected coupled-cluster amplitude equations by means of perturbation theory. In Section 14.6.3, we examine the CCSD(T) approximation, in which the CCSD energy is improved upon by adding triples corrections in a perturbative fashion. Finally, in Section 14.6.4, we compare numerically the different hybrid and nonhybrid methods developed in the present chapter and in Chapter 13.

14.6.1 PERTURBATION ANALYSIS OF THE COUPLED-CLUSTER HIERARCHY

The close relationship between the coupled-cluster and perturbation theories can be understood from an inspection of the equations that determine the amplitudes in the two approaches. Thus, in perturbation theory, we calculate the amplitudes *recursively* from the amplitude equations of order N_{PT} (14.3.17):

$$\varepsilon_\mu t_\mu^{(n)} = -\langle\mu|[\hat{\Phi}^{\mathrm{T}}]^{(n)}|\mathrm{HF}\rangle, \quad n \le N_{\mathrm{PT}} \tag{14.6.1}$$

In coupled-cluster theory, we determine the amplitudes *self-consistently* from the projected equations (13.2.48)

$$\varepsilon_{\mu_n} t_{\mu_n} = -\langle\mu_n|\Phi^{\mathrm{T}^{[N_{\mathrm{CC}}]}}|\mathrm{HF}\rangle, \quad n \le N_{\mathrm{CC}} \tag{14.6.2}$$

after truncating the cluster operator at some excitation level N_{CC}:

$$\hat{T}^{[N_{\mathrm{CC}}]} = \hat{T}_1 + \hat{T}_2 + \cdots + \hat{T}_{N_{\mathrm{CC}}} \tag{14.6.3}$$

In coupled-cluster theory, the level of approximation is established by discarding in (14.6.2) all excitations of rank greater than N_{CC}. In perturbation theory, the level of approximation is established by selecting the order in the fluctuation potential N_{PT} to which the equations are solved. Thus, no excitations are discarded as such, but only those amplitudes that contribute to the given order in the fluctuation potential will be nonzero in the equations. Specifically, to order N_{PT}

in the fluctuation potential, we find that excitations of rank greater than $N_{PT} + 1$ do not contribute to the wave function and their amplitudes are therefore zero; see the discussion in Section 14.3.2.

We now wish to establish to what order N_{PT} in the fluctuation potential the coupled-cluster wave function and the coupled-cluster energy of order N_{CC} are correct. We begin by recalling that the rank-n operator $\hat{T}_n$ is of order $n - 1$ and greater in the fluctuation potential. From this observation, we proceed to determine the perturbation order of the projected coupled-cluster equations (14.6.2). We introduce the notation

$$R_n^N = \varepsilon_{\mu_n} t_{\mu_n} + \langle \mu_n | \hat{\Phi}^{T^{[N]}} | HF \rangle \tag{14.6.4}$$

Expanding the transformed fluctuation potential in orders of the cluster operator, we obtain

$$R_n^N = \varepsilon_{\mu_n} t_{\mu_n} + \langle \mu_n | \hat{\Phi} | HF \rangle + {}^1R_n^N + {}^2R_n^N + {}^3R_n^N + {}^4R_n^N \tag{14.6.5}$$

Inserting the expansion of $\hat{T}^{[N]}$ (14.6.3) in this expression and using (13.2.65), we find that each of the last four terms in (14.6.5) contains contributions only from certain cluster ranks:

$$^1R_n^N = \sum_{s=1}^{N} \langle \mu_n | [\hat{\Phi}, \hat{T}_s] | HF \rangle \qquad \text{nonzero contributions: } n - 1 \le s \le n + 2 \tag{14.6.6}$$

$$^2R_n^N = \frac{1}{2} \sum_{s,t=1}^{N} \langle \mu_n | [[\hat{\Phi}, \hat{T}_s], \hat{T}_t] | HF \rangle \qquad \text{nonzero contributions: } n \le s + t \le n + 2 \tag{14.6.7}$$

$$^3R_n^N = \frac{1}{6} \sum_{s,t,u=1}^{N} \langle \mu_n | [[[\hat{\Phi}, \hat{T}_s], \hat{T}_t], \hat{T}_u] | HF \rangle \qquad \begin{array}{l} \text{nonzero contributions:} \\ n + 1 \le s + t + u \le n + 2 \end{array} \tag{14.6.8}$$

$$^4R_n^N = \frac{1}{24} \sum_{s,t,u,v=1}^{N} \langle \mu_n | [[[[\hat{\Phi}, \hat{T}_s], \hat{T}_t], \hat{T}_u], \hat{T}_v] | HF \rangle \qquad \begin{array}{l} \text{nonzero contributions:} \\ s + t + u + v = n + 2 \end{array} \tag{14.6.9}$$

Combining the information in (14.6.6)–(14.6.9) with the observation that rank-s operators $\hat{T}_s$ are of order $s - 1$ and higher in the perturbation, we conclude that each term ${}^iR_n^N$ is of order $n - 1$ and higher in the perturbation – for example, for ${}^2R_n^N$, the lowest-order contribution occurs for $s + t = n$, which gives the order $n - 2 + 1$, where -2 comes from the cluster operators and $+1$ from the fluctuation potential. Returning to the complete expression R_n^N in (14.6.5), we further note that the first term $\varepsilon_{\mu_n} t_{\mu_n}$ is of order $n - 1$ and that the second term $\langle \mu_n | \hat{\Phi} | HF \rangle$ is also of order $n - 1$ for the nonvanishing excitation ($n = 2$). Thus, we conclude that R_n^N is of order $n - 1$ and higher in the fluctuation potential. It should be recalled, however, that, because of the Brillouin theorem, the first-order amplitudes $\hat{T}_1$ are of order 2 and higher in the fluctuation potential. Therefore, we may more precisely state that R_n^N is of order $n - 1$ and higher except for R_1^N, which is of order 2 and higher in the fluctuation potential.

Next, let us examine the relationship between the projected coupled-cluster equations of different orders. Inspection of (14.6.5)–(14.6.9) reveals the following recursive relationships:

$$R_n^{N+1} = R_n^N, \qquad n \le N - 2 \tag{14.6.10}$$

$$R_{N-1}^{N+1} = R_{N-1}^N + \langle \mu_{N-1} | [\hat{\Phi}, \hat{T}_{N+1}] | HF \rangle = R_{N-1}^N + O(N+1) \tag{14.6.11}$$

$$R_N^{N+1} = R_N^N + \langle \mu_N | [\hat{\Phi}, \hat{T}_{N+1}] | HF \rangle + \langle \mu_N | [[\hat{\Phi}, \hat{T}_1], \hat{T}_{N+1}] | HF \rangle = R_N^N + O(N+1) \tag{14.6.12}$$

$$R_{N+1}^{N+1} = \varepsilon_{\mu_{N+1}} t_{\mu_{N+1}} + \langle \mu_{N+1} | \hat{\Phi}^{T^{[N+1]}} | HF \rangle = O(N) \tag{14.6.13}$$

In these expressions, $O(N)$ represents unspecified terms of order N and higher in the fluctuation potential. Thus, we see that only the two highest projections R_{N-1}^N and R_N^N are directly affected when we go from excitation level N to $N + 1$. The correction terms added to these equations are both of order $N + 1$ and higher in the perturbation. Also, the new set of equations (14.6.13) are of order N and higher in the fluctuation potential. It should be recalled, however, that the projected equations are coupled. Thus, a correction of order k added to any term R_n^N in the projected equations will affect all the remaining terms in orders higher than k. The propagation of terms may be illustrated by the addition of excitation level $N + 1$. The operator $\hat{T}_{N+1}$ is of order N and higher and couples directly to R_N^{N+1} and R_{N-1}^{N+1}, giving terms of order $N + 1$; see (14.6.11) and (14.6.12). Subsequently, these excitation levels couple to R_{N-2}^{N+1} and R_{N-3}^{N+1}, giving terms of order $N + 2$. The propagation then continues to the lower excitation levels.

Let us first consider the CCSD wave function. From the preceding discussion, we note that the singles and doubles equations are of orders 2 and 1 in the perturbation:

$$R_1^2 = O(2) \tag{14.6.14}$$

$$R_2^2 = O(1) \tag{14.6.15}$$

These equations should be interpreted to indicate that the double amplitudes are at least linear in the fluctuation potential and that the singles are at least quadratic in the fluctuation potential. The order equations (14.6.14) and (14.6.15) should not be interpreted as giving any direct indication of the accuracy of the singles and doubles amplitudes. To establish their accuracy, we go to the next level of coupled-cluster theory (CCSDT) and determine the lowest orders of the corrections introduced. Referring to (14.6.11)–(14.6.13), we find

$$R_1^3 = R_1^2 + O(3) \tag{14.6.16}$$

$$R_2^3 = R_2^2 + O(3) \tag{14.6.17}$$

$$R_3^3 = O(2) \tag{14.6.18}$$

Whereas the singles and doubles projections are affected directly by third-order terms, the new triples equations are of second order. Higher-order corrections may be taken into account by considering the CCSDTQ equations, but the corrections thus introduced will be of higher orders in the fluctuation potential. We conclude that the CCSD wave function is correct only to first order (because of the second order error in the triples), but that the singles and doubles amplitudes are correct to second order. Moreover, since the CCSD energy is calculated from the singles and doubles amplitudes, we conclude that the energy is correct to third order in the fluctuation potential; see (14.3.34). It is noteworthy that the CCSD energy is correct to two orders higher in the perturbation than is the wave function itself. In Table 14.5, we have summarized these results.

Let us now consider the accuracy of the CCSDT wave function. Going from CCSDT to CCSDTQ, the projected equations are affected in the following manner:

$$R_1^4 = R_1^3 \tag{14.6.19}$$

$$R_2^4 = R_2^3 + O(4) \tag{14.6.20}$$

$$R_3^4 = R_3^3 + O(4) \tag{14.6.21}$$

$$R_4^4 = O(3) \tag{14.6.22}$$

Table 14.5 The order in the fluctuation potential to which the coupled-cluster model is correct. In general, the wave function of order N is correct to order $N-1$ and the energy to order $[3N/2]$

	CCSD	CCSDT	CCSDTQ
Singles	2	4	5
Doubles	2	3	5
Triples	1*	3	4
Quadruples	2*	2*	4
Quintuples	3*	3*	3*
n-tuples	$n-2$*	$n-2$*	$n-2$*
Wave function	1	2	3
Energy	3	4	6

*Amplitudes zero in the wave function.

There are no direct modifications of the singles equations (14.6.19), but R_1^3 is indirectly affected by modifications in the doubles and triples. These amplitudes are corrected by fourth-order terms, implying that the corrections in the singles are of fifth order in the fluctuation potential. We conclude that the CCSDT singles are correct to fourth order. The CCSDT doubles and triples amplitudes are both correct to third order since the CCSDTQ corrections are of fourth order. Finally, the neglect of quadruples at the CCSDT level introduces errors of third order in the perturbation, implying that the wave function itself is correct only to second order. The CCSDT energy depends linearly on the doubles and quadratically on the singles. Since the doubles are correct to third order and the singles to fifth order, we conclude that the CCSDT energy is correct to fourth order. These results are summarized in Table 14.5.

Proceeding in the same manner for CCSDTQ, we note the relationships:

$$R_1^5 = R_1^4 = R_1^3 \tag{14.6.23}$$

$$R_2^5 = R_2^4 \tag{14.6.24}$$

$$R_3^5 = R_3^4 + O(5) \tag{14.6.25}$$

$$R_4^5 = R_4^4 + O(5) \tag{14.6.26}$$

$$R_5^5 = O(4) \tag{14.6.27}$$

At this stage, the doubles equations R_2^4 are no longer directly affected (in the sense of new terms introduced) by higher-order corrections. The doubles amplitudes are therefore affected only indirectly through the appearance of the triples and quadruples in R_2^4. From (14.6.25) and (14.6.26), we conclude that the CCSDTQ triples and quadruples are correct to fourth order and hence that the CCSDTQ doubles are correct to fifth order in the perturbation. Consequently, the CCSDTQ energy is correct to sixth order, although the wave function is correct to only third order as quintuple and higher excitations have been neglected; see Table 14.5 for details. A similar analysis reveals that the CCSDTQ5 energy is correct to seventh order. In general, we may state that the coupled-cluster wave function of order N is correct to order $N-1$ and its energy to order $[3N/2]$.

It should be emphasized that an order analysis of the coupled-cluster model can serve only as a crude indication of the overall quality of the coupled-cluster wave function and energy. If anything, we would expect the coupled-cluster model to be more accurate and more flexible than indicated by a strict perturbation analysis. This expectation is based on the observation that, for example, the CCSD wave function – as well as being correct to first order in the fluctuation potential – has been constructed so as to satisfy the Schrödinger equation exactly when projected against the singles and doubles subspaces. This requirement makes the CCSD wave function more accurate and useful than the MP1 wave function, although both are correct to first order in the fluctuation potential. In particular, the CCSD singles and doubles amplitudes are both correct to second order in the fluctuation potential, whereas the MP1 singles (which are zero) and doubles are correct only to first order. The reason why the CCSD wave function is classified as correct only to first order is the lack of triples, which from a perturbational point of view enter the wave function to second order. Similarly, we expect the CCSD energy to be more accurate and useful than the MP3 energy, although both contain errors of fourth order in the fluctuation potential.

To illustrate the convergence of the coupled-cluster and perturbation series, we have in Figure 14.8 plotted the coupled-cluster and Møller–Plesset errors in the correlation energy for the water molecule [18]. The errors are relative to FCI in the cc-pVDZ basis, on a linear scale to the left and on a logarithmic scale to the right.

For this particular system, the convergence of the two methods is much the same, although the logarithmic plot reveals that the coupled-cluster calculations are consistently of higher quality than the corresponding Møller–Plesset calculations. In either case, there is a dramatic improvement from a first-order treatment (which corresponds to the Hartree–Fock approximation) to the next level of theory – MP2 and CCSD. The improvement from MP2 to MP3 is small but brings the Møller–Plesset energy in line with the CCSD energy, which is correct to third order in the perturbation. Further improvements in the wave functions produce only minor improvements in the energy on a linear scale but we note the almost perfect linear reduction in the error of the coupled-cluster energy on a logarithmic scale.

The calculations on water in Figure 14.8 are typical of a situation where the Hartree–Fock wave function provides a good zero-order approximation to the true wave function. In such situations, the first perturbation corrections give large improvements in the energy and the differences between the coupled-cluster and Møller–Plesset energies are rather small. As a different example, we have in Figure 14.9 plotted the errors in the correlation energy of water using the same basis but at the stretched geometry $2R_{ref}$, where the Hartree–Fock wave function provides a poor representation of the true wave function. The different convergence patterns of the two approaches are now evident,

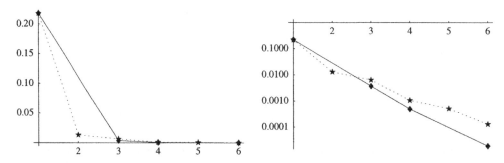

Fig. 14.8. The errors in the cc-pVDZ correlation energy of the coupled-cluster (full line) and Møller–Plesset (dotted line) models relative to FCI for the water molecule at the reference geometry of Table 5.5, plotted against the order in the fluctuation potential to which the energies are correct (in E_h). On the left, the plot is on a linear scale; on the right, on a logarithmic scale. For both approaches, the first-order energy is equal to the Hartree–Fock energy.

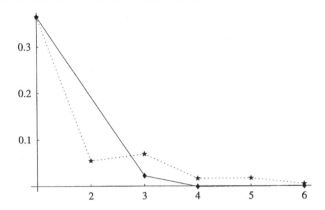

Fig. 14.9. The errors in the cc-pVDZ correlation energy of the coupled-cluster (full line) and Møller–Plesset (dotted line) models relative to FCI for the water molecule at the stretched geometry of Table 5.5, plotted against the order in the fluctuation potential to which the energies are correct (in E_h). In both approaches, the first-order energy is equal to the Hartree–Fock energy.

even on a linear scale. In particular, we note the much smoother convergence of the coupled-cluster energy than of the Møller–Plesset energy. Only at sixth order in the perturbation (MP6 and CCSDTQ) are the calculated energies identical on the linear scale of the plot. No logarithmic plot has been attempted in this case since the errors are in some cases negative.

14.6.2 ITERATIVE HYBRID METHODS

Although the CCSD wave function provides a reasonably accurate description of the electronic system, higher accuracy is often desired. Within the coupled-cluster hierarchy of approximations, we may improve on our description by going to the CCSDT level of theory. Unfortunately, the CCSDT model is usually too expensive to be applied to any but the smallest systems, the evaluation of the energy scaling as n^8, which should be compared with the cost of n^6 for the CCSD model. Clearly, some approach is needed that allows us to improve the description of the system for a more modest increase in the computational cost. There are several methods that achieve this goal, scaling with the system as n^7 and being correct to fourth order in the fluctuation potential.

To improve on the CCSD description, we must take into account the main effects of the connected triples, which are neglected at the CCSD level but are known from perturbation theory to make a fourth-order contribution to the energy – see (14.3.74):

$$E_{\mathrm{T}}^{(4)} = \langle \bar{t}_2^{(1)} | [\hat{\Phi}, \hat{T}_3^{(2)}] | \mathrm{HF} \rangle \tag{14.6.28}$$

We shall include the effects of triples by combining the coupled-cluster and Møller–Plesset models. Let us begin by reviewing the form of the coupled-cluster equations for the *coupled-cluster singles (CCS)*, CCSD and CCSDT wave functions. Introducing the following notation for the right-hand sides of the coupled-cluster amplitude equations

$$\bar{R}_n^N = R_n^N - \varepsilon_{\mu_n} t_{\mu_n} \tag{14.6.29}$$

the projected coupled-cluster equations may be written in the form

$$\varepsilon_{\mu_n} t_{\mu_n} = -\bar{R}_n^N \tag{14.6.30}$$

Expanding (14.6.4), we obtain for the CCS right-hand side (in the tilde notation of Section 13.7.3 for T1-transformed fluctuation potential)

$$\overline{R}_1^1 = \langle \mu_1 | \tilde{\hat{\Phi}} | \text{HF} \rangle \tag{14.6.31}$$

for the CCSD right-hand side

$$\overline{R}_1^2 = \overline{R}_1^1 + \langle \mu_1 | [\tilde{\hat{\Phi}}, \hat{T}_2] | \text{HF} \rangle \tag{14.6.32}$$

$$\overline{R}_2^2 = \langle \mu_2 | \tilde{\hat{\Phi}} | \text{HF} \rangle + \langle \mu_2 | [\tilde{\hat{\Phi}}, \hat{T}_2] | \text{HF} \rangle + \tfrac{1}{2} \langle \mu_2 | [[\tilde{\hat{\Phi}}, \hat{T}_2], \hat{T}_2] | \text{HF} \rangle \tag{14.6.33}$$

and, finally, for the CCSDT right-hand side

$$\overline{R}_1^3 = \overline{R}_1^2 + \langle \mu_1 | [\tilde{\hat{\Phi}}, \hat{T}_3] | \text{HF} \rangle \tag{14.6.34}$$

$$\overline{R}_2^3 = \overline{R}_2^2 + \langle \mu_2 | [\tilde{\hat{\Phi}}, \hat{T}_3] | \text{HF} \rangle \tag{14.6.35}$$

$$\overline{R}_3^3 = \langle \mu_3 | [\tilde{\hat{\Phi}}, \hat{T}_2] | \text{HF} \rangle + \tfrac{1}{2} \langle \mu_3 | [[\tilde{\hat{\Phi}}, \hat{T}_2], \hat{T}_2] | \text{HF} \rangle$$

$$+ \langle \mu_3 | [\tilde{\hat{\Phi}}, \hat{T}_3] | \text{HF} \rangle + \langle \mu_3 | [[\tilde{\hat{\Phi}}, \hat{T}_2], \hat{T}_3] | \text{HF} \rangle \tag{14.6.36}$$

Going from CCS to CCSD, the main additional cost arises from the last two terms in the doubles right-hand side (14.6.33), which scale as n^6 – see Sections 13.7.8 and 13.7.9. The remaining terms in the CCSD equations scale as the fifth or lower powers of the size of the system. Likewise, going from CCSD to CCSDT, the extra cost arises mainly from the last two terms in (14.6.36), which scale as n^8. As we shall show shortly, all other terms in the CCSDT equations scale as the seventh or lower powers of the size of the system. We also make the observation that, at each level, the terms of highest cost – the last two terms in (14.6.33) and the last two terms in (14.6.36) – all describe higher-order adjustments of the 'top-level' excitations – the doubles in CCSD theory and the triples in CCSDT theory. As such, they are not required if we wish to describe the top-level excitations in an approximate manner.

These observations suggest the following strategy for simplifying the coupled-cluster equations. To reduce the cost of the calculations while retaining some approximate description of the top-level excitations, we discard from R_N^N all terms of orders N and higher in the fluctuation potential, associating as before order $n - 1$ with the rank-n cluster operator $\hat{T}_n$. The resulting equations are solved in the usual manner, yielding an approximate coupled-cluster wave function that we shall refer to as CCN, where N is the rank of the approximated top-level excitations. For instance, we obtain CC2 (cost n^5) from CCSD (cost n^6) by approximating the doubles equations to first order in the fluctuation potential [19], and we obtain CC3 (cost n^7) from CCSDT (cost n^8) by approximating the triples equations to second order in the fluctuation potential [20].

In setting up the CC2 and CC3 approximations, we treat $\hat{T}_1$ as a zero-order operator, in agreement with the general rule for cluster operators but in disagreement with Møller–Plesset theory, where singles are of second order because of the Brillouin theorem. The special treatment of singles in Møller–Plesset theory occurs since the Hartree–Fock reference state has been optimized with respect to the orbitals, making the orbital relaxations due to the fluctuation potential rather small. For a reference state that has not been variationally optimized with respect to the orbitals, the singles amplitudes would be much larger and enter to zero order in the perturbational treatment. By also treating the singles as zero-order parameters in the Hartree–Fock case, we ensure that any large adjustments needed in the orbitals may be fully incorporated in our perturbational treatment at any stage. This may be important in situations where the single-reference dominance is weak and also when properties other than energies are considered. In particular, it turns out that, for the treatment of time-dependent properties, it is essential to treat the singles as zero-order parameters. All things considered, for a general, unbiased treatment of the singles, we classify $\hat{T}_1$ as a zero-order operator, recognizing the key role played by the singles as orbital-relaxation parameters.

Let us consider the CC2 approximation in greater detail. The CC2 equations are obtained from their CCSD counterparts (14.6.32) and (14.6.33) by treating the doubles equations to first order. The CC2 equations may therefore be written in the form

$$\varepsilon_{\mu_1} t_{\mu_1} = -\overline{R}_1^1 - \langle\mu_1|[\hat{\tilde{\Phi}}, \hat{T}_2]|\text{HF}\rangle \tag{14.6.37}$$

$$\varepsilon_{\mu_2} t_{\mu_2} = -\langle\mu_2|\hat{\tilde{\Phi}}|\text{HF}\rangle \tag{14.6.38}$$

Solving (14.6.38) explicitly for the doubles, we may write the CC2 equations in the form

$$\varepsilon_{\mu_1} t_{\mu_1} = -\overline{R}_1^1 - \langle\mu_1|[\hat{\tilde{\Phi}}, \hat{Q}_2]|\text{HF}\rangle \tag{14.6.39}$$

where

$$\hat{Q}_2 = -\sum_{\mu_2} \varepsilon_{\mu_2}^{-1} \langle\mu_2|\hat{\tilde{\Phi}}|\text{HF}\rangle \hat{\tau}_{\mu_2} \tag{14.6.40}$$

Similarly, the CC2 wave function may be written as

$$|\text{CC2}\rangle = \exp(\hat{T}_1 + \hat{Q}_2)|\text{HF}\rangle \tag{14.6.41}$$

where the only free parameters are the singles amplitudes, obtained by solving the modified CCS equations (14.6.39).

The CC2 wave function and energy are correct to the same orders in the fluctuation potential as the MP2 wave function and energy. However, we would expect the CC2 wave function to be slightly more accurate and more robust than the MP1 wave function since the CC2 wave function has the additional merit of fulfilling the CCSD amplitudes equations to first order. The improvements are expected to be small, however. A direct comparison is given after the discussion of the CC3 approximation.

In Table 14.6, we have listed the order in the fluctuation potential to which the amplitudes of the different wave-function approximations are correct. We have also listed the overall order of accuracy of each wave function, as well as the number of electrons for which each approximation recovers the FCI solution. The CC2 wave function is intermediate in quality between the MP1 and CCSD wave functions. In Table 14.7, we compare the energies of the various approximations (in a perturbational sense) as well as the overall cost of the approximation and the number of electrons for which the energy reproduces the FCI value. The CC2 energy is similar to the MP2 energy in both cost and accuracy.

Following the same procedure as for CC2, we obtain the CC3 approximation from CCSDT by retaining, in the triples equations, only those terms that are of second order in the fluctuation potential:

$$\varepsilon_{\mu_1} t_{\mu_1} = -\overline{R}_1^2 - \langle\mu_1|[\hat{\tilde{\Phi}}, \hat{T}_3]|\text{HF}\rangle \tag{14.6.42}$$

$$\varepsilon_{\mu_2} t_{\mu_2} = -\overline{R}_2^2 - \langle\mu_2|[\hat{\tilde{\Phi}}, \hat{T}_3]|\text{HF}\rangle \tag{14.6.43}$$

$$\varepsilon_{\mu_3} t_{\mu_3} = -\langle\mu_3|[\hat{\tilde{\Phi}}, \hat{T}_2]|\text{HF}\rangle \tag{14.6.44}$$

Table 14.6 The wave function at different levels of approximation. The first four rows give the orders to which the different amplitudes are correct; the fifth row gives the order to which the wave function is correct; the last row indicates the number electrons treated exactly at each level

	MP1	CC2	CCSD	MP2	CC3	CCSDT
Singles	1*	2	2	2	3	4
Doubles	1	1	2	2	3	3
Triples	1*	1*	1*	2	2	3
Quadruples	2*	2*	2*	2*	2*	2*
Wave function	1	1	1	2	2	2
Exact system	1	1	2	1	2	3

*Amplitudes zero in the wave function.

Table 14.7 The energy at different levels of approximation. The first row gives the order to which the energy is correct; the second row indicates the cost of the method; the last row indicates the number of electrons for which the treatment is exact

	MP2	CC2	MP3	CCSD	MP4	CCSD(T)	CC3	CCSDT
Order	2	2	3	3	4	4	4	4
Cost	n^5	n^{5*}	n^6	n^{6*}	n^7	n^7	n^{7*}	n^{8*}
Exact system	1	1	1	2	1	2	2	3

*Iterative in the time-consuming steps.

Solving for the triples in (14.6.44) and introducing the notation

$$\hat{Q}_3 = -\sum_{\mu_3} \varepsilon_{\mu_3}^{-1} \langle \mu_3 | [\hat{\tilde{\Phi}}, \hat{T}_2] | \mathrm{HF} \rangle \hat{\tau}_{\mu_3} \tag{14.6.45}$$

we may write the CC3 wave function in the form

$$|\mathrm{CC3}\rangle = \exp(\hat{T}_1 + \hat{T}_2 + \hat{Q}_3)|\mathrm{HF}\rangle \tag{14.6.46}$$

where the singles and doubles amplitudes are obtained from the modified CCSD equations

$$\varepsilon_{\mu_1} t_{\mu_1} = -\overline{R}_1^2 - \langle \mu_1 | [\hat{\tilde{\Phi}}, \hat{Q}_3] | \mathrm{HF} \rangle \tag{14.6.47}$$

$$\varepsilon_{\mu_2} t_{\mu_2} = -\overline{R}_2^2 - \langle \mu_2 | [\hat{\tilde{\Phi}}, \hat{Q}_3] | \mathrm{HF} \rangle \tag{14.6.48}$$

The CC3 wave function may be optimized in the same manner as the CCSD wave function, correcting in each iteration the CCSD right-hand sides by the terms given in (14.6.47) and (14.6.48). The additional cost scales as n^7.

Let us consider the quality of the CC3 wave function and energy. We recall that the CCSDT singles are correct to fourth order in the fluctuation potential and that the CCSDT doubles and triples are correct to third order; see Table 14.6. In the CC3 model, we have reduced the accuracy of the triples to second order. The terms containing $\hat{Q}_3$ in (14.6.47) and (14.6.48) therefore introduce fourth-order errors in the CCSDT singles and the doubles. These errors do not affect the overall quality of the doubles (since these are correct only to third order in CCSDT) but affect the singles, reducing their accuracy to third order. In conclusion, the CC3 wave function is correct to third order in the singles and doubles and to second order in the triples; see Table 14.6. Comparing with CCSD, we have improved the accuracy in the singles, doubles and triples by one order in the fluctuation potential. The overall accuracy of the CC3 wave function is of second order (the same as for CCSDT) because the quadruples are neglected in both approximations. The accuracy of the CC3 energy is of fourth order, again the same as for the CCSDT energy; see Table 14.7.

14.6.3 NONITERATIVE HYBRID METHODS: THE CCSD(T) MODEL

In the development of the CC3 model, we replaced the CCSDT triples equations by approximate equations where all terms of order higher than 2 in the fluctuation potential were discarded. We then proceeded in the same manner as for CCSDT, iterating the (approximate) CCSDT equations until convergence. The resulting wave function and energy are correct to the same order in the fluctuation potential as CCSDT.

However, if all we wish to accomplish by our post-CCSD treatment of the electronic system is to improve the accuracy of the energy from third to fourth order in the fluctuation potential, then we are overdoing it with the CC3 approximation. In a perturbational sense, we may achieve the same improvement in the energy by not iterating the CC3 equations. Even simpler, we may obtain the necessary corrections to the CCSD energy directly from perturbation theory by examining the lowest-order terms that contain connected triples. In particular, let us identify all terms in CCPT that involve connected triples to fourth and fifth orders.

The connected triples in (14.3.74) give one fourth-order contribution

$$E_T^{(4)} = \langle \bar{t}_2^{(1)} | [\hat{\Phi}, \hat{T}_3^{(2)}] | \mathrm{HF} \rangle \tag{14.6.49}$$

To fifth order, several contributions arise. From a consideration of excitation levels, the first two terms in (14.3.75) are seen not to contain contributions from connected triples. The remaining two terms give rise to the following fifth-order connected triples contributions:

$$E_T^{(5)} = \langle \bar{t}_1^{(2)} | [\hat{\Phi}, \hat{T}_3^{(2)}] | \mathrm{HF} \rangle + \langle \bar{t}_2^{(2)} | [\hat{\Phi}, \hat{T}_3^{(2)}] | \mathrm{HF} \rangle + \langle \bar{t}_3^{(2)} | [\hat{\Phi}, \hat{T}_3^{(2)}] | \mathrm{HF} \rangle + \langle \bar{t}_4^{(2)} | [\hat{\Phi}, \hat{T}_3^{(2)}] | \mathrm{HF} \rangle$$
$$+ \langle \bar{t}_3^{(2)} | [\hat{\Phi}, \hat{T}_2^{(2)}] | \mathrm{HF} \rangle + \tfrac{1}{2} \langle \bar{t}_3^{(2)} | [[\hat{\Phi}, \hat{T}_2^{(1)}], \hat{T}_2^{(1)}] | \mathrm{HF} \rangle \tag{14.6.50}$$

The connected triples thus give a fourth-order contribution when projected against the doubles space and fifth-order contributions when projected against the singles, doubles, triples and quadruples space. The $E_T^{(4)}$ term in (14.6.49) and the first two terms of $E_T^{(5)}$ in (14.6.50) are the only fourth- and fifth-order connected triples contributions that arise from projection onto the singles and doubles space. These three contributions are all included in the *CCSD(T) correction* to the CCSD energy [21]

$$\Delta E^{\mathrm{CCSD(T)}} = \langle \bar{t} | [\hat{\Phi}, {}^*\hat{T}_3^{(2)}] | \mathrm{HF} \rangle \tag{14.6.51}$$

In this expression, the triples correspond to the second-order amplitudes (14.4.52) calculated from the CCSD amplitudes rather than from the first-order CCPT amplitudes

$$ {}^*\hat{T}_3^{(2)} = \hat{T}_3^{(2)} + O(3) \tag{14.6.52}$$

and the amplitudes in $\langle \bar{t} |$ are the CCSD amplitudes. The terms that originate from the projection on the singles and doubles space in (14.6.49) and (14.6.50) are included in (14.6.51) since

$$\bar{t}_2 = \bar{t}_2^{(1)} + \bar{t}_2^{(2)} + O(3) \tag{14.6.53}$$

$$\bar{t}_1 = \bar{t}_1^{(2)} + O(3) \tag{14.6.54}$$

The substitution (14.6.52) modifies the energy to fifth and higher orders, whereas those in (14.6.53) and (14.6.54) give terms of orders 6 and higher.

We have thus arrived at a new approximation to the FCI energy; see Table 14.7. The CCSD(T) energy [21]

$$E^{\text{CCSD(T)}} = E^{\text{CCSD}} + \Delta E^{\text{CCSD(T)}} \tag{14.6.55}$$

is similar to the CC3 energy but less expensive since the most expensive computational step is not iterated; see Table 14.7. The CCSD(T) approximation is therefore the preferred one for the calculation of energies, although for other molecular properties – in particular, time-dependent properties – the CCSD(T) approach is less useful than CC3 since no wave function is associated with this approximation. In the *CCSD[T] method* (also known as the *CCSD + T(CCSD) method*), the triples correction is slightly different from that of the CCSD(T) method and is obtained from (14.6.51) by including in $\langle \bar{t} |$ only the doubles amplitudes [22].

Let us consider the evaluation of the CCSD(T) energy correction for a closed-shell system. The multipliers in (14.6.51) are the CCSD singles and doubles amplitudes

$$\langle \bar{t} | = \sum_{ai} \bar{t}_i^a \left\langle \frac{a}{i} \right| + \frac{1}{2} \sum_{abij} \bar{t}_{ij}^{ab} \left\langle \frac{ab}{ij} \right| \tag{14.6.56}$$

However, because of the biorthogonal representation, the bra amplitudes in (14.6.56) are not identical to the ket amplitudes obtained from the CCSD calculation:

$$|t\rangle = \sum_{ai} t_i^a \left| \frac{a}{i} \right\rangle + \frac{1}{2} \sum_{abij} t_{ij}^{ab} \left| \frac{ab}{ij} \right\rangle \tag{14.6.57}$$

From (13.7.56) and (13.7.57), we establish the following relationships

$$\bar{t}_i^a = 2 t_i^a \tag{14.6.58}$$

$$\bar{t}_{ij}^{ab} = 4 t_{ij}^{ab} - 2 t_{ji}^{ab} \tag{14.6.59}$$

which determine the coefficients of the bra state used in the calculation of the CCSD(T) energy correction (14.6.51). The triples amplitudes $^*\hat{T}_3^{(2)}$ of the CCSD(T) calculation are obtained in the usual way according to (14.4.52)

$$^*t_{ijk}^{abc(2)} = -P_{ijk}^{abc} \frac{\sum_d t_{ij}^{ad} g_{ckbd} - \sum_l t_{il}^{ab} g_{cklj}}{\varepsilon_a + \varepsilon_b + \varepsilon_c - \varepsilon_i - \varepsilon_j - \varepsilon_k} \tag{14.6.60}$$

where the amplitudes are those of the CCSD state.

Expanding the bra state in (14.6.51), we may now write the CCSD(T) correction in the form

$$\Delta E^{\text{CCSD(T)}} = \sum_{ai} \bar{t}_i^a \left\langle \frac{a}{i} \right| [\hat{\Phi}, {}^*\hat{T}_3^{(2)}] \left| \text{HF} \right\rangle + \frac{1}{2} \sum_{abij} \bar{t}_{ij}^{ab} \left\langle \frac{ab}{ij} \right| [\hat{\Phi}, {}^*\hat{T}_3^{(2)}] \left| \text{HF} \right\rangle \tag{14.6.61}$$

The second term corresponds to the triples part of the MP4 energy (14.4.67) and may be evaluated as described in Section 14.4.4. The first term has no counterpart at the MP4 level and must be evaluated separately, following the same procedure as for (14.4.71). The final expression for the CCSD(T) energy correction may be written as

$$\Delta E^{\text{CCSD(T)}} = \sum_{ai} \bar{t}_i^a {}^*T_i^a + \sum_{abij} \bar{t}_{ij}^{ab} {}^*T_{ij}^{ab} \tag{14.6.62}$$

where

$$*T_i^a = \sum_{cdkl} ({}^*t_{ikl}^{acd^{(2)}} - {}^*t_{lki}^{acd^{(2)}})L_{kcld} \tag{14.6.63}$$

$$*T_{ij}^{ab} = \sum_{cdk} \left({}^*t_{ijk}^{acd^{(2)}} L_{bckd} - {}^*t_{kji}^{acd^{(2)}} g_{kdbc}\right) - \sum_{ckl} \left({}^*t_{ikl}^{abc^{(2)}} L_{kjlc} - {}^*t_{lki}^{abc^{(2)}} g_{kjlc}\right) \tag{14.6.64}$$

and the amplitudes are obtained from (14.6.58)–(14.6.60). The computational cost of the CCSD(T) correction scales as O^3V^4, the same as for MP4.

14.6.4 HYBRID AND NONHYBRID METHODS COMPARED

We have developed three hybrid approximations to the FCI wave function and energy: CC2, CC3 and CCSD(T). The characteristics of these methods are summarized in Tables 14.6 and 14.7, where comparisons are made with the coupled-cluster and Møller–Plesset methods. In Figure 14.10, we have, for the methods in Table 14.7, plotted the cc-pVDZ errors relative to FCI for the water molecule with bond lengths R_{ref} and $2R_{\text{ref}}$. The two plots represent situations where the Hartree–Fock dominance in the FCI wave function is strong (weight 0.941) and weak (weight 0.589).

There is a steady improvement in the energy as we go through the sequence of methods MP2, CC2, MP3, CCSD, MP4, CCSD(T), CC3 and CCSDT. The only exception occurs at the stretched geometry, where the MP3 energy is considerably higher than those of the second-order MP2 and CC2 methods. As discussed in Section 14.5, such an oscillatory behaviour of the perturbation series is frequent for systems not dominated by a single configuration. As expected, the convergence of the correlation energy is slower at the stretched geometry, in agreement with what we would expect for schemes based on the dominance of a single electronic configuration.

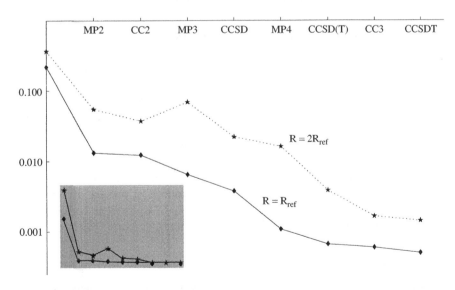

Fig. 14.10. The errors in the cc-pVDZ correlation energy of water at various levels of theory (in E_h). In the main plot, the absolute values of the errors relative to FCI are plotted on a logarithmic scale. In the inset, the errors are shown on a linear scale. The geometries are those of Table 5.5, with bond lengths R_{ref} and $2R_{\text{ref}}$.

Most of the correlation energy is recovered at the MP2 and CC2 levels, which are similar in quality and cost except that CC2 is iterative. Indeed, the attraction of the CC2 method relative to MP2 lies not so much in the calculation of the energy as in the calculation of various molecular (in particular, time-dependent) properties.

Going to the third-order methods MP3 and CCSD, we note a significant improvement at the CCSD level. Indeed, its performance is clearly better than that of the MP3 method, although both methods are correct to the same order in the fluctuation potential – illustrating again that we cannot judge the merits of a particular method based only on an order analysis. Although the MP3 and CCSD models are correct to the same order in the fluctuation potential, the CCSD wave function is a better approximation to the FCI solution since it fulfils the additional requirements imposed by the projected coupled-cluster equations. In particular, whereas the CCSD energy is exact (relative to FCI) for a two-electron system, the MP3 is exact only for a single electron (like Hartree–Fock). In Tables 14.6 and 14.7, we have, for each method, listed the number of electrons for which they reproduce the FCI solution.

Next we turn our attention to the more accurate fourth-order methods. As for the third-order methods, the noniterative MP4 method performs poorer than the iterative ones. Again, this behaviour is related to the overall quality of the wave function: the noniterative MP4 method of cost n^7 is exact (relative to FCI) for one electron only. The pseudo-iterative CCSD(T) method of cost n^7 (in the noniterative step) is exact for two electrons, as is the iterative CC3 method of the same cost. Finally, the iterative CCSDT method of cost n^8 is exact for three electrons.

Comparing the CCSD(T) and CC3 methods with CCSD and CCSDT, we find that both hybrid methods represent a considerable improvement on the CCSD treatment. We have clearly been quite successful in incorporating the main effects of the triples amplitudes in the CCSD(T) and CC3 models. For the calculation of energies and static properties, CCSD(T) is the method of choice since the correction is obtained in a noniterative fashion. The CC3 method has some advantages for the calculation of time-dependent properties.

14.7 Multiconfigurational perturbation theory

Except for the general theory of Section 14.1, the perturbation theory developed so far has been restricted to systems dominated by a single electronic configuration. For multiconfigurational systems, the standard Møller–Plesset theory fails. First, it may not be obvious which determinant should be chosen as the zero-order state. Second, even if a unique choice can be made, the lower-order corrections will usually be inaccurate and the higher-order ones divergent. To treat such systems, we must develop a more general perturbation theory, where the zero-order state is multiconfigurational. In this section, we examine the most important such theory, *CAS perturbation theory (CASPT)*, where the zero-order state is taken to be a CASSCF wave function [23–28]. The importance of CASPT stems from the fact that, at present, it represents the only generally applicable method for the *ab initio* calculation of dynamical correlation effects of open- and closed-shell multiconfigurational electronic systems.

14.7.1 THE ZERO-ORDER CASPT HAMILTONIAN

Based on the success of Møller–Plesset theory for systems dominated by a single electronic configuration, it is natural to enquire whether the zero-order Hamiltonian operator of Møller–Plesset theory can be extended to multiconfigurational systems. We recall that, in Møller–Plesset theory, we use the Fock operator in the canonical representation as the zero-order Hamiltonian – see

Section 14.2.1. Extending the Fock operator in the orbital basis (10.4.2) to multiconfigurational CASSCF wave functions $|0\rangle$, we obtain the *CASSCF Fock operator* [26]

$$\hat{f} = \sum_{pq} f_{pq} E_{pq} = \tfrac{1}{2} \sum_{pq} \sum_{\sigma} \langle 0|[a_{q\sigma}^{\dagger}, [a_{p\sigma}, \hat{H}]]_{+}|0\rangle E_{pq} \qquad (14.7.1)$$

In Møller–Plesset theory, $\hat{f}$ can be brought to a diagonal form with the single-reference zero-order state as an eigenstate. We shall now examine the CASSCF Fock operator and see how it may be modified to constitute a zero-order Hamiltonian with the CASSCF state as an eigenstate.

The Fock matrix may be partitioned following the division of the orbitals into inactive, active and secondary spaces. From (10.8.38) and from the discussion in Section 12.5.1, we note that the inactive–secondary elements of the Fock matrix f_{ia} are proportional to the corresponding elements of the CASSCF orbital gradient

$$4f_{ia} = 4(^{I}F_{ia} + {}^{A}F_{ia}) = {}^{o}E_{ia}^{(1)} \qquad (14.7.2)$$

and therefore vanish for an optimized state. Since the CASSCF Fock matrix is symmetric, the secondary–inactive elements f_{ai} vanish as well. The remaining inactive–active and active–secondary elements of the Fock matrix, on the other hand, are not directly related to the corresponding elements of the CASSCF orbital gradient and do not vanish for the optimized wave function. Finally, the three diagonal blocks of the Fock matrix may be separately diagonalized to obtain canonical orbitals by carrying out redundant intraspace (inactive–inactive, active–active and secondary–secondary) rotations; see the discussion in Section 12.2.6. These results are summarized in Fig. 14.11, where the Fock matrix is given in the canonical representation of an optimized CASSCF state.

In general, the Fock operator (14.7.1) does not have the CASSCF wave function $|0\rangle$ as an eigenfunction, $\hat{f}|0\rangle$ being a sum of single excitations from $|0\rangle$:

$$\hat{f}|0\rangle = \sum_{pq} f_{pq} E_{pq} |0\rangle \qquad (14.7.3)$$

However, the zero-order Hamiltonian operator $\hat{H}_0$ must have $|0\rangle$ as an eigenstate. To achieve this, we choose

$$\hat{H}_0 = E^{(0)}|0\rangle\langle 0| + \hat{P}\hat{f}\hat{P} \qquad (14.7.4)$$

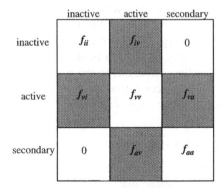

Fig. 14.11. The Fock matrix of an optimized CASSCF state in the canonical representation, where the three diagonal blocks are diagonal.

where $\hat{P}$ is the projector

$$\hat{P} = 1 - |0\rangle\langle 0| \tag{14.7.5}$$

and $E^{(0)}$ the zero-order energy, chosen as the expectation value of the CASSCF Fock matrix

$$E^{(0)} = \langle 0|\hat{f}|0\rangle \tag{14.7.6}$$

This choice of $E^{(0)}$ ensures that, when $|0\rangle$ reduces to a single-determinant closed-shell state, the zero-order Hamiltonian (14.7.4) reduces to the Møller–Plesset Fock operator.

The operator (14.7.4) may be used to generate the perturbation expansion. However, this choice of zero-order operator is inconvenient since the corrections would then, already to first order, involve the full set of determinants belonging to the FCI expansion. To see this, consider the first-order correction to the wave function given in (14.1.42) for $n = 1$:

$$\mathbf{C}^{(1)} = -\mathbf{P}(\mathbf{H}_0 - E^{(0)}\mathbf{1})^{-1}\mathbf{PUC}^{(0)} \tag{14.7.7}$$

In the determinantal representation, the zero-order operator (14.7.4) becomes a sparse matrix $\mathbf{H}_0$ which is not block diagonal. Since the inverse of such a matrix is a full matrix, the first-order correction (14.7.7) will include contributions from all determinants in the FCI wave function, even though $\mathbf{UC}^{(0)}$ is nonzero only for determinants that are at most doubly excited with respect to the zero-order state.

To avoid long perturbation vectors, we impose a diagonal block structure on $\mathbf{H}_0$. This is accomplished by defining the zero-order CASPT operator as [25]

$$\hat{H}_0 = E^{(0)}|0\rangle\langle 0| + \hat{P}_{\text{K}}\hat{f}\hat{P}_{\text{K}} + \hat{P}_{\text{SD}}\hat{f}\hat{P}_{\text{SD}} + \hat{P}_{\text{TQ}}\hat{f}\hat{P}_{\text{TQ}} + \cdots \tag{14.7.8}$$

in terms of the orthogonal projectors

$$1 = |0\rangle\langle 0| + \hat{P}_{\text{K}} + \hat{P}_{\text{SD}} + \hat{P}_{\text{TQ}} + \cdots \tag{14.7.9}$$

where $\hat{P}_{\text{K}}$ projects onto the part of the CAS space that is orthogonal to $|0\rangle$, $\hat{P}_{\text{SD}}$ projects onto the space spanned by the single and double excitations from $|0\rangle$ that are not contained in the CAS space, $\hat{P}_{\text{TQ}}$ onto the space spanned by the triple and quadruple excitations orthogonal to the CAS space, and so on. Since the excitations that refer only to the CAS space can be described fully in terms of the determinants of the CAS wave function, they have been excluded from $\hat{P}_{\text{SD}}$, $\hat{P}_{\text{TQ}}$, and so on. In Fig. 14.12, we have illustrated the structure of the resulting zero-order CASPT Hamiltonian matrix (14.7.8) in Fock space. The block-diagonal structure imposed by the projectors ensures that the corrections to the wave function truncate at a finite level of excitation. For example, the first-order correction contains at most double excitations and the second-order correction no higher than quadruple excitations.

14.7.2 SIZE-EXTENSIVITY IN CASPT

To set up a size-extensive perturbation theory, the zero-order Hamiltonian must be additively separable. We shall here show that, for multiplicatively separable zero-order wave functions, the CASSCF Fock operator and the zero-order energy are both additively separable but that the presence of projection operators nevertheless makes the zero-order CASPT Hamiltonian $\hat{H}_0$

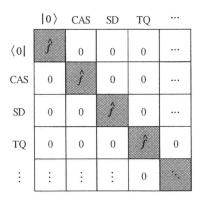

Fig. 14.12. The block-diagonal structure of the zero-order CASPT Hamiltonian matrix.

nonseparable [29]. To keep things simple, we shall consider the zero-order operator (14.7.4) rather than the more complicated CASPT Hamiltonian (14.7.8). We assume that separate CASSCF calculations have been carried out for the noninteracting subsystems A and B, giving the normalized reference wave functions

$$|0_A\rangle = \hat{\psi}_A|\text{vac}\rangle \qquad (14.7.10)$$

$$|0_B\rangle = \hat{\psi}_B|\text{vac}\rangle \qquad (14.7.11)$$

and furthermore that an active space has been chosen for the compound system AB such that the compound wave function is multiplicatively separable:

$$|0_{AB}\rangle = \hat{\psi}_{AB}|\text{vac}\rangle = \hat{\psi}_A\hat{\psi}_B|\text{vac}\rangle \qquad (14.7.12)$$

The separability of the Fock operator now follows from the additive separability of the Hamiltonian and from the multiplicative separability of the wave function (14.7.12)

$$
\begin{aligned}
\hat{f}_{AB} &= \tfrac{1}{2}\sum_{pq}\sum_{\sigma}\langle 0_{AB}|[a_{q\sigma}^{\dagger},[a_{p\sigma},\hat{H}_{AB}]]_+|0_{AB}\rangle E_{pq} \\
&= \tfrac{1}{2}\sum_{p_A q_A}\sum_{\sigma}\langle 0_A|[a_{q_A\sigma}^{\dagger},[a_{p_A\sigma},\hat{H}_A]]_+|0_A\rangle E_{p_A q_A} + \tfrac{1}{2}\sum_{p_B q_B}\sum_{\sigma}\langle 0_B|[a_{q_B\sigma}^{\dagger},[a_{p_B\sigma},\hat{H}_B]]_+|0_B\rangle E_{p_B q_B} \\
&= \hat{f}_A + \hat{f}_B
\end{aligned}
\qquad (14.7.13)
$$

and the separability of the zero-order energy then follows trivially:

$$E_{AB}^{(0)} = \langle 0_{AB}|\hat{f}_{AB}|0_{AB}\rangle = \langle 0_A|\hat{f}_A|0_A\rangle + \langle 0_B|\hat{f}_B|0_B\rangle = E_A^{(0)} + E_B^{(0)} \qquad (14.7.14)$$

The zero-order Hamiltonian for the compound system can now be written as

$$
\begin{aligned}
\hat{H}_{0AB} &= \hat{P}_{AB}\hat{f}_{AB}\hat{P}_{AB} + E_{AB}^{(0)}|0_{AB}\rangle\langle 0_{AB}| \\
&= \hat{f}_A + \hat{f}_B - (\hat{f}_{AB} - E_{AB}^{(0)})|0_{AB}\rangle\langle 0_{AB}| - |0_{AB}\rangle\langle 0_{AB}|(\hat{f}_{AB} - E_{AB}^{(0)})
\end{aligned}
\qquad (14.7.15)
$$

On the other hand, the sum of the zero-order Hamiltonians of the noninteracting subsystems is given by

$$\hat{H}_{0A} + \hat{H}_{0B} = \hat{P}_A \hat{f}_A \hat{P}_A + \hat{P}_B \hat{f}_B \hat{P}_B + E_A^{(0)} |0_A\rangle\langle 0_A| + E_B^{(0)} |0_B\rangle\langle 0_B|$$

$$= \hat{f}_A + \hat{f}_B - (\hat{f}_A - E_A^{(0)})|0_A\rangle\langle 0_A| - |0_A\rangle\langle 0_A|(\hat{f}_A - E_A^{(0)})$$

$$- (\hat{f}_B - E_B^{(0)})|0_B\rangle\langle 0_B| - |0_B\rangle\langle 0_B|(\hat{f}_B - E_B^{(0)}) \qquad (14.7.16)$$

Comparing (14.7.15) and (14.7.16), we find that $\hat{H}_{0AB}$ is not equal to $\hat{H}_{0A} + \hat{H}_{0B}$. A similar argument would show that the zero-order CASPT Hamiltonian (14.7.8) is not additively separable.

For a single-determinant closed-shell Hartree–Fock state, the zero-order CASPT operator (14.7.8) reduces to the closed-shell Fock operator. In this limit, the operators $\hat{f}_A$ and $\hat{f}_B$ have $|0_A\rangle$ and $|0_B\rangle$ as eigenfunctions:

$$\hat{f}_A |0_A\rangle = E_A^{(0)} |0_A\rangle \qquad (14.7.17)$$

$$\hat{f}_B |0_B\rangle = E_B^{(0)} |0_B\rangle \qquad (14.7.18)$$

The nonseparable contributions to the zero-order CASPT Hamiltonian (14.7.15) then vanish and $\hat{H}_{0AB}$ becomes separable:

$$\hat{H}_{0AB} = \hat{f}_A + \hat{f}_B = \hat{H}_{0A} + \hat{H}_{0B} \qquad (14.7.19)$$

For a CASSCF reference wave function dominated by a single determinant, we therefore expect the contributions to the energy from the terms that are not size-extensive to be small. Conversely, for a reference wave function in which several determinants have large weights, we expect larger nonseparable contributions.

14.7.3 THE CASPT WAVE FUNCTION AND ENERGY

In this subsection, we consider *the first-order CASPT (CASPT1) wave-function correction* and *the second-order CASPT (CASPT2) energy correction*. We adopt the *internally contracted scheme*, in which the CASPT1 correction is written as a linear combination of all single and double excitations from $|0\rangle$ [26]:

$$|0^{(1)}\rangle = \sum_{pq} C_{pq}^{(1)} E_{pq} |0\rangle + \sum_{pq \geq rs} C_{pqrs}^{(1)} e_{pqrs} |0\rangle \qquad (14.7.20)$$

Whereas the indices q and s in the summations are occupied (i.e. inactive or active), p and r are noninactive (i.e. active or secondary). In (14.7.20), no terms with all indices active are included. As shown below, such terms do not contribute to the first-order wave function with $\hat{H}_0$ in (14.7.8) chosen as the zero-order Hamiltonian. Similarly, the triples and higher excitations from $|0\rangle$ do not contribute.

The contracted configurations in (14.7.20) constitute a nonorthogonal basis. Whereas (14.1.23)–(14.1.25) are independent of the basis and need not be modified because of nonorthogonality, the expansion in (14.1.42) must be modified. For the CASPT1 wave-function correction, we obtain (see Exercise 14.7)

$$\mathbf{C}^{(1)} = -\mathbf{P}(\mathbf{H}_0 - E^{(0)}\mathbf{S})^{-1} \mathbf{PUC}^{(0)} \qquad (14.7.21)$$

where

$$S_{ij} = \langle i | j \rangle \tag{14.7.22}$$

$$\mathbf{P} = \mathbf{1} - \mathbf{C}_0 \mathbf{C}_0^{\mathrm{T}} \tag{14.7.23}$$

and $|i\rangle$ refers to one of the contracted configurations $E_{pq}|0\rangle$ or $e_{pqrs}|0\rangle$. The first- and second-order CASPT energy corrections may be calculated in the same manner as before:

$$E^{(1)} = \mathbf{C}^{(0)\mathrm{T}} \mathbf{U} \mathbf{C}^{(0)} \tag{14.7.24}$$

$$E^{(2)} = \mathbf{C}^{(1)\mathrm{T}} \mathbf{U} \mathbf{C}^{(0)} \tag{14.7.25}$$

In the CASPT1 wave function (14.7.21), the components belonging to the CAS space become

$$C_k^{(1)} = - \sum_l [\mathbf{P}(\mathbf{H}_0 - E^{(0)}\mathbf{S})^{-1}]_{kl} \langle l | \hat{P} \hat{U} | 0 \rangle \tag{14.7.26}$$

where we have used the zero-order Hamiltonian (14.7.8) and where k and l both refer to the determinants in the CAS space. The matrix elements

$$\langle l | \hat{P} \hat{U} | 0 \rangle = \langle l | \hat{P} \hat{H} | 0 \rangle \tag{14.7.27}$$

vanish since we have assumed that $|0\rangle$ is a CASSCF wave function, for which the CI gradient (12.2.26) is equal to zero. The components belonging to the CAS space (14.7.26) therefore vanish in the first-order correction (14.7.21). Moreover, since the elements of $\mathbf{PUC}^{(0)}$ that correspond to triples and higher excitations from $|0\rangle$ are zero, the first-order corrections (14.7.21) vanish for these components as well.

The use of internally contracted configurations in CASPT leads to a rather complicated formalism, involving, for example, three-particle density matrices over active orbitals already for the CASPT2 energy. A more straightforward but computationally more demanding approach would be to span the wave-function corrections in Slater determinants, redefining the projectors in (14.7.9) so that, for example, $\hat{P}_{\mathrm{SD}}$ corresponds to the space of single and double excitations from the full set of CAS determinants rather than from $|0\rangle$. The internally contracted and determinantal formulations give similar energy corrections.

14.7.4 SAMPLE CASPT CALCULATIONS

To illustrate the behaviour of the CASPT method, we have, on the left in Figure 14.13, plotted the CASPT2/cc-pVDZ dissociation curve of the water molecule, calculated at a fixed HOH bond angle of 110.565°, using a (3,0,1,2) valence CASSCF reference state. The energies have been obtained using the internally contracted scheme. For comparison, the corresponding FCI and MP2 curves have been included. On the right in Figure 14.13, we have plotted the CASPT2 and MP2 energies relative to the FCI energy in the same basis.

Close to the equilibrium geometry, the errors of the MP2 and CASPT2 energies are similar. However, whereas the MP2 description deteriorates markedly for large separations, the CASPT2 description is uniform, the error decreasing only slightly as the molecule dissociates. The CASPT2 curve in Figure 14.13 should also be compared with the CASSCF curve in Figure 5.15. Although the CASSCF model also provides a uniform description of the dissociation, its error is more than a factor of 10 larger than that of CASPT2.

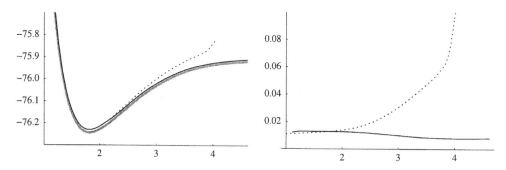

Fig. 14.13. The CASPT2 dissociation of the C_{2v} water molecule in the cc-pVDZ basis (in atomic units) for a fixed HOH bond angle of 110.565°. On the left, we have plotted the CASPT2 potential-energy curve (black line) together with the corresponding FCI (grey line) and MP2 (dotted line) curves; on the right, we have plotted the difference between the CASPT2 and FCI energies (black line) and the MP2 and FCI energies (dotted line).

In Table 14.8, we have listed the CASPT and MPPT energies calculated up to order 10 for the water molecule at the geometries R_{ref} and $2R_{\text{ref}}$. The CASPT results have been calculated using the uncontracted, determinantal scheme. In the CASPT calculations, a (3,0,1,2) valence CAS reference state is employed, with the inactive $1a_1$ CAS orbital frozen; in the MPPT calculations, the canonical $1a_1$ Hartree–Fock orbital has been kept frozen.

Owing to the single-configuration character of the wave function at R_{ref} and its multiconfigurational character at $2R_{\text{ref}}$, the convergence patterns are rather different at the two geometries. At R_{ref}, the CASPT and MPPT series behave in the same manner, both converging to within 10^{-5} E$_\text{h}$ at order 10. At $2R_{\text{ref}}$, the CASPT series is strongly divergent: although CASPT2 recovers

Table 14.8 The convergence of the CASPT and MPPT series for the water molecule in the cc-pVDZ basis at the reference geometries R_{ref} and $2R_{\text{ref}}$. The listed energies are the errors (in E$_\text{h}$) relative to the FCI energy (in the same basis and with the same frozen orbitals)

	CASPT		MPPT	
	$R_{\text{ref}}^{\text{a}}$	$2R_{\text{ref}}^{\text{b}}$	$R_{\text{ref}}^{\text{c}}$	$2R_{\text{ref}}^{\text{d}}$
1	0.16091	0.13212	0.21574	0.36221
2	0.01262	0.00798	0.01335	0.05491
3	0.00419	0.00442	0.00645	0.06908
4	0.00019	−0.00084	0.00107	0.01602
5	0.00036	0.11878	0.00051	0.01665
6	0.00001	1.17522	0.00013	0.00434
7	0.00005	27.45409	0.00007	0.00060
8	0.00000	416.56579	0.00001	−0.00051
9	0.00001	7758.23440	0.00001	−0.00208
10	0.00000	129958.90000	0.00000	−0.00168

[a]FCI energy: −76.23875 E$_\text{h}$.
[b]FCI energy: −75.95022 E$_\text{h}$.
[c]FCI energy: −76.23977 E$_\text{h}$.
[d]FCI energy: −75.94992 E$_\text{h}$.

as much as 80% of the correlation energy, the series diverges and the higher-order corrections have no physical meaning. By contrast, the corresponding MPPT series at $2R_{ref}$ appears to converge. However, to orders higher than those reported in the table, the MPPT series diverges, just as for the HF molecule at the stretched geometry in Figure 14.6.

The amount of data available on the convergence of multireference perturbational schemes for truly multiconfigurational states is rather limited, but the behaviour in Table 14.8 appears to be typical of this rather simple zero-order Hamiltonian. Although the lowest-order corrections give an improved energy (CASPT2 is successfully used to describe correlation effects), the series ultimately diverges. It is perhaps not surprising that the requirements for accurate low-order perturbation corrections as in CASPT2 are less demanding than the requirements for a convergent series, where considerations must be given not only to the ability of the zero-order Hamiltonian to represent the physical system but also to the convergence properties of the series.

References

[1] S. T. Epstein, *The Variation Method in Quantum Chemistry*, Academic Press, 1974.
[2] T. Helgaker and P. Jørgensen, in S. Wilson and G. H. F. Diercksen (eds), *Methods in Computational Molecular Physics*, Plenum, 1992, p. 353.
[3] E. A. Hylleraas, *Z. Phys.* **65**, 209 (1930).
[4] K. A. Brueckner, *Phys. Rev.* **97**, 1353 (1955); *Phys. Rev.* **100**, 36 (1955).
[5] C. Møller and M. S. Plesset, *Phys. Rev.* **46**, 618 (1934).
[6] R. J. Bartlett and D. M. Silver, *Phys. Rev.* **A10**, 1927 (1974).
[7] J. S. Binkley and J. A. Pople, *Int. J. Quantum Chem.* **9**, 229 (1975).
[8] J. A. Pople, J. S. Binkley and R. Seeger, *Int. J. Quantum Chem.* **S10**, 1 (1976).
[9] R. J. Bartlett and I. Shavitt, *Chem. Phys. Lett.* **50**, 190 (1977).
[10] G. D. Purvis and R. J. Bartlett, *J. Chem. Phys.* **68**, 2114 (1978).
[11] R. Krishnan and J. A. Pople, *Int. J. Quantum Chem.* **14**, 91 (1978).
[12] R. Krishnan, M. J. Frisch and J. A. Pople, *J. Chem. Phys.* **72**, 4244 (1980).
[13] T. Kato, *Perturbation Theory for Linear Operators*, Springer-Verlag, 1966.
[14] T. H. Schucan and H. A. Weidenmuller, *Ann. Phys. (NY)* **73**, 108 (1972).
[15] J. Olsen, O. Christiansen, H. Koch and P. Jørgensen, *J. Chem. Phys.* **105**, 5082 (1996).
[16] O. Christiansen, J. Olsen, P. Jørgensen, H. Koch and P.-Å. Malmqvist, *Chem. Phys. Lett.* **261**, 369 (1996).
[17] G. Arfken, *Mathematical Methods for Physicists*, Academic Press, 1970.
[18] J. Olsen, P. Jørgensen, H. Koch, A. Balkova and R. J. Bartlett, *J. Chem. Phys.* **104**, 8007 (1996).
[19] O. Christiansen, H. Koch and P. Jørgensen, *Chem. Phys. Lett.* **243**, 409 (1995).
[20] H. Koch, O. Christiansen, P. Jørgensen, A. M. Sanchez de Merás and T. Helgaker, *J. Chem. Phys.* **106**, 1808 (1997).
[21] K. Raghavachari, G. W. Trucks, J. A. Pople and M. Head-Gordon, *Chem. Phys. Lett.* **157**, 479 (1989).
[22] M. Urban, J. Noga, S. J. Cole and R. J. Bartlett, *J. Chem. Phys.* **83**, 4041 (1985).
[23] B. O. Roos, P. Linse, P. E. M. Siegbahn and M. R. A. Blomberg, *Chem. Phys.* **66**, 197 (1982).
[24] K. Wolinski, H. L. Sellers and P. Pulay, *Chem. Phys. Lett.* **140**, 225 (1987).
[25] K. Wolinski and P. Pulay, *J. Chem. Phys.* **90**, 3647 (1989).
[26] K. Andersson, P.-Å. Malmqvist, B. O. Roos, A. J. Sadlej and K. Wolinski, *J. Phys. Chem.* **94**, 5483 (1990).
[27] K. Andersson, P.-Å. Malmqvist and B. O. Roos, *J. Chem. Phys.* **96**, 1218 (1992).
[28] H.-J. Werner, *Mol. Phys.* **89**, 645 (1996).
[29] P.-Å. Malmqvist, private communication.

Further reading

K. Andersson and B. O. Roos, Multiconfigurational second-order perturbation theory, in D. R. Yarkony (ed.), *Modern Electronic Structure Theory*, World Scientific, 1995, p. 55.

S. T. Epstein, *The Variation Method in Quantum Chemistry*, Academic Press, 1974.

Exercises

EXERCISE 14.1

In Section 14.1, we developed *Rayleigh–Schrödinger perturbation theory (RSPT)*, where the energy and the wave-function corrections are obtained in a noniterative form. In this exercise, we develop *Brillouin–Wigner perturbation theory (BWPT)*, where the corrections are generated iteratively.

1. Consider the Schrödinger equation

$$\mathbf{HC} = E\mathbf{C} \tag{14E.1.1}$$

with the partitioning of the Hamiltonian matrix

$$\mathbf{H} = \mathbf{H}_0 + \mathbf{U} \tag{14E.1.2}$$

and the zero-order equation

$$\mathbf{H}_0 \mathbf{C}^{(0)} = E^{(0)} \mathbf{C}^{(0)} \tag{14E.1.3}$$

where $\mathbf{C}^{(0)}$ is the zero-order state. Show that, in the intermediate normalization

$$\mathbf{C} = \mathbf{C}^{(0)} + \mathbf{PC} \tag{14E.1.4}$$

$$\mathbf{P} = \mathbf{1} - \mathbf{C}^{(0)} \mathbf{C}^{(0)^\mathrm{T}} \tag{14E.1.5}$$

the Schrödinger equation (14E.1.1) can be written in the form

$$(E\mathbf{1} - \mathbf{H}_0)\mathbf{PC} = \mathbf{UC} - (E - E^{(0)})\mathbf{C}^{(0)} \tag{14E.1.6}$$

2. Show that (14E.1.6) can be written in the form

$$E = E^{(0)} + \mathbf{C}^{(0)^\mathrm{T}} \mathbf{UC} \tag{14E.1.7}$$

$$\mathbf{C} = \mathbf{C}^{(0)} + (E\mathbf{1} - \mathbf{H}_0)^{-1} \mathbf{PUC} \tag{14E.1.8}$$

3. Show that (14E.1.7) and (14E.1.8) can be iterated to give

$$E = E^{(0)} + \sum_{m=0}^{\infty} \mathbf{C}^{(0)^\mathrm{T}} \mathbf{U}[(E\mathbf{1} - \mathbf{H}_0)^{-1} \mathbf{PU}]^m \mathbf{C}^{(0)} \tag{14E.1.9}$$

$$\mathbf{C} = \mathbf{C}^{(0)} + \sum_{m=1}^{\infty} [(E\mathbf{1} - \mathbf{H}_0)^{-1} \mathbf{PU}]^m \mathbf{C}^{(0)} \tag{14E.1.10}$$

These equations are equivalent to the Schrödinger equation; no approximations have been introduced. In BWPT, the expansions (14E.1.9) and (14E.1.10) are truncated to a given order and the resulting equations are then solved for E and $\mathbf{C}$. For example, the second-order BWPT equations are given by

$$E = E^{(0)} + \mathbf{C}^{(0)^\mathrm{T}} \mathbf{U} \mathbf{C} \tag{14E.1.11}$$

$$\mathbf{C} = \mathbf{C}^{(0)} + (E\mathbf{1} - \mathbf{H}_0)^{-1} \mathbf{P} \mathbf{U} \mathbf{C}^{(0)} \tag{14E.1.12}$$

These equations are solved iteratively. From an initial guess of the energy, we obtain the wave function from (14E.1.12) and the new energy from (14E.1.11). With this energy, an improved wave function is obtained, and the procedure is continued until convergence.

4. In the following, let the reference state be denoted by 0 and let i and j denote zero-order states orthogonal to the reference state. Show that the second-order BWPT equations become

$$C_0 = 1 \tag{14E.1.13}$$

$$C_i = [(E\mathbf{1} - \mathbf{H}_0)^{-1}]_{ii} H_{i0} \tag{14E.1.14}$$

and that the energy is given by

$$E = H_{00} + \sum_i H_{0i} C_i \tag{14E.1.15}$$

5. Show that the equations in Exercise 14.1.4 correspond to the eigenvalue problem

$$\mathbf{M} \begin{pmatrix} 1 \\ \mathbf{c} \end{pmatrix} = E \begin{pmatrix} 1 \\ \mathbf{c} \end{pmatrix} \tag{14E.1.16}$$

with

$$M_{00} = H_{00} \tag{14E.1.17}$$

$$M_{0i} = M_{i0} = H_{0i} \tag{14E.1.18}$$

$$M_{ij} = [\mathbf{H}_0]_{ij} \tag{14E.1.19}$$

The second-order BWPT equations thus correspond to an eigenvalue problem where the matrix elements between two states orthogonal to the reference state are approximated by the corresponding elements of the zero-order Hamiltonian.

6. Is the second-order BWPT energy size-extensive?

EXERCISE 14.2

In this exercise, we consider the evaluation of the energy as an expectation value of the RSPT wave function.

1. Consider the wave function correct to nth order in RSPT:

$$|0_n\rangle = \sum_{i=0}^{n} |0^{(i)}\rangle \tag{14E.2.1}$$

Show that the energy calculated as the expectation value

$$E = \frac{\langle 0_n | \hat{H} | 0_n \rangle}{\langle 0_n | 0_n \rangle} \tag{14E.2.2}$$

is correct to order $2n + 1$, illustrating a close connection between odd-order energies and expectation values.

2. As a special case, consider the wave function correct to first order

$$|0_1\rangle = |0^{(0)}\rangle + |0^{(1)}\rangle \tag{14E.2.3}$$

Show that the expectation value (14E.2.2) may then be expressed in terms of the RSPT corrections as

$$\frac{\langle 0_1|\hat{H}|0_1\rangle}{\langle 0_1|0_1\rangle} = E^{(0)} + E^{(1)} + \frac{E^{(2)} + E^{(3)}}{1 + \langle 0^{(1)}|0^{(1)}\rangle} \tag{14E.2.4}$$

demonstrating explicitly that it is correct to third order in the perturbation.

EXERCISE 14.3

In this exercise, we apply RSPT with the Møller–Plesset partitioning of the Hamiltonian to our HeH$^+$ STO-3G system. All numbers quoted are in atomic units.

1. As discussed in Exercise 11.11, three CSFs with $M = 0$ can be obtained from the two Hartree–Fock orbitals:

$$|1\rangle = a_{1\alpha}^{\dagger} a_{1\beta}^{\dagger}|\text{vac}\rangle \tag{14E.3.1}$$

$$|2\rangle = \frac{1}{\sqrt{2}}(a_{1\alpha}^{\dagger} a_{2\beta}^{\dagger} + a_{2\alpha}^{\dagger} a_{1\beta}^{\dagger})|\text{vac}\rangle \tag{14E.3.2}$$

$$|3\rangle = a_{2\alpha}^{\dagger} a_{2\beta}^{\dagger}|\text{vac}\rangle \tag{14E.3.3}$$

In this representation, the Hamiltonian is given by (see Exercise 11.11)

$$\mathbf{H} = \begin{pmatrix} -2.84189825 & 0.00000000 & 0.14536406 \\ 0.00000000 & -1.75145864 & -0.29741518 \\ 0.14536406 & -0.29741518 & -0.57640710 \end{pmatrix} \tag{14E.3.4}$$

and the Hartree–Fock orbital energies are (see Exercise 10.4)

$$\varepsilon_1 = -1.63255846 \tag{14E.3.5}$$

$$\varepsilon_2 = -0.17271555 \tag{14E.3.6}$$

The nuclear-repulsion energy, which is included in the Hamiltonian (14E.3.4), is given by

$$h_{\text{nuc}} = 1.36631061 \tag{14E.3.7}$$

In the Møller–Plesset partitioning, set up the zero-order and perturbed Hamiltonians.

2. Determine the zero- and first-order energy corrections.

3. Determine the second- and third-order energy corrections.

4. Determine the fourth- and fifth-order energy corrections.

5. Calculate the total RSPT energies up to fifth order and compare these with the FCI energy of -2.85152628 E$_\text{h}$, obtained by diagonalization of the Hamiltonian $\mathbf{H}$.

6. Compare the total RSPT energies up to third order and the expectation value of $\mathbf{C}^{(0)} + \mathbf{C}^{(1)}$.

7. Consider the following approximation to the first-order wave-function correction:

$$\mathbf{C}^{(1)} = \begin{pmatrix} 0.00 \\ 0.00 \\ -0.05 \end{pmatrix} \tag{14E.3.8}$$

From this correction, calculate the second-order energy correction from the standard expression and from the Hylleraas functional

$$E^{(2)} = \mathbf{C}^{(0)^\mathrm{T}} \mathbf{U} \mathbf{C}^{(1)} \tag{14E.3.9}$$

$$J_\mathrm{H}^{(2)} = \mathbf{C}^{(1)^\mathrm{T}} (\mathbf{H}_0 - E_0 \mathbf{1}) \mathbf{C}^{(1)} + 2\mathbf{C}^{(1)^\mathrm{T}} \mathbf{U} \mathbf{C}^{(0)} \tag{14E.3.10}$$

Compare these energies with the exact second-order energy correction of Exercise 14.3.3.

EXERCISE 14.4

Verify the operator identity (14.4.14):

$$E_{ai}E_{bj}E_{ck} + E_{ai}E_{bk}E_{cj} + E_{aj}E_{bi}E_{ck} + E_{aj}E_{bk}E_{ci} + E_{ak}E_{bi}E_{cj} + E_{ak}E_{bj}E_{ci} = 0 \tag{14E.4.1}$$

EXERCISE 14.5

In this exercise, we investigate the convergence of perturbation theory for a two-state system with the following matrix representation of the Hamiltonian ($\beta > \alpha$):

$$\mathbf{H} = \begin{pmatrix} \alpha & \delta \\ \delta & \beta \end{pmatrix} \tag{14E.5.1}$$

1. Determine a closed-form expression for the energies of this Hamiltonian.

2. Let the diagonal of $\mathbf{H}$ define the zero-order system and the off-diagonal elements the perturbation:

$$\mathbf{H}_0 = \begin{pmatrix} \alpha & 0 \\ 0 & \beta \end{pmatrix} \tag{14E.5.2}$$

$$\mathbf{U} = \begin{pmatrix} 0 & \delta \\ \delta & 0 \end{pmatrix} \tag{14E.5.3}$$

By expanding the expression for the energy in Exercise 14.5.1 in orders of δ, determine the nth-order energy for the lowest state.

3. For fixed α and β, determine the interval of δ for which the expansion converges.

EXERCISE 14.6

In this exercise, we examine how RSPT expansions are affected by a universal diagonal shift of the zero-order Hamiltonian

$$\mathbf{H}_0(s) = \mathbf{H}_0 + s\mathbf{1} \tag{14E.6.1}$$

$$\mathbf{U}(s) = \mathbf{U} - s\mathbf{1} \tag{14E.6.2}$$

where s is a scalar. The energy and wave-function correction are written in the form:

$$E = \sum_{n=0}^{\infty} E^{(n)}(s) \tag{14E.6.3}$$

$$\mathbf{C} = \sum_{n=0}^{\infty} \mathbf{C}^{(n)}(s) \tag{14E.6.4}$$

1. Show that

$$\mathbf{C}^{(0)}(s) = \mathbf{C}^{(0)}(0) \tag{14E.6.5}$$

$$E^{(0)}(s) = E^{(0)}(0) + s \tag{14E.6.6}$$

$$E^{(1)}(s) = E^{(1)}(0) - s \tag{14E.6.7}$$

The sum of the zero- and first-order energies is thus independent of s:

$$E^{(0)}(s) + E^{(1)}(s) = E^{(0)}(0) + E^{(1)}(0) \tag{14E.6.8}$$

2. Show that, for the higher-order corrections,

$$\mathbf{C}^{(n)}(s) = \mathbf{C}^{(n)}(0), \qquad n \geq 1 \tag{14E.6.9}$$

$$E^{(n)}(s) = E^{(n)}(0), \qquad n > 1 \tag{14E.6.10}$$

A universal diagonal shift of the zero-order Hamiltonian thus affects only the zero- and first-order energy corrections. In particular, the convergence of the perturbation expansion is not affected by the shift.

EXERCISE 14.7

In Section 14.1.1, we derived general RSPT expressions for the perturbed energy (14.1.14) and the perturbed wave function (14.1.8) in the operator form ($n > 0$)

$$E^{(n)} = \langle 0^{(0)} | \hat{U} | 0^{(n-1)} \rangle \tag{14E.7.1}$$

$$(\hat{H}_0 - E^{(0)}) | 0^{(n)} \rangle = -\hat{U} | 0^{(n-1)} \rangle + \sum_{k=1}^{n} E^{(k)} | 0^{(n-k)} \rangle \tag{14E.7.2}$$

assuming the intermediate normalization (14.1.10):

$$\langle 0^{(0)} | 0^{(n)} \rangle = 0 \tag{14E.7.3}$$

A matrix formulation of these equations was developed in Section 14.1.2, using an orthonormal basis for the expansion of the perturbed wave functions (14.1.28). In this exercise, this matrix formulation is generalized to the expansion of the perturbed states

$$| 0^{(n)} \rangle = \sum_{i} C_i^{(n)} | i \rangle \tag{14E.7.4}$$

in a nonorthogonal basis, whose overlap elements are different from the Kronecker delta:

$$S_{ij} = \langle i | j \rangle \neq \delta_{ij} \tag{14E.7.5}$$

1. Show that the zero-order equations and the condition for intermediate normalization become

$$\mathbf{H}_0\mathbf{C}^{(0)} = E^{(0)}\mathbf{S}\mathbf{C}^{(0)} \tag{14E.7.6}$$

$$\mathbf{C}^{(0)^{\mathrm{T}}}\mathbf{S}\mathbf{C}^{(n)} = 0 \tag{14E.7.7}$$

2. Show that the nth-order corrections satisfy the equations

$$E^{(n)} = \mathbf{C}^{(0)^{\mathrm{T}}}\mathbf{U}\mathbf{C}^{(n-1)} \tag{14E.7.8}$$

$$(\mathbf{H}_0 - E^{(0)}\mathbf{S})\mathbf{C}^{(n)} = -\mathbf{U}\mathbf{C}^{(n-1)} + \sum_{k=1}^{n} E^{(k)}\mathbf{S}\mathbf{C}^{(n-k)} \tag{14E.7.9}$$

3. Show that the nth-order correction to the wave function (14E.7.9) may be written as

$$\mathbf{C}^{(n)} = -\mathbf{P}(\mathbf{H}_0 - E^{(0)}\mathbf{S})^{-1}\mathbf{P}\left(\mathbf{U}\mathbf{C}^{(n-1)} - \sum_{k=1}^{n-1} E^{(k)}\mathbf{S}\mathbf{C}^{(n-k)}\right) \tag{14E.7.10}$$

where the generalized projection matrix is given by

$$\mathbf{P} = \mathbf{1} - \mathbf{S}\mathbf{C}^{(0)}\mathbf{C}^{(0)^{\mathrm{T}}} \tag{14E.7.11}$$

4. In multiconfigurational perturbation theory, the overlap matrix takes the form

$$\mathbf{S} = \begin{pmatrix} \mathbf{1} & \mathbf{0} \\ \mathbf{0} & \mathbf{T} \end{pmatrix} \tag{14E.7.12}$$

where the unit matrix refers to the subspace that spans the zero-order state. Show that the generalized projection matrix (14E.7.11) then reduces to the usual projector

$$\mathbf{P} = \mathbf{1} - \mathbf{C}^{(0)}\mathbf{C}^{(0)^{\mathrm{T}}} \tag{14E.7.13}$$

Solutions

SOLUTION 14.1

1. Inserting (14E.1.2) in (14E.1.1) and rearranging, we obtain

$$(E\mathbf{1} - \mathbf{H}_0)\mathbf{C} = \mathbf{U}\mathbf{C} \tag{14S.1.1}$$

Next, we introduce (14E.1.4):

$$(E\mathbf{1} - \mathbf{H}_0)\mathbf{P}\mathbf{C} = \mathbf{U}\mathbf{C} - (E\mathbf{1} - \mathbf{H}_0)\mathbf{C}^{(0)} \tag{14S.1.2}$$

Using (14E.1.3), we arrive at (14E.1.6).

2. To obtain the expression for the energy (14E.1.7), we project (14E.1.6) from the left by $\mathbf{C}^{(0)^{\mathrm{T}}}$, noting that the left-hand side of (14E.1.6) then vanishes. Next, to arrive at the corresponding expression for the wave function (14E.1.8), we project (14E.1.6) from the left by $\mathbf{P}$ and obtain

$$(E\mathbf{1} - \mathbf{H}_0)\mathbf{P}\mathbf{C} = \mathbf{P}\mathbf{U}\mathbf{C} \tag{14S.1.3}$$

or equivalently

$$\mathbf{PC} = (E\mathbf{1} - \mathbf{H}_0)^{-1}\mathbf{PUC} \tag{14S.1.4}$$

Inserting this expression into (14E.1.4), we obtain (14E.1.8).

3. Iterating (14E.1.8) once, we obtain

$$\mathbf{C} = \mathbf{C}^{(0)} + (E\mathbf{1} - \mathbf{H}_0)^{-1}\mathbf{PUC}^{(0)} + (E\mathbf{1} - \mathbf{H}_0)^{-1}\mathbf{PU}(E\mathbf{1} - \mathbf{H}_0)^{-1}\mathbf{PUC} \tag{14S.1.5}$$

If this process is continued indefinitely, we arrive at (14E.1.10). The iterated expression for the energy (14E.1.9) is obtained by substituting (14E.1.10) into (14E.1.7).

4. Let p be a general index such that $p = 0$ for the reference state and $p > 0$ for all other states. In the intermediate normalization, the elements of $\mathbf{C}^{(0)}$ and $\mathbf{P}$ are then given by

$$C_p^{(0)} = \delta_{p0} \tag{14S.1.6}$$

$$P_{pq} = \delta_{pq} - \delta_{p0}\delta_{q0} \tag{14S.1.7}$$

Introducing these expressions into (14E.1.12) and carrying out some simple algebra, we obtain

$$C_p = \delta_{p0} + [(E\mathbf{1} - \mathbf{H}_0)^{-1}]_{pp}(U_{p0} - \delta_{p0}U_{00}) \tag{14S.1.8}$$

assuming that the zero-order Hamiltonian is diagonal. This equation is equivalent to (14E.1.13) and (14E.1.14), noting the relationship

$$U_{i0} = H_{i0} \tag{14S.1.9}$$

which again follows from the assumption of a diagonal zero-order Hamiltonian. For the second-order energy (14E.1.11), we obtain

$$E = E^{(0)} + \mathbf{C}^{(0)\mathrm{T}}\mathbf{UC} = E^{(0)} + U_{00} + \sum_i U_{0i}C_i = H_{00} + \sum_i H_{0i}C_i \tag{14S.1.10}$$

where we have used (14S.1.6) and (14S.1.9).

5. With the definition of the elements of $\mathbf{M}$ in (14E.1.17)–(14E.1.19), we write the BWPT expression for the energy (14E.1.15) in the form

$$M_{00} + \sum_i M_{0i}C_i = E \tag{14S.1.11}$$

Next, the equation for the coefficients (14E.1.14) can be written in the form

$$[\mathbf{H}_0\mathbf{C}]_i + H_{i0} = EC_i \tag{14S.1.12}$$

which corresponds to

$$M_{i0} + \sum_j M_{ij}C_j = EC_i \tag{14S.1.13}$$

Combining (14S.1.11) and (14S.1.13), we obtain the eigenvalue problem (14E.1.16).

6. According to (14E.1.16), second-order BWPT is equivalent to CI theory, where the off-diagonal Hamiltonian matrix elements involving only states orthogonal to the reference state are approximated by the zero-order Hamiltonian. We have previously shown that the CI method is not in general size-extensive; in general, therefore, BWPT is not size-extensive either.

SOLUTION 14.2

1. From Section 4.2.2, we recall that, if the wave function $|0\rangle$ has an error $|\delta\rangle$, then the energy calculated as an expectation value is correct to second order in $|\delta\rangle$. Since $|0_n\rangle$ has an error of order $n + 1$, then the energy (14E.2.2) has an error of order $2n + 2$ and is therefore correct to order $2n + 1$.

2. Expanding the numerator of the expectation value, we obtain

$$\langle 0_1|\hat{H}|0_1\rangle = \langle 0^{(0)}|\hat{H}|0^{(0)}\rangle + 2\langle 0^{(0)}|\hat{H}|0^{(1)}\rangle + \langle 0^{(1)}|\hat{H}|0^{(1)}\rangle \tag{14S.2.1}$$

The first two contributions are easily identified from (14.1.13) and (14.1.14):

$$\langle 0^{(0)}|\hat{H}|0^{(0)}\rangle = \langle 0^{(0)}|\hat{H}_0|0^{(0)}\rangle + \langle 0^{(0)}|\hat{U}|0^{(0)}\rangle = E^{(0)} + E^{(1)} \tag{14S.2.2}$$

$$\langle 0^{(0)}|\hat{H}|0^{(1)}\rangle = \langle 0^{(0)}|\hat{U}|0^{(1)}\rangle = E^{(2)} \tag{14S.2.3}$$

The third contribution to (14S.2.1) may be rewritten in the form

$$\langle 0^{(1)}|\hat{H}|0^{(1)}\rangle = \langle 0^{(1)}|\hat{H}_0 - E^{(0)}|0^{(1)}\rangle + E^{(0)}\langle 0^{(1)}|0^{(1)}\rangle + \langle 0^{(1)}|\hat{U} - E^{(1)}|0^{(1)}\rangle + E^{(1)}\langle 0^{(1)}|0^{(1)}\rangle \tag{14S.2.4}$$

Identifying the first term in this expression from (14.1.8) and (14.1.14) and the third term from (14.1.26)

$$\langle 0^{(1)}|\hat{H}_0 - E^{(0)}|0^{(1)}\rangle = -\langle 0^{(1)}|\hat{U}|0^{(0)}\rangle = -E^{(2)} \tag{14S.2.5}$$

$$\langle 0^{(1)}|\hat{U} - E^{(1)}|0^{(1)}\rangle = E^{(3)} \tag{14S.2.6}$$

we obtain

$$\langle 0^{(1)}|\hat{H}|0^{(1)}\rangle = (E^{(0)} + E^{(1)})\langle 0^{(1)}|0^{(1)}\rangle - E^{(2)} + E^{(3)} \tag{14S.2.7}$$

Substituting (14S.2.2), (14S.2.3) and (14S.2.7) into (14S.2.1) and dividing by

$$\langle 0_1|0_1\rangle = 1 + \langle 0^{(1)}|0^{(1)}\rangle \tag{14S.2.8}$$

we arrive at (14E.2.4).

SOLUTION 14.3

1. In the Møller–Plesset partitioning of the Hamiltonian, the zero-order operator (including the nuclear repulsion) is given by:

$$\hat{H}_0 = \varepsilon_1 E_{11} + \varepsilon_2 E_{22} + h_{\text{nuc}} \tag{14S.3.1}$$

Evaluating the matrix elements for the states (14E.3.1)–(14E.3.3), we obtain the following zero-order Hamiltonian matrix

$$\mathbf{H}_0 = \begin{pmatrix} 2\varepsilon_1 + h_{\text{nuc}} & 0 & 0 \\ 0 & \varepsilon_1 + \varepsilon_2 + h_{\text{nuc}} & 0 \\ 0 & 0 & 2\varepsilon_2 + h_{\text{nuc}} \end{pmatrix}$$

$$= \begin{pmatrix} -1.89880632 & 0.00000000 & 0.00000000 \\ 0.00000000 & -0.43896341 & 0.00000000 \\ 0.00000000 & 0.00000000 & 1.02087951 \end{pmatrix} \tag{14S.3.2}$$

The perturbation is consequently represented by the matrix

$$\mathbf{U} = \mathbf{H} - \mathbf{H}_0 = \begin{pmatrix} -0.94309194 & 0.00000000 & 0.14536406 \\ 0.00000000 & -1.31249524 & -0.29741518 \\ 0.14536406 & -0.29741518 & -1.59728660 \end{pmatrix} \tag{14S.3.3}$$

2. The zero-order Møller–Plesset state (i.e. the Hartree–Fock determinant) is represented by

$$\mathbf{C}^{(0)} = \begin{pmatrix} 1 \\ 0 \\ 0 \end{pmatrix} \tag{14S.3.4}$$

from which we obtain the following zero- and first-order energy corrections:

$$E^{(0)} = \mathbf{C}^{(0)^{\mathrm{T}}} \mathbf{H}_0 \mathbf{C}^{(0)} = -1.89880632 \tag{14S.3.5}$$

$$E^{(1)} = \mathbf{C}^{(0)^{\mathrm{T}}} \mathbf{U} \mathbf{C}^{(0)} = -0.94309194 \tag{14S.3.6}$$

3. First, we calculate the first-order wave-function correction from (14.1.42), obtaining

$$\mathbf{C}^{(1)} = -(\mathbf{H}_0 - E^{(0)}\mathbf{1})^{-1}(\mathbf{U}\mathbf{C}^{(0)} - E^{(1)}\mathbf{C}^{(0)}) = \begin{pmatrix} 0.00000000 \\ 0.00000000 \\ -0.04978757 \end{pmatrix} \tag{14S.3.7}$$

which is orthogonal to $\mathbf{C}^{(0)}$. Next, the second- and third-order energies are obtained from (14.1.63) and (14.1.64):

$$E^{(2)} = \mathbf{C}^{(0)^{\mathrm{T}}} \mathbf{U} \mathbf{C}^{(1)} = -0.00723732 \tag{14S.3.8}$$

$$E^{(3)} = \mathbf{C}^{(1)^{\mathrm{T}}} (\mathbf{U} - E^{(1)}\mathbf{1}) \mathbf{C}^{(1)} = -0.00162162 \tag{14S.3.9}$$

4. According to (14.1.42), the second-order correction to the wave function is given by

$$\mathbf{C}^{(2)} = -(\mathbf{H}_0 - E^{(0)}\mathbf{1})^{-1}(\mathbf{U}\mathbf{C}^{(1)} - E^{(1)}\mathbf{C}^{(1)} - E^{(2)}\mathbf{C}^{(0)}) = \begin{pmatrix} 0.00000000 \\ -0.01014327 \\ -0.01115557 \end{pmatrix} \tag{14S.3.10}$$

which gives us the following fourth- and fifth-order energy corrections (14.1.65) and (14.1.66):

$$E^{(4)} = \mathbf{C}^{(1)^{\mathrm{T}}} (\mathbf{U} - E^{(1)}\mathbf{1}) \mathbf{C}^{(2)} - E^{(2)} \mathbf{C}^{(1)^{\mathrm{T}}} \mathbf{C}^{(1)} = -0.00049560 \tag{14S.3.11}$$

$$E^{(5)} = \mathbf{C}^{(2)^{\mathrm{T}}} (\mathbf{U} - E^{(1)}\mathbf{1}) \mathbf{C}^{(2)} - 2E^{(2)} \mathbf{C}^{(1)^{\mathrm{T}}} \mathbf{C}^{(2)} - E^{(3)} \mathbf{C}^{(1)^{\mathrm{T}}} \mathbf{C}^{(1)} = -0.00017467 \tag{14S.3.12}$$

5. The total RSPT energies up to fifth order are listed in Table 14S.3.1. The convergence towards the FCI energy is monotonic, with a reduction in the error by a factor of 3–4 in each iteration.

Table 14S.3.1 The total RSPT energies up to fifth order and their differences relative to the FCI energy and to the expectation value of the first-order wave function (E_h)

n	E_{RSPT}	$E_{\mathrm{RSPT}} - E_{\mathrm{FCI}}$	$E_{\mathrm{RSPT}} - E(\mathbf{C}^{(0)} + \mathbf{C}^{(1)})$
1	-2.84189825	0.00962802	0.00883704
2	-2.84913558	0.00239070	0.00159971
3	-2.85075720	0.00076908	-0.00002191
4	-2.85125280	0.00027348	-0.00051751
5	-2.85142747	0.00009881	-0.00069218

6. The expectation value of the first-order energy $\mathbf{C}^{(0)} + \mathbf{C}^{(1)}$ is given by

$$E(\mathbf{C}^{(0)} + \mathbf{C}^{(1)}) = \frac{(\mathbf{C}^{(0)} + \mathbf{C}^{(1)})^{\mathrm{T}}\mathbf{H}(\mathbf{C}^{(0)} + \mathbf{C}^{(1)})}{1 + \mathbf{C}^{(1)\mathrm{T}}\mathbf{C}^{(1)}} = -2.85073529 \tag{14S.3.13}$$

As seen from Table 14S.3.1, the third-order RSPT energy is close to the expectation value of $\mathbf{C}^{(0)} + \mathbf{C}^{(1)}$, in agreement with the results of Exercise 14.2.

7. Carrying out the calculations with the approximate first-order state (14E.3.8), we obtain the following energies:

$$E^{(2)} = -0.00726820 \tag{14S.3.14}$$

$$J_{\mathrm{H}}^{(2)} = -0.00723719 \tag{14S.3.15}$$

Comparing with the exact second-order energy correction of -0.00723732 obtained from the exact first-order state in Exercise 14.3.3, we find that, whereas the usual expression gives an error of -0.00003088, the Hylleraas functional gives an error of only 0.00000013. Note that the Hylleraas functional gives a positive error.

SOLUTION 14.4

Expanding each excitation operator in creation and annihilation operators and collecting terms, we obtain

$$E_{ai}E_{bj}E_{ck} + E_{ai}E_{bk}E_{cj} + E_{aj}E_{bi}E_{ck} + E_{aj}E_{bk}E_{ci} + E_{ak}E_{bi}E_{cj} + E_{ak}E_{bj}E_{ci}$$

$$= \sum_{\sigma\tau\delta} a_{a\sigma}^{\dagger}a_{i\sigma}a_{b\tau}^{\dagger}a_{j\tau}a_{c\delta}^{\dagger}a_{k\delta} + \sum_{\sigma\tau\delta} a_{a\sigma}^{\dagger}a_{i\sigma}a_{b\tau}^{\dagger}a_{k\tau}a_{c\delta}^{\dagger}a_{j\delta} + \sum_{\sigma\tau\delta} a_{a\sigma}^{\dagger}a_{j\sigma}a_{b\tau}^{\dagger}a_{i\tau}a_{c\delta}^{\dagger}a_{k\delta}$$

$$+ \sum_{\sigma\tau\delta} a_{a\sigma}^{\dagger}a_{j\sigma}a_{b\tau}^{\dagger}a_{k\tau}a_{c\delta}^{\dagger}a_{i\delta} + \sum_{\sigma\tau\delta} a_{a\sigma}^{\dagger}a_{k\sigma}a_{b\tau}^{\dagger}a_{i\tau}a_{c\delta}^{\dagger}a_{j\delta} + \sum_{\sigma\tau\delta} a_{a\sigma}^{\dagger}a_{k\sigma}a_{b\tau}^{\dagger}a_{j\tau}a_{c\delta}^{\dagger}a_{i\delta}$$

$$= -\sum_{\sigma\tau\delta} a_{a\sigma}^{\dagger}a_{b\tau}^{\dagger}a_{c\delta}^{\dagger} \begin{vmatrix} a_{i\sigma} & a_{i\tau} & a_{i\delta} \\ a_{j\sigma} & a_{j\tau} & a_{j\delta} \\ a_{k\sigma} & a_{k\tau} & a_{k\delta} \end{vmatrix} = 0 \tag{14S.4.1}$$

The determinant vanishes since the indices σ, τ and δ cannot all be different.

SOLUTION 14.5

1. In the intermediate normalization, the eigenvalue problem becomes

$$\begin{pmatrix} \alpha & \delta \\ \delta & \beta \end{pmatrix} \begin{pmatrix} 1 \\ x \end{pmatrix} = E \begin{pmatrix} 1 \\ x \end{pmatrix} \tag{14S.5.1}$$

which is equivalent to the following two equations:

$$\alpha + \delta x = E \tag{14S.5.2}$$

$$\delta + \beta x = Ex \tag{14S.5.3}$$

Eliminating x from these equations, we obtain

$$(E - \alpha)(E - \beta) - \delta^2 = 0 \tag{14S.5.4}$$

which has the two solutions

$$E = \frac{\alpha + \beta}{2} \pm \frac{1}{2}\sqrt{(\beta - \alpha)^2 + 4\delta^2} \tag{14S.5.5}$$

2. We first rewrite the lowest solution in (14S.5.5) in the form:

$$E = \frac{\alpha + \beta}{2} - \frac{\beta - \alpha}{2}\sqrt{1 + \left(\frac{2\delta}{\beta - \alpha}\right)^2} \tag{14S.5.6}$$

Introducing the Taylor expansion

$$\sqrt{1 + x} = \sum_{n=0}^{\infty}(-1)^{n-1}\frac{(2n-3)!!}{2^n n!}x^n \tag{14S.5.7}$$

where the double factorial function is defined in (6.5.10), we arrive at the following expansion of the energy (14S.5.6)

$$E = \frac{\alpha + \beta}{2} - \frac{\beta - \alpha}{2}\left[1 + \sum_{n=1}^{\infty}(-1)^{n-1}\frac{(2n-3)!!}{2^n n!}\frac{(2\delta)^{2n}}{(\beta - \alpha)^{2n}}\right]$$

$$= \alpha + \sum_{n=1}^{\infty}(-1)^n\frac{(2n-3)!!2^{n-1}}{n!}\frac{\delta^{2n}}{(\beta - \alpha)^{2n-1}} \tag{14S.5.8}$$

which gives (14.5.7)–(14.5.9).

3. The expansion of the energy converges if the Taylor expansion (14S.5.7) converges. Since this expansion converges for $|x| < 1$, the expansion of the energy converges whenever the condition

$$\frac{4\delta^2}{(\beta - \alpha)^2} < 1 \tag{14S.5.9}$$

is satisfied. This condition is equivalent to (14.5.11).

SOLUTION 14.6

1. We first note that, since $\mathbf{H}_0$ and $\mathbf{H}_0 + s\mathbf{1}$ have the same eigenvectors, the zero-order state is not affected by the shift:

$$\mathbf{C}^{(0)}(s) = \mathbf{C}^{(0)}(0) = \mathbf{C}^{(0)} \tag{14S.6.1}$$

From (14.1.33) and (14.1.41), we then obtain

$$E^{(0)}(s) = \mathbf{C}^{(0)\mathrm{T}}\mathbf{H}_0(s)\mathbf{C}^{(0)} = \mathbf{C}^{(0)\mathrm{T}}\mathbf{H}_0\mathbf{C}^{(0)} + s\mathbf{C}^{(0)\mathrm{T}}\mathbf{C}^{(0)} = E^{(0)}(0) + s \tag{14S.6.2}$$

$$E^{(1)}(s) = \mathbf{C}^{(0)\mathrm{T}}\mathbf{U}(s)\mathbf{C}^{(0)} = \mathbf{C}^{(0)\mathrm{T}}\mathbf{U}\mathbf{C}^{(0)} - s\mathbf{C}^{(0)\mathrm{T}}\mathbf{C}^{(0)} = E^{(1)}(0) - s \tag{14S.6.3}$$

2. From (14.1.42), we find that the first-order wave function is unaffected by the shift:

$$\mathbf{C}^{(1)}(s) = -\mathbf{P}[\mathbf{H}_0(s) - E^{(0)}(s)\mathbf{1}]^{-1}\mathbf{P}\mathbf{U}(s)\mathbf{C}^{(0)}$$

$$= -\mathbf{P}(\mathbf{H}_0 + s\mathbf{1} - E^{(0)}\mathbf{1} - s\mathbf{1})^{-1}\mathbf{P}(\mathbf{U} - s\mathbf{1})\mathbf{C}^{(0)}$$

$$= -\mathbf{P}(\mathbf{H}_0 - E^{(0)}\mathbf{1})^{-1}\mathbf{P}\mathbf{U}\mathbf{C}^{(0)}$$

$$= \mathbf{C}^{(1)}(0) \tag{14S.6.4}$$

We now proceed by induction, assuming that the relation

$$\mathbf{C}^k(s) = \mathbf{C}^{(k)}(0) \tag{14S.6.5}$$

holds for $k \leq m$. We then obtain from (14.1.41) and (14.1.42)

$$E^{(m+1)}(s) = \mathbf{C}^{(0)\mathrm{T}}\mathbf{U}(s)\mathbf{C}^{(m)}(s) = \mathbf{C}^{(0)\mathrm{T}}\mathbf{U}\mathbf{C}^{(m)}(0) - s\mathbf{C}^{(0)\mathrm{T}}\mathbf{C}^{(m)}(0) = E^{(m+1)}(0) \quad (14\mathrm{S}.6.6)$$

$$\mathbf{C}^{(m+1)}(s) = -\mathbf{P}[\mathbf{H}_0(s) - E_0(s)\mathbf{1}]^{-1}\mathbf{P}\left[\mathbf{U}(s)\mathbf{C}^{(m)}(s) - \sum_{k=1}^{m} E^{(k)}(s)\mathbf{C}^{(m+1-k)}(s)\right]$$

$$= -\mathbf{P}[\mathbf{H}_0(0) - E_0(0)\mathbf{1}]^{-1}\mathbf{P}\left[\mathbf{U}(0)\mathbf{C}^{(m)}(0) - \sum_{k=1}^{m} E^{(k)}(0)\mathbf{C}^{(m+1-k)}(0)\right]$$

$$= \mathbf{C}^{(m+1)}(0) \quad (14\mathrm{S}.6.7)$$

In (14S.6.7), the level shifts in $\mathbf{H}_0(s)$ and $\mathbf{U}(s)$ are cancelled by the same shifts in $E^{(0)}(s)$ and $E^{(1)}(s)$ – see (14E.6.6) and (14E.6.7). Since (14S.6.5) holds for $m = 1$, the induction is complete.

SOLUTION 14.7

1. The zero-order equation is given by

$$\hat{H}_0|0^{(0)}\rangle = E^{(0)}|0^{(0)}\rangle \quad (14\mathrm{S}.7.1)$$

Inserting the expansion (14E.7.4) in (14S.7.1) and projecting from the left by $\langle j|$, we obtain

$$\sum_k \langle j|\hat{H}_0|k\rangle C_k^{(0)} = E^{(0)} \sum_k \langle j|k\rangle C_k^{(0)} \quad (14\mathrm{S}.7.2)$$

which, in matrix notation, is equivalent to (14E.7.6). Likewise, expanding $\mathbf{C}^{(n)}$ and $\mathbf{C}^{(0)}$ in (14E.7.3), we obtain the condition

$$\sum_{jk} C_j^{(0)}\langle j|k\rangle C_k^{(n)} = 0 \quad (14\mathrm{S}.7.3)$$

which is equivalent to (14.E.7.7).

2. Relation (14E.7.8) is obtained by inserting the expansions for the wave-function corrections in (14E.7.1) and expressing the result in matrix notation. Likewise, expanding the corrections in (14E.7.2) in $|k\rangle$, projecting from the left by $\langle j|$ and introducing matrix notation, we arrive at (14E.7.9).

3. From (14E.7.8), it follows that the right-hand side of (14E.7.9) is orthogonal to $\mathbf{C}^{(0)}$. We may then write (14E.7.9) in the form

$$\mathbf{C}^{(n)} = -(\mathbf{H}_0 - E^{(0)}\mathbf{S})^{-1}\left(\mathbf{U}\mathbf{C}^{(n-1)} - \sum_{k=1}^{n} E^{(k)}\mathbf{S}\mathbf{C}^{(n-k)}\right)$$

$$= -(\mathbf{H}_0 - E^{(0)}\mathbf{S})^{-1}\left(\mathbf{U}\mathbf{C}^{(n-1)} - \mathbf{C}^{(0)\mathrm{T}}\mathbf{U}\mathbf{C}^{(n-1)}\mathbf{S}\mathbf{C}^{(0)} - \sum_{k=1}^{n-1} E^{(k)}\mathbf{S}\mathbf{C}^{(n-k)}\right)$$

$$= -(\mathbf{H}_0 - E^{(0)}\mathbf{S})^{-1}(\mathbf{1} - \mathbf{S}\mathbf{C}^{(0)}\mathbf{C}^{(0)\mathrm{T}})\left(\mathbf{U}\mathbf{C}^{(n-1)} - \sum_{k=1}^{n-1} E^{(k)}\mathbf{S}\mathbf{C}^{(n-k)}\right) \quad (14\mathrm{S}.7.4)$$

Since, for all $\mathbf{X}$, $(\mathbf{H}_0 - E^{(0)}\mathbf{S})^{-1}\mathbf{PX}$ is orthogonal to $\mathbf{C}^{(0)}$, we may write this equation in the form (14E.7.10).

4. From (14E.7.12), we obtain

$$\mathbf{SC}^{(0)} = \mathbf{C}^{(0)} \tag{14S.7.5}$$

which shows that (14E.7.11) reduces to (14E.7.13) for this particular metric.

15 CALIBRATION OF THE ELECTRONIC-STRUCTURE MODELS

In *ab initio* electronic-structure calculations, approximate solutions are obtained to the molecular electronic Schrödinger equation. The errors made in such calculations arise from the truncation of the one-electron space discussed in Chapters 6–8 and from the approximate treatment of the N-electron space discussed in Chapters 10–14. In the present chapter, we shall make an extensive assessment of the methods, presenting investigations and comparisons of calculations made using standard basis sets and wave functions. Clearly, the usefulness of such investigations increases if they are performed for a large variety of molecular systems, comprising a number of different prototypical systems, and if the results are subjected to an appropriate statistical analysis. With this in mind, we shall examine the performance of the different hierarchies of one- and N-electron models to establish what levels are sufficient to achieve a given precision. In particular, we shall investigate what accuracy is obtainable in calculations of molecular equilibrium geometries, dipole moments, atomization energies, reaction enthalpies and conformational barriers.

15.1 The sample molecules

The calculations and statistical analysis in this chapter will be carried out for the 20 molecules listed in Table 15.1. These molecules contain a variety of chemical bonds between hydrogen and first-row atoms as well as single, double and triple bonds between two first-row atoms. We note, however, that the investigation concerns only closed-shell molecules containing first-row atoms. Therefore, the results do not necessarily carry over to open-shell molecules or to molecules containing heavier elements such as transition-metal compounds.

The methods employed in this chapter all require, to some extent at least, the wave function to be dominated by the Hartree–Fock configuration, with small contributions from static correlation. To see how well these methods work for the molecules in Table 15.1, we shall here probe their Hartree–Fock dominance by examining their natural-orbital occupation numbers. In Table 15.2, we have listed the natural-orbital occupation numbers most relevant for this purpose – namely, the lowest occupation number of the orbitals that are occupied in the dominant (usually Hartree–Fock) configuration $n_{\min}^{\mathrm{occ}}$ and the highest occupation number of the orbitals not present in this configuration $n_{\max}^{\mathrm{vir}}$. These occupation numbers have been obtained from the variational CCSD(T) one-electron density matrix (see Section 13.5.3), calculated in the cc-pVQZ basis at the optimized geometries with all electrons correlated.

Table 15.1 The molecules used in the statistical investigations of molecular equilibrium bond lengths R_e and bond angles Θ_e, dipole moments μ_e, atomization energies D_e and reaction enthalpies $\Delta_r H_e^\circ(0K)$. The molecules included in the statistical investigation of a given property are marked with a cross

	R_e	Θ_e	μ_e	D_e	$\Delta_r H_e^\circ(0K)$
H_2	×			×	×
CH_2	×	×	×	×	×
CH_4	×			×	×
H_2O	×	×	×	×	×
NH_3	×	×	×	×	×
HF	×		×	×	×
N_2	×			×	×
F_2	×			×	×
CO	×		×	×	×
HCN	×		×	×	×
HNC	×		×		
C_2H_2	×			×	×
C_2H_4	×	×		×	×
CH_2O	×	×	×	×	×
HNO			×	×	×
N_2H_2	×	×			
HOF	×	×	×		
H_2O_2	×		×		
CO_2	×			×	×
O_3	×	×	×	×	×

Table 15.2 Natural-orbital occupation numbers calculated from variational densities at the all-electron CCSD(T)/cc-pVQZ level

	n_{min}^{occ}	n_{max}^{vir}
H_2	1.964	0.020
CH_4	1.954	0.023
NH_3	1.952	0.027
HF	1.962	0.028
H_2O	1.957	0.029
CO	1.932	0.065
HNC	1.923	0.066
C_2H_2	1.908	0.068
CO_2	1.933	0.069
HCN	1.911	0.072
N_2	1.917	0.074
CH_2	1.895	0.077
CH_2O	1.910	0.077
H_2O_2	1.915	0.078
C_2H_4	1.896	0.078
HOF	1.905	0.092
N_2H_2	1.893	0.097
HNO	1.891	0.107
F_2	1.883	0.118
O_3	1.806	0.218

The O_3 molecule has a large static contribution to the correlation energy, with a virtual occupation number as large as 0.22. Moreover, with virtual occupation numbers greater than 0.09, F_2, HNO, N_2H_2 and HOF have significant static contributions as well. Only for H_2, CH_4, NH_3, HF and H_2O is the static contribution negligible, the virtual occupation numbers being smaller than 0.03. Our sample thus comprises molecules with a varying degree of Hartree–Fock dominance and should in this respect be representative of the systems that are encountered in problems of chemical interest.

15.2 Errors in quantum-chemical calculations

Before we consider the different properties analysed in the present chapter, it is appropriate to discuss in general terms the errors that arise in quantum-chemical calculations. Thus, in Section 15.2.1, we discuss the sources of errors in quantum-chemical calculations, introducing the concepts of apparent and intrinsic errors. Next, in Section 15.2.2, we review the measures of errors that we shall employ in our statistical investigations in this chapter.

15.2.1 APPARENT AND INTRINSIC ERRORS

In electronic-structure calculations, it is important to distinguish between the different types of errors introduced by the approximations made in the one- and N-electron spaces. In the present chapter, in particular, we shall distinguish between the following kinds of errors: the apparent error, the basis-set error, the N-electron error and the intrinsic N-electron error [1]. The relationships between these errors are illustrated in Figure 15.1. Thus, in a given calculation, the *apparent error* is simply the error relative to the true solution to the Schrödinger equation as obtained by carrying out an FCI calculation in a complete basis. The apparent error is often referred to as the computational error or simply the error of the calculation. For a given N-particle model, the *basis-set error* is the difference between the result in a given one-electron basis and the result obtained in a complete basis for the same N-electron model. Similarly, in a given one-electron basis, the

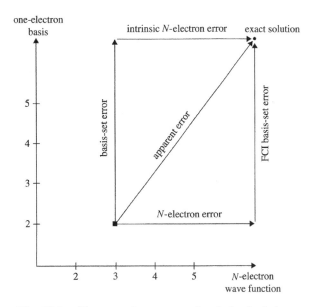

Fig. 15.1. The errors in quantum-chemical calculations.

N-electron error is the difference between the result obtained with a given *N*-electron model and the result that would be obtained using the FCI wave function in the same basis set. As illustrated in Figure 15.1, we may think of the apparent error as the combined effect of the basis-set error and the *N*-electron error. It should be noted that the *N*-electron error depends on the basis set employed in the calculation. Of interest is also the *N*-electron error in the limit of a complete basis, which we shall refer to as the *intrinsic N-electron error* or simply the *intrinsic error*.

The errors arising from the truncation of the one-electron space and from the approximate treatment in the *N*-electron space are often oppositely directed, giving considerable scope for cancellation of errors, in particular at low levels of theory. Although we must often rely on the cancellation of errors in electronic-structure calculations, it is particularly difficult in this case because the errors have different physical origins. Specific studies must therefore be carried out in each individual case to justify any cancellation.

After the intrinsic and apparent errors of the standard models and basis sets have been established, this knowledge may be used in the design of *balanced calculations*, where the quality of the basis set matches the quality of the *N*-electron model. Consider the calculation depicted in Figure 15.2. From basis 2 to basis 3, a significant improvement in the calculated property is observed compared with the intrinsic error of the model; beyond basis 4, the basis-set change is small compared with the intrinsic error. Thus, little improvement is observed beyond basis 4, which is said to represent a *balanced level of description*. Beyond this basis, the accuracy of the description cannot be substantially improved upon because the basis-set error is small compared with the intrinsic error of the wave-function model. The calculation is balanced in the sense that

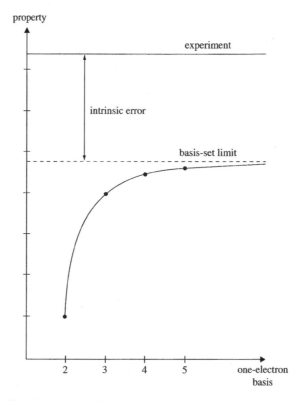

Fig. 15.2.   The dependence of the calculated property on the quality of the basis.

a higher accuracy requires improvements in the correlation description as well as in the basis. Indeed, the use of a larger basis in the same N-electron model does not improve the results, it only makes the calculation more expensive.

15.2.2 STATISTICAL MEASURES OF ERRORS

To quantify the errors in the calculations, we shall employ several statistical measures. Let the calculated property for a given model and for a given basis set be denoted by P_i^{calc} and let the corresponding experimental value be P_i^{exp}. The error is then given by

$$\Delta_i = P_i^{\mathrm{calc}} - P_i^{\mathrm{exp}} \tag{15.2.1}$$

The *mean error* $\overline{\Delta}$, the *standard deviation* Δ_{std}, the *mean absolute error* $\overline{\Delta}_{\mathrm{abs}}$ and the *maximum absolute error* Δ_{max} are then given by

$$\overline{\Delta} = \frac{1}{n} \sum_{i=1}^{n} \Delta_i \tag{15.2.2}$$

$$\Delta_{\mathrm{std}} = \sqrt{\frac{1}{n-1} \sum_{i=1}^{n} (\Delta_i - \overline{\Delta})^2} \tag{15.2.3}$$

$$\overline{\Delta}_{\mathrm{abs}} = \frac{1}{n} \sum_{i=1}^{n} |\Delta_i| \tag{15.2.4}$$

$$\Delta_{\mathrm{max}} = \max_i |\Delta_i| \tag{15.2.5}$$

Each measure characterizes a specific aspect of the performance of the model and the basis set. The standard deviation Δ_{std} characterizes the distribution of errors $\Delta_i - \overline{\Delta}$ about a mean value; $\overline{\Delta}$ and Δ_{std} thus quantify systematic and nonsystematic errors. The mean absolute error $\overline{\Delta}_{\mathrm{abs}}$ represents the typical magnitude of the error in the calculations and Δ_{max} gives the largest error. Graphically, the mean error and standard deviation can be represented in terms of the *normal distribution*

$$\rho(P) = N_{\mathrm{c}} \exp\left[-\frac{1}{2} \left(\frac{P - \overline{\Delta}}{\Delta_{\mathrm{std}}} \right)^2 \right] \tag{15.2.6}$$

where N_{c} is a normalization constant. In the following sections, we shall frequently plot such normal distributions to illustrate the performance of the different models and basis sets.

15.3 Molecular equilibrium structures: bond distances

In this section, we investigate the accuracy of *ab initio* electronic-structure predictions of bond distances [2]. In Section 15.3.1, we review the experimental data on which the investigation is based. Next, in Sections 15.3.2–15.3.5, we consider the statistical measures of errors, piecing together a picture of the performance of the standard models with respect to the calculation of equilibrium bond distances. After a discussion of higher-order effects at the CCSDT level in Section 15.3.6 and core correlation in Section 15.3.7, we attempt to rationalize the behaviour of the different models in Section 15.3.8. We conclude our discussion in Section 15.3.9, which contains a table of bond distances, calculated using the most important N-electron models in a large one-electron basis.

15.3.1 EXPERIMENTAL BOND DISTANCES

The experimental bond lengths are listed in Table 15.3. All distances are equilibrium values R_e except for HNO, for which only the vibrationally averaged distances R_0 are available. Since R_e and R_0 may be rather different – in particular, for the anharmonic bonds to hydrogen – the HNO distances have been excluded from the statistical sample.

For the rigid molecules in the sample, the experimental uncertainties (where available) are 0.3 pm or smaller. For the nonrigid molecules NH_3 and H_2O_2, the accurate measurement of bond distances is more difficult. Indeed, for NH_3 (for which we have used an old measurement that recent theoretical studies suggest is the most accurate), the uncertainty is as large as 0.6 pm; for H_2O_2, the uncertainty is unknown but we note that a recent experimental study has corrected previous experimental distances by about 2 pm [2].

Since the bond lengths of larger molecules are usually less accurately determined, the molecules in Table 15.3 constitute a good sample for investigating the performance of the standard electronic-structure methods, at least for molecules containing only first-row atoms.

15.3.2 MEAN ERRORS AND STANDARD DEVIATIONS

In the present subsection, we consider the mean errors and standard deviations of bond distances calculated using the Hartree–Fock, MP2, MP3, MP4, CCSD, CCSD(T) and CISD models in the cc-pVDZ, cc-pVTZ and cc-pVQZ sets with all electrons correlated. The mean errors and standard deviations are listed in Table 15.4 and plotted in Figure 15.3. As discussed in Section 15.3.1, the sample comprises all bonds in Table 15.3 except those of HNO.

From Figure 15.3, we see that the bonds usually lengthen with improvements in the N-electron description and shorten with improvements in the one-electron basis. Going from the cc-pVDZ basis to the cc-pVTZ basis, the bonds contract by about 0.8 pm at the Hartree–Fock level and by 1.6 pm at the correlated levels. From cc-pVTZ to cc-pVQZ, the contraction is much smaller – of the order of 0.1 pm for all methods. Clearly, for most purposes, the cc-pVTZ basis gives results sufficiently close to the basis-set limit; the cc-pVDZ basis, on the other hand, gives large errors.

For all basis sets, the bonds lengthen in the sequence Hartree–Fock, CISD, MP3, CCSD, MP2, CCSD(T), MP4. Moreover, relative to experiment, the Hartree–Fock bonds are always too short and the MP4 bonds always too long. We also note that, even though improvements in the correlation treatment tend to lengthen the bonds, the Møller–Plesset series oscillates, with bond lengths increasing in the order MP3, MP2, MP4.

Since improvements in the one- and N-electron descriptions affect bond lengths in opposite directions, there is considerable scope for cancellation of errors. Thus, at the cc-pVDZ level, the CISD bond lengths are quite accurate in the mean, with an error of just $+0.3$ pm. However, as the one-electron description improves, the CISD bonds shorten and the error is -1.7 pm at the cc-pVQZ level. The MP3 model behaves similarly, with mean errors of $+0.5$ and -1.2 pm at the cc-pVDZ and cc-pVQZ levels, respectively. By comparison, the CCSD(T) error is $+1.7$ pm at the cc-pVDZ level but only -0.1 pm at the cc-pVQZ level.

Clearly, the CISD and MP3 models are not sufficiently flexible to yield accurate bond lengths with large basis sets and should not be used for the calculation of molecular structures. The good

Table 15.3 Experimental equilibrium bond lengths R_e in pm

Molecule	Bond	Bond length
H_2	R_{HH}	74.144[a]
HF	R_{FH}	91.680(8)[a]
H_2O	R_{OH}	95.72[b]
HOF	R_{OH}	96.57(16)[c]
H_2O_2	R_{OH}	96.7[d]
HNC	R_{NH}	99.40(8)[e]
NH_3	R_{NH}	101.1(6)[f]
N_2H_2	R_{NH}	102.9(1)[g]
C_2H_2	R_{CH}	106.215(17)[h]
HNO	R_{NH} (R_0)	106.2(8)[i]
HCN	R_{CH}	106.501(8)[j]
C_2H_4	R_{CH}	108.1(2)[k]
CH_4	R_{CH}	108.58(10)[l]
N_2	R_{NN}	109.768(5)[a]
CH_2O	R_{CH}	110.05(20)[k]
CH_2	R_{CH}	110.7(2)[m]
CO	R_{CO}	112.832[a]
HCN	R_{CN}	115.324(2)[j]
CO_2	R_{CO}	115.995[n]
HNC	R_{CN}	116.89(2)[e]
C_2H_2	R_{CC}	120.257(9)[h]
CH_2O	R_{CO}	120.33(10)[k]
HNO	R_{NO} (R_0)	121.1(6)[i]
N_2H_2	R_{NN}	124.7(1)[g]
O_3	R_{OO}	127.17(2)[o]
C_2H_4	R_{CC}	133.4(2)[k]
F_2	R_{FF}	141.193[a]
HOF	R_{FO}	143.50(31)[c]
H_2O_2	R_{OO}	145.56[d]

[a] K. P. Huber and G. H. Herzberg, *Constants of Diatomic Molecules*, Van Nostrand Reinhold, 1979.

[b] A. R. Hoy, I. M. Mills and G. Strey, *Mol. Phys.* **24**, 1265 (1972). See also A. R. Hoy and P. R. Bunker, *J. Mol. Spectrosc.* **74**, 1 (1979): $R_{OH} = 95.78$.

[c] L. Halonen and T.-K. Ha, *J. Chem. Phys.* **89**, 4885 (1988).

[d] G. Pelz, K. M. T. Yamada and G. Winnewisser, *J. Mol. Spectrosc.* **159**, 507 (1993).

[e] R. A. Creswell and A. G. Robiette, *Mol. Phys.* **36**, 869 (1978).

[f] J. L. Duncan and I. M. Mills, *Spectrochim. Acta* **20**, 523 (1964). See also W. S. Benedict and E. K. Plyler, *Can. J. Phys.* **35**, 1235 (1957): $R_{NH} = 101.24$, $\Theta_{HNH} = 106.7$.

[g] J. Demaison, F. Hegelund and H. Bürger, *J. Mol. Struc.* **413**, 447 (1997).

[h] A. Baldacci, S. Ghersetti, S. C. Hurlock and K. N. Rao, *J. Mol. Spectrosc.* **59**, 116 (1976).

[i] F. W. Dalby, *Can. J. Phys.* **36**, 1336 (1958).

[j] S. Carter, I. M. Mills and N. C. Handy, *J. Chem. Phys.* **97**, 1606 (1992). See also G. Winnewisser, A. G. Maki and D. R. Johnson, *J. Mol. Spectrosc.* **39**, 149 (1971): $R_{CN} = 115.321(5)$, $R_{CH} = 106.549(24)$.

[k] J. L. Duncan, *Mol. Phys.* **28**, 1177 (1974). See also K. Yamada, T. Nakagawa, K. Kuchitsu and Y. Morino, *J. Mol. Spectrosc.* **38**, 70 (1971): $R_{CO} = 120.3(3)$, $R_{CH} = 109.9(9)$, $\Theta_{HCH} = 116.5(12)$.

[l] D. L. Gray and A. G. Robiette, *Mol. Phys.* **37**, 1901 (1979). See also L. S. Bartell and K. Kuchitsu, *J. Chem. Phys.* **68**, 1213 (1978): $R_{CH} = 108.62(24)$; E. Hirota, *J. Mol. Spectrosc.* **77**, 213 (1979): $R_{CH} = 108.70(7)$.

[m] H. Petek, D. J. Nesbitt, D. C. Darwin, P. R. Ogilby, C. B. Moore and D. A. Ramsay, *J. Chem. Phys.* **91**, 6566 (1989).

[n] G. Graner, C. Rossetti and D. Bailly, *Mol. Phys.* **58**, 627 (1986).

[o] T. Tanaka and Y. Morino, *J. Mol. Spectrosc.* **33**, 538 (1970).

Table 15.4 Errors relative to experiment in the calculated bond distances (pm). The calculations have been carried out with all electrons correlated

		HF	MP2	MP3	MP4	CCSD	CCSD(T)	CISD
$\overline{\Delta}$	cc-pVDZ	−1.80	1.35	0.48	1.83	1.07	1.68	0.26
	cc-pVTZ	−2.63	−0.12	−1.06	0.33	−0.63	0.01	−1.43
	cc-pVQZ	−2.74	−0.23	−1.19	0.27	−0.79	−0.12	−1.66
Δ_{std}	cc-pVDZ	2.25	0.75	0.98	0.82	0.76	0.80	1.24
	cc-pVTZ	2.23	0.69	0.93	0.60	0.66	0.24	1.34
	cc-pVQZ	2.28	0.62	1.03	0.49	0.78	0.20	1.51
$\overline{\Delta}_{abs}$	cc-pVDZ	1.94	1.35	0.87	1.83	1.19	1.68	0.95
	cc-pVTZ	2.63	0.56	1.06	0.47	0.64	0.20	1.44
	cc-pVQZ	2.74	0.51	1.19	0.38	0.80	0.16	1.67
Δ_{max}	cc-pVDZ	7.49	3.16	2.72	3.85	1.99	4.56	3.86
	cc-pVTZ	8.28	1.61	3.84	1.80	2.59	0.45	5.16
	cc-pVQZ	8.45	1.66	4.21	1.51	3.04	0.61	5.69

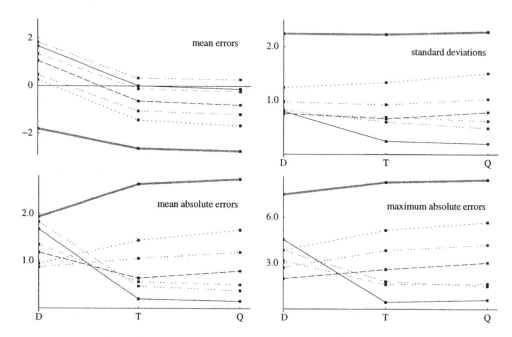

Fig. 15.3. Errors relative to experiment in the calculated bond distances (pm). To distinguish between the different models, we use thick grey lines for Hartree–Fock, dotted lines for CISD, dashed lines for CCSD and full lines for CCSD(T). For the MP2, MP3 and MP4 models, we use lines of grouped dots where the number of dots in each group corresponds to the order of the perturbation theory. The bond distances have been calculated in the cc-pVXZ basis sets with all electrons correlated.

performance at the cc-pVDZ level is fortuitous and does not allow for an improvement in the one-electron description. It does explain, however, the success of the CISD model in the 1970s for the calculation of molecular structures, using basis sets of polarized double-zeta quality.

At the cc-pVTZ level, two models stand out from the others: the MP2 model with a mean error of −0.12 pm and the CCSD(T) model with a mean error of +0.01 pm. Again, there is a certain element of cancellation of errors in the calculations since, at the cc-pVQZ level, the MP2 and CCSD(T) bond distances are less accurate than at the cc-pVTZ level (mean errors -0.23 and -0.12 pm, respectively). Although the MP4 model is less accurate than the MP2 model for all basis sets in Table 15.4, we expect MP4 to be slightly more accurate than MP2 in the basis-set limit.

Having discussed the mean errors in the bond distances, it is appropriate to consider the standard deviations so as to characterize more fully the distribution of errors in the calculations. Only for MP2, MP4 and CCSD(T) does the standard deviation decrease monotonically with improvements in the basis – see Figure 15.3. For MP3 and CCSD, it decreases from cc-pVDZ to cc-pVTZ but then increases as we go to cc-pVQZ. For the CISD model, the standard deviation increases monotonically; for the Hartree–Fock model, it is always large.

15.3.3 NORMAL DISTRIBUTIONS

In Figure 15.4, we have, for each basis and each N-electron model, plotted the normal distributions for the calculated bond distances. Although we make no claim that the errors in the calculated bond distances are indeed normally distributed, these plots neatly summarize the performance of the various levels of theory.

The Hartree–Fock model is characterized by broad distributions centred off the origin and its performance does not improve with the basis set. The Møller–Plesset distances, by contrast, are characterized by sharper distributions located closer to the origin. The poor performance of the MP3 model relative to MP2 and MP4 is evident from these plots. The progression of the MP4 distribution as the basis improves is more satisfactory than that of MP2 theory but only slightly so. Indeed, considering the much higher cost of the MP4 model, the improvement is rather disappointing.

The performance of the CCSD model is likewise disappointing, being intermediate between MP2 and MP3. Clearly, the CCSD model is not well suited to the calculation of bond distances – only with the inclusion of the triples at the CCSD(T) level does the coupled-cluster model yield satisfactory results. Indeed, at the cc-pVTZ and cc-pVQZ levels, the CCSD(T) model performs excellently, with sharply peaked distributions close to the origin.

From these investigations, it appears that the inclusion of doubles amplitudes at the MP2 level yields satisfactory bond distances, but that the inclusion of doubles to higher orders as in MP3, CISD and CCSD without the simultaneous incorporation of triples, as in MP4 and CCSD(T), yields distances in poorer agreement with the true solution. The CISD method performs less satisfactory than any other correlated method, with no improvement as the basis set is increased.

15.3.4 MEAN ABSOLUTE DEVIATIONS

With a few exceptions, the plots of $\overline{\Delta}_{\mathrm{abs}}$ in Figure 15.3 are similar to what we would obtain by plotting the absolute values of the mean values $|\overline{\Delta}|$, confirming the systematic nature of the errors obtained in *ab initio* calculations. From the $\overline{\Delta}_{\mathrm{abs}}$ plots, the different behaviour of the cc-pVDZ set on the one hand and of the cc-pVTZ and cc-pVQZ sets on the other hand is evident. Among the correlated methods, the CISD and MP3 models perform best at the cc-pVDZ level and worst at the cc-pVTZ and cc-pVQZ levels.

At this point, let us comment on the relative performance of the Møller–Plesset approximations. In the cc-pVDZ basis, the absolute mean errors relative to the uncorrelated Hartree–Fock errors are 70%, 45% and 94% at the MP2, MP3 and MP4 levels, respectively. Thus, in this basis, the improvements on the uncorrelated description are small and MP3 performs better than MP2 and

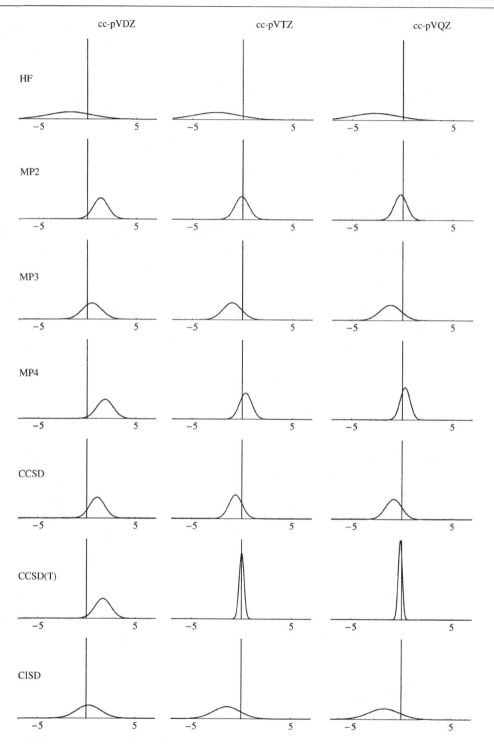

Fig. 15.4. Normal distributions of the errors in the calculated bond distances (pm). For ease of comparison, all distributions have been normalized to one and plotted against the same horizontal and vertical scales.

MP4. At the cc-pVTZ level, the situation is reversed and the errors relative to Hartree–Fock are 21%, 40% and 18%, respectively. Finally, in the cc-pVQZ basis, the errors at the MP2, MP3 and MP4 levels are 19%, 43% and 14%. These examples demonstrate clearly the oscillation of the Møller–Plesset series and the inherent inability of the small cc-pVDZ basis to recover correlation effects, indicating that, at the cc-pVDZ level, any comparison of correlated models with experiment should be treated with caution as it may give a false impression of the performance of the models. The cc-pVTZ basis, on the other hand, yields satisfactory results for the bond distances (as compared with the basis-set limit) and should be sufficient for most purposes.

15.3.5 MAXIMUM ERRORS

The maximum absolute errors in Figure 15.3 are important in representing worst-case errors of the different models and basis sets. For the Hartree–Fock and MP2 models in the cc-pVQZ basis, the largest errors occur for F_2 and are 8.5 and 1.7 pm, respectively. For the MP3, MP4, CCSD and CISD models, O_3 gives the largest errors of 4.2, 1.5, 3.0 and 5.7 pm, respectively. Finally, the largest CCSD(T) error is only 0.6 pm and occurs for the R_{OO} distance in H_2O_2, while the second largest error of 0.5 pm occurs of the R_{OO} distance in O_3. As discussed in Section 15.3.9, the discrepancy for H_2O_2 may in fact arise from an error in the experiment.

Although these numbers are indicative of the largest intrinsic errors of the standard methods, we should keep in mind that they are based on a rather small sample of molecules, containing elements from only the first row of the periodic table. Larger errors may occur for other systems and, in particular, for systems containing heavier atoms. For example, for ferrocene, the Hartree–Fock wave function overestimates the vertical cyclopentadienyl–iron distance by 21 pm, MP2 underestimates the same distance by 19 pm, while CCSD and CCSD(T) give distances within 1–2 pm of the experimental bond length [3]. This example shows that, although less accurate for molecules containing first-row atoms, the CCSD model appears to be more robust than the MP2 model for molecules containing heavier atoms. As already noted, the maximum errors in Figure 15.3 may arise from errors in the experimental measurements rather than from errors in the calculations, in particular for the most accurate methods. We shall return to this point shortly.

15.3.6 THE CCSDT AND CCSD(T) MODELS

At the cc-pVQZ level, the CCSD(T) bond distances differ from the experimental ones by just a few tenths of a picometer. To see if the errors can be reduced further, we here consider calculations where, in the cc-pVQZ basis, the correlation treatment is extended to the CCSDT model and where, for the CCSD(T) model, the basis is extended to cc-pV5Z [4]. Since these calculations are demanding, they have been carried out for only six molecules – see Table 15.5. We first note that, even though the bond lengths in general shorten from cc-pVQZ to cc-pV5Z, there are exceptions to this rule – in the HF molecule, for example, the bond distance lengthens.

The CCSDT and CCSD(T) bond lengths are very similar – the largest difference being 0.06 pm for N_2 in the cc-pVQZ basis. Except for the small cc-pVDZ basis, the CCSD(T) bonds are always longer than the CCSDT bonds. The effect of the triples is thus overestimated by the CCSD(T) model – by as little as 0.4% in HF and as much as 9.6% in N_2. In general, however, the CCSD(T) model provides a good approximate description of the effect of triple excitations on bond distances, at least for molecules containing only first-row atoms.

From Table 15.5, we see that the differences between the CCSDT and CCSD(T) bond distances are almost the same at the cc-pVTZ and cc-pVQZ levels, suggesting that we should be able to

Table 15.5 Calculated equilibrium bond distances (in pm) for six small molecules. The distances that in the cc-pVXZ basis are longer than those in the cc-pV(X − 1)Z basis have been shaded. The numbers marked by an asterisk have been obtained as discussed in the text

Molecule	Model	cc-pVDZ	cc-pVTZ	cc-pVQZ	cc-pV5Z
HF	HF	90.15	89.79	89.69	89.70
	CCSD	91.87	91.41	91.26	91.27
	CCSD(T)	91.96	91.62	91.52	91.55
	CCSDT	91.97	91.62	91.52	91.55*
H_2O	HF	94.63	94.06	93.96	93.96
	CCSD	96.44	95.55	95.36	95.36
	CCSD(T)	96.58	95.78	95.62	95.64
	CCSDT	96.58	95.78	95.62	95.64*
CH_2	HF	110.71	109.57	109.46	109.44
	CCSD	112.68	110.28	110.36	110.18
	CCSD(T)	112.80	110.42	110.52	110.34
	CCSDT	112.80	110.40	110.50	110.32*
N_2	HF	107.73	106.71	106.56	106.54
	CCSD	111.23	109.36	109.09	108.95
	CCSD(T)	111.84	110.06	109.81	109.68
	CCSDT	111.80	110.00	109.75	109.62*
CO	HF	111.01	110.45	110.20	110.18
	CCSD	113.79	112.56	112.19	112.02
	CCSD(T)	114.41	113.25	112.89	112.72
	CCSDT	114.44	113.24	112.88	112.71*
F_2	HF	134.77	132.91	132.75	132.67
	CCSD	143.18	139.23	138.86	138.59
	CCSD(T)	145.75	141.36	141.11	140.88
	CCSDT	145.76	141.33	141.08	140.85*

predict the CCSDT/cc-pV5Z bond distances to within 0.01 pm by adding the cc-pVQZ differences to the CCSD(T)/cc-pV5Z values. The predicted bond distances are marked with an asterisk in the table. A similar estimate of the full triples correction based on the differences between CCSD and CCSDT in a smaller basis is impossible since these much larger corrections are more sensitive to the basis set. In F_2, for example, the correction from CCSD to CCSDT differs by as much as 0.12 pm in the cc-pVTZ and cc-pVQZ basis sets.

From Table 15.5, we conclude that the CCSDT/cc-pVQZ level provides a balanced description where the basis-set and N-electron errors cancel – the mean absolute deviation from experiment is 0.11 pm, with a maximum deviation of 0.2 pm in CH_2. Except for CO, the bonds are shorter than experiment. At the cc-pV5Z level, the bonds are reduced further and all bonds are now shorter than experiment (by about 0.2 pm). Since the CCSD(T) bonds are longer than the CCSDT bonds, the agreement with experiment is in fact slightly better for CCSD(T) than for CCSDT.

15.3.7 THE EFFECT OF CORE CORRELATION ON BOND DISTANCES

We now examine the contribution to the bond lengths from the correlation of the core electrons. As the core-correlation energy to a large extent is independent of the geometry, it is expected to have a small effect on geometries. This was demonstrated in Section 8.3.1, where correlation of the core reduced the bond distance of BH by just 0.2 pm. The contributions discussed here are therefore important only for applications that require high precision.

In the calculations discussed up to now, we have correlated all electrons, even though the optimizations have been carried out in the cc-pVXZ family of basis sets, designed for valence correlation. This approach was taken for technical reasons since, for most of the wave functions studied, the programs did not have the capability of freezing the core orbitals in the course of the geometry optimization. At the MP2 level, however, the optimized geometries were also determined with a frozen core. Since the MP2 geometries agree closely with those obtained with the CCSD(T) method, we expect the MP2 model to predict accurately the changes arising from core correlation.

To examine the effect of core correlation, we have in Table 15.6 listed the statistical data that describe the differences in the MP2 bond distances between all-electron cc-pCVXZ calculations and frozen-core cc-pVXZ calculations. Core correlation is seen to shorten the bonds, with an average reduction of about 0.20 pm for the cc-pCVTZ and cc-pCVQZ sets. The cc-pCVDZ basis is too small to give a reliable description of core correlation.

Since our investigations have been carried out in the cc-pVXZ basis sets with all electrons correlated, it is of interest to examine how accurately the core contraction is described in such calculations. From Table 15.6, we see that use of the cc-pVTZ basis leads to a significant overestimation of the contraction, with a largest overestimation of 0.48 pm (for the CH bond in HCN). An inspection of the numbers underlying the statistical data in Table 15.6 reveals that, whereas the core contributions to bonds that do not involve hydrogen atoms are accurately described by the cc-pVTZ and, in particular, the cc-pVQZ sets (with mean absolute deviations of 0.07 and 0.02 pm, respectively), the core contributions to bonds to a hydrogen atom are poorly described (with mean absolute deviations of 0.20 and 0.10 pm, respectively).

In practice, the core contribution is usually included as an additive correction. First, the equilibrium geometry is determined in a sufficiently accurate frozen-core cc-pVXZ calculation. Next, the core correction is estimated by comparing all-electron and frozen-core calculations in the cc-pCVYZ and cc-pVYZ sets where $Y < X$. This approach is acceptable since the core correction is small and need only be determined to a low accuracy.

15.3.8 TRENDS IN THE CONVERGENCE TOWARDS EXPERIMENT

From the discussion in this section, some general trends in the convergence of the calculated bond distances towards the experimental values may be discerned. In general, bonds shorten with improvements in the basis set and lengthen with improvements in the correlation treatment. These generalizations, however, gloss over some interesting details in the dependency of the bond lengths on the correlation treatment. Thus, among the methods that introduce correlation only through the inclusion of double excitations – that is, MP2, MP3, CCSD and CISD – the simplest treatment (MP2) gives the longest bond distances. Any further improvement in the treatment

Table 15.6 The effect of core correlation on bond distances (in pm) at the MP2 level. The statistical measures are based on the molecules in Table 15.1

	all cc-pCVXZ – valence cc-pVXZ			all cc-pCVXZ – all cc-pVXZ		
	DZ	TZ	QZ	DZ	TZ	QZ
$\overline{\Delta}$	−0.13	−0.20	−0.19	−0.06	0.13	0.05
Δ_{std}	0.07	0.11	0.09	0.06	0.13	0.05
$\overline{\Delta}_{abs}$	0.13	0.20	0.19	0.06	0.14	0.06
Δ_{max}	0.30	0.36	0.34	0.22	0.48	0.14

of the doubles contracts the bonds back towards the Hartree–Fock limit. The magnitude of this contraction depends on the nature of the improvement in the correlation description as illustrated in Figure 15.5, where the bond-length changes have been calculated from the $\overline{\Delta}$ of the cc-pVQZ basis.

In general terms, we may rationalize these observations as follows. First, the inclusion of doubles at the MP2 level stretches the bonds since a new type of 'interaction' among the electrons (parametrized by means of doubles amplitudes) is introduced. The bonds are stretched by the 'repulsive' nature of this interaction. Second, any further refinement in the treatment of the interaction (by relaxation of the amplitudes) reduces its overall 'repulsive' character, allowing the bonds to contract somewhat. By this argument, we expect a simple treatment of triples to stretch the bonds further, and any relaxation of the wave function in the presence of the triples to contract the bonds again. However, since the triples are less important than the doubles, we expect these effects to be much smaller than for the doubles.

The simplest treatment of the triples occurs at the MP4 and CCSD(T) levels, the two methods differing in that, at the CCSD(T) level, the doubles are fully relaxed (in the absence of the triples), whereas no such relaxation occurs at the MP4 level. In agreement with this observation, we find that the MP4 and CCSD(T) models both lengthen the bonds (relative to MP3 and CCSD) but that the MP4 bonds are the longest since neither the doubles nor the triples have been fully relaxed. Finally, relaxation of the triples at the CCSDT level contracts the bonds somewhat relative to CCSD(T).

The smallness of the CCSD(T) error at the cc-pVQZ level arises from a cancellation of errors – the contraction that would occur upon relaxation of the triples and upon extension of the basis beyond cc-pVQZ is approximately balanced by the stretching that would occur upon the introduction of quadruples and higher amplitudes. The same cancellation of errors is observed in MP2 theory, where the contraction that occurs upon relaxation of the doubles is balanced by the stretching that occurs upon introduction of higher-order amplitudes. It is unknown whether this cancellation of errors is fortuitous or systematic (and would thus also occur for CCSDT(Q) and higher-order wave functions).

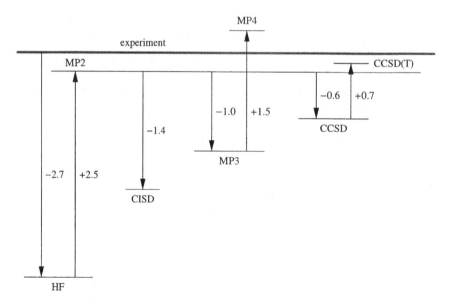

Fig. 15.5. The relationships between the calculated equilibrium bond distances for the standard models (in pm).

15.3.9 SUMMARY

The calculations presented in this section have shown that, whereas improvements in the correlation treatment in general increase the bond length, improvements in the one-electron treatment reduce it. The coupled-cluster hierarchy converges smoothly, with a reduction in the error by a factor of 3–5 at each level. Thus, whereas the intrinsic error (based on $\overline{\Delta}$ and $\overline{\Delta}_{abs}$ in Figure 15.3) at the Hartree–Fock level is about -3 pm, it is reduced to about -0.8 pm at the CCSD level and to about -0.2 pm at the CCSD(T) level. In contrast, the Møller–Plesset hierarchy oscillates, with typical errors of about 0.5 pm at the MP2 level, 1.2 pm at the MP3 level and 0.4 pm at the MP4 level. Even for the small molecules in this study, the CISD model is inadequate, with an intrinsic error of -1.7 pm. Based on these observations, we may state that, in terms of computational accuracy and cost, the only successful models are Hartree–Fock, MP2, CCSD and CCSD(T).

In Table 15.7, we have collected the equilibrium bond distances of the molecules in Table 15.1, calculated using the core–valence cc-pCVQZ basis set, correlating all the electrons in the system. For the CCSD(T) model, a comparison with experiment shows that, for 22 of the 29 bond lengths, the difference is less than or equal to 0.1 pm. This error is less than the intrinsic error of the CCSD(T) model and arises from a cancellation of errors.

The largest difference (1.0 pm) occurs for the R_{NH} bond in HNO, for which only a vibrationally averaged experimental geometry is available. The next largest differences occur for O_3 and H_2O_2.

Table 15.7 Calculated and experimental equilibrium bond lengths R_e (in pm). The calculations have been carried out in the cc-pCVQZ basis with all electrons correlated

Molecule	Bond	HF	MP2	CCSD	CCSD(T)	Experiment
H_2	R_{HH}	73.4	73.6	74.2	74.2	74.1
HF	R_{FH}	89.7	91.7	91.3	91.6	91.7
H_2O	R_{OH}	94.0	95.7	95.4	95.7	95.7
HOF	R_{OH}	94.5	96.6	96.2	96.6	96.6
H_2O_2	R_{OH}	94.1	96.2	95.8	96.2	96.7
HNC	R_{NH}	98.2	99.5	99.3	99.5	99.4
NH_3	R_{NH}	99.8	100.8	100.9	101.1	101.1
N_2H_2	R_{NH}	101.1	102.6	102.5	102.8	102.9
C_2H_2	R_{CH}	105.4	106.0	106.0	106.2	106.2
HNO	R_{NH}	103.0	104.8	104.8	105.2	106.2 (R_0)
HCN	R_{CH}	105.7	106.3	106.3	106.6	106.5
C_2H_4	R_{CH}	107.4	107.8	107.9	108.1	108.1
CH_4	R_{CH}	108.2	108.3	108.5	108.6	108.6
N_2	R_{NN}	106.6	110.8	109.1	109.8	109.8
CH_2O	R_{CH}	109.3	109.8	109.9	110.1	110.1
CH_2	R_{CH}	109.5	110.1	110.5	110.7	110.7
CO	R_{CO}	110.2	113.2	112.2	112.9	112.8
HCN	R_{CN}	112.3	116.0	114.6	115.4	115.3
CO_2	R_{CO}	113.4	116.4	115.3	116.0	116.0
HNC	R_{CN}	114.4	117.0	116.2	116.9	116.9
C_2H_2	R_{CC}	117.9	120.5	119.7	120.4	120.3
CH_2O	R_{CO}	117.6	120.6	119.7	120.4	120.3
HNO	R_{NO}	116.5	121.5	119.7	120.8	121.1 (R_0)
N_2H_2	R_{NN}	120.8	124.9	123.6	124.7	124.7
O_3	R_{OO}	119.2	127.6	124.1	126.6	127.2
C_2H_4	R_{CC}	131.3	132.6	132.5	133.1	133.4
F_2	R_{FF}	132.7	139.5	138.8	141.1	141.2
HOF	R_{FO}	136.2	142.0	141.2	143.3	143.5
H_2O_2	R_{OO}	138.4	144.3	143.1	145.0	145.6

For O_3, the bond is 0.6 pm too short at the CCSD(T) level, reflecting the multiconfigurational character of this system. For H_2O_2, the calculated R_{OO} and R_{OH} differ from experiment by 0.6 and 0.5 pm, respectively. However, H_2O_2 is a molecule dominated by a single determinant and there is no reason to assume that the accuracy achieved by the CCSD(T) model for this particular molecule is significantly lower than for the other such molecules in the table. More likely, the discrepancies for H_2O_2 are related to the difficulties associated with determining experimentally equilibrium geometries of nonrigid molecules. Of the remaining bonds in the table, the two largest errors (0.3 pm for R_{CC} in C_2H_4 and 0.2 pm for R_{OF} in HOF) are just outside and inside the experimental error bars, respectively.

Clearly, the CCSD(T) model is capable of providing structures of high quality, often surpassing that of experimental investigations. The MP2 model is less accurate but more generally applicable, being considerably less expensive. The CCSD model appears to be rather less successful, being more expensive but usually less accurate than MP2 for simple organic molecules, but apparently more robust for molecules with a more complicated electronic structure such as ferrocene. The Hartree–Fock model is suitable only for preliminary calculations, in which quantitative accuracy is not sought.

The basis-set requirements are different for the different models. Whereas useful calculations can be carried out in the cc-pVDZ basis for the Hartree–Fock model, the correlated models require at least the cc-pVTZ basis. For high-precision work at the CCSD(T) level, the cc-pVQZ basis is mandatory and some account must be taken of core correlation.

15.4 Molecular equilibrium structures: bond angles

Having considered equilibrium bond lengths in Section 15.3, we now turn our attention to bond angles, proceeding in much the same manner as for bond lengths. In Section 15.4.1, we discuss the experimental equilibrium bond angles; in Section 15.4.2, the statistical analysis of the different computational models is presented. Our conclusions and calculated bond angles for the most important N-electron models in large one-electron basis sets are given in Section 15.4.3.

15.4.1 EXPERIMENTAL BOND ANGLES

In Table 15.8, we have listed the experimental equilibrium bond angles for the molecules in Table 15.1. We note that, for some of the molecules, the bond angles are not independent of one another. In C_2H_4, for example, the two HCC bond angles and the HCH bond angle on each carbon atom add up to $360°$. In such cases, we have chosen to use the smallest bond angle – that is, the HCH angle rather than the HCC angle in C_2H_4.

Except for the vibrationally averaged HNO bond angle, the angles in Table 15.8 are equilibrium angles, which correspond directly to those calculated theoretically. Since the vibrational corrections may be significant, the HNO bond angle has been excluded from the analysis. In addition, we have excluded the H_2O_2 bond angle since, according to the discussion in Section 15.3.9, the experimental measurement is in error. Indeed, we shall later find that the experimental HOO bond angle exceeds the best calculated angle by $2.3°$, which is an order of magnitude more than for the other bond angles in Table 15.8.

Having excluded the bond angles in HNO and H_2O_2 from our analysis, we are left with a rather small set of eight bond angles for our statistical study. Moreover, as seen from Table 15.8, the experimental uncertainties for the remaining bond angles (where available) are rather large – between $0.2°$ and $0.5°$. Indeed, we shall later find that, in the basis-set limit, the

Table 15.8 Molecular equilibrium bond angles (in degrees). The bond angles marked with an asterisk are excluded from the statistical analysis. For references, see Table 15.3

Molecule	Bond angle	Experiment
H_2O	HOH	104.52[b]
HOF	HOF	97.54(50)[c]
H_2O_2	HOO*	102.32[d]
NH_3	HNH	106.7[f]
N_2H_2	HNN	106.3(2)[g]
HNO	HNO* (Θ_0)	108.5(8)[i]
C_2H_4	HCH	117.37(33)[k]
CH_2O	HCH	116.30(25)[k]
CH_2	HCH	102.4(4)[m]
O_3	OOO	116.78(33)[o]

MP2 and CCSD(T) bond angles are nearly all within these uncertainties, making it difficult to distinguish between these models in our analysis. In spite of these difficulties, we shall be able to make a meaningful comparison of the different *ab initio* models, establishing their typical convergence patterns and errors.

15.4.2 CALCULATED BOND ANGLES

In Figure 15.6 we have, for the eight calculated equilibrium bond angles in the test set, plotted the mean errors, standard deviations, mean absolute errors and maximum absolute errors relative

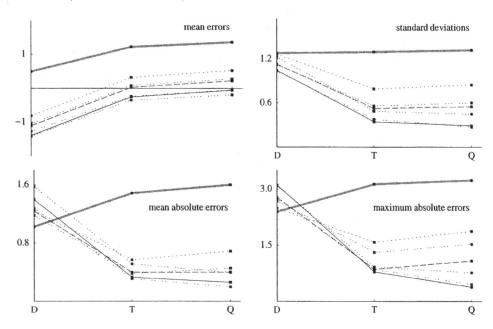

Fig. 15.6. Errors in the calculated bond angles relative to the experimental values in Table 15.8 for the Hartree–Fock, MP2, MP3, MP4, CCSD, CCSD(T) and CISD models (in degrees). For line patterns, see Figure 15.3.

to the experimental values in Table 15.8. Since the signs of the errors are somewhat arbitrary, depending on our choice of independent bond angles as discussed in Section 15.4.1, the mean errors should be interpreted with some caution.

The first thing to note about the plots in Figure 15.6 is that, except for the opposite signs of the mean errors, they resemble closely those for bond distances in Figure 15.3, indicating that the computational methods behave in much the same manner with respect to bond angles and bond distances. This similarity should perhaps be expected, noting that bond angles and bond distances are two sets of parameters that describe the equilibrium geometrical structure of the molecular systems.

Comparing the mean errors in Figures 15.3 and 15.6, we note a striking coupling between the errors in bond lengths and bond angles – at all levels of theory, any reduction in bond lengths (relative to experiment) leads to a corresponding increase in bond angles. This behaviour may be rationalized by noting that, whenever two bonds AX and XB contract, the atoms A and B bonded to X will come into closer contact with each other, which is compensated for by opening up the bond angle AXB. Whatever the true explanation, we note that, in the plot of mean errors in the bond angles, the order of the methods is the exact reverse (the mirror image) of that for the bond distances.

These similarities notwithstanding, some differences exist between bond distances and bond angles. Thus, for the bond angles, the differences among the correlated models are smaller than for bond distances. The CCSD and MP3 plots, for instance, are now almost indistinguishable. More important, for bond angles, the MP2 model appears to perform as well as the CCSD(T) model. However, since the experimental uncertainties are typically $0.3°$, their performances cannot be distinguished.

For bond angles, the cc-pVDZ basis appears to be even less adequate than for bond distances. In particular, at the cc-pVDZ level, the standard deviation is $1.0–1.3°$ for all models. As a curiosity, in this small basis, the Hartree–Fock model has the smallest maximum error ($2.4°$ for the bond angle in HOF), while CCSD(T) has the largest maximum error ($-3.1°$ for the HNH bond angle in NH_3). This peculiar situation serves as a reminder that, except for calibration and extrapolation, correlated calculations should not be carried out in a basis as small as cc-pVDZ. On the other hand, already at the cc-pVTZ level, the calculations are balanced in the sense that the Hartree–Fock model has the largest maximum error ($+3.1°$ for the bond angle in HOF) and CCSD(T) the smallest maximum error ($-0.8°$ for the bond angle in H_2O).

At the cc-pVQZ level, the largest CCSD(T) error occurs for the O_3 bond angle, which is overestimated by $0.4°$. Although this error is not much larger than the experimental uncertainty ($0.3°$), it is probably genuine and related to the multiconfigurational electronic structure of O_3 – indeed, a priori, we would expect the CCSD(T) model to exhibit the largest error for O_3. The fact that this does not happen for cc-pVTZ may perhaps be taken as an indication that the CCSD(T) calculations are not fully balanced at the cc-pVTZ level.

In Figure 15.7, we have plotted the normal distributions of the errors in the calculated bond angles. The main difference from the plots for bond distances in Figure 15.4 is that, for bond angles, the CCSD(T) distributions are less sharply peaked and similar to those of the MP2 model, reflecting the poorer quality of the experimental reference data for bond angles than for bond distances.

15.4.3 SUMMARY

The calculated bond angles follow the pattern observed for bond distances in Section 15.3. Whereas the intrinsic Hartree–Fock error is typically $1.6°$, this error is reduced to about $0.4°$

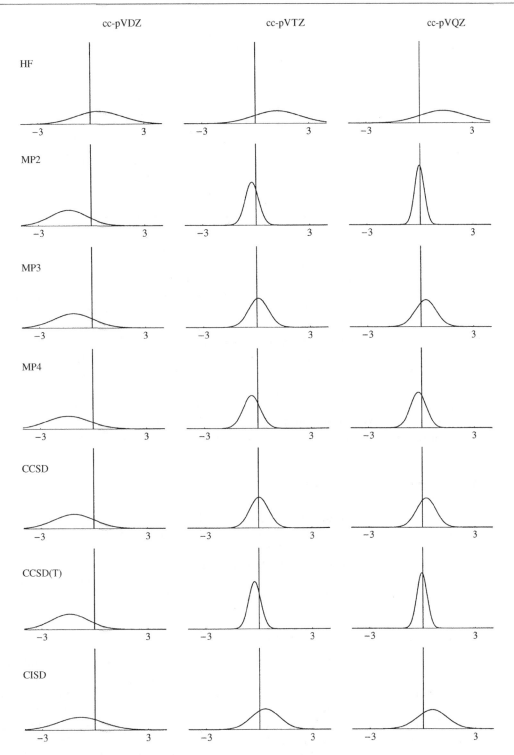

Fig. 15.7. Normal distributions of the errors in the calculated bond angles (in degrees) relative to the experimental values in Table 15.8.

Table 15.9 Calculated and experimental equilibrium bond angles (in degrees). The calculations have been carried out in the cc-pCVQZ basis with all electrons correlated

Molecule	Bond angle	HF	MP2	CCSD	CCSD(T)	Experiment
H_2O	HOH	106.2	104.1	104.5	104.2	104.5
HOF	HOF	100.8	98.0	98.6	97.8	97.5
H_2O_2	HOO	103.1	99.8	100.7	100.0	102.3
NH_3	HNH	107.9	106.7	106.6	106.4	106.7
N_2H_2	HNN	108.1	105.9	106.6	106.2	106.3
HNO	HNO	109.4	107.8	108.3	108.1	108.5 (Θ_0)
C_2H_4	HCH	116.7	117.3	117.0	117.1	117.4
CH_2O	HCH	116.0	116.4	116.4	116.4	116.3
CH_2	HCH	103.7	102.2	102.0	102.0	102.4
O_3	OOO	119.3	116.8	117.8	117.2	116.8

for the CCSD model. Because of the errors in the experimental measurements, the intrinsic error of the CCSD(T) model cannot be established with a high degree of certainty but it is probably smaller than 0.2° and perhaps as small as 0.1°, which is smaller than that of most experimental measurements. Indeed, this is the accuracy that we would expect if we assume that the CCSD(T) model is able to reproduce all interatomic distances (not only the bond distances) to within 0.1 pm. For bond angles, the basis-set requirements are the same as for bond distances, with a minimal requirement of cc-pVTZ for correlated calculations.

In Table 15.9, we have listed the bond angles calculated using the Hartree–Fock, MP2, CCSD and CCSD(T) models in the cc-pCVQZ basis with all electrons correlated. With the exception of H_2O_2, the CCSD(T) angles compare well with the experimental ones. In view of the documented high quality of the CCSD(T) bond angles, we suspect that the discrepancy for H_2O_2 arises from an error in the experimental measurement. All things considered, we conclude that the ability of the different N-electron models to describe bond angles is similar to that for bond distances.

15.5 Molecular dipole moments

In this section, we consider the equilibrium dipole moments of the polar molecules in Table 15.1. Because of the lack of accurate experimental equilibrium dipole moments, the analysis of the performance of the different models becomes more speculative than for the geometries, atomization energies and reaction enthalpies. Still, by carrying out a careful analysis of the data, we shall see that it is possible to determine the probable errors in the calculated dipole moments for the most important standard models: Hartree–Fock, MP2, CCSD and CCSD(T).

15.5.1 EXPERIMENTAL DIPOLE MOMENTS

In Table 15.10, we have listed, in order of increasing polarity, the experimental dipole moments of the polar molecules in Table 15.1. For the equilibrium dipole moment μ_e, there are experimental values only for CO, NH_3, H_2O and HF; for the dipole moment in the vibrational ground state μ_0, there are measurements for all molecules except HNO, CH_2 and HOF. However, a comparison of μ_e and μ_0 (where both exist) shows that the vibrational correction to μ_e may be quite large (-0.013 D for CO, -0.089 D for NH_3, $+0.007$ D for H_2O, and $+0.023$ D for HF), making it impossible to use the experimental μ_0 for analysing the calculated μ_e. Indeed, for a satisfactory

Table 15.10 Experimental and calculated dipole moments (in D) for the polar molecules in Table 15.1. The calculated dipole moments have been obtained at the CCSD(T) level in the aug-cc-pVQZ basis; the predicted dipole moments have been generated from the CCSD(T) values as explained in Section 15.5.3

	Experimental		Theoretical	
	μ_0	μ_e	Calculated μ_e	Predicted μ_e
CO	0.1097(1)[a]	0.123(2)[a]	0.12	0.13(1)
O_3	0.5324(24)[b]	–	0.56	0.52(4)
NH_3	1.471932(7)[c]	1.561(5)[d]	1.52	1.52(1)
HNO	–	–	1.69	1.69(2)
CH_2	–	–	1.69	1.69(1)
H_2O_2	2.2[e]	–	1.75	1.74(2)
HF	1.826178(3)[f]	1.803(2)[f]	1.80	1.80(1)
H_2O	1.8546(6)[g]	1.8473(10)[g]	1.85	1.85(1)
HOF	–	–	1.89	1.89(1)
CH_2O	2.3321(5)[h]	–	2.39	2.38(2)
HCN	2.985188(3)[i]	–	3.02	3.01(2)
HNC	3.05(1)[j]	–	3.11	3.11(1)

[a]W. L. Meerts, F. H. De Leeuw and A. Dymanus, *Chem. Phys.* **22**, 319 (1977).
[b]M. Lichtenstein, J. J. Gallagher and S. A. Clough, *J. Mol. Spectrosc.* **40**, 10 (1971).
[c]K. Tanaka, H. Ito and T. Tanaka, *J. Chem. Phys.* **87**, 1557 (1987).
[d]M. D. Marshall, K. C. Izgi and J. S. Muenter, *J. Chem. Phys.* **107**, 1037 (1997).
[e]D. R. Lide, (ed.), *Handbook of Chemistry and Physics*, 72nd edn, CRC Press, 1991.
[f]J. M. Brown, J. Demaison, A. Dubrulle, W. Hüttner and E. Tiemann, *Landolt-Börnstein: Molecular Constants Mostly from Microwave, Molecular Beam and Electron Resonance Spectroscopy*, K.-H. Hellwege and A. M. Hellwege (eds), Vol. II/14, Subvol. b, Springer-Verlag, 1983.
[g]S. A. Clough, Y. Beers, G. P. Klein and L. S. Rothman, *J. Chem. Phys.* **59**, 2254 (1973).
[h]B. Fabricant, D. Krieger and J. S. Muenter, *J. Chem. Phys.* **67**, 1576 (1977).
[i]W. L. Ebenstein and J. S. Muenter, *J. Chem. Phys.* **80**, 3989 (1984).
[j]G. L. Blackman, R. D. Brown, P. D. Godfrey and H. I. Gunn, *Nature* **261**, 395 (1976).

comparison with experiment, we would have to carry out a vibrational averaging of the calculated dipole moments. Such an averaging is expensive – in particular, for molecules other than diatomics – and has not been attempted.

In the present section, we shall therefore proceed differently, comparing our calculated dipole moments with a set of *predicted dipole moments*, obtained from our best calculated values (obtained at the CCSD(T) level in the aug-cc-pVQZ basis) by adding corrections for higher-order correlation effects and basis-set incompleteness. The best calculated and predicted dipole moments are both listed in Table 15.10. Note that the uncertainties in the predicted μ_e (mostly 0.01 or 0.02 D) are smaller than the differences between the experimental μ_e and μ_0, substantiating further our assumption that a comparison with the predicted μ_e is more satisfactory than a direct comparison with the experimental μ_0.

15.5.2 CALCULATED DIPOLE MOMENTS

The calculations in Table 15.11 have been carried out at the Hartree–Fock, MP2, CCSD and CCSD(T) levels (all electrons correlated) in the aug-cc-pVXZ basis sets with $X \leq 4$. Augmented basis sets have been used to ensure a flexible description of the outer valence region. Each dipole moment has been calculated at the geometry optimized using the same wave function.

To examine basis-set saturation, we have carried out additional calculations in the corresponding cc-pVXZ and cc-pCVXZ sets. First, we examine the effects of the core-correlating orbitals,

Table 15.11 Equilibrium dipole moments μ_e (in D) calculated at the geometry optimized using the same wave function. For CO, the sign is relative to the indicated polarity of the molecule

	HF	MP2	CCSD	CCSD(T)		
	aug-cc-pVQZ	aug-cc-pVQZ	aug-cc-pVQZ	aug-cc-pVDZ	aug-cc-pVTZ	aug-cc-pVQZ
C^-O^+	−0.054	0.265	0.084	0.083	0.120	0.118
O_3	0.837	0.486	0.642	0.532	0.554	0.559
NH_3	1.544	1.517	1.530	1.541	1.513	1.521
HNO	1.877	1.708	1.718	1.674	1.683	1.694
CH_2	1.904	1.829	1.733	1.637	1.689	1.695
H_2O_2	1.863	1.764	1.782	1.747	1.742	1.750
HF	1.884	1.811	1.813	1.799	1.797	1.800
H_2O	1.936	1.864	1.870	1.859	1.845	1.853
HOF	2.024	1.875	1.901	1.914	1.883	1.888
CH_2O	2.703	2.395	2.440	2.397	2.385	2.392
HCN	3.262	3.029	3.059	3.005	3.003	3.017
HNC	2.956	3.271	3.104	3.072	3.107	3.110

comparing the (all-electron) cc-pVXZ and cc-pCVXZ calculations. At the quadruple-zeta level, the correction from the core-correlating orbitals is 0.006 D for H_2O_2 but mostly smaller than 0.001 D. At the double- and triple-zeta levels, the correction is larger but never larger than 0.006 and 0.016 D, respectively, and mostly three times smaller. Henceforth, we shall ignore these corrections in our discussion of dipole moments.

Next, comparing the cc-pVXZ and aug-cc-pVXZ dipole moments, we find the mean absolute changes due to augmentation are 0.08, 0.04 and 0.02 D at the double-, triple- and quadruple-zeta levels, respectively. The effect of augmentation is therefore significant but decreases with the cardinal number. Mostly, the dipole moments are reduced but the mean changes are only 0.00, −0.01 and −0.01 D at the double-, triple and quadruple-zeta levels – the largest changes (+0.33, +0.12 and +0.05 D) occur for CH_2O at the MP2 level of approximation. Augmentation with a second set of diffuse functions does not affect the calculated dipole moments significantly. In the following, we shall discuss only the calculations carried out at the singly augmented aug-cc-pVXZ level.

15.5.3 PREDICTED DIPOLE MOMENTS

The predicted dipole moments in Table 15.10 were obtained in the following manner. For each dipole moment calculated at the CCSD(T) level in the aug-cc-pVQZ basis, we first added a correction for the connected quadruples, obtained by assuming that the ratio between the quadruples and triples corrections is the same as the ratio between the triples and doubles corrections. Obviously, this procedure can give only a crude estimate of the quadruples correction, but it may at least serve as a rough indication of the error in the calculated dipole moments.

In general, the estimated quadruples correction is small, typically −0.005 D. However, for O_3, the correction is as large as −0.035 D, indicating that higher-order excitations are important for this molecule (as expected from the presence of several important configurations in the wave function). For H_2O_2, the correction is −0.013 D, but for all other molecules it is smaller than 0.01 D in magnitude. Except for CO, for which a large positive correction of 0.009 D is obtained, all corrections are negative. For each corrected dipole moment, the uncertainty was set equal to the magnitude of the correction.

After the application of this correction, the effect of basis-set incompleteness was considered. Comparing the CCSD(T) dipole moments of the three basis sets in Table 15.11, we note that the differences are rather small. Indeed, in no case was it deemed necessary to adjust the correlation-corrected dipole moment, although the uncertainties were adjusted. The uncertainties in Table 15.10 therefore reflect both basis-set errors and errors in the correlation treatment.

Let us now compare the predicted equilibrium dipole moments with experiment. Of the four experimental values in Table 15.10, the dipole moments for CO, HF and H_2O are within 0.01 D of the predicted value. For NH_3, on the other hand, the dipole moment differs by as much as 0.04 D from the predicted value. It has been argued, however, that the experimental μ_e of NH_3 is too large by several hundredths of a debye, suggesting that this discrepancy arises from an error in the experimental rather than predicted dipole moment [5].

15.5.4 ANALYSIS OF THE CALCULATED DIPOLE MOMENTS

In Figure 15.8 we have, for the calculated dipole moments in the aug-cc-pVXZ basis sets, plotted the mean errors, the standard deviations, the mean absolute errors and the maximum absolute errors. The plots indicate that the calculated dipole moments depend in a systematic manner on the cardinal number and the correlation treatment. In general, the dipole moments are reduced as we improve the correlation treatment. Indeed, with the exceptions of CO and HNC, the dipole moment is always reduced as we go from Hartree–Fock to CCSD and then on to CCSD(T) – see Table 15.11. The MP2 dipole moments are less systematic but are usually slightly smaller than the CCSD numbers. At the aug-cc-pVQZ level, the mean absolute errors are 0.17 D for the Hartree–Fock model, 0.05 D for the MP2 model, 0.04 D for the CCSD model and 0.01 D for

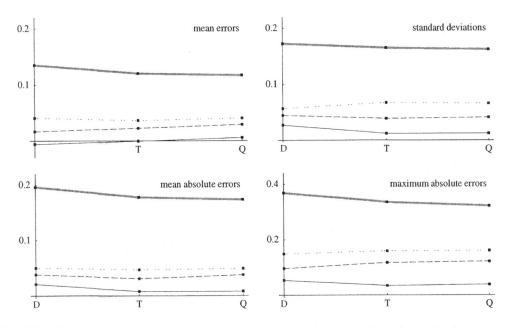

Fig. 15.8. The mean errors, standard deviations, mean absolute errors and maximum absolute errors of the aug-cc-pVXZ dipole moments relative to the predicted dipole moments in Table 15.10 (in D). The Hartree–Fock, MP2, CCSD and CCSD(T) models use the line patterns described in the legend of Figure 15.3.

the CCSD(T) model. These values may be taken as indicative of the typical intrinsic errors of these models. The corresponding maximum absolute errors are 0.32, 0.16, 0.12 and 0.04 D for the Hartree–Fock, MP2, CCSD and CCSD(T) models, respectively.

From Figure 15.8, we see that the calculated dipole moments are more sensitive to the choice of the N-electron model than to the choice of cardinal number for the basis set. Even in the smallest basis, the CCSD(T) model is considerably more accurate than the CCSD model. Thus, for the CCSD(T) model, the smallest basis (aug-cc-pVDZ) gives a mean error of -0.006 D and the largest basis (aug-cc-pVQZ) a mean error of $+0.006$ D, the corresponding mean absolute errors being 0.02 and 0.008 D. In comparison, for the CCSD model, the mean errors are 0.02 and 0.03 D at the aug-cc-pVDZ and aug-cc-pVQZ levels, respectively, and the mean absolute errors 0.04 D in both cases. Whereas, at the CCSD and CCSD(T) levels, the dipole moment increases with the cardinal number, at the Hartree–Fock level, the change is in the opposite direction. The performance of the MP2 model appears to be less systematic.

Of the molecules listed in Table 15.11, the weakly polar systems CO and O_3 deserve special attention. The dipole moment of CO is very small – about 0.13 D. Moreover, at the Hartree–Fock level, the polarity is reversed, giving an incorrect sign of the dipole moment. The O_3 molecule is special in that the change in the dipole moment from CCSD to CCSD(T) is large (0.08 D), suggesting that the contributions from a full relaxation of the triples at the CCSDT level and from the quadruples at the CCSDTQ level may be substantial. From a comparison of the Hartree–Fock, CCSD and CCSD(T) levels (0.837, 0.642 and 0.559 D, respectively), it appears that the true dipole moment of O_3 is smaller than 0.559 D and probably close to 0.53 D. The problems encountered with O_3 arise from the multiconfigurational character of this molecule.

15.5.5 SUMMARY

For the calculation of dipole moments, the aug-cc-pVXZ basis sets should be used for a flexible description of the outer valence region. In general, however, the quality of the calculations depends more on the correlation treatment than on the cardinal number, the aug-cc-pVTZ basis being sufficient for most applications. Whereas the Hartree–Fock model is typically in error by 0.1–0.2 D, the introduction of correlation at the MP2 and CCSD levels reduces the errors to about 0.05 and 0.04 D, respectively. The CCSD(T) errors are small – typically smaller than 0.01 D. At the coupled-cluster level, the dipole moment usually increases with the cardinal number; at the Hartree–Fock level, the change is typically in the opposite direction but small.

15.6 Molecular and atomic energies

We now begin a series of four sections devoted to different aspects of the electronic energy: total energies in the present section, atomization energies in Section 15.7, reaction enthalpies in Section 15.8 and conformational barriers in Section 15.9. Although atomization energies, reaction enthalpies and conformational barriers are all differences of total energies, their calculation requires quite different levels of theory for convergence and agreement with experimental measurements. To understand the different requirements of these properties, we shall in this section consider the total energy itself. Since, for the total energy, there are no experimental measurements to guide us, we shall proceed differently from the other sections in this chapter, analysing the calculated energies with respect to extrapolated limits, and also comparing the different contributions to the energy with one another so as to build up an understanding of the relative importance of these contributions.

15.6.1 THE TOTAL ELECTRONIC ENERGY

To study the total electronic energy, calculations have been performed at the Hartree–Fock, MP2, CCSD and CCSD(T) levels of theory for the 20 closed-shell molecules in Table 15.1 and their constituent atoms. The molecular calculations were carried out at the spin-restricted all-electron CCSD(T)/cc-pCVQZ geometries. As seen from Tables 15.7 and 15.9, these geometries are close to the true equilibrium geometries. The open-shell atomic calculations were carried out in the same basis sets, using spin-unrestricted wave functions.

For each atomic and molecular system, valence-electron calculations were carried out in the cc-pVXZ basis sets with $X \leq 6$, and all-electron calculations in the cc-pCVXZ basis sets with $X \leq 5$. In addition, for each system, we have generated a *cc-pcV6Z energy*, obtained from the valence-electron cc-pV6Z energy by adding the difference between the all-electron cc-pCV5Z and valence-electron cc-pV5Z energies. The cc-pcV6Z energies, which should be close to the all-electron cc-pCV6Z energies, are listed in Table 15.12.

Total electronic energies were discussed in Section 8.4. From this discussion, we recall that, at the Hartree–Fock level, the cc-pV6Z energies should be in error relative to the basis-set limit by less than 1 mE_h, probably by about 0.1 mE_h. The energies in Table 15.12 may therefore be taken as the true Hartree–Fock energies of these systems at the CCSD(T)/cc-pCVQZ geometry, the uncertainty being in the last quoted digit. However, we also learned in Section 8.4 that the correlation energy converges much more slowly, with an error of about 10 mE_h remaining at

Table 15.12 The total electronic energy (E_h). The systems have been listed in order of increasing absolute energy

	HF	MP2		CCSD		CCSD(T)	
	pV6Z	pcV6Z	pcV(56)Z	pcV6Z	pcV(56)Z	pcV6Z	pcV(56)Z
H	−0.5000	−0.500	−0.500	−0.500	−0.500	−0.500	−0.500
H_2	−1.1336	−1.168	−1.168	−1.174	−1.174	−1.174	−1.174
C	−37.6937	−37.821	−37.824	−37.840	−37.842	−37.843	−37.845
$CH_2(^1A_1)$	−38.8960	−39.102	−39.106	−39.125	−39.127	−39.131	−39.133
CH_4	−40.2171	−40.487	−40.491	−40.504	−40.507	−40.512	−40.515
N	−54.4045	−54.569	−54.572	−54.583	−54.586	−54.587	−54.589
NH_3	−56.2249	−56.543	−56.548	−56.551	−56.554	−56.560	−56.564
O	−74.8189	−75.045	−75.050	−75.059	−75.063	−75.064	−75.067
H_2O	−76.0674	−76.423	−76.429	−76.424	−76.428	−76.434	−76.439
C_2H_2	−76.8556	−77.304	−77.311	−77.311	−77.316	−77.330	−77.336
C_2H_4	−78.0707	−78.546	−78.553	−78.565	−78.570	−78.583	−78.588
HNC	−92.9003	−93.379	−93.387	−93.382	−93.388	−93.403	−93.409
HCN	−92.9157	−93.407	−93.415	−93.406	−93.411	−93.427	−93.433
F	−99.4163	−99.715	−99.721	−99.723	−99.728	−99.729	−99.734
HF	−100.0708	−100.448	−100.455	−100.445	−100.451	−100.454	−100.460
N_2	−108.9930	−109.521	−109.530	−109.513	−109.519	−109.535	−109.542
N_2H_2	−110.0497	−110.616	−110.625	−110.621	−110.627	−110.643	−110.650
CO	−112.7908	−113.301	−113.310	−113.299	−113.305	−113.319	−113.326
CH_2O	−113.9234	−114.478	−114.487	−114.482	−114.488	−114.502	−114.509
HNO	−129.8498	−130.454	−130.464	−130.452	−130.459	−130.476	−130.484
H_2O_2	−150.8523	−151.533	−151.544	−151.531	−151.539	−151.554	−151.563
HOF	−174.8230	−175.524	−175.536	−175.520	−175.530	−175.544	−175.553
CO_2	−187.7252	−188.574	−188.588	−188.555	−188.566	−188.590	−188.601
F_2	−198.7734	−199.500	−199.513	−199.495	−199.505	−199.518	−199.529
O_3	−224.3661	−225.410	−225.427	−225.364	−225.377	−225.419	−225.432

the sextuple-zeta level. Without introducing interelectronic coordinates in the wave function, it is impossible to calculate directly the correlation energy to within 1 mE_h using standard basis sets.

Fortunately, using the extrapolation scheme developed in Section 8.4, it is possible to estimate the basis-set limit rather accurately, reducing the error in the calculated electronic energy by an order of magnitude. Assume that we have calculated the total energy at the all-electron cc-pCVXZ and cc-pCVYZ levels, with $X < Y$:

$$E_X = E_X^{HF} + E_X^{cor} \tag{15.6.1}$$

$$E_Y = E_Y^{HF} + E_Y^{cor} \tag{15.6.2}$$

As described in Section 8.4.3, the extrapolated energy is now given as

$$E_{XY} = E_Y^{HF} + \frac{X^3 E_X^{cor} - Y^3 E_Y^{cor}}{X^3 - Y^3} \tag{15.6.3}$$

The cc-pCV(XY)Z energies obtained in this manner are used extensively both in this section and in the subsequent sections on atomization energies and reaction enthalpies. Since atomization energies and reaction enthalpies are simple linear combinations of total molecular and atomic energies, the linear extrapolation (15.6.3) can be carried out either before or after the combination of atomic and molecular energies into atomization energies and reaction enthalpies, without affecting the results.

In Section 8.4, we found that, for molecules containing only first-row atoms, the cc-pCV(56)Z energies are within 1 mE_h of the basis-set limit (as established by the R12 model). However, since the all-electron cc-pCV6Z energies are not available to us, we have in Table 15.12 instead listed the cc-pcV(56)Z energies, obtained from the valence cc-pV(56)Z energies by adding a core correction equal to the difference between the all-electron cc-pCV(Q5)Z and valence-electron cc-pV(Q5)Z energies. The resulting extrapolated MP2, CCSD and CCSD(T) energies in Table 15.12 should be within 1–2 mE_h of the all-electron basis-set limits, making them suitable for accurate predictions of atomization and reaction energies.

15.6.2 CONTRIBUTIONS TO THE TOTAL ELECTRONIC ENERGY

From Table 15.12, it is immediately apparent that the total energies are dominated by the Hartree–Fock contribution, which increases in magnitude with increasing nuclear charges of the molecular systems. However, to analyse the contributions to the total electronic energy, it is better to express it differently. In Table 15.13, we have divided the electronic energy by the number of electrons present in the system. Moreover, rather than giving the total energies, we have listed separately the Hartree–Fock contribution, the total correlation contribution, the MP2 doubles contribution, the CCSD singles-and-doubles contribution and the CCSD(T) triples contribution. We see that the Hartree–Fock contribution is not only much larger than the correlation contribution, but also differs more among the molecules.

Apart from the hydrogen atom, the least strongly correlated system is H_2, whose correlation energy per electron is only -20.4 mE_h. For the remaining molecules, the mean correlation energy per electron is -37.2 mE_h. The most weakly correlated molecules are CH_2 and CH_4, with correlation energies per electron of only -29.7 and -29.8 mE_h, respectively. By contrast, the most strongly correlated molecules are the electron-rich systems F_2 and O_3, whose correlation energies per electron are -42.0 and -44.4 mE_h, respectively.

With a mean correlation energy per electron of -29.5 mE_h, the atoms are in general less strongly correlated than the molecules. Thus, for carbon and nitrogen, the CCSD(T) correlation energy per

Table 15.13 The contributions to the total electronic energy per electron (mE$_h$). The atomic and molecular systems have been listed in order of increasing magnitude of the Hartree–Fock energy per electron. All energies have been calculated at the cc-pcV6Z level

	HF	Correlation	D/MP2	SD/CCSD	T/CCSD(T)
H	−500.0	0.0	0.0	0.0	0.0
H$_2$	−566.8	−20.4	−17.1	−20.4	0.0
CH$_4$	−4 021.7	−29.8	−27.4	−29.0	−0.8
CH$_2$($^1A_1$)	−4 862.0	−29.7	−26.2	−28.9	−0.8
C$_2$H$_4$	−4 879.4	−32.4	−30.2	−31.2	−1.1
C$_2$H$_2$	−5 489.7	−34.3	−32.5	−32.9	−1.4
NH$_3$	−5 622.5	−33.9	−32.3	−32.9	−1.0
C	−6 282.3	−25.1	−21.7	−24.7	−0.5
HNC	−6 635.7	−36.3	−34.8	−34.8	−1.5
HCN	−6 636.8	−36.9	−35.7	−35.4	−1.5
N$_2$H$_2$	−6 878.1	−37.5	−36.0	−36.1	−1.5
CH$_2$O	−7 120.2	−36.6	−35.3	−35.3	−1.3
H$_2$O	−7 606.7	−37.1	−36.2	−36.1	−1.0
N	−7 772.1	−26.3	−24.0	−25.9	−0.5
N$_2$	−7 785.2	−39.2	−38.3	−37.6	−1.6
CO	−8 056.5	−38.2	−37.1	−36.7	−1.5
HNO	−8 115.6	−39.6	−38.4	−38.1	−1.5
H$_2$O$_2$	−8 380.7	−39.5	−38.4	−38.2	−1.3
CO$_2$	−8 533.0	−39.8	−39.2	−38.2	−1.6
O	−9 352.4	−31.0	−28.9	−30.5	−0.6
O$_3$	−9 348.6	−44.4	−44.2	−42.1	−2.3
HOF	−9 712.4	−40.6	−39.6	−39.3	−1.3
HF	−10 007.1	−38.9	−38.4	−38.0	−0.9
F	−11 046.3	−35.3	−33.9	−34.7	−0.6
F$_2$	−11 043.0	−42.0	−41.1	−40.6	−1.3

electron is only −25.1 and −26.3 mE$_h$, respectively. Not surprisingly, the electron-richer oxygen and fluorine atoms are more strongly correlated, with correlation energies per electron of −31.0 and −35.3 mE$_h$, respectively.

The difference between the atoms and molecules is perhaps best seen by comparing the correlation energy per electron in N and N$_2$ (−26.3 and −39.2 mE$_h$, respectively), in O and O$_3$ (−31.0 and −44.4 mE$_h$), and in F and F$_2$ (−35.3 and −42.0 mE$_h$). Clearly, the open-shell electronic systems of the atoms are considerably less strongly correlated than the closed-shell molecular systems, where all electrons are paired. It is noteworthy that, for O and O$_3$ and also for F and F$_2$, the Hartree–Fock energy *increases* as the molecules are formed (by +3.8 and +3.3 mE$_h$, respectively), implying that O$_3$ and F$_2$ are bonded by correlation. Clearly, the calculation of atomization energies will turn out to be a difficult task, requiring an accurate description of electron correlation.

Examining the various contributions to the correlation energy per electron in Table 15.13, we find that it is dominated by the doubles correction, which in the molecules is about 30 times larger than the triples correction and in the atoms about 50 times larger. We note that, in weakly correlated systems, the MP2 correction is smaller than the CCSD correction; conversely, in strongly correlated systems, the CCSD correction is smaller than the MP2 correction. Also, the MP2 correction is significantly smaller (relative to the CCSD correction) in atoms than in molecules. Per electron, the triples energy is about −0.5 mE$_h$ in the atoms and −1.3 mE$_h$ in the molecules. The largest triples correction per electron of −2.3 mE$_h$ is found in O$_3$.

The relative importance of the different contributions to the total electronic energy is illustrated in Table 15.14. Except for H$_2$, the Hartree–Fock energy constitutes more than 99% of the total energy

Table 15.14 Contributions to the extrapolated CCSD(T) energy (%)

	HF	Correlation	SD		T	
			Valence	Core	Valence	Core
H_2	96.52	3.48	100.00	0.00	0.00	0.00
CH_4	99.27	0.73	78.30	19.11	2.44	0.14
C_2H_4	99.34	0.66	74.60	21.91	3.30	0.19
C_2H_2	99.38	0.62	72.27	23.65	3.86	0.23
CH_2	99.39	0.61	74.05	23.33	2.39	0.22
NH_3	99.40	0.60	79.51	17.60	2.76	0.13
HCN	99.45	0.55	73.43	22.42	3.94	0.22
HNC	99.46	0.54	73.16	22.75	3.88	0.21
N_2H_2	99.46	0.54	76.44	19.69	3.69	0.18
CH_2O	99.49	0.51	76.22	20.28	3.33	0.17
N_2	99.50	0.50	74.35	21.55	3.89	0.20
HNO	99.51	0.49	77.05	19.08	3.70	0.17
H_2O	99.51	0.49	80.36	16.85	2.68	0.11
O_3	99.53	0.47	77.44	17.37	5.02	0.17
CO	99.53	0.47	74.05	22.09	3.67	0.19
H_2O_2	99.53	0.47	79.15	17.48	3.22	0.14
CO_2	99.54	0.46	75.32	20.71	3.79	0.17
HOF	99.58	0.42	79.37	17.39	3.11	0.14
C	99.60	0.40	61.77	36.27	1.69	0.26
HF	99.61	0.39	80.81	16.81	2.29	0.09
F_2	99.62	0.38	79.65	17.17	3.05	0.13
N	99.66	0.34	66.43	31.71	1.70	0.16
O	99.67	0.33	73.25	24.92	1.70	0.13
F	99.68	0.32	77.72	20.48	1.70	0.10
H	100.00	0.00	–	–	–	–

and again its dominance is most pronounced for the atoms. (Note that the large relative correlation energies of H_2 and of the hydrocarbons do not imply that they are strongly correlated but arise from small Hartree–Fock energies.) The correlation energy is dominated by the valence doubles part, which amounts to 70–80% of the total correlation energy, the core doubles contributing only about 20%. The triples contribute less than 2% in all atoms but more than 2% in the molecules and as much as 5% in O_3. The triples core contribution is very small, less than 0.3% in all systems.

15.6.3 BASIS-SET CONVERGENCE

Having studied the total energy and its different contributions in the basis-set limit, we now turn our attention to the basis-set dependence of the energy. For a first impression of the different contributions to the energy and their basis-set dependence, we have in Figure 15.9 plotted the Hartree–Fock, MP2, CCSD and CCSD(T) energies of C_2H_4 and CO_2 as functions of the cardinal number. From Section 15.6.2, we recall that C_2H_4 and CO_2 are rather different in the sense that C_2H_4 is a weakly correlated system where the MP2 correction is smaller than the CCSD correction, whereas CO_2 is strongly correlated, the MP2 correction being larger than the CCSD correction. Nonetheless, the plots in Figure 15.9 are rather similar and typical of all molecules in our sample.

From Figure 15.9, we first note that the basis-set dependence of the Hartree–Fock model is not only smaller than for the correlated models but also small compared with the difference between the correlated and uncorrelated energies. On the other hand, the differences between the various

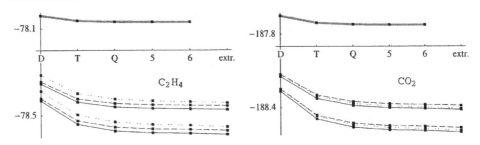

Fig. 15.9. The electronic energies (in E_h) of C_2H_4 and CO_2, plotted as functions of the cardinal number for Hartree–Fock (thick grey line), MP2 (dotted line), CCSD (dashed line) and CCSD(T) (full line). For each molecule, both the valence-electron and all-electron correlated total energies have been plotted.

correlated models are rather small – they all converge in much the same manner and the differences between the basis sets are large compared with the differences between the models – at least for small cardinal numbers.

In Table 15.15, we have listed the mean and maximum absolute errors relative to the basis-set limit for the Hartree–Fock energy and for the correlation contributions to the total energy. The errors, which are based on the molecular and atomic systems in Table 15.12, confirm our expectations that the Hartree–Fock energy converges considerably faster than the correlation energy. We also note that, for a given allowed absolute error in the energy, the triples correction converges significantly faster than the doubles correction. At the cc-pCVQZ level, for example, the maximum absolute error in the CCSD(T) triples correction is only 2.5 mE_h, compared with 39.8 mE_h in the singles-and-doubles correction and 5.0 mE_h in the Hartree–Fock energy. Even at the cc-pCVTZ level, the triples error is quite small, the mean error being 2.4 mE_h and the maximum absolute error 6.3 mE_h. These results indicate that it is unnecessary to calculate the CCSD(T) triples correction in the same basis as the CCSD energy, leading to large savings in the calculations.

In Table 15.16, we have listed the proportion of the energy that is recovered by the different basis sets. Even the smallest sets recover nearly 100% of the Hartree–Fock energy. However, the remaining error may still be important since the differential energies that are of interest in chemical problems often constitute no more than a few parts per million of the total electronic energy. For instance, at the quadruple-zeta level, where 100.0% of the Hartree–Fock energy has been recovered, the maximum error in Table 15.15 is 5.0 mE_h.

Table 15.15 Mean and maximum absolute errors (mE_h) relative to the cc-pV6Z energy for the Hartree–Fock method and relative to the cc-p(c)V(56)Z correlation energies for the correlated methods

		All-electron cc-pCVXZ				Valence-electron cc-pVXZ		
		HF	MP2	SD	T	MP2	SD	T
$\overline{\Delta}$	DZ	39.6	140.8	127.6	8.2	108.6	95.7	7.9
	TZ	9.3	52.6	43.4	2.4	43.4	34.4	2.5
	QZ	2.0	22.9	17.1	0.9	19.9	14.0	1.0
	5Z	0.2	11.8	8.4	0.4	10.2	6.8	0.5
	6Z	0.0	7.5	5.5	0.2	5.9	3.9	0.3
Δ_{max}	DZ	97.9	310.0	280.7	21.6	245.7	216.4	20.9
	TZ	22.5	118.0	98.3	6.3	101.6	81.8	6.9
	QZ	5.0	52.5	39.8	2.5	47.3	34.3	3.0
	5Z	0.6	27.0	19.6	1.1	24.3	16.8	1.4
	6Z	0.0	16.7	12.5	0.6	14.1	9.7	0.8

Table 15.16 The proportion of the basis-set limit energy (%) recovered by the basis sets, calculated relative to the cc-pV6Z energy for the Hartree–Fock model and relative to the cc-p(c)V(56)Z correlation energy for the correlated models.

	All-electron cc-pCVXZ				Valence-electron cc-pVXZ		
	HF	MP2	SD	T	MP2	SD	T
DZ	99.9	68.5	72.1	44.7	69.3	73.7	44.1
TZ	100.0	88.4	90.7	83.4	88.0	90.8	81.6
QZ	100.0	95.0	96.4	93.8	94.5	96.3	92.6
5Z	100.0	97.4	98.2	97.2	97.2	98.2	96.7
6Z	100.0	98.3	98.8	98.6	98.4	99.0	98.1

Turning our attention to the correlation energy, we find that the cc-pCVDZ set usually recovers about 70% of the doubles energy but only about 50% of the triples energy; for cc-pCVTZ, the corresponding proportions are about 90% and 85%. Recalling our recent discussion of the errors in Table 15.15, it is perhaps surprising that larger sets are needed for the triples than for the doubles to recover a given proportion of the correlation energy. Clearly, the small errors in the triples energy in Table 15.15 do not arise from a more rapid rate of convergence – the triples converge like the doubles – but rather from the smallness of the correction.

Let us finally point out an interesting relation between the changes in the energy with improvements in the basis and the energy itself. In Figure 15.10, the change in the energy from the cc-pCVDZ basis to the basis-set limit has been plotted against the basis-set limit energy – for the Hartree–Fock energy, the CCSD singles-and-doubles energy and the CCSD(T) triples energy. The plotted points lie along the diagonal, indicating that the changes in the energy with the cardinal number are proportional to the total (Hartree–Fock or correlation) energy itself. In short, if we increase the cardinal number, the energy is reduced most for systems of low energy.

As we shall see in Sections 15.7 and 15.8, this observation may help us predict the changes that occur in atomization energies and reaction enthalpies as we increase the cardinal number. Here, we note that it may also be used to rationalize the bond contraction that occurs as we increase the cardinal number – see Section 15.3.2. In a molecular system, the electronic energy (omitting the nuclear–nuclear part) is always lowered as we contract bonds from the equilibrium geometry. Therefore, when we increase the cardinal number, the lowering of the electronic energy will be larger for the contracted system, leading to a shortening of the equilibrium bond lengths.

15.6.4 CCSDT CORRECTIONS

Although our investigations are chiefly concerned with the Hartree–Fock, MP2, CCSD and CCSD(T) models, we shall briefly consider also the CCSDT model – both in this section and

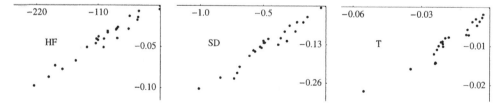

Fig. 15.10. The energy change from cc-pCVDZ to the basis-set limit plotted against the basis-set limit for the Hartree–Fock energy, the CCSD singles-and-doubles energy and the CCSD(T) triples energy (in E_h).

in the section on atomization energies. Because of the high cost of the model, the CCSDT study has been restricted to CH_2, H_2O, HF, N_2, F_2 and CO. Likewise, the cardinal number has been restricted to $2 \leq X \leq 5$ for the valence-electron cc-pVXZ calculations and to $2 \leq X \leq 4$ for the all-electron cc-pCVXZ calculations. In addition, a set of CCSDT/cc-pcV5Z energies has been generated in the same manner as the cc-pcV6Z energies in Section 15.6.1. The core–valence energies will be used in our study of atomization energies; in the present section, we consider only the valence-electron energies.

In Table 15.17, we have listed the valence-electron CCSDT energies for the six selected molecules. The main point of interest is the difference between the CCSDT and CCSD(T) models. In most cases, the CCSDT model changes the energy only very little relative to the CCSD(T) model. The largest change occurs for the CH_2 molecule, where the CCSDT model reduces the energy by about 0.8 mE_h and where the CCSD(T) triples correction constitutes only 88% of the total CCSDT triples correction.

For the atoms, the shift in the energy from CCSD(T) to CCSDT is negative. Except for CH_2, the molecular shifts are small and positive for the larger basis sets but negative for the smaller sets. Owing to the smallness of these shifts, we expect the changes in the atomization energies and the reaction enthalpies to be small as well. However, the changes should be largest for the atomization energies since no cancellation usually occurs between the atoms and the molecule. In the quintuple-zeta basis, for example, the changes from CCSD(T) to CCSDT are 0.53 mE_h for N_2 and -0.31 mE_h for N.

15.6.5 MOLECULAR VIBRATIONAL CORRECTIONS

Up to this point, we have discussed only the nonrelativistic electronic contribution to the total energy within the Born–Oppenheimer approximation $E(\mathbf{R})$. The total energy of a molecular system also has a vibrational contribution. For the vibrational ground state, the total energy is obtained by adding the *zero-point vibrational energy (ZPVE)* E_{ZPV} to the electronic equilibrium energy $E(\mathbf{R}_e)$:

$$E_0 = E(\mathbf{R}_e) + E_{ZPV} \tag{15.6.4}$$

Our purpose here is to outline how ZPVEs may be obtained, without giving a detailed account of vibrational theory. We begin by discussing the evaluation of ZPVEs in the *harmonic approximation*, where we replace the true molecular potential-energy surface by a harmonic surface.

Expanding the surface in a set of internal coordinates $\mathbf{R}$ about the minimum $\mathbf{R}_e$, we obtain

$$E(\mathbf{R}) = E(\mathbf{R}_e) + \tfrac{1}{2} \sum_{ij} F_{ij} R_i R_j \tag{15.6.5}$$

where the F_{ij} are the *quadratic force constants*. In a set of coordinates that diagonalizes the force-constant matrix $\mathbf{F}$ in a suitable metric, the vibrational energy becomes

$$E(\mathbf{v}) = \sum_i \left(v_i + \tfrac{1}{2} d_i \right) \omega_i \tag{15.6.6}$$

where ω_i are the *harmonic frequencies* (obtained by diagonalizing $\mathbf{F}$), d_i the *degeneracies* of the vibrational levels, and v_i the *vibrational quantum numbers*. In the ground state, all v_i are zero, yielding the following expression for the ZPVE in the harmonic approximation:

$$E_{ZPV}^{hrm} = \tfrac{1}{2} \sum_i d_i \omega_i \tag{15.6.7}$$

Table 15.17 Breakdown of the valence-electron CCSDT energies at the all-electron CCSD(T)/cc-pCVQZ geometries

		Total energy (E_h)		Energy corrections (mE_h)		
		CCSDT	HF	HF $\to$ SD	SD $\to$ (T)	(T) $\to$ T
C	DZ	-37.76069	-37.68654	-72.91	-0.93	-0.31
	TZ	-37.78123	-37.69157	-87.16	-2.04	-0.46
	QZ	-37.78700	-37.69331	-90.87	-2.36	-0.46
	5Z	-37.78868	-37.69368	-92.09	-2.47	-0.44
CH$_2$($^1A_1$)	DZ	-39.02274	-38.88109	-138.05	-2.89	-0.72
	TZ	-39.06218	-38.89237	-164.19	-4.81	-0.81
	QZ	-39.07269	-38.89514	-171.42	-5.34	-0.78
	5Z	-39.07578	-38.89588	-173.63	-5.53	-0.75
N	DZ	-54.47863	-54.39111	-86.76	-0.63	-0.13
	TZ	-54.51503	-54.40069	-111.75	-2.28	-0.32
	QZ	-54.52516	-54.40372	-118.30	-2.81	-0.33
	5Z	-54.52825	-54.40444	-120.51	-2.99	-0.31
O	DZ	-74.91006	-74.79217	-117.03	-0.75	-0.11
	TZ	-74.97425	-74.81176	-159.29	-2.91	-0.29
	QZ	-74.99388	-74.81729	-172.58	-3.69	-0.32
	5Z	-75.00041	-74.81878	-177.34	-3.99	-0.30
H$_2$O	DZ	-76.24121	-76.02680	-211.21	-3.04	-0.16
	TZ	-76.33228	-76.05716	-267.40	-7.65	-0.08
	QZ	-76.35981	-76.06482	-285.98	-8.99	-0.01
	5Z	-76.36899	-76.06708	-292.44	-9.52	0.04
F	DZ	-99.52765	-99.37524	-151.45	-0.88	-0.08
	TZ	-99.62054	-99.40552	-211.25	-3.59	-0.18
	QZ	-99.65045	-99.41377	-231.86	-4.63	-0.19
	5Z	-99.66080	-99.41605	-239.52	-5.06	-0.17
HF	DZ	-100.22823	-100.01946	-206.76	-1.92	-0.09
	TZ	-100.33838	-100.05807	-273.93	-6.36	-0.03
	QZ	-100.37316	-100.06774	-297.61	-7.83	0.02
	5Z	-100.38511	-100.07049	-306.26	-8.43	0.07
N$_2$	DZ	-109.27538	-108.95407	-309.36	-11.87	-0.08
	TZ	-109.37357	-108.98339	-371.97	-18.50	0.28
	QZ	-109.40397	-108.99100	-393.20	-20.18	0.41
	5Z	-109.41366	-108.99268	-400.70	-20.82	0.53
CO	DZ	-113.05488	-112.74923	-294.53	-10.67	-0.46
	TZ	-113.15561	-112.78027	-358.27	-16.93	-0.14
	QZ	-113.18789	-112.78877	-380.59	-18.53	-0.00
	5Z	-113.19807	-112.79053	-388.52	-19.13	0.11
F$_2$	DZ	-199.09773	-198.68573	-402.72	-9.01	-0.27
	TZ	-199.29609	-198.75212	-526.01	-17.97	0.01
	QZ	-199.35884	-198.76835	-569.69	-20.86	0.06
	5Z	-199.38084	-198.77282	-586.13	-22.04	0.14

However, the true surface is anharmonic. An improved treatment begins with the *quartic force field*, which contains cubic and quartic terms as well as the quadratic ones:

$$E(\mathbf{R}) = E(\mathbf{R}_e) + \frac{1}{2}\sum_{ij} F_{ij}R_iR_j + \frac{1}{6}\sum_{ijk} F_{ijk}R_iR_jR_k + \frac{1}{24}\sum_{ijkl} F_{ijkl}R_iR_jR_kR_l \qquad (15.6.8)$$

Here, the F_{ijk} and F_{ijkl} are the *cubic and quartic force constants*, respectively. Second-order perturbation theory now gives the following expression for the vibrational energy

$$E(\mathbf{v}) = \sum_i \left(v_i + \tfrac{1}{2}d_i\right)\omega_i + \sum_{i \geq j}\left(v_i + \tfrac{1}{2}d_i\right)\left(v_j + \tfrac{1}{2}d_j\right)X_{ij} \tag{15.6.9}$$

where the X_{ij} are the *anharmonic constants* and the summations are over the distinct harmonic frequencies. We shall not discuss the second-order perturbation method here, noting only that it breaks down whenever there are close-lying vibrational states. The ZPVE is now separated into a harmonic term and an anharmonic term

$$E_{\text{ZPV}} = E_{\text{ZPV}}^{\text{hrm}} + E_{\text{ZPV}}^{\text{anh}} \tag{15.6.10}$$

where the harmonic term is given by (15.6.7) and the anharmonic term by

$$E_{\text{ZPV}}^{\text{anh}} = \tfrac{1}{4}\sum_{i \geq j}d_i d_j X_{ij} \tag{15.6.11}$$

In principle, therefore, it is straightforward to generate the harmonic frequencies and anharmonic constants from electronic-structure calculations although the large number of quartic constants – for example, 3060 in a system of seven atoms – makes this approach cumbersome for large molecules.

It is less straightforward to extract the harmonic frequencies and anharmonic constants from experimental data since the force constants are not observables. Instead, the force constants are obtained from fits to the vibrational levels. From an initial set of force constants, the harmonic frequencies ω_i and anharmonic constants X_{ij} are calculated, and the vibrational energy levels are predicted from (15.6.9). The differences between the predicted and experimental energy levels are next used to modify the force constants. This procedure is repeated until convergence.

In Table 15.18, we have listed the available theoretical and experimental ZPVEs for the molecules in Table 15.1. The harmonic and anharmonic energies are quoted separately except for CH_2, where this separation is impossible due to the breakdown of perturbation theory. The theoretical ZPVEs have been obtained from quartic CCSD(T) force fields, calculated in a basis of at least triple-zeta quality. Where both theoretical and experimental data are available, the difference in the ZPVEs is always less than 0.30 kJ/mol – highly accurate ZPVEs can thus be generated from *ab initio* calculations.

The harmonic approximation overestimates the ZPVEs by about 1%; the anharmonic corrections are most important for molecules containing hydrogen atoms, the largest correction of 2.3 kJ/mol occurring for CH_4. Comparing the harmonic and anharmonic energies, we conclude that a treatment more elaborate than the quartic force field (15.6.8) is unnecessary since the corrections would be considerably smaller than the errors in the calculated electronic energies, noting that the vibrational energy contributes only about 0.03% to the total energy. To correct the experimental atomization energies in Section 15.7 and reaction enthalpies in Section 15.8 for vibrational contributions, we shall use the experimental ZPVEs where available and the theoretical ZPVEs otherwise. Comparing the experimental and theoretical ZPVEs, we estimate the errors in the ZPVEs to be at most 0.5 kJ/mol.

15.6.6 RELATIVISTIC CORRECTIONS

The electronic energies discussed so far have been obtained from approximate solutions to the nonrelativistic Schrödinger equation. To calculate relative energies such as atomization energies and reaction enthalpies to an accuracy of 1 kJ/mol or better, we must also take into account relativistic

Table 15.18 Harmonic, anharmonic and total zero-point vibrational energies (kJ/mol)

Molecule	Method	Harmonic	Anharmonic	Total
F_2	experiment[a]	5.48	−0.03	5.45
H_2	experiment[a]	26.33	−0.36	25.96
HF	experiment[a]	24.75	−0.27	24.48
O_3	experiment[b]	17.59	−0.23	17.35
	CCSD(T)/cc-pVTZ[c]	17.49	−0.36	17.13
HOF	exp. (empirical potential)[d]	36.34	−0.45	35.89
$CH_2(^1A_1)$	variational calculation[e]	–	–	43.22
HNO	CCSD(T)/cc-pVQZ[f]	36.51	−0.72	35.79
N_2	experiment[a]	14.11	−0.04	14.06
H_2O	experiment[g]	56.36	−0.92	55.44
CO	experiment[a]	12.98	−0.04	12.94
HNC	experiment[h]	41.20	−0.59	40.62
	CCSD(T)/ANO(TZP)[h]	40.81	−0.41	40.40
NH_3	CCSD(T) cc-pVQZ[i]	90.57	−1.59	88.98
HCN	experiment[h]	42.00	−0.39	41.61
	CCSD(T)/ANO(TZP)[h]	41.94	−0.40	41.55
CH_2O	experiment[j]	70.41	−1.26	69.14
	CCSD(T)/cc-pVTZ[k]	70.08	−1.14	68.93
CO_2	experiment[l]	30.48	−0.19	30.29
	CCSD(T)/cc-pVTZ[m]	30.29	−0.17	30.11
C_2H_2	experiment[n]	69.74	−0.94	68.80
CH_4	experiment[o]	118.21	−2.28	115.93
	CCSD(T)/cc-pVQZ[p]	117.74	−2.09	115.64
C_2H_4	CCSD(T)[q]	133.80	−1.58	132.21

[a]K. P. Huber and G. Herzberg, *Constants of Diatomic Molecules*, Van Nostrand Reinhold, 1979.
[b]A. Barbe, C. Secroun and P. Jouve, *J. Mol. Spectrosc.* **49**, 171 (1974).
[c]T. J. Lee and G. E. Scuseria, *J. Chem. Phys.* **93**, 489 (1990).
[d]L. Halonen and T.-K. Ha, *J. Chem. Phys.* **89**, 4885 (1988).
[e]P. Jensen and P. R. Bunker, *J. Chem. Phys.* **89**, 1327 (1988).
[f]C. E. Dateo, T. J. Lee and D. W. Schwenke, *J. Chem. Phys.* **101**, 5853 (1994).
[g]A. R. Hoy, I. M. Mills and G. Strey, *Mol. Phys.* **24**, 1265 (1972).
[h]See T. J. Lee, C. E. Dateo, B. Gazdy and J. M. Bowman, *J. Phys. Chem.* **97**, 8937 (1993). The experimental value for HCN is based on an experimental quartic force field of S. Carter, I. M. Mills and N. C. Handy, *J. Chem. Phys.* **99**, 4379 (1993). The experimental value for HNC is given by R. A. Creswell and A. G. Robiette, *Mol. Phys.* **36**, 869 (1978).
[i]J. M. L. Martin, T. J. Lee and P. R. Taylor, *J. Chem. Phys.* **97**, 8361 (1992).
[j]D. E. Reisner, R. W. Field, J. L. Kinsey and H.-L. Dai, *J. Chem. Phys.* **80**, 5968 (1984).
[k]J. M. L. Martin, T. J. Lee and P. R. Taylor, *J. Mol. Spectrosc.* **160**, 105 (1993).
[l]J.-L. Teffo, O. N. Sulakshina and V. I. Perevalov, *J. Mol. Spectrosc.* **156**, 48 (1992).
[m]J. M. L. Martin, P. R. Taylor and T. J. Lee, *Chem. Phys. Lett.* **205**, 535 (1993).
[n]B. C. Smith and J. S. Winn, *J. Chem. Phys.* **89**, 4638 (1988).
[o]D. L. Gray and A. G. Robiette, *Mol. Phys.* **37**, 1901 (1979).
[p]T. J. Lee, J. M. L. Martin and P. R. Taylor, *J. Chem. Phys.* **102**, 254 (1995).
[q]J. M. L. Martin and P. R. Taylor, *Chem. Phys. Lett.* **248**, 336 (1996). The vibrational contributions have been obtained from a CCSD(T) force field, with the harmonic constants extrapolated to the basis-set limit and the anharmonic constants calculated in the cc-pVTZ basis.

corrections. A fully relativistic treatment based on the Dirac equation and its extension to many electrons is beyond the scope of this book. Fortunately, for molecules without heavy elements, we need only include the main relativistic corrections: the first-order spin–orbit correction and the first-order one-electron spin-free corrections. In this subsection, we first show how the spin–orbit correction may be obtained from experimental data – this treatment is convenient as the presence of spin means that nonrelativistic programs cannot easily be used to evaluate the spin–orbit correction. Next, we discuss the spin-free corrections, which are easily evaluated in the same manner as for example the molecular dipole moment.

First-order spin–orbit splittings occur only for open-shell systems for which several couplings are possible between the spin angular momentum and the orbital angular momentum of the electrons. Thus, among the systems in Table 15.12, we need only consider the spin–orbit coupling for the carbon, oxygen and fluorine atoms; there are no first-order spin–orbit corrections for the high-spin atoms hydrogen and nitrogen nor for the closed-shell molecular systems.

In a nonrelativistic calculation on an atom of orbital angular momentum L and spin S, we obtain the term energy $E(^{2S+1}L)$. The spin–orbit operator splits this term into levels $^{2S+1}L_J$, whose energies may be expressed as

$$E(^{2S+1}L_J) = E(^{2S+1}L) + \Delta E_{SO}(^{2S+1}L_J) \tag{15.6.12}$$

The first-order splittings are obtained by constructing and diagonalizing the spin–orbit matrix $\mathbf{H}_{SO}$ in the space of the $(2S+1)(2L+1)$ degenerate components of the nonrelativistic ^{2S+1}L atom, producing a number of levels $^{2S+1}L_J$ of different energies and with $|L - S| \leq J \leq |L + S|$. The corrections $\Delta E_{SO}(^{2S+1}L_J)$ are thus the eigenvalues of $\mathbf{H}_{SO}$. In the following, we shall see how these corrections may be obtained from tabulated atomic level energies.

Since the $(2S+1)(2L+1)$ degenerate wave functions may be chosen to be real and since the expectation value of a Hermitian imaginary operator vanishes for real wave functions, we obtain for the spin–orbit operator

$$\mathrm{Tr}\,\mathbf{H}_{SO} = 0 \tag{15.6.13}$$

Furthermore, since the corrections $\Delta E_{SO}(^{2S+1}L_J)$ may be obtained by diagonalizing $\mathbf{H}_{SO}$ and since the trace of a matrix is invariant to unitary transformations, they must satisfy the relation

$$\sum_J (2J + 1)\Delta E_{SO}(^{2S+1}L_J) = 0 \tag{15.6.14}$$

where $2J + 1$ is the degeneracy of the level $^{2S+1}L_J$. Summing (15.6.12) over all states in the degenerate zero-order space and using (15.6.14), we may express $E(^{2S+1}L)$ as a weighted sum of the level energies:

$$E\left(^{2S+1}L\right) = \frac{\sum_J (2J + 1)E(^{2S+1}L_J)}{\sum_J (2J + 1)} \tag{15.6.15}$$

Substituting this result in (15.6.12), we arrive at the following expression for the spin–orbit correction of the lowest atomic level:

$$\Delta E_{SO}\left(^{2S+1}L_{J_{\min}}\right) = E\left(^{2S+1}L_{J_{\min}}\right) - \frac{\sum_J (2J + 1)E(^{2S+1}L_J)}{\sum_J (2J + 1)} \tag{15.6.16}$$

Table 15.19 contains the ground-state spin–orbit corrections $\Delta E_{SO}(^{2S+1}L_{J_{\min}})$ for the atoms in our sample molecules, calculated from (15.6.16) using the experimental energies $E(^{2S+1}L_J)$ tabulated by Moore [6]. For hydrogen and nitrogen, there are no first-order corrections.

The spin–orbit interaction is important as it splits otherwise degenerate states. However, for the total electronic energy, other relativistic corrections are more important. For light atoms, in

Table 15.19 First-order relativistic corrections (mE$_h$). The spin−orbit correction has been obtained from experimental data as described in the text; the one-electron scalar corrections have been calculated at the all-electron CCSD(T)/cc-pCVQZ level

System	Spin−orbit correction	One-electron scalar corrections		
		Mass-velocity	Darwin	Total
F_2	0.00	−837.24	662.29	−174.95
H_2	0.00	−0.08	0.07	−0.01
HF	0.00	−417.57	330.39	−87.18
O_3	0.00	−770.65	613.27	−157.38
HOF	0.00	−674.86	535.11	−139.75
$CH_2(^1A_1)$	0.00	−77.49	62.48	−15.01
HNO	0.00	−403.65	321.98	−81.67
N_2	0.00	−294.76	235.98	−58.78
H_2O	0.00	−255.87	203.69	−52.18
CO	0.00	−334.19	266.71	−67.48
H_2O_2	0.00	−512.66	408.05	−104.60
N_2H_2	0.00	−293.89	235.40	−58.49
HNC	0.00	−224.13	179.90	−44.23
NH_3	0.00	−146.49	117.38	−29.11
HCN	0.00	−224.35	180.06	−44.29
CH_2O	0.00	−333.31	266.11	−67.21
CO_2	0.00	−589.67	470.11	−119.56
C_2H_2	0.00	−154.11	124.26	−29.85
CH_4	0.00	−76.93	62.07	−14.86
C_2H_4[a]	0.00	−151.49	121.79	−29.69
H $(^2S_{1/2})$	0.00	−0.03	0.02	−0.01
C $(^3P_0)$	−0.13	−77.90	62.76	−15.14
N $(^4S_{3/2})$	0.00	−147.82	118.32	−29.50
O $(^3P_2)$	−0.36	−257.29	204.69	−52.60
F $(^2P_{3/2})$	−0.61	−418.63	331.13	−87.50

[a]CCSD(T)/cc-pCVTZ results.

particular, the main corrections arise from two scalar (i.e. spin-free) operators: the mass-velocity operator H_{MV} and the one-electron Darwin operator H_{1D}:

$$H_{MV} = -\frac{1}{8c^2} \sum_i \nabla_i^4 \qquad (15.6.17)$$

$$H_{1D} = \frac{\pi}{2c^2} \sum_{il} Z_l \delta(\mathbf{r}_{il}) \qquad (15.6.18)$$

In Table 15.19, we have listed the first-order corrections arising from these scalar operators

$$E_{MVD} = \langle H_{MV} + H_{1D} \rangle \qquad (15.6.19)$$

as calculated at the all-electron CCSD(T)/cc-pCVQZ level using the variational density matrices of Section 13.5.3. Even though the two corrections partly cancel, their combined contribution is still as large as about 10% of the total correlation energy. However, unlike the correlation energy, this scalar relativistic correction does not change much in the course of molecular rearrangements and reactions.

Consider, for example, the dissociation of the flourine molecule at the all-electron CCSD(T)/cc-pCVQZ level. For this molecule, the correction from the mass-velocity and Darwin operators is

-174.95 mE$_h$, which may be compared with the total correlation energy of -720 mE$_h$. For the two separate atoms, the total relativistic correction is -175.00 mE$_h$ and the correlation energy -602 mE$_h$. Thus, whereas electron correlation makes a large contribution of 118 mE$_h$ to the dissociation energy, the scalar relativistic correction is only -0.05 mE$_h$. Indeed, for this particular dissociation, the spin–orbit correction of 1.22 mE$_h$ is larger than the mass-velocity and Darwin corrections. The reason for the very nearly constant contribution from the mass-velocity and Darwin operators (15.6.17) and (15.6.18) is not hard to understand: both corrections arise primarily from the core electrons, which are largely unaffected by chemical rearrangements.

There are other first-order relativistic corrections to the Hamiltonian operator. From Exercise 2.2, we recall the two-electron Darwin operator and the spin–spin contact operator:

$$H_{2D} = -\frac{\pi}{c^2} \sum_{i>j} \delta(\mathbf{r}_i - \mathbf{r}_j) \tag{15.6.20}$$

$$H_{SSC} = -\frac{8\pi}{3c^2} \sum_{i>j} \delta(\mathbf{r}_i - \mathbf{r}_j) \mathbf{s}_i \cdot \mathbf{s}_j \tag{15.6.21}$$

In addition, there exists a two-electron operator that couples the spins of the electrons in a dipolar fashion as well as an operator that couples their orbital angular momenta. In general, the two-electron relativistic operators are less important than the one-electron mass-velocity and Darwin operators. For the neon atom, for example, we obtain the following first-order one- and two-electron corrections in the cc-pVDZ basis using a valence-electron FCI wave function:

$$\langle FCI|H_{MV}|FCI \rangle = -632.9 \text{ mE}_h \tag{15.6.22}$$

$$\langle FCI|H_{1D}|FCI \rangle = 496.6 \text{ mE}_h \tag{15.6.23}$$

$$\langle FCI|H_{2D}|FCI \rangle = -7.1 \text{ mE}_h \tag{15.6.24}$$

$$\langle FCI|H_{SSC}|FCI \rangle = 14.2 \text{ mE}_h \tag{15.6.25}$$

The two-electron relativistic corrections are at least an order of magnitude smaller than the one-electron corrections. In accordance with Exercise 2.2, we note that the contribution from the two-electron Darwin operator is exactly minus one-half of the contribution from the spin–spin contact operator.

15.6.7 SUMMARY

The nonrelativistic energy of a closed-shell molecule containing only first-row atoms is dominated by the Hartree–Fock contribution, which constitutes as much as 99.5% of the total energy. Electron correlation reduces this energy further, with a contribution of about 0.5% from the connected doubles and about 0.02% from the connected triples. By contrast, the vibrational correction increases the energy, contributing about -0.03% – that is, about as much as the triples but in the opposite direction. Although the relativistic corrections are significant – amounting to about 0.05% of the total energy for the molecules considered here – their contribution changes little in the course of chemical processes and may be neglected except in calculations aiming for a very high accuracy. The nonrelativistic small contributions to the energy – the singles-and-doubles correlation energy, the triples correlation energy and the zero-point vibrational energy – depend sensitively on the molecular and electronic structure and must be taken into account in accurate calculations of molecular properties, as we shall see in the remaining sections of this chapter.

15.7 Atomization energies

In this section, we examine the accuracy in electronic-structure calculations of the *atomization energy* – that is, the difference between the energy of the constituent atoms of a molecular system and the energy of the molecule itself. Since atomization energies represent differences in the energies of systems containing different numbers of paired electrons, they pose a difficult challenge for quantum chemistry, requiring a high degree of flexibility in the description of the short-range interactions. As such, the calculation of atomization energies represents a particularly stringent test of the quality of the different N-electron models [7,8].

15.7.1 EXPERIMENTAL ATOMIZATION ENERGIES

The calorimetric measurement of heats of formation and atomization energies is a classical discipline of chemistry and a number of compilations of such data exist. In Table 15.20, we have listed the experimental atomization energies for the molecules in Table 15.1.

The experimental atomization energy D_0 of a molecule represents the difference between the relativistic energy of the constituent atoms in their electronic ground-state level $^{\text{rel}}E^A(^{2S+1}L_J)$ and the relativistic energy of the molecule in its vibronic ground state $^{\text{rel}}E_0$:

$$D_0 = \sum_A {}^{\text{rel}}E^A(^{2S+1}L_J) - {}^{\text{rel}}E_0 \tag{15.7.1}$$

By contrast, the nonrelativistic equilibrium atomization energy D_e calculated theoretically represents the difference between the nonrelativistic atomic ground-state term energy $E^A(^{2S+1}L)$ and the nonrelativistic molecular equilibrium ground-state electronic energy $E(\mathbf{R}_e)$:

$$D_e = \sum_A E^A(^{2S+1}L) - E(\mathbf{R}_e) \tag{15.7.2}$$

As discussed in Sections 15.6.5 and 15.6.6, the most important corrections that must be considered to obtain the 'experimental' equilibrium atomization energies D_e from D_0 are the zero-point vibrational correction to the molecular energy, the spin–orbit corrections to the atomic energies, and the relativistic mass-velocity and one-electron Darwin corrections to the atomic and molecular energies:

$$D_0 = D_e - E_{\text{ZPV}} + \sum_A \Delta E_{\text{SO}}^A(^{2S+1}L_J) + \Delta E_{\text{MVD}} \tag{15.7.3}$$

$$\Delta E_{\text{MVD}} = \sum_A E_{\text{MVD}}^A - E_{\text{MVD}} \tag{15.7.4}$$

In Table 15.20, we have listed the vibrational and relativistic corrections for the atomization energies considered in this section, obtained from the ZPVEs collected in Table 15.18 and the relativistic corrections in Table 15.19. From these corrections, we have calculated a set of 'experimental' equilibrium atomization energies – see Exercise 15.1. The vibrational corrections to the atomization energies are sizeable – typically 4–5%, and larger than 7% for NH_3 and CH_4. The relativistic corrections are smaller but cannot be neglected in accurate work.

For our statistical analysis, we exclude those molecules for which either the experimental atomization energy or the vibrational correction is unknown: H_2O_2, N_2H_2 and HNC. In addition, we have excluded HOF, which has the largest experimental uncertainty and for which we shall later see that the experimental value must be in error. Except for O_3 (1.7 kJ/mol), C_2H_2 (1 kJ/mol), CH_2 (2.2 kJ/mol) and HCN (2.6 kJ/mol), the uncertainty in D_e for the other molecules is smaller than

Table 15.20 Experimental atomization energies D_0, molecular zero-point vibrational energies E_{ZPV}, atomic spin–orbit corrections $\sum_A \Delta E_{SO}^A$, scalar one-electron relativistic corrections ΔE_{MVD} and estimated nonrelativistic equilibrium atomization energies D_e (kJ/mol)

Molecule	$D_0{}^a$	E_{ZPV}	$\sum_A \Delta E_{SO}^A$	ΔE_{MVD}	D_e
F_2	154.56(60)	5.45	−3.222	−0.12	163.35
H_2	432.07(1)	25.96	0.000	−0.01	458.04
HF	566.22(71)^b	24.48	−1.611	−0.85	593.16
O_3	595.02(172)	17.35	−2.796	−1.07	616.24
HOF	635.53(421)	35.89	−2.543	−0.92	674.88
$CH_2{}^c$	713.11(215)	43.22	−0.354	−0.37	757.06
HNO	823.63(32)^d	35.79	−0.932	−1.15	861.50
N_2	941.64(20)	14.06	0.000	−0.58	956.28
H_2O	917.78(15)	55.44	−0.932	−1.13	975.28
CO	1071.79(51)	12.94	−1.286	−0.68	1086.70
H_2O_2	1055.46	–	−1.864	−1.59	–
N_2H_2	1153.71(2090)	–	0.000	−1.38	–
HNC	–	40.62	−0.354	−1.09	–
NH_3	1157.83(42)	88.98	0.000	−1.08	1247.88
HCN	1269.85(262)^e	41.61	−0.354	−0.94	1312.75
CH_2O	1494.73(67)^f	69.14	−1.286	−1.42	1566.58
CO_2	1597.92(50)	30.29	−2.218	−2.03	1632.46
C_2H_2	1627.16(100)^g	68.80	−0.708	−1.17	1697.84
CH_4	1642.24(57)	115.93	−0.354	−0.81	1759.33
C_2H_4	2225.53(71)	132.21	−0.708	−1.37^h	2359.82

^{a}Unless otherwise indicated, the data are taken from M. W. Chase Jr, C. A. Davies, J. R. Downey Jr, D. J. Frurip, R. A. McDonald and A. N. Syverud, *JANAF Thermochemical Tables*, 3rd edn, *J. Phys. Chem. Ref. Data* **4**, Monograph 9, 1998.

^{b}K. P. Huber and G. Herzberg, *Constants of Diatomic Molecules*, Van Nostrand Reinhold, 1979.

^{c}We have used $\Delta_f H^O$ (0K) = 392.5 ± 2.1 kJ/mol for the $\tilde{X}^3 B_1$ ground state, obtained by photodissociation from methane by K. E. McCulloh and V. H. Dibeler [*J. Chem. Phys.* **64**, 4445 (1976)]. In the same work, photodissociation from ketene gives the slightly different result of 390.8 ± 1.7 kJ/mol. C. C. Hayden, D. M. Neumark, K. Shobatake, R. K. Sparks and Y. T. Lee [*J. Chem. Phys.* **76**, 3607 (1982)] obtained 393.7 ± 2.5 kJ/mol. To arrive at D_0 for the $\tilde{a}^1 A_1$ state, we have further used $T_0(\tilde{a}^1 A_1)$ = 3147 ± 5 cm^{-1} from P. Jensen and P. R. Bunker [*J. Chem. Phys.* **89**, 1327 (1988)].

^{d}R. N. Dixon, *J. Chem. Phys.* **104**, 6905 (1996).

^{e}Obtained by adding D_0(H−CN) = 523.25(179) kJ/mol [G. P. Morley, I. R. Lambert, M. N. R. Ashfold, K. N. Rosser and C. M. Western, *J. Chem. Phys.* **97**, 3157 (1992)] to D_0(CN) = 746.60(192) kJ/mol [Y. Huang, S. A. Barts and J. B. Halpern, *J. Phys. Chem.* **96**, 425 (1992)].

^{f}D. L. Baulch, R. A. Cox, P. J. Crutzen, R. F. Hampson Jr, J. A. Kerr, J. Troe and R. T. Watson, *J. Phys. Chem. Ref. Data* **11**, 327 (1982); see also J. B. Pedley, R. D. Naylor and S. D. Kirby, *Thermodynamical Data of Organic Compounds*, 2nd edn, Chapman & Hall, 1986.

$^g\Delta_f H^O$ (0K) for C_2H_2 in ref. a is in error. The present data are taken from D. D. Wagman, W. H. Evans, V. B. Parker, R. H. Schumm, I. Halow, S. M. Bailey, K. L. Churney and R. L. Nuttall, *J. Phys. Chem. Ref. Data* **11**, (1982), supplement 2.

^{h}Calculated at the CCSD(T)/cc-pCVTZ level. The sum of the mass-velocity and Darwin corrections is −29.694 mE_h for C_2H_4 (see Table 15.19), −15.095 mE_h for C and −0.007 mE_h for H.

1 kJ/mol – ignoring the uncertainty in the vibrational and relativistic corrections, which should be significantly smaller than 1 kJ/mol.

Our final statistical sample for the atomization energy thus consists of the 16 molecules ticked in Table 15.1. These molecules should be well suited for testing the performance of the standard electronic-structure models; for the four molecules omitted from the statistical analysis, the calculations presented here should serve as accurate predictions of the equilibrium atomization energies.

15.7.2 STATISTICAL ANALYSIS OF ATOMIZATION ENERGIES

In the present subsection, we examine the statistical measures of error in the calculated atomization energies. For the sample molecules in Table 15.1, the atomization energies have been calculated as described for total energies in Section 15.6.1. The statistical error measures are plotted in Figure 15.11 and listed in Table 15.21. Since the Hartree–Fock errors are considerably larger than those at the correlated levels, they have been omitted in Figure 15.11.

From Figure 15.11 and Table 15.21, we see that the atomization energy behaves in a systematic manner with respect to both the N-electron treatment and the one-electron treatment. Thus, with few exceptions, D_e increases with the cardinal number of the basis set – the only exceptions occur at the Hartree–Fock level, where D_e decreases from cc-pCV5Z to cc-pcV6Z for F_2 and from cc-pCVTZ to cc-pCVQZ for CO and CO_2. The reason for this systematic behaviour is that improvements in the basis favour systems of low total energy – that is, the molecule rather than its constituent atoms.

Likewise, the atomization energy increases in the sequence Hartree–Fock, CCSD, CCSD(T), MP2. While there are no exceptions – for any basis or any molecule – to the rule that D_e increases in the sequence Hartree–Fock, CCSD, CCSD(T), the MP2 model is less systematic. Mostly, the MP2 model gives the largest atomization energy, but in a few cases it falls between the HF and CCSD energies or between the CCSD and CCSD(T) energies.

From Table 15.21, we note that the intrinsic error in the Hartree–Fock atomization energy is very large, on average about -420 kJ/mol for the molecules in the sample. Indeed, a closer scrutiny of the individual atomization energies reveals that, for two of the molecules (F_2 and O_3), the Hartree–Fock model does not even predict a stable system, giving a negative atomization energy. For the remaining molecules, the model underestimates the atomization energy by typically 40% and sometimes by as much as 60% – see Table 15.22, which contains the relative contributions to the equilibrium atomization energies. Clearly, the Hartree–Fock model cannot be used even for a qualitative investigation of atomization energies, being unable to describe the changes that occur as electron pairs are separated.

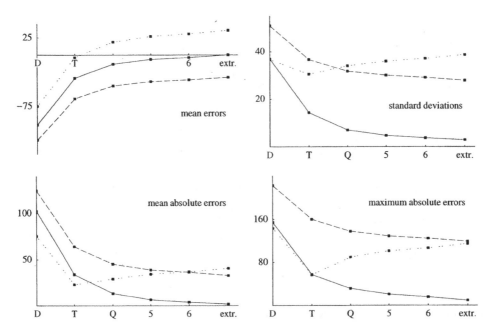

Fig. 15.11. Errors relative to experiment in all-electron calculations of atomization energies (kJ/mol) in the cc-pCVXZ basis sets at the MP2 level (dotted line), CCSD level (dashed line) and CCSD(T) level (full line).

Table 15.21 Errors relative to experiment in atomization energies (kJ/mol), calculated with all electrons correlated. The cc-pcV6Z energies have been obtained as described in Section 15.6.1

		HF	MP2	CCSD	CCSD(T)
$\overline{\Delta}$	cc-pCVDZ	−450.1	−76.2	−125.4	−103.3
	cc-pCVTZ	−426.1	−4.7	−65.1	−34.9
	cc-pCVQZ	−423.7	17.7	−46.1	−14.3
	cc-pCV5Z	−423.2	26.1	−39.8	−7.4
	cc-pcV6Z[a]	−423.1	29.7	−37.3	−4.7
Δ_{std}	cc-pCVDZ	187.3	36.8	51.2	37.4
	cc-pCVTZ	179.6	30.4	37.0	14.7
	cc-pCVQZ	179.1	33.8	32.1	7.3
	cc-pCV5Z	179.3	35.7	30.3	4.9
	cc-pcV6Z[a]	179.3	36.8	29.4	3.8
$\overline{\Delta}_{abs}$	cc-pCVDZ	450.1	76.2	125.4	103.3
	cc-pCVTZ	426.1	23.3	65.1	34.9
	cc-pCVQZ	423.7	28.6	46.1	14.3
	cc-pCV5Z	423.2	33.7	39.8	7.4
	cc-pcV6Z[a]	423.1	36.2	37.3	4.7
Δ_{max}	cc-pCVDZ	901.9	144.7	223.3	155.7
	cc-pCVTZ	857.8	55.2	161.4	58.8
	cc-pCVQZ	855.1	87.7	139.0	32.6
	cc-pCV5Z	854.8	99.3	129.9	21.9
	cc-pcV6Z[a]	854.4	104.5	125.7	17.0

[a]cc-pV6Z for the Hartree–Fock model.

Table 15.22 Relative contributions (%) to the atomization energy at the all-electron CCSD(T)/cc-pcV6Z level

	HF	SD	T
F_2	−97.8	177.1	20.6
H_2	76.6	23.4	0.0
HF	68.6	29.9	1.6
O_3	−39.7	121.6	18.1
HOF	35.0	59.7	5.3
$CH_2(^1A_1)$	70.2	28.7	1.1
HNO	38.8	56.2	5.1
N_2	50.9	44.9	4.3
H_2O	67.1	31.4	1.5
CO	67.4	29.4	3.2
H_2O_2	50.2	46.4	3.4
N_2H_2	51.2	45.4	3.5
HNC	63.8	33.2	3.0
NH_3	67.6	31.0	1.3
HCN	63.8	33.2	3.0
CH_2O	69.0	28.9	2.2
CO_2	63.5	32.8	3.6
C_2H_2	72.6	25.3	2.1
CH_4	78.2	21.1	0.7
C_2H_4	76.2	22.5	1.3

Turning our attention to the correlated models, we find that the CCSD(T) model is the only model capable of reproducing the experimental atomization energies to chemical accuracy – that is, to better than 5 kJ/mol (see Table 15.21). However, this accuracy is obtained only for large basis sets. At the cc-pCVDZ level, the error is typically −103 kJ/mol. At the cc-pCVTZ level, it is reduced to −35 kJ/mol, but this is still unacceptable for most applications. Indeed, only at the cc-pCVQZ level is the error sufficiently small (−14 kJ/mol) to make the results useful for qualitative if not quantitative investigations. For higher precision, calculations at the cc-pCV5Z level (error −7 kJ/mol) or at the cc-pcV6Z level (error −5 kJ/mol) should be sufficient for the most demanding applications. In short, although chemical accuracy is obtainable with the CCSD(T) model, the basis-set requirements are so high that the calculations are possible only for rather small systems.

Their high cost notwithstanding, it is gratifying to see that the CCSD(T) calculations converge smoothly, with respect to both mean errors and standard deviations. Indeed, the smooth but slow convergence of the atomization energy is directly related to the difficulties experienced by determinantal models in describing short-range electronic interactions. As we shall see in Section 15.7.3, it is possible to carry out simple but accurate extrapolations of the atomization energy, reducing the basis-set requirements to the cc-pCVQZ level for quantitative investigations.

It is likewise gratifying to see that, for most molecules, the CCSD(T) model is sufficient for the calculation of atomization energies. Thus, at the cc-pcV6Z level, the largest error (−17.0 kJ/mol) occurs for O_3 – as we would expect from the multiconfigurational electronic structure of this molecule. The second largest error (−6.8 kJ/mol) occurs for HCN, but – as we shall see in Section 15.7.3 – extrapolation reduces this error to less than −1.7 kJ/mol. Clearly, the chief difficulty faced in CCSD(T) calculations of atomization energies arises from the truncation of the one-electron basis rather than from the shortcomings of the N-electron model.

Having discussed the CCSD(T) model at some length, let us briefly consider the simpler MP2 and CCSD models. In relative terms, these models do not behave badly at all. As seen from Figure 15.11 and Table 15.21, the basis-set dependence is much like that of CCSD(T), except that MP2 overestimates the atomization energy by 3% while CCSD underestimates it by about 5%. Unfortunately, in absolute terms, these errors translate to 30–40 kJ/mol, making the calculations rather useless except in preliminary or exploratory investigations. As a curiosity, we note that, at the cc-pCVTZ level, the MP2 model is very accurate in the mean, with an error of only −4.7 kJ/mol. However, this small error arises from a rather unsystematic cancellation of errors in the one- and N-electron treatments, as evidenced by the standard deviation of 30 kJ/mol and the mean absolute error of 23 kJ/mol. For the same reason, in the cc-pCVTZ basis, the largest MP2 error occurs for CO_2 rather than for O_3 as in the CCSD and CCSD(T) calculations.

The normal distributions of the errors in the correlated atomization energies are plotted in Figure 15.12. These plots illustrate strikingly the superiority of the CCSD(T) model with respect to the calculation of atomization energies. Whereas the progression of the CCSD(T) distribution with the cardinal number is satisfactory with respect to both width and position, the MP2 and CCSD distributions do not improve substantially, remaining broad and off-centre for all cardinal numbers.

15.7.3 EXTRAPOLATION OF ATOMIZATION ENERGIES

For the accurate calculation of atomization energies, the slow convergence of the dynamical correlation energy with respect to the cardinal number is a severe problem. Thus, in Section 15.7.2, we found that, even in the largest basis sets developed, the errors in the atomization energies are

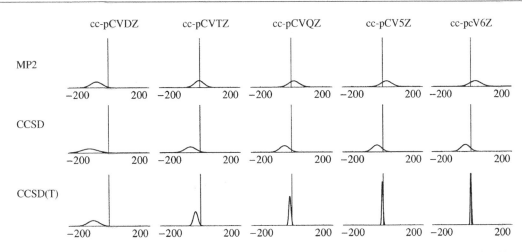

Fig. 15.12. Normal distributions of the errors in the atomization energies (kJ/mol), calculated with all electrons correlated.

larger than those of most experimental measurements. In particular, for the cc-pCVQZ, cc-pCV5Z and cc-pcV6Z basis sets, the mean all-electron CCSD(T) errors for the 16 sample molecules are -14.3, -7.4 and -4.7 kJ/mol, respectively. Recalling that 1 mE$_h$ = 2.625 kJ/mol, we find that the corresponding errors in the all-electron CCSD(T) correlation energies in Table 15.15 (relative to the *extrapolated* limits) are 47.3, 23.1 and 15.0 kJ/mol. The ratios between the errors in the correlation and atomization energies are -3.1 to -3.3 for the three basis sets, indicating that the atomization energy converges in the same manner as the total correlation energy – that is, as the third inverse power of the cardinal number, X^{-3}. Although the errors in the atomization energy are somewhat smaller than those in the total correlation energy (because of a cancellation of errors between the molecules and atoms), the slow convergence arises for the same reason – namely, from the difficulties faced by determinantal expansions in describing the short-range electronic interactions.

It is noteworthy that the X^{-3} dependence of the errors in the atomization energies has been established using experimental reference energies, indicating that, in the basis-set limit, the CCSD(T) and experimental energies are very similar. The X^{-3} dependence enables us to extrapolate the atomization energies from the simple formula (15.6.3). In Table 15.23, we compare the statistical measures of errors for the calculated and extrapolated all-electron CCSD(T) atomization energies. The extrapolations have been carried out as described in Section 15.6.1.

Table 15.23 Errors in calculated and extrapolated all-electron CCSD(T) atomization energies relative to experiment (kJ/mol)

	cc-pCVXZ				cc-pCV(XY)Z			
	T	Q	5	6	DT	TQ	Q5	56
$\overline{\Delta}$	-34.9	-14.3	-7.4	-4.7	-16.2	-1.1	-0.6	-0.9
Δ_{std}	14.7	7.3	4.9	3.8	10.9	4.4	3.0	2.8
$\overline{\Delta}_{abs}$	34.9	14.3	7.4	4.7	16.2	2.6	1.6	1.4
Δ_{max}	58.8	32.6	21.9	17.0	42.5	15.4	11.0	10.7

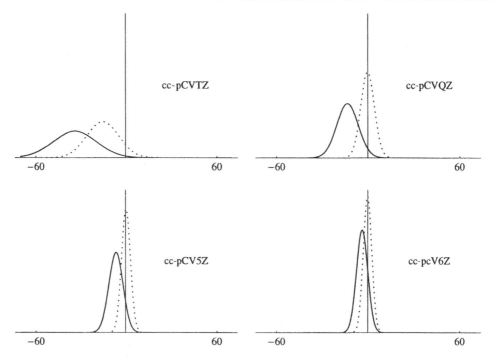

Fig. 15.13. Normal distributions of the errors in the calculated all-electron CCSD(T) atomization energies (kJ/mol) without extrapolation (full line) and with extrapolation (dotted line).

The improvement in the atomization energies is quite remarkable at all levels. Thus, at the cc-pCVQZ level, the mean error is reduced from -14.3 to -1.1 kJ/mol and the mean absolute error from 14.3 to 2.6 kJ/mol; at the cc-pcV(56)Z level, the mean and mean absolute errors are -0.9 and 1.4 kJ/mol, respectively. Excluding from our sample O_3 with a large intrinsic error of about 10 kJ/mol, we find that the extrapolation (15.6.3) gives a mean absolute error *relative to experiment* of 0.9 kJ/mol at the cc-pCV(Q5)Z level. Furthermore, based on the observation that the cc-pCV(Q5)Z extrapolation in Section 8.4.3 reduces the error in the correlation energy of N_2 to 2.2 kJ/mol and that the errors in the atomization energies appear to be 3.0–3.5 times smaller at each level, we expect the *basis-set error* in the cc-pCV(Q5)Z atomization energies to be no larger than 0.7 kJ/mol – that is, similar to the experimental uncertainties. In Figure 15.13, we have plotted the normal distributions of the errors in the extrapolated atomization energies, superimposed on the corresponding distributions for the unextrapolated atomization energies. For each basis set, the improvement in position and width of the distribution upon extrapolation is well illustrated.

The basis-set requirements of the different contributions to the atomization energy are illustrated in Table 15.24, which contains the errors relative to the cc-pV6Z energy for the Hartree–Fock contribution and relative to the extrapolated limits for the correlation contributions. The convergence of the atomization energy is governed by the doubles, whose contribution is converged to chemical accuracy only at the cc-pcV6Z level. For the Hartree–Fock and triples contributions, convergence is already obtained at the cc-pCVTZ level. In general, therefore, there is no need to calculate the expensive triples correction to the atomization energy in the same basis as the doubles contribution.

Table 15.24 Statistical measures of the basis-set errors in the atomization energies (kJ/mol) relative to the cc-pV6Z energy for the Hartree–Fock model and relative to the cc-pcV(56)Z correlation energies for the correlated models

		HF	MP2–HF	CCSD–HF	CCSD(T)–CCSD
$\overline{\Delta}$	cc-pCVDZ	−28.0	−88.8	−68.1	−11.6
	cc-pCVTZ	−3.1	−38.5	−29.9	−2.9
	cc-pCVQZ	−0.6	−17.5	−12.5	−1.2
	cc-pCV5Z	−0.1	−9.1	−6.3	−0.5
	cc-pcV6Z[a]	0.0	−5.4	−3.8	−0.3
Δ_{std}	cc-pCVDZ	9.0	29.7	26.2	6.0
	cc-pCVTZ	2.2	14.0	11.6	1.4
	cc-pCVQZ	0.8	6.7	5.1	0.7
	cc-pCV5Z	0.2	3.6	2.6	0.3
	cc-pcV6Z[a]	0.0	2.1	1.6	0.2
$\overline{\Delta}_{abs}$	cc-pCVDZ	28.0	88.8	68.1	11.6
	cc-pCVTZ	3.3	38.5	29.9	2.9
	cc-pCVQZ	0.8	17.5	12.5	1.2
	cc-pCV5Z	0.2	9.1	6.3	0.5
	cc-pcV6Z[a]	0.0	5.4	3.8	0.3
Δ_{max}	cc-pCVDZ	47.5	131.2	113.2	27.9
	cc-pCVTZ	7.4	57.8	46.8	6.9
	cc-pCVQZ	2.1	27.0	20.4	3.1
	cc-pCV5Z	0.4	14.1	10.4	1.5
	cc-pcV6Z[a]	0.0	8.3	6.4	0.8

[a]cc-pV6Z for the Hartree–Fock model.

15.7.4 CORE CONTRIBUTIONS TO ATOMIZATION ENERGIES

In the calculations of atomization energies discussed up to now, we have correlated the full set of electrons in the cc-pCVXZ basis sets. Such calculations are rather expensive, however, and significant savings would be obtained if instead the calculations could be carried out in the smaller cc-pVXZ basis sets, correlating only the valence electrons. Unfortunately, the errors arising from the neglect of core correlation in such valence-electron calculations would be unacceptable, at least for quantitative work.

In Table 15.25, we have listed the core-correlation contributions to the atomization energies, calculated as the differences between the all-electron cc-pCV5Z and valence-electron cc-pV5Z energies at the CCSD(T) level. The core contributions are all positive but vary considerably among the molecules – for a few molecules, it is smaller than 1 kJ/mol but for C_2H_2 it is as large as 10.8 kJ/mol.

Clearly, the core contribution to the atomization energy cannot be neglected if high accuracy is required – that is, for CCSD(T) calculations in basis sets of quadruple-zeta quality or higher.

Table 15.25 The core contribution to the CCSD(T) atomization energy (kJ/mol), calculated as the difference between the all-electron cc-pCV5Z and valence-electron cc-pV5Z atomization energies

F_2	H_2	HF	O_3	HOF	CH_2	HNO	N_2	H_2O	CO
0.1	0.0	0.8	1.4	1.1	1.9	2.9	4.8	1.9	4.8

H_2O_2	N_2H_2	HNC	NH_3	HCN	CH_2O	CO_2	C_2H_2	CH_4	C_2H_4
2.3	4.5	7.0	3.2	7.9	6.1	8.6	10.8	5.4	10.3

On the other hand, in calculations carried out at the triple-zeta level using a model without triples correction, the total errors relative to experiment are in any case so large (see Table 15.21) that there is little point in adding core corrections. The importance of the core correlation is illustrated also in Figure 15.14, where we have plotted the normal distributions of the CCSD(T) errors relative to experiment with and without core correlation for the molecules in the statistical sample.

In Section 15.3.7, we discussed the core corrections to equilibrium structures and considered a simple additive scheme to account for these corrections. Proceeding in the same manner, we first calculate the atomization energy in a large valence basis cc-pVXZ, correlating only the valence electrons. Next, we add a correction for the core correlation, obtained as the difference between all-electron and valence-electron energies calculated in a basis of a lower cardinal number. Such a procedure may often enable calculations to be carried out with a cardinal number one higher than otherwise possible, without compromising the total accuracy since the core contributions are at least two orders of magnitude smaller than the total atomization energies and need only be calculated to a relative accuracy of a few per cent.

In Table 15.26, we have listed the statistical data that describe the errors in the core corrections calculated at different levels of theory relative to the core contributions in Table 15.25. The corrections calculated at the cc-p(C)VQZ level are similar to those obtained at the cc-p(C)V5Z level, the largest difference being only 0.4 kJ/mol and the typical error 0.2 kJ/mol. Since the cc-pCVQZ basis is smaller than the cc-pV5Z basis (see Table 8.13) and since, for the molecules considered here, the number of core electrons is small, the all-electron cc-pCVQZ calculations are less expensive than the corresponding valence-electron cc-pV5Z calculations. In short, we may estimate the core correction reliably by reducing the cardinal number by one, at a cost no higher than that of the original valence-electron calculation.

The data in Table 15.26 also show that, if the core corrections are estimated by reducing the cardinal number by two rather than one, errors as large as 2.7 kJ/mol are introduced. Considering

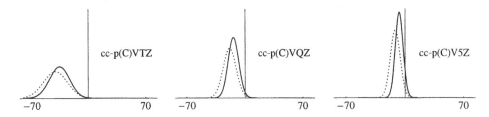

Fig. 15.14. The errors relative to experiment in CCSD(T) atomization energies (kJ/mol) calculated with core correlation (full line) and without core correlation (dotted line).

Table 15.26 Errors in the core-correlation contributions to the atomization energies relative to the CCSD(T)/cc-pCV5Z level (kJ/mol)

	CCSD(T)				MP2			
	DZ	TZ	QZ	5Z	DZ	TZ	QZ	5Z
$\overline{\Delta}$	0.0	0.4	−0.1	0	0.7	−0.4	−1.1	−1.0
Δ_{std}	1.3	1.2	0.2	0	1.8	1.4	0.8	0.7
$\overline{\Delta}_{abs}$	1.0	1.0	0.2	0	1.5	1.1	1.1	1.0
Δ_{max}	2.6	2.7	0.4	0	5.0	3.6	2.6	2.1

the high cost of the original valence-electron cc-pV5Z calculations, such an error is unacceptable. Also, it appears difficult to estimate the core corrections reliably at a simpler level of theory such as MP2 – even at the quintuple-zeta level, the MP2 core corrections differ from the CCSD(T) corrections by as much as 2.1 kJ/mol.

15.7.5 CCSDT CORRECTIONS

For large basis sets, the CCSD(T) model gives atomization energies close to the experimental values, particularly when extrapolation is carried out for the correlation part – see Table 15.23. The cc-pCV(56)Z extrapolation, for example, gives a mean error and a standard deviation of −0.9 and 2.8 kJ/mol, respectively. It is thus of some interest to examine what happens when the triples treatment is extended to the full CCSDT model. Since such calculations are demanding, they have been performed only for the six molecules in Table 15.27. For details of the calculations, we refer to Section 15.6.4.

The CCSDT and CCSD(T) atomization energies are similar, the largest difference being 2.7 kJ/mol for the cc-pCVQZ calculation on N_2. At the CCSD(T) level, the triples correction is underestimated by 11% for CH_2 but overestimated for the other molecules, by as little as 3.2% for F_2 and as much as 7.1% for N_2. In general, therefore, the CCSD(T) model provides a useful approximate treatment of the connected triples for atomization energies. The difference between the CCSDT and CCSD(T) atomization energies is quite stable with respect to the cardinal number

Table 15.27 CCSDT atomization energies (kJ/mol). The shaded CCSDT/cc-pCV5Z energies have been obtained by adding to the CCSD(T)/cc-pCV5Z energies the difference between the CCSDT/cc-pCVQZ and CCSD(T)/cc-pCVQZ energies

Molecule	Model	cc-pCVDZ	cc-pCVTZ	cc-pCVQZ	cc-pCV5Z
F_2	HF	−169.8	−153.8	−155.3	−155.6
	CCSD	92.3	116.8	121.7	124.5
	CCSD(T)	112.2	146.7	153.4	156.9
	CCSDT	112.5	145.8	152.4	155.9
HF	HF	380.9	401.2	404.4	405.5
	CCSD	526.7	566.0	577.5	581.4
	CCSD(T)	529.6	573.6	586.1	590.4
	CCSDT	529.6	573.2	585.5	589.8
$CH_2(^1A_1)$	HF	515.2	528.4	530.3	530.9
	CCSD	687.4	731.7	743.1	746.5
	CCSD(T)	692.8	739.3	751.3	754.9
	CCSDT	693.9	740.3	752.3	755.9
N_2	HF	452.4	479.6	482.4	482.7
	CCSD	812.1	873.5	896.5	905.1
	CCSD(T)	841.2	911.8	936.3	945.6
	CCSDT	840.6	909.4	933.7	942.9
H_2O	HF	620.5	645.5	650.3	651.9
	CCSD	869.0	930.4	949.2	955.7
	CCSD(T)	875.3	943.3	963.5	970.5
	CCSDT	875.4	942.7	962.7	969.7
CO	HF	711.5	729.0	730.7	730.2
	CCSD	988.9	1026.0	1041.7	1046.6
	CCSD(T)	1013.4	1058.6	1075.5	1080.9
	CCSDT	1013.4	1057.0	1073.5	1078.9

except for the cc-pCVDZ basis, which is too small for an accurate calculation of the triples correction.

It is noteworthy that the inclusion of the full triples correction does not improve the agreement with the experimental atomization energies, indicating that there is an element of error cancellation at the CCSD(T) level which is absent at the CCSDT level. For better agreement with experiment, other corrections must also be included – most important, the contributions from the connected quadruples.

15.7.6 SUMMARY

Our calculations have shown that the errors in the atomization energies are dominated by the correlation treatment. For small molecules containing only first-row atoms, atomization energies may be calculated to chemical accuracy. However, to achieve this level of accuracy, the calculations must be carried out at the CCSD(T) level in a basis of sextuple-zeta quality and corrections must be made for core correlation. In the cc-pcV6Z calculations, the errors relative to experiment are typically less than 5 kJ/mol and are reduced substantially by extrapolation of the correlation contribution. In Table 15.28, we have listed the resulting atomization energies. The extrapolated numbers are expected to be close to the basis-set limit.

The Hartree–Fock model underestimates atomization energies severely – it predicts negative atomization energies for F_2 and O_3 and the relative stability is not correctly predicted for compounds with similar atomization energies. The reason for the failure of the Hartree–Fock model is its inability to describe the changes that occur as electron pairs are broken upon atomization. The Hartree–Fock model therefore cannot be used for any but the crudest estimates of atomization energies.

Although the MP2 model represents a significant improvement on the Hartree–Fock description, it is still in error by several per cent relative to experiment. Except for H_2, CH_2, CH_4 and possibly NH_3, the atomization energies are overestimated in the basis-set limit. Somewhat surprisingly, the largest error of 113 kJ/mol occurs for CO_2 rather than for O_3, which has an error of 110 kJ/mol. Large errors occur also for CO (59 kJ/mol) and HCN (51 kJ/mol). The MP2 model gives the largest errors for electron-rich bonds and atoms (oxygen and fluorine). In contrast, the singly bonded molecules containing carbon, nitrogen and hydrogen (NH_3 and CH_4) are rather accurately described. The relative stability of the compounds is not always correctly predicted and the MP2 model appears to be useful only for qualitative investigations.

For all molecules except H_2, the CCSD model gives atomization energies that are too small. The CCSD errors are as large as the MP2 errors but of opposite sign and more systematic, predicting the correct relative stability in all cases except for O_3. The largest error occurs for O_3 (-120 kJ/mol) and all molecules containing triple bonds exhibit errors of about 40 kJ/mol. The molecules containing only carbon, nitrogen and hydrogen and only single bonds are again most accurately described.

The CCSD(T) model reduces the CCSD errors by more than an order of magnitude, increasing the atomization energies as the correlation treatment is improved. This model gives highly accurate atomization energies, typically within 1 kJ/mol of experiment after extrapolation – for all bonds and atoms. Relaxation of the triples at the full CCSDT level generally lowers the atomization energy by 1–2 kJ/mol, indicating an element of error cancellation in the CCSD(T) model. For CCSD(T) in Table 15.28, differences larger than 2.3 kJ/mol occur for only two systems: O_3 (-10.7 kJ/mol) and HOF (-12.0 kJ/mol). The corresponding experimental uncertainties are 1.7 and 4.2 kJ/mol. For O_3, the difference probably arises from an error in the calculated energy, related

Table 15.28 Calculated and experimental electronic atomization energies (kJ/mol).

	HF	MP2		CCSD		CCSD(T)		Experiment
	pV6Z	pcV6Z	pcV(56)Z	pcV6Z	pcV(56)Z	pcV6Z	pcV(56)Z	
F_2	−155.3	182.8	185.4	126.1	128.0	158.8	161.1	163.4(06)
H_2	350.8	439.8	440.7	457.7	458.1	457.7	458.1	458.0(00)
HF	405.7	611.6	613.8	582.5	583.9	591.6	593.3	593.2(07)
O_3	−238.2	718.8	726.6	490.5	496.1	599.2	605.5	616.2(17)
HOF	230.4	690.3	695.0	624.2	627.5	659.1	662.9	674.9(42)
$CH_2(^1A_1)$	531.1	738.1	740.8	747.7	749.4	756.2	757.9	757.1(22)
HNO	331.6	890.7	897.2	812.2	816.8	855.4	860.4	861.5(03)
N_2	482.9	1003.6	1010.6	908.9	913.9	949.6	954.9	956.3(02)
H_2O	652.3	992.2	996.1	957.6	960.2	972.7	975.5	975.3(02)
CO	730.1	1141.2	1145.8	1048.8	1052.3	1083.2	1086.9	1086.7(05)
H_2O_2	562.9	1159.5	1166.0	1082.6	1087.2	1121.1	1126.1	−
N_2H_2	631.8	1253.4	1261.6	1191.6	1197.2	1234.2	1240.2	−
HNC	793.1	1282.2	1288.8	1205.2	1209.9	1242.8	1247.8	−
NH_3	841.2	1243.4	1248.6	1227.3	1230.7	1243.8	1247.4	1247.9(04)
HCN	833.5	1356.6	1363.5	1266.5	1271.3	1305.9	1311.0	1312.8(26)
CH_2O	1078.2	1605.4	1611.5	1529.4	1533.8	1563.4	1568.0	1566.6(07)
CO_2	1033.4	1736.9	1745.2	1567.2	1573.6	1626.5	1633.2	1632.5(05)
C_2H_2	1229.1	1736.1	1742.5	1656.5	1661.0	1692.2	1697.1	1697.8(10)
CH_4	1374.1	1749.0	1753.1	1744.3	1747.0	1756.6	1759.4	1759.3(06)
C_2H_4	1793.9	2372.4	2379.3	2324.0	2328.9	2355.7	2360.8	2359.8(07)

to the absence of connected quadruples in the CCSD(T) model. For HOF, on the other hand, the discrepancy is most likely caused by an error in the experiment. Consequently, the extrapolated CCSD(T) value of 662.9 kJ/mol in Table 15.28 probably represents the best estimate of the equilibrium atomization energy D_e of this molecule.

15.8 Reaction enthalpies

Although chemical reactions come in many different forms, in the present section we shall restrict ourselves to an important special class of reactions – namely, reactions among closed-shell molecules. Since such reactions are *isogyric*, conserving the number of unpaired electrons, we expect the accurate calculation of enthalpies of such reactions to be simpler than the accurate calculation of atomization energies, requiring less stringent levels of theory for agreement with experiment. For nonisogyric reactions, with different numbers of electron pairs in the reactants and the products, we expect the accuracy of the calculated reaction enthalpies to be similar to that of the atomization energies in Section 15.7.

15.8.1 EXPERIMENTAL REACTION ENTHALPIES

In Table 15.29, we have listed the reaction enthalpies of the 17 isogyric reactions examined in the present section. At $T = 0K$, the total enthalpy of a reaction may be decomposed into a nonrelativistic electronic equilibrium contribution, a vibrational contribution and a relativistic correction:

$$\Delta_r H^\circ(0K) = \Delta_r H_e^\circ(0K) + \Delta_r H_v^\circ(0K) + \Delta_r H_{rel}^\circ(0K) \qquad (15.8.1)$$

Table 15.29 Experimental reaction enthalpies (kJ/mol). The reactions used for the statistical analysis are given in bold

	Reaction	$\Delta_r H^\circ(0K)$	$\Delta_r H_v^\circ(0K)$		$\Delta_r H_{rel}^\circ(0K)$	$\Delta_r H_e^\circ(0K)$
			Harm.	Anharm.		
R1	$CO + H_2 \rightarrow CH_2O$	$+9.1(08)$	31.1	-0.9	0.7	-21.8
R2	$HNC \rightarrow HCN$	$-$	0.8	0.2	-0.2	$-$
R3	$H_2O + F_2 \rightarrow HOF + HF$	$-129.4(43)$	-0.8	0.2	0.5	-129.4
R4	$N_2 + 3H_2 \rightarrow 2NH_3$	$-77.8(06)$	88.1	-2.0	1.6	-165.4
R5	$N_2H_2 \rightarrow N_2 + H_2$	$-220.0(209)$	$-$	$-$	-0.8	$-$
R6	$C_2H_2 + H_2 \rightarrow C_2H_4$	$-166.3(12)$	37.7	-0.3	0.2	-203.9
R7	$CO_2 + 4H_2 \rightarrow CH_4 + 2H_2O$	$-151.6(08)$	95.2	-2.5	1.0	-245.3
R8	$CH_2O + 2H_2 \rightarrow CH_4 + H_2O$	$-201.2(09)$	51.5	-1.2	0.5	-251.9
R9	$CO + 3H_2 \rightarrow CH_4 + H_2O$	$-192.0(08)$	82.6	-2.1	1.2	-273.8
R10	$HCN + 3H_2 \rightarrow CH_4 + NH_3$	$-234.0(27)$	87.8	-2.4	0.9	-320.3
R11	$H_2O_2 + H_2 \rightarrow 2H_2O$	-348.0	$-$	$-$	0.7	$-$
R12	$HNO + 2H_2 \rightarrow H_2O + NH_3$	$-387.8(06)$	57.8	-1.1	1.0	-445.6
R13	$C_2H_2 + 3H_2 \rightarrow 2CH_4$	$-361.1(13)$	87.7	-2.5	0.4	-446.7
R14	$CH_2 + H_2 \rightarrow CH_4$	$-497.1(22)$	46.8	$-$	0.4	-544.2
R15	$F_2 + H_2 \rightarrow 2HF$	$-545.8(12)$	17.7	-0.1	1.6	-564.9
R16	$2CH_2 \rightarrow C_2H_4$	$-799.3(31)$	45.8	$-$	0.6	-845.7
R17	$O_3 + 3H_2 \rightarrow 3H_2O$	$-862.1(17)$	72.5	-1.5	2.3	-935.5

where the total reaction enthalpy $\Delta_r H^\circ(0K)$ is obtained from experiment. To arrive at the electronic contribution $\Delta_r H_e^\circ(0K)$ studied in the present section, we must subtract the vibrational and relativistic contributions. The total reaction enthalpies and their contributions collected in Table 15.29 have been obtained from Tables 15.18 and 15.20 in the following manner

$$\Delta_r H^\circ(0K) = -\sum_P D_0^P + \sum_R D_0^R \tag{15.8.2}$$

$$\Delta_r H_v^\circ(0K) = \sum_P E_{ZPV}^P - \sum_R E_{ZPV}^R \tag{15.8.3}$$

$$\Delta_r H_e^\circ(0K) = -\sum_P D_e^P + \sum_R D_e^R \tag{15.8.4}$$

$$\Delta_r H_{rel}^\circ(0K) = -\sum_P \Delta E_{MVD}^P + \sum_R \Delta E_{MVD}^R \tag{15.8.5}$$

where P and R label the products and reactants. As seen from Table 15.29, the vibrational contributions are large and cannot be ignored if we wish to make meaningful comparisons with experiment. Indeed, for the reaction $CO + H_2 \rightarrow CH_2O$, the vibrational contribution dominates the reaction, making it endothermic rather than exothermic, as we would expect from a consideration of the electronic contribution alone. For many purposes, an account of the harmonic vibrational contribution may be sufficient, although in a few cases the anharmonic contribution is sizeable.

At $T = 0K$, the electronic contribution to the reaction enthalpy may be calculated simply as the difference between the ground-state electronic energies of the products and reactants at their equilibrium geometries:

$$\Delta_r H_e^o(0K) = \sum_P E^P(\mathbf{R}_e) - \sum_R E^R(\mathbf{R}_e) \qquad (15.8.6)$$

As the reactants and products are all closed-shell molecules, there are no first-order spin–orbit corrections – the only first-order relativistic corrections are the mass-velocity and Darwin corrections (15.8.5), which are similar in magnitude to the anharmonic corrections but of opposite sign. The non-Born–Oppenheimer corrections may be assumed to be small for the reactions considered here.

In Table 15.29, the reactions have been listed and labelled in the order of increasing exothermicity of the *electronic contribution* to the reaction enthalpy $\Delta_r H_e^o(0K)$. Note that this order differs from the one we would obtain if the reactions were sorted in order of increasing total reaction enthalpy $\Delta_r H^o(0K)$. Moreover, in our discussions, we shall refer to all reactions in Table 15.29 as exothermic, even though (as noted above) R1 is endothermic because of a large positive vibrational contribution. Finally, in this section, we shall use the terms (electronic) *reaction enthalpies* and *reaction energies* interchangeably for the electronic contribution $\Delta_r H_e^o(0K)$.

For our statistical analysis, we have excluded all reactions involving HNC, N_2H_2 and H_2O_2, for which no experimental equilibrium atomization energies exist, and HOF, for which the experimental equilibrium atomization energy is probably in error. This leaves us with a set of 13 reactions (shown in bold in Table 15.29) on which our statistical analysis of the performance of the Hartree–Fock, MP2, CCSD and CCSD(T) models is based.

15.8.2   STATISTICAL ANALYSIS OF REACTION ENTHALPIES

The equilibrium reaction enthalpies have been calculated as described in Section 15.6.1. The statistical measures of errors in the calculated reaction enthalpies have been plotted in Figure 15.15 and listed in Table 15.30.

The first thing to note about the enthalpies is that the exothermicity in general increases with the cardinal number as improvements in the basis usually favour systems of low energy – that is, the reaction products. In Section 15.7.2, a similar behaviour was observed for the atomization energy, which nearly always increases with the cardinal number. Still, in some cases, increased exothermicity does not follow from an increased cardinal number – for R5 and R13, in particular, the opposite happens.

Concerning the N-electron models, we note that the Hartree–Fock model has a surprisingly small intrinsic mean error of about -5 kJ/mol. However, the inadequacy of the Hartree–Fock model for the calculation of reaction enthalpies becomes apparent when we consider the mean absolute errors, which are almost an order of magnitude larger than the mean errors, and also the standard deviations, which are about 70 kJ/mol for all basis sets. Apparently, the Hartree–Fock enthalpies are more or less randomly distributed about the experimental enthalpies. In Figure 15.16, we have plotted the normal distributions of the errors in the calculated electronic reaction enthalpies. The broad distributions of the Hartree–Fock errors are here well illustrated.

In discussing the Hartree–Fock electronic reaction enthalpies, it should be noted that the errors plotted in Figures 15.15 and 15.16 are *absolute errors*. The *relative errors* are more acceptable, as may be seen from Table 15.31, where we have decomposed the CCSD(T) enthalpies into contributions from the Hartree–Fock determinant, from the singles and doubles amplitudes, and from the triples amplitudes. Disregarding the weakly exothermic reaction R1, the mean absolute relative error in the Hartree–Fock enthalpies is about 10%, and the largest error of 21% occurs for the ozone reaction R17. Also, except for R8 and R9, the Hartree–Fock model predicts the

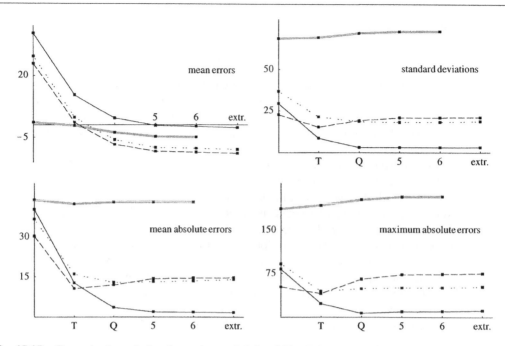

Fig. 15.15. Errors in the calculated reaction enthalpies (kJ/mol) in the cc-pCVXZ basis sets at the Hartree–Fock level (thick grey line), the all-electron MP2 level (dotted line), the all-electron CCSD level (dashed line) and the all-electron CCSD(T) level (full line).

Table 15.30 Errors relative to experiment in reaction enthalpies (kJ/mol), calculated with all electrons correlated. The cc-pcV6Z enthalpies have been obtained as described in Section 15.6.1

		HF	MP2	CCSD	CCSD(T)
$\overline{\Delta}$	cc-pCVDZ	0.9	27.7	24.8	36.9
	cc-pCVTZ	−0.3	3.0	0.9	12.0
	cc-pCVQZ	−3.1	−6.0	−7.9	2.7
	cc-pCV5Z	−4.7	−9.2	−10.7	−0.3
	cc-pcV6Z[a]	−5.0	−9.5	−11.0	−0.7
Δ_{std}	cc-pCVDZ	68.5	36.6	22.6	29.3
	cc-pCVTZ	69.1	21.3	15.4	8.7
	cc-pCVQZ	71.6	18.6	19.2	3.2
	cc-pCV5Z	72.6	18.1	20.8	3.1
	cc-pcV6Z[a]	72.8	18.2	20.9	3.1
$\overline{\Delta}_{abs}$	cc-pCVDZ	43.8	36.5	30.2	40.3
	cc-pCVTZ	42.3	16.0	10.5	12.7
	cc-pCVQZ	43.0	12.9	12.0	3.6
	cc-pCV5Z	43.0	13.2	14.3	2.0
	cc-pcV6Z[a]	43.0	13.5	14.5	1.8
Δ_{max}	cc-pCVDZ	186.4	91.6	51.6	82.5
	cc-pCVTZ	192.3	44.4	40.5	23.4
	cc-pCVQZ	202.3	48.8	65.2	7.0
	cc-pCV5Z	206.6	50.0	72.8	9.4
	cc-pcV6Z[a]	207.2	50.5	73.6	10.1

[a]cc-pV6Z for the Hartree–Fock model.

correct exothermic order of the reactions – see Table 15.36. All things considered, therefore, the Hartree–Fock model works much better for reaction enthalpies than for atomization energies; see Table 15.22 – at least for isogyric reactions such as those studied here.

Having seen that the Hartree–Fock enthalpies are in error by about 10% relative to experiment, we now turn our attention to the correlated models. From Figures 15.15 and 15.16, we see that the inclusion of correlation improves the reaction enthalpies significantly. The MP2 and CCSD

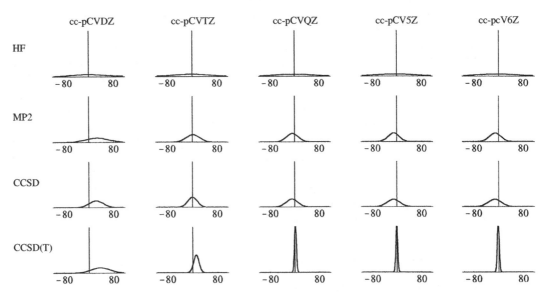

Fig. 15.16. Normal distributions of the errors in the reaction enthalpies (kJ/mol), calculated with all electrons correlated.

Table 15.31 Relative contributions (%) to the CCSD(T) reaction energy at the all-electron cc-pcV6Z level

	HF	SD	T
R1	−12.0	113.8	−1.8
R2	64.2	32.9	2.9
R3	116.7	−13.6	−3.1
R4	89.3	15.3	−4.6
R5	116.6	−15.5	−1.1
R6	104.0	−2.1	−2.0
R7	98.9	7.9	−6.9
R8	98.4	4.2	−2.6
R9	89.4	13.2	−2.6
R10	102.5	0.8	−3.3
R11	106.7	−4.4	−2.3
R12	103.3	−0.7	−2.6
R13	104.2	−1.7	−2.5
R14	90.7	8.6	0.7
R15	108.7	−6.1	−2.5
R16	86.8	11.5	1.8
R17	120.8	−14.1	−6.7

models behave similarly, with an absolute mean error of about 14 kJ/mol. Although this represents only a few per cent of the total enthalpy, it is still large compared with the experimental uncertainties; see Table 15.29. The exothermicity is usually overestimated in both the MP2 and CCSD approximations, which, at the cc-pcV6Z level, have mean errors of -10 and -11 kJ/mol, respectively. Comparing Table 15.31 with Table 15.22, we note that the doubles contributions are less important for reaction enthalpies than for atomization energies, presumably because of the isogyric nature of the reactions considered here.

For chemical accuracy – that is, for errors smaller than 5 kJ/mol – we must take into account the effects of the connected triples. As can be seen in Figure 15.15, the inclusion of connected triple excitations at the CCSD(T) level reduces the exothermicity of the reactions relative to the MP2 and CCSD levels. However, for quantitative agreement with experiment, basis sets of at least quadruple-zeta quality are needed. Thus, at the all-electron cc-pCVQZ level, the CCSD(T) model gives a mean error and a standard deviation of 2.7 and 3.2 kJ/mol, respectively, and the relative errors are 1–2%. As the basis increases, the exothermicity increases and, at the cc-pcV6Z level, the mean error and the standard deviation are -0.7 and 3.1 kJ/mol, respectively.

The reason for the overestimation of the exothermicity at the CCSD(T) level is probably the lack of connected quadruples in the wave function. Indeed, an inspection of Table 15.36 reveals that the largest discrepancy occurs for the ozone reaction R17, whose exothermicity is overestimated by as much as 10.1 kJ/mol at the cc-pcV6Z level; for the remaining 12 reactions in the statistical sample, the mean error and the standard deviation are -0.1 and 1.4 kJ/mol, respectively, with a largest deviation of -2.4 kJ/mol for R16.

15.8.3 EXTRAPOLATION AND CONVERGENCE TO THE BASIS-SET LIMIT

From our discussion of atomization energies, we recall that extrapolation improves the agreement with experiment significantly. In this subsection, we shall investigate how the same extrapolation scheme works for reaction enthalpies.

We first note that the reaction enthalpies in general converge faster than the atomization energies. At the all-electron CCSD(T) level, for example, the mean errors for $2 \leq X \leq 6$ are -103.3, -34.9, -14.3, -7.4 and -4.7 kJ/mol for the atomization energies, compared with only 36.9, 12.0, 2.7, -0.3 and -0.7 kJ/mol for the enthalpies – see Tables 15.21 and 15.30. Apparently, in the description of short-range electronic interactions, there is a considerable cancellation of errors in the enthalpies, arising from the conservation of the number of paired electrons. In particular, the CCSD(T) enthalpies are already within chemical accuracy in the cc-pCVQZ basis.

Nevertheless, as for atomization energies, we expect the basis-set convergence of the enthalpies to be dominated by the errors in the correlation treatment. For this reason, we shall here attempt to carry out extrapolations of the correlation part of the enthalpy, applying the procedure described in Section 15.6.1. In Table 15.32, we have listed the usual statistical measures of errors for the calculated and extrapolated all-electron CCSD(T) reaction enthalpies. In Figure 15.17, we have plotted the normal distributions of the errors in the calculated reaction enthalpies, with and without extrapolation.

Although the cc-pCV(DT)Z enthalpies are considerably more accurate than the cc-pCVTZ enthalpies, beyond the quadruple-zeta level, there is little or no improvement upon extrapolation. This occurs since the errors in the description of the short-range correlation of the closed-shell products and reactants cancel even for rather small basis sets. Thus, for cardinal numbers greater than four, the main errors in the correlation treatment arise no longer from the basis set but from the CCSD(T) model itself. We note that the intrinsic error of the CCSD(T) model is similar for

Table 15.32 Errors in calculated and extrapolated all-electron CCSD(T) reaction enthalpies relative to experiment (kJ/mol)

	cc-pCVXZ				cc-pCV(XY)Z			
	T	Q	5	6	DT	TQ	Q5	56
$\overline{\Delta}$	12.0	2.7	−0.3	−0.7	2.0	−2.1	−1.7	−1.2
Δ_{std}	8.7	3.2	3.1	3.1	4.3	3.4	3.6	3.2
Δ_{abs}	12.7	3.6	2.0	1.8	3.7	2.3	2.1	1.7
Δ_{max}	23.4	7.0	9.4	10.1	10.8	12.7	13.1	11.1

reaction enthalpies and atomization energies – compare the cc-pcV(56)Z errors in Tables 15.23 and 15.32.

In Table 15.33, we compare the basis-set convergence of the different contributions to the calculated reaction enthalpies. The errors have been calculated relative to the cc-pcV6Z level for the full set of 17 reactions in Table 15.29. The convergence of the Hartree–Fock contribution is surprisingly slow. Moreover, the triples errors are considerably smaller than the Hartree–Fock and CCSD errors, suggesting that it is unnecessary to carry out the triples calculation in a basis larger than cc-pCVTZ, giving considerable savings in the calculation of reaction enthalpies.

15.8.4 CORE CONTRIBUTIONS TO REACTION ENTHALPIES

Since the core correlation is largely unaffected by chemical reactions, it is worthwhile to investigate whether its contribution to the reaction enthalpies can be obtained at less demanding levels of theory, reducing the overall cost of the calculations.

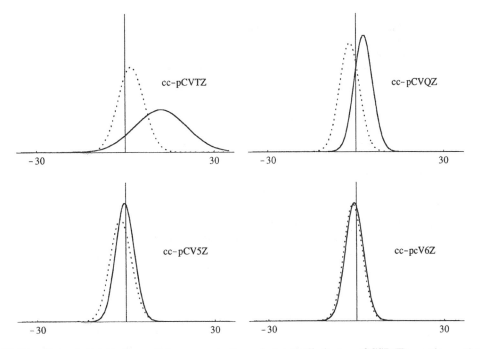

Fig. 15.17. Normal distributions of the errors in the calculated all-electron CCSD(T) reaction enthalpies (kJ/mol) without extrapolation (full line) and with extrapolation (dotted line).

Table 15.33 Statistical measures of the basis-set errors (kJ/mol) relative to the cc-pV6Z enthalpy (for the Hartree–Fock contribution) and to the extrapolated enthalpy (for the correlation contributions)

		HF	MP2–HF	CCSD–HF	CCSD(T)–CCSD
$\overline{\Delta}$	cc-pCVDZ	5.8	24.8	23.8	1.3
	cc-pCVTZ	3.9	6.7	6.3	0.5
	cc-pCVQZ	1.5	1.8	1.5	0.2
	cc-pCV5Z	0.2	0.4	0.5	0.0
	cc-pcV6Z[a]	0.0	0.4	0.4	0.0
Δ_{std}	cc-pCVDZ	12.3	23.4	21.4	2.2
	cc-pCVTZ	5.5	6.8	5.6	0.8
	cc-pCVQZ	1.9	2.3	1.6	0.3
	cc-pCV5Z	0.3	0.9	0.7	0.1
	cc-pcV6Z[a]	0.0	0.6	0.5	0.1
$\overline{\Delta}_{abs}$	cc-pCVDZ	10.3	27.1	25.7	1.9
	cc-pCVTZ	5.0	7.9	6.9	0.7
	cc-pCVQZ	1.9	2.5	1.8	0.3
	cc-pCV5Z	0.3	0.8	0.6	0.1
	cc-pcV6Z[a]	0.0	0.6	0.5	0.1
Δ_{max}	cc-pCVDZ	26.2	75.0	73.9	5.4
	cc-pCVTZ	14.9	21.0	19.3	2.2
	cc-pCVQZ	4.9	6.2	4.6	0.9
	cc-pCV5Z	0.9	2.2	2.3	0.4
	cc-pcV6Z[a]	0.0	1.5	1.5	0.2

[a] cc-pV6Z for the Hartree–Fock model.

Table 15.34 The core contribution to the CCSD(T) reaction enthalpy (kJ/mol), calculated as the difference between the all-electron cc-pCV5Z and valence-electron cc-pV5Z reaction enthalpies

R1	R2	R3	R4	R5	R6	R7	R8	R9	R10
−1.3	−0.9	0.0	−1.6	−0.4	0.5	−0.5	−1.1	−2.4	−0.7

R11	R12	R13	R14	R15	R16	R17
−1.5	−2.2	0.0	−3.5	−1.6	−6.5	−4.2

In Table 15.34, we have listed the core-correlation corrections to the CCSD(T) reaction enthalpies. Except for R6, the significant corrections are negative, increasing the exothermicity of the reactions. Although many of the corrections are small, this is not always so and a few are too large to be ignored in high-precision work. In passing, we note that, since the corrections increase the exothermicity, the inclusion of core correlation does not necessarily improve the agreement with experiment – it merely improves the agreement with the CCSD(T) limit. This is also evident from Figure 15.18, where we compare the normal distributions of the enthalpies calculated with and without the application of core corrections. Nevertheless, for accurate work and for work aimed at establishing the basis-set limit or to estimate higher-order correlation effects, the core corrections cannot be neglected.

Let us investigate whether the core correction can be reliably estimated in a basis smaller than that needed for the dominant valence-correlation contribution to the reaction enthalpy. From Section 15.7.4, we recall that the core correction to the atomization energy may be estimated

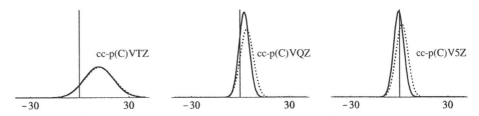

Fig. 15.18. The errors relative to experiment in CCSD(T) reaction enthalpies (kJ/mol) calculated with core correlation included (full line) and without core correlation included (dotted line).

Table 15.35 Errors in the core-correlation contributions to the reaction enthalpies relative to the CCSD(T)/cc-pCV5Z level (kJ/mol)

	CCSD(T)				MP2			
	DZ	TZ	QZ	5Z	DZ	TZ	QZ	5Z
$\overline{\Delta}$	0.4	1.2	0.3	0	0.6	1.3	0.4	0.2
Δ_{std}	0.9	1.2	0.3	0	1.1	1.2	0.4	0.4
$\overline{\Delta}_{abs}$	0.7	1.5	0.3	0	0.9	1.5	0.5	0.2
Δ_{max}	2.9	3.0	1.0	0	3.3	3.1	1.4	1.1

reliably in a basis of cardinal number one less than for the valence energy. In Table 15.35, we have listed the statistical data that summarize the errors in the calculated core corrections relative to the core contributions in Table 15.34. Comparing with Table 15.26 for atomization energies, we find that the errors are quite similar and that we may proceed in the same manner, estimating the all-electron cc-pCVXZ core correction as the difference between the all-electron cc-pCV(X−1)Z and valence-electron cc-pV(X−1)Z enthalpies. Since the cc-pCV(X−1)Z basis is smaller than the cc-pVXZ basis, this approach makes it possible to include the core correction at a cost that is no higher than that of the valence-electron calculation. Unlike for atomization energies, where the MP2 and CCSD(T) core corrections differ considerably, it appears that the MP2 and CCSD(T) core corrections are similar for reaction enthalpies.

15.8.5 SUMMARY

For the study of isogyric reactions, the Hartree–Fock model works reasonably well, producing enthalpies that are in qualitative agreement with experiment. Although the errors are often 10% or larger, the Hartree–Fock model predicts the correct order of exothermicity for all reactions in Table 15.36 except R8 and R9. In disagreement with CCSD(T), R1 is predicted to be endothermic. (However, with the vibrational correction included, R1 becomes endothermic also at the CCSD(T) level, in agreement with experiment – see Table 15.29.) In general, the vibrational contributions are large (often considerably larger than the correlation contributions) and must be estimated to make the predictions of reaction enthalpies meaningful.

For quantitative agreement with experiment, it is necessary to take into account the effects of electron correlation. At the MP2 and CCSD levels of theory, the mean absolute error is reduced from 16% at the Hartree–Fock level to about 4%. At the CCSD(T) level, the error is reduced by a further factor of 4 to about 1%. In absolute terms, the mean absolute errors are 14, 15 and

Table 15.36 Calculated and experimental electronic reaction enthalpies (kJ/mol)

	HF	MP2		CCSD		CCSD(T)		Experiment
	pV6Z	pcV6Z	pcV(56)Z	pcV6Z	pcV(56)Z	pcV6Z	pcV(56)Z	
R1	2.7	−24.4	−25.0	−22.8	−23.4	−22.4	−23.0	−21.8(08)
R2	−40.5	−74.6	−74.7	−61.3	−61.3	−63.1	−63.2	−
R3	−139.1	−126.8	−127.2	−122.9	−123.3	−119.2	−119.5	−129.4(43)
R4	−147.1	−163.7	−164.4	−172.4	−173.1	−164.8	−165.5	−165.4(06)
R5	−201.9	−190.0	−189.6	−175.0	−174.8	−173.2	−172.8	−
R6	−214.1	−196.5	−196.1	−209.8	−209.7	−205.7	−205.6	−203.9(12)
R7	−242.0	−237.2	−237.3	−261.3	−261.3	−244.6	−244.7	−245.3(08)
R8	−246.5	−256.1	−256.3	−257.0	−257.2	−250.4	−250.6	−251.9(09)
R9	−243.8	−280.6	−281.3	−279.9	−280.5	−272.8	−273.7	−273.8(08)
R10	−329.3	−316.1	−316.0	−331.8	−332.0	−321.2	−321.5	−320.3(27)
R11	−390.9	−385.1	−385.6	−374.9	−375.1	−366.5	−366.8	−
R12	−460.3	−465.3	−466.0	−457.2	−457.8	−445.6	−446.3	−445.6(06)
R13	−466.6	−442.4	−441.6	−458.8	−458.6	−447.7	−447.4	−446.7(13)
R14	−492.2	−571.1	−571.6	−538.8	−539.5	−542.7	−543.4	−544.2(22)
R15	−616.0	−600.5	−601.4	−581.1	−581.7	−566.7	−567.3	−564.9(12)
R16	−731.8	−896.2	−897.7	−828.6	−830.1	−843.3	−844.9	−845.7(31)
R17	−1142.7	−938.4	−939.7	−1009.1	−1010.1	−945.6	−946.6	−935.5(17)

2 kJ/mol, respectively, at the MP2, CCSD and CCSD(T) levels. Thus, for quantitative agreement with experiment, the connected triples cannot be neglected.

For all models, the exothermicity increases with improvements in the basis. All models overestimate the exothermicity of the reactions, with mean errors of −10, −11 and −1 kJ/mol at the MP2, CCSD and CCSD(T) levels of theory, respectively. The CCSD(T) mean error is dominated by the ozone reaction R17, whose error is −11.1 kJ/mol. The remaining 12 errors are all within chemical accuracy, and 11 reactions have an error smaller than 2.0 kJ/mol.

For basis-set convergence, the calculations should be carried out at the cc-pCVQZ level, although the cc-pCV(DT)Z numbers are also quite reliable. Beyond quadruple-zeta, there is little improvement relative to experiment. The CCSD(T) triples correction can be calculated in a smaller basis than the Hartree–Fock and CCSD enthalpies. For most purposes, the cc-pCVTZ level should be sufficient for the triples correction, and even the cc-pCVDZ correction may be useful.

The contributions from the connected quadruples and the relaxation of the connected triples appear to be about 1–2 kJ/mol – that is, of the same order of magnitude as the relativistic and anharmonic vibrational corrections. The core corrections are usually somewhat larger but can be accurately estimated from all-electron calculations in a basis of cardinal number one less than that needed for the valence electrons.

15.9 Conformational barriers

In the preceding sections, we discussed the energy differences associated with atomizations and chemical reactions. In the present section, we consider the smaller differences associated with conformational changes [10]: the barrier to linearity of water in Section 15.9.1, the inversion barrier of ammonia in Section 15.9.2 and the torsional barrier of ethane in Section 15.9.3. All barriers have been studied at the Hartree–Fock, MP2, CCSD, CCSD(T) and CCSDT levels of theory in the cc-pVXZ, aug-cc-pVXZ and cc-pCVXZ basis sets, with the valence electrons correlated in the valence sets and all electrons correlated in the core–valence sets. For water and ammonia, the calculations have been carried out at the geometries optimized at the all-electron CCSD(T)/cc-pCVQZ level; for ethane, the CCSD(T)/cc-pCVTZ geometries were used.

15.9.1   THE BARRIER TO LINEARITY OF WATER

We begin our study of conformational barriers by considering the barrier to linearity of the water molecule – that is, the difference in energy between the linear and bent conformations of the molecule. Due to the rehybridization of oxygen as the molecule becomes linear, the barrier is high. Indeed, from an accurate spectroscopic potential, the electronic part of the barrier has been estimated at 131.2 kJ/mol [11] – that is, almost as large as the electronic part of the binding energy of F_2 (163.4 kJ/mol). The calculated barriers are listed in Table 15.37.

The barrier is dominated by the Hartree–Fock contribution, which in the basis-set limit is equal to 134.5 kJ/mol. It is noteworthy that the small correlation correction goes from positive to negative as we increase the cardinal number – see Figure 15.19. Since the Hartree–Fock model overestimates the barrier, correlation improves it only for the larger basis sets. Comparing the correlated models, we find that the MP2 correction is too large and that the CCSD and CCSD(T) corrections are very similar. Full optimization of the triples at the CCSDT level hardly changes the CCSD(T) results. In view of the small triples correction, the higher connected excitations can be neglected. The convergence of the doubles correction is very slow, however, making it difficult to establish the basis-set limit to within 0.1 kJ/mol. The triples correction, on the other hand, is more stable, especially for the larger basis sets and for the augmented sets.

The calculated barrier converges faster if the basis is augmented with diffuse functions, the aug-cc-pV$(X-1)$Z results being similar to the cc-pVXZ results. Extrapolating the valence-correlation correction at the cc-pV(56)Z and aug-cc-pV(56)Z levels, we obtain the barriers 133.2 and 133.4 kJ/mol, respectively. We take 133.4 kJ/mol as the basis-set limit, using the difference of 0.2 kJ/mol as an indication of the error. The valence-correlation correction is thus about $-1.1(2)$ kJ/mol – that is, only 0.8% of the total barrier.

The inclusion of core correlation reduces the barrier further. From the all-electron cc-pCVXZ and valence-electron cc-pVXZ calculations with $X \leq 5$, we obtain the CCSD(T) core corrections

Table 15.37   The electronic barrier to linearity of water (kJ/mol)

Basis	XZ	HF	MP2	CCSD	CCSD(T)	CCSDT
cc-pVXZ (val)	DZ	146.0	150.2	152.3	153.4	153.5
	TZ	137.7	139.0	141.2	141.9	141.9
	QZ	135.7	134.3	136.8	137.0	137.0
	5Z	134.5	131.5	134.5	134.4	134.5
	6Z	134.5	130.9	134.0	133.9	–
aug-cc-pVXZ (val)	DZ	138.2	137.4	140.4	140.6	140.6
	TZ	134.9	133.3	136.4	136.3	136.3
	QZ	134.6	131.5	134.5	134.3	134.4
	5Z	134.5	131.0	134.1	134.0	–
	6Z	134.5	130.7	133.9	133.7	–
cc-pCVXZ (all)	DZ	145.6	150.0	152.0	153.2	153.2
	TZ	137.8	137.9	140.2	140.9	141.0
	QZ	135.7	132.9	135.6	135.8	135.9
	5Z	134.5	130.2	133.3	133.2	–

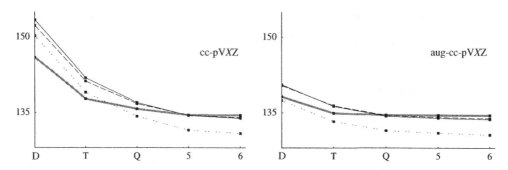

Fig. 15.19. The electronic barrier to linearity of water (kJ/mol) calculated at the Hartree–Fock level (thick grey line), the valence-electron MP2 level (dotted line), the valence-electron CCSD level (dashed line) and the valence-electron CCSD(T) level (full line).

−0.2, −1.0, −1.2 and −1.2 kJ/mol. The core correction is thus as large as the valence correction – at least for large basis sets. Assuming additivity of the two correlation corrections, we obtain 132.2(2) kJ/mol as our final estimate of the barrier to linearity of water, to which electron correlation contributes merely −2.3 kJ/mol (1.7%). The calculated barrier is thus 1 kJ/mol higher than the experimental estimate. Non-Born–Oppenheimer and relativistic corrections have been estimated to increase the barrier by about 0.6 kJ/mol, increasing this discrepancy further [10].

15.9.2   THE INVERSION BARRIER OF AMMONIA

We now consider the inversion barrier of ammonia – that is, the difference in energy between the planar and pyramidal conformations of the molecule. Again, since nitrogen undergoes rehybridization when the molecule becomes planar, we would also expect this barrier to be large. However, from spectroscopic measurements, the barrier of ammonia has been determined to be 24.2(1) kJ/mol [12]. From theoretical studies, the zero-point vibrational contribution has been estimated to be 2.9 kJ/mol [10,13,14], giving an electronic barrier of 21.2 kJ/mol – that is, about six times smaller than that of water. The calculated inversion barriers are listed in Table 15.38.

Table 15.38   The electronic inversion barrier of ammonia (kJ/mol)

Basis	XZ	HF	MP2	CCSD	CCSD(T)	CCSDT
cc-pVXZ (val)	DZ	30.1	33.6	34.8	35.9	35.9
	TZ	22.0	24.8	25.9	26.7	26.6
	QZ	20.6	22.5	23.6	24.2	24.2
	5Z	19.6	20.7	22.1	22.6	–
	6Z	19.4	20.2	21.7	22.2	–
aug-cc-pVXZ (val)	DZ	21.1	22.2	23.8	24.3	24.4
	TZ	19.6	21.1	22.7	23.2	23.2
	QZ	19.3	20.4	21.8	22.3	–
	5Z	19.3	20.1	21.6	22.1	–
cc-pCVXZ (all)	DZ	29.8	33.5	34.7	35.7	35.8
	TZ	22.2	24.2	25.3	26.2	26.1
	QZ	20.7	21.7	22.8	23.5	–
	5Z	19.6	19.9	21.3	21.9	–

As for water, the Hartree–Fock model provides the dominant contribution to the barrier, with an estimated basis-set limit of 19.3 kJ/mol. The correlation contribution is small and positive at all levels but decreases with the cardinal number – see Figure 15.20. At the cc-pCV5Z level, this reduction leads to a near-zero MP2 correlation correction. By contrast, the CCSD correction of 1.7 kJ/mol is close to the CCSD(T) correction of 2.3 kJ/mol. The triples correction is quite stable, in particular for the augmented sets. Moreover, a full optimization of the triples at the CCSDT level hardly changes the CCSD(T) results. From the smallness of the triples correction, we expect the contributions from higher connected excitations to be negligible.

Augmentation with diffuse functions improves the convergence of the calculations. Thus, from Table 15.38, we note that the aug-cc-pV5Z barrier is close to the cc-pV6Z barrier. Extrapolating the correlation contribution, we obtain the same valence-electron CCSD(T) limits at the aug-cc-pV(Q5)Z and cc-pV(56)Z levels: 21.9 kJ/mol, with a valence-correlation contribution of 2.6 kJ/mol. The core correction is smaller but significant: -0.2, -0.5, -0.8 and -0.8 kJ/mol for $X \leq 5$. Adding this correction to the extrapolated valence-correlated barrier, we obtain a final inversion barrier of 21.1(1) kJ/mol, with a correlation contribution of 1.8 kJ/mol (8%). This barrier compares well with the experimental estimate of 21.2 kJ/mol, keeping in mind that non-Born–Oppenheimer and relativistic corrections increase the calculated barrier by about 0.1 kJ/mol [10].

15.9.3 THE TORSIONAL BARRIER OF ETHANE

The torsional barrier of ethane represents the difference between the energies of the eclipsed and staggered conformations. Unlike for water and ammonia, the conformational rearrangement of ethane does not involve any rehybridization. Instead, the barrier arises from weak interactions between the methyl groups. We therefore expect the barrier to be smaller and perhaps easier to calculate, requiring less demanding levels of theory for agreement with experiment. The experimental barrier is 12.1 kJ/mol [15]. In the harmonic approximation, the zero-point vibrational contribution has been estimated to be -0.7 kJ/mol [16], giving an electronic barrier of 11.4 kJ/mol. The calculated barriers are listed in Table 15.39.

Although the change in the barrier from double- to triple-zeta is quite large, the Hartree–Fock barrier appears to be converged to within 0.1 kJ/mol already at the triple-zeta level. Moreover, for high cardinal numbers, there are no differences between the cc-pVXZ, aug-cc-pVXZ and cc-pCVXZ

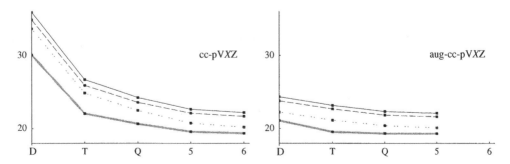

Fig. 15.20. The electronic inversion barrier of ammonia (kJ/mol) calculated at the Hartree–Fock level (thick grey line), the valence-electron MP2 level (dotted line), the valence-electron CCSD level (dashed line) and the valence-electron CCSD(T) level (full line).

Table 15.39 The electronic rotational barrier of ethane (kJ/mol)

Basis	XZ	HF	MP2	CCSD	CCSD(T)	CCSDT
cc-pVXZ (val)	DZ	13.8	13.6	13.1	13.2	13.2
	TZ	12.7	12.2	11.8	11.7	11.7
	QZ	12.7	11.9	11.5	11.4	–
	5Z	12.7	11.9	11.6	11.5	–
aug-cc-pVXZ (val)	DZ	13.5	12.7	12.3	12.2	12.2
	TZ	12.7	12.1	11.7	11.6	–
	QZ	12.7	11.8	11.5	11.4	–
cc-pCVXZ (all)	DZ	13.8	13.7	13.2	13.2	13.2
	TZ	12.7	12.1	11.8	11.7	11.7
	QZ	12.7	11.9	11.6	11.5	–
	5Z	12.7	12.0	–	–	–

sets, which all yield a barrier of 12.7 kJ/mol. The Hartree–Fock model thus overshoots the barrier by about 11% – in relative terms, slightly more than for the inversion barrier of ammonia.

The correlated models are more dependent on the quality of the basis set – see Figure 15.21. At the CCSD(T) level, for example, the barrier changes by −1.5 kJ/mol from cc-pVDZ to cc-pVTZ and by a further −0.3 kJ/mol as we go on to cc-pVQZ. From cc-pVQZ to cc-pV5Z, the barrier changes by only 0.04 kJ/mol, suggesting that the cc-pVQZ barrier is within 0.1 kJ/mol of the basis-set limit. We take 11.5 kJ/mol as the valence-electron CCSD(T) basis-set limit of the barrier in ethane, with a correlation contribution of −1.2 kJ/mol. Full optimization of the triples correction does not change the CCSD(T) barrier. At the CCSD(T) level, the core correction is less than 0.1 kJ/mol and positive.

Concerning the two other correlated methods, we note that, whereas the MP2 model recovers only about 60% of the correlation contribution to the barrier, the CCSD model recovers about 90%, the connected triples lowering the barrier by about 0.1 kJ/mol. The effect of higher connected excitations should therefore be negligible. The relativistic and non-Born–Oppenheimer corrections have also been shown to be negligible [10]. Our final estimate of the electronic contribution to the torsional barrier is therefore 11.5(1) kJ/mol, in good agreement with the experimental value of 11.4 kJ/mol.

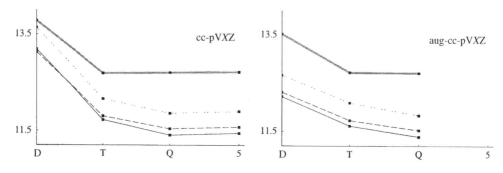

Fig. 15.21. The electronic rotation barrier of ethane (kJ/mol) calculated at the Hartree–Fock level (thick grey line), the valence-electron MP2 level (dotted line), the valence-electron CCSD level (dashed line) and the valence-electron CCSD(T) level (full line).

15.9.4 SUMMARY

The conformational rearrangements of water, ammonia and ethane studied in this section exhibit a number of features common to all symmetry-allowed conformational processes dominated by the Hartree–Fock description. In all three cases, the Hartree–Fock barrier is less than 3 kJ/mol from the true electronic barrier. For processes involving rehybridization, diffuse functions improve the convergence significantly. Although the Hartree–Fock contribution is poorly reproduced at the double-zeta level, for larger aug-cc-pVXZ sets, the barriers are within 0.4 kJ/mol of the basis-set limit. Thus, even in modest basis sets, the Hartree–Fock model reproduces the electronic conformational barriers to within a few kilojoules per mole.

For conformational barriers, the valence-correlation contribution is sensitive to the quality of the basis. For the barrier to linearity of water, for example, the correlation correction changes sign as we go from cc-pVQZ to cc-pV5Z. In general, therefore, the correlation contribution should be calculated only in large basis sets. The core correction is sizeable (as large as the valence correction in water) and cannot be neglected in studies aiming for high accuracy. However, the core-correlation contribution is more stable than the valence contribution and may be estimated in a basis of cardinal number one less than that used for valence correlation.

The MP2 model is inadequate for the study of conformational processes – overestimating the correlation correction by almost an order of magnitude in water and underestimating it by a factor of 3 in ammonia. The inadequacy of the MP2 model arises from its inability to describe molecular systems away from the equilibrium geometry. By contrast, the CCSD model is stable, recovering the bulk of the correlation correction in all three cases; the connected triples contribute less than 0.5 kJ/mol, with no significant differences between CCSD(T) and CCSDT. Because of the smallness of the triples correction, it may be calculated in a smaller basis set than the CCSD correction. Higher-order connected excitations may be neglected.

15.10 Conclusions

In this chapter, we have carried out systematic investigations, comparing calculations carried out using the standard models of *ab initio* theory with experimental measurements of geometrical structures, dipole moments, atomization energies, reaction enthalpies and conformational barriers. In addition, we investigated van der Waals systems in Section 8.5. Let us now take stock of these calculations to see where the best calculations – namely, those carried out at the CCSD(T) level – stand *vis-à-vis* the best experiments.

For molecular equilibrium structures, we found in Sections 15.3 and 15.4 that the CCSD(T) model is correct to within 0.1–0.2 pm for bond lengths and to 0.1–0.2° for bond angles. Most experimental structures are confirmed, but a few differences occur. For HNO, only a vibrationally averaged structure is available, differing significantly from the structure calculated by us – the experimental NH bond length is 1.0 pm too long and the bond angle 0.4° too open. For the nonrigid molecule H_2O_2, the experimental structure is inaccurate, the bonds being too long by about 0.5 pm and the bond angle too open by 2.3°. Finally, for the multiconfigurational O_3 molecule, the calculations are in error: the bond length is too short by 0.6 pm and the bond angle too open by 0.4°. In short, the CCSD(T) model is competitive with experimental measurements, failing only for O_3 and refuting the experiment for H_2O_2.

For the dipole moments in Section 15.5, experimental equilibrium values are available only for CO, NH_3, HF and H_2O. Except for NH_3, the experimental values are reproduced to within 0.01 D. For NH_3, the experimental dipole moment of 1.56 D is larger than the CCSD(T) dipole moment of 1.52 D, probably because of an error in the analysis of the experimental data. For the remaining polar molecules in our sample, vibrationally averaged experimental dipole moments exist for five systems and no measurements for three systems. In short, the CCSD(T) model appears to be the most reliable source of equilibrium dipole moments of small systems, providing dipole moments correct to within 0.01–0.02 D.

Let us now consider the atomization energies discussed in Section 15.7. Experimental measurements exist for 17 of the 20 molecules in our sample. Upon basis-set extrapolation of the calculated atomization energies, the agreement with experiment is usually good – that is, the errors are typically 1–2 kJ/mol but 2.3 kJ/mol for F_2. However, for two systems the discrepancy with experiment is large. First, the calculated atomization energy of HOF is 662.9 kJ/mol, whereas the experimental measurement gives 674.9(42) kJ/mol. With no indication that the CCSD(T) model fails for this system, we put this discrepancy down to an error in the experiment. Second, the atomization energy of O_3 is underestimated by 10.7 kJ/mol at the CCSD(T) level, which gives 605.5 kJ/mol compared with the experimental value of 616.2(17) kJ/mol. Unlike for HOF, this discrepancy arises from a deficiency of the CCSD(T) model, which is unable to provide a sufficiently accurate description of the multiconfigurational electronic structure of O_3. All things considered, however, the CCSD(T) model does an excellent job *vis-à-vis* experiment: 15 experimental atomization energies are reproduced, one experimental atomization energy is refuted (HOF), one calculated atomization energy is in error by 11 kJ/mol (O_3), and three atomization energies are predicted (H_2O_2, N_2H_2, HNC).

The enthalpies of the isogyric reactions studied in Section 15.8 are somewhat easier to calculate than the atomization energies, requiring no extrapolation for convergence. For three reactions, no experimental measurements are available. For the remaining ones, the reaction enthalpy is reproduced to within 2 kJ/mol for 11 reactions and large discrepancies are obtained only for the two reactions involving HOF and O_3, as expected from our calculations of atomization energies. Again, the CCSD(T) model appears to be an excellent alternative to experimental measurements.

Concerning the conformational energy barriers studied in Section 15.9, they are all well reproduced by calculations. For H_2O, we obtain an electronic barrier to linearity of 132.2 kJ/mol, compared with the experimental barrier of 131.2 kJ/mol. For the inversion of NH_3, the experimental and calculated electronic barriers are 21.2 and 21.1 kJ/mol, respectively. Finally, for the rotation of ethane, the experimental and calculated electronic barriers are 11.4 and 11.5 kJ/mol, respectively. Clearly, the theoretical calculations compare well with the measurements, again providing a useful alternative to experiment.

Finally, we recall the two van der Waals systems of Section 8.5. For the neon dimer, we obtained a CCSD(T) interaction energy of -132 μE_h, in good agreement with the experimental energy of -134 μE_h. For the water dimer, we obtained at the CCSD(T) level an interaction energy of -7.9 mE_h, again in agreement with the experimental value of $-8.6(11)$ mE_h.

All evidence thus indicates that, for small systems consisting of first-row atoms, we are today in a position to provide accurate theoretically calculated values for a variety of molecular properties, often confirming the experimental measurements, sometimes challenging or even refuting them. In some cases, our calculations break down, notably for systems not dominated by a single electronic configuration. All things considered, however, we are well prepared to carry out accurate, quantitative work on chemical systems and processes that involve first-row atoms.

The last decade has seen a tremendous and perhaps unexpected development of *density-functional methods*. These methods combine the simplicity and general applicability of the Hartree–Fock method with the accuracy of correlated models that include the effects of the connected doubles and triples. As such, they are invaluable to the practising chemist. On the other hand, the density-functional methods contain empirical elements and cannot be systematically refined towards the exact solution, making it difficult to establish, for a given chemical problem, the trustworthiness of the calculations or even to detect its failure. In the future, new developments may introduce these essential elements of *ab initio* theory into density-functional theory as well. At present, however, the only theory that can be said to provide a fully independent, accurate alternative to experiment is wave-function based *ab initio* theory – in particular, as realized in the CCSD(T) model.

Still, it is worth recalling that chemistry usually involves atoms heavier than neon (the heaviest atom considered by us) and systems containing more than eight atoms (the largest system considered by us). Although nonrelativistic chemistry works quite well for many atoms heavier than neon, heavy-atom chemistry requires more sophisticated models, where relativistic effects are accounted for. At present, much work is directed towards the development of *relativistic computational models*. Although these models have not yet reached the advanced state of development that characterizes the standard nonrelativistic methods, they will no doubt do so in the near future.

With regard to large systems, the simple Hartree–Fock model can already be applied to systems containing several thousand atoms, owing to the recent development and realization of *linear-scaling techniques*. At present, these techniques are being introduced into the simplest correlated levels MP2 and CCSD. In the near future, we shall no doubt also see activity directed towards the implementation of the CCSD(T) model for large systems, thereby significantly extending the area of applicability of these quantitative models, raising hopes for an interaction between experiment and theory that is unhampered by the usual limitations on the size of the systems and the often annoying need to replace the real chemical problems by model problems.

However, to achieve this goal, we shall probably have to find different solutions to the basis-set problem so as to reduce the size of the basis needed for convergence to the basis-set limit. The obvious candidates here are the *explicitly correlated methods*, elements of which must be introduced into the standard models in a black-box manner, to enable the user to approach the basis-set limit of, for example, the CCSD(T) model at a cost smaller than presently possible. Again, there is enough work going on in this direction for the prospects to appear bright and exciting.

Finally, the standard black-box coupled-cluster methods developed today for the accurate description of electronic-structure work well only for systems dominated by a single configuration. In our comparisons with experiment, we invariably found that the CCSD(T) model fails for O_3 – not badly, but enough to make the results useless for high-precision work. Clearly, a significant improvement would be observed at the CCSDT(Q) level, but apart from the very high cost of such a procedure, this is not the way to go if we wish to solve the *multiconfigurational problem*. Thus, our aim must be to generalize the coupled-cluster model so that it becomes applicable also to degenerate configurations, making it possible to study not only molecules close to equilibrium but also the full potential-energy surface, including transition states and regions characterized by avoided crossings. Only then can the coupled-cluster model be said to constitute a truly chemical method, applicable to the study of chemical reactions and dynamics as well as energetics. For this important development to occur, we must introduce into coupled-cluster theory some elements of MCSCF theory, the great challenges being to avoid complicated algebra and to retain the black-box hierarchical structure of the coupled-cluster model, which works so well for systems dominated by a single configuration.

References

[1]   T. H. Dunning Jr, K. A. Peterson and D. E. Woon, in P. v. R. Schleyer, N. L. Allinger, T. Clark, J. Gasteiger, P. A. Kollman, H. F. Schaefer III and P. R. Schreiner (eds), *Encyclopedia of Computational Chemistry*, Vol. 1, Wiley, 1998, p 88.
[2]   T. Helgaker, J. Gauss, P. Jørgensen and J. Olsen, *J. Chem. Phys.* **106**, 6430 (1997).
[3]   H. Koch, P. Jørgensen and T. Helgaker, *J. Chem. Phys.* **104**, 9528 (1996).
[4]   A. Halkier, P. Jørgensen, J. Gauss and T. Helgaker, *Chem. Phys. Lett.* **274**, 235 (1997).
[5]   A. Halkier and P. R. Taylor, *Chem. Phys. Lett.* **285**, 133 (1998).
[6]   C. E. Moore, *Atomic Energy Levels, Vol.* **1**, *Circular 467*, National Bureau of Standards (1949).
[7]   J. M. L. Martin, *Chem. Phys. Lett.* **259**, 669 (1996).
[8]   J. M. L. Martin and P. R. Taylor, *J. Chem. Phys.* **106**, 8620 (1997).
[9]   A. Halkier, T. Helgaker, P. Jørgensen, W. Klopper and J. Olsen, *Chem. Phys. Lett.* **302**, 437 (1999).
[10] A. G. Császár, W. D. Allen and H. F. Schaefer III, *J. Chem. Phys.* **108**, 9751 (1998).
[11] O. L. Polyansky, P. Jensen and J. Tennyson, *J. Chem. Phys.* **101**, 7651 (1994); *J. Chem. Phys.* **105**, 6490 (1996).
[12] V. Špirko and W. P. Kraemer, *J. Mol. Spectrosc.* **133**, 331 (1989).
[13] J. M. L. Martin, T. J. Lee and P. R. Taylor, *J. Chem. Phys.* **97**, 8361 (1992).
[14] T. J. Lee, R. B. Remington, Y. Yamaguchi and H. F. Schaefer III, *J. Chem. Phys.* **89**, 408 (1988).
[15] R. Fantoni, K. van Helvoort, W. Knippers and J. Reuss, *Chem. Phys.* **110**, 1 (1986).
[16] W. D. Allen, A. L. L. East and A. G. Császár, in J. Laane, M. Dakkouri, B. van der Veken and H. Oberhammer (eds), *Structures and Conformations of Non-Rigid Molecules*, Kluwer, 1993, p. 343.

Further reading

T. J. Lee and G. E. Scuseria, Achieving chemical accuracy with coupled-cluster theory, in S.R. Langhoff (ed.), *Quantum-Mechanical Electronic Structure Calculations with Chemical Accuracy*, Kluwer, 1995, p. 47.
K. Raghavachari and L. A. Curtiss, Evaluation of bond energies to chemical accuracy by quantum chemical techniques, in D. R. Yarkony (ed.), *Modern Electronic Structure Theory*, World Scientific, 1995, p. 991.
H. F. Schaefer III, J. R. Thomas, Y. Yamaguchi, B. J. DeLeeuw and G. Vacek, The chemical applicability of standard methods in *ab initio* molecular quantum mechanics, in D. R. Yarkony (ed.), *Modern Electronic Structure Theory*, World Scientific, 1995, p. 3.
P. R. Taylor, Accurate calculations and calibration, in B. O. Roos (ed.), *Lecture Notes in Quantum Chemistry*, Lecture Notes in Chemistry Vol. 58, Springer-Verlag, 1992, p. 325.

Exercises

EXERCISE 15.1

In this exercise, we determine the nonrelativistic equilibrium atomization energy D_e of BH, correcting the experimental atomization energy D_0 for the effects of vibrational motion and relativity. Useful conversion factors are $1E_h = 27.211606$ eV $= 219\,474.6$ cm$^{-1} = 2625.5$ kJ/mol.

1. From the constants $\omega_1 = 2366.9$ cm^{-1} and $X_{11} = -49.39$ cm^{-1} [16], determine the zero-order vibrational energy E_{ZPV}.

2. The ^2P ground state of boron exhibits first-order spin–orbit splitting, the energy of the $^2P_{3/2}$ level being 16 cm^{-1} above the ground-state level $^2P_{1/2}$ [6]. Determine the first-order spin–orbit correction $\Delta E_{SO}(^2P_{1/2})$.

3. At the all-electron CCSD(T)/cc-pCVTZ level, the first-order mass-velocity and one-electron Darwin contributions to the energy (mE$_h$) are: -35.671 and 28.870 for BH at $R_e = 2.3289a_0$; -35.741 and 28.914 for B; -0.031 and 0.024 for H. Determine ΔE_{MVD} for BH.

4. From the experimental atomization energy of $D_0 = 3.42$ eV and the vibrational and relativistic corrections, calculate the experimental equilibrium atomization energy D_e in E_h and kJ/mol units.

Solutions

SOLUTION 15.1

1. The zero-point vibrational energy (15.6.10) is calculated from the harmonic and anharmonic expressions (15.6.7) and (15.6.11):

$$E_{ZPV} = \tfrac{1}{2}\omega_1 + \tfrac{1}{4}X_{11} = 0.005336 \; E_h \tag{15S.1.1}$$

2. According to (15.6.16), the first-order spin–orbit correction to the ground-state energy of the boron atom is given by

$$\Delta E_{SO}(^2P_{1/2}) = E(^2P_{1/2}) - \frac{2E(^2P_{1/2}) + 4E(^2P_{3/2})}{2+4}$$

$$= \frac{2}{3}[E(^2P_{1/2}) - E(^2P_{3/2})] = -0.000049 \; E_h \tag{15S.1.2}$$

3. From (15.7.4), we obtain

$$\Delta E_{MVD} = E^B_{MVD} + E^H_{MVD} - E^{BH}_{MVD} = -0.033 \, mE_h \tag{15S.1.3}$$

4. The equilibrium atomization energy is obtained from (15.7.3) as

$$D_e = D_0 - \Delta E_{SO}(^2P_{1/2}) + E_{ZPV} - \Delta E_{MVD}$$

$$= 0.131100 \; E_h = 344.2 \; kJ/mol \tag{15S.1.4}$$

where we have retained more digits than the experimental D_0 warrants, in order to quantify the relativistic correction.

LIST OF ACRONYMS

LFF	local FF [box]	421
MCSCF	multiconfigurational SCF [model]	176,598
MDn	McMurchie–Davidson [type] n [two-electron integral scheme]	380
MO	molecular orbital	144
MOC	minimal operation-count [method]	560
MPn	nth-order Møller–Plesset [perturbation theory]	193,747
MPPT	Møller–Plesset perturbation theory	192,724,739
MP4(SDQ)	MP4 singles-doubles-and-quadruples [perturbation theory]	769
MRCI	multireference CI [model]	183
MRSDCI	multireference singles-and-doubles CI [model]	183,526
NEO	norm-extended optimization	619
NF	near-field [box]	418
NN	nearest neighbour [box]	418
OCC	orbital-optimized coupled-cluster [model]	699
OCCD	OCC doubles [model]	701
ON	occupation-number [vector, operator]	1,6
OSn	Obara–Saika [type] n [two-electron integral scheme]	386
QCI	quadratic CI [model]	702
QCISD	QCI singles-and-doubles [model]	702
RAS	restricted active space	526,599
RASSCF	RAS self-consistent field [model]	599
RFF	remote FF [box]	421
RHF	restricted Hartree–Fock [model]	170,434
RPA	random-phase approximation	501
RSPT	Rayleigh–Schrödinger perturbation theory	724,728
Rn	Rys [type] n [two-electron integral scheme]	397
SCF	self-consistent field [model]	169,448
STO	Slater-type orbital	226
TZ	triple-zeta [basis set]	300
UHF	unrestricted Hartree–Fock [model]	170,435
ZPVE	zero-point vibrational energy	847

INDEX

References to main entries are indicated by ¶